SpringerWienNewYork

Fortschritte der Chemie
organischer Naturstoffe

Progress in the Chemistry
of Organic Natural Products

Founded by L. Zechmeister

Editors:
A. D. Kinghorn, Columbus, OH
H. Falk, Linz
J. Kobayashi, Sapporo

Honorary Editor:
W. Herz, Tallahassee, FL

Editorial Board:
V. M. Dirsch, Vienna
S. Gibbons, London
N. H. Oberlies, Greensboro, NC
Y. Ye, Shanghai

91

Fortschritte der Chemie
organischer Naturstoffe

Progress in the Chemistry
of Organic Natural Products

Naturally Occurring Organohalogen Compounds –
A Comprehensive Update

Author:
G. W. Gribble

SpringerWienNewYork

Prof. A. Douglas Kinghorn, College of Pharmacy,
Ohio State University, Columbus, OH, USA

em. Univ.-Prof. Dr. H. Falk, Institut für Organische Chemie,
Johannes-Kepler-Universität, Linz, Austria

Prof. Dr. J. Kobayashi, Graduate School of Pharmaceutical Sciences,
Hokkaido University, Sapporo, Japan

This work is subject to copyright.
All rights are reserved, whether the whole or part of the material is concerned, specifically those of translation, reprinting, re-use of illustrations, broadcasting, reproduction by photocopying machines or similar means, and storage in data banks.

© 2010 Springer-Verlag/Wien
Printed in Germany

SpringerWienNewYork is a part of
Springer Science + Business Media
springer.at

Product Liability: The publisher can give no guarantee for the information contained in this book. This also refers to that on drug dosage and application thereof. In each individual case the respective user must check the accuracy of the information given by consulting other pharmaceutical literature. The use of registered names, trademarks, etc. in this publication does not imply, even in the absence of a specific statement, that such names are exempt from the relevant protective laws and regulations and therefore free for general use.

Cover photos: Both photos were taken by Professor Joseph R. Pawlik, University of North Carolina, Wilmington, Center for Marine Science Wilmington, NC 28409, USA. Sponge on left: Agelas clathrodes. Sponge on right: Aplysina archeri. Photos taken on coral reefs, 10–20 m depth, Bahamas Islands.

Library of Congress Catalog Card Number 2009934003

Typesetting: SPI, Chennai

Printed on acid-free and chlorine-free bleached paper
SPIN: 12674480

With 41 coloured Figures

ISSN 0071-7886
ISBN 978-3-211-99322-4   e-ISBN 978-3-211-99323-1
DOI 10.1007/978-3-211-99323-1
SpringerWienNewYork

Dedicated to the memory of my mother, Jane Adena Gribble, 1917–2007.

Brief CV – Gordon W. Gribble

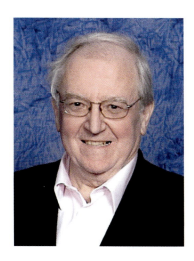

Gordon W. Gribble is a native of San Francisco, California, and completed his undergraduate education at the University of California at Berkeley, in 1963. He earned a Ph.D. in organic chemistry at the University of Oregon, in 1967. Following a Postdoctoral Fellowship at the University of California, Los Angeles, he joined the faculty of Dartmouth College, in 1968, where has been Full Professor of Chemistry since 1980. He served as Department Chair from 1988 to 1991. In 2005, he was named to the newly endowed Chair as "The Dartmouth Professor of Chemistry." Dr. Gribble has published 320 papers in natural product synthesis, synthetic methodology, heterocyclic chemistry, natural organohalogen compounds, and synthetic triterpenoids. Since 1995, he has coedited the annual series *Progress in Heterocyclic Chemistry*, and the 2nd edition of *Palladium in Heterocyclic Chemistry*, coauthored with Jack Li, was published in 2008. Dr. Gribble has had a long-standing interest in organic chemical toxicity, chemical carcinogenesis, environmental chemistry, and naturally occurring organohalogen compounds. As an award winning home winemaker for the past 30 years, he has a strong interest in the chemistry of wine and winemaking.

Acknowledgements

The author is grateful to Gettysburg College for a sabbatical leave in Gettysburg, Pennsylvania, during 2006–2007 when the majority of this review was prepared, and is especially appreciative to the late Professor Richard Moore (University of Hawai'i) who suggested *Fortschritte der Chemie organischer Naturstoffe* as a forum for this review and the preceding one in Volume 68. A special thanks goes to Ms. Wendy Berryman who typed the manuscript and drew all of the structures, and to Professors Heinz Falk and Douglas Kinghorn for their enormous assistance with the preparation of the manuscript. The author also acknowledges the inspiration and encouragement by three late outstanding scientists and pioneers in the war against "chemophobia" and the public's irrational fear of chlorine: Professors Gordon Edwards (San Jose State University), Fredrick Stare (Harvard School of Public Health), and Thomas Jukes (University of California, Berkeley). A special personal acknowledgement goes to the late Mr. Jack Grant for his friendship during my time in Gettysburg.

Contents

List of Contributors xv

1 Introduction 1

2 Origins 3
   2.1 Marine Environment 3
   2.2 Terrestrial Environment 5
   2.3 Extraterrestrial Environment 8

3 Occurrence 9
   3.1 Simple Alkanes 9
      3.1.1 Chloromethane 9
      3.1.2 Dichloromethane 12
      3.1.3 Trichloromethane 13
      3.1.4 Tetrachloromethane (Carbon Tetrachloride) 15
      3.1.5 Bromomethane 15
      3.1.6 Other Simple Bromoalkanes 17
      3.1.7 Mixed Bromochloromethanes 18
      3.1.8 Iodomethanes 19
      3.1.9 Other Simple Iodoalkanes 19
      3.1.10 Mixed Iodomethanes 20
      3.1.11 Simple Alkenes 20
      3.1.12 Simple Alkynes 22
      3.1.13 Simple Organofluorines 23
      3.1.14 Other Simple Organochlorines 24
   3.2 Simple Functionalized Acyclic Organohalogens 25
   3.3 Simple Functionalized Cyclic Organohalogens 27
      3.3.1 Cyclopentanes 27
      3.3.2 Cyclitols and Benzoquinones 28

3.4	Terpenes	32
	3.4.1 Monoterpenes	32
	3.4.2 Sesquiterpenes	38
	3.4.3 Diterpenes	60
	3.4.4 Higher Terpenes	86
3.5	Steroids	92
3.6	Marine Nonterpenes: C_{15} Acetogenins	96
3.7	Iridoids	104
3.8	Lipids and Fatty Acids	105
3.9	Fluorine-Containing Carboxylic Acids	124
3.10	Prostaglandins	127
3.11	Furanones	130
3.12	Amino Acids and Peptides	134
3.13	Alkaloids	174
3.14	Heterocycles	177
	3.14.1 Pyrroles	177
	3.14.2 Indoles	197
	3.14.3 Carbazoles	217
	3.14.4 Indolocarbazoles	217
	3.14.5 Carbolines	218
	3.14.6 Quinolines and Other Nitrogen Heterocycles	220
	3.14.7 Benzofurans and Related Compounds	226
	3.14.8 Pyrones and Chromones	227
	3.14.9 Coumarins and Isocoumarins	227
	3.14.10 Flavones and Isoflavones	231
	3.14.11 Carbohydrates	231
3.15	Polyacetylenes	231
	3.15.1 Terrestrial Polyacetylenes and Derived Thiophenes	231
	3.15.2 Marine Polyacetylenes	232
3.16	Enediynes	232
3.17	Macrolides and Polyethers	234
3.18	Naphthoquinones, Higher Quinones, and Related Compounds	249
3.19	Tetracyclines	253
3.20	Aromatics	254
3.21	Simple Phenols	256
	3.21.1 Terrestrial	256
	3.21.2 Marine	265
3.22	Complex Phenols	270
	3.22.1 Diphenylmethanes and Related Compounds	270
	3.22.2 Diphenyl Ethers	273
	3.22.3 Tyrosines	281
	3.22.4 Depsides	314
	3.22.5 Depsidones	315
	3.22.6 Xanthones	317
	3.22.7 Anthraquinones and Related Compounds	319

	3.22.8 Griseofulvin and Related Compounds	322
	3.22.9 Miscellaneous Fungal Metabolites and Other Complex Phenols	322
3.23	Glycopeptides	328
3.24	Orthosomycins	333
3.25	Dioxins and Dibenzofurans	337
3.26	Humic Acids	345

4 Biohalogenation 349
    4.1 Introduction 349
    4.2 Chloroperoxidase 349
    4.3 Bromoperoxidase 355
    4.4 Halogenases, Other Haloperoxidases and Peroxidases 356
    4.5 Myeloperoxidase 360
    4.6 Abiotic Processes 361
    4.7 Biofluorination 361
    4.8 Biosynthesis 362

5 Biodegradation 367

6 Natural Function 369

7 Significance 375

8 Outlook 377

References 379

Author Index 507

Subject Index 577

Listed in PubMed

List of Contributors

Gribble, Prof. Dr. G. W., Dartmouth College, Department of Chemistry, 6128 Burke Laboratory, Hanover, NH 03755, USA
email: Gordon.W.Gribble@Dartmouth.EDU

1 Introduction

The previous survey documented 2,448 naturally occurring organohalogen compounds, both biogenic and abiotic (*1*). In the intervening years an additional 2,266 compounds have been identified from a myriad of natural sources, which include chlorine-, bromine-, iodine-, and fluorine-containing organic compounds. The organization herein follows that used earlier so as to provide continuity. Numerous reviews covering natural organohalogens, both very general and highly specialized, have appeared since 1994 (*2–72*). Others will be cited in the appropriate section.

While the furor over "chlorine" has abated to some extent, there remains an underlying "chlorophobia" – an irrational fear of chlorine and organochlorine compounds. It is hoped that the present review along with the references cited herein will help to balance the need to regulate persistent organic organohalogen pollutants (e.g., "POPs") against the clearly demonstrated important role of organohalogens – both natural and anthropogenic – in our society and in the environment.

G.W. Gribble, *Naturally Occurring Organohalogen Compounds – A Comprehensive Update*, Progress in the Chemistry of Organic Natural Products, Vol. 91,
DOI 10.1007/978-3-211-99323-1_1, © Springer-Verlag/Wien 2010

2 Origins

The ubiquitous abundance of the four halides (Table 2.1) has resulted in the evolution of organohalogens in all regions of our earth, both biogenic and abiotic (*73–77*).

2.1 Marine Environment

As will be seen, most naturally occurring organohalogen compounds are unique to individual marine organisms and are not widely dispersed in the environment. However, the more volatile haloalkanes, which have several marine sources, are important contributors to the atmosphere. The salinity of Earth's early ocean was probably twice that of the present value (*77*), and sea-salt spray is the major atmospheric source of reactive halogens (Cl_2, Br_2, BrCl, HOCl, HOBr) that are subsequently converted to chlorine oxide and bromine oxide. This atmosphere chemistry is exceedingly complex and beyond the scope of this review (*78–85*). The formation of reactive chlorine and bromine in sea-salt aerosols, for which there is compelling evidence, may explain the low ozone concentrations that are often observed above the oceans (*86–94*). The importance of bromine oxide to both tropospheric and stratospheric bromine-ozone chemistry has been stressed (*95–100*). Not surprising is the observation of similar bromine oxide-ozone interactions over the Dead Sea (*101*), and both bromine oxide and chlorine oxide chemical reactions with ozone over the Great Salt Lake (*102*), with concomitant ozone depletion in both areas. Although less studied, iodine in the marine boundary layer is well known and can involve the photolysis of marine biogenic organoiodine compounds (*103, 104*). Moreover, it appears that the global aerosol load has a major contribution from marine organohalogen aerosols with the inevitable formation of reactive halogens (*105*).

As abundantly illustrated in the first survey, marine organisms produce and sequester an enormous number of organohalogens. It is estimated that more than 15,000 marine natural products of all types have been described (*106*). This author

G.W. Gribble, *Naturally Occurring Organohalogen Compounds – A Comprehensive Update*, Progress in the Chemistry of Organic Natural Products, Vol. 91,
DOI 10.1007/978-3-211-99323-1_2, © Springer-Verlag/Wien 2010

Table 2.1 Distribution of halides/mg kg^{-1} in the environment

Halide	Oceans (73, 74)	Sedimentary rocks (66, 74)	Fungi (75)	Wood pulp (218)	Plants (74, 76)
Cl$^-$	19,000	10–320		70–2100	200–10,000
Br$^-$	65	1.6–3	100		
I$^-$	0.05	0.3			
F$^-$	1.4	270–740			

has determined from the published literature (1998–2005) that 15–20% of all newly discovered marine natural products are organohalogens (107). Given the salinity of the world's oceans, which occupy more than 70% of the earth's surface and over 90% of the volume of the crust (108), it is not surprising that organohalogens are plentiful in the 500,000 estimated species of marine organisms spread over 30 phyla (109). This figure includes 100,000 marine invertebrates (110), 80,000 molluscs (111), 15,000 sponges (112), and 4,000 species of bryozoa (moss animals) (113).

Perhaps due to their accessibility (and visibility!), sponges – the simplest and earliest multicellular organisms that evolved about one billion years ago (114) – have been widely examined for their chemical content, and new sponge species are still being discovered (115). However, to acquire significant quantities of biologically active sponge metabolites, it is necessary to develop "farming" methods (116, 117) or to employ cell culture and gene cluster tactics (118). A major sponge research area has been to explore the now well-established sponge-bacteria symbiosis (119–123). Such studies of sponges include *Aplysina cavernicola* (124–126), *Aplysina aerophoba* (126–128), *Theonella swinhoei* (128), *Rhopaloeides odorabile* (129, 130), and *Xestospongia muta* and *X. testudinaria* (131), all of which have associated active bacterial communities that may produce the metabolite.

Even older than sponges are cyanobacteria (blue-green algae, Fig. 2.1), which date back 2.8 billion years (132). As will be seen in Chap. 3 (Occurrence), the 2,000 species of cyanobacteria produce a multitude of organohalogen and other metabolites (133–135), which are often highly toxic to humans (136–138). The cyanobacterium *Oscillatoria spongeliae* is a common symbiont of the sponges *Dysidea herbacea* and *Dysidea granulosa* (Fig. 2.2) (139–144), but the actual producer of the organohalogen metabolites remains uncertain.

As will be presented in Chap. 3 (Occurrence), other marine organisms such as molluscs (145), sea hares (146), mussels (147), bryozoans (148), tunicates, and soft corals (149) produce a myriad of organohalogen metabolites. Interestingly, symbiotic bacteria can also be associated with these organisms (123). Marine phytoplankton (150) and macroalgae (151, 152) are rich sources of organohalogens, particularly volatile haloalkanes. Two relatively new areas for ocean exploration are marine bacteria and fungi (108, 153–155, 178). Finally, as more remote and deeper regions of the oceans are explored, new marine species are being discovered; for example, the new genus, *Osedax*, of marine worms (156).

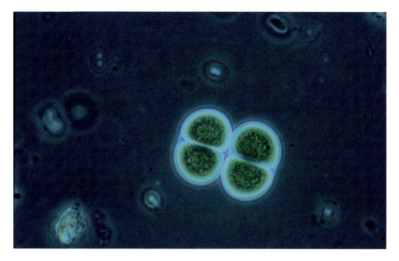

Fig. 2.1 *Chroococcus turgidus*, a species of cyanobacteria, which are prolific producers of organohalogens (Photo: A. D. Wright)

Fig. 2.2 *Dysidea granulosa*, a cyanobacterium containing sponge (Photo: F. J. Schmitz)

2.2 Terrestrial Environment

Organohalogens are present in many terrestrial environments: sediments, soils, plants, fungi, lichen, volcanoes, biomass combustion, bacteria, insects, and higher organisms. The high concentration and dispersal of chloride in minerals, soils, and plants results in a multitude of both biogenic and abiotic organochlorine compounds in these terrestrial environments (*37, 157–167*). Humic forest lake sediments (*168*) and peatlands (*169–171*) contain large quantities of organohalogens, including organoiodines that form during humification in peatlands (*171*). Peatlands that comprise 2% of the earth's continental surface are a major reservoir of organically bound iodine in the terrestrial environment (*171*), accumulating 280–1,000 million tons of organochlorines during the postglacial period (*169*). Likewise, 91% of the bromine found in peat is organically bound (*170*).

X-ray absorption spectroscopy has revealed the formation of organochlorine compounds from chloride and chloroperoxidase in "weathering" plant material (*172–174*). Moreover, this technique has uncovered the bromide-to-organobromine conversion in environmental samples (*174*). In addition to chloroperoxidase mediated chlorination, the abiotic chlorination in soils and sediments involving the alkylation of halides during Fe(III) oxidation of natural organic phenols in soils and sediments has been discovered (*175–177*).

Other terrestrial organisms frequently contain organohalogens. Lichens dating back 400 million years are a rich source of chlorinated phenolics (*1, 178, 179*). The even older fungi, which date back one billion years (*180*), perhaps number 1.5 million species of which only 70,000 are described (*181*). Basidiomycetes fungi are ubiquitous producers of organohalogens (*182*), and fungi, bacteria, and lichen engage in symbiosis (*183, 184*). Tundra fungi under snow-cover (*185*) and insect pathogenic fungi (*186*) are of recent interest and undoubtedly will yield novel natural products. Slime molds (myxomycetes) (*187*) and bryophytes (liverworts, mosses, hornworts) (*188, 189*) possess a rich assortment of natural products, including organohalogens, but terrestrial bacteria remain king of the biosynthesizers (*190*).

Volcanoes have been comparatively little studied for their chemical content. However, a few studies have provided some astonishing results as will be described later in Chap. 3 (Occurrence) (*1*). The origin of volcanic organohalogens may simply be a result of the halides present in sediments and minerals reacting with organic matter within the volcano at the high temperatures and pressures during eruptions and outgassing. The four halides are known to be entombed in rocks and sediments (*190–199*), and also in the ocean mantle (*200, 201*). Volcanic emissions invariably contain massive quantities of HCl and HF (*1*). Recent studies of Mt. Etna (*202–205*), Mt. Pinatubo (*206*), Soufriere Hills (*207, 208*), Popocatépetl (*209*), Villarrica (*210*), Satsuma-Iwojima (*211*), Sakurajima (*212*), and Laki (*213*) confirm their ubiquity of gaseous HCl and HF. Reactive bromine (BrO) and iodine emissions are also reported (*206, 208, 211, 212*). The largest known point source of both HCl and HF is the 1997 Mt. Etna eruption, with emission rates of 8.6 and 2.2 kg s^{-1}, respectively (*203*). Newer detection techniques such as solar occultation spectroscopy (*203*) and remote infrared spectroscopy (*214*) will obviously lessen the hazards of sampling. The role that these halogens play in depleting ozone has been discussed (*215*).

Only two studies of organohalogen volcanic gases were reported since the last survey, those of Vulcano (Fig. 2.3) (*216*) and Kuju, Satsuma-Iwojima, Mt. Etna,

Fig. 2.3 A volcano on Stromboli, an island in the Tyrrhenian Sea off the north coast of Sicily (Photo: F. M. Schwandner)

and Vulcano (*217*). These results will be described in Chap. 3 (Occurrence). The mysterious chemistry that occurs in volcanoes has been addressed with regard both to halogen (*218*) and hydrocarbon formation (*219*). An interesting personal account of the Kamchatka hot springs, which are a rich source of organohalogens (*1, 218*), has appeared (*220*). A novel volcanic source of HCl stems from the heating and evaporation of seawater by molten lava from the Hawaiian volcano Kilauea, which has been in continuous eruption since 1986, leading to highly acidic plumes ("acid rain") estimated at 3–30 tons of HCl daily (*221*).

A related pyrolytic source of HCl, HBr, and low molecular weight haloalkanes is biomass combustion. Human controlled fires may date back 790,000 years (*222*), but natural forest and grass fires presumably date from the time vegetation first appeared on earth (350–400 million years ago), and continue unabated today (*223–225*). Recent massive fires include those in Indonesia 1997 (*226*), Northern Alberta 1998 (*227*), Alaska 2004 (*228*), and Russia 2002–2005 (*229*). In Canada alone some 10,000 forest fires occur annually (*230*), and forest fires have plagued the Western United States for decades (*225, 231*). The 1988 Yellowstone fire, which burned more than 3 months, consumed 600,000 ha (*225*). Interestingly, a model study revealed that heating a mixture of methane, hydrogen chloride, and oxygen forms haloalkanes, chlorinated aromatics, dioxins, and many other organochlorines (*232*), indicating the plausibility of finding such compounds in volcanic plumes and biomass combustion fires. Cigarette smoke contains 30–66 mg kg^{-1} of unidentified organochlorines *(233)*.

2.3 Extraterrestrial Environment

Although both HCl *(234)* and HF *(235)* are present in interstellar space, it came as a stunning surprise when meteorites were found to contain organohalogen compounds *(236)*. The several earlier claims of meteoritic organochlorines were only cautiously advanced and perhaps even viewed with skepticism *(237–240)*. In various forms, chlorine has been detected in and around Io, Jupiter's largest moon *(241–245)*.

3 Occurrence

3.1 Simple Alkanes

No other class of natural organohalogens has the wide diversity of sources, as do the simple halokanes – marine, terrestrial biogenic, terrestrial abiotic, biomass combustion, and volcanoes. However, since the previous survey, only a few newly discovered natural simple halogenated alkanes have been reported.

3.1.1 Chloromethane

Chloromethane is the most abundant organohalogen – anthropogenic or natural – in the atmosphere. The myriad natural sources of CH_3Cl dwarf the anthropogenic contribution (Table 3.1). Subsequent to the previous survey (*1*) a number of new natural sources of CH_3Cl have been identified, and other reviews have appeared (*42, 246, 247*).

3.1.1.1 Marine

Laboratory cultures of marine phytoplankton (*Phaeodactylum tricornutum*, *Phaeocystis* sp., *Thalassiosira weissflogii*, *Chaetoceros calcitrans*, *Isochrysis* sp., *Porphyridium* sp., *Synechococcus* sp., *Tetraselmis* sp., *Prorocentrum* sp., and *Emiliana huxleyi*) produce CH_3Cl, but in relatively insignificant amounts (*248, 249*). Similarly, low production of CH_3Cl was observed from several macroalgae (*Fucus vesticulosus*, *Enteromorpha compressa*, *Ulva lactuca*, and *Corallina officinalis*) but not for two others (*250*). Another laboratory study of micro- and macroalgae failed to detect CH_3Cl in 58 species, although CH_3Br and CH_3I were observed (*251*). An extensive study of 30 species of polar macroalgae revealed the release of significant amounts of CH_3Cl in only *Gigartina skottsbergii* and *Gymnogongrus antarcticus* (*152*). In contrast to these studies, ten species of salt marsh (notably

G.W. Gribble, *Naturally Occurring Organohalogen Compounds – A Comprehensive Update*, Progress in the Chemistry of Organic Natural Products, Vol. 91,
DOI 10.1007/978-3-211-99323-1_3, © Springer-Verlag/Wien 2010

Table 3.1 Sources and estimated amounts[a,b] of chloromethane/Gg y^{-1}

Source	Best estimate	Range[c]	References
Biomass combustion	910	650–1,120	(285, 286)
Biomass combustion	611 ± 38	–	(295)
Biomass combustion	515	226–904	(283)
Savanna fires	420	–	(285)
Asian tropical plants[c]	910	820–8,200	(268)
Oceanic	650	325–1,300	(253)
Oceanic	–	200–400	(300)
Salt marshes	170	65–440	(252)
Wood rotting fungi	160	43–470	(271, 295)
Industrial incineration	162	–	(296)
Coal combustion	105	5-205	(287)
Conifer forest floor	84.7	38.7–130.8	(267)
Leaf litter	–	75–2,500	(281)
Fossil fuel combustion	75	5–145	(296)
Wetlands	48	6–270	(270)
Waste incineration	32	15–75	(296)
Other industry	7	–	(296)
Peatlands	5.5	0.9–43.3	(267)
Macroalgae	0.14	–	(250)
Volcanoes	0.074 ± 0.045	–	(216)
Volcanoes	0.012	–	(217)

[a] 1 Gg (gigagram) = 10^9 g ≅ 1,000 tons
[b] For excellent compilations of these data and discussions of the missing CH$_3$Cl, see (279, 280, 287, 295, 297)
[c] Some of these ranges (uncertainties) were taken from those cited in (280)

Salicornia sp., *Batis maritima*, and *Frankenia grandifolia*) produce large amounts of CH$_3$Cl, perhaps constituting the largest natural terrestrial source (Table 3.1) (252).

Chloromethane has a strong atmospheric presence, particularly over the oceans and ice packs (253–259). A new technique has been reported for the determination of CH$_3$Cl in ice cores (260). The measured CH$_3$Cl concentration of 528 ± 26 pptv in pre-industrial and/or early industrial ice cores is similar both to present day concentrations in the remote atmosphere and to concentrations measured from ice cores dating back 300 years. These results support previous conclusions that the CH$_3$Cl concentration has remained relatively constant over the past few hundred years. Interesting is the finding that a strong source of CH$_3$Cl is warm coastal areas, such as tropical islands (261).

3.1.1.2 Terrestrial Biogenic

Both new and preexisting terrestrial sources of CH$_3$Cl have been reported since the previous survey. Rice paddies (262–266), peatlands (Fig. 3.1) (267), tropical plants (268), shrublands (269), wetlands (270), woodrot fungi (271, 272), root fungi (273), forest leaf litter (274), and coastal wetlands (275) are significant and, in some cases, major sources of atmospheric CH$_3$Cl (Table 3.1). Higher plants such as potato

3.1 Simple Alkanes

Fig. 3.1 A peatbog in Ireland that is a rich source of natural organohalogens (Photo: H. Falk)

tubers (*Solanum tuberosum*) (*276, 277*) and the saltwort (*Batis maritima*) (*277*) (also mentioned in Sect. 3.1.1.1) yield CH_3Cl, as do other terrestrial plants (*278–280*). It is quite possible that the "missing" CH_3Cl source may have its origin in tropical green plants (*278, 280*).

3.1.1.3 Terrestrial Abiotic

It is proposed that the CH_3Cl missing source (vide supra) may be the abiotic methylation of chloride in plants and soils (*280, 281*). This methylation by plant pectin in senescent and dead leaves efficiently produces CH_3Cl and shows a positive correlation with temperature. Plants studied include Norway maple, horse chestnut, cherry, oak, beech, a *Eucalyptus* sp., and a salt marsh (*Batis maritima*) (*281*). This important study complements that of *Myneni* (*172–174*) and *Keppler* et al. (*175, 176*), cited earlier, and *Öberg* (*298, 299*).

3.1.1.4 Biomass Combustion

Biomass combustion is a major global source and may even be the major source of atmospheric CH_3Cl (*223, 282–288*). These studies indicate that CH_3Cl production is maximized in low intensity fires, from incomplete combustion, and by increased chloride concentration. It is estimated that six billion tons of biomass are consumed by fire per year (*282*). The actual mechanism of CH_3Cl formation during biomass combustion may be similar to the abiotic formation of CH_3Cl in plants (*281*).

3.1.1.5 Volcanic Emissions

As described previously, volcanoes liberate a myriad of organic chemicals including CH_3Cl (*1*) and two recent studies confirm the earlier findings (*216, 217*). Thus, Vulcano, Mt. Etna, Kuju, and Satsuma-Iwojima all emit CH_3Cl from both the fumaroles and the lava gas.

3.1.1.6 Biogenesis

In addition to the abiotic mechanism suggested for CH_3Cl formation (vide supra), there is compelling evidence for biosynthetic pathways (*289–292*). The salt marsh plant *Batis maritima* contains the enzyme methyl chloride transferase that catalyzes the synthesis of CH_3Cl from *S*-adenosine-L-methionine and chloride (*291*). This protein has been purified and expressed in *E. coli*, and seems to be present in other organisms such as white rot fungi (*Phellinus pomaceus*), red algae (*Endocladia muricata*), and the ice plant (*Mesembryanthemum crystallium*), each of which is a known CH_3Cl producer (*291, 292*).

A vexing factor in understanding CH_3Cl sources is the observation that the bacterial conversion of CH_3Br to CH_3Cl with chloride has been reported in a biotranshalogenation S_N2 reaction (*293*). Since bromide is a better nucleophile than chloride in aqueous media (*294*), the reverse biotranshalogenation reaction is plausible.

3.1.2 Dichloromethane

Dichloromethane is a widely used industrial and academic laboratory solvent. New natural sources are recognized subsequent to the previous review, although the amounts are small compared to industrial emissions (Table 3.2). These include estimates of biomass combustion (*256, 283, 286*), oceanic sources (*250, 253, 256, 275, 302*), wetlands (*275*), and volcanoes (*216, 217*). Macroalgae (*Desmarestia*

Table 3.2 Sources and estimated amounts[a,b] of dichloromethane/Gg y^{-1}

Source	Best estimate	Range[c]	References
Industry[c]	650	–	(*259*)
Biomass combustion	60	50–70	(*256*)
Oceanic	190	100–290	(*256*)
Macroalgae	0.32	–	(*250*)
Volcanoes	0.021 ± 0.013	–	(*216*)

[a]The latest estimates are provided
[b]See (*253, 283, 286, 287, 301*)
[c]See also (*301*)

3.1 Simple Alkanes

Table 3.3 Sources and estimated amounts[a] of trichloromethane/Gg y^{-1}

Source	Best estimate	Range	References
Oceanic	360	210–510	(256)
Soil/fungi	235	110–450	(256, 259)
Termites	100	10–100	(327)
Peatlands	4.7	0.1–151.9	(267)
Biomass combustion	2	0.9–4	(283)
Rice paddies	23	7.7–50	(304)
Volcanoes	0.095	0.067–0.12	(216)
Microalgae	23	7.9–49	(304)
Macroalgae (non-kelp)	0.25	–	(250)
Macroalgae (non-kelp)	10	–	(308)
Macroalgae (non-kelp)	0.06	–	(306)
Macroalgae (kelp)	0.17	–	(306)
Macroalgae	0.84	0.009–3.1	(304)
Industry	66	40–100	(256, 259)

[a]See also (253, 286, 287, 303–305)

antarctica, *Lambia antarctica*, *Laminaria saccharina*, and *Neuroglossum ligulatum*) release substantial amounts of CH$_2$Cl$_2$, surpassing bromoform in some cases (*302*). Interestingly, CH$_2$Cl$_2$ forms when CH$_4$ and HCl are heated in the presence of oxygen (*232*).

3.1.3 Trichloromethane

Trichloromethane (chloroform) is a widely used industrial and academic laboratory solvent. In contrast to CH$_2$Cl$_2$, CHCl$_3$ has a multitude of natural sources, both biogenic and abiotic (*1*), and several excellent reviews are available (*303–305*). Noteworthy is the estimate that greater than 90% of atmospheric CHCl$_3$ is of natural origin (Table 3.3).

3.1.3.1 Marine

Trichloromethane is produced by brown seaweeds (*Laminaria digitata*, *Laminaria saccharina*, *Fucus serratus*, *Pelvetia canalicuta*, *Ascophyllum nodosum*), red seaweeds (*Gigartina stellata*, *Corallina officinalis*, *Polysiphonia lanosa*), and green seaweeds (*Ulva lactuca*, *Enteromorpha* sp., *Cladophora albida*) (*306*). Similarly, the macroalga *Eucheuma denticulatum*, which is cultivated and harvested on a large scale for carrageenan production, produces CHCl$_3$ (*307*), as do *Hypnea spinella*, *Falkenbergia hillebrandii*, and *Gracilara cornea* along with seven indigenous macroalgae inhabiting a rock pool (*308*). These studies show increased CHCl$_3$ production with increased light intensity, presumably when photosynthesis is at a maximum. Trichloromethane is also produced by the brown alga *Fucus vesiculosus*, the green algae *Cladophora glomerata*, *Enteromorpha ahlneriana*,

Enteromorpha flexuosa, and *Enteromorpha intestinalis*, and the diatom *Pleurosira laevis* (*309*). Other studies observe $CHCl_3$ in *Fucus serratus*, *Fucus vesiculosis*, *Corallina officinalis*, *Cladophora pellucida*, and *Ulva lactuca* (*250*), and *Desmarestia antarctica, Lambia antarctica, Laminaria saccharina, Neuroglossum ligulatum* (*302*). The yields of $CHCl_3$ often vary widely in these studies. Microalgae are also emitters of $CHCl_3$, as first found with laboratory cultures of *Porphyridium purpureum* and *Dunaliella tertiolecta* (*310, 311*). Oceanic atmospheric trichloromethane measurements provide the estimate of 320–360 Gg y^{-1} for ocean emissions (*253, 256, 259, 287, 312*) (Table 3.3). A new source of $CHCl_3$ arises from the sediments of salt lakes that harbor halobacteria capable of biosynthesizing volatile chloroalkanes (*312*).

3.1.3.2 Terrestrial Biogenic

Subsequent to the seminal work of *Asplund* and *Grimvall* (*313*), it is now well established that $CHCl_3$, trichloroacetic acid, and other simple organochlorides are naturally produced in soil, perhaps involving both biogenic and abiotic pathways (*278, 303–305, 314–323*). A study with $Na^{37}Cl$ demonstrated isotopic enrichment in the $CHCl_3$ (*316*), and numerous worldwide remote forest sites (spruce, beech, Douglas fir, grasslands) all generate trichloromethane, a process believed to involve a microbial enzymatic origin (*317–322*). One particular spruce forest was found to liberate $CHCl_3$ at a rate of 12 µg m^{-2} day^{-1} (*319*). Another forest site with a rich humic layer emits 24 µg m^{-2} day^{-1} (*321*). As will be presented in later sections, $CHCl_3$ formation is believed to involve chloroperoxidase-mediated chlorination of phenolic structures in soil humic acid, followed by ring rupture, and degradation to afford both $CHCl_3$ and trichloroacetic acid, along with other organic products (*317, 324*). The laboratory chlorination of resorcinol yielding $CHCl_3$ is an undergraduate experiment (*325*).

Other terrestrial biogenic sources of $CHCl_3$ exist. The fungi *Mycena metata*, *Peniophora pseudopini*, and *Caldariomyces fumago* produce 0.07–70 µg L^{-1} culture per day for the latter fungus and 0.7–40 ng L^{-1} culture per day for the first two fungi. The fungi *Agaricus arvensis*, *Bjerkandera* sp. BOS55, and *Phellinus pini* produce $CHCl_3$, but only in incidental cases (*326*). Nevertheless, the authors of this latter study conclude that fungi are important sources of soil air $CHCl_3$. Coastal wetlands and grassland areas in Tasmania produce $CHCl_3$, and a major contributor is the eucalypt soil-plant material site, probably due to the high chloride content of eucalyptus leaves (*275*). Irish peatlands are significant sources of $CHCl_3$ (*267*), as is leaf litter from aspen and willow trees (*274*). It is estimated that termites worldwide produce less than 15% (<100 Gg y^{-1}) of the atmospheric $CHCl_3$. Six termite species (*Coptotermes lacteus, Amitermes laurensis, Nasutitermes magnus, Nasutitermes triodiae, Drepanotermes perniger, Tumulitermes pastinator*) produce $CHCl_3$ within their mounds, and in the mound of one species, *Coptotermes lacteus*, the $CHCl_3$ concentration is 1,000 times higher than the ambient concentration (*327*).

3.1 Simple Alkanes

3.1.3.3 Biomass Combustion

The combustion of biomass, which invariably contains carbon and chloride, produces $CHCl_3$, although the amount is much less than that recorded for CH_3Cl and CH_2Cl_2 *(283, 286)* and is a minor emission source of $CHCl_3$ (<1%) *(303–305)* (Table 3.3). The high temperature (700°C) reaction of methane, HCl, and oxygen furnishes trichloromethane *(232)*.

3.1.3.4 Volcanic Emissions

Following the pioneering studies of Isidorov *(218)*, other investigations find $CHCl_3$ in several Italian (Vulcano, Mt. Etna) and Japan (Kuju, Satsuma-Iwojima) volcanoes *(216, 217)*, but the amounts are relatively small *(216)*.

3.1.4 Tetrachloromethane (Carbon Tetrachloride)

Tetrachloromethane (carbon tetrachloride, CCl_4) is a toxic industrial chemical with several natural sources *(1)*. The solfataras and hydrothermal vents of Kamchatka *(328)* and the thermal springs in Ashkhabad (Turkmenia) and Tskhaltubo (Georgia) *(329)* emit CCl_4. The carbonaceous black shales from Central Asia contain CCl_4 *(330)*, consistent with earlier studies of similar abiotic sources *(218)*. Volcanic emissions contain CCl_4 *(216, 217)*, and one study determined a global volcanic emission rate of 0.00341 Gg y^{-1} *(216)*. A larger global emission rate is estimated for biomass combustion of 3 Gg y^{-1} on the average *(283)*. Thermolysis of a mixture of CH_4, HCl, and O_2 also produces CCl_4 *(232)*. Forest soil has been reported to emit CCl_4 at low levels or not at all *(274, 278, 318, 319)*, but it has been suggested that the presence of CCl_4 in these studies "is probably due to an equilibrium with atmospheric concentrations of anthropogenic origin" *(278)*. The commercially important seaweed *Eucheuma denticulatum*, which is used to make the food thickener carrageenan, emits CCl_4 *(307)*. Other studies of marine algae and salt lakes are inconclusive with regard to CCl_4 *(308, 312, 392)*.

3.1.5 Bromomethane

Unlike other simple haloalkanes, bromomethane (methyl bromide, CH_3Br) has very large natural and anthropogenic sources. Indeed, CH_3Br is an outstanding pesticide (e.g., soil fumigant) for which there are no suitable alternatives *(331–334)*. Although CH_3Br is now banned in the United States because of its presumed toxicity *(335)*, it is still used by some farmers for selective applications *(334)*.

Table 3.4 Distribution of halides/mg kg^{-1} in the environment

Halide	Oceans (73, 74)	Sedimentary rocks (66, 74)	Fungi (75)	Wood pulp (218)	Plants (74, 76)
Cl$^-$	19,000	10–320		70–2,100	200–10,000
Br$^-$		65	1.6–3	100	
I$^-$		0.05	0.3		
F$^-$		1.4	270–740		

Table 3.5 Sources and estimated amounts[a] of bromomethane/Gg y^{-1}

Source	Best estimate	Range	References
Oceanic	60	–	(531)
Biomass burning	20	10–50	(285)
Salt marshes	14	7–29	(252)
Macroalgae	0.056	–	(250)
Rice paddies	–	0.5–0.9	(266)
Litter decomposition (fungi)	1.7	0.5–5.2	(350)
Wetlands	4.6	–	(270)
Peatlands	0.9	0.1–3.3	(267)
Automobiles	1.5	–	(352)
Creosote bush (*Larrea tridentata*)	0.2	–	(269)
Brassica plants[b]	7	–	(347)
Fumigation	47	–	(42)
Phytoplankton	–	2.6–47.0	(340)
Volcanoes	0.00098 ± 0.00047		(216)

[a]See also (288, 353)
[b]Rapeseed and cabbage

In previous years some 20,000 metric tons of CH$_3$Br was used annually in the US as a soil pesticide (331).

Despite the low abundance of bromide relative to chloride in sediments, soil, and the ocean (Table 3.4), the ease with which nature can manipulate bromide (reduction potentials: $E°$ = 1.09 V for Br$^-$ vs. 1.36 V for Cl$^-$, 0.54 V for I$^-$, and 2.87 V for F$^-$ (2671)) results in a multitude of natural organobromine compounds especially in the oceans (41, 45) (Table 3.5).

3.1.5.1 Marine

Perhaps because of its volatility (boiling point, 4°C), CH$_3$Br has not been identified in marine organisms (i.e., algae) as frequently as bromoform and other bromoalkanes. Nevertheless, CH$_3$Br is released from both Antarctic and Arctic cultivated macroalgae, including brown algae, red algae, and green algae (152, 336–338). Several macroalgae from the north coast of Norfolk, England, yield CH$_3$Br (250). Methyl bromide is also found in cultures of marine microalgae (248, 249, 251, 311, 339, 340). The several reported atmospheric measurements of CH$_3$Br are consistent with oceanic sources (252, 254, 255, 259, 341–346). Methyl bromide originates from the vegetation zones of coastal salt marshes (252), and CH$_3$Br is

3.1 Simple Alkanes

supersaturated over part of the northeast Atlantic due in large measure to the phytoplankton *Phaeocystis* (*343*).

3.1.5.2 Terrestrial Biogenic

In addition to the emission of CH_3Br from coastal salt marshes (vide supra) (*252*), which could be considered a terrestrial source, shrublands and wetlands near coastal sites also emit CH_3Br (*269, 270, 275*). Extraordinary is the observation that higher plants (e.g., rapeseed, mustard, cabbage, broccoli, turnip, radish, alyssum, etc.) produce significant amounts of CH_3Br from natural soil bromide, estimated to be 6.6 Gg y^{-1} CH_3Br from rapeseed worldwide (*347*). It might be noted that plants are also a sink for both CH_3Br (*348*) and bromide (*347, 349*). Peatlands are a source of CH_3Br (*267*) as are rice paddies (*262–266*) and fungi (*273*). Wood-rotting fungi are another source of CH_3Br, but of low significance (*350*) (Table 3.5).

3.1.5.3 Terrestrial Abiotic

The abiotic bromination involving the alkylation of bromide during Fe(III) oxidation of natural organic phenols in organic matter is a pathway to CH_3Br that has been demonstrated in the laboratory (*175*).

3.1.5.4 Biomass Combustion

Where there is halide, carbon, oxygen, and fire there will be organohalogens. Such is the case with bromide giving rise to CH_3Br in forest fires (*223, 284, 285, 288*). Indeed, vegetation fires are a major source of CH_3Br (*285*).

3.1.5.5 Volcanic Emissions

Studies of volcanoes (Vulcano, Mt. Etna, Kuju, and Satsuma-Iwojima) have found CH_3Br in the fumarolic emissions (*216, 217*).

3.1.6 Other Simple Bromoalkanes

3.1.6.1 Marine

Marine algae biosynthesize and emit several other simple bromoalkanes, including dibromomethane (CH_2Br_2), tribromomethane (bromoform, $CHBr_3$), bromoethane

(CH$_3$CH$_2$Br), and 1,2-dibromoethane (BrCH$_2$CH$_2$Br) (*1*). Indeed, CHBr$_3$ is a promiscuous marine metabolite and is invariably produced by both macro- and microalgae (*152, 250, 302, 306–309, 336–339, 342, 344–346, 354–370*), and is the major contributor of organic bromine to the atmosphere. Bromoform may supply reactive bromine (e.g., BrO) to the upper troposphere and lower stratosphere for reaction with and destruction of ozone (*371, 372*). Macroalgae may contribute 70% of the world's CHBr$_3$ (*373*). An estimate of a global emission rate of 220 Gg y^{-1} (*50–390*) is proposed for CHBr$_3$ (*373*). Studies of brown, red, and green macroalgae from the polar region show that CHBr$_3$ is released by all 30 species examined (*152*). Dibromomethane frequently accompanies CHBr$_3$ in marine algae emissions, although usually with lower release rates (*152, 250, 302, 306, 308, 309, 336, 337, 339, 342, 345, 346, 354–366, 368–370*). In one polar algae study cited above, only 12 of 30 species were reported to emit CH$_2$Br$_2$ (*152*). Likewise, bromoethane is only occasionally reported as a marine algae volatile (*152, 250, 336, 337, 345*). The previous survey listed 1,2-dibromoethane as a tentative marine algae metabolite (*1*), but subsequent independent studies clearly establish 1,2-dibromoethane (BrCH$_2$CH$_2$Br, **1**) as a bona fide natural product (*359, 362*). This organobromine, which has anthropogenic sources (*359*), is emitted by several algae species (*152, 250, 336, 337, 359, 360, 362*).

BrCH$_2$CH$_2$Br	C$_4$H$_9$Br	CHBr$_2$CHBr$_2$	CBrCl$_3$
1	**2**	**3**	**4**

3.1.6.2 Volcanic Emissions

Several simple organobromines are emitted from Mt. Etna, Vulcano, Kuju, and Satsuma-Iwojima, including CH$_2$Br$_2$, CHBr$_3$, CH$_3$CH$_2$Br, each of which has large biogenic (marine algae) contributions, and the new compound bromobutane (**2**, isomer unknown) (*217, 216*). 1,1,2,2-Tetrabromoethane (**3**) is present in carbonaceous black shale (*330*).

3.1.7 Mixed Bromochloromethanes

The mixed bromochloromethanes, chlorobromomethane (CH$_2$BrCl), chlorodibromomethane (CHBr$_2$Cl), and bromodichloromethane (CHBrCl$_2$), often accompany CHBr$_3$ and CH$_2$Br$_2$ in marine algae (*1*). Both macro- and microalgae produce CH$_2$BrCl (*152, 250, 309, 336, 337, 346, 358, 365*), CHBr$_2$Cl (*152, 250, 306–309, 336, 342, 344, 345, 346, 354–362, 364, 365, 368, 373*), and CHBrCl$_2$ (*152, 250, 302, 306, 309, 336, 337, 342, 346, 356, 357–362, 365*). Volcanic emissions of bromochloromethanes include the known CH$_2$BrCl, CHBr$_2$Cl, and CHBrCl$_2$, along with the previously unreported CBrCl$_3$ (**4**) (*217*).

3.1 Simple Alkanes

3.1.8 Iodomethanes

3.1.8.1 Marine

The widely used organic chemical reagent, iodomethane (CH_3I, methyl iodide), has a large biogenic source in worldwide marine algae (*1*). Like CH_3Br and $CHBr_3$, CH_3I is often detected in emissions from algae and in the oceanic atmosphere (*152, 248–251, 255, 302, 306–311, 338, 339, 342, 344–346, 356, 357, 359, 361, 365*). Diiodomethane also has numerous marine algae sources (*152, 307, 309, 336, 337, 339, 342, 344, 345, 356, 357, 359–362, 365, 367, 369*), but iodoform (CHI_3) has not been described in nature following its report in *Asparagopsis taxiformis* (*1*). Diiodomethane is a more significant source of iodine in the atmosphere than CH_3I (*365*).

3.1.8.2 Terrestrial

Iodomethane has several terrestrial biogenic and abiotic sources (Table 3.6). It is emitted from volcanoes (*216, 217*), fungi (*273*), wetlands (*275*), peatlands (*267*), rice paddies (*262–266, 374*), and oat plants (*374*). Biomass combustion also accounts for some CH_3I (*284, 285, 288*). The abiotic soil source cited earlier can also produce CH_3I (*175*).

3.1.9 Other Simple Iodoalkanes

Given the dearth of iodide in the ecosystem relative to chloride and bromide (Table 3.4), it is perhaps surprising that several other alkyl iodides are found in the environment.

3.1.9.1 Marine

Marine algae are a rich source of iodoethane (CH_3CH_2I) (*152, 250, 309, 336, 337, 339, 344, 357, 365, 367*), 1-iodopropane ($CH_3CH_2CH_2I$) (*152, 309, 342, 356, 357, 365, 367*), 2-iodopropane (($CH_3)_2CHI$) (*152, 308, 309, 356, 357, 365, 367*),

Table 3.6 Sources and estimated amounts[a] of iodomethane/Gg y^{-1}

Source	Best estimate	Range	References
Oceanic	–	128–335	(*375*)
Biomass burning	< 10	–	(*285*)
Rice paddies	–	16–29	(*266*)
Peatlands	1.4	0.1–12.8	(*267*)
Macroalgae	–	0.00092–0.011	(*365*)
Macroalgae	0.28	–	(*250*)

[a]See also (*42*)

1-iodobutane (CH$_3$(CH$_2$)$_3$I) (*152, 307–309, 357, 361, 365, 367*), 2-iodobutane (CH$_3$CH$_2$CH(CH$_3$)I) (*152, 307, 309, 357, 365*), and the new 1-iodo-2-methylpropane (**5**) (*152, 365*).

3.1.9.2 Volcanic Emissions

An examination of the emissions from Mt. Etna, Vulcano, Kuju, and Satsuma-Iwojima has revealed the presence of ethyl iodide (*217*).

3.1.10 Mixed Iodomethanes

By unknown biogenetic mechanisms (perhaps nucleophilic substitution reactions?), marine algae produce several iodomethanes containing chlorine or bromine (*1*). Recent studies confirm the oceanic presence of chloroiodomethane (CH$_2$ClI) (*152, 250, 307–309, 336, 337, 339, 342, 344, 345, 355–357, 359, 360, 362, 365, 367, 369*), bromoiodomethane (BrCH$_2$I) (*339, 344, 345*), dibromoiodomethane (CHBr$_2$I) (*345*), and the new dichloroiodomethane (Cl$_2$CHI) (**6**) (*345*). Volcanic emissions from Mt. Etna, Vulcano, Kuju, and Satsuma-Iwojima are reported to contain CH$_2$ClI (*217*).

3.1.11 Simple Alkenes

While the carbon–carbon double bond is a common functional group in complex natural products, it is far rarer in simple natural compounds.

3.1.11.1 Marine

The chemically productive red algae *Asparagopsis taxiformis* and *Asparagopsis armata*, which are prized by Hawaiians for flavor and aroma ("limu kohu", supreme seaweed) (*1*), contain the novel (*E*)-1,2-dibromoethylene (**7a**), (*Z*)-1,2-dibromoethylene (**7b**), and tribromoethylene (**8**) (*364*). Trichloroethylene (TCE) (*301, 307–309, 312, 361, 376, 377*) and tetrachloroethylene (PERC) (*307–309, 312, 376, 377*) continue to be found in marine algae in concentrations larger than anticipated, but their origin remains controversial (*253, 278, 301, 307–309, 376–378*). Initially, TCE and PERC were found in 27 species of macroalgae (*376*).

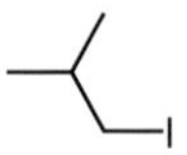

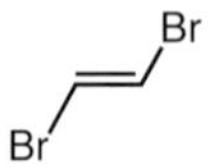

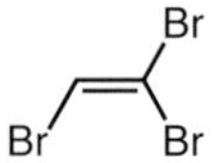

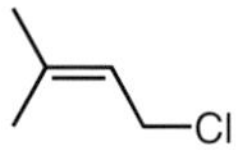

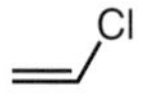

3.1.11.2 Terrestrial Biogenic

One of the most intriguing newly discovered natural organohalogens is 1-chloro-3-methyl-2-butene (**9**) from a secretion of a male flying fox (*Pteropus giganteus*) (*380*). Since this compound is known to be a powerful lachrymator (personal experience), it may function as a chemical defensive agent ("allomone") for the fox.

3.1.11.3 Terrestrial Abiotic

Vinyl chloride is obviously essential for the manufacture of poly(vinyl chloride) polymer (PVC), and is a known human liver carcinogen. Therefore, it is surprising that vinyl chloride (**10**) ($CH_2 = CHCl$) is produced abiotically in soil, during the oxidative degradation of soil matter (*381*), similar to the proposed abiotic soil formation of CH_3Cl (vide supra). Model experiments with catechol (1,2-dihydroxy-benzene), Fe(III), and chloride yield both vinyl chloride and CH_3Cl. It might be noted that CH_3CCl_3, TCE, and PERC are not formed under these conditions (*381*). The latter observations are consistent with field studies of spruce forests (*319, 322*); for a review, see (*278*). Both TCE and PERC are reported in biomass fires (*283*).

3.1.11.4 Volcanic Emissions

Several chlorinated and brominated alkenes are present in emissions of Mt. Etna, Vulcano, Kuju, and Satsuma-Iwojima (*217*). In addition to vinyl chloride (**10**), TCE, PERC, and 1,1-dichloroethylene, compounds **11–57** were detected in these emissions. For many of these compounds, exact structures remain unknown.

11 **12** **13** (or isomer) **14, 15** **16**

17-20 **21-23** **24, 25** **26, 27** **28-30** **31**

32 **33** **34, 35** **36-38** **39**

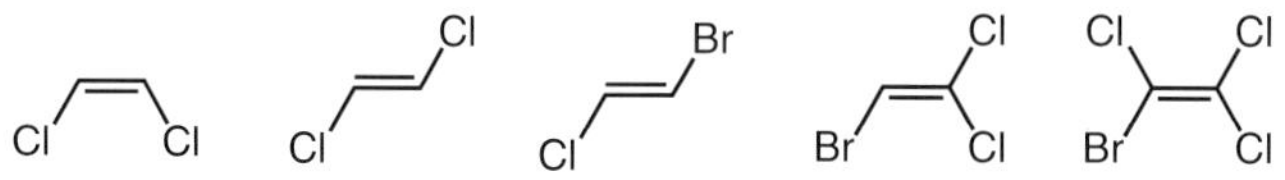

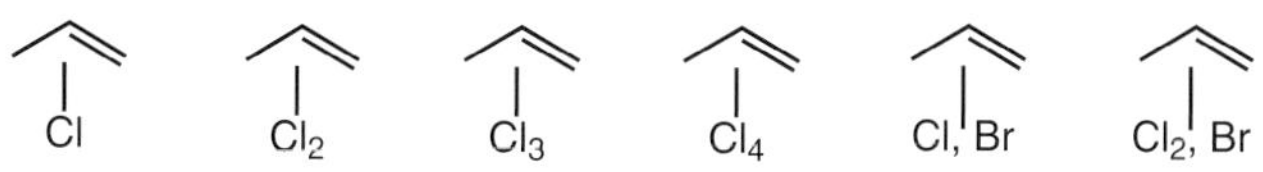

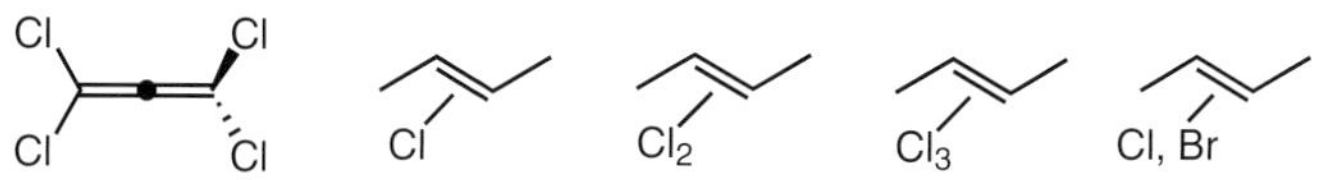

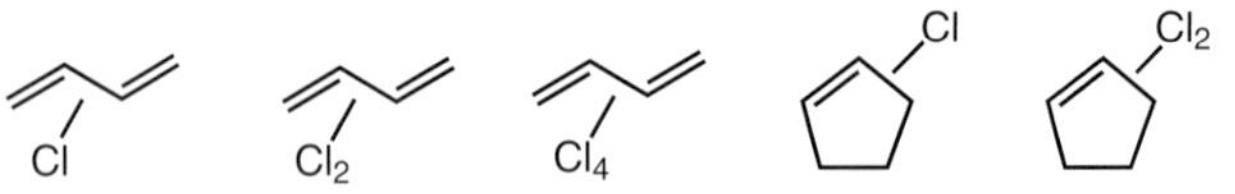

40, 41 **42, 43** **44** **45-47** **48, 49**

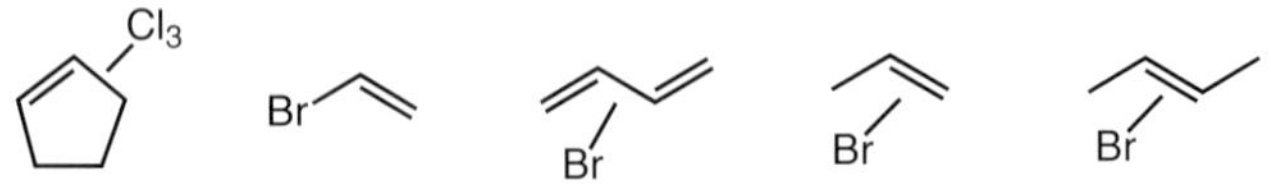

50 **51** **52, 53** **54-56** **57**

3.1.12 Simple Alkynes

Although acetylenes are widely used in organic synthesis, and are present both in pharmaceuticals and in interstellar space, simple halogenated acetylenes are virtually unknown in nature.

3.1.12.1 Terrestrial Abiotic

In model experiments with catechol, Fe(III), and chloride, and in soil emission studies, it is found that chloroethyne (chloroacetylene) (**58**) is produced (*382*). The natural formation of **58** parallels that of vinyl chloride, which is also found in these experiments. The in vitro and in vivo mechanisms are unknown, but the authors propose the path shown in Scheme 3.1 (*382*). Both chloroethyne and vinyl chloride are emitted from three soil types (coastal salt marsh, peatland, and a deciduous forest).

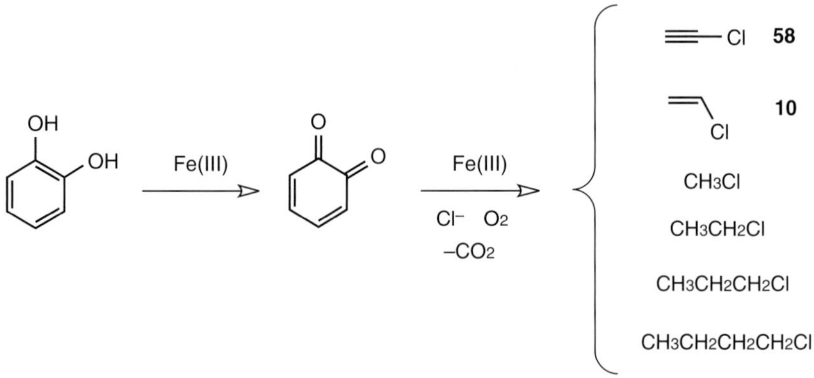

Abiotic formation of organochlorines from catechol (*382*).

Scheme 3.1

3.1.12.2 Volcanic Emissions

A study of Mt. Etna, Vulcano, Kuju, and Satsuma-Iwojima volcanoes reveals the presence of several halogenated alkynes **59–77** (*217*).

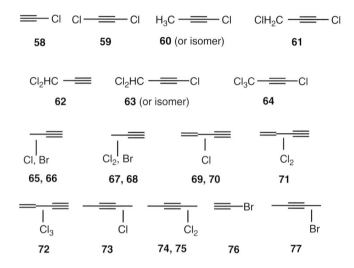

3.1.13 Simple Organofluorines

Following the early reports of chlorofluorocarbons (CFCs) in volcanic and drill well emissions (*1*), more recent work confirms these studies (*216, 217*), although the emission rates of $CFCl_3$ and CF_2Cl_2 are very small compared to the anthropogenic sources of these compounds (*383*). Tetrafluoromethane occurs in natural fluorites and granites (*384, 385*) and has been detected in natural gas (*385*). The global emission rate from cold degassing from the Earth's crust is negligible at 0.0001–0.01 Gg y^{-1} compared to anthropogenic emissions (*383*), but, because it has the very long atmospheric lifetime of 200,000 years, about half of the present CF_4 is from natural sources (*384, 385*). In this study, tetrafluoroethylene was also found (*384*), which had been previously found in volcanic emissions (*1*). Further study of fluorites, plutonites, and other rocks reveals the presence of CF_4, CF_2Cl_2, $CFCl_3$, and SF_4, along with CHF_3 and the previously unreported CF_3Cl (**78**) (*386*). Both $CFCl_3$ and CF_2Cl_2 are emitted from the Kamchatka solfataras and hydrothermal vents (*328*) and other thermal springs (*329*). The novel $CF_3CF_2CF_2H$ (**79**) is found in carbonaceous black shales (*330*).

CF_3Cl $CF_3CF_2CF_2H$ CCl_3CCl_3 CCl_3CHCl_2

78   79   80   81

3.1.14 Other Simple Organochlorines

Another controversial compound is 1,1,1-trichloroethane (methyl chloroform, CH_3CCl_3), which may or may not have significant natural sources (*253, 286, 287*). A few studies reveal emissions of CH_3CCl_3 from marine algae (*307*) and halobacteria (*312*), but studies of biomass burning (*283, 286, 387*) and forest soil emissions (*318*) do not indicate a significant natural source for CH_3CCl_3. The carbonaceous black shales from Central Asia are reported to contain the novel CCl_3CCl_3 (**80**) and CCl_3CHCl_2 (**81**) (*330*). The meteorites, Orgueil and Cold Bokkeveld, have yielded the long-chain chloroalkanes, 1-chlorododecane (**82**), 1-chlorotridecane (**83**), 1-chlorotetradecane (**84**), 1-chloropentadecane (**85**), 1-chlorohexadecane (**86**), 1-chloroheptadecane (**87**), and 1-chlorooctadecane (**88**) (*388*). The closely related long-chain chloroalkanes, 1-chlorononadecane (**89**), 1-chloroicosane (**90**), 1-chlorohenicosane (**91**), 1-chlorodocosane (**92**), 1-chlorotricosane (**93**), 1-chlorotetracosane (**94**), 1-chloropentacosane (**95**), 1-chlorohexacosane (**96**), 1-chloroheptacosane (**97**), 1-chlorooctacosane (**98**), and 1-chlorononacosane (**99**), are found in three salt marsh plants (*Suaeda vera, Sarcocornia fruticosa, Halimione portulacoides*) (*389*).

$n\text{-}C_{12}H_{25}Cl$	$n\text{-}C_{13}H_{26}Cl$	$n\text{-}C_{14}H_{29}Cl$	$n\text{-}C_{15}H_{31}Cl$	$n\text{-}C_{16}H_{33}Cl$	$n\text{-}C_{17}H_{35}Cl$
82	**83**	**84**	**85**	**86**	**87**
$n\text{-}C_{18}H_{37}Cl$	$n\text{-}C_{19}H_{39}Cl$	$n\text{-}C_{20}H_{41}Cl$	$n\text{-}C_{21}H_{43}Cl$	$n\text{-}C_{22}H_{45}Cl$	$n\text{-}C_{23}H_{47}Cl$
88	**89**	**90**	**91**	**92**	**93**
$n\text{-}C_{24}H_{49}Cl$	$n\text{-}C_{25}H_{51}Cl$	$n\text{-}C_{26}H_{53}Cl$	$n\text{-}C_{27}H_{55}Cl$	$n\text{-}C_{28}H_{57}Cl$	$n\text{-}C_{29}H_{59}Cl$
94	**95**	**96**	**97**	**98**	**99**

An examination of several marine worms (*Marenzellaria viridis, Polydora socialis, Scolelepsis squamata, Spiophanes bombyx,* and *Streblospio benedicti*) has tentatively identified alkyl and alkenyl halides **100–105** (*390*).

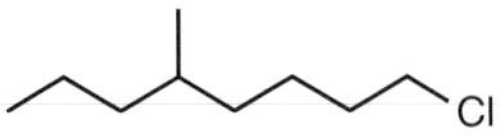

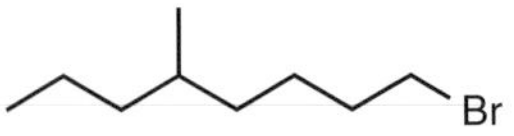

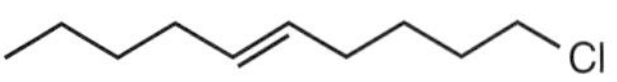

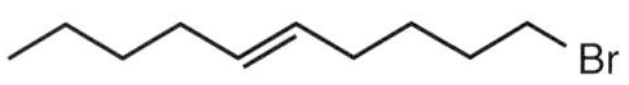

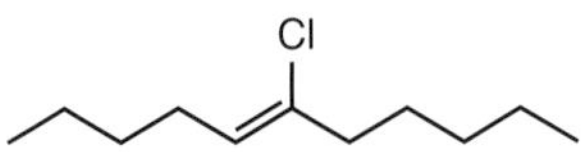

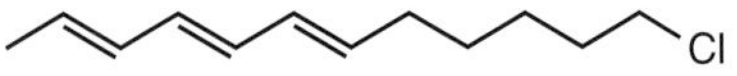

3.2 Simple Functionalized Acyclic Organohalogens

Several simple functionalized natural organohalogens do not reasonably fit into categories such as terpenes, alkaloids, or fatty acids (*1*) and are therefore included here.

The prolific red seaweed *Asparagopsis taxiformis* has afforded eight new halogenated carboxylic acids **106–113** (*391*) in addition to several already described (*1*). In some cases the double bond stereochemistry or the exact location of the halogens has not been established. Another study of this seaweed has identified the two heavily brominated enol esters **114** and **115**, the structures of which were confirmed by synthesis (*392*). These compounds, which are aldose reductase inhibitors, are two of the most heavily halogenated natural compounds known. This brings to over 100 the number of halogenated compounds found in this alga (*1*). An Antarctic collection of the red alga *Delisea fimbriata* yielded two new brominated acetates **116** and **117**, in addition to four brominated furanones and two bromooctenones that were previously known from this alga (*393*). Five Antarctic sponges (*Phorbas glaberrima*, *Kirkpatrickia variolosa*, *Artemisina apollinis*, *Halichondria* sp., and *Leucetta antarctica*) contain the known 1,1,2-tribromooct-1-en-3-one (*394*). The Okinawan alga *Wrangelia* sp. has afforded the simple tribromoacetamide (**118**), which displays potent biofilm inhibition against *Rhodospirillum salexigens* and cytotoxicity towards P388 leukemia (*395*). An unidentified fungus on the surface of the red alga *Gracillaria verrucosa* has furnished the novel brominated keto esters **119** and **120** (*396*).

106 X = Br
107 X = I

108 X = Br
109 X = I

110

111

112

113

114 R = CBr_3
115 R = $CHBr_2$

116

117

118

119

120

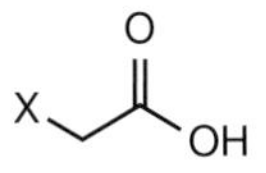

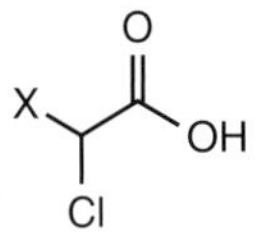

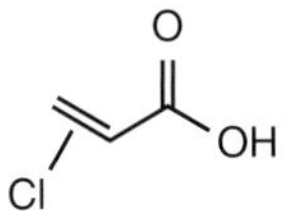

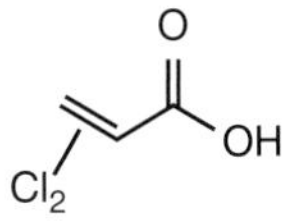

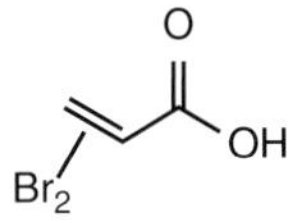

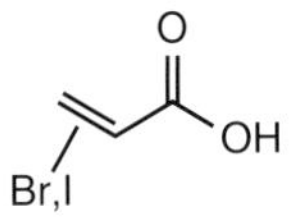

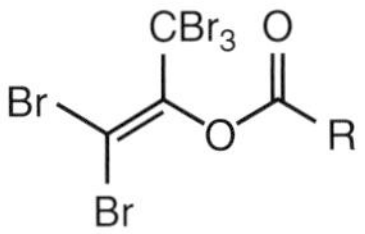

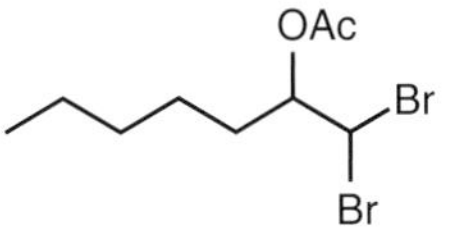

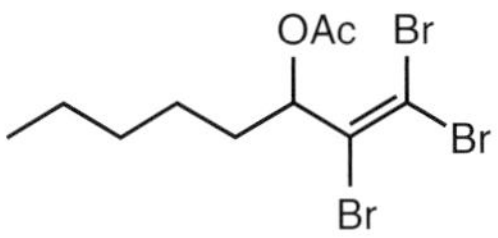

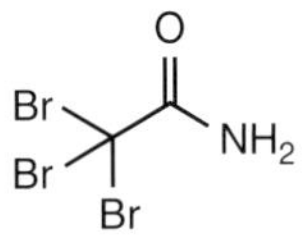

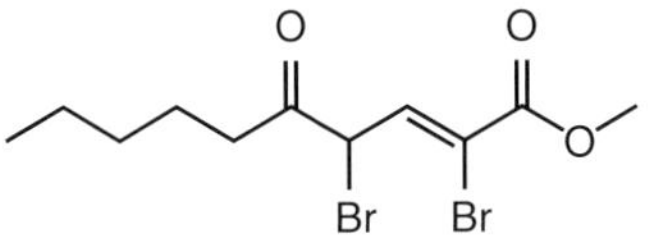

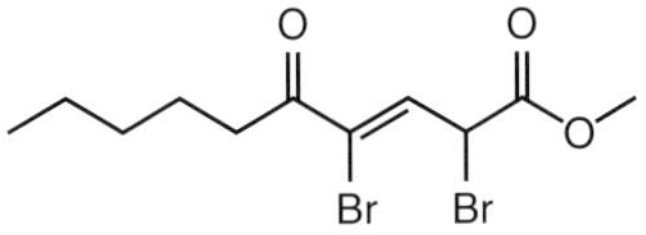

Pinicoloform (**121**), which has antibiotic and cytotoxic activities, was isolated from the fungus *Resinicium pinicola* (*397*). The edible wild milk cap (*Lactarius* spp.) contains 1-chloro-5-heptadecyne (**122**) (*398*). In addition to containing the toxic alkaloid coniine, the Yemenese plant *Aloe sabaea* has afforded the novel *N*-4′-chlorobutylbutyramide (**123**), which is the first report of a chlorinated compound in the Aloeaceae family (*399*). Several polychlorinated acetamides (**124**–**128**) were characterized from the cyanobacterium *Microcoleus lyngbyaceus* (*400*). The novel fosfonochlorin (chloroacetylphosphonic acid) (**129**) was isolated from four fungi (*Fusarium avenaceum*, *Fusarium oxysporum*, *Fusarium tricinctum*, and *Talaromyces flavus*) (*401*). The remarkable previously known bromoester **A** [2-octyl 4-bromo-3-oxobutanoate], which is present in mammalian cerebrospinal fluid and is involved in REM-sleep (*1, 402*), has been synthesized and investigated further in comparison with synthetic analogues (*403, 404*).

122

121 (pinicoloform) **123** **129** (fosfonochlorin)

124 R = H
125 R = Cl

128

126 R = Cl
127 R = H

A

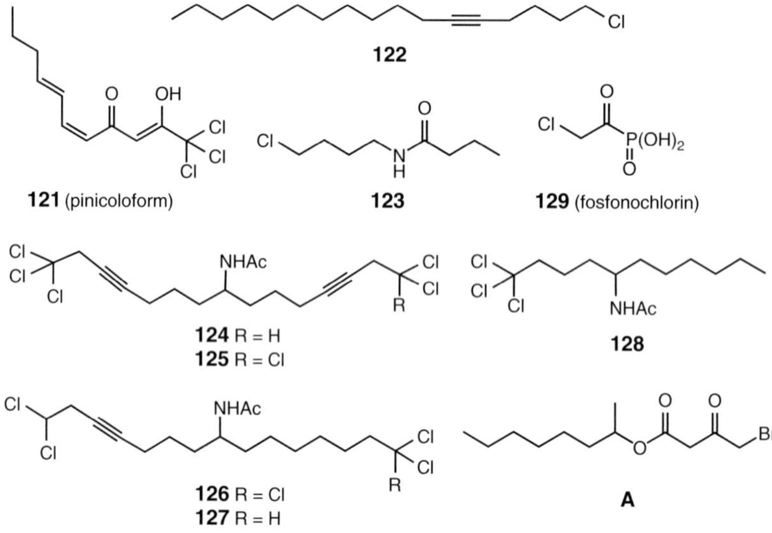

All three chloroacetic acids (chloroacetic acid [MCA], dichloroacetic acid [DCA], and trichloroacetic acid [TCA]) are naturally occurring (*1*), with TCA being identified in the environment most frequently (reviews: (*278, 405–408*)). However, these chlorinated acetic acids also have anthropogenic sources. The major source of natural TCA appears to be the enzymatic (chloroperoxidase) or abiotic degradation of humic and fulvic acids, which ultimately leads to chloroform and TCA. Early studies (*409*) and subsequent work confirm both a biogenic and an abiotic pathway. Model experiments with soil humic and fulvic acids, chloroperoxidase, chloride, and hydrogen peroxide show the formation of TCA, chloroform, and other chlorinated compounds (*317, 410–412*). Other studies reveal an abiotic source of TCA (*412, 413*).

3.3 Simple Functionalized Cyclic Organohalogens

3.3.1 Cyclopentanes

The previously known cryptosporiopsinol (**B**) (*1*) is also found in the marine fungus *Coniothyrium* sp. living on the sponge *Ectyplasia perox* (*414*), and its biosynthesis has been investigated (*415, 416*). These results reveal that chlorination occurs early in the sequence and a ring contraction from an isocoumarin seems to be involved. An asymmetric synthesis of (–)-cryptosporiopsin (**130**), the antipode of the previously described natural product (*1*), has been reported (*417*). The known (+)-cryptosporiopsin was found in cultures of the fungus *Pezicula livida* (*418*). The chlorinated isonitrile MR566A (**131**) is a melanin synthesis inhibitor produced by *Trichoderma harzianum* (*419, 420*). This same fungus found on the sponge *Halichondria okadai* has yielded trichodenones B (**132**) and C (**133**), which show significant cytotoxicity against P388 leukemia (*421*); total syntheses have established their stereochemistry (*422*). Metabolite **134** was isolated from a culture of the ascomycete A23-98 (*423*). The diastereomeric cyclopentenones VM 4798-1a (**135**) and VM 4798-1b (**136**) were obtained from the fungus *Dasyscyphus* sp. A47-98 found growing on tree bark (*424*). The pentachlorinated cyclopentenone A11-99-1 (**137**) was isolated from cultures of the ascomycete *Mollisia melaleuca* and displays inhibition of human TNF-α promoter activity and synthesis (*425*). The Madagascan plant *Samadera madagascariensis* contains the quassinoid 2-chlorosamaderine A (**138**), which has a 2-chlorocyclopentenone ring (*426*).

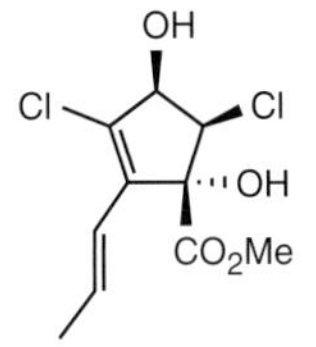

B (cryptosporiopsinol) 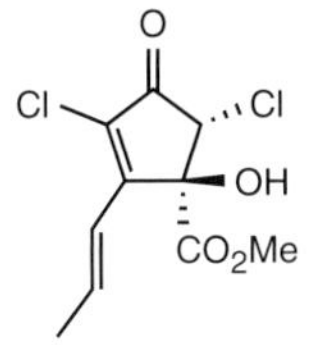   **130** ((–)-cryptosporiopsin) 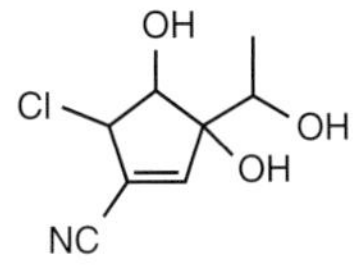   **131** (MR566A) 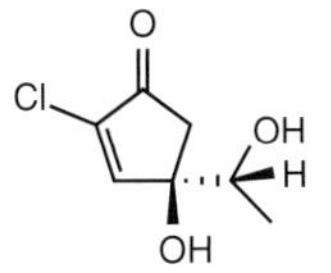   **132** (trichodenone B)

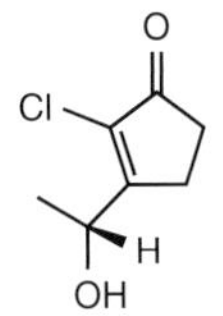

133 (trichodenone C) 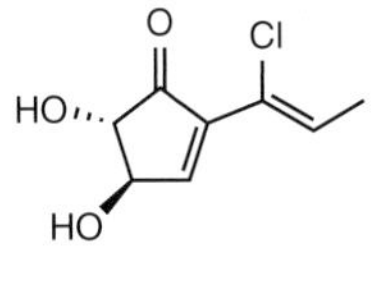   **134** 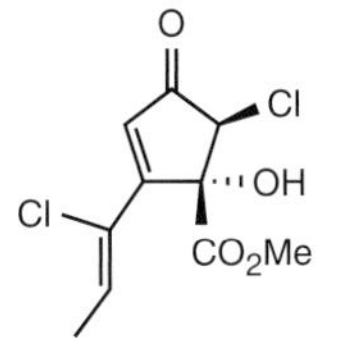   **135** (VM 4798-1a) 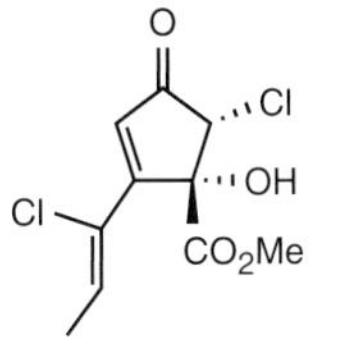   **136** (VM 4798-1b)

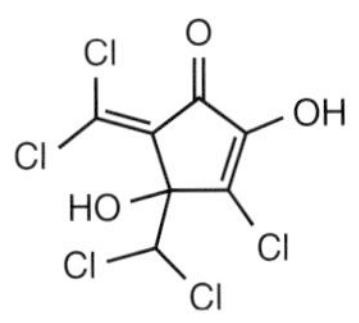

137 (A11-99-1) 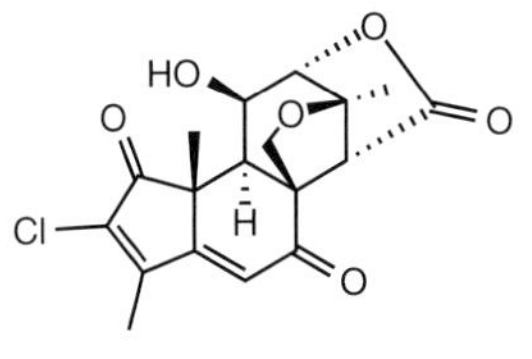   **138** (2-chlorosamaderine A)

3.3.2 Cyclitols and Benzoquinones

The halogenated natural products in this section are mainly fungal metabolites that are cyclohexene derived via a shikimate or polyketide pathway (*1*). The simple inhibitor of rat brain neutral sphingomyelinase activity, chlorogentisylquinone (**139**), is produced by the marine fungus FOM-8108 found on beach sand in Japan (*427*). The structure of the cytotoxic (+)-pericosine A (**140**), which was isolated from a marine fungus (*Periconia byssoides*) in a sea hare (*Aplysia kurodai*) (*428*), was revised by total synthesis (*429, 430*). Because of its pronounced antitumor activity, the previously reported Maui acorn worm (*Ptychodera*) (+)-bromoxone (**C**) (*1*) has been of synthetic interest (*431–433*; review, *434*). The antipode (−)-mycorrhizin A (**141**) of the previously reported (+)-mycorrhizin A was isolated from the fungi *Pezicula carpinea* and *Pezicula livida* (*418*). Some "forced" brominated metabolites of mycorrhizin A, chloromycorrhizin A, and the related lachnumons were characterized from cultures of *Lachnum papyraceum* (e.g., **D**) (*435, 436*), but these are not counted as natural products. Chlovalicin (**142**) from *Sporothrix* sp. inhibits interleukin 6 (*437, 438*), and a *Cytospora* sp. produces cytosporin B (**143**) an antagonist of angiotensin II (*439*).

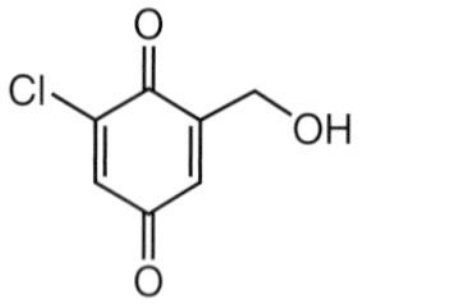

139 (chlorogentisylquinone) 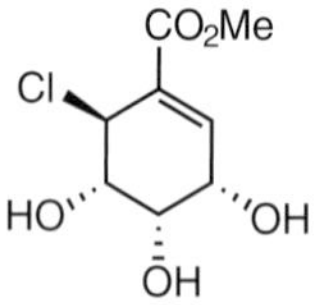   **140** (pericosine A) 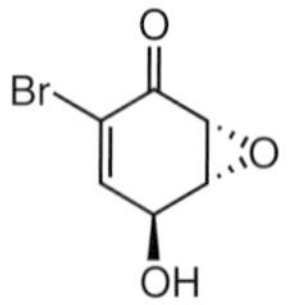   **C** (bromoxone) 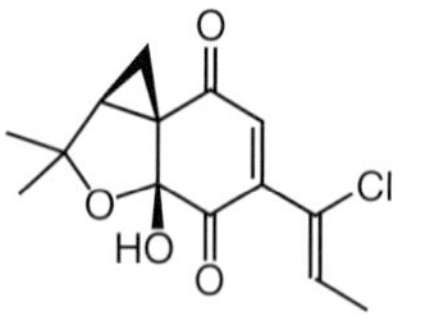   **141** ((−)-mycorrhizin A)

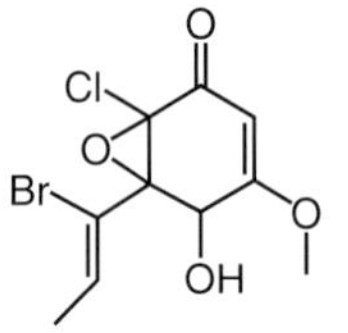

D (lachnumon B2) 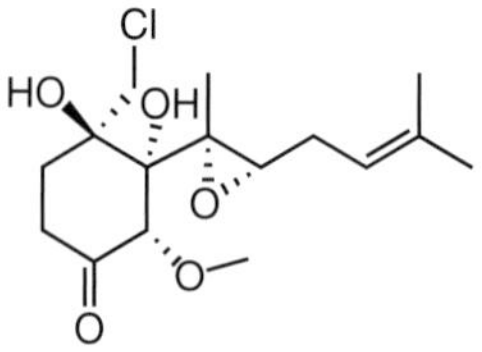   **142** (chlovalicin) 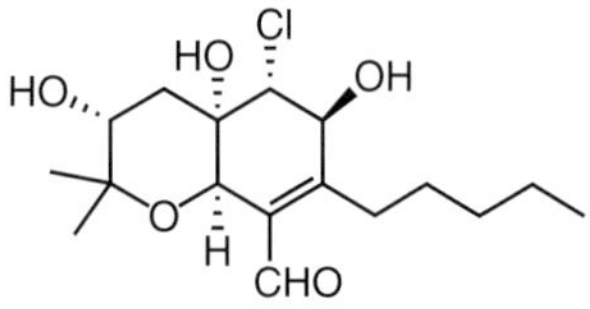   **143** (cytosporin B)

Several new fungal metabolites ("azaphilones", having an affinity for nitrogen nucleophiles) related to the well-known sclerotiorin (*1*) have been reported in recent years. Studies of the fungus *Talaromyces luteus* have uncovered luteusins A–E (**144–148**) (*440–443*). Luteusins A and B were originally named TL-1 and TL-2, and the stereochemistry of C and D was later revised (*443*). These metabolites have monoamine oxidase inhibitory properties. The related fungus *Talaromyces helicus* has furnished helicusins A–D (**149–152**) (*444*).

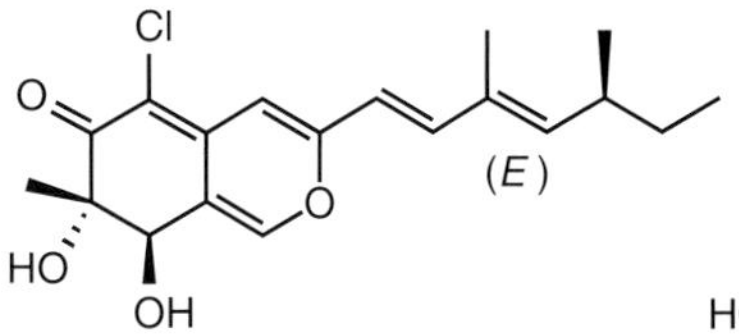

3.3 Simple Functionalized Cyclic Organohalogens

144 (luteusin A)
145 (luteusin B) (Z)

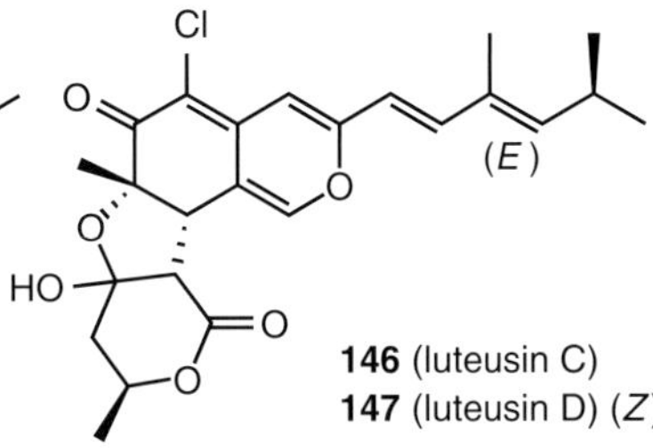

146 (luteusin C)
147 (luteusin D) (Z)

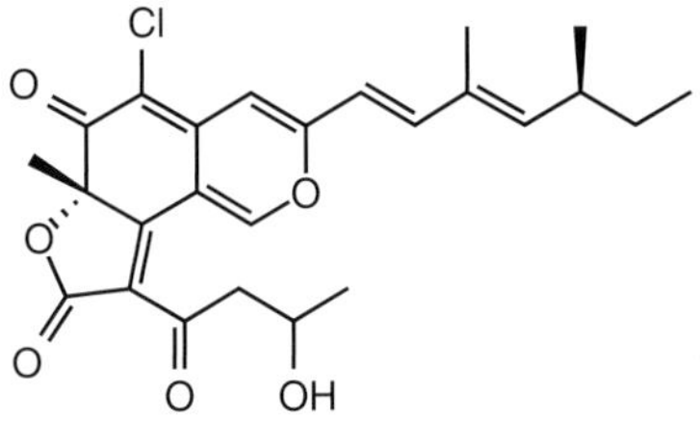

148 (luteusin E)

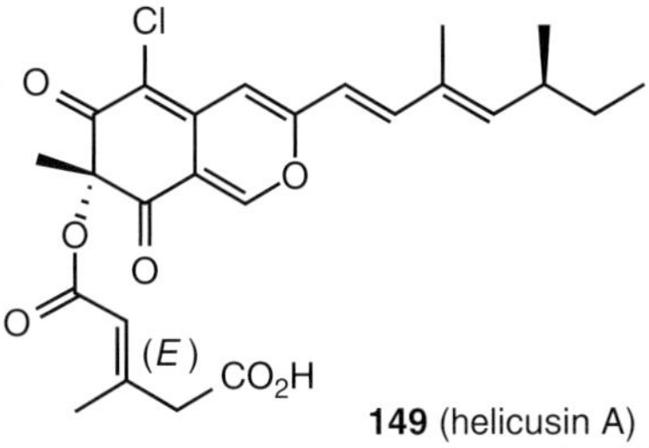

149 (helicusin A)
150 (helicusin B) (Z)

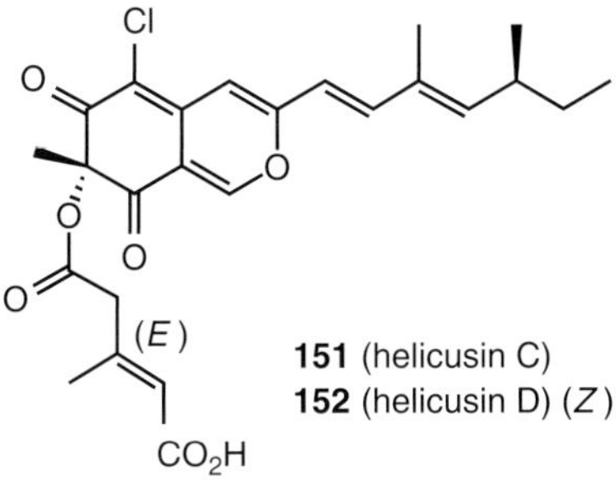

151 (helicusin C)
152 (helicusin D) (Z)

A full account and structures are reported for isochromophilones I (**E**) and II (**F**) from *Penicillium multicolor* FO-2338 (*445, 446*). The latter study also confirms the structure of sclerotiorin. The ACAT inhibitors isochromophilones III-VI (**153–156**) are found in cultures of *Penicillium multicolor* FO-3216 (*447*) and isochromophilones VII and VIII (**157, 158**) are produced by *Penicillium* sp. FO-4164 (*448*). These azaphilones inhibit cholesteryl ester transfer protein (CETP), which promotes the exchange and transfer of neutral lipids between plasma lipoproteins, and is involved in atherosclerosis. Thus, CETP is a logical target for anti-atherosclerotic drugs (*449*). A study of *Penicillium sclerotiorum* has reported 5-chloroisorotiorin (**159**), which may be isochromophilone III (**153**) (*450*). Unnatural brominated azaphilones ("forced metabolites") are produced when bromide is added to the cultures (*451*). The novel GABA-containing isochromophilone IX (**160**) is found in cultures of *Penicillium* sp. (*452*).

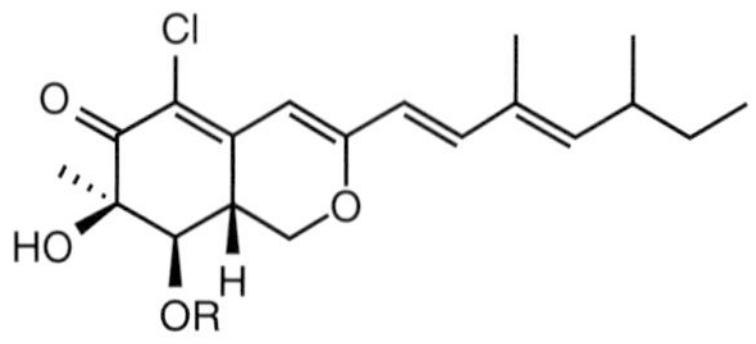

153 R = H (isochromophilone III)
154 R = Ac (isochromophilone IV)

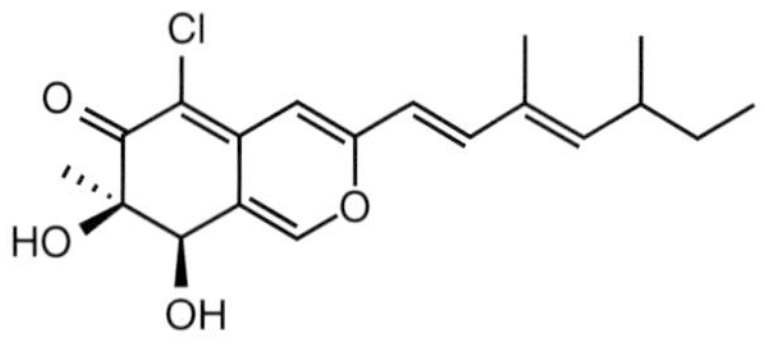

155 (isochromophilone V)

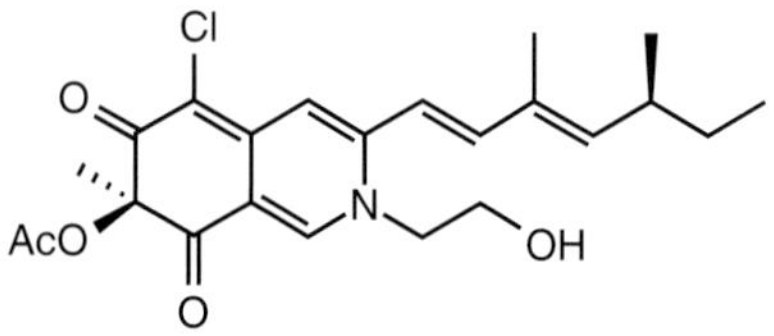

156 (isochromophilone VI)

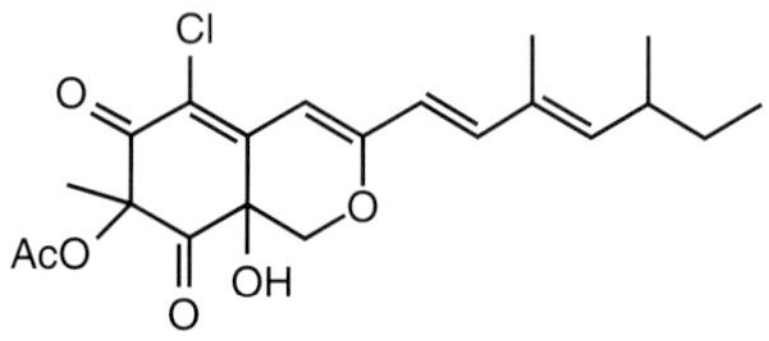

157 (isochromophilone VII)

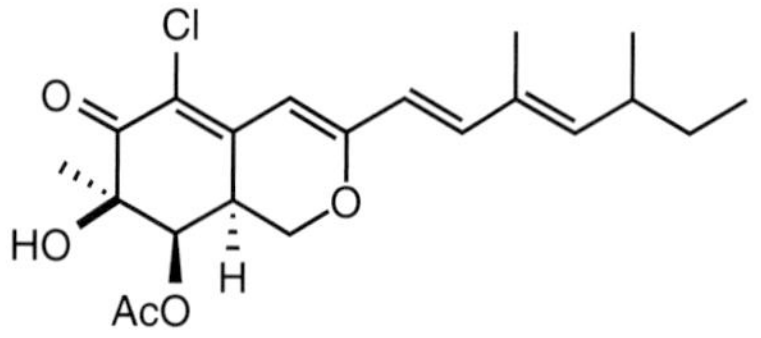

158 (isochromophilone VIII)

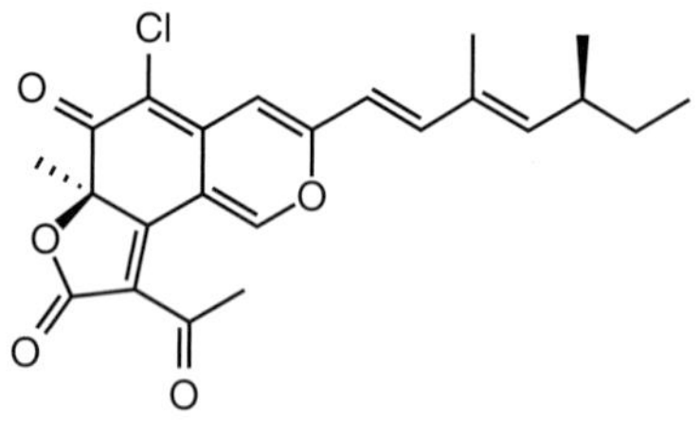

159 (5-chloroisorotiorin)

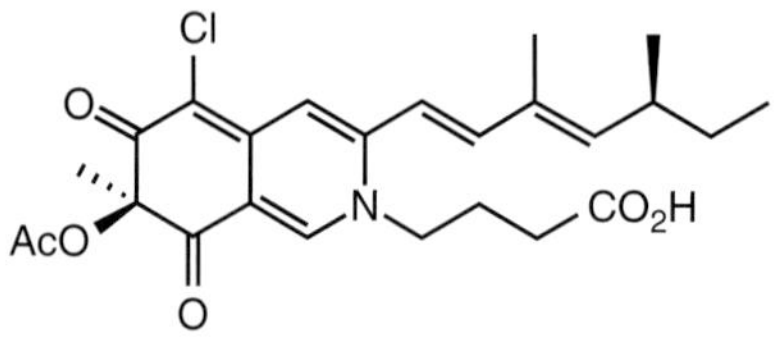

160 (isochromophilone IX)

A new family of azaphilones, the RP-1551s (**161–168**), is produced by *Penicillium* sp. SPC-21609 and they inhibit the binding of PDGF to its receptor (*453*). RP-1551-7 is identical to luteusin A (**144**). RP-1551-1 (**161**) and RP-1551-6 (**166**) are diastereomers and are different stereoisomers from luteusins C (**146**) and D (**147**).

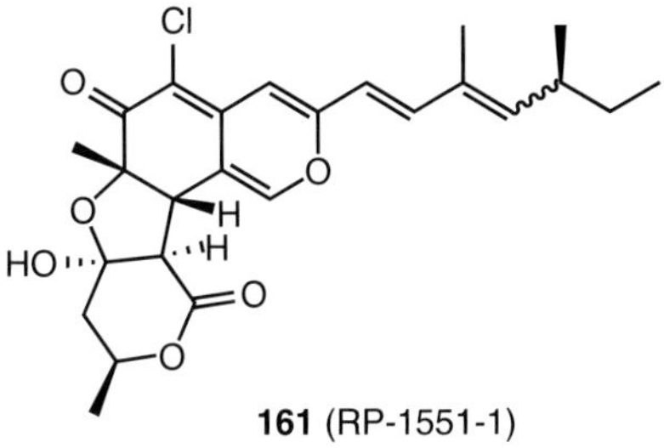

3.3 Simple Functionalized Cyclic Organohalogens

161 (RP-1551-1)

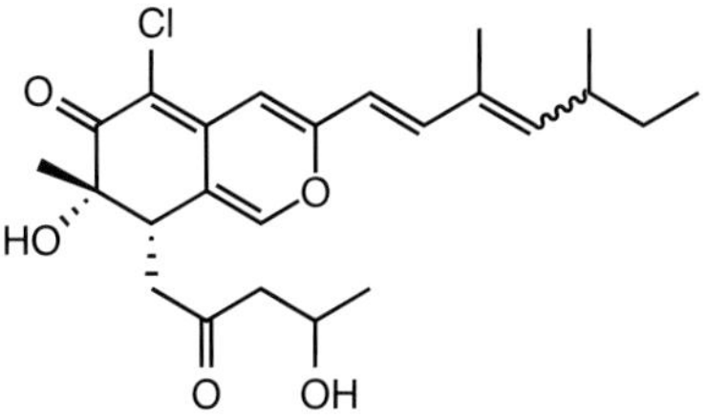

162 (RP-1551-2)

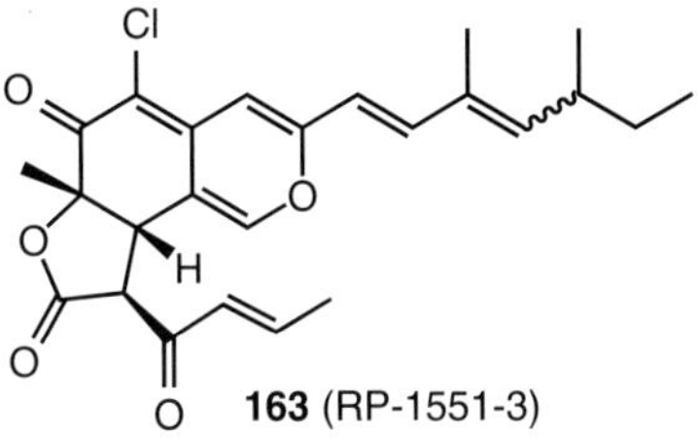

163 (RP-1551-3)

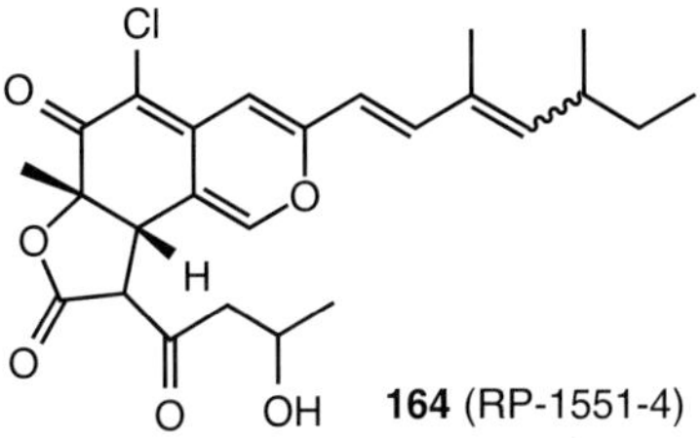

164 (RP-1551-4)

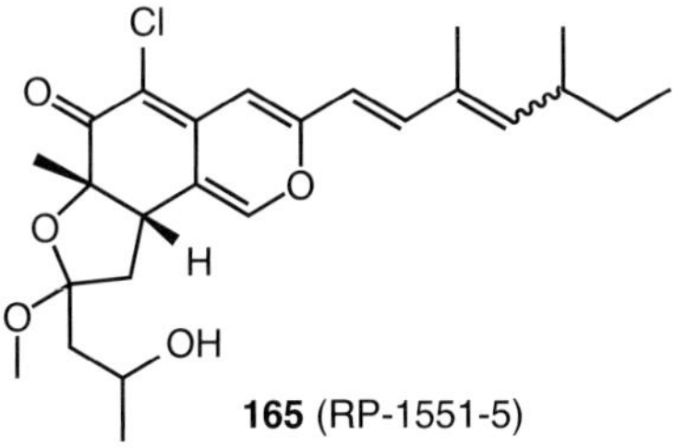

165 (RP-1551-5)

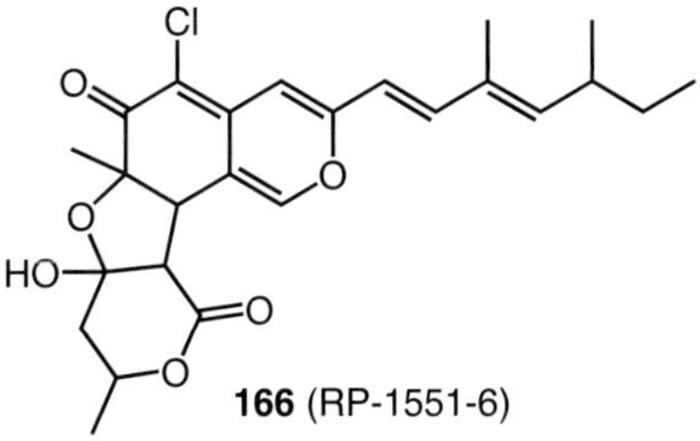

166 (RP-1551-6)

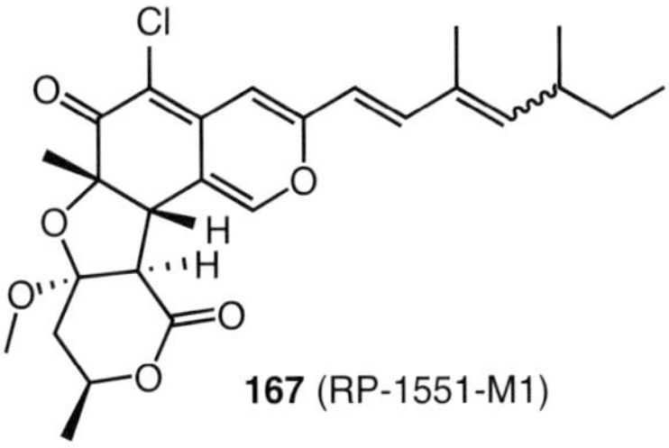

167 (RP-1551-M1)

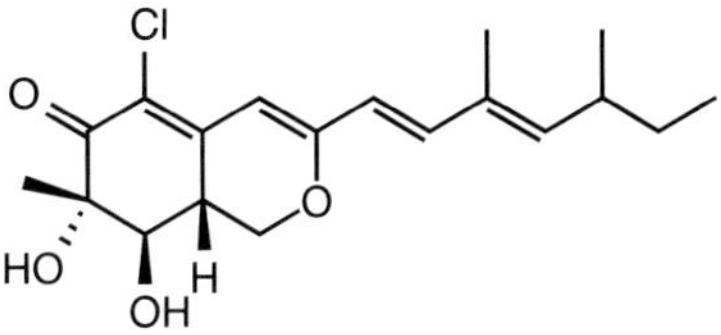

168 (RP-1551-M2)

The novel 8-*O*-methylsclerotiorinamine (**169**) was isolated from a strain of *Penicillium multicolor* and is a strong antagonist of the Grb2-SH2 domain (*454*). A *Fusarium* sp. fungus has yielded the cyclopeptide chlorofusin (**170**), which is a p53-MDM2 antagonist (*455*). The biosynthesis of chlorofusin involves an acetogenic origin coupled with an aminodecanoic acid piece (*456*). A total synthesis has established the absolute configuration (*2657*). Fluostatin E (**171**) is a minor member of the fluostatin family of metabolites from *Streptomyces* sp. (*457, 458*). The epoxide (fluostatin C) corresponding to **171** is also isolated; the authors cannot exclude fluostatin E as an artifact arising from HCl ring opening of fluostatin C.

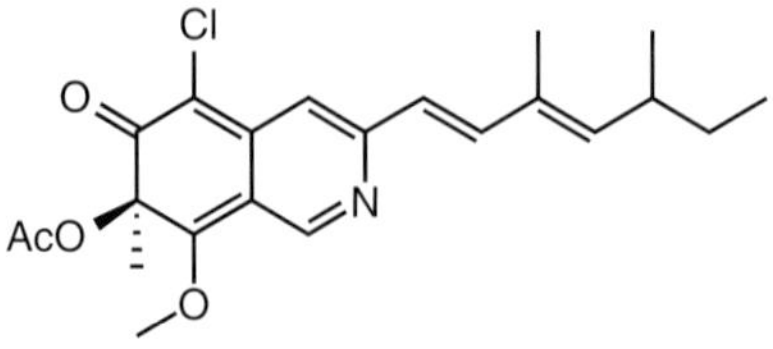

169 (8-*O*-methylsclerotiorinamine)

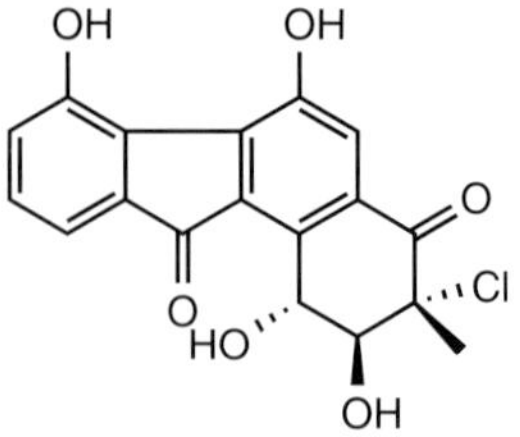

171 (fluostatin E)

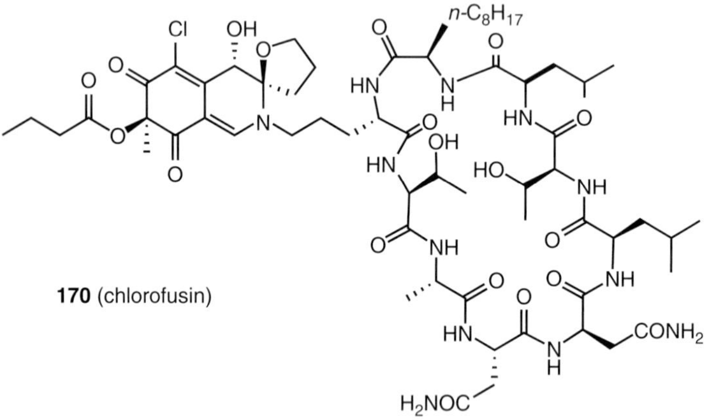

170 (chlorofusin)

3.4 Terpenes

The first survey of natural organohalogens documented 570 halogenated terpenes (*1*). The present update describes many additional new members of this important class of marine and terrestrial natural products.

3.4.1 Monoterpenes

3.4.1.1 Acyclic Monoterpenes

A review on halogenated monoterpenes has appeared (*459*). The structure of the previously known plocamenone (incorrectly named plocamenone A in (*1*)) from a *Plocamium* red alga has been revised to **G** (*460*). A collection of *Plocamium cartilagineum* from the Antarctic has yielded the new monoterpene **172** along with two known halogenated compounds (*461*). The previously isolated **H** (now named halomon) from *Portieria hornemannii* (*1*) has been fully characterized (*462*) and synthesized from myrcene (*463*). Halomon has broad range activity against human cancer cell lines and has undergone clinical evaluation. A subsequent study of this seaweed afforded isohalomon (**I**), which was isolated previously but with

3.4 Terpenes

undefined stereochemistry (*1*), and the new metabolites **173** and **174**. In addition, the stereochemistry of the known metabolites **J** and **K** was established (*464*). A Portuguese collection of *Plocamium cartilagineum* contains the novel **175** and **176** (*465*); the former is the first natural halogenated dimethyloctadiene with a (Z)-alkene.

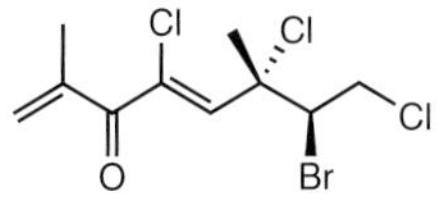

G (plocamenone) 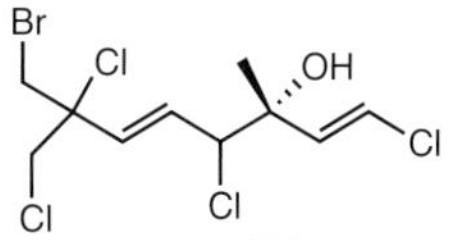   **172** 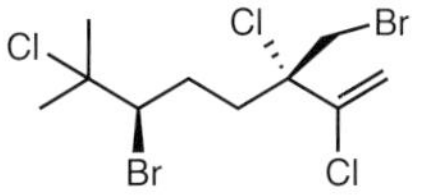   **H** (halomon)

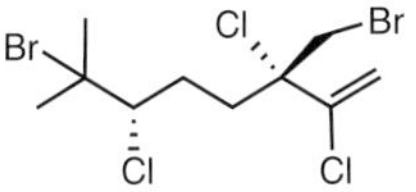

I (isohalomon) 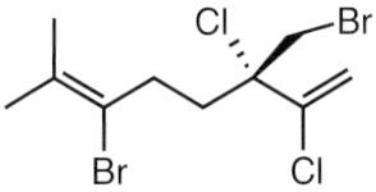   **173** 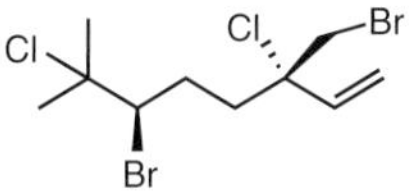   **174** 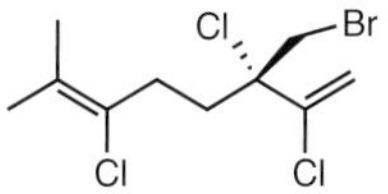   **J**

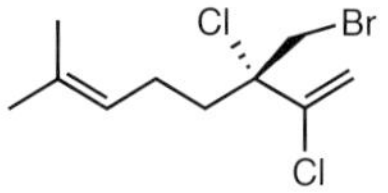

K 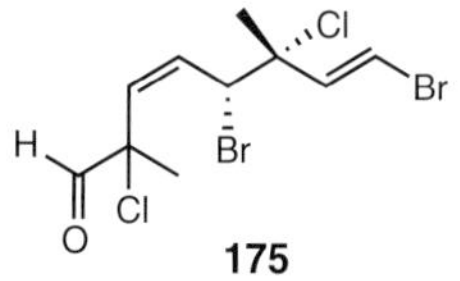   **175** 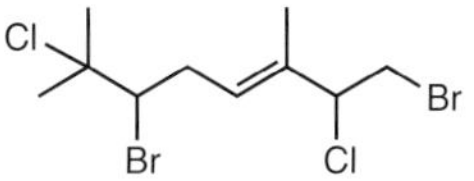   **176**

The Spanish sea hare *Aplysia punctata* contains the four novel acetates **177–180**, which are perhaps biotransformation products of dietary algae compounds (*466*). A study of the Antarctic red alga *Pantoneura plocamioides* uncovered the novel pantoneurotriols (**181, 182**) and **183**, in addition to establishing the stereochemistry of the previously isolated **172** (*467*). A Tasmanian collection of *Plocamium costatum* yielded the new **184** and three known halogenated monoterpenes, some of which deter barnacle larvae settlement (*468*). A detailed survey of six samples of *Plocamium hamatum* from the Great Barrier Reef identified 11 known halogenated monoterpenes, the occurrence of which differed significantly between locations. In this study the previously reported monoterpene **L** (*1*) was obtained for the first time in pure form (*469*). A collection of *Plocamium cartilagineum* from Tasmania uncovered two new polyhalogenated monoterpenes **185** and **186** (*470*).

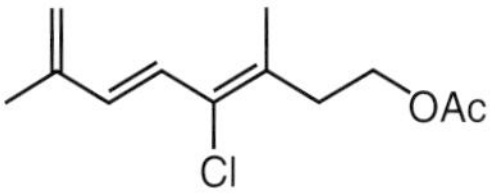

177 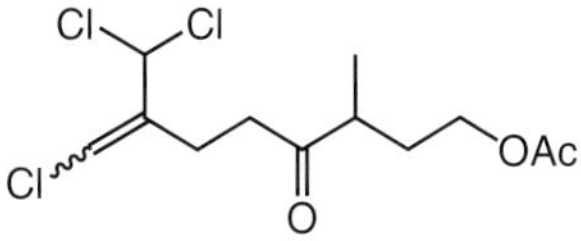   **178** (*Z*)   **180**
         179 (*E*)

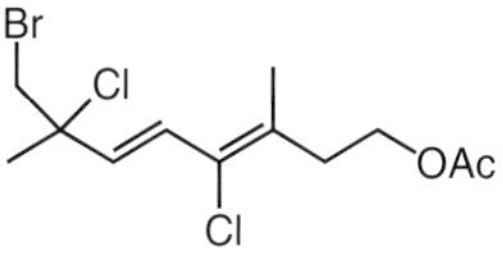

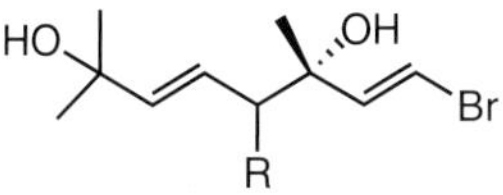

181 R = β-OH   **183** R = β-Cl   **184**
182 R = α-OH   **172** R = α-Cl

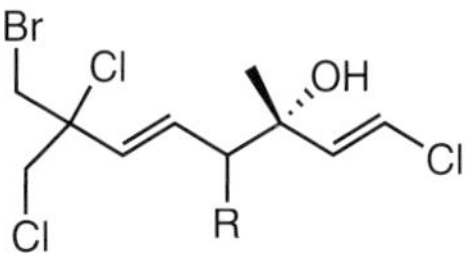

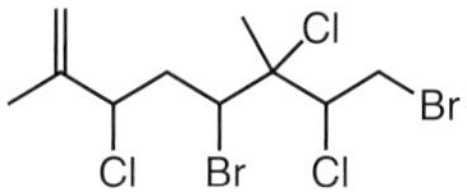

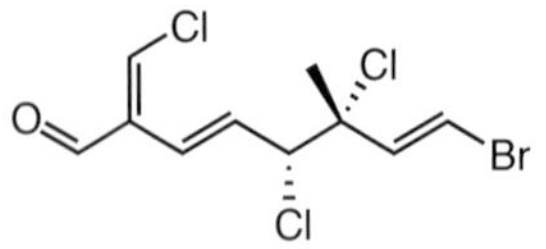

185 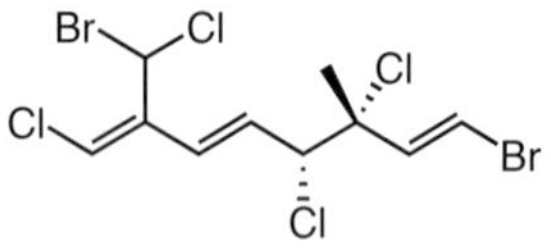 **186** 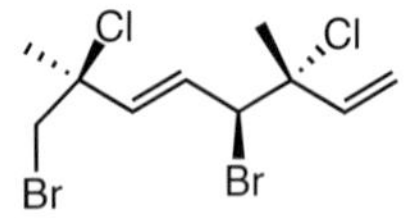 **L**

Three plocamenols A–C (**187–189**) were isolated from a Chilean collection of *Plocamium cartilagineum* (*471*), and this seaweed also yielded prefuroplocamioid (**190**) (*472*), **191** (*473*), **192**, and **193** (*474*). The alga *Plocamium corallorhiza* from Cape Town, South Africa, has afforded plocoralides A–C (**194–196**), which display some cytotoxicity toward esophageal cancer cells (*475*). Anverene (**197**) was found in an Antarctic collection of *Plocamium cartilagineum* (*476*). The Madagascar red alga *Portieria hornemannii* has yielded the new monoterpenes **198–200** along with halomon and two other known compounds (*477*). The known marine alga metabolite **M** (*1*) is found in several fish, monk seals, hooded seals, and harp seals, presumably as a bioaccumulative compound (*478*). It is also present in Norwegian predatory bird eggs (white-tailed eagle, osprey, goshawk, golden eagle, and merlin) (*479*).

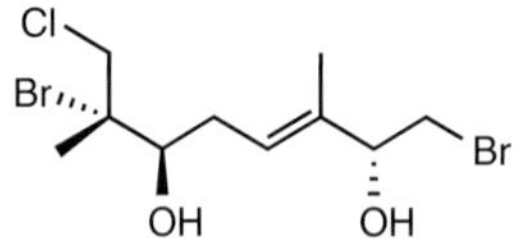

187 (plocamenol A) 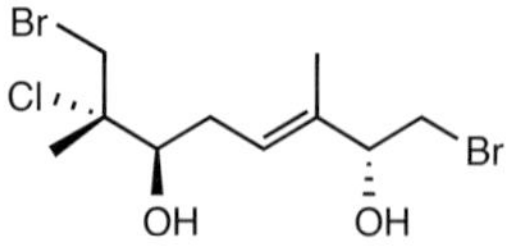   **188** (plocamenol B) 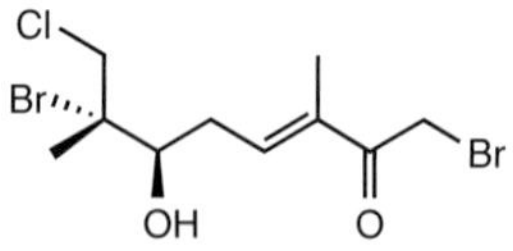   **189** (plocamenol C)

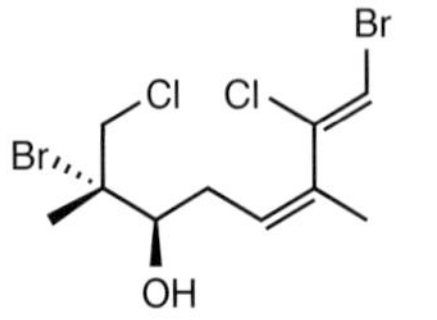

190 (prefuroplocamioid) 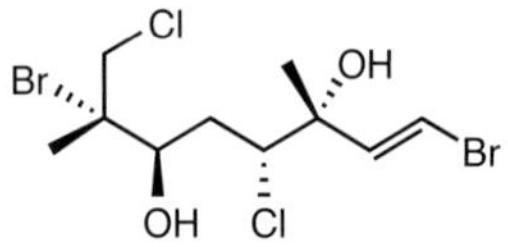   **191** 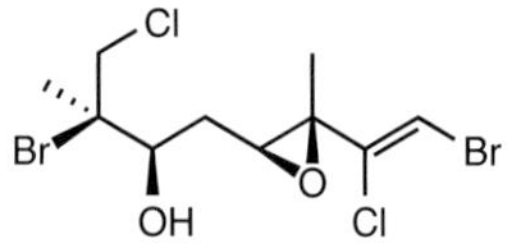   **192**

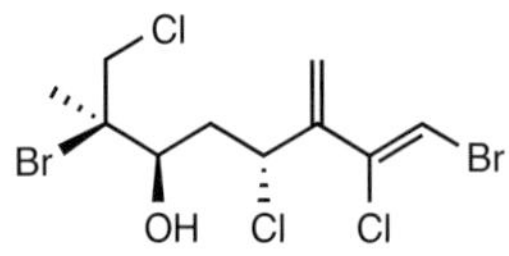

193 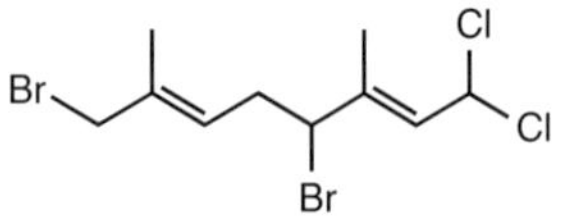   **194** (plocoralide A) 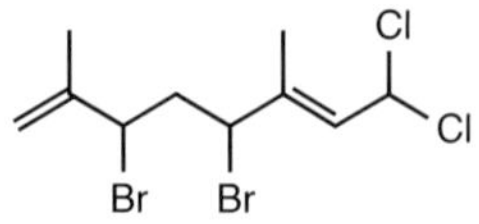   **195** (plocoralide B)

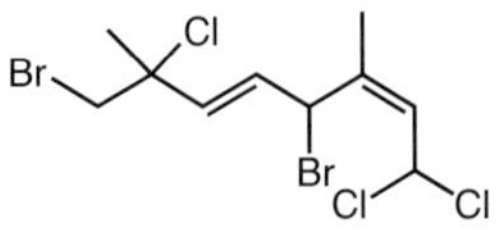

196 (plocoralide C) 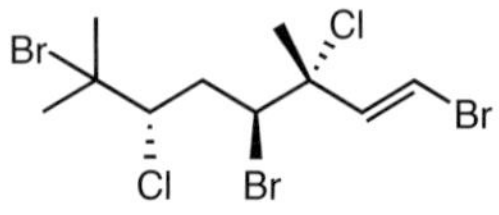   **197** (anverene) 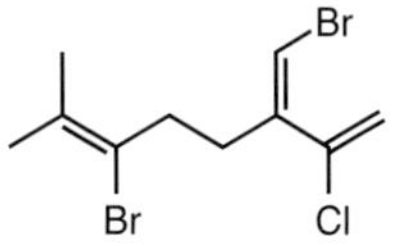   **198**

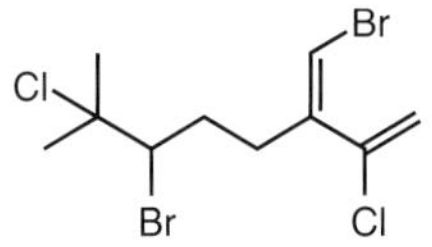

3.4 Terpenes

199, 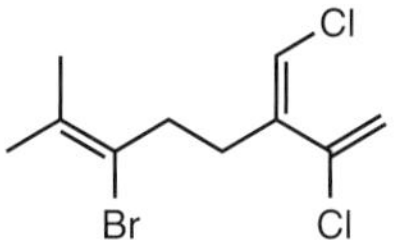 **200**, 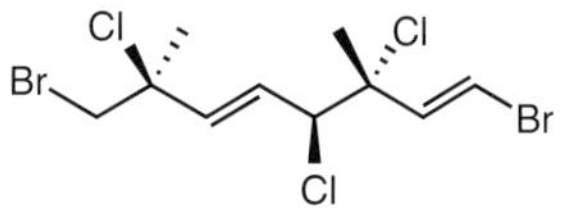 **M (or isomer)**

Many polyhalogenated monoterpenes have potent biological activity (*1*). In addition to cytotoxic activity, several compounds display insect repellent and antifeedant activity, and selective insect cell toxicity (*480*). To acquire sufficient quantities of these and other target metabolites for biological evaluation, the laboratory cultivation of marine algae – "bioprocess engineering" – is under intense exploration (*481–483*).

3.4.1.2 Alicyclic Monoterpenes

The halogenated acyclic marine monoterpenes are often considered to be the biogenetic precursors of the alicyclic monoterpenes that are presented in this section. Many of the preceding algae species also contain cyclic monoterpenes. As was the case in preceding sections only newly characterized compounds are numbered and the reader is referred to the first survey for structures of previously isolated compounds (*1*).

The halomon-containing red alga *Portieria hornemannii* contains the new cyclic trichloro metabolite **201** (*462*), and another study of this seaweed has furnished **202–205** (*464*). The prolific *Plocamium cartilagineum* from the Portuguese coast produces epoxide **205**, which is the first natural alicyclic polyhalogenated epoxy-monoterpene to be isolated (*484*). The absolute configuration of the previously known **N** (*1*) was determined (*469*). A Guam collection of *Portieria hornemannii* contains the novel apakaochtodenes A (**206**) and B (**207**) (*485*), which are effective feeding deterrents toward herbivorous reef fish (*486*). The Japanese red seaweed *Carpopeltis crispata* contains the new ochtodenes **208–212** (*487*). Several known antifeedant alicyclic monoterpenes are found in the sea hare *Aplysia punctata* that are apparently diet derived and used for chemical defense (*488*). An alicyclic polyhalogenated monoterpene (as yet unidentified) has been detected in fish, seal, and birds (*478, 479, 489*). The known monoterpenes mertensene and violacene, and some synthetic derivatives, found in *Plocamium cartilagineum* have insecticidal activity (*490*).

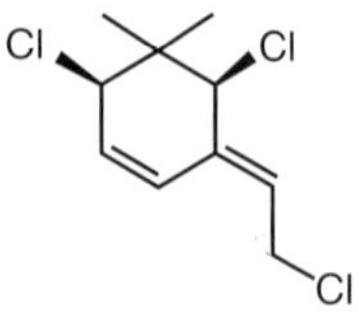

201 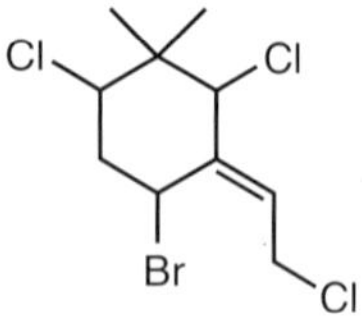 **202** 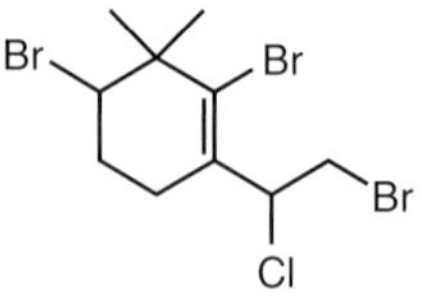 **203** 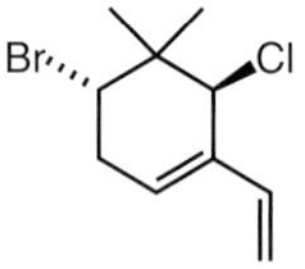 **204**

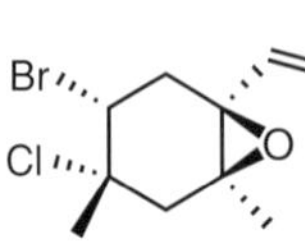

205 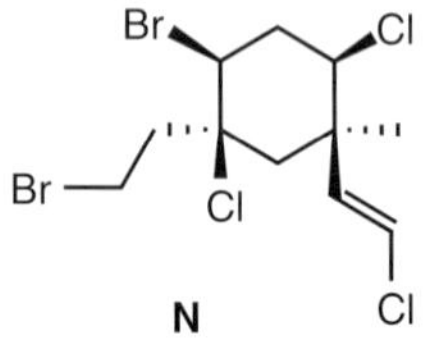 **N** 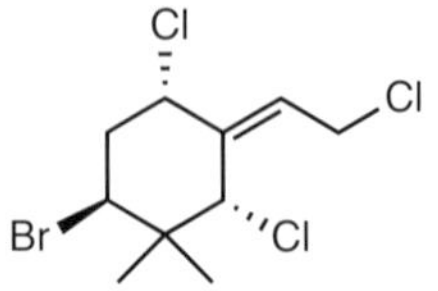 **206** (apakaochtodene A)

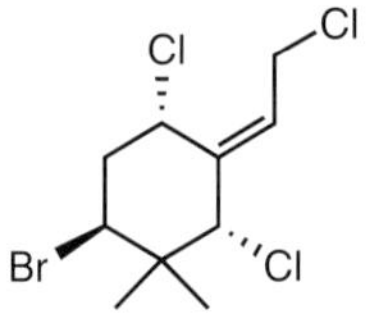

207 (apakaochtodene B) 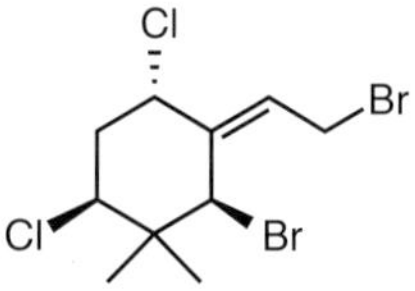 **208** 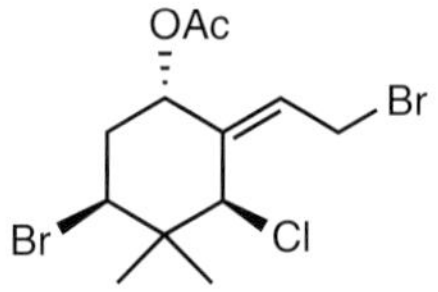 **209**

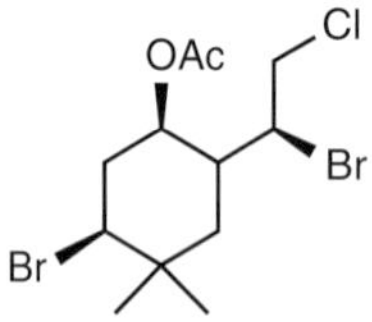

210 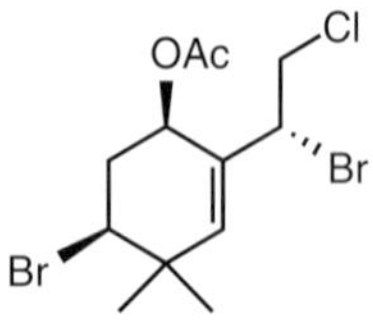 **211** 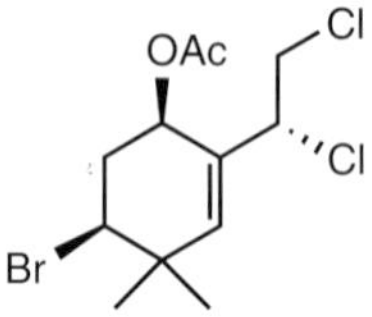 **212**

Two different Okinawan collections of *Portieria hornemanni* have yielded the novel cyclohexadienones **213–217** (*491*). The previously known *Portieria hornemanni* metabolite **O** has been characterized from the cyanobacterium *Lyngbya majuscula* (*492*). Several cyclic monoterpene ethers are seaweed metabolites, presumably derived by cyclization of a proximate hydroxyl group (i.e., "neighboring group participation"). For example, the Antarctic *Pantoneura plocamioides* has yielded pantofuranoids A–F (**218–223**) (*493*) and pantoisofuranoids A–C (**224–226**) (*494*). The Antarctic *Plocamium cartilagineum* contains furoplocamioids A–C (**227–229**), which possess the unusual bromochlorovinyl moiety (*495*), and also **230a** and **230b** (*473*).

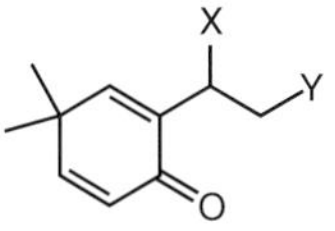

3.4 Terpenes

213 X = Y = Br
214 X = Cl, Y = Br
215 X = Y = Cl

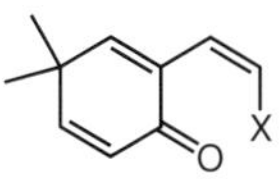

216 X = Br
217 X = Cl

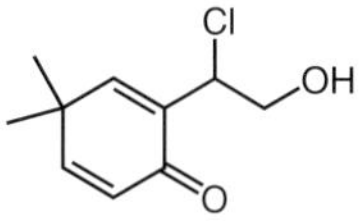

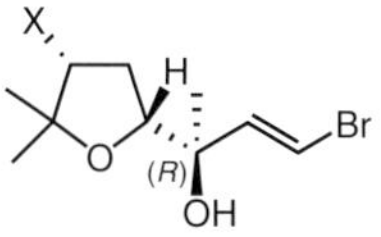

218 X = Br (pantofuranoid A)
222 X = OH (pantofuranoid E)

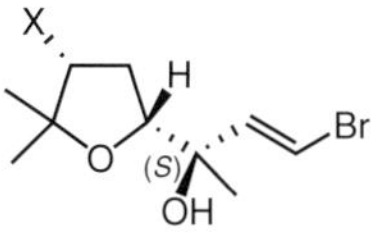

219 X = Br (pantofuranoid B)
221 X = OH (pantofuranoid D)

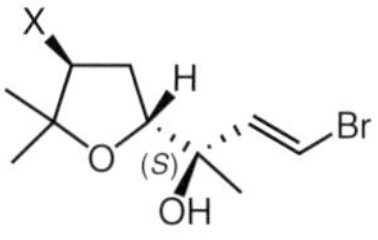

220 X = Br (pantofuranoid C)
223 X = OH (pantofuranoid F)

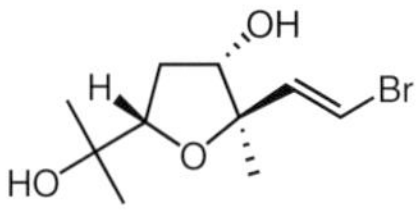

224 (pantoisofuranoid A)

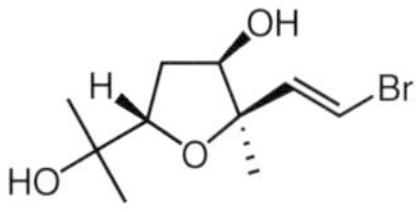

225 (pantoisofuranoid B)

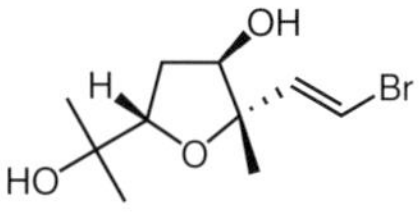

226 (pantoisofuranoid C)

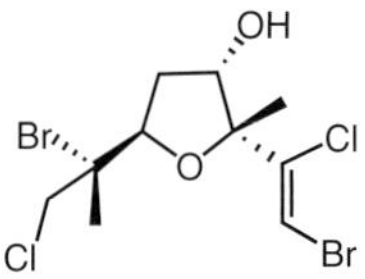

227 (furoplocamioid A)

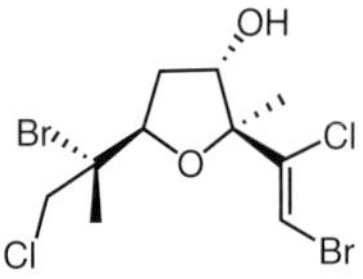

228 (furoplocamioid B)

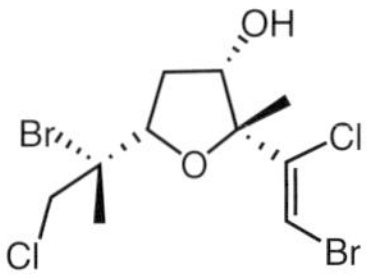

229 (furoplocamioid C)

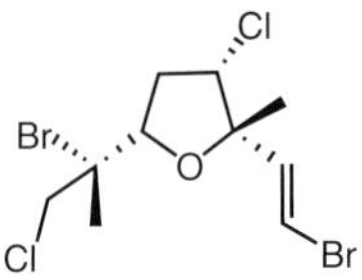

230a

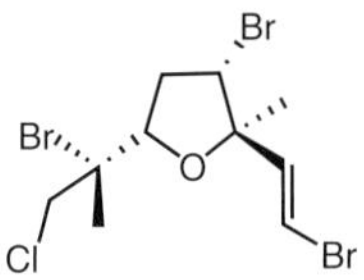

230b

The pyran metabolites, pantopyranoids A–C (**231**–**233**), have been isolated from the Antarctic alga *Pantoneura plocamioides* (*494*). This seaweed and *Plocamium cartilagineum* contain plocamiopyranoid (**234**) and **235**, and pantoneurines A (**236**) and B (**237**) (*496*). The Pakistani herb *Mentha longifolia* has yielded the novel chlorinated menthone longifone **238** (*497*), one of the few known terrestrial halogenated monoterpenes.

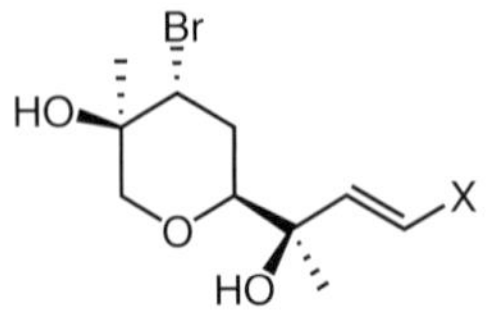

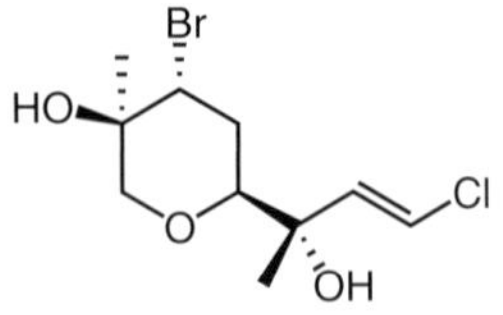

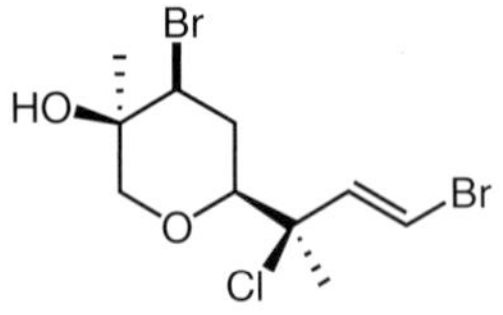

231 X = Br (pantopyranoid A)
233 X = Cl (pantopyranoid C)
232 (pantopyranoid B)
234 (plocamiopyranoid)

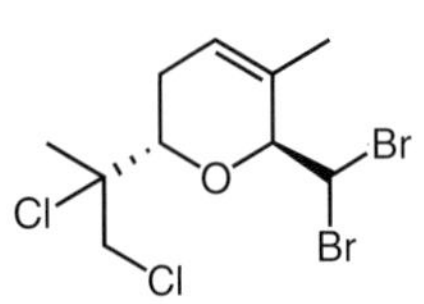

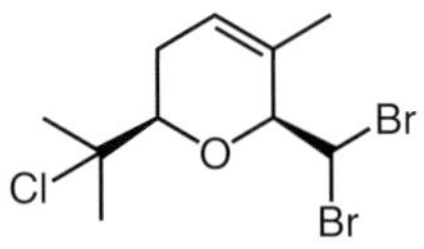

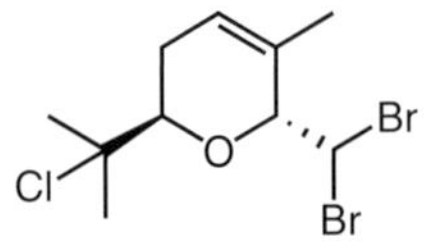

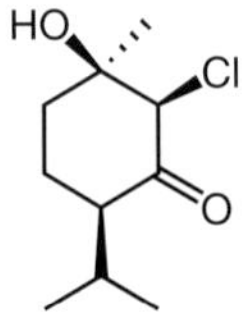

235
236 (pantoneurine A)
237 (pantoneurine B)
238

3.4.2 Sesquiterpenes

In addition to the myriad natural marine halogenated sesquiterpenes (vide infra), the terrestrial plant kingdom is also a major source of halogenated (chlorinated) sesquiterpenes, most of which possess the guaianolide skeleton (*498*).

3.4.2.1 Terrestrial Sesquiterpene Lactones

Although the first survey listed 45 natural chlorinated sesquiterpene lactones, several such compounds were omitted in that coverage (*1*) and are described here. The novel sesquiterpene lactone chlorochrymorin (**239**) was isolated from *Chrysanthemum morfolium* (*499*), and the chlorohydrin graminichlorin (**240**) is found in *Liatris graminifolia* (*500*). The antibacterial AA-57 (**241**), which is related to pentalenolactone, is produced by a *Streptomyces* sp. (*501*). The plant *Eupatorium chinense* var. *simplicifolium* has yielded eupachifolin D (**242**) (*502*) (side-chain double bond stereochemistry revised (*518*)), and the new guaianolide andalucin (**243**) was characterized from *Artemisia lanata* (*503*). The previously known chlorohyssopifolins (*1*) have been studied for cytostatic activity, and the presence of one and even two chlorine atoms amplifies this activity (*504*).

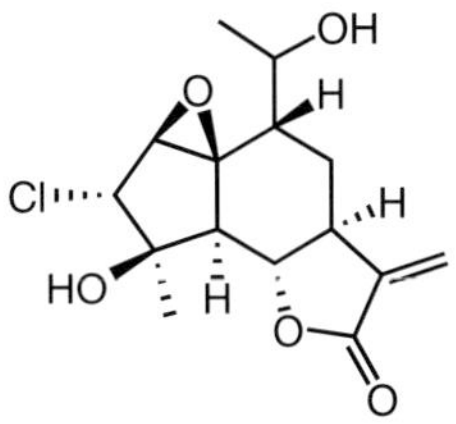

3.4 Terpenes

239 (chlorochrymorin)

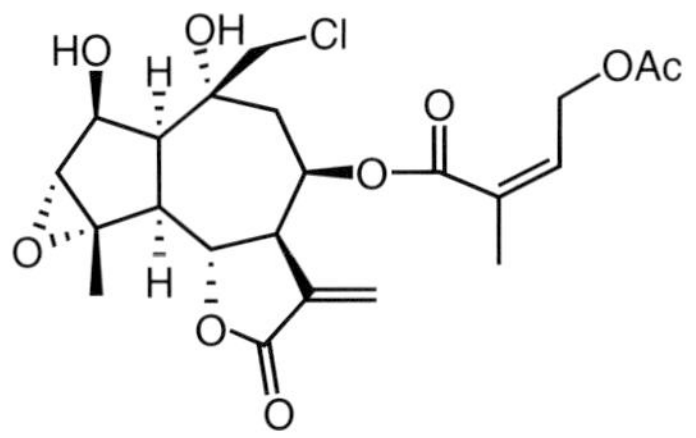

240 (graminichlorin)

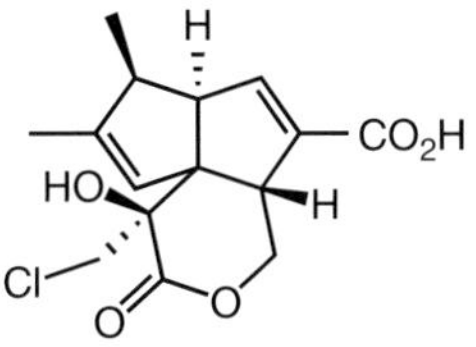

241 (AA-57)

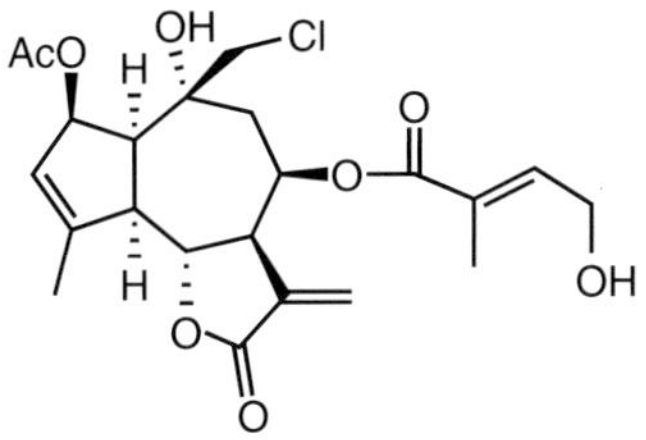

242 (eupachifolin-D)

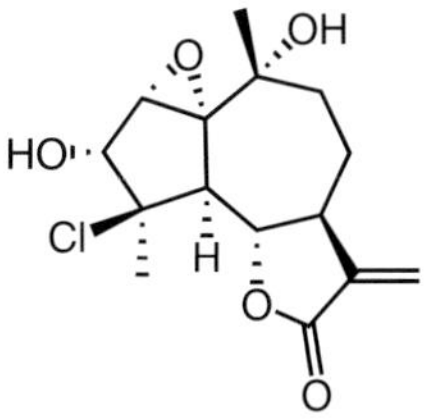

243 (andalucin)

It should be noted that at least some of these chlorohydrin sesquiterpenes could be artifacts (505) formed by epoxide ring opening during isolation. It is essential that acid and acid-forming reagents (e.g., $CHCl_3$) be avoided during isolation of these compounds when epoxides might also be present, and that investigators be cognizant of this potential problem. Unless otherwise indicated, the following new compounds were isolated in the absence of chlorinated solvents.

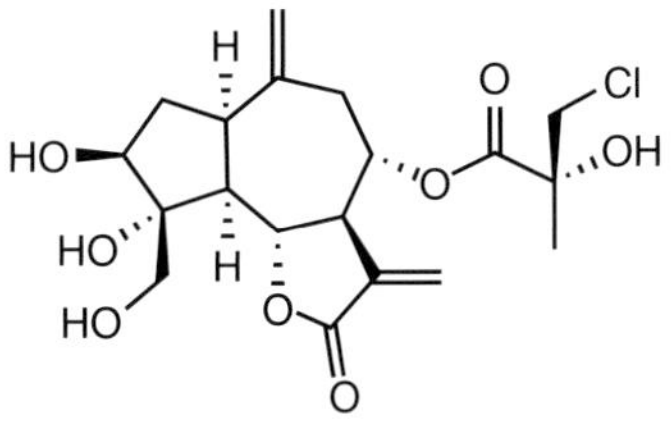

244 (epicebellin J)

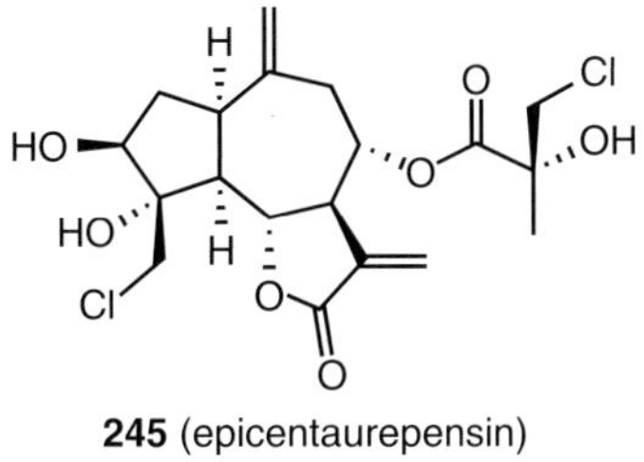

245 (epicentaurepensin)

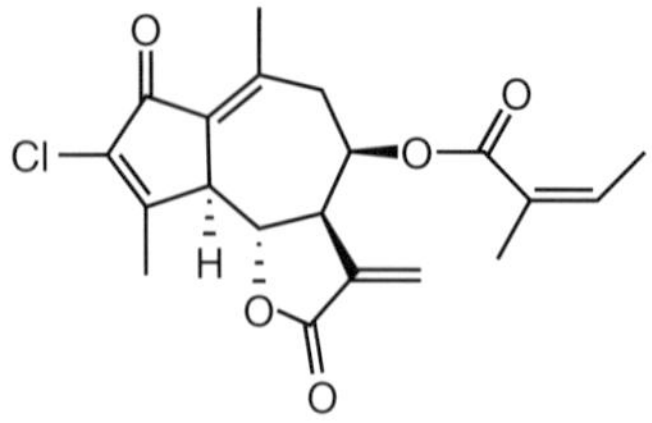

246

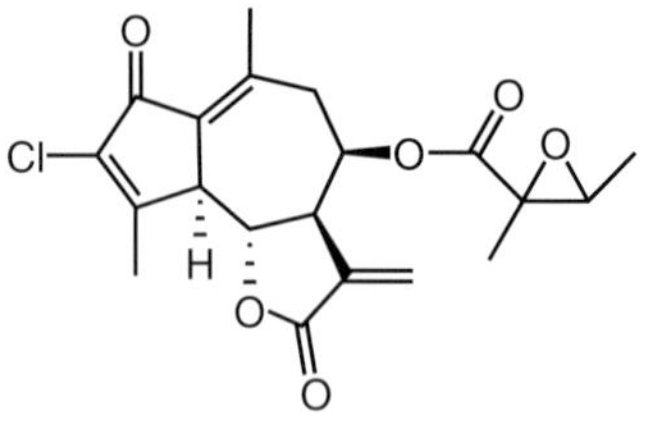

247

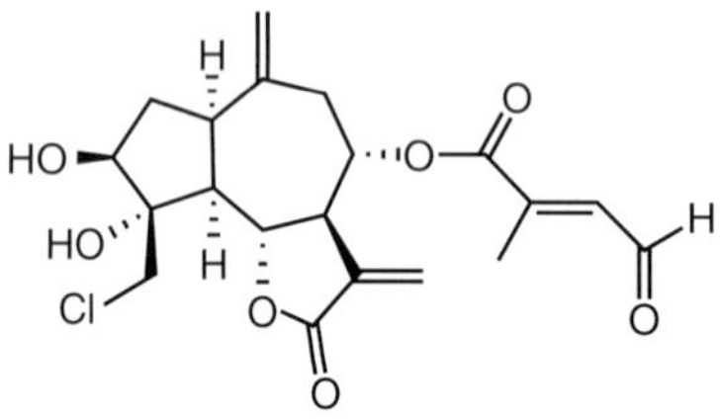

248

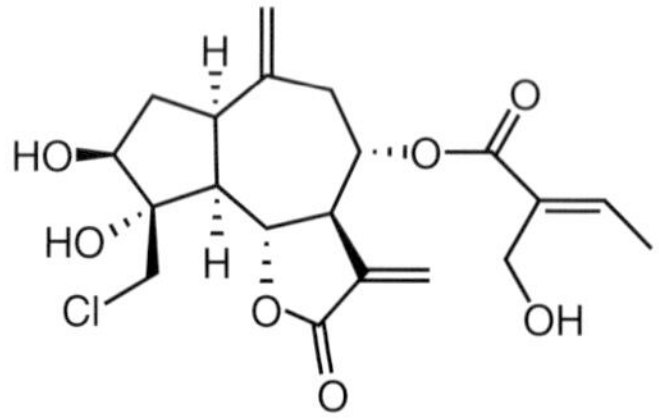

249

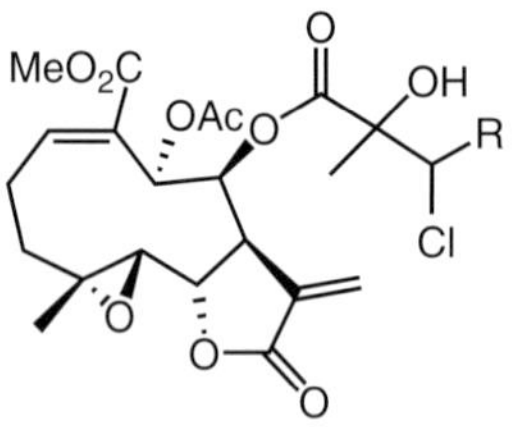

250 R = H
251 R = Me

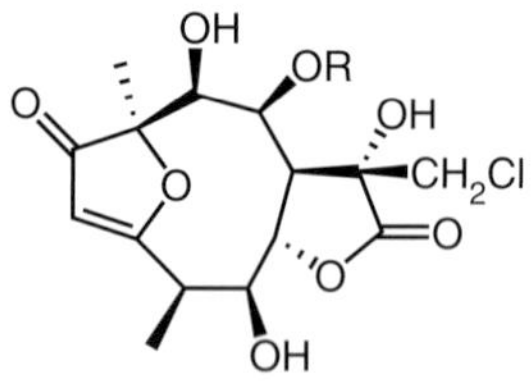

252 R = tiglyl
253 R = angelyl

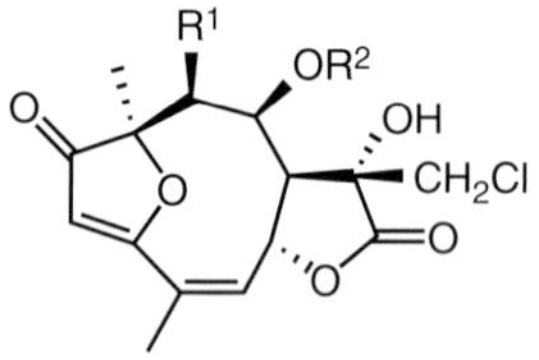

254 R^1 = OH, R^2 = tiglyl
255 R^1 = H, R^2 = 5-OAc-tiglyl

The new epicebellin J (**244**) was isolated from *Centaurea glatifolia* along with several known guaianolides (*506*). *Centaurea conifera* has yielded the C-17 epimer of the previously known chlorohyssopifolin A (centaurepensin) (**245**), and the previously described chlorohyssopifolin A and chlorojanerin (*507*). The high altitude Argentinean plant *Stevia sanguinea* contains the new **246** and **247** (*508*). Another study of *Centaurea scoparia* has identified the new **248** and **249** (*509*). The eastern India medicinal plant *Enhydra fluctuans* has yielded two new chlorinated melampolides **250** and **251** (*510*). The South American plant *Bejaranoa balansae* contains the novel furanoheliangolides **252–254**, and *Bejaranoa semistriata* has afforded **255** (*511*).

The plant *Achillea clusiana* from the mountains of Bulgaria contains the new 2-*epi*-chloroklotzchin (**256**), which is the first report of a halogenated sesquiterpene lactone from *Achillea* genus (*512*). Chloroform was used to process the plant. The Egyptian medicinal plant *Ambrosia maritima*, which is still used to treat renal colic and other aliments, has afforded 11β-hydroxy-13-chloro-11,13-dihydrohymenin (**257**) (*513*). Eupaglehnins E (**258**) and F (**259**) are novel germacranolides isolated

3.4 Terpenes

from the Japanese plant *Eupatorium glehni* (*514, 515*). In addition to containing the new guaianolide **260**, a Greece collection of *Achillea ligustica* has uncovered the seco-tanapartholide **261** (*516*). The first chlorinated sesquiterpene lactone glucoside to be isolated is **262**, 13-chloro-3-*O*-β-D-glucopyranosylsolstitialin, from *Leontodon palisae* (*517*).

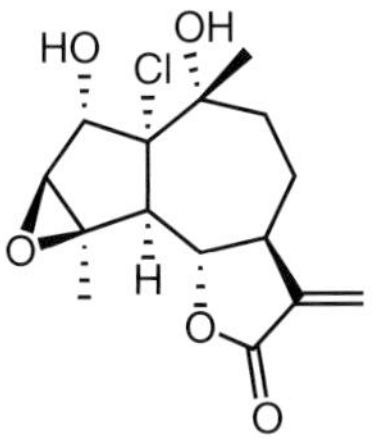

256 (2-*epi*-chloroklotzchin) 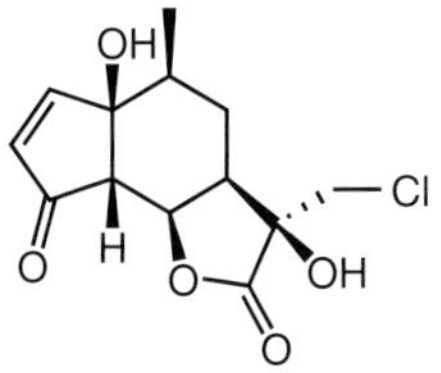
257 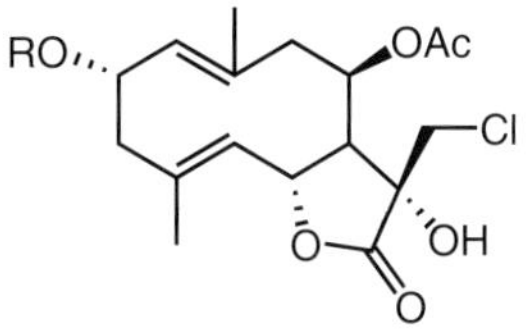
258 R = H (eupaglehnin E)
259 R = Ac (eupaglehnin F)

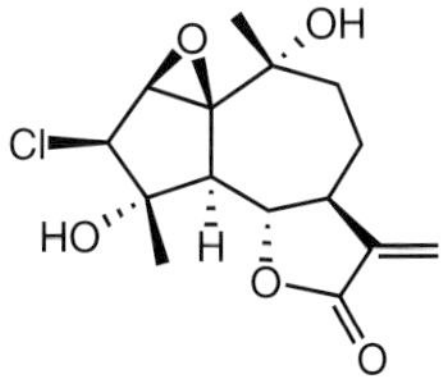

260
261 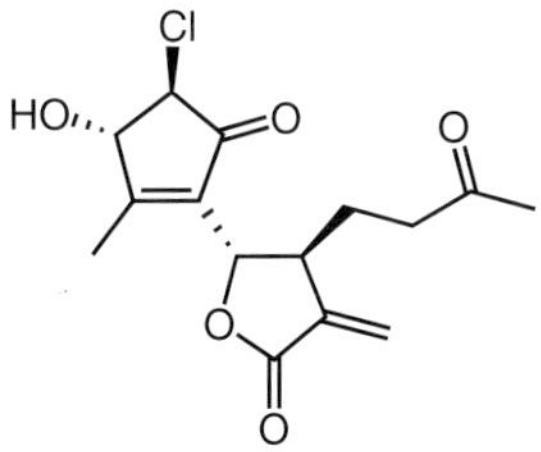
262 R = β-D-glucose 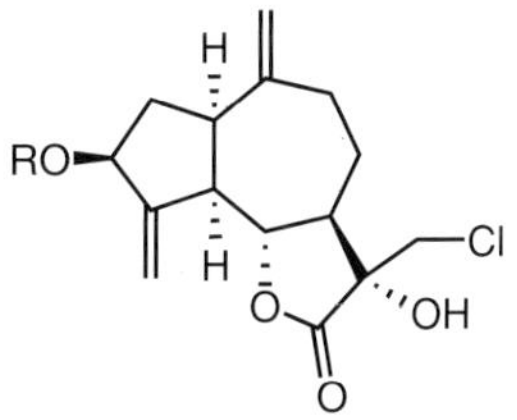

The Chinese *Eupatorium chinense* has afforded ten new sesquiterpenoids, three of which are chlorinated, eupachinilides C (**263**), E (**264**), and F (**265**) (*518*). The Chinese medicinal plant *Eupatorium lindleyanum* contains the chlorinated guaianes eupalinilides A (**266**), D (**267**), E (**268**), and H (**269**), amongst other non-chlorinated eupalinilides and nine known sesquiterpenoids (*519*). The Oregon coastal perennial plant *Artemisia suksdorfii* contains four novel chlorinated sesquiterpene lactones **270**–**273** (*520*).

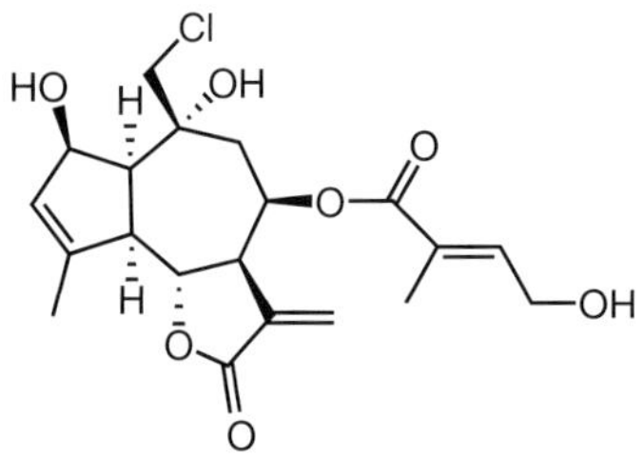

263 (eupachinilide C)
264 R = H (eupachinilide E) 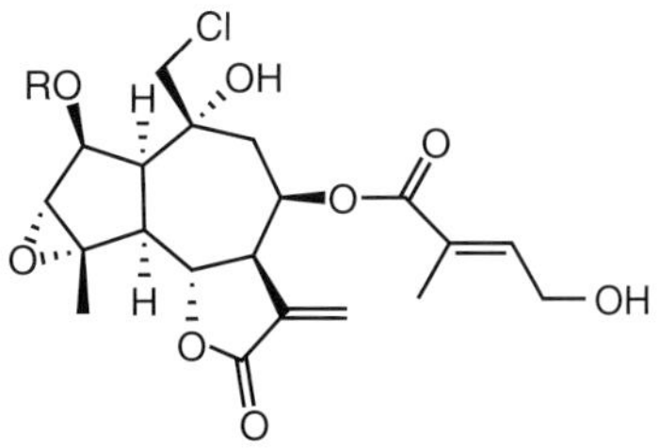
265 R = Ac (eupachinilide F)

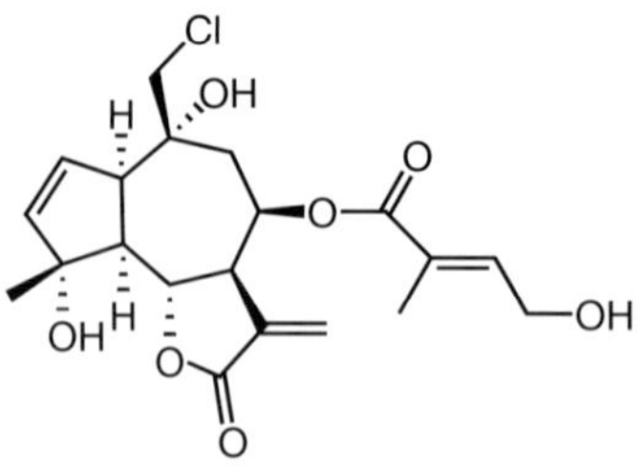

266 (eupalinilide A)

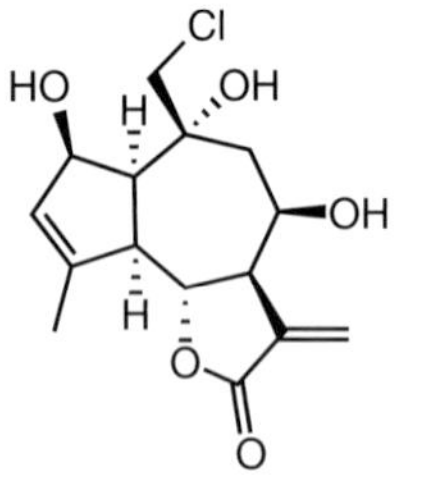

267 (eupalinilide D)

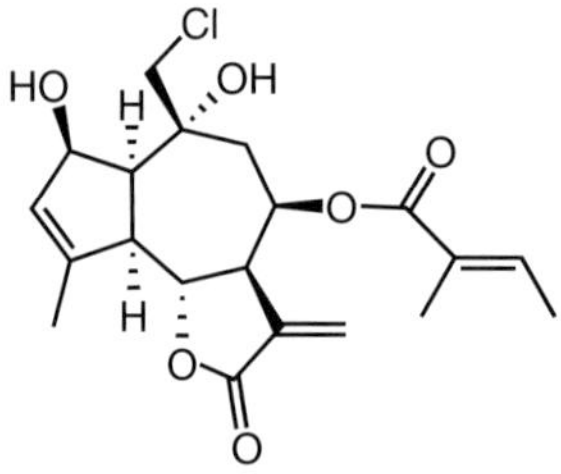

268 (eupalinilide E)

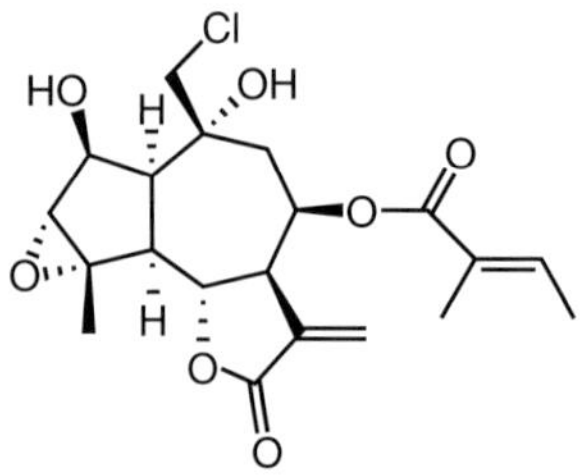

269 (eupalinilide H)

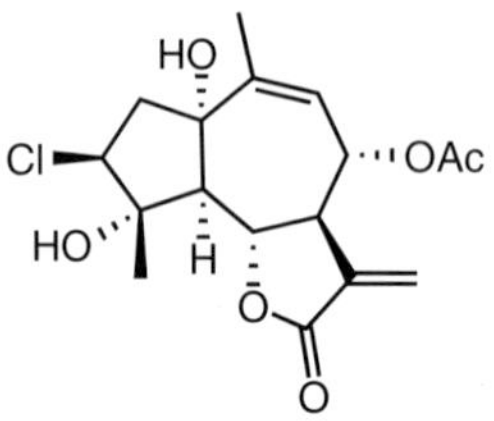

270

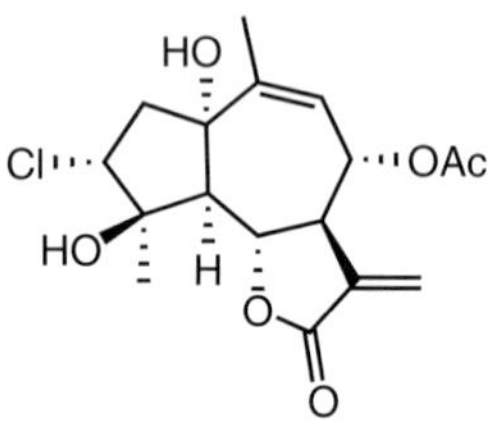

271

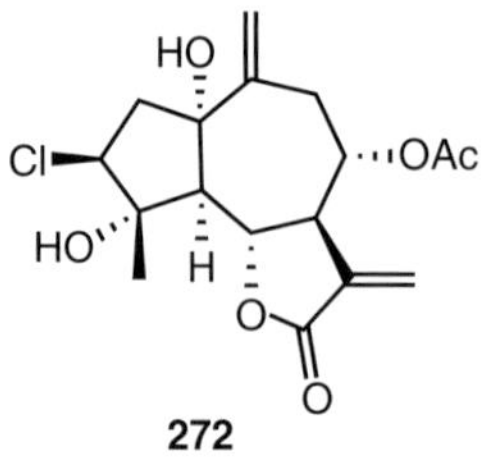

272

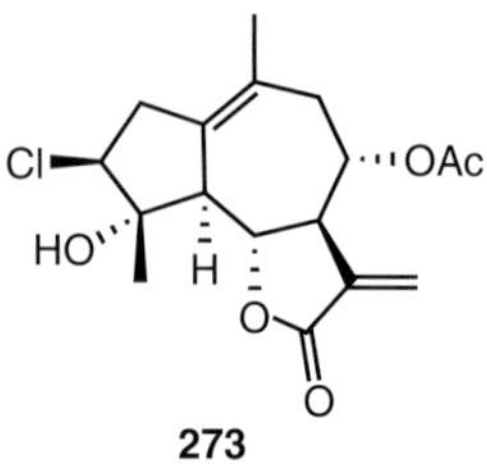

273

The Balkan Peninsula plant *Achillea depressa* contains the previously discussed **260** and its novel hydroxy derivative **274** (*521*), which is apparently a diastereomer of the known bibsanin (*1*). *Centaurea acaulis* from Algeria has afforded 14-chloro-10β-hydroxy-10(14)-dihydrozaluzanin D (**275**) (*522*). The widely distributed medicinal herbaceous perennial plant *Cynara scolymus* contains the new cynarinin B (**276**) as one of nine related sesquiterpenoids (*523*).

3.4 Terpenes

Examination of the Montenegro *Achillea clavennae* reveals the presence of three new chlorine-containing guaianolides **277–279** in addition to several known analogues (*524*). The first investigation of the Chinese medicinal plant *Vernonia chinensis* has uncovered the new chlorinated sesquiterpene lactones vernchinilides A (**280**), B (**281**), C (**282**) and E (**283**) (*525*). Vernchinilides B and E exhibit potent cytotoxic activity against the P-388 and A-549 cell lines. The structurally similar vernolide C (**284**) was found in the Cambodian traditional medicinal plant *Vernonia cinera* (fever, colic, malaria) (*526*). Indeed, vernolide C could be identical with vernchinilide A.

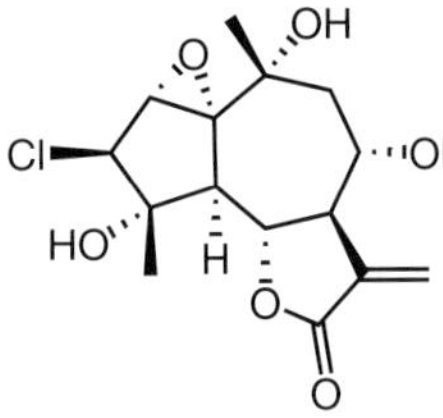

274

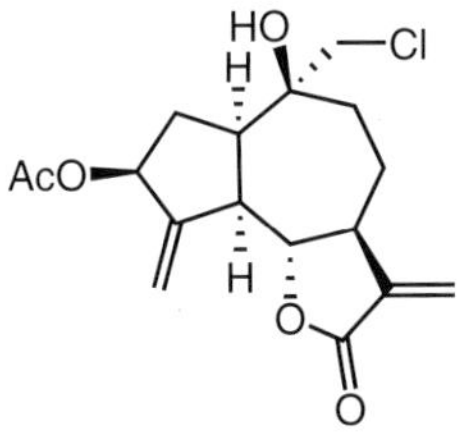

275

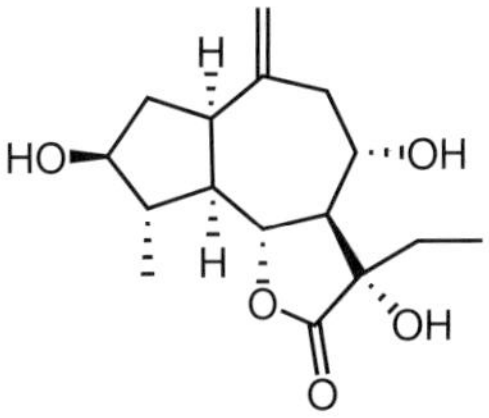

276 (cynarinin B)

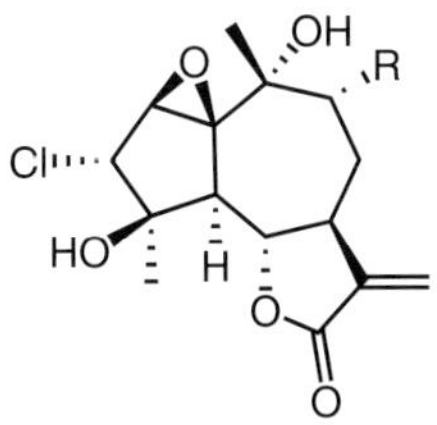

277 R = H
278 R = OAc

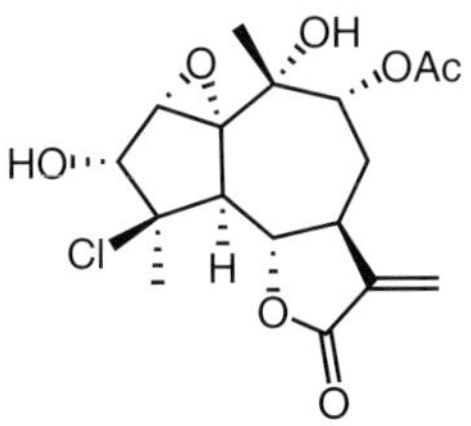

279 (9α-acetoxyanadalucin)

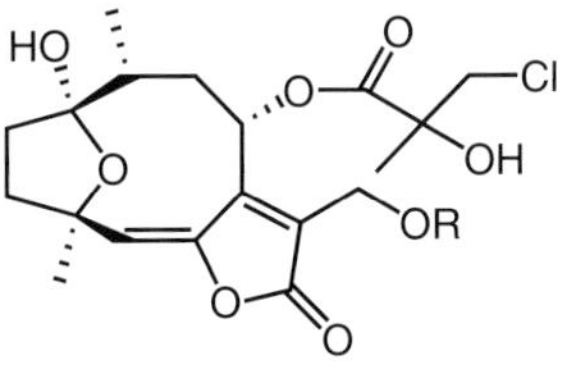

280 R = Ac (vernchinilide A)
282 R = H (vernchinilide C)

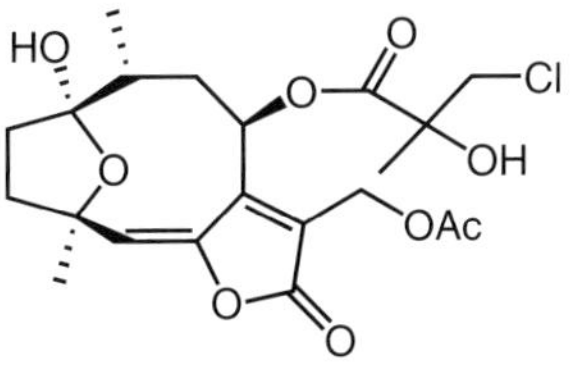

281 R = Ac (vernchinilide B)

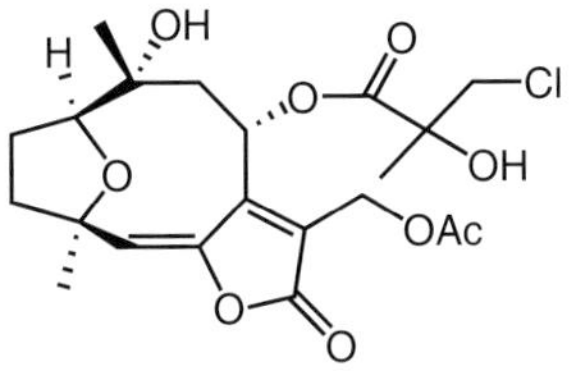

283 R = Ac (vernchinilide E)

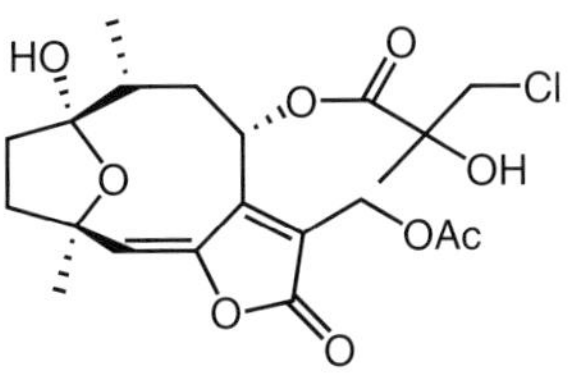

284 (vernolide C)

3.4.2.2 Indanone Sesquiterpenes

A review on the isolation, chemistry, and biochemistry of the bracken fern (*Pteridium aquilinum*) carcinogen ptaquiloside (**P**) has been published (*527*). This fern and others contain the pterosins that were summarized previously (*1*). No new examples were reported in the interim.

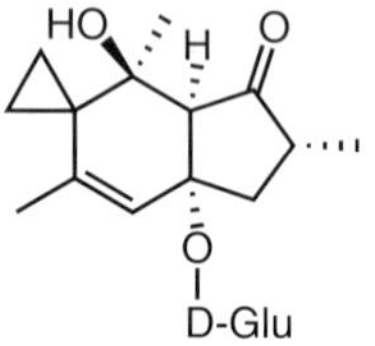

P (ptaquiloside)

3.4.2.3 Other Terrestrial Sesquiterpenes

The toxic plant *Illicium tashiroi* is the source of many novel sesquiterpenes and one new chlorine derivative, 12-chloroillifunone C (**285**) (*528*). A subsequent investigation of this plant revealed 12-chloroillicinone E (**286**) and (2*R*)-12-chloro-2,3-dihydroillicinone E (**287**) (*529*). The latter metabolite increases choline acetyltransferase activity and thus may find use in the treatment of Alzheimer's disease. The common Pakistani weed *Pluchea arguta* has yielded 3,4-di-*epi*-3′-chloro-2′-hydroxyarguticinin (**288**) (*530*), which is a diastereomer of a compound previously reported in this plant (*1*), and *Pluchea carolonesis* from Haiti contains the eudesmane **289** (*531*). Five novel chlorinated bisabolanes **290**–**294** were characterized from the Himalayan plant *Cremanthodium discoideum*, the genus of which is used as a Tibetan traditional herbal medicine for the treatment of fever, pain, inflammation, and other ailments. Compound **290** shows antibacterial activity against *Bacillus acidilatici* and *Bacillus subtilis* (*532*). The related bisabolane **295** was isolated from the roots of *Ligularia cymbulifera* (*533*). As with the previous studies in this section, no chloroform or HCl was employed in the isolation process, which might otherwise convert the corresponding epoxides to these chlorohydrins. The fungus *Phomopsis* sp., which was found growing on the plant *Adenocarpus foliolosus*, produces the sesquiterpene acid **296** (*534*).

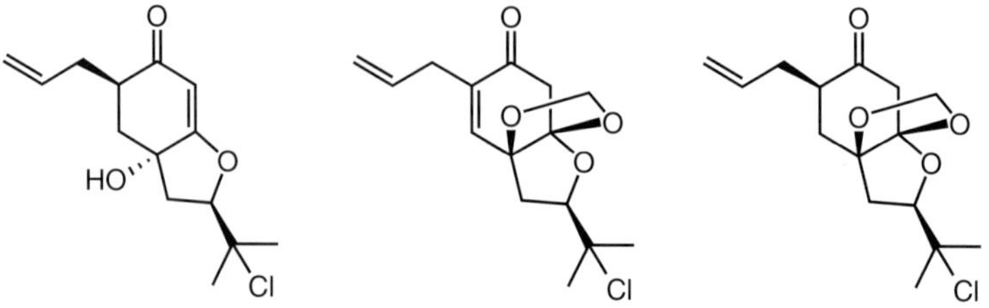

285 (12-chloroillifunone C)   **286** (12-chloroillicinone E)   **287**

3.4 Terpenes

Fig. 3.2 *Laurencia subopposita*, an example of the widespread red alga genus *Laurencia* (Photo: W. Fenical)

3.4.2.4 Marine Sesquiterpenes

The first survey covered more than 200 halogenated marine sesquiterpenes (*1*), and these extraordinarily structurally diverse natural products continue to be discovered from marine organisms. The red algal genus *Laurencia* (*Rhodomelaceae*, *Ceramiales*) is a large genus comprising at least 140 species distributed throughout the world's oceans, but mainly in warm waters. *Laurencia* is a treasure trove of halogenated (i.e., brominated) metabolites, and the morphology of this genus is of great interest (Fig. 3.2) (*535–538*).

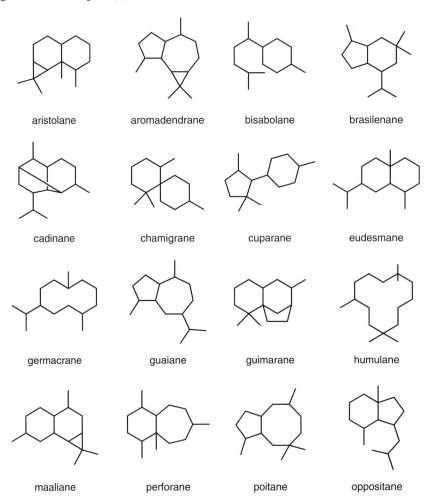

Sesquiterpene skeletons from *Laurencia* spp. (*553*).

Scheme 3.2

3.4 Terpenes

The diversity of ring systems found in *Laurencia* is shown in Scheme 3.2 (*553*), although the present organization of marine sesquiterpenes does not follow these categories, but rather continues the previous one (*1*).

Monocyclic and Other Simple Sesquiterpenes

The new red algal species *Laurencia mariannensis* from the Great Barrier Reef provides the novel sesquiterpene **297**, along with the known pacifenol and deoxy-prepacifenol, which are now fully characterized by NMR for the first time (*539*). The Philippine *Laurencia majuscula* has furnished 13 novel halogenated sesquiterpenes **298–310**, of which the major components are the majapolenes A (**298**, **299**) (two diastereomers), which are also found in *Laurencia caraibica* (*540*). Most of these compounds occur as inseparable diastereomers. A collection of *Laurencia majuscula* from the South China Sea has yielded the cedrene-type sesquiterpene majusin (**311**) (*541*). A new sesquiterpene dichloroimine, stylotellane A (**312**), was isolated from the sponge *Stylotella aurantium* (Fig. 3.3) (*542*).

Feeding experiments with carbon-14 reagents revealed the incorporation of both cyanide and thiocyanate into **312**. Three new dichloroimines were isolated from the sponge *Axinyssa* sp., axinyssimides A–C (**313–315**), and possess strong larval settlement inhibitory activity against the infamous barnacle *Balanus amphitrite* (*543*). The Australian sponge *Ulosa spongia* has furnished the new carbonimide dichlorides ulosins A (**316**) and B (**317**) (*544*), and **316** was isolated independently from the sponge *Stylotella aurantium* along with the new **318** (*545*). This sponge has also yielded the novel stylotellane D (**319**) (*546*). Biosynthetic labeling studies have been performed with these dichloroimine sesquiterpenes (*542, 547, 548*).

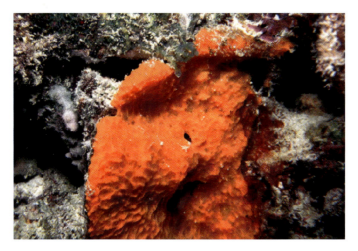

Fig. 3.3 *Stylotella aurantium*, a sponge that produces the dichloroimine stylotellane A (**312**) (Photo: A. Flowers)

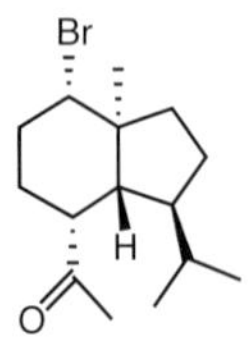

297

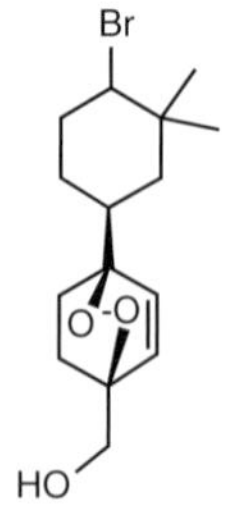

298, 299 (majapolene A)

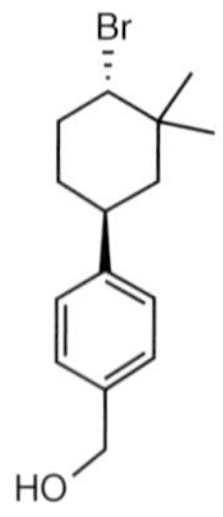

300 (majapolene B)

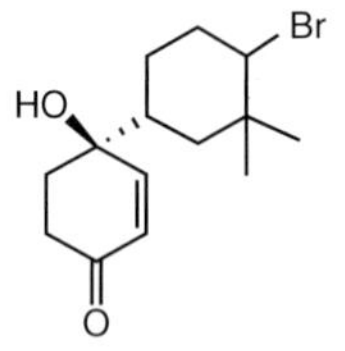

301, 302 (majapolene)

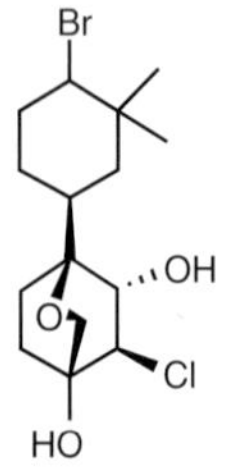

303, 304 (majapol A)

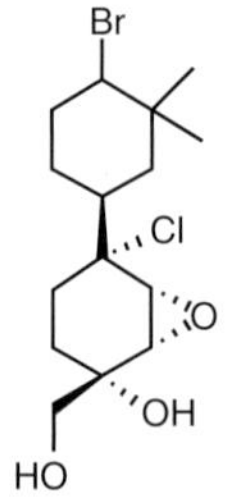

305, 306 (majapol B)

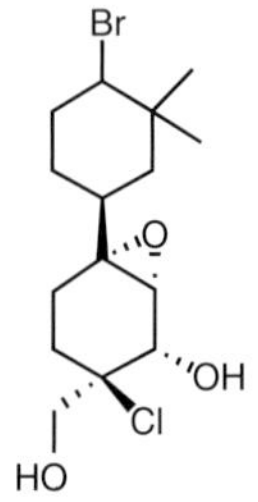

307, 308 (majapol C)

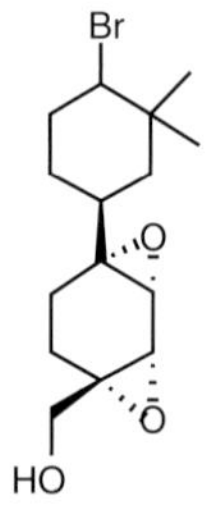

309, 310 (majapol D)

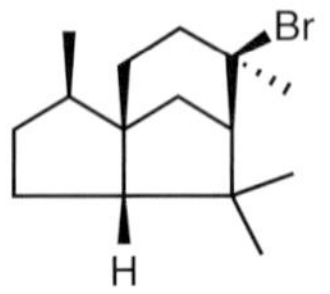

311 (majusin)

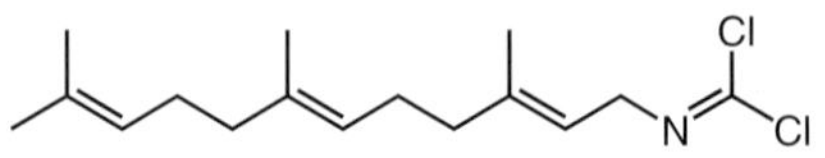

312 (stylotellane A)

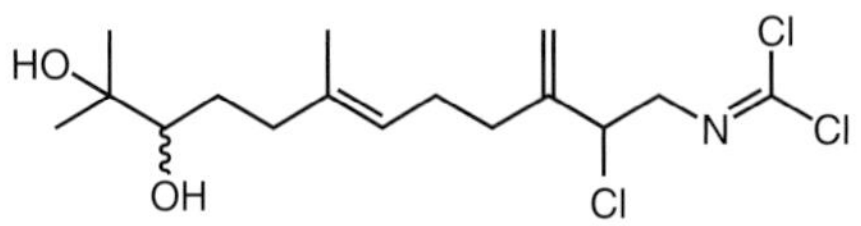

314 (axinyssimide B)
315 (axinyssimide C)

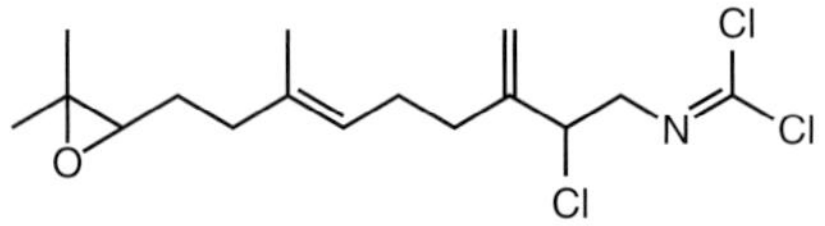

313 (axinyssimide A)

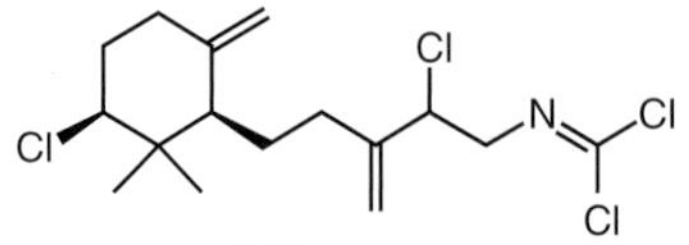

318

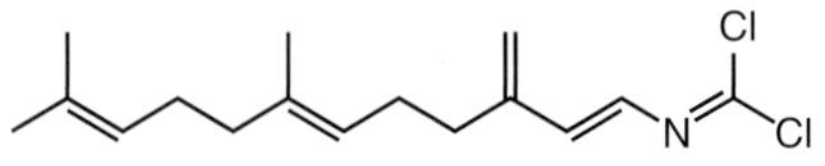

316 (ulosin A)

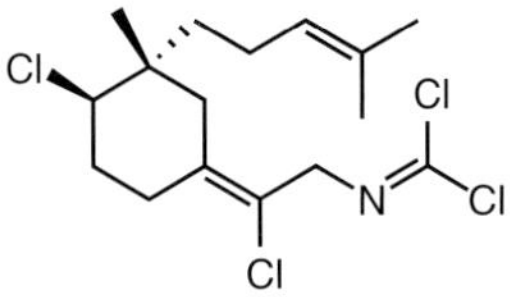

3.4 Terpenes

319 (stylotellane D)

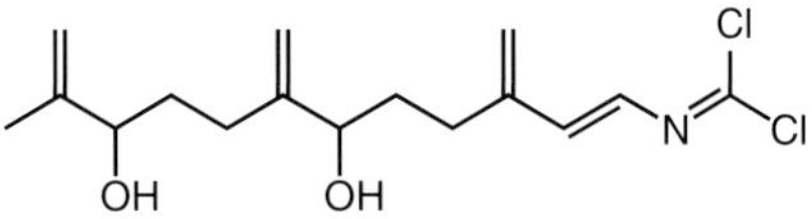

317 (ulosin B)

Laurencia seaweeds continue to be a rich source of brominated sesquiterpenes. The Okinawan *Laurencia luzonensis* contains the new isopalisol (**320**), luzonensol (**321**), luzonensol acetate (**322**), luzonensin (**323**), and triene bromohydrin **324** (*549*). A collection of *Laurencia obtusa* from Turkey has furnished the β-snyderol analogues **325** and **326** and ketone **327** (*550*), whereas *Laurencia scoparia* from Brazil contains the novel β-bisabolenes **328**–**330**, the first β-bisabolenes from the genus *Laurencia* (*551*). Bromocyclococanol (**331**) from *Laurencia obtusa* in Cuba, has a novel fused cyclopropane-cyclopentane ring system (*552*). The authors propose a biogenesis involving an interesting cyclopropyl carbinyl cation intermediate (**332**).

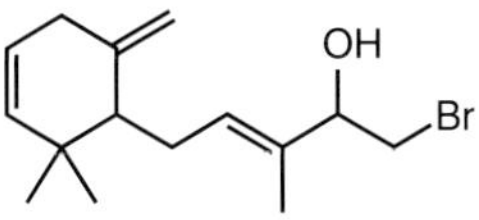

320 (isopalisol)

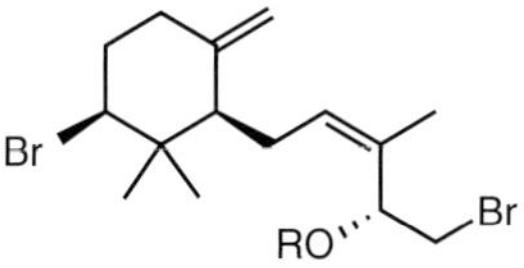

321 R = H (luzonensol)
322 R = Ac (luzonensol acetate)

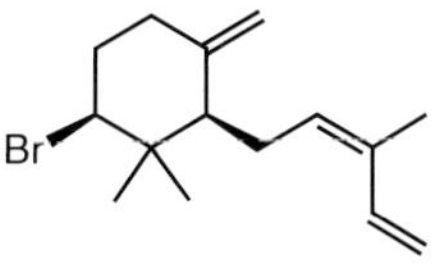

323 (luzonensin)

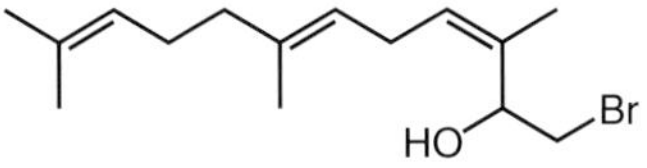

324

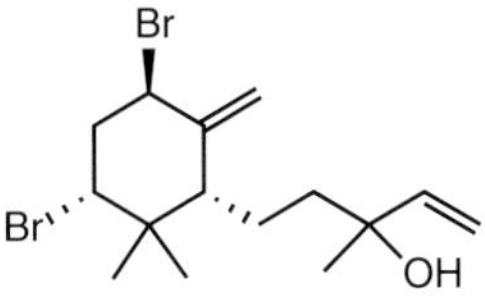

325

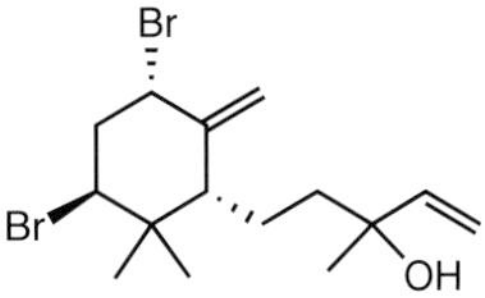

326

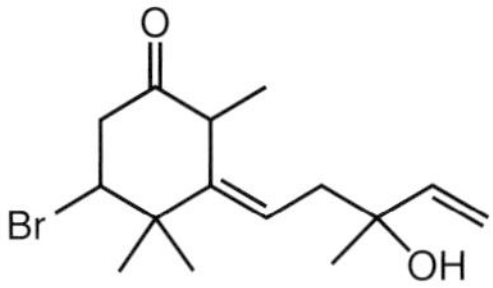

327

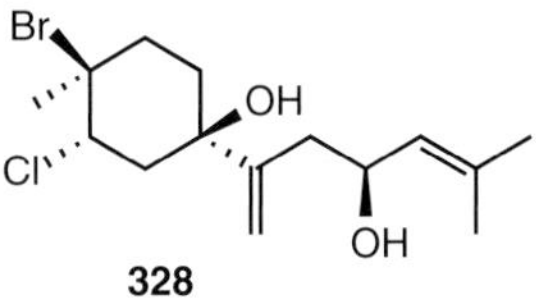

328

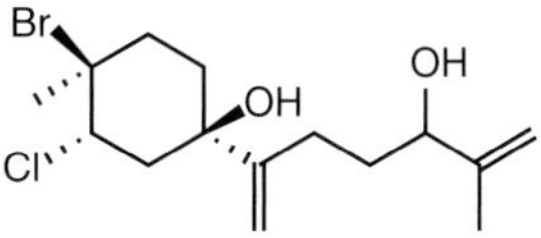

329

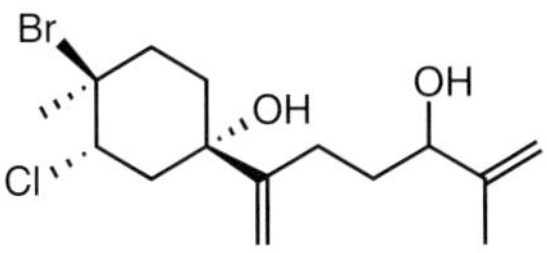

330

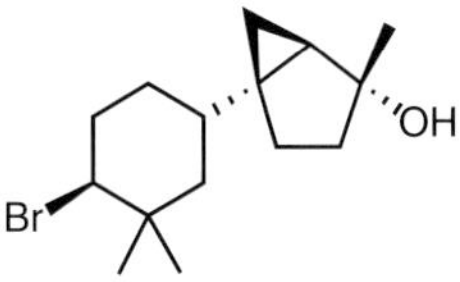

331 (bromocyclococanol)

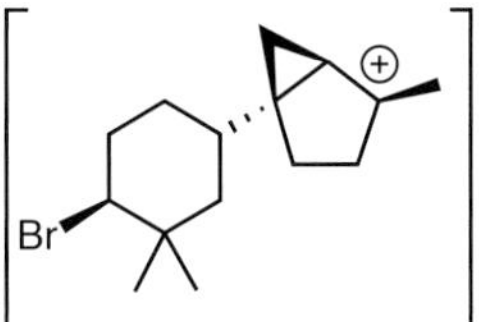

332

Sea hares and nudibranchs feed on seaweeds to acquire their chemicals for defense. The common sea hare *Aplysia dactylomela* from Spain contains puertitol-B acetate (**333**) (*554*), whereas this animal collected off the coast of South Africa affords the new algoane (**334**), 1-deacetoxyalgoane (**335**), 1-deacetoxy-8-deoxyalgoane (**336**), and ibhayinol (**337**) (*555, 556*). Investigation of this sea hare from La Palma has uncovered the new **338** along with two other halogenated compounds shown in section "Eudesmane and Other Types" (*557*).

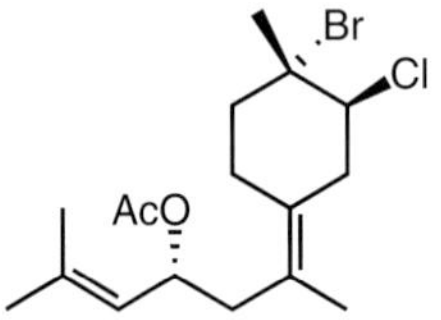

333 (puertitol-B acetate)

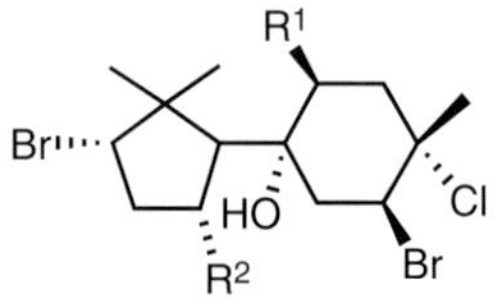

334 R^1 = OAc, R^2 = OH (algoane)
335 R^1 = H, R^2 = OH
336 R^1 = R^2 = H

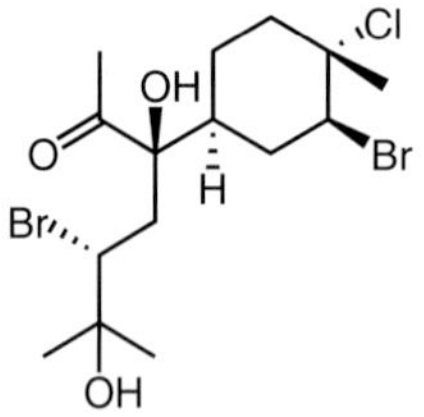

338

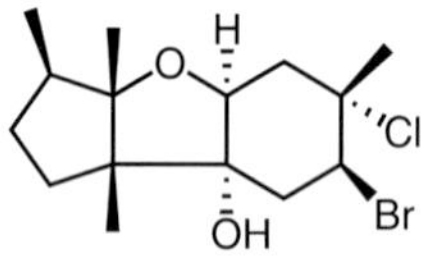

337 (ibhayinol)

An Okinawan collection of *Laurencia luzonensis* has yielded five new sesquiterpenes, luzonenone (**339**), luzofuran (**340**), 3,4-epoxypalisadin B (**341**), 1,2-dehydro-3,4-epoxypalisadin B (**342**), and 15-hydroxypalisadin A (**343** (*558*). In addition, the relative stereochemistry of luzonensol (**321**) (vide supra) (*549*) was assigned by conversion to the known palisadin B (**P**) (*1*). The novel fused bisabolene aldingenins A (**344**), B (**345**), C (**346**), and D (**347**) were isolated from *Laurencia aldingensis* (*559, 560*). The interesting chlorotriol **348** was found in a Turkish *Laurencia obtusa* (*561*).

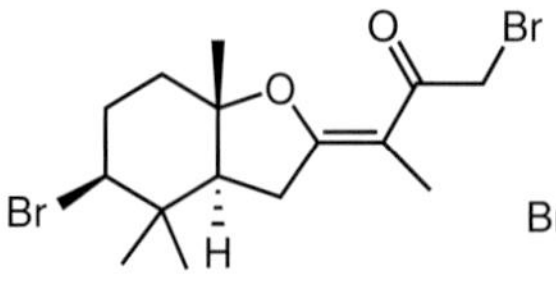

339 (luzonenone) 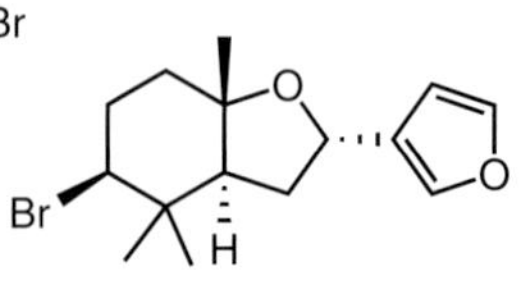   **340** (luzofuran) 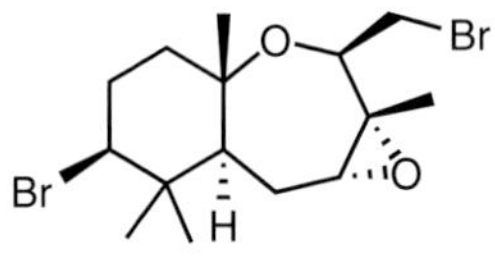   **341** (3,4-epoxypalisadin B)

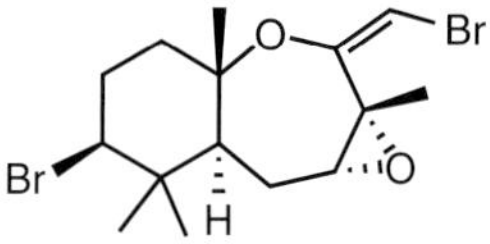

3.4 Terpenes

342

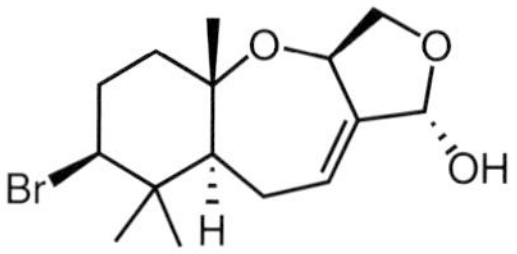

343 (15-hydroxypalisadin A)

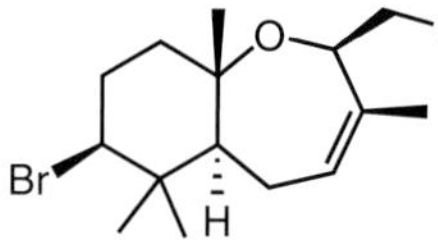

P (palisadin B)

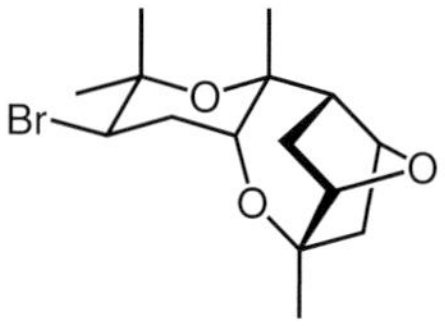

344 (aldingenin A)

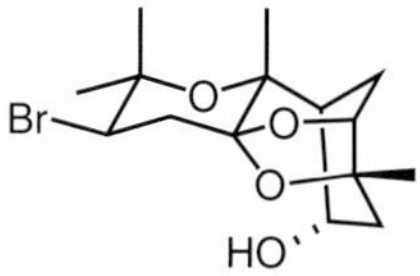

345 (aldingenin B)

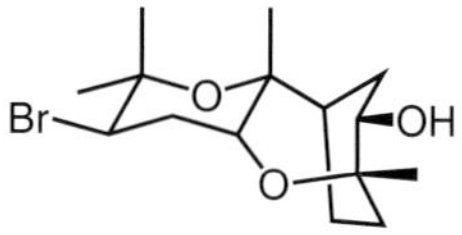

346 (aldingenin C)

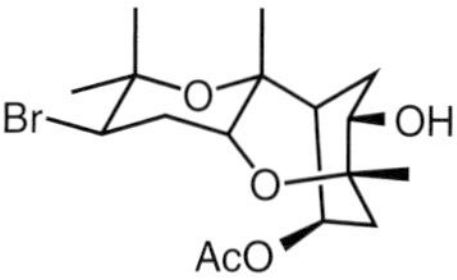

347 (aldingenin D)

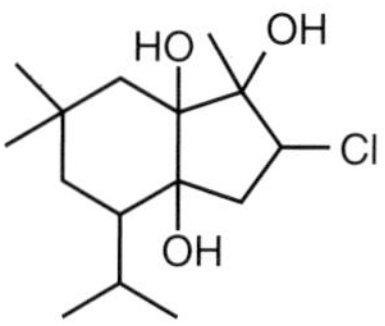

348

Chamigrene and Related Types

The halogenated spiro-chamigrene, and related metabolites represent a huge class of marine natural products, mainly from *Laurencia* seaweeds. The initial survey documented 85 examples (*1*).

Many of these chamigrenes are found in sea hares, presumably from their diet of algae, and the new 10-bromo-β-chamigren-4-one (**349**) was isolated from an *Aplysia* sp. (*562*). The Hawaiian red alga *Laurencia cartilaginea* has yielded the new ma'ilione (**350**) and *allo*-isoobtusol (**351**) (*563*). The latter is a diastereomer of isoobtusol. However, this assignment has been questioned and *allo*-isoobtusol should be reassigned as **Q** and renamed as cartilagineol (*564*), a correction that has now been confirmed (*565*). An Australian collection of *Laurencia rigida* contains the new (−)-10α-bromo-9β-hydroxy-α-chamigrene (**352**), rigidol (**353**), and (+)-(10S)-10-bromo-β-chamigrene (**354**) (*566*), for which the latter metabolite was subjected to detailed NMR analysis (*567*). It should be noted that extensive NMR studies have been performed on several known halogenated chamigrenes (e.g., prepacifenol epoxide, johnstonol, pacifenediol, pacifidiene, pacifenol, etc.) (*568–570*). Furthermore, dynamic NMR conformational analysis studies have been described with the polyhalogenated α-chamigrenes (*571, 572*). *Laurencia claviformis*, which is endemic to Easter Island, has afforded the new claviol (**355**) in addition to a suite of known halogenated chamigrenes (*573*). The Oahu red seaweed *Laurencia nidifica* contains the new **356** and **357**, along with ten known halogenated chamigranes (*574*). Tribrominated ma'iliohydrin (**358**) was isolated from a Philippine *Laurencia* sp. (*575*). Mailione (**359**) and isorigidol (**365**), which are found in *Laurencia scoparia*, were subjected to X-ray crystallography (*576*). It is not clear if **350** = **359** from the data provided (*563, 576*). *Laurencia mariannensis* contains 9-deoxyelatol (**360**) (*2658*).

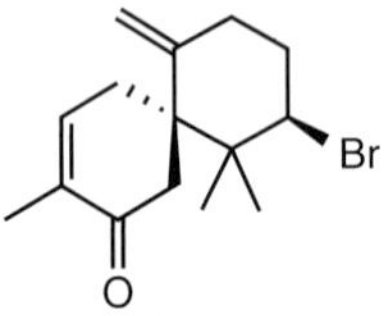

349 (10-bromo-β-chamigren-4-one) 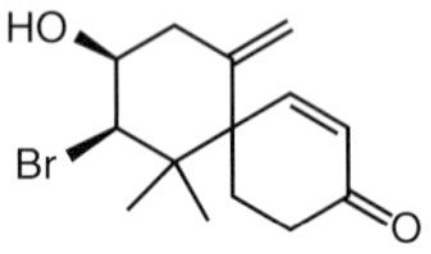   **350** (ma'ilione) 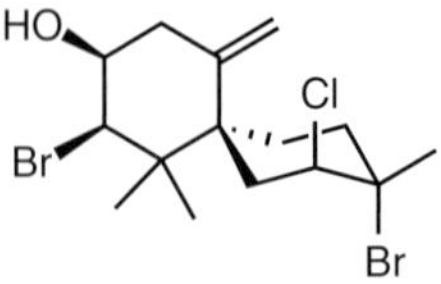   **351** (*allo*-isoobtusol)

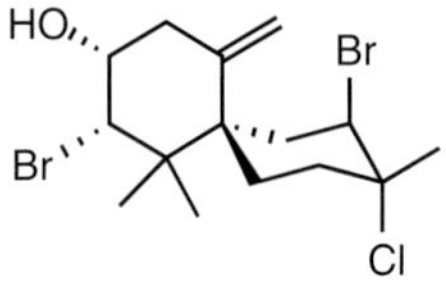

Q (cartilagineol = "*allo*-isoobtusol") 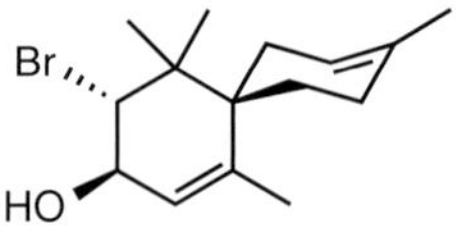   **352** 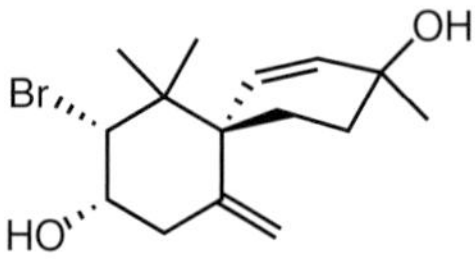   **353** (rigidol)

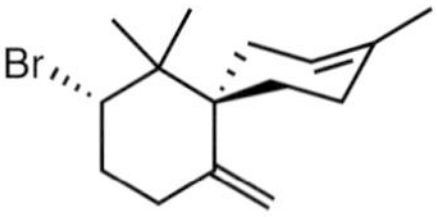

354 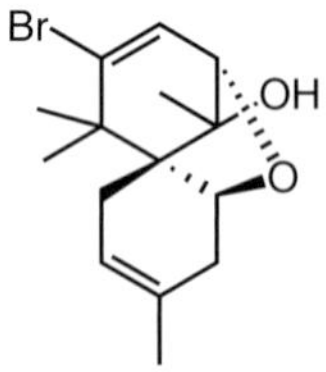   **355** (claviol) 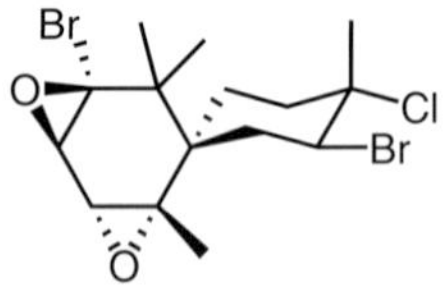   **356**

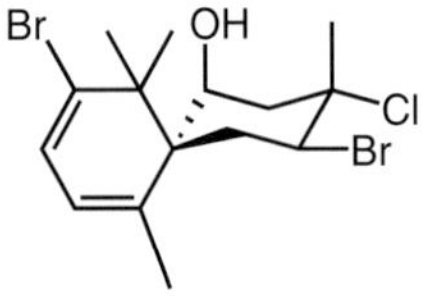

357 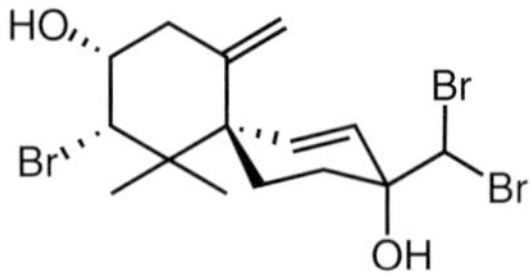   **358** (ma'iliohydrin) 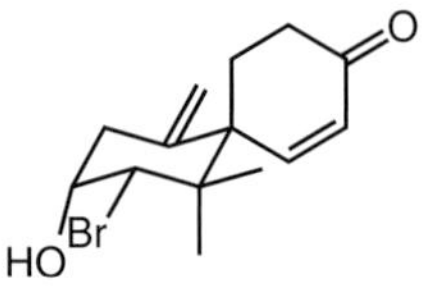   **359** (mailione)

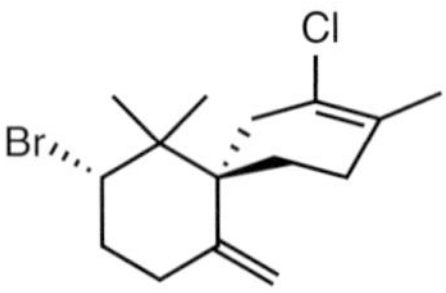

360 (9-deoxyelatol)

Whereas Okinawan *Laurencia cartilaginea* and *Laurencia concreta* yielded no halogenated metabolites, *Laurencia majuscula* from these waters afforded (6*R*,9*R*,10*S*)-10-bromo-9-hydroxychamigra-2,7(14)-diene (**361**) (*577*), which

3.4 Terpenes

appears to be an epimer of the known deschloroelatol (*1*). A Malaysian *Laurencia pannosa* contains the new pannosanol (**362**) and pannosane (**363**) (*578*). The new 5-acetoxy-2,10-dibromo-3-chloro-7,8-epoxy-α-chamigrene (**364**) is found in both *Laurencia filiformis* and the sea hare *Aplysia parvula* from Tasmania (*579*). The NMR spectra of the known 2,10-dibromo-3-chloro-7-chamigrene are assigned for the first time. A Brazilian collection of *Laurencia scoparia* has yielded three new halogenated chamigrenes, isorigidol (**365**), (+)-3-(Z)-bromomethylidene-10β-bromo-β-chamigrene (**366**), and (–)-3-(E)-bromomethylidene-10β-bromo-β-chamigrene (**367**) (*580*). A South China Sea collection of *Laurencia majuscula* afforded the simple 8-bromo-chamigren-1-en (**368**) (*581*). Oxachamigrene (**369**) and 5-acetoxyoxachamigrene (**370**) were found in *Laurencia obtusa* from Cuba, metabolites that are proposed to arise from a γ-bisabolene (*582*). The sea hare *Aplysia dactylomela* contains the novel and highly strained aplydactone (**371**) (*583*). The authors propose a biosynthesis of **371** via a formal (2 + 2) cycloaddition from **349**, which is also found in this sea hare (*562*). Sea hares from La Palma Island have furnished the new compounds **372–374** (*584*). It should be noted that several of these chamigrenes have both cytotoxic and antibacterial activity (*584–586*). The chemical diversity of halogenated chamigrenes within four Japanese *Laurencia* species has been compared and contrasted (*587*).

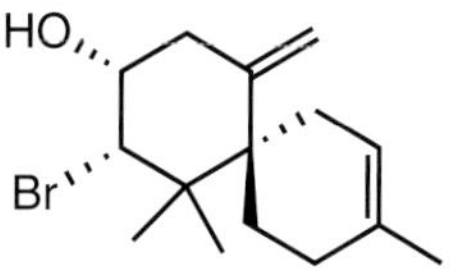

361

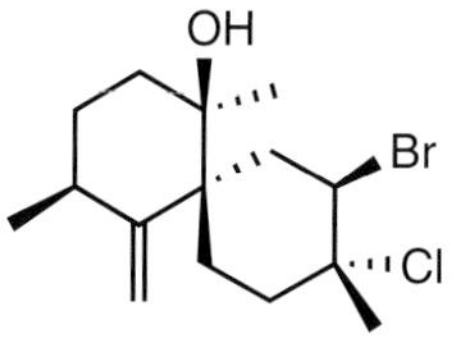

362 (pannosanol)

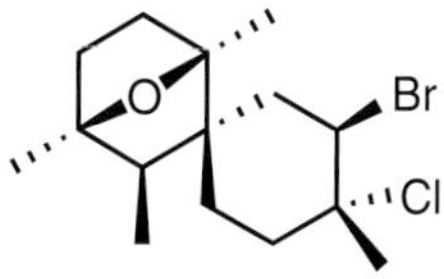

363 (pannosane)

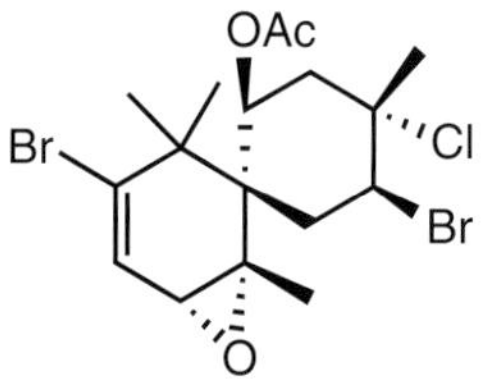

364

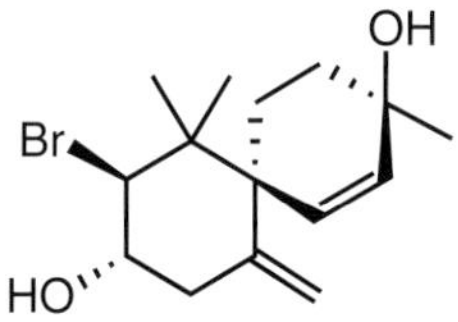

365 (isorigidol)

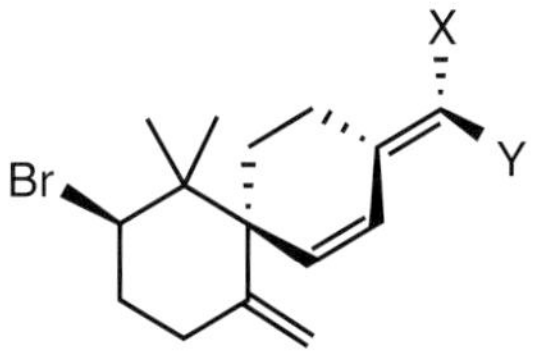

366 X = H, Y = Br
367 X = Br, Y = H

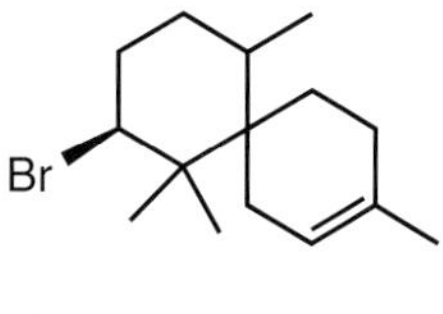

368

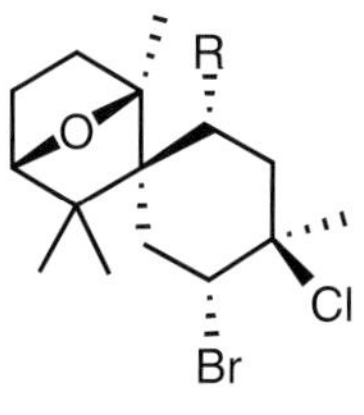

369 R = H (oxachamigrene)
370 R = OAc

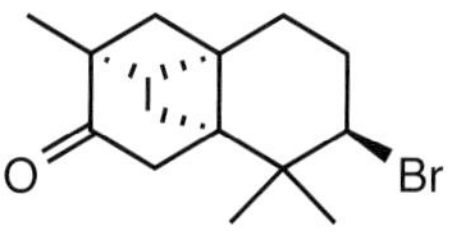

371 (aplydactone)

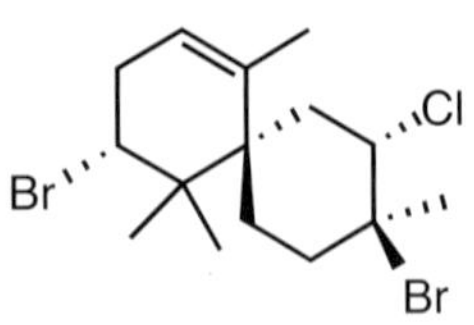

372 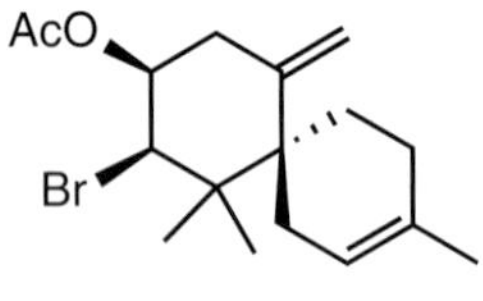 **373** (acetyldeschloroelatol) 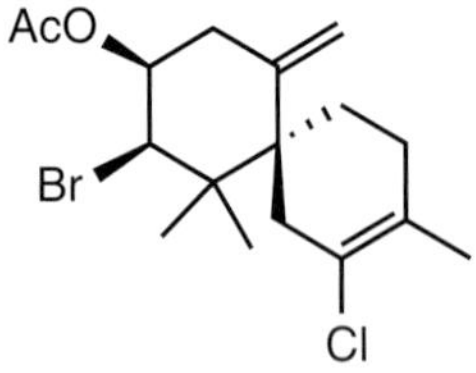 **374** (acetylelatol)

Eudesmane and Other Types

Nearly 40 halogenated eudesmanes and related halogenated sesquiterpenes were documented in the first survey (*1*), and terrestrial eudesmanes of all types are well represented with some 100 known examples (*588*).

The sea hare *Aplysia dactylomela* contains lankalapuols A (**375**) and B (**376**), which interestingly have opposite absolute configurations (*589*). The tropical green alga *Neomeris annulata* has yielded the three novel brominated sesquiterpenes **377–379**, which are effective feeding deterrents (*590*), and an Okinawan collection of *Laurencia intricata* contains itomanol (**380**), which is a diastereomer of lankalapuol A (**375**) (*591*). The Elba Island *Laurencia microcladia* has afforded the novel 6,8-cycloeudesmanes calenzanol (**381**) (*553, 592*) and **382** (*592, 593*), which feature the new sesquiterpene skeleton, calenzanane (*553, 592*). A study of *Laurencia obtusa* from the Aegean Sea, Greece, has revealed the presence of four new perforatone analogs **383–386** (*594*), and, from a different location in the Aegean Sea, the new perforenol B (**387**) and **388** (*595*). The sea hare *Aplysia punctata* contains the new perforatol (**389**) (*596*).

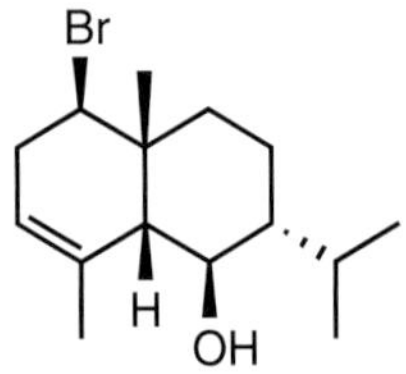

375 (lankalapuol A) 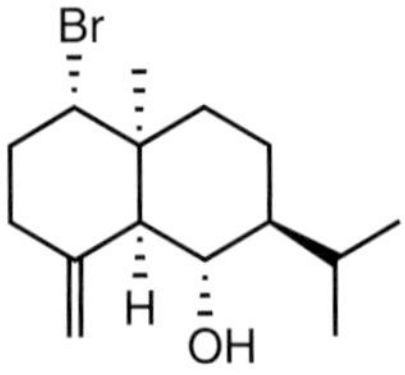   **376** (lankalapuol B) 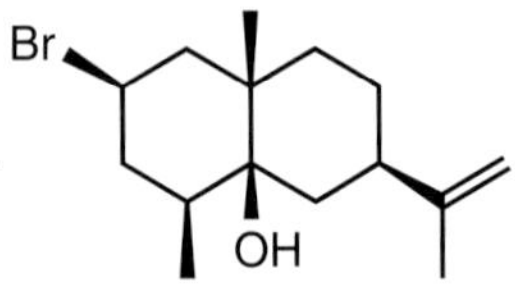   **377**

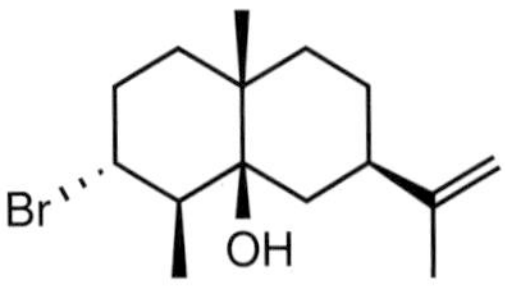

378 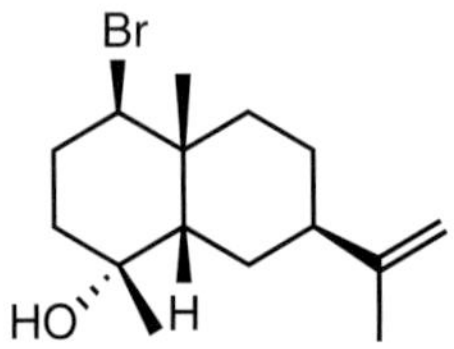   **379** 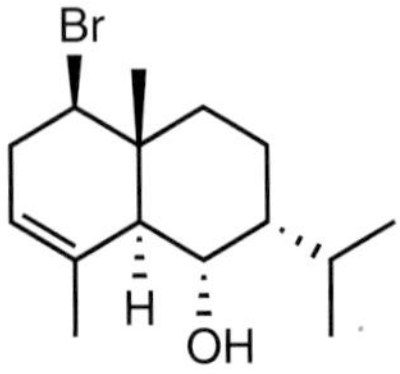   **380** (itomanol)

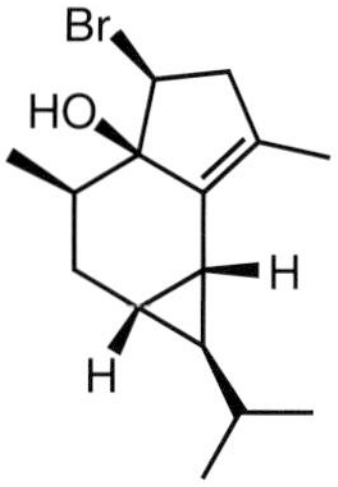

3.4 Terpenes

381 (calenzanol) 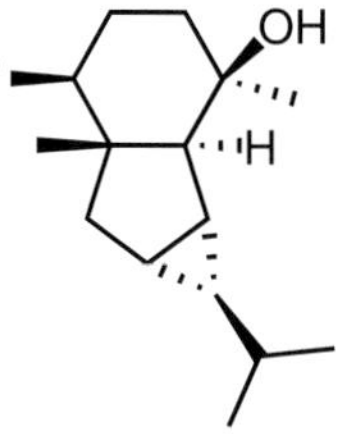 **382** 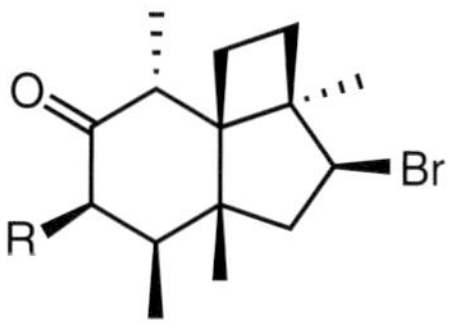 **383** R = Br
384 R = OMe
385 R = H

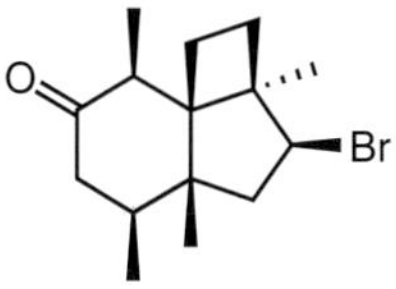

386 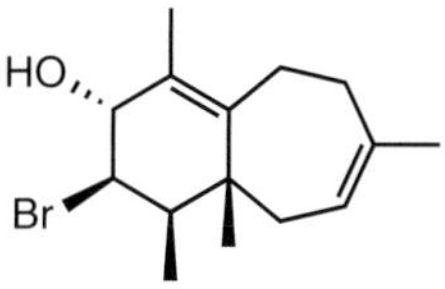 **387** (perforenol B)

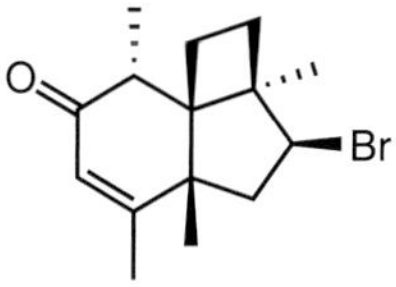

388 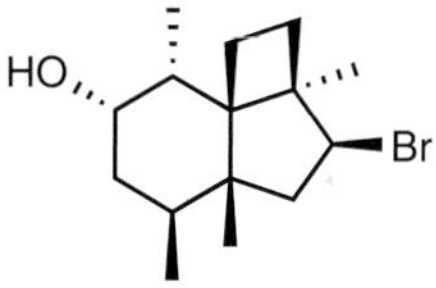 **389** (perforatol)

An examination of the red seaweed *Laurencia obtusa* from Symi Island in the Aegean Sea has uncovered the brasilanes **390–392** (*597*). The soft coral *Paralemnalia thyrsoides* from Taiwan has afforded the chlorinated norsesquiterpenoid paralemnolin A (**393**) (*598*), and the Australian sponge *Euryspongia* sp. provides the sesquiterpene quinone (*E*)-chlorodeoxyspongiaquinone (**394**) and related hydroquinone (*E*)-chlorodeoxyspongiaquinol (**395**) (*599*). Several new sesquiterpene chlorohydrins and carbonimidic dichlorides (*600*) have been found in the sponge *Stylotella aurantium* (Fig. 3.3), **396–398** (*545, 546*), and from the nudibranch *Reticulidia fungia* (Fig. 3.4), reticulidins A (**399**) and B (**400**) (*601*). A biosynthetic pathway to these sesquiterpene dichloroimines involving farnesyl isocyanide and isothiocyanate is supported by labeling experiments (*548*). The sea hare *Aplysia dactylomela* contains caespitenone (**401**) and 8-acetylcaespitol (**402**) (*554*), and *Aplysia punctata* has yielded punctatol (**403**) (*596*). The former mollusc also contains deschlorobromocaespitol (**404**) and furocaespitanelactol (**405**) (*557*).

Fig. 3.4 *Reticulidia fungia*, a nudibranch collected in Manza, Okinawa, that contains reticulidins A and B (**399** and **400**) (Photo: J. Tanaka)

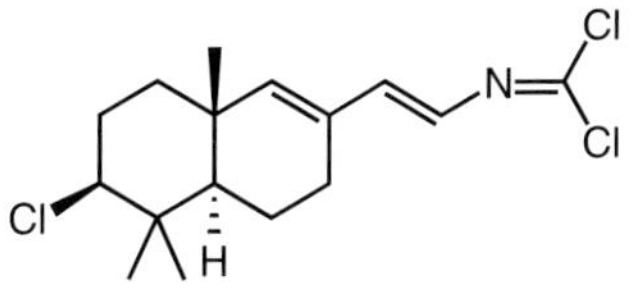

3.4 Terpenes

398 (deoxyisoreticulidine B)

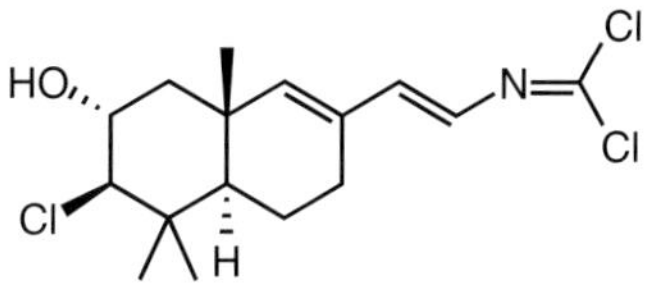

399 (reticulidin A)

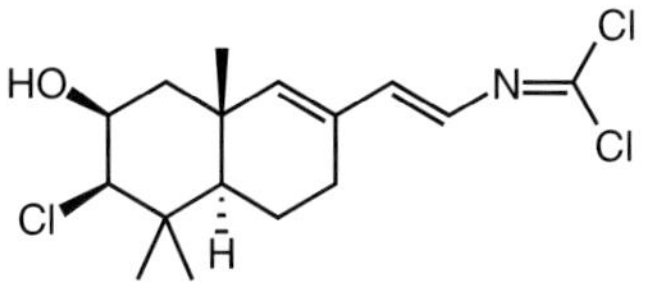

400 (reticulidin B)

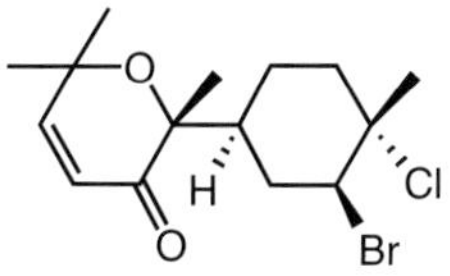

401 (caespitenone)

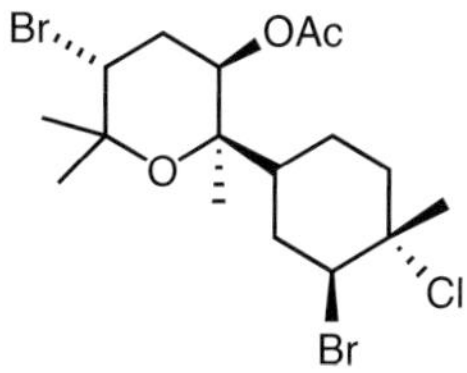

402 (8-acetylcaespitol)

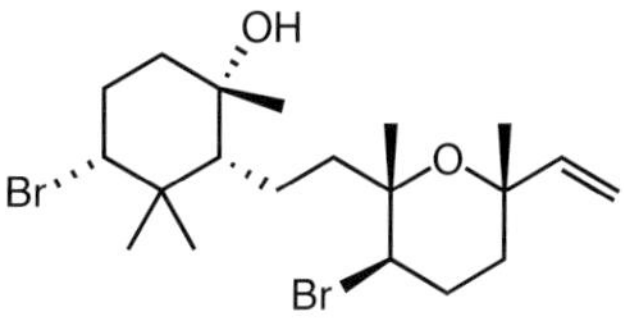

403 (punctatol)

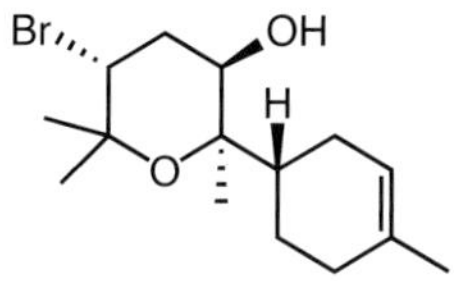

404 (deschlorobromocaespitol)

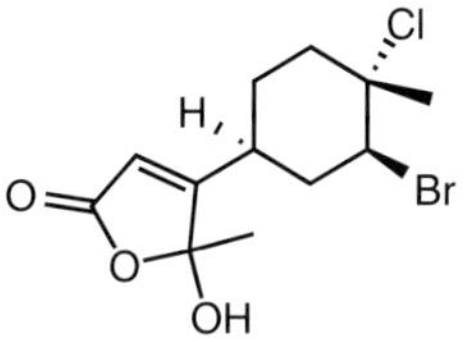

405 (furocaespitanelactol)

Cuparene, Laurene, and Other Aromatic Types

Some 50 marine aromatic halogenated sesquiterpenes were documented in the first survey (*1*). In the interim a number of new examples have been reported, mainly from *Laurencia* red algae.

A study of *Laurencia tristicha* from the South China Sea has discovered the hydroxylated aplysins, 10-hydroxyepiaplysin (**406**) and 10-hydroxyaplysin (**407**) (*602*), and 4-bromo-1,1-epoxylaur-11-ene (**408**), which was previously synthesized but not found naturally (*603*). *Laurencia microcladia* from the North Aegean Sea has yielded the new **409** and **410**, which exhibit significant cytotoxicity against two lung cancer cell lines (*604*). This red alga also contains the dimeric cyclolaurane **411** (*595*). An East China Sea collection of *Laurencia okamurai* has led to the isolation of the novel laureperoxide (**412**) and 10-bromoisoaplysin (**413**)

(*605*), and "3β-hydroxyaplysin" and "laurokamurene A" (*606*), which would appear to be identical to 10-hydroxyaplysin (**407**) and **410**, respectively. Specimens of the sea hare *Aplysia kurodai* from the Sea of Japan have afforded the new laurinterol acetate (**414**) (*607*). It should be noted that syntheses of these halogenated cuparane and asplysin sesquiterpenes are known (*608, 609*). The previously known laurinterol, isolaurinterol, aplysinal, and aplysin (*1*) show pronounced cytotoxicity against the A549, SK-OV-3, SK-MEL-2, XF498, and HT15 cell lines (*610*).

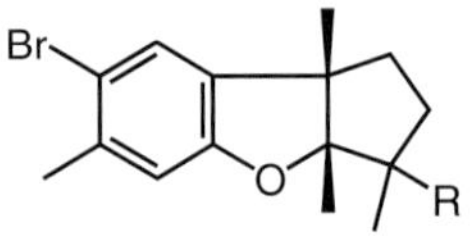

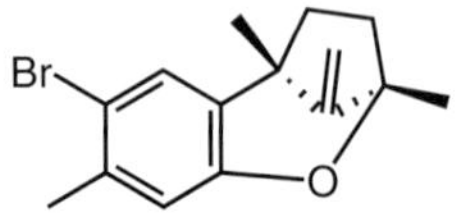

406 R = α–OH (10-hydroxyepiaplysin)   **408** (4-bromo-1,10-epoxylaur-11-ene)
407 R = β–OH (10-hydroxyaplysin)

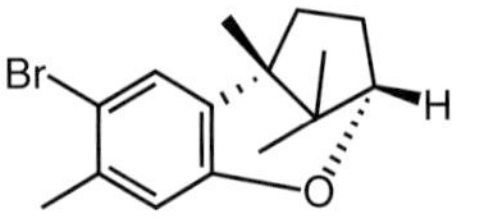

409 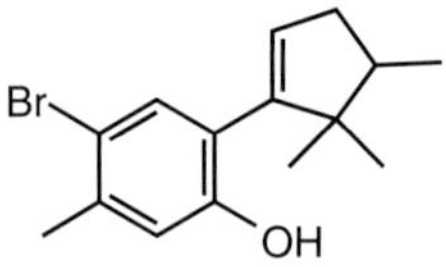   **410** 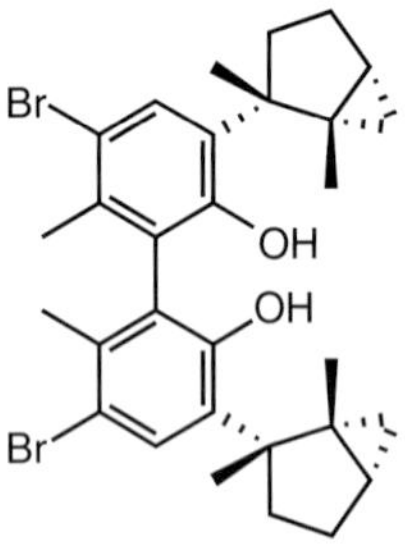   **411**

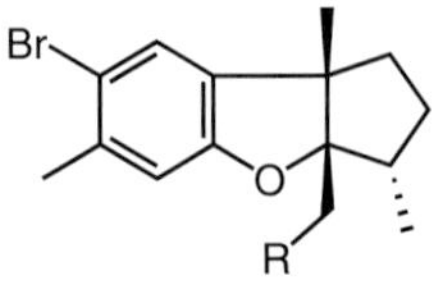

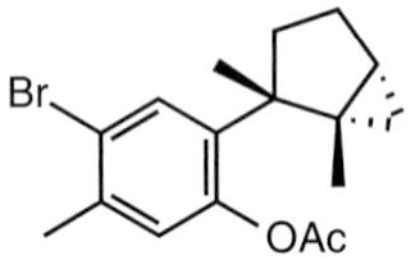

412 R = OOH (laureperoxide)   **414** (laurinterol acetate)
413 R = Br (10-bromoisoaplysin)

The green alga *Cymopolia barbata* from Cuba contains the new prenylated hydroquinones 3′-methoxy-7-hydroxycymopol (**415**), 3-hydroxycymopolone (**416**), 3,7-dihydroxycymopolone (**417**), 7-hydroxycymopochromanone (**418**), 7-hydroxycymopochromenol (**419**), 6-hydroxycymopochromenol (**420**) (*611*), and a Jamaican collection of this alga yielded 7-hydroxycymopol (**421**) (*612*). The latter compound was previously described as a synthetic intermediate (*613*). The structurally similar known brominated cacoxanthenes from the sponge *Cacospongia* are found in the blubber of monk seal, in commercial fish samples and mussels (*489*). The sponge *Spirastrella hartmani* from Martinique has yielded the two halogenated heliananes **422** and **423** (*614*). The New Zealand sponge *Hamigera tarangaensis* produces hamigeran A (**424**), hamigeran B (**425**), 4-bromohamigeran B (**426**), hamigeran C (**427**), hamigeran D (**428**), and hamigeran E (**429**) (*615*). The structure of hamigeran E was revised from that reported earlier (*616*), and the structures of **424**, **425**, and **426** have been confirmed by total synthesis (*617, 618*).

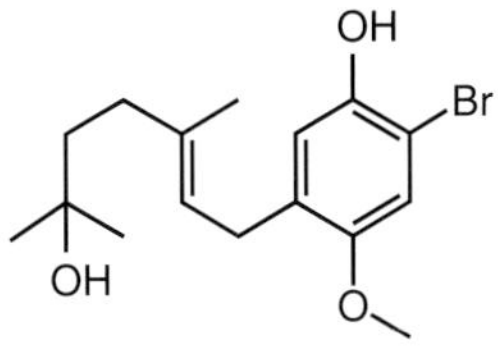

3.4 Terpenes

415

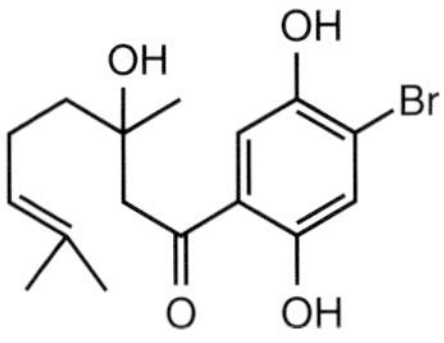

416

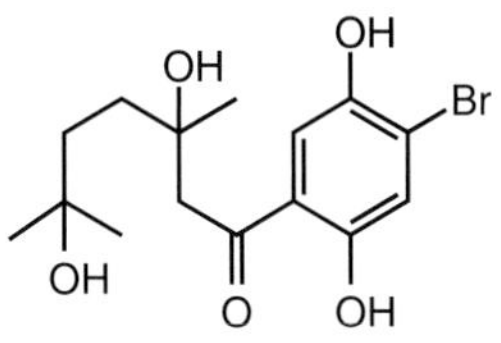

417

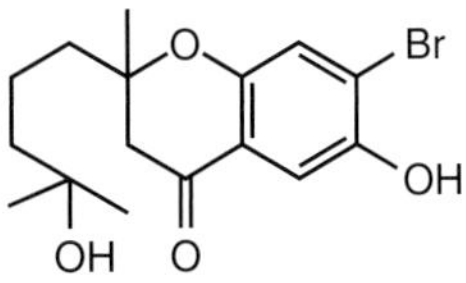

418

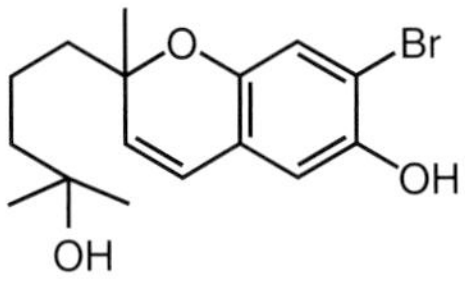

419

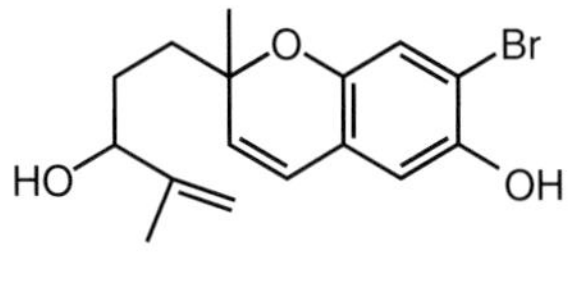

420

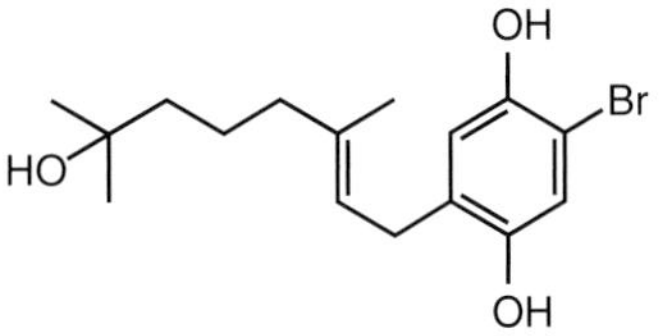

421

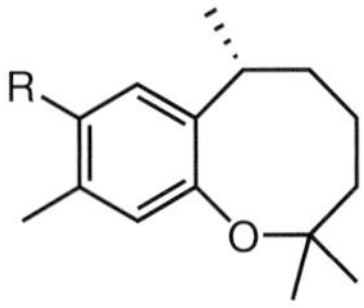

422 R = Cl
423 R = Br

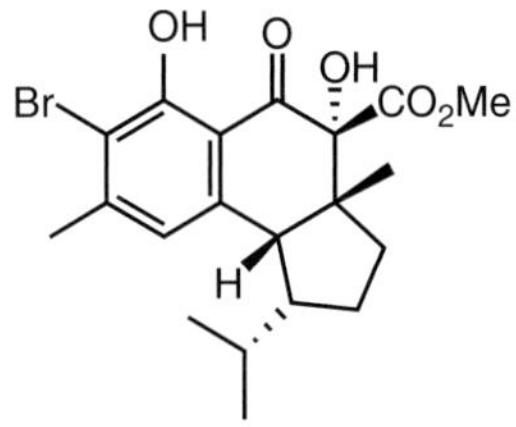

424 (hamigeran A)

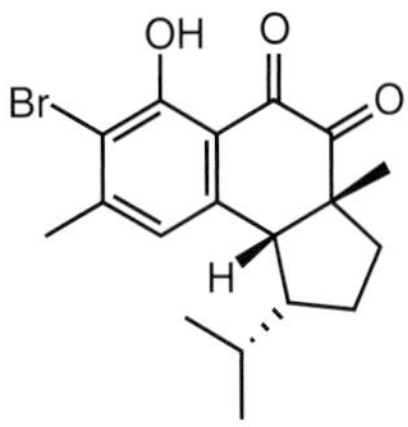

425 (hamigeran B)

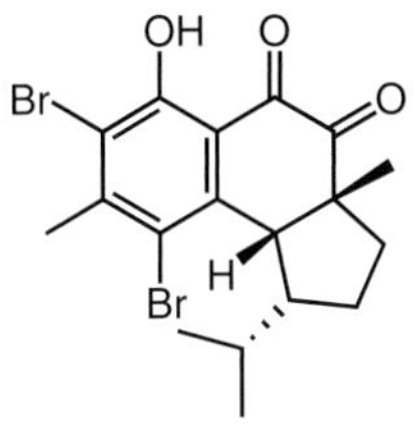

426 (4-bromohamigeran B)

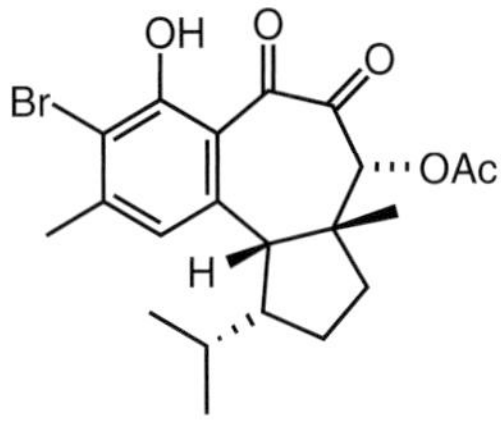

427 (hamigeran C)

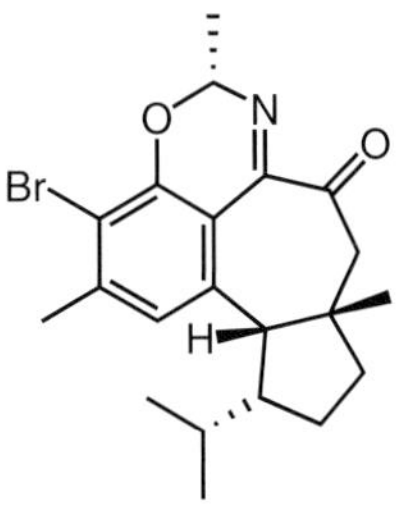

428 (hamigeran D)

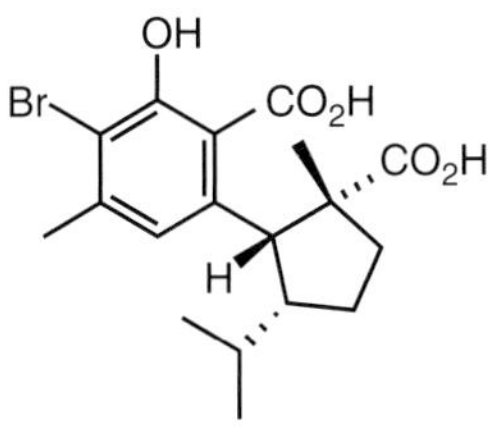

429 (hamigeran E)

The Hawaiian sponge *Hyrtios* sp. from Oahu has furnished 21-chloropuupehenol (**430**), while the same sponge from Maui contains molokinenone (**431**) (*619*). The absolute configuration of these drimane-phenolic metabolites has been assigned as shown based on that of puupehenone (*620*). The South Georgia Island soft coral *Alcyonium paessleri* produces several novel illudalane sesquiterpenoids, including the chlorinated alcyopterosins A (**432**), D (**433**), K (**434**), and L (**435**) (*621*).

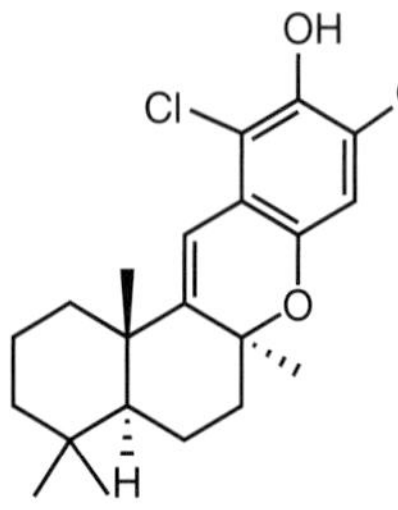

430 (21-chloropuupehenol) 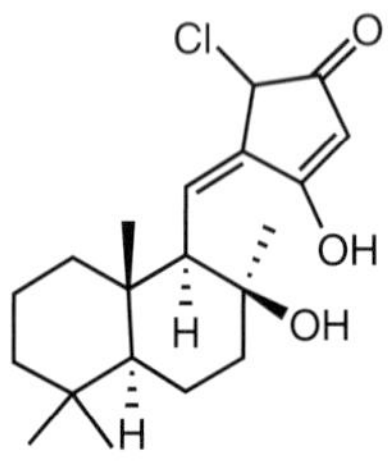   **431** (molokinenone) 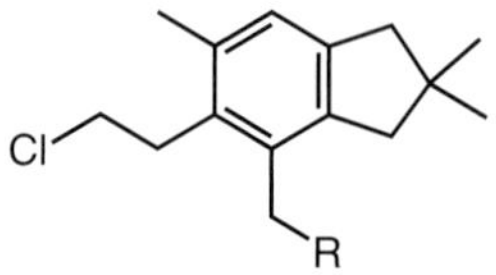   **432** R = H (alcyopterosin A)
                                                         433 R = OH (alcyopterosin D)

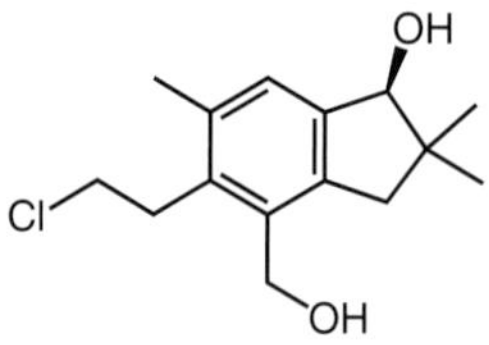

434 (alcyopterosin K) 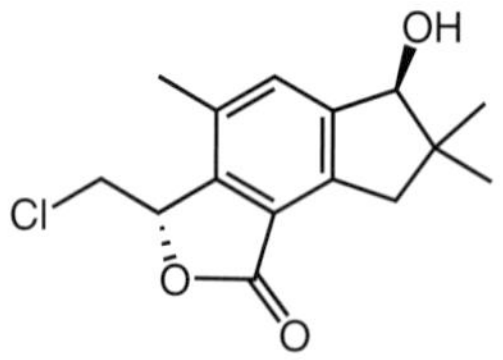   **435** (alcyopterosin L)

3.4.3 Diterpenes

3.4.3.1 Terrestrial Diterpenes

As was illustrated in the first survey, all known halogenated terrestrial diterpenes are chlorohydrins (*1*), and that continues to be mainly the case. Obviously, one must be alert to the possibility of artifact formation from ring opening of the corresponding epoxide during isolation. Many nonhalogenated terrestrial diterpenoids also continue to be isolated (*622*).

The Brazilian plant *Vellozia bicolor* contains the isopimarane diterpene 12-chloroillifunone C (**436**). The corresponding epoxide, which is also found in this plant, is not converted to **436** under the isolation conditions (*623*). Teuracemin (**437**), a novel *neo*-clerodane diterpene, was isolated from *Teucrium racemosum* and is the 7-hydroxy derivative of the known tafricanin A (*624*). Examination of fresh plant material revealed the presence of **437**. The new *neo*-clerodane ajugarin-I chlorohydrin (**438**) has been characterized from the Indian plant *Ajuga parviflora*

3.4 Terpenes

(*625*). The wood resin of *Excoecaria agallocha* has furnished several labdane-type diterpenes including the chlorinated excoercarin F (**439**), which is the first example of a chlorine-containing metabolite from this "shore plant" (*626*). Another investigation of this mangrove plant from India revealed the labdanes agallochins A (**440**), B (**441**), and C (**442**) in the roots (*627*). Another mangrove plant, *Bruguiera gymnorrhiza*, from China contains the *ent*-kaurane **443** (*628*).

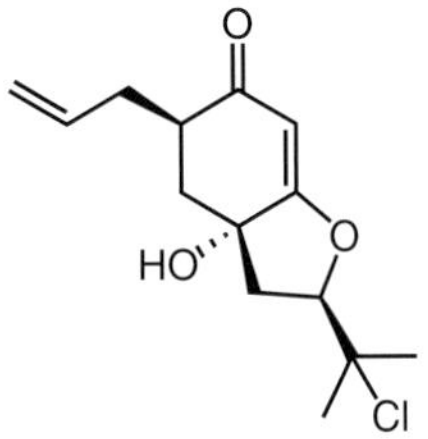

436 (12-chloroillifunone C) 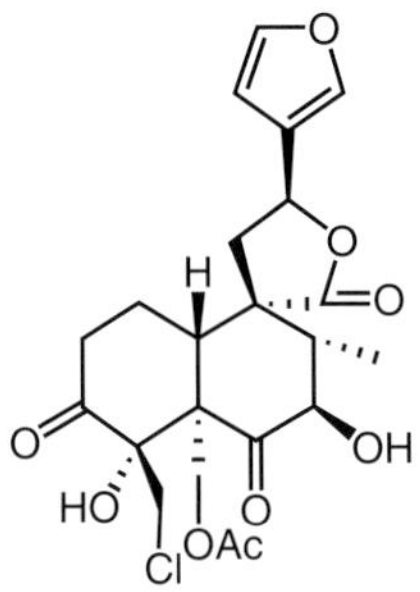   **437** (teuracemin) 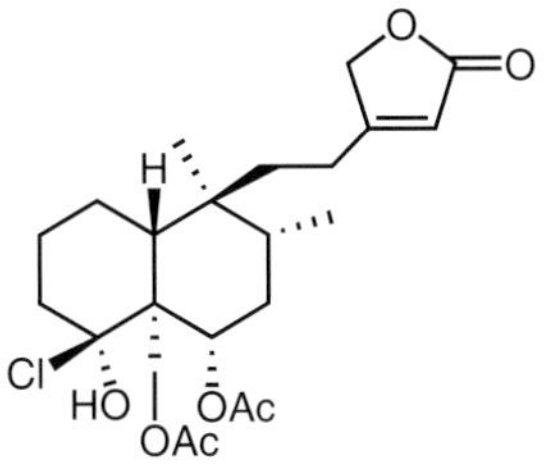   **438**

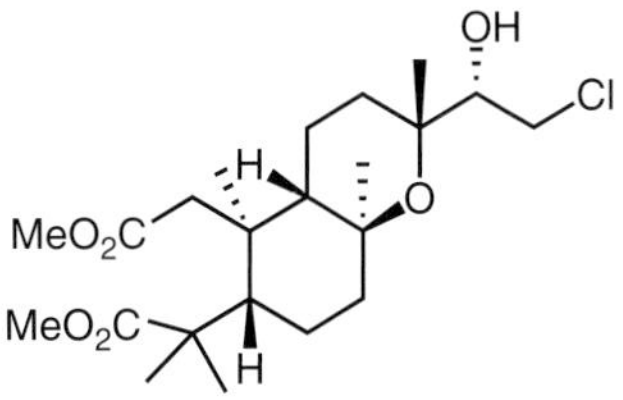

439 (excoercarin F) 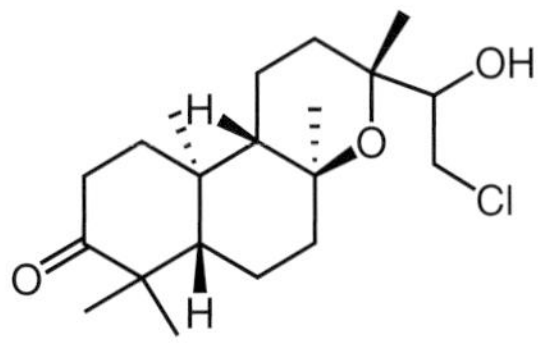   **440** (agallochin A)

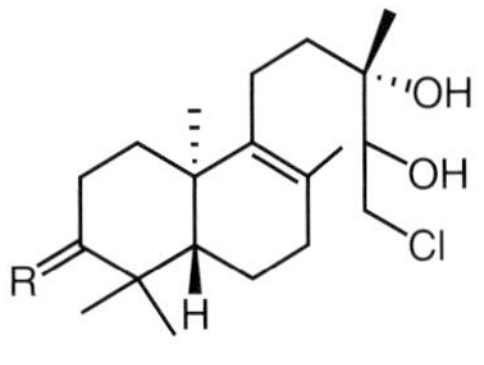

441 R = O (agallochin B)
442 R = βOH, H (agallochin C)

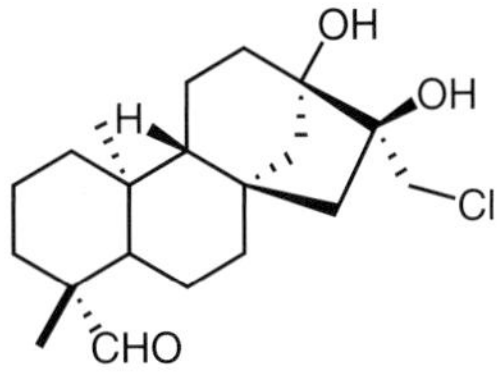

443

The novel chloroenone quassinoid eurycolactone B (**444**) was characterized from the roots of *Eurycoma longifolia* from Malaysia (*629*). This is the first halogenated quassinoid discovered in a plant. A series of norditerpene dilactones, including the chlorinated rakanmakilactones E (**445**), G (**446**), and **447**, were isolated from the leaves of *Podocarpus macrophyllus* from Japan (*630, 631*). These represent the first halogenated norditerpene dilactones found in the Podocarpaceae.

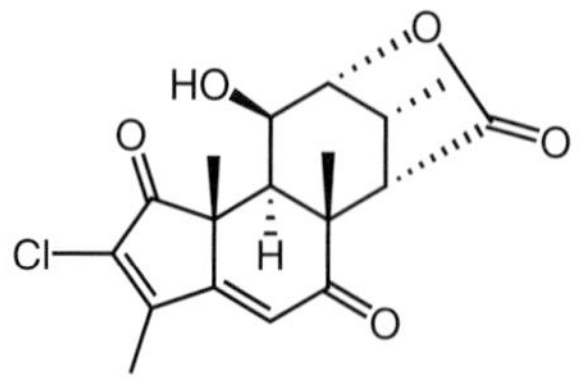

444 (eurycolactone B)

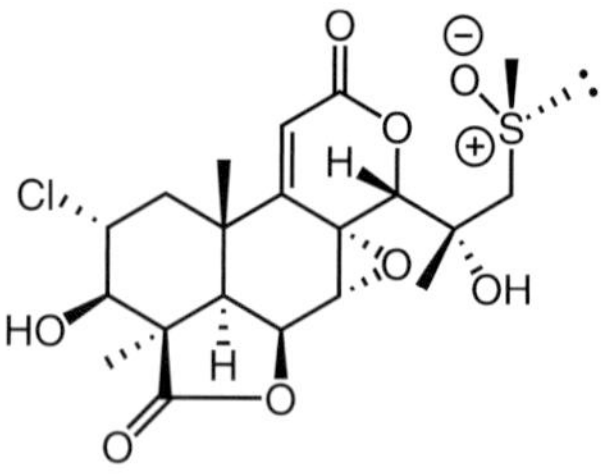

445 (rakanmakilactone E)

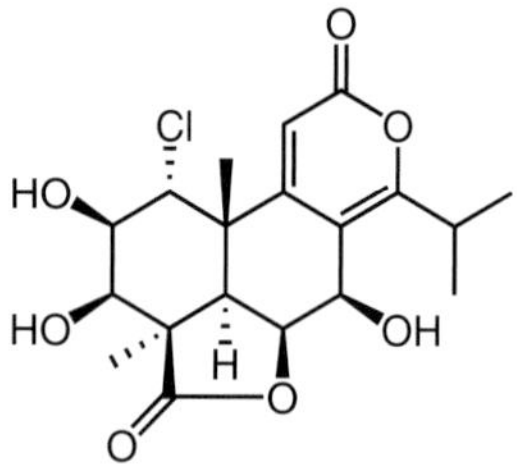

446 (rakanmakilactone G)

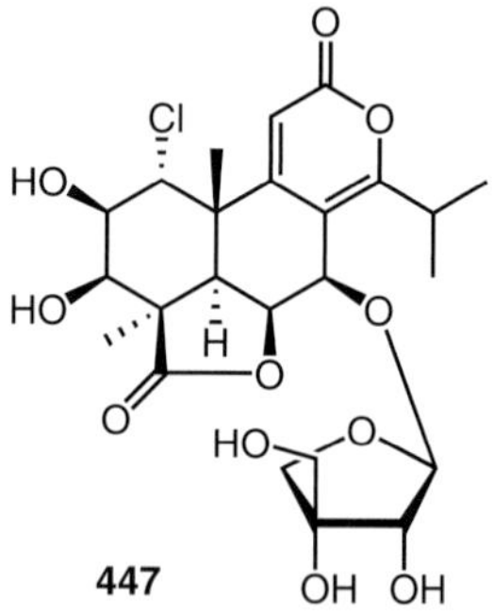

447

3.4.3.2 Marine Diterpenes

In contrast to the small number of known halogenated (chlorinated) terrestrial diterpenes (vide supra), the number of marine diterpenes is very large, and more than 130 were documented in the initial survey (*1*).

Diterpenes of Aplysia

Sea hares of genus *Aplysia* continue to be the source of new halogenated and nonhalogenated diterpenes, and a review of diterpenes from marine opisthobranch molluscs has appeared (*632*). The Tenerife *Aplysia dactylomela* contains isopinnatol B (**448**) and dactylopyranoid (**449**) (*554*). *Aplysia punctata* from Sardinia has afforded the novel neopargueroldione (**450**), which may arise from the previously known isoparguerol, and deacetylparguerol (**451**) (*596*). Most probably all of these brominated diterpenes originate from the *Laurencia* and other algae diet of the sea hare.

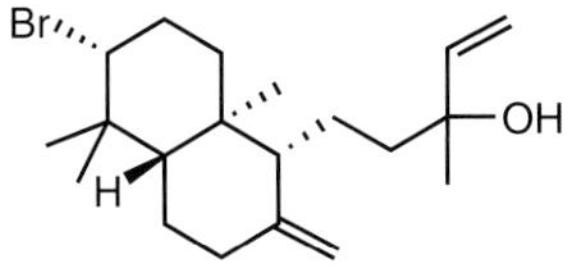

448 (isopinnatol B)

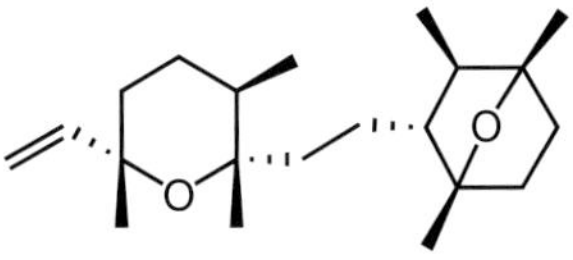

449 (dactylopyranoid)

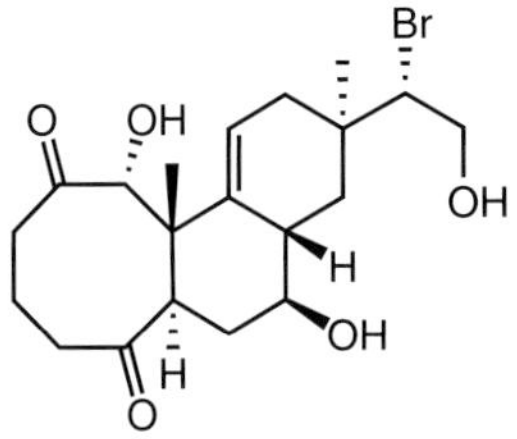

450 (neopargueroldione)

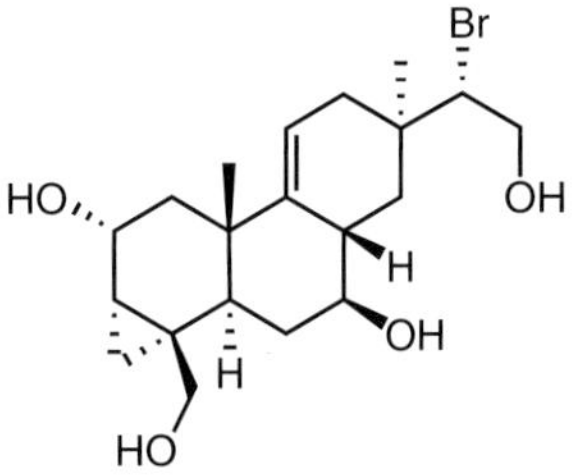

451 (deacetylparguerol)

Diterpenes of Laurencia

Laurencia red seaweeds produce a large and varied assortment of halogenated (mainly brominated) diterpenes (*1*), and this trend continues with the present survey. Five new parguerenes, **452–456**, are found in the Southern Australian *Laurencia filiformis*, along with a plausible biogenic precursor **R** (*633*). A collection of *Laurencia nipponica* from Russian waters in the Sea of Japan has identified the new parguerones **457** and **458**, where the former metabolite appears to be a diastereomer of **455** (*634*). The new *Laurencia japonensis* species contains anhydroaplysiadiol (**459**) along with the known aplysiadiol and a halogenated chamigrene (*635*). A collection of *Laurencia paniculata* from the Arabian Gulf furnished the *ent*-labdane paniculatol (**460**), which contains an unusual tetrahydropyran ring and is closely related to the known *ent*-isoconcinndiol isolated from *Aplysia kurodai* (*636*). The closely related *ent*-labdanes **461** and **462** were found in an Okinawan *Laurencia* sp. collection, and are the first labdane bromoditerpenoids to be functionalized at C-1 (*637*). Two collections of *Laurencia* sp. in different locations in Japanese waters have yielded **463**, closely related to paniculatol (**460**) (*638*).

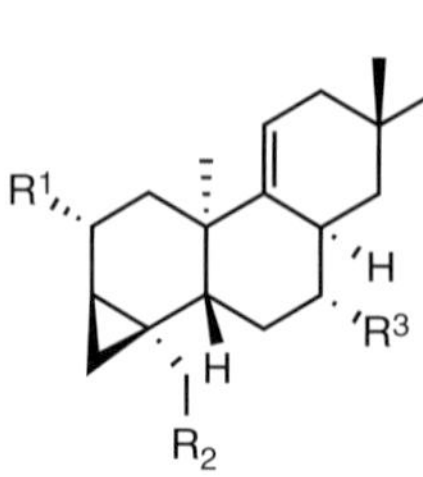

452 $R^1 = R^3 = R^4 = OAc, R^2 = H$
453 $R^1 = R^3 = OAc, R^2 = H, R^4 = OH$
454 $R^1 = OAc, R^2 = R^3 = H, R^4 = OH$
455 $R^1 = R^2 = R^3 = H, R^4 = OH$

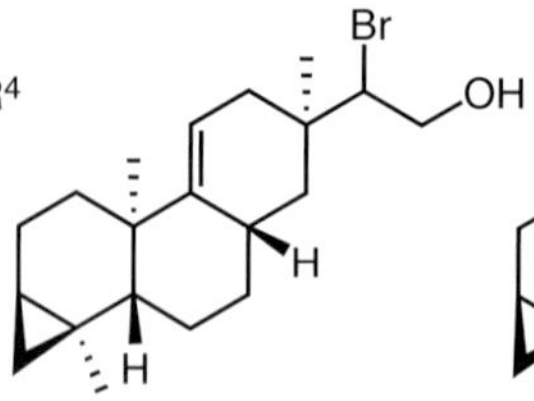

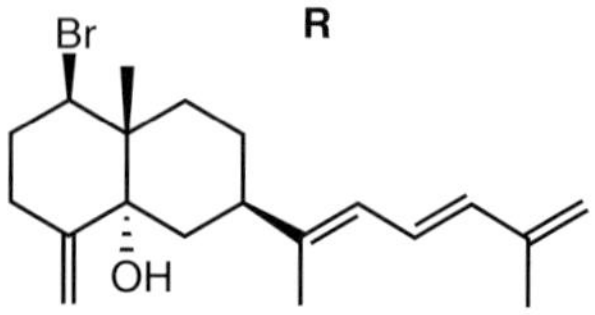

459 (anhydroaplysiadiol)

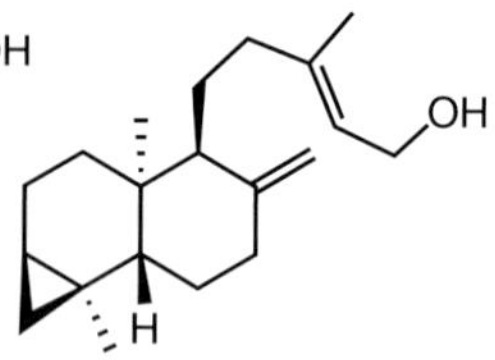

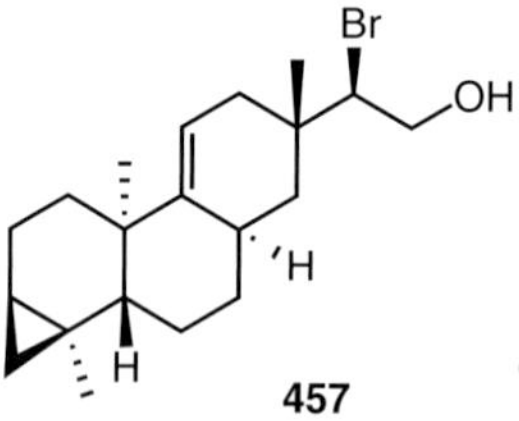

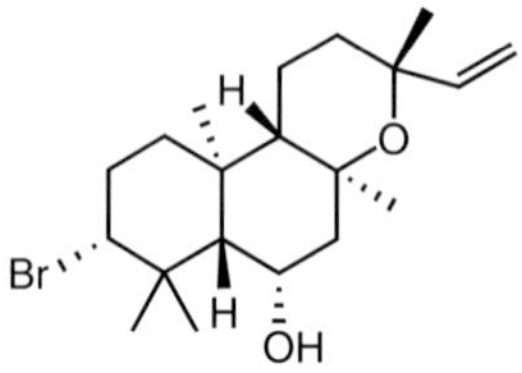

460 (paniculatol)

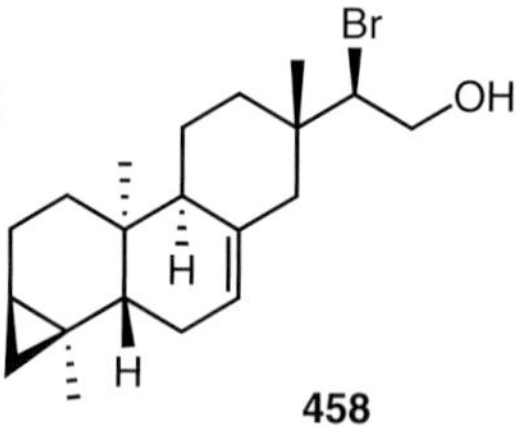

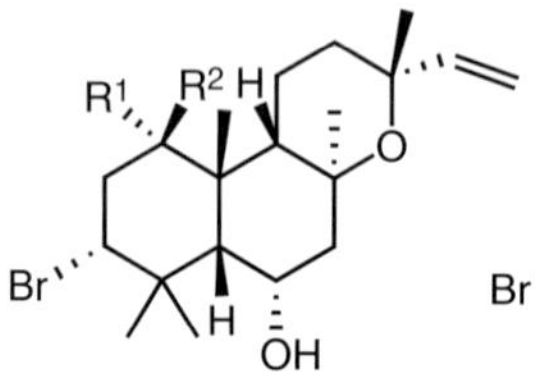

461 $R^1 = OAc, R^2 = H$
462 $R^1 = R^2 = O$

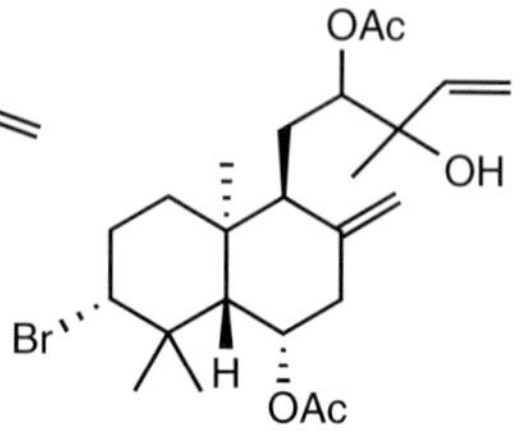

The Okinawan *Laurencia luzonensis* contains 3-bromobarekoxide (**464**), a novel seven-membered ring diterpene (*549, 639*). Equally unprecedented are the labdanes **465** and **466** from *Laurencia obtusa* gathered in the Ionean Sea (*640*). This source of *Laurencia obtusa* has also yielded the new prevezols A (**467**) and B (**468**), which are marginally related to the known obtusadiol and rogioldiol A (vide infra) (*641*). This red alga also contains the new prevezols C-E (**469–471**) and neorogioldiol B (**472**) (*642*). Prevezol B (**468**) was revised (*642*) from the original assignment (*641*).

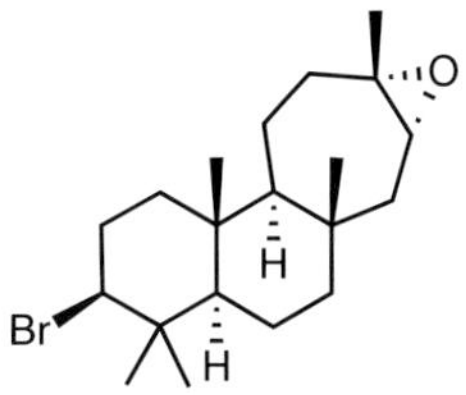

3.4 Terpenes

464 (3-bromobarekoxide)

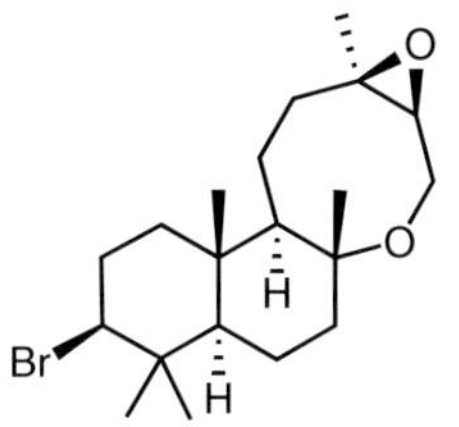

465

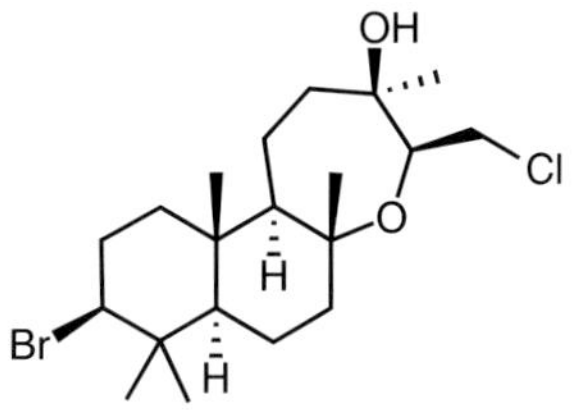

466

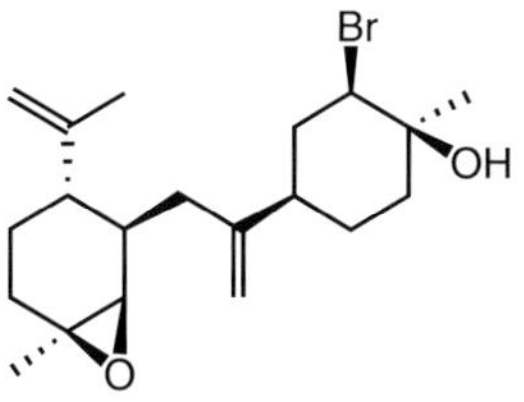

467 (prevezol A)

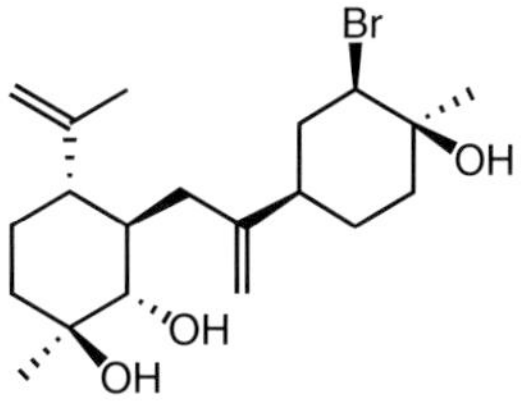

468 (prevezol B)

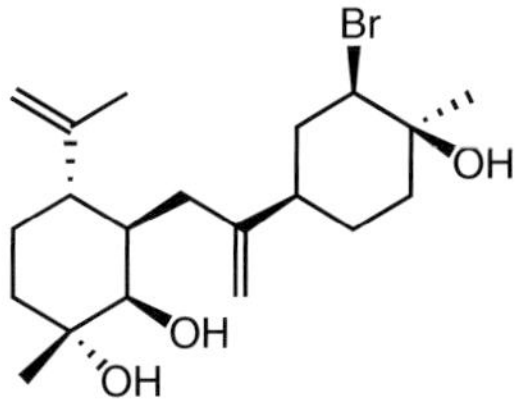

469 (prevezol C)

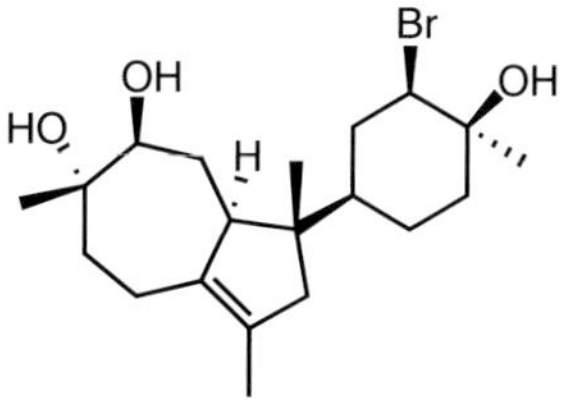

470 (prevezol D)

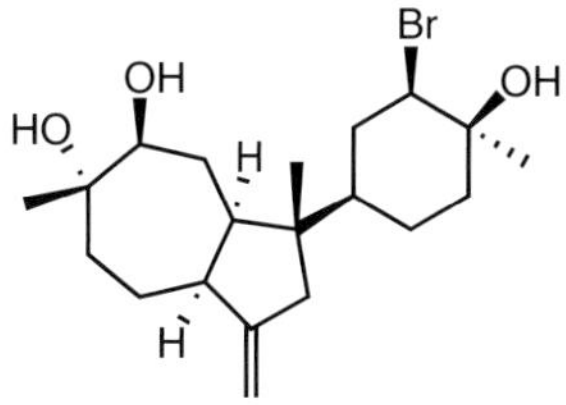

471 (prevezol E)

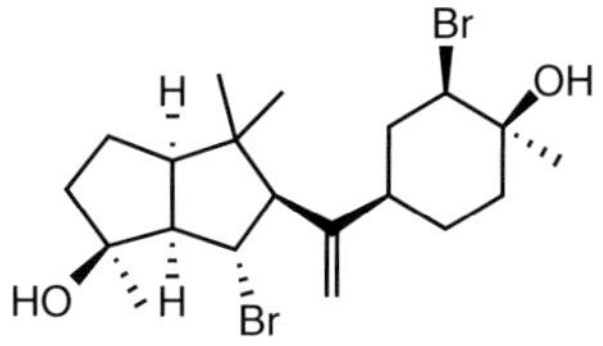

472 neorogioldiol B)

The neoirieane diterpene neoirietetraol (**473**) is found in the new *Laurencia yonaguniensis* species (*643*), and the novel luzodiol (**474**) was isolated from *Laurencia luzonensis* found in Okinawa (*558*). A study of *Laurencia microcladia* from the coast of Tuscany has yielded rogioldiol A (**475**), rogiolal (**476**), isorogiolal **477** (*644*), rogioldiols B (**478**), and C (**479**) (*645*). Further studies of this seaweed identified neorogioldiol (**480**), rogioldiol D (**481**), and **482** (*646*).

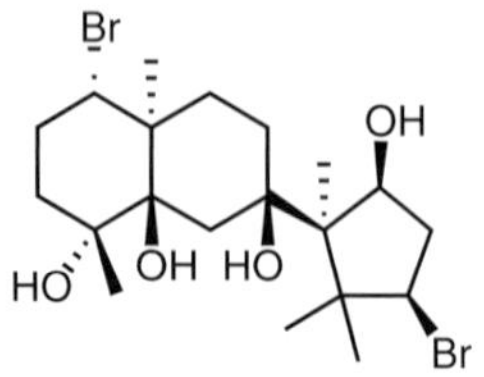

473 (neoirietetraol)

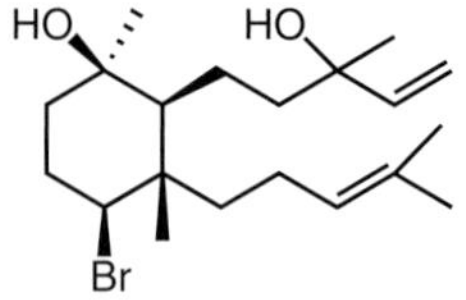

474 (luzodiol)

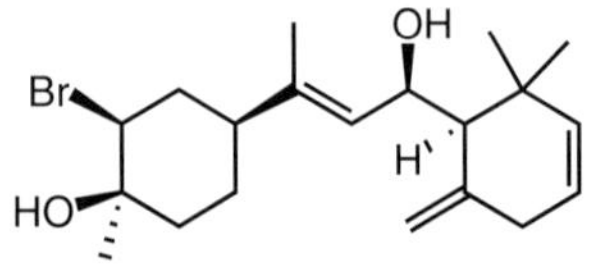

475 (rogioldiol A)

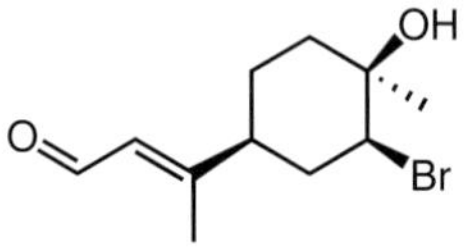

476 (rogiolal)

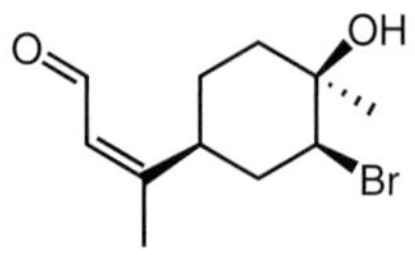

477 (isorogiolal)

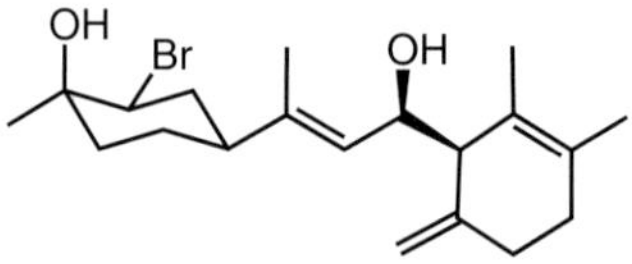

478 (rogioldiol B)

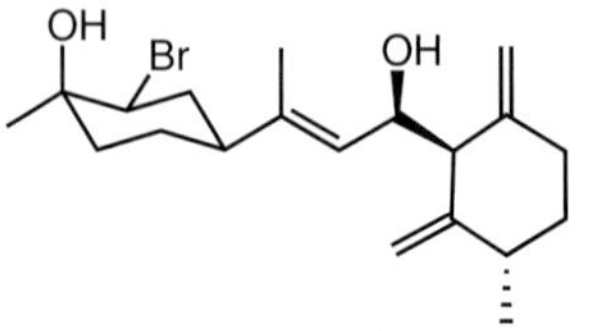

479 (rogioldiol C)

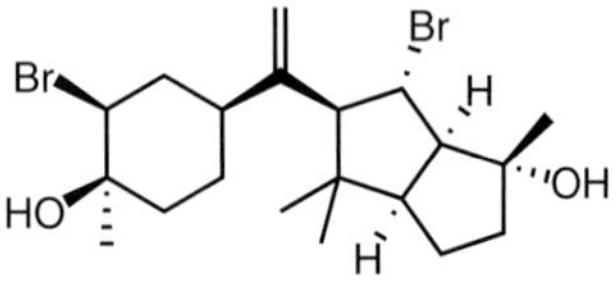

480 (neorogioldiol)

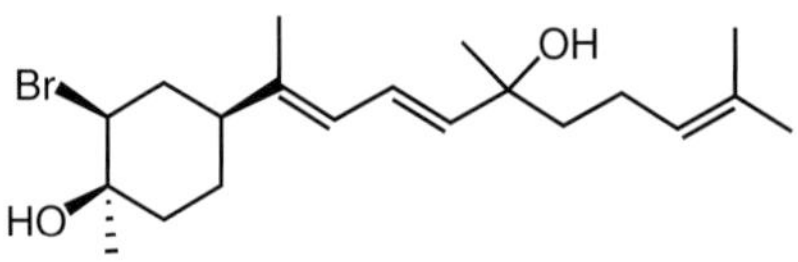

481 (rogioldiol D)

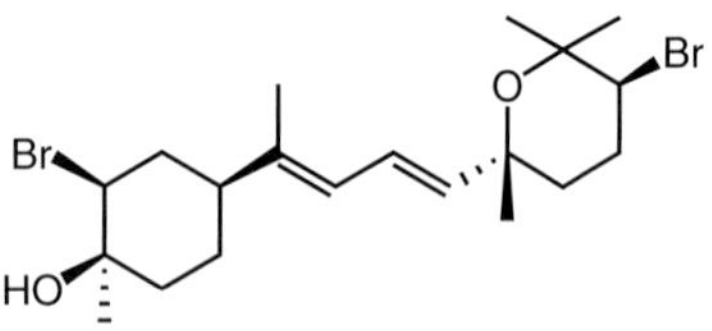

482

Sphaerococcus and Other Red Algae Diterpenes

The Mediterranean red alga *Sphaerococcus coronopifolius*, which was seen to be a rich source of novel bromine-containing diterpenes in the first survey (*1*), has furnished some new examples. A Naples collection of this seaweed has afforded norsphaerol (**483**), and sphaerolabdadiene-3,14-diol (**484**) and bromosphaerone (**485**) were characterized from a Morocco version of this alga (*648*). The novel and unprecedented iodinated diterpenes tasihalides A (**486**) and B (**487**) were isolated from a *Symploca* cyanobacterium associated with an unidentified red alga (*649*). The Fijian red alga *Callophycus serratus* produces the nine novel bromophycolides A–I (**488**–**496**), which contain a diterpene-benzoate skeleton (*650, 651*).

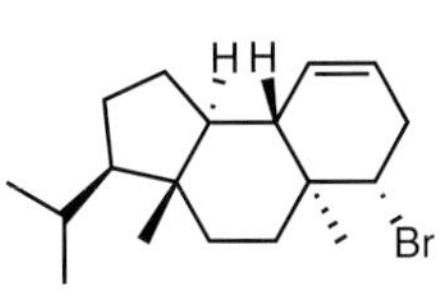

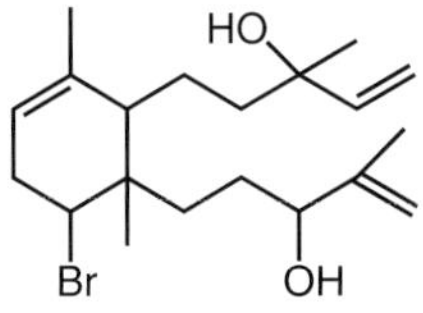

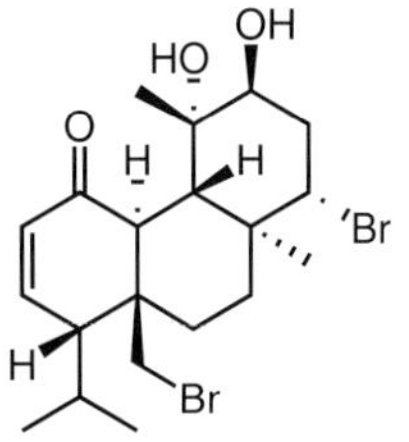

483 (norsphaerol)   **484** (sphaeroladadiene-3,14-diol)   **485** (bromosphaerone)

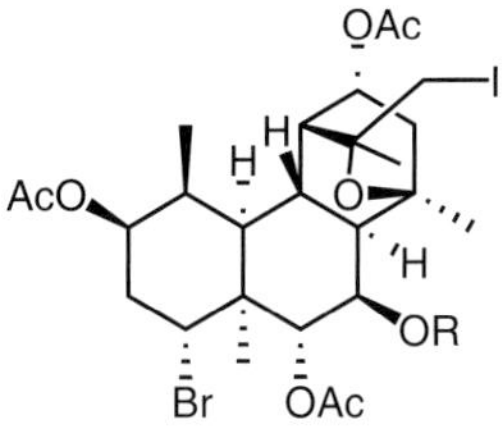

486 R = H (tasihalide A)
487 R = Ac (tashalide B)

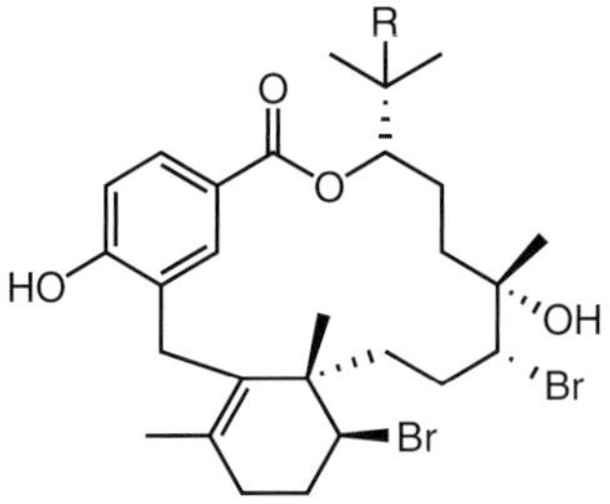

488 R = Br (bromophycolide A)
490 R = OH (bromophycolide C)

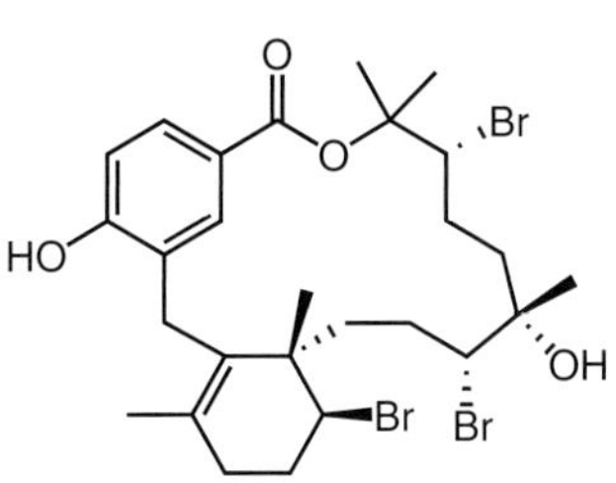

489 (bromophycolide B)

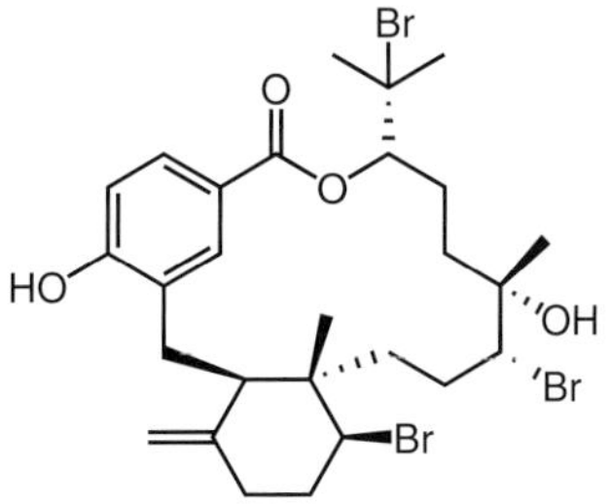

491 (bromophycolide D)

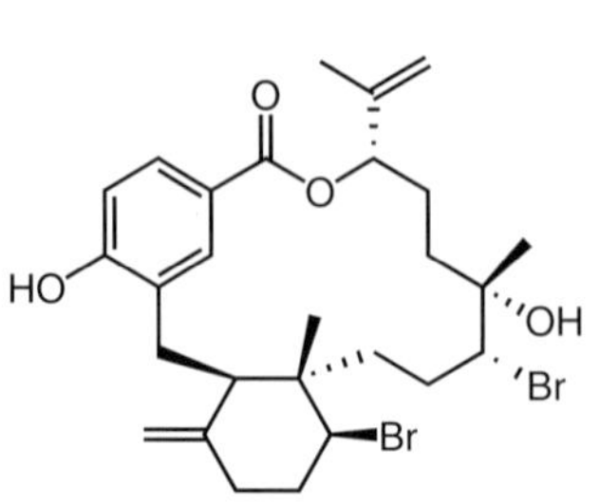

492 (bromophycolide E)

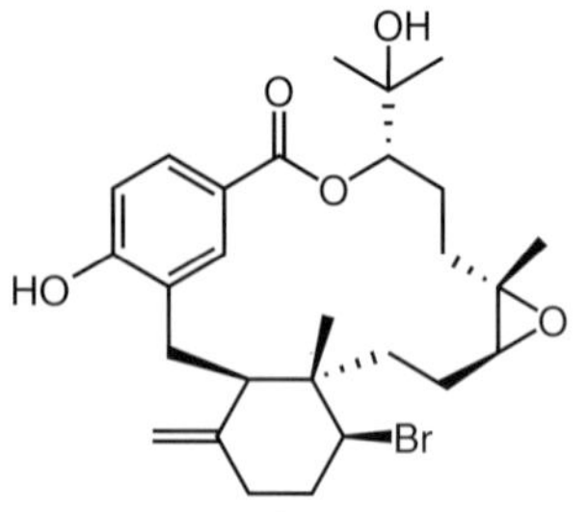

493 (bromophycolide F)

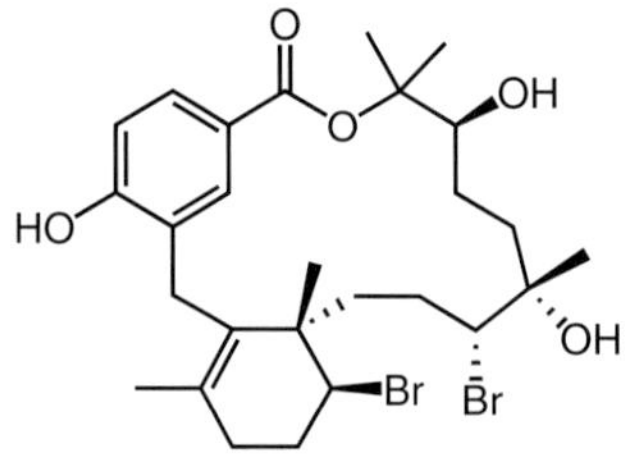

494 (bromophycolide G)

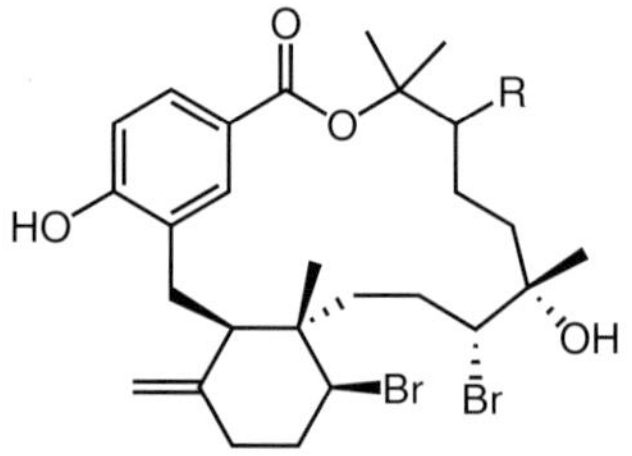

495 R = α-Br (bromophycolide H)
496 R = β-OH (bromophycolide I)

Sponge Diterpenes

Relatively few sponge diterpenes are known and these are typified by the isocyano kalihinanes (*600, 652*), such as the kalihinols that were presented in the first survey (*1, 653, 654*). The absolute configuration of kalihinol A has been determined (*655*) and it is correct as shown in the first survey (*1*). Sponges of genus *Acanthella* from the Pacific Ocean are the producers of the kalihinanes. Investigations of *Acanthella cavernosa* in Pacific waters south of Tokyo (Yakushima Island) have uncovered the new kalihinenes X (**497**), Y (**498**), and Z (**499**) (*656*), and a total synthesis of kalihinene X (**497**) has established its relative and absolute configuration (*657*). A collection from a slightly different location revealed 10β-formamidokalihinol-A (**500**), 10β-formamidokalihinol-E (**501**), 10β-formamido-5-isocyanatokalihinol-A (**502**), and 10β-formamido-5β-isothiocyanatokalihinol-A (**503**) (*658*). The Yakushima sample also yielded kalihipyran B (**504**) (*659*). All of these metabolites display potent antifouling activity against larvae of the barnacle *Balanus amphitrite*, suggesting a natural function for these compounds in the sponge. An Okinawan collection of *Acanthella* sp. contains the new Δ^9-kalihinol Y (**505**) and 10-epikalihinol I (**506**) (*660*). This paper also describes the powerful antimalarial activity of kalihinol A. The Philippines sponge *Phakellia pulcherrima*, which is

3.4 Terpenes

in the same family as *Acanthella cavernosa*, contains several known kalihinols, including **505** (*661*).

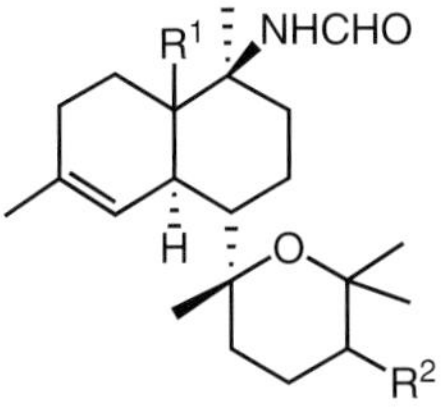

497 R¹ = α-H, R² = β-Cl (kalihinene X)
498 R¹ = β-H, R² = β-Cl (kalihinene Y)
499 R¹ = α-H, R² = α-Cl (kalihinene Z)

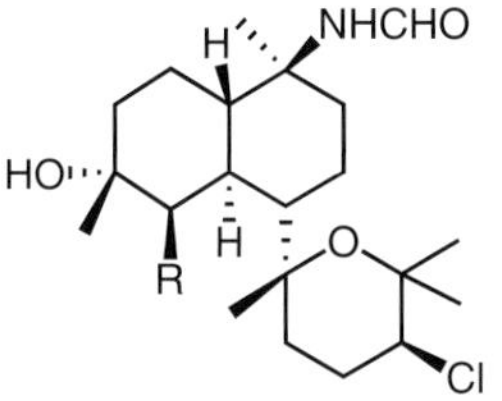

500 R = NC
502 R = NCO
503 R = NCS

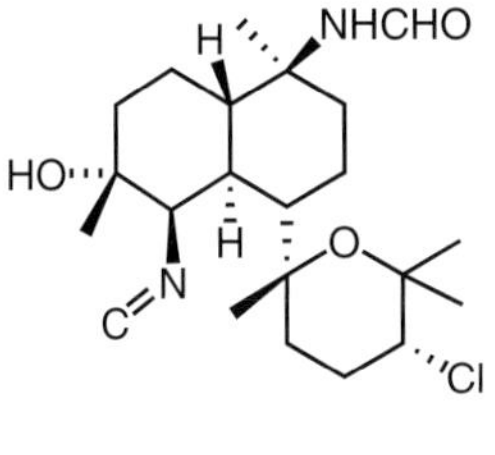

501

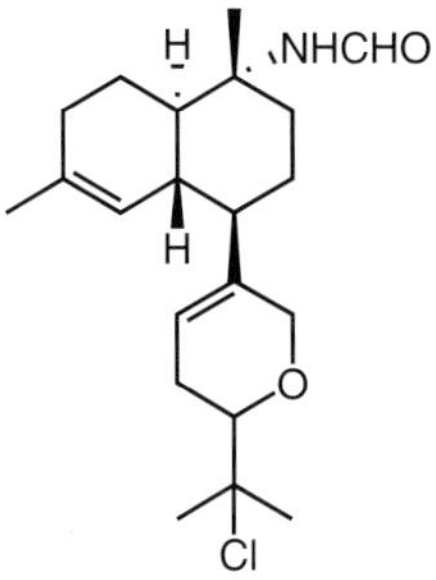

504 (kalihipyran B)

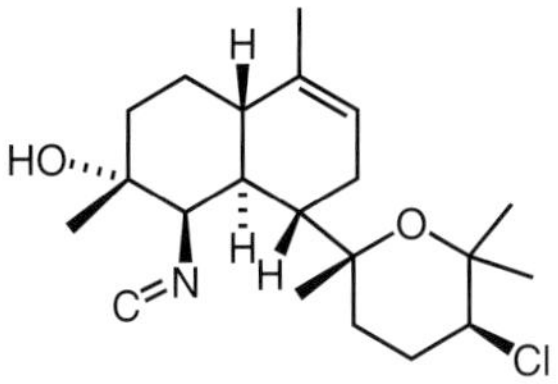

505 (Δ^9-kalihinol Y)

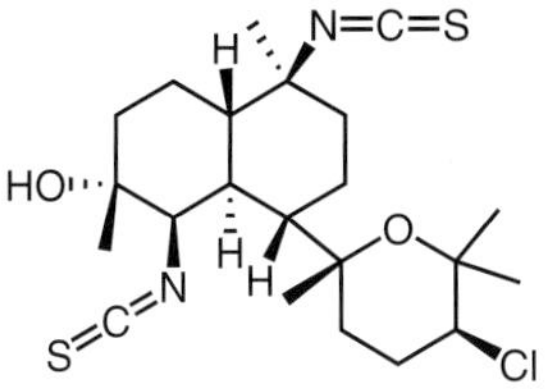

506 (10-epikalihinol I)

The first report of isocyanide diterpenes occurring in nudibranchs (sea slugs) has appeared, which describes the known kalihinol A and kalihinol E, along with nonchlorinated metabolites, in *Phyllidiella pustulosa* from the South China Sea (*662*). This supports the notion that nudibranchs feed on sponges and thereby acquire metabolites for their own purposes, giving new meaning to the term "lazy slugs". A nudibranch from South Africa, *Chromodoris hamiltoni*, contains the novel hamiltonins A–D (**507–510**) (*663*). The Mediterranean dorid nudibranch *Doris verrucosa* has afforded several novel diterpene isocopalane verrucosins, including the chlorinated verrucosins-7 (**511**) and -9 (**512**) (*664*).

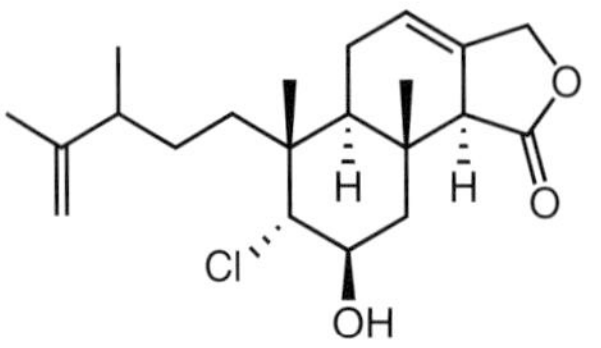

507 (hamiltonin A)

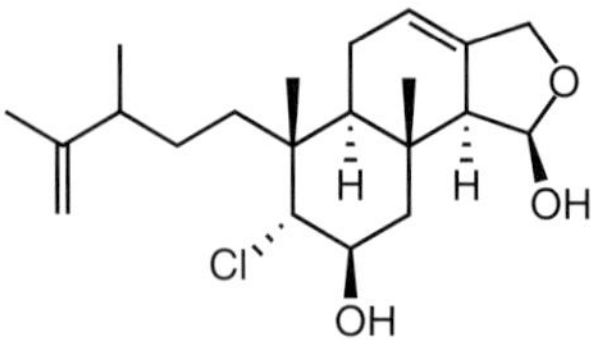

508 (hamiltonin B)

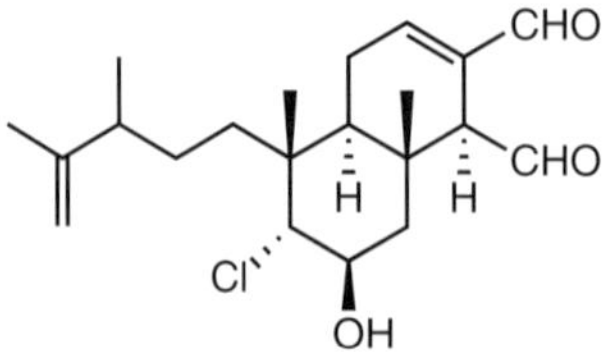

509 (hamiltonin C)

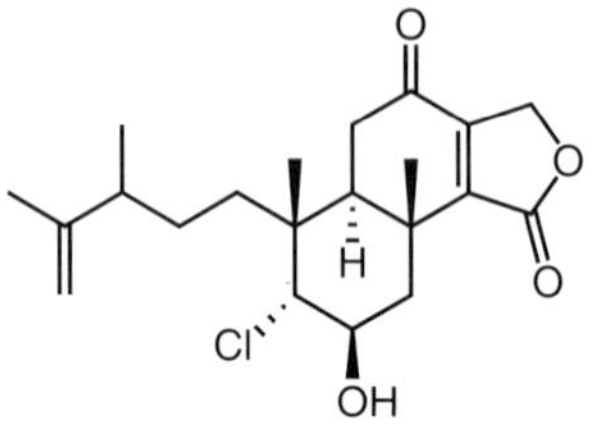

510 (hamiltonin D)

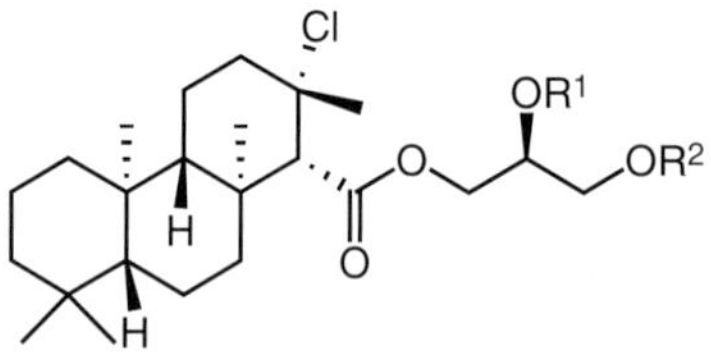

511 R^1 = H, R^2 = Ac (verrucosin-7)
512 R^1 = Ac, R^2 = H (verrucosin-9)

Gorgonian Diterpenes

Gorgonians produce the largest complement of chlorine-containing marine metabolites – more than 50 were illustrated in the first survey (*1*) – and many more nonchlorinated gorgonian diterpenes are known (*665–667*). There is evidence to indicate that these gorgonian diterpenoids are feeding deterrents to reef fishes. Gorgonian corals can achieve densities of up to 20 colonies per square meter on the reef (*668, 669*).

The stereochemistry of the diterpenoid praelolide, which was originally isolated from *Plexaureides praelonga* from the South China Sea (*1*) and subsequently from the Indian Ocean gorgonian *Gorgonella umbraculum* (*670*) and the Taiwanese *Junceella fragilis* and *Junceella juncea* (*671*), has now been confirmed as **S** (*670, 671*). Likewise the structure of junceellin, which was incorrect in the first survey (*1*), is corrected as **T** (*670, 671*). The stereochemistry of the previously known solenolides C and D are proposed to be revised as **U** and **V** (*672*). Several new chlorinated briareins have been identified from the common Caribbean gorgonian *Briareum asbestinum*, including briareins B (**513**), C (**514**), D (**515**), E (**516**), F (**517**), G (**518**), and J (**519**) (*673*). Although briarein B was isolated some years previously, its structure was not positively established at that time (*674, 675*).

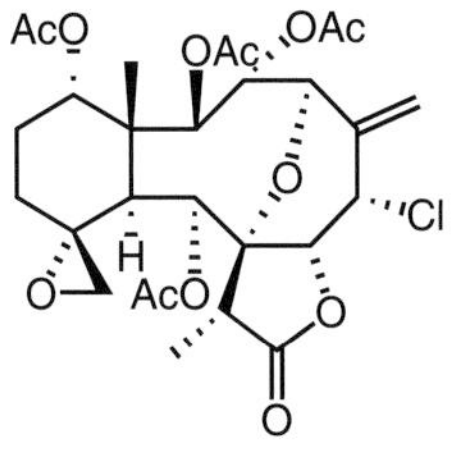

3.4 Terpenes

S (praelolide)

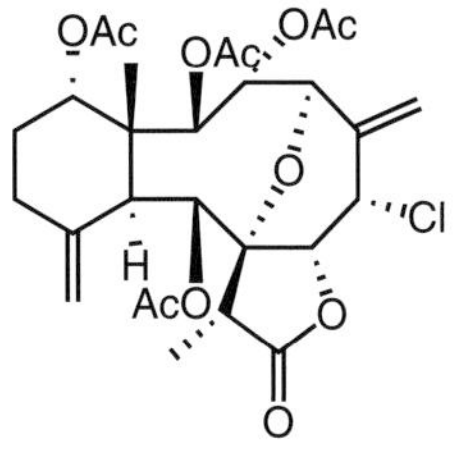

T (junceellin)

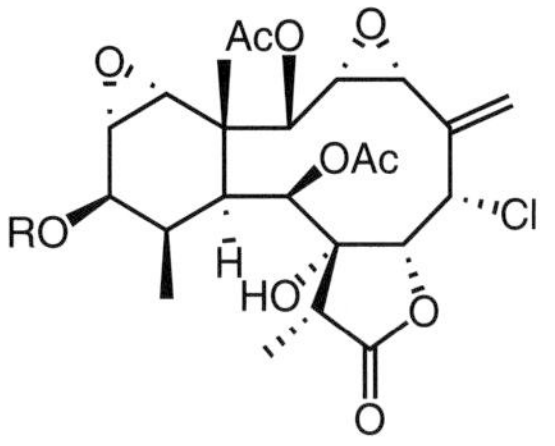

U R = H (solenolide C)
V R = Ac (solenolide D)

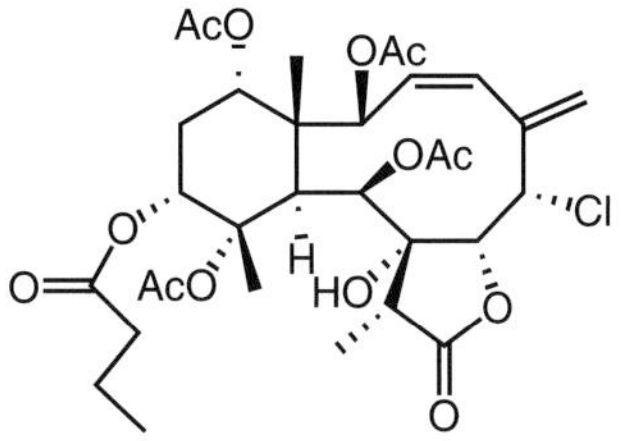

513 (briarein B)

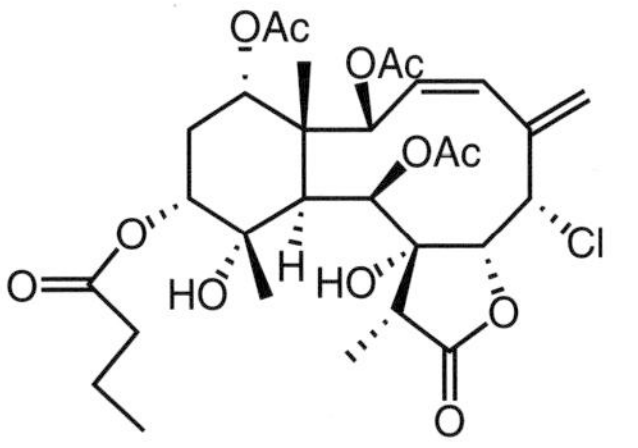

514 (briarein C)

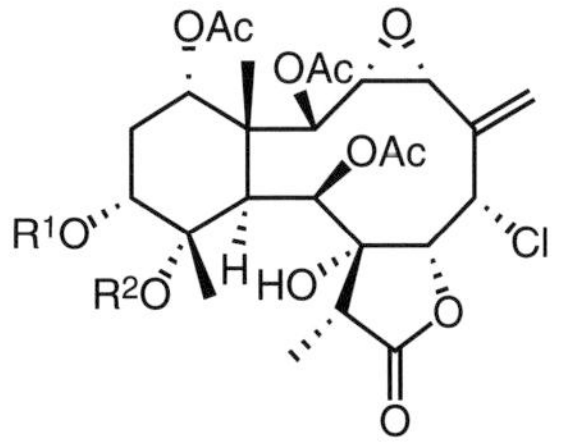

515 R¹ = R² = Ac (briarein D)
516 R¹ = Ac, R² = H (briarein E)
518 R¹ = COPr, R² = Ac (briarein G)

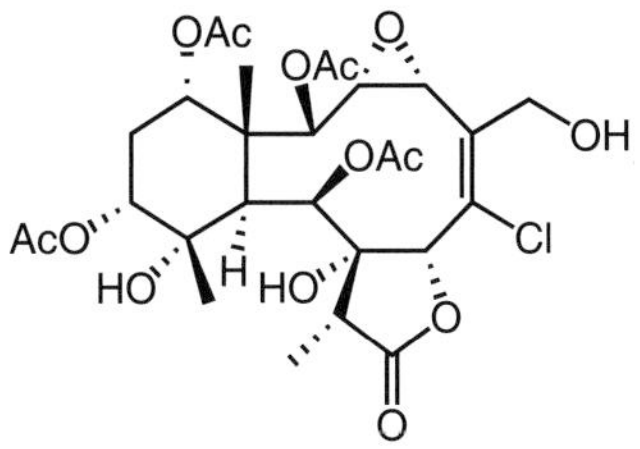

517 (briarein F)

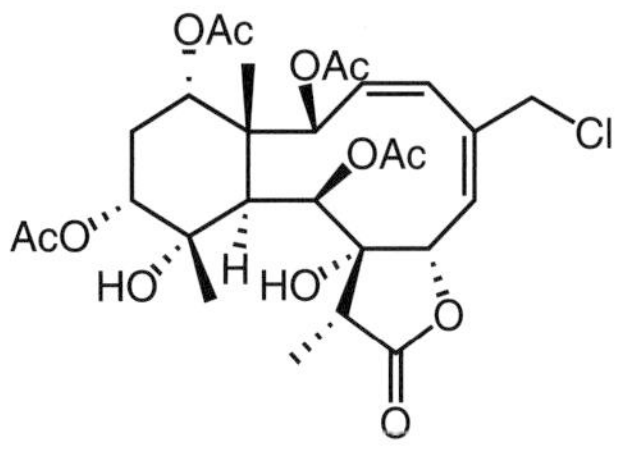

519 (briarein J)

A Bahamian collection of *Briareum asbestinum* (Fig. 3.5) has provided the new 11-hydroxybrianthein V (**520**), 11-hydroxybrianthein U (**521**), 11-hydroxybrianthein Y (**522**), 3,4-dihydro-11-hydroxybrianthein V (**523**), and 3,4-dihydro-

Fig. 3.5 *Briareum asbestinum*, a Caribbean gorgonian soft coral that produces numerous chlorinated diterpenes such as the briareins (**513–519**) and briantheins (**520–524**) (Photo: W. Fenical)

11-hydroxybrianthein U (**524**) (*676*), whereas a study of *Briareum excavatum* from Taiwan yielded a series of excavatolides, one of which is chlorinated, excavatolide A (**525**) (*677*). This latter gorgonian collection also contains seven chlorinated briaexcavatolides E (**526**), F (**527**), G (**528**), H (**529**), I (**530**), J (**531**) (*678*), and M (**532**) (*679*).

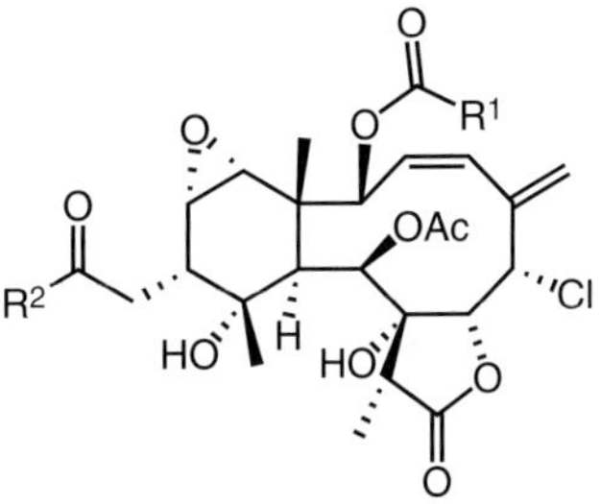

3.4 Terpenes

520 R¹ = R² = n-Pr
521 R¹ = Me, R² = n-Pr
522 R¹ = n-Pr, R² = Me

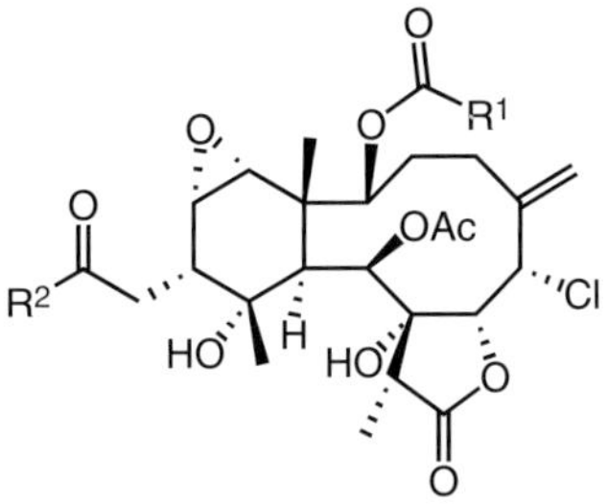

523 R¹ = R² = n-Pr
524 R¹ = Me, R² = n-Pr

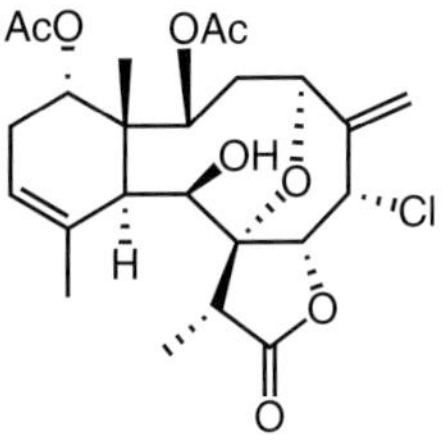

525 (excavatolide A)

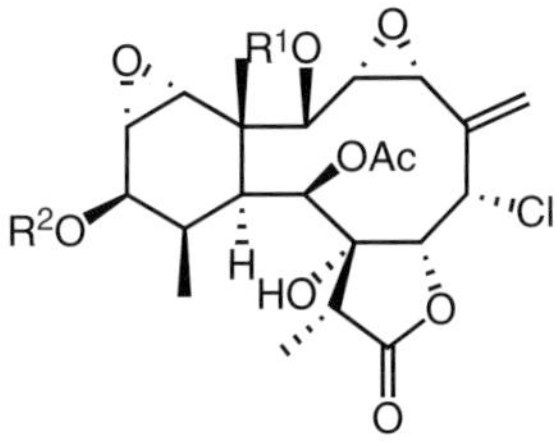

526 R¹ = R² = Ac (briaexcavatolide E)
527 R¹ = H, R² = COCH$_2$CH$_2$CH$_3$ (briaexcavatolide F)
528 R¹ = R² = H (briaexcavatolide G)

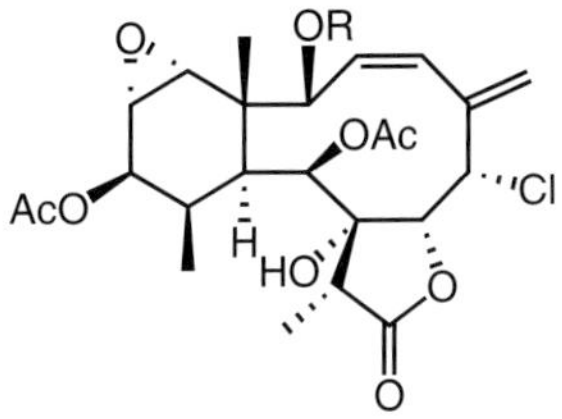

529 R = H (briaexcavatolide H)
530 R = Ac (briaexcavatolide I)

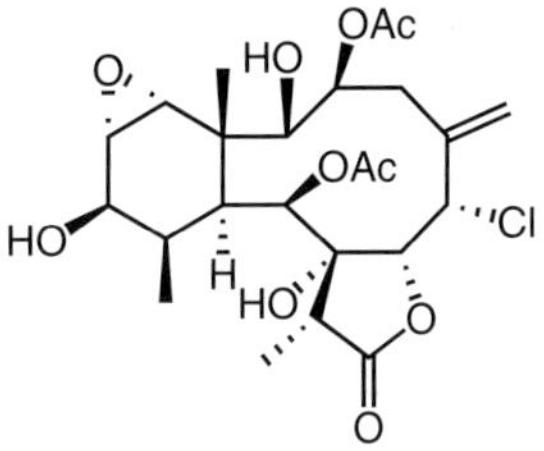

531 (briaexcavatolide J)

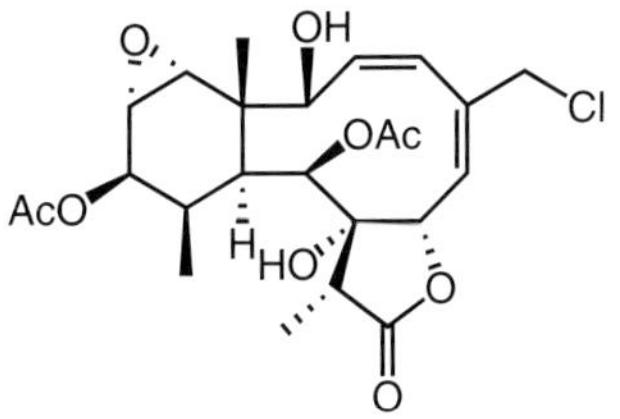

532 (briaexcavatolide M)

The Micronesian soft coral *Briareum stechei* (Fig. 3.6) has furnished a series of milolides, including 16-chloromilolide B (**533**), milolide C (**534**), 4-hydroxymilolide C (**535**), milolide D (**536**), milolide E (**537**) (*680*), and milolide L (**538**) (*681*).

Fig. 3.6 *Briareum stechei*, a Western Pacific octocoral that produces the milolides (**533–538**) (Photo: F. J. Schmitz)

The stereochemistry of solenolide C (**U**) is also revised in this study (*680*). The new briviolides B (**539**) and C (**540**) were characterized from a Japanese collection of *Briareum* sp. (*682*). A study of octocorals from Pohnpei and Ant atoll in Micronesia led to the novel nui-inoalides A–D (**541–544**) (*683*). The absolute configuration of juncin E (**W**) was also established by these researchers.

533 (16-chloromilolide B)

534 (milolide C)

535 (4-hydroxymilolide C)

536 R^1 = H, R^2 = COC_3H_7 (milolide D)
537 R^1 = Ac, R^2 = COC_5H_{11} (milolide E)

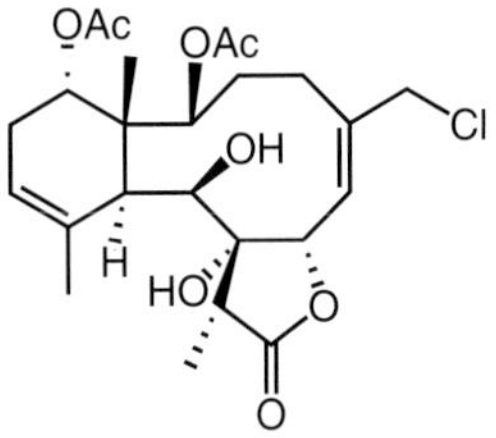

3.4 Terpenes

538 (milolide L)

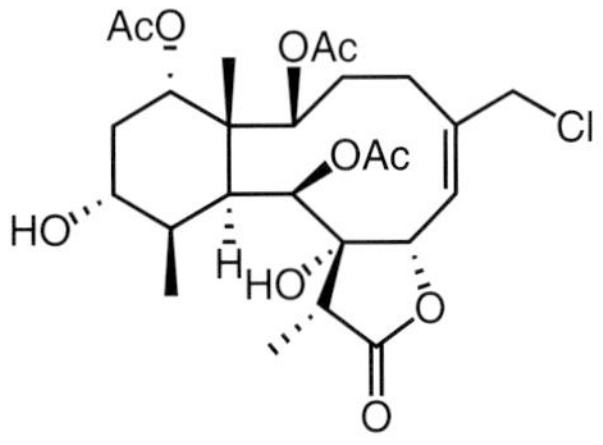

539 (briviolide B)

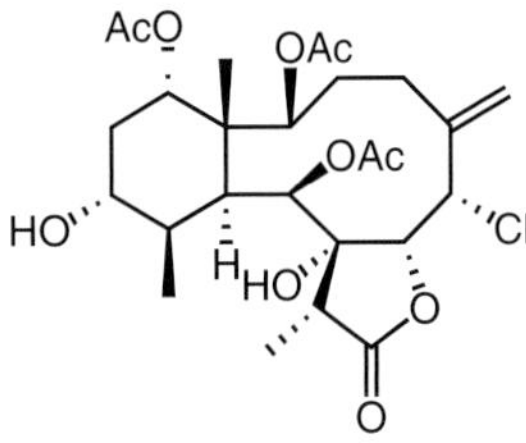

540 (briviolide C)

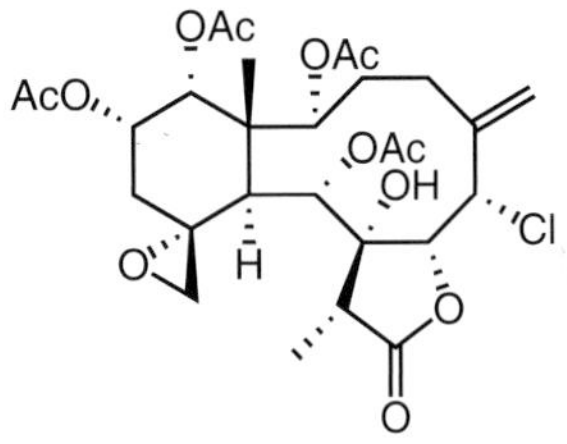

541 (nui-inoalide A)

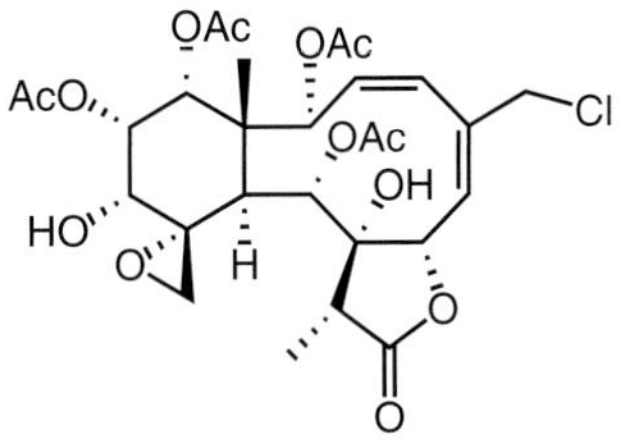

542 (nui-inoalide B)

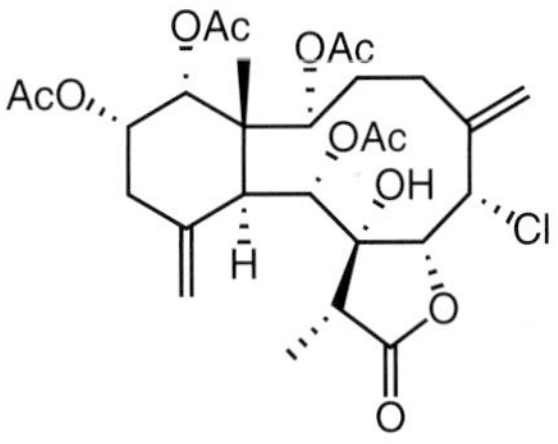

543 (nui-inoalide C)

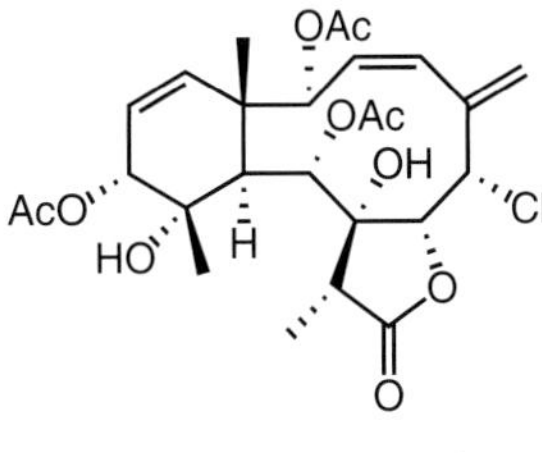

544 (nui-inoalide D)

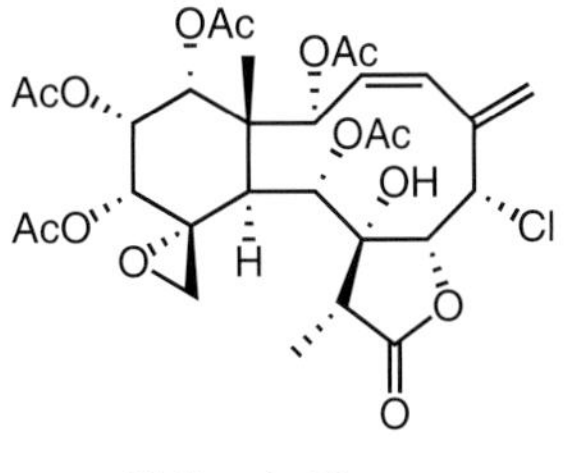

W (juncin E)

Another prolific gorgonian is *Junceella*, and these animals contributed 15 chlorinated briarane diterpenoids to the first survey. A series of juncins A–F from the Red Sea *Junceella juncea* were presented earlier, although C and F were not precisely defined (*1*). The Indian Ocean *Junceella juncea* contains the new juncins G (**545**) and H (**546**), along with the antipodes of the previously known gemmacolides A (**547**) and B (**548**) (*684*). The same research group isolated chlorinated juncins L (**549**) and M (**550**) from this collection of *Junceella juncea* (*685*). Chemical extraction of this octocoral living in Taiwan waters furnished juncin N (**551**) (*686*), and the chlorine-containing juncins O (**552**) and P (**553**) were characterized from a South China Sea *Junceella juncea* (*687*), which also afforded juncins R (**554**), S (**555**), and ZI (**556**), along with seven new non-chlorinated briaranes (*688*). The Taiwanese *Junceella juncea* is also the source of juncenolides A (**557**) (*689*), F (**558**), and G (**559**) (*690*).

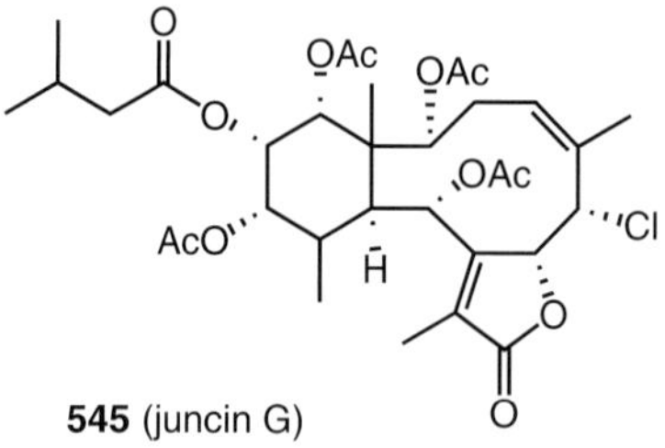

545 (juncin G)

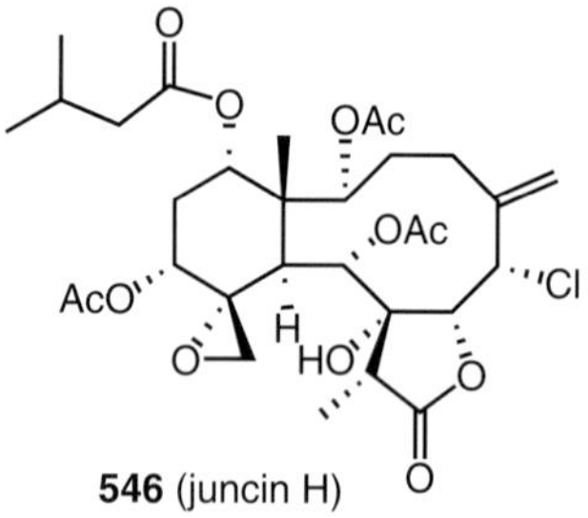

546 (juncin H)

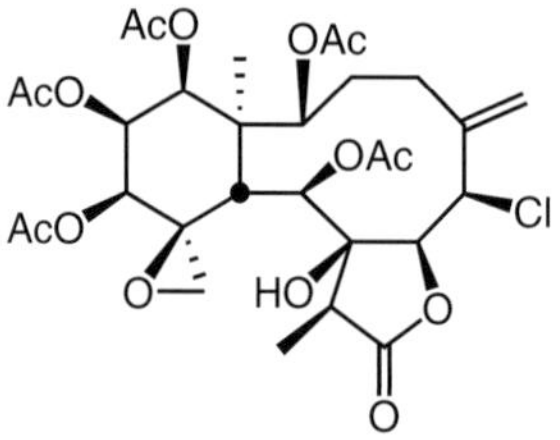

547 (gemmacolide A)

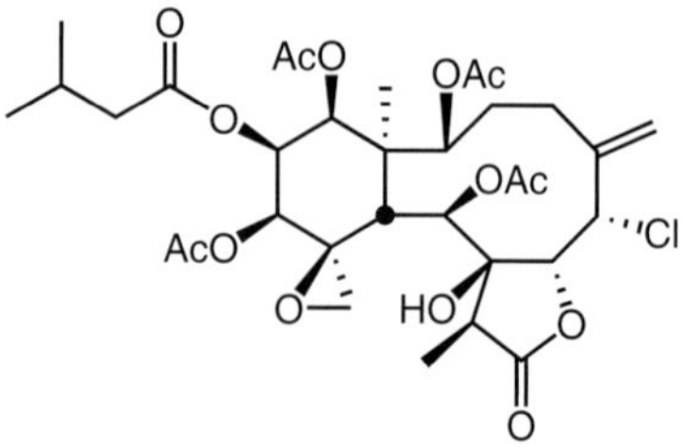

548 (gemmacolide B)

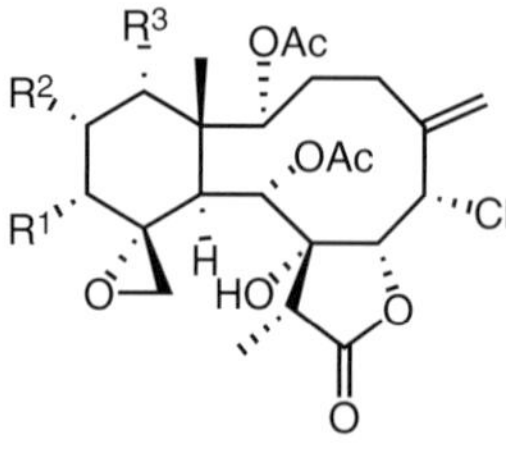

549 $R^1 = R^2 = OCOCH_2CH(CH_3)_2$, $R^3 = OAc$ (juncin L)
550 $R^1 = R^3 = OCOCH_2CH(CH_3)_2$, $R^2 = H$ (juncin M)

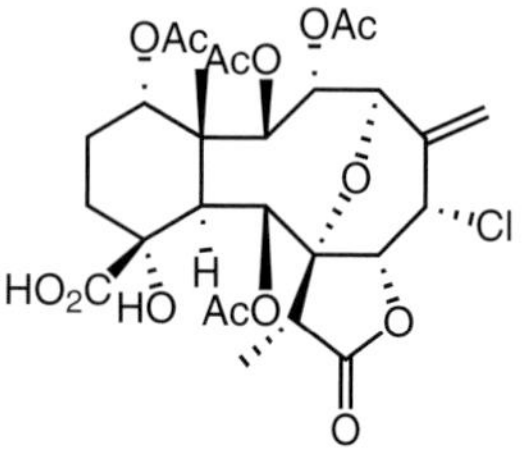

551 (juncin N)

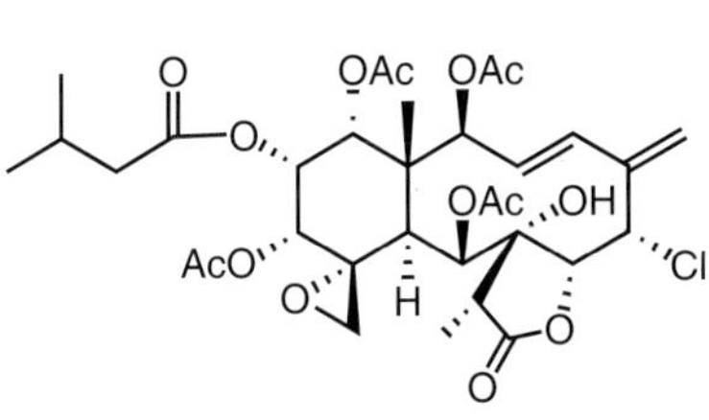

3.4 Terpenes

552

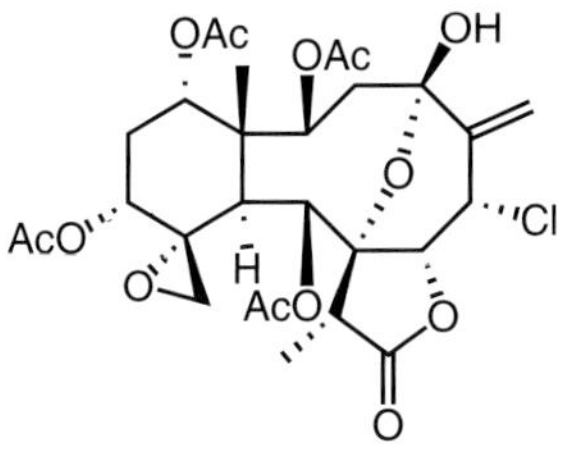

553 (juncin P)

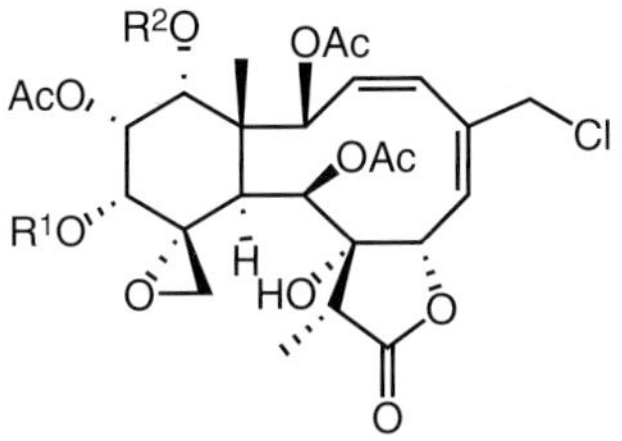

554 R¹ = Ac, R² = COCH$_2$CH(CH$_3$)$_2$ (juncin R)
555 R² = Ac, R¹ = COCH$_2$CH(CH$_3$)$_2$ (juncin S)

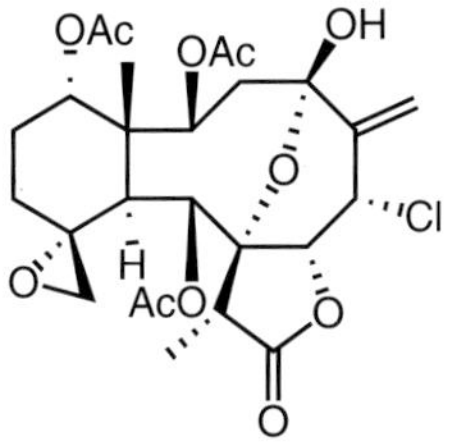

556 (juncin Zl)

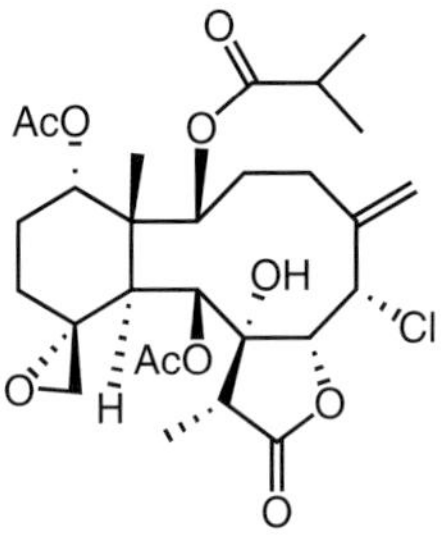

557 (juncenolide A)

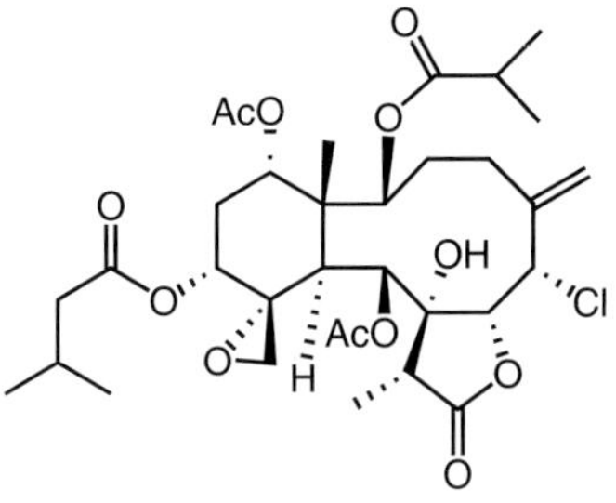

558 (juncenolide F)

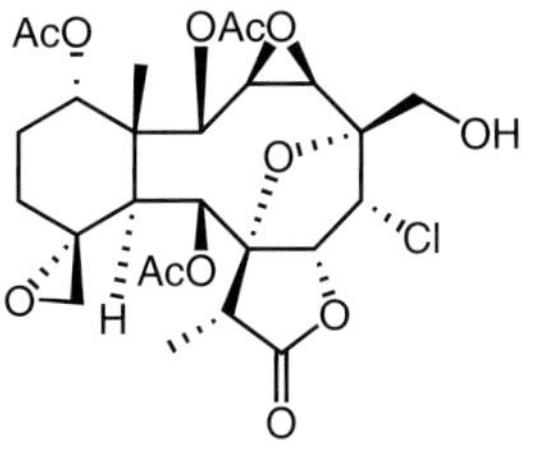

559 (juncenolide G)

Junceella fragilis from Indonesia contains the novel antipode (+)-junceelloide A (**560**) of the known (−)-junceelloide A (drawn incorrectly in (*1*)) (*691*). A collection of *Junceella fragilis* from the South China Sea yielded junceellonoid A (**561**) (*692*) and junceellonoids C–E (**562–564**) (*693*), and a Papuan sample of this coral led to (−)-2-deacetyljunceellin (**565**) and (−)-3-deacetyljunceellin (**566**) (absolute configuration shown) (*694*).

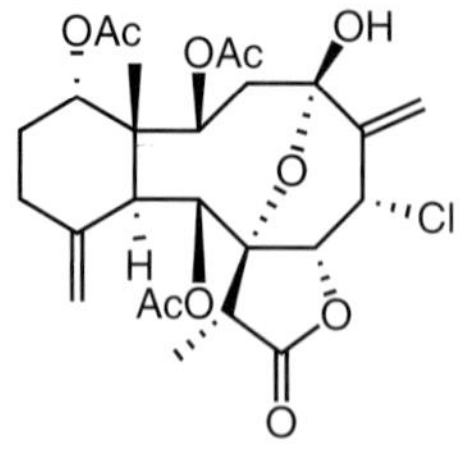

560 ((+)-junceellolide A)

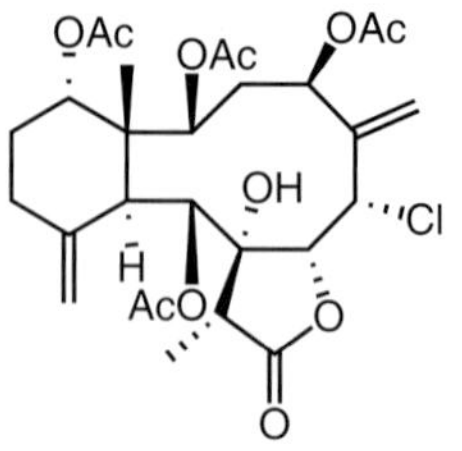

561 (junceellonoid A)

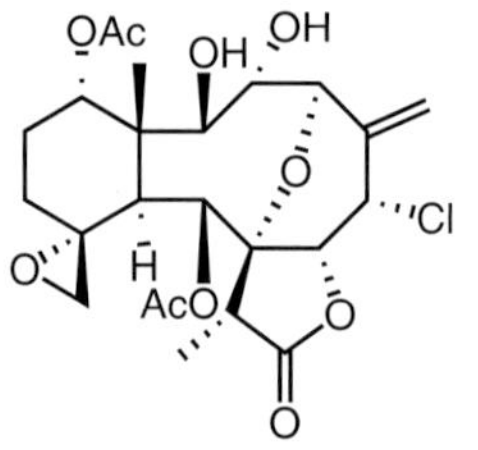

562 (junceellonoid C)

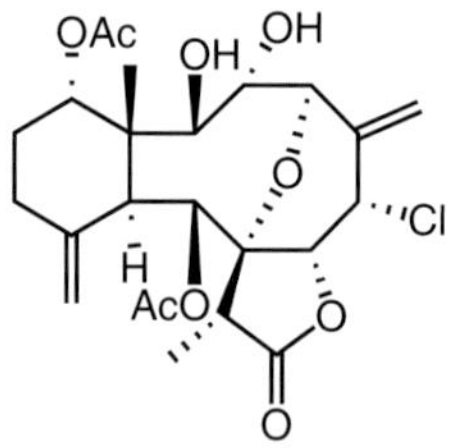

563 (junceellonoid D)

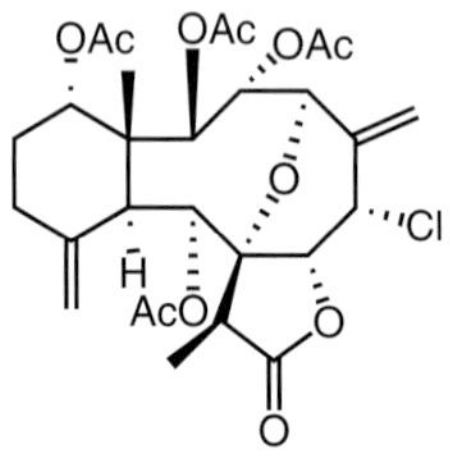

564 (junceellonoid E)

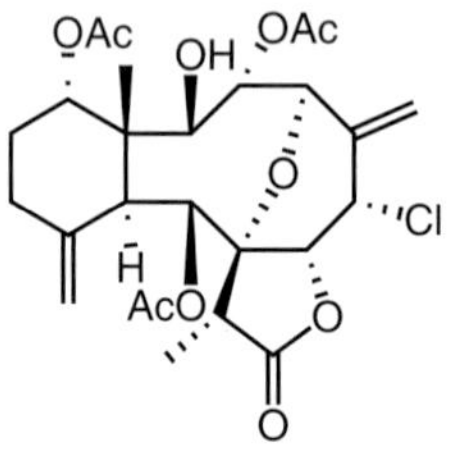

565 ((−)-2-deacetyljunceellin)

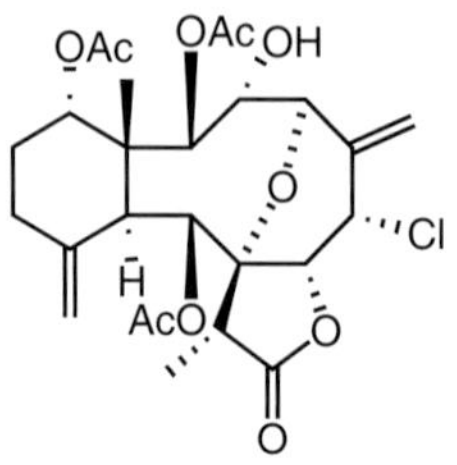

566 ((−)-3-deacetyljunceellin)

The common Caribbean octocoral *Erythropodium caribaeorum* has yielded several additional chlorinated diterpenes since the first survey (*1*).

3.4 Terpenes

Fig. 3.7 *Erythropodium caribaeorum*, a common Caribbean octocoral that produces chlorinated diterpenes such as the erythrolides and aquariolides (**567–574**) (Photo: W. Fenical)

The structure of the novel erythrolide K (**567**), which was isolated from *Erythropodium caribaeorum* (Fig. 3.7) collected in Tobago, was confirmed by synthesis from the known erythrolide A (*695*). A Jamaican source of this soft coral has afforded the new **568** and **569** in addition to six known erythrolides (*696*). These two new compounds are acetyl derivatives of erythrolides E and I. The Tobagoan *Erythropodium caribaeorum* also contains the three new chlorinated erythrolides L (**570**), P (**571**), and Q (**572**) (*697*). A survey of *Erythropodium caribaeorum* from Dominica has revealed the new erythrolides R (**573**), T (**574**), U (**575**), V (**576**), and aquariolides B (**577**) and C (**578**) (*698*). Aquariolide A (**579**) was earlier isolated from cultured (aquarium grown) *Erythropodium caribaeorum* (*699*). On the basis of these studies, the authors suggest a biogenesis of: briaranes to erythranes (i.e., erythrolides) then to aquarianes (i.e., aquariolides), involving sequential di-π-methane and vinyl cyclopropane rearrangements (*699*), the first transformation of which was mentioned previously (*1*).

567 (erythrolide K)

568 R = Ac
569 R = COCH$_2$OAc

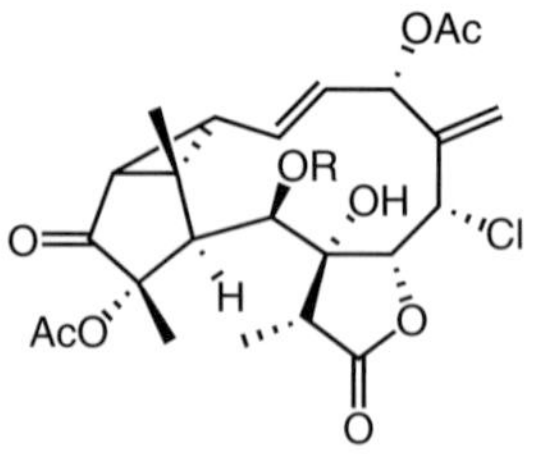

570 R = COCH$_2$OAc (erythrolide L)

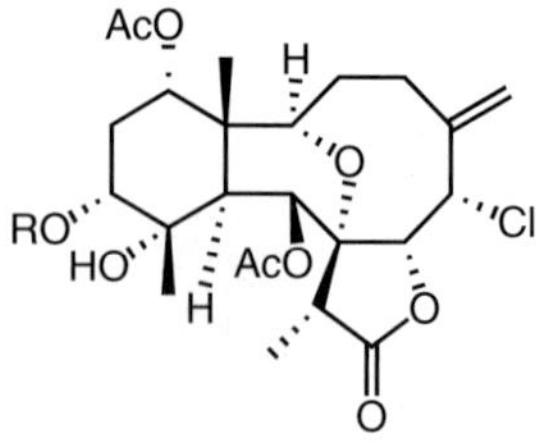

571 R = H   (erythrolide P)
572 R = Ac (erythrolide Q)

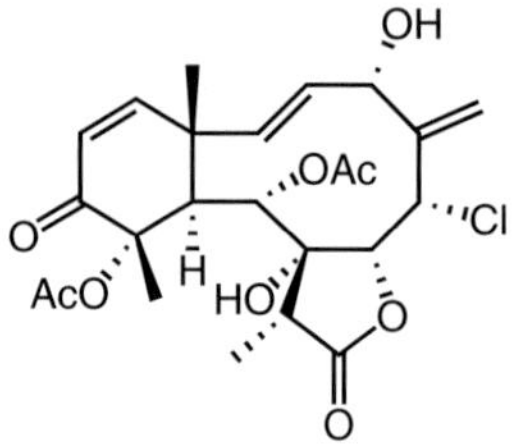

573 (erythrolide R)

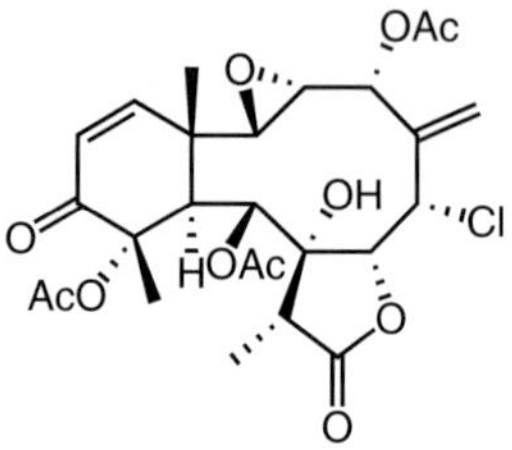

574 (erythrolide T)

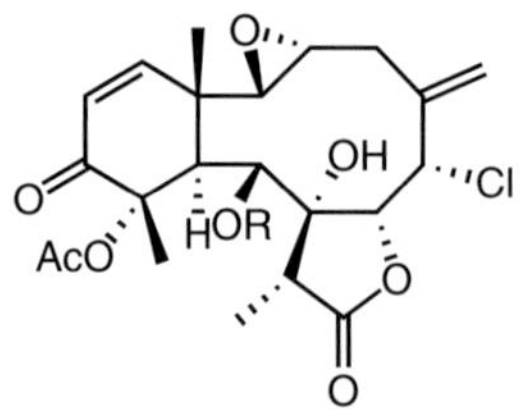

575 R = COCH$_2$OH (erythrolide U)

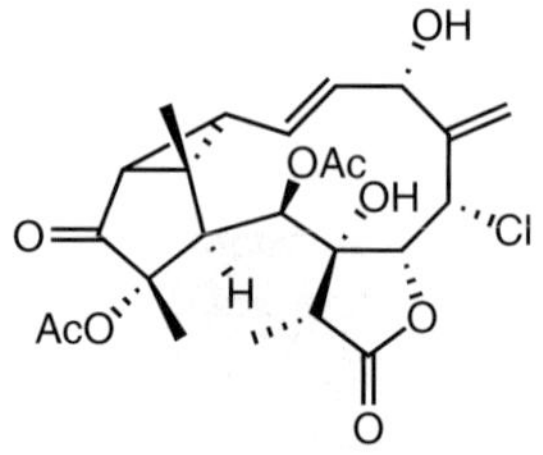

576 (erythrolide V)

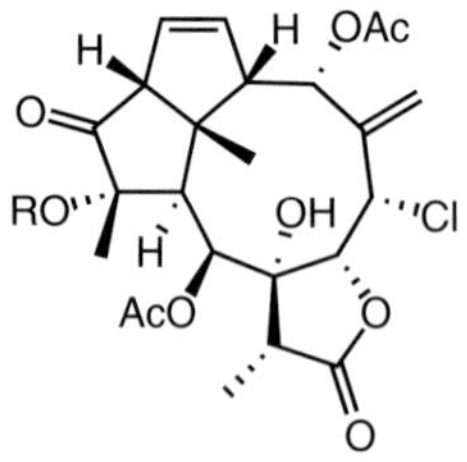

577 R = Me (aquariolide B)
578 R = Ac (aquariolide C)

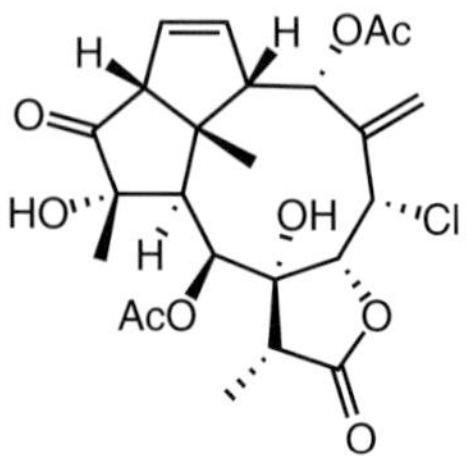

579 (aquariolide A)

The Indian Ocean gorgonian *Gorgonella umbraculum* has yielded the new umbraculolides C (**580**) and D (**581**) (*700*). The Okinawan sea whip *Ellisella* sp. (Fig. 3.8) furnished the four new briaranes **582–585**, and the sea pen *Pteroeides* sp. (Fig. 3.9) was likewise found to contain the novel **586** and **587** (*701*). Renillins A (**588**) and B (**589**) were isolated from the sea pansy *Renilla reniformis* (*702*). These

3.4 Terpenes

Fig. 3.8 *Ellisella* sp., a briarane-containing sea whip, collected in Alor, Indonesia (Photo: J. Tanaka)

new compounds with an unprecedented oxygenation pattern deterred feeding by the predatory blue crab, *Callinectes similis*, and two nonchlorinated renillins were deterrents to the predatory mummichog fish, *Fundulus heteroclitus*. The soft coral *Pachyclavularia violacea* has furnished pachyclavulide D (**590**) along with three nonchlorinated analogs (*703*). The sponge *Psammaplysilla purpurea* contains bis (deacetyl)solenolide D (**591**) (*704*). Examination of the Pohnpei octocoral

Fig. 3.9 *Pteroeides* sp., a sea pen from Flores, Indonesia, that contains the novel diterpenes **586** and **587** (Photo: J. Tanaka)

Eleutherobia sp. led to several known briareins. This study also found that minabien-6 is identical to 11-hydroxyptilosarcenone and that minabein-4 and nui-inoalide D are the same except they are epimeric at C-2 (*705*).

580 (umbraculolide C)

581 (umbraculolide D)

582

583

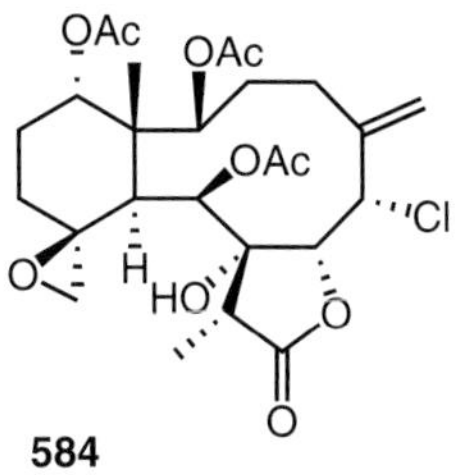

3.4 Terpenes

584

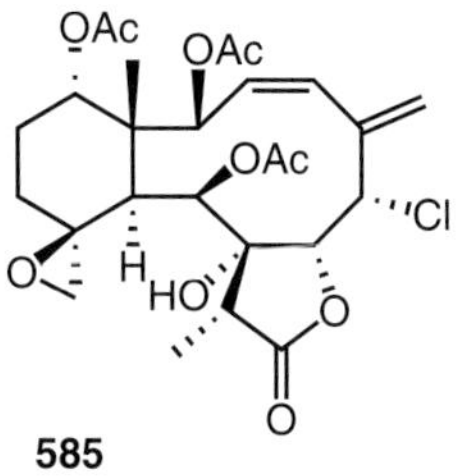

585

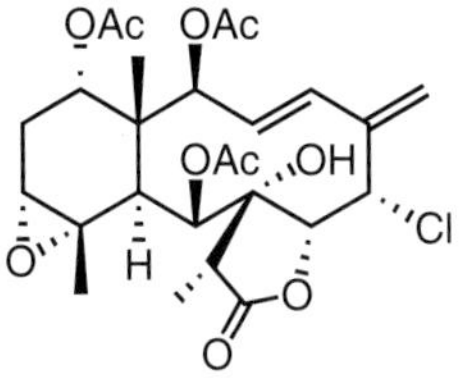

586

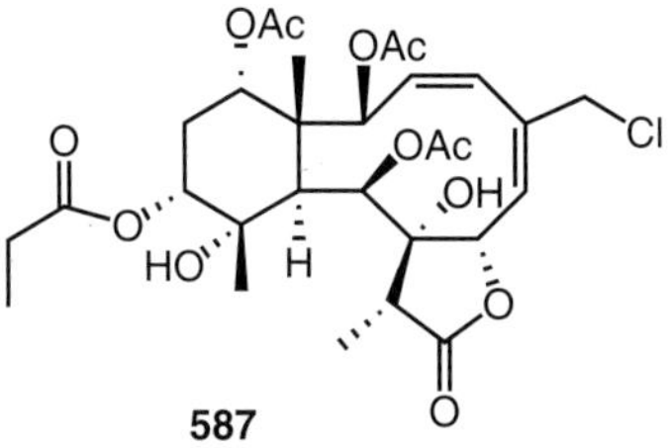

587

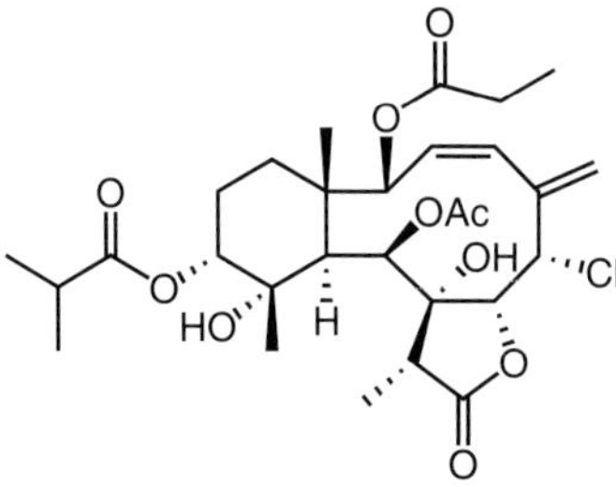

588 (renillin A)

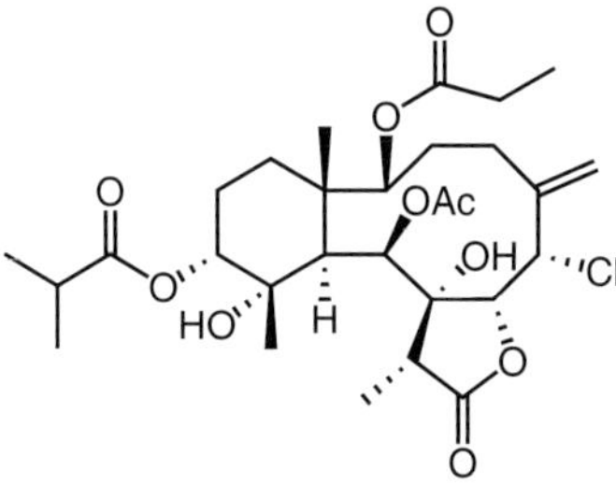

589 (renillin B)

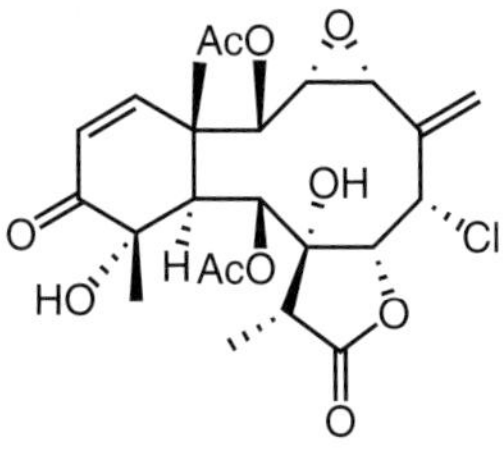

590 (pachyclavulide D)

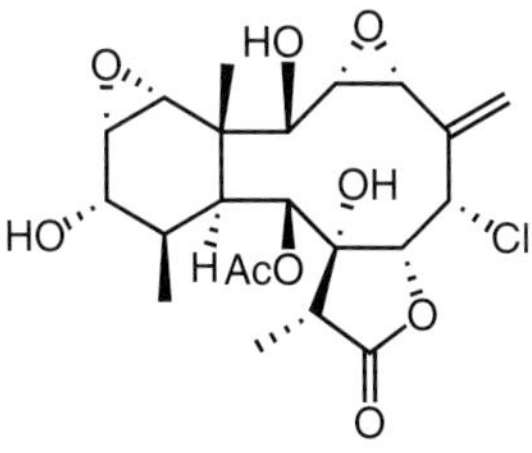

591

New diterpenoids of the dolabellane class have been reported, such as clavinflol B (**592**) from the Taiwanese soft coral *Clavularia inflata* (*706*). This metabolite has comparable cytotoxicity against the KB cell line to doxorubicin. A sea whip of the genus *Eunicea* has yielded the cembrane **593** (*707*). Both **592** and **593** are considered to be natural since no chlorinated solvents were used in the isolation process, and in both metabolites the chlorine is attached to the less substituted carbon, opposite to what is expected for acid-induced epoxide ring opening. A Kenyan soft coral, *Sinularia erecta*, contains the norcembrane sinularectin (**594**) (*708*).

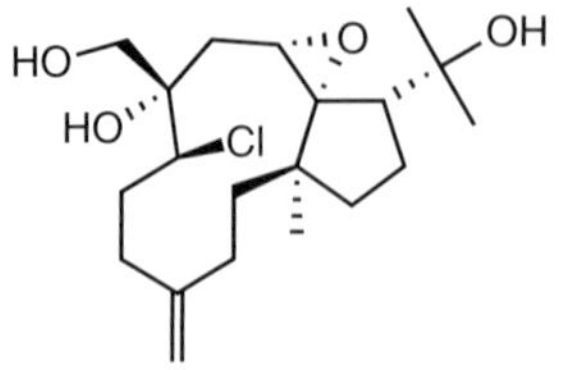

592 (clavinflol B)

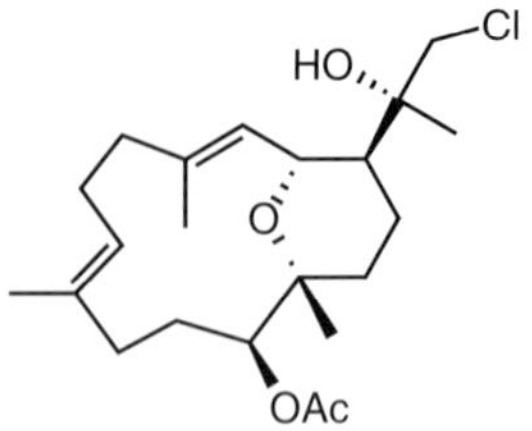

593

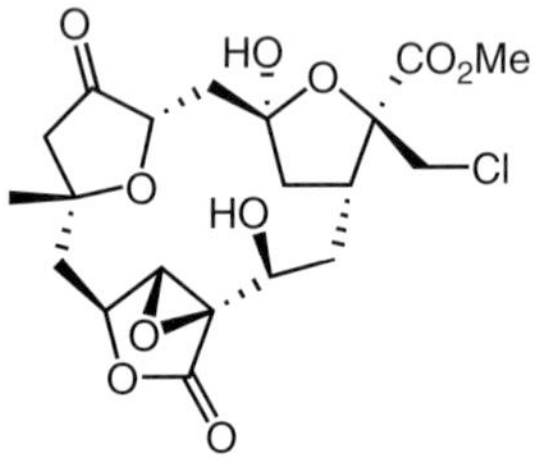

594 (sinularectin)

The New Caledonian ascidian *Lissoclinum voeltzkowi* has yielded several cytotoxic labdane diterpenes. Dichlorolissoclimide was described earlier (*1*) and the related chlorolissoclimide (**595**) was isolated subsequently (*709*). The C-7 hydroxy stereochemistry was more recently revised as shown for both **595** and dichlorolissoclimide (*710*). The antiproliferative activity of these compounds on a non-small-cell bronchopulmonary carcinoma cell line has been investigated (*711*). The lissoclimides are believed to be involved in human food poisoning from the consumption of oysters contaminated by *Lissoclinum voeltzkowi* (*709*). An Okinawan *Lissoclinum* sp. has yielded an array of chlorinated lissoclimide-type diterpenoids, the haterumaimides (*712–715*). This collection includes haterumaimides A–E (**596–600**) (*712*). Both C (**598**) and D (**599**) were detected in the animal and are not considered to be artifacts of B (**597**). Haterumaimides F-I (**601–604**) are also present, and both collections also produced the known chloro- (**595**) and dichlorolissoclimides (*713*). Heating G (**602**) and treating H (**603**) with *p*-toluenesulfonic acid resulted in no conversion to H (**603**) or I (**604**), respectively, supporting the natural origin of H and I. Haterumaimides J (**605**) and K (**606**) (*714*), and N (**607**), O (**608**), and P (**609**) (*715*) complete this metabolite collection. Only one compound in this set, haterumaimide Q, is not chlorinated. Some of the haterumaimides have sub-nanogram cytotoxicity and a structure–activity relationship is known; e.g., a chlorine atom at C-2 is essential for maximum activity (*715*). Haterumaimides L (**610**) and M (**611**), and 3β-hydroxychlorolissoclimide (**612**) were isolated from the molluscs *Pleurobranchus albiguttatus* (**610–612**) and *Pleurobranchus forskalii* (**610, 611**) from the Philippines (*716*). The mechanism of cytotoxicity ascribed to the chlorolissoclimides seems to involve protein synthesis inhibition (*717*).

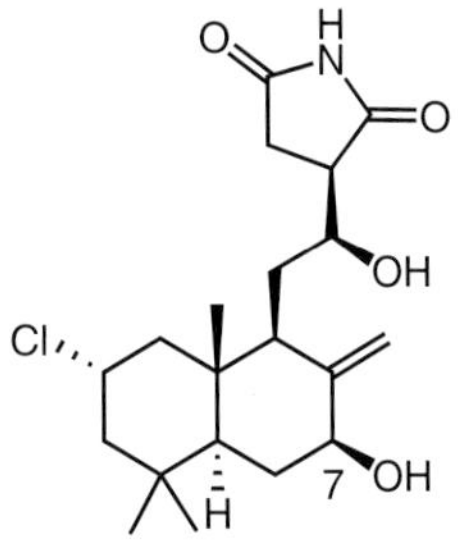

3.4 Terpenes

595 (chlorolissoclimide)
596 (haterumaimide A) 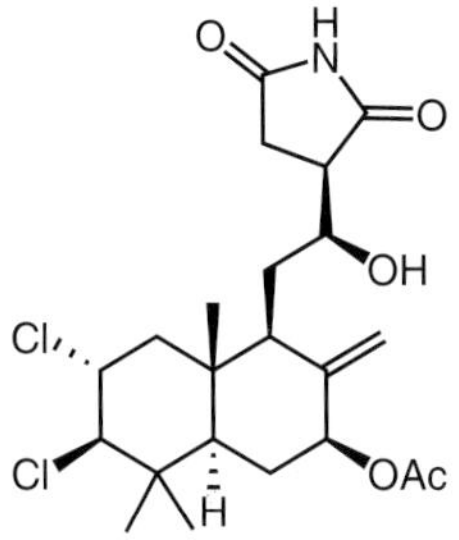
597 (haterumaimide B) 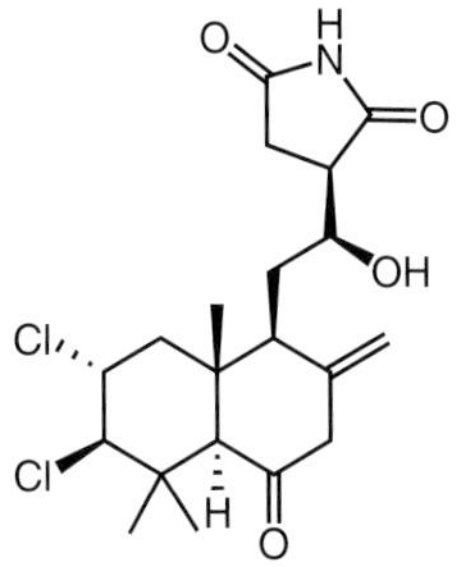
598 (haterumaimide C) 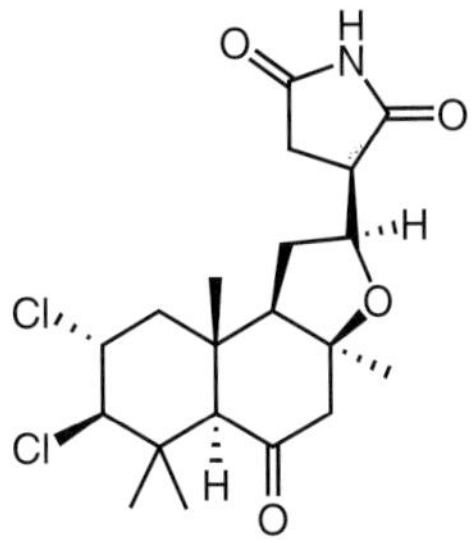
599 (haterumaimide D) 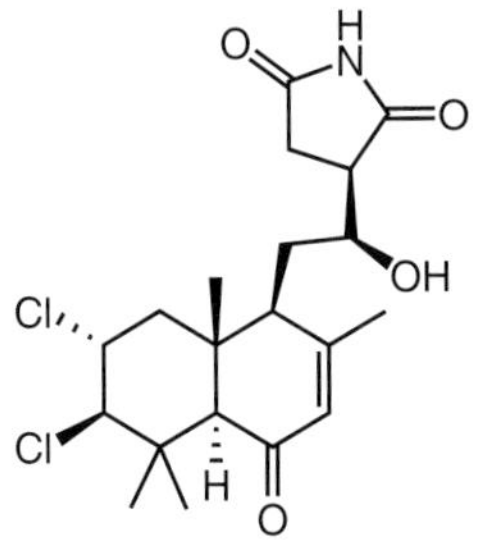
600 (haterumaimide E) 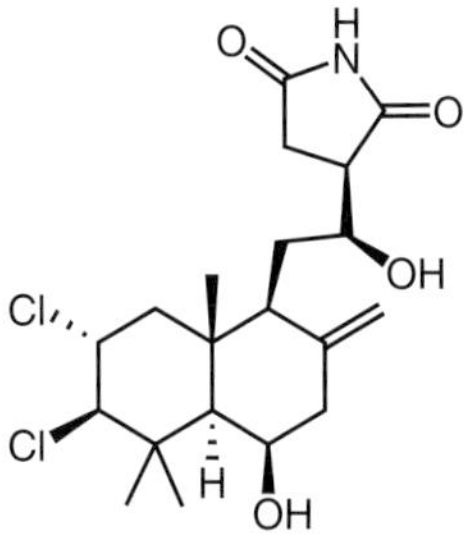
601 (haterumaimide F) 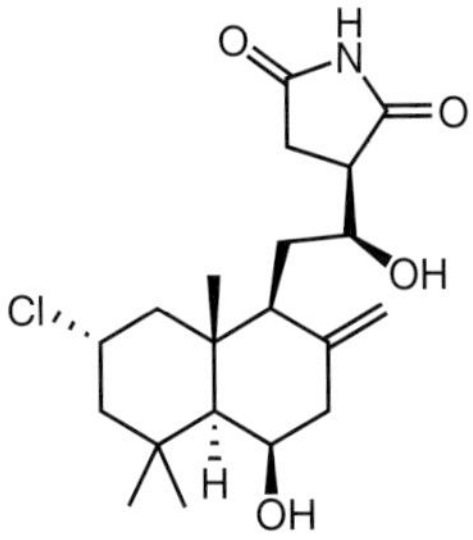
602 (haterumaimide G) 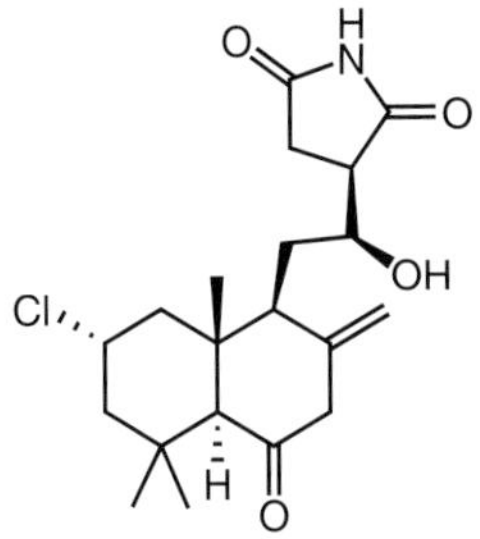
603 (haterumaimide H) 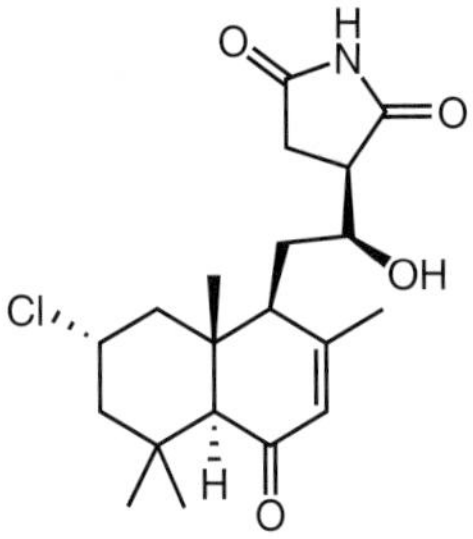
604 (haterumaimide I) 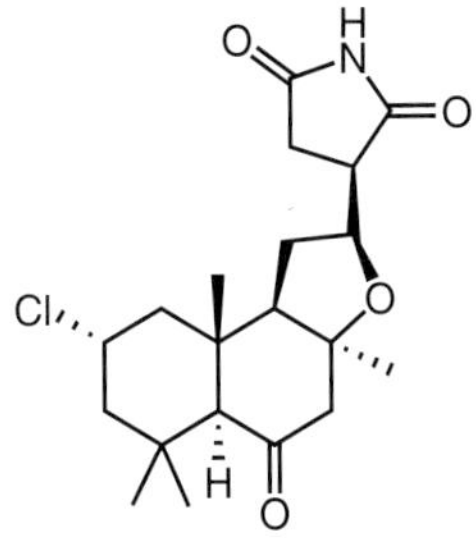
605 R = H (haterumaimide J) 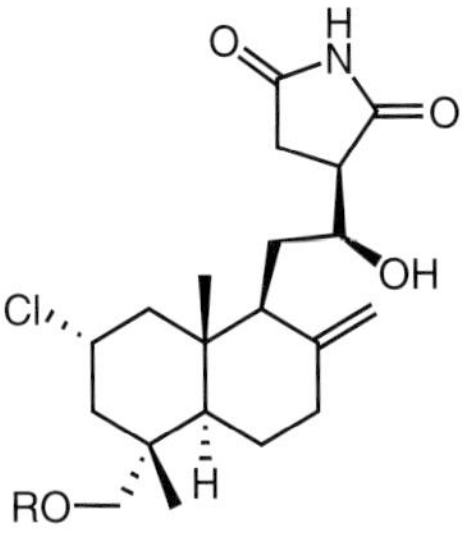
606 R = Ac (haterumaimide K)
607 (haterumaimide N)

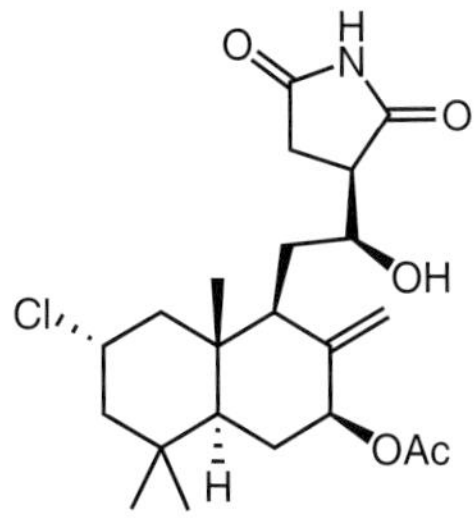

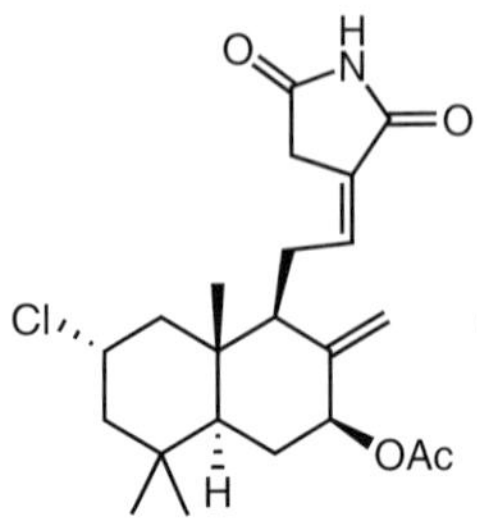

608 (haterumaimide O) 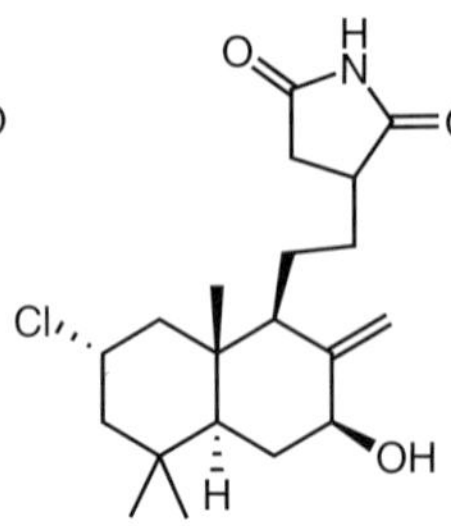   **609** (haterumaimide P) 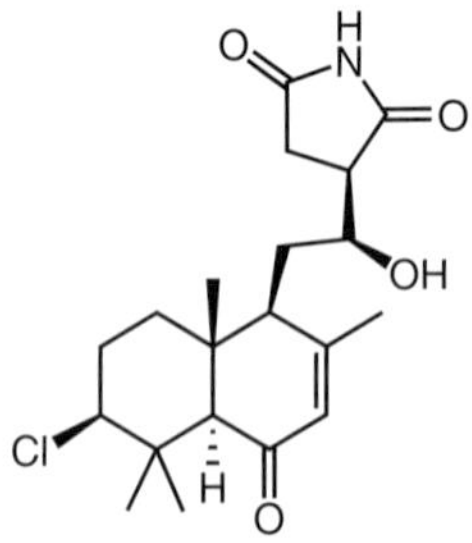   **610** (haterumaimide L)

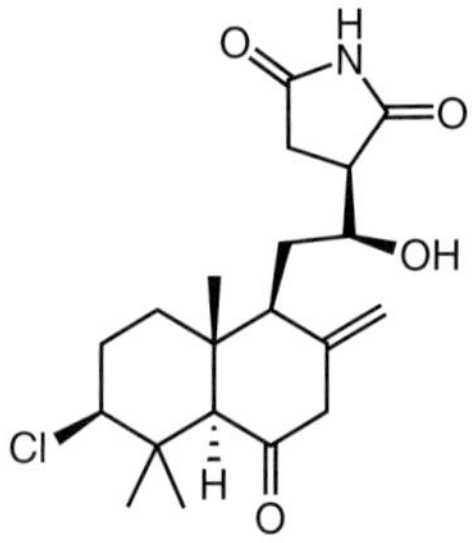

611 (haterumaimide M) 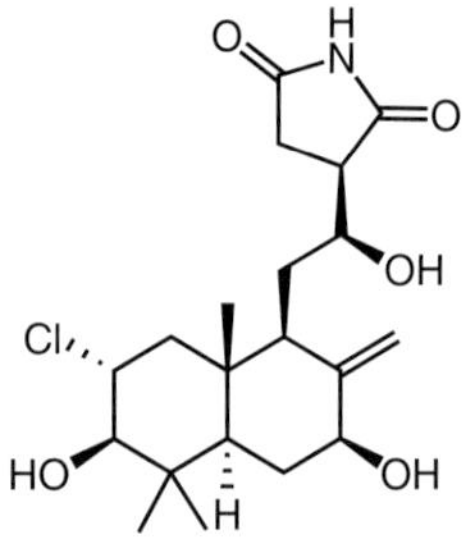   **612**

3.4.4 Higher Terpenes

Although the numbers are relatively small, several new halogenated triterpenes and other higher terpenes have been described since the earlier review (*1*). Accounts of marine polyether triterpenes (*718*) and heterocyclic triterpenes (*719*) have been published. The novel pentacyclic triterpene squalene-derived enshuol (**613**) is found in the new species *Laurencia omaezakiana* from central Japan Pacific waters (*720*). The first study of *Laurencia* living in Vietnamese waters has led to callicladol (**614**) from *Laurencia calliclada* (*721*). The Canary Islands *Laurencia viridis*, also a new species, has yielded several novel brominated polyether squalene-derived metabolites (*722–724*). These include thyrsenols A (**615**) and B (**616**) (*722*), isodehydrothrysiferol (**617**) and 10-epidehydrothyrsiferol (**618**) (*723*), and dehydrovenustatriol (**619**), 15,16-dehydrovenustatriol (**620**), 16-hydroxydehydrothyrsiferol (**621**), and 10-*epi*-15,16-dehydrothyrsiferol (**622** (*724*). The Canary Islands *Laurencia pinnatifida* contains dehydrothyrsiferol (**623**) (*725*). Two later collections of *Laurencia viridis* from around the Canary Islands revealed the presence of dioxepandehydrothyrsiferol (**624**), 16-*epi*-hydroxydehydrothyrsiferol (**625**) (*726*), clavidol (**626**), and 3-*epi*-dehydrothyrsiferol (**627**) (*727*).

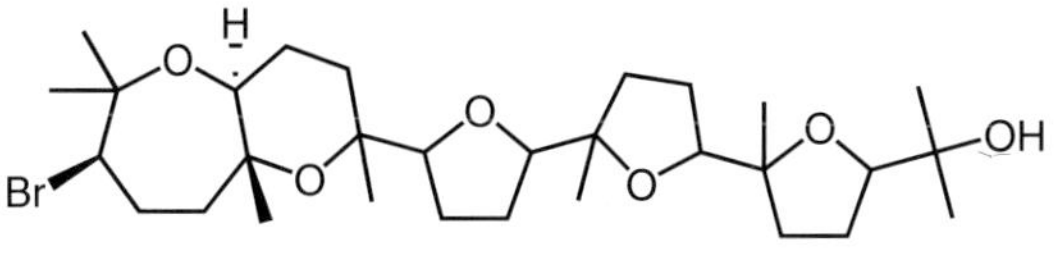

3.4 Terpenes

613 (enshuol)

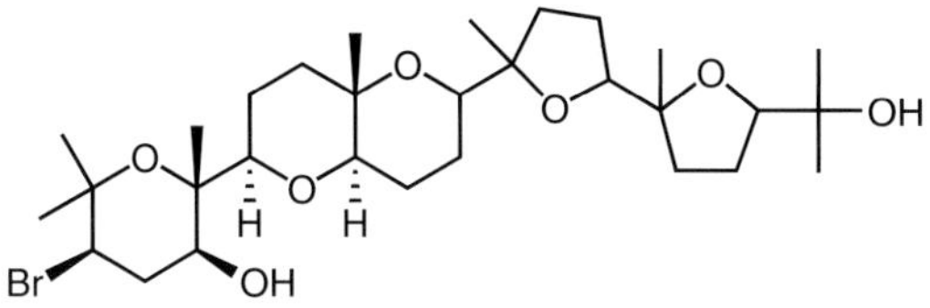

614 (callicladol)

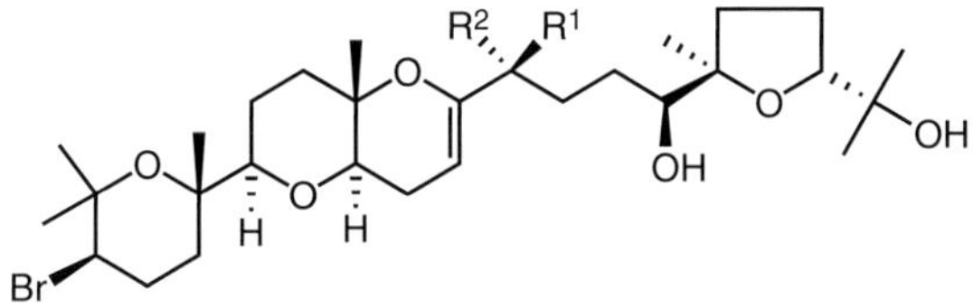

615 R^1 = OH, R^2 = CH$_2$OH (thyrsenol A)
616 R^1 = CH$_2$OH, R^2 = OH (thyrsenol B)

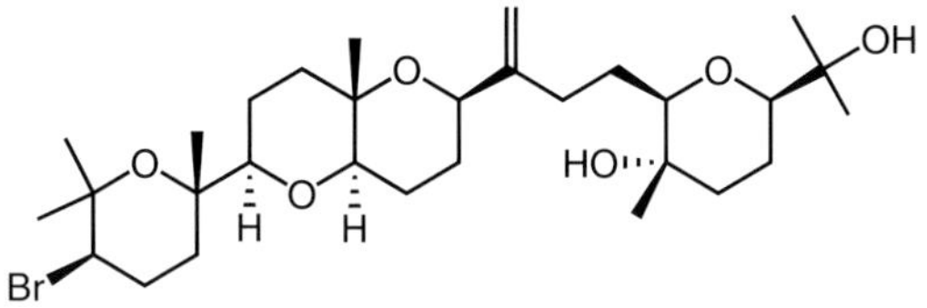

617 (isodehydrothyrsiferol)

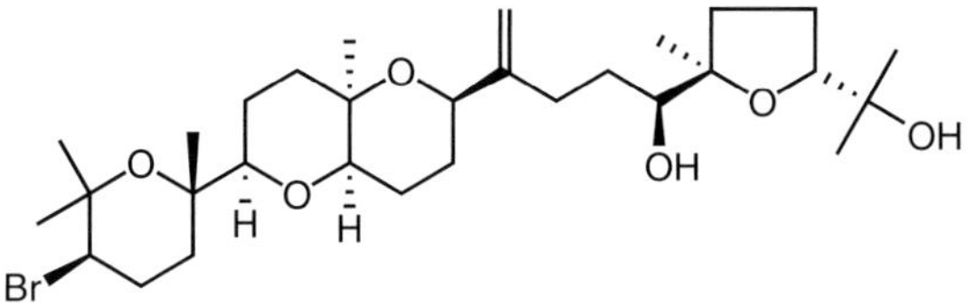

618 (10-epidehydrothyrsiferol)

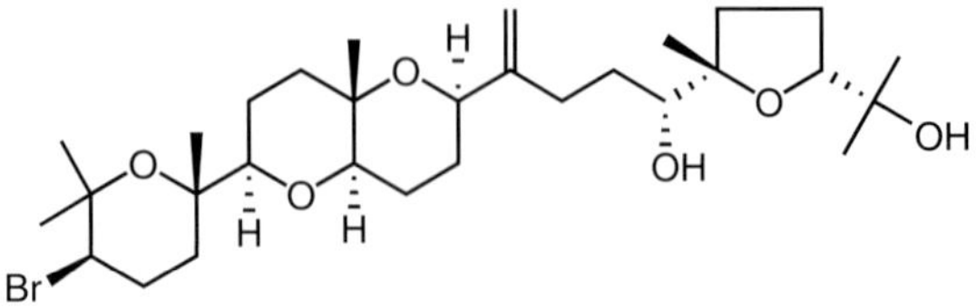

619 (dehydrovenustatriol)

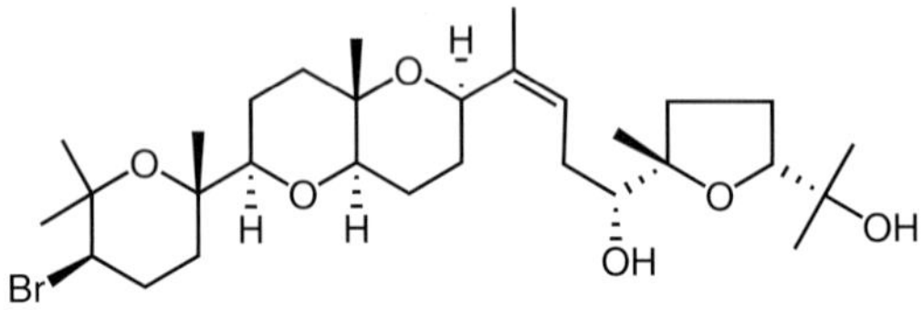

620 (15,16-dehydrovenustatriol)

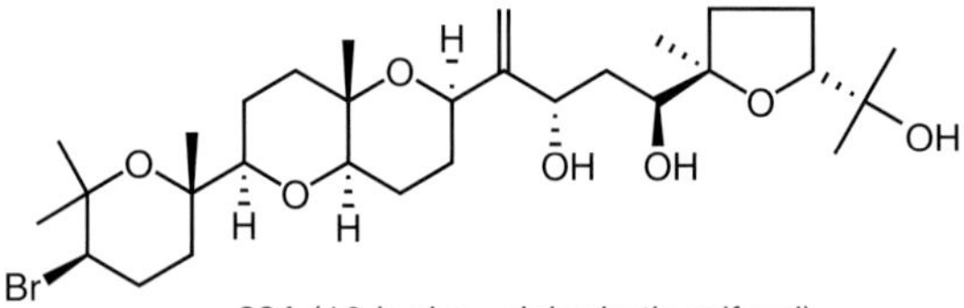

621 (16-hydroxydehydrothyrsiferol)

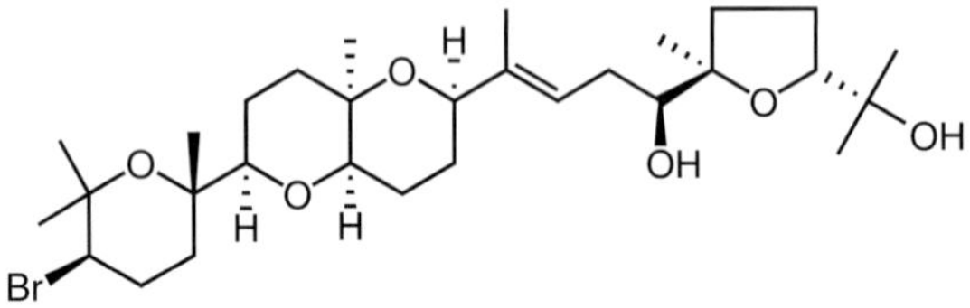

622 (10-*epi*-15,16-dehydrothyrsiferol)

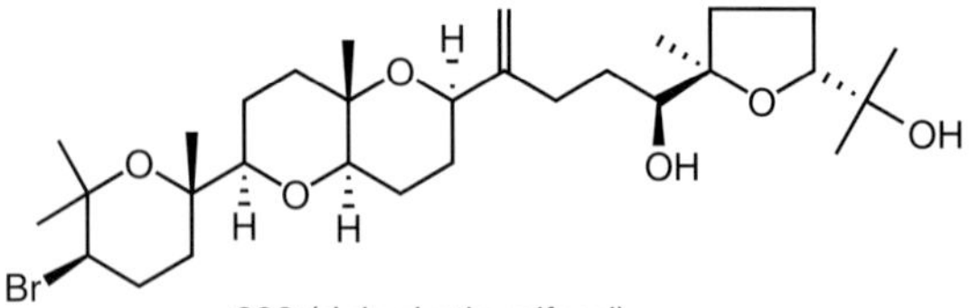

623 (dehydrothyrsiferol)

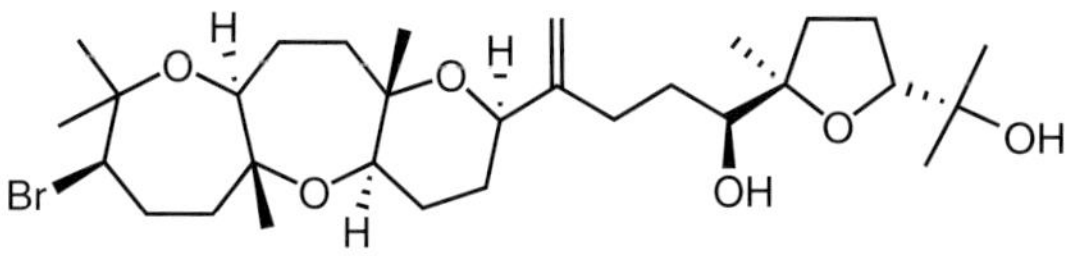

3.4 Terpenes

624 (dioxepandehydrothyrsiferol)

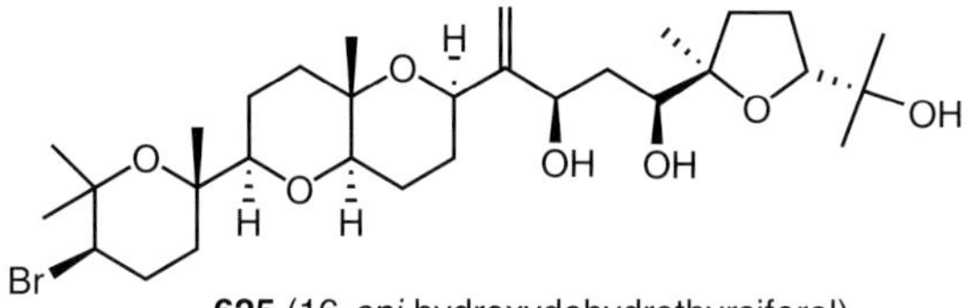

625 (16-*epi*-hydroxydehydrothyrsiferol)

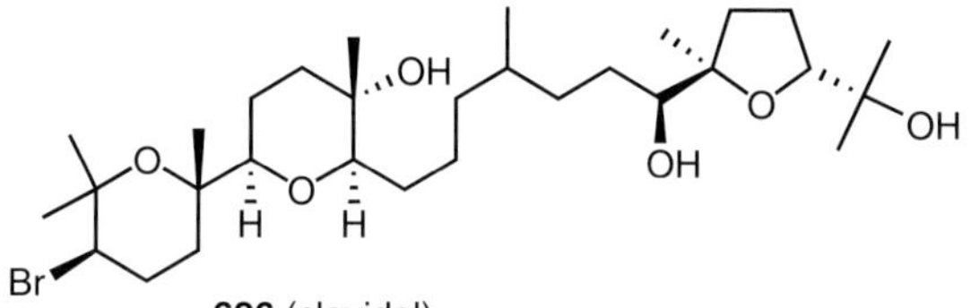

626 (clavidol)

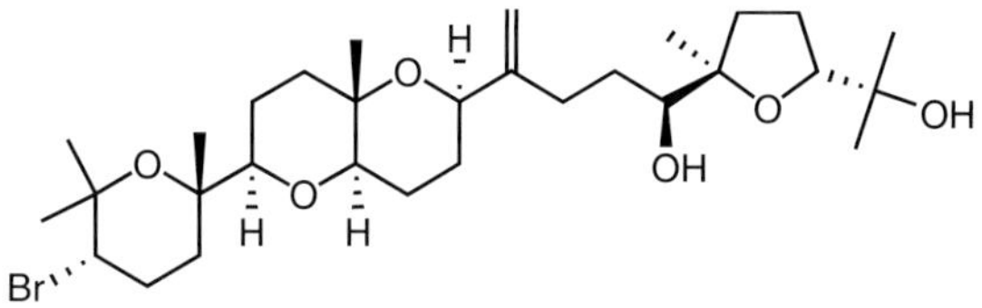

627 (3-*epi*-dehydrothyrsiferol)

The sea hare *Dolabella auricularia* has furnished aurilol (**628**), which is cytotoxic (*728*) and for which the structure has now been fully assigned by total synthesis (*729*). The Indian Ocean red alga *Chondria armata*, a member of the *Laurencia* family, contains armatols A–F (**629–634**) (*730*).

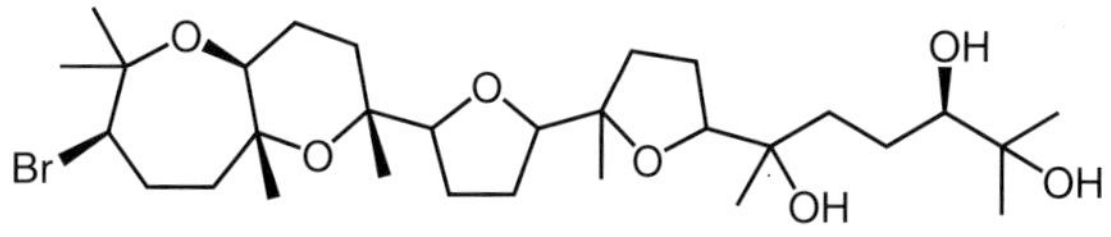

628 (aurilol)

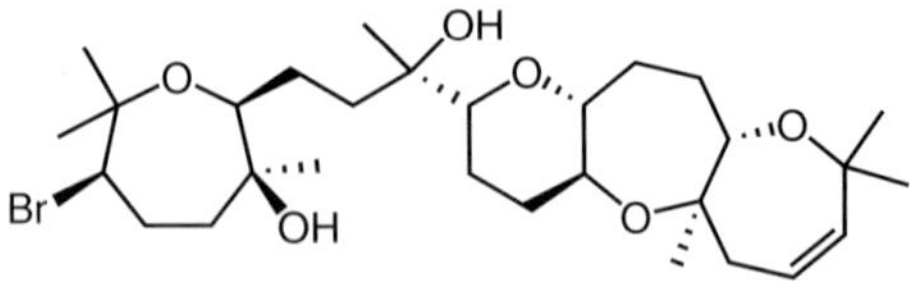

629 (armatol A)

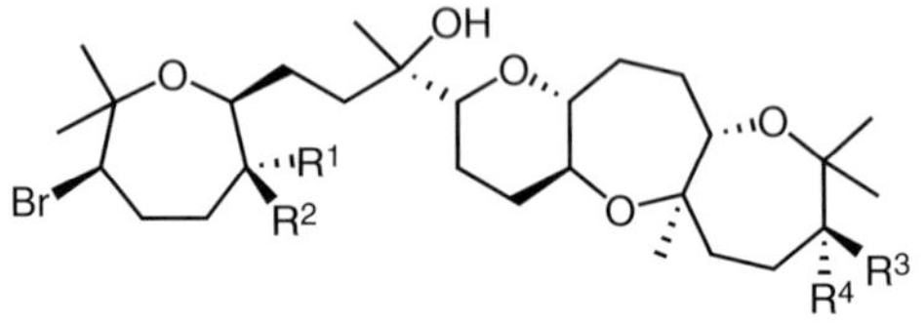

630 R¹ = Me, R² = OH, R³ = H, R⁴ = Br (armatol B)
631 R¹ = OH, R² = Me, R³ = H, R⁴ = Br (armatol C)
632 R¹ = Me, R² = OH, R³ = Br, R⁴ = H (armatol D)
633 R¹ = OH, R² = Me, R³ = Br, R⁴ = H (armatol E)

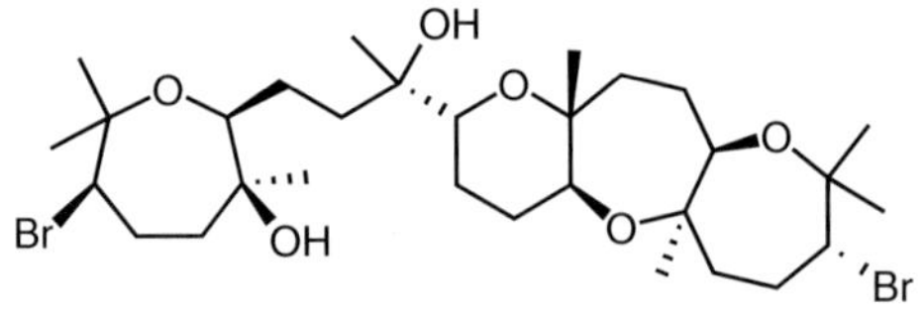

634 (armatol F)

Despite their stereochemical complexity, a few brominated polyethers have succumbed to total synthesis. In addition to aurilol (**628**), discussed above (*728*), the previously described thyrsiferol and thyrsiferyl 23-acetate (*1*) have been synthesized (*731*), as has (unnatural) 7,11-*epi*-thyrsiferol (*732*). Several other synthetic studies are known (*733*). The cytotoxicity of thyrsiferyl 23-acetate and some of the other *Laurencia* polyether terpenoids has generated considerable interest into the biological mechanisms and possible drug development (*734–738*). A marine fungus of the genus *Fusarium* found on driftwood in a Bahamas mangrove habitat produces the halogenated sesterterpenes neomangicols A (**635**) and B (**636**), which have some activity against several human cancer cell lines (*740*). These compounds are the first natural halogenated sesterterpenes. A subsequent study tentatively identified this fungus as *Fusarium heterosporum* (*741*). The medicinal terrestrial plant *Turraea pubescens* has yielded turrapubesin A (**637**) (*742*), and the Okinawan sponge *Ircinia* sp. contains the new furanosesterterpenes **638** and **639** (*743*). An *Aspergillus* sp. culture has yielded ICM0301C (**640**) and ICM0301D (**641**), along with several nonchlorinated analogs (*744, 745*).

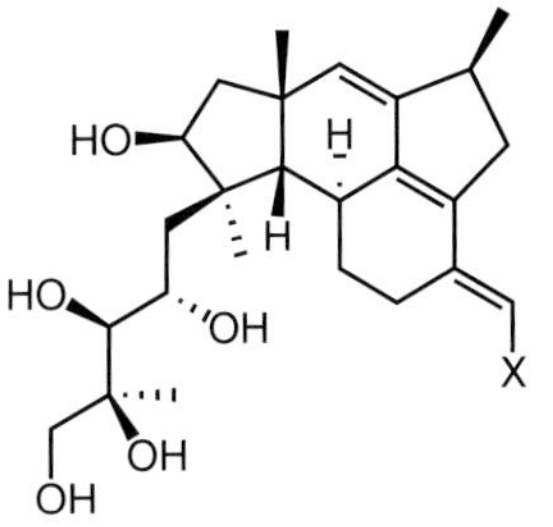

3.4 Terpenes

635 X = Cl (neomangicol A)
636 X = Br (neomangicol B)

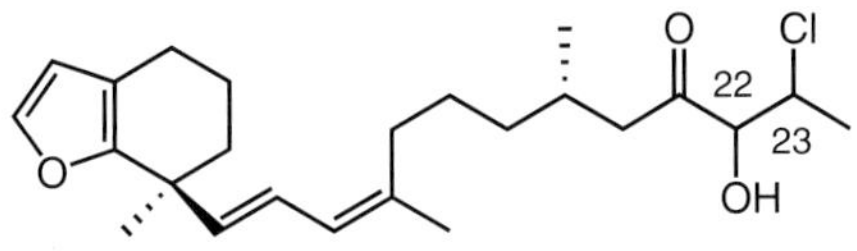

638 (22, 23 *anti*)
639 (22, 23 *syn*)

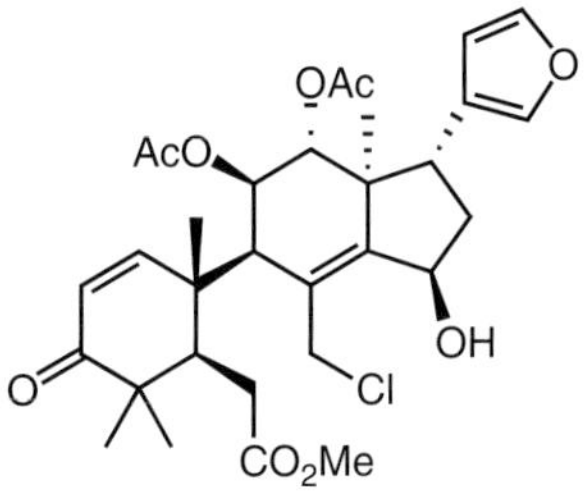

637 (turrapubesin)

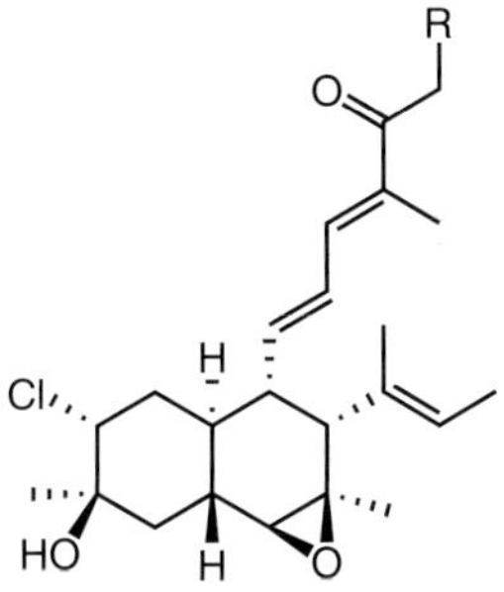

640 R = Me (ICM0301C)
641 R = H (ICM0301D)

The Chinese plant *Amoora yunnanensis* contains dammaranes **642** and **643** (*746*), and oleanane **644** was isolated as a triacetate from *Mentha villosa* (*747*). It is conceivable that **644** is an artifact formed by HCl acting on the corresponding unsaturated carboxylic acid, since this type of acid-catalyzed lactone formation is well known (*748*).

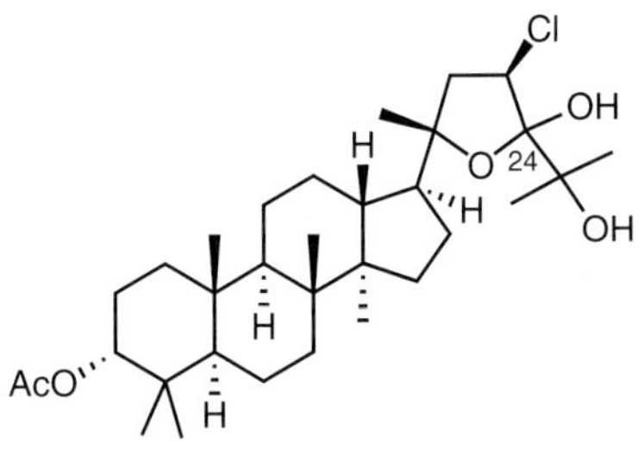

642, 643 (isomeric at C-24)

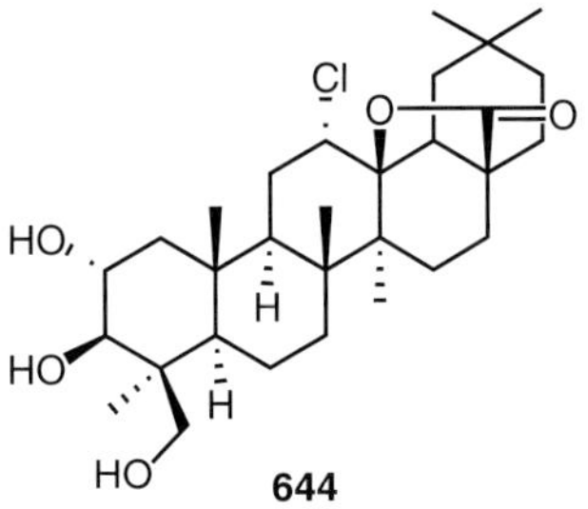

644

3.5 Steroids

Because most natural halogenated steroids are chlorohydrins, which are usually accompanied by the corresponding epoxide, one must ensure that the former are not artifacts formed during the isolation process.

The structure of physalin H (**645**) from the plant *Physalis angulata* has been revised to the chlorohydrin shown (*749*). The Argentinian *Jaborosa sativa* has afforded the new jaborosalactone T (**646**) (*750*), and *Jaborosa runcinata* contains jaborosalactones 3 (**647**) and 6 (**648**) (*751*). Another Argentina collection of *Jaborosa odonelliana* revealed jaborosalactone 10 (**649**), which was present in plants collected in December but not in April (*752*). The Argentinian *Jaborosa bergii* contains chlorohydrins **650–652** in addition to nonchlorinated withanolides (*753*). As a group these steroids and the corresponding chlorohydrins display interesting biological activity. Several withanolides induce quinone reductase (*754*) and inhibit the growth of human cancer cell lines (*755*). Physalin H (**645**) has potent leishmanicidal activity (*756*).

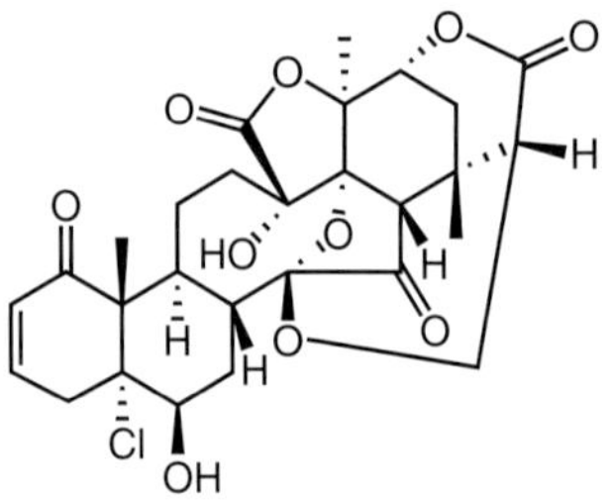

645 (physalin H)

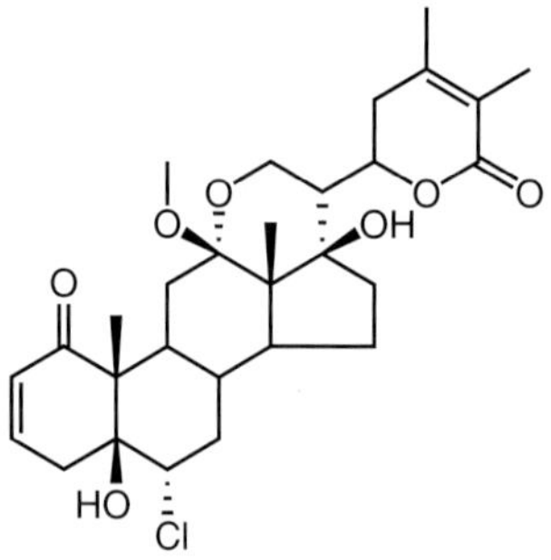

646 (jaborosalactone T)

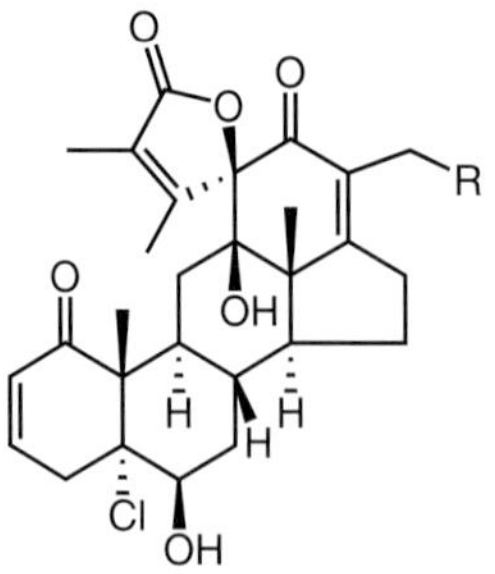

647 R = H (jaborosalactone 3)
648 R = OH (jaborosalactone 6)

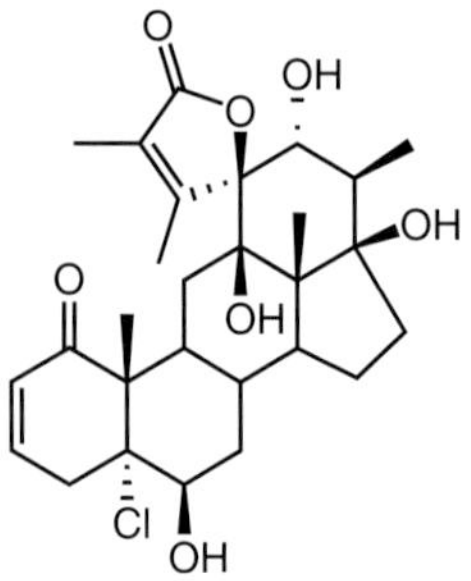

649 (jaborosalactone 10)

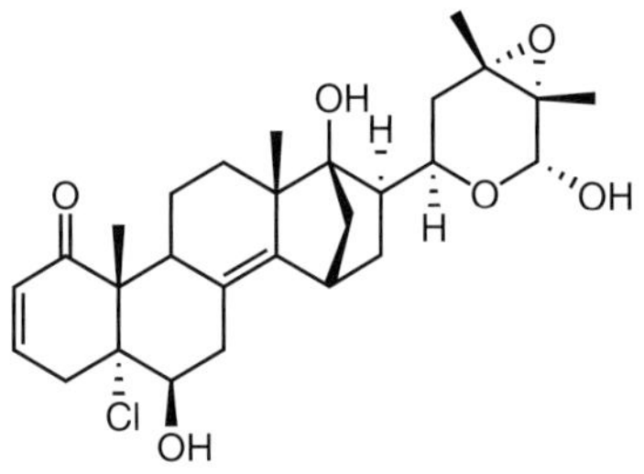

3.5 Steroids

650 (jaborosalactol 21)

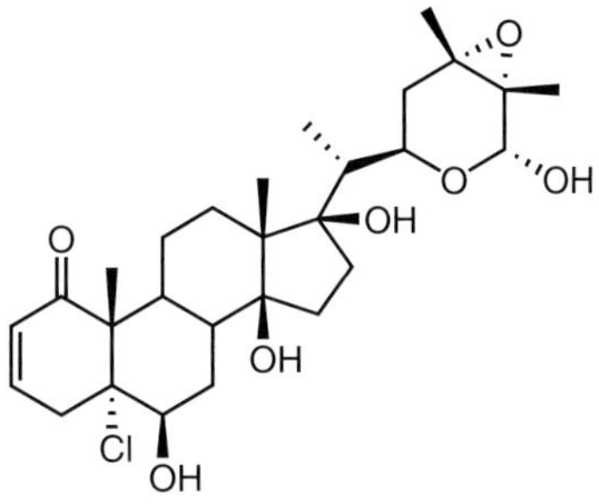

651 (jaborosalactol 23)

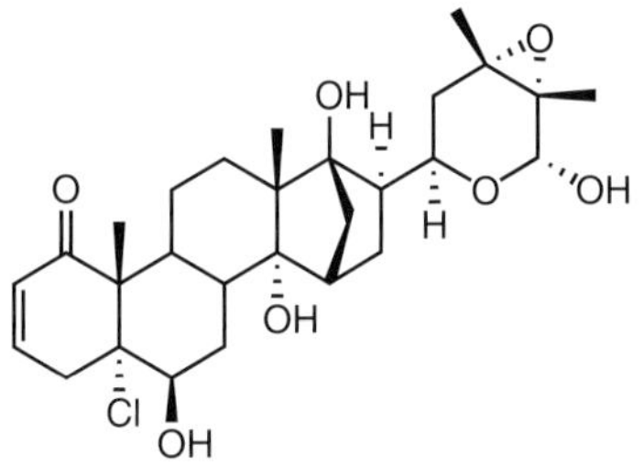

652

Several new halogenated marine steroids have been characterized. An Okinawan marine sponge of *Xestospongia* sp. contains aragusteroketal C (**653**), which has nanogram activity against KB cells (*757*). Since methanol was not used in the isolation procedure, this compound, along with the corresponding nonchlorinated epoxide, are rare examples of natural dimethylketals. Yonarasterols G (**654**), H (**655**), and I (**656**) were isolated from the Okinawan soft coral *Clavularia viridis* (*758*). The epoxide corresponding to chlorohydrin **655** was heated in methanol for 3 days in the presence of NaCl with and without silica gel, but remained unaffected. Another Okinawan sponge, *Terpios hoshinota*, contains nakiterpiosin (**657**) and nakiterpiosinone (**658**), novel mixed bromochloro nor-steroids that show potent cytotoxicity against P388 leukemia cells (*759, 760*). It might be noted that CH_2Cl_2 was not employed in the isolation and purification process.

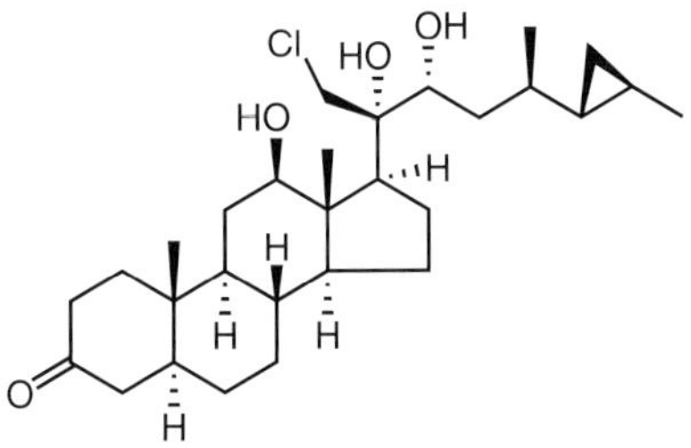

653 (aragusteroketal C)

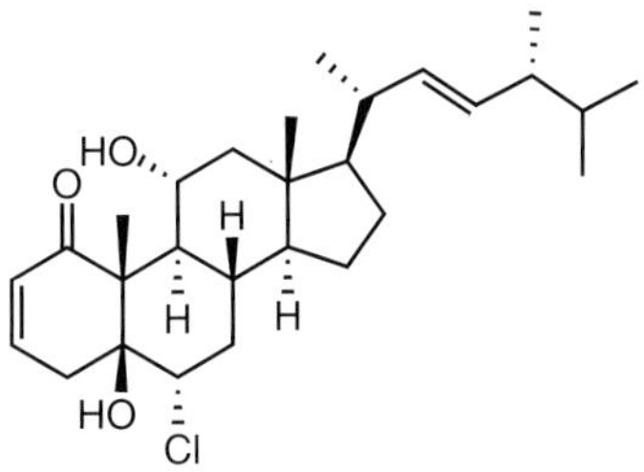

654 (yonarasterol G)

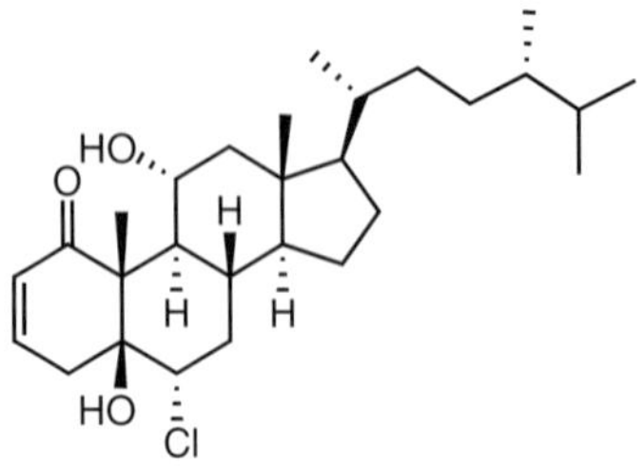

655 (yonarasterol H)

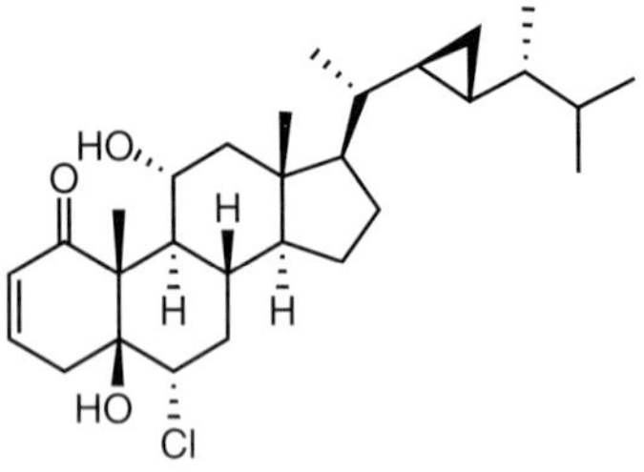

656 (yonarasterol I)

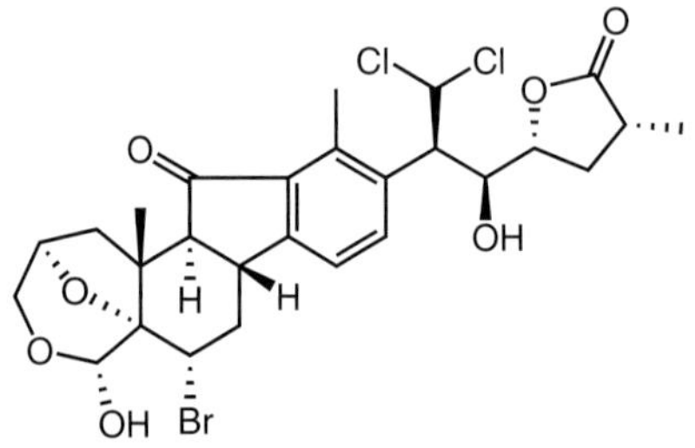

657 (nakiterpiosin)

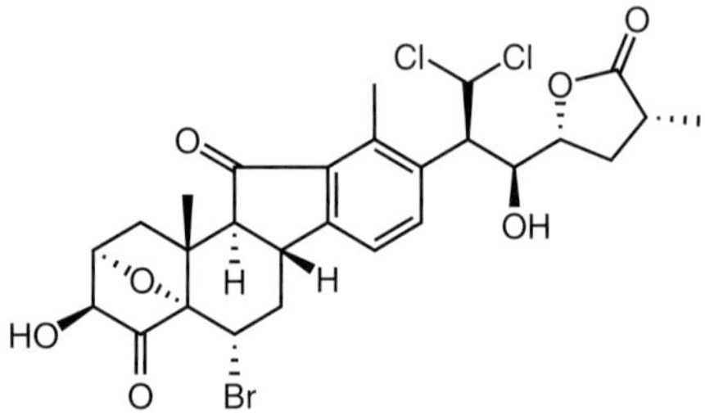

658 (nakiterpiosinone)

The polychlorinated androstanes clionastatins A (**659**) and B (**660**) were isolated from the burrowing sponge *Cliona nigricans* (Fig. 3.10) collected in two locations along the Italian coast (*761*). These unique metabolites have good cytotoxic activity against murine and human cancer cell lines. The eastern Pacific octocoral *Carijoa multiflora* has yielded the unusual chlorinated pregnanes **661** and **662** in a chloroform-free isolation process (*762*).

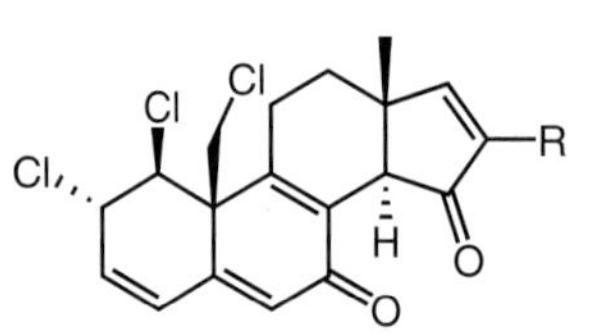

659 R = H (clionastatin A)
660 R = Cl (clionastatin B)

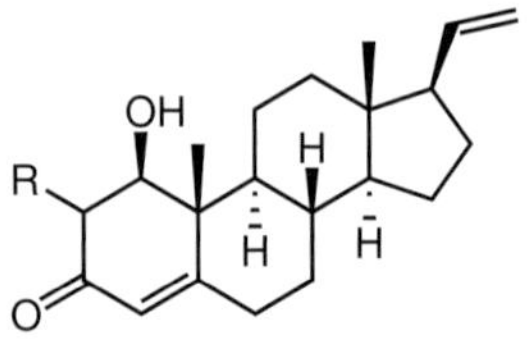

661 R = α-Cl
662 R = β-Cl

Another group of natural chlorinated steroids are the products of enzymatic (or anthropogenic) chlorination of cholesterol, estrone, and other natural steroids. Thus, the well known myeloperoxidase–H_2O_2–chloride system in white blood cells (*763, 764*) targets cholesterol leading to at least three chlorohydrins (*765–769*), the structures of which have now been confirmed (**663–665**) (*769*). These studies also show that the same chlorohydrins are produced when HOCl and cholesterol are allowed to react. Interestingly, three chlorinated estrones have been identified in wastewater effluents treated with hypochlorous acid (*770*), although these products are not considered to be naturally occurring for the present survey.

3.5 Steroids

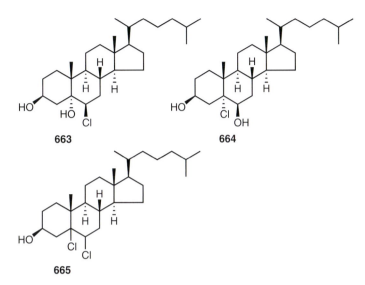

Fig. 3.10 *Cliona nigricans*, a boring sponge found in the Ligurian Sea, Italy, that contains the clionastatins A and B (**659** and **660**). The brown color is due to symbiotic zooxanthellae (Photo: C. Cerrano)

3.6 Marine Nonterpenes: C_{15} Acetogenins

When chemists began to explore the oceans for novel natural products, halogenated C_{15} acetogenins were an unknown class of compounds. But since the initial discovery of laurencin from *Laurencia glandulifera* in 1965, a large number of these compounds have been found – 130 examples in the first survey (*1*). This number continues to grow, especially as produced by the prolific *Laurencia* red algae. The biological properties and synthesis of allenic natural products, of which many are bromoallene C_{15} acetogenins, have been published (*771*).

A collection of *Laurencia elata* from the coast of Victoria has provided the pyrano[3,2-*b*]pyranyl vinyl acetylene elatenyne (**666**) (*772*), which is related to the known (Z)-dactomelyne (*1*). Japonenynes A (**667**), B (**668**), and C (**669**), which possess a furo[3,2-*b*]pyranyl framework, were isolated from *Laurencia japonensis* (*773*). Compound **669** may be an isolation (methanol) artifact although it is isolated as a single compound. The report of "aplysiallene" from the sea hare *Aplysia kurodai* (*774*) is erroneous and this compound is actually a known bromoallene (*775*) described earlier (*1*). The Vietnamese *Laurencia pannosa* contains pannosallene (**670**), which is closely related to the known laurallene (*776*). During this investigation the authors discovered that their earlier proposed structure of epilaurallene must be incorrect. A new isomer of pannosallene, nipponallene (**671**), along with the novel neonipponallene (**672**) was isolated from *Laurencia nipponica* collected off the Russian shore of the Sea of Japan (*777*). The sea hare *Aplysia parvula* has yielded aplyparvunin (**673**), which has potent fish toxicity (*778*). *Laurencia intricata* has furnished itomanallenes A (**674**) and B (**675**); the former is an epimer of the known neolaurallene (*591*). Chinzallene (**676**) was characterized from a Japanese *Laurencia* sp. (*638*). The stereochemistry of chinzallene is not fully established.

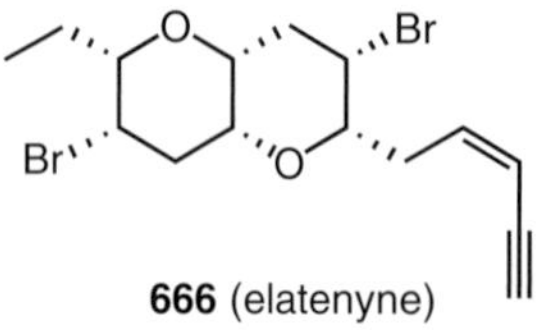

666 (elatenyne)

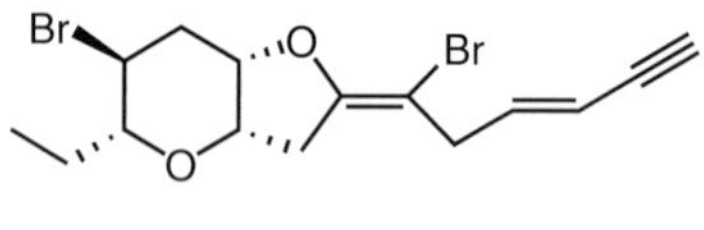

667 (japonenyne A)

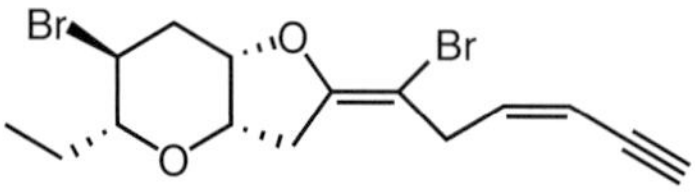

668 (japonenyne B)

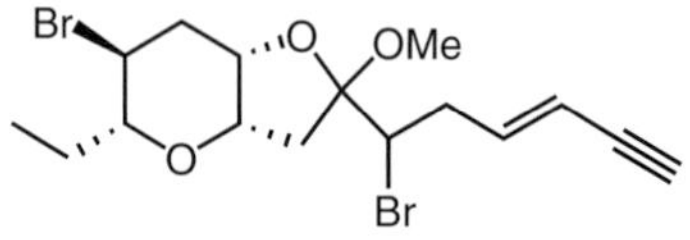

669 (japonenyne C)

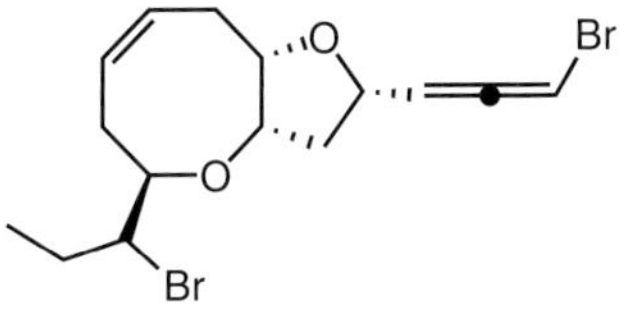

3.6 Marine Nonterpenes: C_{15} Acetogenins

670 (pannosallene)

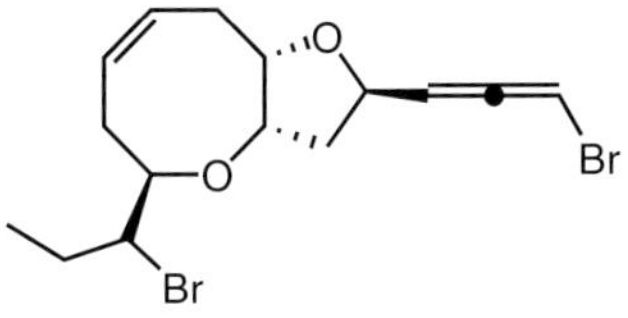

671 (nipponallene)

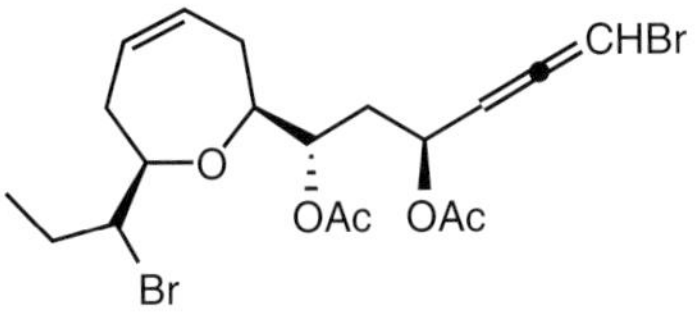

672 (neonipponallene)

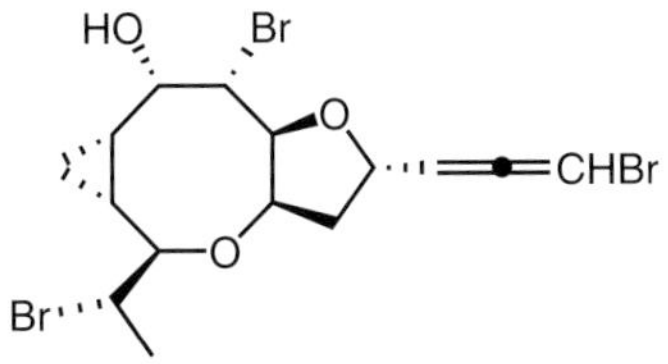

673 (aplyparvunin)

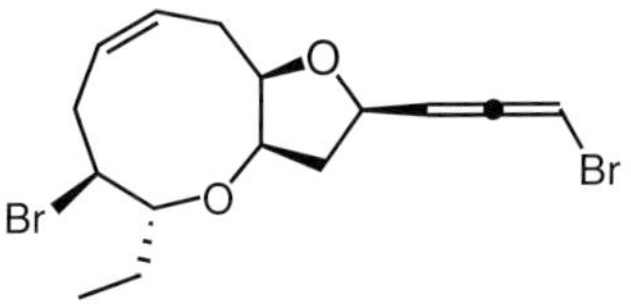

674 (itomanallene A)

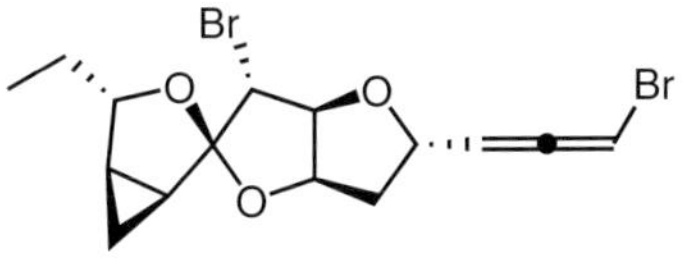

676 (chinzallene)

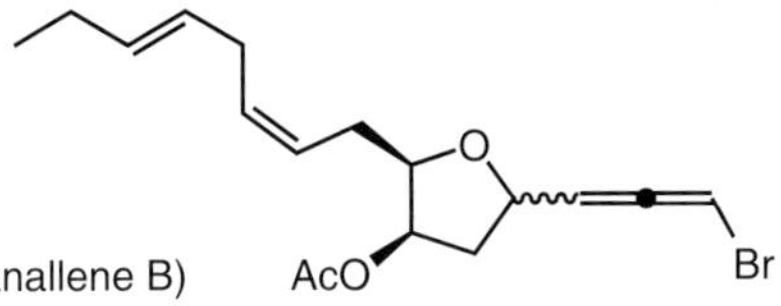

675 (itomanallene B)

The C_{15} acetogenin (E)-dihydrorhodophytin (**677**), an isomer of the previously known (Z)-isomer (*1*), is found in *Laurencia pinnatifida* from the Canary Islands (*779*). An Easter Island variety of *Laurencia claviformis* has afforded (3Z)-13-epipinnatifidenyne (**678**) (*780*). Likewise, an epimer of the previously known laurencienyne, 13-epilaurencienyne (**679**), is found in *Laurencia obtusa* from the Aegean coast (*781*). This same seaweed and locale yielded laurencienyne B (**680**), the *cis* isomer of laurencienyne (*782*), and the acetate **681** (*783*). This Aegean Sea *Laurencia obtusa* has also provided (3Z)-13-epilaurencienyne (**682**), (3E)-13-epipinnatifidenyne (**683**) (revised in (*785*)), **684**, and **685** (*784*). The (Z)-diastereomers **682** and **685** showed very potent insecticidal activity. Three enantiomers of known compounds were identified in the sea hare *Aplysia dactylomela*, (−)-(3E,6R,7R)-pinnatifidenyne (**686**), (+)-(3E,6R,7R)-obtusenyne (**687**), and (+)-(3Z,6R,7R)-obtusenyne (**688**) (*785*).

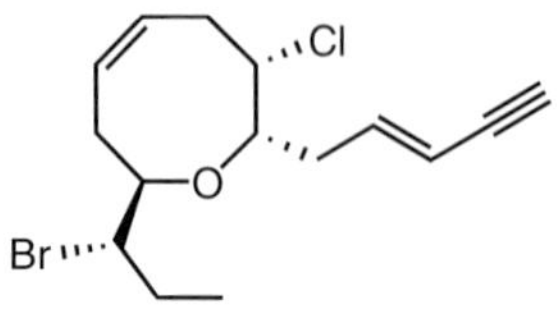

677 ((3*E*)-dihydrorhodophytin)

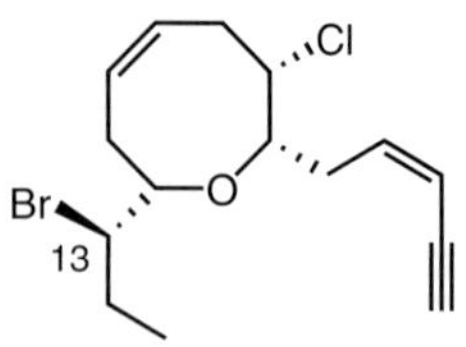

678 ((3*Z*)-13-*epi*-pinnatifidenyne)

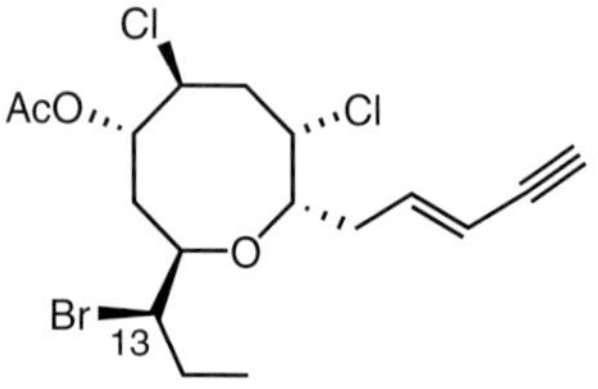

679 (13-*epi*-laurencienyne)

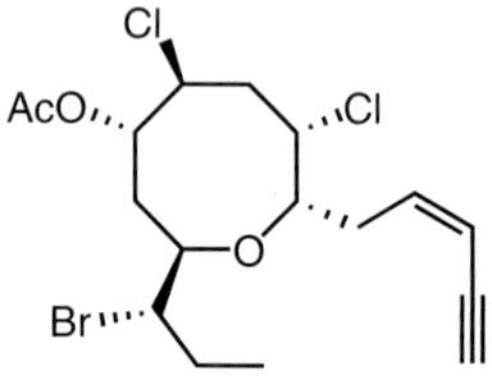

680 (laurencienyne B)

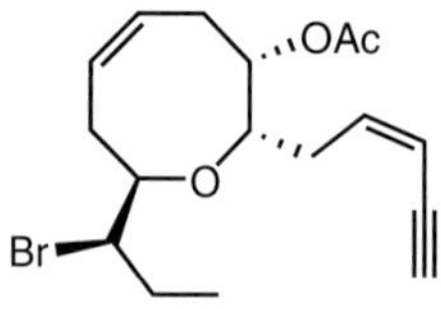

681

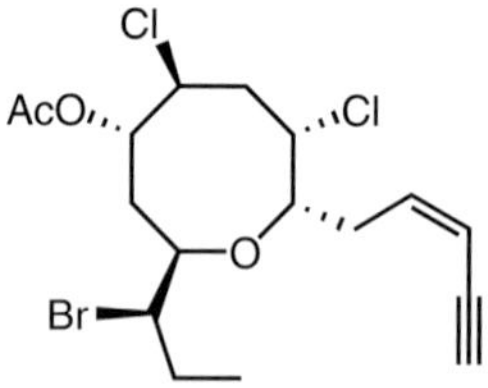

682 ((3*Z*)-13-*epi*-laurencienyne)

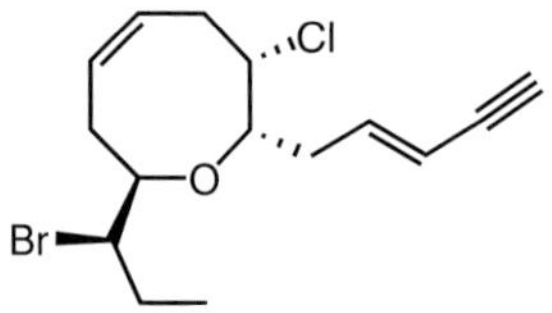

683 ((3*E*)-13-*epi*-pinnatifidenyne)

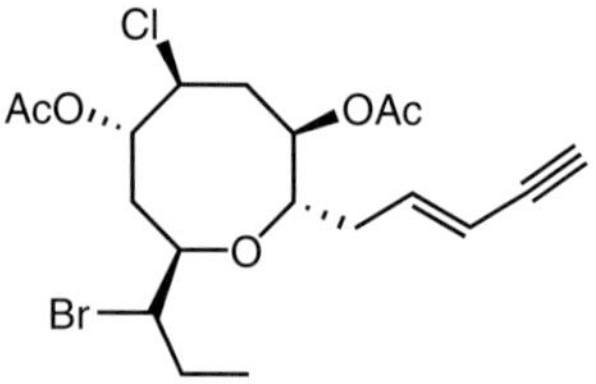

684

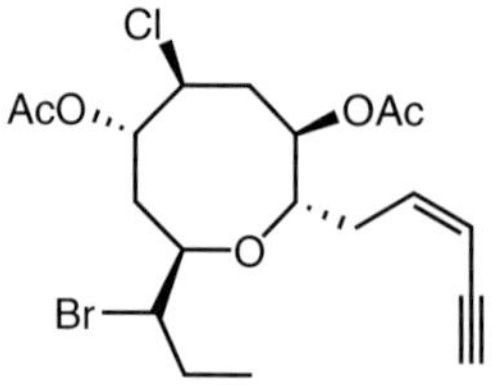

685

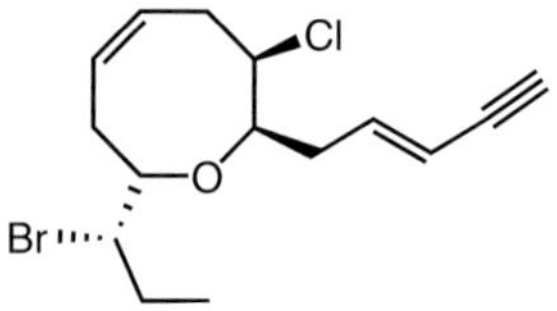

686 ((3*E*,6*R*,7*R*)-pinnatifidenyne)

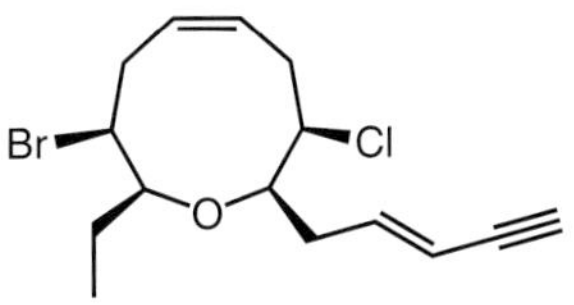

3.6 Marine Nonterpenes: C_{15} Acetogenins

687 ((3E,6R,7R)-obtusenyne)

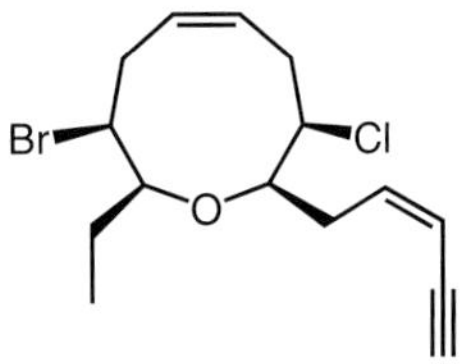

688 ((3Z,6R,7R)-obtusenyne)

Laurencia obtusa from the western coast of Ireland has afforded scanlonenyne (**689**), the first reported study of *Laurencia* red algae from Irish waters (*786*). A Japanese *Laurencia* sp. contains the new bisezakyne-A (**690**) and -B (**691**) (*787*). The red alga *Ptilonia magellanica* is the source of pyranosylmagellanicus A–C (**692–694**) and the linear **695**, a possible biogenetic precursor (*788*).

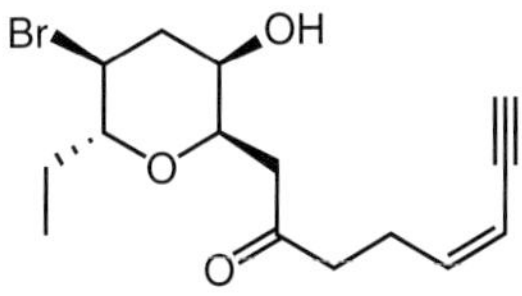

689 (scanlonenyne)

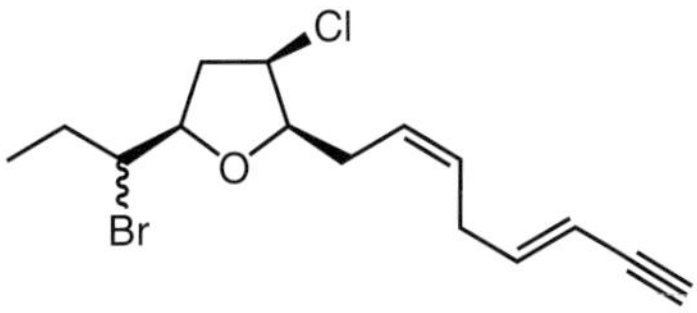

690 (bisezakyne-A)

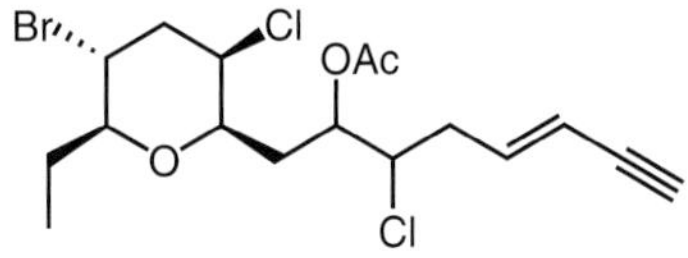

691 (bisezakyne-B)

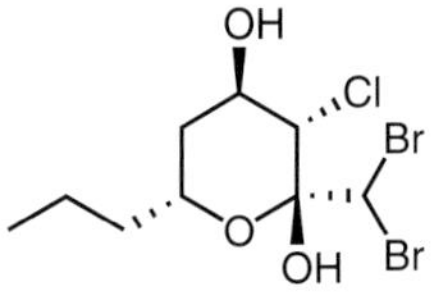

692 (pyranosylmagellanicus A)

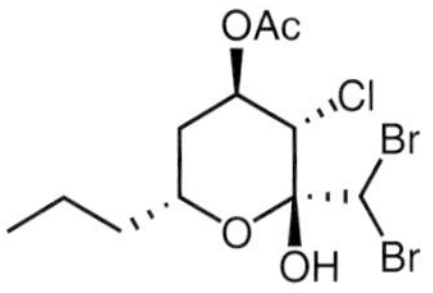

693 (pyranosylmagellanicus B) 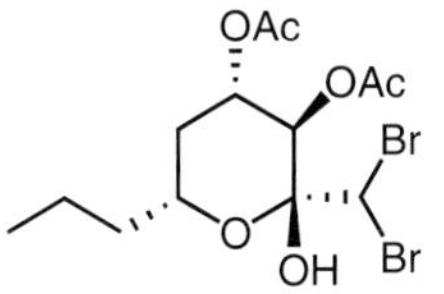 **694** (pyranosylmagellanicus C) 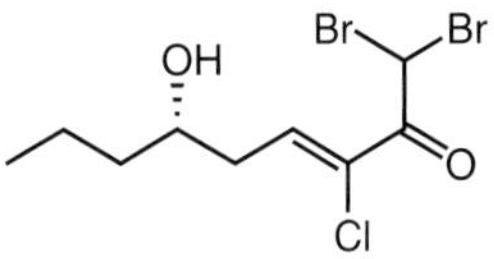 **695**

Several new halogenated bicyclic acetogenins of the laurefucin type have been discovered in marine organisms since the first survey (*1*). Thus, the Coral Sea red seaweed *Dasyphila plumariodes* contains the new isolaurefucin methyl ether (**696**) (*789*). Neoisoprelaurefucin (**697**) was characterized from a Japanese

Laurencia nipponica (*790*), and the structure and absolute configuration were confirmed by total synthesis (*791*). This new compound is a stereoisomer of the known (3Z)-isoprelaurefucin. A Malaysian *Laurencia pannosa* has yielded (3Z)-chlorofucin (**698**) (*578*), and (3Z)-bromofucin (**699**), which is also a new C_{15}-acetogenin, is found in a South African sea hare, *Aplysia parvula* (*792*). The (3E)-neoisoprelaurefucin (**700**) was found in *Laurencia obtusa* collected in Turkish waters (*557*).

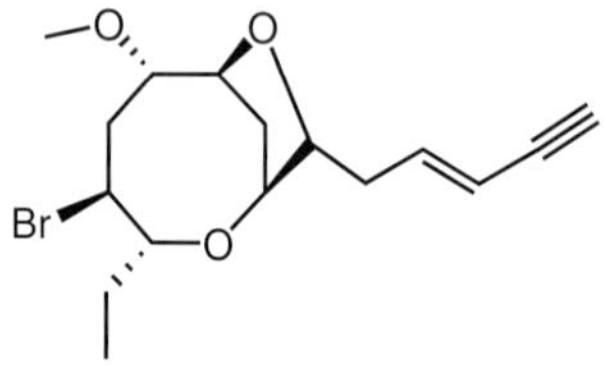

696

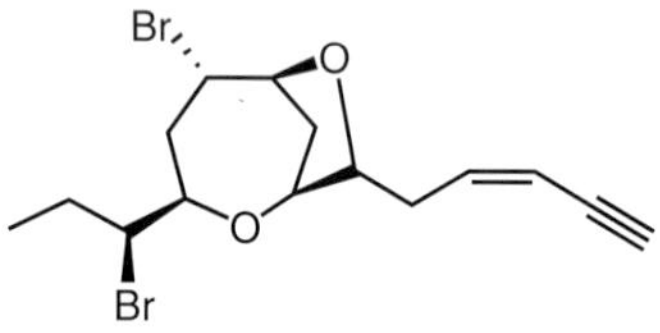

697 (neoisoprelaurefucin)

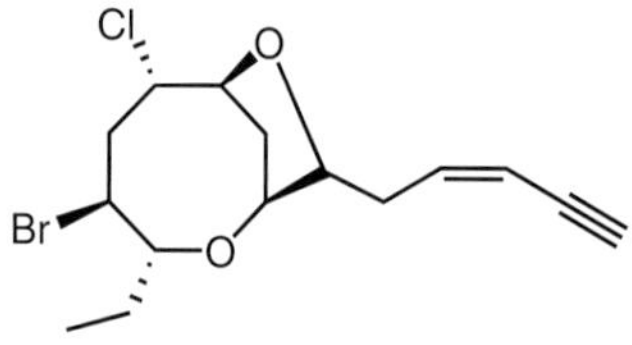

698 R = Cl ((3Z)-chlorofucin)
699 R = Br ((3Z)-bromofucin)

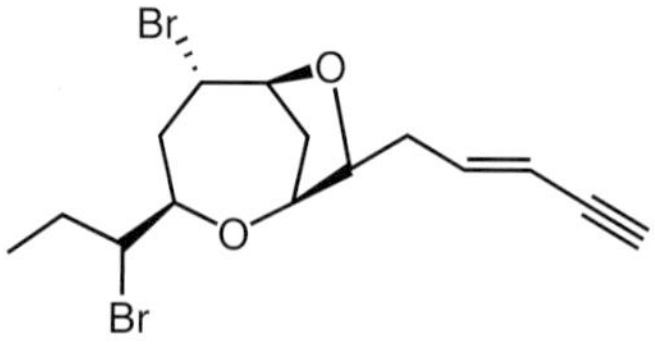

700 ((3E)-neoisoprelaurefucin)

The sea hare *Aplysia dactylomela* from the Canary Islands contains dactylallene (**701**), which is highly toxic to the mosquito fish (*Gambusia affinis*) at 10 ppm and deters feeding by the golden fish (*Carassius auratus*) at low concentrations (*793*). Dactylallene is a stereoisomer of the known obtusallene II. Obtusallene IV (**702**) was isolated from *Laurencia obtusa* collected in Turkish waters (*794*). This study also describes the conformational properties of several obtusallenes as does a subsequent investigation, which reports the isolation of five new obtusallenes from *Laurencia obtusa*, V–IX (**703**–**707**) (*795*). An unrecorded Malaysian *Laurencia* species has afforded the novel lembyne-A (**708**) and lembyne-B (**709**), the former of which is a (Z)-diastereomer of the known *cis*-maneonene C, while the latter is a stereoisomer of isomaneonene A (*796*). The related (12E)-lembyne A (**710**) has been isolated from an Okinawan *Laurencia mariannensis* (*577*). This metabolite appears to be a stereoisomer of *cis*-maneonene C.

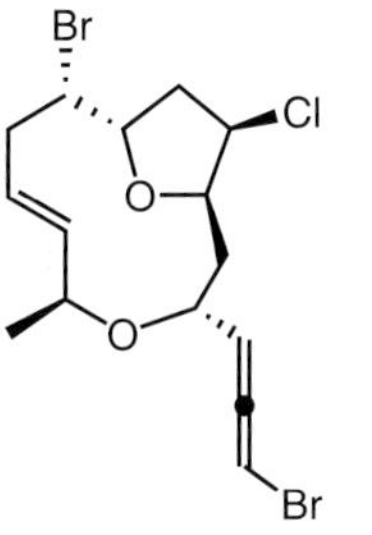

3.6 Marine Nonterpenes: C_{15} Acetogenins

701 (dactylallene)

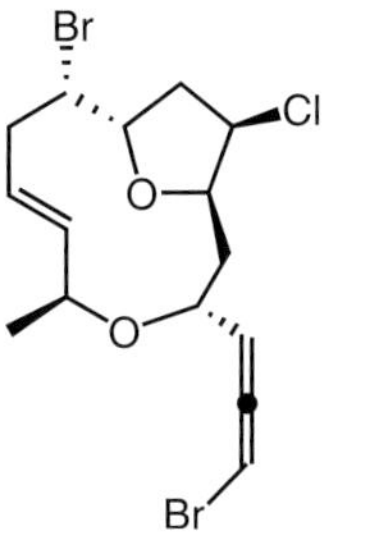

702 (obtusallene IV)

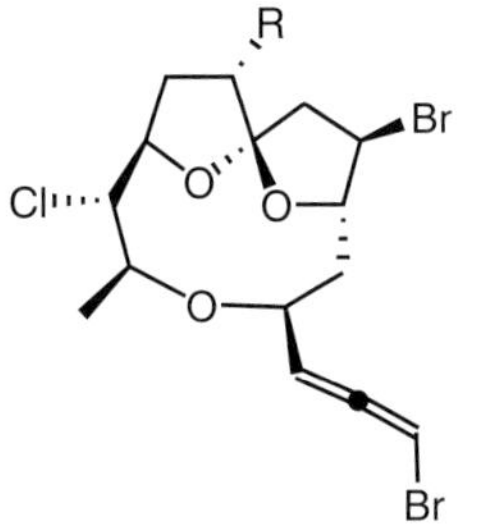

703 R = Br (obtusallene V)
704 R = H (obtusallene VI)

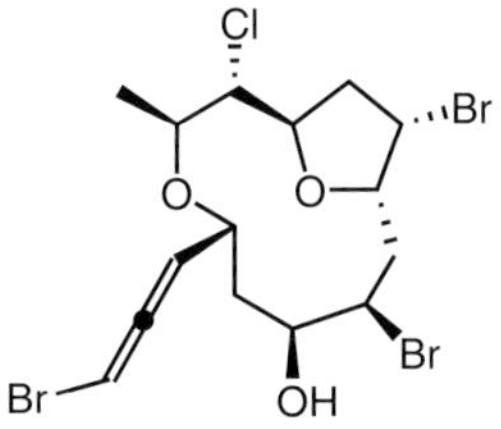

705 (obtusallene VII)

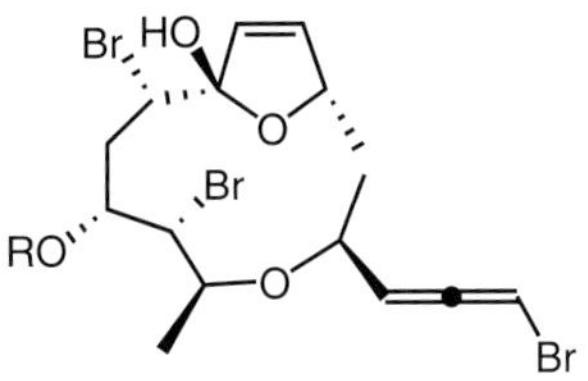

706 R = H (obtusallene VIII)
707 R = Ac (obtusallene IX)

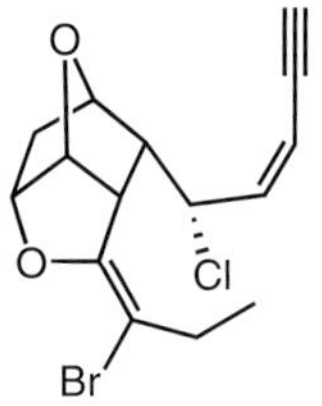

708 (lembyne-A)

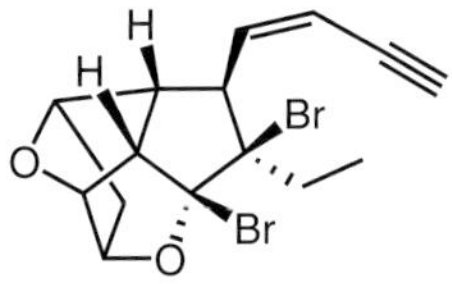

709 (lembyne-B)

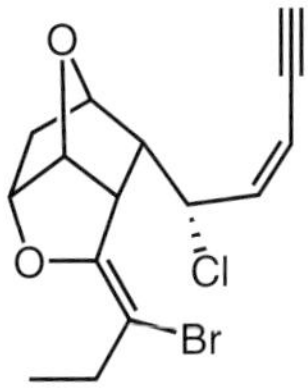

710 ((12E)-lembyne-A)

Only a few of the many reported total syntheses of the halogenated C_{15} acetogenins are listed here. The first total synthesis of (+)-isolaurepinnacin (**X**) confirms the proposed structure, notably the (*S*)-configuration at C-3, and corrects the rotation of natural **X** as being dextrorotatory (*797*). Total syntheses of (–)-*trans*-kumausyne (**Y**) (*798*) and (–)-kumausallene (**Z**) (*799*) support the proposed structures and absolute configurations of these compounds. Several syntheses of (+)-obtusenyne (**AA**) have been described (*800*), and the absolute configurations were established for "Norte's obtusenynes" (**BB**) and (**CC**) (*801*). Several syntheses of (+)-laurencin (**DD**) have been described (*802*), as has the first total synthesis of (+)-(Z)-laureatin (**EE**), which confirms its absolute configuration (*803*). Likewise, the first total syntheses of (+)-rogioloxepane A (**FF**) (*804*) and (–)-isolaurallene (**GG**) (*805*) validate the proposed structures of these *Laurencia* metabolites. Although the antipode (**HH**) was synthesized earlier, the first total synthesis of (+)-laurenyne (**II**) was described later (*806*). A second total synthesis of (+)-rogioloxepane A (**FF**) (*807*) and syntheses of both (+)-(3E)- (**JJ**) and (+)-(3Z)-pinnatifidenyne (**KK**) (*808*) confirm the proposed structures. An asymmetric synthesis of (–)-panacene (**LL**) has corrected its relative configuration as shown (*809*). In contrast to these and other successful syntheses, some synthetic efforts expose incorrectly proposed structures for natural products. Total syntheses of elatenyne (**MM**) and **NN** reveal that the proposed structures for these pyrano[3,2-*b*]pyrans are probably incorrect (*810*). The authors suggest a 2,2'-bifuranyl core for **MM** and **NN**. Unfortunately, space does not permit a full presentation of the many other elegant total syntheses of the halogenated C_{15} acetogenins.

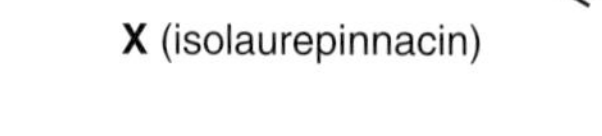

X (isolaurepinnacin)

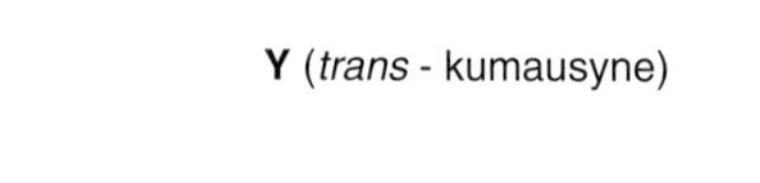

Y (*trans* - kumausyne)

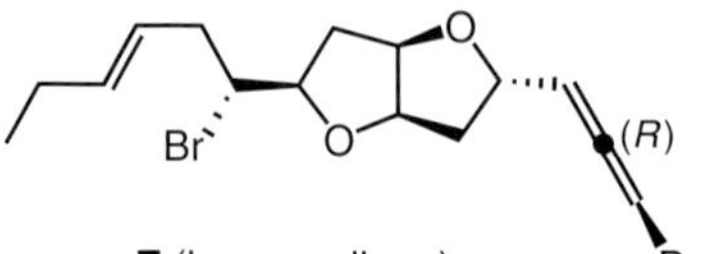

Z (kumausallene)

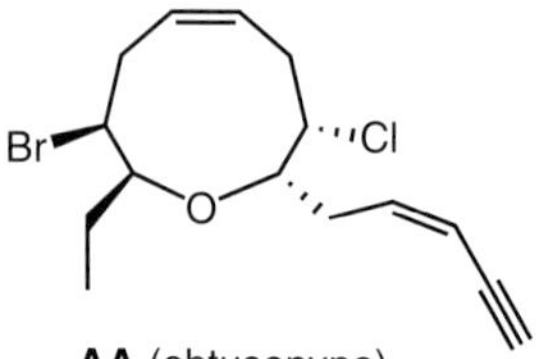

AA (obtusenyne)

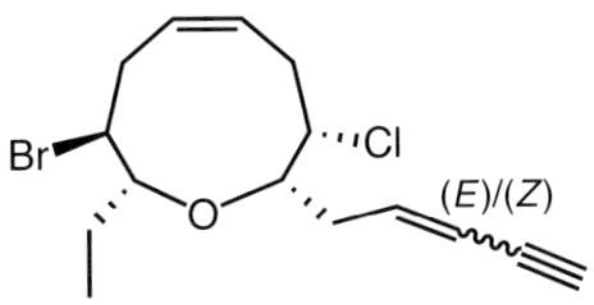

3.6 Marine Nonterpenes: C_{15} Acetogenins

BB ((E)-isomer)
CC ((Z)-isomer)

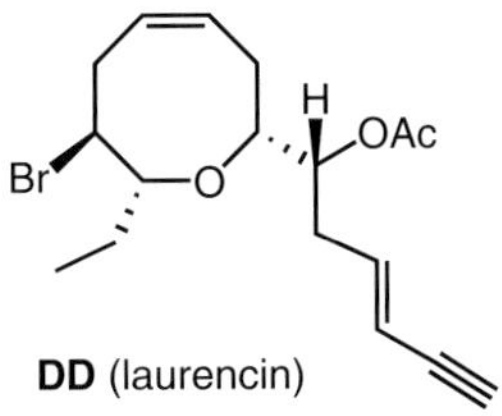

DD (laurencin)

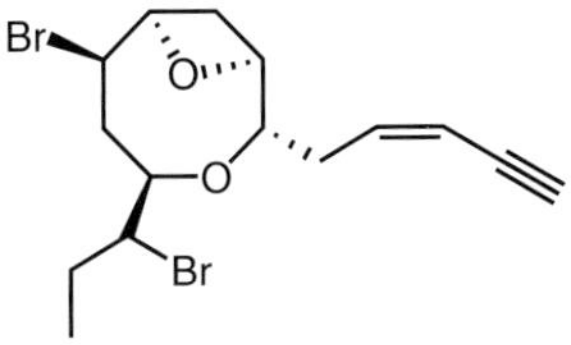

EE ((Z)-laureatin)

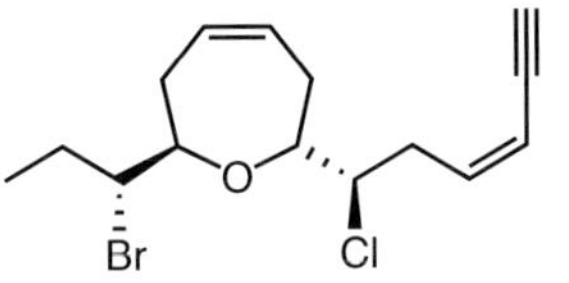

FF (rogioloxepane A)

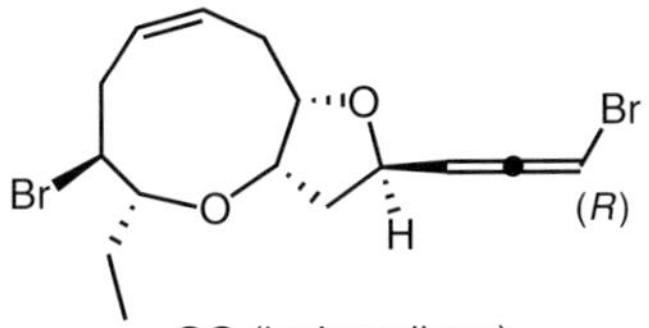

GG (isolaurallene)

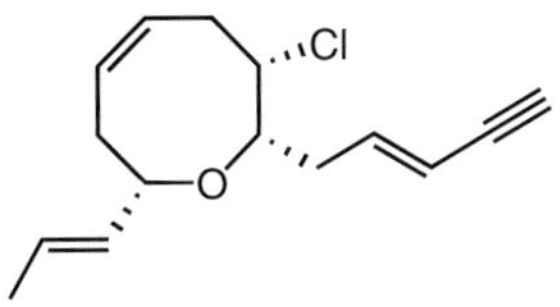

HH

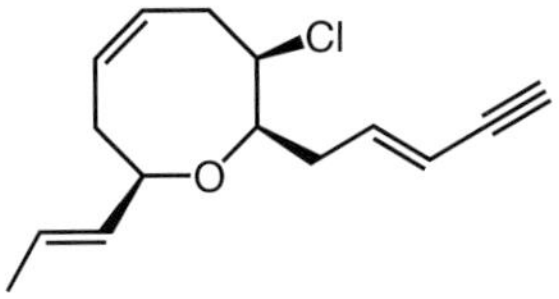

II (laurenyne)

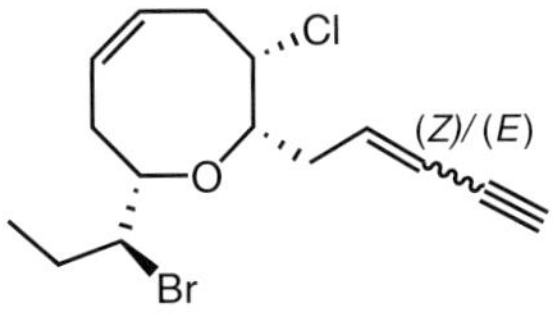

JJ ((E)-pinnatifidienyne)
KK ((Z)-pinnatifidienyne)

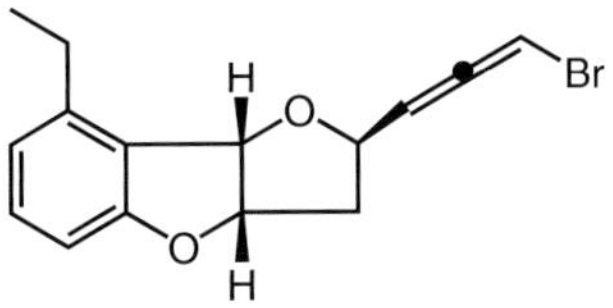

LL (panacene)

MM (elatenyne)

NN

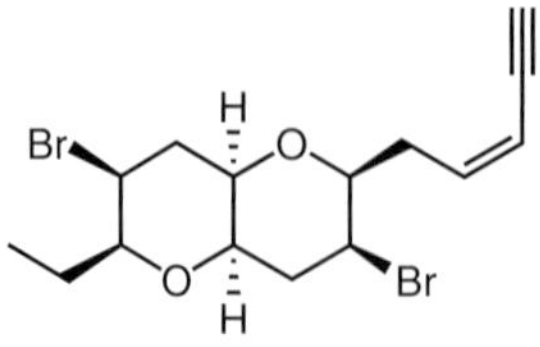

3.7 Iridoids

Iridoids are a large group of plant metabolites that are mevalonate-derived in origin and isoprenoid in carbon skeleton (*811*). A dozen chlorinated iridoids were cited in the first review (*1*), all of which are chlorohydrins. Phloyoside II (**711**) was isolated from the roots of *Phlomis younghusbandii* from Tibet (*812*). The South American shrub *Mentzelia cordifolia* contains the novel mentzefoliol (**712**) and glucosylmentzefoliol (**713**) in addition to the known 7-chlorodeutziol (*813*). The full paper describing the isolation, characterization of the previously reported glutinoside has appeared (*814*). Stegioside I (**714**) is found in *Physostegia virginiana* ssp. *virginiana*, and is a dehydroxylinarioside (*815*). The Okinawan plant *Premia subscandens* has furnished the four novel 10-*O*-acyl derivatives of the known asystasioside E (**715–718**) (*816*). Likewise, **719** and **720** from *Calalpae fructus* are 4-hydroxybenzoyl esters of known iridoid glucosides (*817*). Urphoside B (**721**), which is closely related to **719**, was isolated from a Turkish collection of *Veronica pectinata* var. *glandulosa* (*818*). Similarly, piscroside A (**722**) is a methoxy analogue of **720** and was characterized from the roots of the Chinese plant *Neopicrorhiza scrophulariiflora* (*819*). Globularioside (**723**), with a unique beta-chlorine atom, has been isolated from the Moroccan plant *Globularia alypum* (*820*). A collection of the parasitic plant *Cistanche tubulosa* has yielded kankanoside C (**724**) and kankanol (**725**) along with several other new nonchlorinated iridoids (*821*). A ^{13}C NMR analysis of iridoids is available (*822*).

711 (phloyoside II)

712 R = H (mentzefoliol)
713 R = Glc

714 (stegioside I)

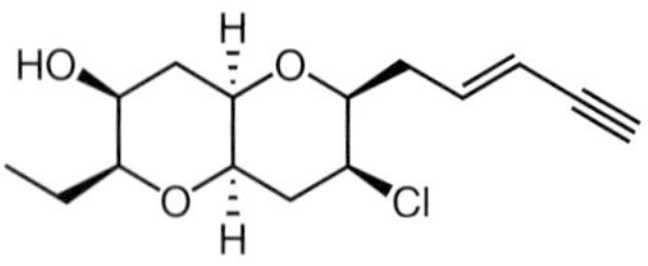

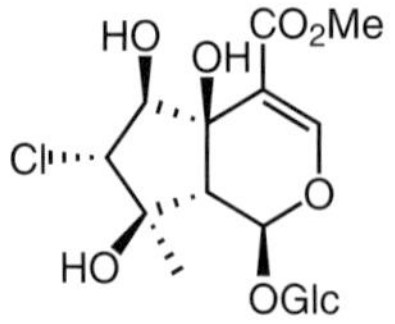

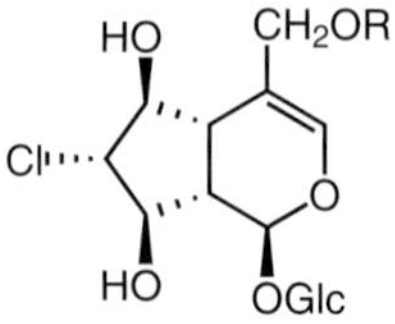

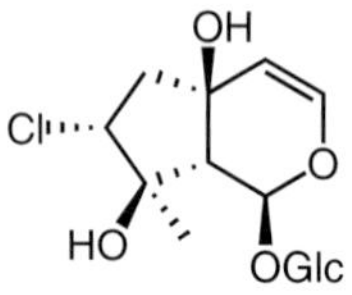

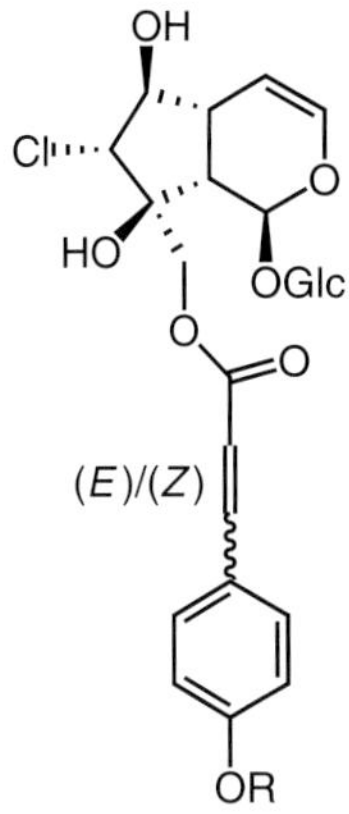

715 R = Me, (*E*)
716 R = Me, (*Z*)
717 R = H, (*E*)
718 R = H, (*Z*)

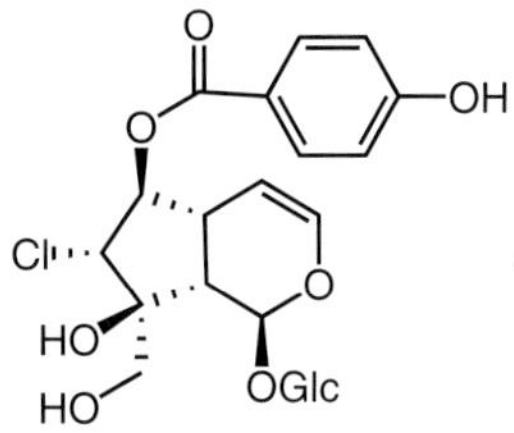

719

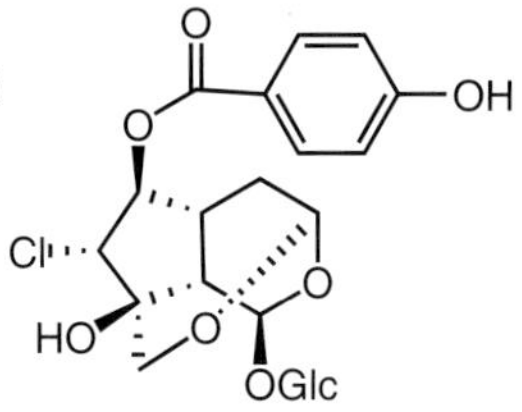

720

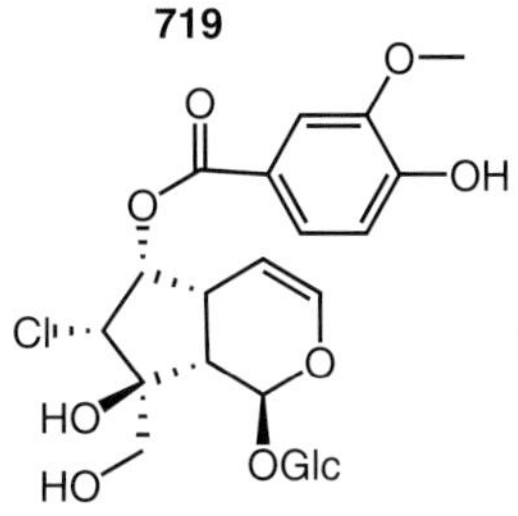

721 (urphoside B)

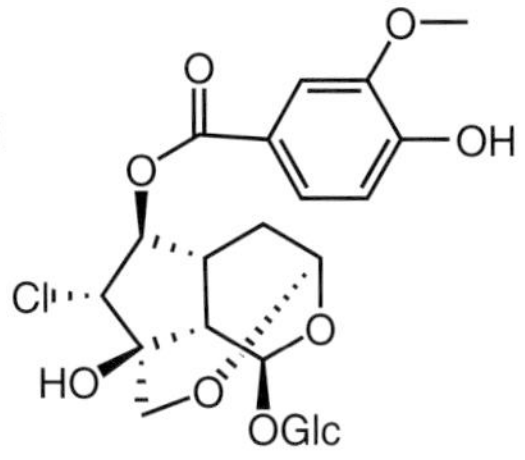

722 (piscroside A)

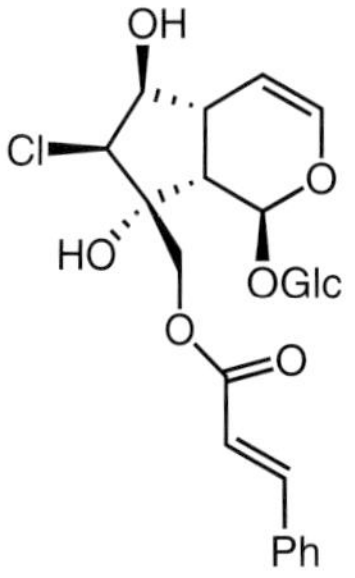

723 (globularioside)

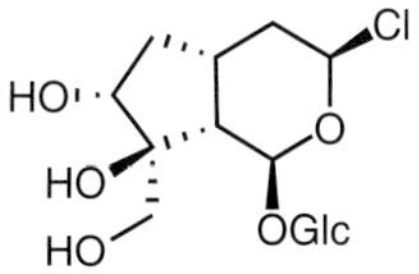

724 (kankanoside C)

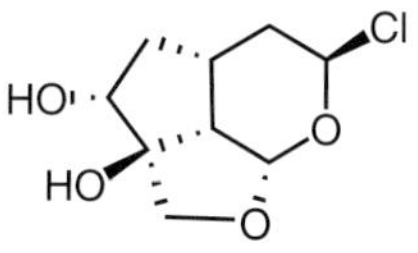

725 (kankanol)

3.8 Lipids and Fatty Acids

The evaluation of halogenated lipids and fatty acids is rendered difficult because many examples of chlorinated fatty acids are indisputably man-made and do not have clear natural sources. The order of presentation follows that adopted in the first survey. As before, only newly isolated and characterized compounds are numbered. Studies of the toxic (contaminated) edible mussel *Mytilus galloprovincialis* from the Adriatic Sea have led to the isolation of three novel chlorosulfolipids **726–728** (*823–825*). These cytotoxic compounds are found in the digestive glands of the animals and are antiproliferative against several cell lines (J774, WEHI164, P388).

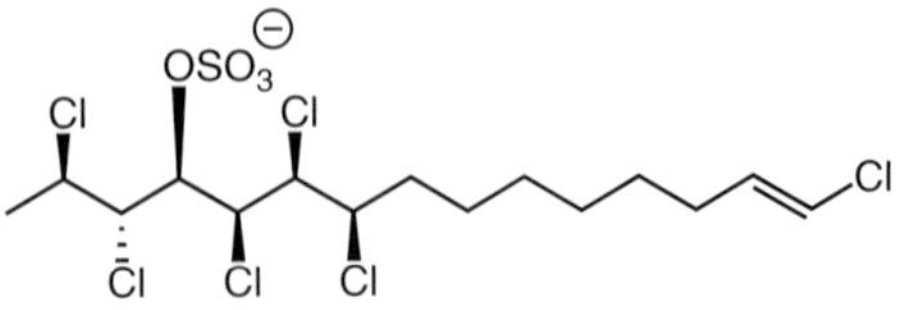

Halogenated fatty acids have both anthropogenic and natural sources (*826–830*), and the distinction is not always unambiguous, particularly with chlorinated fatty acids (*826–829*). Nevertheless, an abundance of natural halogenated fatty acids is beyond dispute (*830*). Several new members of the class of bromine-containing fatty acids, which numbered 14 in the first survey (*1*), have been identified from both marine and terrestrial sources. An Indonesian sponge, *Oceanapia* sp., has furnished the two novel bromo acids **729** and **730** (*831*), and the Australian sponge *Amphimedon terpenensis* contains 6-bromo-(5E,9Z)-tetracosadienoic acid (**731**) and 6-bromo-(5E,9Z)-pentacosadienoic acid (**732**) (*832, 833*). In addition to the latter two bromoacids, the Caribbean sponge *Agelas* (Fig. 3.11) has afforded **733** and **734** (*834*). The sea anemone *Stoichactis helianthus* contains 6-bromo-(5E,9Z)-heneicosadienoic acid (**735**) and 6-bromo-(5E,9Z)-docosadienoic acid (**736**) (*835*). The phospholipid extracts of both the sea anemone *Condylactis gigantea* and the zoanthid *Palythoa caribaeorum* furnished the novel 6-bromo-(5E,9Z)-eicosadienoic acid (double bond stereochemistry assumed) (**737**) (*836*).

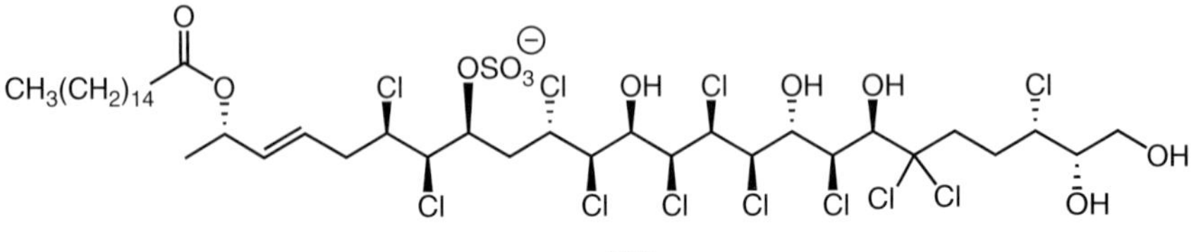

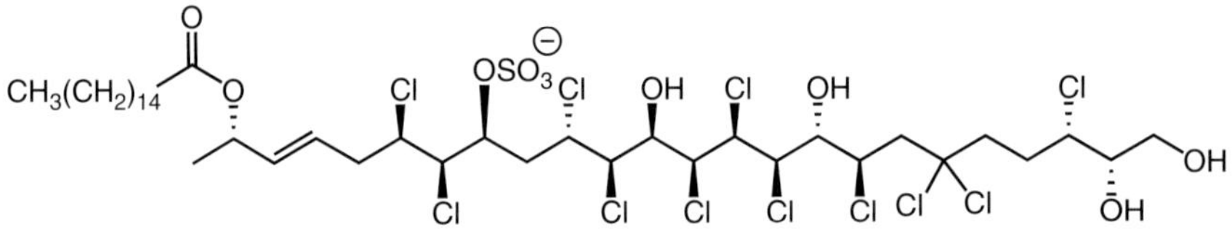

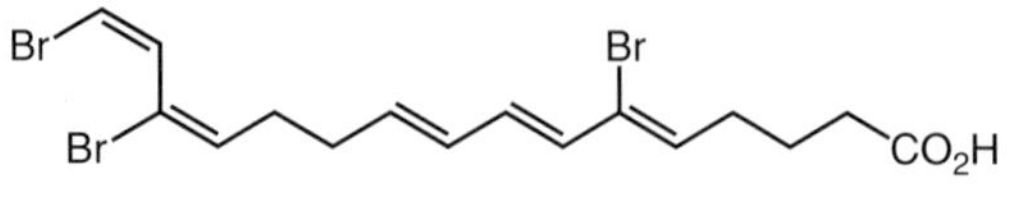

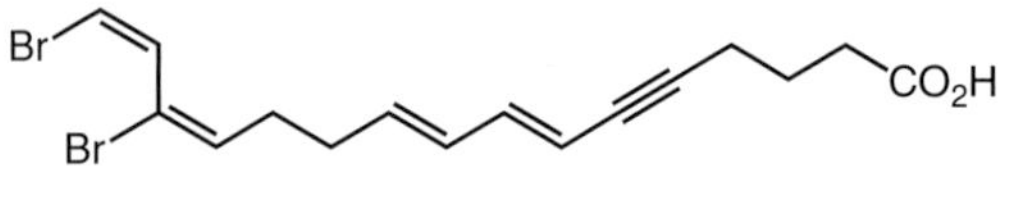

3.8 Lipids and Fatty Acids

CH₃(CH₂)ₙ ~~~~ Br ~~~~ CO₂H

731 $n = 13$
732 $n = 14$
735 $n = 10$
736 $n = 11$
737 $n = 9$

(CH₃)₂CH(CH₂)ₙ ~~~~ Br ~~~~ CO₂H

733 $n = 12$
734 $n = 13$

Fig. 3.11 *Agelas* sp. An example of a very common marine sponge that produces a wide variety of halogenated metabolites such as the brominated fatty acids **733** and **734** (Photo: J. R. Pawlik)

Sponges of genus *Xestospongia* are rich suppliers of brominated fatty acids. An Okinawan sampling of this sponge yielded 14 new brominated fatty acids (**738–751**), along with three previously identified examples (*837*). A study of this sponge from the Indian Ocean characterized the novel **752–754** (*838*). The new xestosterol esters **755** and **756** were discovered in *Xestospongia testudinaria* from Australia, which had previously been found to contain xestosterol (*839*).

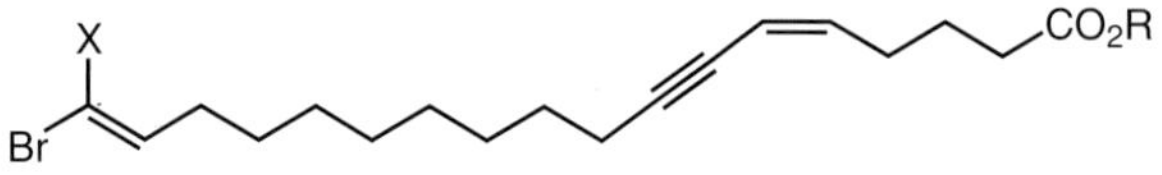

738 R = H, X = Br
739 R = X = H
740 R = Me, X = H
741 R = Me, X = Br

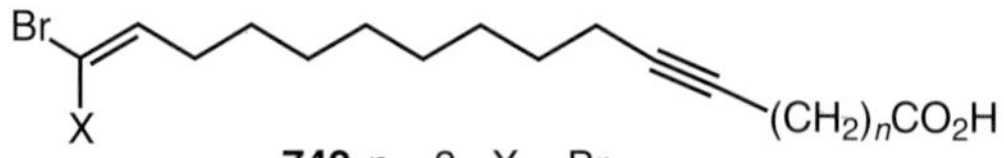

742 n = 3, X = Br
743 n = 3, X = H
744 n = 5, X = Br
745 n = 5, X = H

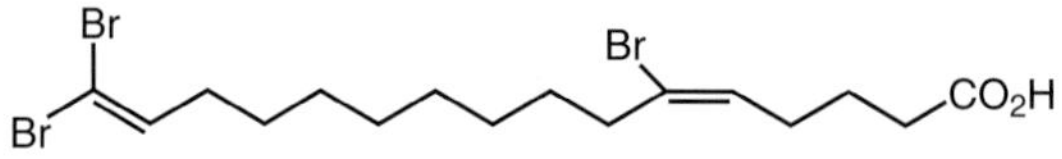

746

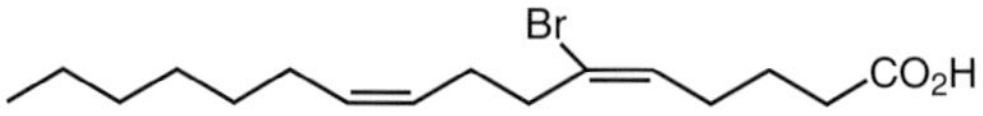

747

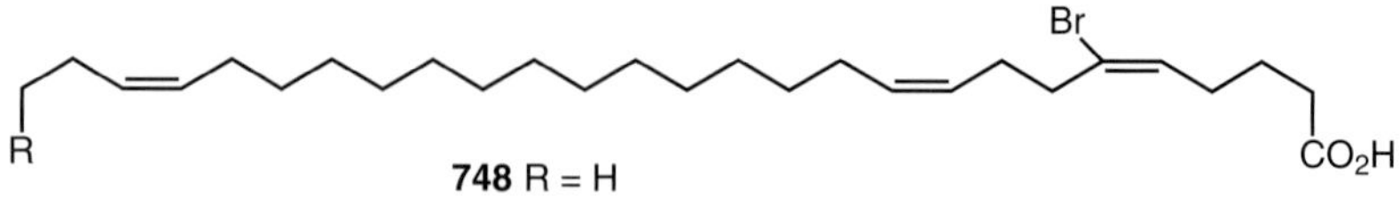

748 R = H
749 R = Me

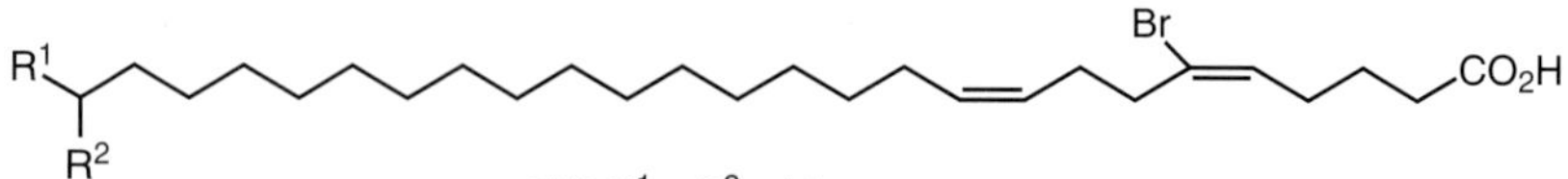

750 R^1 = R^2 = Me
751 R^1 = H, R^2 = *i*-Pr

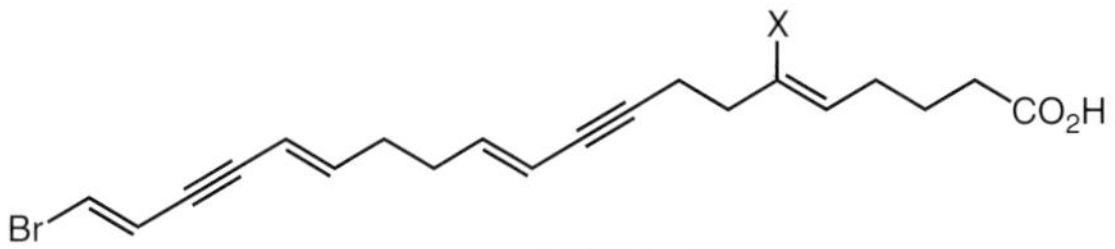

3.8 Lipids and Fatty Acids

752 X = H
753 X = Br

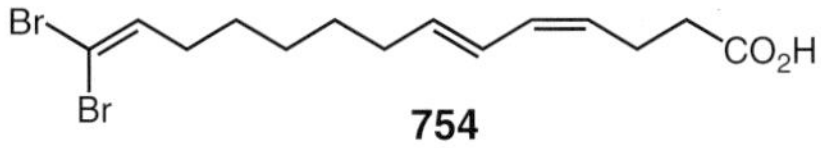

754

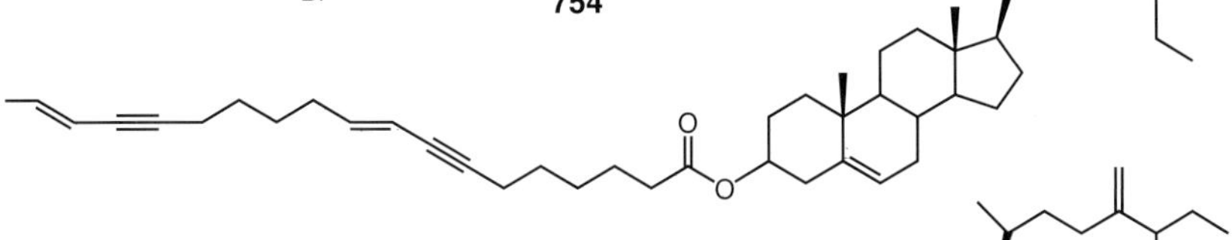

755

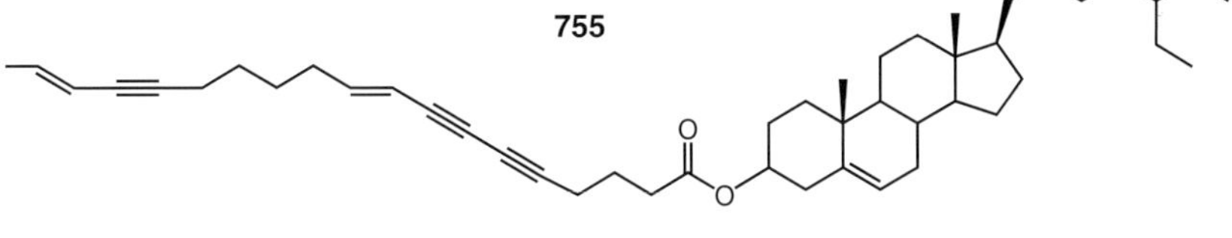

756

Surrounding a fresh water lake in Central Asia, which has a high salt content (up to 5,800 ppm), is the lichen *Acorospora gobiensis*. This lichen yielded the novel bromo acids **757** and **758** (*840*). Further study of this lichen and others collected around Lake Issyk-Kul in Central Asia (*Cladonia furcata, Lecanora fructulosa, Leptogium saturninum, Parmelia linctina, Parmelia comtseliadalis, Peltigera canina*, and *Xanthoria* sp.) uncovered six additional new brominated fatty acids, **759–764** (*841*). Another study of these and other lichens around this lake led to the first natural bromoallenic fatty acids **765** and **766** (*842*). The new lichens examined were *Rhizoplaca peltata, Xanthoparmelia camtschadalis, Xanthoparmelia tinctina*, and *Xanthoria elegans*.

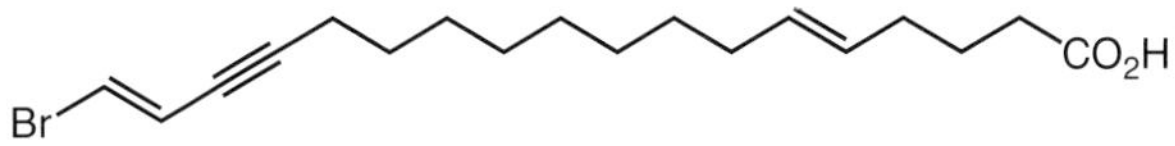

757

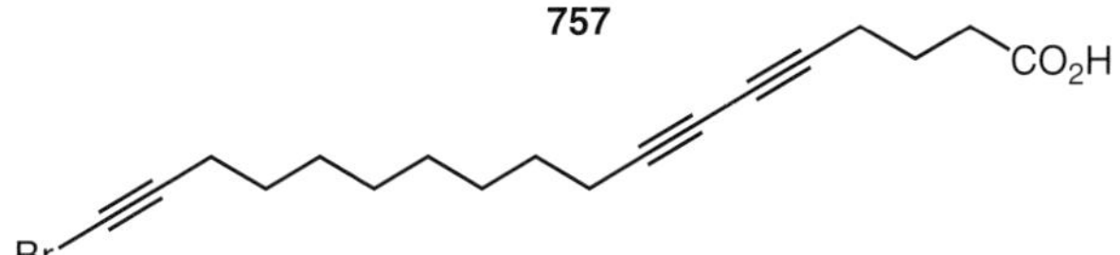

758

759

760

761

762

763

764

765

766

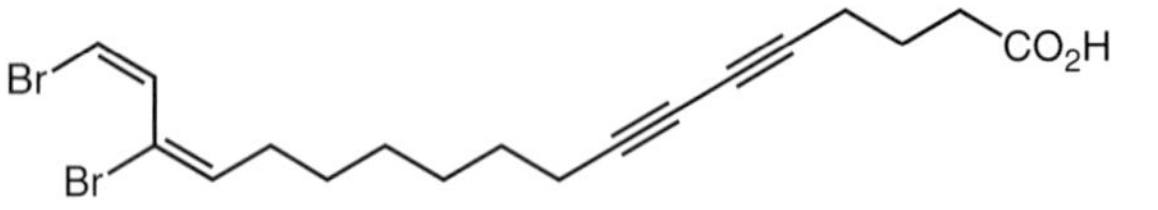

A collection of the red alga *Plocamium cartilagineum* from Corsica and the Maltese Islands has yielded the four new halogenated homosesquiterpenic fatty acids **767–770** (*843*). The Pohnpei sponge *Dysidea fragilis* contains the novel (4*E*)-

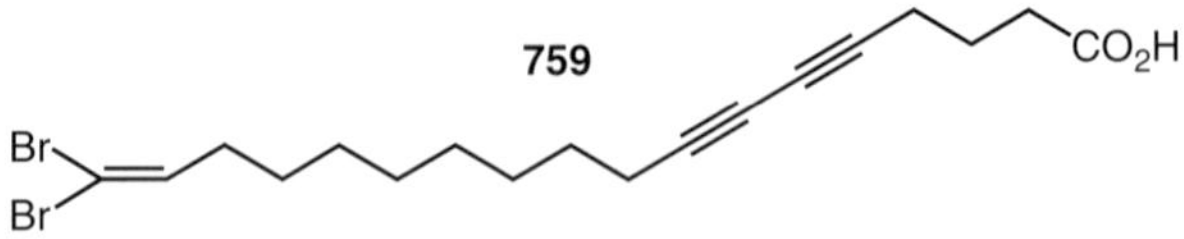

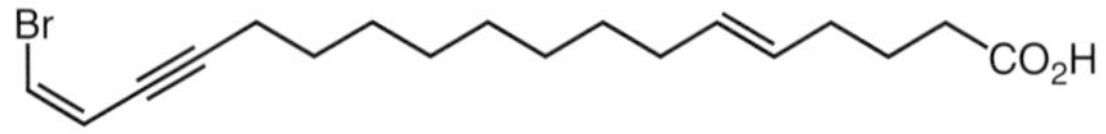

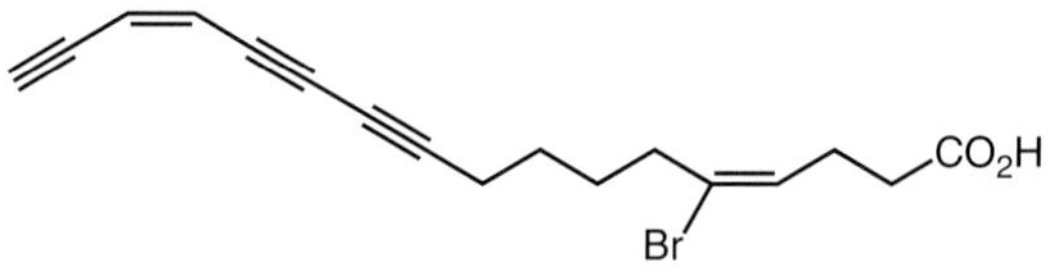

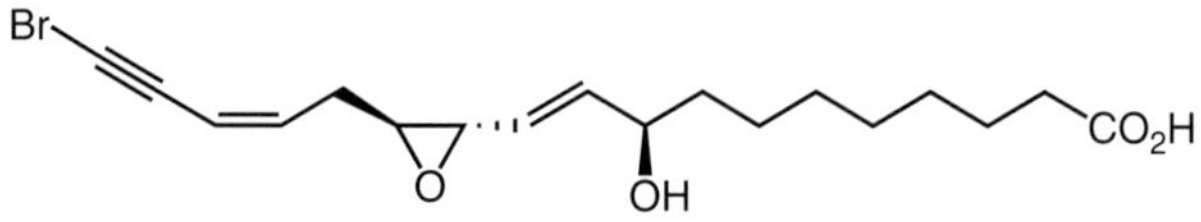

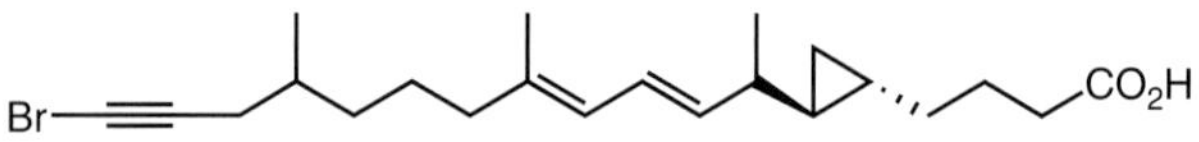

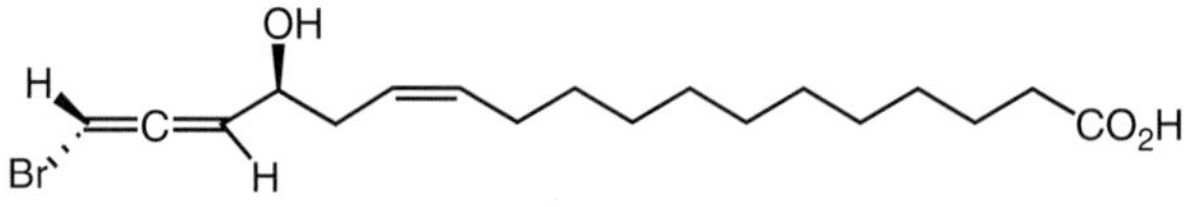

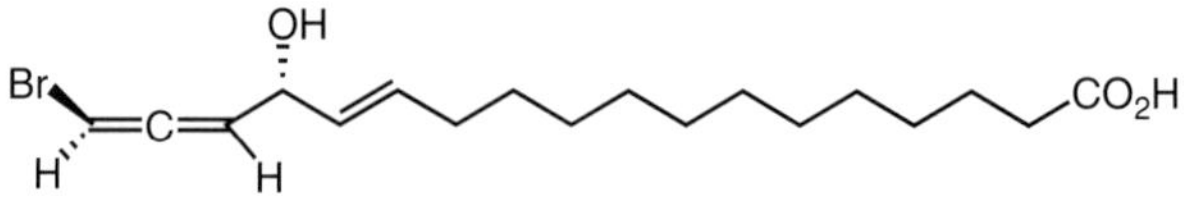

3.8 Lipids and Fatty Acids

(*S*)-antazirine (**771**) and (4*Z*)-antazirine (**772**) (*844*). Majusculoic acid (**773**) is a novel metabolite isolated from a Bahamian cyanobacterial mat microbial community (*845*). A further cyclopropane fatty acid, grenadadiene (**774**), was found in the marine blue-green alga (cyanobacterium) *Lyngbya majuscula* from Grenada (*846*), a seaweed implicated in causing contact dermatitis ("swimmer's itch") (*847, 848*). Another genus of cyanobacteria, *Nostoc*, can also cause severe health problems (*849*).

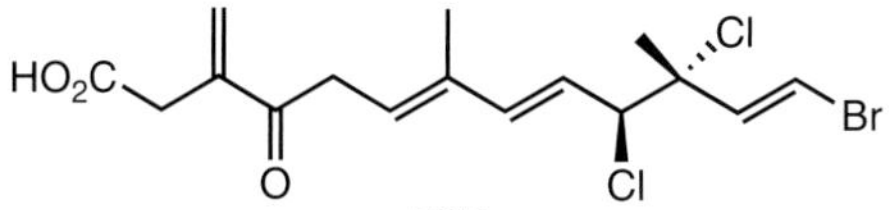

767

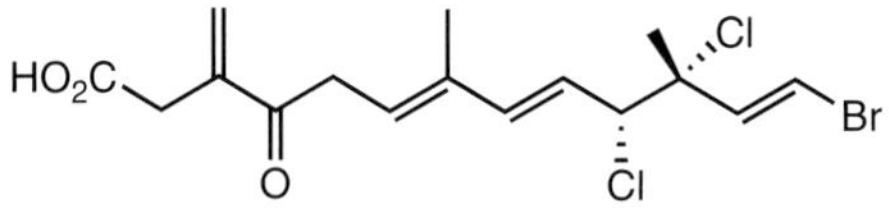

768

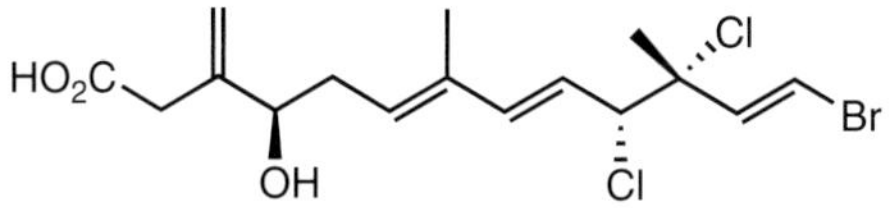

769

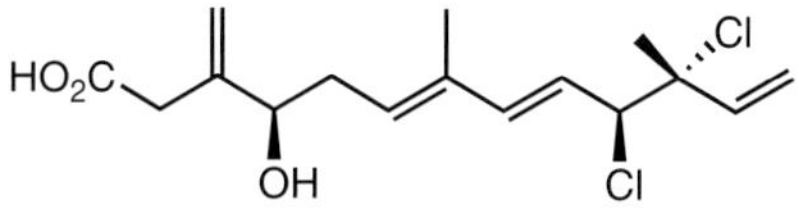

770

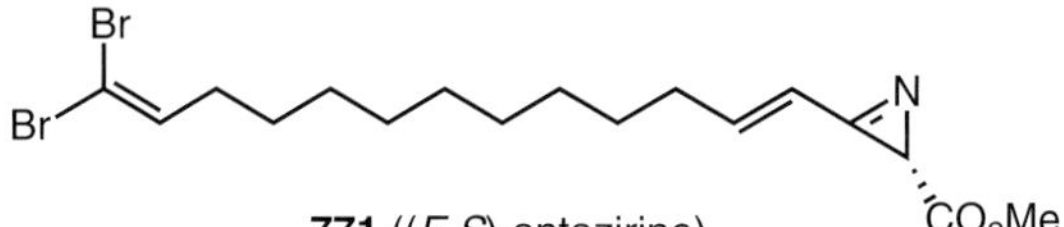

771 ((*E,S*)-antazirine)

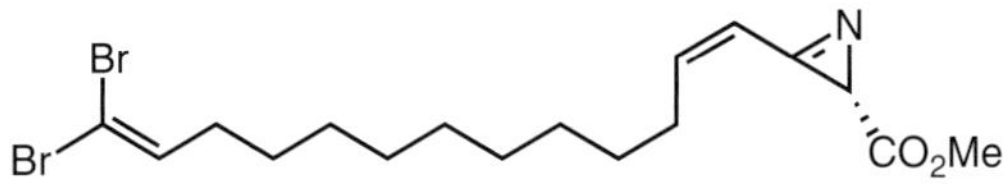

772 ((*Z,S*)-antazirine)

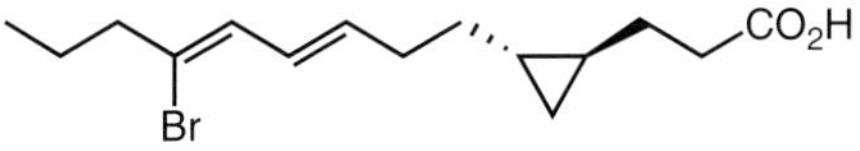

773 (majusculoic acid)

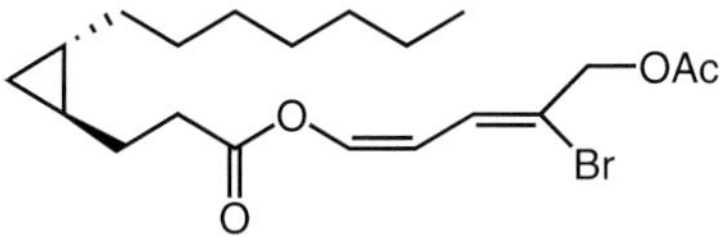

774 (grenadadiene)

In addition to producing grenadadiene (**774**), *Lyngbya majuscula* is an amazingly prolific source of diverse fatty acid metabolites (*1*). An Okinawan collection of this cyanobacterium has yielded the novel malyngamide I (**775**), and this study led to the revision of the previously reported stylocheilamide as the acetate **OO** (*850*). The latter metabolite was originally isolated from the sea hare *Stylocheilus longicauda*, which feeds on *Lyngbya majuscula* (*1*). An assemblage of this blue-green alga from Curacao contains malyngamides K (**776**) and L (**777**) (*851*), a Madagascan sample yielded malyngamides Q (**778**) and R (**779**) (*852*), and a Puerto Rican specimen afforded malyngamide T (**780**) (*853*). The Hawaiian sea hare *Stylocheilus longicauda*, which feeds on *Lyngbya majuscula*, contains malyngamides O (**781**) and P (**782**) (*854*), and the New Zealand sea hare *Bursatella leachii* has yielded malyngamide S (**783**), which displays some antiinflammatory and cytotoxic activity (*855*). This animal is also known to feed on *Lyngbya majuscula*. The Hawaiian red alga *Gracilaria coronopifolia*, known as the source of the toxic aplysiatoxin, contains malyngamides M (**784**) and N (**PP**), the latter of which is a revised structure of deacetoxystylocheilamide (*856*), a compound described earlier (*1*). Malyngamide M is the first natural aromatized malyngamide. The authors suggest that an associated cyanobacterium actually produces the malyngamides.

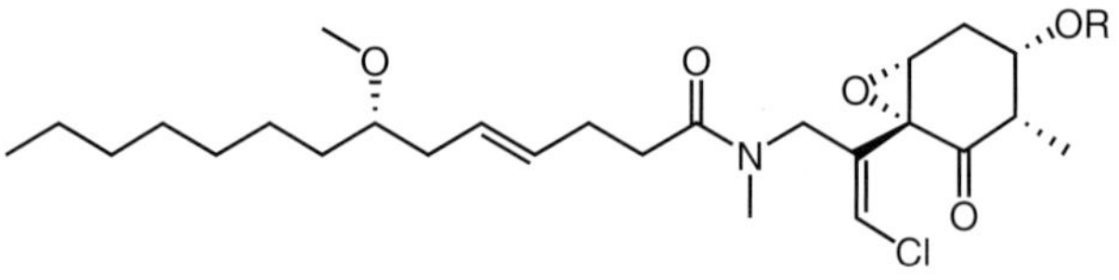

775 R = H (malyngamide I)
OO R = Ac (stylocheilamide)

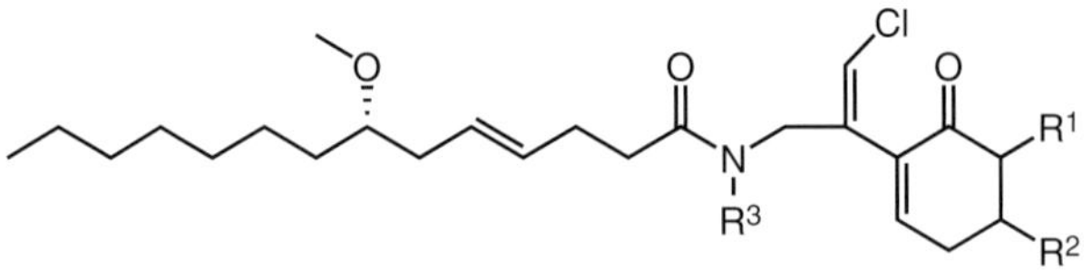

776 $R^1 = R^2 = R^3 = H$ (malyngamide K)
777 $R^1 = R^3 = Me$, $R^2 = OH$ (malyngamide L)

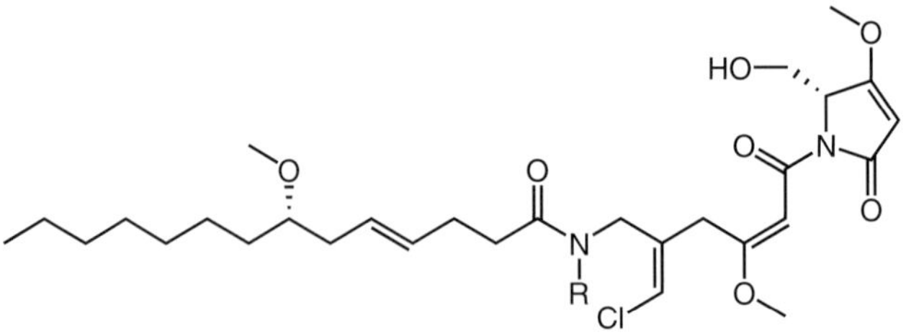

778 R = H (malyngamide Q)
779 R = Me (malyngamide R)

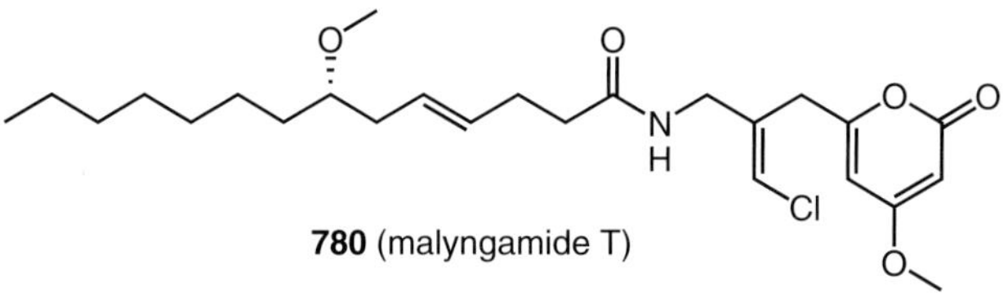

3.8 Lipids and Fatty Acids

780 (malyngamide T)

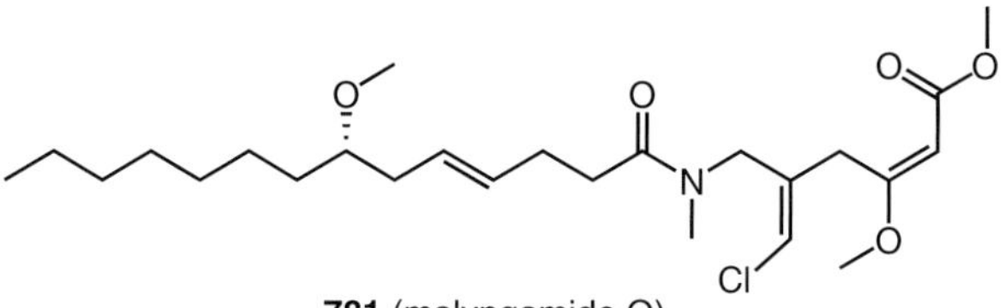

781 (malyngamide O)

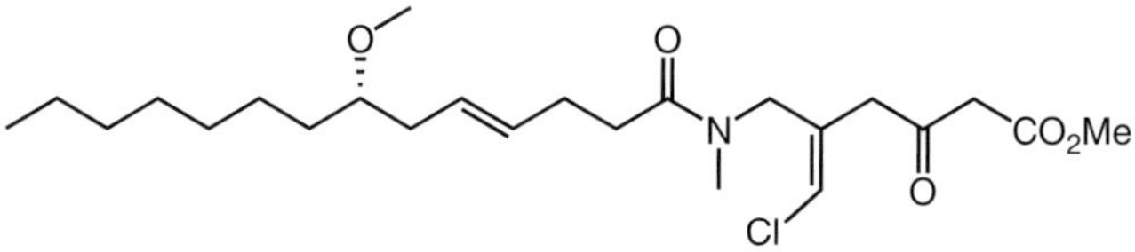

782 (malyngamide P)

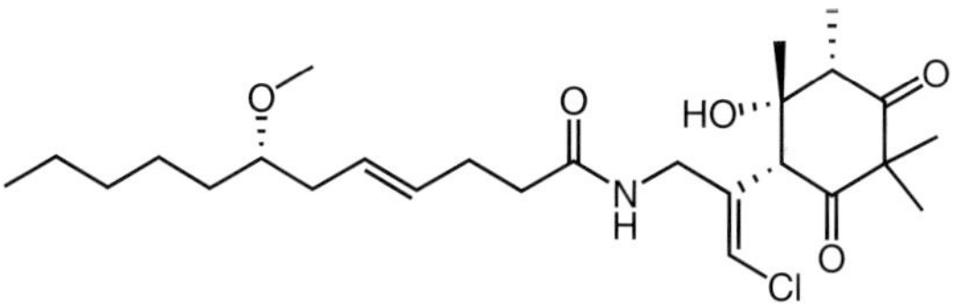

783 (malyngamide S)

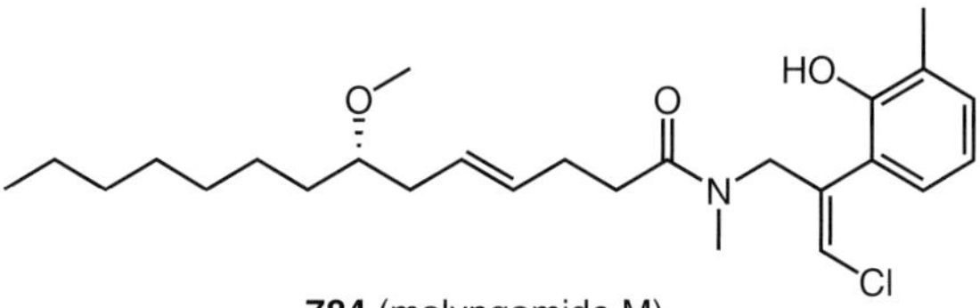

784 (malyngamide M)

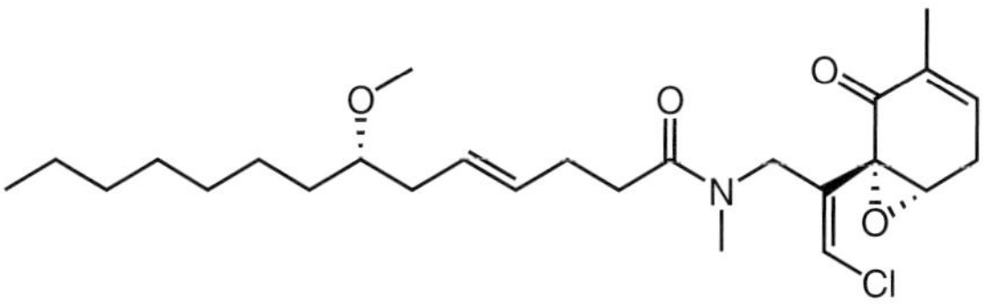

PP (malyngamide N = deacetoxystylocheilamide)

The Hawaiian *Lyngbya majuscula* has afforded isomalyngamides A (**785**) and B (**786**) (*857*), isomeric with the known malyngamide A, and the previously unreported malyngamide B (**787**) (*858*). The novel pitiamide A (**788**) was isolated from a mixed cyanobacterial sample of *Lyngbya majuscula* and *Microcoleus* sp. growing on the hard coral *Porites cylindra* from Guam (*859*). The structure and absolute configuration were confirmed by total synthesis (*860*). A homologue, pitiamide B, with one additional methylene group remains unidentified (*859*). A Jamaican strain of *Lyngbya majuscula* has yielded jamaicamides A (**789**), B (**790**), and C (**791**) (*861, 862*). The sea hare *Stylocheilus longicauda* from Oahu contains the unprecedented makalika ester (**792**) and makalikone ester (**793**) (*863*). The South African red alga *Gracilaria verrucosa* has furnished the two chlorohydrins **794** and **795** (*864*), and the related α-chloro divinyl ethers, maracens A (**796**), B (**797**), C (**798**), and D (**799**), were isolated from *Sorangium cellulosum* and display some activity against mycobacteria, related to the cause of tuberculosis (*865*).

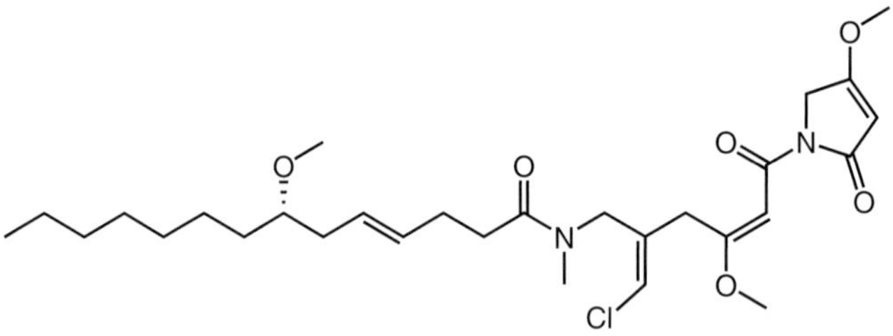

785 (isomalyngamide A)

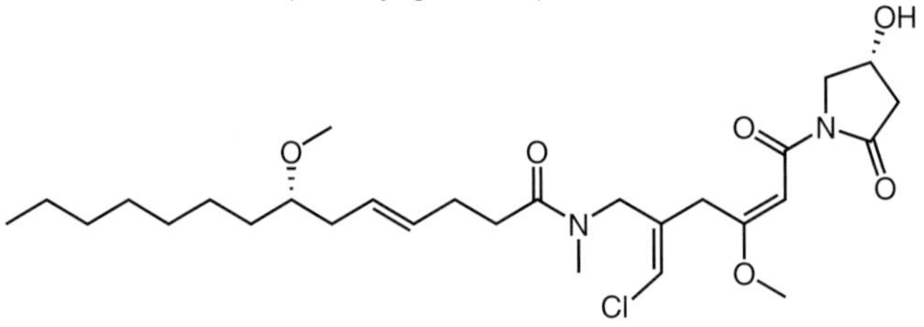

786 (isomalyngamide B)

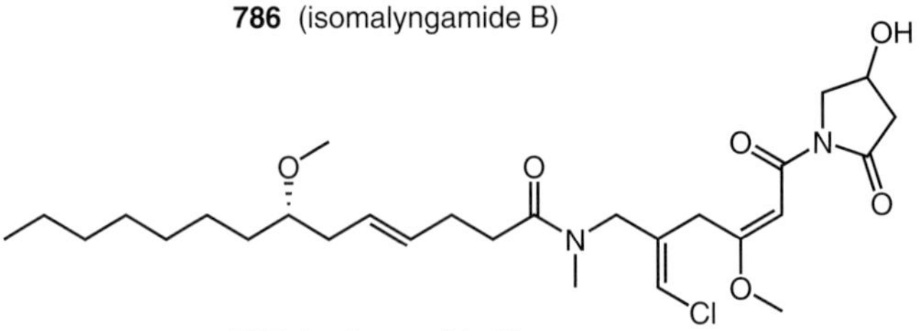

787 (malyngamide B)

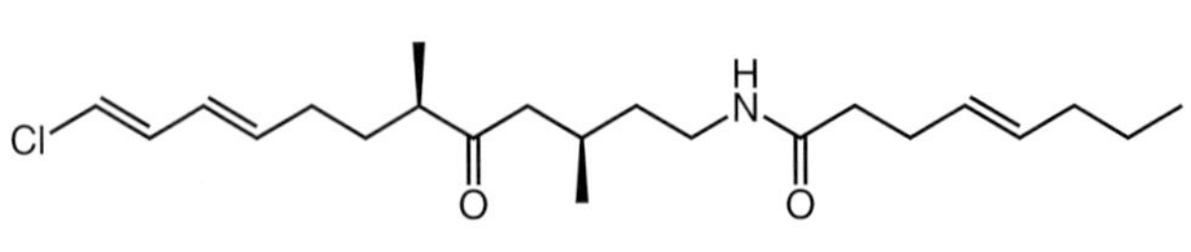

788 (pitiamide A)

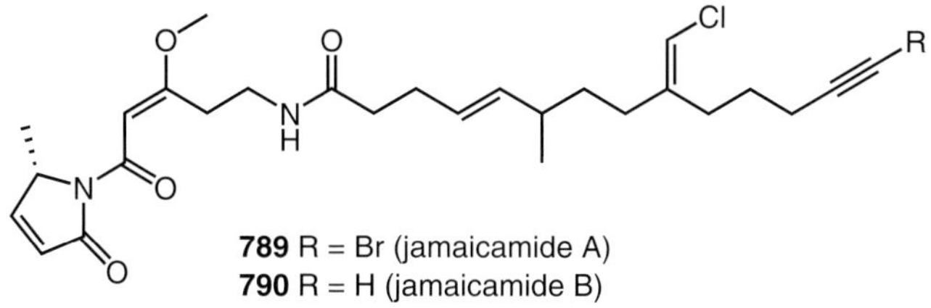

3.8 Lipids and Fatty Acids

789 R = Br (jamaicamide A)
790 R = H (jamaicamide B)

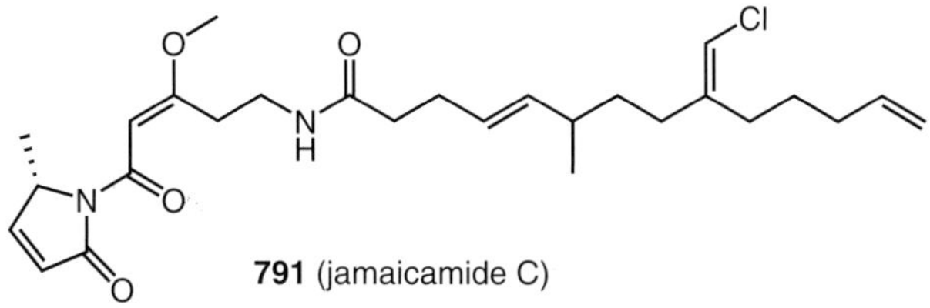

791 (jamaicamide C)

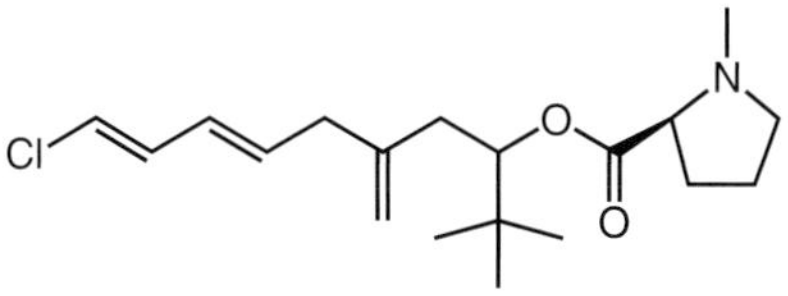

792 (makalika ester)

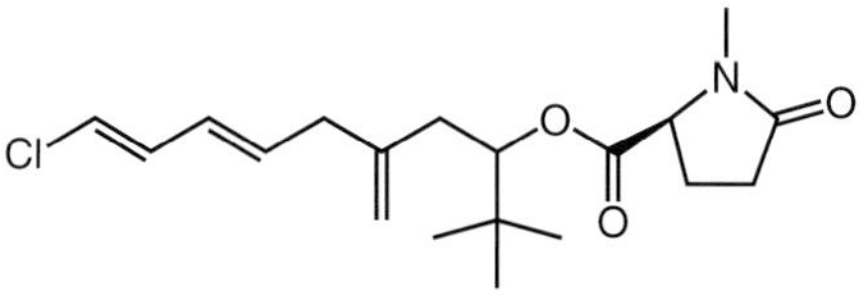

793 (makalikone ester)

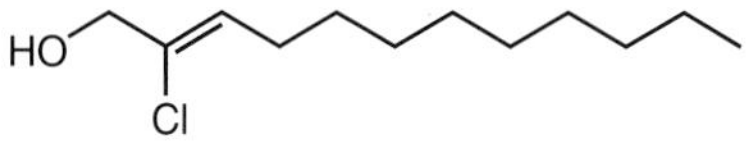

794

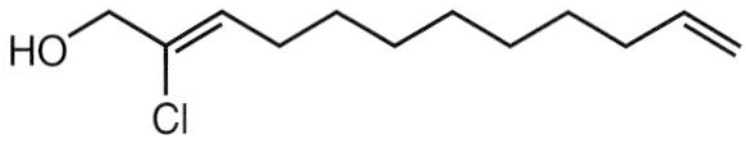

795

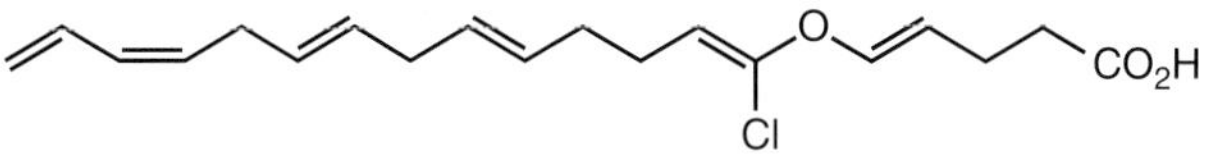

796 (maracen A)

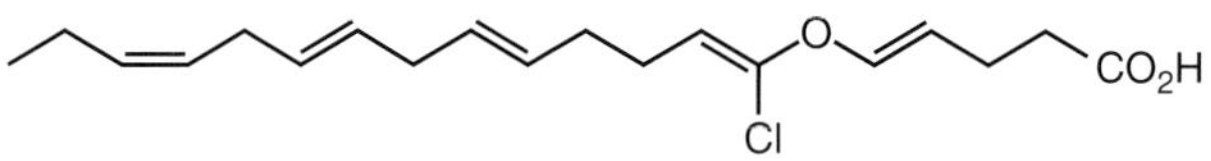

797 (maracen B)

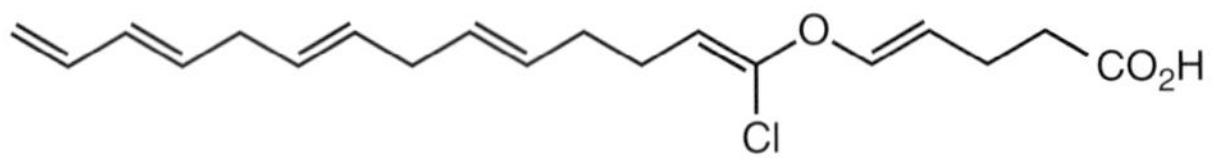

798 (maracen C)

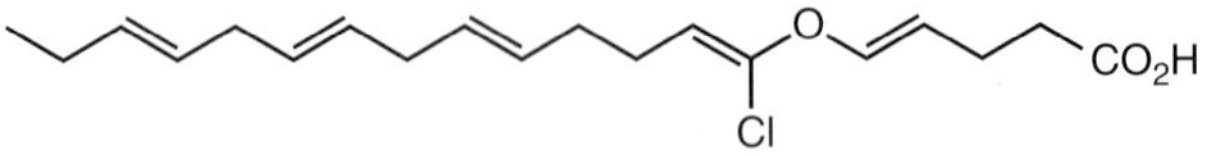

799 (maracen D)

Blue-green algae are not the sole producers of halogenated fatty acid metabolites; several marine sponges and associated fungi have furnished new examples. The fungus *Gymnasella dankaliensis* from the sponge *Halichondria japonica* has supplied the novel gymnastatins A–H, most of which are chlorinated (A, **800**; B, **801**; C, **802**; D, **803**; E, **804**; F, **805**; G, **806**) (*866–868*). Gymnastatin A (**800**) has been synthesized (*869*). Several of these metabolites have pronounced cytotoxic and cytostatic activity. It should be noted that gymnastatins A, D, and E are each mixtures of two hemiacetals. The related aranochlors A (**807**) and B (**808**) were isolated from the fungus *Pseudoarachniotus roseus* (*870*). It seems possible that these hemiacetals also each exist as two epimers.

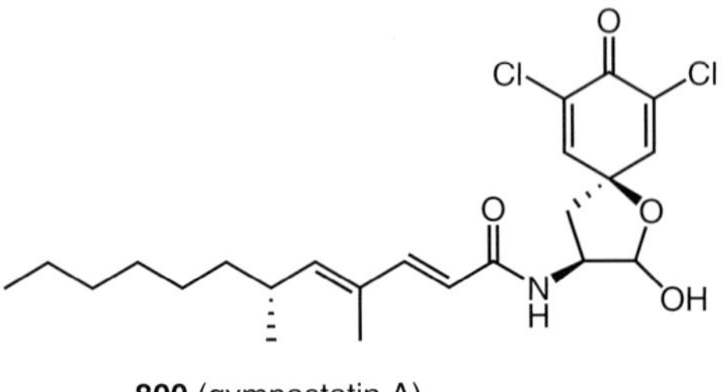

800 (gymnastatin A)

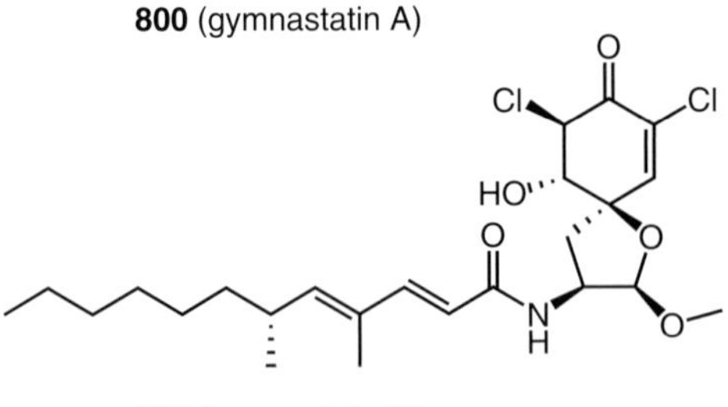

801 (gymnastatin B)

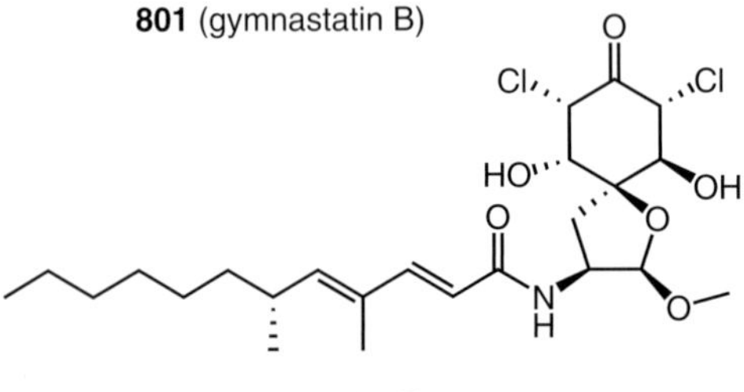

802 (gymnastatin C)

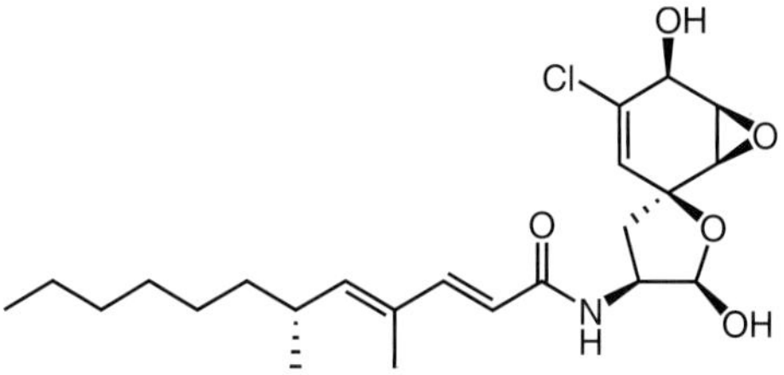

3.8 Lipids and Fatty Acids

803 (gymnastatin D)

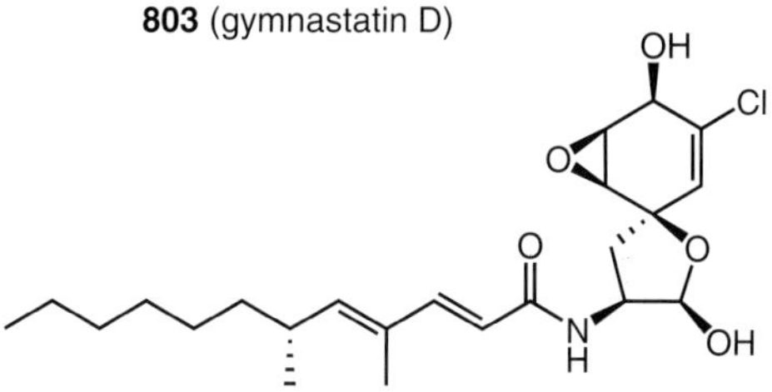

804 (gymnastatin E)

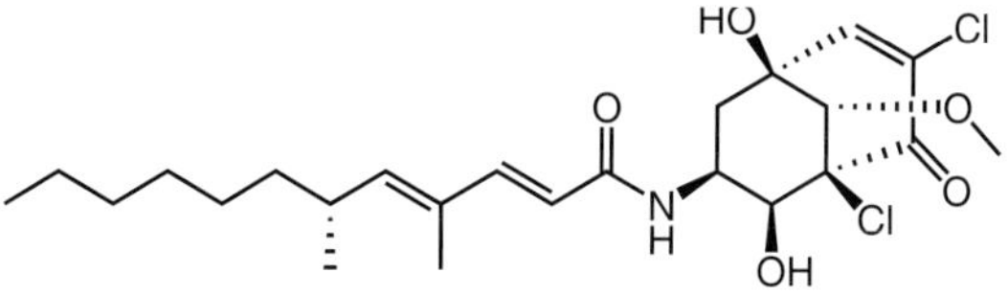

805 (gymnastatin F)

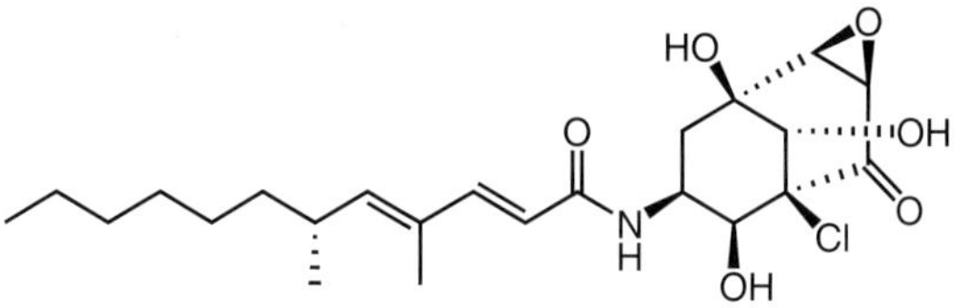

806 (gymnastatin G)

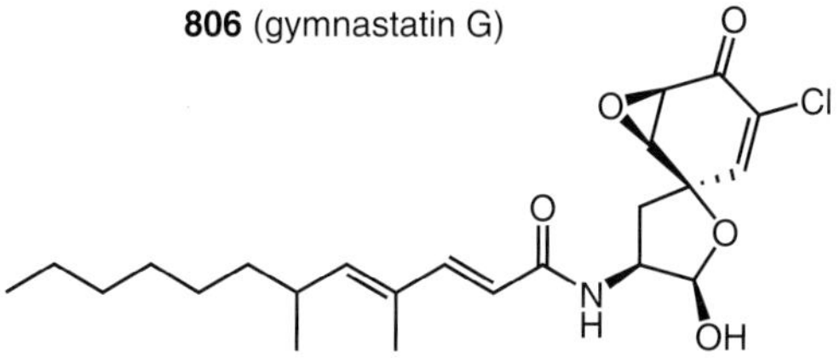

807 (aranochlor A)

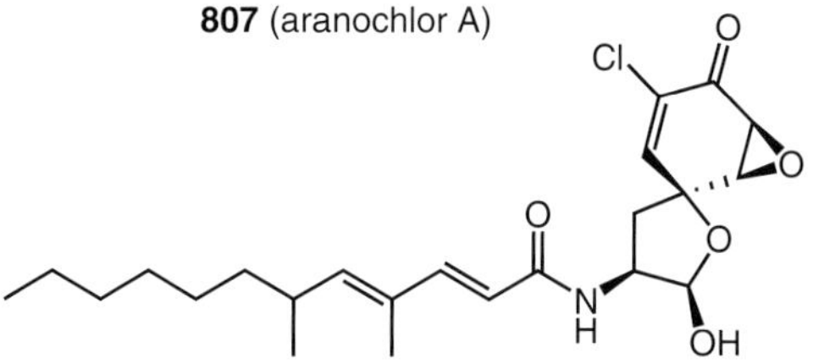

808 (aranochlor B)

The sponge metabolites aurantosides A (**QQ**) and B (**RR**) were described in the first survey (*1*); subsequently, several new aurantosides have been isolated. Thus, aurantoside C (**809**) is found in the sponge *Homophymia conferta* and has the same absolute configuration as aurantosides A and B (*871*). The stereochemistry about the terminal double bond in the latter two compounds was revised as shown (**QQ**, **RR**) in a report that described the isolation of the new aurantosides D (**810**), E (**811**), and F (**812**), which have both antifungal and cytotoxic activity, from the sponge *Siliquariaspongia japonica* (*872*). The Papua New Guinea sponge *Theonella swinhoei* has afforded aurantosides G (**813**), H (**814**), and I (**815**) (*873*). A series of related tetramic acid glycosides, rubrosides A–H (**816–823**), was characterized from the Japanese sponge *Siliquariaspongia japonica* (*874*). Several of these rubrosides have antifungal (*Aspergillus fumigatus*, *Candida albicans*) and cytotoxic (P388) activity.

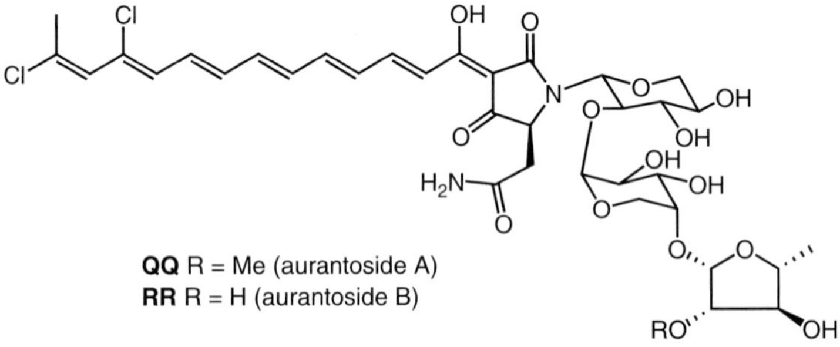

QQ R = Me (aurantoside A)
RR R = H (aurantoside B)

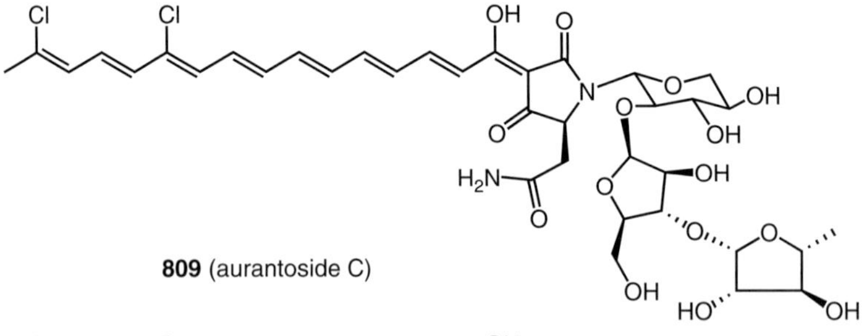

809 (aurantoside C)

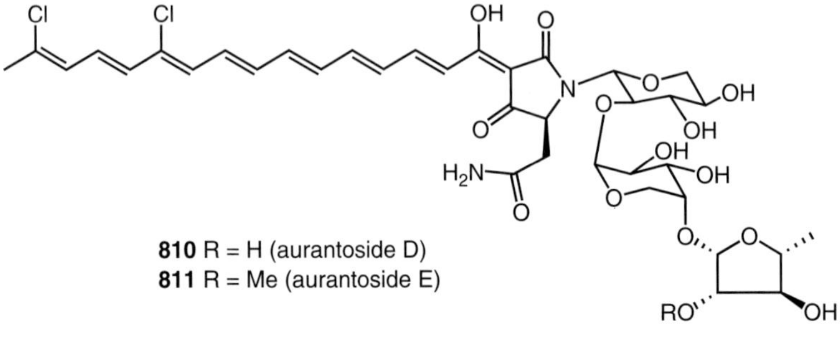

810 R = H (aurantoside D)
811 R = Me (aurantoside E)

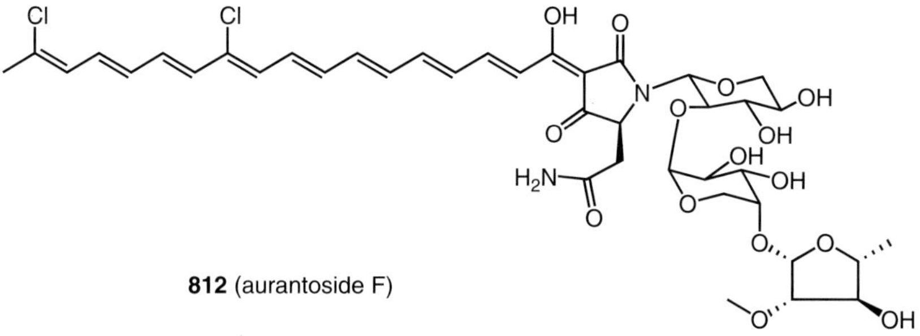

3.8 Lipids and Fatty Acids

812 (aurantoside F)

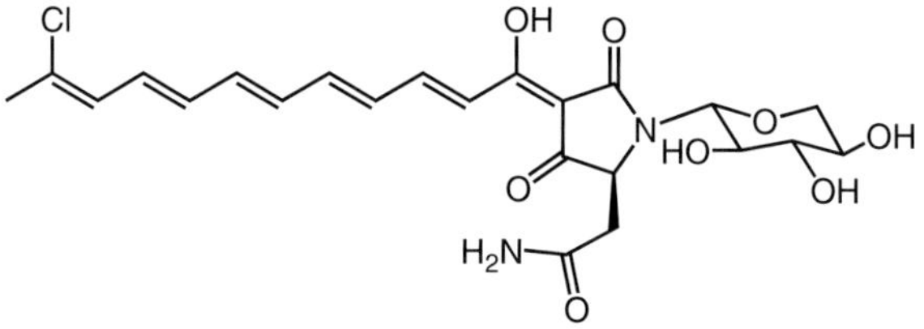

813 (aurantoside G)

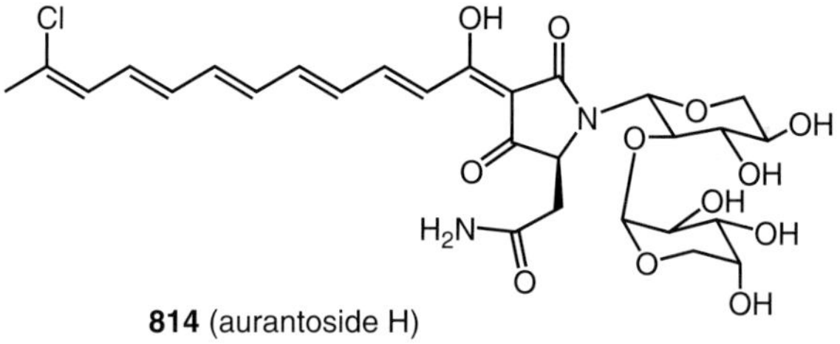

814 (aurantoside H)

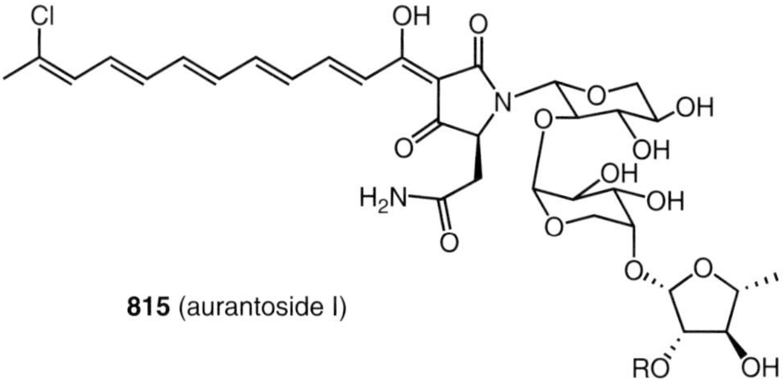

815 (aurantoside I)

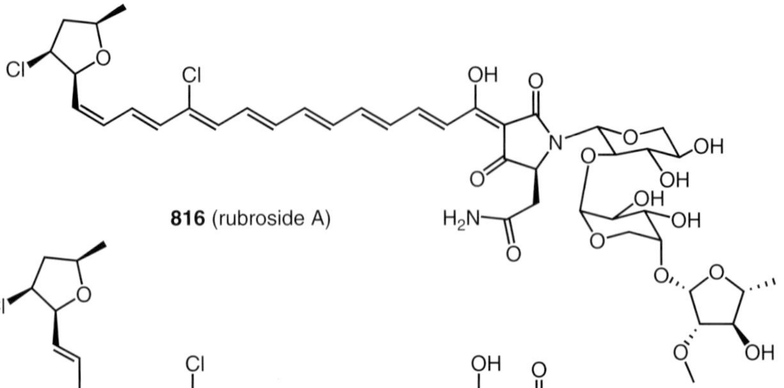

816 (rubroside A)

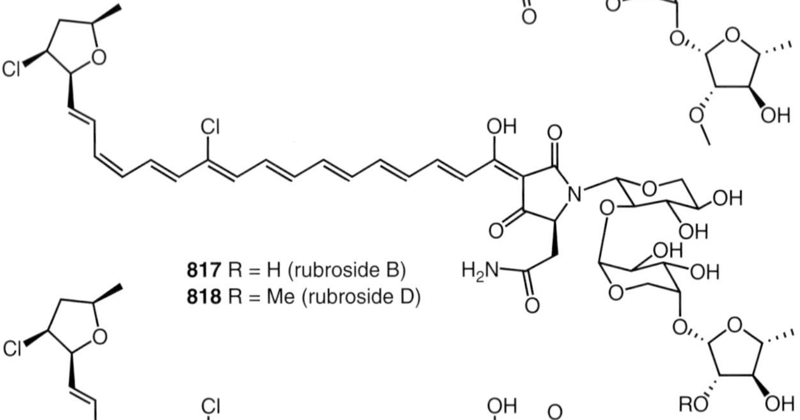

817 R = H (rubroside B)
818 R = Me (rubroside D)

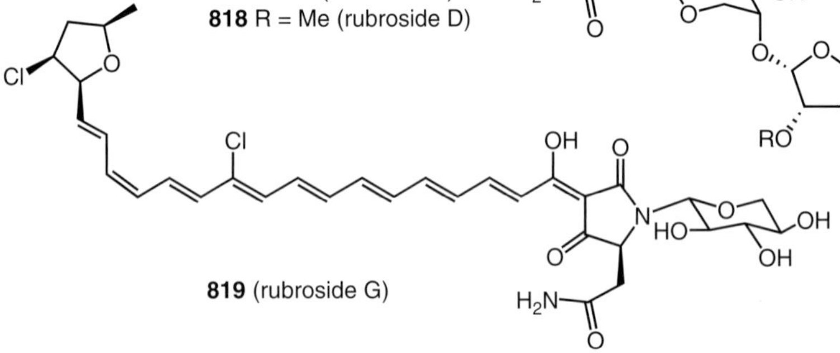

819 (rubroside G)

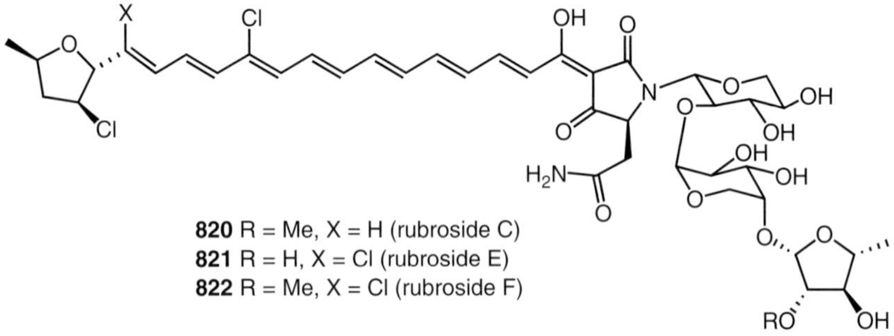

820 R = Me, X = H (rubroside C)
821 R = H, X = Cl (rubroside E)
822 R = Me, X = Cl (rubroside F)

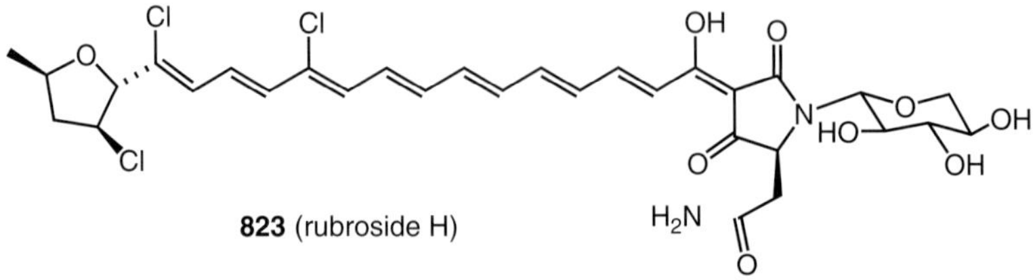

823 (rubroside H)

3.8 Lipids and Fatty Acids 121

Fig. 3.12 *Theonella* cf. *swinhoei*, a sponge collected at Bitung, Indonesia, that contains the bitungolides A–D (**824–827**) (Photo: J. Tanaka)

The novel polyketides, bitungolides A–D (**824–827**), were characterized from the Indonesian sponge *Theonella* cf. *swinhoei* (Fig. 3.12), compounds that inhibit dual-specificity phosphatase VHR (*875*). The first chlorine-containing compound found in a Caribbean *Plakoris* sponge (*Plakoris simplex*) is plakortether C (**828**), along with several non-halogenated analogs (*876*). In addition to the known clathrynamide A, the Okinawan sponge *Psammoclemma* sp. has afforded the new (6*E*)-clathrynamide A (**829**) (*877*). Moreover, the absolute stereochemistry of clathrynamide A was established as shown for **829**. The marine bacterium *Pseudoalteromonas* sp. F-420 produces korormicin analog **830** (*878*).

824 (bitungolide A)

825 (bitungolide B)

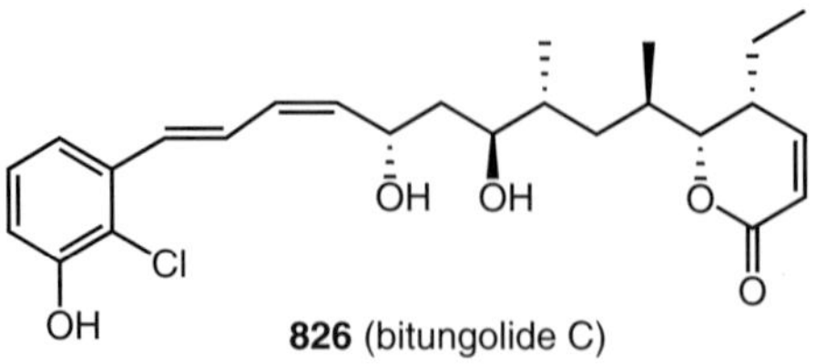

826 (bitungolide C)

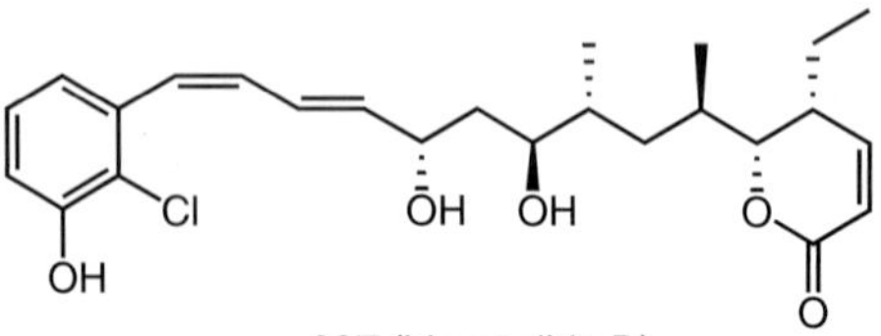

827 (bitungolide D)

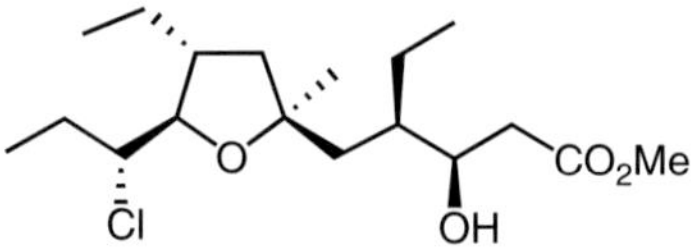

828 (plakortether C)

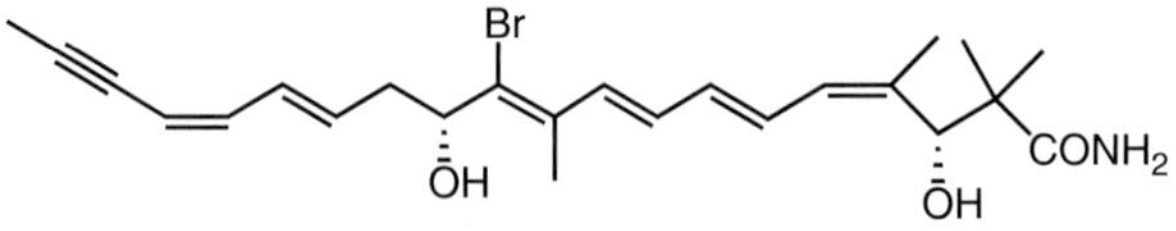

829 ((6*E*)-clathrynamide A)

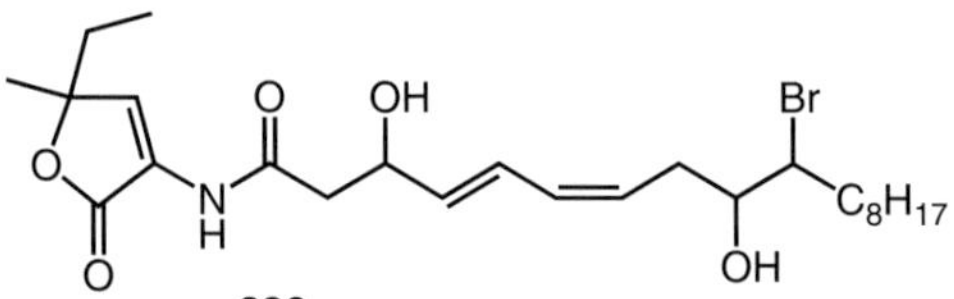

830

The myxobacterium *Chondromyces crocatus* contains the novel chondrochlorens A (**831**) and B (**832**) (*879*). A marine *Streptomyces* species has afforded the manumycin antibiotics chinikomycins A (**833**) and B (**834**) (*880*). While inactive in antiviral, antimicrobial, and phytotoxicity screens, these chinikomycins display some cytotoxic activity.

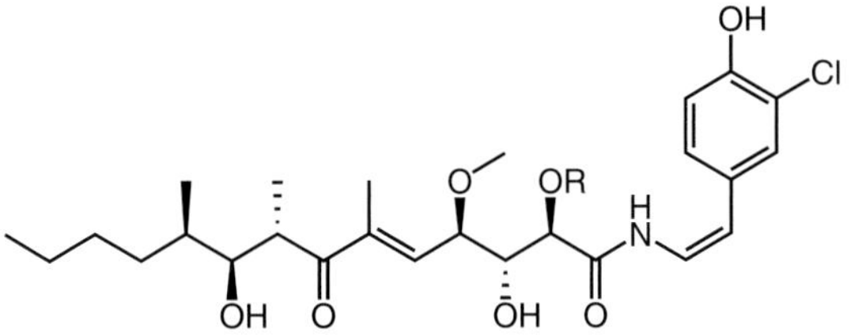

3.8 Lipids and Fatty Acids

831 R = Me (chondrochloren A)
832 R = Et (chondrochloren B)

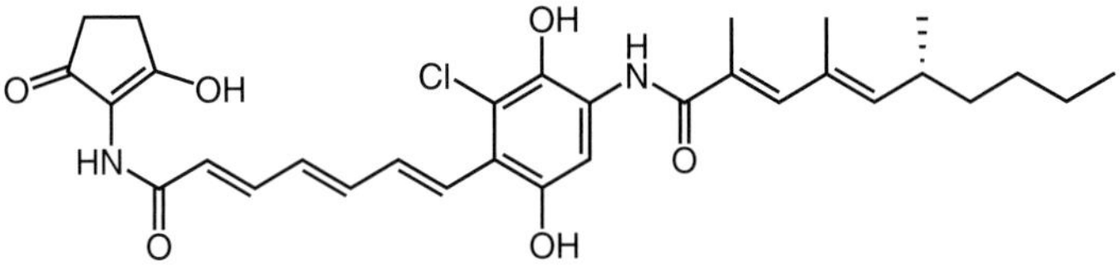

833 (chinikomycin A)

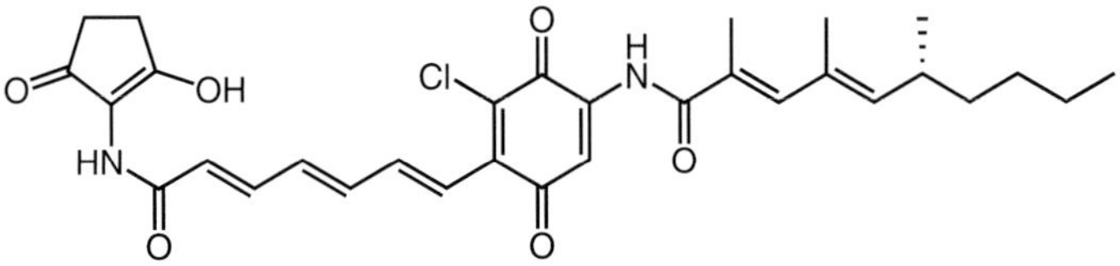

834 (chinikomycin B)

The absolute configuration of the previously described enacyloxin IIa (formerly named enacyloxin II) (*1*) has now been partially established as **SS** (*881*), following earlier structural studies (*882*). This same bacterium *Frateuria* sp. W-315 produces enacyloxin IVa (**835**) (*883*). Several studies on the biological activity of enacyloxin IIa reveal that it inhibits protein biosynthesis (*884–887*). Two iodolactones found in the thyroid gland of dogs were presented in the first survey (*1*). The novel 2-iodohexadecanal (**836**) is present in the horse, dog, and rat thyroid (*888*). Studies indicate that **836** serves as a "mediator of some of the regulatory actions of iodide on the thyroid gland" (*889, 890*). The aforementioned iodolactones (*1*) seem to have a different role than **836** in the thyroid gland (*891, 892*).

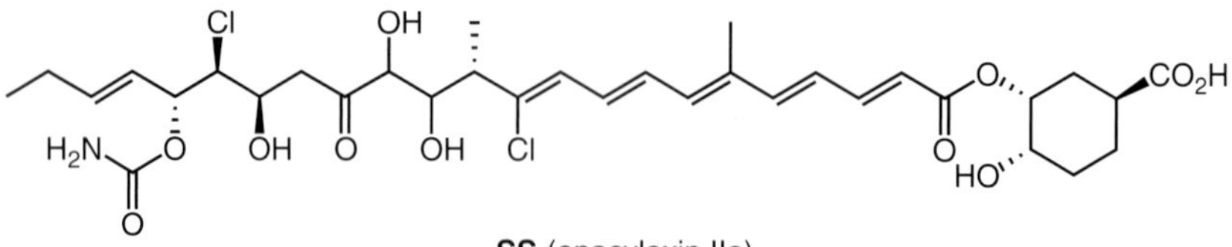

SS (enacyloxin IIa)

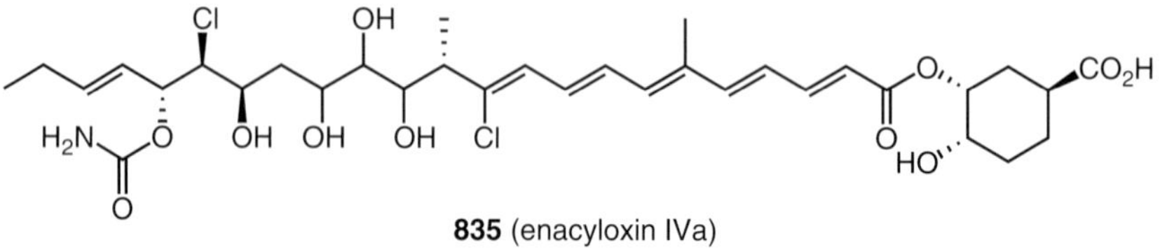

835 (enacyloxin IVa)

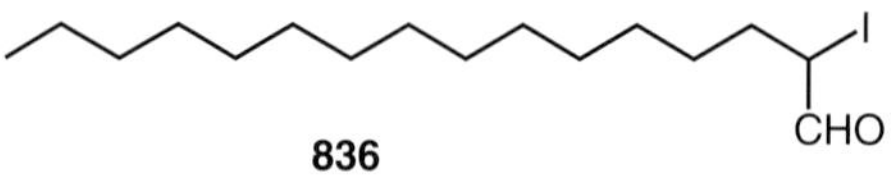

836

3.9 Fluorine-Containing Carboxylic Acids

The infamous fluoroacetic acid and the equally toxic naturally occurring even-numbered ω-fluorinated fatty acids were discussed in detail earlier (*1*), and several reviews are available (*34, 44, 66*). Although not counted as being natural in the earlier survey (*1*), 4-fluorothreonine (**837**) is now considered to be a bona fide natural metabolite of *Streptomyces cattleya* (*893*), the stereochemistry of which has been confirmed by synthesis (*894*). In addition to the five ω-fluorinated fatty acids presented earlier (*1*), new studies of the seed oil of *Dichapetalum toxicarium* have uncovered 16-fluoropalmitoleic acid (**838**), 18-fluorostearic acid (**839**), 18-fluorolinoleic acid (**840**), 20-fluoroarachidic acid (**841**), 20-fluoroeicosenoic acid (**842**), 18-fluoro-9,10-epoxystearic acid (**843**) (*895*), (Z)-16-fluorohexadec-7-enoic acid (**844**), (Z)-18-fluorooctadec-9-enoic acid (**845**), and (Z)-20-fluoroicos-9-enoic acid (**846**) (*896*).

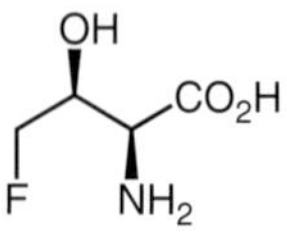

837

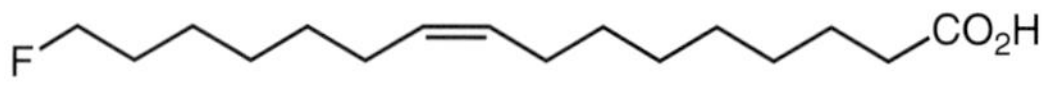

838 (16-fluoropalmitoleic acid)

3.9 Fluorine-Containing Carboxylic Acids

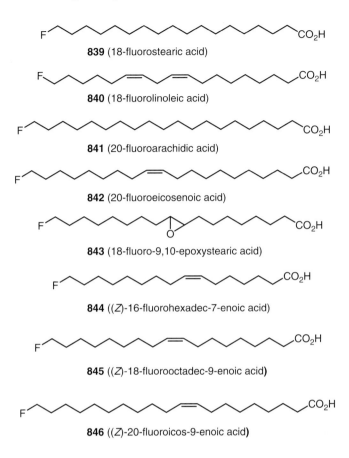

839 (18-fluorostearic acid)

840 (18-fluorolinoleic acid)

841 (20-fluoroarachidic acid)

842 (20-fluoroeicosenoic acid)

843 (18-fluoro-9,10-epoxystearic acid)

844 ((Z)-16-fluorohexadec-7-enoic acid)

845 ((Z)-18-fluorooctadec-9-enoic acid)

846 ((Z)-20-fluoroicos-9-enoic acid)

Extensive and elegant biosynthetic studies on these fluorine-containing metabolites have revealed some additional natural organofluorines produced by *Streptomyces cattleya*, and, more importantly, provide a comprehensive understanding of the biosynthesis of fluoroacetate and 4-fluorothreonine (*895, 897–913*). A summary of this biosynthesis is shown in Scheme 3.3 and reviews are available (*914, 915*).

The first step in this pathway involves S_N2 displacement by fluoride on S-adenosine-L-methionine (SAM) catalyzed by the newly discovered enzyme fluorinase (*905–910*), which also can function as a chlorinase (*912*). Fluorinase has been isolated and characterized, and the gene has been cloned (*916*). Both 5'-fluoro-5'-deoxyadenosine (**847**) and 5'-fluoro-5'-deoxy-D-ribose-1-phosphate (**848**) have been identified as intermediates (*905–908*). Fluoroacetaldehyde (**850**) is the immediate precursor, presumably via fluororibulose-1-phosphate (**849**) (*915*), to both fluoroacetate and 4-fluorothreonine (**837**) (*901*). The requisite enzymes fluoroacetaldehyde dehydrogenase (*902*) and L-threonine transaldolase-PLP (*903*) have been isolated and purified. The steps from **848** to **850** remain to be established but are based on known biochemistry. The pronounced toxicity of fluoroacetic acid

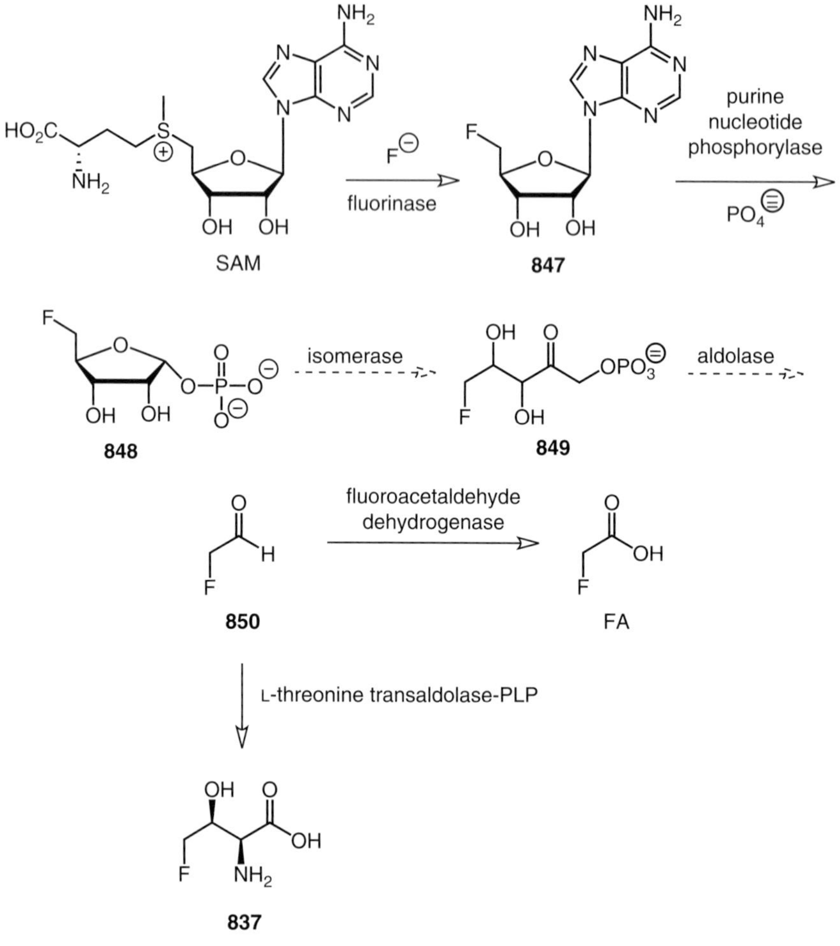

Scheme 3.3 Proposed abbreviated biosynthesis of fluoroacetic acid and 4-fluorothreonine (*909*, *915*, *2395*).

(fluoroacetate) is still of major concern, especially with the realization that it is a metabolite of several fluorinated drugs, pesticides, and other industrial chemicals, and thus may pose an environmental threat to aquatic organisms (*917*). A study of genetically modified ruminal bacteria, with a gene encoding fluoroacetate dehalogenase, indicates promise in protecting sheep against fluoroacetate poisoning (*918*). The mechanism of fluoroacetate toxicity, which is known to involve conversion to fluorocitrate, has been shown to entail further transformation to 4-hydroxy-(*E*)-aconitate, which then binds to aconitase (*919*). While it has been known as an

3.10 Prostaglandins

anthropogenic atmospheric pollutant for a long time, trifluoroacetic acid (**851**) is now considered to have (unknown) natural sources with an ocean concentration of about 200 ng L^{-1} (*920*).

$$CF_3CO_2H$$

851

3.10 Prostaglandins

The first survey identified 15 marine halogenated prostaglandins, some of which display striking biological activity (*1*), and a review is available on the occurrence, biological activity, and biogenesis of these interesting organohalogen compounds (*921*).

A practical racemic synthesis of the known chlorovulone II from the Okinawan soft coral *Clavularia viridis* has been accomplished (*922*). This coral has more recently afforded the new prostanoids **852–856** (*923*), **857–871** (*924*), and, from a Taiwanese collection, **872, 873** in addition to **857** and **858** (*925*). The absolute configuration of the previously known punaglandin 8 (**852**, X = Cl) was determined as shown (*923*). This soft coral also contains several non-halogenated possible biosynthetic precursors to these halogenated metabolites (*926*).

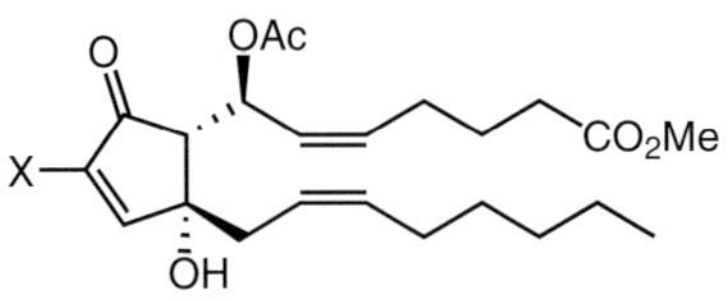

852 X = I
853 X = Br

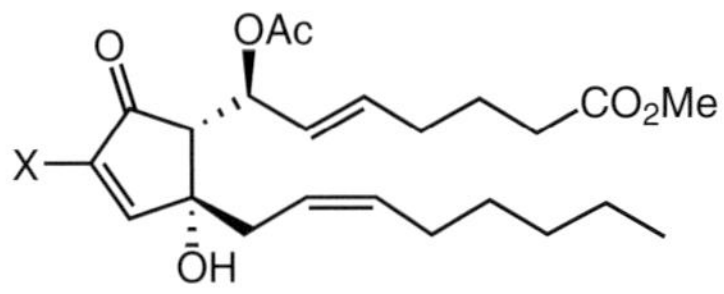

854 X = I
855 X = Br
856 X = Cl

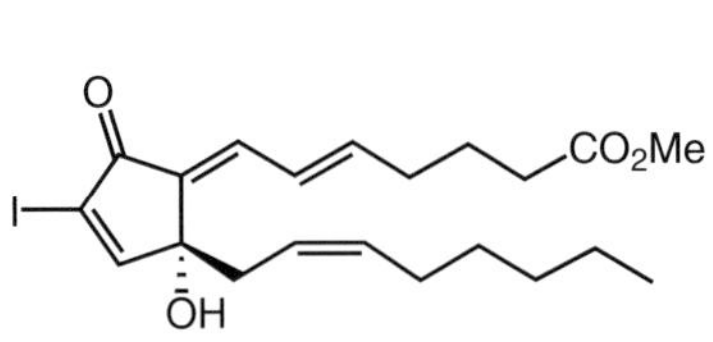

857 (iodovulone II)

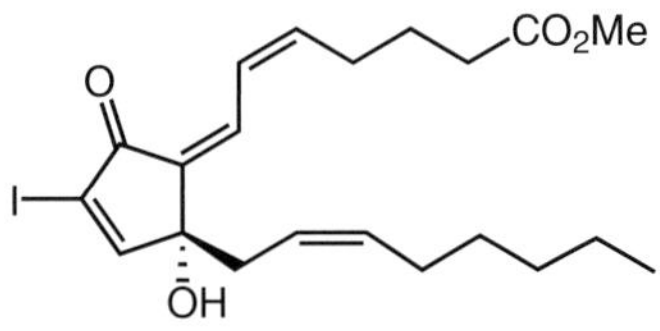

859 (iodovulone IV)

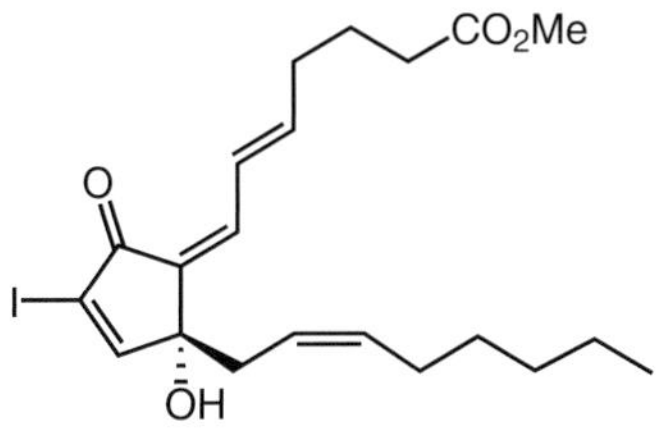

858 (iodovulone III)

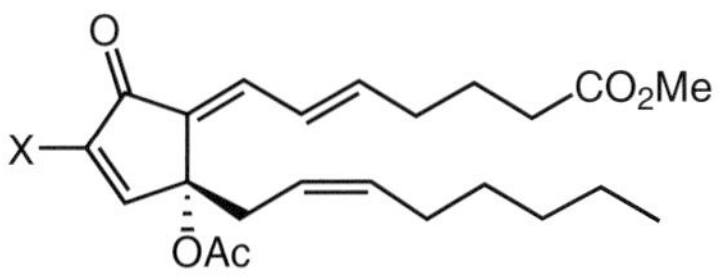

860 X = I   (12-*O*-acetyliodovulone II)
861 X = Br (12-*O*-acetylbromovulone II)
862 X = Cl (12-*O*-acetylchlorovulone II)

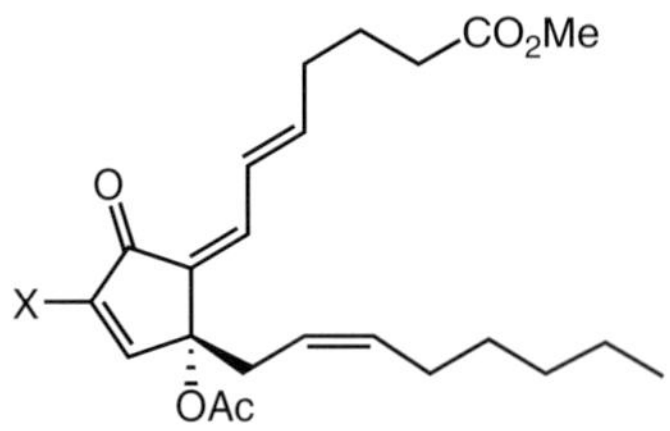

863 X = I (12-*O*-acetyliodovulone III)
864 X = Br (12-*O*-acetylbromovulone III)
865 X = Cl (12-*O*-acetylchlorovulone III)

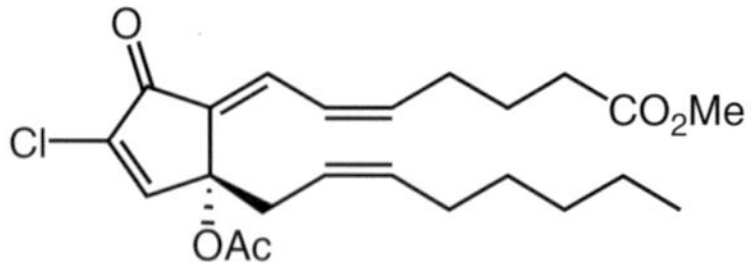

866 (12-*O*-acetylchlorovulone I)

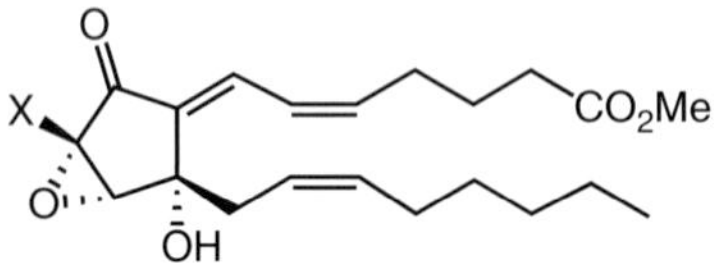

870 X = I (10,11-epoxyliodovulone I)
871 X = Br (10,11-epoxybromovulone I)

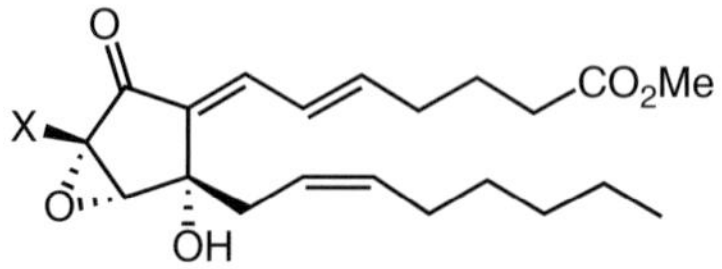

867 X = I (10,11-epoxyliodovulone II)
868 X = Br (10,11-epoxybromovulone II)
869 X = Cl (10,11-epoxychlorovulone II)

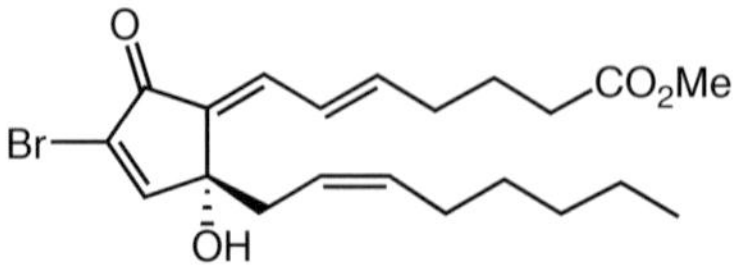

872 (bromovulone II)

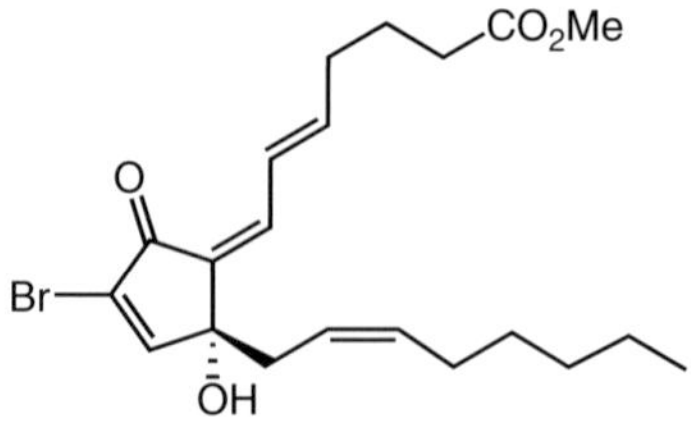

873 (bromovulone III)

A Red Sea collection of the soft corals *Dendrophyllia* sp., *Dendronephthya* sp. (red variety), *Dendronephthya* sp. (yellow variety), and *Tubipora musica* revealed the eight new brominated oxylipins **874–881** (*927*). The brown alga *Eisenia bicyclis*, which was gathered around the coast of Japan, has afforded eiseniachlorides A–C (**882–884**), eiseniaiodides A (**885**) and B (**886**), and **887** (*928*). The eastern Pacific octocoral *Carijoa multiflora* contains the novel prostanoid carijenone (**888**) (*929*). Both natural punaglandins and synthetic analogs can function as Michael reaction acceptors to inhibit ubiquitin isopeptidase activity, which may represent a target for anticancer agents (*930, 931*).

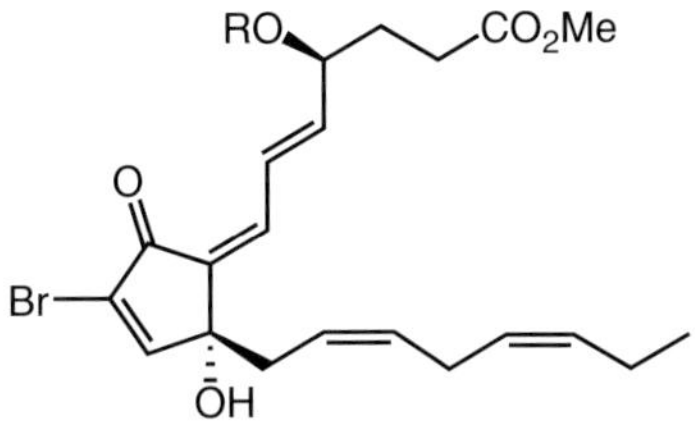

3.10 Prostaglandins

874 R = H
875 R = α-D-glucopyranosyl

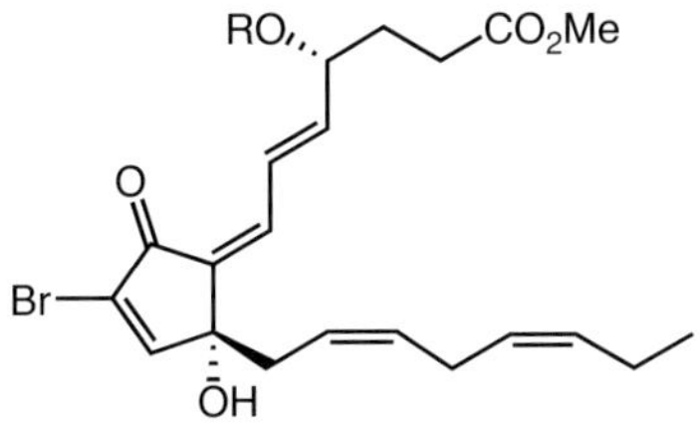

876 R = H
877 R = β-D-glucopyranosyl

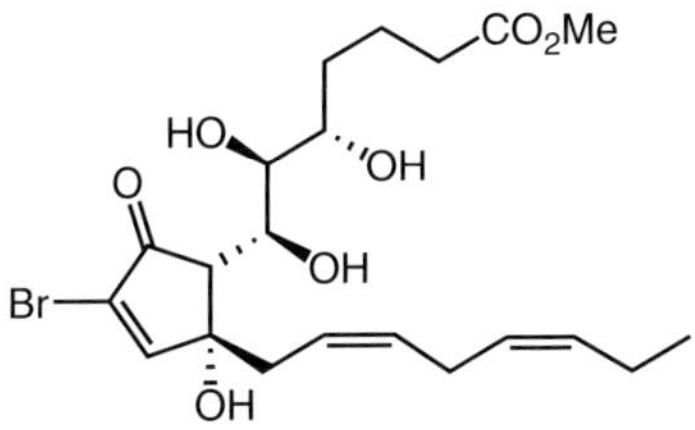

878

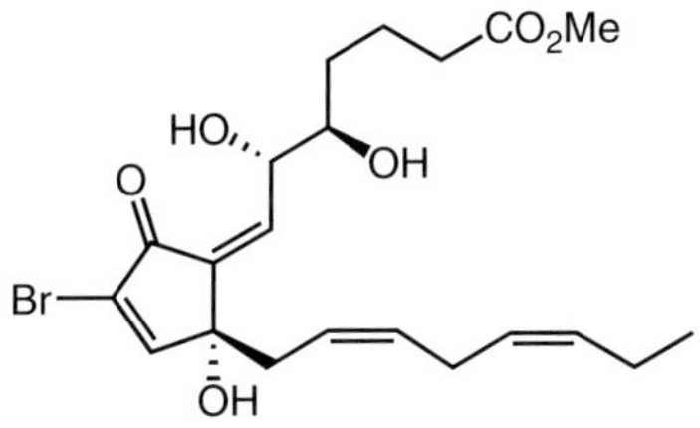

879

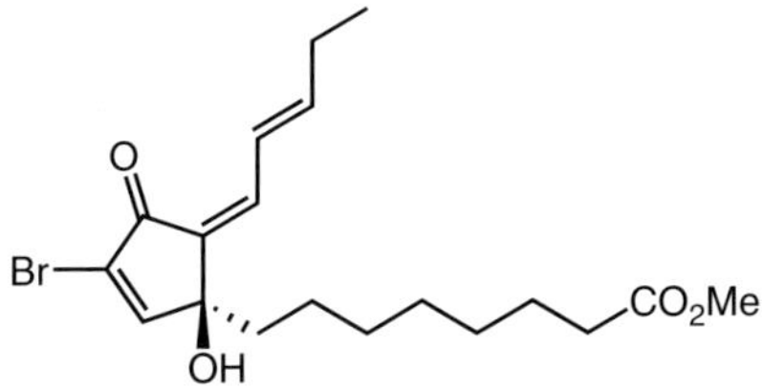

880

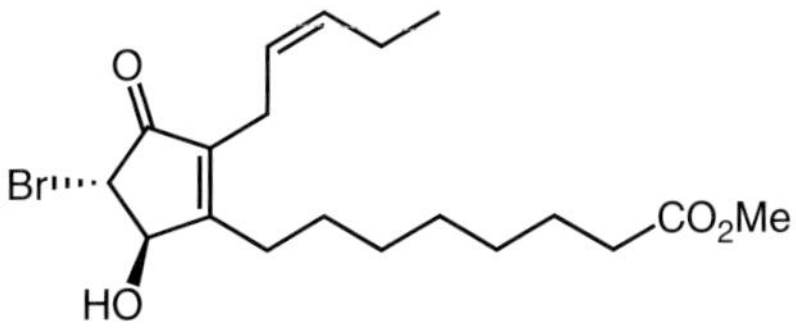

881

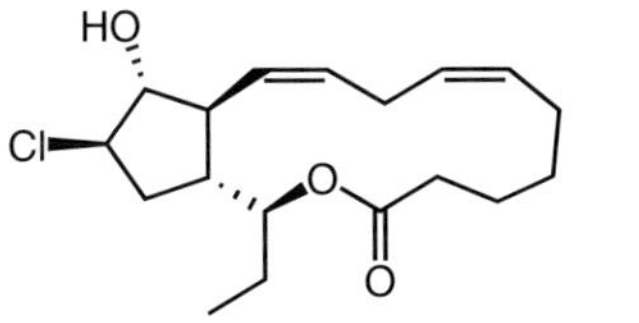

882 (eiseniachloride A)

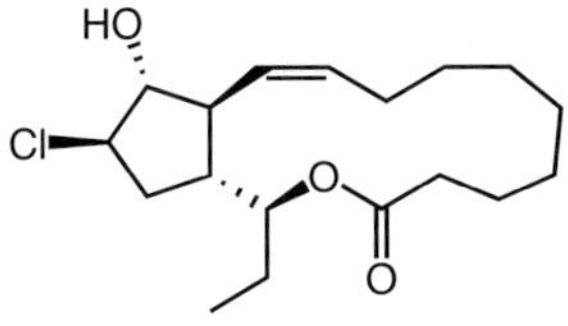

883 (eiseniachloride B)

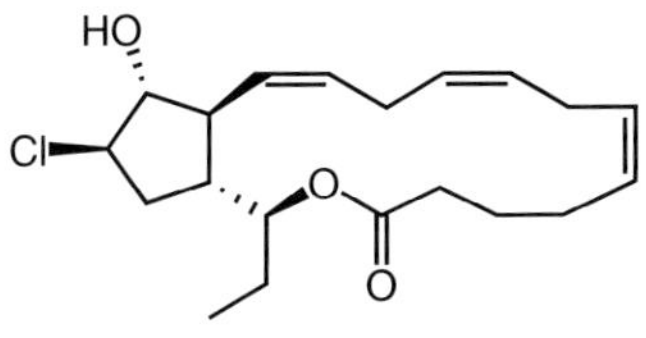

884 (eiseniachloride C)

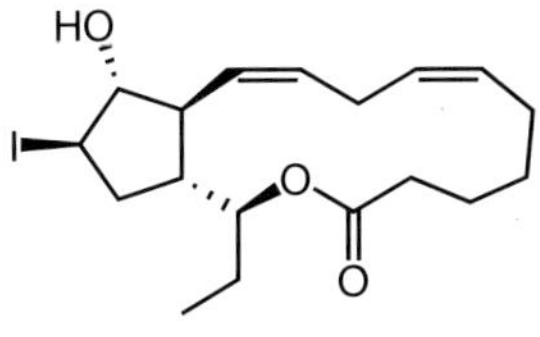

885 (eiseniaiodide A)

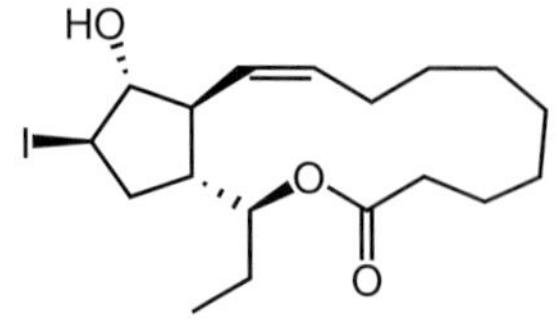

886 (eiseniaiodide B)

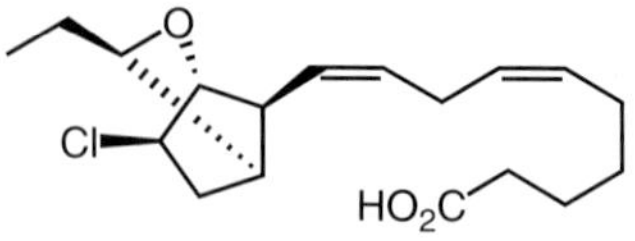

887

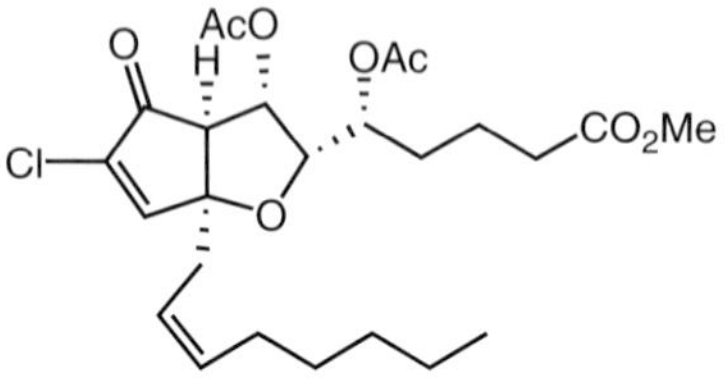

888 (carijenone)

3.11 Furanones

Red algae of the genus *Delisea* enriched the first survey with 40 mainly brominated furanones, some of which have powerful antibacterial activity (*1*). The absolute configuration of the previously known *Delisea pulchra* furanone **TT**, and others by extension, has been determined (*932*). While the only new examples of these heavily brominated furanones appear to be the pulchralides A–C (**889–891**) from an Antarctic collection of *Delisea pulchra* (*476*), these compounds have been extensively studied from a biological standpoint (*933–945*). Most notably, these halogenated furanones inhibit bacterial colonization (*933–942, 944*), and this quorum-sensing inhibitory activity may lead to drugs for the treatment of bacterial infections. These furanones also display feeding deterrence to herbivores (*943*) and show cytotoxic, antimicrobial, and antiplasmodial activity (*945*). The sea hare *Aplysia parvula* feeds on *Delisea pulchra* to acquire halogenated furanones for chemical defense (*946*).

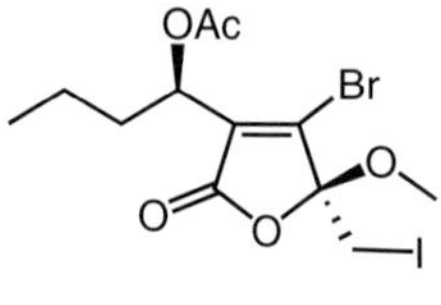

TT

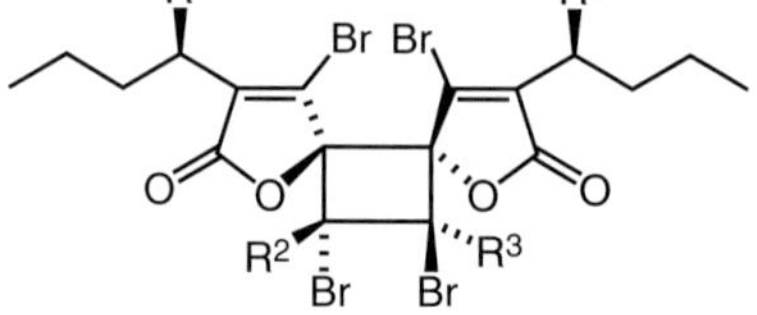

889 R = R¹ = OAc, R² = R³ = H (pulchralide A)
890 R = R¹ = R² = R³ = H (pulchralide B)
891 R = OAc, R¹ = R² = R³ = H (pulchralide C)

3.11 Furanones

Six new halogenated rubrolides I (**892**), J (**893**), K (**894**), L (**895**), M (**896**), and N (**897**) were characterized from the Spanish ascidian *Synoicum blochmanni* (*947*). Several of these compounds exhibit significant cytotoxicity against these tumor cell lines: HT-29, MEL-28, P-388, and A-549, with rubrolide M (**895**) showing the greatest activity. A New Zealand variety of *Synoicum* sp. has provided rubrolide O as a mixture of Z (**898**) and E (**899**) isomers, which display some antiinflammatory activity (*948*). The previously known rubrolides C and E have been efficiently synthesized (*949*).

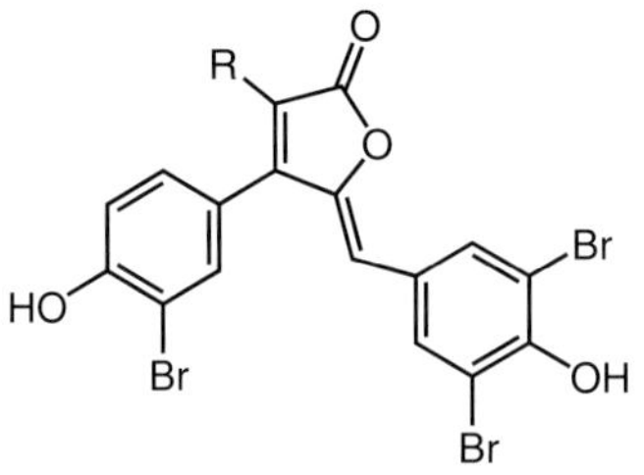

892 R = Cl (rubrolide I)
893 R = H (rubrolide J)

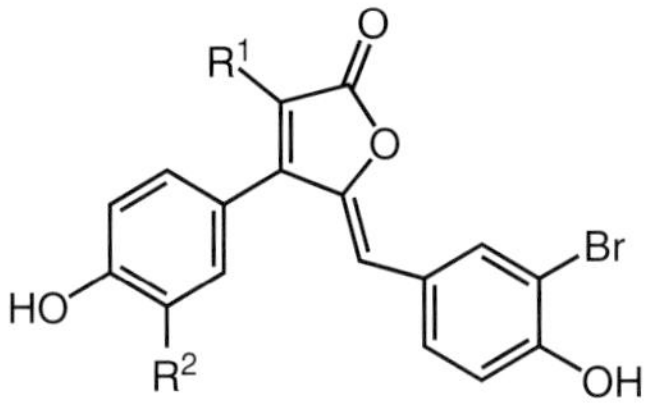

894 R¹ = Cl, R² = Br (rubrolide K)
895 R¹ = Cl, R² = H (rubrolide M)
896 R¹ = Br, R² = Cl (rubrolide N)

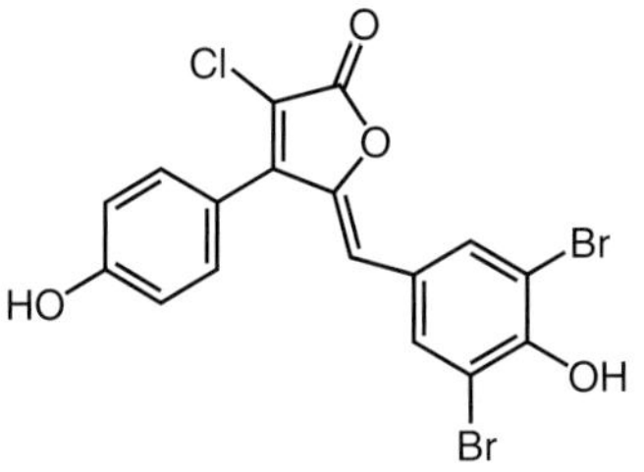

897 (rubrolide L)

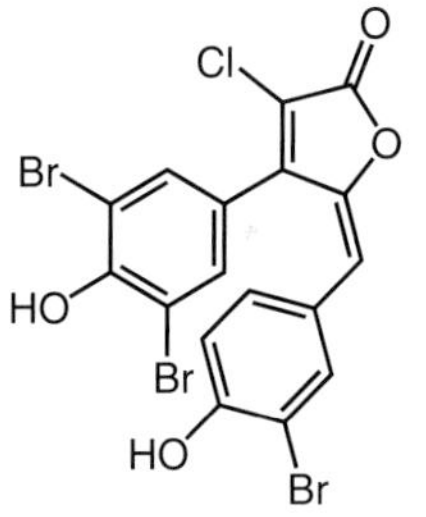

899 ((E)-rubrolide O)

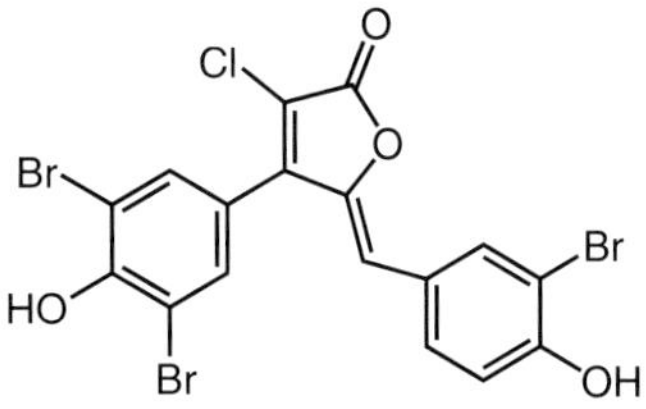

898 ((Z)-rubrolide O)

An Indonesian ascidian of the genus *Botryllus* contains the novel cadiolides A (**900**) and B (**901**) (*950*). The tetraphenolic bis-spiroketals prunolides A (**902**, **903**) and B (**904**, **905**) were characterized from an Australian ascidian *Synoicum prunum* (*951*).

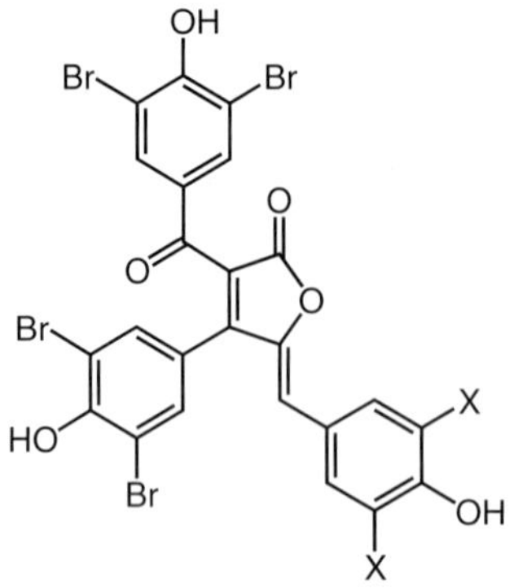

900 X = H (cadiolide A)
901 X = Br (cadiolide B)

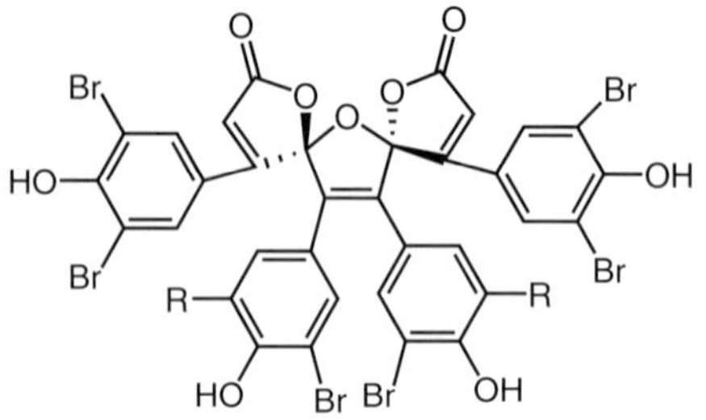

902, 903 R = Br (prunolide A; racemic mixture)
904, 905 R = H (prunolide B; racemic mixture)

The ant-cultivated fungus *Lepiota* sp. produces the antibacterial lepiochlorin (**906**), which is racemic, perhaps due to ring-opening tautomerism (*952*). The mushroom *Clitocybe flaccida* (Fig. 3.13), which is repugnant to the banana slug (*Ariolimax columbianus*) (Fig. 3.14), secretes clitolactone (**907**) (*953*). Control experiments clearly demonstrate the potent antifeedant properties of clitolactone to these slugs (Fig. 3.15). The Asian shrub *Prinsepia utilis* affords lactone **908** (*954*). An *Aspergillus* sp. fungus from the sponge *Jaspis* cf. *coriacea* has yielded chlorocarolides A (**909**) and B (**910**) (*955*). Another marine-derived fungus, *Aspergillus ostianus*, from Pohnpei has provided chlorinated furanones **911** and **912** (*956*). A related pyrone from this organism is shown later in Sect. 3.14.8 (Pyrones and Chromones).

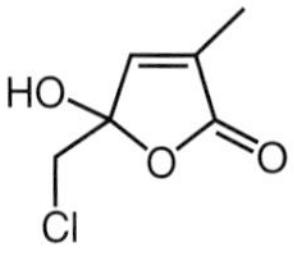

906 (lepiochlorin) 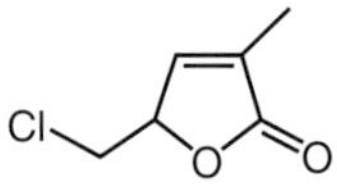   **907** (clitolactone) 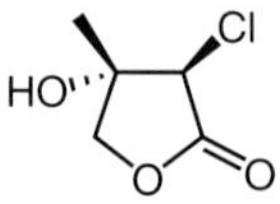   **908**

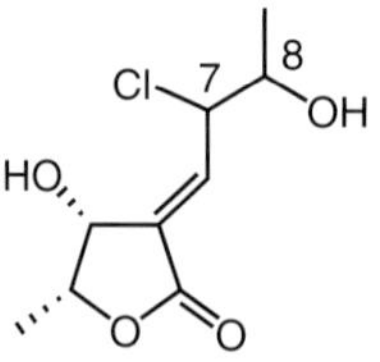

909 ((7*R*,8*R*)-chlorocarolide A)
910 ((7*S*,8*S*)-chlorocarolide B)

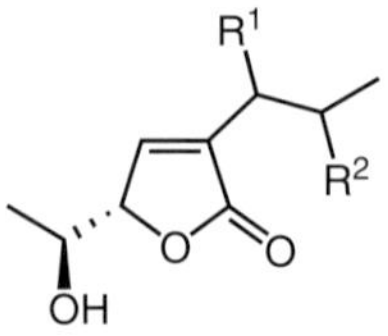

911 R^1 = Cl, R^2 = OH
912 R^1 = OH, R^2 = Cl

3.11 Furanones

Fig. 3.13 *Clitocybe flaccida*, the mushroom that contains the antifeedant clitolactone (**907**) (Photo: W. F. Wood)

Fig. 3.14 The banana slug (*Ariolimax columbianus*) feeding on the mushroom *Russula roscea* (= *R. sanguinea*) (Photo: W. F. Wood)

Fig. 3.15 The banana slug (*Ariolimax columbianus*) tasting and being repelled by clitolactone-treated lettuce (Photo: W. F. Wood)

3.12 Amino Acids and Peptides

Halogenated amino acids and peptides represent an enormous class of natural products. The first survey counted nearly 100 such examples of marine and terrestrial bacterial origin (*1*). One of the first halogenated natural products to be characterized, chloramphenicol, continues to draw attention (*2656*). The new analog 3′-*O*-acetylchloramphenicol (**913**), which is a possible intermediate in the biosynthesis of chloramphenicol, has been isolated from *Streptomyces venezuelae* (*957*). The antifungal acrodontiolamide (**914**) is produced by *Acrodontium salmoneum* (*958, 959*). The dichloromethyl ether moiety is unique amongst natural products and is expected to be a potent alkylating agent, assuming that this structure is correct. An enantioselective synthesis of (−)-(1*R*,2*R*)-chloramphenicol has been reported (*960*), and the genes required for its biosynthesis by *Streptomyces venezuelae* have been identified (*961, 962*), including those required for the dichloroacetyl unit (*962*). Synthetic derivatives of bactobolin (*1*), which also contains a dichloromethyl group, show less activity than the natural product in various cytotoxicity and antibacterial assays (*963*).

Mushrooms are a source of simple chlorinated amino acids (*1*), and several new examples are known. Thus, *Amanita vergineoides* has furnished (2*S*,4*Z*)-2-amino-5-chloro-4-pentenoic acid (**915**) (*964*), and a full account of the isolation of (2*S*)-2-

amino-5-chloro-4-hydroxy-5-hexenoic acid from *Amanita gymnopus* (*1*) and other *Amanita* species has been published (*965, 966*). Likewise, a full report on the isolation of 2-amino-5-chloro-5-hexenoic acid from *Amanita miculifera* (*1*) has appeared (*967*). Interestingly, the related amino acid **915** was isolated as a racemate from *Amanita castanopsidis* (*968*). The biosynthesis of the known armentomycin (2-amino-4,4-dichlorobutyric acid) from *Streptomyces armentosus* var. *armentosus* was studied using radiolabelling and the results support a pathway of pyruvate to acetyl-CoA and perhaps dichloropyruvate, followed by condensation and conversion to armentomycin along known amino acid pathways (*969*).

Cyanobacteria blooms can pose an extremely serious threat to human health (*970–972*), and some of the causative toxins contain halogen. The fresh water toxic cyanobacterium *Oscillatoria agardhii* produces oscillaginin A (**916**), which features the novel 3-amino-10-chloro-2-hydroxydecanoic acid, and is the source of the microcystins, which are heptatoxins (*973*). The prolific cyanobacterium *Lyngbya majuscula* from Curacao has furnished the novel barbamide (**917**) (*974*) and dechlorobarbamide (**918**) (*975*). Extensive biosynthetic studies show that the amino acids leucine, cysteine, and phenylalanine are involved in barbamide production (*976–982*). The chlorination of leucine is of great interest and may involve a radical mechanism (*976, 980–983*).

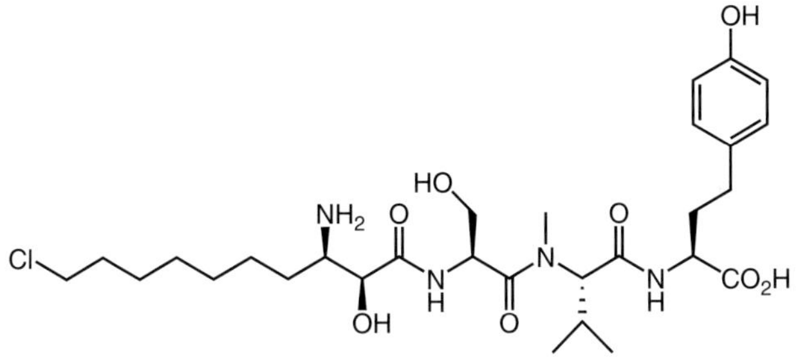

916 (oscillaginin A)

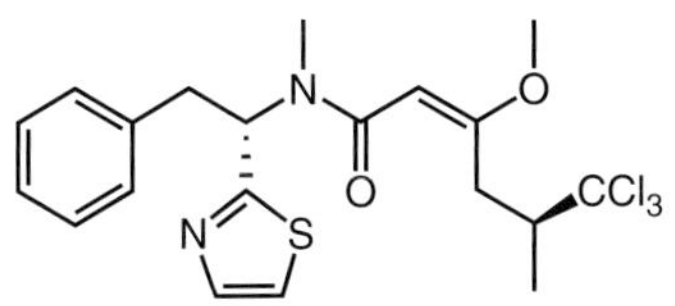

917 (barbamide) 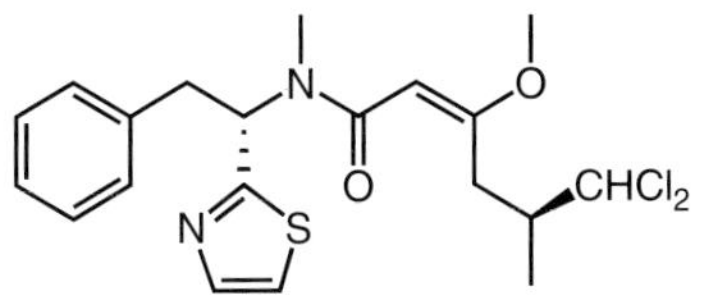   **918**

The marine sponge *Dysidea herbacea*, perhaps in association with its cyanobacterial symbiont *Oscillatoria spongeliae*, is responsible for furnishing some 20 polychlorinated amino acid-derived metabolites (*1*). A Papua New Guinea sample of *Dysidea herbacea*, which contains *Oscillatoria spongeliae*, has yielded the novel herbamide A (**919**) (*984*), whereas a collection of this sponge from the Great Barrier

Reef provided **920–922**, and the absolute configuration of the latter metabolite was established as shown (*985*). *Dysidea fragilis* from the South China Sea has yielded dysamide D (**923**) (*986*), and *Dysidea chlorea* from Micronesia afforded 12 new polychlorinated diketopiperazines, dysamides I-T (**924–935**) (*987*). In addition, this study (*987*) confirmed the structure of dysamide E (**936**) (*988*). Based on previous assignments the absolute configurations of **924–936** are believed to be those indicated. A Pacific Ocean collection of *Dysidea* sp. provided dysamide U (**937**), which is the first trichlorinated member of the diketopiperazine family to be identified (*989*).

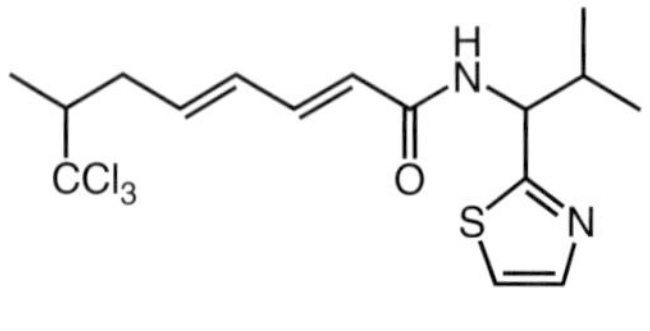

919 (herbamide A)

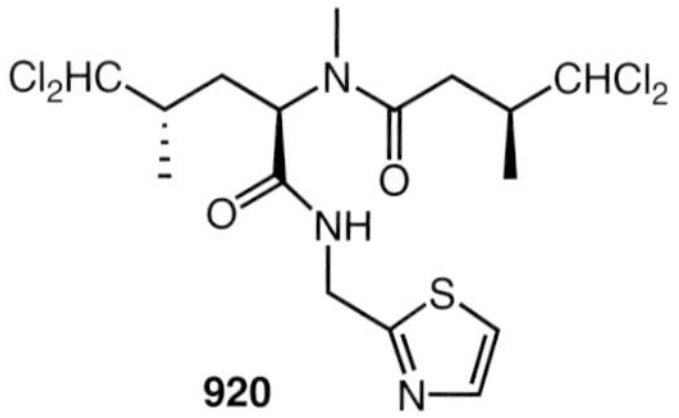

920

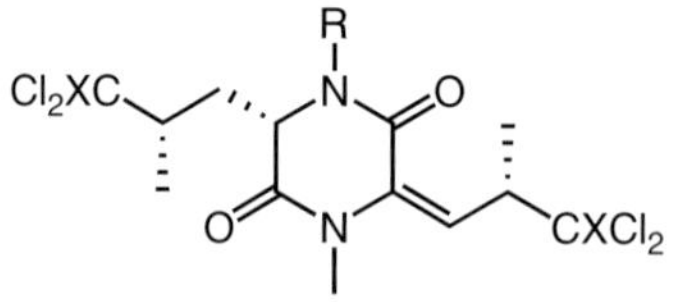

921 R = Me, X = H
922 R = H, X = Cl

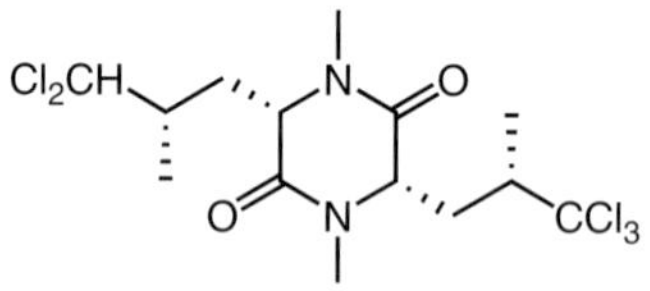

923 (dysamide D)

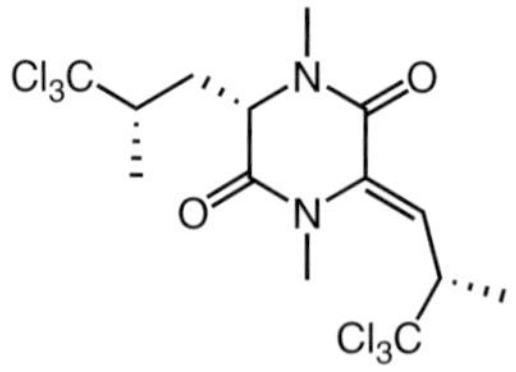

924 (dysamide I)

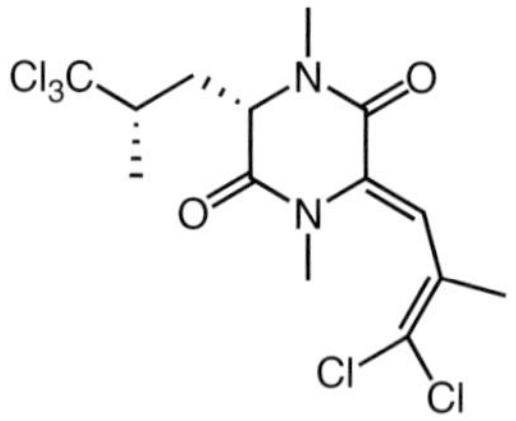

925 (dysamide J)

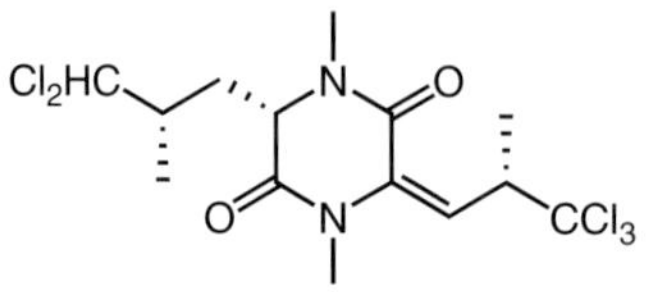

926 (dysamide K)

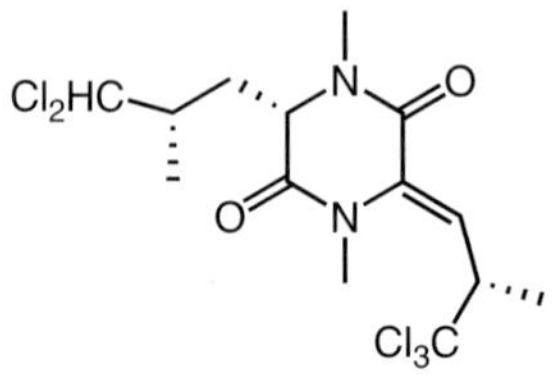

927 (dysamide L)

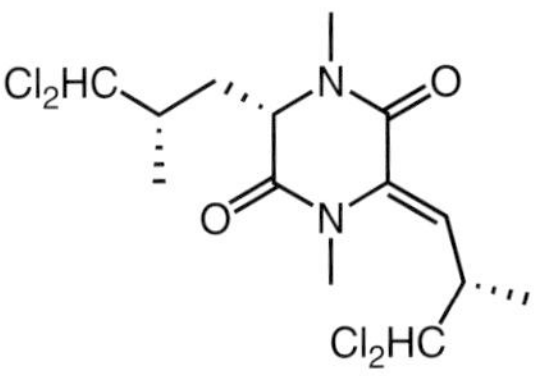

3.12 Amino Acids and Peptides

928 (dysamide M)

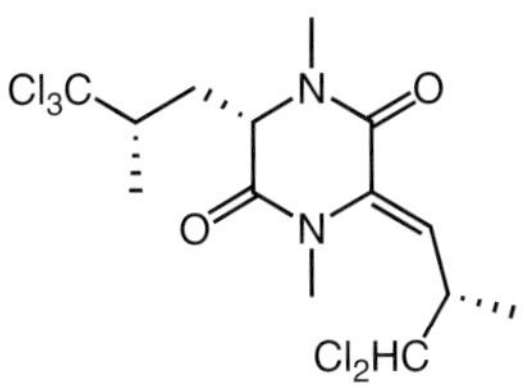

929 (dysamide N)

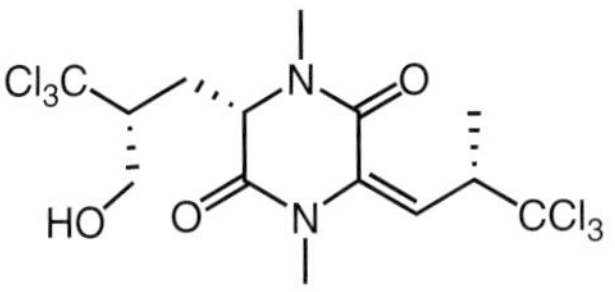

930 (dysamide O)

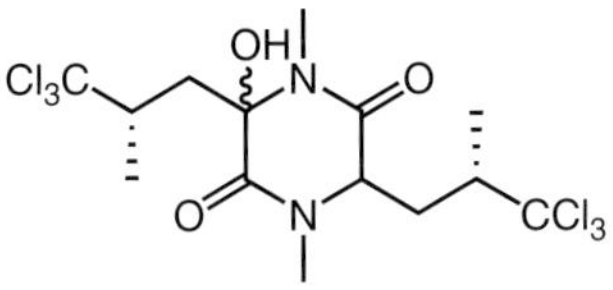

931 (dysamide P)

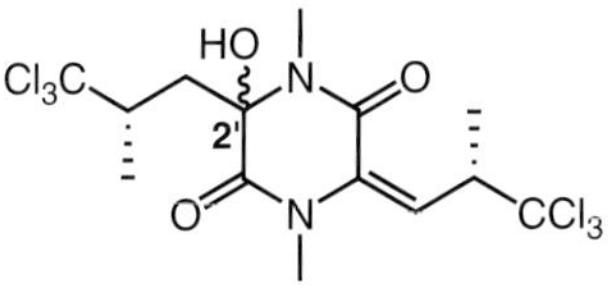

932, 933 (dysamide Q and R)
(epimers at C-2')

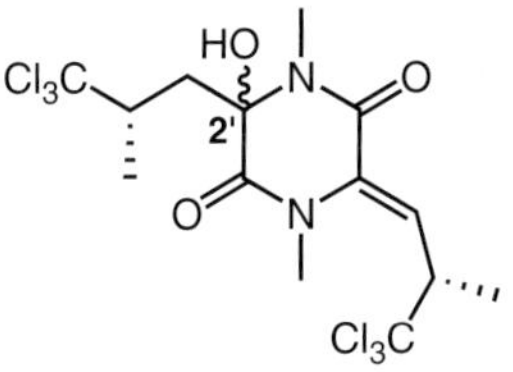

934, 935 (dysamide S and T)
(epimers at C-2')

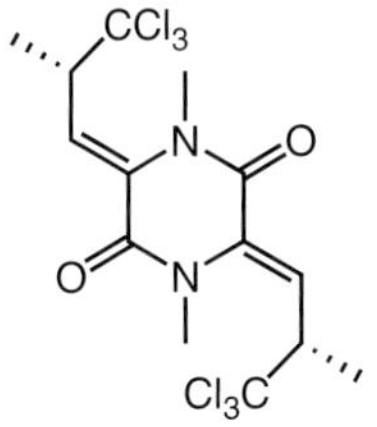

936 (dysamide E)

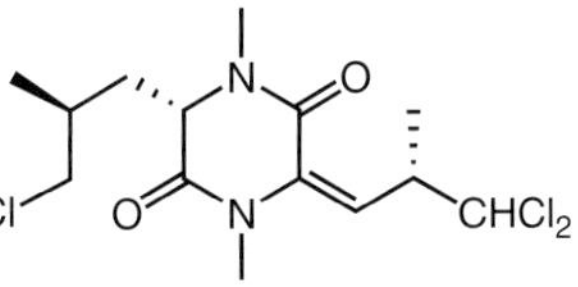

937 (dysamide U)

The simple herbacic acid (**938**) was isolated from *Dysidea herbacea* from the Great Barrier Reef, and may be a precursor to more complex trichloromethyl metabolites (*990*). Another collection of *Dysidea* sp. from Australia's Great Barrier Reef yielded five new metabolites (**939**–**943**) for which the absolute stereochemistry was determined by correlation with (–)-(S)-4,4,4-trichloro-3-methylbutanoic acid (*991*). *Dysidea herbacea* from the Great Barrier Reef contains (–)-neodysidenin (**944**), which is an isomer of the well-known and often isolated dysidenin.

This new metabolite belongs to the L-series of trichloroleucine peptides and is a rare example of a non-N-methylated trichloroleucine amino acid (*992*). Another sample of this sponge from the same locale has yielded the new thiazoles **945** and **946**, which are also related to dysidenin (*993*). The Panamanian *Lyngbya majuscula* has afforded the new dysidenamide (**947**), pseudodysidenin (**948**), and nordysidenin (**949**), which is the first report of dysidenin-like compounds from a free-living cyanobacterium (*994*).

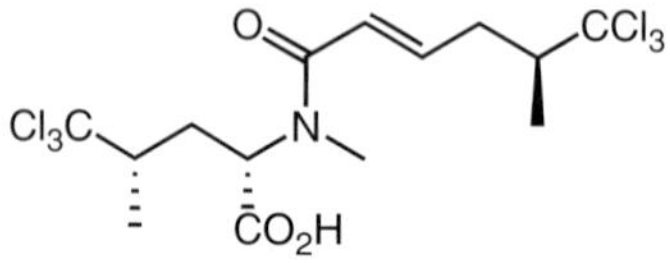

938 (herbacic acid)

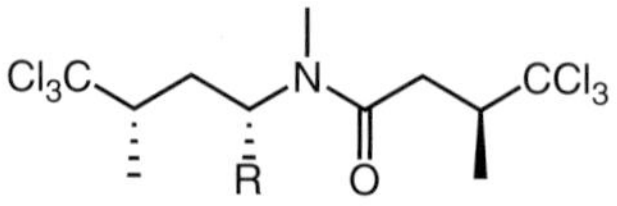

939 R = CO$_2$Me
940 R = CO$_2$H

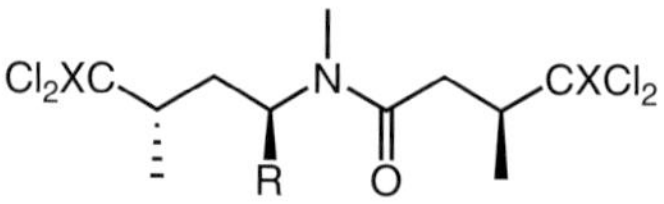

941 X = Cl, R = CO$_2$Me
942 X = H, R = CO$_2$Me
943 X = Cl, R = CO$_2$H

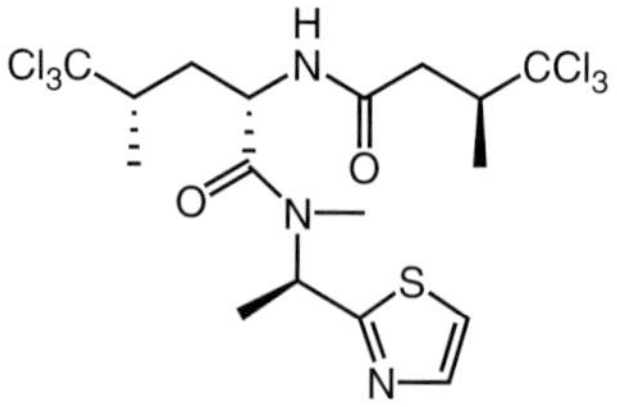

944 (neodysidenin)

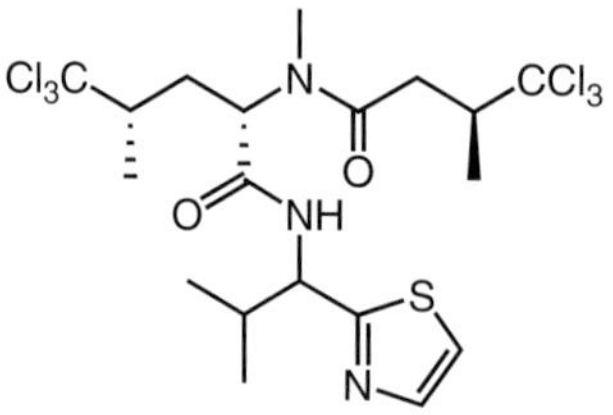

945

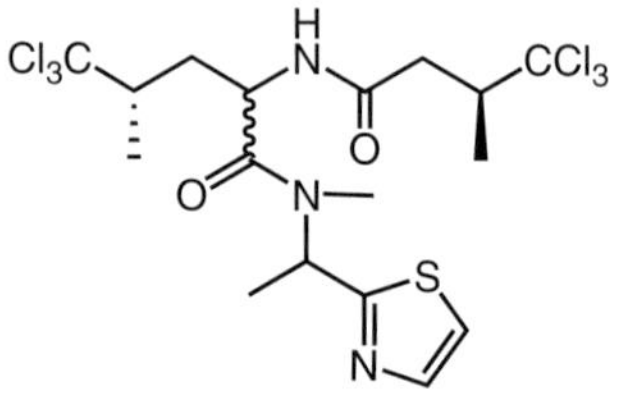

946

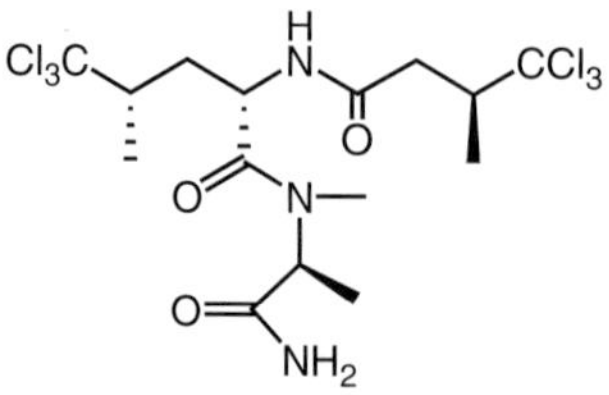

947 (dysidenamide)

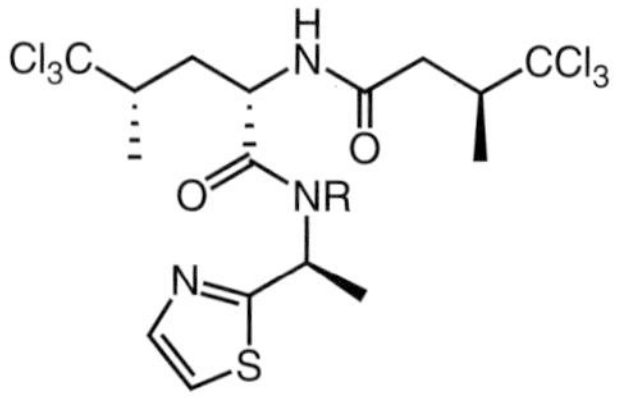

948 R = Me (pseudodysidenin)
949 R = H (nordysidenin)

3.12 Amino Acids and Peptides

A study of a Philippines *Dysidea* sp. has yielded the novel proline analogs of dysidenin, dysideaprolines A–F (**950–955**) and barbaleucamides A (**956**) and B (**957**), which are reminiscent of barbamide (*995*). Also from the Philippines was isolated the novel pyrrolidone **958** from the nudibranch *Asteronotus cespitosus* (*996*), which is the first example of a *Dysidea*-type polychlorinated metabolite found in a carnivorous mollusc. An Indonesian collection of *Dysidea* sp. has furnished dysithiazolamide (**959**), having the suggested absolute configuration shown (*997*).

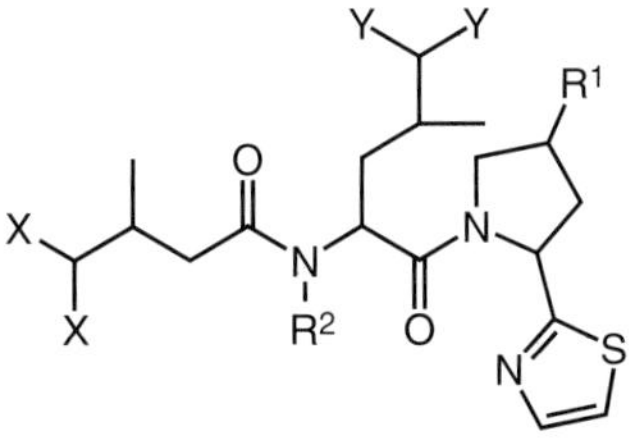

950 R^1 = H, R^2 = Me, X = Y = Cl (dysideaproline A)
951 R^1 = R^2 = Me, X = Y = Cl (dysideaproline B)
952 R^1 = R^2 = H, X = Y = Cl (dysideaproline C)
953 R^1 = H, R^2 = Me, X = H, Y = Cl (dysideaproline D)
954 R^1 = H, R^2 = Me, X = Cl, Y = H (dysideaproline E)
955 R^1 = H, R^2 = Me, X = Cl, Y = H,Cl (dysideaproline F)

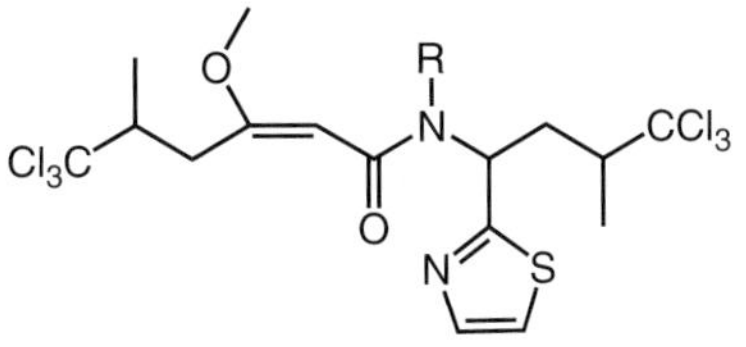

956 R = H (barbaleucamide A)
957 R = Me (barbaleucamide B)

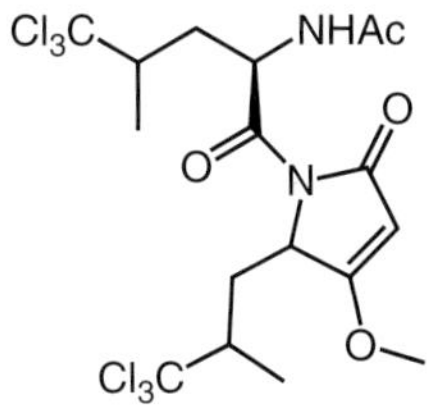

958

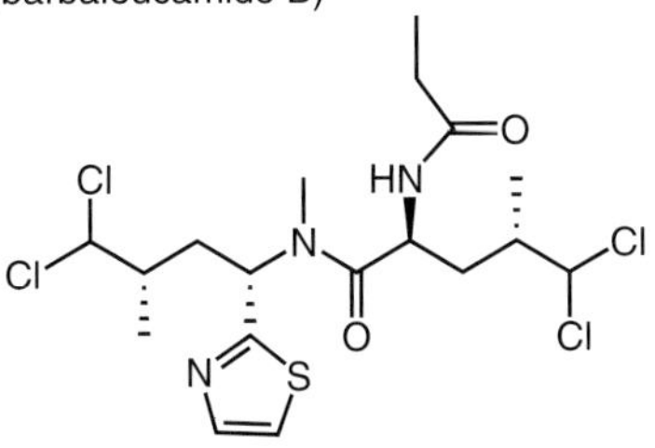

959 (dysithiazolamide)

The Red Sea sponge *Lamellodysidea herbacea* contains the new dysidamides D–H (**960–965**) and ring-opened analogs **966** and **967** (*998*). As has been pointed out several times (*999*), the determination of absolute stereochemistry of the *Dysidea* polychlorinated peptides has been difficult and revisions are not uncommon. The X-ray crystal structure of a zinc chelate of dechlorinated dysidenin has confirmed its absolute configuration as (5*S*,13*S*) as shown in **UU** (*999*).

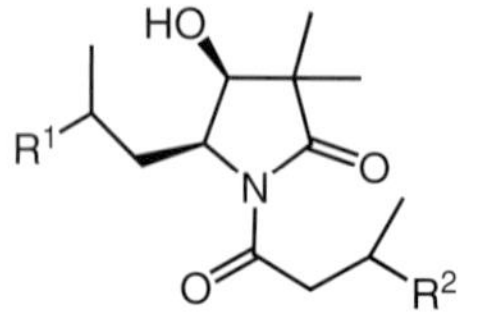

960 R¹ = CCl₃, R² = CHCl₂ (dysidamide D)
961 R¹ = CHCl₂, R² = CCl₃ (dysidamide E)

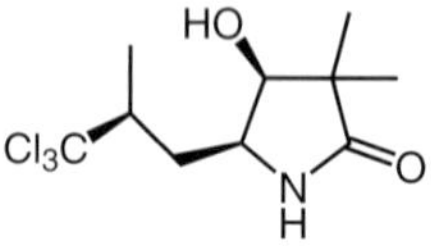

962 (dysidamide F)

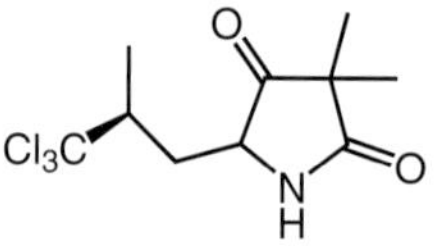

963 (dysidamide G)
964 (5-epidysidamide G)

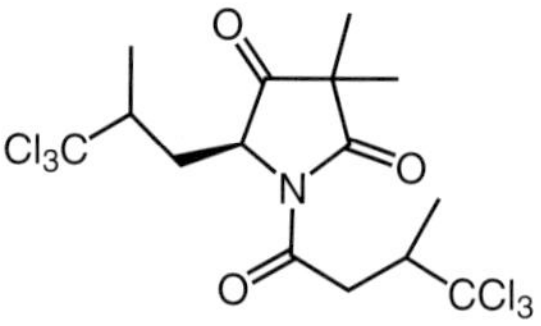

965 (dysidamide H)

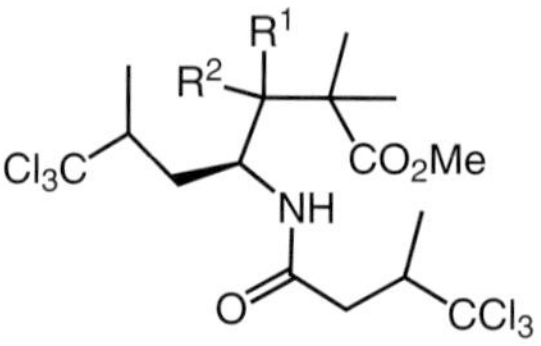

966 R¹ = H, R² = OH
967 R¹ = R² = O

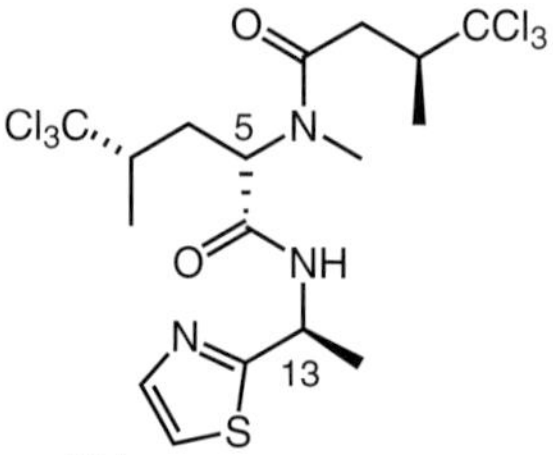

UU

Dysidenin is a strong inhibitor of iodide transport in dog thyroid, and the trichloromethyl group is recognized by the binding site (*1000, 1001*). Interestingly, the configuration at C-5 has no influence on this biological activity. The relationship between *Dysidea* and other sponges and the cyanobacterium *Oscillatoria spongeliae* continues to be studied (*1002, 1003*), revealing that only certain strains of this cyanobacterium are capable of producing either polychlorinated metabolites or polybrominated diphenyl ethers but not both (*1003*).

Several other, more complex cysteine-derived polychlorinated peptide metabolites are known to arise in marine organisms. Dolabellin (**968**) was characterized from the Japanese sea hare *Dolabella auricularia* (*1004*). The absolute

3.12 Amino Acids and Peptides

configuration was established by chemical degradation and a total synthesis of dolabellin. A Guamanian strain of the cyanobacterium *Lyngbya majuscula* has yielded lyngbyabellins A (**969**) (*1005*) and B (**970**) (*1006*). The latter metabolite was independently isolated from a collection of this alga in Florida (*1007*). Lyngbyabellin A has been synthesized (*1008*) and the absolute configuration of lyngbyabellin B is as shown (*1006*). A Palauan variety of *Lyngbya* sp. contains lyngbyabellin C (**971**) (*1009*), and the structurally related hectochlorin (**972**) was isolated from a Panamanian sample of *Lyngbya majuscula* and its absolute configuration was determined (*1010*). Hectochlorin has been synthesized (*1011*) and has potent inhibitory action against the fungus *Candida albicans* and displays inhibition of cell growth (*1010*). The Thai sea hare *Bursatella leachii* contains deacetylhectochlorin (**973**), which is more potent than hectochlorin against human carcinoma cell lines (*1012*).

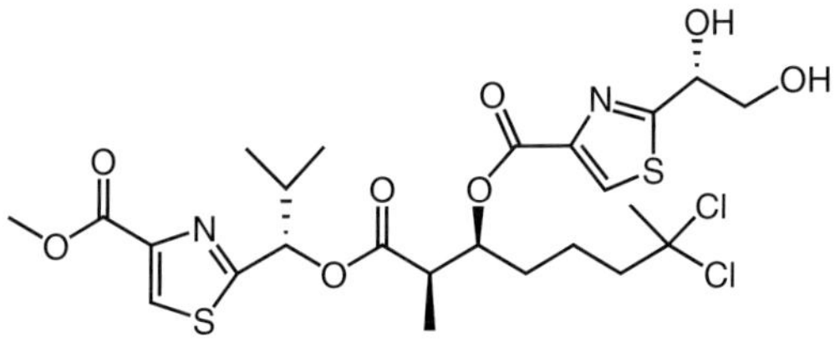

968 (dolabellin)

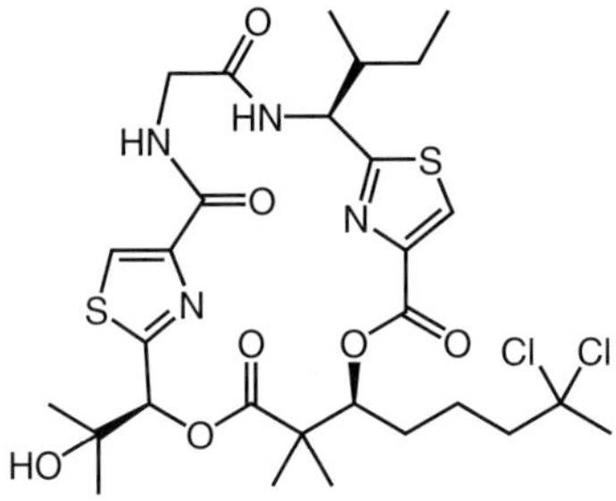

969 (lyngbyabellin A)

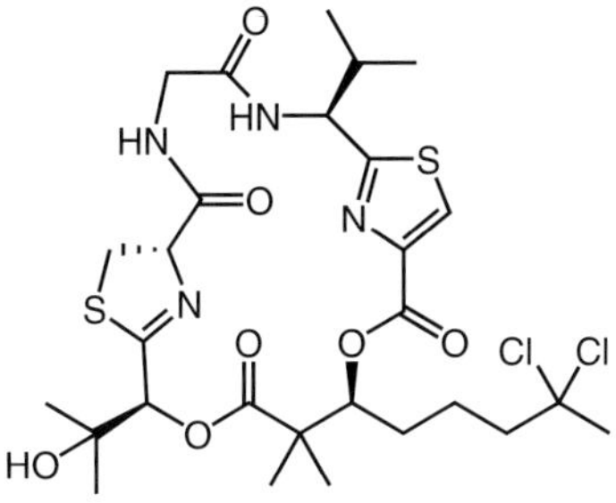

970 (lyngbyabellin B)

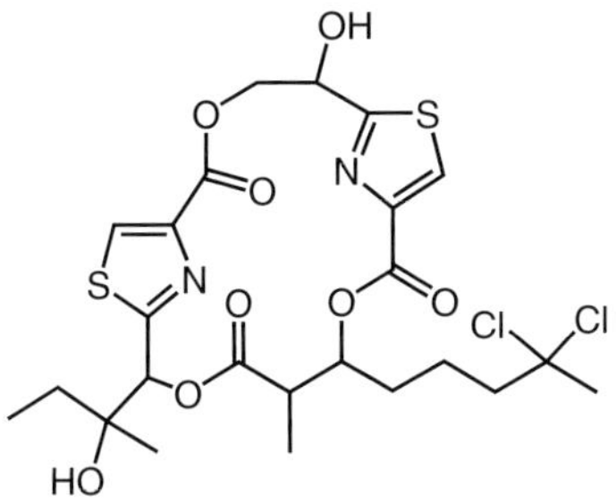

971 (lyngbyabellin C)

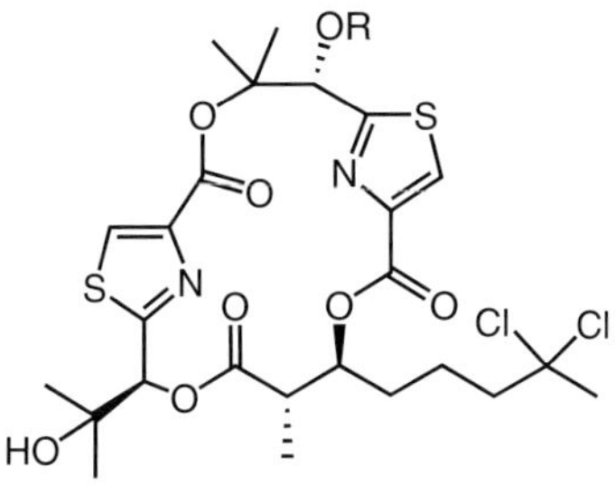

972 R = Ac (hectochlorin)
973 R = H (deacetylhectochlorin)

Additional collections of Guamanian *Lyngbya* sp. led to the discovery of lyngbyabellin D (**974**), having the absolute configuration shown and which displays good cytotoxicity against the KB cell line (*1013*). Lyngbyabellins E-I (**975–979**), which exhibit significant cytotoxicity against cancer cell lines, were characterized from *Lyngbya majuscula* from Papua New Guinea (*1014*). The absolute configurations of E (**975**) and H (**978**) were established via degradation products.

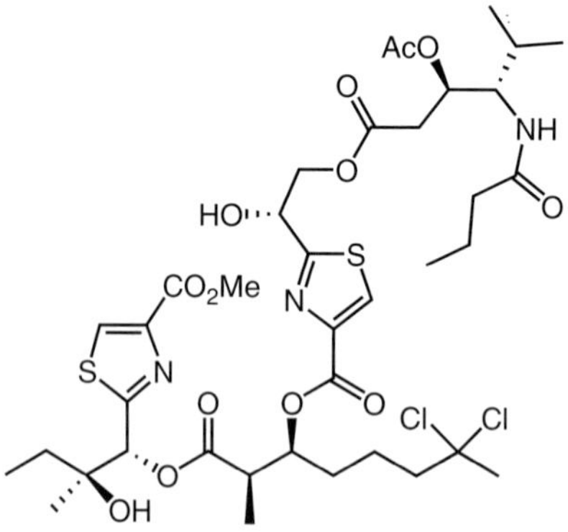

974 (lyngbyabellin D)

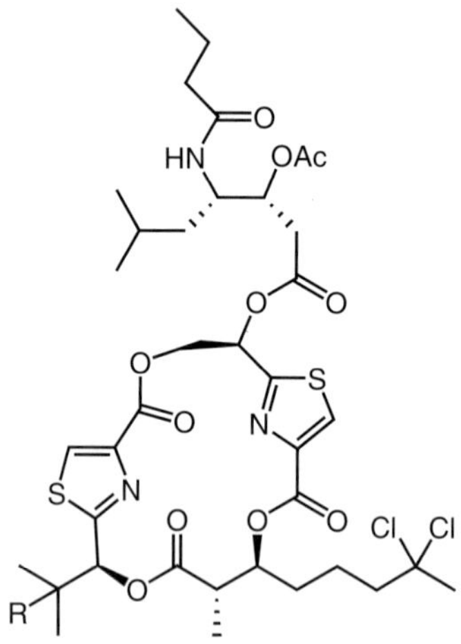

975 R = OH (lyngbyabellin E)
978 R = H (lyngbyabellin H)

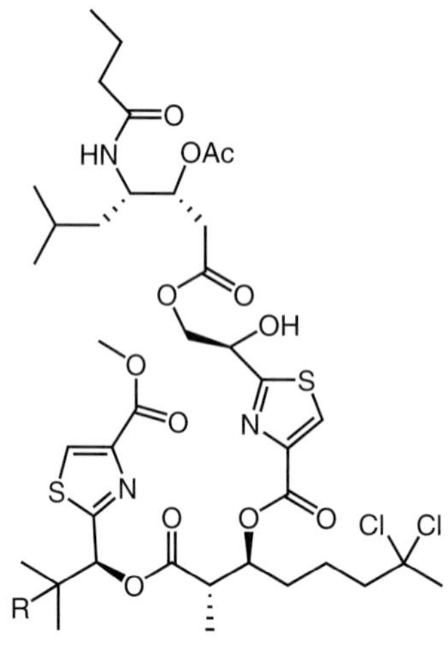

976 R = OH (lyngbyabellin F)
979 R = H (lyngbyabellin I)

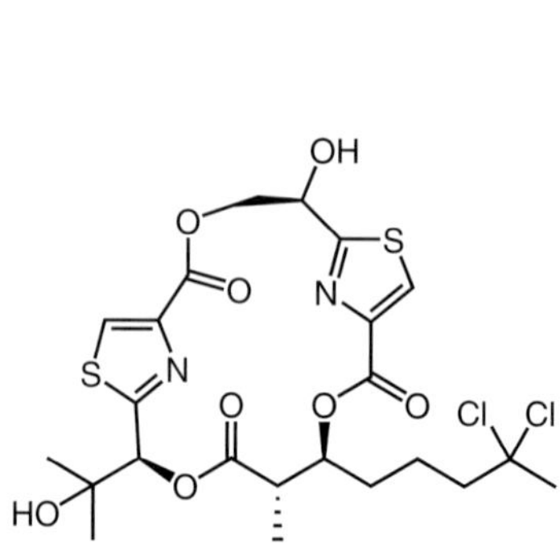

977 (lyngbyabellin G)

3.12 Amino Acids and Peptides

The chlorophenolic peptide resormycin (**980**), which was isolated from cultures of *Streptomyces platensis*, displays herbicidal and fungicidal activities (*1015, 1016*). Kaitocephalin (**981**) is produced by the fungus *Eupenicillium shearii* and is a glutamate receptor antagonist (*1017, 1018*). The original structure has been slightly revised (*1019*) and confirmed by total synthesis (*1019–1022*). The unique zinc-containing antibiotic transvalencin A (**982**) was isolated from *Nocardia transvalensis* found in a human clinical patient (*1023, 1024*).

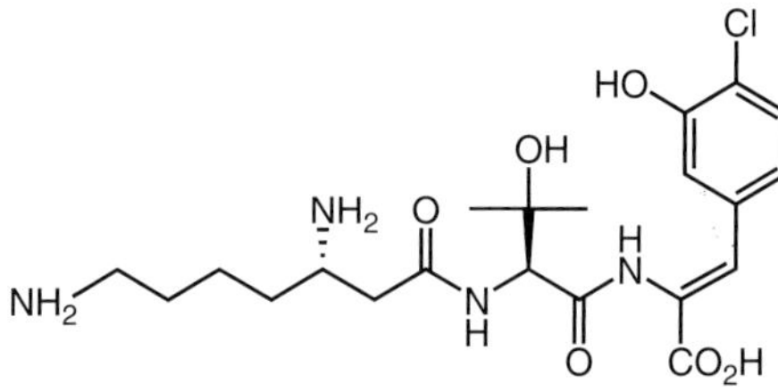

980 (resormycin)

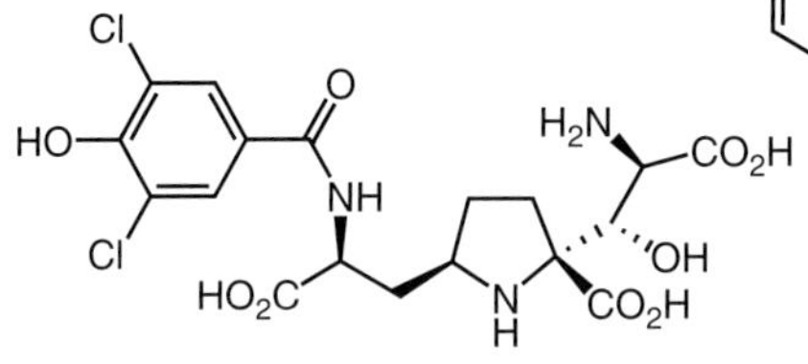

981 (kaitocephalin)

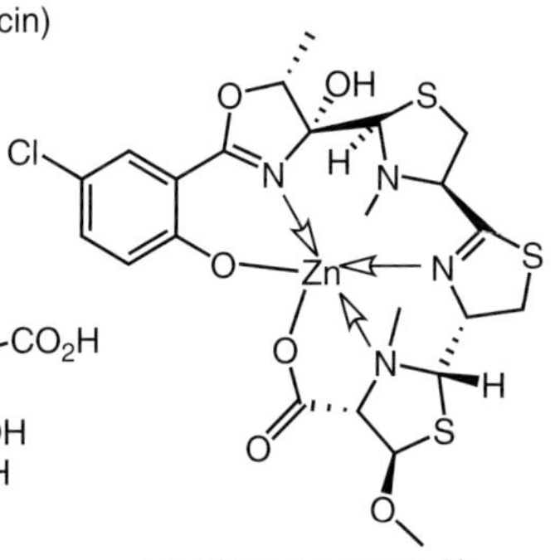

982 (transvalencin A)

A series of proline-containing cyclopentapeptides was disclosed in the first review (*1*), and subsequent work has shown that islanditoxin and cyclochlorotine are identical metabolites; the latter is the correct structure (*1025*). A new member of this group, astin I (**983**), has been isolated from *Aster tataricus* (*1026*), which is the source of several previously known astins (*1*). A full account of the structure determination of the known astins A–C has appeared (*1027*), and the antitumor activity of the astins seems to be related to the conformation of the dichloroproline unit in astins A–C (*1028, 1029*). A total synthesis of astin G, which is the only non-chlorinated astin, has been reported (*1030*). A cyclic pentapeptide related to the astins is destruxin-A4 chlorohydrin (**984**), which was found in the fungal culture OS-F68576 (*1031*). This chlorohydrin induces erythropoietin gene expression. The marine-derived fungus *Beauveria felina* has afforded [β-Me-Pro] destruxin E chlorohydrin (**985**) (*1032*).

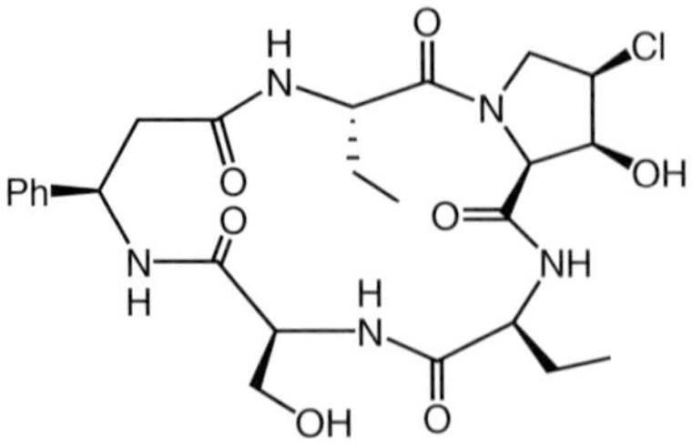

983 (astin I)

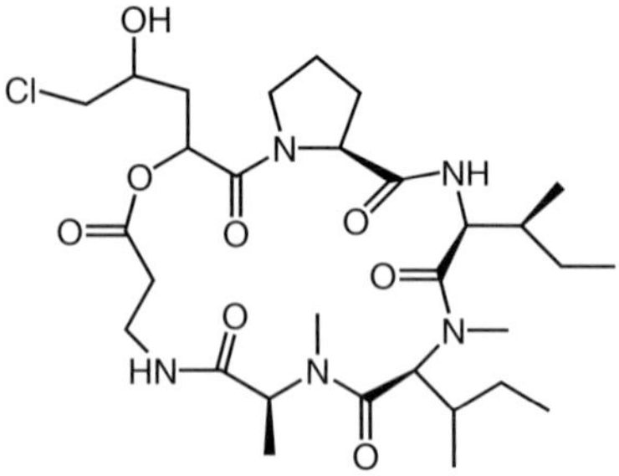

984 (destruxin-A4 chlorohydrin)

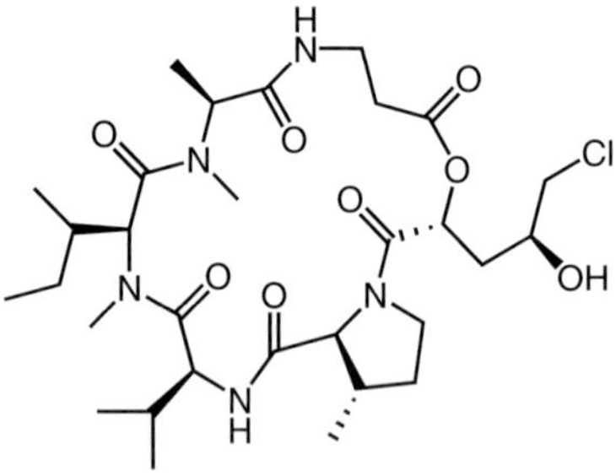

985

The freshwater blue-green alga *Microcystis aeruginosa* is the source of novel peptides, several of which contain chlorine. Aeruginosin 98-A (**986**) was isolated from *Microcystis aeruginosa* (NIES-98), and this compound inhibits trypsin, plasmin, and thrombin (*1033*). The cyanobacterium *Oscillatoria agardhii* produces glycopeptides aeruginosins 205A (**987**) and 205B (**988**), which are also potent inhibitors of trypsin and thrombin (*1034*). A Japanese bloom of *Microcystis aeruginosa* (NIES-299) yielded microginins 299-A (**989**) and 299-B (**990**), which are leucine aminopeptidase inhibitors. The absolute stereochemistries are indicated (*1035*). From another strain of *Microcystis aeruginosa* (NIES-478) there were isolated micropeptins 478-A (**991**) and 478-B (**992**), which inhibit plasmin but not trypsin, thrombin, papain, chymotrypsin, or elastase at 10 μg mL^{-1} (*1036*).

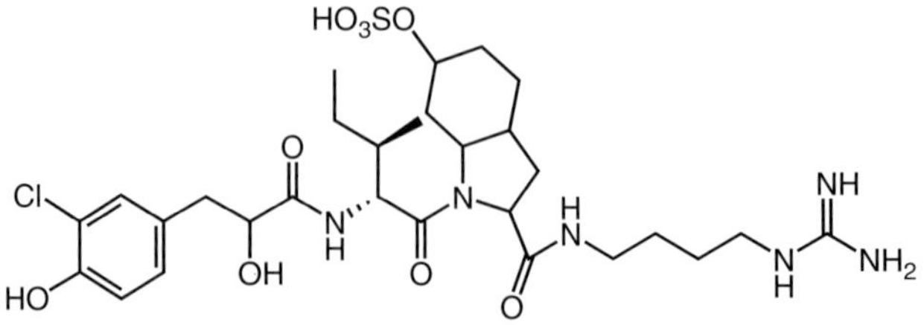

986 (aeruginosin 98-A)

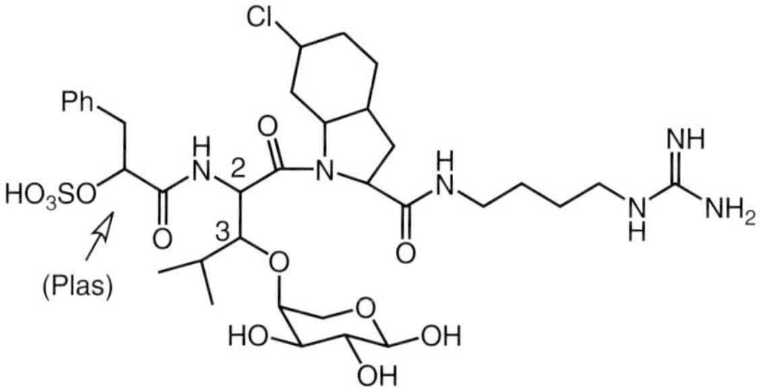

987 (2*R*,3*S*) L-Plas (aeruginosin 205A)
988 (2*S*,3*R*) D-Plas (aeruginosin 205B)

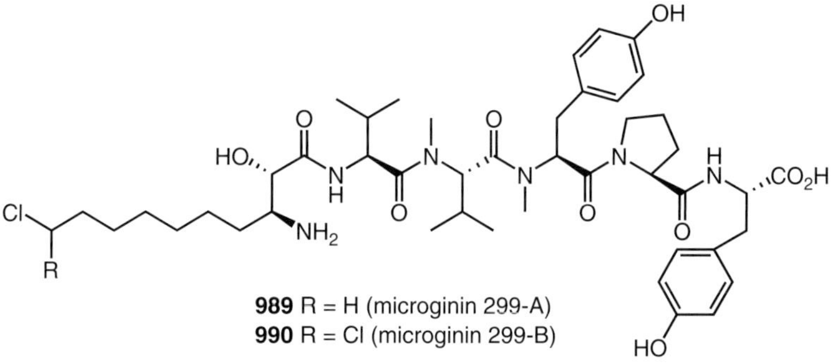

989 R = H (microginin 299-A)
990 R = Cl (microginin 299-B)

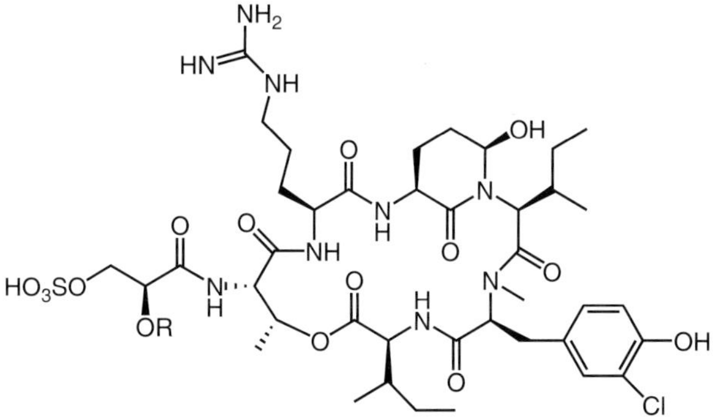

991 R = H (micropeptin 478-A)
992 R = SO^{3}H (micropeptin 478-B)

The new chlorinated microginins 99-A (**993**), 99-B (**994**), and 299-D (**995**) were isolated from two blooms of *Microcystis aeruginosa* (*1037*). Structurally similar to aeruginosin 98-A (**986**), the new aeruginosins 98-C (**996**), 101 (**997**), 89-A (**998**), and 89-B (**999**) were isolated from different strains of *Microcystis aeruginosa* (*1038*). The absolute configurations are as shown. A subsequent investigation of this blue-green alga revealed the presence of the new microginins 91-A (**1000**), 91-B (**1001**), 91-D (**1002**), and 91-E (**1003**) (*1039*). The total synthesis of the non-chlorinated aeruginosin 298-A has corrected a stereocenter; thus, D-leucine and not L-leucine is incorporated in this compound (*1040*), which may apply to other aeruginosins.

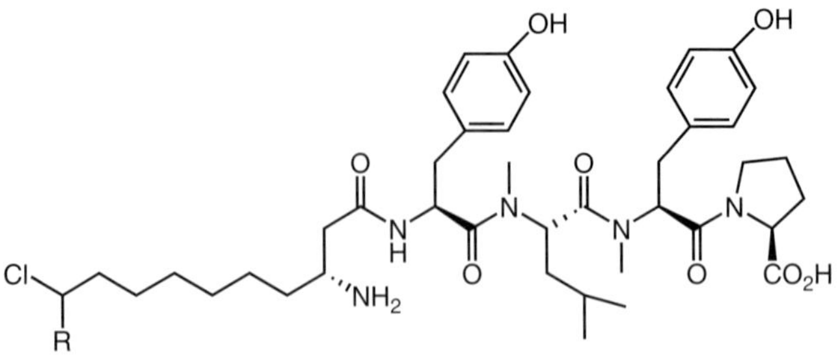

993 R = H (microginin 99-A)
994 R = Cl (microginin 99-B)

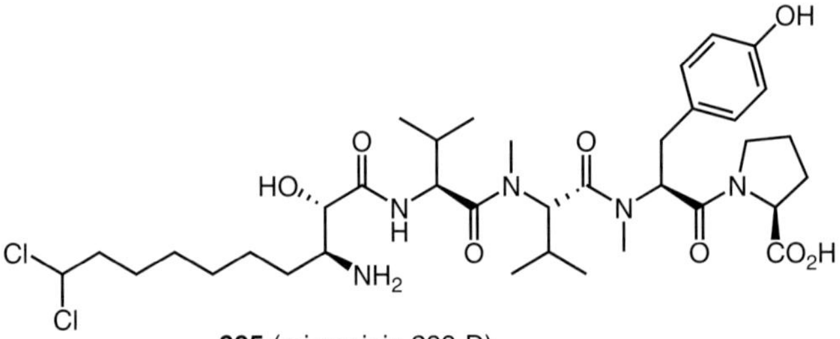

995 (microginin 299-D)

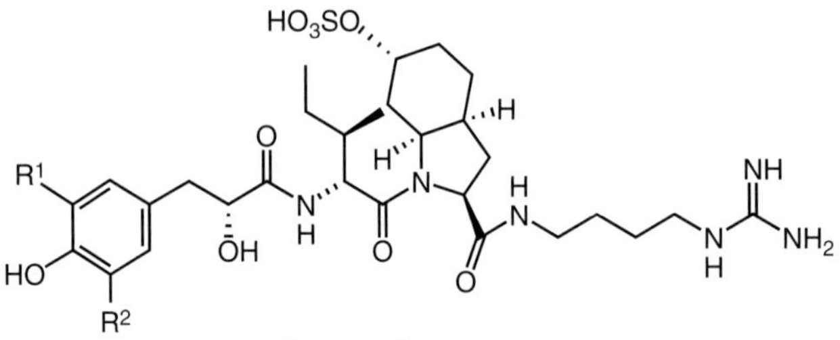

996 R^1 = Br, R^2 = H (aeruginosin 98-C)
997 R^1 = R^2 = Cl (aeruginosin 101)

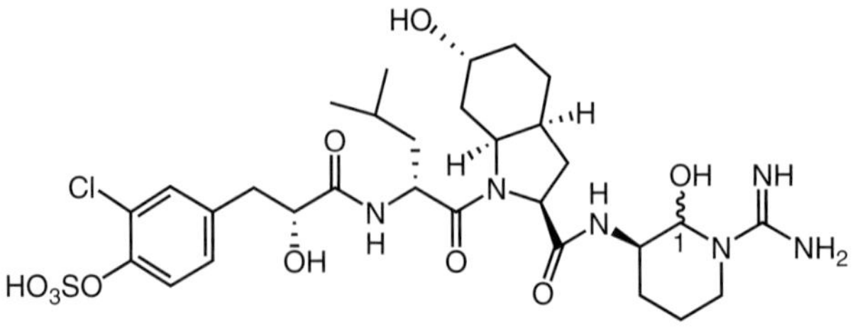

998 (aeruginosin 89-A)
999 (aeruginosin 89-B)   (epimeric at C-1)

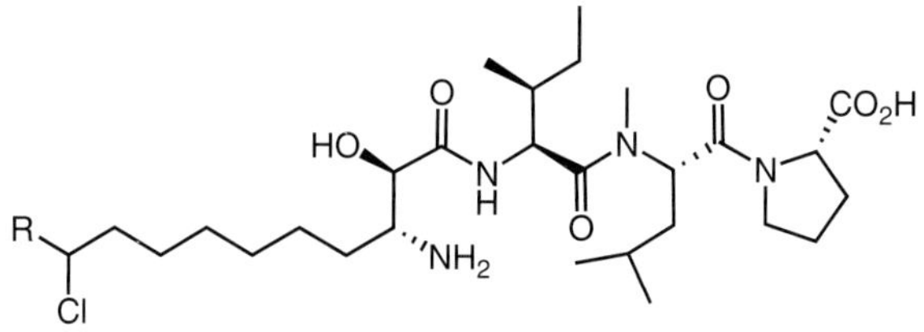

3.12 Amino Acids and Peptides   147

1000 R = H (microginin 91-A)
1001 R = Cl (microginin 91-B)

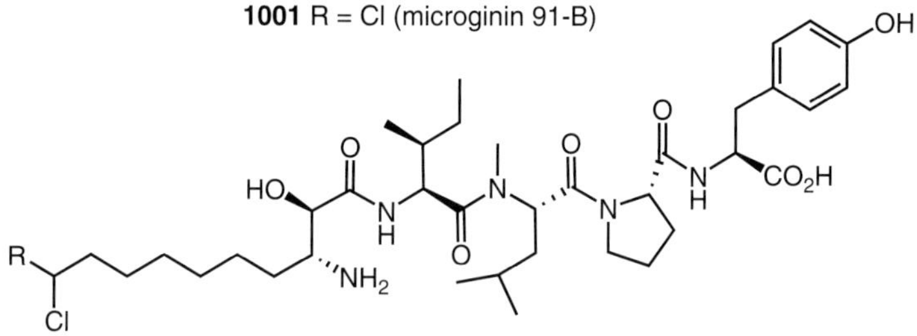

1002 R = H (microginin 91-D)
1003 R = Cl (microginin 91-E)

Another cultivation of *Microcystis aeruginosa* NIVA Cya 43 yielded the new cyanopeptolin 954 (**1004**), which is a chymotrypsin inhibitor (*1041*). The newly isolated chlorodysinosin A (**1005**), from a *Dysidea* sponge, has been synthesized and its structure confirmed (*1042*).

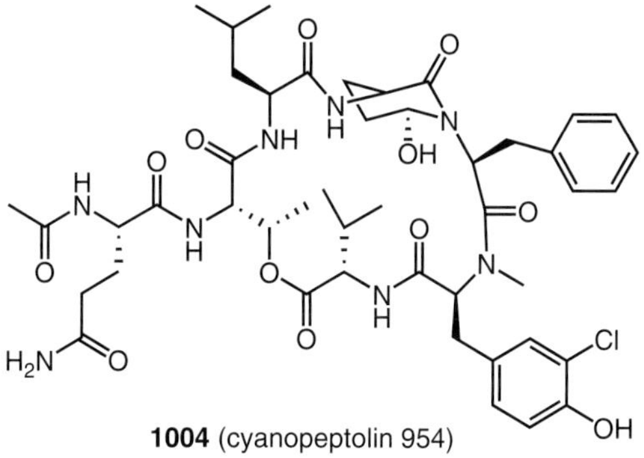

1004 (cyanopeptolin 954)

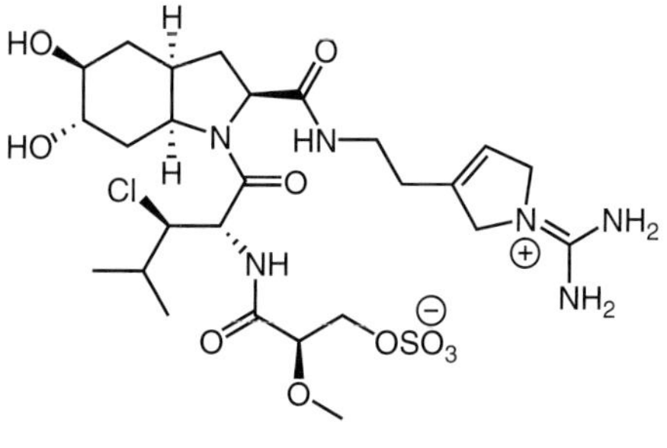

1005 (chlorodysinosin A)

A disease of oats in North America is caused by the fungus *Cochliobolus victoriae* (*Helminthosporium victoriae*) and the major causal agent is victorin C. After considerable work on degradation products (*1043, 1044*) the structure of victorin C (**1006**) was finally established (*1045*). Additional study of this fungus has afforded the minor victorins B, D, and E (**1007–1009**) and victoricine (**1010**) (*1046*). The victorin binding protein from oats has been identified (*1047, 1048*). The fungal pathogen *Periconia circinata*, which causes milo disease of grain, produces peritoxins A (**1011**) and B (**1012**), and periconins A (**1013**) and B (**1014**) (*1049*).

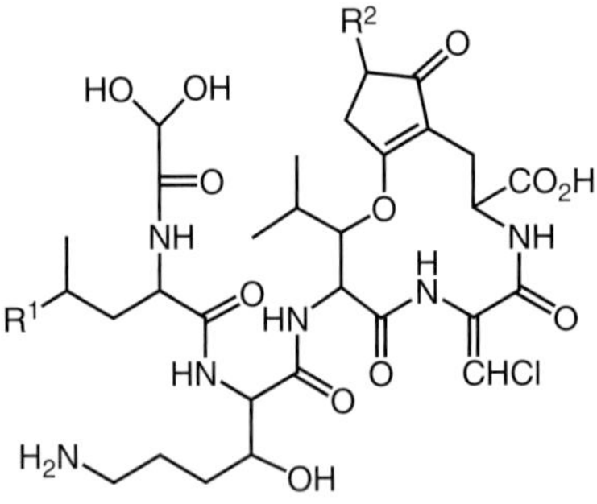

1006 $R^1 = CHCl_2$, $R^2 = OH$ (victorin C)
1007 $R^1 = CH_2Cl$, $R^2 = OH$ (victorin B)
1008 $R^1 = CHCl_2$, $R^2 = H$ (victorin D)
1009 $R^1 = CCl_3$, $R^2 = OH$ (victorin E)

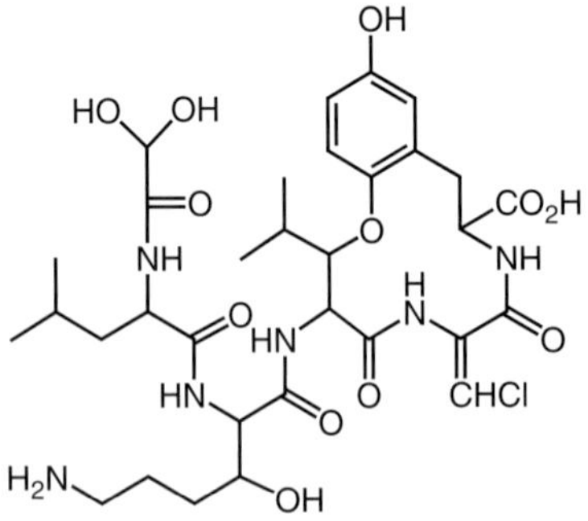

1010 (victoricine)

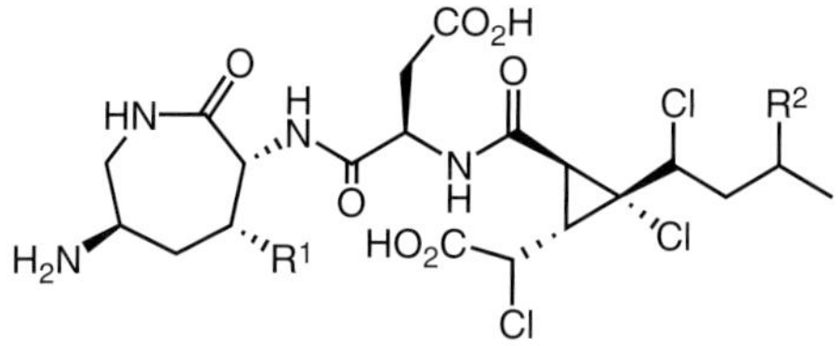

3.12 Amino Acids and Peptides

1011 $R^1 = R^2 = OH$ (peritoxin A)
1012 $R^1 = OH$, $R^2 = H$ (peritoxin B)
1013 $R^1 = H$, $R^2 = OH$ (periconin A)

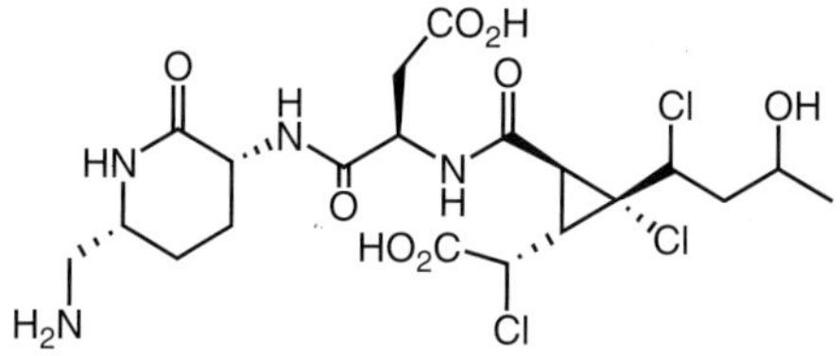

1014 (periconin B)

The monamycins are a group of 15 antibiotic cyclodepsipeptides from *Streptomyces jamaicensis*. After much structural elucidation work on the hydrolysis and degradation products (*1050, 1051*), the structures of the six chlorine-containing monamycins G_1 (**1015**), G_2 (**1016**), G_3 (**1017**), H_1 (**1018**), H_2 (**1019**), and I (**1020**) were determined (*1052*).

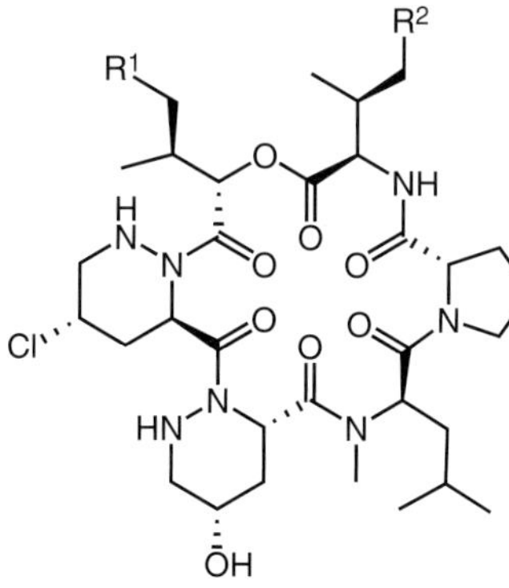

1015 $R^1 = R^2 = H$, $R^3 = Me$ (monamycin G_1)
1016 $R^1 = R^3 = H$, $R^2 = Me$ (monamycin G_2)
1017 $R^1 = Me$, $R^2 = R^3 = H$ (monamycin G_3)
1018 $R^1 = R^3 = Me$, $R^2 = H$ (monamycin H_1)
1019 $R^1 = H$, $R^2 = R^3 = Me$ (monamycin H_2)
1020 $R^1 = R^2 = R^3 = Me$ (monamycin I)

Fig. 3.16 *Auletta* cf. *constricta*, a brownish sponge that contains jasplakinolide, which is related to jasplakinolides B and C (**1023** and **1024**)–to make sure which of the two organisms of the picture are addressed (Photo: P. Crews)

Some 15 halogenated tryptophan-derived peptides were described in the first account (*1*), and several new examples have been discovered in the interim. Chondramides A–D, two of which are the chlorinated B (**1021**) and D (**1022**), were isolated from a strain of myxobacteria (*Chondromyces crocatus*), and possess antifungal and cytostatic activity (*1053, 1054*). The chondramides, which are structurally related to jasplakinolide (= jaspamide) but which can be produced in large quantities by fermentation, exhibit antiproliferative activity against carcinoma cell lines (*1055*). Two new jasplakinolides, B (**1023**) and C (**1024**) (Fig. 3.16), have been characterized from the Vanuatu sponge *Jaspis splendans* (*1056*). The well-known jasplakinolide has been the object of biological studies (e.g., actin cytoskeleton disruptor) (*1057, 1058*), conformational studies with and without lithium (*1059*), and synthesis and biological evaluation of analogs (*1060, 1061*). A new celenamide, celenamide E (**1025**), was found in the Patagonian sponge *Cliona chilensis* (*1062*). This metabolite may be the biosynthetic precursor of the known celenamides A–C (*1*), as it contains an unusual *N*-terminal dehydroamino acid.

3.12 Amino Acids and Peptides

1021 R = OMe (chondramide B)
1022 R = H (chondramide D)

1023 R = =O (jasplakinolide B)
1024 R = OH (jasplakinolide C)

1025 (celenamide E)

Fig. 3.17 *Psammocinia* aff. *bulbosa*, a Papua New Guinea sponge that produces cyclocinamide A (**1026**) (Photo: P. Crews)

A *Psammocinia* sp. (Fig. 3.17) sponge from Papua New Guinea has yielded the potent cytotoxic cyclocinamide A (**1026**) (*1063*). The novel 6-chlorotryptophan derivatives, microsclerodermins C (**1027**) and D (**1028**), were characterized from a Philippines *Theonella* sp. sponge (*1064*). The related dehydromicrosclerodermins C (**1029**) and D (**1030**) are found in the sponge *Theonella cupola* from Okinawa (*1065*).

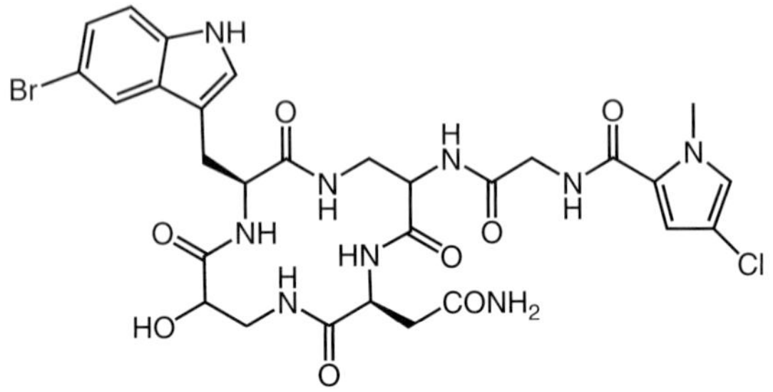

1026 (cyclocinamide A)

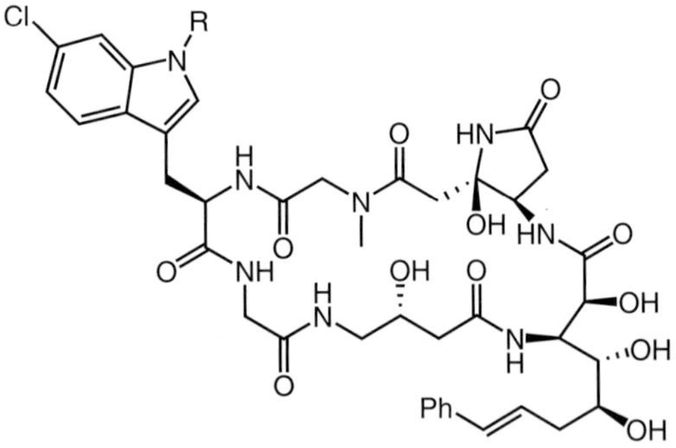

1027 R = CONH2 (microsclerodermin C)
1028 R = H (microsclerodermin D)

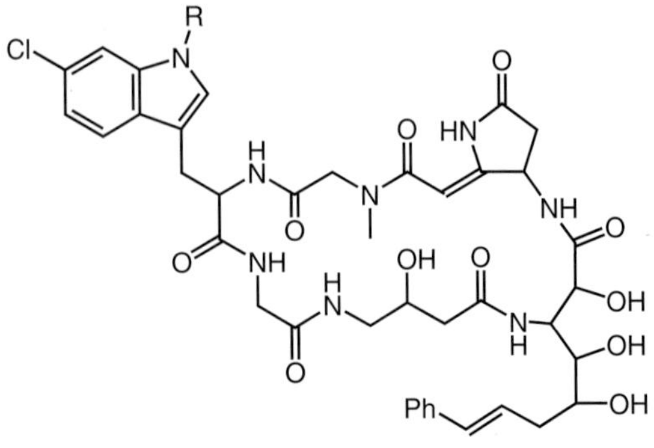

1029 R = CONH$_2$ (dehydromicrosclerodermin C)
1030 R = H (dehydromicrosclerodermin D)

3.12 Amino Acids and Peptides

The stereochemistry of the modified tryptophan amino acids in the previously known konbamide and keramamide A sponge metabolites (*1*) has been determined to be L for both 2-bromo-5-hydroxytryptophan and 6-chloro-5-hydroxy-*N*-methyltryptophan, respectively (*1066*). However, based on synthetic studies, doubt has been raised as to the structure of konbamide (*1067*). A series of investigations of Okinawan *Theonella* sp. sponges has uncovered the new halogenated keramamides, E (**1031**), H (**1032**) (*1068*), L (**1033**) (*1069*), M (**1034**), and N (**1035**) (*1070*).

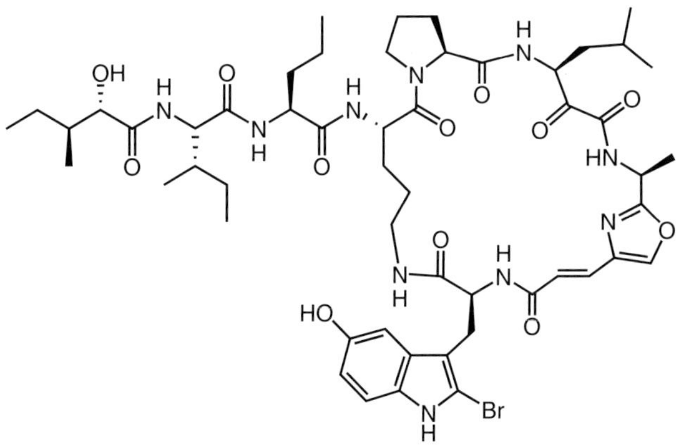

1031 (keramamide E)

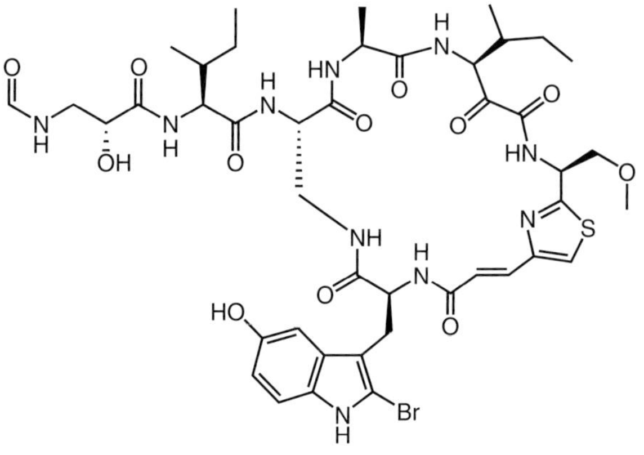

1032 (keramamide H)

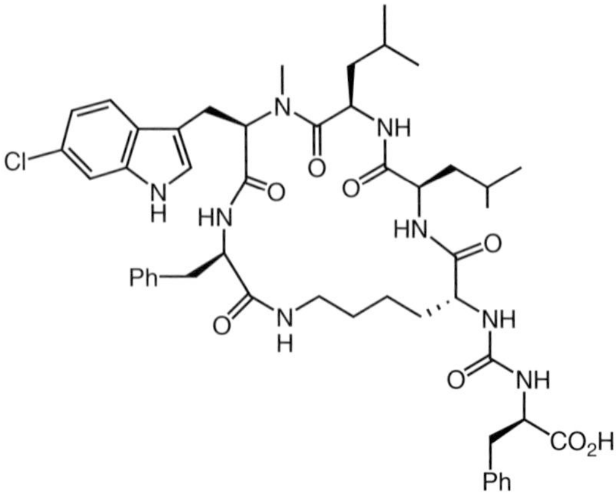

1033 (keramamide L)

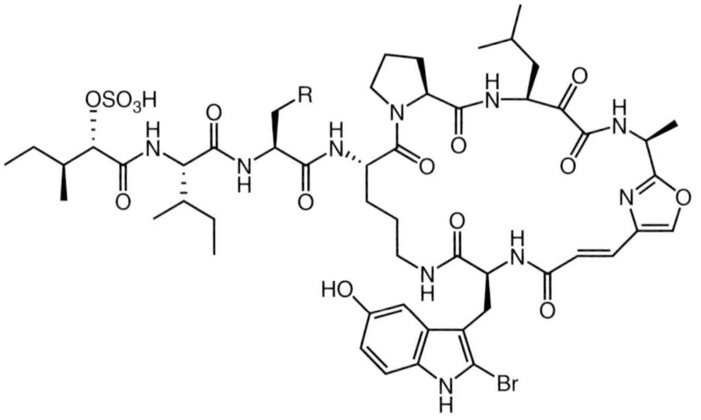

1034 R = Me (keramamide M)
1035 R = Et (keramamide N)

The polychlorinated cyclic hexadepsipeptides kutznerides 1–9 (**1036–1044**), which contain both the novel 6,7-dichlorohexahydropyrrolo[2,3-β]indole core and several unusual amino acids, are found in the actinomycete *Kutzneria* sp. 744 inhabiting the roots of *Picea abies* (*1071, 1072*). These compounds show moderate activity against root-rotting fungi. Another chlorinated hexahydropyrrolo[2,3-b] indole cyclohexapeptide is the dimeric chloptosin (**1045**) isolated from a *Streptomy-*

3.12 Amino Acids and Peptides

ces strain (*1073*). This unique metabolite induces apoptosis and shows strong antimicrobial activity against Gram-positive bacteria including methicillin-resistant *Staphylococcus aureus*.

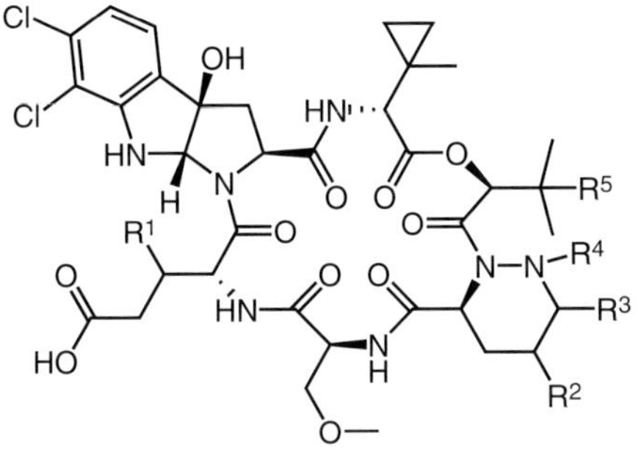

1036 $R^1 = \alpha$-OH, $R^2 = R^3 = R^4 = H$, $R^5 = Me$ (kutzneride 1)
1037 $R^1 = \alpha$-OH, $R^2 = Cl$, $R^3 = R^4 = H$, $R^5 = Me$ (kutzneride 2)
1038 $R^1 = \beta$-OH, $R^2 = R^3 = R^4 = H$, $R^5 = Me$ (kutzneride 3)
1039 $R^1 = \beta$-OH, $R^2 = H$, $R^3 = R^4 = \pi$ bond, $R^5 = Me$ (kutzneride 4)
1040 $R^1 = \beta$-OH, $R^2 = R^3 = R^4 = R^5 = H$ (kutzneride 5)
1041 $R^1 = \beta$-OH, $R^2 = OH$, $R^3 = R^4 = \pi$ bond, $R^5 = Me$ (kutzneride 6)
1042 $R^1 = \alpha$-OH, $R^2 = R^3 = R^4 = R^5 = H$ (kutzneride 7)
1043 $R^1 = \beta$-OH, $R^2 = Cl$, $R^3 = R^4 = H$, $R^5 = Me$ (kutzneride 8)
1044 $R^1 = \alpha$-OH, $R^2 = H$, $R^3 = R^4 = \pi$ bond, $R^5 = Me$ (kutzneride 9)

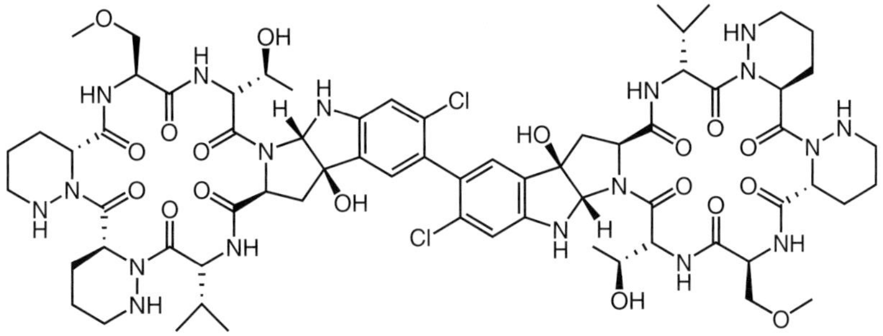

1045 (chloptosin)

One major recent development in the area of natural products is the discovery and subsequent medicinal application of the toxic peptides from cone snails. These *Conus* peptides, several of which contain a 6-bromotryptophan amino acid, are finding utility for the treatment of neuropathic pain and other neurological conditions (*1074–1078*). For example, ω-conopeptide MVIIA (Ziconotide, trade name Prialt) has been approved by the US FDA since 2004 for the treatment of severe pain. It is estimated that the 500–700 species of cone snails (*Conus* genus) contain more than 50,000 distinct toxins, since the venom in each snail consists of 40–200 individual peptides with a specific biological action (*1074, 1077, 1080*). However, relatively few of these toxic peptides have been characterized. The venomous worm-hunting cone snail *Conus imperialis* contains the heptapeptide **1046** in which 6-bromotryptophan is present, and *Conus radiatus* produces a 33-amino acid peptide **1047** that also contains 6-bromotryptophan (*1081, 1988*). The venom from this latter snail has also furnished the octapeptide bromocontryphan (**1048**) (*1082*). Numerous conotoxins contain 6-bromotryptophan, such as a 31-amino acid peptide (**1049**) and others from *Conus textile* (*1083, 1984–1987*), and several peptides from other *Conus* species (*1084*), including *Conus delessertii* (*1989*) and *Conus monile* (*1990*).

Pca–cys–gly–gln–ala–Brtrp–cys–NH$_2$

1046

[Pca = pyroglutamic acid]

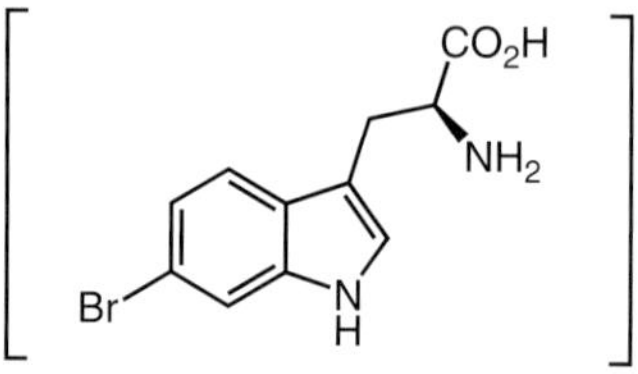

Brtrp

1047 (33-amino acid peptide)   **1049** (31-amino acid peptide)

gly–cys–hyp–D-trp–glu–pro–Brtrp–cys–NH$_2$

1048 (bromocontryphan)   [hyp = *trans*-4-hydroxyproline]

(**1050-1054** are not shown)

Other marine organisms contain peptides and proteins with 6-bromotryptophan. The ascidian *Phallusia mammillata* has a polypeptide morulin Pm (**1050**) that includes 6-bromotryptophan (*1085*). In fact, this amino acid is the major residue in the peptide. Similarly, the solitary ascidian *Styela clava* produces styelin D (**1051**), another 6-bromotryptophan-containing 32-amino acid peptide (*1086, 1087*). Styelin D shows excellent activity against marine bacteria and human pathogens (*1087*). Three cathelicidins, HFIAP-1 (**1052**), HFIAP-2 (**1053**), and HFIAP-3 (**1054**), each containing one or two 6-bromotryptophans, were isolated from the Atlantic hagfish (*Myxine glutinosa*) (*1088*). This report suggests that the role of 6-bromotryptophan in these and other peptides (vide supra) is to block proteolytic degradation. Thus, the large bromine atom makes the peptide a poor fit for chymotrypsin, which normally would cleave tryptophan residues (*1088*).

One of the more interesting and structurally complex indole-containing peptides is the previously described diazonamide (A and B) from the ascidian *Diazona*

3.12 Amino Acids and Peptides

chinensis, which is now named *Diazona angulata* (*1*). However, as shown by synthesis, the originally proposed structure of the diazonamides (*1*) is incorrect (*1089*). Reevaluation of the data, X-ray analysis, and biogenetic considerations led to structure **VV** for diazonamide A (*1090, 1091*), which has been confirmed by total synthesis (*1092–1094*). Diazonamide A and synthetic analogs continue to find interest as potential anticancer agents (*1095*).

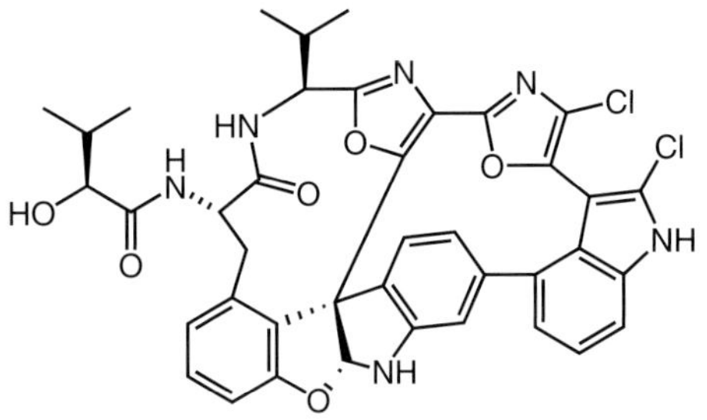

VV (diazonamide A)

Although iodine-containing natural products are exceedingly rare, the previously described geodiamolides A–F are a group of marine sponge cyclic peptides containing chlorine, bromine, and iodine (*1*). Several new examples are known, such as geodiamolide G (**1055**) from a *Cymbastela* sp. Papua New Guinea sponge (*1096*), H (**1056**) and I (**1057**) from the sponge *Geodia* sp. in Trinidad (*1097*), and from a Papua New Guinea *Cymbastela* sp. J (**1058**), K (**1059**), L (**1060**), M (**1061**), N (**1062**), O (**1063**), P (**1064**), and R (**1065**) (*1098*). Five of these new metabolites contain iodine, and several are cytotoxic (*1096–1098*). The related geodiamolide TA (**1066**) was isolated from the sponge *Hemiasterella minor*, and it has the same configuration as the geodiamolides (*1099*). It is also quite cytotoxic against P388. Related to the geodiamolides (i.e., D) is the iodinated neosiphoniamolide A (**1067**) from the New Caledonian sponge *Neosiphonia superstes* (*1100*).

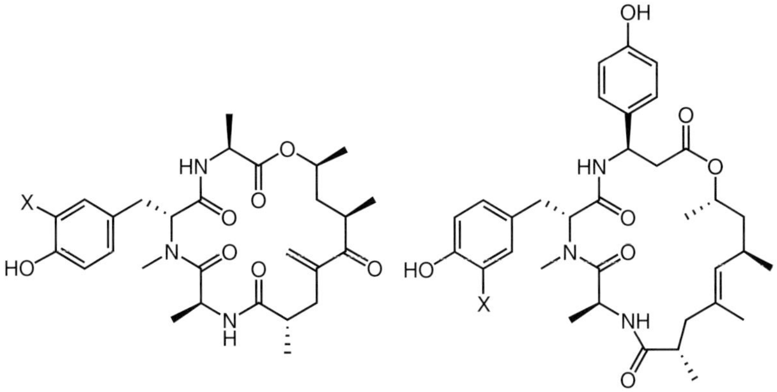

1055 X = I (geodiamolide G)
1058 X = Br (geodiamolide J)
1059 X = Cl (geodiamolide K)

1056 X = I (geodiamolide H)
1057 X = Br (geodiamolide I)

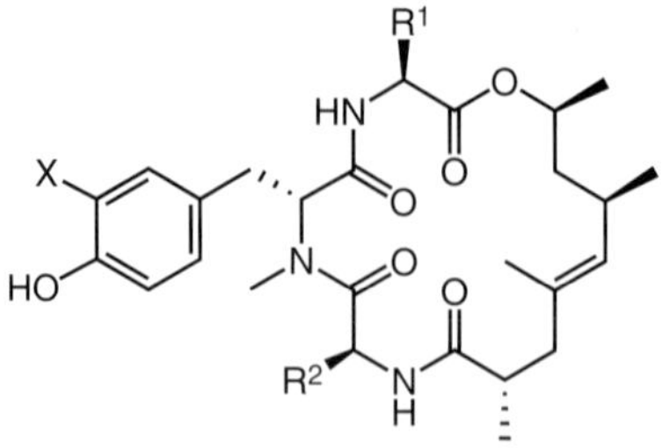

1060 X = I, R^1 = CH$_2$OH, R^2 = Me (geodiamolide L)
1061 X = Br, R^1 = CH$_2$OH, R^2 = Me (geodiamolide M)
1062 X = Cl, R^1 = CH$_2$OH, R^2 = Me (geodiamolide N)
1063 X = I, R^1 = Me, R^2 = CH$_2$OH (geodiamolide O)
1064 X = Br, R^1 = Me, R^2 = CH$_2$OH (geodiamolide P)
1065 X = I, R^1 = CH$_2$OH, R^2 = H (geodiamolide R)
1066 X = I, R^1 = CHMe$_2$, R^2 = Me (geodiamolide TA)
1067 X = I, R^1 = CHMe$_2$, R^2 = H (neosiphoniamolide A)

Closely related to the geodiamolides are the seragamides A–F (**1068–1073**) isolated from the Okinawan sponge *Suberites japonicus* (Fig. 3.18) (*1101*). Seragamide A promotes G-actin polymerization and stabilizes F-actin filaments. A different group of cyclic depsipeptides, miuraenamides A (**1074**) and B (**1075**), are produced by the halophilic myxobacterial strain, SMH-27-4, and are potent and selective inhibitors of the phytopathogenic *Phytophthora* sp. (*1102*). Interest in the previously known chlorine-containing pepticinnamin E, which is an inhibitor of farnesyl-protein transferase isolated from a *Streptomyces* strain (*1*), has extended to its total synthesis, and the synthesis of diastereomers (*1103*) and compound libraries (*1104–1106*).

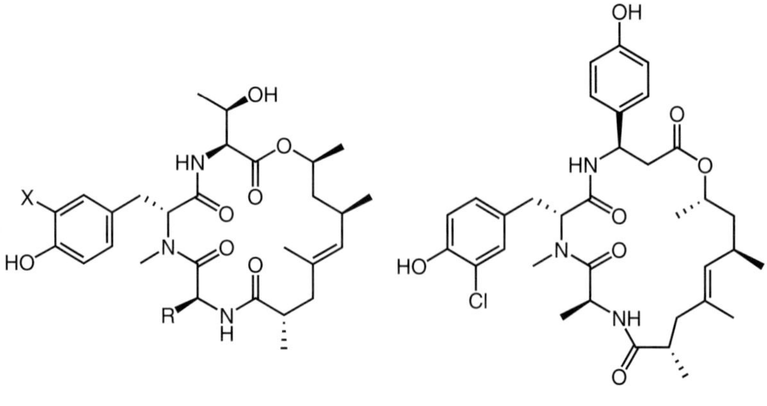

1068 X = I, R = Me (seragamide A)
1069 X = Br, R = Me (seragamide B)
1070 X = Cl, R = Me (seragamide C)
1071 X = I, R = H (seragamide D)
1072 X = I, R = CH$_2$OH (seragamide E)

1073 (seragamide F)

3.12 Amino Acids and Peptides

1074 X = Br (miuraenamide A)
1075 X = I (miuraenamide B)

Fig. 3.18 *Suberites japonicus*, a sponge collected in Zampa, Okinawa, that contains seragamides A–F (**1068–1073**) (Photo: J. Tanaka)

The extraordinarily biologically active and clinically promising cryptophycins, which were discovered in a *Nostoc* sp. terrestrial blue-green alga (*1*), continue to be isolated from this genus of filamentous cyanobacteria (*1107–1109*). The ecology of *Nostoc* has been reviewed (*849*). The previously recorded cryptophycins A (**WW**) and C (**XX**) (*1*) have been shown by total synthesis to have the absolute configuration corresponding to the D-series of 3-chloro-*O*-methyltyrosine (*1110*). A study of *Nostoc* sp. GSV 224 has uncovered 22 new chlorine-containing cryptophycins (which are now designated by numbers): cryptophycin-30 (**1076**), -28 (**1077**), -16 (**1078**), -23 (**1079**), -31 (**1080**), -17 (**1081**), -45 (**1082**), -175 (**1083**), -46 (**1084**), -29 (**1085**), -21 (**1086**), -176 (**1087**), -40 (**1088**), -326 (**1089**), -38 (**1090**), and -18 (**1091**), -49 (**1092**), -50 (**1093**), -54 (**1094**), -19 (**1095**), -26 (**1096**), and -327 (**1097**) (*1111–1113*).

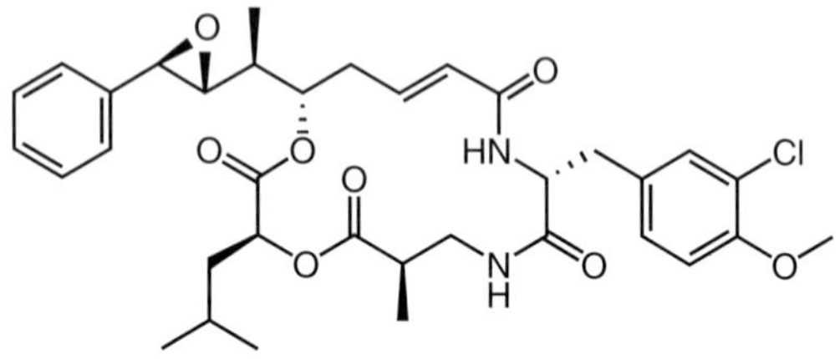

WW (cryptophycin A)

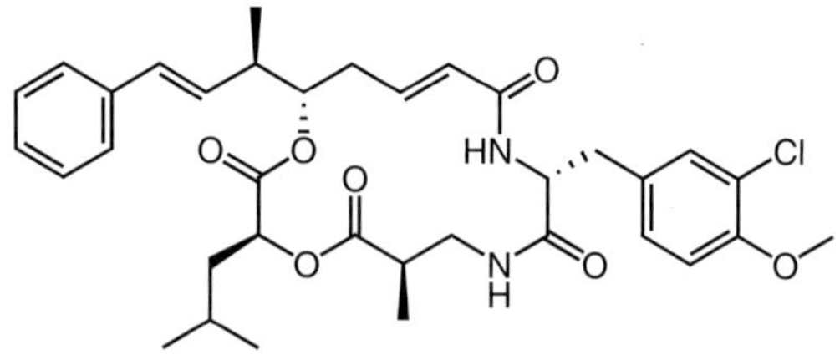

XX (cryptophycin C)

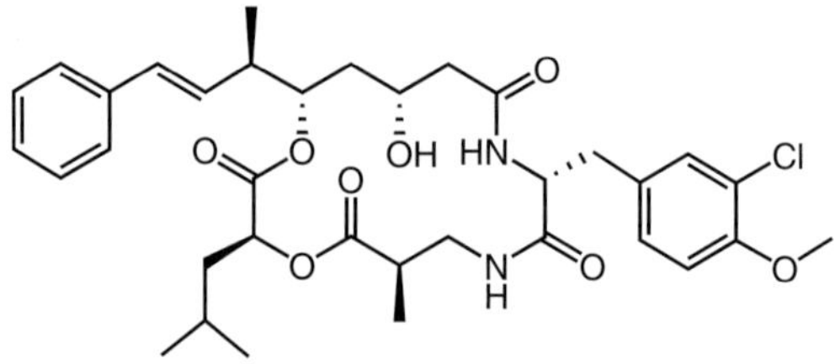

1076 (cryptophycin-30)

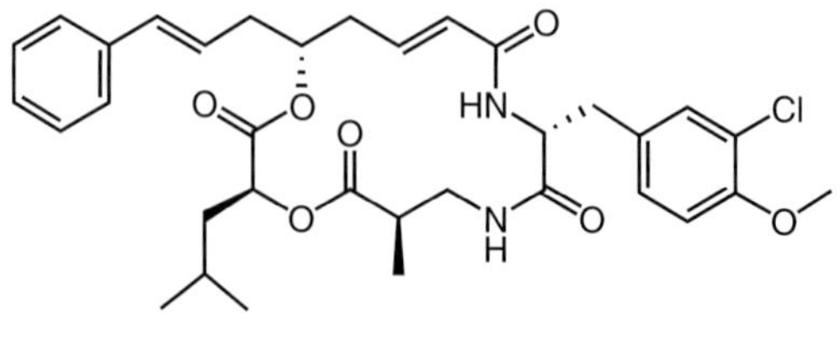

1077 (cryptophycin-28)

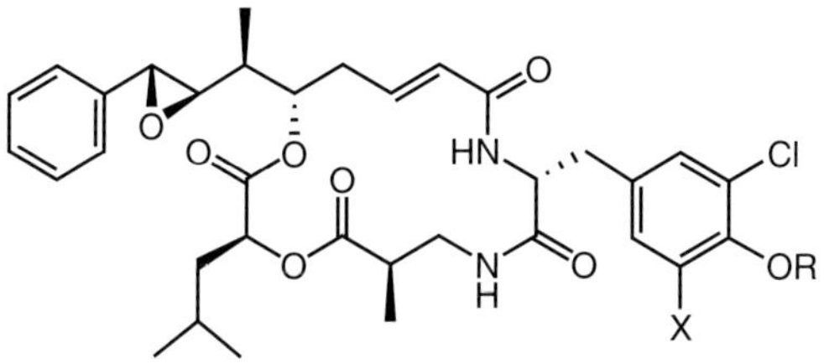

1078 R = H, X = H (cryptophycin-16)
1079 R = H, X = Cl (cryptophycin-23)
1080 R = Me, X = Cl (cryptophycin-31)

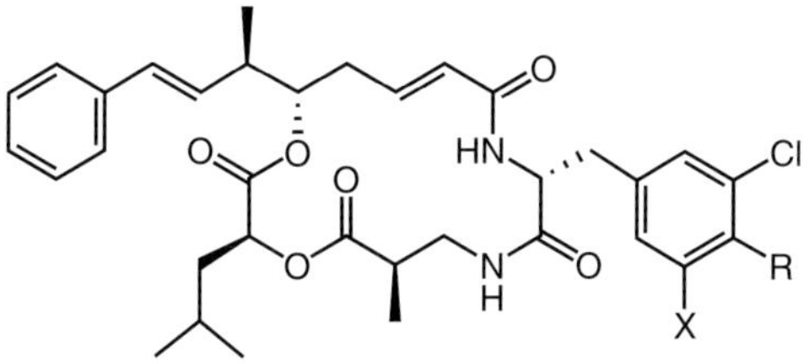

1081 R = H, X = H (cryptophycin-17)
1082 R = H, X = Cl (cryptophycin-45)
1083 R = Me, X = Cl (cryptophycin-175)

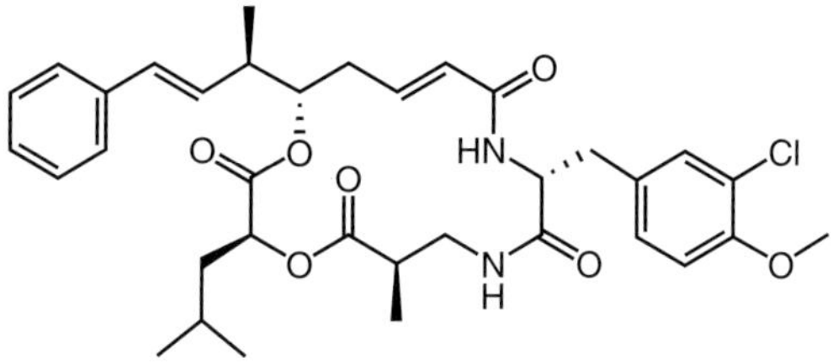

1084 (cryptophycin-46)

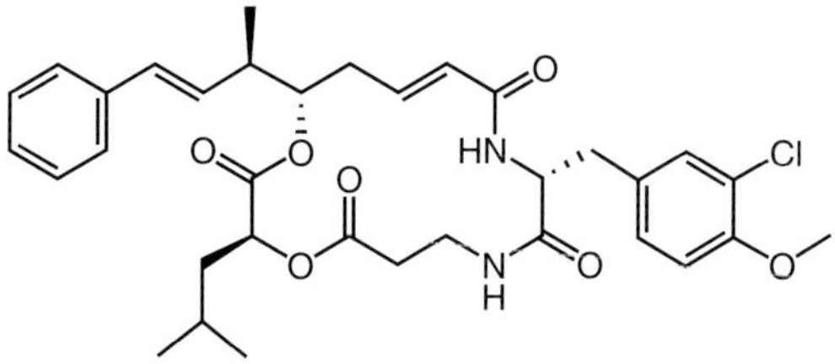

1085 (cryptophycin-29)

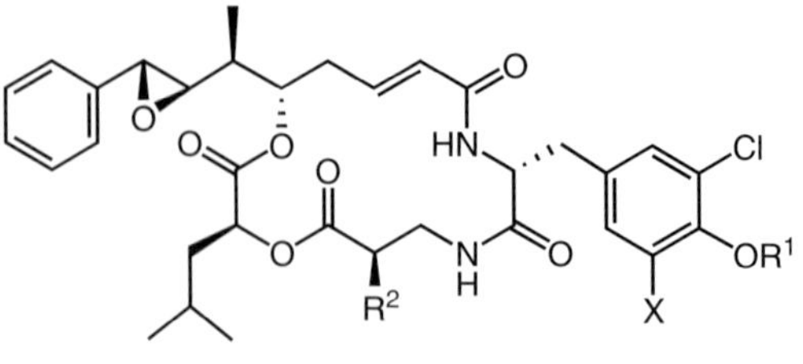

1086 R¹ = Me, R² = X = H (cryptophycin-21)
1087 R¹ = R² = X = H (cryptophycin-176)
1088 R¹ = R² = Me, X = H (cryptophycin-40)
1089 R¹ = Me, R² = H, X = Cl (cryptophycin-326)

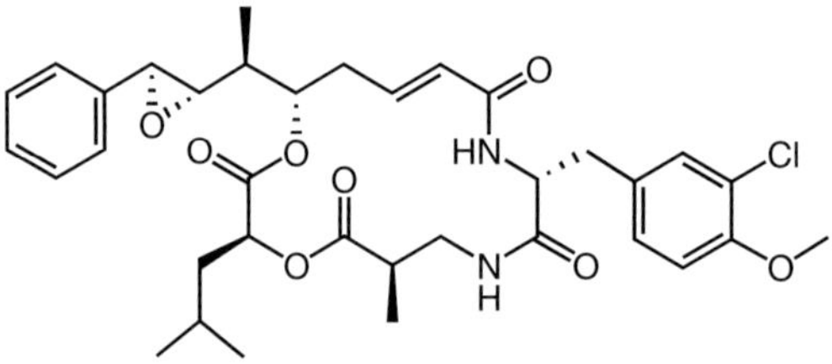

1090 (cryptophycin-38)

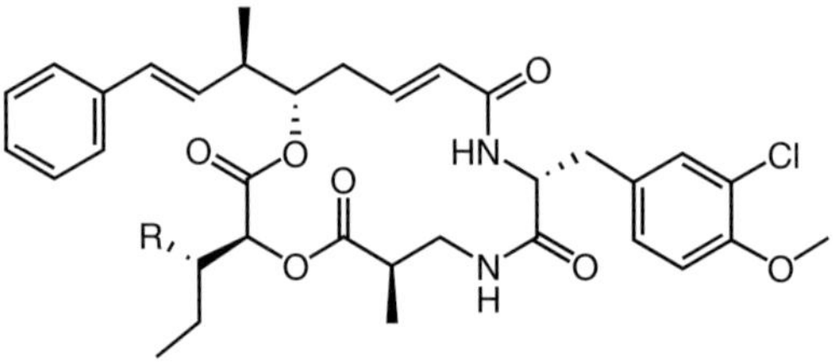

1091 R = Me (cryptophycin-18)
1092 R = H (cryptophycin-49)

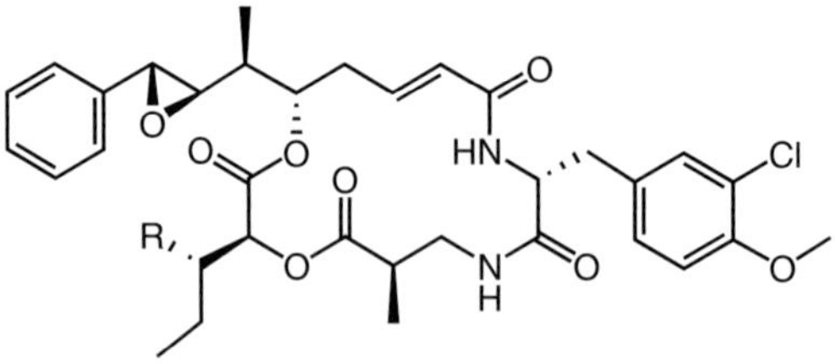

1093 R = H (cryptophycin-50)
1094 R = Me (cryptophycin-54)

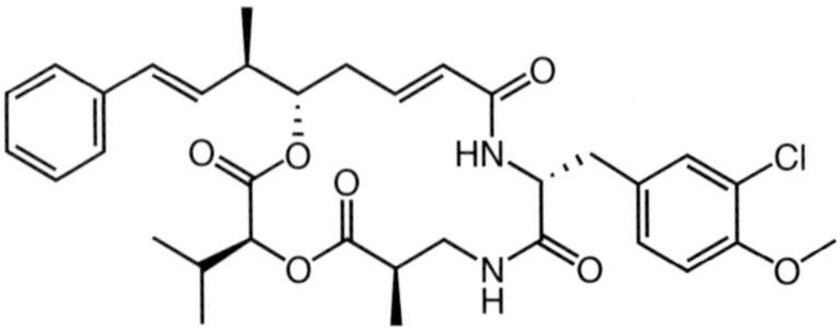

1095 (cryptophycin-19)

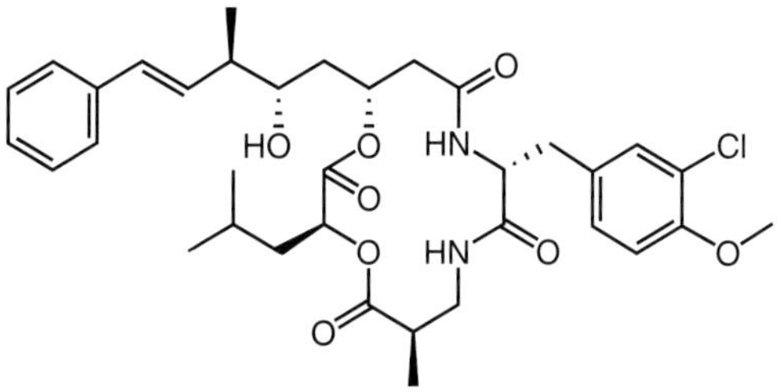

3.12 Amino Acids and Peptides

1096 (cryptophycin-26)

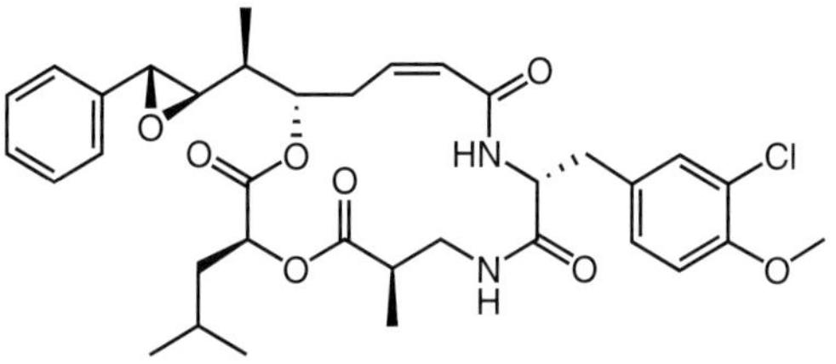

1097 (cryptophycin-327)

The striking cytotoxic activity of some cryptophycins, comparable or superior to taxol and vincristine in some cell lines, has generated intense synthetic interest (*1107, 1109, 1114, 1115*). A synthetic analog, cryptophycin-52 (the C6 *gem*-dimethyl analog of cryptophycin 1 (= cryptophycin A); not shown), has been selected for clinical evaluation (*1116–1119*). Unfortunately, neurotoxicity may preclude further development of cryptophycin-52 (*1118, 1119*). The simple fungal metabolite (–)-xylariamide A (**1098**), which resembles the "right-half" of the cryptophycins, is produced by the terrestrial fungus *Xylaria* sp. (*1120*). The structure of **1098** has been confirmed by synthesis (*1121*).

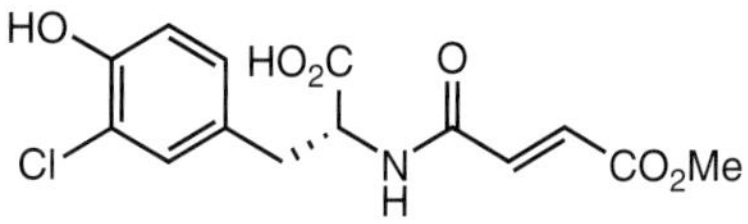

1098 (xylariamide A)

Perthamide B (**1099**) is a novel cyclic octapeptide found in a *Theonella* sp. sponge near Perth, Australia (*1122*). A collection of the Japanese sponge *Halichondria cylindrata* has afforded halicylindramides A–E (**1100–1103**), and the absolute configurations are shown (*1123, 1124*). Each compound features a 4-bromophenylalanine residue.

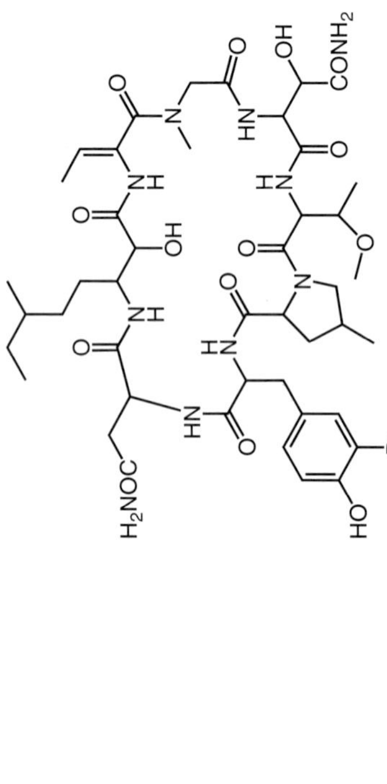

1099 (perthamide B)

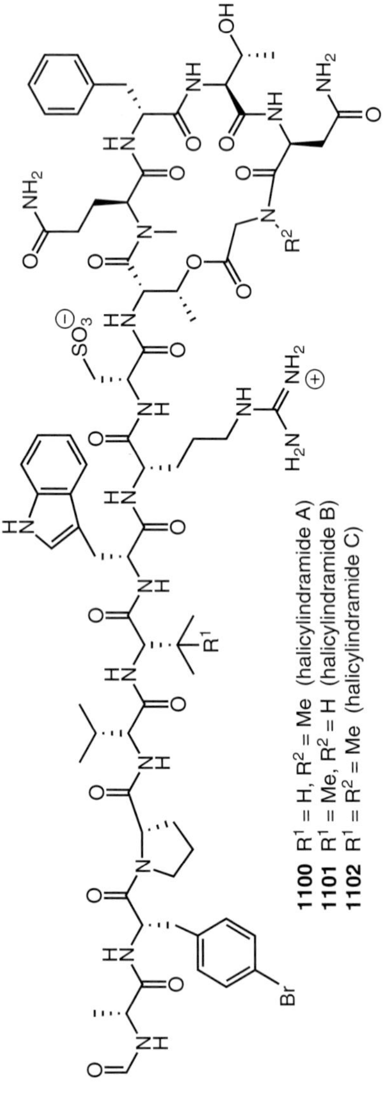

1100 R^1 = H, R^2 = Me (halicylindramide A)
1101 R^1 = Me, R^2 = H (halicylindramide B)
1102 R^1 = R^2 = Me (halicylindramide C)

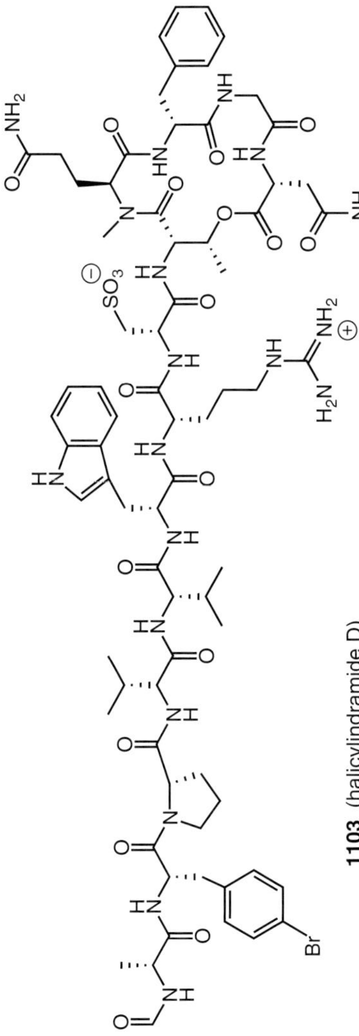

3.12 Amino Acids and Peptides

1103 (halicylindramide D)

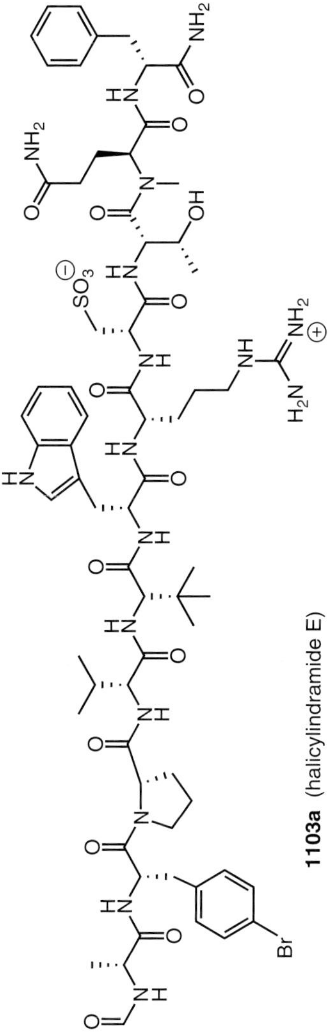

1103a (halicylindramide E)

Theonellamides A–E (**1104–1108**), which are analogs of the previously isolated theonellamide F (*1*), were characterized from the Japanese sponge *Theonella* sp. (*1125*). The related theopalauamide (= P951) (**1109**) was isolated from *Theonella swinhoei* found in both Palau and Mozambique (*1126, 1127*). It should be noted that a minor structural correction (misplaced methyl group) has been reported for the previously known theonegramide (*1128*).

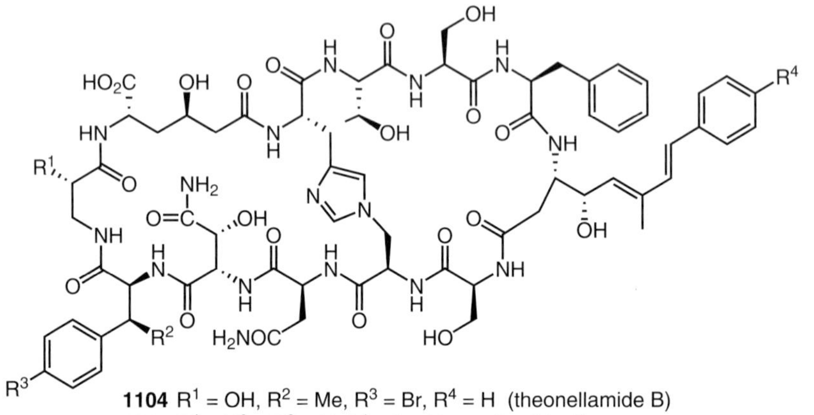

1104 R^1 = OH, R^2 = Me, R^3 = Br, R^4 = H (theonellamide B)
1105 R^1 = R^2 = R^3 = H, R^4 = Br (theonellamide C)

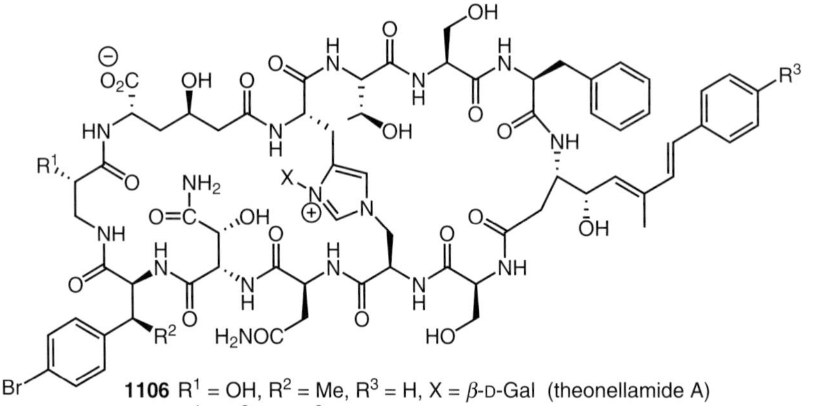

1106 R^1 = OH, R^2 = Me, R^3 = H, X = β-D-Gal (theonellamide A)
1107 R^1 = R^2 = H, R^3 = Br, X = β-L-Ara (theonellamide D)
1108 R^1 = R^2 = H, R^3 = Br, X = β-D-Gal (theonellamide E)

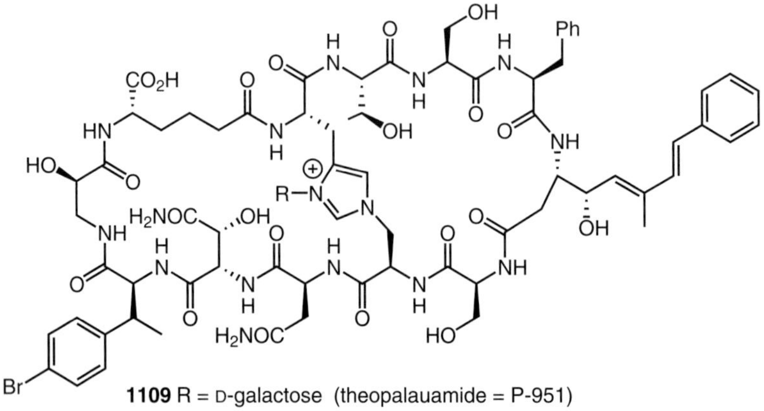

1109 R = D-galactose (theopalauamide = P-951)

3.12 Amino Acids and Peptides 167

Studies of *Theonella swinhoei* sponges from Papua New Guinea and Indonesia revealed the presence of the chloroleucine-containing cyclolithistide A (**1110**) (*1129*). Anabaenopeptilide 90B (**1111**) is a cyclic depsipeptide produced by the cyanobacterium *Anabaena* strain 90 (*1130*). An Indonesian collection of the sponge *Sidonops microspinosa* yielded microspinosamide (**1112**), a novel HIV-inhibitory cyclic depsipeptide (*1131*). Structurally similar to cyclolithistide A (**1110**) are phoriospongins A (**1113**) and B (**1114**) found in the Australian sponges *Phoriospongia* sp. and *Callyspongia bilamellata* (*1132*).

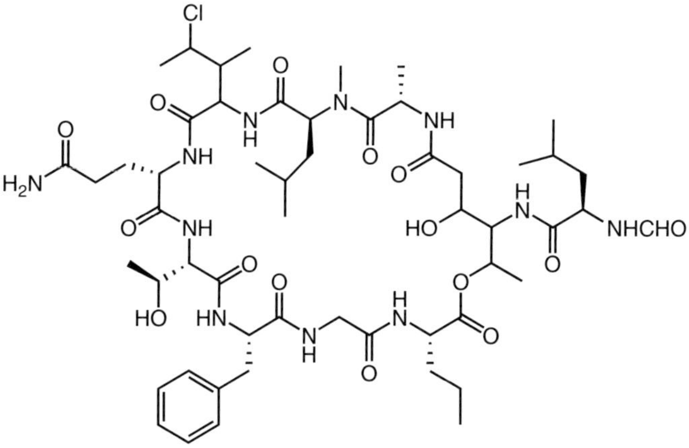

1110 (cyclolithistide A)

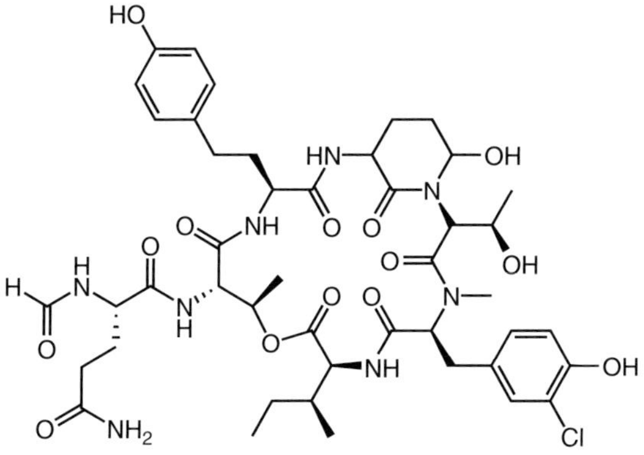

1111 (anabaenopeptilide 90B)

1112 (microspinosamide)

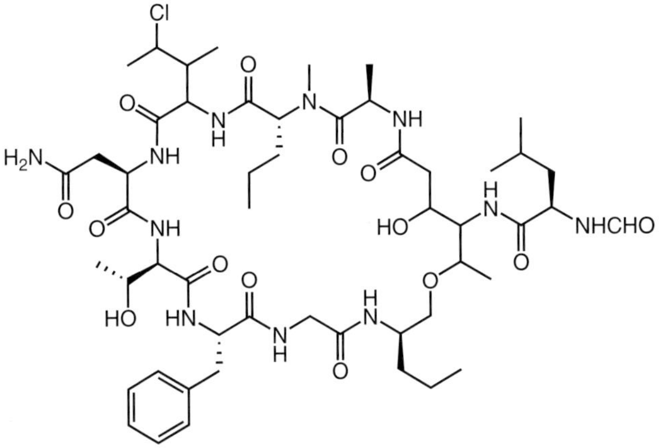

3.12 Amino Acids and Peptides

1113 (phoriospongin A)

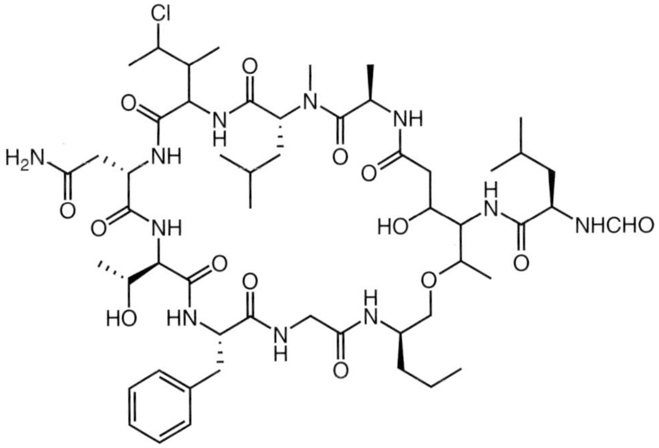

1114 (phoriospongin B)

The prevalent *Microcystis aeruginosa* has afforded the plasmin inhibitors micropeptins 478-A (**1115**) and 478-B (**1116**) (*1133*). The terrestrial cyanobacterium *Scytonema hofmanni* PCC 7110 gives rise to scyptolins A (**1117**) and B (**1118**), which contain the 3-chloro-*N*-methyltyrosine residue (*1134*). The binding of **1117** to pancreatic elastase has been determined by X-ray crystallography (*1135*). Myriastramide B (**1119**) was isolated from the Philippine sponge *Myriastra clavosa* and features a novel chlorinated ether moiety (*1136*).

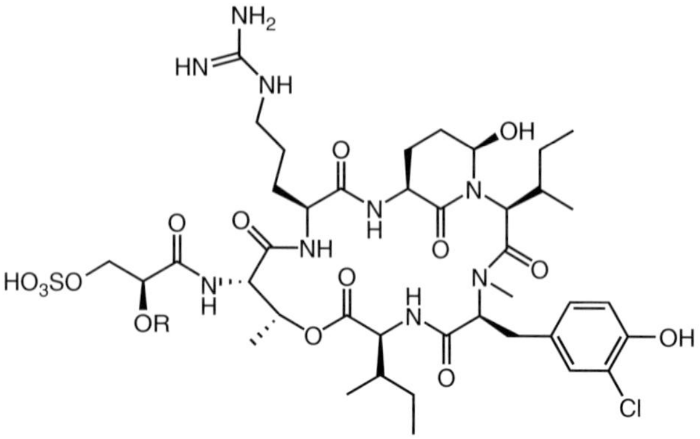

1115 R = H (micropeptin 478-A)
1116 R = SO₃H (micropeptin 478-B)

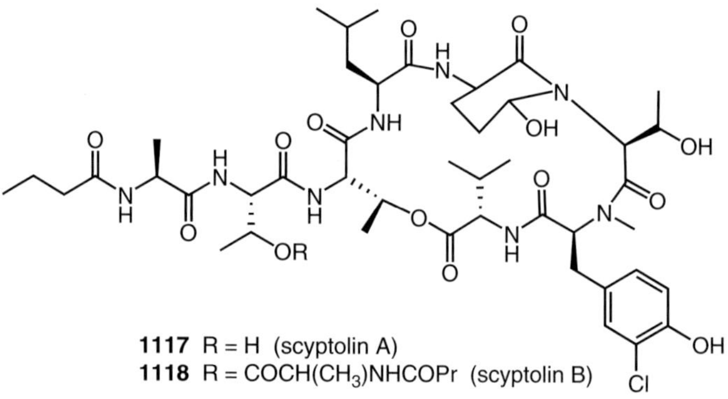

1117 R = H (scyptolin A)
1118 R = COCH(CH₃)NHCOPr (scyptolin B)

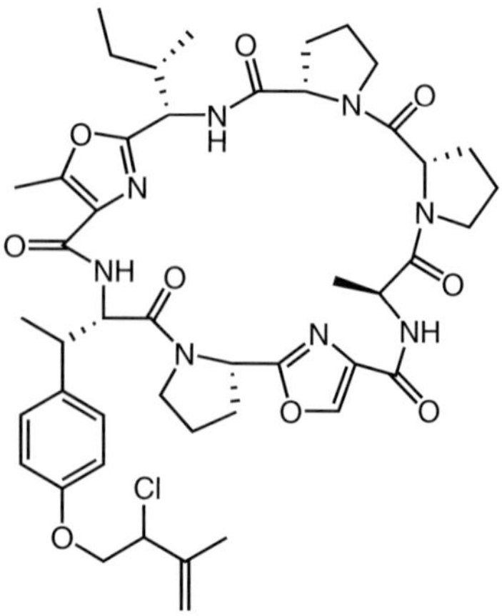

1119 (myriastramide B)

3.12 Amino Acids and Peptides

The novel peptide takaokamycin (**1120**) was isolated from a *Streptomyces* sp. culture (*1136*), but it subsequently became clear that this antibiotic was identical to hormaomycin (= **1120**) (*1137*), which was isolated independently from *Streptomyces griseoflavus* (*1138, 1139*). The structure of hormaomycin was later confirmed (*1140*), and it has been synthesized (*1141*). Two chlorinated actinomycins, Z_3 (**1121**) and Z_5 (**1122**), were isolated from cultures of *Streptomyces fradiae* and are more active than the non-chlorinated actinomycin D (*1142*). *Streptomyces iakyrus* provides the new actinomycin G_2 (**1123**), which is the major component of this family of actinomycins and the most biologically active (*1143*). Although no new examples of syringomycins and syringtoxins have been described, the synthesis and study of synthetic analogs reveal the importance of 4-chlorothreonine residues for antibiotic activity (*1144*). The biological chlorination of threonine in syringomycin involves a non-haem halogenase, SyrB2 (*1145, 1146*), and a mechanism is presented in Chap. 4 (Biohalogenation).

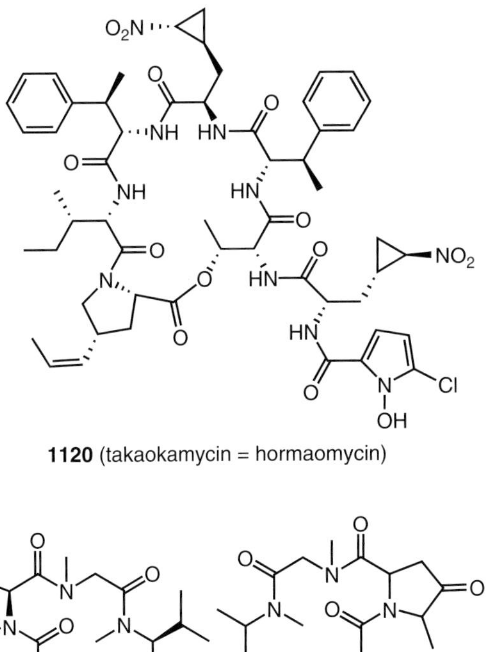

1120 (takaokamycin = hormaomycin)

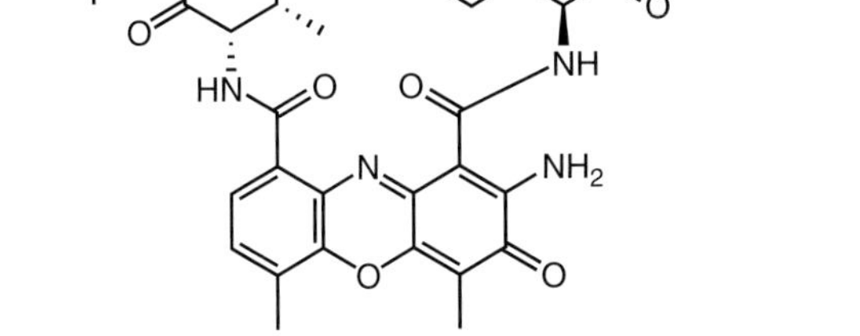

1121 R = OH (actinomycin Z_3)
1122 R = H (actinomycin Z_5)

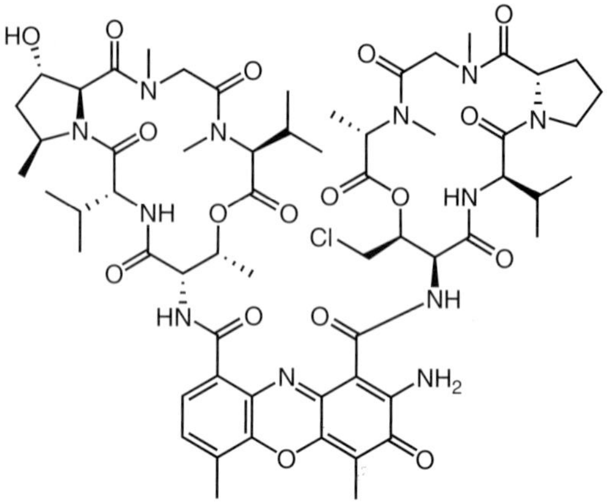

1123 (actinomycin G$_2$)

An exciting development in the area of halogenated natural products is the isolation and characterization of salinosporamide A (**1124**) from a new genus of marine bacteria, *Salinospora* (*1147*) (subsequently renamed *Salinispora*). Salinosporamide A displays potent cytotoxicity against a number of human cell lines (HCT-116 colon, NCI-H226 non-small cell lung, SK-MEL-28 melanoma, and others) and is now in phase I clinical trials (as NPI-0052) for the treatment of cancer (*1148*). Additional studies of *Salinispora tropica* yielded salinosporamides C (**1125**) (*1149*), F (**1126**), I (**1127**), and J (**1128**) (*1150*). Several degradation products and non-chlorinated salinosporamides were also isolated in both studies. Cytotoxicity data indicate that the chloroethyl substituent is crucial for activity (*1149–1151*). The mechanism of action of salinosporamide A seems to involve nucleophilic addition of a threonine in the 20S proteasome to the lactone carbonyl group followed by attack on the chloroethyl group leading to a cyclic ether and irreversible binding (*1152*). As anticipated, synthetic interest in the salinosporamides has been intense (*1153, 2655*).

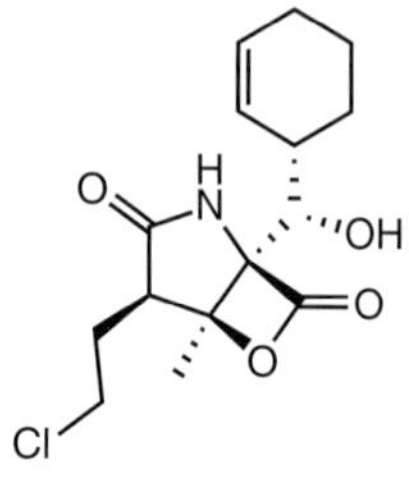

1124 (salinosporamide A) 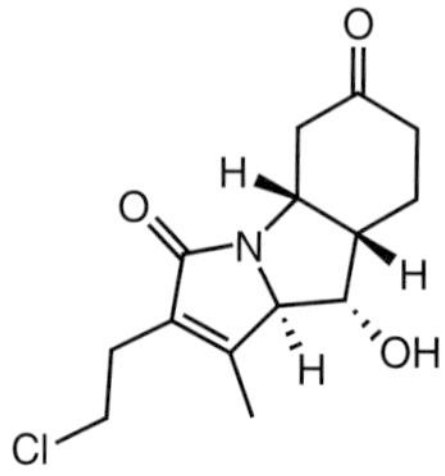   **1125** (salinosporamide C) 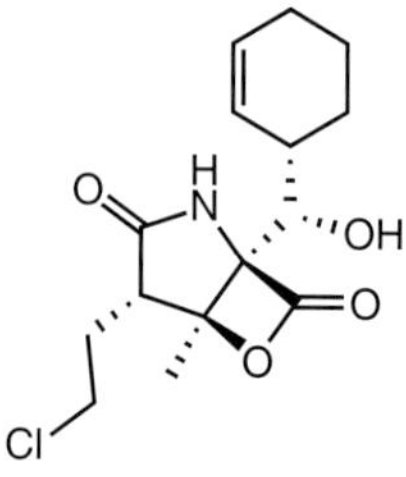   **1126** (salinosporamide F)

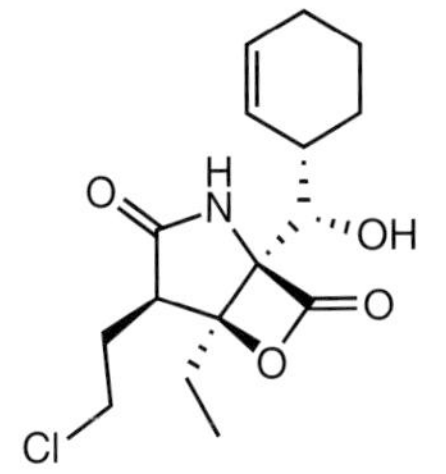

3.12 Amino Acids and Peptides

1127 (salinosporamide I) 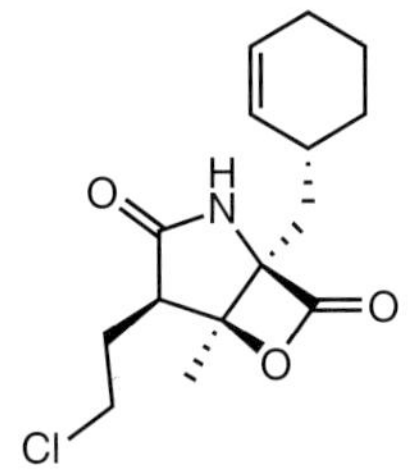   **1128** (salinosporamide J)

A marine-derived fungus, *Trichoderma virens*, has yielded the novel chlorinated trichodermamide B (**1129**), which displays significant cytotoxicity towards HCT-116 human colon carcinoma (*1154*). Interestingly, trichodermamide A is devoid of both chlorine and biological activity in all of the assays tested. A *Helicomyces* fungal strain (No. 19353) produces the gluconeogenesis inhibitors FR225659 (**1130**), FR225656 (**1131**), and the related **1132–1134** (*1155–1158*).

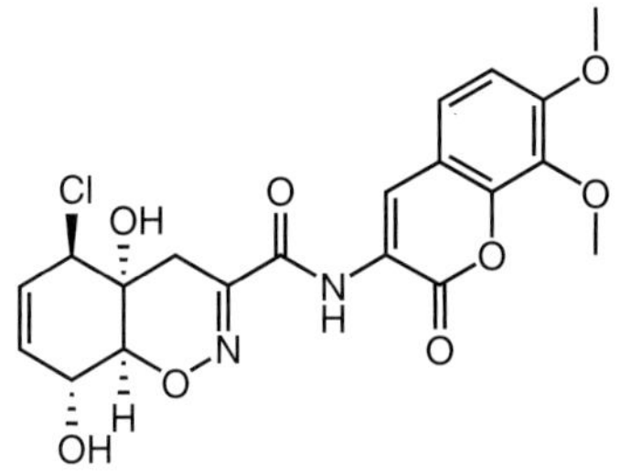

1129 (trichodermamide B)

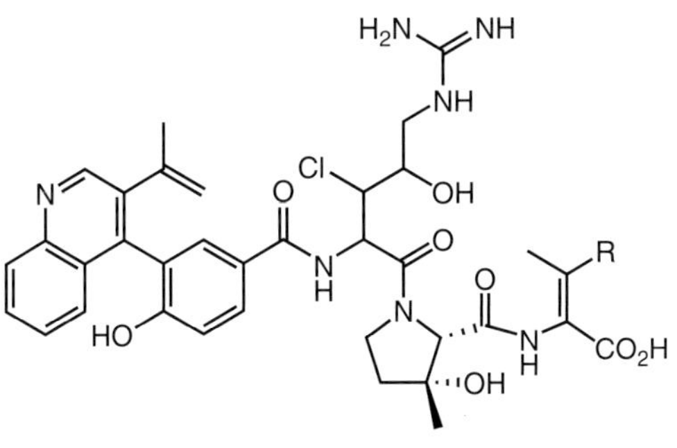

1130 R = Me (FR225659)
1131 R = Et (FR225656)

1132 R¹ = OH, R² = H, R³ = Me
1133 R¹ = OMe, R² = OH, R³ = Et
1134 R¹ = OMe, R² = OH, R³ = Me

3.13 Alkaloids

This section and the section on alkaloids in the first survey (*1*) are artificially small since many halogenated alkaloids are presented in the sections on pyrroles, indoles, carbolines, tyrosines, and other nitrogen heterocycles. It might be noted that the very large number of brominated alkaloids that are obviously tyrosine-derived are now included in Sect. 3.22.3 (Tyrosines).

The previously known novel frog alkaloid epibatidine (*1*), continues to be of pharmacological (*1159, 1160*) and synthetic interest (*1161*), including the synthesis of many analogs (*1162*). The new clolimalongine (**1135**), a hasubanan type alkaloid related to the previously described chlorine-containing acutumine (*1*), was characterized from *Limacia oblonga* (*1163*). Two new epimers of known alkaloids are dauricumine (**1136**) and dauricumidine (**1137**) isolated from plant cultures of *Menispermum dauricum* (*1164*). The known acutumine (*1*), which is an epimer of **1136**, incorporates ¹⁴C-labelled L-tyrosine (*1165*), and it along with **1136** and **1137** incorporate ³⁶Cl when this radiolabel is fed to the roots (*1164*). An extraction of the Brazilian plant *Senecio selloi* yielded 18-hydroxyjaconine (**1138**) (*1166*). Plants of this genus are infamous for their armament of poisonous pyrrolizidine alkaloids; five chlorinated examples were cited earlier (*1*).

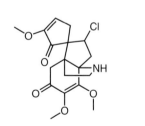

1135 (clolimalongine)   1136 R = Me (dauricumine)
                          1137 R = H (dauricumidine)

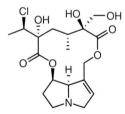

1138 (18-hydroxyjaconine)

3.13 Alkaloids

Petrosamine B (**1139**), an isomer of the known petrosamine (*1*), was characterized from the Australian sponge *Oceanapia* sp. (*1167*). The structure of the tetracyclic brominated alkaloid pantherinine (*1*) has been confirmed by total synthesis (*1168*). The novel aporphine alkaloids romucosine F (**1140**) from *Annona purpurea* (*1169*) and romucosine B (**1141**) from *Rollinia mucosa* (*1170*) have antiplatelet aggregation activity. Interestingly, synthetic halogenated boldine (*1171*) and protoberberine alkaloids (*1172*) show enhanced biological activity (monoamine receptor selectivity and cytotoxicity, respectively) over their non-halogenated counterparts. For boldine the order of activity is I > Br > Cl, and it is suggested that both increased lipophilicity and interaction with aromatic residues are involved in the mode of action (*1171*).

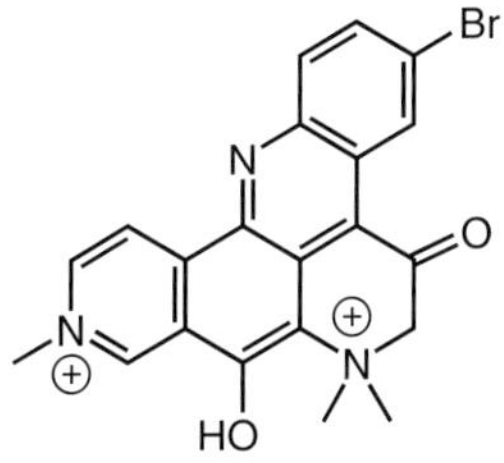

1139 (petrosamine B)

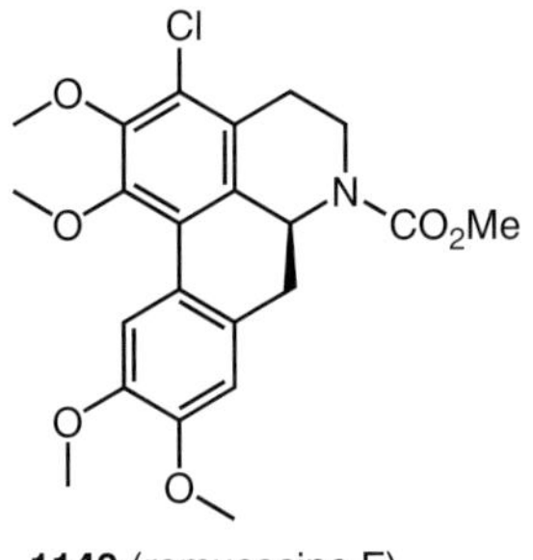

1140 (romucosine F) 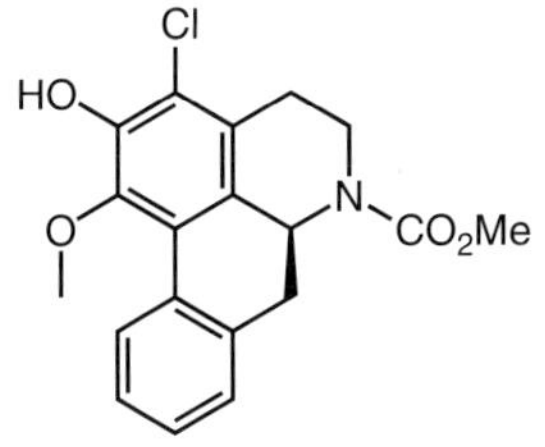   **1141** (romucosine B)

A Papua New Guinea sponge, *Pseudoceratina* sp., contains the unusual alkaloids ceratamines A (**1142**) and B (**1143**), and a biogenesis involving histidine and tyrosine is proposed (*1173*). The fermentation broth from *Aspergillus fischeri* var. *thermomutatus* has yielded CJ-12662 (**1144**) and UK-88051 (**1145**) (*1174*). The former metabolite was confirmed by X ray spectroscopy and partial synthesis. A marine-derived *Streptomyces* sp. produces the novel pyrrolizidine 5-chlorobohemamine C (**1146**), which was shown not to be an isolation artifact (*1175*). The Chinese medicinal plant *Huperzia serrata* has furnished 2-chlorohyperzine E (**1147**) (*1176*).

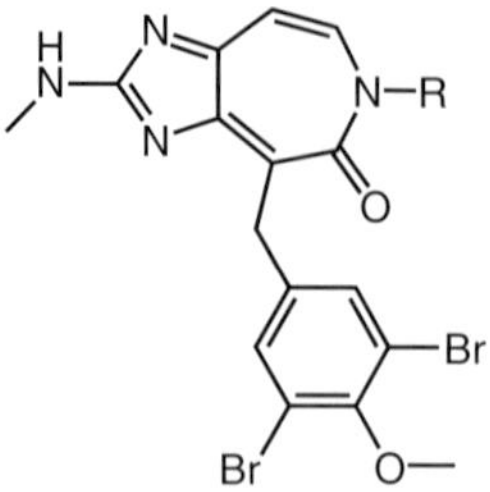

1142 R = Me (ceratamine A)
1143 R = H (ceratamine B)

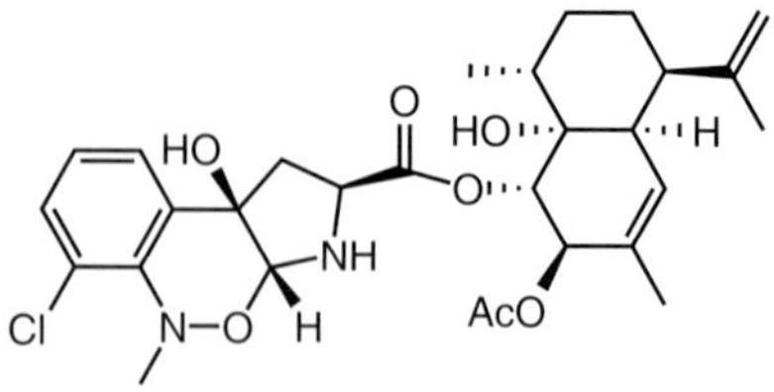

1144 (CJ-12662)

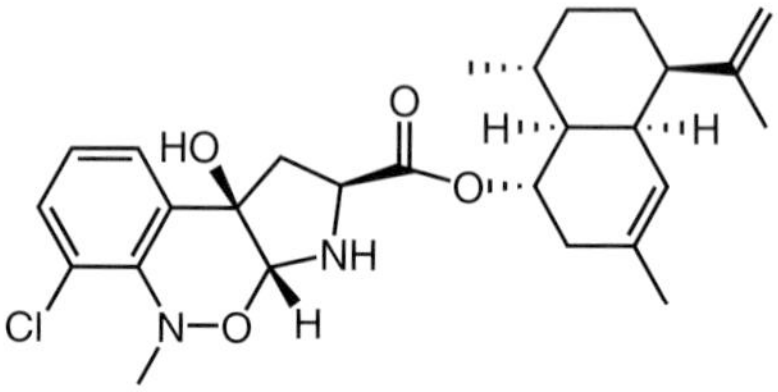

1145 (UK-88051)

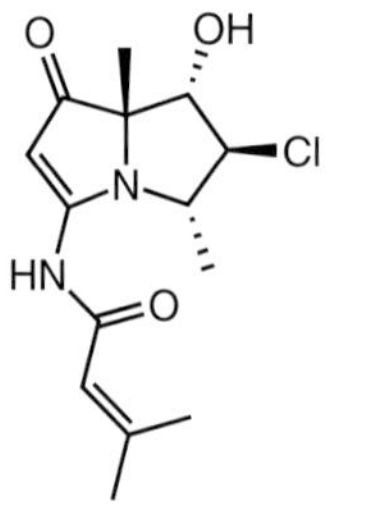

1146 (5-chlorobohemamine C)

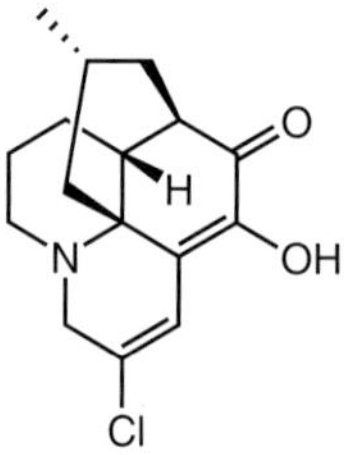

1147 (2-chlorohyperzine E)

Three novel related marine alkaloids, halichlorine (**1148**) from the sponge *Halichondria okadai* (*1177, 1178*) and pinnaic acid (**1149**) and tauropinnaic acid (**1150**) from the bivalve *Pinna muricata* (*1179*), have been the objects of much synthetic interest in view of their pronounced biological activity (inhibition of the vascular cell adhesion molecule-1) (*1180*). Synthesis of these alkaloids led to both revision and confirmation of the original structures (*1181, 1182*). The syntheses of the previously known chlorine-containing cylindricines have been reviewed (*1183*).

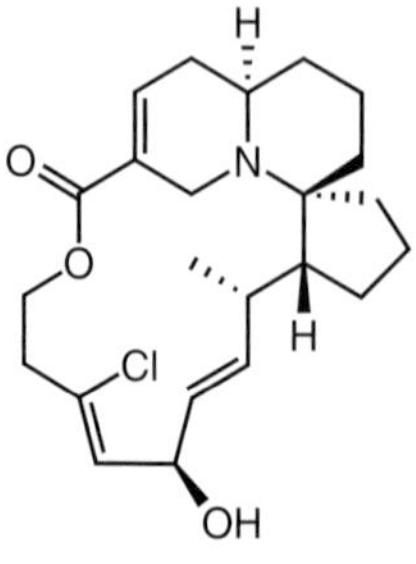

1148 (halichlorine)

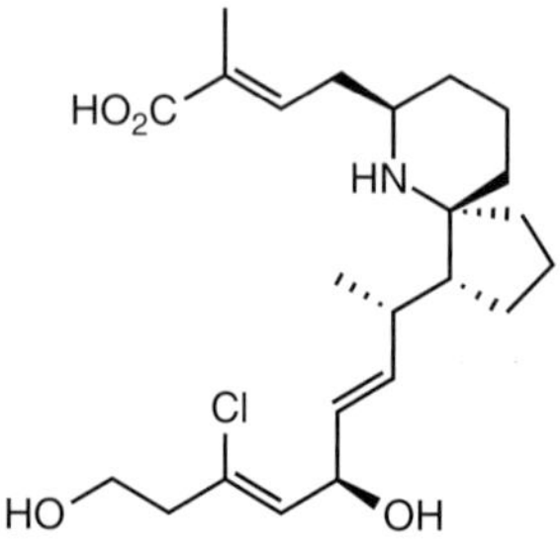

1149 R = OH (pinnaic acid)
1150 R = NHCH$_2$CH$_2$SO$_3$H (tauropinnaic acid)

3.14 Heterocycles

3.14.1 Pyrroles

The abundance of proline (and therefore pyrroles) in the biosphere, coupled with the enormous reactivity of pyrrole towards electrophilic substitution (i.e., halogenation), portends an abundance of naturally occurring halogenated pyrroles (*2668*). The first survey documented more than 70 such compounds (*1*), including the prototypical pyoluteorin and pyrrolnitrin, which continue to receive attention. The latter metabolite is produced by both *Pseudomonas* bacteria and, as recently discovered, *Enterobacter agglomerans* (*1184*). Pyrrolnitrin is active against a wide range of bacteria and fungi (*1184*), such as *Mycobacterium tuberculosis* (*1185*). The biological activity of pyrrolnitrin involves blocking the electron-transport system of the respiratory chain (*1186*), and this metabolite seems to play an important role in the biocontrol of pathogenic fungi (*1187*). Pyrrolnitrin is also used clinically to treat dermatophytosis (*1188*).

The biosynthesis of pyrrolnitrin has been extensively investigated for 40 years and the current state of affairs is summarized in Chap. 4 (Biohalogenation) (*1189*). Noteworthy is that the chlorination of tryptophan by tryptophan 7-halogenase is the first step in the sequence (*1190, 1191*). Although less well studied, the biosyntheses of other members of this family (pyoluteorin, dioxapyrrolomycin, and pentabromopseudilin) will be briefly presented in Chap. 4 (Biohalogenation). A culture of *Streptomyces fumanus* has yielded the new pyrrolomycin G (**1151**), H (**1152**), I (**1153**), and J (**1154**) (*1192*). The absolute configuration of G and H was determined as (*S*). A compound missed in the earlier survey is pentachloropseudilin (**1155**) (*1193*), the chlorine analog of the known pentabromopseudilin (*1*).

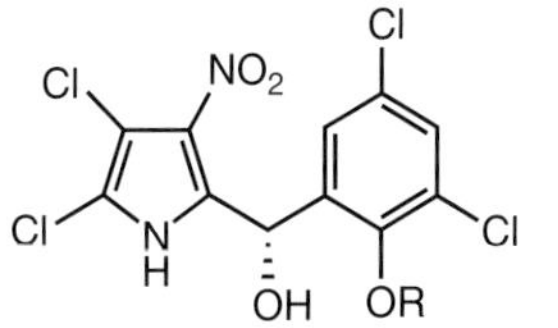

1151 R = H   (pyrrolomycin G)
1152 R = Me (pyrrolomycin H)

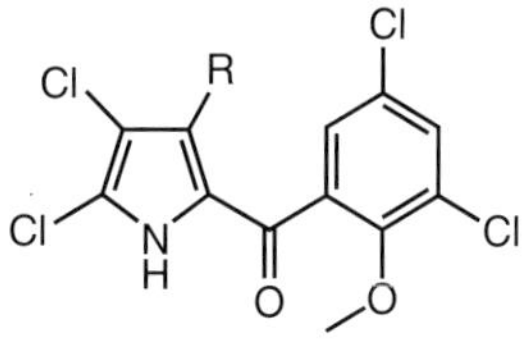

1153 R = H   (pyrrolomycin I)
1154 R = Cl (pyrrolomycin J)

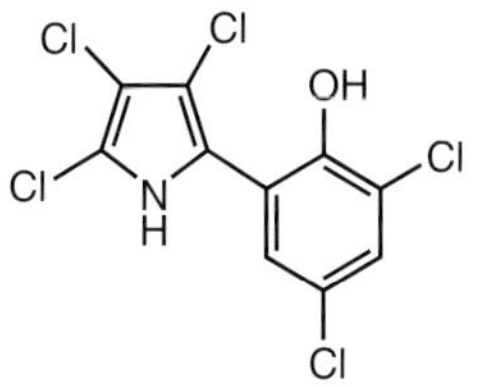

1155 (pentachloropseudilin)

A *Streptomyces* sp. has yielded the novel TAN-876 A (**1156**) and TAN-876 B (**1157**), where the former compound is a unique example of the chromeno[2,3-*b*] pyrrole ring system (*1194*). These metabolites exhibit strong antibacterial activity against Gram-positive and Gram-negative bacteria and fungi. The pyralomicins 1a (**1158**), 1c (**1159**), 1d (**1160**), 1b (**1161**), 2a (**1162**), 2c (**1163**), and 2b (**1164**) were isolated from cultures of *Microtetraspora* (formerly *Actinomadura*) *spiralis* (*1195–1198*). Cultures of *Streptomyces armeniacus* produce streptopyrrole (**1165**) (*1199*), and *Streptomyces rimosus* has afforded an additional six streptopyrroles **1166–1171** (*1200*).

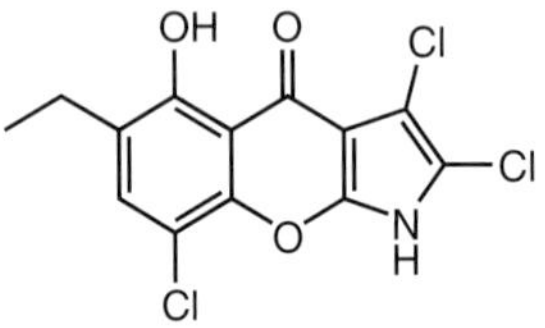

1156 (TAN-876 A)

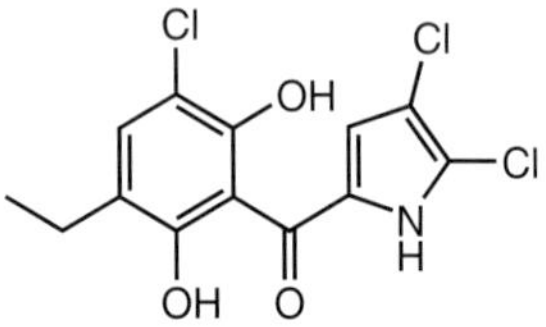

1157 (TAN-876 B)

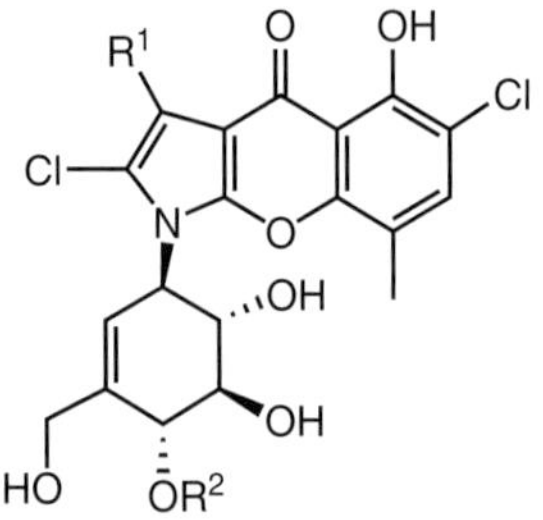

1158 R^1 = H, R^2 = Me (pyralomicin 1a)
1159 R^1 = R^2 = H (pyralomicin 1c)
1160 R^1 = Cl, R^2 = H (pyralomicin 1d)

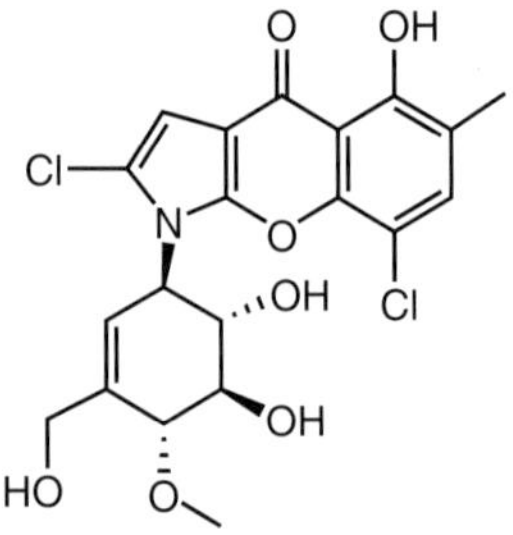

1161 (pyralomicin 1b)

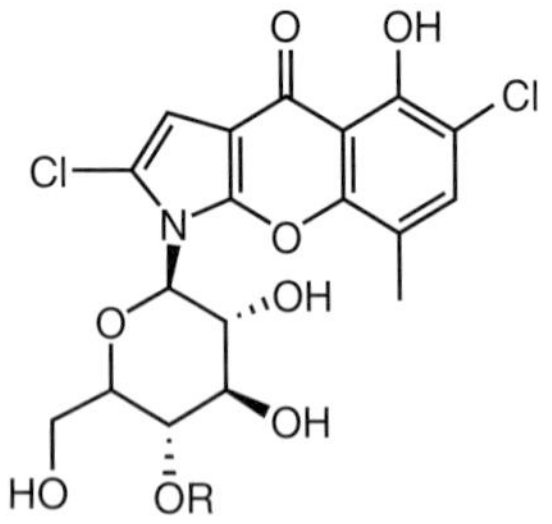

1162 R = Me (pyralomicin 2a)
1163 R = H (pyralomicin 2c)

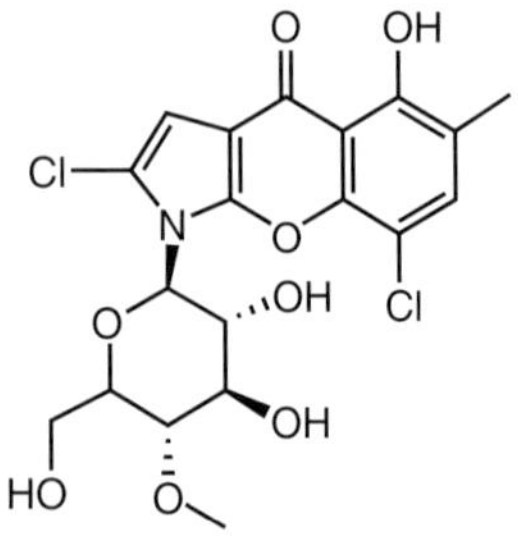

1164 (pyralomicin 2b)

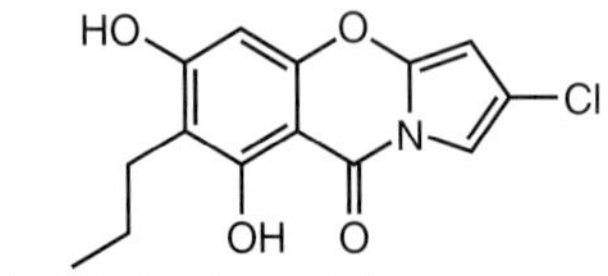

1165 (streptopyrrole)

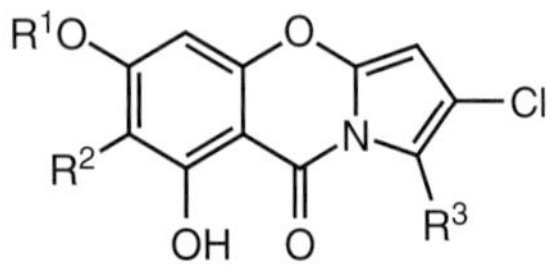

1166 R^1 = R^3 = H, R^2 = Et
1167 R^1 = H, R^2 = Pr, R^3 = Cl
1168 R^1 = Me, R^2 = Pr, R^3 = H
1169 R^1 = R^3 = H, R^2 = Bu
1170 R^1 = Me, R^2 = Et, R^3 = H
1171 R^1 = Me, R^2 = Pr, R^3 = Cl

3.14 Heterocycles

The sodium sulfamate salt **1172** of the previously known marine acorn worm metabolite 2,3,4-tribromopyrrole was isolated from *Saccoglossus kowalevskii* (*1201*). This compound may serve as the stable, non-toxic, and non-volatile precursor to 2,3,4-tribromopyrrole, which is probably the actual deterrent to predators. The Tasmanian bryozoan *Bugula dentata* has afforded the new bipyrroles tambjamines G-J (**1173–1176**) (*1202*). Several new analogs of the polybrominated 2,2'-bipyrrole that was described earlier (*1*) have been isolated from seabird eggs (gulls, petrel, albatross, puffin, bald eagle) (*1203*). As the isolated amounts were too small to be identified, total synthesis verified the two major compounds as **1177** and **1178** (*1204*). Further confirmation was provided by X-ray crystallography (*1205*). Subsequent studies reveal that **1177**, **1178** and other analogs (**1179–1184**) are ubiquitous in the marine environment, being present in zooplankton, fish, seabirds, seal, porpoise, dolphin, and whale (*1206–1209*). Porpoise and whale blubber also contain the less heavily halogenated **1181–1184**, which have not yet been fully characterized (*1209*). Given their structural similarity to polychlorinated biphenyls (PCBs), it is not surprising that these polyhalogenated bipyrroles bioaccumulate in the food chain.

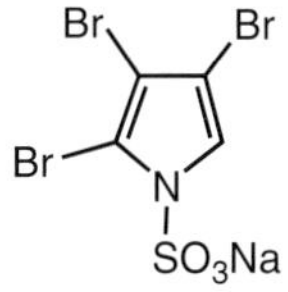

1172

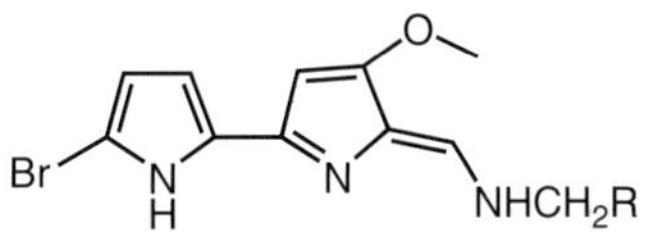

1173 R = CH_3 (tambjamine G)
1174 R = CH_2CH_3 (tambjamine H)
1175 R = $CH(CH_3)_2$ (tambjamine I)
1176 R = $CH(CH_3)CH_2CH_3$ (tambjamine J)

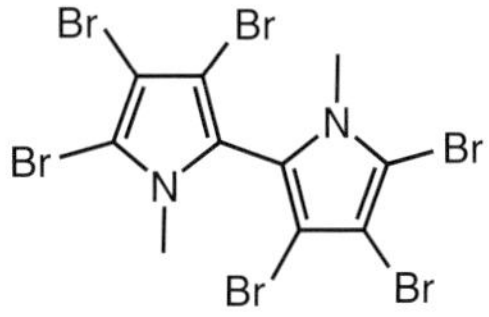

1177

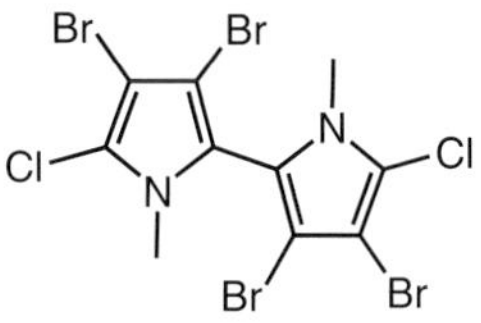

1178

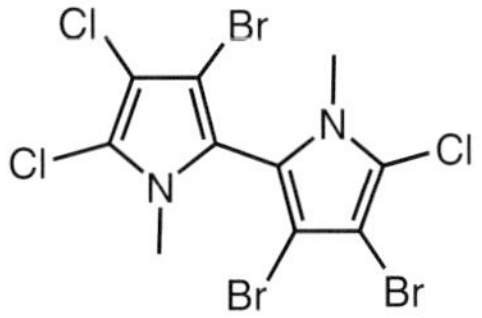

1179 (or isomer)

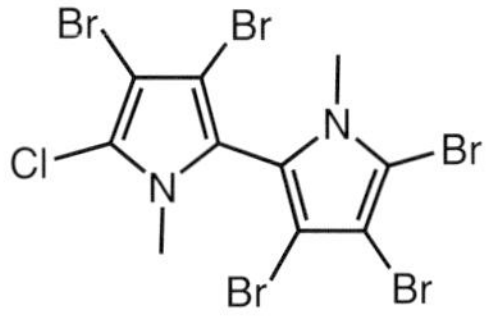

1180 (or isomer)

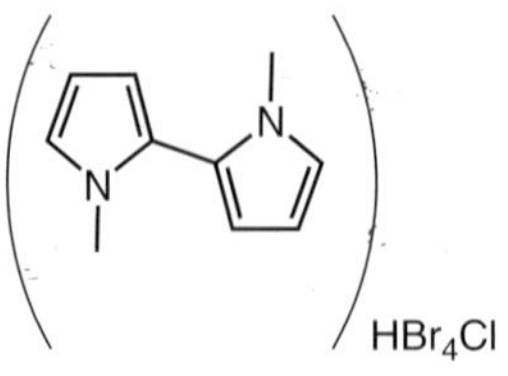

(1181) HBr₄Cl

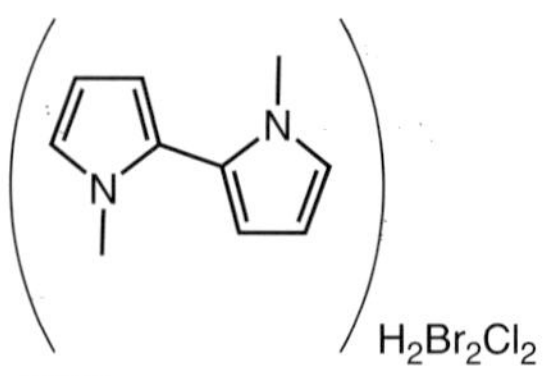

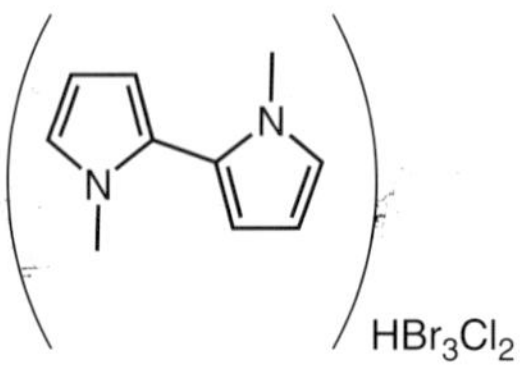

(1182) H₂Br₂Cl₂

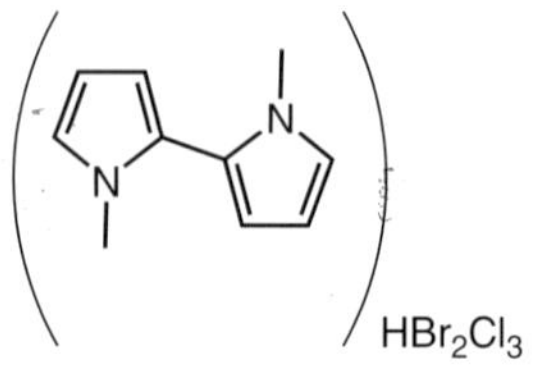

(1183) HBr₃Cl₂

(1184) HBr₂Cl₃

An investigation of marine samples (penguin and skua eggs, in whale and seal blubber, deep sea fish, and human milk) from Antarctica and Africa uncovered "Q1", a novel heptachloro-1'-methyl-1,2'-bipyrrole (**1185**) (*1210–1213*), which was confirmed by total synthesis (*1214*). The discovery of Q1 in human milk samples from women of the Faeroe Islands who consume whale blubber indicates that Q1 is the first natural bioaccumulative compound to move up the food chain to humans, in levels up to 230 µg kg^{-1} (*1213*). Examination of worldwide marine samples shows that Q1 is widespread in the environment, particularly in marine mammals and birds, but also in a green turtle and a python in Australia (*1215*). Q1 is also present in the eggs of five different Norwegian predatory birds (white-tailed sea eagle, golden eagle, merlin, osprey, and goshawk) (*479*). The environmental occurrence and structure of Q1 are reviewed (*1216, 1217*). The highest concentration measured to date is 14 mg kg^{-1} in an Australian dolphin (*1217*). As might be expected for a nonplanar PCB-like compound, Q1 has low biological activity and only modest affinity as an AHR ligand (*1218*). Three brominated derivatives (**1186–1188**) of Q1 have been isolated from the blubber of nine New England marine mammals (species of dolphin, porpoise, whale, seal, and a squid) (*1219, 1220*). Concentrations of **1186–1188** in a squid (*Loligo pealei*) were measured up to 2.7 mg kg^{-1} (*1219*). Marine samples from Australia and Antarctica (melon-headed whale, pygmy sperm whale, common dolphin, bottlenose dolphin) contain 22 new polyhalogenated Q1 analogs in addition to **1186** and **1187** (*1221*). These are **1189–1210**, with varying numbers of bromines and chlorines and different isomers. An important complementary finding is that Q1, several brominated analogs (**1186–1188**), and the new **1211** were discovered in archived whale oil collected in 1921 from the last voyage of the whaling ship *Charles W. Morgan*, a sample of oil that predates large-scale industrial manufacture of organohalogen compounds (*1222*). This oil was found not to contain PCBs, DDT, or DDE. Equally noteworthy is that

3.14 Heterocycles

radiocarbon analysis of both Q1 (**1185**) and **1178** reveals that these compounds originate naturally and are not anthropogenic (petroleum-derived) (*1223*). It is highly likely that Q1 and brominated Q1s have been isolated and misidentified in the past. A case in point could be the "halogenated naphthols" isolated from a white-sided dolphin (*1224*).

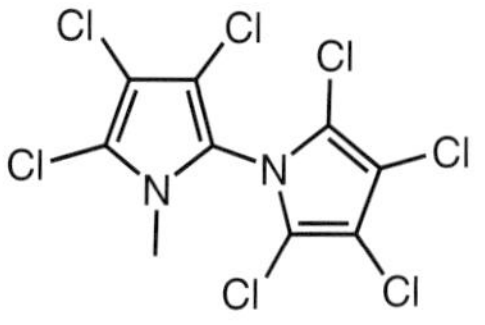

1185 (Q1) — octachloro bipyrrole	**1186** / Br_6Cl
1187 / Br_7	**1188** / HBr_5Cl
1189-1192 / $BrCl_6$	**1193-1197** / Br_2Cl_5
1198-1202 / Br_3Cl_4	**1203-1206** / Br_4Cl_3
1207-1210 / Br_5Cl_2	**1211** / HBr_6

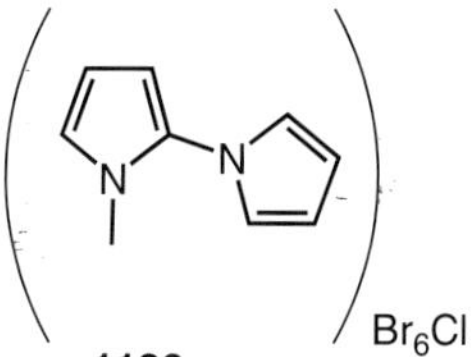

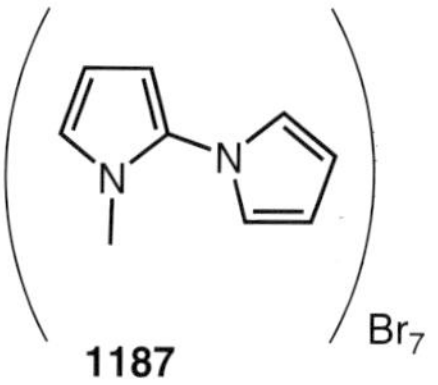

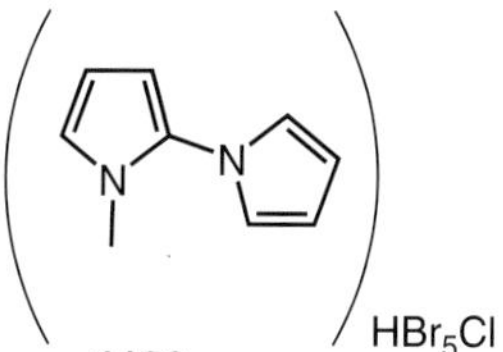

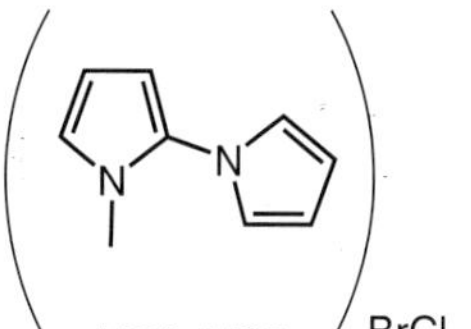

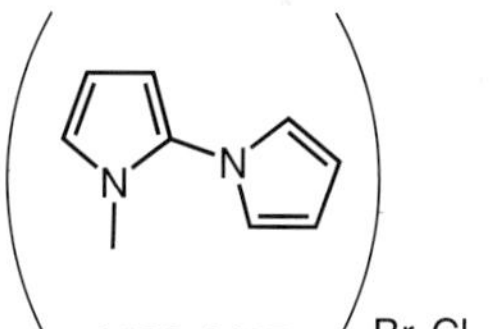

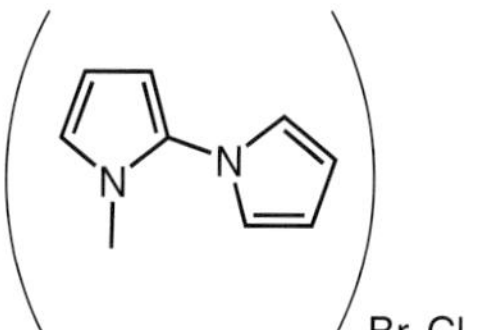

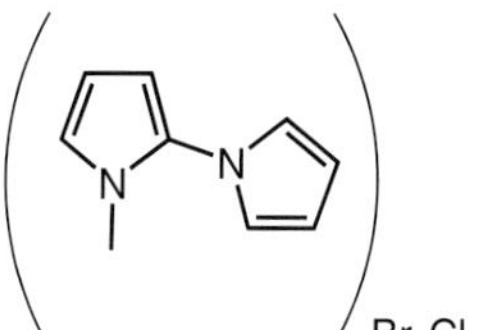

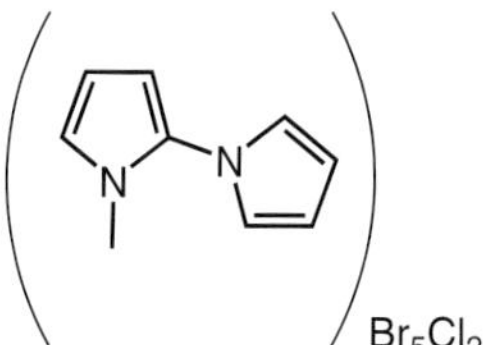

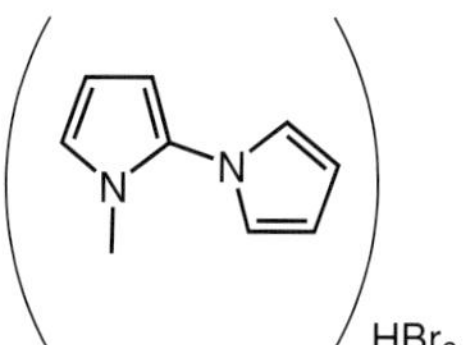

A *Streptomyces* strain isolated from Brazilian (*Maytenus aquifolia*) and South African (*Putterlickia retrospinosa, Putterlickia verrucosa*) plants has furnished celastramycin A (**1212**) (*1225*). A new isomer of the known rumbrin (*1*) was isolated from an Australian soil ascomycete, *Gymnoascus reessii*, and named (12E)-isorumbrin (**1213**) (*1226*). Somewhat earlier, the three related auxarconjugatins A, B, and (3'Z)-A (**1214–1216**) were characterized from an Arizona soil microorganism *Auxarthron conjugatum* (*1227*). A basidiomycete fungus from a New Zealand forest, *Chamonixia pachydermis*, produces pachydermin (**1217**) (*1228*).

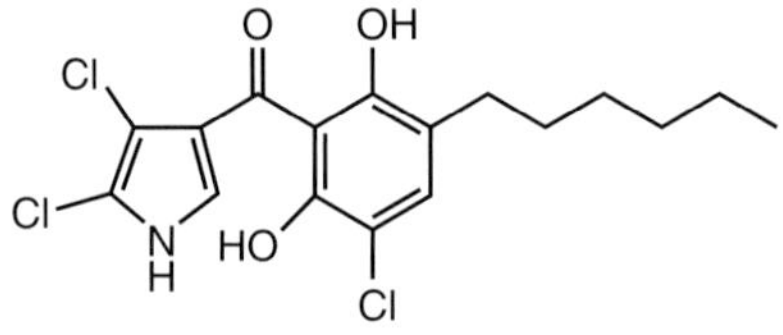

1212 (celastramycin A)

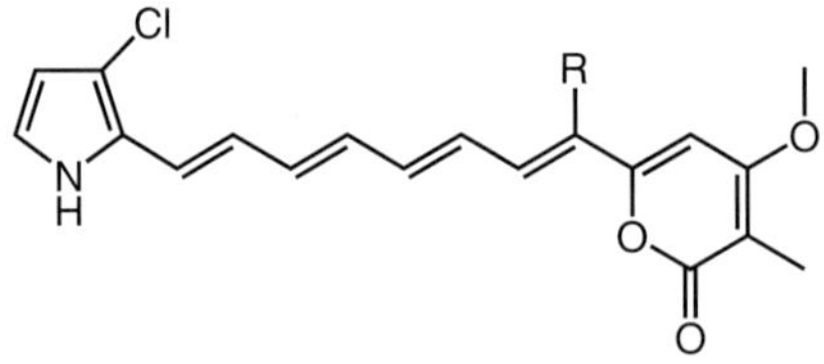

1213 R = Me ((12E)-isorumbrin)
1214 R = H (auxarconjugatin A)

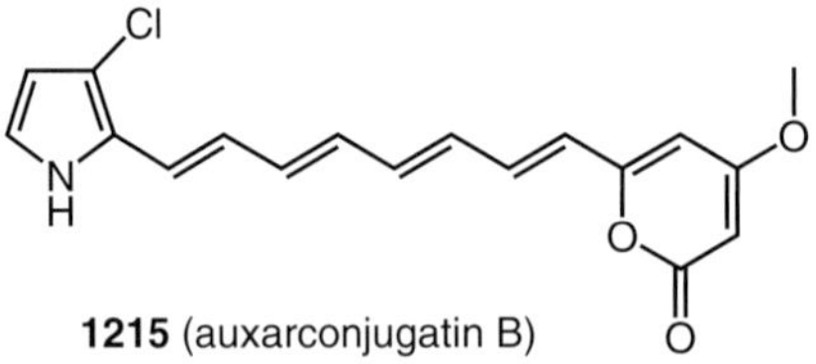

1215 (auxarconjugatin B)

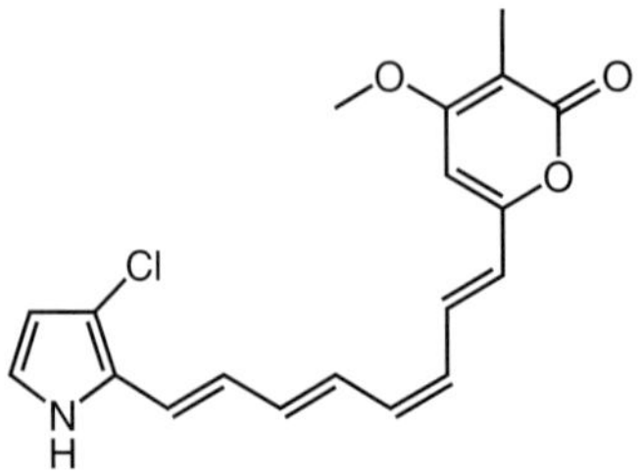

1216 (3'-*cis*-auxarconjugatin A)

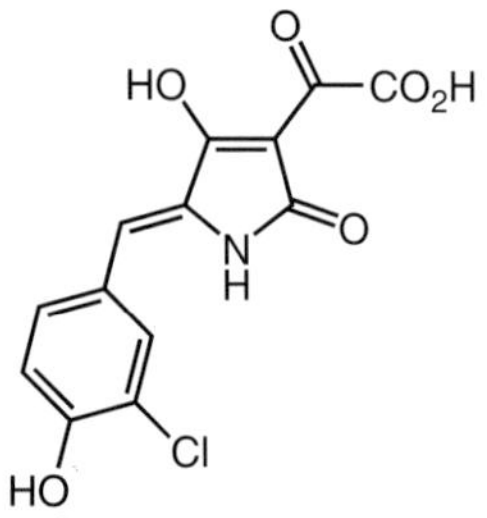

1217 (pachydermin)

3.14 Heterocycles

Decatromicins A (**1218**) and B (**1219**) are produced by an *Actinomadura* sp. and are active against Gram-positive bacteria including methicillin-resistant *Staphylococcus aureus* (*1229, 1230*). These compounds are closely related to pyrrolosporin A (**1220**) from *Micromonospora* sp. (*1231, 1232*). The ascidian *Polycitor africanus* from Madagascar has afforded the new polycitone B (**1221**) (*1233*), which is related to the known polycitone A (*1*), a potent inhibitor of retroviral reverse transcriptases and cellular DNA polymerases (*1234*). The known polycitrin B was synthesized for the first time (*1235*).

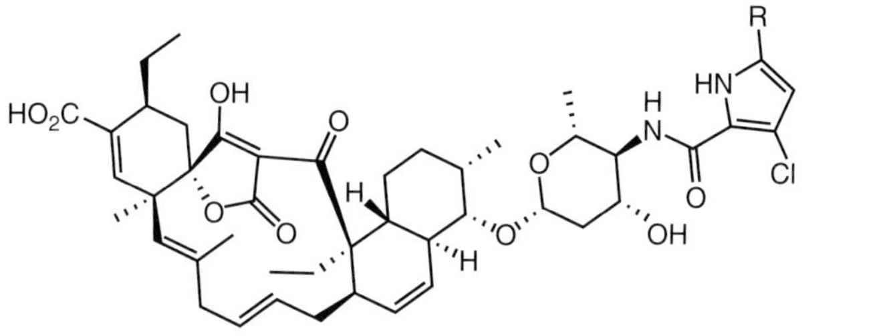

1218 R = H (decatromicin A)
1219 R = Cl (decatromicin B)

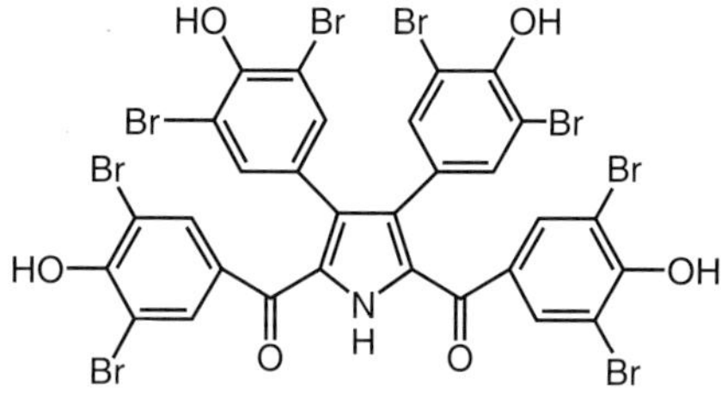

1221 (polycitone B)

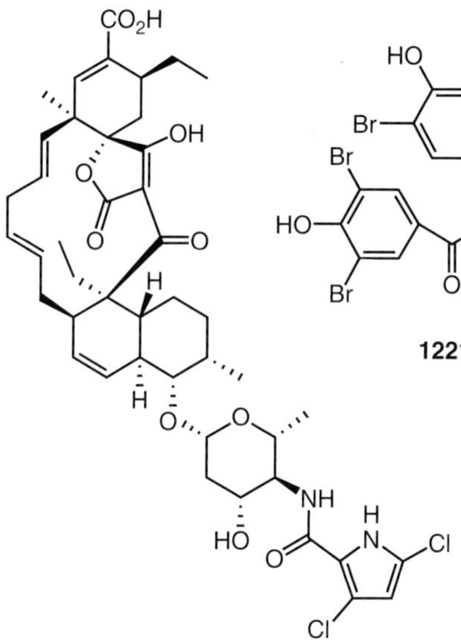

1220 (pyrrolosporin A)

Most of the known natural brominated pyrrole alkaloids are found in sponges, and several new examples were isolated since the first survey (*1*). Reviews are available that discuss the occurrence and syntheses of these metabolites (*1236–1238*). The Papua New Guinea sponge *Agelas nakamurai* has yielded the new simple pyrroles **1222** and **1223** (*1239*). The dibromo analog (**1224**) of **1223** along with enantiomeric lactams **1225/1226**, which were separated by chiral HPLC, and racemic longamide, previously isolated as the (+)-isomer (vide infra), were all characterized from the Japanese sponge *Homaxinella* sp. (*1240*). Interestingly, lactam ester (+)-(*S*)-**1225** was isolated from *Agelas ceylonica* collected in India (*1241*), and (+)-(*S*)-longamide (**1227**) was first described from *Agelas longissima* (*1242*). The (racemic) ethyl ester of methyl esters **1225/1226** is known as hanishin (**1228/1229**), and was isolated from the Red Sea (Hanish Islands) sponge *Acanthella carteri*, along with amide **1230** (*1243*), an isomer of **1222**. A debromo analog (**1231**) of **1225** was isolated from the Indian sponge *Axinella tenuidigitata* (*1244*). Longamide B (**1232**) was found in the Caribbean sponge *Agelas dispar* (*1245*). Debromolongamide **1233** was isolated from the Micronesian *Axinella carteri* (*1246*), and the same compound as "mukanadin C" was found in *Agelas nakamurai* (*1247*). Total syntheses of hanishin (**1228**), longamide B (**1232**), and longamide B methyl ester (**1225**) show that the *levorotary* enantiomers of these natural products have the (*S*)-configuration (*1267*).

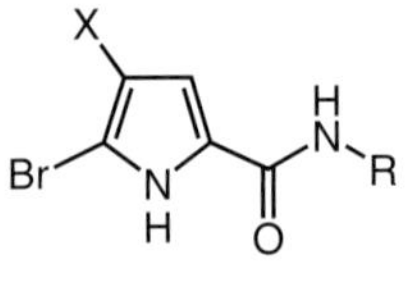

1222 R = X = H
1223 R = CH$_2$OMe, X = H
1224 R = CH$_2$OMe, X = Br

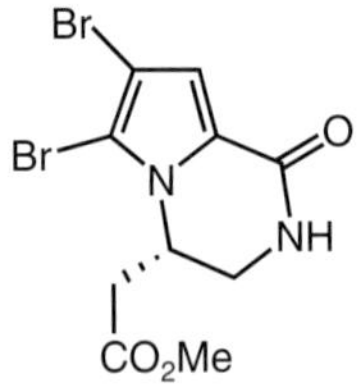

1225

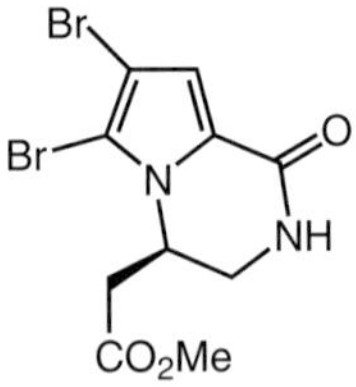

1226

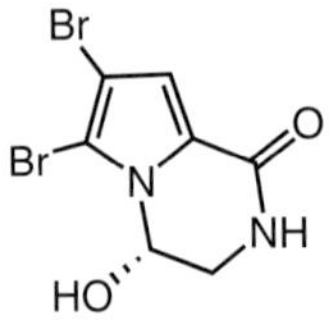

1227 (longamide)

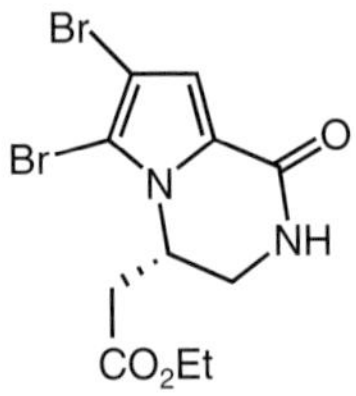

1228 (hanishin)

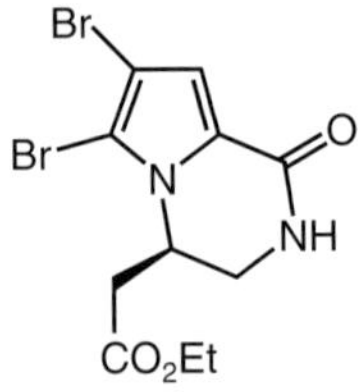

1229 (hanishin)

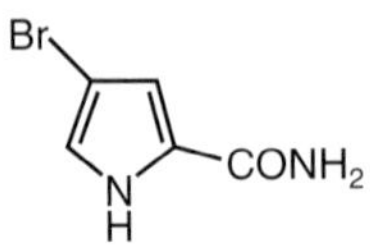

1230

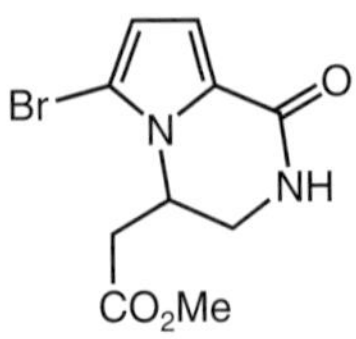

1231

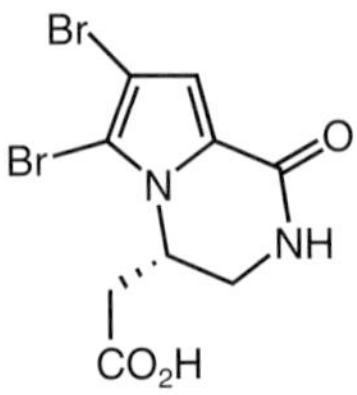

1232 (longamide B)

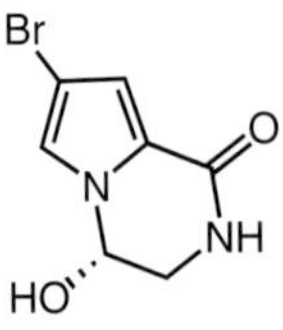

1233 (debromolongamide = mukanadin C)

3.14 Heterocycles

The aforementioned Caribbean collection of *Agelas dispar* affords the novel clathramides C (**1234**) and D (**1235**) (*1245*), which are demethylated examples of the earlier isolated clathramides A (**1236**) and B (**1237**) (*1248*). The zwitterionic alkaloid agelongine (**1238**) was isolated from the Caribbean sponge *Agelas longissima* and displays antiserotonergic activity (*1249*). This sponge and three others from the Caribbean (*Agelas conifera*, *Agelas clathrodes*, and *Agelas dispar*) have afforded dispacamide (**1239**) and its monobromo analog (dispacamide B) **1240** (*1250*). These two metabolites are the first of many new bromopyrroles to be isolated that are related to oroidin. These same four *Agelas* sponges produce dispacamides C (**1241**) and D (**1242**) (*1251*). The latter bromopyrrole was also isolated from *Agelas nakamurai* as "mukanadin A" with the new mukanadin B (**1243**) (*1247*). Along with the latter compound, mukanadin D (**1244**) was found in the Jamaican sponge *Didiscus oxeata* (*1252*).

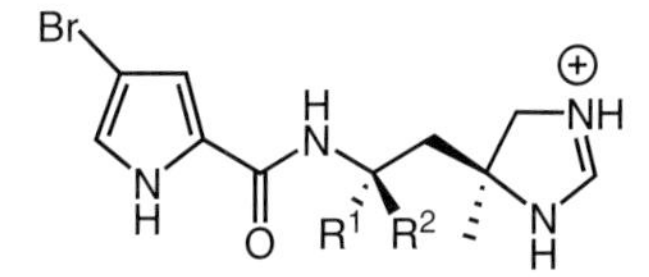

1234 R^1 = H, R^2 = $CO_2^{\ominus}$ (clathramide C)
1235 R^1 = $CO_2^{\ominus}$, R^2 = H (clathramide D)

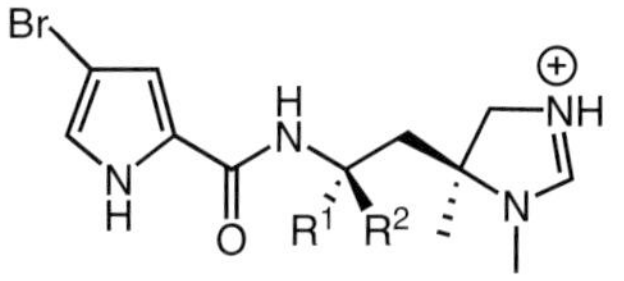

1236 R^1 = H, R^2 = $CO_2^{\ominus}$ (clathramide A)
1237 R^1 = $CO_2^{\ominus}$, R^2 = H (clathramide B)

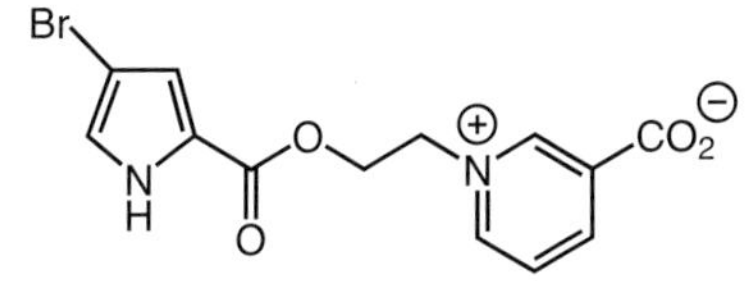

1238 (agelongine)

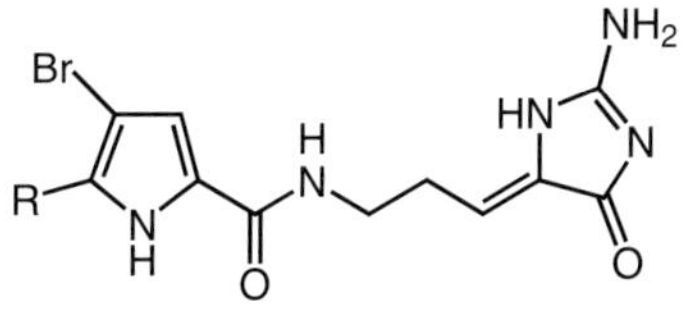

1239 R = Br (dispacamide)
1240 R = H (dispacamide B)

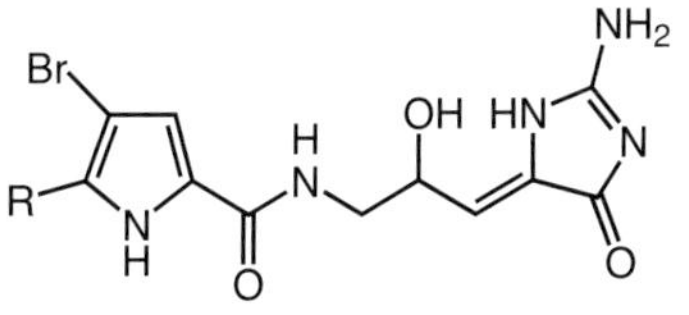

1241 R = Br (dispacamide C)
1242 R = H (dispacamide D)

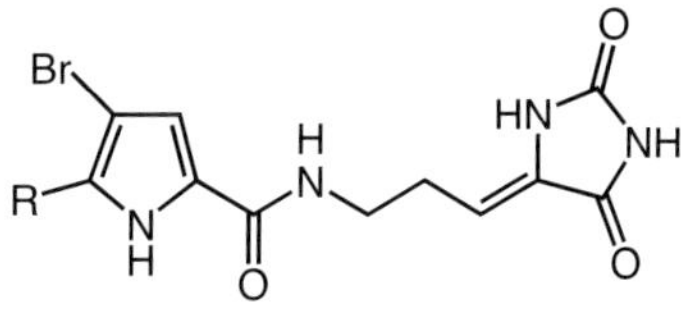

1243 R = H (mukanadin B)
1244 R = Br (mukanadin D)

The antifouling sponge metabolite pseudoceratidine (**1245**) was characterized from the Japanese *Pseudoceratina purpurea* (*1253*). This spermidine derivative has excellent larval settlement and metamorphosis inhibitory activity against the barnacle *Balanus amphitrite* (ED_{50} = 8.0 µg cm^{-3}), and is the first example of an antifouling spermidine derivative. The four tauroacidins A (**1246/1247**) and B (**1248/1249**), with tyrosine kinase inhibitory activity, were isolated from the

Okinawan sponge *Hymeniacidon* sp. (*1254*). The closely related taurodispacamide A (**1250**) is found in the Mediterranean sponge *Agelas oroides* (*1255*). This compound exhibits good antihistaminic activity. A Florida collection of *Agelas wiedenmayeri* contains the bromopyrrole homoarginine **1251**, which may be a biosynthetic precursor to hymenidin and oroidin derivatives (*1256*). The decarboxylated version of **1251**, laughine (**1252**), was isolated from the Dominican sponge *Eurypon laughlini* (*1257*). The arginine (**1253**) and lysine (**1254**) analogs of **1251** were found in the Bahamanian sponge *Stylissa caribica* (*1258*). A total synthesis of **1251** confirms its structure (*1259*). Sventrin (**1255**) (*N*-methyloroidin) is present in the Bahamanian sponge *Agelas sventres*, and is a feeding deterrent to the reef fish *Thalassoma bifasciatum* (*1260*). Four analogs of known bromopyrrole alkaloids (**1256–1259**) were isolated from the Corsican sponge *Axinella verrucosa* (*1261*). Thus, compound **1256** is 9-hydroxymukanadin B, **1257** is 9-methoxydispacamide B, **1258** is 2-debromotaurodispacamide A, and **1259** is the 2-debromo analog of **1224**.

1245 (pseudoceratidine)

1246 R = Br (9*S*-tauroacidin A)
1247 R = Br (9*R*-tauroacidin A)
1248 R = H (9*S*-tauroacidin B)
1249 R = H (9*R*-tauroacidin B)

1250 (taurodispacamide A)

1251 R = CO_2H
1252 R = H (laughine)

1253

1254

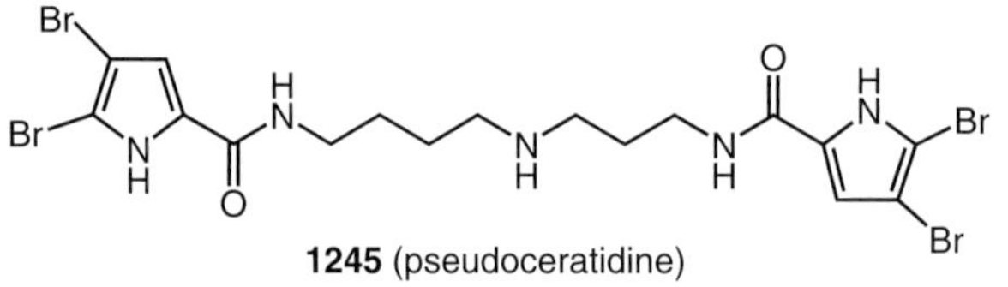

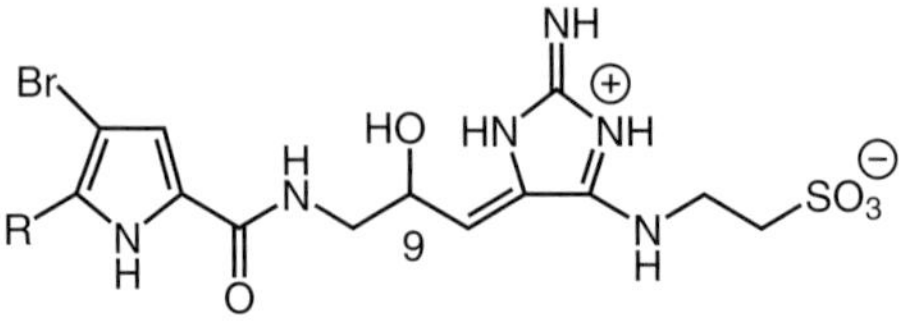

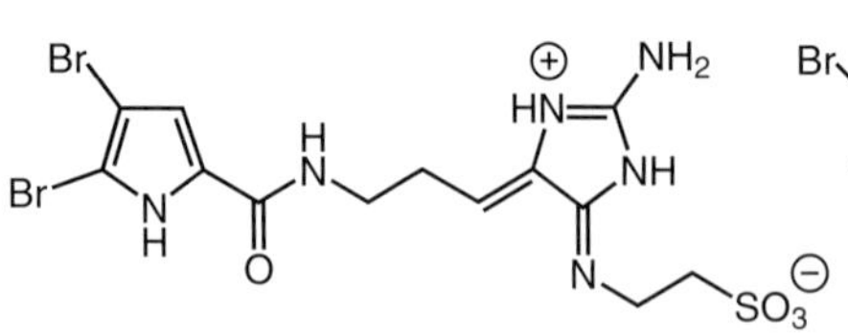

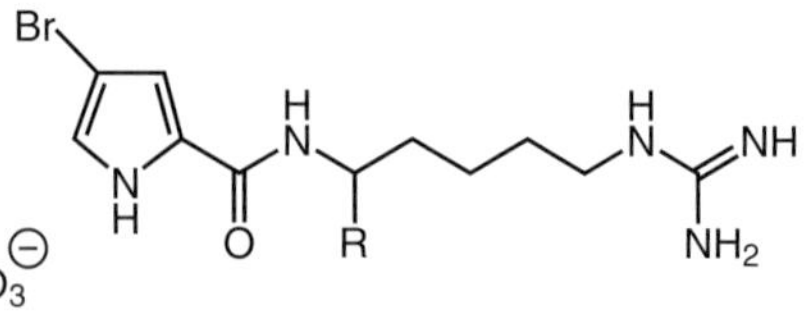

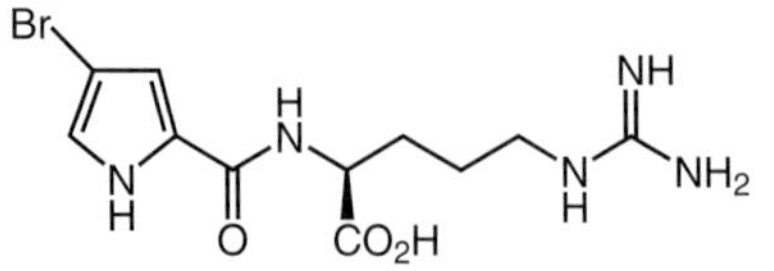

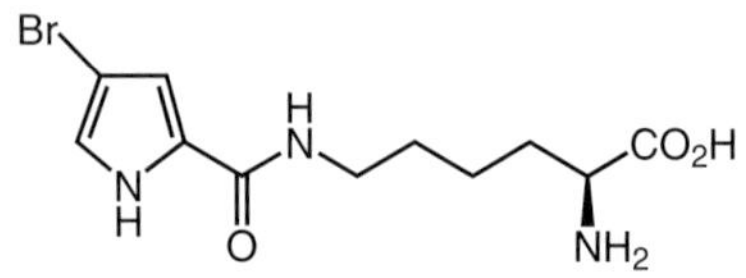

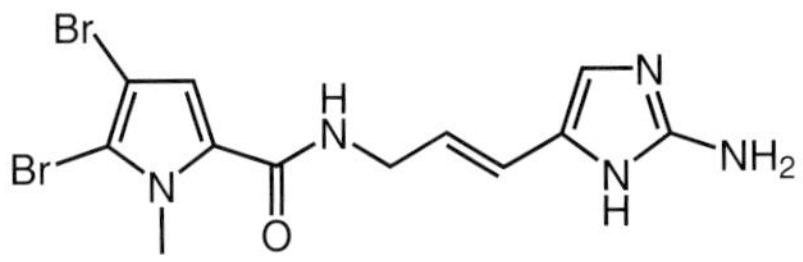

3.14 Heterocycles

1255 (sventrin)

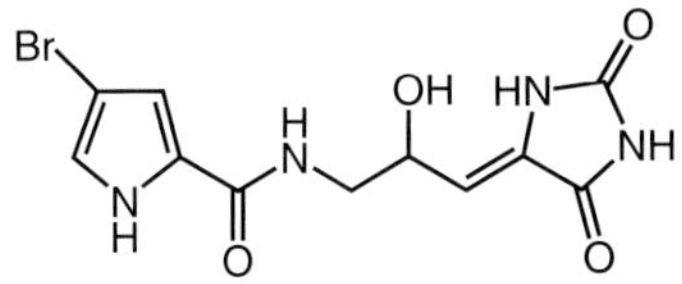

1256

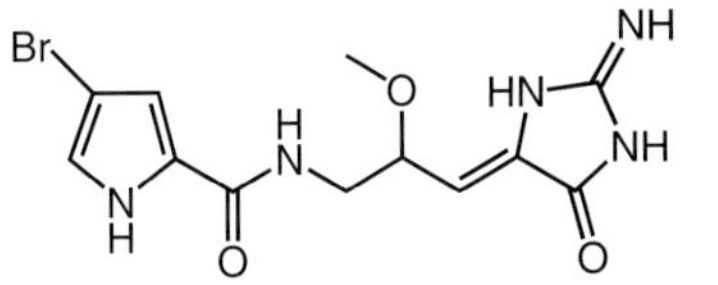

1257

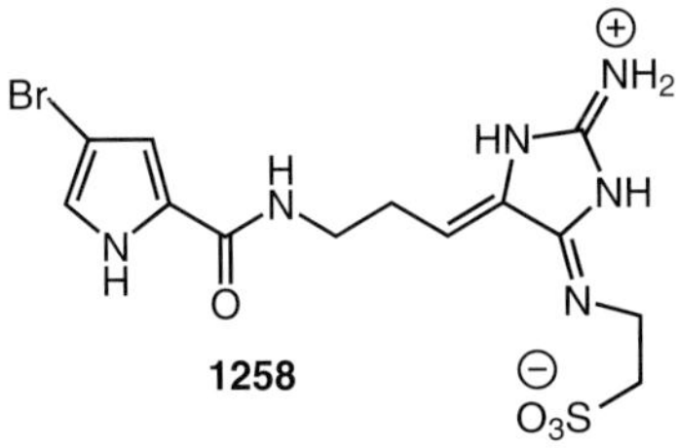

1258

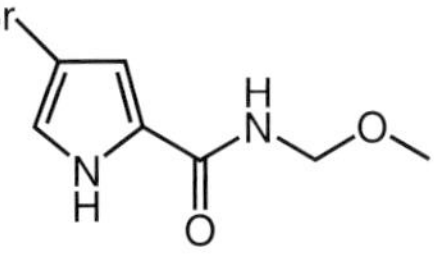

1259

The cyclized slagenins A–C (**1260–1262**) were discovered in the sponge *Agelas nakamurai* living in Okinawan waters (*1262*). Slagenins A and B are active in the L1210 murine leukemia screen. The absolute configurations of **1260–1262** were established by total synthesis (*1263*). A Japanese collection of *Agelas mauritiana* yielded mauritiamine (**1263**) (*1264*). This novel oroidin dimer inhibits barnacle growth (*Balanus amphitrite*). Cyclooroidin (**1264**) was isolated from the Mediterranean *Agelas oroides* (*1255*), and ugibohlin (**1265**) was found in the Philippines sponge *Axinella carteri* (*1265*). Total synthesis confirms the assigned structure of **1264** and establishes its absolute configuration as shown (*1266*). Ageladine A (**1266**) was isolated from the sponge *Agelas nakamurai*, and is a potent inhibitor of matrix metalloproteinases (*1268*). Ageladine A has been synthesized (*1269, 1270*).

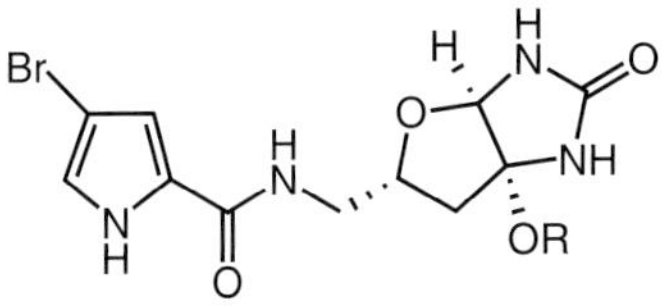

1260 R = H (slagenin A)
1261 R = Me (slagenin B)

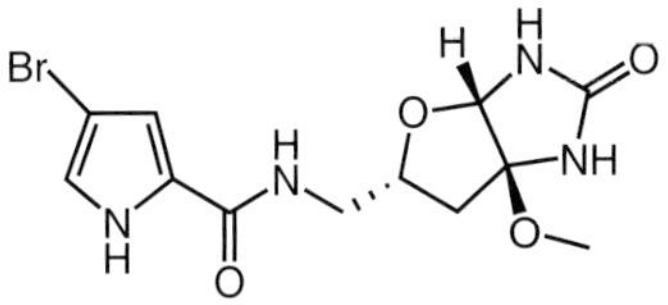

1262 (slagenin C)

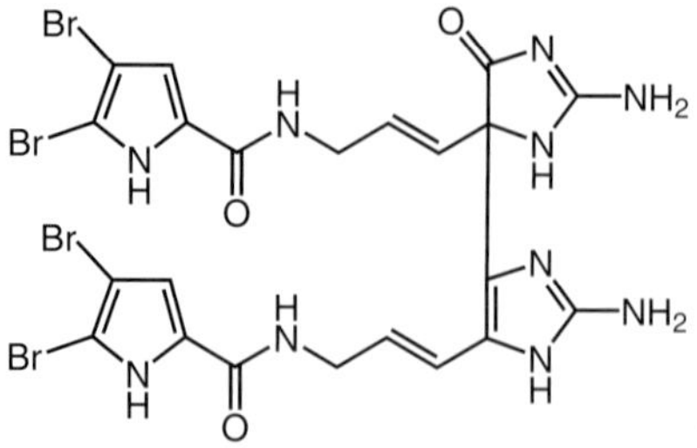

1263 (mauritiamine)

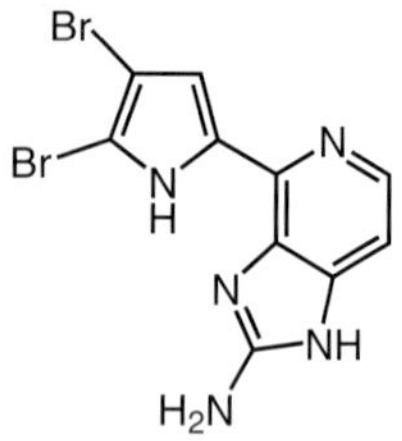

1266 (ageladine A)

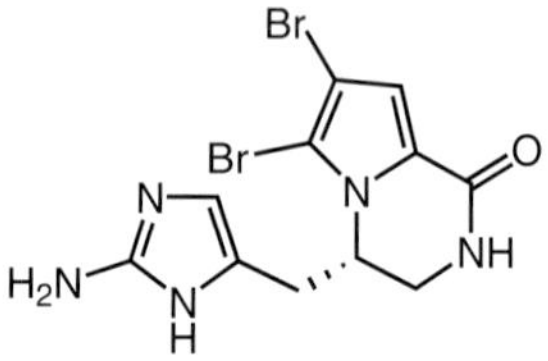

1264 (S - cyclooroidin)

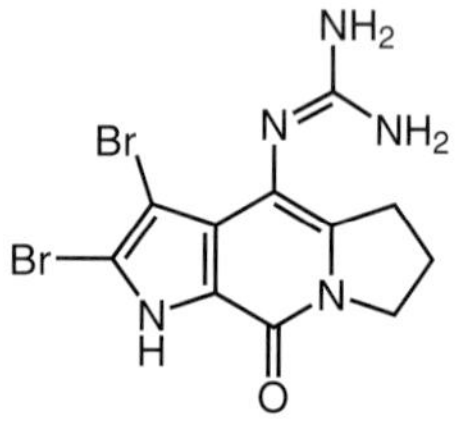

1265 (ugibohlin)

Agesamides A (**1267**) and B (**1268**) were isolated from an Okinawan sponge *Agelas* sp. (*1271*), and the interesting oxocyclostylidol (**1269**) was found in the Bahamian sponge *Stylissa caribica*, and is structurally related to cyclooroidin (*1272*). Some new examples of the hymenialdisine-axinohydantoin bromopyrrole class have been discovered since the first survey (*1*). The prolific sponge *Axinella carteri* from Indonesia contains 3-bromohymenialdisine (**1270**) (*1273*). The Palauan sponge *Stylotella aurantium* has furnished the (10*E*)-diastereomer of hymenialdisine (**1271**) (*1274*), and the (10*Z*)-diastereomer of axinohydantoin (**1272**) was also isolated from the sponge *Stylotella aurantium* (*1275*). An Okinawan sponge *Hymeniacidon* sp. has afforded the new spongiacidins A (**1273**) and B (**1274**), along with **1272** ("spongiacidin D") (*1276*). Spongiacidin A is the (10*E*)-diastereomer of **1270**. The Indonesian sponge *Stylissa carteri* yielded 2-debromostevensine (**1275**) and 2-debromohymenin (**1276**) (*1277*). Syntheses of several axinohydantoins (*1278*) and hymenialdisines (*1279*) have been reported.

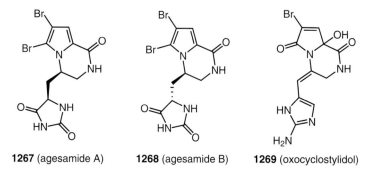

1267 (agesamide A)   **1268** (agesamide B)   **1269** (oxocyclostylidol)

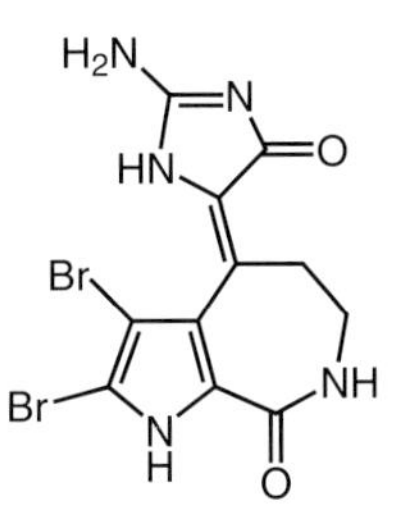

3.14 Heterocycles

1270 (3-bromohymenialdisine) 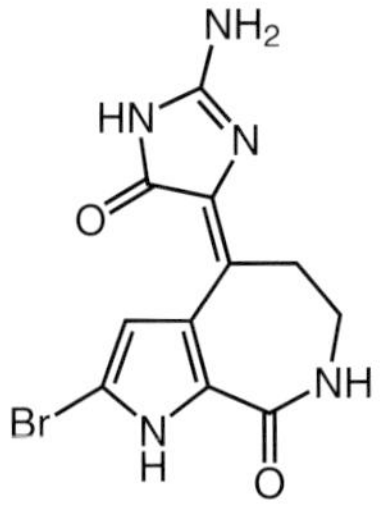 **1271** ((E)-hymenialdisine) 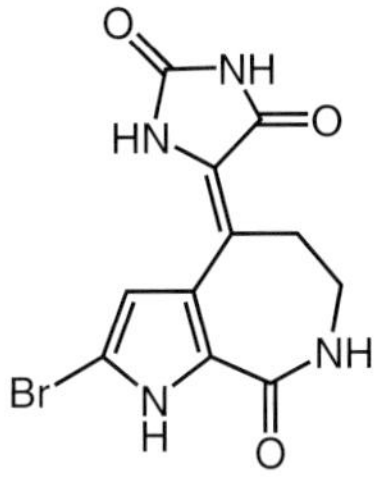 **1272** ((Z)-axinohydantoin)

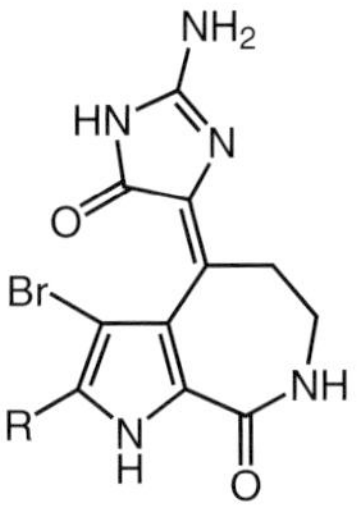

1273 R = Br (spongiacidin A)
1274 R = H (spongiacidin B)
1275 (debromostevensine) 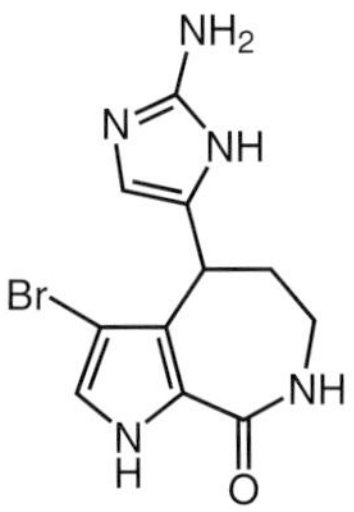 **1276** (debromohymenin) 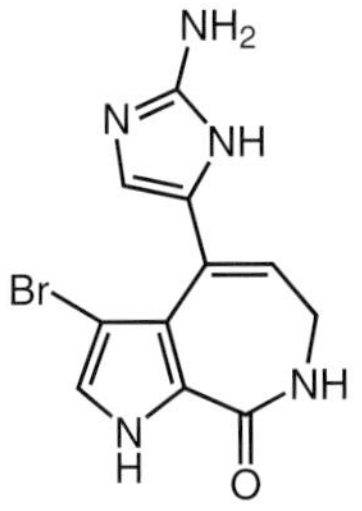

Latonduines A (**1277**) and B (**1278**), which have a novel ring system, were isolated from the Indonesian sponge *Stylissa carteri* and the structures confirmed by total synthesis (*1280*). Some new phakellin-type bromopyrroles have been characterized. The Indian Ocean sponge *Phakellia mauritiana* contains dibromophakellstatin (**1279**), which is the principal antineoplastic component of this sponge and shows activity against ovarian, brain, kidney, lung, colon, and melanoma human cell lines (*1281*). The sponge *Stylissa caribica* from the Bahamas produces *N*-methyldibromoisophakellin (**1280**), which displays excellent feeding deterrent activity against the common reef fish *Thalassoma bifasciatum* (*1282*). This metabolite is more active than oroidin in this assay. The related monobromoisophakellin (**1281**) was identified in an *Agelas* sp. sponge from the Bahamas (*1283*). Syntheses of dibromophakellstatin (*1284–1286*) and dibromoisophakellin (*1284*) confirm the proposed structures. The Japanese sponge *Axinella brevistyla* produces four new pyrrole-derived alkaloids (*1287*). In addition to the simple 3-bromomaleimide (**1282**) and 3,4-dibromomaleimide (**1283**), *N*-methylmanzacidin C (**1284**) and 12-chloro-11-hydroxydibromoisophakellin (**1285**) were characterized from this sponge. The two new phakellin alkaloids, (–)-7-*N*-methyldibromophakellin (**1286**) and (–)-7-*N*-methylmonobromophakellin (**1287**), are present in the Papua New Guinea sponge *Agelas* sp. (*1288*).

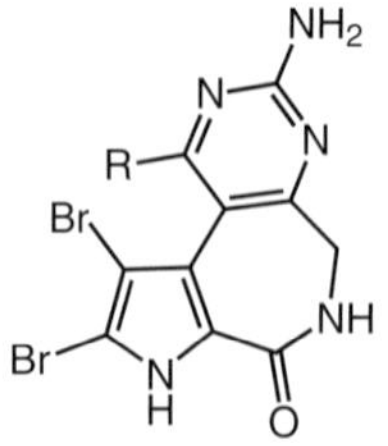

1277 R = H (latonduine A)
1278 R = CO₂H (latonduine B)

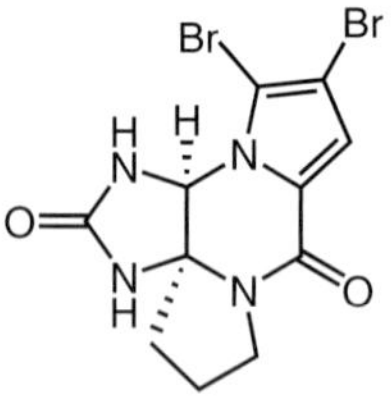

1279 (dibromophakellstatin)

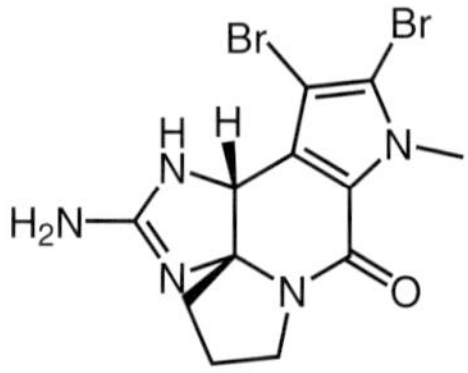

1280 (*N*-methyldibromoisophakellin)

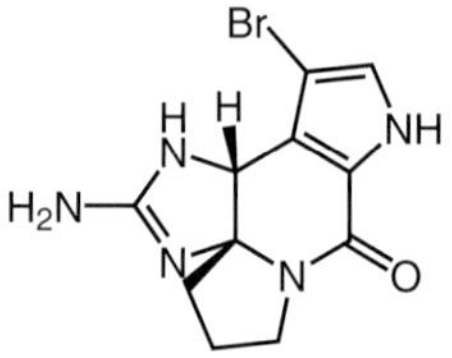

1281 (monobromoisophakellin)

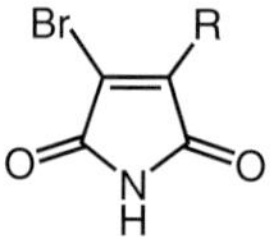

1282 R = H
1283 R = Br

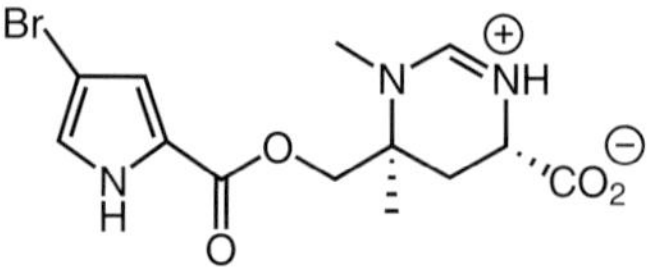

1284 (*N*-methylmanzacidin C)

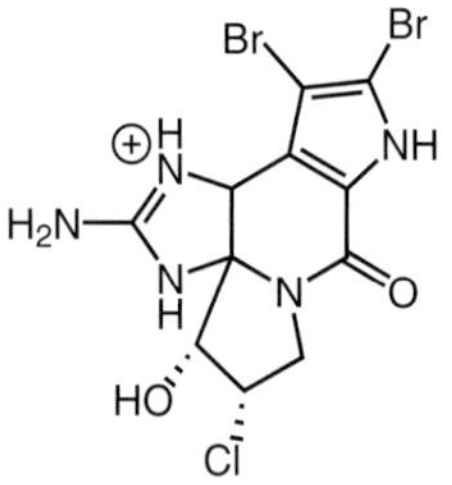

1285

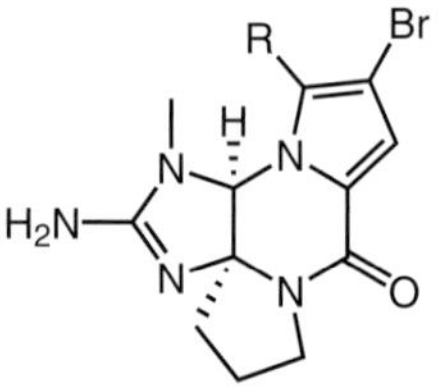

1286 R = Br
1287 R = H

The Pohnpei sponge *Astrosclera willeyana* has furnished seven new *N*-methylageliferins: *N*(1′)-methylageliferin (**1288**), *N*(1),*N*(1′)-dimethylageliferin (**1289**), *N*(1′)-methylisoageliferin (**1290**), *N*(1),*N*(1′)-dimethylisoageliferin (**1291**), *N*(1′)-methyl-2-bromoageliferin (**1292**), *N*(1′)-methyl-2′-bromoageliferin (**1293**), and *N*(1′)-methyl-2,2′-dibromoageliferin (**1294**) (*1289*). The Indonesian sponge *Agelas nakamurai* contains the sceptrin-related nakamuric acid (**1295**) and the corresponding methyl ester, which is considered to be an isolation artifact (*1290*). The new bromosceptrin (**1296**) was characterized from the Florida sponge *Agelas*

3.14 Heterocycles

conifera (*1291*). The ageliferin and sceptrin families of bromopyrroles have been of synthetic interest (*1292*, *1293*).

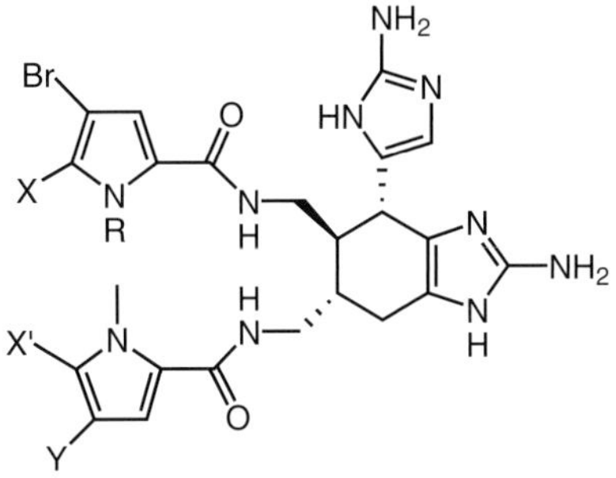

1288 R = X = X' = H, Y = Br
1289 R = Me, X = X' = H, Y = Br
1290 R = X = Y = H, X' = Br
1291 R = Me, X = Y = H, X' = Br
1292 R = X' = H, X = Y = Br
1293 R = X = H, X' = Y = Br
1294 R = H, X = X' = Y = Br

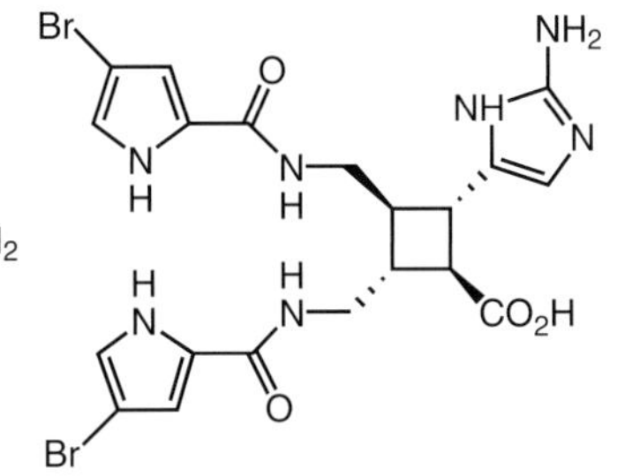

1295 R = H (nakamuric acid)

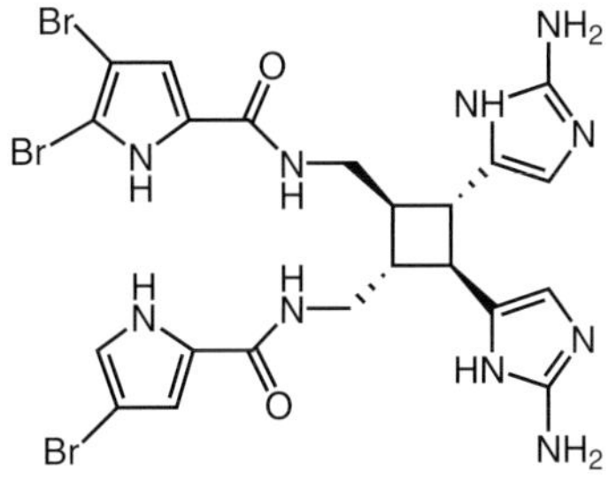

1296 (bromosceptrin)

The new agelastatins C (**1297**) and D (**1298**) were discovered in the Indian Ocean sponge *Cymbastela* sp. (*1294*). The absolute configurations are as shown. The known agelastatin A, which has been the object of synthesis (*1295–1298*), was also isolated in this study and shown to have insecticidal activity against larvae of beet army worm and corn rootworm (*1294*). A series of nagelamides A–H (**1299–1306**) was characterized in the Okinawan sponge *Agelas* sp., along with the new 9,10-dihydrokeramadine (**1307**) (*1299*). All of the nagelamides display antibacterial activity and nagelamide G (**1305**) inhibits protein phosphatase 2A. The structurally complex stylissadines A (**1308**) and B (**1309**) were isolated from the sponge *Stylissa caribica* collected in the Bahamas (*1300*). The stylissadines appear to be dimers of massadine (**1314**).

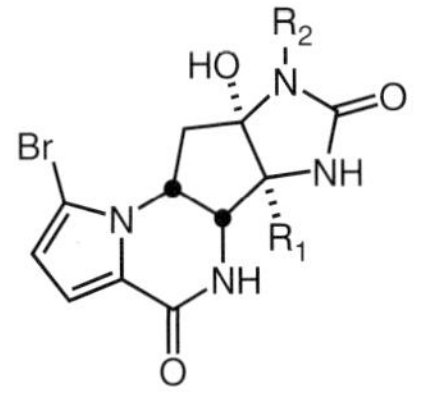

1297 R_1 = OH, R_2 = Me (agelastatin C)
1298 R_1 = R_2 = H (agelastatin D)

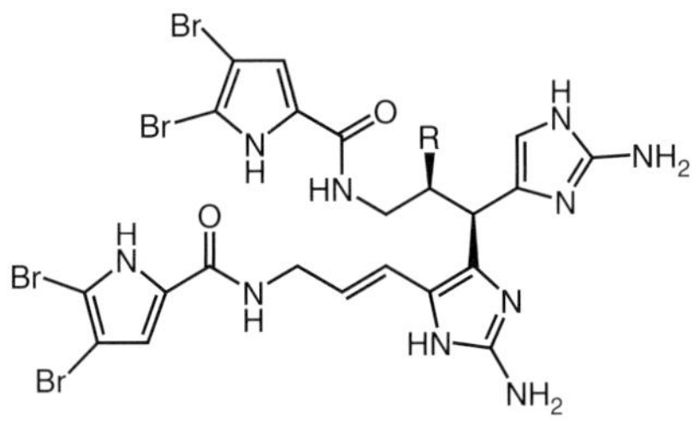

1299 R = H (nagelamide A)
1300 R = OH (nagelamide B)

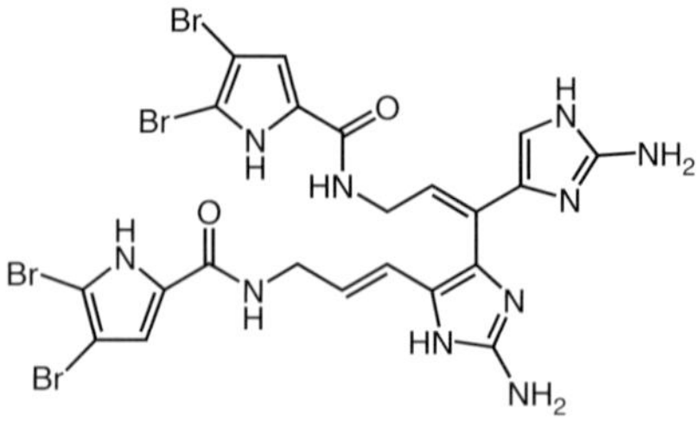

1301 (nagelamide C)

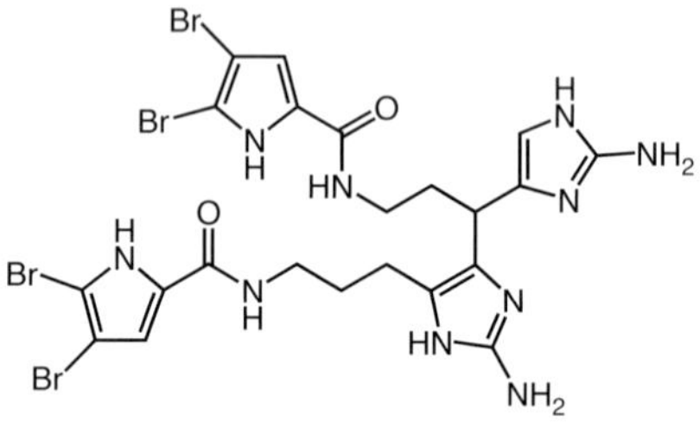

1302 (nagelamide D)

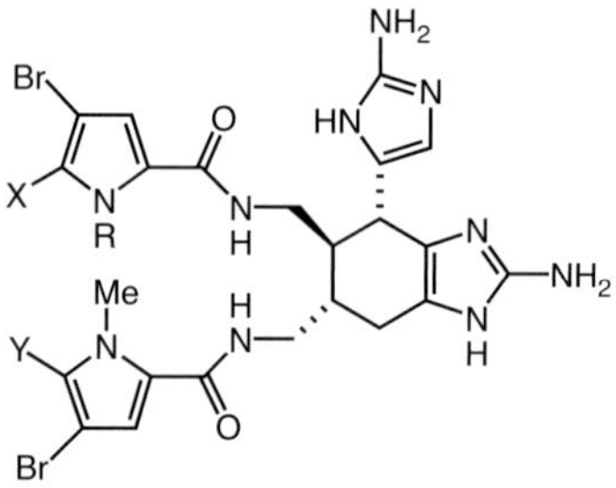

1303 X = Y = H (nagelamide E)
1304 X = Br, Y = H (nagelamide F)
1305 X = Y = Br (nagelamide G)

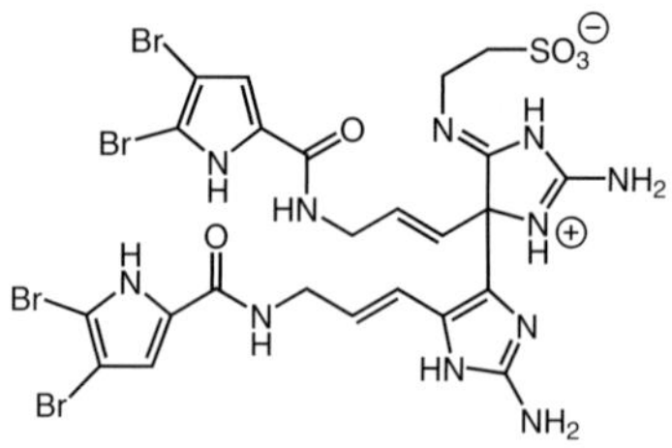

1306 (nagelamide H)

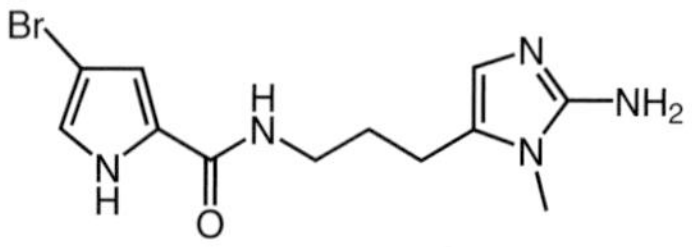

1307 (9,10-dihydrokeramadine)

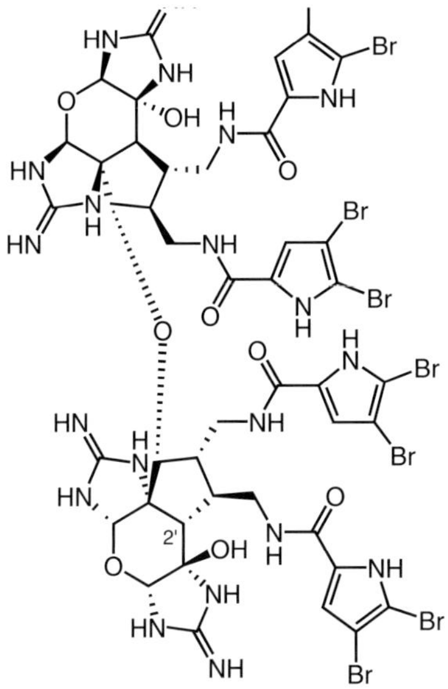

3.14 Heterocycles

1308, 1309 (stylissadines A and B) (epimeric at C-2')

An Australian sponge *Axinella* sp. has afforded the novel axinellamines A–D (**1310–1313**), which exhibit activity against *Helicobacter pylori* (*1301*). The similar massadine (**1314**) from the Japanese sponge *Stylissa* aff. *massa* inhibits geranylgeranyltransferase type I and the fungus *Candida albicans* (*1302*). Like many of these 2-aminoimidazole alkaloids, **1310–1314** were isolated as acid salts; trifluoroacetic acid salts in these cases.

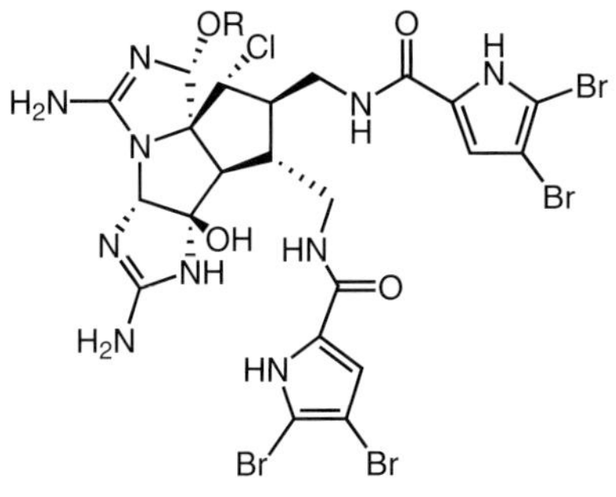

1310 R = H (axinellamine A)
1312 R = Me (axinellamine C)

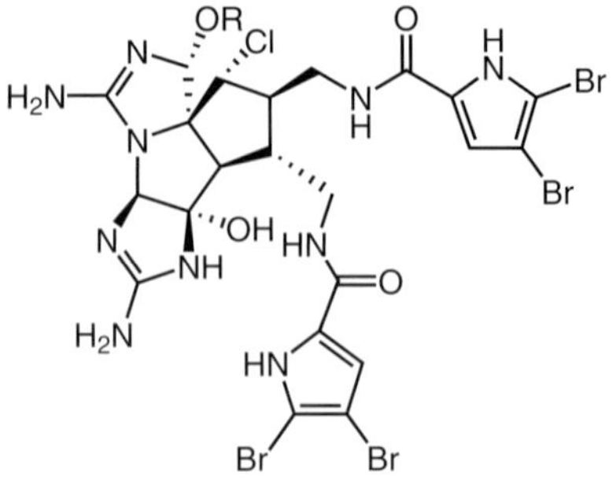

1311 R = H (axinellamine B)
1313 R = Me (axinellamine D)

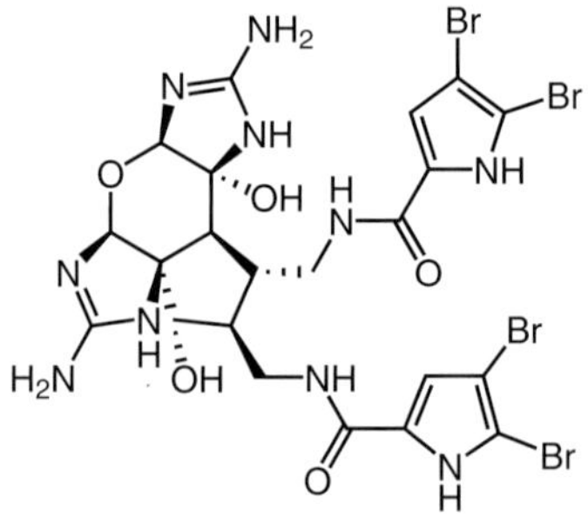

1314 (massadine)

The stunningly complex and intricate palau'amine (**1315**) was isolated from the South Pacific sponge *Stylotella agminata* (renamed as *Stylotella aurantium*) (*1303, 1304*), along with 4-bromopalau'amine (**1316**) and 4,5-dibromopalau'amine (**1317**) (*1304*). This sponge has also yielded styloguanidine (**1318**), 3-bromostyloguanidine (**1319**), and 2,3-dibromostyloguanidine (**1320**), which are potent antifouling compounds against barnacles (*1305*). These three compounds were also identified in a previous examination of *Stylotella aurantium* as "isopalau'amines" (*1304*). The related konbu'acidin A (**1321**), having cyclin dependent kinase activity, was isolated from the Okinawan sponge *Hymeniacidon* sponge (*1306*).

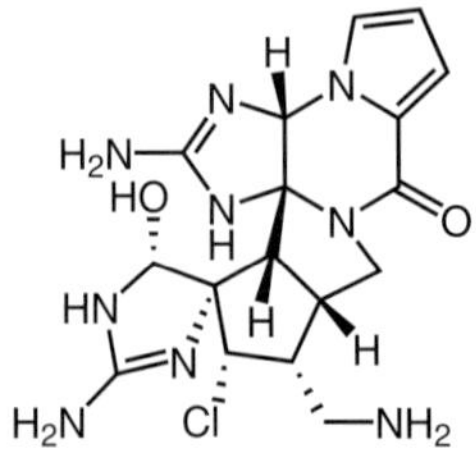

1315 (palau'amine)

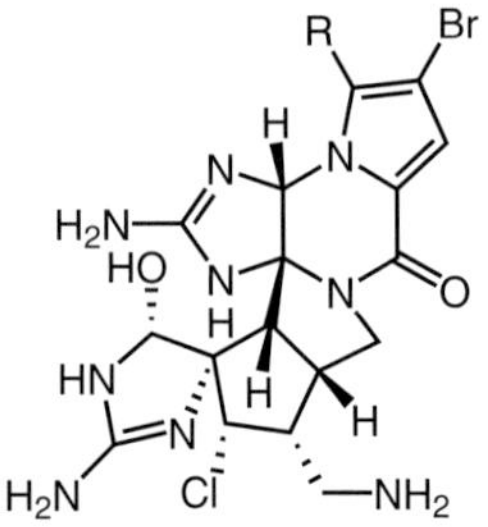

1316 R = H
1317 R = Br

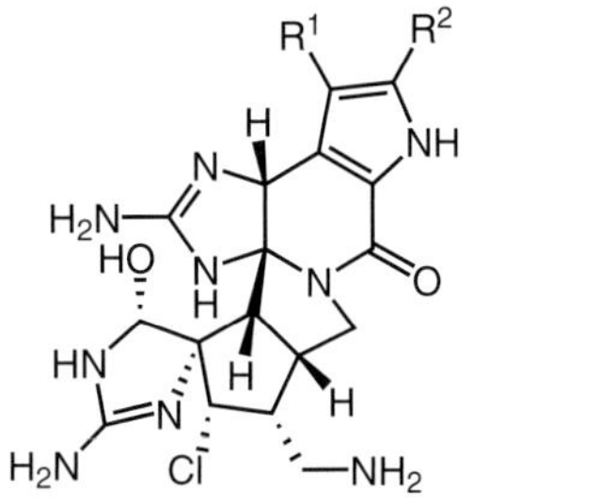

3.14 Heterocycles

1318 R¹ = R² = H (styloguanidine)
1319 R¹ = Br, R² = H
1320 R¹ = R² = Br

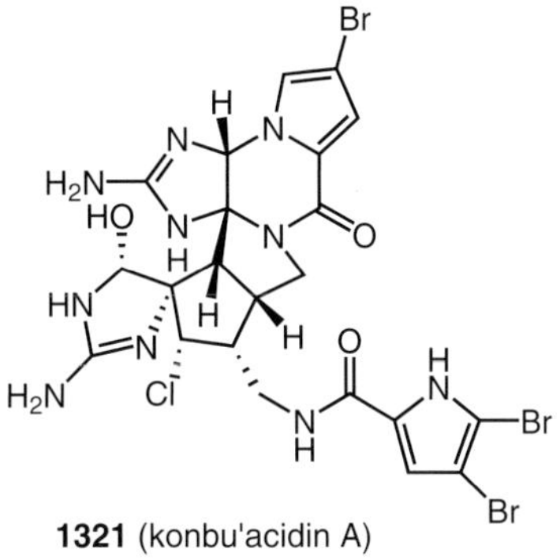

1321 (konbu'acidin A)

Three independent studies present evidence that palau'amine, styloguanidine, and derivatives need to be revised at three stereocenters (i.e. (12R,7S,20S) (*1307–1309*)). Based on these studies the revised structure of palau'amine is shown as **YY**, and the other structures (**1316–1321**) may need to be reconsidered as well. The Australian sponge *Stylissa flabellata* contains stylissadines A (**1308**) and B (**1309**), which were initially named "flabellazoles A and B", respectively (*1307*). This study also uncovered the new konbu'acidin B (**1322**), and reported that **1308** and **1309** are the most potent natural product P2X$_7$ antagonists to be isolated to date (*1307*). The Bahamaian sponge *Stylissa caribica* contains tetrabromostyloguanidine (**1323**) (*1308*), and *Stylissa carteri* has yielded carteramine A, which appears to be **1323** (same optical rotation) (*1309*). Related to nagelamides A–H (**1299–1300**), the new nagelamide J (**1324**) was isolated from an Okinawan *Agelas* sponge (*2659*).

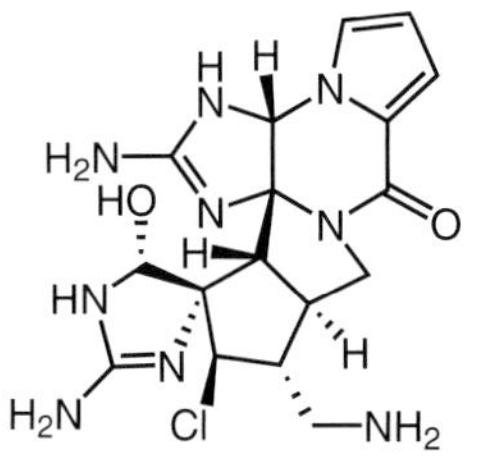

YY (palau'amine) (revised)

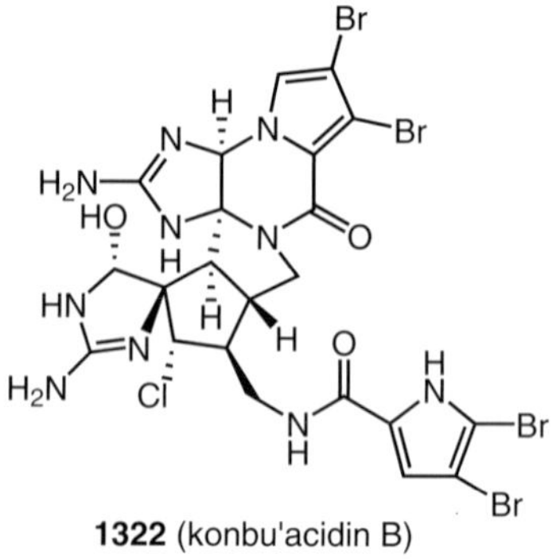

1322 (konbu'acidin B)

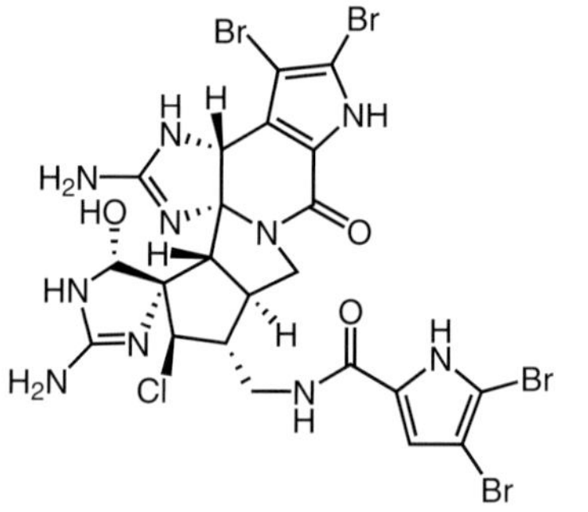

1323 (tetrabromostyloguanidine) (= carteramine A?)

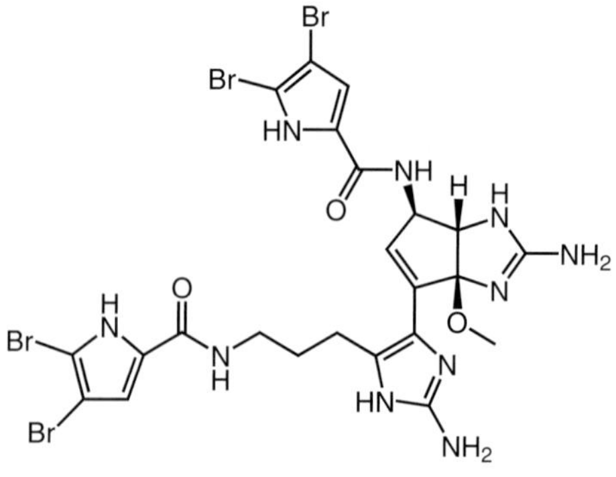

1324 (nagelamide J)

3.14.2 Indoles

Like pyrrole, the enormous reactivity of indole guarantees a large number of natural halogenated indoles. The earlier survey documented nearly 200 halogenated indoles, not including halogenated carbazoles, carbolines, and related fused indoles (*1*). Simple halogenated indoles, both previously known and new, continue to be identified in natural sources. The Palauan ascidian *Distaplia regina* contains 3,6-dibromoindole (*1310*), which was earlier misidentified (*1*). The previously known 1-methyl-2,3,5-tribromoindole was identified in the Indian red alga *Nitophyllum marginata* (*1311*), and the marine acorn worm metabolite 3-chloroindole (*1*) has now been found in the mushroom *Hygrophorus paupertinus* (*1312*). The common oyster *Crassostrea virginica* contains two dibromoindoles and one tribromoindole, which are not yet identified (*1313*). In addition to the previously reported 3,6- and 4,6-dibromoindoles (*1*), both the open ocean and sediments from the North and Baltic Seas contain 4-bromoindole (**1325**), 5-bromoindole (**1326**), 6-bromoindole (**1327**), and 3,4-dibromoindole (**1328**) (*1314*). Three new bromoindoles, 3,5,6-tribromoindole (**1329**), 1-methyl-3,5,6-tribromoindole (**1330**), and 2,3,6-tribromoindole (**1331**) were isolated from the red alga *Laurencia similis* collected off the coast of Hainan Island, China (*1315*). The muricid gastropod *Drupella fragum*, a predator of Madreporaria corals, contains in its mid-intestinal gland novel brominated hydroxyindoles, 6-bromo-5-hydroxyindole (**1332**), 6-bromo-4,5-dihydroxyindole (**1333**), and 5-bromo-4,7-dihydroxyindole (**1334**) (*1316*). Indole **1332** is comparable to BHT and superior to α-tocopherol for antioxidative activity, and the structure of **1332** is confirmed by synthesis. The unusual sulfate-sulfamate indoles ancorinolates A (**1335**) and C (**1336**) were isolated from the sponge *Ancorina* sp. (*1317*). These indoles show weak HIV-inhibitory activity. The Formosan red alga *Laurencia brongniartii* has yielded three new sulfur-containing polybromoindoles, **1337–1339**, and two related dimeric polybromoindoles, **1340** and **1341** (*1318*).

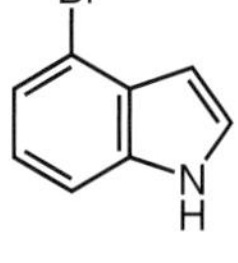

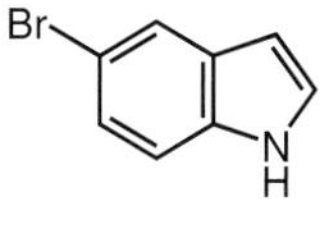

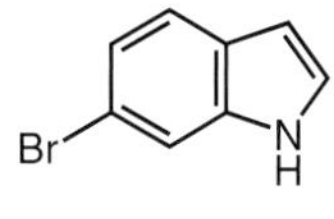

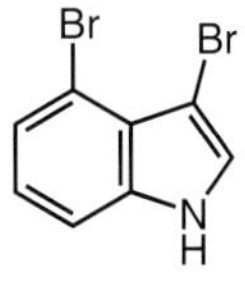

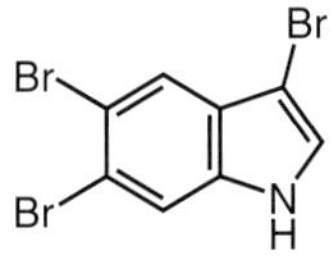

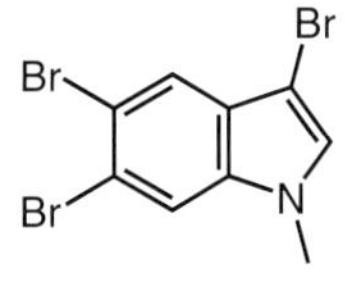

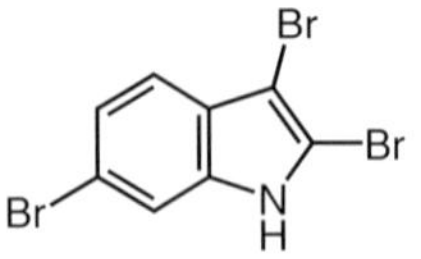

1331

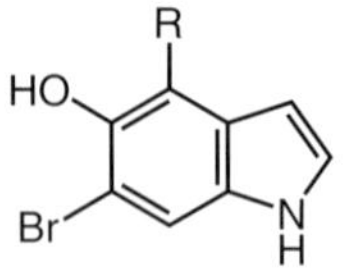

1332 R = H
1333 R = OH

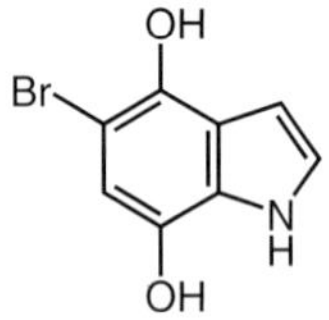

1334

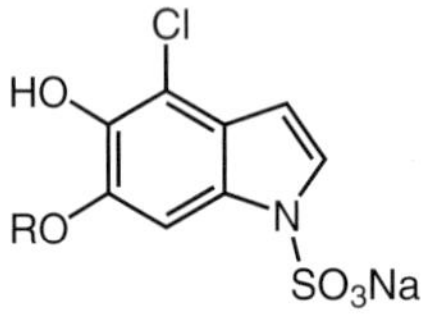

1335 R = SO₃Na (ancorinolate A)
1336 R = H (ancorinolate C)

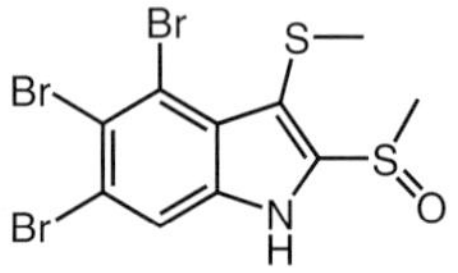

1337

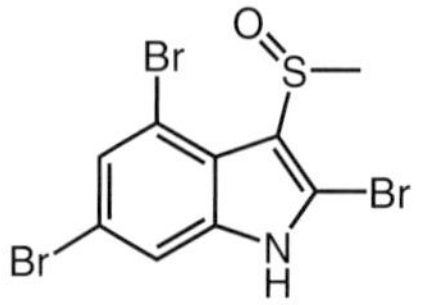

1338

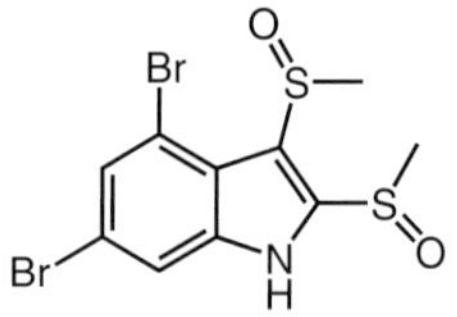

1339

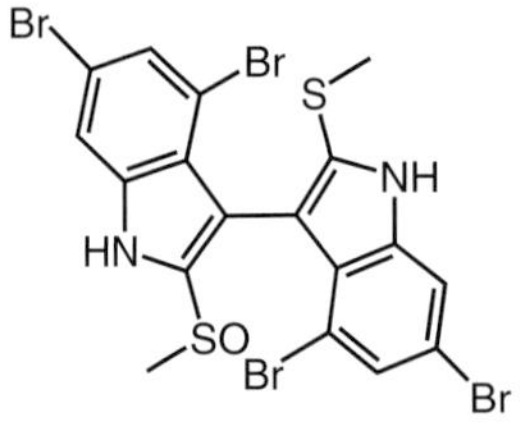

1340

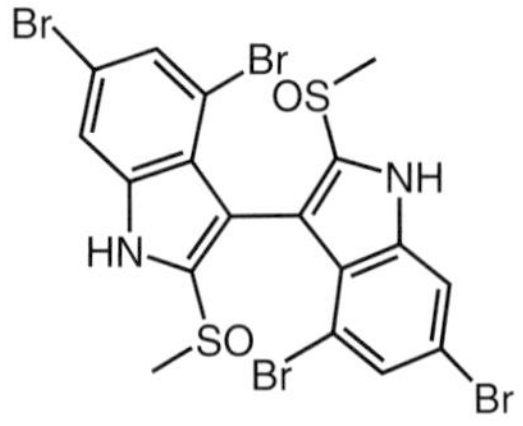

1341

Several new brominated tryptamines and tryptophans have been described. The Tasmanian bryozoan *Amathia convoluta* contains convolutindole A (**1342**) (*1319*), and the North Sea bryozoan *Flustra foliacea* (Fig. 3.19) has afforded deformylflustrabromine B (**1343**) (*1320*), **1344**, **1345**, and deformylflustrabromine (**1346**) (*1321*). A Philippine sponge *Smenospongia* sp. contains 5-bromotryptophan (**1347**), 5-bromoabrine (**1348**), 5,6-dibromoabrine (**1349**), and 5-bromoindole-3-acetic acid (**1350**) (*1322*). A study of the sponge *Thorectandra* sp. has furnished 5-bromo-*N*,*N*-dimethyltryptophan (**1351**), 5-bromohypaphorine (**1352**), and aply-sinopsin **1353** (*1323*), while the Papua New Guinea sponge *Smenospongia* sp. produces methyl 6-bromoindole-3-carboxylate (**1354**) (*1323*). The novel iodinated

3.14 Heterocycles

plakohypaphorines A–F (**1355–1360**) were isolated from the Caribbean sponge *Plakortis simplex* (*1324, 1325*). These novel compounds are the first examples of naturally occurring iodine-containing indoles. The Pacific Coast snail *Calliostoma canaticulatum* secretes the disulfide-linked dimer of 6-bromo-2-mercaptotryptamine (**1361**) that repels the predatory starfish *Pycnopodia helianthoides* (*1326*).

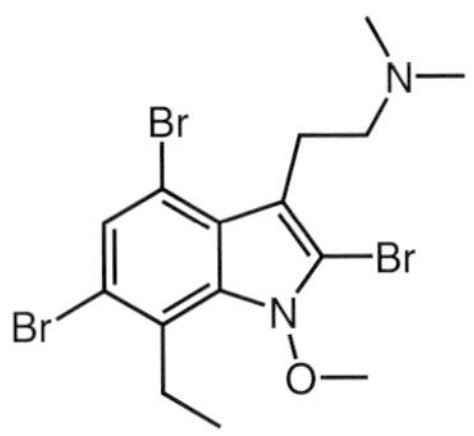

1342 (convolutindole A)

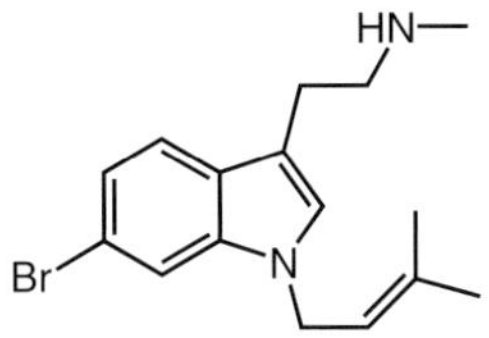

1343 (deformylflustrabromine B)

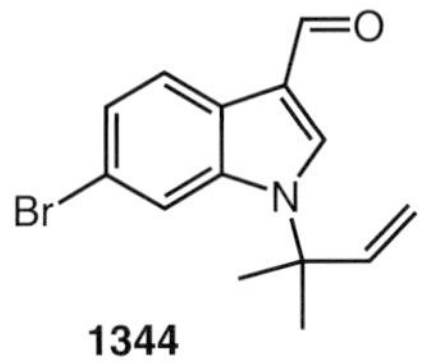

1344

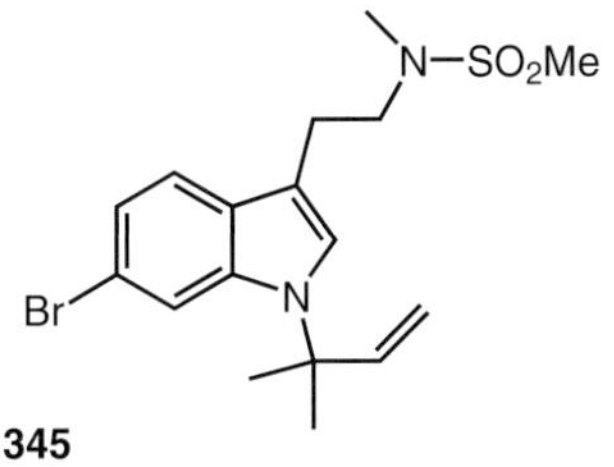

1345

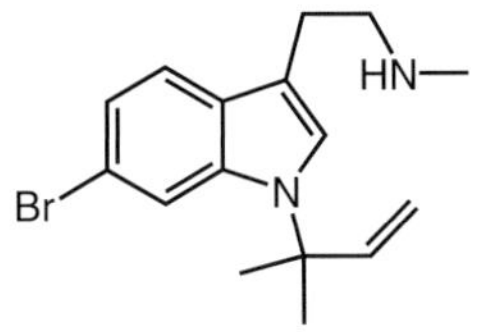

1346 (deformylflustrabromine)

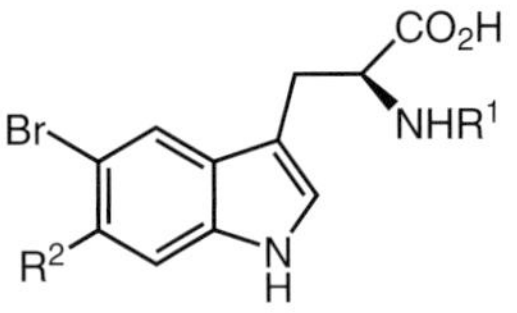

1347 R^1 = R^2 = H
1348 R^1 = Me, R^2 = H (5-bromoabrine)
1349 R^1 = Me, R^2 = Br (5,6-dibromoabrine)

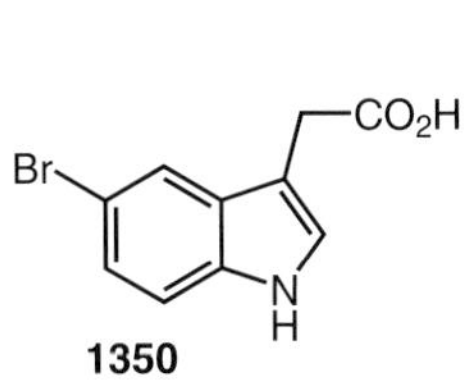

1350

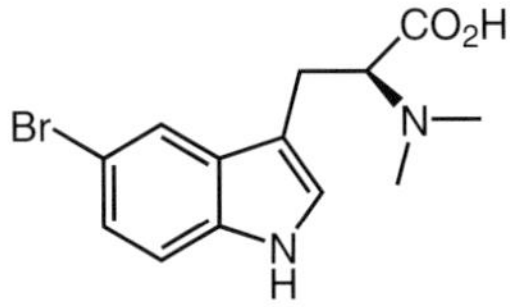

1351

1352 (5-bromohypaphorine)

1353 (aplysinopsin)

1354

1355 R¹ = R² = H, R³ = I (plakohypaphorine A)
1356 R¹ = H, R² = R³ = I (plakohypaphorine B)
1357 R¹ = R³ = I, R² = H (plakohypaphorine C)
1358 R¹ = I, R² = I, R³ = H (plakohypaphorine D)
1359 R¹ = I, R² = I, R³ = I (plakohypaphorine E)
1360 R¹ = I, R² = H, R³ = Cl (plakohypaphorine F)

1361

Fig. 3.19 *Flustra foliacea*, a North Sea bryozoan and a producer of many bromotryptamines and brominated indole alkaloids, such as **1343–1346** (Photo: A. D. Wright)

3.14 Heterocycles

Barettin (**1362**) was isolated from the Swedish sponge *Geodia baretti* (*1327*), and the structure was subsequently revised (*1328*) after a synthesis of the proposed structure proved it incorrect (*1329*). Dihydrobarettin (**1363**) is also found in this sponge (*1330*). The three novel amino acid derivatives **1364–1366** were identified in the New Caledonian ascidian *Leptoclinides debius* (*1331*). The latter metabolite features the rare amino acid enduracididine. Four new bromotryptamine peptides, alternatamides A (**1367**), B (**1368**), C (**1369**), and D (**1370**), were characterized from the bryozoan *Amathia alternata* collected along the North Carolina coast (*1332*). The absolute stereochemistry of the previously known chelonin B (*1*) has been determined as (S) by total synthesis (*1333*). The sponge *Hyrtios erecta* contains the new (Z)-5,6-dibromo-2′-demethylaplysinopsin (**1371**) and (E)-5,6-dibromo-2′-demethylaplysinopsin (**1372**) (*1334*). The New Zealand ascidian *Pycnoclavella kottae* has furnished the four kottamides A–D (**1373–1376**), which display antiinflammatory, antimetabolic, and antitumor activity to varying degrees (*1335*). This same marine animal also contains kottamide E (**1377**), which incorporates an unusual 1,2-dithiolane moiety (*1336*).

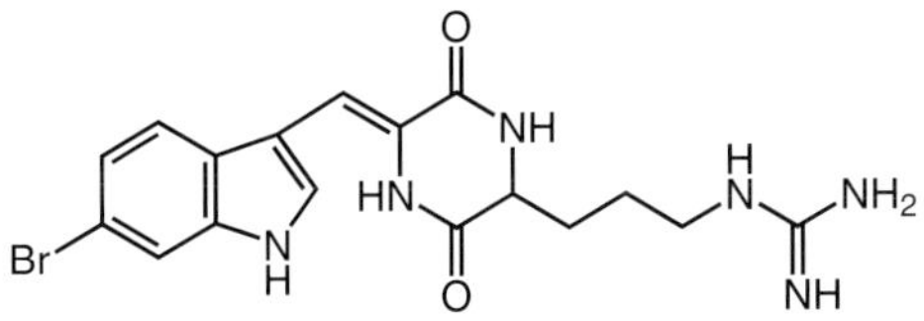

1362 (barettin)

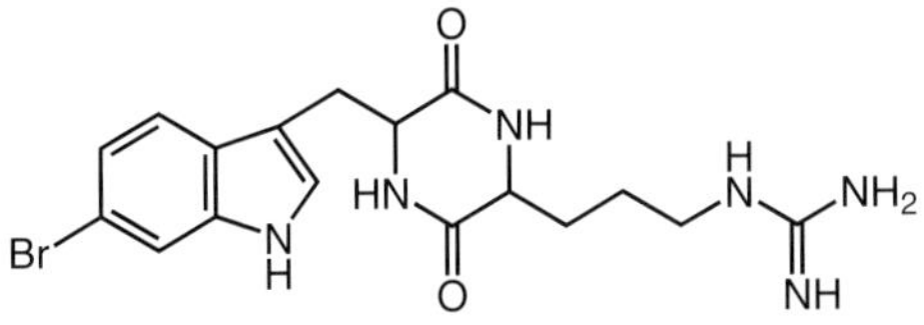

1363 (8,9-dihydrobarettin)

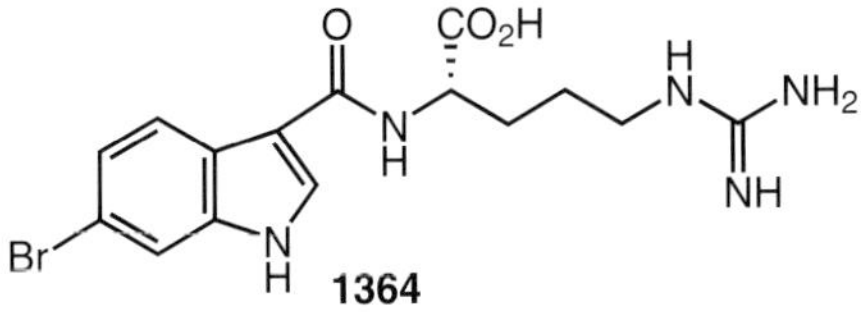

1364

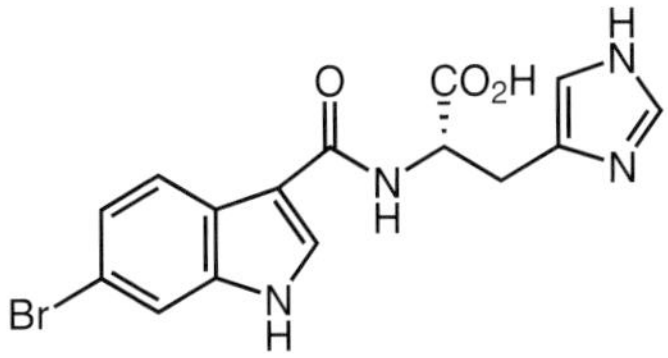

1365

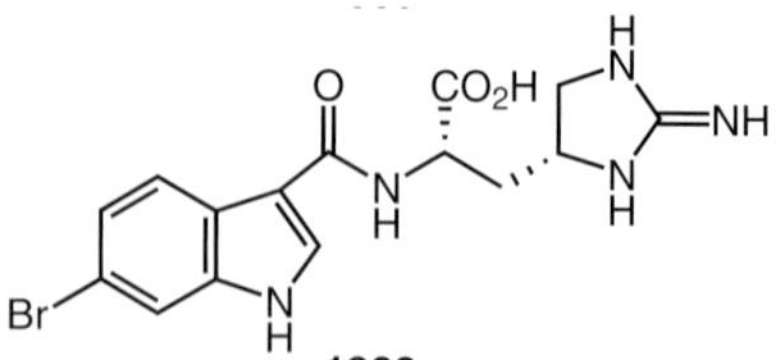

1366

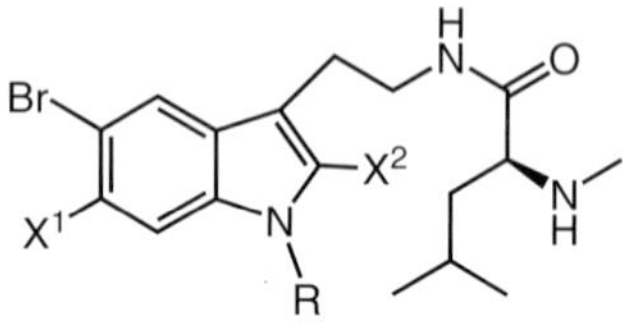

1367 R = Me, X¹ = X² = Br (alternatamide A)
1368 R = H, X¹ = X² = Br (alternatamide B)
1369 R = X¹ = H, X² = Br (alternatamide C)
1370 R = X² = H, X¹ = Br (alternatamide D)

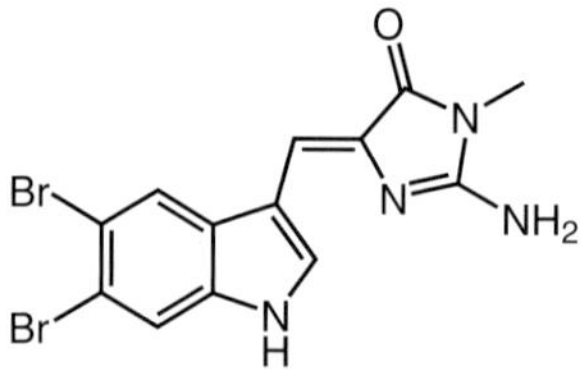

1371

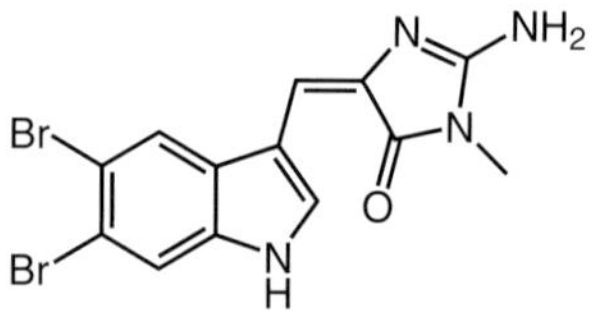

1372

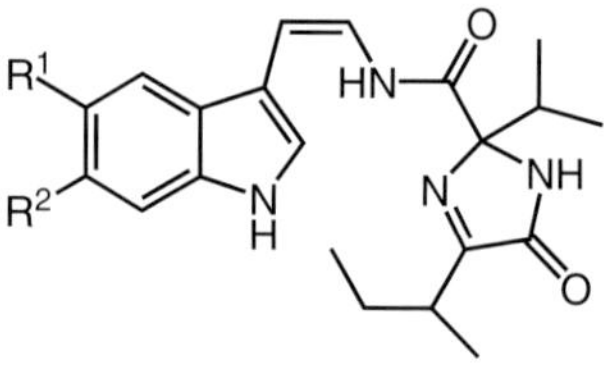

1373 R¹ = R² = Br (kottamide A)
1374 R¹ = Br, R² = H (kottamide B)
1375 R¹ = H, R² = Br (kottamide C)

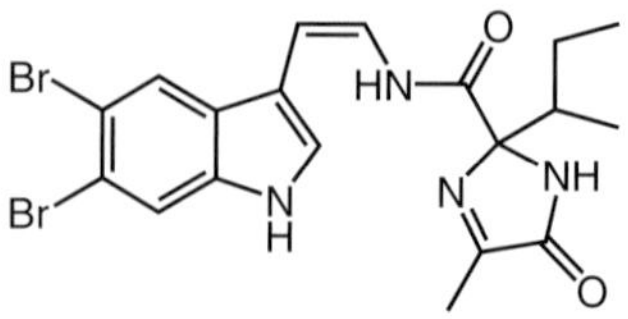

1376 (kottamide D)

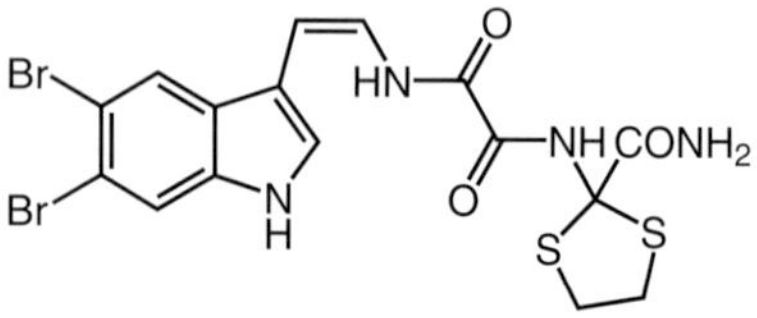

1377 (kottamide E)

3.14 Heterocycles

An Antarctica collection of the sponge *Psammopemma* sp. has yielded the new 4-hydroxyindole alkaloids, psammopemmins A (**1378**), B (**1379**), and C (**1380**), which embody the unique 2-bromopyrimidine unit (*1337*). The related meridianins B (**1381**), C (**1382**), D (**1383**), E (**1384**), and F (**1385**) were found in the tunicate *Aplidium meridianum* collected at 100 m near the South Georgia Islands (*1338, 1339*). These protein kinase inhibitors have been synthesized (*1340*). The sponge *Discodermia polydiscus* has afforded 6-hydroxydiscodermindole (**1386**) (*1341*).

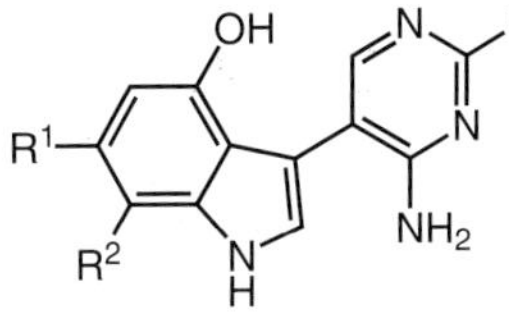

1378 $R^1 = R^2 = H$ (psammopemmin A)
1379 $R_1 = H, R^2 = Br$ (psammopemmin B)
1380 $R^1 = Br, R^2 = H$ (psammopemmin C)

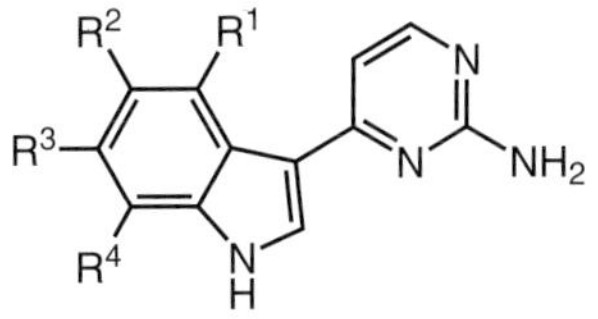

1381 $R^1 = OH, R^2 = R^4 = H, R^3 = Br$ (meridianin B)
1382 $R^1 = R^3 = R^4 = H, R^2 = Br$ (meridianin C)
1383 $R^1 = R^2 = R^4 = H, R^3 = Br$ (meridianin D)
1384 $R^1 = OH, R^2 = R^3 = H, R^4 = Br$ (meridianin E)
1385 $R^1 = R^4 = H, R^2 = R^3 = Br$ (meridianin F)

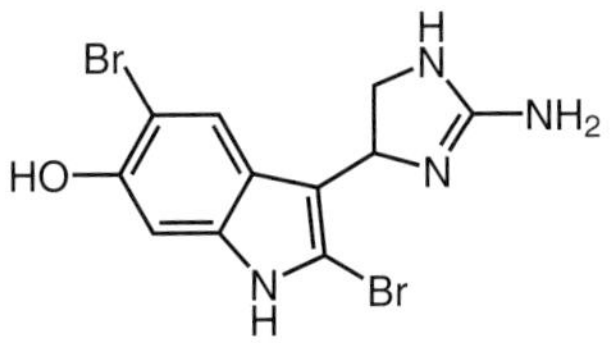

1386 (6-hydroxydiscodermindole)

Numerous halogenated bis-indole natural products have been described since the early discovery of Tyrian purple, the dibrominated analog of indigo (*1*). A new example of the topsentin family has been isolated, isobromotopsentin (**1387**), from the deep water Australian sponge *Spongosorites* sp. (*1342*). The Caribbean mangrove ascidian *Didemnum conchyliatum* (Fig. 3.20) has provided four new didemnimides, two of which, didemnimides B (**1388**) and D (**1389**), contain bromine

(*1343*). The latter is a potent feeding deterrent against mangrove-specific carnivorous fish. The structure of **1388** is confirmed by total synthesis (*1344*). The Korean sponge *Spongosorites genitrix* has afforded the new bromodeoxytopsentin (**1390**) and isobromodeoxytopsentin (**1391**), which display moderate cytotoxicity against human leukemia K-562 (*1345*). The Okinawan tunicate *Rhopalaea* sp. has yielded four rhopaladins, two of which, rhopaladins A (**1392**) and C (**1393**), are brominated (*1346*), and all four rhopaladins have been synthesized (*1347*). The Mediterranean sponge *Rhaphisia lacazei* produces seven new bis-indoles of the topsentin and hamacanthin classes, *cis*-3,4-dihydrohamacanthin B (**1394**), 6′-debromo-*cis*-3,4-dihydrohamacanthin B (**1395**), 6″-debromo-*cis*-3,4-dihydrohamacanthin B (**1396**), *cis*-3,4-dihydrohamacanthin A (**1397**), *trans*-3,4-dihydrohamacanthin A (**1398**), 6′-debromo-*trans*-3,4-dihydrohamacanthin A (**1399**), and 6″-debromo-*trans*-3,4-dihydrohamacanthin A (**1400**) (*1348*). A Korean collection of the sponge *Spongosorites* sp. has afforded (*R*)-6″-debromohamacanthin A (**1401**), (*R*)-6′-debromohamacanthin A (**1402**), dibromodeoxytopsentin (**1403**) (*1349*), (*R*)-6″-debromohamacanthin B (**1404**) (*1350*), (*R*)- and (*S*)-6′-debromohamacanthin B (**1405, 1406**), spongotine A (**1407**), spongotine B (**1408**), and spongotine C (**1409**) (*1351*). The previously reported "(*S*)-6″-debromohamacanthin B" (*1349*) has been reassigned as spongotine B (**1408**) (*1351*). The absolute configuration of spongotine A (**1407**) is established as (*S*) by total synthesis (*2670*) (and assumed for **1408** and **1409**). Some of these bis-indoles have significant antibacterial activity against methicillin-resistant *Staphylococcus aureus* and pathogenic fungi (*1352*). Syntheses of the hamacanthins have been accomplished (*1353, 1354*).

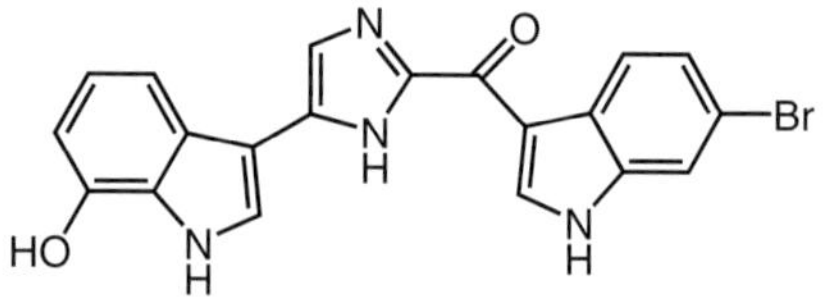

1387 (isobromotopsentin)

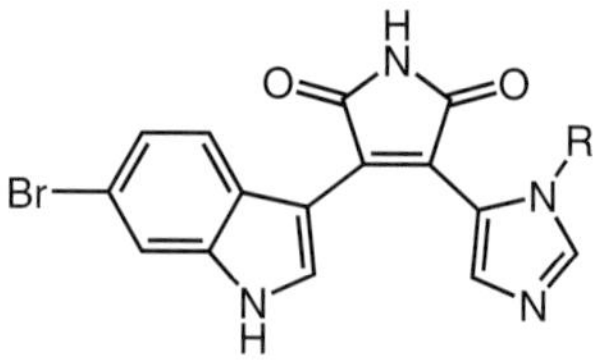

1388 R = H (didemnimide B)
1389 R = Me (didemnimide D)

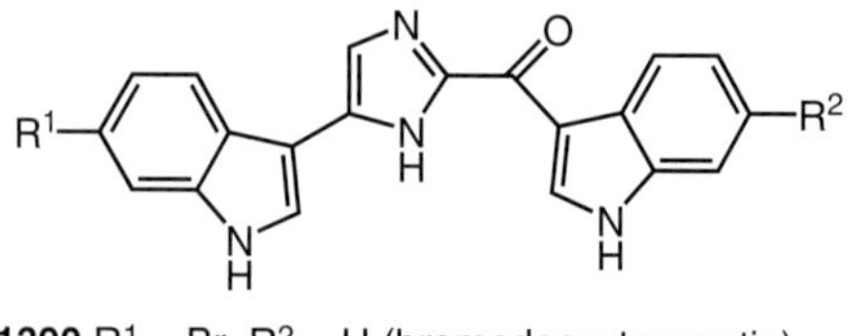

1390 R^1 = Br, R^2 = H (bromodeoxytopsentin)
1391 R^1 = H, R^2 = Br (isobromodeoxytopsentin)

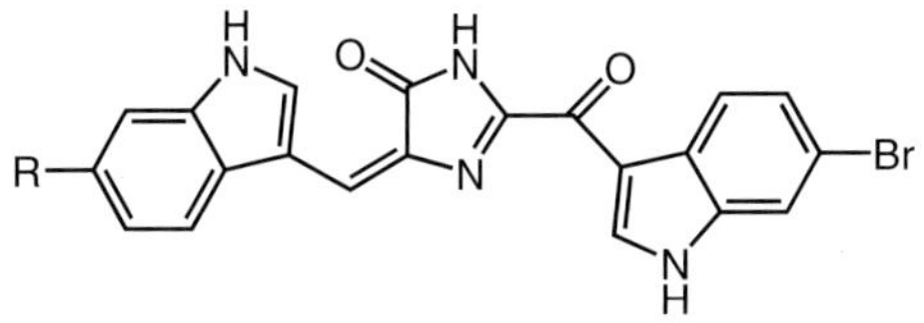

3.14 Heterocycles

1392 R = OH (rhopaladin A)
1393 R = H (rhopaladin C)

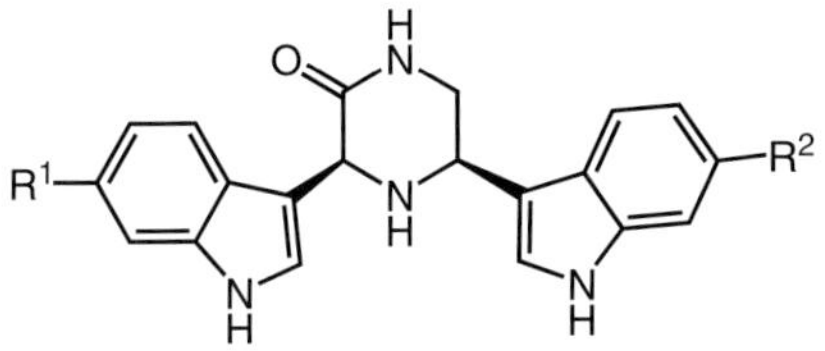

1394 R¹ = R² = Br (*cis* - 3,4-dihydrohamacanthin B)
1395 R¹ = H, R² = Br (6'-debromo-*cis*-3,4-dihydrohamacanthin B)
1396 R¹ = Br, R² = H (6''-debromo-*cis*-3,4-dihydrohamacanthin B)

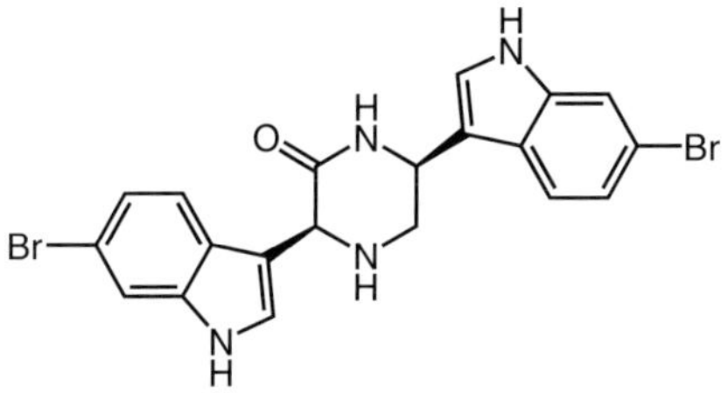

1397 (*cis*-3,4-dihydrohamacanthin A)

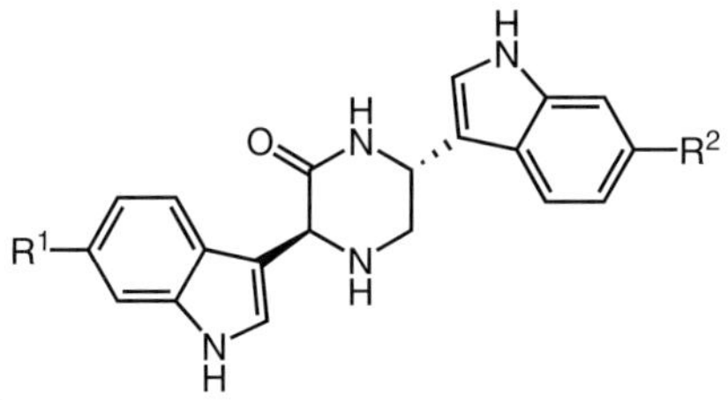

1398 R¹ = R² = Br (*trans*-3,4-dihydrohamacanthin A)
1399 R¹ = H, R² = Br (6'-debromo-*trans*-3,4-dihydrohamacanthin A)
1400 R¹ = Br, R² = H (6''-debromo-*trans*-3,4-dihydrohamacanthin A)

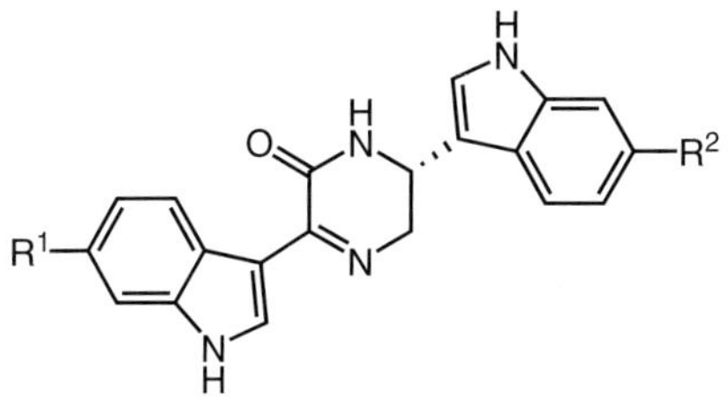

1401 R¹ = Br, R² = H ((*R*)-6''-debromohamacanthin A)
1402 R¹ = H, R² = Br ((*R*)-6'-debromohamacanthin A)

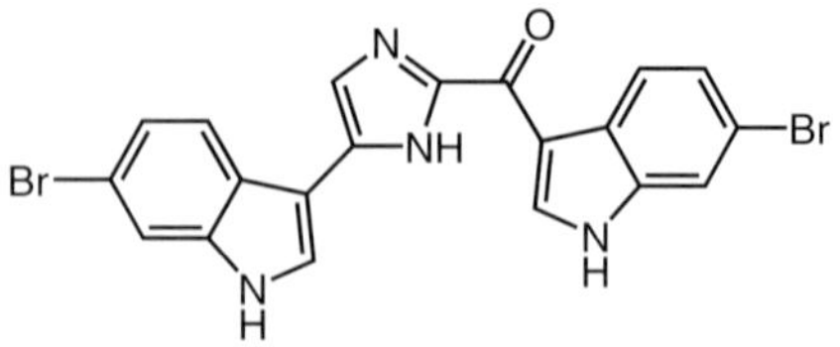

1403 (dibromodeoxytopsentin)

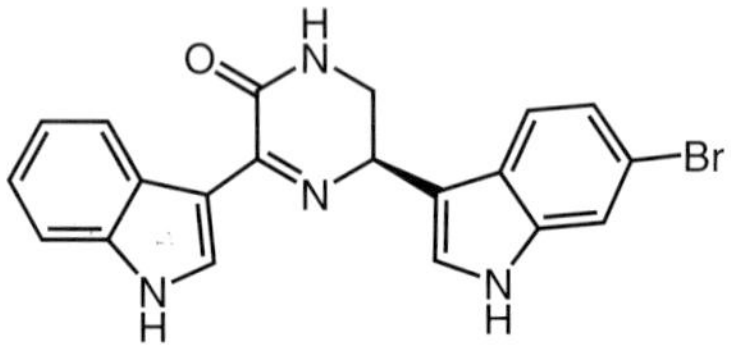

1404 ((*R*)-6''-debromohamacanthin B)

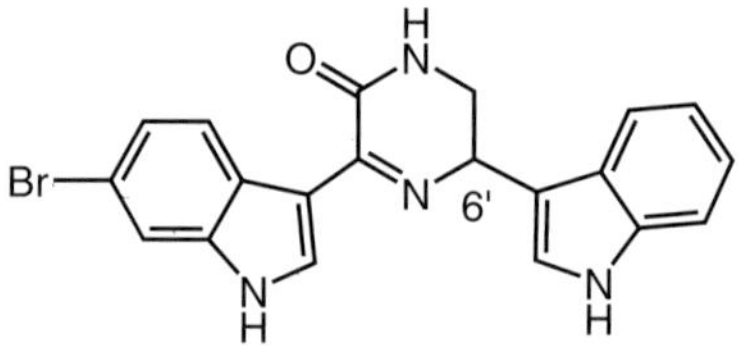

1405 ((*R*)-6'-debromohamacanthin B)
1406 ((*S*)-6'-debromohamacanthin B)

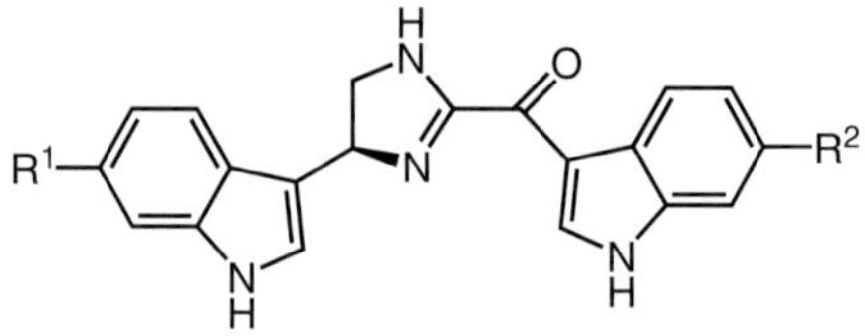

1407 R^1 = Br, R^2 = H (spongotine A)
1408 R^1 = H, R^2 = Br (spongotine B)
1409 R^1 = R^2 = Br (spongotine C)

The deep water New Caledonian sponge *Dragmacidon* sp. contains the novel nortopsentin D (**1410**), which is inactive on KB cancer cells (*1355*). However, a polymethylated synthetic derivative (seven methyl groups) is highly cytotoxic. Another deep-water sponge, *Spongosorites* sp., collected from the southern coast of Australia, has provided dragmacidin E (**1411**) (*1356*). Dragmacidin F (**1412**) was isolated from the Mediterranean sponge *Halicortex* sp. (*1357*). This complex metabolite, as well as other dragmacidins, has yielded to total synthesis (*1358*). One of the few calcareous hard corals to be investigated for secondary metabolites is *Tubastraea* sp. from Japan, and this stony coral has yielded tubastrindole A (**1413**), a novel bis-indole (*1359*).

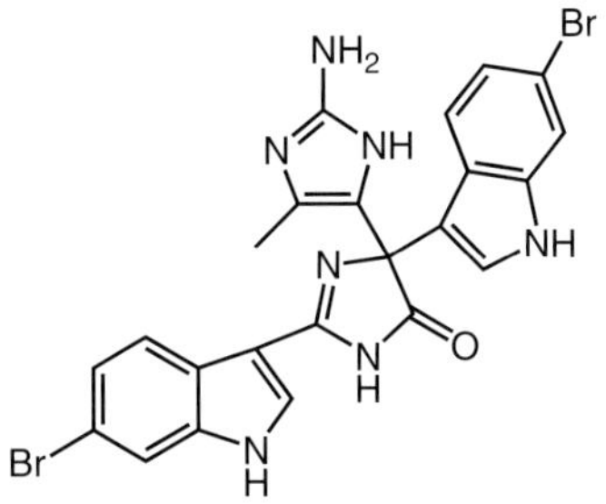

3.14 Heterocycles

1410 (nortopsentin D)

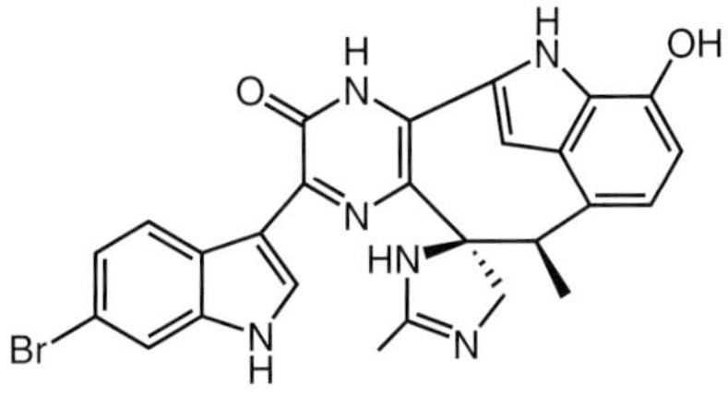

1411 (dragmacidin E)

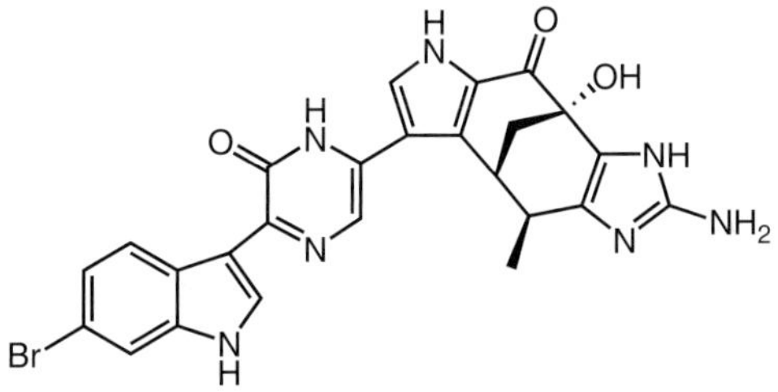

1412 (dragmacidin F)

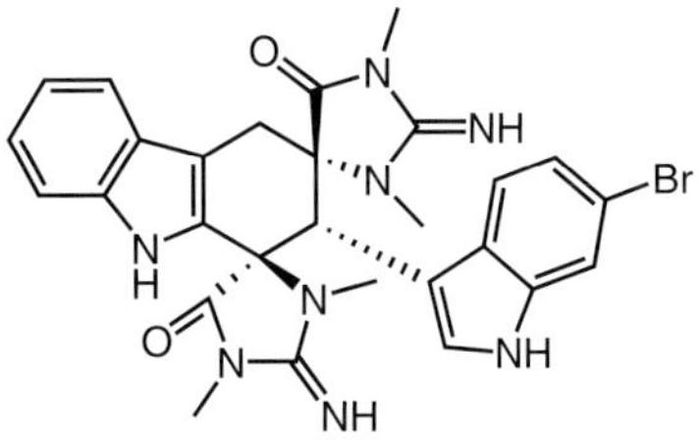

1413 (tubastrindole A)

Several new halogenated polyindole gelliusines were isolated from the deep-water New Caledonian sponge *Orina* sp. (*1360*). These include the racemic (±)-gelliusine C (**1414**, **1415**), (±)-gelliusine D (**1416**, **1417**), (±)-gelliusine E (**1418**, **1419**), and (±)-gelliusine F (**1420**, **1421**). Echinosulfonic acids B

(**1422**) and C (**1423**) along with echinosulfone A (**1424**) are produced by the Southern Australian sponge *Echinodictyum* sp. (*1361*). Echinosulfonic acid A (**1422**, ethoxy in place of methoxy) is probably an artifact produced during storage of the sponge in ethanol. The New Caledonian sponge *Psammoclemma* sp. has afforded echinosulfonic acid D (**1425**) along with echinosulfonic acid B (**1422**) (*1362*). The Papua New Guinea sponge *Coscinoderma* sp. contains coscinamides A (**1426**) and C (**1427**), which are the first reported alkaloids from this genus (*1363*).

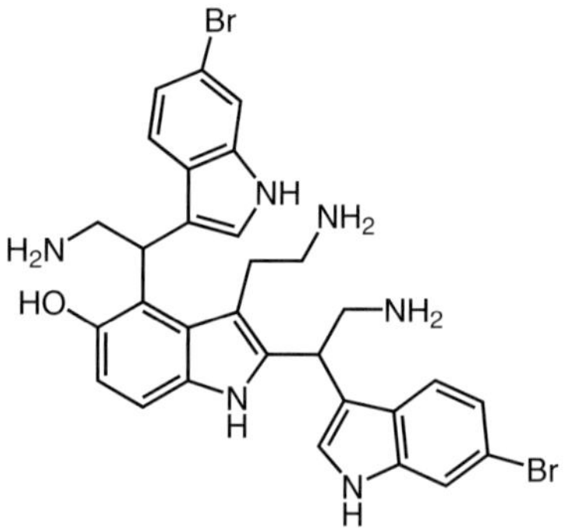

1414, 1415 (gelliusine C)

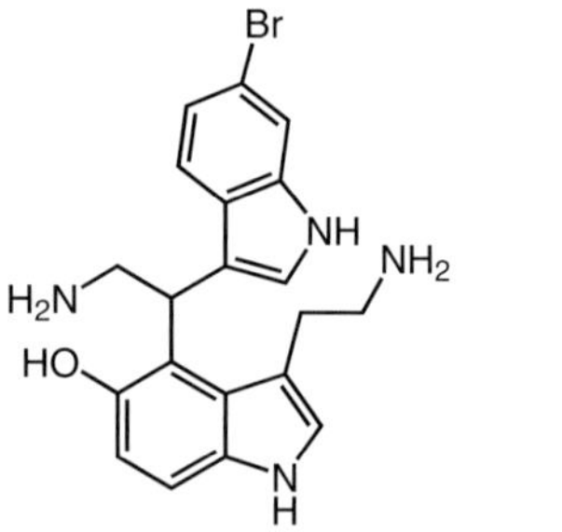

1416, 1417 (gelliusine D)

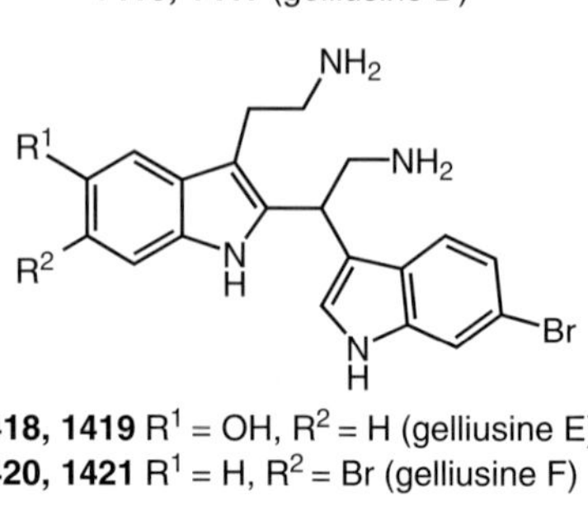

1418, 1419 R^1 = OH, R^2 = H (gelliusine E)
1420, 1421 R^1 = H, R^2 = Br (gelliusine F)

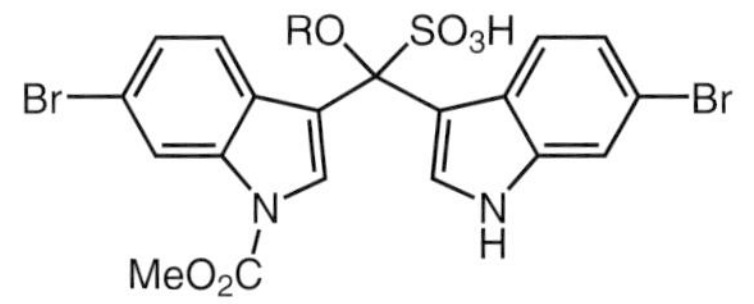

3.14 Heterocycles

1422 R = Me (echinosulfonic acid B)
1423 R = H (echinosulfonic acid C)
R = Et (echinosulfonic acid A)

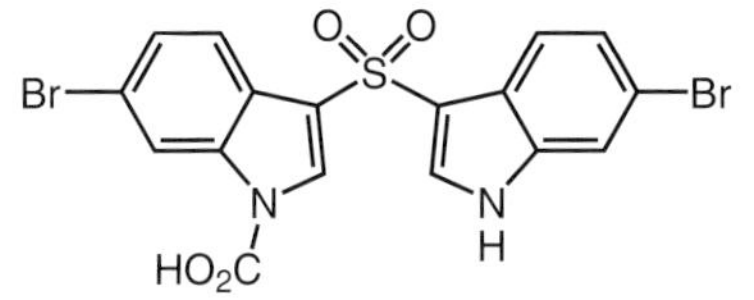

1424 (echinosulfone A)

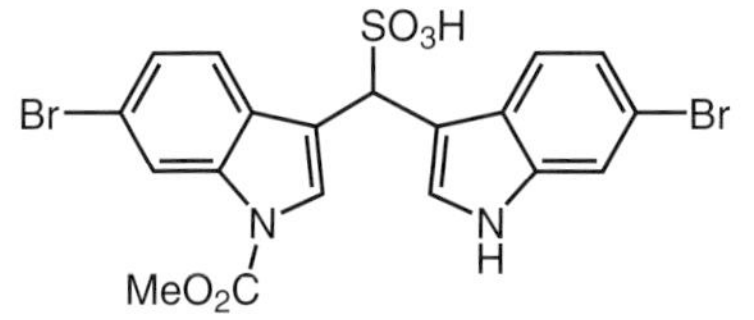

1425 (echinosulfonic acid D)

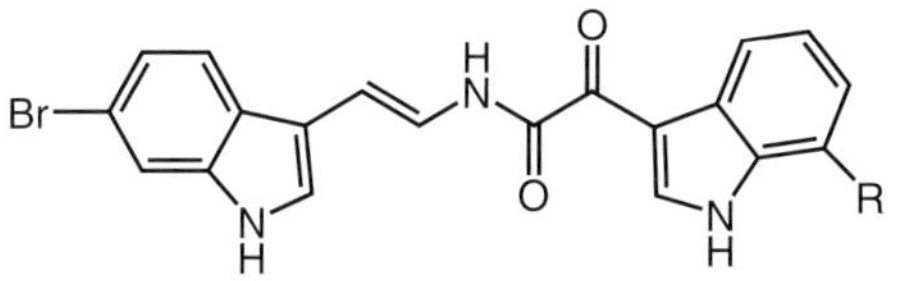

1426 R = H (coscinamide A)
1427 R = OH (coscinamide C)

An Okinawan collection of the red alga *Laurencia brongniartii* yielded the five new polybromoindoles **1428–1432** (*1364*); the latter compound is similar to **1340** and **1341**. Halogenated biindole **1433** was isolated from the South China Sea green alga *Chaetomorpha basiretorsa* (*1365*). The Okinawan sponge *Dictyodendrilla* sp. provided the novel brominated bis-tryptamine dendridine A (**1434**) (*1366*). The colonial Philippine ascidian *Perophora namei* produces the complex fused indole perophoramidine (**1435**), which is the first metabolite to be reported from the genus *Perophora* (*1367*). The Antarctica tunicate *Aplidium cyaneum* (Fig. 3.21) produces aplicyanins A–F (**1436–1440**) (*1368*), which are related to the meridianins (**1381–1385**). Aplicyanins B (**1437**), D (**1439**), E (**1440a**), and F (**1440b**) exhibit cytotoxic and antimiotic activities. Aplicyanin E (**1440a**) was readily acetylated to aplicyanin F (**1440b**).

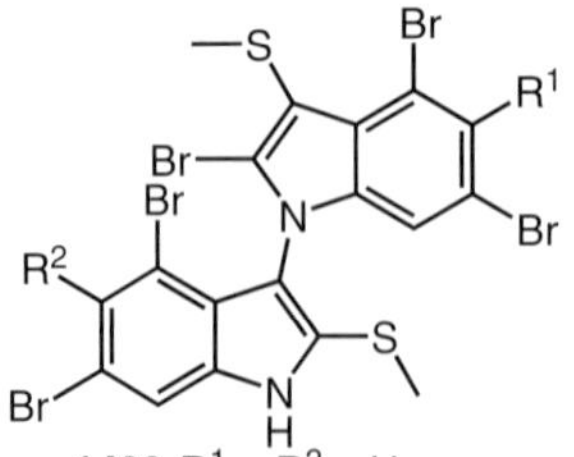

1428 R¹ = R² = H
1429 R¹ = H, R² = Br
1430 R¹ = Br, R² = H
1431 R¹ = R² = Br

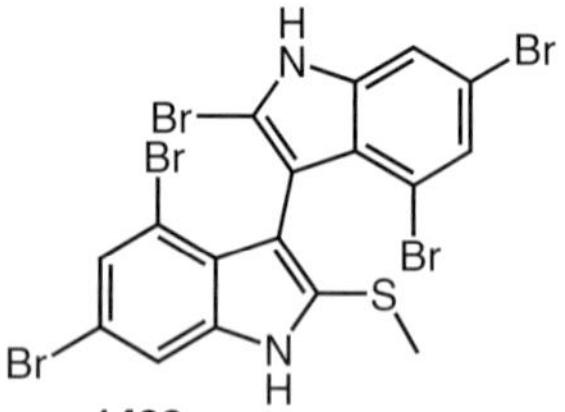

1432

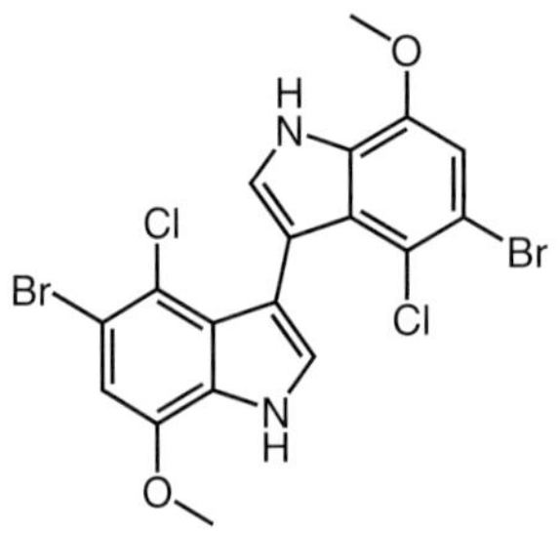

1433

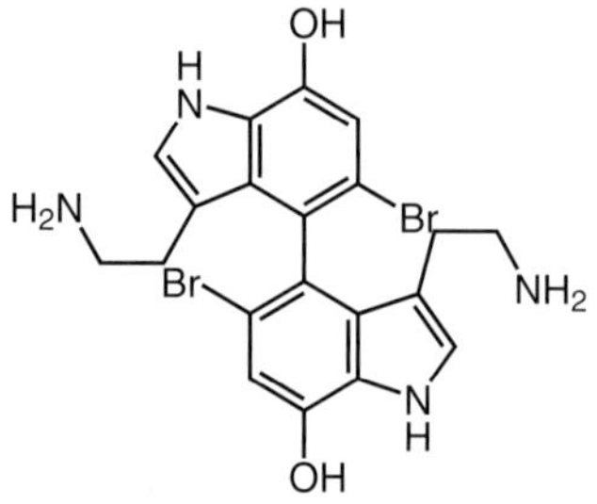

1434 (dendridine A)

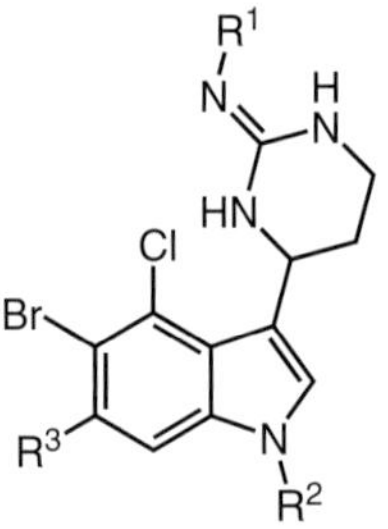

1435 (perophoramidine)

1436 R¹ = R² = R³ = H (aplicyanin A)
1437 R¹ = Ac, R² = R³ = H (aplicyanin B)
1438 R¹ = R³ = H, R² = OMe (aplicyanin C)
1439 R¹ = Ac, R² = OMe, R³ = H (aplicyanin D)
1440a R¹ = H, R² = OMe, R³ = Br (aplicyanin E)
1440b R¹ = Ac, R² = OMe, R³ = Br (aplicyanin F)

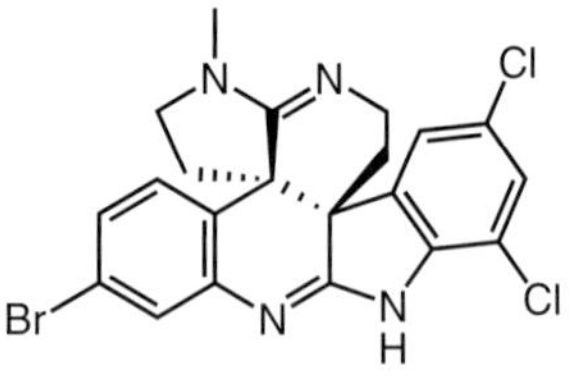

The most famous halogenated bis-indole is Tyrian purple, the dibromo analog of indigo (*1*). This colorful mollusc metabolite, which was the major component of the ancient dye, continues to receive attention (*1369–1374*). Additional studies of Tyrian purple from various molluscs have revealed the presence of 6,6′-dibromoindirubin (**1441**) (*1375*), 6-bromoindigotin (**1442**) (*1375, 1376*), 6-bromoisatin (**1443**) (*1377, 1378*), 6-bromoindoxyl (**1444**) (*1378*), **1445** (*1378*), **1446** (*1378*), 6-bromoindirubin (**1447**) (*1379*), and 6′-bromoindirubin (**1448**) (*1379*). The brominated indirubins are potent and selective kinase inhibitors (*1379*). A direct-exposure

3.14 Heterocycles

mass spectrometry technique has been developed to characterize the constituents of indigo and Tyrian purple extracts (*1380*). A *Streptomyces* sp. has furnished the chlorinated indigo glycosides akashins A–C (**1449–1451**), which display significant antitumor activity against various human cell lines (*1381*).

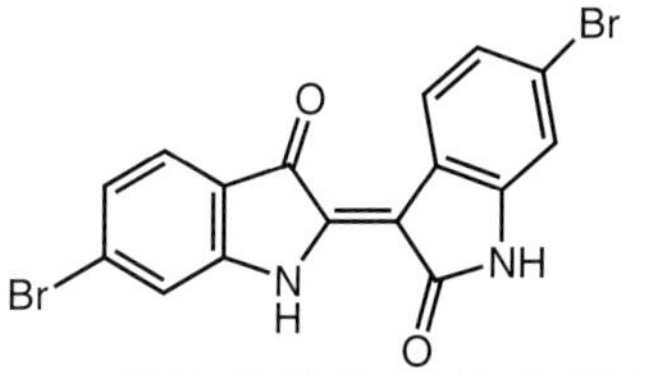

1441 (6,6'-dibromoindirubin)

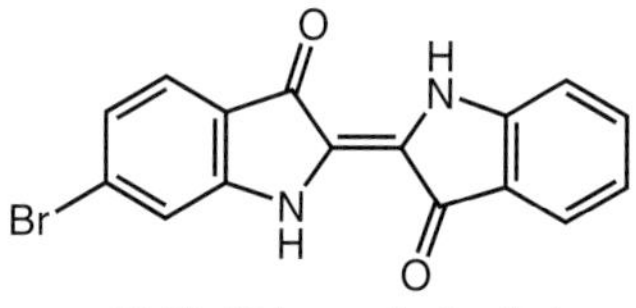

1442 (6-bromoindigotin)

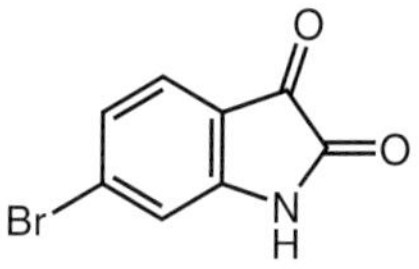

1443

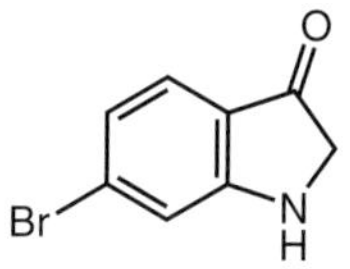

1444

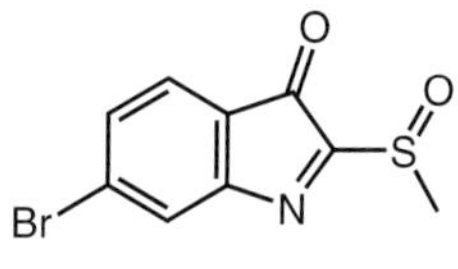

1445

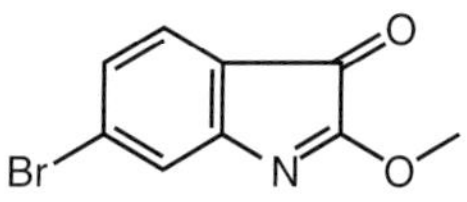

1446

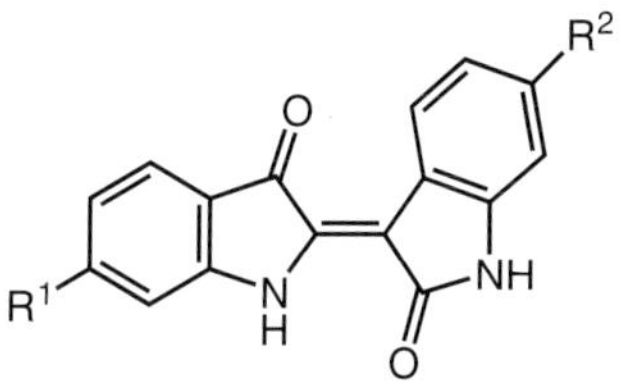

1447 R^1 = H, R^2 = Br (6-bromoindirubin)
1448 R^1 = Br, R^2 = H (6-bromoindirubin)

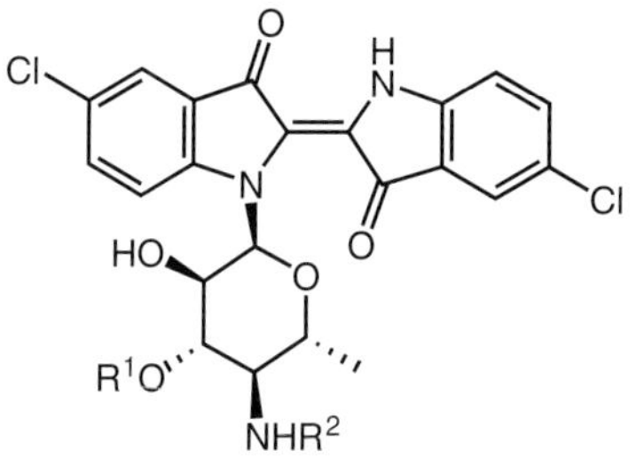

1449 R^1 = R^2 = H (akashin A)
1450 R^1 = H, R^2 = COCH$_3$ (akashin B)

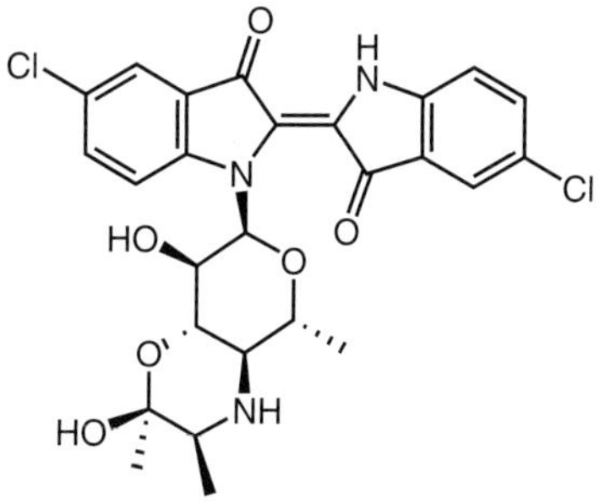

1451 (akashin C)

As illustrated in the first survey, marine bryozoans ("moss animals") are preeminent practitioners of organic synthesis, particularly in the production of halogenated indoles. Thus, the North Sea bryozoan *Securiflustra securifrons* produces securamines A-G (**1452–1458**). Compounds A (**1452**) and B (**1453**) are in equilibrium with their tautomers, securines A and B, as shown (*1382, 1383*). The prolific North Sea bryozoan *Flustra foliacea* (Fig. 3.19) has yielded the new hexahydropyrrolo[2,3-*b*]indole **1459** (*1321*). Total syntheses of the structurally related and previously known flustramines A–C and flustramides A and B have been achieved (*1384–1386*), as has the total synthesis of the known chartelline C (*1387, 1388*).

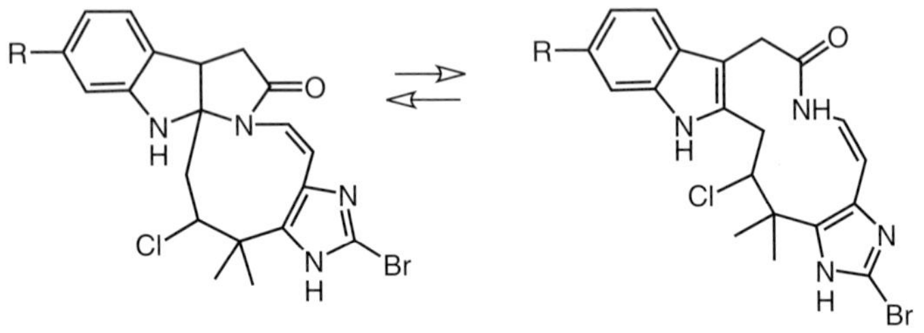

1452 R = H (securamine A)
1453 R = Br (securamine B)

R = H (securine A)
R = Br (securine B)

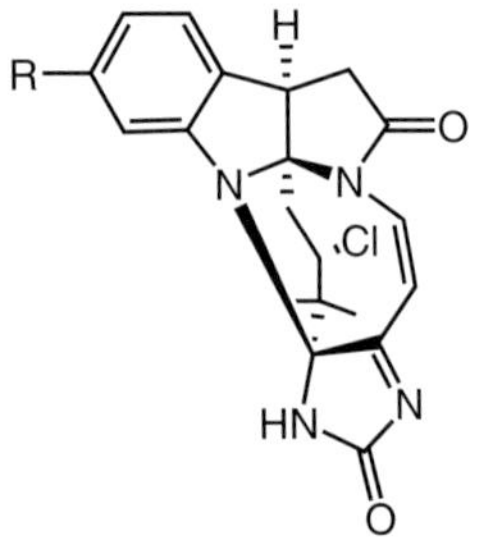

1454 R = Br (securamine C)
1455 R = H (securamine D)

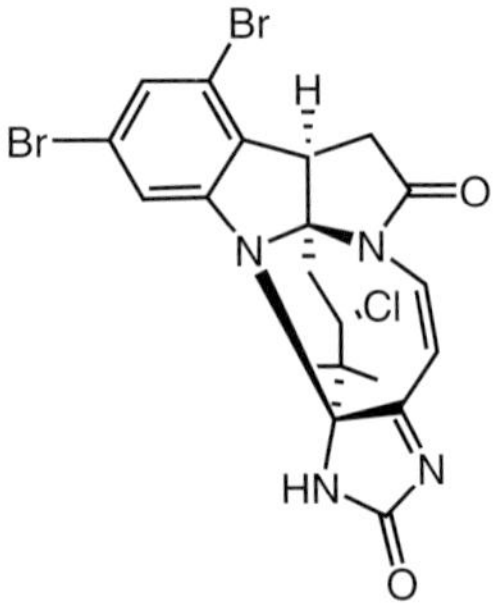

1456 (securamine E)

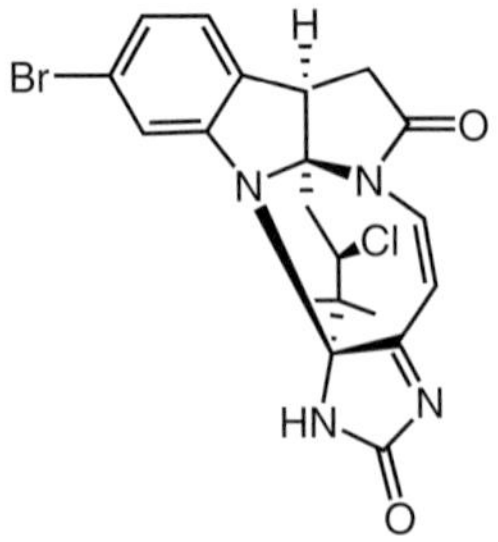

1457 (securamine F)

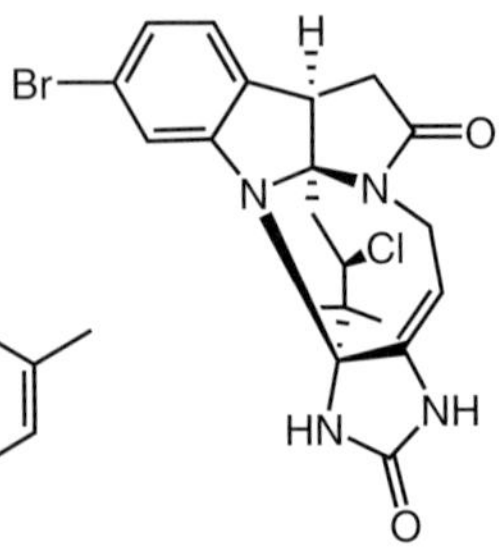

1458 (securamine G)

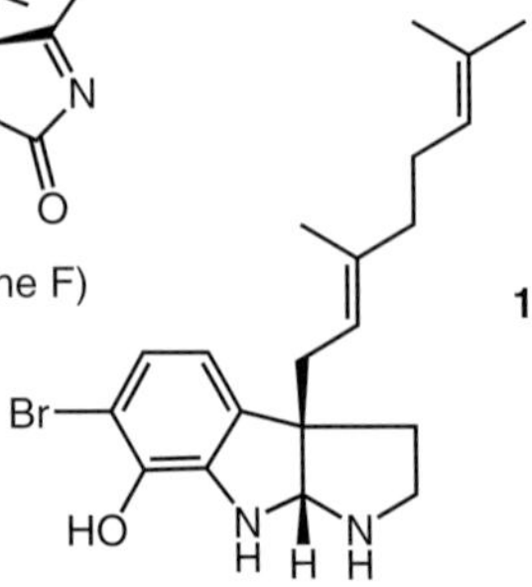

1459

3.14 Heterocycles

Several species of blue-green algae contain four groups of complex halogenated (and nonhalogenated) indoles: hapalindoles, ambiguines, fischerindoles, and welwitindolinones (*1*). New examples of these interesting metabolites have been discovered, such as 12-*epi*-hapalindole G (**1460**) from the blue-green alga *Hapalosiphon laingii* (*1389*) and ambiguine G nitrile (**1461**) from *Hapalosiphon delicatulus*, the first nitrile found in the Stigonemataceae (*1390*). The terrestrial *Fischerella muscicola* contains 3-hydroxy-*N*-methylwelwitindolinone C isonitrile (**1462**) and 3-hydroxy-*N*-methylwelwitindolinone C isothiocyanate (**1463**) (*1391*). The welwitindolinones can reverse P-glycoprotein mediated multiple drug resistance (*1392, 1393*). Total syntheses of the previously known welwitindolinone A and fischerindoles I and G have been accomplished (*1394, 2648*). The first ergoline marine alkaloids, pibocins A (**1464**) and B (**1465**), were isolated from the Far Eastern ascidian *Eudistoma* sp. (*1395, 1396*).

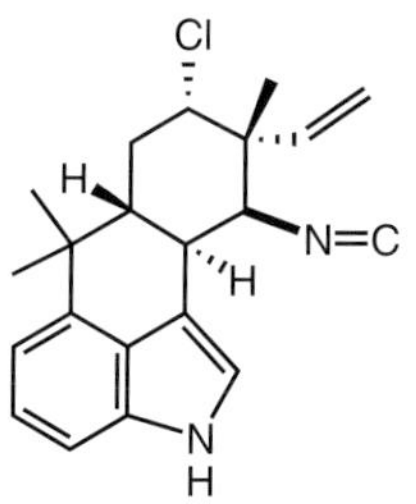

1460 (12-*epi*-hapalindole G)

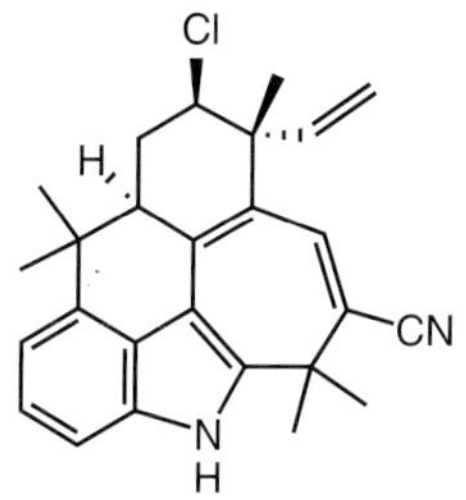

1461 (ambiguine G nitrile)

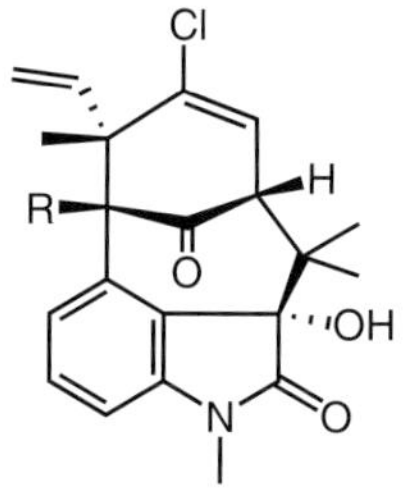

1462 R = NC
1463 R = NCS

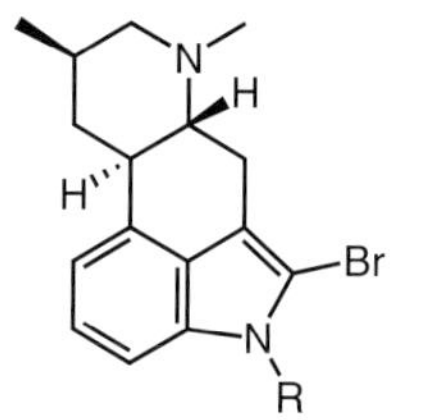

1464 R = H (pibocin A)
1465 R = OMe (pibocin B)

A new halogen-containing member of the penitrem family of indole-diterpenoids, which have insecticidal activity (*1397*), is thomitrem A (**1466**) from *Penicillium crustosum* (*1398*). The novel dichlorinated calmodulin inhibitor, malbrancheamide (**1467**), was characterized from the fungus *Malbranchea aurantiaca* (*1399*). The microbe *Streptomyces rugosporus* produces pyrroindomycin B (**1468**), which is active against both methicillin-resistant *Staphylococcus aureus* and vancomycin-resistant *Enterococci* (*1400*). The Chinese shrub *Acacia confusa* has yielded the unusual chlorotryptamine alkaloid **1469**, which does not appear to be an artifactual dichloromethane adduct (*1401*).

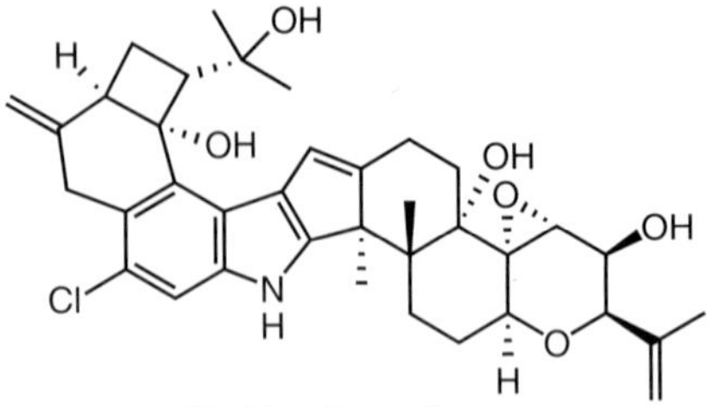

1466 (thomitrem A)

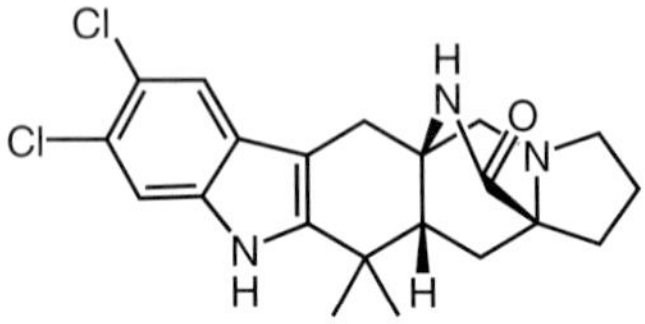

1467 (malbrancheamide)

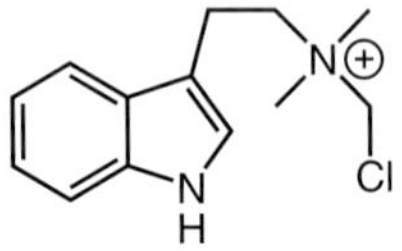

1469

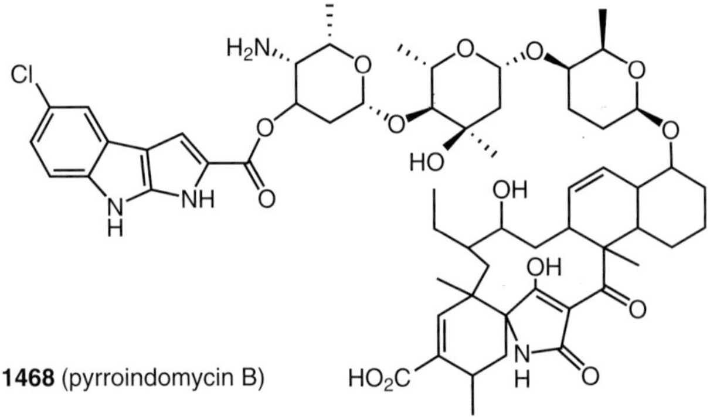

1468 (pyrroindomycin B)

The earlier survey documented a number of indoloquinones and related compounds, the discorhabins, makaluvamines, and batzellines (*1*), which have anticancer activity and are DNA topoisomerase II inhibitors (*1402*). Several new examples of these fused indoles have been uncovered in the interim (*1403, 1404*). The Antarctic sponge *Latrunculia apicalis* contains the novel discorhabdin G (**1470**), which is a feeding deterrent towards the sea star *Perknaster fuscus*, the major Antarctic sponge predator (*1405, 1406*). A South African undescribed latrunculid sponge has afforded 14-bromodiscorhabdin C (**1471**) and 14-bromodihydrodiscorhabdin C (**1472**), which are the first discorhabdins having a 2-bromoindole unit (*1407*). Discorhabdin P (**1473**) is present in the Bahamian sponge *Batzella* sp., and inhibits phosphatase activity (*1408*). The Australian sponges *Latrunculia purpurea*, *Zyzzya massalis*, *Zyzzya fuliginosa*, and *Zyzzya* spp. contain discorhabdin Q (**1474**) (*1409*). Discorhabdins S (**1475**), T (**1476**), and U (**1477**) were isolated from the deep-water Caribbean sponge *Batzella* sp. (*1410*). The newly classified South African sponges *Tsitsikamma pedunculata*, *Tsitsikamma favus*, *Latrunculia bellae*, and *Strongylodesma algoensis* yielded 3-dihydro-7,8-dehydrodiscorhabdin C

3.14 Heterocycles

(**1478**), 14-bromo-3-dihydro-7,8-dehydrodiscorhabdin C (**1479**), discorhabdin V (**1480**), and 14-bromo-1-hydroxydiscorhabdin V (**1481**) (*1411*). The novel dimeric discorhabdin W (**1482**) is present in the New Zealand sponge *Latrunculia* sp. and exhibits potent cytotoxicity against the P388 murine leukemia cell line (*1412*). The total synthesis of the previously known discorhabdin A has been achieved (*1413, 1414*), as have semi-syntheses of discorhabdins P (**1473**) and U (**1477**) (*1415*).

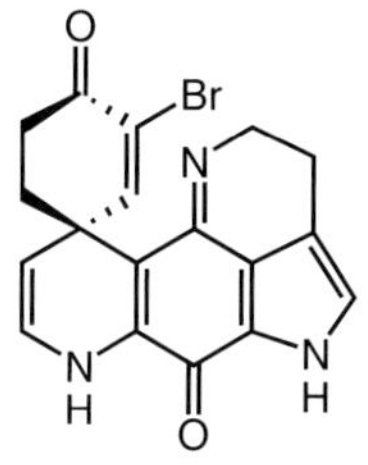

1470 (discorhabdin G)

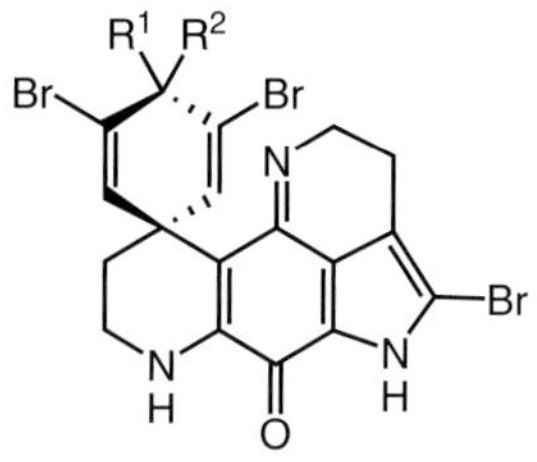

1471 $R^1, R^2 = O$
1472 $R^1 = OH, R^2 = H$

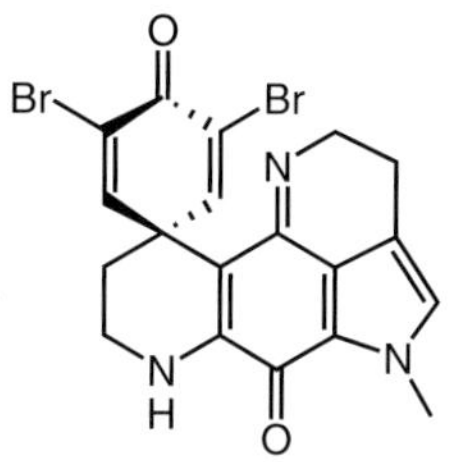

1473 (discorhabdin P)

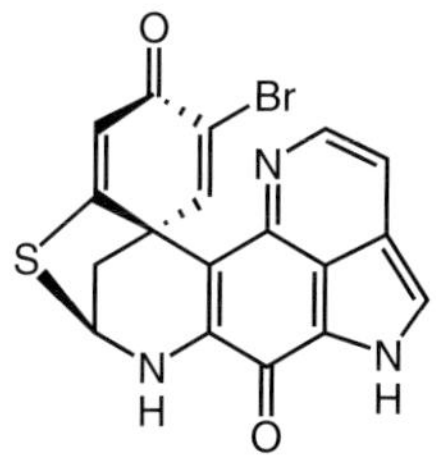

1474 (discorhabdin Q)

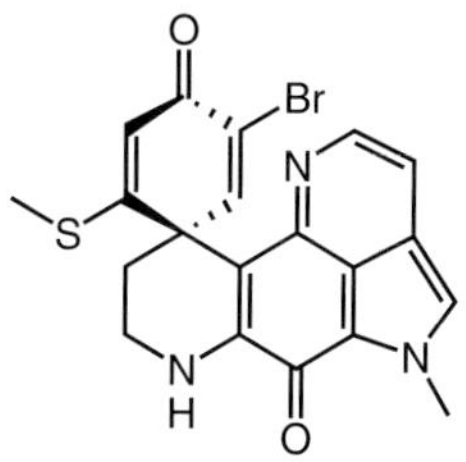

1475 (discorhabdin S)

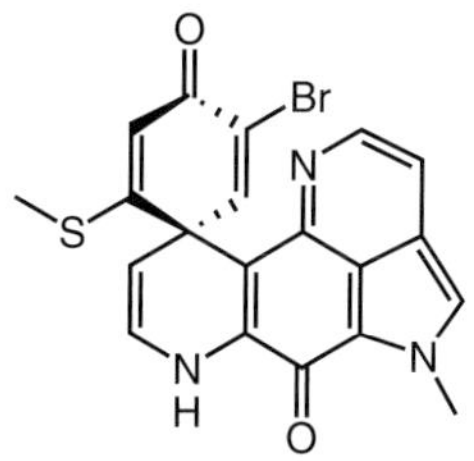

1476 (discorhabdin T)

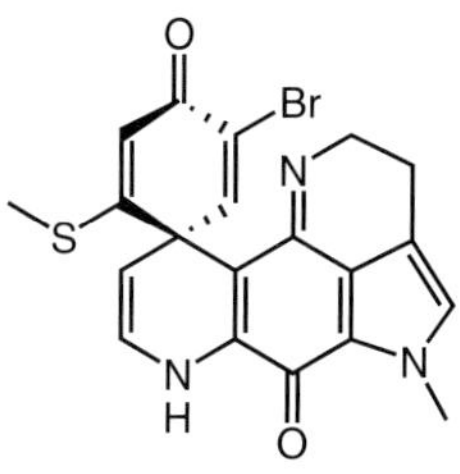

1477 (discorhabdin U)

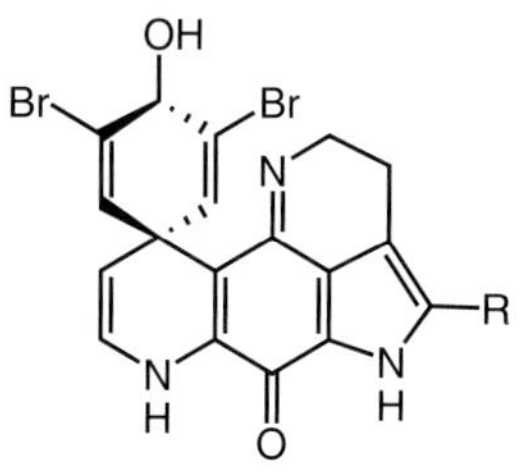

1478 R = H
1479 R = Br

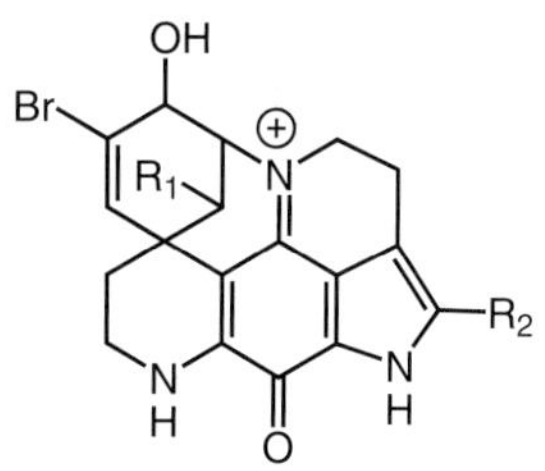

1480 (discorhabdin V)
1481

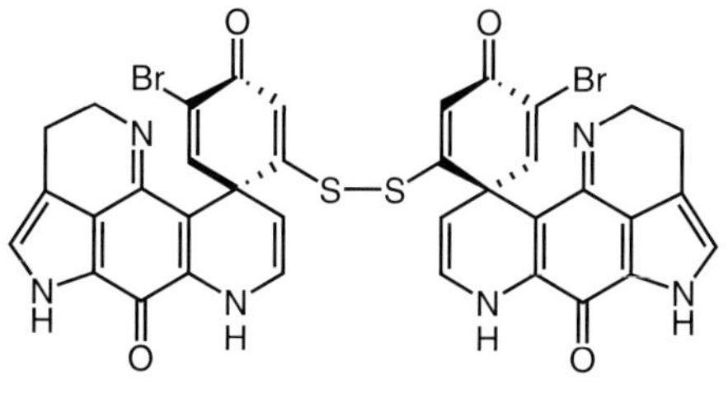

1482 (discorhabdin W)

Related to the discorhabdins are epinardins B–D (**1483–1485**), which were isolated from undetermined deep-water green demosponges from pre-Antarctic Indian Ocean waters (*1416*). Epinardin C (**1485**) is strongly cytotoxic towards doxorubicin-resistant L1210/DX tumor cells. The Philippino sponge *Zyzzya fuliginosa* contains makaluvamine N (**1486**) in addition to five related known compounds (*1417*). A study of this sponge from Papua New Guinea has yielded batzelline D (**1487**) and isobatzelline E (**1488**) (*1418*). Four collections of *Zyzzya* were analyzed for antitumor activity and three nonhalogenated makaluvamines were the most potent (*1419*). A deep-water *Batzella* sponge from the Bahamas has furnished secobatzelline A (**1489**), which is a potent inhibitor of the phosphatase activity of calcineurin and the peptidase function of CPP32 (*1420*). The Jamaican sponge *Smenospongia aurea* contains makaluvamine O (**1490**) (*1421*).

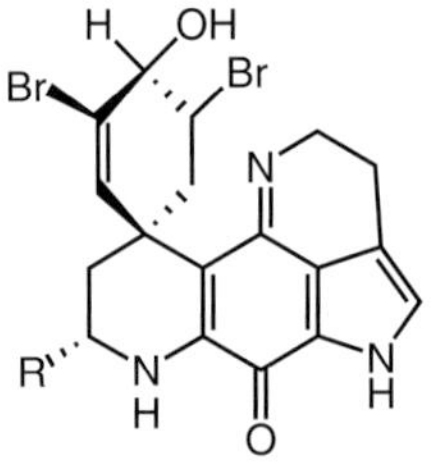

1483 R = H (epinardin B)
1484 R = OMe (epinardin D)

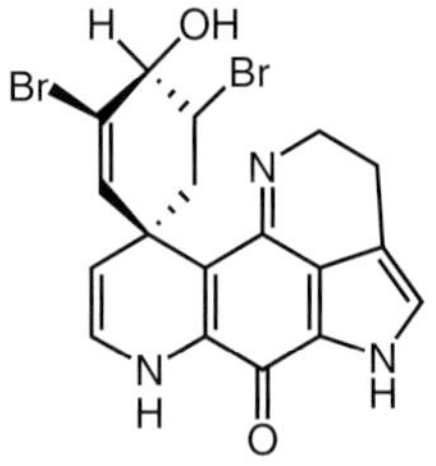

1485 (epinardin C)

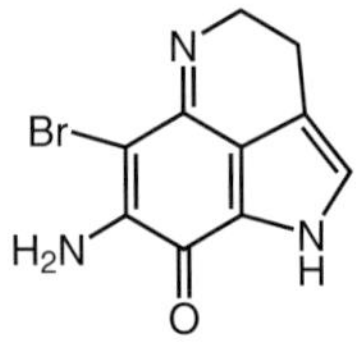

1486 (makaluvamine N)

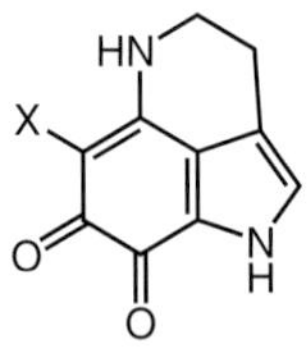

1487 X = Cl (batzelline D)
1490 X = Br (makaluvamine O)

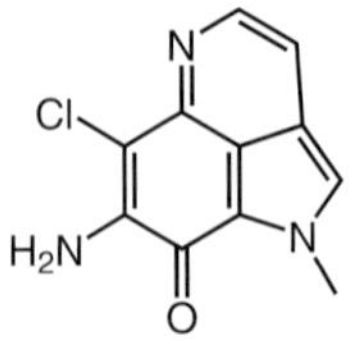

1488 (isobatzelline E)

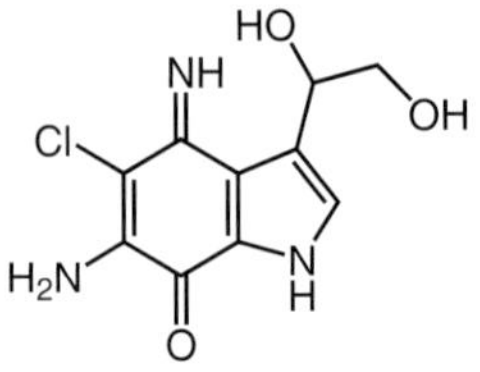

1489 (secobatzelline A)

Several halogenated metabolites with an oxidized indole ring have been discovered since the first survey (*1*). Indisocin (**1491**) was isolated from cultures of the actinomycete *Nocardia blackwellii* and displays strong antimicrobial activity against a range of both Gram-positive and Gram-negative bacteria and fungi (*1422*). The Floridian bryozoan *Amathia convoluta* produces convolutamydines A–E (**1492–1496**) (*1423–1425*), and total syntheses of convolutamydines A and B

3.14 Heterocycles

established their absolute configuration (*1426–1428*). The Indian Ocean sponge *Iotrochota purpurea* contains matemone (**1497**), which inhibits division of sea-urchin eggs (*1429*). Cynthichlorine (**1498**) was isolated from the Moroccan tunicate *Cynthia savignyi* and displays both antifungal and antibacterial activity (*1430*).

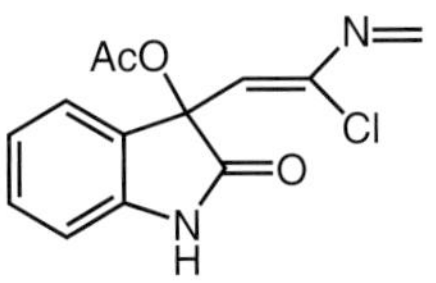

1491 (indisocin)

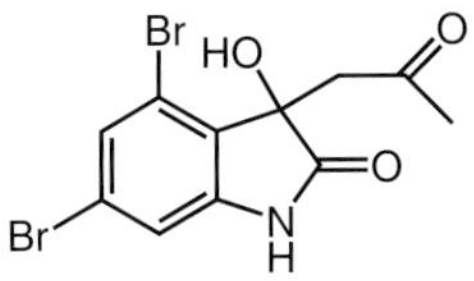

1492 (convolutamydine A)

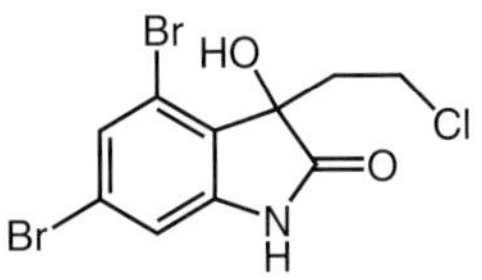

1493 (convolutamydine B)

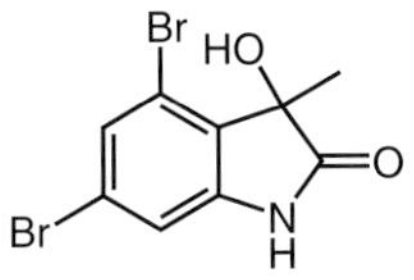

1494 (convolutamydine C)

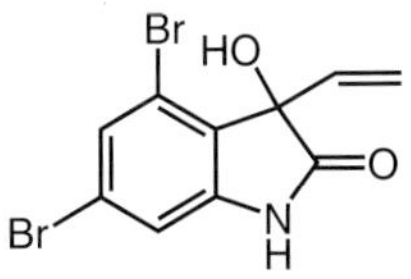

1495 (convolutamydine D)

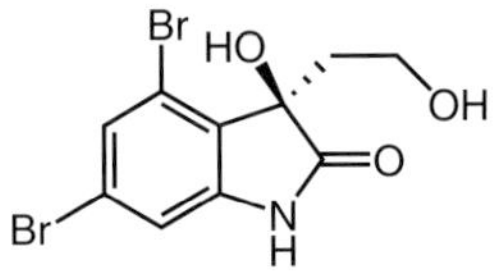

1496 (convolutamydine E)

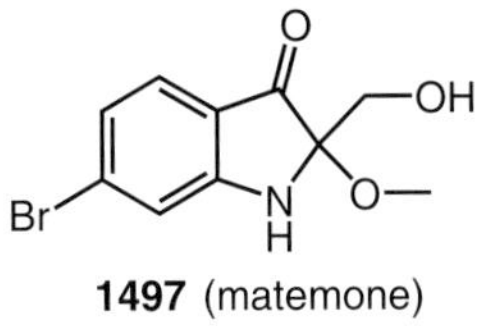

1497 (matemone)

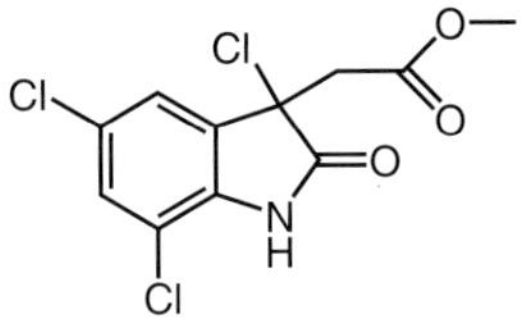

1498 (cynthichlorine)

3.14.3 Carbazoles

Newly discovered halogenated carbazoles will be presented in a future volume.

3.14.4 Indolocarbazoles

Although no new natural halogenated indolocarbazoles were reported following the 1996 survey (*1*), an enormous effort has focused on the discovery of synthetic analogs of the chlorine-containing rebeccamycin (*1431, 1432*), comprising fluoro-indolocarbazoles (*1432, 1433*), sugar analogs (*1434*), and others (*1435, 1436*), including combinatorial biosynthesis (*1437*). A synthetic rebeccamycin analog is in phase II trials for metastatic renal cell cancer (*1438*). The biosynthesis of rebeccamycin is also of interest (*1439–1441*).

3.14.5 Carbolines

The previous survey presented more than 20 halogenated carbolines from ascidians of genus *Eudistoma*, *Ritterella*, and others (*1*). The Mariana Islands ascidian *Didemnum* sp. has afforded didemnolines A (**1499**) and C (**1500**), along with two nonhalogenated analogs (*1442*). A Western Australia *Eudistoma* sp. ascidian contains 19-bromoisoeudistomin U (**1501**) (*1443*), and the Australian ascidian *Pseudodistoma aureum* provided eudistomin V (**1502**) (*1444*). A Palau *Eudistoma gilboverde* has afforded 2-methyleudistomin D (**1503**), 2-methyleudistomin J (**1504**), and 14-methyleudistomidin C (**1505**) (*1445*). The latter metabolite displays excellent cytotoxicity towards four different human tumor cell lines. The Palauan sponge *Plakortis nigra*, which was collected at a depth of 380 feet, contains plakortamines A–D (**1506–1509**). These compounds are active against the HCT-116 human colon cell line with **1507** being the most active (*1446*). Total syntheses of several eudistomins are known (*1447, 1448*).

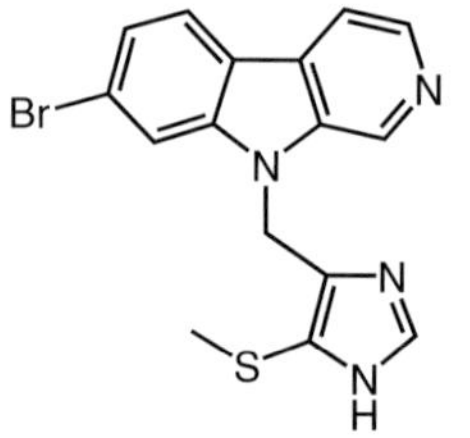

1499 (didemnoline A)

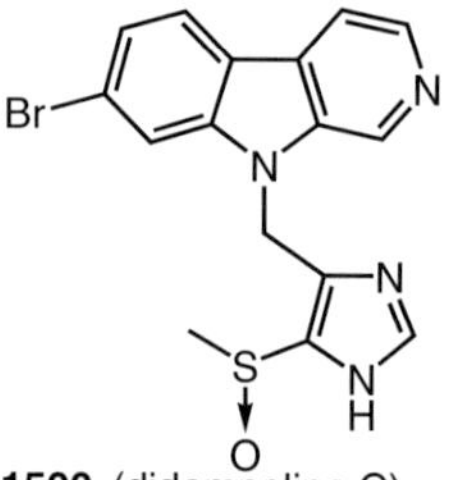

1500 (didemnoline C)

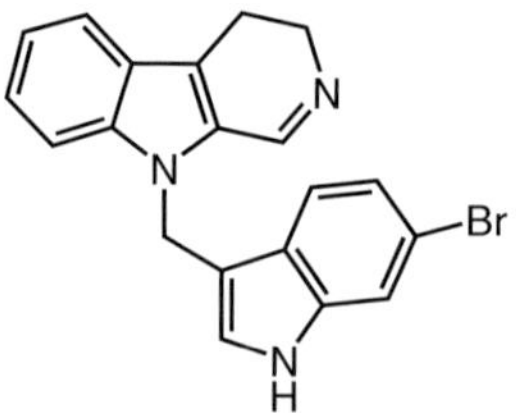

1501 (19-bromoisoeudistomin U)

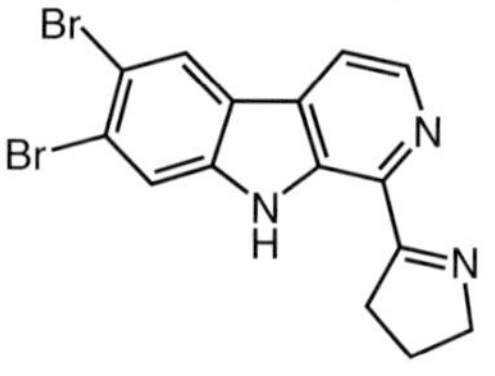

1502 (eudistomin V)

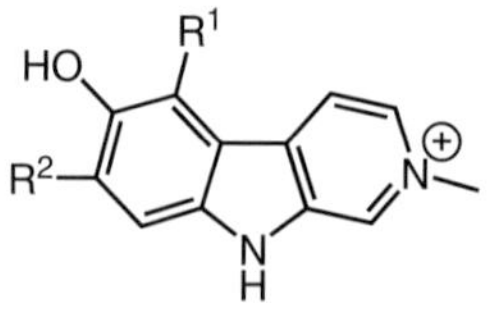

1503 R^1 = Br, R^2 = H (2-methyleudistomin D)
1504 R^1 = H, R^2 = Br (2-methyleudistomin J)

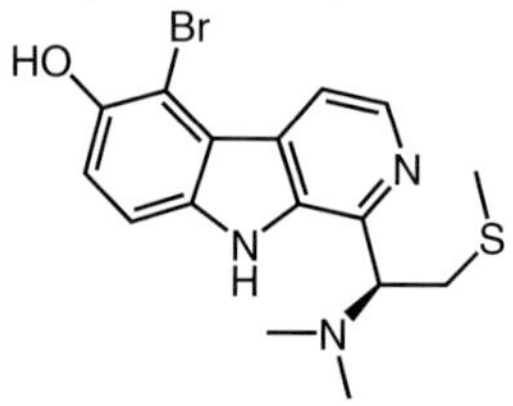

1505 (14-methyleudistomidin C)

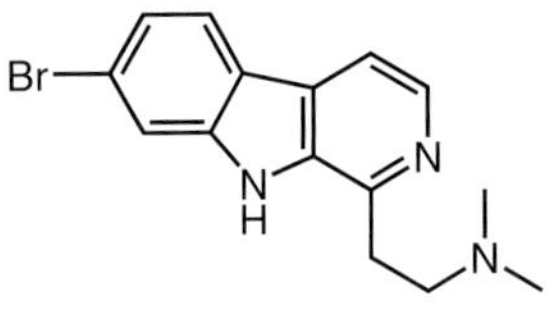

3.14 Heterocycles

1506 (plakortamine A)

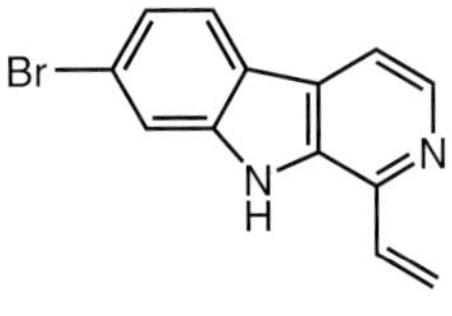

1507 (plakortamine B)

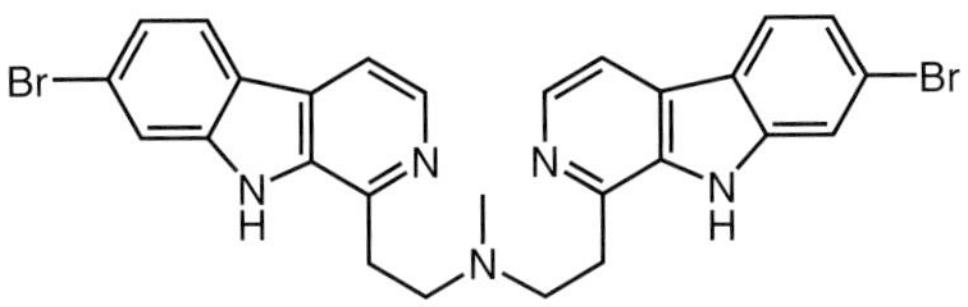

1508 (plakortamine C)

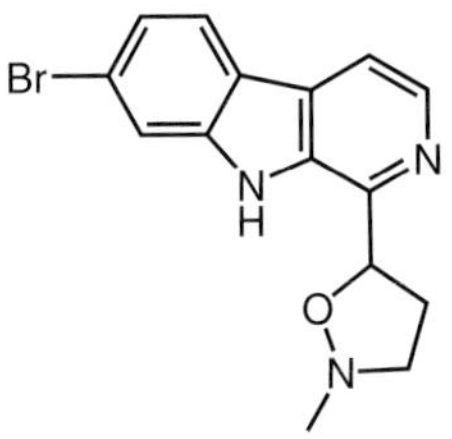

1509 (plakortamine D)

Studies of the sponge *Fascaplysinopsis reticulata* and the tunicate *Didemnum* sp. (Fig. 3.20) have identified several brominated derivatives of the previously known nonbrominated fascaplysin and reticulatine, including 3-bromofascaplysin (**1510**), 14-bromoreticulatine (**1511**), and 14-bromoreticulatate (**1512**) (*1449*). Further investigation of these organisms uncovered 10-bromofascaplysin (**1513**), 3,10-dibromofascaplysin (**1514**), 3-bromohomofascaplysin B (**1515**), 3-bromohomofascaplysin B-1 (**1516**), 3-bromohomofascaplysin C (**1517**), 7,14-dibromoreticulatine (**1518**), 14-bromoreticulatol (**1519**), 3-bromosecofascaplysin A (**1520**), and 3-bromosecofascaplysin B (**1521**) (*1450*). The fresh water cyanobacterium *Nostoc* 78-12A has provided nostocarboline (**1522**), which was synthesized for structural confirmation (*1451, 1452*). Nostocarboline is a potent butyrylcholinesterase inhibitor, comparable to the Alzheimer's disease drug galanthamine. This β-carboline and some derivatives are also potent algicides (*1452*).

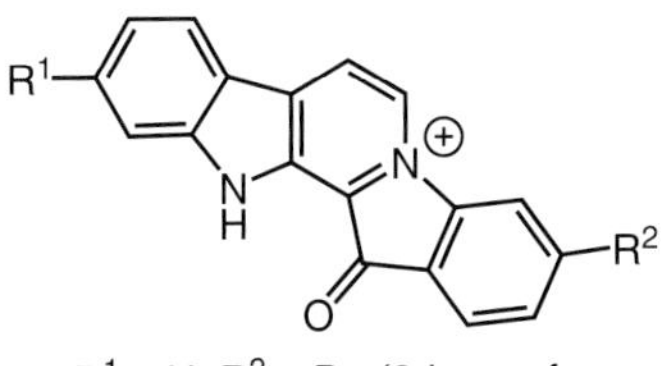

1510 R^1 = H, R^2 = Br (3-bromofascaplysin)
1513 R^1 = Br, R^2 = H (10-bromofascaplysin)
1514 R^1 = R^2 = Br (3,10-bromofascaplysin)

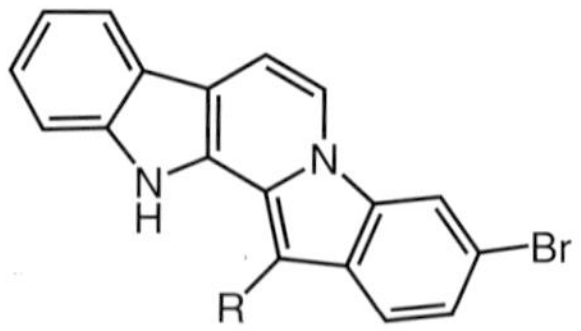

1515 R = COCO$_2$Me (10-bromohomofascaplysin B)
1516 R = COCO$_2$Et (3-bromohomofascaplysin B-1)
1517 R = CHO (3-bromohomofascaplysin C)

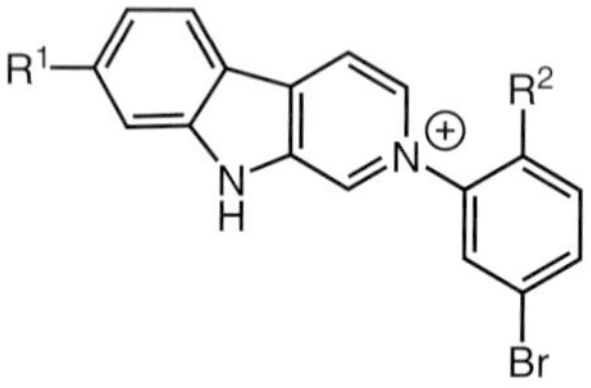

1511 R^1 = H, R^2 = CO$_2$Me (14-bromoreticulatine)
1512 R^1 = H, R^2 = CO$_2$ (14-bromoreticulatate)
1518 R^1 = Br, R^2 = CO$_2$Me (7,14-dibromoreticulatine)
1519 R^1 = H, R^2 = OH (14-bromoreticulatol)

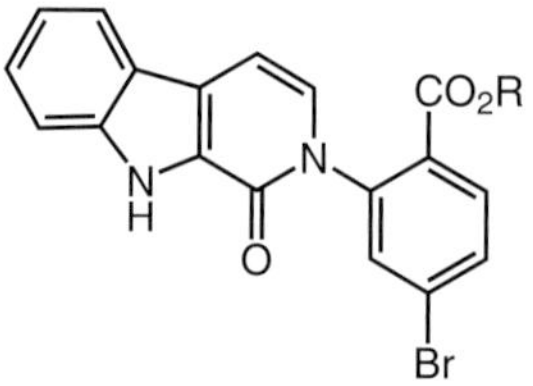

1520 R = Me (3-bromosecofascaplysin A)
1521 R = H (3-bromosecofascaplysin B)

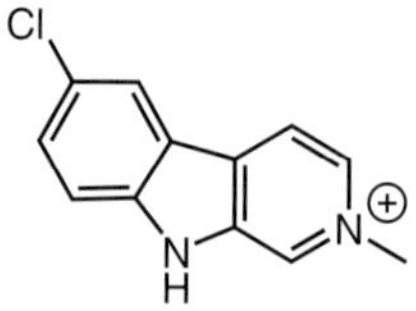

1522 (nostocarboline)

3.14.6 *Quinolines and Other Nitrogen Heterocycles*

Unlike π-excessive nitrogen heterocycles (pyrroles, indoles), π-deficient nitrogen heterocycles are much less reactive towards electrophilic halogenation, and relatively few halogenated π-deficient heterocycles are found naturally (*1*). The Thai spiny herb *Acanthus ilicifolius* contains several benzoxazinoid glucosides, including the chlorine-containing **1523** (*1453*). This medicinal plant is distributed in the mangroves of southern Thailand. The *N*-hydroxy derivative, **1524**, of **1523** is found in the medicinal mangrove plant *Acanthus ebracteatus* (*1454*). The sponge *Hyrtios erecta* has furnished the novel quinolones **1525** and **1526** (*1334*). A collection of a Puerto Rican *Lyngbya majuscula* cyanobacterium has yielded **1527** (*853*), and three novel tetrahydroquinolines, **1528–1530**, were characterized from the red alga *Rhodomela confervoides* (*1455*). The Palau bryozoan *Caulibugula intermis*

3.14 Heterocycles

Fig. 3.20 *Didemnum* sp., a tunicate that contains the fascaplysin alkaloids **1510–1521** (Photo: F. J. Schmitz)

contains caulibugulones B (**1531**) and C (**1532**), along with four nonhalogenated analogs (*1456*). These structures were confirmed by chemical interconversion (*1456*) and total synthesis (*1457*), and the compounds display cytotoxicity (*1456*) and potent phosphatase inhibitory activity (*1457*). The amide of the previously known virantmycin (*1*), benzastatin C (**1533**), is produced by *Streptomyces nitrosporeus* (*1458, 1459*). This compound is a potent free radical scavenger. The relative and absolute stereochemistry of (–)-virantmycin have been established by synthesis (*1460, 1461*). The New Zealand bryozoan *Euthyroides episcopalis* contains the novel quinone methides, euthyroideones A–C (**1534–1536**) (*1462*). Chlorodesnkolbisine (**1537**) was characterized from the African traditional medicine plant *Teclea nobilis* (*1463*). This chlorohydrin is detected in crude hexane extracts of the plant and is excluded as an artifact. Likewise, the new acridone alkaloid A6 (**1538**), which is found in several *Ruta* plants (*Ruta bracteosa*, *Ruta macrophylla*, *Ruta chalepensis*), is not formed when the corresponding allylic alcohol (gravacridonol) is heated with HCl, and, therefore, **1538** is judged not to be artifactual (*1464*).

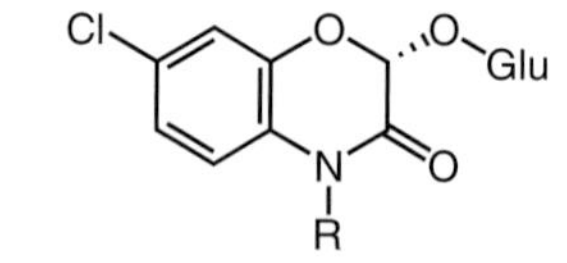

1523 R = H
1524 R = OH

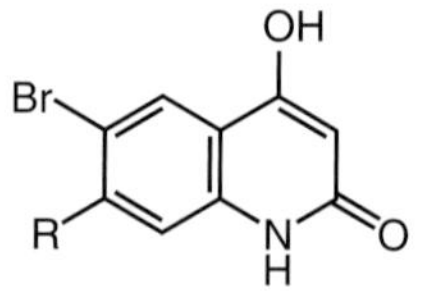

1525 R = H
1526 R = Br

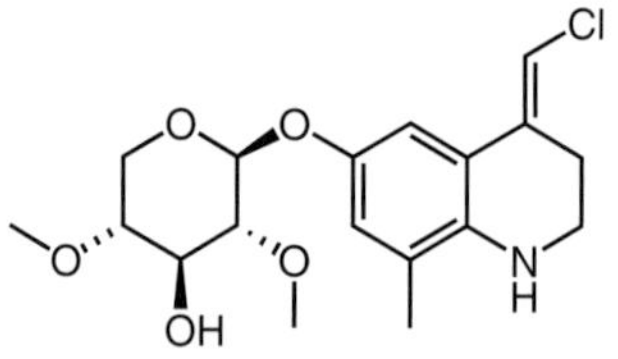

1527

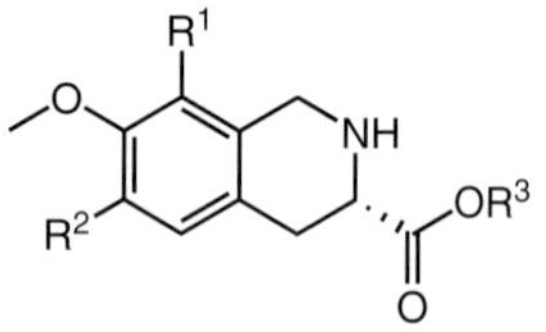

1528 R^1 = Br, R^2 = OH, R^3 = H
1529 R^1 = Br, R^2 = OH, R^3 = Me
1530 R^1 = OH, R^2 = Br, R^3 = Me

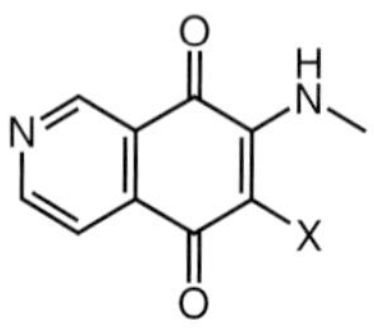

1531 X = Br (caulibugulone B)
1532 X = Cl (caulibugulone C)

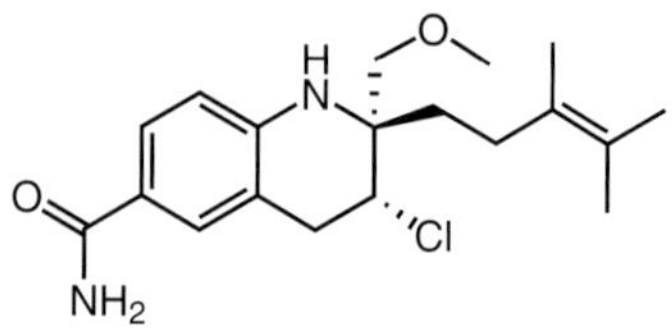

1533 (benzastatin C)

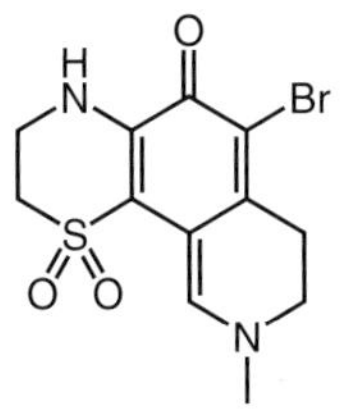

3.14 Heterocycles

1534 (euthyroideone A) 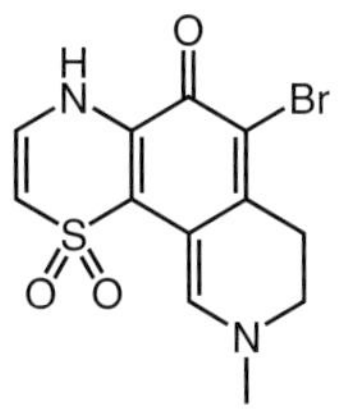 **1535** (euthyroideone B) 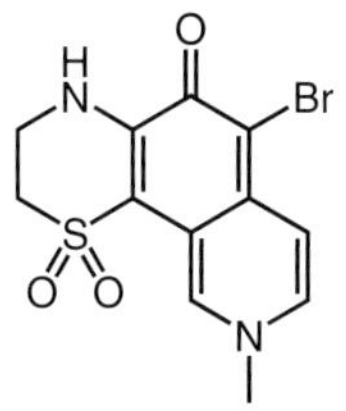 **1536** (euthyroideone C)

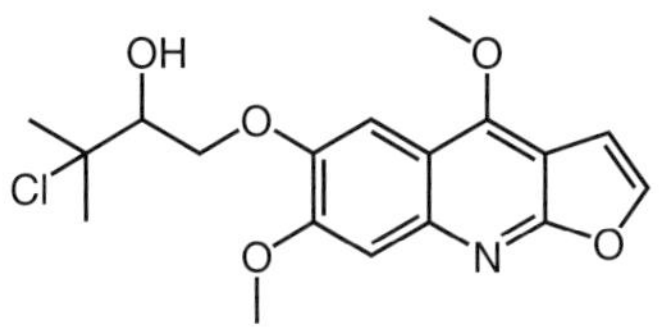

1537 (chlorodesnkolbisine) 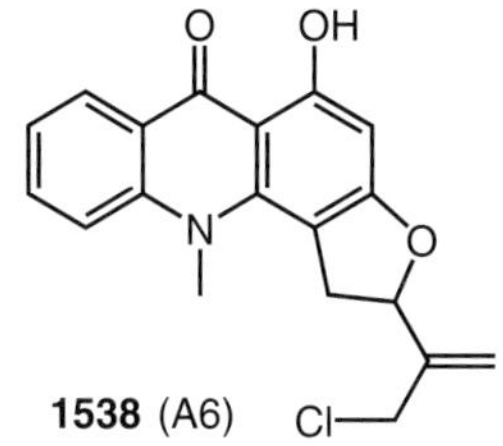 **1538** (A6)

The simple auxin inhibitor, 4-chloro-6,7-dimethoxy-2-benzoxazolinone (**1539**), was isolated from maize (*Zea mays*) (*1465*), as was the related 5-chloro-6-methoxy-2-benzoxazolinone (**1540**), which causes growth inhibition of crabgrass, ryegrass, lettuce, and oats (*1466*). The egg masses of three muricid molluscs (*Trunculariopsis trunculus*, *Ceratostoma erinaceum*, *Trophon geversianus*) contain 2,4,5-tribromo-1*H*-imidazole (**1541**) (*1467*), and the Indian medicinal plant *Jatropha curcas* has afforded chlorinated imidazole **1542** (*1468*). The structure of the *Streptomyces griseoluteus* metabolite 593A is incorrectly shown in the first survey ((*1*), compound "**1553**") and should be revised as shown **ZZ** (*1469*). The extraordinary cyclic *N*-bromoimide **1543** is claimed to be produced by the sponge *Rhaphisia pallida* (*1470*). The novel antifungal antibiotics atpenins A4 (**1544**) and A5 (**1545**) are produced by a *Penicillium* sp. and possess unique chloroalkane side chains (*1471*). X-ray crystallography supports the 2-hydroxypyridine tautomer.

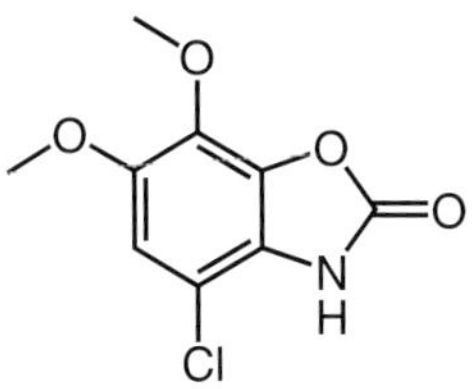

1539 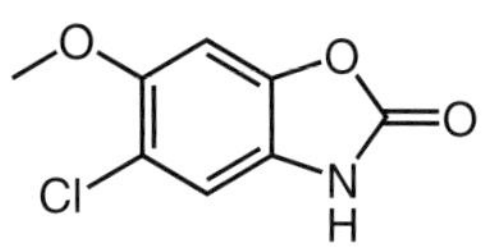 **1540** 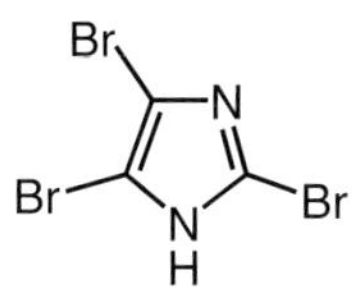 **1541**

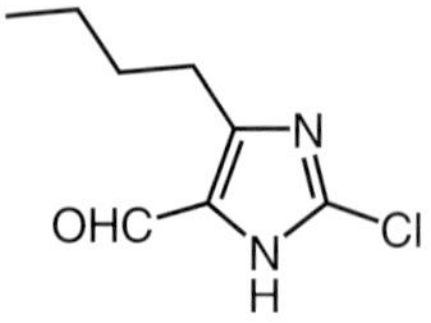

1542

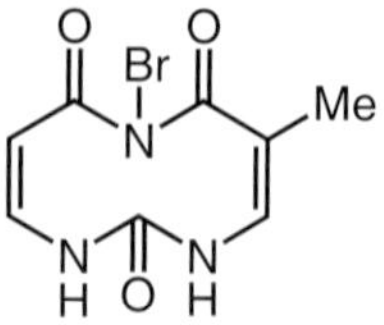

1543

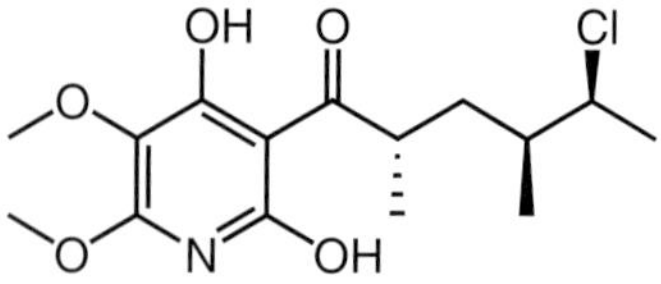

1544 (atpenin A4)

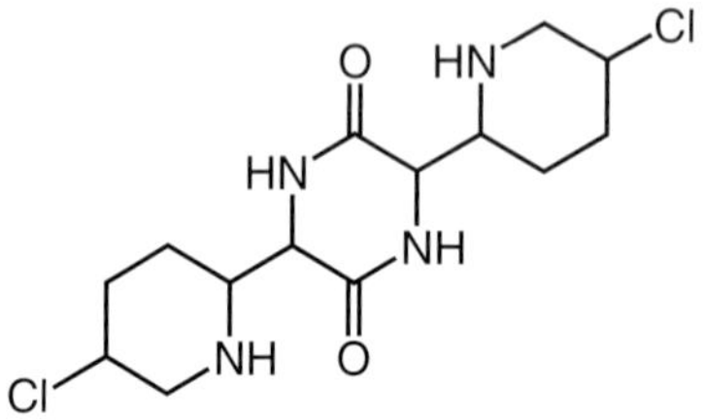

ZZ (593A)

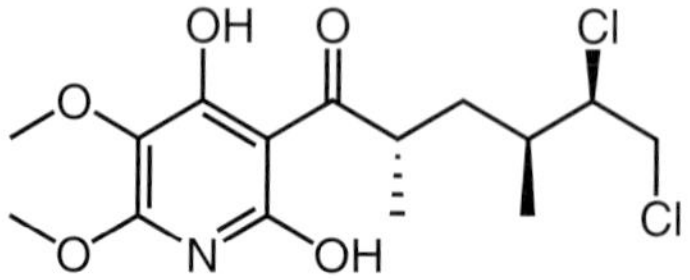

1545 (atpenin A5)

As reported in the first survey, several halogenated nucleic acid bases and nucleosides have been isolated from natural sources. These compounds are generally thought to arise from the action of myeloperoxidase on halide in the presence of DNA (*1*). New examples of halogenated nucleic acid bases and nucleosides include 5-chlorocytosine (**1546**) and 5-chlorodeoxycytidine (**1547**) from salmon sperm (*1472*). Kumusine (**1548**) was isolated from the Indonesian sponge *Theonella* sp. (*1473*), and as "trachycladine A" from the Western Australian sponge *Trachycladus laevispirulifer* (*1474*), and later from the sponge *Theonella cupola* (*1475*). The ascidian *Didemnum voeltzkowi* contains 5'-deoxy-3-bromotubercidin (**1549**) along with the previously known iodo analog (*1476*). The sponge *Phakellia fusca* has yielded five 5-fluorouracil derivatives **1550–1554** (*1477*). The action of myeloperoxidase on human inflammatory tissue produces 5-chlorouracil (**1555**) (*1478, 1479*), 5-bromouracil (**1556**) (*1478, 1480*), and 8-chloroadenine (**1557**) (*1481*), each of which is considered as being natural. It has been suggested that these halogenated nucleic acid bases, which are products of inflammation, may exert cytotoxic and mutagenic effects (*1478–1480, 1482, 1483*). Thus, the incorporation of 5-bromouracil into DNA results in mutagenesis (*1482*).

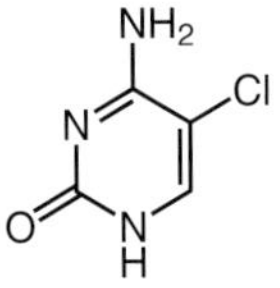

3.14 Heterocycles

1546 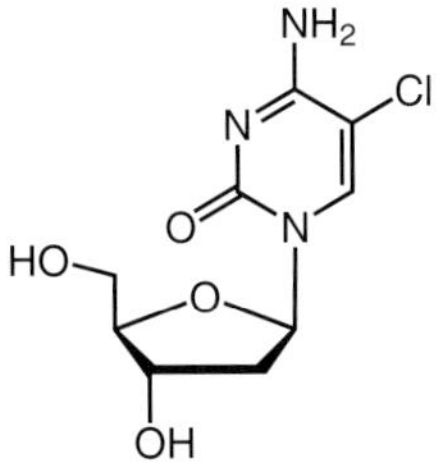 **1547** 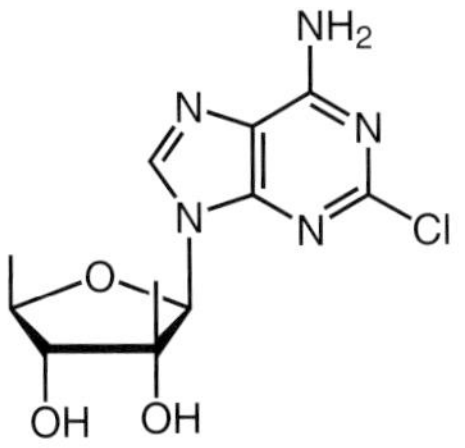 **1548** (kumusine)
1549 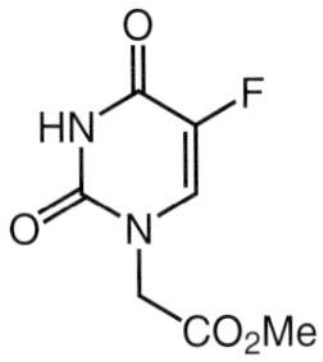 **1550** 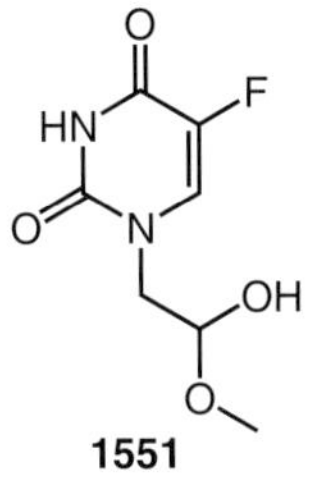 **1551** 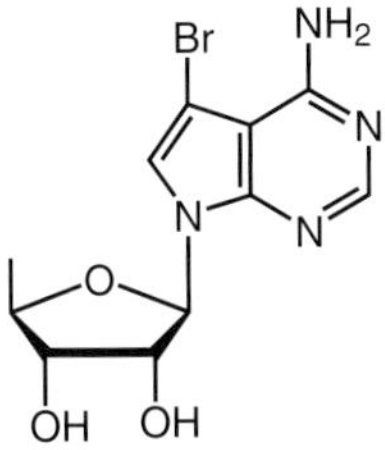
1552 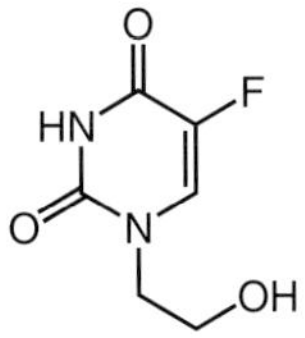 **1553** 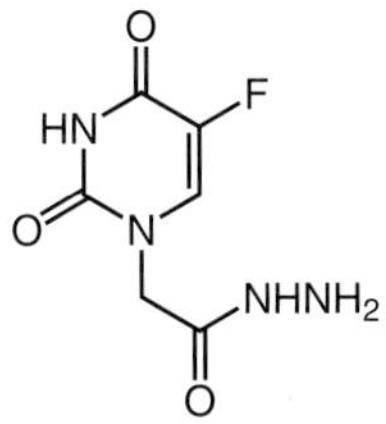 **1554** 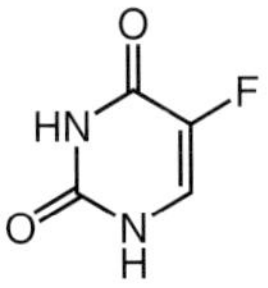
1555 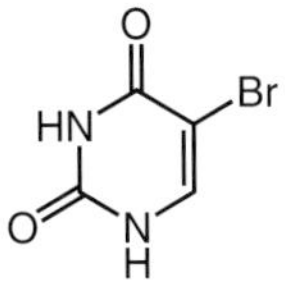 **1556** 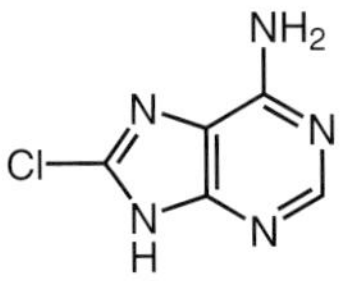 **1557** 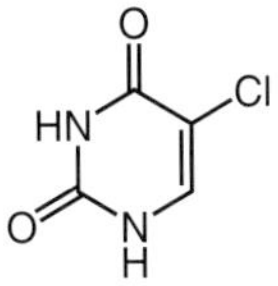

While there are no newly reported naturally occurring benzodiazepines since the first survey (*1*), the interest in this area remains high. The current status of research and clinical implications has been reviewed (*1484*). It is noted that natural benzodiazepines, including seven known halogenated examples (*1*), are found in soil, plants, animal and human tissues, and are chemically identical to their pharmaceutical counterparts. The endogenous formation of benzodiazepines by plant cells (*Artemisia dracunculus* and *Solanum tuberosum*) has been demonstrated for delorazepam, temazepam, and diazepam (*1485*).

3.14.7 Benzofurans and Related Compounds

The novel heterocycles pterulinic acids (**1558** and **1559**) and pterulone (**1560**) were isolated from a *Pterula* sp. fungus (*1486, 1487*). These compounds are inhibitors of NADH:ubiquinone oxidoreductase (complex I). Pterulone B (**1561**) was characterized from cultures of *Pterula* sp. 82168 living on wood (*1488*). The wood-rotting fungus *Mycena galopus* has yielded the chlorinated 2,3-dihydro-1-benzoxepins **1562–1564** (*1489*). The two aldehydes **1565** and **1566** appear to be minor components in *Mycena galopus*, and several of these metabolites (**1559, 1563–1566**) have been synthesized (*1490*). The novel dimeric polybrominated benzofurans, iantherans A (**1567**) and B (**1568**) are produced by the Australian sponge *Ianthella* sp., and display Na,K-ATPase inhibitory activity (*1491, 1492*). The aurones **1569** and **1570** were isolated from the brown alga *Spatoglossum variabile*, collected along the coast of Pakistan (*1493*). These metabolites are rare examples of the halogen atom residing on an unactivated benzene ring.

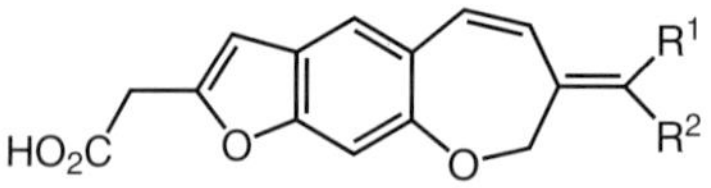

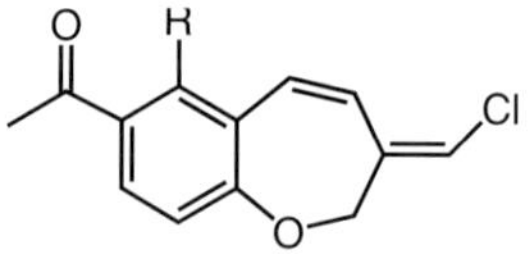

1558 R^1 = Cl, R^2 = H ((E)-pterulinic acid; major)
1559 R^1 = H, R^2 = Cl ((Z)-pterulinic acid; minor)
1560 R = H (pterulone)
1562 R = OH

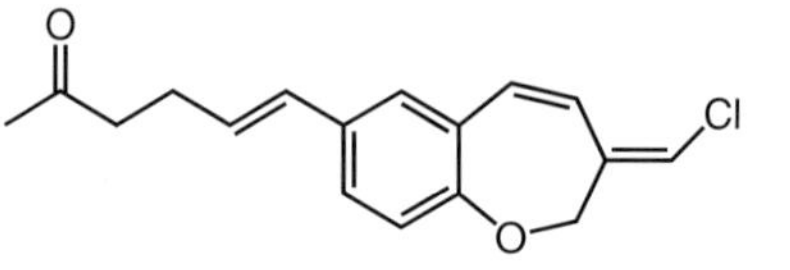

1561 (pterulone B)

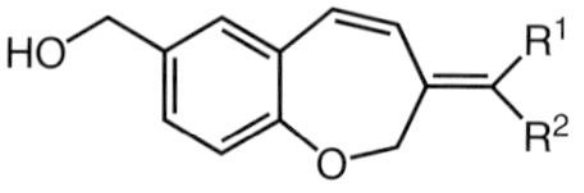

1563 R^1 = Cl, R^2 = H
1564 R^1 = H, R^2 = Cl

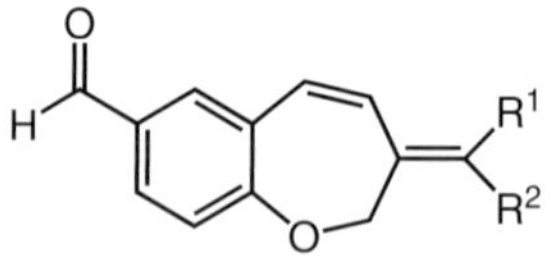

1565 R^1 = Cl, R^2 = H
1566 R^1 = H, R^2 = Cl

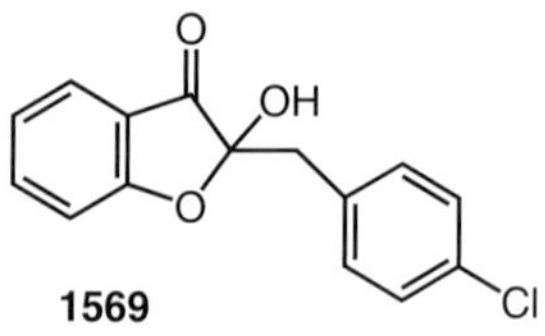

1569

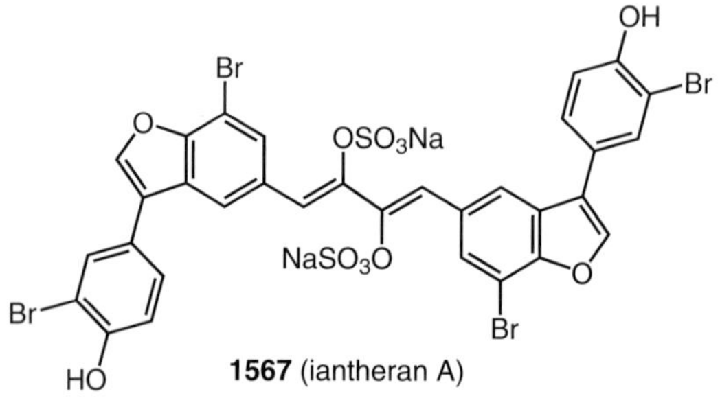

1567 (iantheran A)

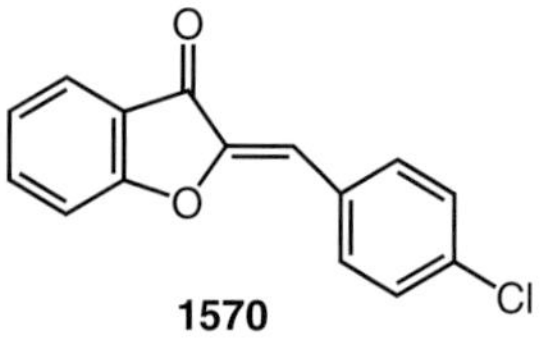

1570

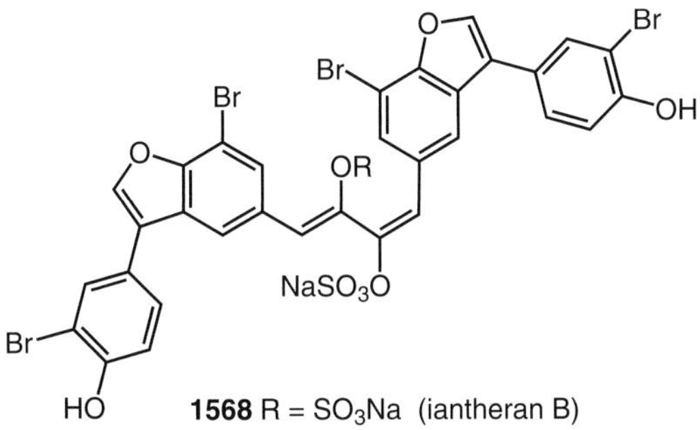

1568 R = SO_3Na (iantheran B)

3.14.8 *Pyrones and Chromones*

The medicinal plant *Goniothalamus amuyon* from Taiwan contains the pyrone 8-chlorogoniodiol (**1571**) with the absolute configuration shown (*1494*). A Taiwanese collection of withered wood of *Aguilaria sinensis* ("Agarwood"), which is used as incense and for medicinal purposes, has yielded the novel chromone **1572** (*1495*). Cultures of *Aspergillus candidus* F1484 produce the antifungal compound F1484 (**1573**) (*1496*).

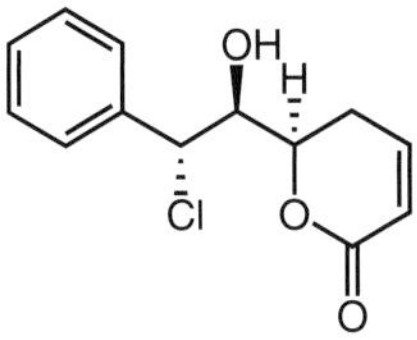

1571 (8-chlorogoniodiol)

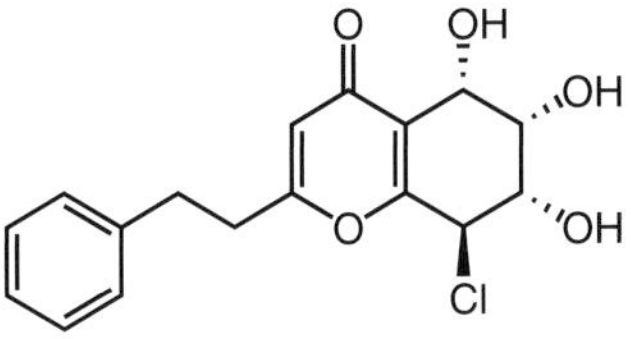

1572

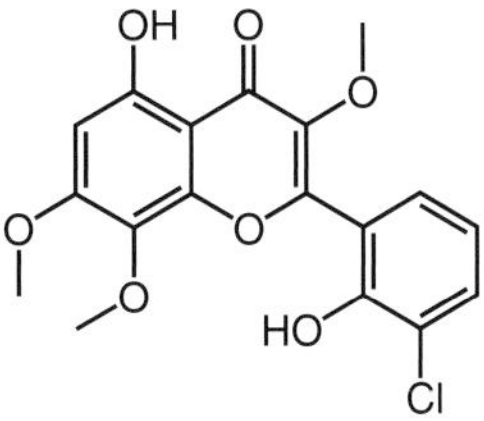

1573 (F1484)

3.14.9 *Coumarins and Isocoumarins*

Cylindrocarpon olidum, a fungus isolated from the root knot nematode *Meloidogyne incognita*, contains the antifungal 8-chlorocannabiorcichromenic acid (**1574**) (*1497*).

Bark from the Indonesian medicinal plant *Aegle marmelos* has provided chloromarmin (**1575**), having the absolute configuration shown (*1498*). The novel aminocoumarin simocyclinone D8 (**1576**) was isolated from cultures of *Streptomyces antibioticus* Tü 6040 and displays antibiotic activity on Gram-positive bacteria and cytotoxicity against various tumor cell lines (*1499, 1500*). A biosynthetic study of **1576** in an oxygen-18 rich atmosphere reveals the incorporation of four oxygen atoms (*1501*). The well-studied clorobiocin (chlorobiocin, RP 18,631) is one of many aminocoumarins from *Streptomyces* strains (*1502*), and a biosynthesis has been proposed (*1503*). The Western United States plant *Harbouria trachypleura* ("whiskbroom parsley") has yielded (±)-trachypleuranin B (**1577**), confirmed by total synthesis (*1504*). The rare folk medicinal Colombian herb *Niphogeton ternata* contains the new psoralen **1578** (8-methoxysaxalin) (*1505*), and the Tanzanian medicinal plant *Mondia whitei* ("Mbombongazi") has afforded 5-chloropropacin (**1579**), which is the first chlorinated example of the coumarinolignan family (*1506*).

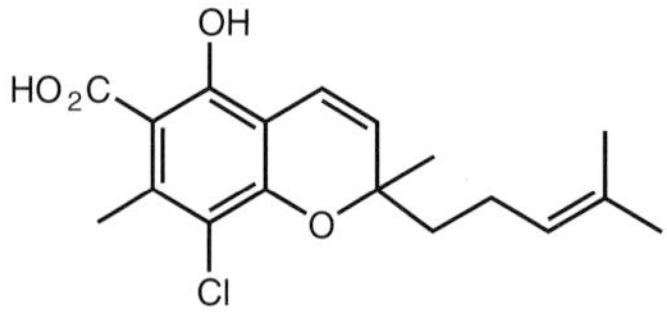

1574

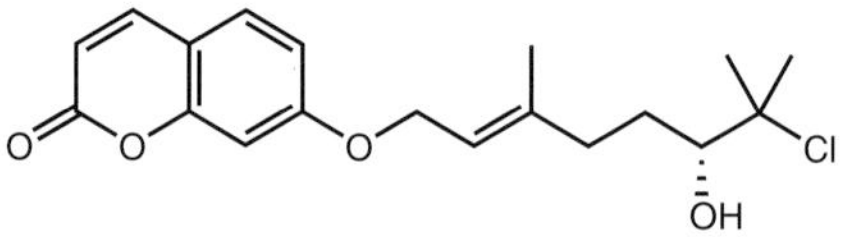

1575 (chloromarmin)

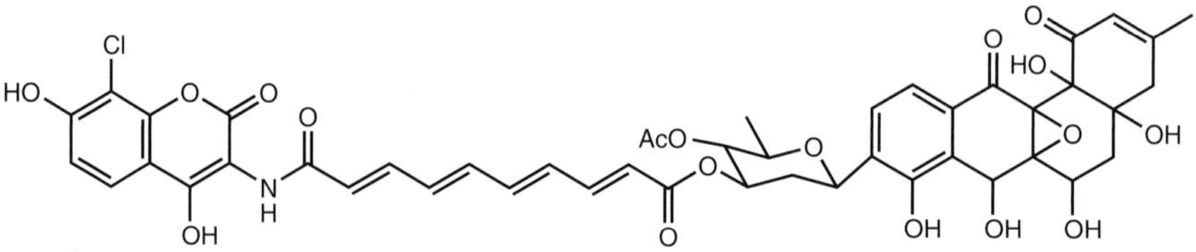

1576 (simocyclinone D8)

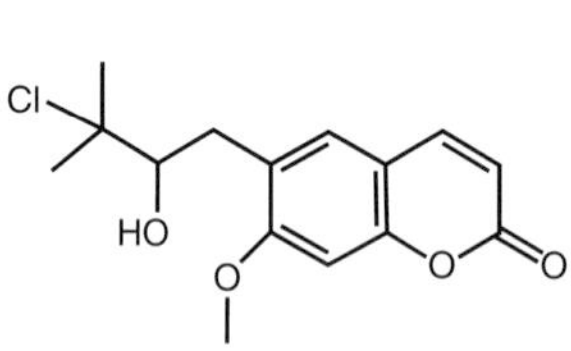

1577 (trachypleuranin-B)

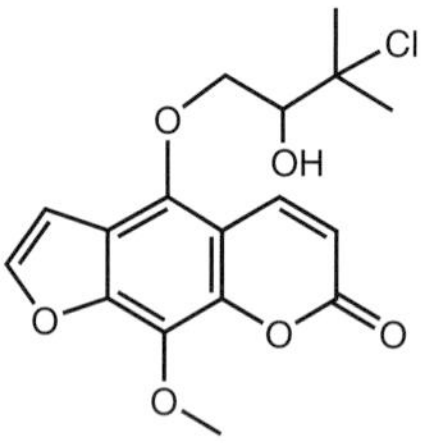

1578 (8-methoxysaxalin)

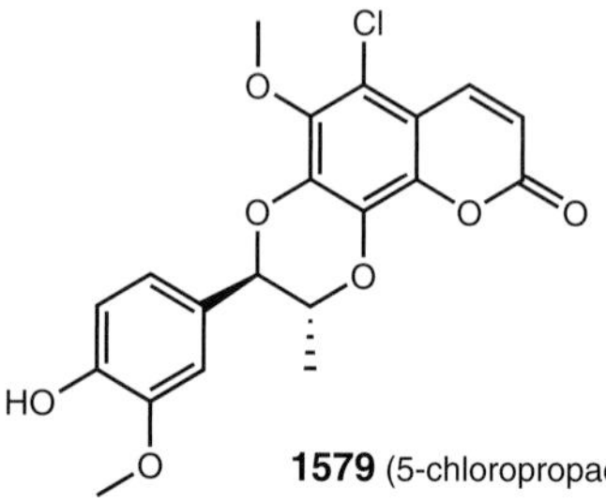

1579 (5-chloropropacin)

3.14 Heterocycles

Several new chlorinated isocoumarins have been isolated since the first review (*1*), and an early review of naturally occurring isocoumarins is available (*1507*). The methyl ester (**1580**) of the notorious ochratoxin A was isolated from *Aspergillus ochraceus*, and is not considered being an artifact (*1508*). The wood-rotting fungus *Heterobasidion annosum* (= *Fomes annosus*) yielded isocoumarin **1581** (*1509*), and **1582–1584** were characterized from the ascomycete *Lachnum papyraceum* (*1510*). The fungus *Plectophomella* sp. has yielded the two mellein derivatives, 5-chloro-6-hydroxymellein (**1585**) and 5-chloro-4,6-dihydroxymellein (**1586**) (*1511*). The fungus *Periconia macrospinosa* is the source of several metabolites, which are of biosynthetic interest (*415, 416*), and this organism also produces the new **1587** (*416*). The absolute configuration of the previously known bromine-containing isocoumarin, hiburipyranone, has been established as (*R*) by total synthesis (*1512*). The cheese-associated fungus *Penicillium nalgiovense* produces dichlorodiaportin (**1588**) (*1513*), while the related methylated diaportins **1589** and **1590** are found in the cultured lichen mycobiont of *Graphis* sp. from a Philippines tree (*1514*). Avicennin A (**1591**) is a novel isocoumarin isolated from a mangrove entophytic fungus in the South China Sea (*1515*).

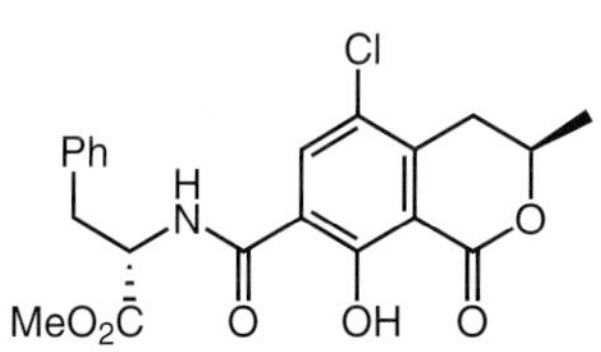

1580

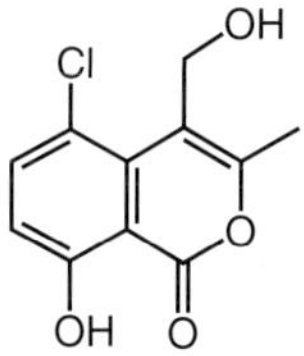

1581

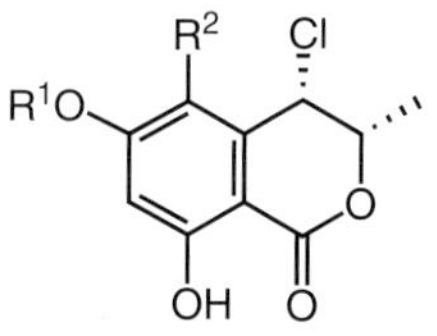

1582 R^1 = H, R^2 = H
1583 R^1 = H, R^2 = OH
1584 R^1 = Me, R^2 = H

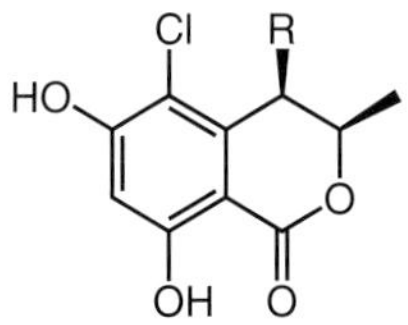

1585 R = H
1586 R = OH

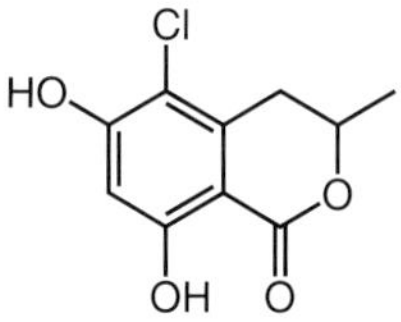

1587

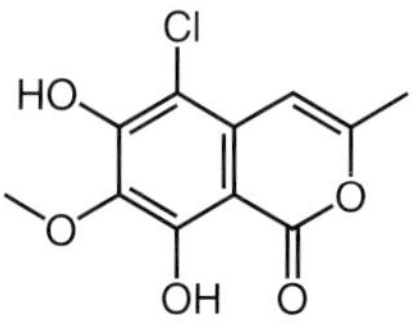

1591 (avicennin A)

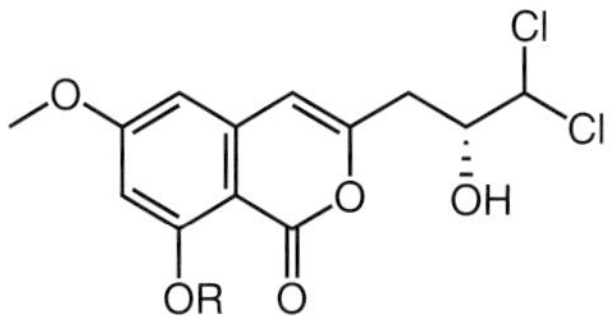

1588 R = H (dichlorodiaportin)
1589 R = Me

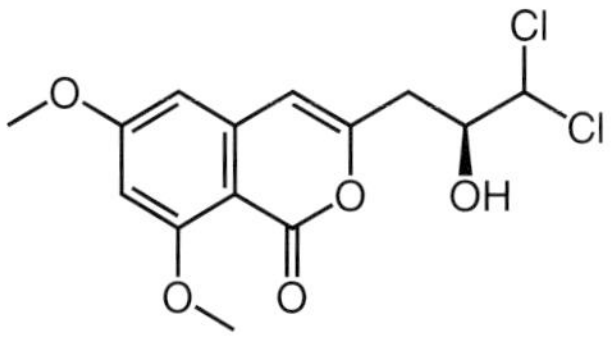

1590 ((10*S*)-8-methyldichlorodiaportin)

The novel chaetochiversins A (**1592**) and B (**1593**) were identified in the fungus *Chaetomium chiversii* living in association with the Sonoran desert plant *Ephedra fasciculata* (*1516*). The absolute configuration of **1592** was established by synthesis from the known radicicol. Tricyclic TMC-264 (**1594**) was isolated from the fungus *Phoma* sp. TC 1674 (*1517, 1518*), and graphislactone G (**1595**) was identified in the endophytic fungus *Cephalosporium acremonium* IFB-E007 residing in the roots of *Trachelospermum jasminoides* (*1519*).

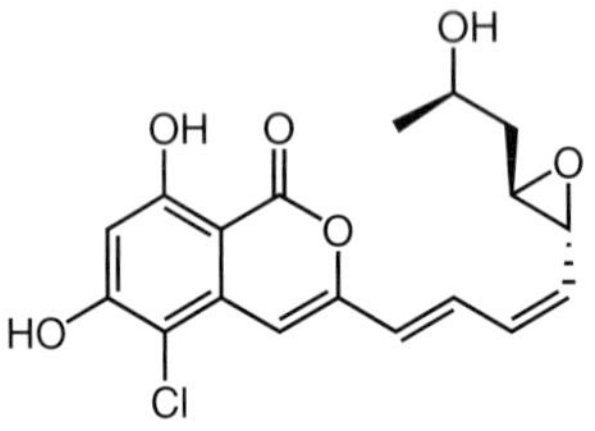

1592 (chaetochiversin A)

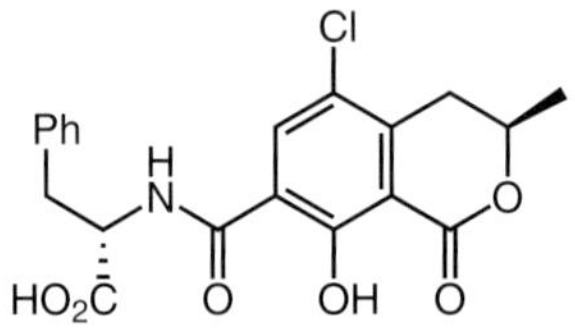

AAA (ochratoxin A)

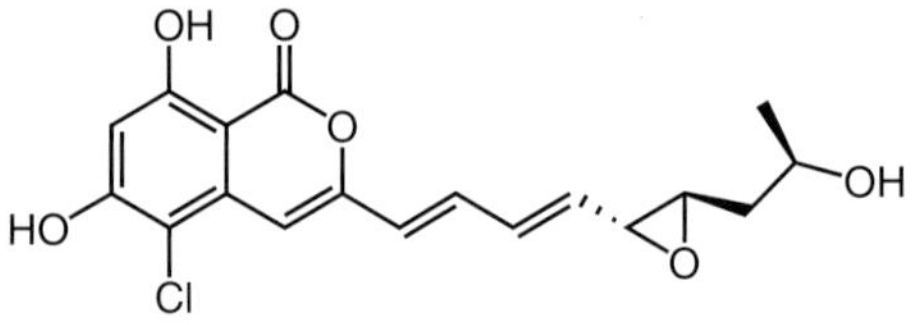

1593 (chaetochiversin B)

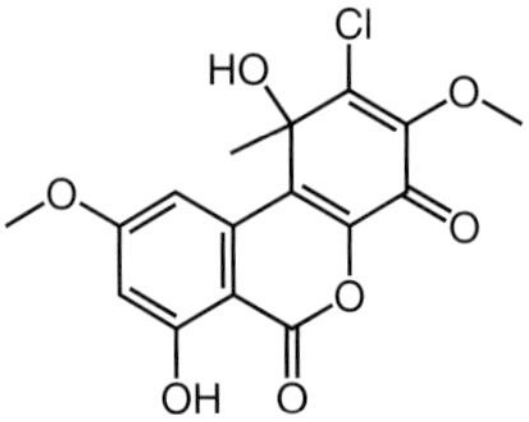

1594 (TMC-264)

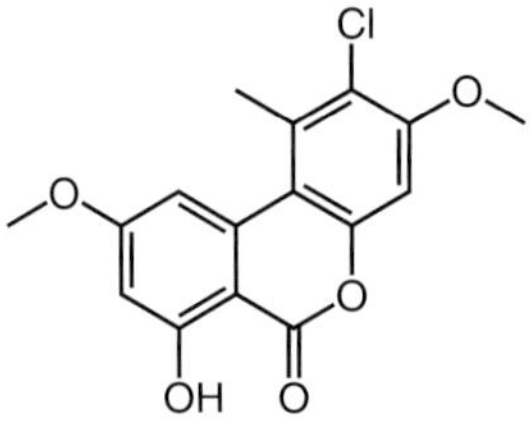

1595 (graphislactone G)

The mycotoxin ochratoxin A (**AAA**) (*1*), which is a possible human carcinogen, continues to receive extensive attention due to its presence in a myriad of foods and beverages (*1520, 1521*) and its well-established toxicity (teratogenicity, mutagenicity, immunotoxicity, genotoxicity, and carcinogenicity) (*1522–1524*). Major sources of ochratoxin A are grapes, must, and wine (*1525–1533*), cereals (*1534*), beer (*1535, 1536*), dried fruit (*1537*), roasted coffee (*1538*), and cocoa products and chocolate (*1539*).

3.14.10 Flavones and Isoflavones

The bromine analog of chlorflavonin, CJ-19,784 (**1596**), is produced by the fungus *Acanthostigmella* sp. in the absence of added bromide ion in the culture medium (*1540*). This metabolite inhibits the growth of the pathogenic fungi *Candida albicans*, *Cryptococcus neoformans*, and *Aspergillus fumigatus*. The roots of the Turkish traditional plant *Rumex patientia* contain 6-chlorocatechin (**1597**), which is the first reported natural halogenated flavan-3-ol (*1541*).

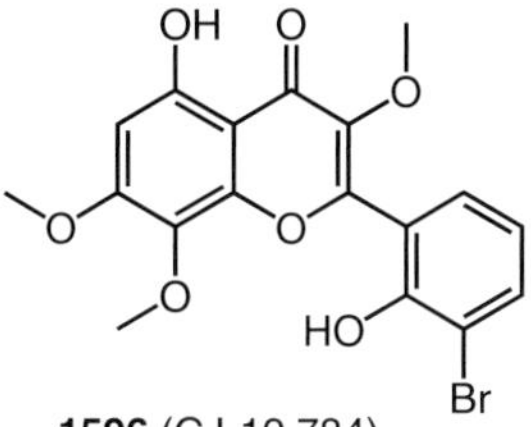

1596 (CJ-19,784) 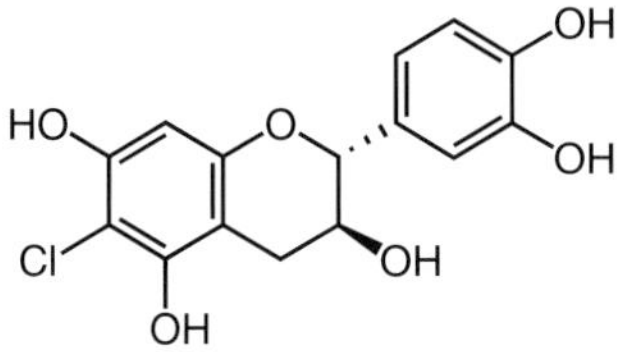   **1597** (6-chlorocatechin)

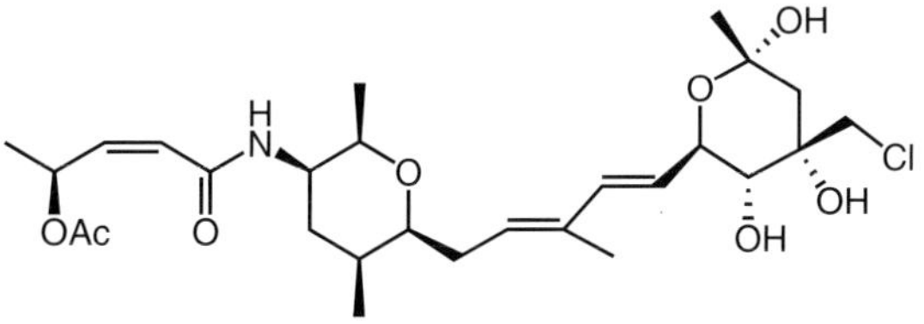

1598 (FR 901463)

3.14.11 Carbohydrates

Despite the wide spread use of the artificial sweetener, "Splenda", which is a synthetic chlorinated carbohydrate, Nature has provided very few halogenated carbohydrates. The antitumor metabolite FR 901463 (**1598**) was isolated from a *Pseudomonas* sp. along with two nonchlorinated epoxides. FR 901463 is not an isolation artifact, being present in the culture medium prior to extraction and isolation (*1542–1544*).

3.15 Polyacetylenes

3.15.1 Terrestrial Polyacetylenes and Derived Thiophenes

The Colombian medicinal plant *Niphogeton ternata* contains the new polyacetylene **1599** (*1505*).

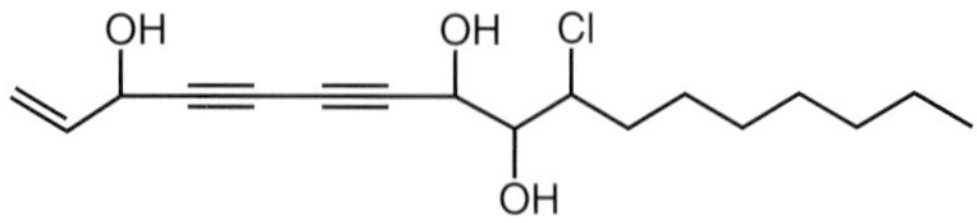

1599

3.15.2 Marine Polyacetylenes

The reader is referred to the section on fatty acids (**3.8**), which includes brominated fatty acids containing multiple acetylene groups.

3.16 Enediynes

Although only a few new members of the extraordinary enediyne class of natural products have been discovered since the first survey (*1*), the powerful biological activity of these natural products, several of which contain halogen, has spurred intensive investigation of their biological activity (*1545–1548*). For example, the calicheamicin family of enediyne antitumor antibiotics continues to be a highlight in this area of natural products (*1549, 1550*). Similarly, the previously known C-1027 is of great interest (*1551*) with regard to its biosynthesis (*1552*), synthesis (*1553*), analog preparation (*1554*), mechanism of action (*1555*), and biological activity (*1556*). The absolute configuration of the C-1027 chromophore (**BBB**) has been established as shown (*1557*), and the new neoC-1027 chromophore (**1600**) was characterized from *Streptomyces globisporus* (*1558*).

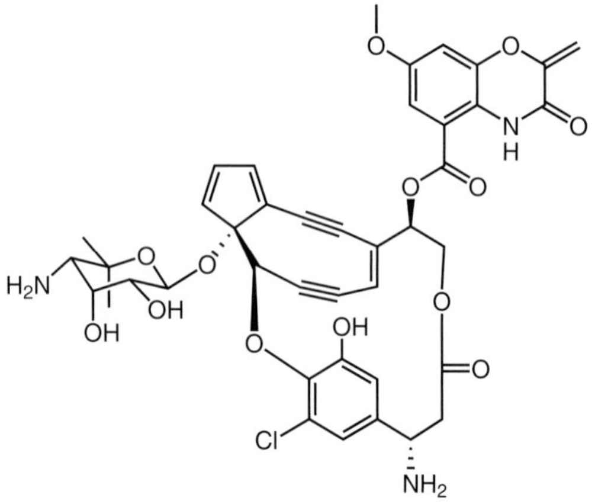

BBB (C-1027 chromophore)

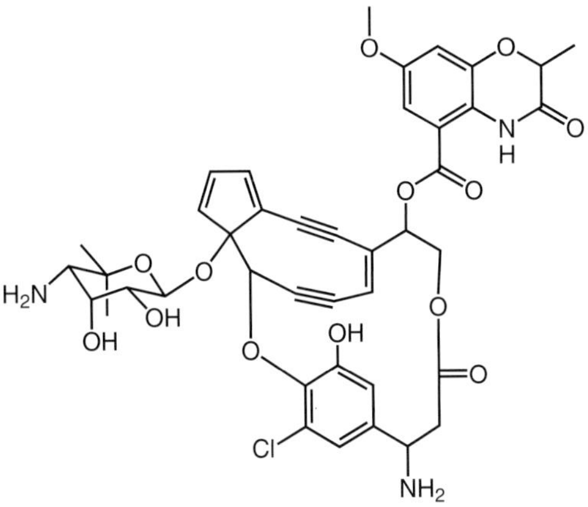

3.16 Enediynes

1600 (neoC-1027 chromophore)

The previously known kedarcidin chromophore (*1*) is revised to **CCC** (*1559*), and the mechanism of action of this enediyne has been studied (*1560*). The new maduropeptin chromophore **1601** was isolated from *Actinomadura madurae*, which is associated with a protein of 14 amino acids (*1561–1563*). The non-protein associated enediyne N1999A2 (**1602**) was characterized from *Streptomyces* sp. AJ 9493 (*1564*), and confirmed by synthesis (*1565, 1566*).

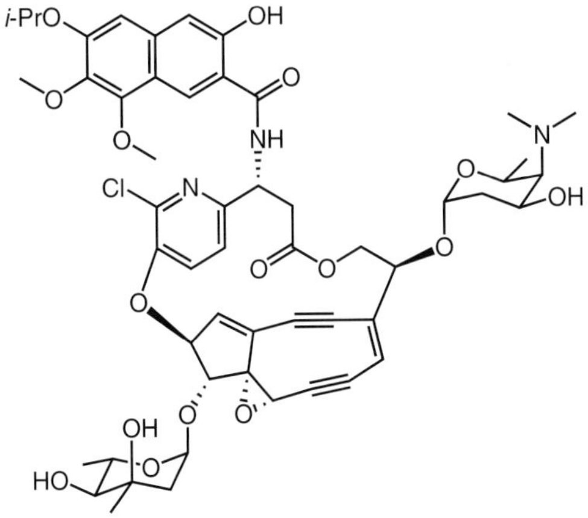

CCC (kedarcidin chromophore)

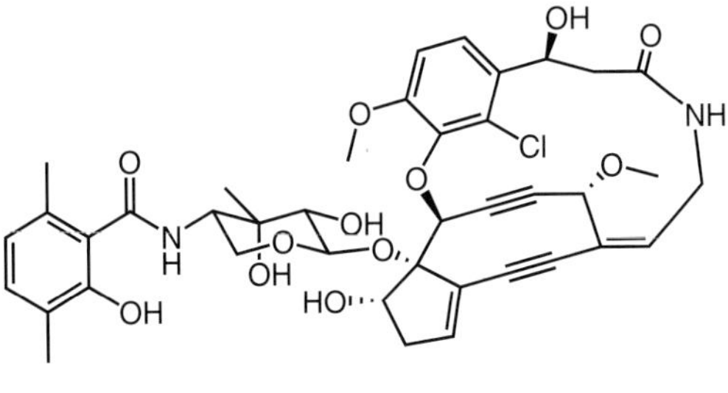

1601 (maduropeptin apoprotein methanol adduct)

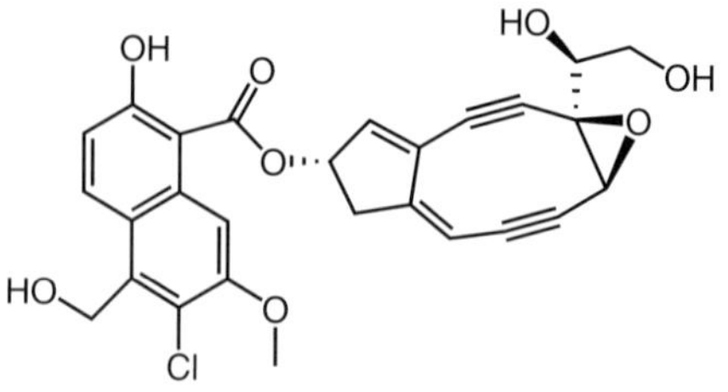

1602 (N1999-A2)

3.17 Macrolides and Polyethers

The large and diverse group of naturally occurring biologically active macrolides includes a number of halogenated examples (*1*). The chlorine-containing maytansinoids, which were once promising anticancer agents, nevertheless continue to receive attention (*1567*), and semisynthetic maytansines show promise as new anticancer agents (*1568*). The two new maytansinoids 2'-N-demethylmaytanbutine (**1603**) and maytanbicyclinol (**1604**) were isolated from the Kenyan plant *Maytenus buchananii* (*1569*), and the new ansamitocin ansamitocinoside P-2 (**1605**) is produced by *Actinosynnema pretiosum* ssp. *auranticum* (*1570*).

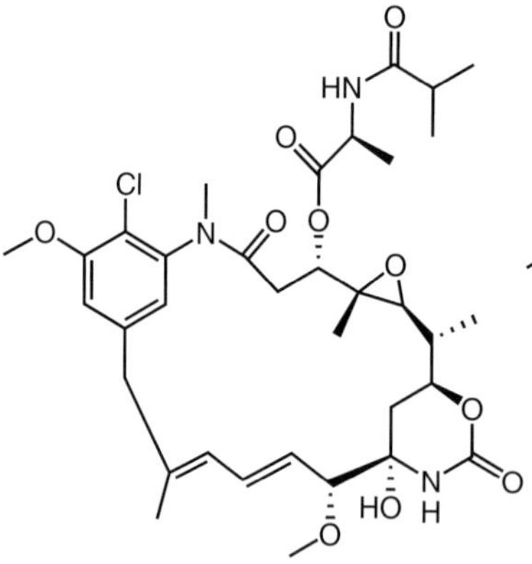

1603 (2'-*N*-demethylmaytanbutine) 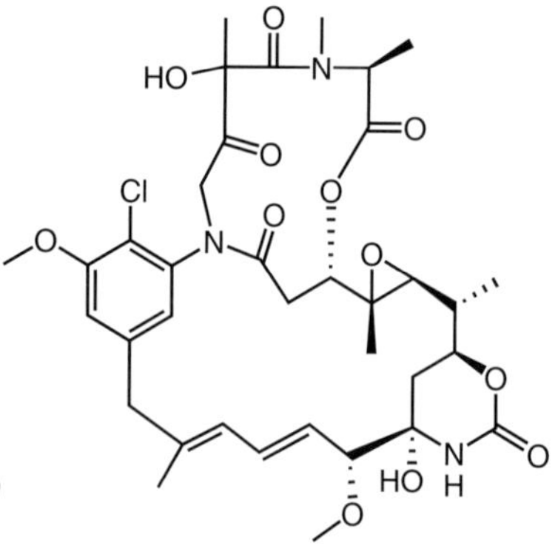　　**1604** (maytanbicyclinol)

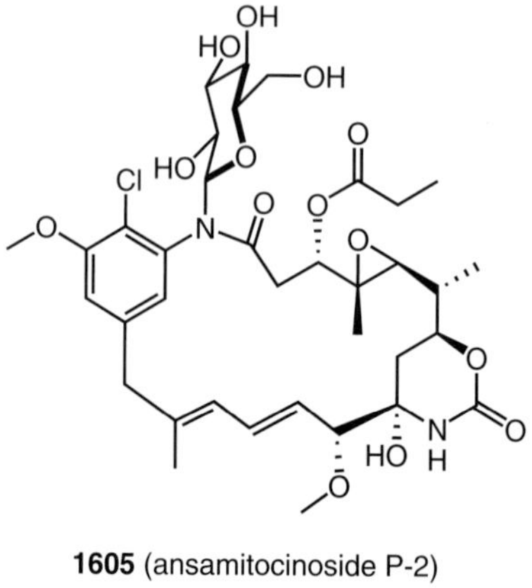

1605 (ansamitocinoside P-2)

3.17 Macrolides and Polyethers

The previously known macrolide radicicol (= monorden A) (*1*) continues to be of interest for its multiplicity of biological activities, such as suppression of oncogene transformation (*1571, 1572*) and inhibition of the human malaria parasite *Plasmodium falciparum* (*1573*). The former activity has been of particular scrutiny, and several novel HSP90 synthetic radicicol analogs are promising anticancer agents (*1574–1576*). The new radicicol **1606** was isolated from the mycoparasite *Humicola fuscoatra* from *Aspergillus flavus* (*1577*). Two groups independently isolated a series of the new pochonins (*1578*) and monordens (*1579, 1580*), which are identical in some cases. Thus, whereas the fungus *Pochonia chlamydosporia* var. *catenulata* produces pochonins A (**1607**), B (**1608**), C (**1609**), D (**1610**), E (**1611**), and tetrahydromonorden (= tetrahydroradicicol) (**1612**) (*1578*), the fungus *Humicola* sp. FO-2942 yields monorden C (= pochonin A, **1607**), monorden D (= pochonin D, **1610**), and monorden E (**1613**) (*1579, 1580*). Diversity-oriented synthesis of the pochonins in a search for ATPase and kinase inhibitors has been reported (*1581*).

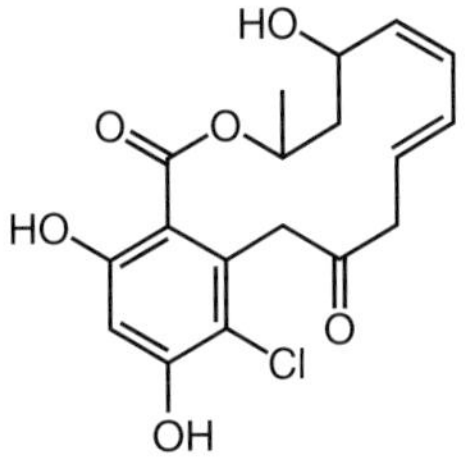

1606

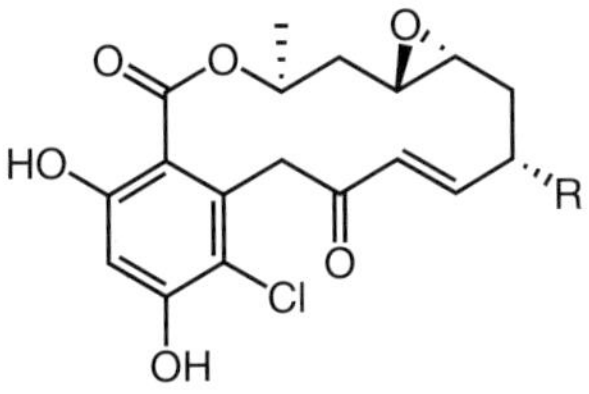

1607 R = H (pochonin A)
1608 R = OH (pochonin B)

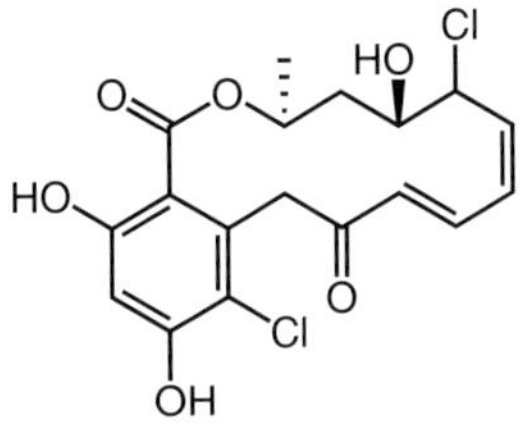

1609 (pochonin C)

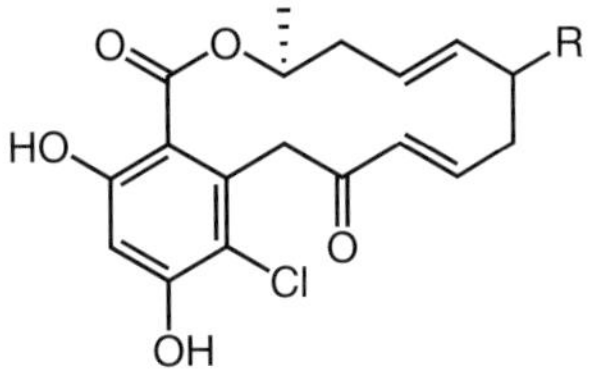

1610 R = H (pochonin D)
1611 R = OH (pochonin E)

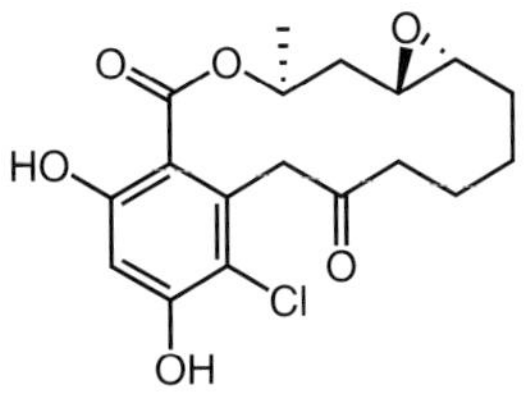

1612 (tetrahydromonorden)

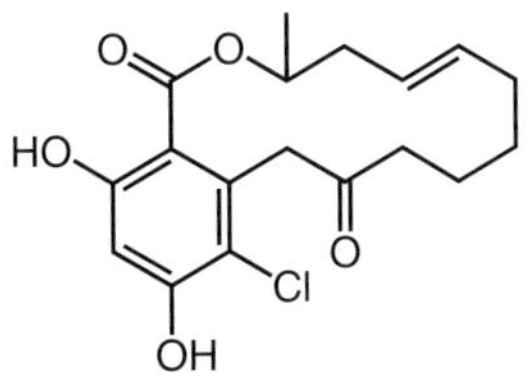

1613 (monorden E)

The previously known and notorious aplysiatoxin (a cause of "swimmer's itch") (*1*) is a protein kinase C inhibitor and has been the object of a structure–activity study (*1582*). The causative agents of a red alga *Gracilaria coronopifolia* poisoning episode in Hawaii are reported to be manauealides A (**1614**) and B (**1615**), compounds that may be associated with a cyanobacterium (*1583*). Manauealide B is an isomer of aplysiatoxin. The novel metacyclophane floresolide C (**1616**) is found in the Indonesian ascidian *Aplidium* sp. (Fig. 3.21) (*1584*). The two macrolides, sporolides A (**1617**) and B (**1618**), were characterized from the marine actinomycete *Salinispora tropica* and have the absolute stereochemistry shown (*1585*).

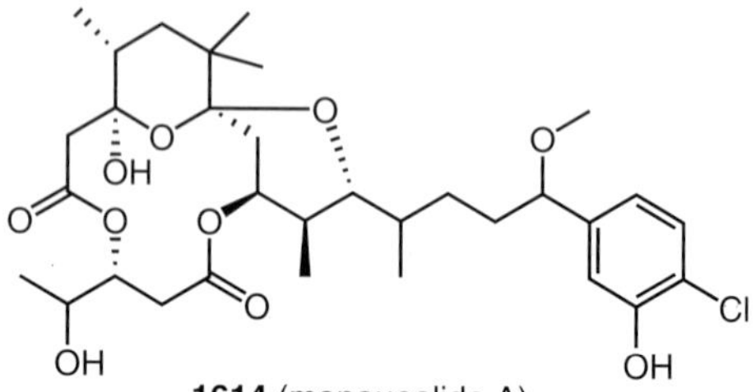

1614 (manauealide A)

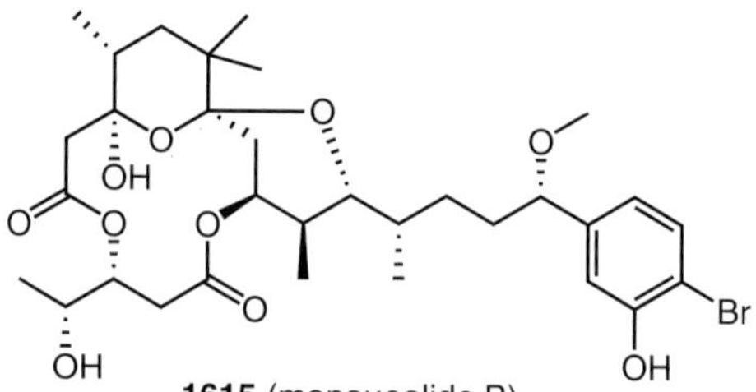

1615 (manauealide B)

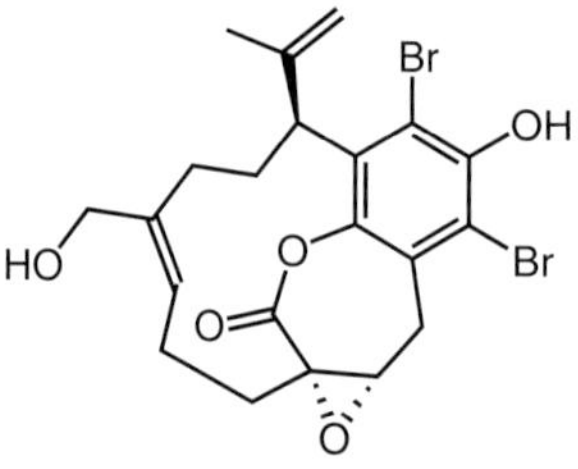

1616 (floresolide C)

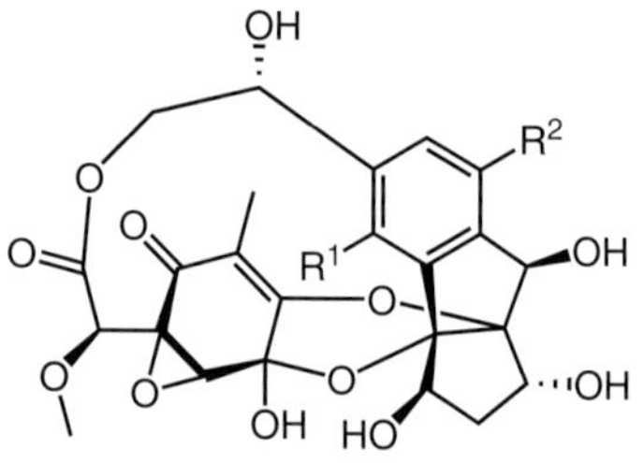

1617 R^1 = Cl, R^2 = H (sporolide A)
1618 R^1 = H, R^2 = Cl (sporolide B)

The extraordinarily complex and biologically important altohyrtins and spongistatins (*1*) have been the object of intense synthetic efforts that have clarified previous stereochemical ambiguities (*1586–1588*). The strain of *Actinoplanes deccanensis* produces lipiarmycins A3 (**1619**), A4 (**1620**) (*1589*), B3 (**1621**), and B4 (**1622**) (*1590*). A group of very similar (or identical) metabolites to the lipiarmycins was isolated from both *Micromonospora echinospora* subsp. *armeniaca* called clostomicins A, B_1 (= lipiarmycin A_3, **1619**), B_2 (= lipiarmycin B_3, **1621**), C, and D (structures undetermined but each has two chlorines) (*1591, 1592*) and from *Dactylosporangium aurantiacum* subsp. *hamdenensis* named tiacumicins B (= lipiarmycin A_3, **1619**), C (= lipiarmycin B_3, **1621**), D (**1623**), E (**1624**), and F (**1625**, = clostomicin A) (*1593, 1594*).

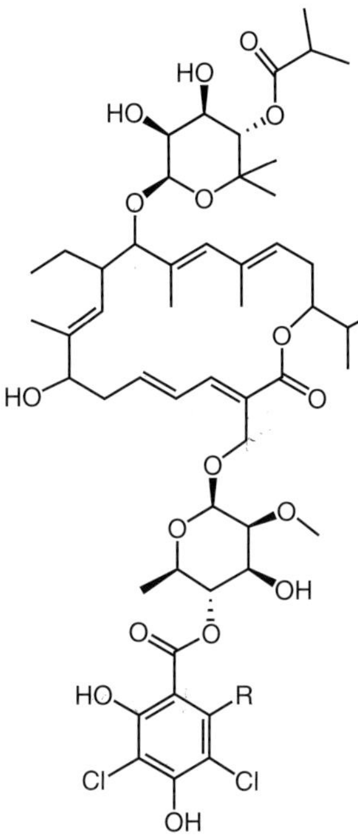

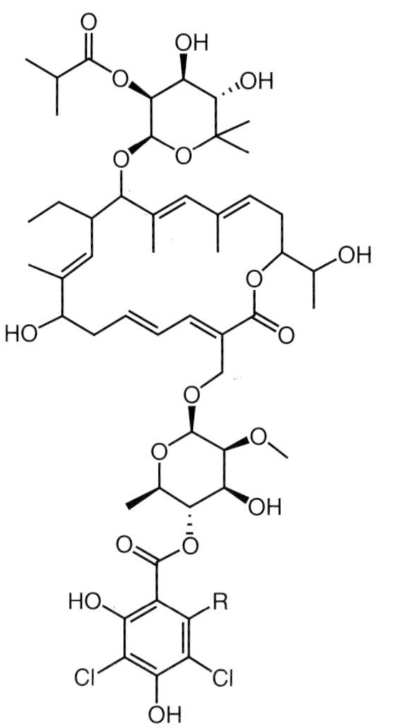

3.17 Macrolides and Polyethers

1619 R = Et (lipiarmycin A$_3$)
1620 R = Me (lipiarmycin A$_4$)

1621 R = Et (lipiarmycin B$_3$ = tiacumicin C)
1622 R = Me (lipiarmycin B$_4$)

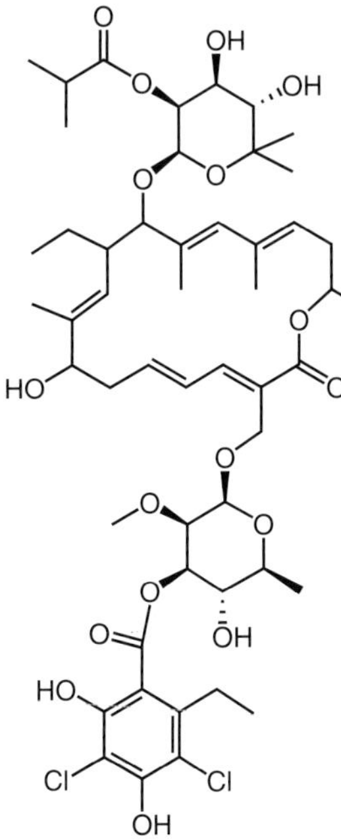

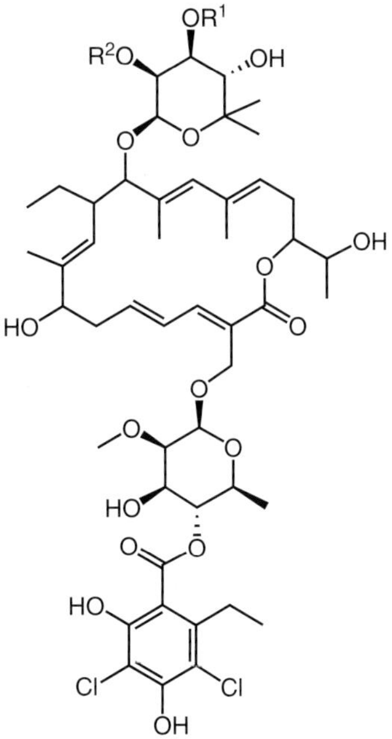

1623 (tiacumicin D)

1624 R^1 = H, R^2 = COEt (tiacumicin E)
1625 R^1 = COi-Pr, R^2 = H (tiacumicin F)

The highly complex chlorinated pyrrole-containing macrolide colubricidin A (**1626**) is produced in cultures of an unidentified *Streptomyces* species (*1595*). This metabolite displays excellent activity against Gram-positive bacteria. The Dominican sponge *Spirastrella coccinea* produces spirastrellolide A (**1627**), which is a potent and selective inhibitor of protein phosphatase 2A (*1596*) (revised later (*1597*)).

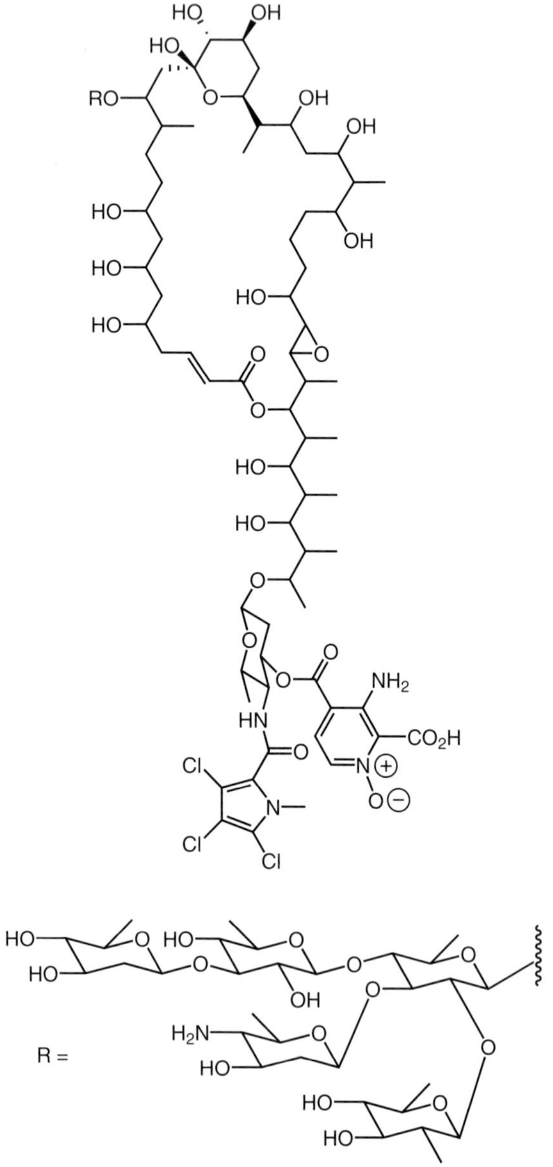

1626 (colubricidin A)

3.17 Macrolides and Polyethers

1627 (spirastrellolide A)

Fig. 3.21 This Indonesian ascidian *Aplidium* sp. produces the novel metacyclophane floresolide C (**1616**) (Photo: J. Tanaka)

The Japanese sea hare *Dolabella auricularia* contains the cytotoxic aurisides A (**1628**) and B (**1629**) (*1598*), which have the absolute stereostructures shown as confirmed by total synthesis (*1599*). Both compounds display excellent cytotoxicity against HeLa S_3 cells (0.17–1.2 µg mL^{-1}). Structurally similar to the aurisides are callipeltosides A (**1630**), B (**1631**), and C (**1632**), which were isolated from the sponge *Callipelta* sp. collected in New Caledonia (*1600, 1601*) and are the targets of several total syntheses (*1586, 2652*). Closely related to the callipeltosides are phorbasides A (**1633**) and B (**1634**) from a Western Australian *Phorbas* sp. sponge (*1602*). The first example of a cyanobacterium glycoside macrolide to be isolated is lyngbyaloside (**1635**) from a Papua New Guinea *Lyngbya bouillonii* (*1603*).

Another Palauan sample of this blue-green alga identified the related bromine-containing lyngbyaloside B (**1636**) (*1604*).

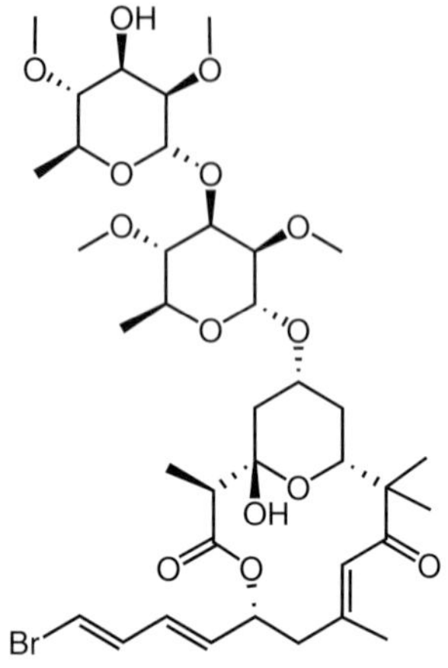

1628 (auriside A)

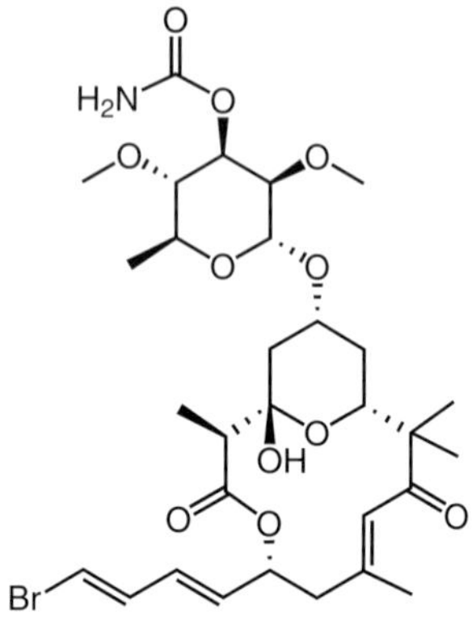

1629 (auriside B)

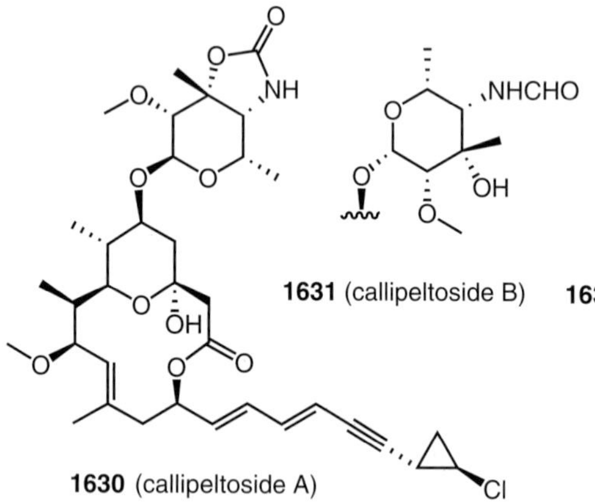

1630 (callipeltoside A)

1631 (callipeltoside B)

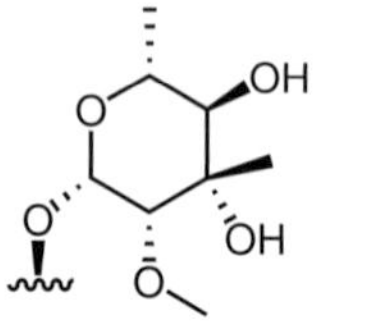

1632 (callipeltoside C)

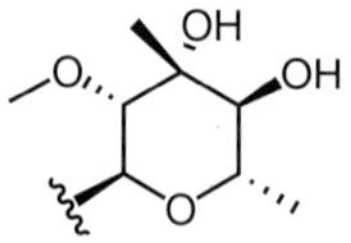

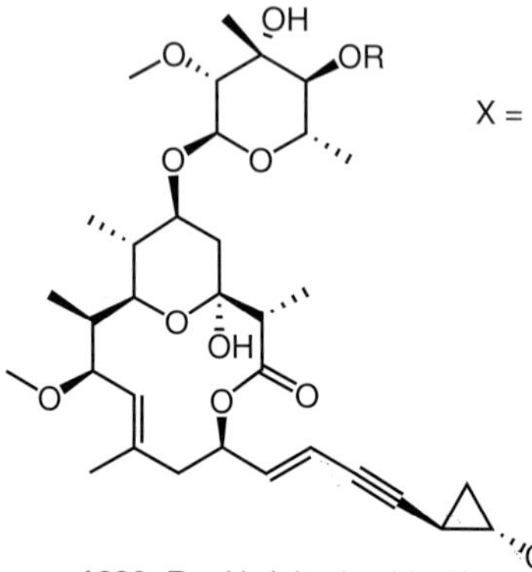

1633 R = H (phorbaside A)
1634 R = X (phorbaside B)

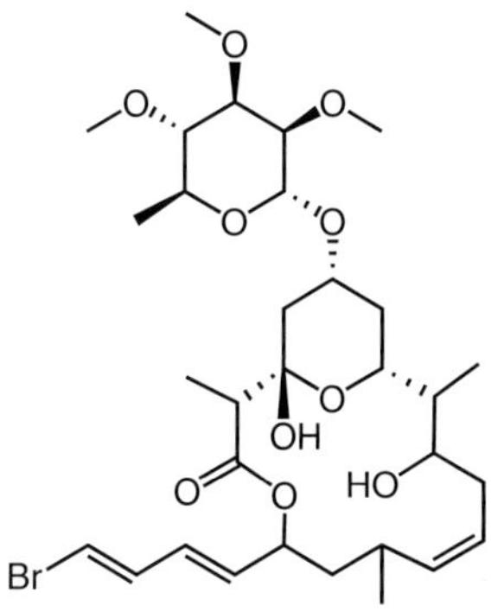

3.17 Macrolides and Polyethers

1635 (lyngbyaloside)

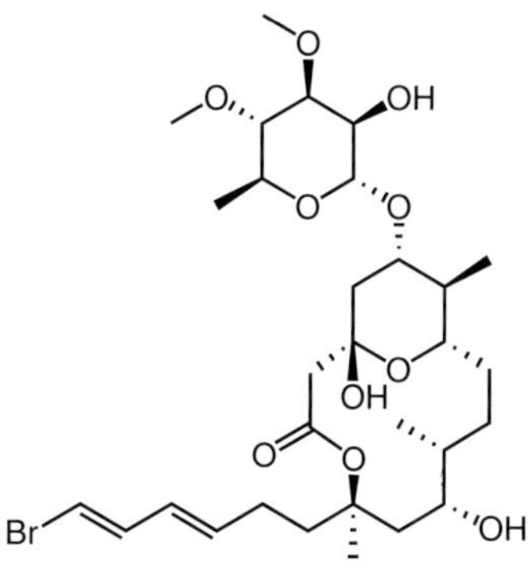

1636 (lyngbyaloside B)

The Indian Ocean sponge *Phorbas* sp. has furnished phorboxazoles A (**1637**) and B (**1638**), which display extraordinary cytostatic activity against human cancer cell lines, comparable to spongistatin 1, and, therefore, are among the most potent cytostatic compounds known (*1605*). Extensive spectral work on model compounds has established the absolute configuration of the phorboxazoles (*1606–1608*), which was confirmed by several elegant total syntheses (*1609–1612*). The sea hare *Dolabella auricularia* from the Gulf of California has afforded dolastatin 19 (**1639**), which is structurally related to the aurisides (*1613*). Total synthesis corrected the stereochemistry and established the absolute configuration of dolastatin 19 (*1614, 2651*). A deep-water (740 feet) Palauan sponge, *Leiodermatium* sp., has yielded leiodelide B (**1640**), along with a non-brominated analog (*1615*). This is the first report of secondary metabolites from the rare genus *Leiodermatium*.

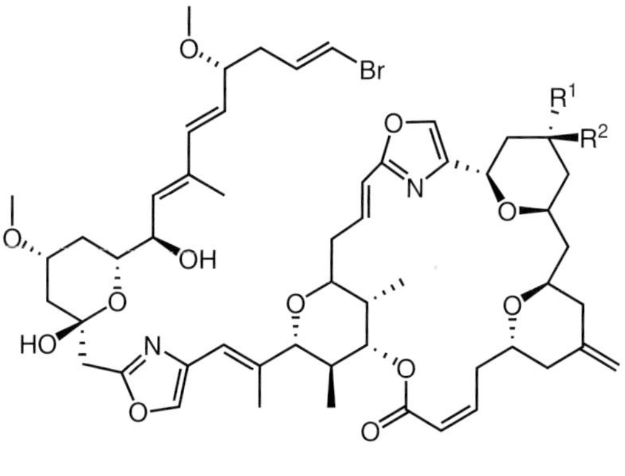

1637 R^1 = OH, R^2 = H (phorboxazole A)
1638 R^1 = H, R^2 = OH (phorboxazole B)

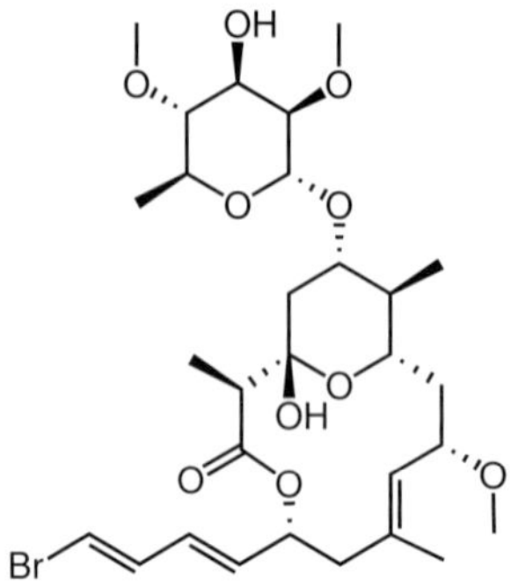

1639 (dolastatin 19)

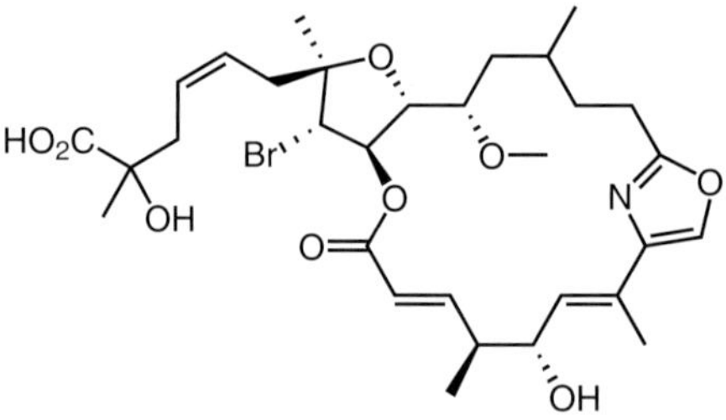

1640 (leiodelide B)

A novel ansamycin, naphthomycin K (**1641**), was isolated from a *Streptomyces* strain of the medicinal plant *Maytenus hookeri* (*1616*). This compound was cytotoxic (P388 and A-549 cell lines) but inactive against *Staphylococcus aureus* and *Mycobacterium tuberculosis*. An *Amycolatopsis* sp. has furnished the ansacarbamitocins A–F, A1, B1 (**1642–1649**), which are similar to the ansamitocins (*1617*).

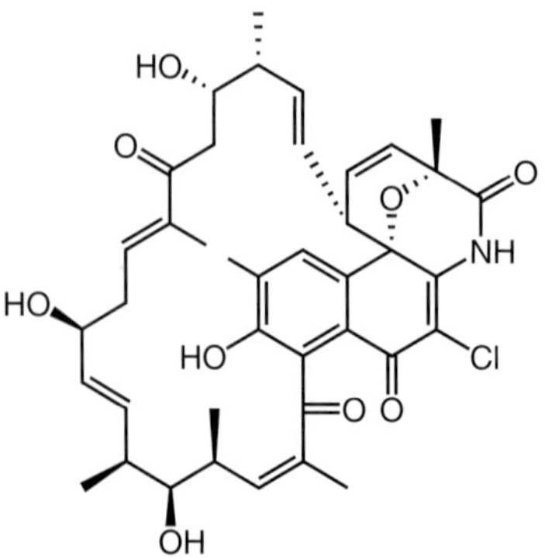

1641 (naphthomycin K)

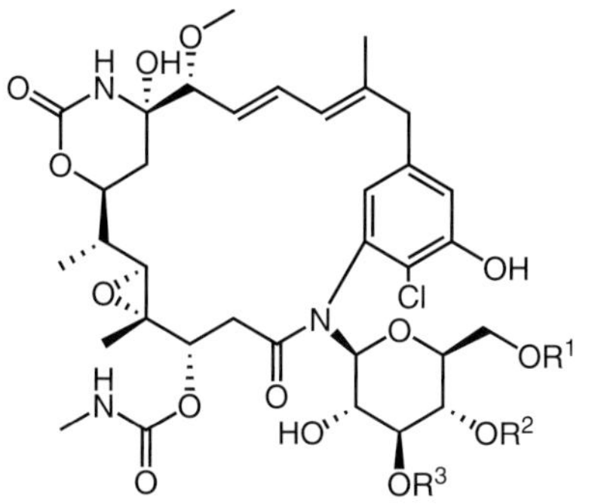

3.17 Macrolides and Polyethers

1642 R^1 = Ac, R^2 = CONH$_2$, R^3 = Me (ansacarbamitocin A)
1643 R^1 = Ac, R^2 = CONH$_2$, R^3 = H (ansacarbamitocin B)
1644 R^1 = H, R^2 = CONH$_2$, R^3 = Me (ansacarbamitocin C)
1645 R^1 = H, R^2 = CONH$_2$, R^3 = H (ansacarbamitocin D)
1646 R^1 = R^2 = H, R^3 = Me (ansacarbamitocin E)
1647 R^1 = R^2 = R^3 = H (ansacarbamitocin F)

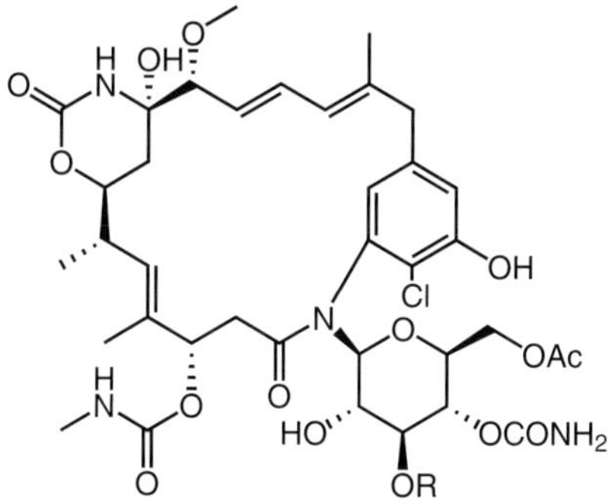

1648 R = Me (ansacarbamitocin A1)
1649 R = H (ansacarbamitocin B1)

An Indonesian marine cyanobacterium, *Phormidium* sp., has provided the toxic phormidolide (**1650**) (*1618*), which is structurally similar to the previously known cyanobacterium metabolite oscillariolide (*1*). Three chlorine-containing macrolides, lytophilippines A–C (**1651–1653**), were isolated from the Red Sea hydroid *Lytocarpus philippinus* ("fireweed") (*1619*), and the Red Sea sponge *Leucetta chagosensis* (Fig. 3.22) contains chagosensine (**1654**) (*1620*). Several novel chlorinated macrolides have been found in both the Okinawan ascidian *Lissoclinum* sp. and the Okinawan sponge *Ircinia* sp. These metabolites are haterumalide B (**1655**) (*1621*) and haterumalides NA-NE (**1656–1660**) (*1622*). Two total syntheses of 15-*epi*-haterumalide NA methyl ester result in a revision of the absolute stereochemistry of haterumalide NA (**1656**) (*1623, 1624*), which was also isolated independently as "oocydin A" from a strain of *Serratia marcescens* growing on the aquatic plant *Rhyncholacis pedicillata* in Venezuela (*1625*) and from the soil bacterium *Serratia plymuthica* in Sweden (*1626*). This revision is depicted for all haterumalides for

Fig. 3.22 *Leucetta chagosensis*, a Red Sea sponge that contains the novel macrolide chagosensine (**1654**) (Photo: T. Rezanka)

convenience. Closely related to the haterumalides are biselides A–E (**1661–1665**), which were characterized from the Okinawan ascidian of the family Didemnidae (*1627, 1628*). A compound FR177391 (**1666**) is produced by *Serratia liquefaciens* and reported to be the enantiomer of haterumalide NA (*1629*).

1650 (phormidolide)

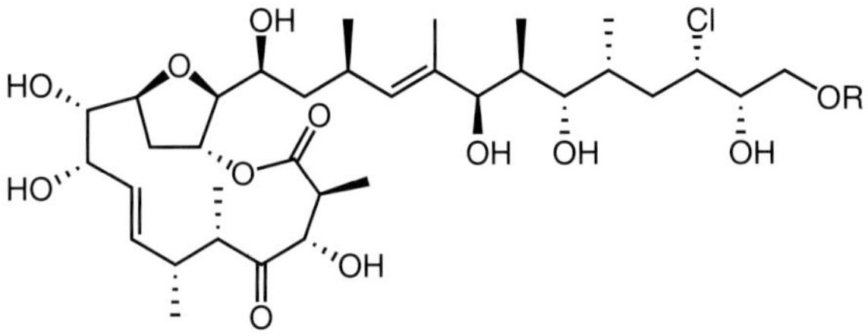

3.17 Macrolides and Polyethers

1651 R = H (lytophilippine A)
1652 R = CO(CH$_2$)$_{14}$CH$_3$ (lytophilippine B)
1653 R = (Z)-CO(CH$_2$)$_7$CH=CH(CH$_2$)$_7$CH$_3$ (lytophilippine A)

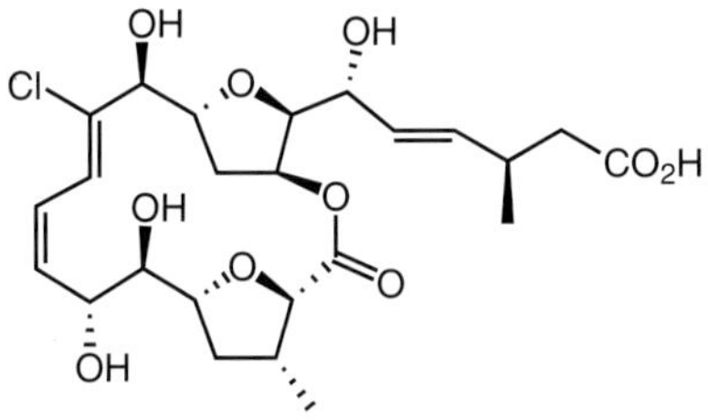

1654 (chagosensine)

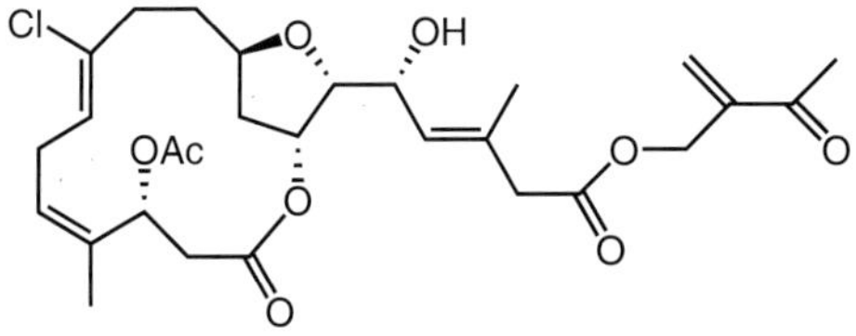

1655 (haterumalide B)

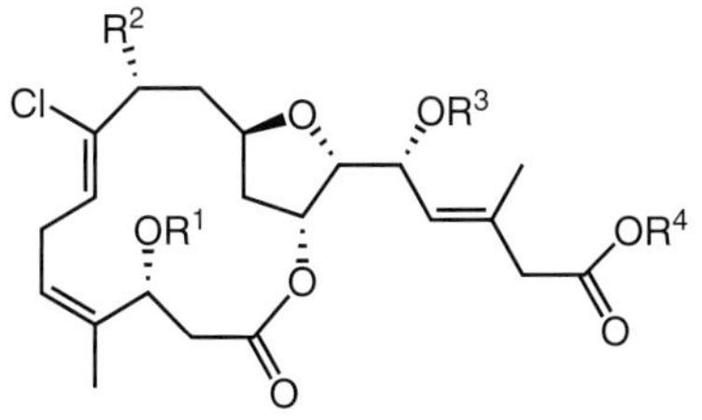

1656 R^1 = Ac, R^2 = R^3 = R^4 = H (haterumalide NA)
1657 R^1 = Ac, R^2 = R^3 = H, R^4 = n-Bu (haterumalide NB)
1658 R^1 = Ac, R^2 = OH, R^3 = H, R^4 = n-Bu (haterumalide NC)
1659 R^1 = Ac, R^2 = OH, R^3 = R^4 = H (haterumalide ND)
1660 R^1 = R^2 = R^3 = R^4 = H (haterumalide NE)

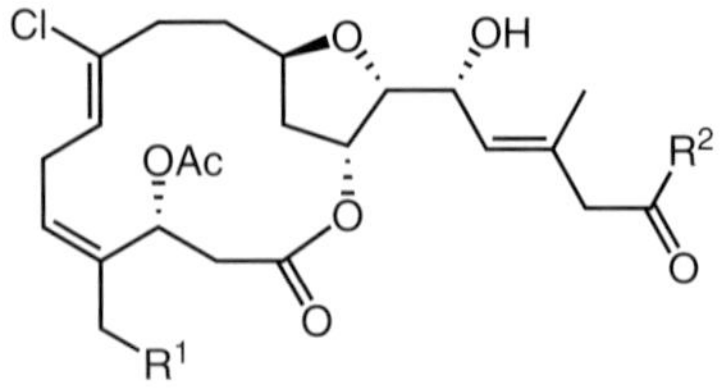

1661 R^1 = OAc; R^2 = OH (biselide A)
1662 R^1 = OAc; R^2 = OCH$_2$C(=CH$_2$)COCH$_3$ (biselide B)
1663 R^1 = R^2 = OH (biselide C)
1664 R^1 = H; R^2 = NHCH$_2$CH$_2$SO$_3$H (biselide D)

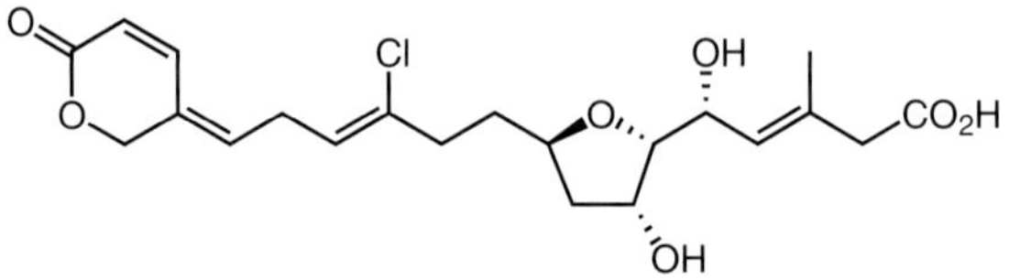

1665 (biselide E)

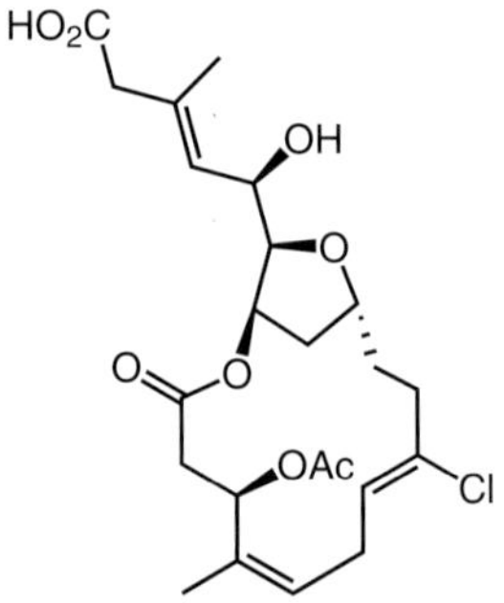

1666 (FR177391)

The Red Sea sponge *Latrunculia corticata* (Fig. 3.23) has afforded latrunculinosides A (**1667**) and B (**1668**), which contain the unusual saccharides β-D-olivose, β-L-digitoxose, α-L-amicetose, and β-D-oliose (*1630*).

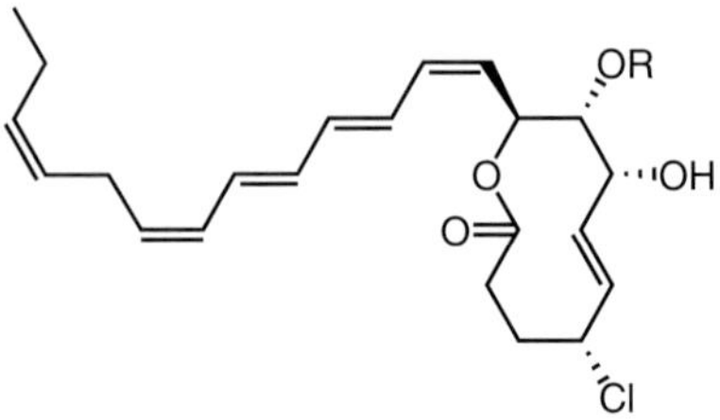

1667 R = β-L-Dig–β-D-Olv–α-L-Ami (latrunculinoside A)
1668 R = β-D-Oli–β-D-Oli (latrunculinoside B)

3.17 Macrolides and Polyethers 247

Fig. 3.23 *Latrunculia corticata*, a Red Sea sponge that contains the latrunculinosides A and B (**1667** and **1668**) (Photo: T. Rezanka)

The Japanese sponge *Discodermia calyx* has yielded the new calyculin, calyculin J (**1669**) (*1631*). The highly toxic (ichthyotoxicity, hemolytic activity) prymnesin-1 (**1670**) and -2 (**1671**), possessing unprecedented structural complexity, were characterized from the red tide alga *Prymnesium parvum* (*1632–1634*).

1669 (calyculin J)

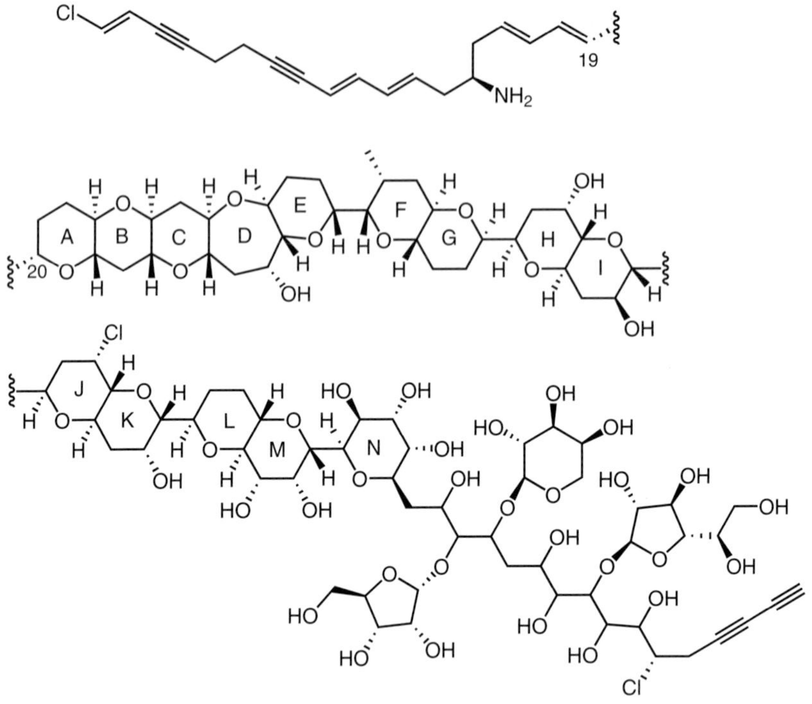

1670 (prymnesin-1)

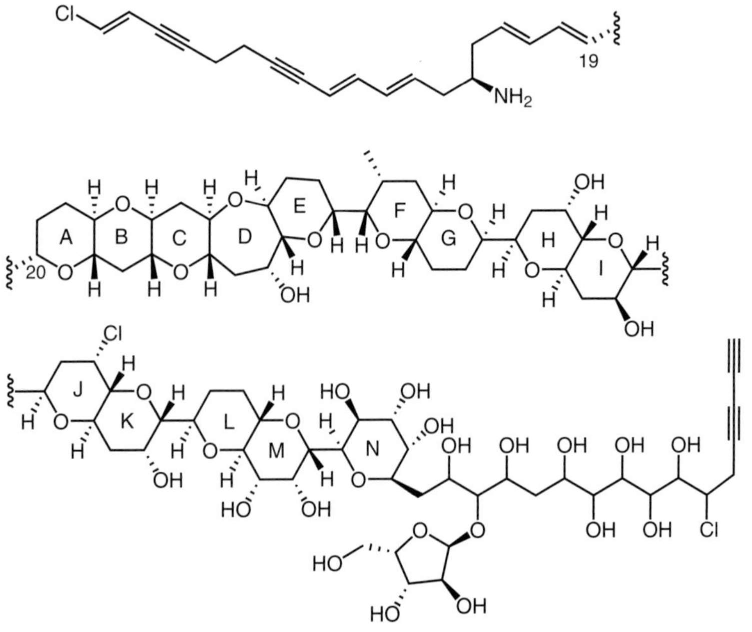

1671 (prymnesin-2)

3.18 Naphthoquinones, Higher Quinones, and Related Compounds

The relatively large group of previously known napyradiomycins and related bacterial metabolites has been augmented by the discovery of A80915-A (**1672**), -B (**1673**), -C (**1674**), and -D (**1675**) from cultures of *Streptomyces aculeolatus* from a Palauan soil sample (*1635*). A deep-sea marine actinomycete has afforded the related **1676–1678**, which exhibit significant antibacterial activity towards drug-resistant *Staphylococcus aureus* and *Enterococcus faecium*, and cytotoxicity toward HCT-116 human colon carcinoma (*1636*). An X-ray structure establishes the absolute configuration of A80915C (**1674**) (*1637*).

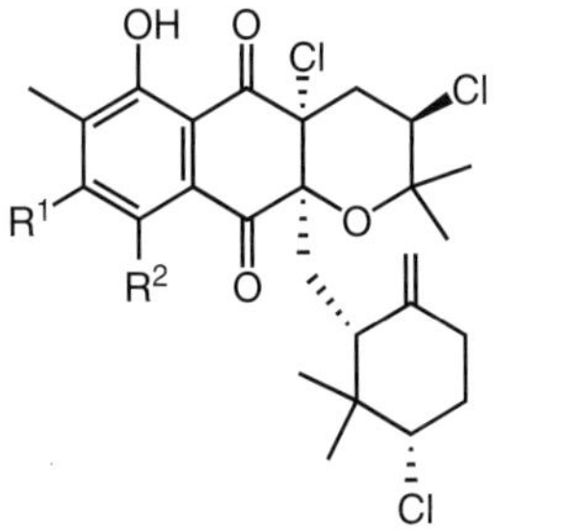

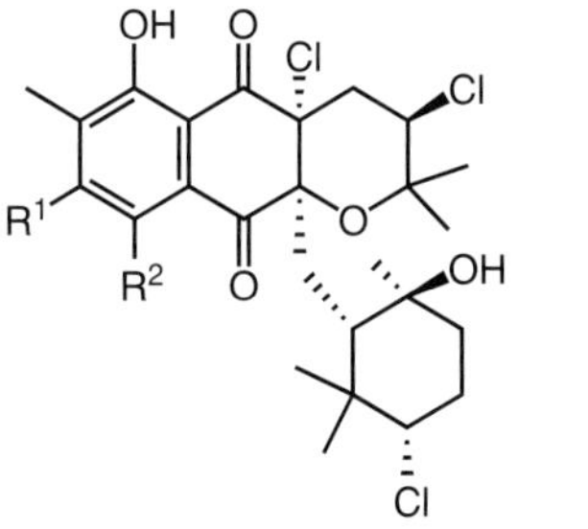

1672 R^1 = OH, R^2 = H (A80915-A)
1673 R^1 = $O^\ominus$, R^2 = $N_2^\oplus$ (A80915-B)
1674 R^1 = OH, R^2 = H (A80915-C)
1675 R^1 = $O^\ominus$, R^2 = $N_2^\oplus$ (A80915-D)

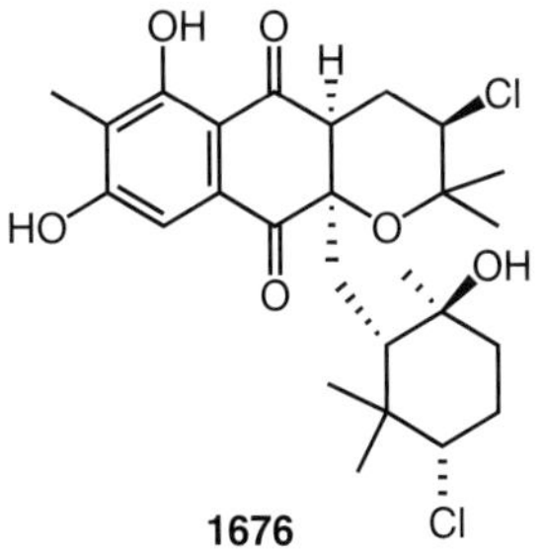

1676

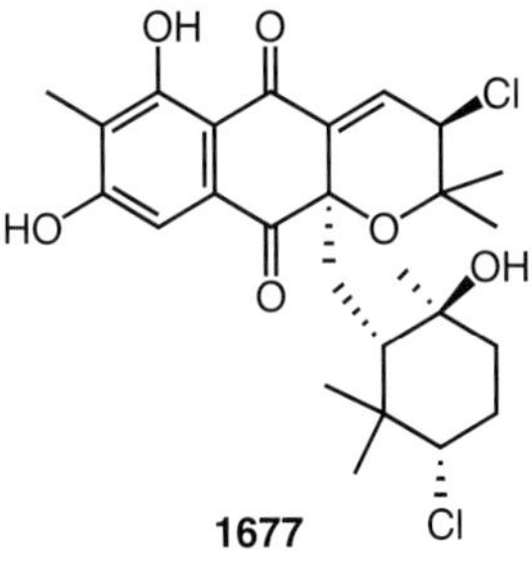

1677

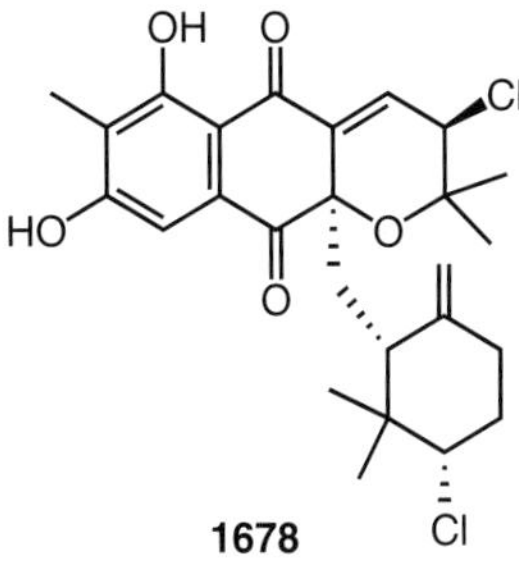

1678

Isomarinone (**1679**), an isomer of the previously known marinone (*1*), was isolated from the same tropical sediment bacterium (*1638*). Another marine-derived bacterium related to the genus *Streptomyces* has yielded the novel azamerone (**1680**) (*1639*). The British Columbian medicinal plant *Moneses uniflora* contains the antibiotic 8-chlorochimaphilin (**1681**), which is more active than chimaphilin (*1640*). Sesame roots (*Sesamum indicum*) have yielded the red chlorinated naphthoquinone chlorosesamone (**1682**) (*1641*). Cultures of *Streptomyces* strain *LL*-A9227 produce chloroquinocin (**1683**), which has some antibacterial activity against Gram-positive bacteria (*1642*). The two xestoquinones **1684** and **1685** were characterized from the Philippino sponge *Xestospongia* sp., and display topoisomerase II activity (*1643*).

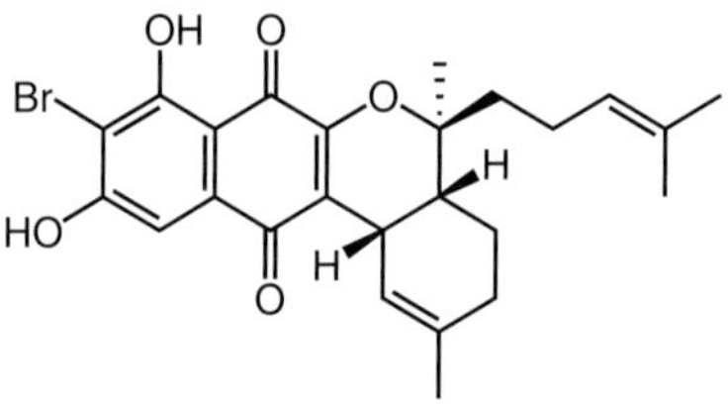

1679 (isomarinone)

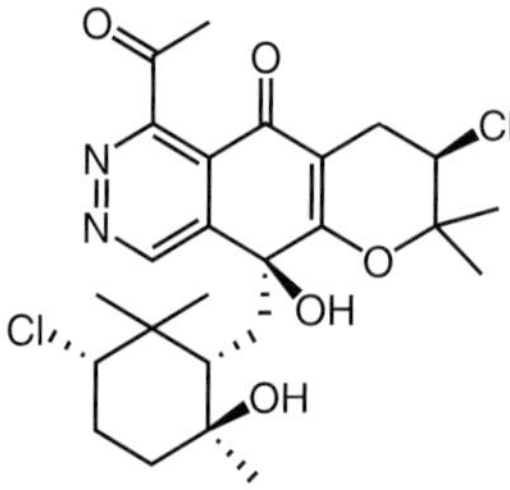

1680 (azamerone)

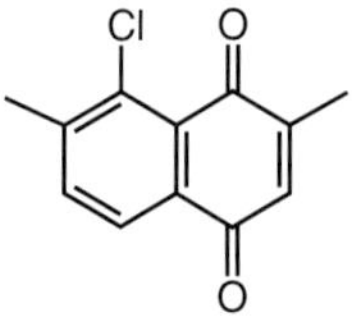

1681 (8-chlorochimaphilin)

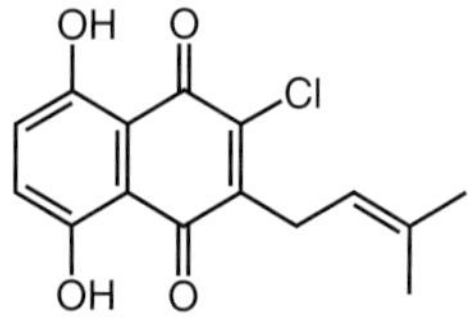

1682 (chlorosesamone)

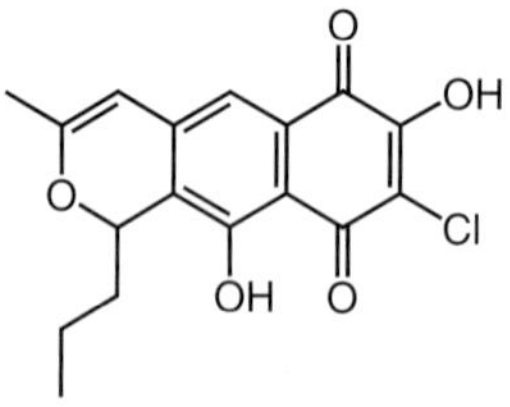

1683 (chloroquinocin)

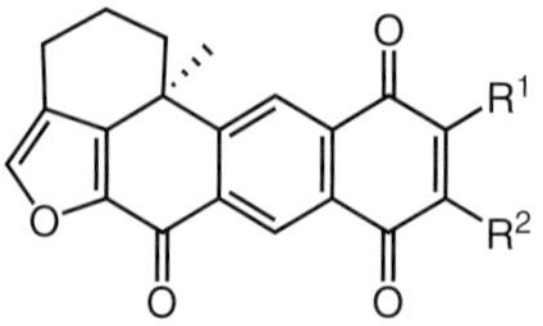

1684 R^1 = Cl, R^2 = OH
1685 R^1 = OH, R^2 = Cl

A synthesis of the chlorinated angucycline antibiotic BE-23254 (**1686**), which was isolated from *Streptomyces* sp. A 23254, has confirmed the structure of this benz[*a*]anthraquinone derivative (*1644, 1939, 1940*). Two detailed examinations of the rare Australian soil actinomycete *Kibdelosporangium* sp. uncovered a series of

3.18 Naphthoquinones, Higher Quinones, and Related Compounds

kibdelones (*1645*) and isokibdelones (*1646*) with novel structures. The former study includes kibdelone A (**1687**), kibdelone A rhamnoside (**1688**), kibdelone B (**1689**), kibdelone B rhamnoside (**1690**), kibdelone C (**1691**), kibdelone C rhamnoside (**1692**) and 13-oxokibdelone A (**1693**) (*1645*), while the latter study includes isokibdelone A (**1694**), isokibdelone A rhamnoside (**1695**), isokibdelone B (**1696**), and isokibdelone C (**1697**) (*1646*).

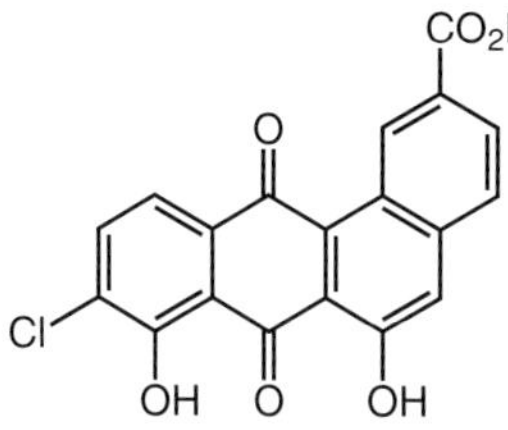

1686 (BE-23254)

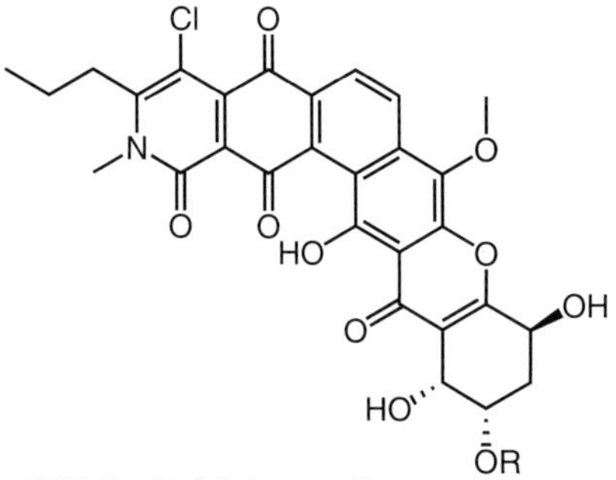

1687 R = H (kibdelone A)
1688 R = α-Rh (kibdelone A rhamnoside)

Rh = α-rhamnose

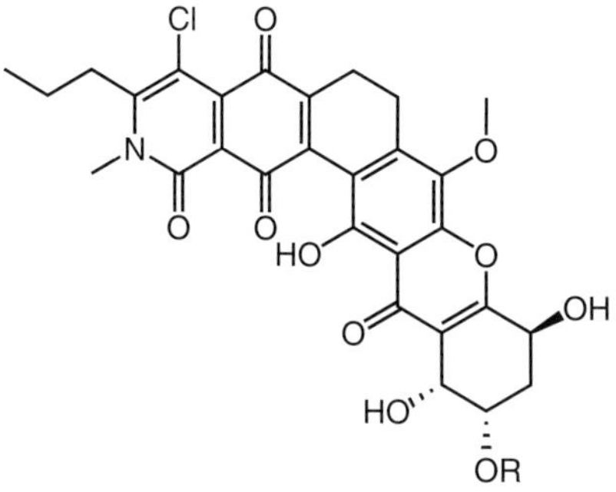

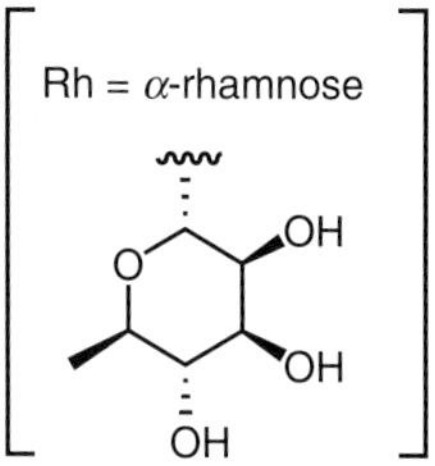

1689 R = H (kibdelone B)
1690 R = α-Rh (kibdelone B rhamnoside)

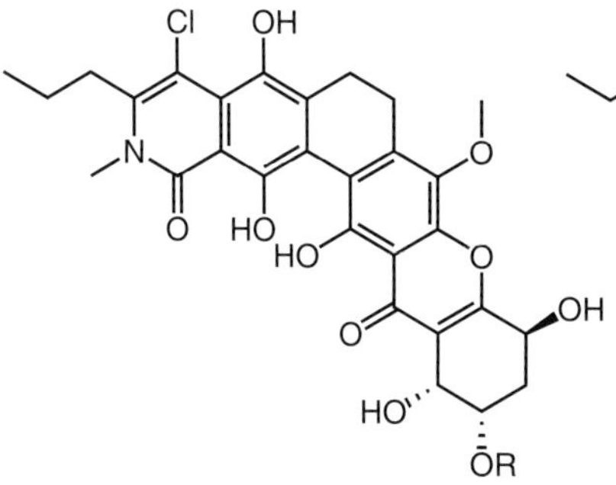

1691 R = H (kibdelone C)
1692 R = α-Rh (kibdelone C rhamnoside)

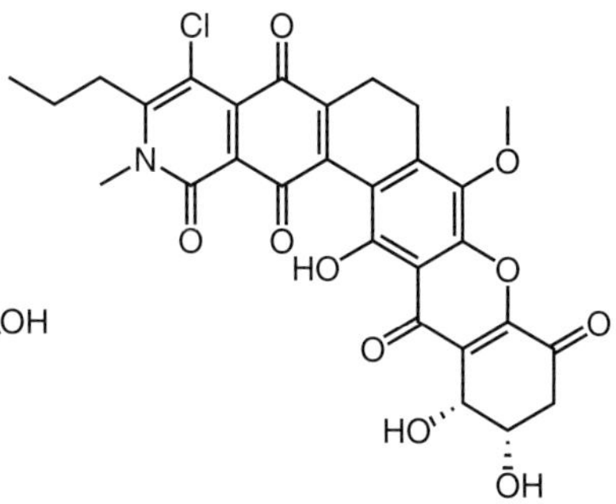

1693 (13-oxokibdelone A)

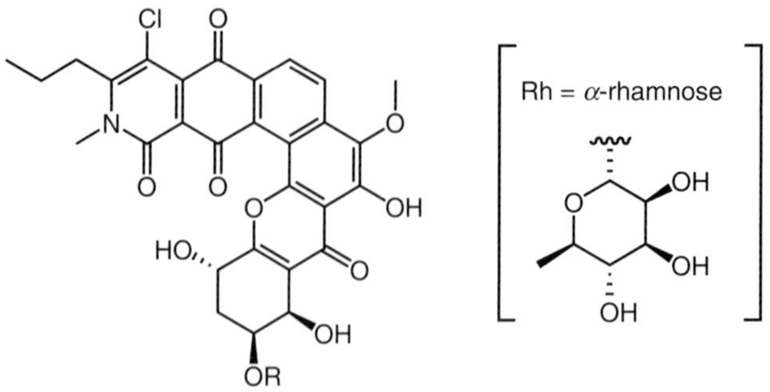

1694 R = H (isokibdelone A)
1695 R = α-Rh (isokibdelone A rhamnoside)

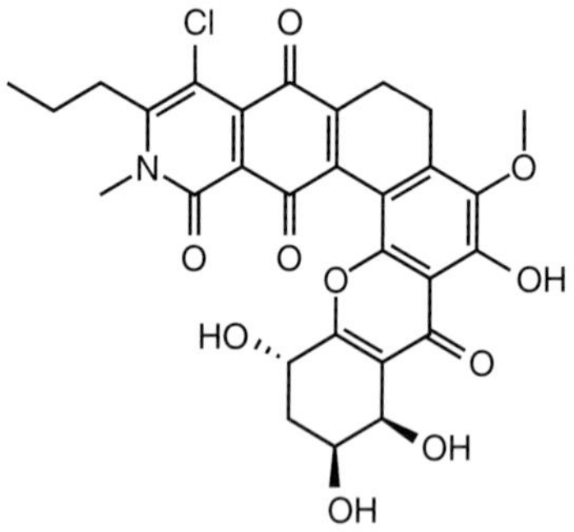

Rh = α-rhamnose

1696 (isokibdelone B)

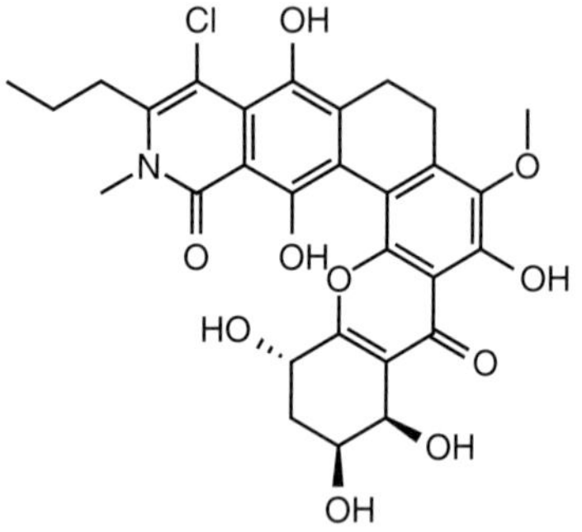

1697 (isokibdelone C)

Four novel antibacterial, antifungal, and herbicidal palmarumycins have been isolated from the West Borneo forest soil microbe *Coniothyrium* sp. (*1647*). These are palmarumycins C_1 (**1698**), C_4 (**1699**), C_7 (**1700**), and C_8 (**1701**), along with several nonchlorinated analogs. The palmarumycins have attracted the interest of synthetic chemists (*434*), as have related naphthalenoid natural products (*2669*). The DNA-cleaving antitumor antibiotics spiroxins A (**1702**), B (**1703**), and E (**1704**) are found in a Vancouver Island soft coral containing an associated fungus LL-37H248 (*1648, 1649*). The absolute configuration of **1702** was determined (*1649*), and a synthesis of the nonchlorinated spiroxin C has been described (*1650*).

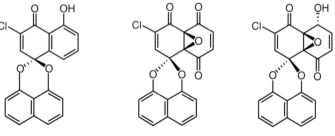

1698 (palmarumycin C_1)   **1699** (palmarumycin C_4)   **1700** (palmarumycin C_7)

1701 (palmarumycin C_8)

1702 R = H, R^1,R^2 = O (spiroxin A)
1703 R = Cl, R^1,R^2 = O (spiroxin B)
1704 R = Cl, R^1 = OH, R^2 = H (spiroxin E)

3.19 Tetracyclines

Although no new halogen-containing tetracyclines have been reported since the first survey (*1*), the gene responsible for the chlorination of tetracycline in *Streptomyces aureofaciens* (Fig. 3.24) has been cloned and the sequence of nucleotides determined (*1651*). The gene product is a 452 amino acid chlorination enzyme.

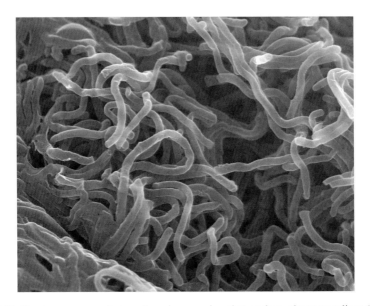

Fig. 3.24 *Streptomyces aureofaciens*, the microorganism that produces the tetracyclines (Photo: T. Rezanka).

3.20 Aromatics

Compared to their more reactive phenol counterparts in electrophilic halogenation, simple unactivated halogenated aromatic rings occur rarely in Nature (*1*). However, several notable examples exist. The high temperatures present in volcanoes and during the formation of meteorites leads to the production of the previously known chlorobenzene and a dichlorobenzene (**1705**) in Orgueil and Cold Bokkeveld meteorites, respectively (*388*), and 1-chloronaphthalene (**1707**) and bromomesitylene (**1708**) in carbonaceous black shales (*330*). Studies of emissions from Vulcano, Mt. Etna, Kuju, and Satsuma Iwojima reveal the presence of chlorobenzene, two dichlorobenzenes (**1705, 1706**), 1,4-dichlorobenzene, fluorobenzene (**1709**), tetrafluorobenzene (**1710**), fluorochlorobenzene (**1711**), chlorostyrenes (**1712–1714**), trichlorobenzenes (**1715, 1716**), chlorotoluenes (**1717–1719**), the previously known bromobenzene, and chloroethylbenzene (**1720**) (*216, 217*). It should be noted that heating (400–950°C) a mixture of methane, hydrogen chloride, and oxygen results in the formation of a plethora of chlorinated aromatics (benzenes, toluenes, xylenes, styrenes, naphthalenes, biphenyls, anisoles, acenaphthylenes, phenanthrenes, fluoranthenes) (*232*). Several chlorinated benzoic acids have natural origins. The meteorites Murray, Murchison, Cold Bokkeveld, and Orgueil contain 4-chlorobenzoic acid (**1721**), 2,4-dichlorobenzoic acid (**1722**), and 2,6-dichlorobenzoic acid (**1723**) to varying degrees (*1652*). These chlorobenzoic acids are also found in remote bogs and sediments, particularly 2,4-dichlorobenzoic acid (**1722**) (*1653*). In all samples, trichloroacetic acid was also detected. Laboratory experiments involving benzoic acid and chloroperoxidase give rise to chlorobenzoic acids, in agreement with their natural production (*1653*). Another biogenic source of 2,4-dichlorobenzoic acid (**1722**) is the terrestrial cyanobacterium *Fischerella ambigua*, which is the first report of this compound from a living organism (*1654*). Along with **1722**, three different tetrachlorobiphenyl carboxylic acids (**1724–1726**) are present in dissolved seawater, and, based on their isomer distribution and global inventory, these compounds are suggested to have a natural source (*1655*). The *de novo* formation of 3,4-dichlorophenylacetic acid (**1727**), 3,4-dichlorobenzoic acid (**1728**), and 3,4,5-trichlorobenzoic acid (**1729**) in a sewage treatment plant were reported, along with several other known chlorinated benzoic acids and phenols (*1656*). Garden compost also produces **1727–1729** and 3,4,5-trichlorophenylacetic acid (**1730**) (*1657*). Acid **1727** has also been identified in fungi (*1656*).

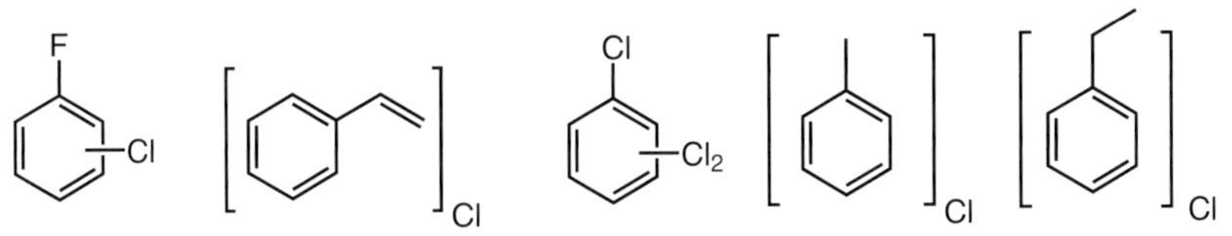

3.20 Aromatics

1711, **1712-1714**, **1715, 1716**, **1717-1719**, **1720**, 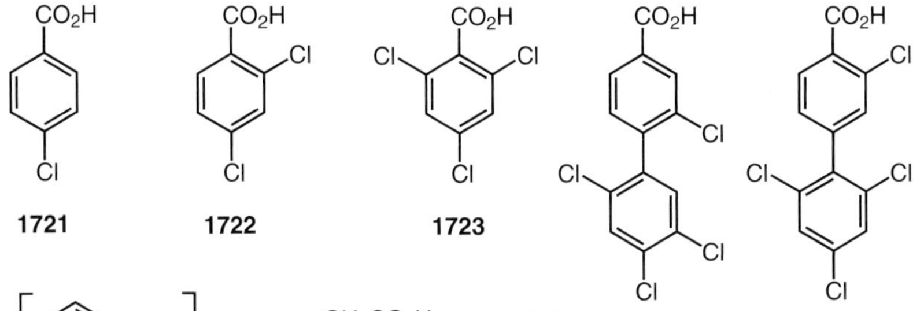 **1721**, **1722**, **1723**, **1724**, **1725**, 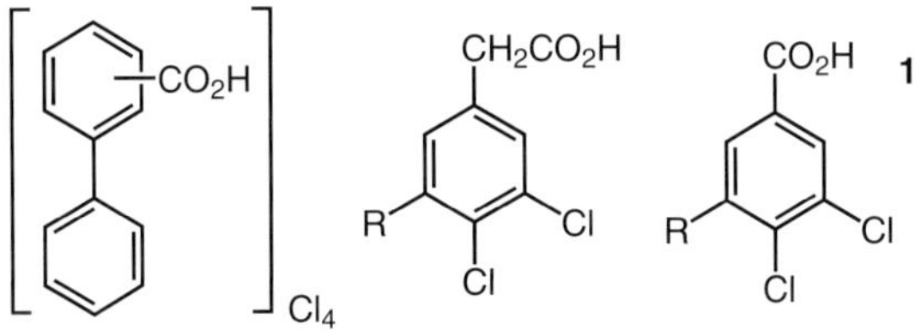 **1726**, **1727** R = H, **1730** R = Cl, **1728** R = H, **1729** R = Cl

The Palauan deep-water (500 m) marine actinomycete *Salinispora pacifica* has yielded cyanosporasides A (**1731**) and B (**1732**), which feature a chlorine on an unactivated benzene ring (*1658*). The authors suggest a novel biosynthesis from an enediyne precursor. Related to the previously known nostocyclophanes (*1*) are the new carbamidocyclophanes A–E, of which A (**1733**), B (**1734**), C (**1735**), and D (**1736**) are chlorinated, being isolated from the Vietnamese cyanobacterium *Nostoc* sp. (*1659*). These compounds exhibit cytotoxicity against MCF-7 (breast) and F1 (amniotic epithelial) human cancer cells.

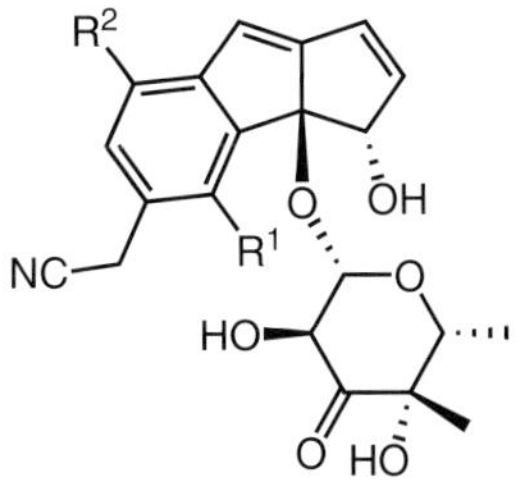

1731 R^1 = Cl, R^2 = H (cyanosporaside A)
1732 R^1 = H, R^2 = Cl (cyanosporaside B)

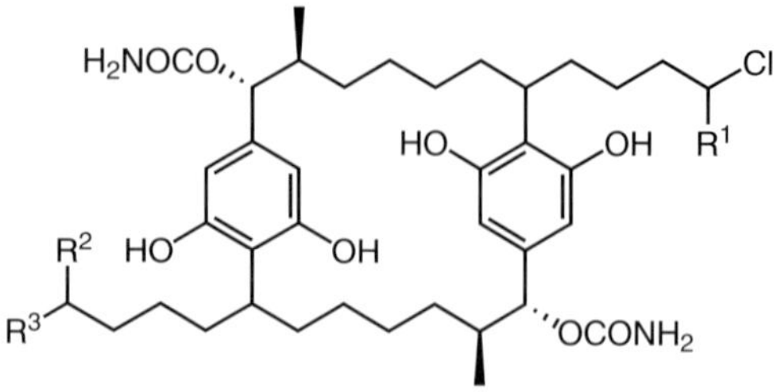

1733 $R^1 = R^2 = R^3 = Cl$ (carbamidocyclophane A)
1734 $R^1 = R^2 = Cl, R^3 = H$ (carbamidocyclophane B)
1735 $R^1 = Cl, R^2 = R^3 = H$ (carbamidocyclophane C)
1736 $R^1 = R^2 = R^3 = H$ (carbamidocyclophane D)

Polycyclic aromatic hydrocarbons (PAH), which are ubiquitous in the environment, including surface waters, undergo facile chlorination by hypochlorite when dissolved in humus-poor water to give a suite of chlorinated PAH (*1660*). It is therefore conceivable that this chlorination can occur under natural conditions, but this is yet to be determined. Another new possible source of natural chlorinated PAH is the reported in vitro reaction of benzo[*a*]pyrene diol epoxide, the ultimate carcinogen of benzo[*a*]pyrene with chloride ion to give chlorohydrin **DDD**, which has been isolated and identified as an intermediate en route to a benzo[*a*]pyrene-DNA adduct (*1661*). However, **DDD** is not considered to be a natural compound at this time.

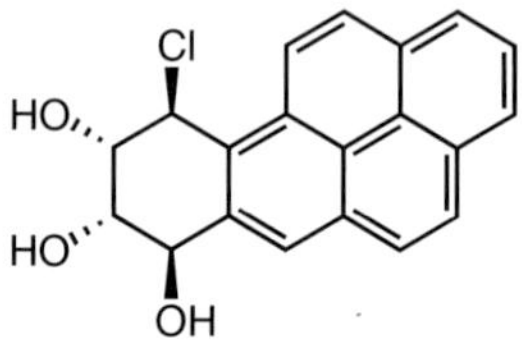

DDD

3.21 Simple Phenols

The enormous reactivity of the phenolic ring towards electrophilic halogenation has led to a multitude of natural halogenated phenols, both terrestrial and marine (*1*).

3.21.1 Terrestrial

The previously reported 2,6-dichlorophenol, which is a sex pheromone of several species of tick (*1*), is also produced by the African tick *Amblyomma hebraeum* (*1662, 1663*). This pheromone has been used to control the African bont tick on cattle in Zimbabwe (*1664*) and the American dog tick (*Dermacentor variabilis*) (*1665*). Neurons in the legs of male ticks (*Amblyomma americanum*) are sensitive to 2,6-dichlorophenol (*1666*). In addition to this well-known chlorophenol, a number of new chlorinated phenols were reported since the first survey, including some that were overlooked by the author (*1*). Oakmoss (*Evernia prunastri*), which contributed bromobenzene to the first survey (*1*), also contains chlororesorcinol **1737** (*1667*). The fungus *Hericium erinaceus* has furnished 4-chloro-3,5-dimethoxybenzoic acid (**1738**) and related esters **1739** and **1740** (*1668*). Another study of this fungus revealed the presence of the related orcinols **1741–1743** (*1669*). Cultures of the basidiomycete *Stropharia* sp. have yielded 3,5-dichloro-4-methoxybenzyl alcohol (**1744**), and 3,5-dichloro-4-methoxybenzaldehyde (**1745**) occurs in *Hypholoma subviride* (*1670*). *Hypholoma elongatum* has provided the new 2,4,6-trichloro-3-methoxyphenol (**1746**), 3,5,6-trichloro-2,4-dimethoxyphenol (**1747**), and 3,4,6-trichloro-2,5-dimethoxyphenol (**1748**) (*1671*). The Japanese mushroom *Russula subnigricans* contains 2,6-dichloro-4-methoxyphenol (**1749**) (*1672*). The basidiomycete strain *Kuehneromyces mutabilis* produces methyl 3,6-dichloro-2-methylbenzoate (**1750**) (*1673*). Chlorinated lactone **1751** was characterized from *Leucoagaricus carneifolia* (*1674*). Metabolites **1744** and **1745** were also isolated from the fungus *Pholiota destruens* (*1675*) and from the American matsutake mushroom *Tricholoma magnivelare* (*1676*).

1737

1738

1739

1740

1741 R = Me
1742 R = CH_2OH
1743 R = CHO

1744 R = CH_2OH
1745 R = CHO

1746

1747

1748

1749

1750

1751

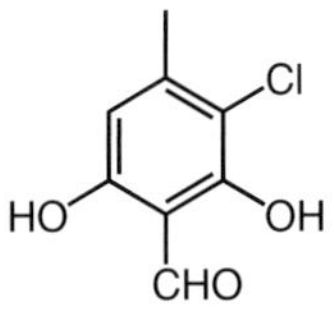

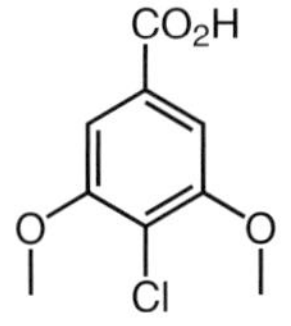

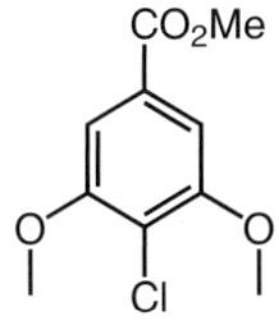

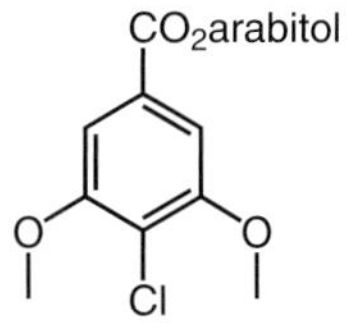

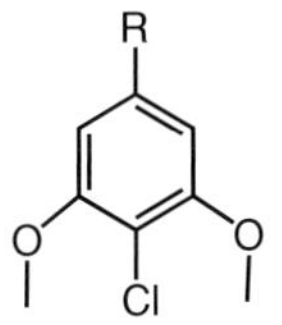

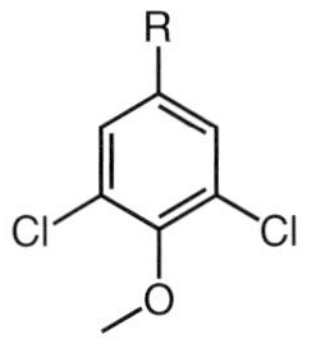

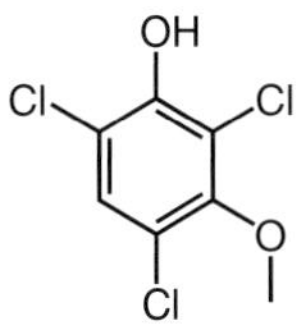

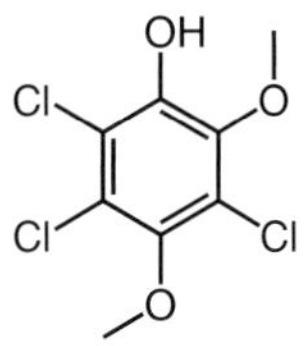

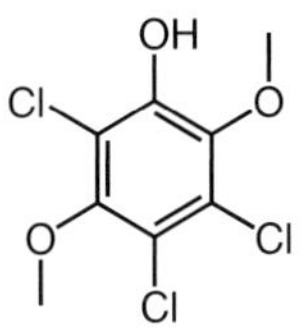

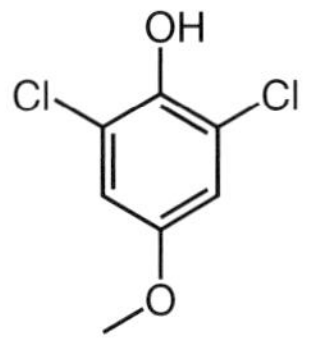

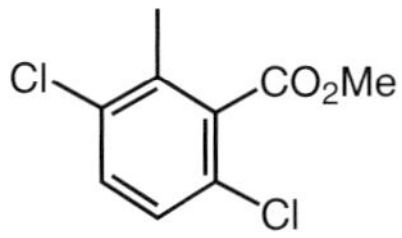

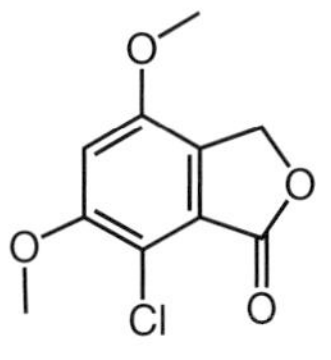

The wood-rotting fungi of genus *Bjerkandera* produce a number of chlorinated phenols and derivatives (*398, 1677*).

Thus, *Bjerkandera* sp. BOS55 has yielded 3-chloro-4-hydroxybenzoic acid (**1752**), 3,5-dichloro-4-hydroxybenzoic acid (**1753**), methyl 3,5-dichloro-4-hydroxybenzoate (**1754**), methyl 3,5-dichloro-4-methoxybenzoate (**1755**), 3-chloro-4-methoxybenzoic acid (**1756**), and 3,5-dichloro-4-methoxybenzoic acid (**1757**) (*1678*). The latter two metabolites and several other new halogenated phenols were discovered in soil around the fungus *Lepista nuda* (wood blewitt): 5-chloro-3,4-dimethoxybenzaldehyde (**1758**), 3-bromo-5-chloro-4-methoxybenzaldehyde (or isomer) (**1759**), 5-bromo-3,4-dimethoxybenzaldehyde (**1760**), 3-bromo-4-methoxybenzaldehyde (**1761**), 4-chloroanisole (**1762**), 2,6-dichloroanisole (**1763**), 2-chloro-1,4-dimethoxybenzene (**1764**), 2,6-dichloro-1,4-dimethoxybenzene (**1765**), a dichlorodimethoxybenzaldehyde isomer (**1766**), and a few other known compounds (*1679*). The simple phenols tetrachlorocatechol (**1767**) and monomethyl ether (**1768**) were isolated from a *Mycena* fungal species (*1680*). *Mycena alcalina* has furnished alcalinaphenols A–C (**1769–1771**) (*1681*). Veratryl chloride (**1772**) (3,4-dimethoxybenzyl chloride) has been reported in *Bjerkandera* sp. BOS55 (*1682*). In view of the enormous reactivity to be expected for this compound (facile S_N1 and S_N2 reactions), this report is surprising. This same fungus and *Bjerkandera fumosa* contain **1773** and **1774** (*1683*). The related trametol (**1775**) was isolated from the fungus *Trametes* sp. (*1684*). The microfungus *Xylaria* sp. contains 3-chloro-4-hydroxyphenylacetamide (**1776**) (*1685*). The chlorinated anisyl metabolites are produced by a wide range of basidiomycete genera including *Mycena, Peniophora, Phellinus, Phylloporia, Bjerandera, Hypholoma,* and *Pholiota* (*1686*). An excellent summary of chlorometabolite production by *Bjerkandera adusta* has been published (*1687*).

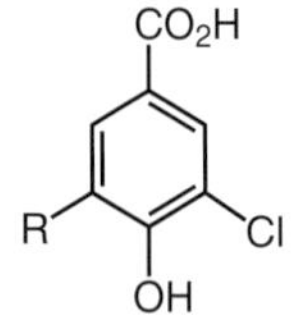

1752 R = H
1753 R = Cl

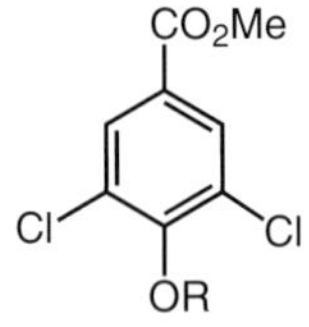

1754 R = H
1755 R = Me

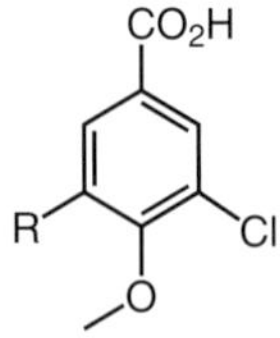

1756 R = H
1757 R = Cl

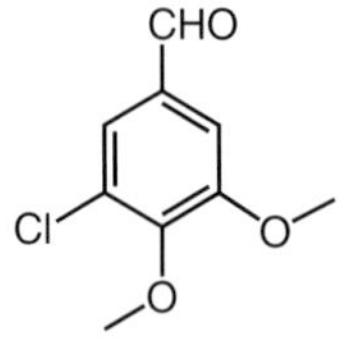

1758

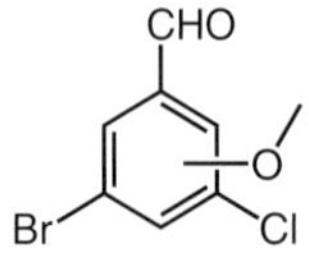

1759

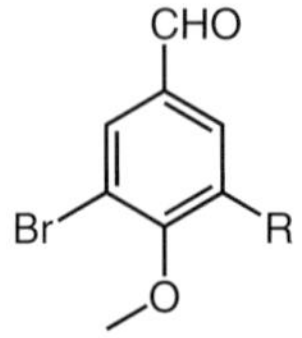

1760 R = OMe
1761 R = H

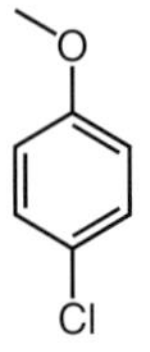

1762

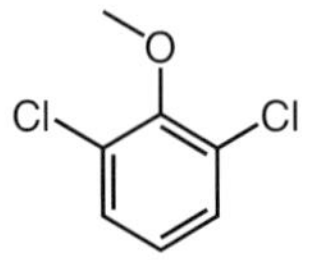

1763

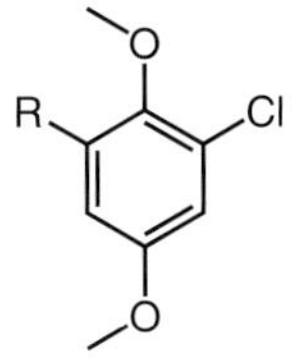

3.21 Simple Phenols

1764 R = H
1765 R = Cl

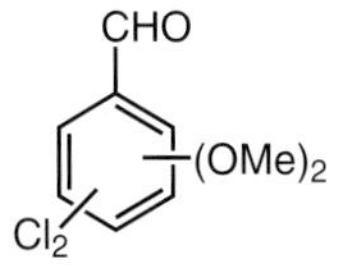

1766

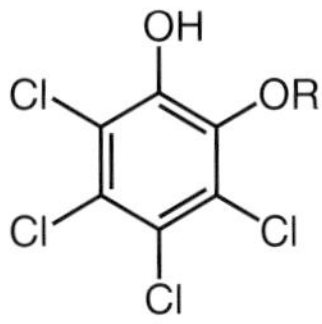

1767 R = H
1768 R = Me

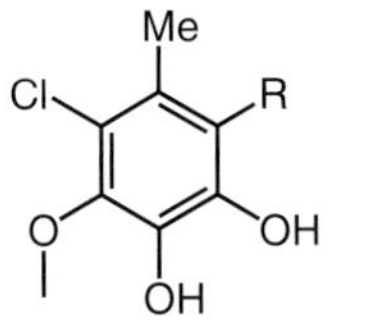

1769 R = H (alcalinaphenol A)
1770 R = OMe (alcalinaphenol B)

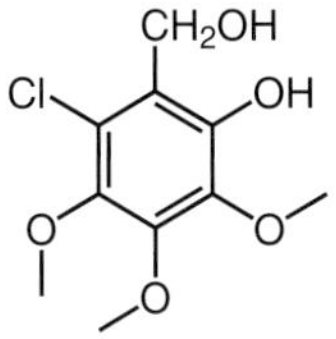

1771 (alcalinaphenol C)

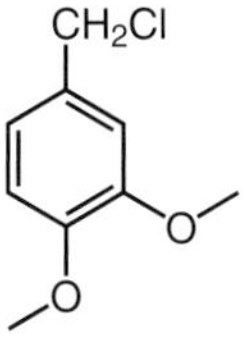

1772

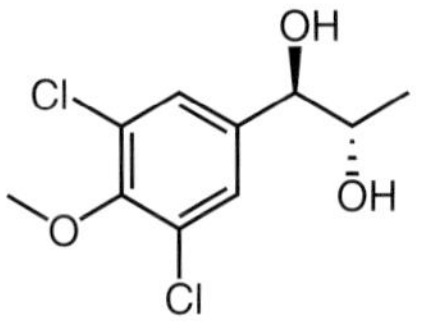

1773 R = Cl
1775 R = H (trametol)

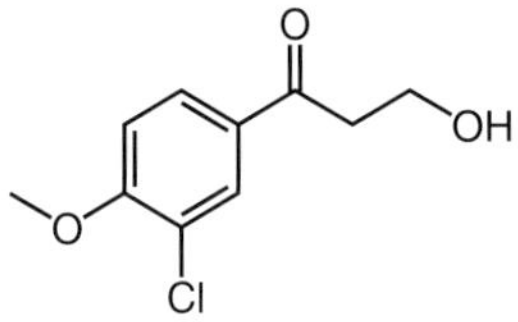

1774

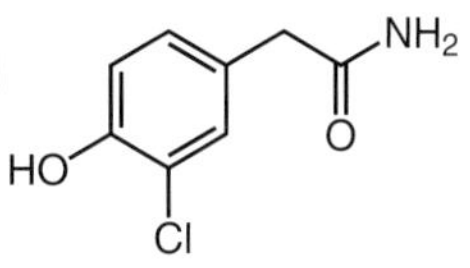

1776

A collection of the carrot truffle, *Stephanospora caroticolor* (Fig. 3.25), from Germany has led to 2-chloro-4-nitrophenol (**1777**), 4-amino-2-chlorophenol (**1778**), and stephanosporin (**1779**) (*1688*). Phenol **1777** is also present in the fruit bodies of *Lindtneria trachyspora* (*1688*). The New Zealand liverwort *Riccardia marginata* has afforded the novel chlorinated bibenzyls **1780–1782** (*1689*). These metabolites have antimicrobial and antifungal activity against *Bacillus subtilis*, *Trichophyton mentagrophytes*, *Candida albicans*, and *Cladosporium resinae*. In response to attack by the pathogenic fungus *Fusarium oxysporum f.* sp. *lilii*, the edible Japanese lily *Lilium maximowiczii* (Fig. 3.26) produces seven chlorinated orcinols **1783–1789** as natural fungicides (*1690*). The Pakistani medicinal lichen (*Usnea longissima*, "Old Man's Beard"), which has been used for pain relief and fever control, contains longissiminone B (**1790**) along with the nonchlorinated analog, which has potent antiinflammatory activity (*1691*). The Canary Islands lichen *Lethariella canariensis* features the new chloroatranol (**1791**), chlorohematommic acid (**1792**), methyl chlorohematommate (**1793**), and ethyl chlorohematommate (**1794**) (*1692*).

1777

1778

1779 (stephanosporin)

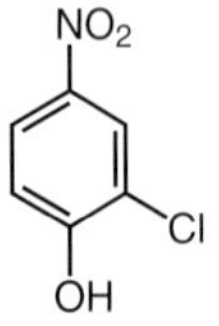

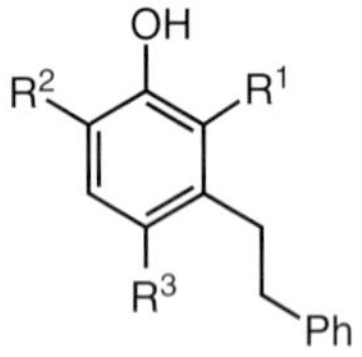

1780 $R^1 = R^2 = R^3 = Cl$
1781 $R^1 = R^2 = Cl$, $R^3 = H$
1782 $R^1 = Cl$, $R^2 = R^3 = H$

1783

1784

1785

1786

1787

1788

1789

1790 (longissiminone B)

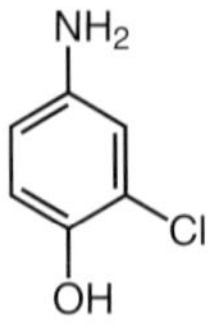

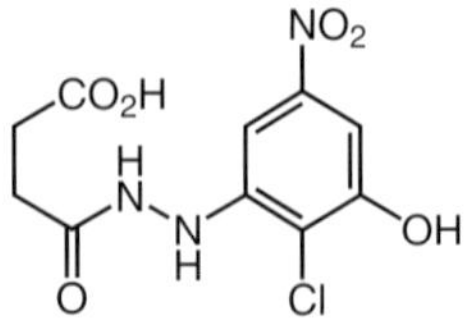

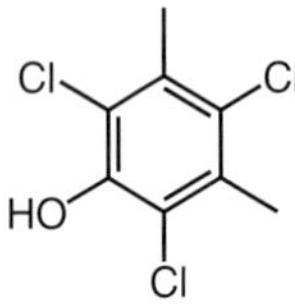

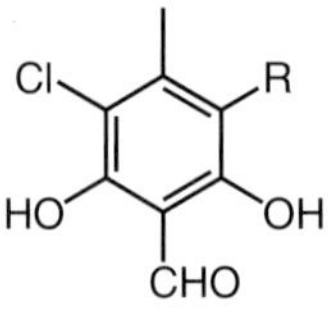

1791 R = H (chloroatranol)
1792 R = CO_2H (chlorohematommic acidl)
1793 R = CO_2Me (methyl chlorohematommate)
1794 R = CO_2Et (ethyl chlorohematommate)

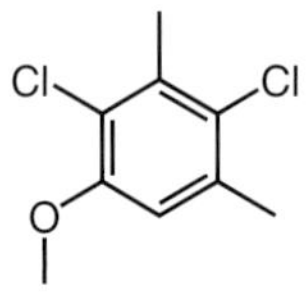

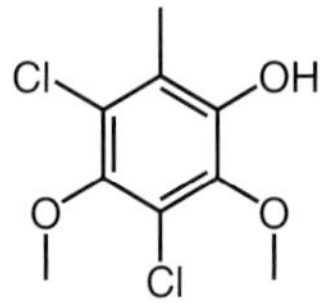

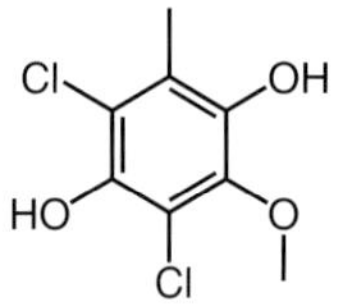

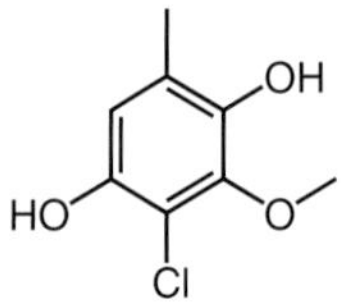

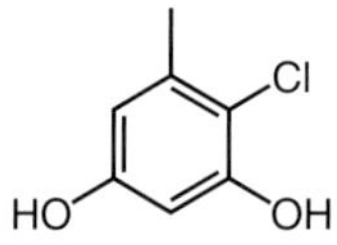

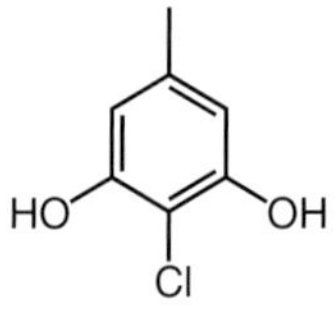

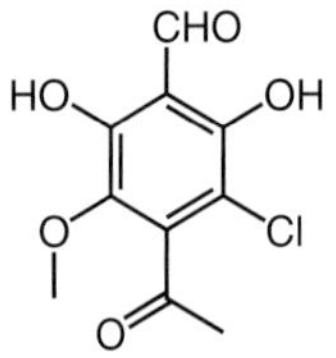

Cultures of *Lentinellus cochleatus* yield chlorostyrene **1795** (*1693*), and the related compound **1796** occurs in *Arnica sachalinensis* (*1694*). The sulfur-oxidizing bacterium *Thialkalivibrio versutus* contains the membrane-bound chloronatronochrome (**1797**) (*1695*). The terrestrial plant *Rumex patientia* from Turkey, which is used in traditional medicine, has yielded the naphthalene glycosides patientosides A (**1798**) and B (**1799**) (*1696*). The Turkish folk medicine plant *Geranium pratense* subsp. *finitimum* contains 6-chloroepicatechin (**1800**) (*1697*). The Indian tree *Gmelina arborea*, which is of commercial importance, has afforded the first bromine-containing lignan **1801** (*1698*).

3.21 Simple Phenols

1795

1796

1798 R = Ac (patientoside A)
1799 R = Cl (patientoside B)

1797 (chloronatronochrome)

1800 (6-chloroepicatechin)

1801

Fig. 3.25 *Stephanospora caroticolor*, the carrot truffle that contains novel chlorophenols **1777** and **1778**, and stephanosporin (**1779**) (Photo: W. Steglich)

Fig. 3.26 *Lilium maximowiczii*, the edible Japanese lily that produces the seven chlorinated fungicides **1783–1789**; the brown portion indicates disease by a *Fusarium* fungus (Photo: K. Monde)

The liverwort genera consisting of 400–500 species have been the source of a large number of chlorinated 'bis-bibenzyls' and related polyphenols (*1699*). The first example appears to be 12-chloroisoplagiochin D (**1802**) found in the Costa Rican liverwort *Plagiochila* sp. (*1700*). In an isolation and identification *tour de force*, a research group has characterized ten chlorinated bridged biphenyls, bazzanins A-J (**1803–1812**), and the novel phenanthrene bazzanin K (**1813**) from the liverwort *Bazzania trilobata* (*1701*). The Japanese liverwort *Herbertus sakuraii* has afforded 2,12-dichloroisoplagiochin D (**1814**), 12,7′-dichloroisoplagiochin D (**1815**), and 12,10′-dichloroisoplagiochin C (**1816**) (*1702, 1703*). The liverwort *Mastigophola diclados* also contains **1802** and **1814** (*1703*). The Taiwanese liverwort *Plagiochila peculiaris* contains bazzanin J and 12-chloroisoplagiochin D (*1704*). Bazzanins L-R (**1817–1823**) and S (**1824**) have been characterized from the liverworts *Lepidozia incurvata* (*1705*) and *Bazzania trilobata* (*1706*), respectively. Several of these bazzanins are optically active, but are not enantiomerically pure in the liverworts (*1706*). The liverwort *Jamesoniella colorata* has furnished the "ring-opened" bis-bibenzyl 6,6′,10,10′,12,12′-hexachloroisoperrottetin (**1825**) (*1707*). That these chlorinated phenols are not isolation artifacts is supported by their presence in the crude liverwort extracts as detected by mass spectrometry (MALD1-TOF and LDI-TOF) (*1708, 1709*). Moreover, a chloroperoxidase enzyme (which will be discussed in detail in Sect. 4.2 (Chloroperoxidase)) has been detected in *Bazzania trilobata* further supporting the natural occurrence of these unusual chlorinated phenols (*1710*).

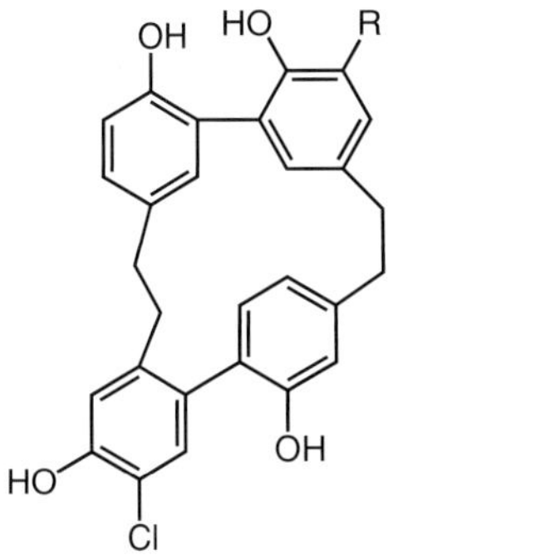

3.21 Simple Phenols

1802 R = H (12-chloroisoplagiochin D)
1812 R = Cl (bazzanin J)

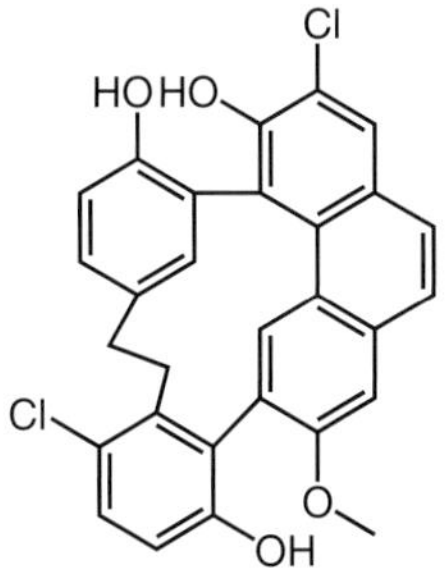

1813 (bazzanin K)

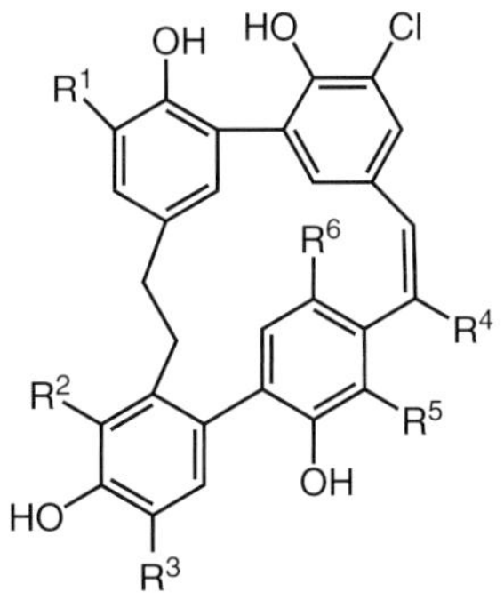

1803 R^1 = H, R^2 = H, R^3 = H, R^4 = H, R^5 = H, R^6 = H (bazzanin A)
1804 R^1 = H, R^2 = H, R^3 = H, R^4 = Cl, R^5 = H, R^6 = H (bazzanin B)
1805 R^1 = H, R^2 = H, R^3 = Cl, R^4 = Cl, R^5 = H, R^6 = H (bazzanin C)
1806 R^1 = H, R^2 = H, R^3 = H, R^4 = Cl, R^5 = Cl, R^6 = H (bazzanin D)
1807 R^1 = H, R^2 = H, R^3 = Cl, R^4 = Cl, R^5 = Cl, R^6 = H (bazzanin E)
1808 R^1 = H, R^2 = Cl, R^3 = H, R^4 = Cl, R^5 = H, R^6 = Cl (bazzanin F)
1809 R^1 = Cl, R^2 = Cl, R^3 = H, R^4 = Cl, R^5 = Cl, R^6 = H (bazzanin G)
1810 R^1 = H, R^2 = Cl, R^3 = H, R^4 = Cl, R^5 = Cl, R^6 = Cl (bazzanin H)
1811 R^1 = Cl, R^2 = H, R^3 = Cl, R^4 = Cl, R^5 = Cl, R^6 = Cl (bazzanin I)

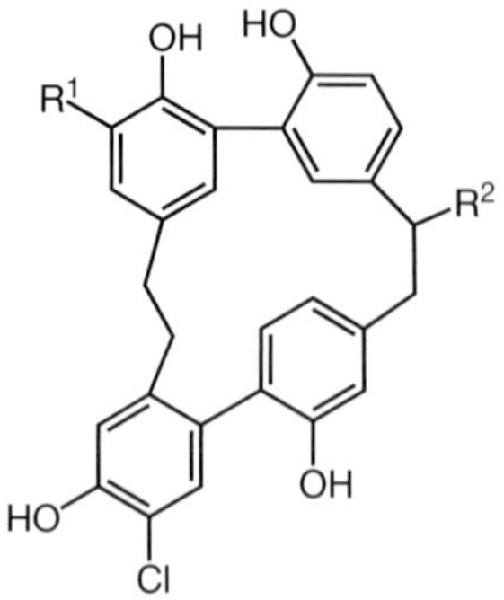

1814 $R^1 = Cl, R^2 = H$
(2,12-dichloroisoplagiochin D)
1815 $R^1 = H, R^2 = Cl$
(12,7'-dichloroisoplagiochin D)

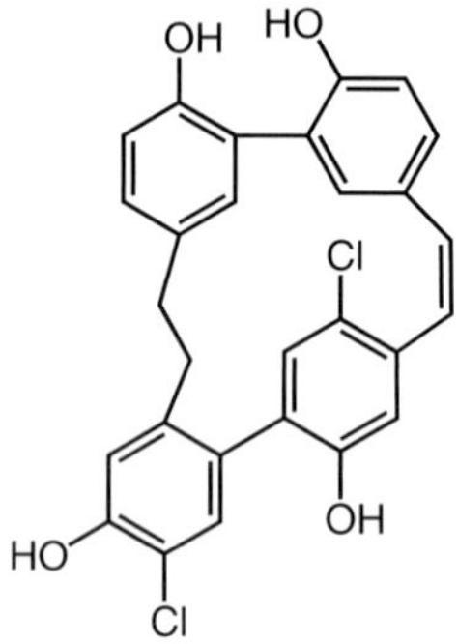

1816 (12,10'-dichloroisoplagiochin C)

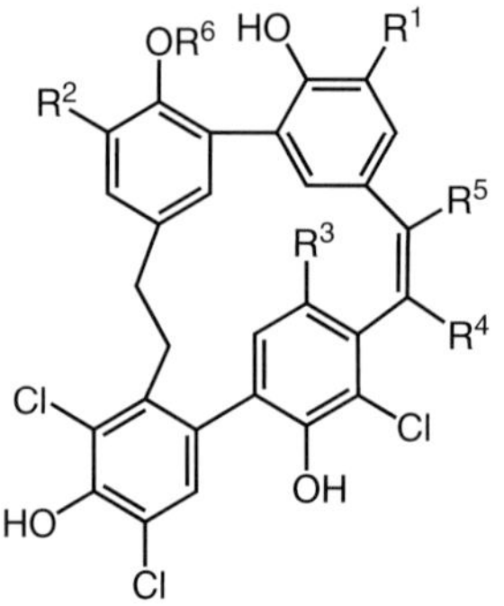

1817 $R^1 = H, R^2 = H, R^3 = H, R^4 = H, R^5 = H, R^6 = Me$ (bazzanin L)
1818 $R^1 = H, R^2 = H, R^3 = H, R^4 = H, R^5 = H, R^6 = H$ (bazzanin M)
1819 $R^1 = H, R^2 = Cl, R^3 = H, R^4 = H, R^5 = H, R^6 = H$ (bazzanin N)
1820 $R^1 = Cl, R^2 = H, R^3 = Cl, R^4 = H, R^5 = H, R^6 = Me$ (bazzanin O)
1821 $R^1 = Cl, R^2 = Cl, R^3 = H, R^4 = H, R^5 = H, R^6 = H$ (bazzanin P)
1822 $R^1 = Cl, R^2 = Cl, R^3 = Cl, R^4 = H, R^5 = H, R^6 = H$ (bazzanin Q)
1823 $R^1 = Cl, R^2 = Cl, R^3 = Cl, R^4 = Cl, R^5 = Cl, R^6 = H$ (bazzanin R)

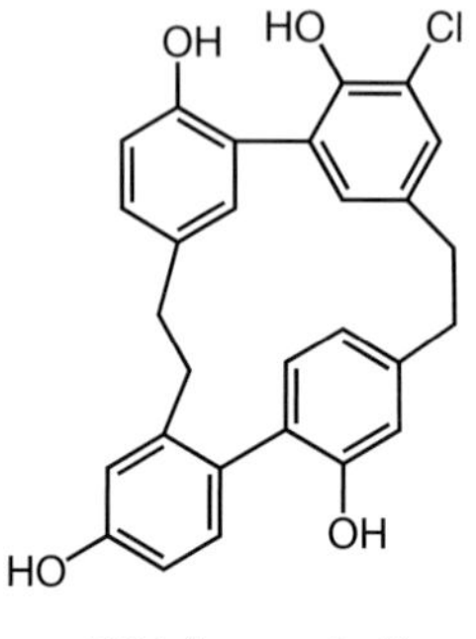

1824 (bazzanin S)

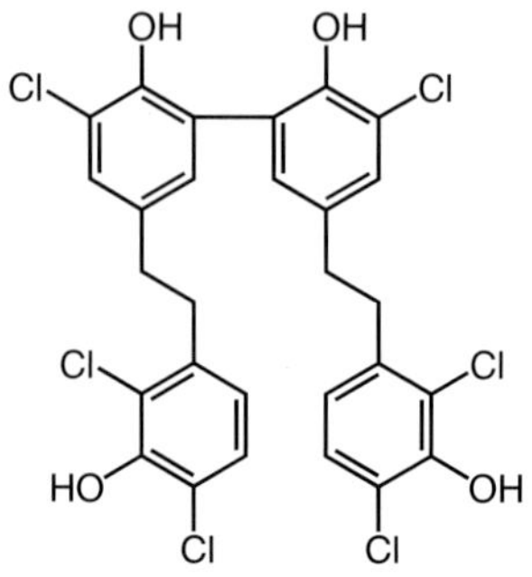

1825

3.21 Simple Phenols

Other sources of chlorophenols are *de novo* formation in a sewage treatment plant (*1656*), composting of organic household waste (*1711*), and production in remote forest soil (*1712*) and by the litter-degrading fungus *Lepista nuda* (*1713*). All composts studied produce a chloromethoxybenzaldehyde in amounts between 5.6 and 73.4 µg kg^{-1} dry matter (*1711*). The chlorophenols detected in rural Douglas fir forest soil are the known 4-chlorophenol, 2,4-, 2,5-, and 2,6-dichlorophenol, and 2,4,5-trichlorophenol, although anthropogenic contributions could not be eliminated (*1712*).

3.21.2 Marine

Most of the known marine-derived halogenated phenols are brominated, in accord with the widespread presence of bromoperoxidase in marine organisms, and 45 simple bromophenols were tabulated in the first survey (*1*). Most of these metabolites were found in red algae and, to a lesser extent, marine acorn worms. The red alga *Polysiphonia lanosa* from Brittany, which is the source of several simple bromophenols, also contains the new rhodomelol (**1826**) and methylrhodomelol (**1827**) (*1714*). A collection of the Senegalese red alga *Polysiphonia ferulacea* has yielded the optically active polysiphenol (**1828**), which is the first 9,10-dihydrophenanthrene found in a marine organism (*1715*). This hindered biphenyl analog is optically active and the absolute configuration was determined from its CD spectrum. Surprisingly, the simple 4-bromophenol (**1829**) was characterized for the first time in the acorn worms *Notomastus lobatus*, *Saccoglossus kowalevskii*, and *Arenicola cristata*, along with the previously known 2,6-dibromophenol and 2,4,6-tribromophenol (*1716, 1717*). The major organobromine metabolite in *Notomastus lobatus* is **1829**. Study of the Indo-Pacific *Dysidea* sp. sponge reveals the presence of a mixture of the new metabolites 2,3-dibromo-5-hydroxyphenol (**1830**) and 3,5-dibromo-2-hydroxyphenol (**1831**) (*1718*). A deep-water (113 m) Bahamian sponge *Aplysina fistularis fulva* has afforded the novel disulfate aplysillin A (**1832**) (*1719*). The sponge *Didiscus* sp. contains 3,5-dibromo-2-methoxybenzoic acid (**1833**) (*1720*). The marine ascidian *Aplidiopsis* sp. from Western Australia has yielded aplidiamine (**1834**), a unique zwitterionic adenine derivative (*1721*). The structure of **1834** was confirmed by synthesis and the original tautomeric structure was reassigned as shown (*1722, 1723*).

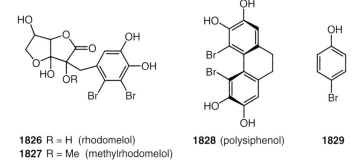

1826 R = H (rhodomelol)
1827 R = Me (methylrhodomelol)

1828 (polysiphenol)

1829

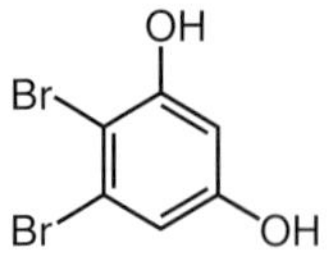

1830

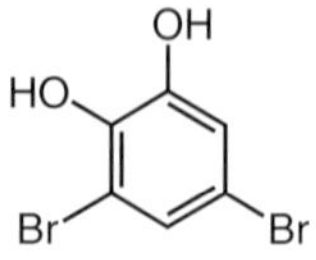

1831

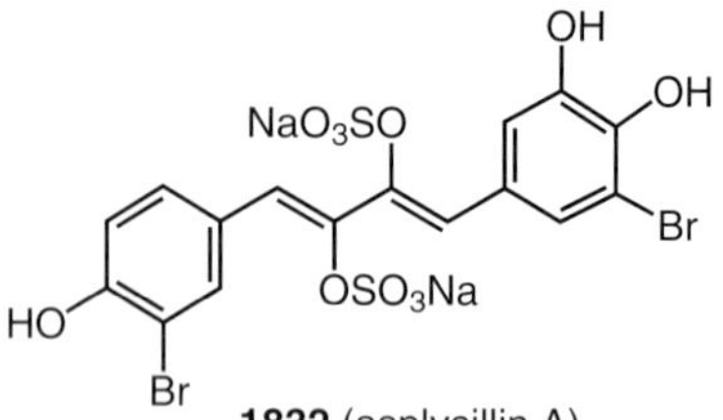

1832 (asplysillin A)

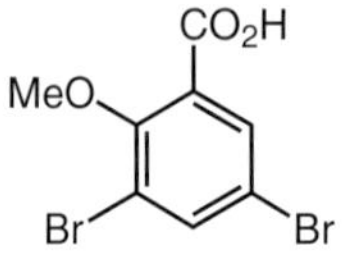

1833

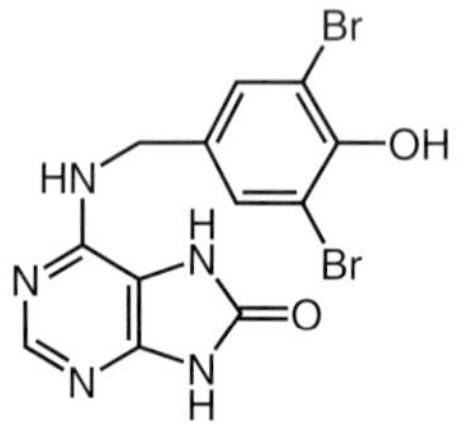

1834 (aplidiamine)

The simple 3,5-dibromo-4-hydroxybenzoic acid (**1835**) and 3,5-dibromo-4-methoxybenzoic acid (**1836**) occur in the green alga *Ulva lactuca* (*1724*) and the Indian Ocean sponge *Psammaplysilla purpurea* (*1725*), respectively. The novel 6-chloro-2,4-dibromophenol (**1837**) was characterized from cultures of the marine bacterium *Pseudoalteromonas luteoviolacea* and displays antibacterial activity against methicillin-resistant *Staphylococcus aureus* and the cystic fibrosis associated pathogen *Burkholderia cepacia* (*1726*). The red alga *Polysiphonia sphaerocarpa* has furnished several previously unreported simple bromophenols, including 2-bromophenol (**1838**), 2,4-dibromoanisole (**1839**), 2,4,6-tribromoanisole (**1840**), and 2-bromo-4-methylphenol (**1841**) (*1727*). Bromoanisole **1840** is ubiquitous in the marine environment (*1728*) and is a compound responsible for the musty aroma of "corked" wine (*1729*). Brominated phenols and anisoles are also found in marine mammals (e.g., Arctic hooded seal and Antarctic Weddell seal) and Antarctic sponges (e.g., *Phorbas glaberrima*) (*394*), and in Norwegian predatory bird eggs (*479*). These compounds are mainly 2,4,6-tribromophenol and 2,4,6-tribromoanisole (**1840**). The flavor and aroma properties of marine bromophenols have been reviewed (*1730*). The Korean red alga *Symphyocladia latiuscula* has provided the new symphyoketone (**1842**), which has radical-scavenging activity (*1731*). Parasitenone (**1843**) was characterized from the Korean marine-derived fungus *Aspergillus parasiticus* and also exhibits radical-scavenging activity (*1732*). A Chinese specimen of the red alga *Rhodomela confervoides* has afforded the new **1844** and **1845** (*1733, 1734*). The isolated dimethyl acetal of **1844** may be an isolation artifact. Further study of this seaweed has uncovered nine new bromophenols, **1846–1854** (*1735*).

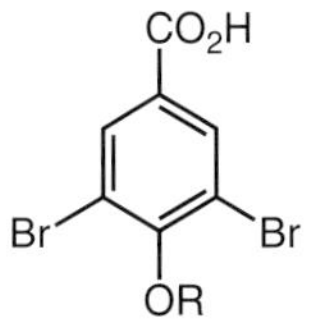

3.21 Simple Phenols

1835 R = H
1836 R = Me

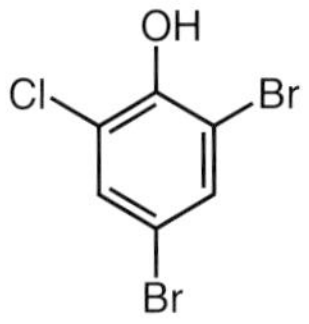

1837

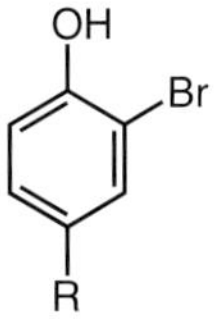

1838 R = H
1841 R = Me

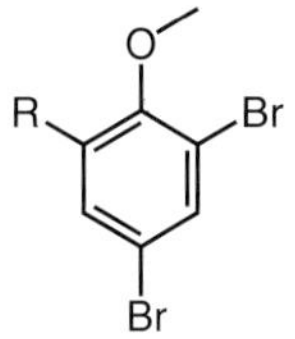

1839 R = H
1840 R = Br

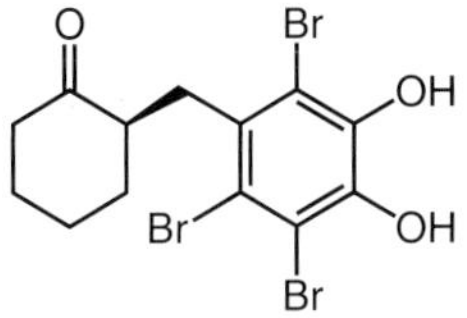

1842 (symphyoketone)

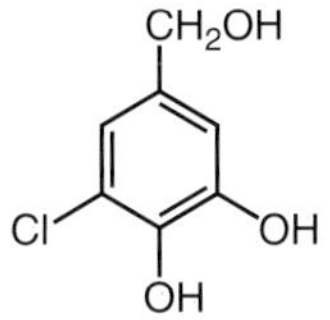

1843 (parasitenone)

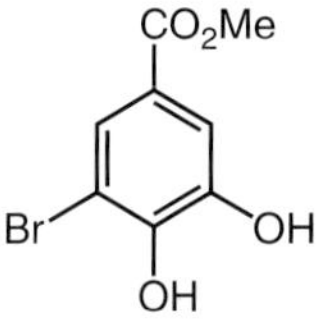

1845

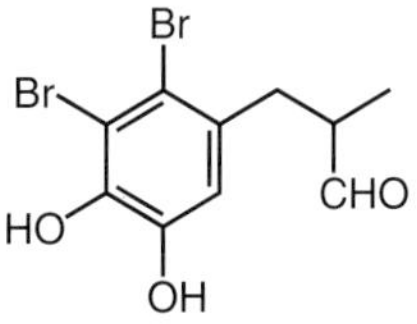

1844

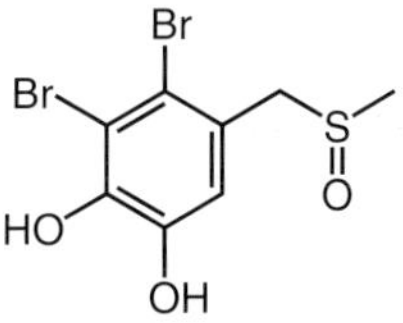

1846

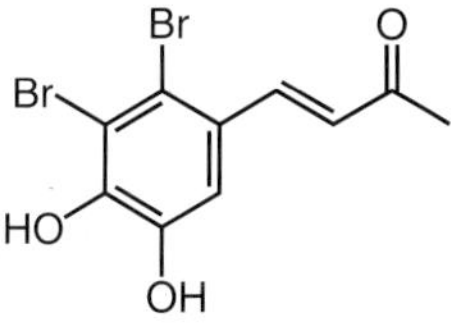

1847

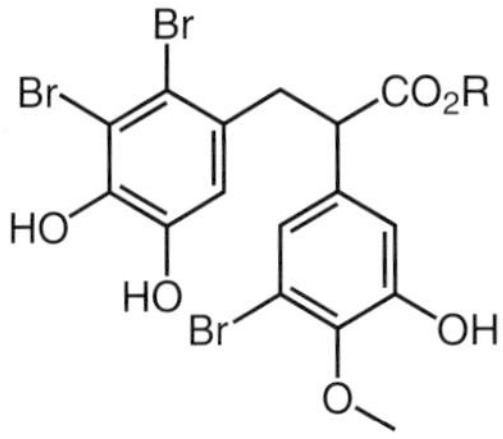

1848 R = H
1849 R = Me

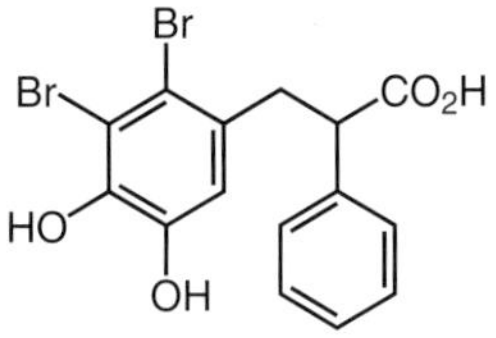

1850

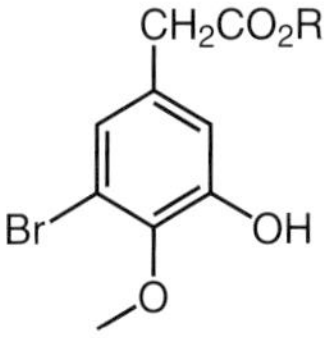

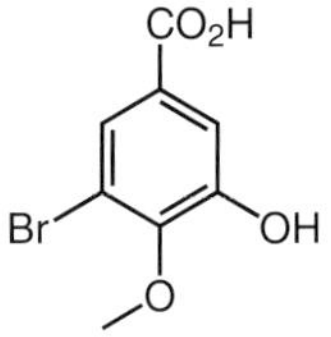

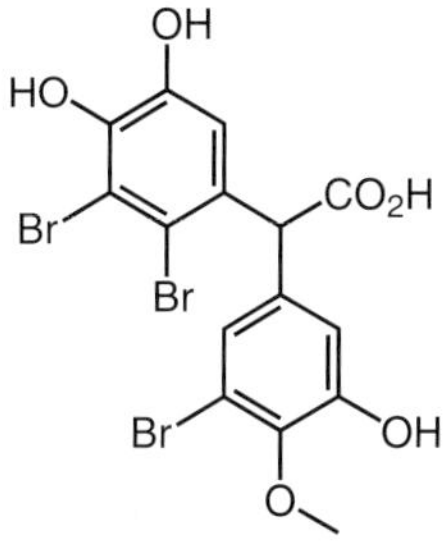

1851 R = H
1852 R = Me

1853

1854

The brown alga *Leathesia nana* from the gulf of the Yellow Sea in China has yielded the new bromophenols **1855** and **1856** (*1736–1738*), and the Chinese red alga *Rhodomela confervoides* contains the five novel brominated catechols **1857–1861**, in addition to several brominated diphenylmethanes discussed in Sect. 3.22.1 (Diphenylmethanes and Related Compounds) (*1739*). The Brazilian red alga *Osmundaria obtusilobu* has yielded the two novel sulfated oligobromophenols **1862** and **1863** (*1740*). The luminous acorn worm *Ptychodera flava* produces 2,3,5,6-tetrabromohydroquinone (**1864**), 2,3,5-tribromohydroquinone (**1865**), and tetrabromo-1,4-benzoquinone (**1866**) (*1741*).

1855

1856

1857

1859 R^1 = Br, R^2 = H
1860 R^1 = Br, R^2 = SO_3H
1861 R^1 = H, R^2 = SO_3H

1858

1864 R = Br
1865 R = H

1866

1862 R^1 = OH, R^2 = H
1863 R^1 = H, R^2 = OH

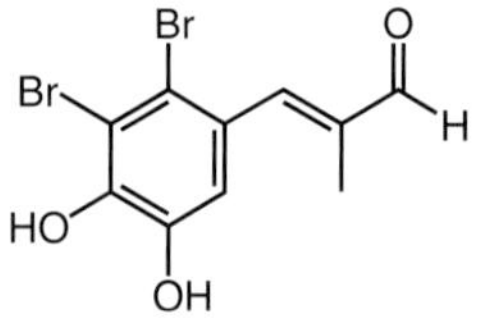

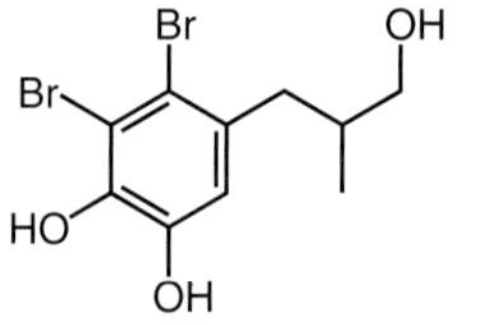

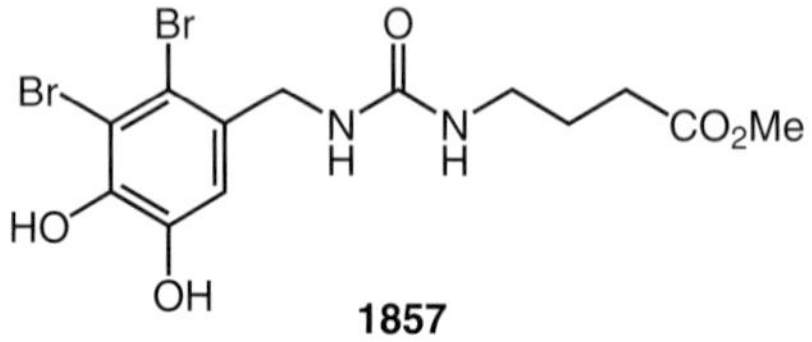

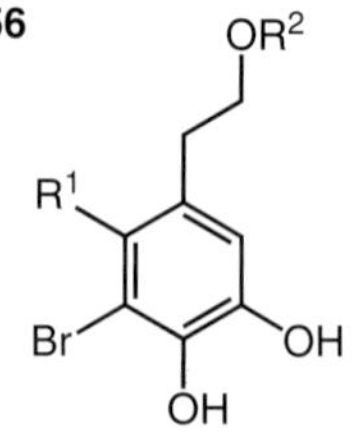

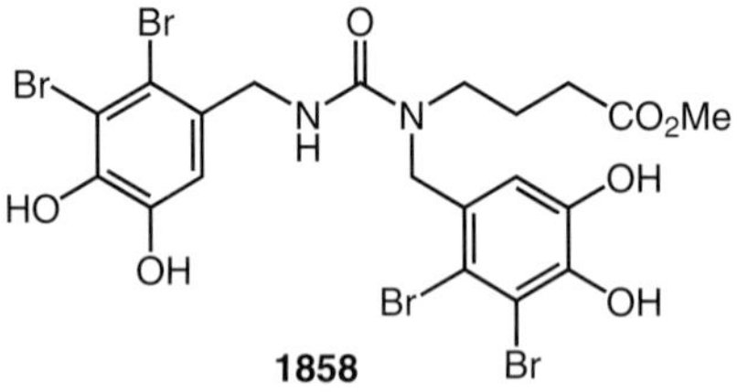

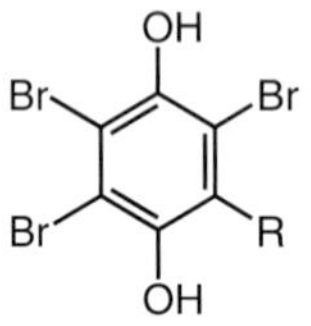

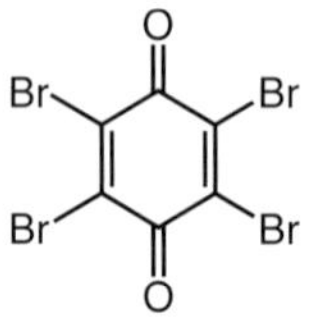

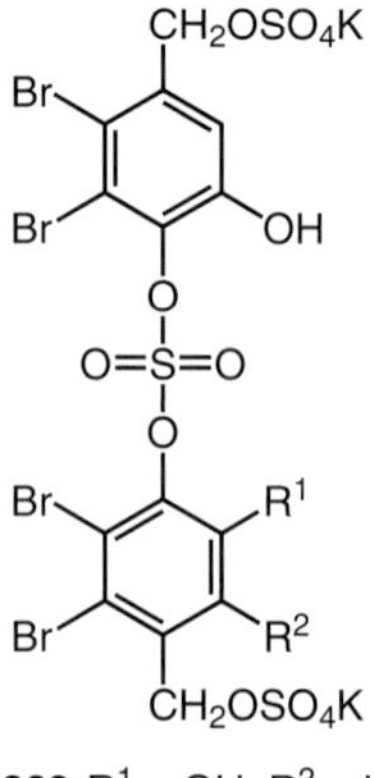

3.21 Simple Phenols

The red alga *Symphyocladia latiuscula* has afforded the new tribromophenols **1867** and **1868** (*1742*), and the red alga *Polysiphonia urceolata* contains urceolatol (**1869**), a novel bromobenzaldehyde dimer having C_2 symmetry (*1743*). The red alga *Rhodomela confervoides* has yielded lanosol-purine metabolite **1870** (*1455*) and lanosol-deoxyguanosine **1871**, along with the new simpler metabolites **1872** and **1873** (*1744*). Poipuol (**1874**) occurs in a Hawaiian *Hyrtios* sp. sponge (*1745*). It is interesting to note that poipuol is a rare halogenated phenol having halogen (chlorine) *meta* to the *ortho*, *para*-directing hydroxy groups.

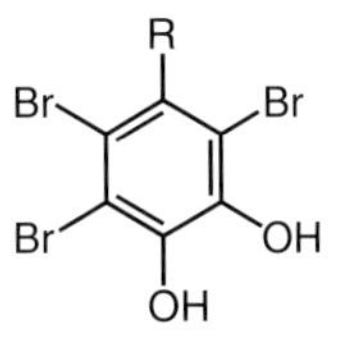

1867 R = Me
1868 R = CHO

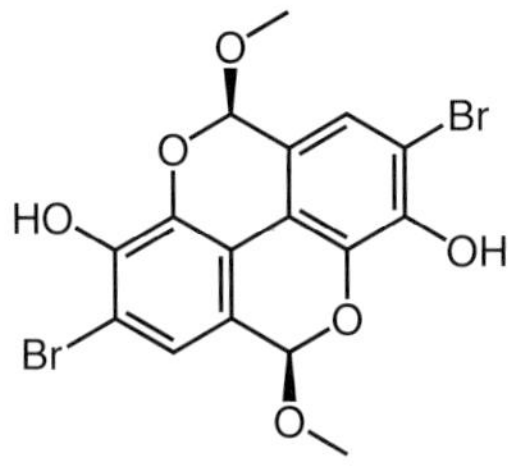

1869 (urceolatol)

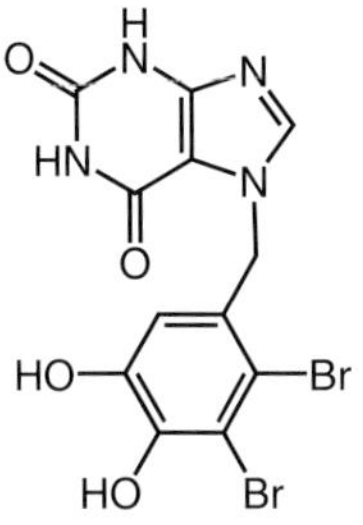

1870

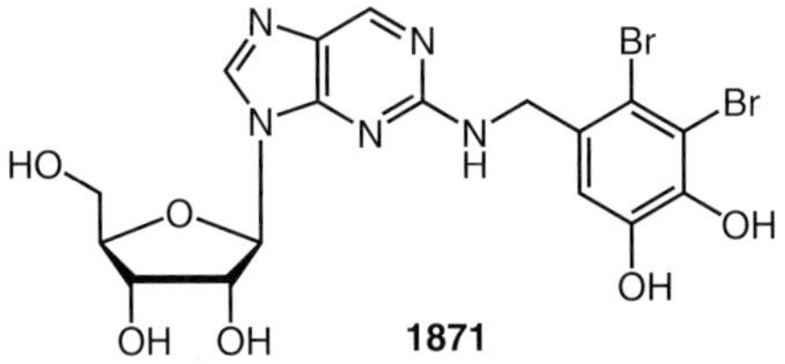

1871

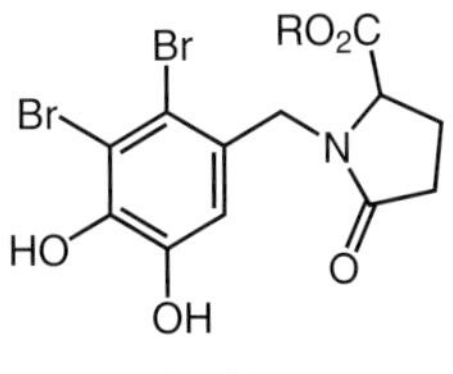

1872 R = Me
1873 R = H

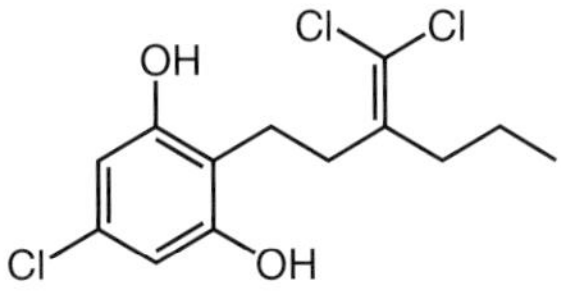

1874 (poipuol)

A study of the natural radiocarbon (^{14}C) in the acorn worm *Saccoglossus bromophenolosus*, which was collected off the Maine coast, revealed that the 2,4-dibromophenol produced by these worms is of recent origin, in contrast to that from petroleum-derived anthropogenic 2,4-dibromophenol (*1746*). Thus, this result combined with the earlier study (*1223*) supports a natural source of 2,4-dibromophenol in these animals. It should be noted that the more recent radiocarbon work utilizes improved methodology (*1746*).

In closing this section, it should be mentioned that simple bromophenols (2-bromophenol, 4-bromophenol, 2,4-dibromophenol, 2,6-dibromophenol, and 2,4,6-

tribromophenol) are ubiquitous in the marine environment, particularly in fresh and saltwater seafood (*1747–1751*). There is compelling evidence for a dietary origin of these compounds, from both marine algae (*1748, 1749, 1751–1753*) and marine polychaetes and bryozoans (*1748–1750*).

3.22 Complex Phenols

3.22.1 *Diphenylmethanes and Related Compounds*

The several known polybrominated diphenylmethanes (*1*) may arise via a pathway analogous to the known dimerization of benzyl alcohols (*1754*). For example, the red algal known metabolite 3,5-dibromo-4-hydroxybenzyl alcohol may condense to give the known thelephenol (**EEE**) via (non-enzymatic) *ipso* electrophilic substitution (Scheme 3.4). These condensations can occur under mild acidic conditions (*1754*).

Several new polybrominated diphenylmethanes have been discovered since the first survey. The Japanese red alga *Odonthalia corymbifera* contains several known bromophenols and the novel diphenylmethane **1875**, which is a potent feeding deterrent towards abalone (*Haliotis discus hannai*) and sea urchin (*Strongylocentrotus intermedius*) (*1755*). Since methanol was used in the isolation process, the actual metabolite may be the corresponding benzylic alcohol. The red alga *Rhodomela confervoides* is a rich source of bromophenols including new brominated diphenylmethanes, such as **1876** (*1733*), **1877** (*1744*), **1878–1880** (*1739*), and **1881** and **1882** (*1455*). The brown alga *Leathesia nana* has yielded **1883–1886** (*1738*), the latter two of which represent very interesting structures.

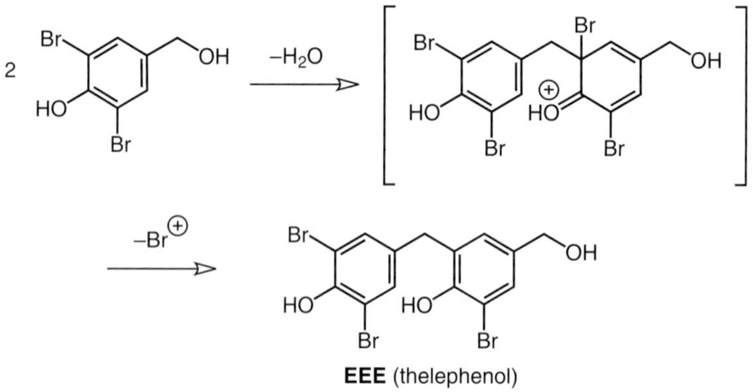

Proposed formation of thelephenol (**EEE**)(*1754*)

Scheme 3.4

3.22 Complex Phenols

The red alga *Symphyocladia latiuscula* has afforded the two heavily brominated **1887** and **1888** (*1742*).

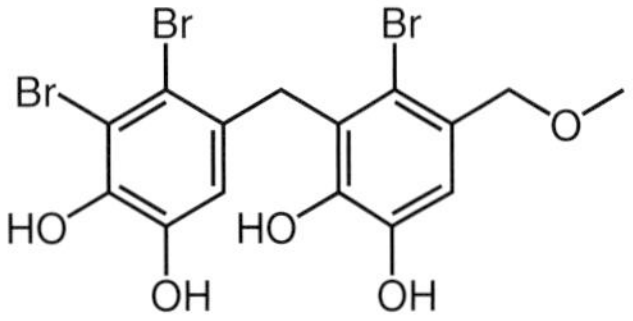

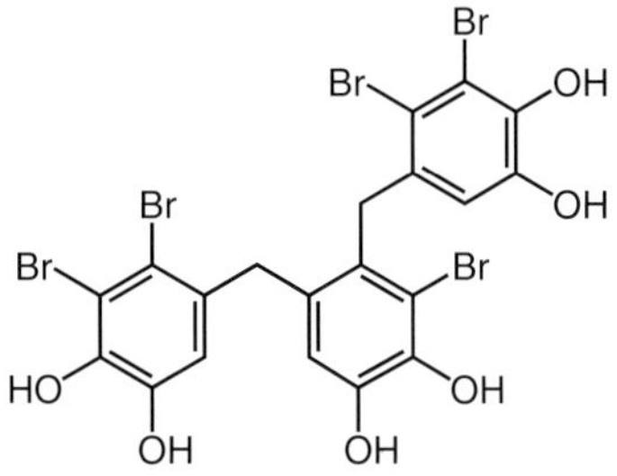

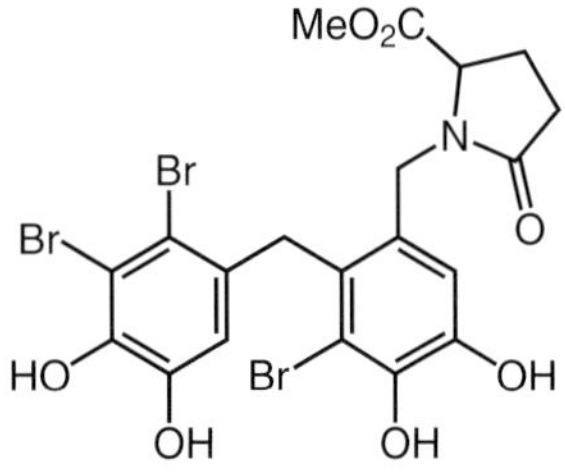

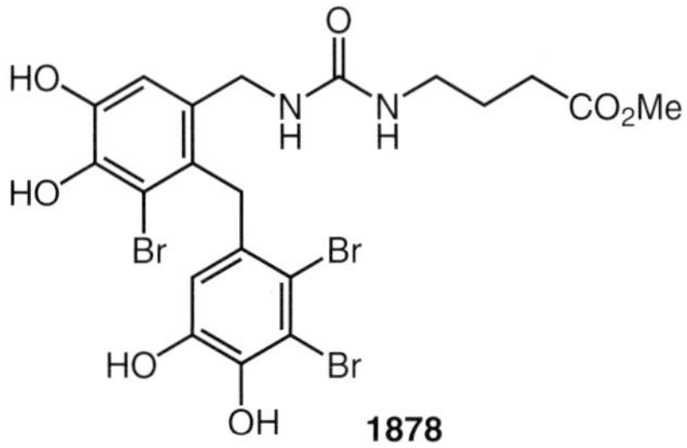

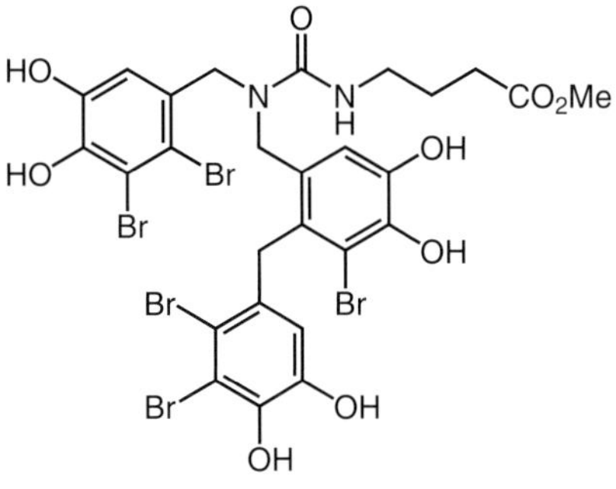

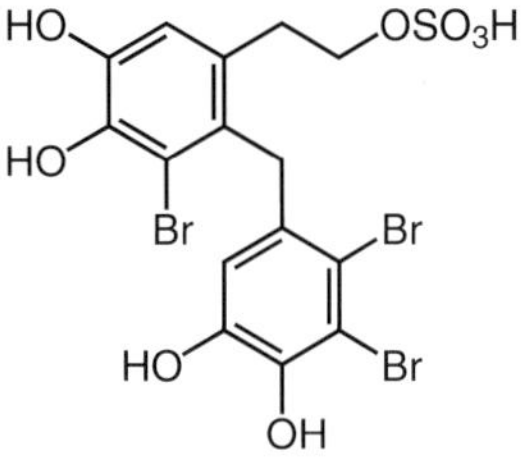

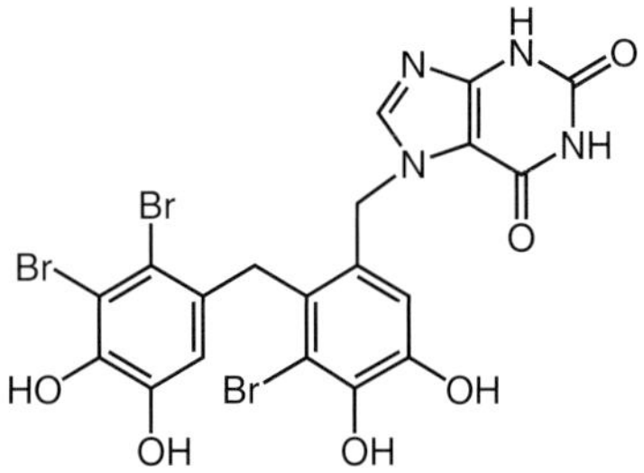

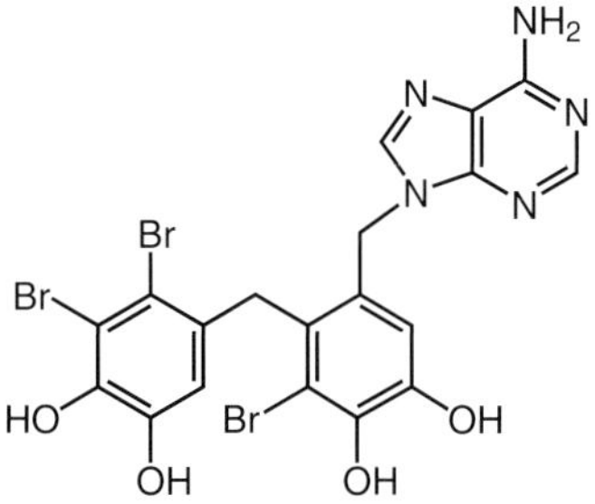

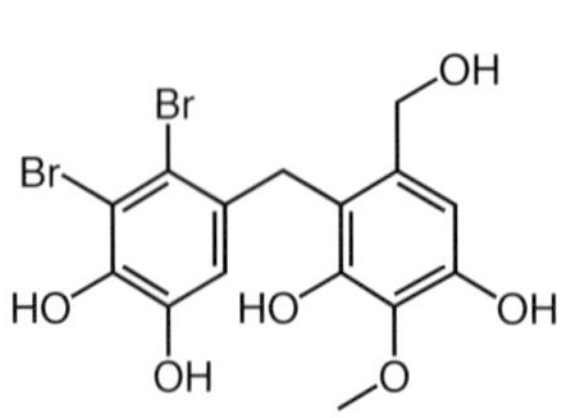

1883

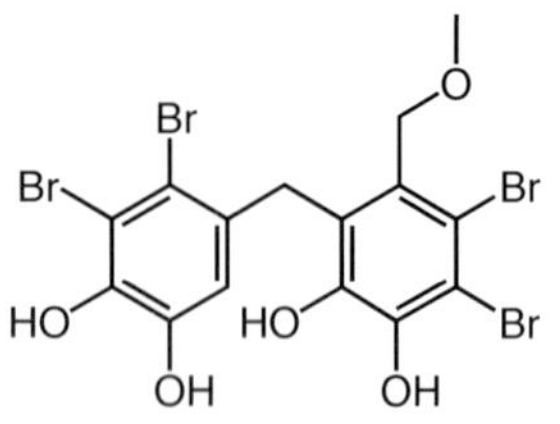

1884

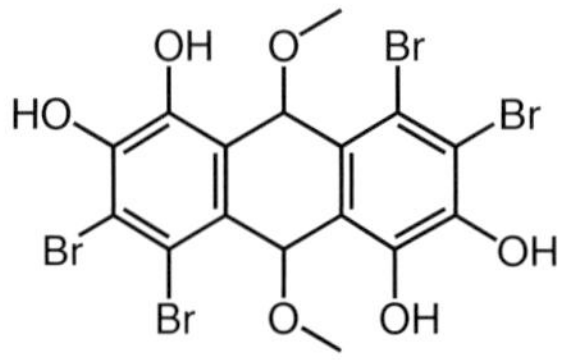

1885

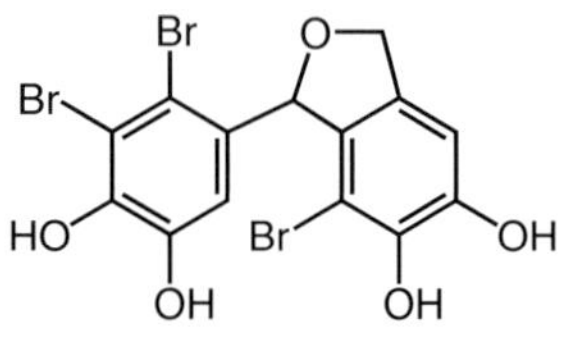

1886

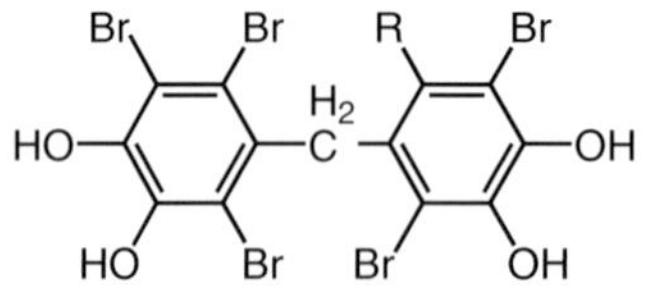

1887 R = Br
1888 R = H

Although not diphenylmethanes, two unique brominated biphenyls were isolated from a marine bacterium and marine mammals. Thus, *Pseudoalteromonas phenolica*, a new marine bacterium, produces MC21-A (**1889**), which appears to be the first naturally occurring polybrominated biphenyl (*1756*). This dimer of 2,4-dibromophenol, 2,2′-dihydroxy-3,3′,5,5′-tetrabromobiphenyl (2,2′-diOH-BB80), has comparable antibacterial activity to vancomycin against methicillin-resistant *Staphylococcus aureus*, and has a higher killing rate than vancomycin. The dimethoxylated version of MC21-A, 2,2′-dimethoxy-3,3′,5,5′-tetrabromobiphenyl (2,2′-diMeO-BB80) (**1890**), is found in several marine mammals, Striped dolphin (*Stenella coeruleoalba*), Bottlenose dolphin (*Tursiops truncatus*), Minke whale (*Balaenoptera acutorostrata*), and Baird's beaked whale (*Berardius bairdii*) (*1757, 1758*). Both of these polybrominated biphenyls (PBBs) are considered to be natural products, as no relevant PBB congener precursor is present in industrial products, no other PBBs are present in the mammalian samples, and the high concentration of **1890** (12–800 ng g^{-1} lipid) represents one of the most abundant compounds analyzed in these samples, which included polybrominated diphenyl ethers (PBDEs), hexabromocyclododecane (HBCDD), and methoxylated PBDEs (*1757, 1758*).

3.22 Complex Phenols

1889 (MC21-A)   **1890**

3.22.2 Diphenyl Ethers

Given the abundance of phenols in the oceans – halogenated or not – and the ease with which they undergo oxidative dimerization (*1759*), it is hardly surprising that halogenated diphenyl ethers are abundant in the marine environment (Fig. 3.27). More than 30 such natural brominated diphenyl ethers were documented in the first survey (*1*), and several new examples have been subsequently identified. It is worth noting that all of the previously identified natural (sponge-derived) brominated diphenyl ethers have at least one additional oxygen atom (hydroxy or methoxy), whereas the industrial fire retardant polybrominated diphenyl ethers do not.

A study of South Pacific marine invertebrates has revealed the new **1891** in the sponge *Dysidea herbacea* (*1760*). The new **1892** was isolated from *Sagaminopteron bilealbum* molluscs feeding on the sponge *Dysidea herbacea* from Guam waters (*1761*). Four samples of *Dysidea* sponges from the Indo-Pacific yielded the

Fig. 3.27 *Aplidium longithorax*, a tunicate that produces polybrominated diphenyl ethers (Photo: F. J. Schmitz)

new polybrominated diphenyl ethers **1893–1896**, in addition to several known analogs (*1718, 1762*). These metabolites inhibit inosine monophosphate dehydrogenase, guanosine monophosphate synthetase, and 15-lipoxygenase. An Indian Ocean collection of *Dysidea herbacea* has afforded the new **1897** (*1763*), and this sponge from West Sumatra, Indonesia, contains **1898–1901**, which show activity against *Bacillus subtilis* and the phytopathogenic fungus *Cladosporium cucumerinum* (*1764*). The novel lanosol-type dimers **1902** and **1903** were isolated along with the known lanosol from the red alga *Odonthalia corymbifera*, and all three bromophenols inactivate α-glucosidase (*1765*). Examination of *Dysidea herbacea* from the Great Barrier Reef reveals the presence of the new polybrominated diphenyl ether **1904** (*1766*). The Palauan sponge *Phyllospongia dendyi* has yielded the new **1905–1907** and the known **1892** (*1767*). Another study of this sponge has uncovered the new **1908** and **1909**, in addition to nine previously identified polybrominated diphenyl ethers (*1768*). The brown alga *Leathesia nana* contains **1910**, which was isolated as the bis-ethoxy ether since ethanol was used in the isolation process (*1738*). The red alga *Symphyocladia latiuscula* contains bis-benzyl ether **1911** (*1742*), and acorn worms of genus *Thelepus* produce the novel bis-benzyl ether **1912** (*1793*). Okinawan crustose coralline red algae have yielded corallinaether (**1913**), along with a novel brominated dibenzofuran described later (*1769*). Further examination of a Great Barrier Reef *Dysidea herbacea* has uncovered the new **1914** (*1770*). The Indonesian sponge *Lamellodysidea herbacea* (Fig. 3.28) has afforded the new **1915–1918** along with ten

Fig. 3.28 *Lamellodysidea herbacea*, a sponge from Sunda Strait, Indonesia, that contains the new diphenyl ethers **1915–1918** and several previously known analogues (Photo: J. Tanaka)

previously known analogs (*1771*). A Solomon Islands sponge *Phyllospongia* sp. has yielded the tribrominated diphenyl ether **1919** (*1772*). This compound, other brominated diphenyl ethers, and related brominated phenolics inhibit various lipoxygenases (*1772*). The methoxylated **1920** occurs both in the red alga *Ceramium tenuicorne* and blue mussels (*Mytilus edulis*) in the Baltic Sea (*1773*). Several other previously known polyhalogenated diphenyl ethers were found in these organisms.

1891

1892

1893 R = H
1894 R = Me

1895

1896

1897

1898

1899

1900 R = H
1901 R = Me

1902

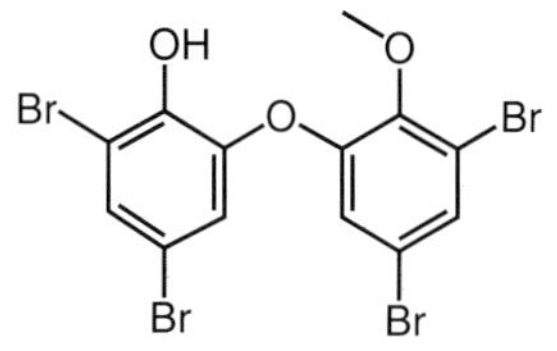

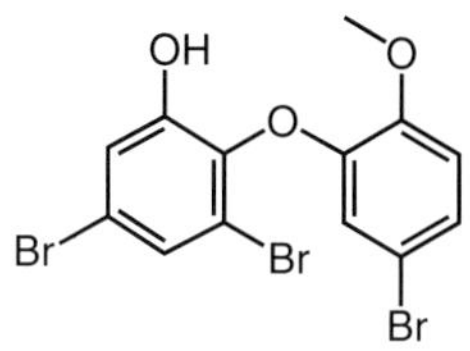

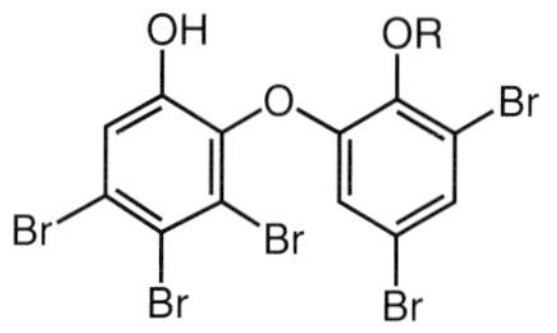

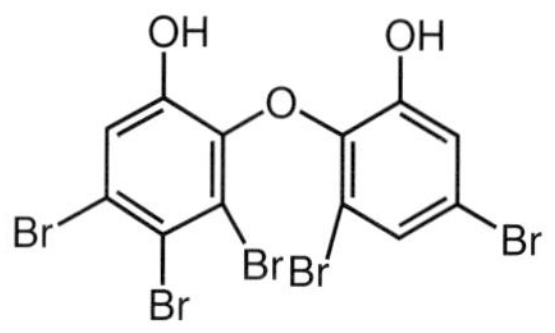

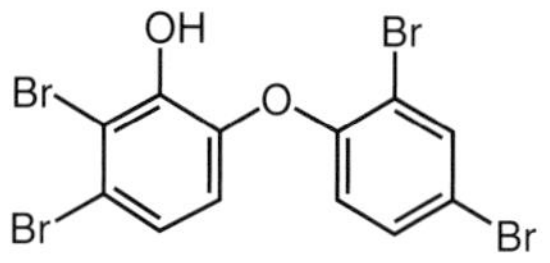

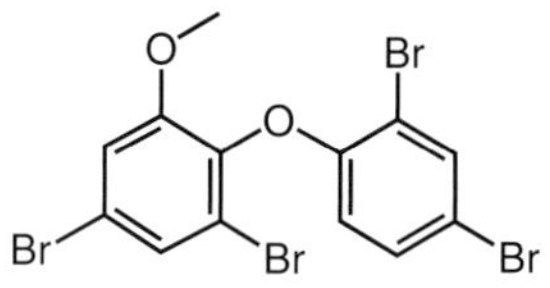

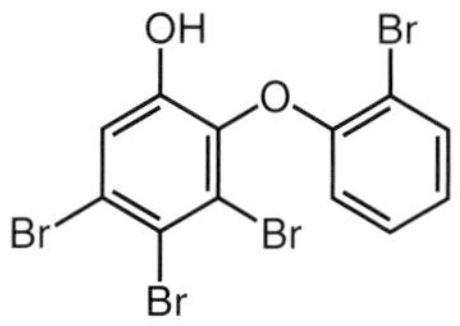

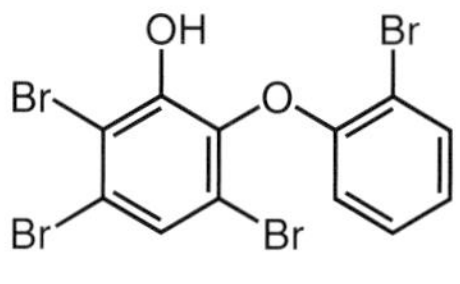

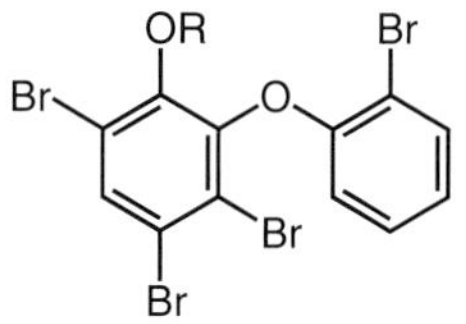

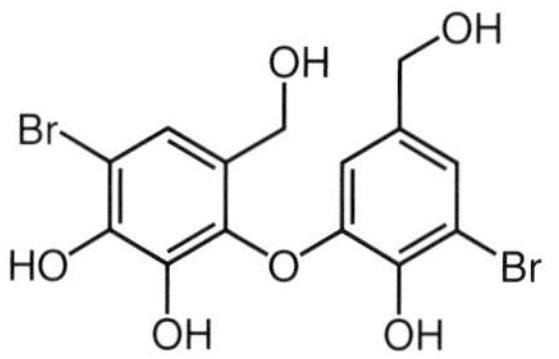

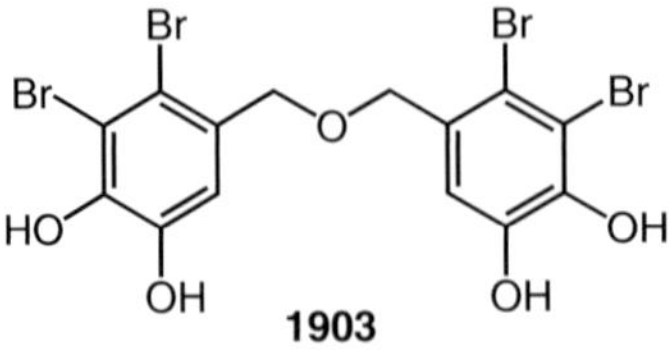

1903
1904 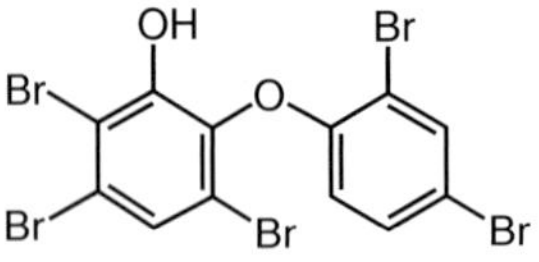
1905 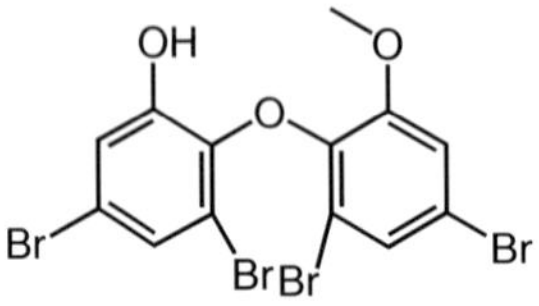
1906 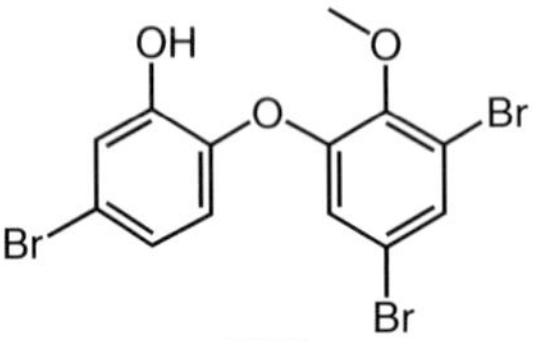
1907 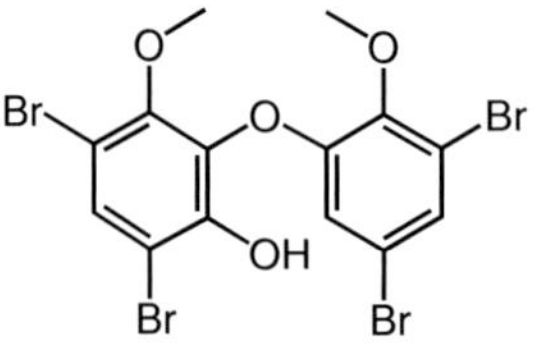
1908 R = Br 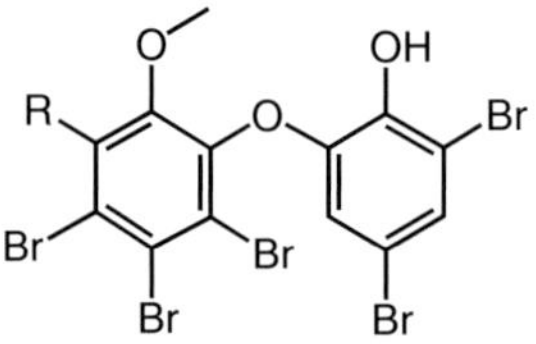
1909 R = H
1910 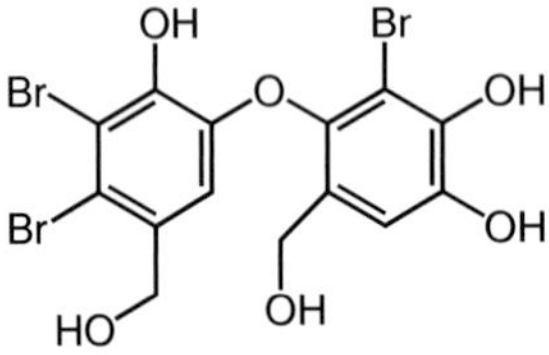
1911 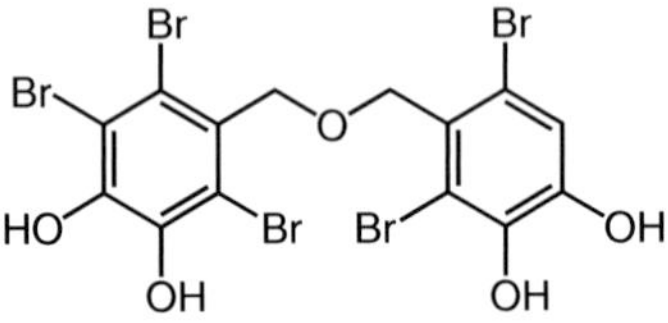
1912 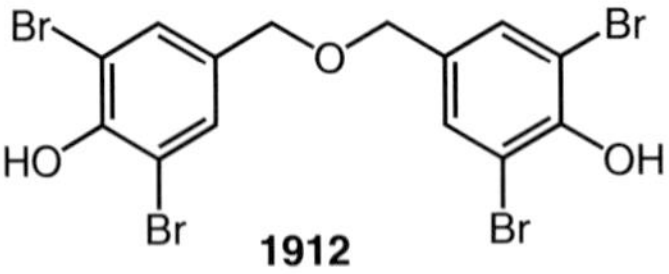
1913 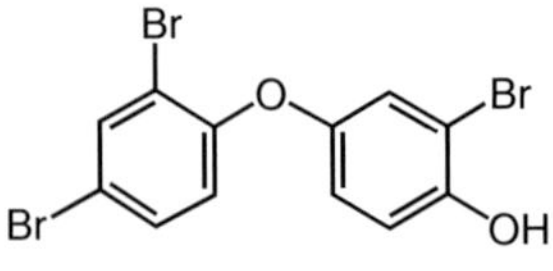
1914 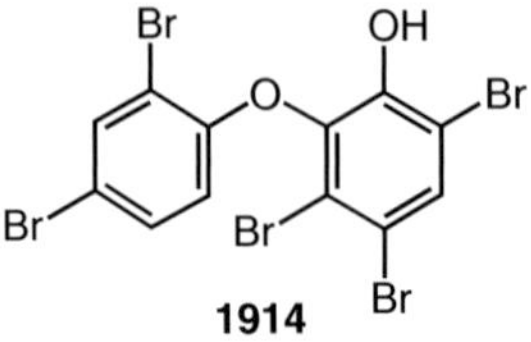
1915

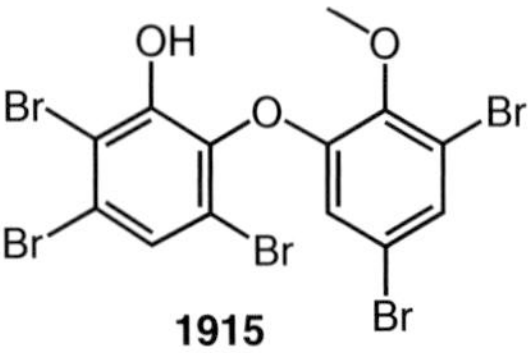

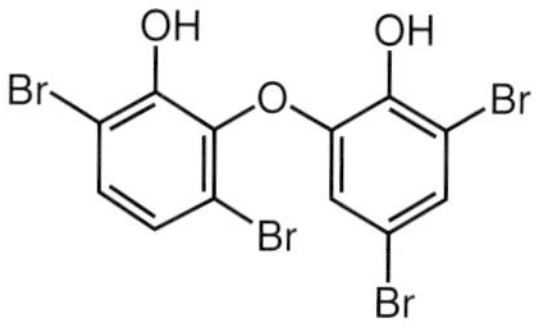

3.22 Complex Phenols

1916

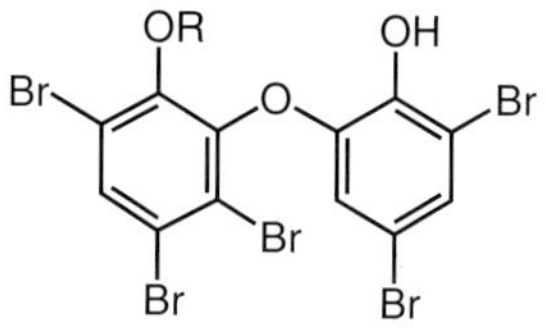

1917 R = H
1918 R = Me

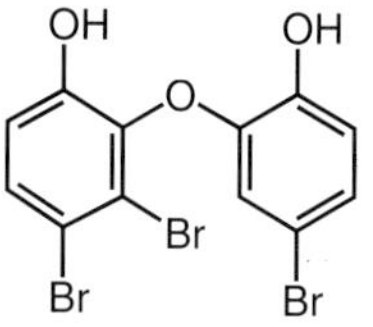

1919

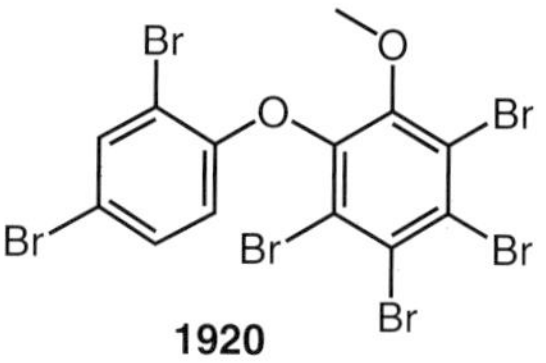

1920

These and other polybrominated diphenyl ethers, along with other halogenated compounds, whether natural marine metabolites or anthropogenic fire retardants (*1774–1776*) have been identified in tunicates (*1777*), nudibranchs (*996, 1778*), cyanobacteria (*1003*) (which may be the actual source of these polybrominated diphenyl ethers), and other marine life such as salmon (*1779, 1780*), other fish (*1781–1783*), several marine mammals (*1215, 1784–1787*), crocodile eggs (*1786*), and human milk from women who consume whale blubber (*1788*). Evidence as to the origin of these polybrominated diphenyl ethers is provided by the observation that some nine halogenated compounds, including polybrominated diphenyl ethers, were discovered in archived whale oil collected in 1921 from the final voyage of the whaling ship *Charles W. Morgan*, obviously predating the large-scale industrial synthesis of brominated fire retardants (*1222*). Noteworthy is that DDT, its metabolites (e.g., DDE), and polychlorinated biphenyls (PCBs) were not detected in this whale oil. Moreover, an analysis of the ^{14}C content of **1897** and two previously described polybrominated diphenyl ethers (*1*) isolated from marine mammals confirms their natural origin (*1223, 1789*). The synthesis of polybrominated diphenyl ethers has been of great interest in view of their biological activity and the need for pure analytical standards (*1790–1792*).

Several other non-marine halogenated diphenyl ethers are newly described. For example, the fungus *Pestalotiopsis* sp. has yielded RES-1214-2 (**1921**) (*1794*), while a fungus of genus *Xylaria* also produces **1921** ("dihydromaldoxin") along with the new isodihydromaldoxin (**1922**) (*1795*). The new methyl dichloroasterrate (**1923**) and methyl chloroasterrate (**1924**) were independently isolated from an *Aspergillus* sp. culture broth (*1796*) and from an unidentified fungal strain B 90911 (*1797*). The corresponding acids **1925** and **1926** were earlier characterized from *Penicillium citrinum* (*1798*). The Brazilian tree *Byrsonima microphylla* contains the novel chlorinated diphenyl ether **1927**, the presence of which in the heartwood was confirmed by HPLC and TLC (*1799*).

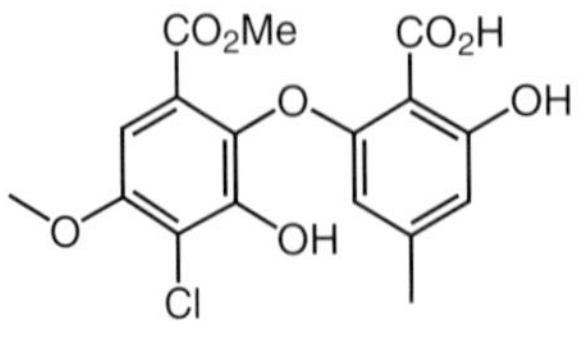

1921 (RES-1214-2)

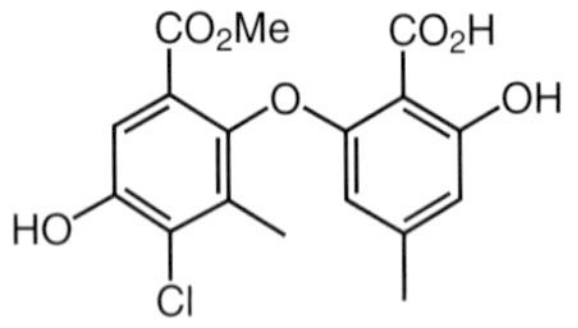

1922 (isodihydromaldoxin)

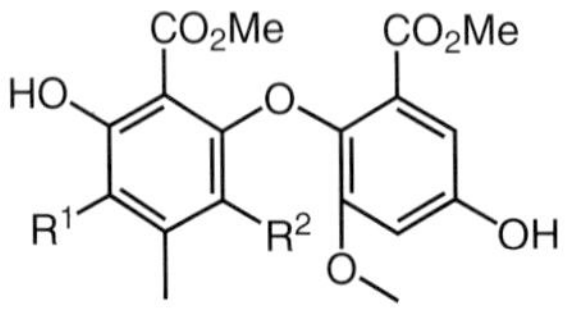

1923 R¹ = R² = Cl (methyl dichloroasterrate)
1924 R¹ = Cl, R² = H (methyl chloroasterrate)

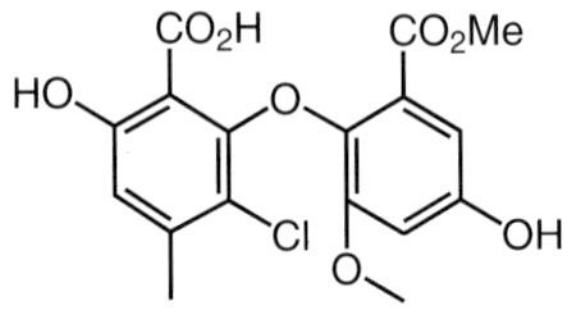

1925

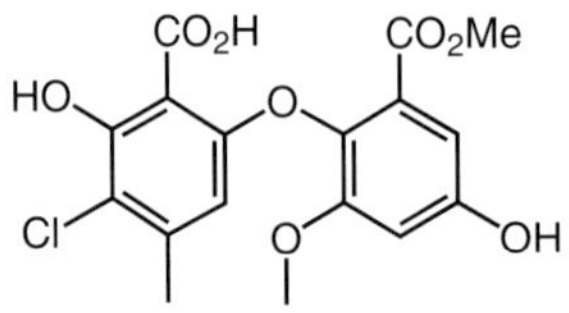

1926

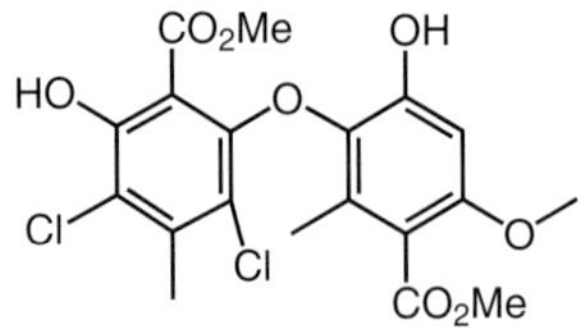

1927

The toxic mushroom *Russula subnigricans* produces the novel polychlorinated phenyl ethers, russuphelins A–F (**1928–1933**), some of which (B, C, D) exhibit cytotoxicity against P388 leukemia cells (*1672, 1800*). Further study of this mushroom uncovered the presence of the optically active russuphelol (**1934**) (*1801*). The terrestrial cyanobacterium *Fischerella ambigua* has afforded ambigol C (**1935**) (*1654*), which is an isomer of the previously described ambigols A and B (*1*).

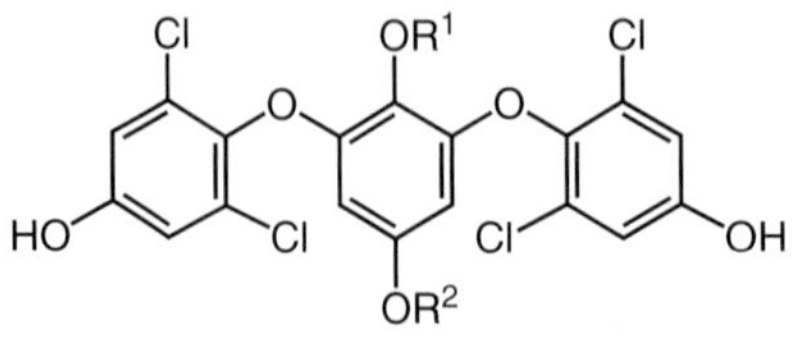

1928 R¹ = R² = Me (russuphelin A)
1929 R¹ = Me, R² = H (russuphelin B)
1930 R¹ = R² = H (russuphelin C)

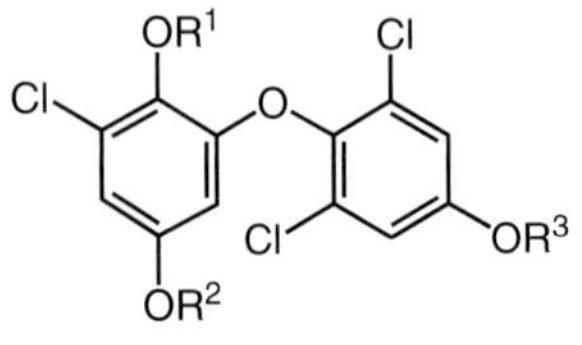

1931 R¹ = R² = Me, R³ = H (russuphelin D)
1932 R¹ = R³ = Me, R² = H (russuphelin E)
1933 R¹ = H, R² = R³ = Me (russuphelin F)

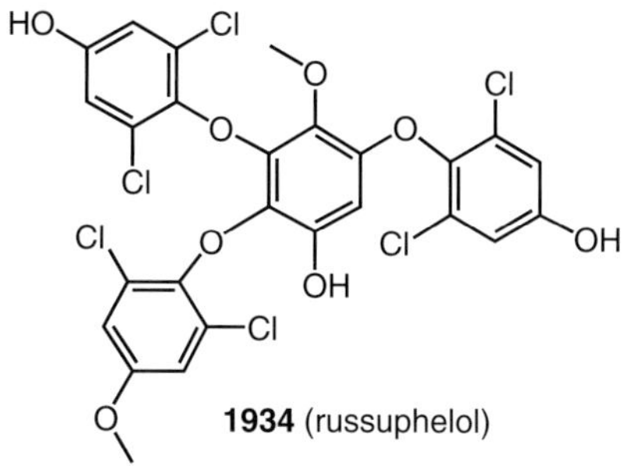

3.22 Complex Phenols

1934 (russuphelol)

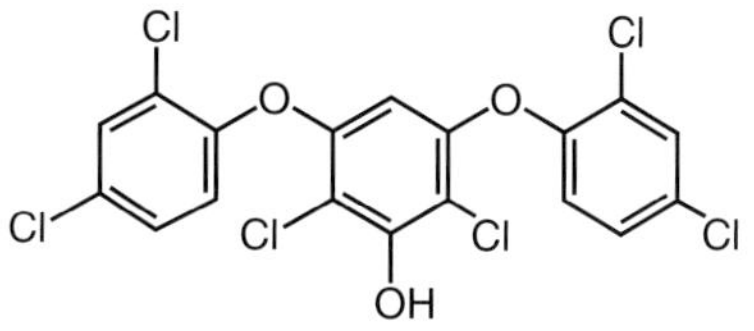

1935 (ambigol C)

The New Zealand brown algae *Sargassum spinuligerum* and *Cystophora torulosa* produce several fucophlorethols including chlorobisfucopentaphlorethol-A (**1936**) (isolated as the peracetate) (*1802*). The brown alga *Carpophyllum angustifolium*, collected in New Zealand, has furnished 45 phloroglucinols including halogenated 2[D']iododiphlorethol (**1937**), 3[A]chlorobifuhalol (**1938**), and 3[A$_4$]chlorodifucol (**1939**) (isolated as peracetates) (*1818*). An examination of the New Zealand brown alga *Cystophora retroflexa* has identified 17 halogenated phlorethols and fucophlorethols, including 12 new compounds, all isolated as peracetates: $2_{[B]}$-bromotriphlorethol-A (**1940**), $2_{[D]}$-bromotriphlorethol-A (**1941**), $2_{[B]},2_{[D]}$-dibromotriphlorethol-A (**1942**), $2_{[D]}$-bromotetraphlorethol-C (**1943**), $3_{[A1]},5_{[A1]}$-dichlorotriphlorethol-A (**1944**), $3_{[A1]},4_{[D]}$-dichlorotriphlorethol-A (**1945**), $3_{[A1]}$-chloro,$4_{[D]}$-bromotriphlorethol-A (**1946**), $2_{[B]},4_{[D]}$-dichlorotriphlorethol-A (**1947**), $2_{[D]},3_{[A1]}$-dibromotriphlorethol-A (**1948**), $3_{[A1]}$-bromo,$2_{[D]}$-chlorotriphlorethol-A (**1949**), $4_{[D]}$-chlorofucotriphlorethol-B (**1950**), and $4_{[D]}$-chlorobisfucotetraphlorethol-A (**1951**) (*1803*). These fascinating polyphenolic phloroglucinols have been reviewed (*1804*).

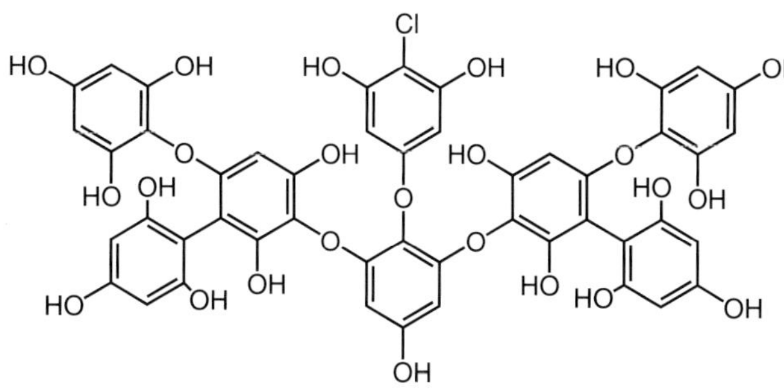

1936

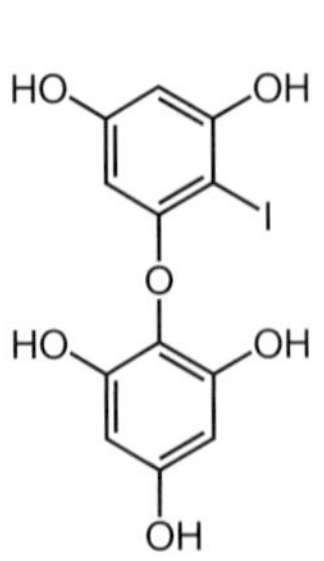

1937

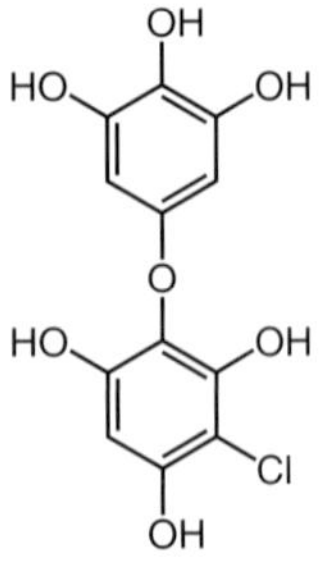

1938

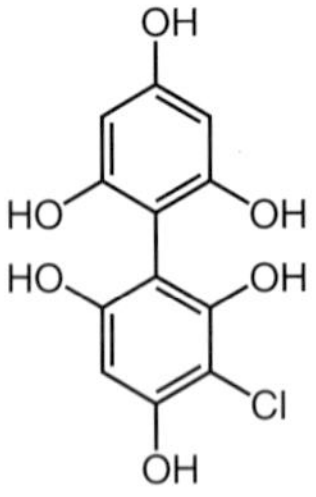

1939

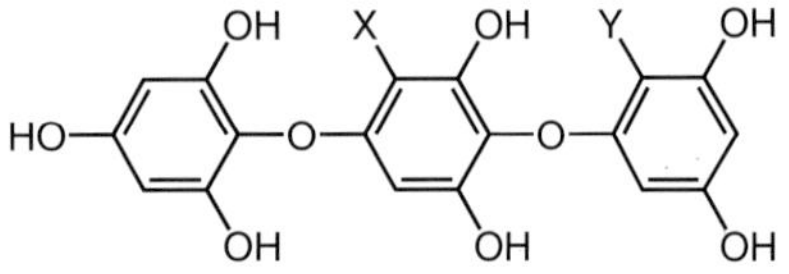

1940 X = Br, Y = H
1941 X = H, Y = Br
1942 X = Y = Br

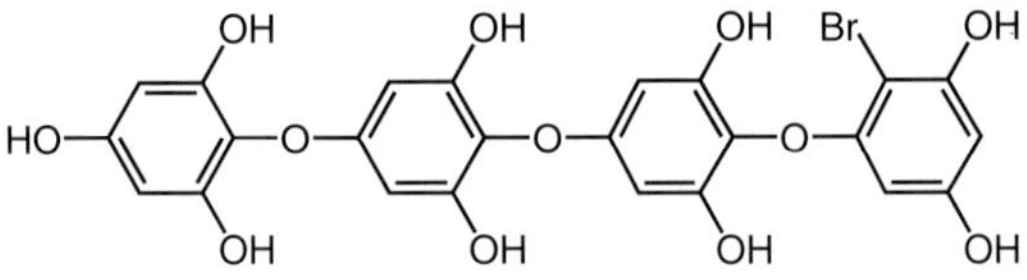

1943

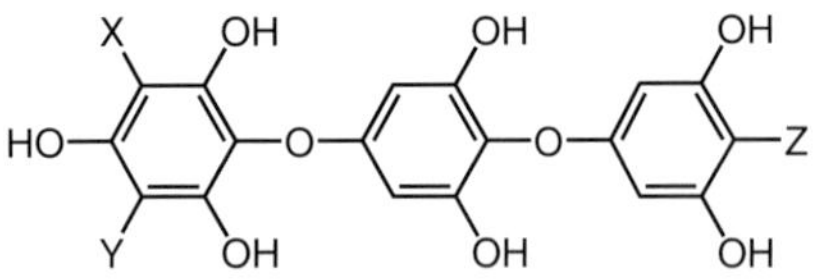

1944 X = Y = Cl, Z = H
1945 X = H, Y = Z = Cl
1946 X = H, Y = Cl, Z = Br

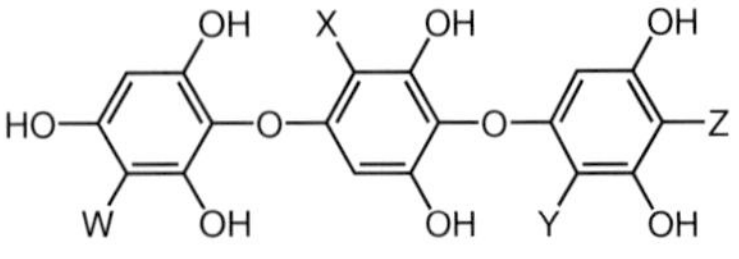

1947 W = Y = H, X = Z = Cl
1948 W = Y = Br, X = Z = H
1949 W = Br, Y = Cl, X = Z = H

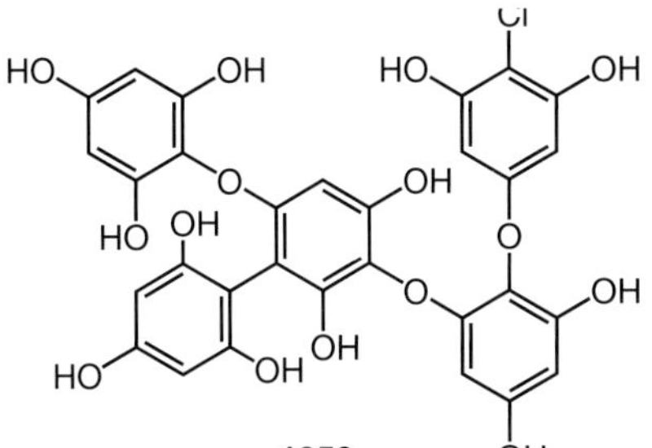

1950

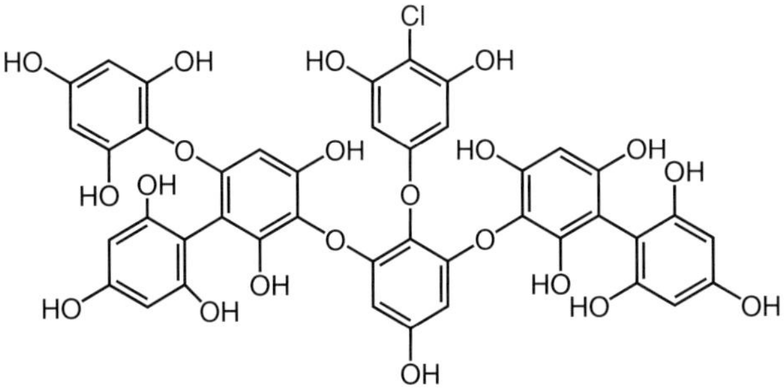

1951

3.22.3 Tyrosines

Although relatively few simple halogenated tyrosines are found naturally, many "transformed" tyrosines are produced by marine organisms and these are covered in Sects. 3.22.3.2–3.22.3.4.

3.22.3.1 Simple Tyrosines, Thyroxine, and Related Compounds

3-Chlorotyrosine, which was previously found to occur in the cuticle of locusts (*1*), is the product of the reaction of tyrosyl residues in albumin (*1805*) and in red blood cells (*1806*) with the human neutrophil myeloperoxidase-hydrogen peroxide-chloride system. The latter study provides evidence that free chlorine gas is involved in this chlorination reaction, rather than hypochlorous acid (*1806*). Furthermore, 3-chlorotyrosine is found in human atherosclerotic tissues, with the highest concentrations present in patients with coronary heart disease, indicating that 3-chlorotyrosine is a specific marker for low-density lipoprotein (LDL) oxidation by myeloperoxidase (MPO) (*1807*). This amino acid forms in dialysis patients as a result of oxidative stress by activated neutrophils. Thus, hemodialysis increases plasma MPO and hypochlorous acid leading to elevated levels of 3-chlorotyrosine (*1808*). This amino acid is also present in high concentrations in cystic fibrosis patients, who have high levels of MPO (*1809*). Consistent with these observations is that MPO-deficient mice fail to generate 3-chlorotyrosine and to kill the fungus *Candida albicans* in vivo (*1810*). Infants who develop chronic lung disease contain high levels of 3-chlorotyrosine, suggesting that MPO and neutrophil oxidants contribute to the pathology of these diseases (*1811*). Likewise, both 3-bromotyrosine and 3,5-dibromotyrosine, which were previously isolated from marine organisms and insects (*1*), appear to be major products of protein oxidation by eosinophil peroxidase (EPO) (*1812*). This EPO-promoted bromination may contribute to the tissue damage that accompanies asthma (*1813*). The red alga *Rhodomela confervoides* has yielded the new bromotyrosine **1952** (*1455*).

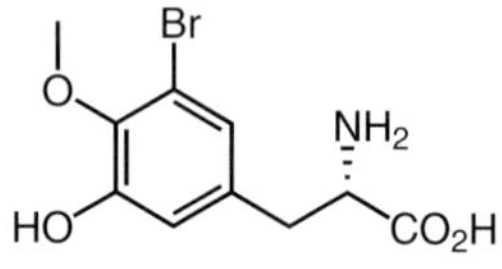

1952

3.22.3.2 Transformed Tyrosines, Tyramines, Phenethylamines and Related Compounds

Tyrosine-derived metabolites in this section do not include spiro-cyclohexadienyl-isoxazolines and related compounds (Sect. 3.22.3.3) or bastadins (Sect. 3.22.3.4), but they do include tyrosine-derived alkaloids that were covered in the Alkaloids section in the first survey (*1*). The prolific bryozoan *Amathia convoluta*, collected in Tasmania, has yielded amathamide G (**1953**) (*1814*), the latest of several amathamide alkaloids from the genus *Amathia* (*1*). A Florida collection of this animal furnished the new convolutamines A–E (**1954–1958**) (*1815*), F (**1959**), and G (**1960**) (*1425*), and a Tasmanian sample of this bryozoan afforded convolutamine H (**1961**) (*1319*). A study of *Amathia convoluta* from the North Carolina coast has yielded volutamides A–E (**1962–1966**) (*1816*). Volutamides B and C reduce feeding by the pinfish (*Lagodon rhomboids*) and the urchin (*Arbacia punctulata*), respectively, and volutamides B and D are toxic toward larvae of the hydroid *Eudendrium carneum*. The New Zealand *Amathia wilsoni* contains the six novel amathaspiramides A–F (**1967–1972**) (*1817*).

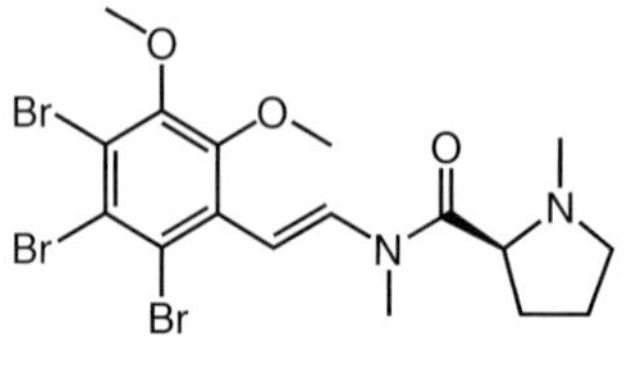

1953 (amathamide G)

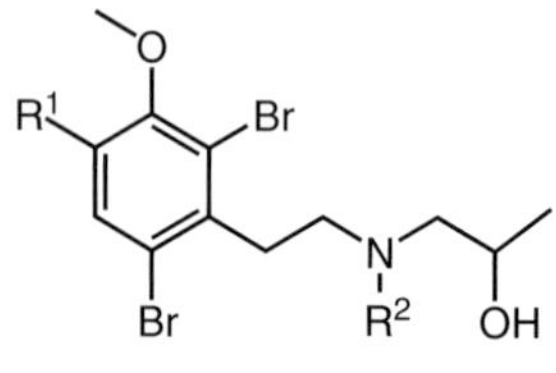

1954 R^1 = Br, R^2 = Me (convolutamine A)
1955 R^1 = H, R^2 = Me (convolutamine B)
1956 R^1 = Br, R^2 = H (convolutamine C)

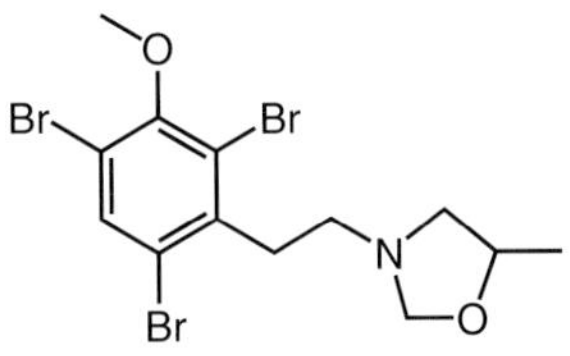

3.22 Complex Phenols

1957 (convolutamine D)

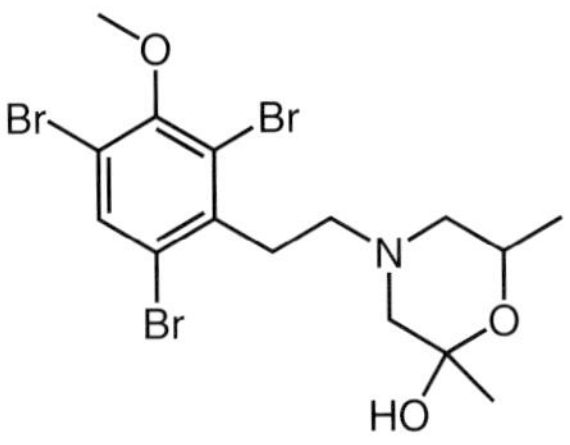

1958 (convolutamine E)

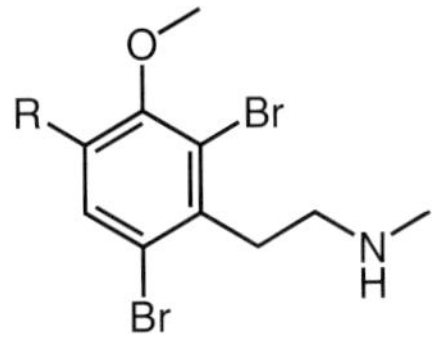

1959 R = Br (convolutamine F)
1960 R = H (convolutamine G)

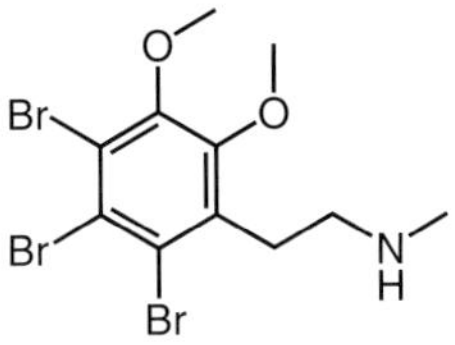

1961 (convolutamine H)

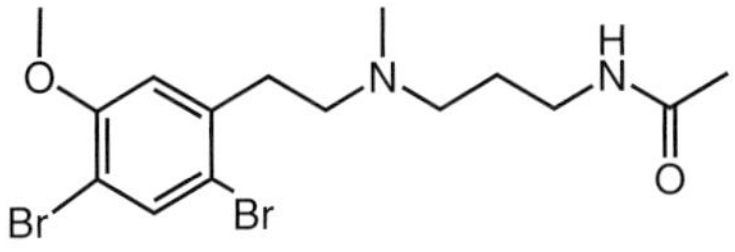

1962 (volutamide A)

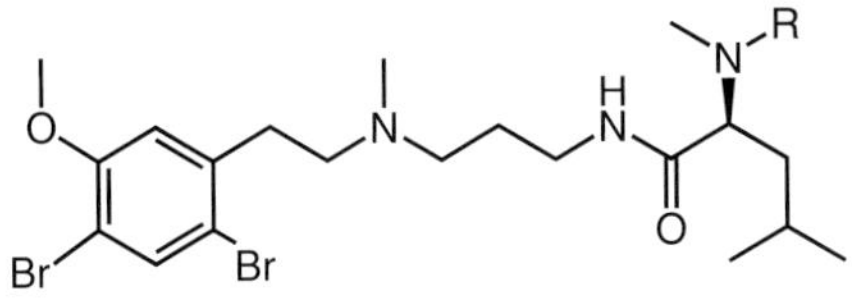

1963 R = H (volutamide B)
1964 R = Me (volutamide C)

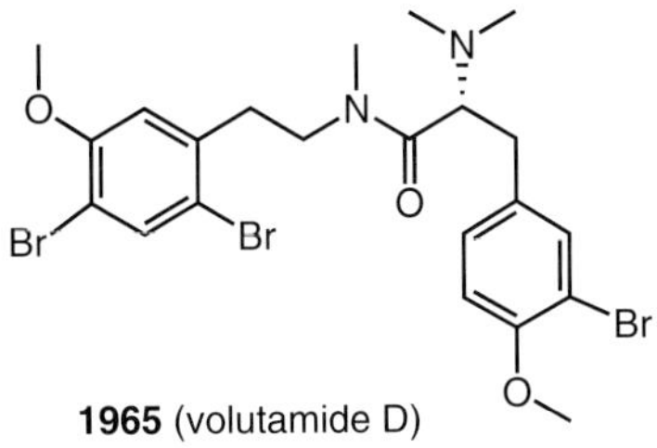

1965 (volutamide D)

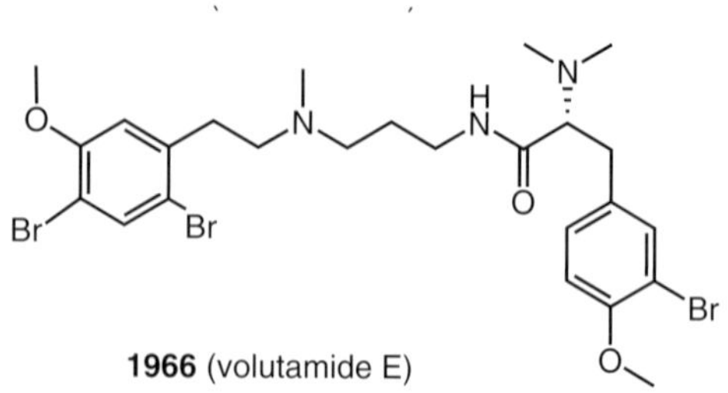

1966 (volutamide E)

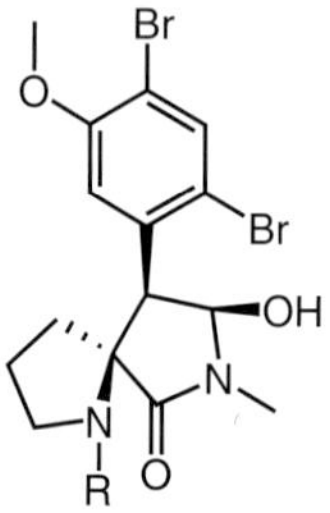

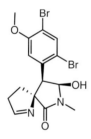

1967 R = Me (amathaspiramide A)
1969 R = H (amathaspiramide C)

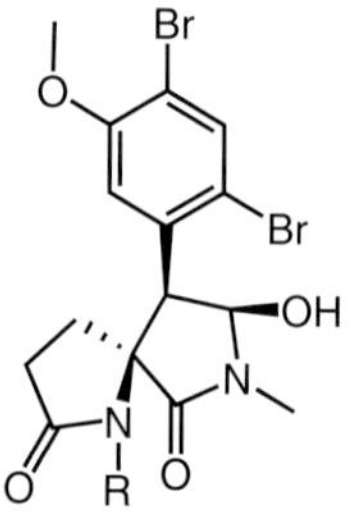

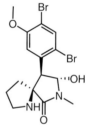

1968 R = Me (amathaspiramide B)
1970 R = H (amathaspiramide D)

1971 (amathaspiramide E)

1972 (amathaspiramide F)

Marine tunicates are also a source of brominated tyrosine derivatives. The colonial ascidian *Aplidium* sp., which was collected in Australia, yielded the novel iodinated tyrosine alkaloids **1973–1975** (*1819*). Collections of *Botryllus* sp. and *Botryllus schlosseri* from the Philippines and the Great Barrier Reef, respectively, have afforded botryllamides A–D (**1976–1979**) (*1820*). A Palauan ascidian *Botrylloides tyreum* produces several new botryllamides, including the brominated botryllamide G (**1980**) (*1821*). The simple brominated tyramines **1981** and **1982** were isolated from the New Zealand ascidian *Cnemidocarpa bicornuta* (*1822*) and an Indonesian *Eudistoma* sp. ascidian (*1823*).

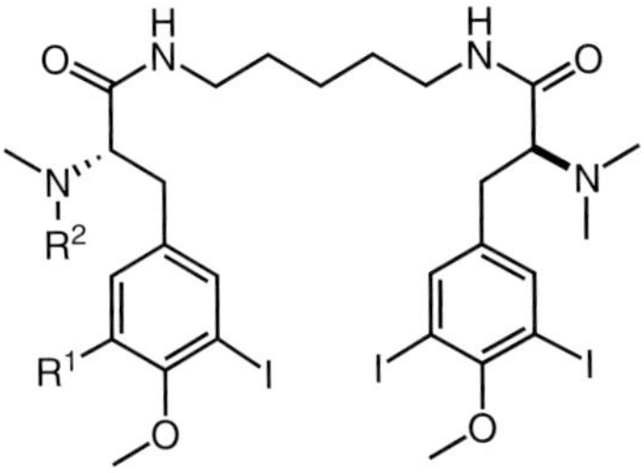

3.22 Complex Phenols

1973 R¹= H, R²= Me
1974 R¹= I, R²= Me
1975 R¹= R²= H

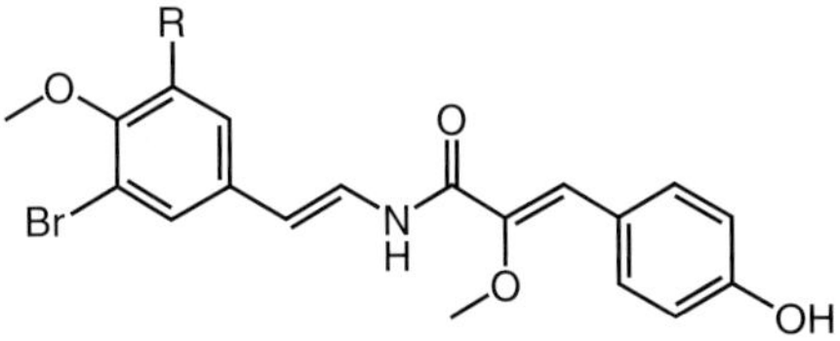

1976 R = Br (botryllamide A)
1978 R = H (botryllamide C)

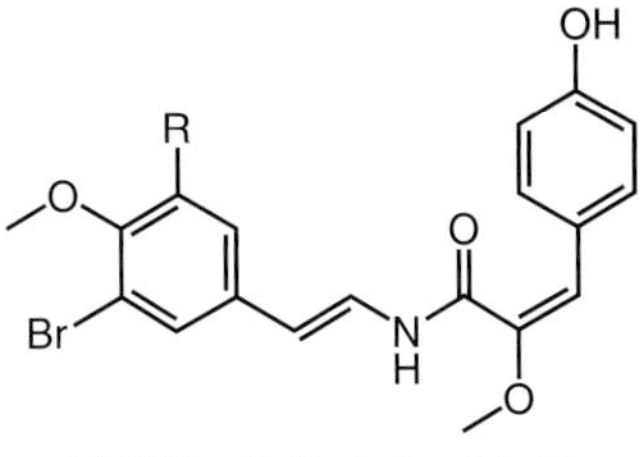

1977 R = Br (botryllamide B)
1979 R = H (botryllamide D)

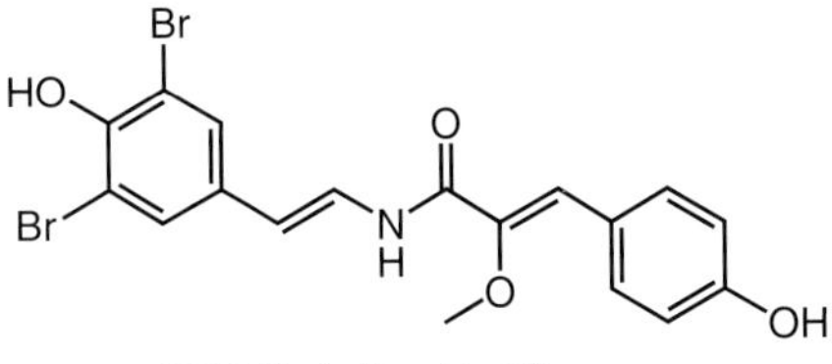

1980 (botryllamide G)

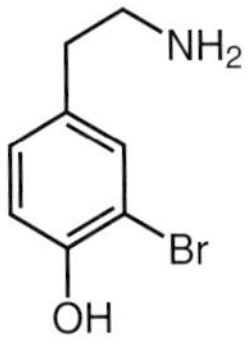

1981

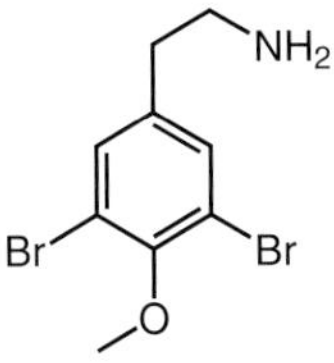

1982

The toxic Japanese gastropod *Turbo marmorata* contains the two toxins, turbotoxins A (**1983**) and B (**1984**), isolated as bis-trifluoroacetates (*1824, 1825*). The turbotoxins A and B show $LD_{99} = 1.0$ and 4.0 mg kg^{-1} in mice. The simple iodinated ammonium salt **1985** is also found in this animal (*1826*). The red alga *Halopytis incurvus* contains the simple brominated phenols **1986** and **1987**, which were isolated as the methyl esters and methyl ethers (*1827*). These presumed degradation products of tyrosine are related to earlier reported brominated metabolites (*1*).

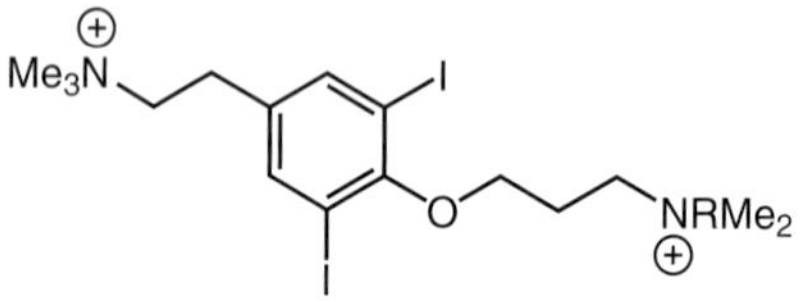

1983 R = Me (turbotoxin A)
1984 R = H (turbotoxin B)

1985

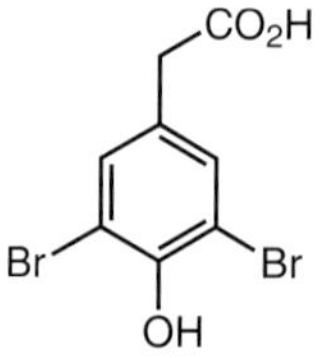

1986

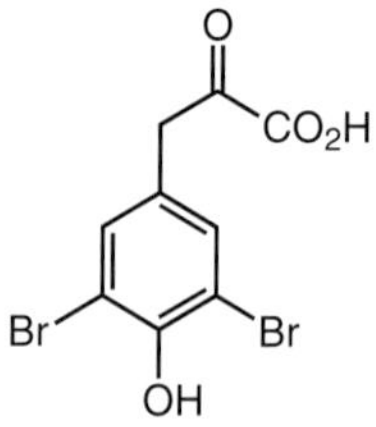

1987

The largest number of brominated tyrosines is found in marine sponges, and more than 100 were documented in the first survey (*1*). The organization in this section is by sponge genus and species, rather than by type of metabolite. The Caribbean sponge *Pseudoceratina crassa* has yielded the new brominated phenylacetonitrile **1988** and imidazole **1989** (*1828*), both of which are close analogs of previously described sponge metabolites, verongamine in the case of **1989** (*1*). In addition to containing several known bromotyrosines, the New Caledonian *Pseudoceratina verrucosa* has afforded pseudoceratinine B (**1990**), in addition to two spirocyclohexadiene isoxazoles reported in the following section (*1829*). A Caribbean *Pseudoceratina* sp. contains 5-bromoverongamine (**1991**), which inhibits the settlement of barnacle larvae at 10 mg cm^{-3} (*1830*). Ceratinamine (**1992**), which was isolated from the Japanese *Pseudoceratina purpurea*, is also an antifouling compound against the barnacle *Balanus amphitrite* and contains the novel cyanoformamide functionality (*1831*). The novel tokaradines A–C (**1993–1995**) are found in the sponge *Pseudoceratina purpurea* (Fig. 3.29) collected in Southern Japan waters (*1832*). These bromotyrosines are lethal to the crab *Hemigrapsus sanguineus* at 20–50 μg g^{-1} (**1993** and **1994**). A Papua New Guinea collection of this sponge yielded the six new psammaplins E-J (**1996–2001**) (*1833*). Psammaplin

3.22 Complex Phenols

Fig. 3.29 *Pseudoceratina purpurea*, a Papua New Guinea sponge that contains several psammaplins such as **1996–2001** (Photo: P. Crews)

F (**1997**) is a potent histone deacetylase inhibitor, and psammaplin G (**1998**) is a potent DNA methyltransferase inhibitor. A Southern Japanese version of *Pseudoceratina purpurea* has yielded pseudoceratins A (**2002**) and B (**2003**) (*1834*).

1988

1989

1990 (pseudoceratinine B)

1991 (5-bromoverongamine)

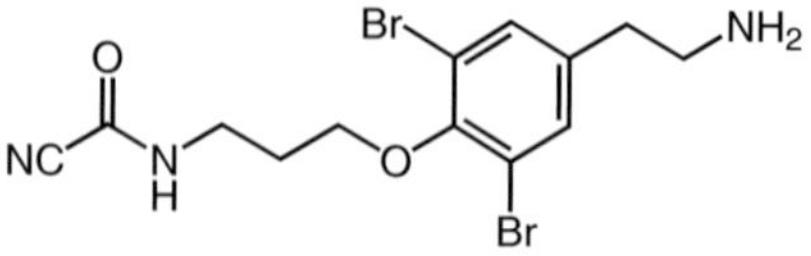

1992 (ceratinamine)

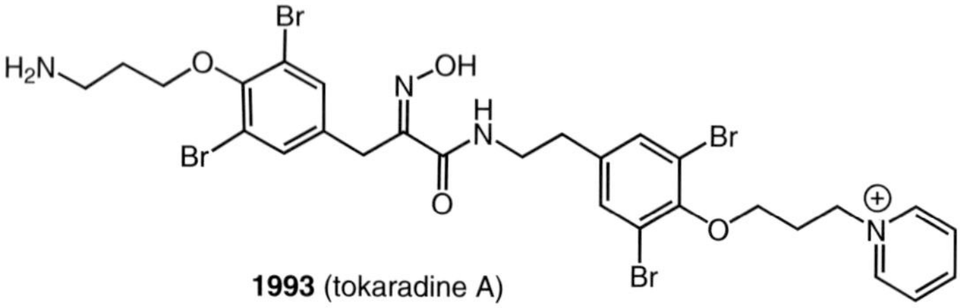

1993 (tokaradine A)

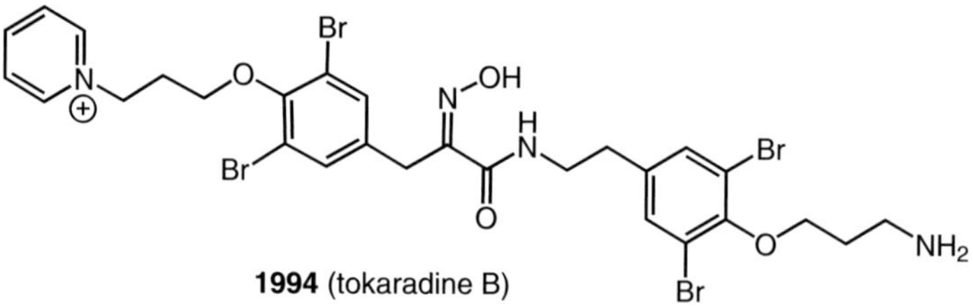

1994 (tokaradine B)

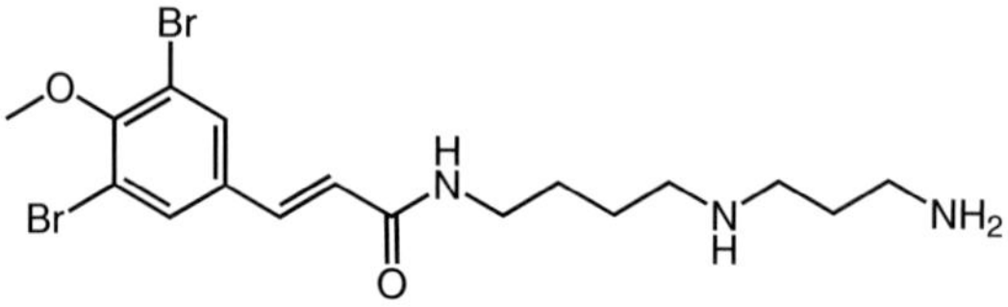

1995 (tokaradine C)

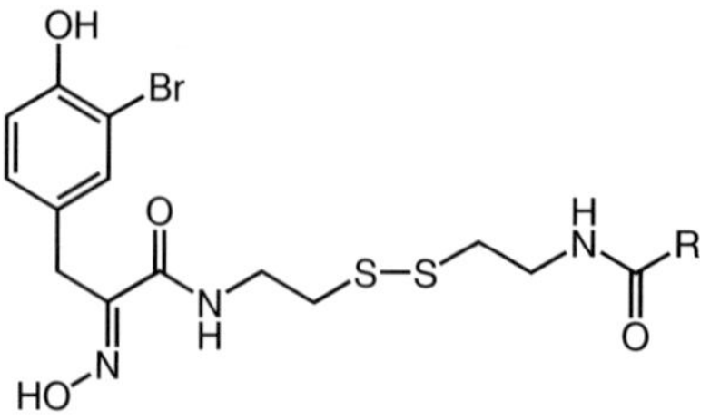

1996 R = CONH$_2$ (psammaplin E)
1997 R = CO$_2$H (psammaplin F)
1999 R = OEt (psammaplin H)

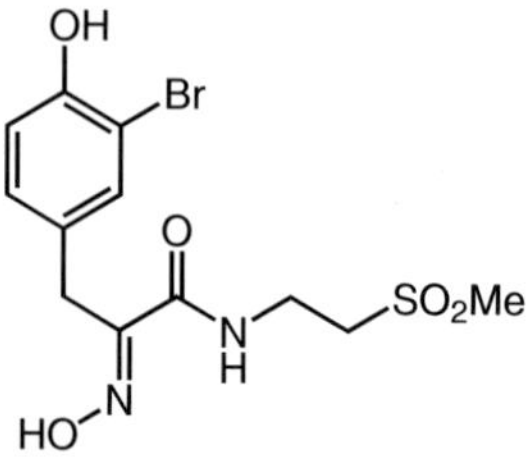

2000 (psammaplin I)

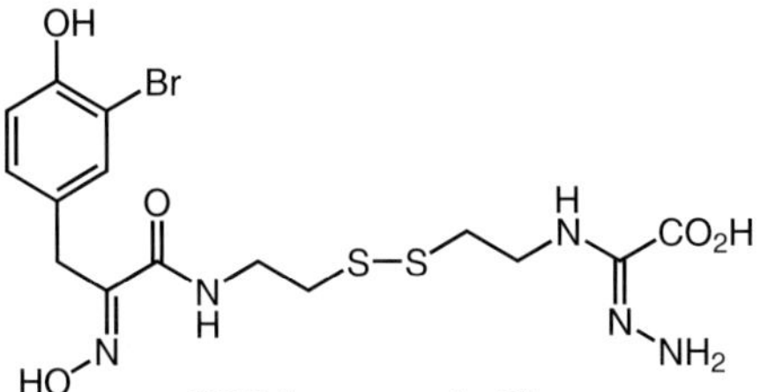

3.22 Complex Phenols

1998 (psammaplin G)

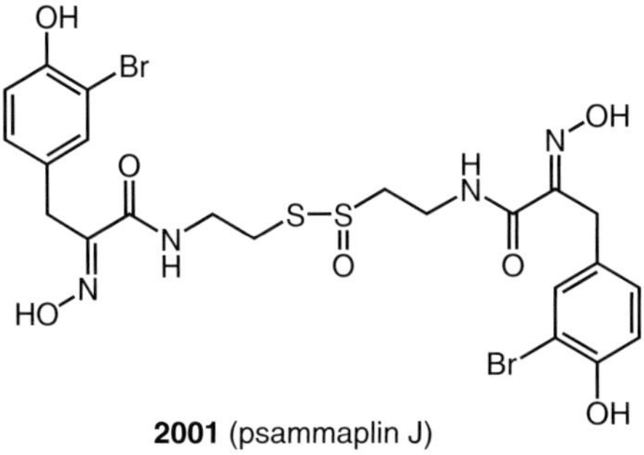

2001 (psammaplin J)

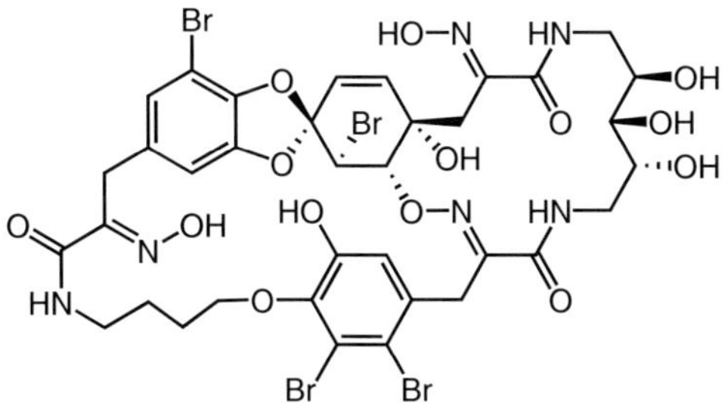

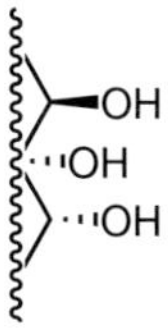

2002 (pseudoceratin A)   **2003** (pseudoceratin B)

Sponges of the genus *Psammaplysilla* have been a rich source of bromotyrosine metabolites (*1*), and that trend continues for the present survey. An examination of the Okinawan *Psammaplysilla purea* has revealed the presence of purealidins M–O (**2004**–**2006**) (*1835*), and purealidin H (**2007**) and lipopurealins D (**2008**) and E (**2009**) (*1836*). Several collections of *Psammaplysilla purpurea* from India have yielded new bromotyrosines and related compounds. These include the simple **2010** and **2011** (*1837*), **2012** (*1838*), **2013** and **2014** (*1839*), purpuramines K (**2015**) and L (**2016**) (*1840*), **2017** (*1841*), and purpurealidins F (**2018**), G (**2019**), and H (**2020**) (*1842*).

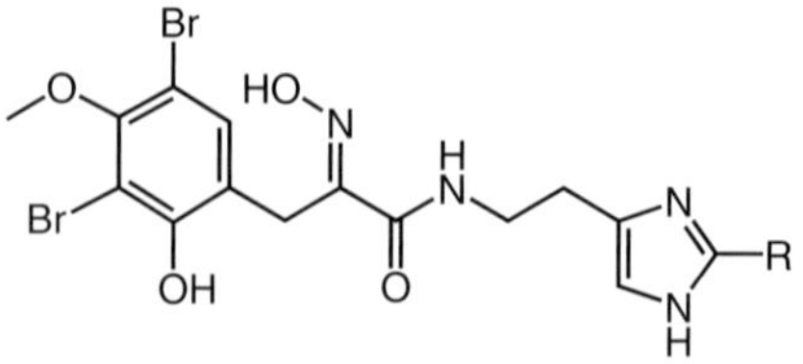

2004 R = NH₂ (purealidin M)
2005 R = H (purealidin N)

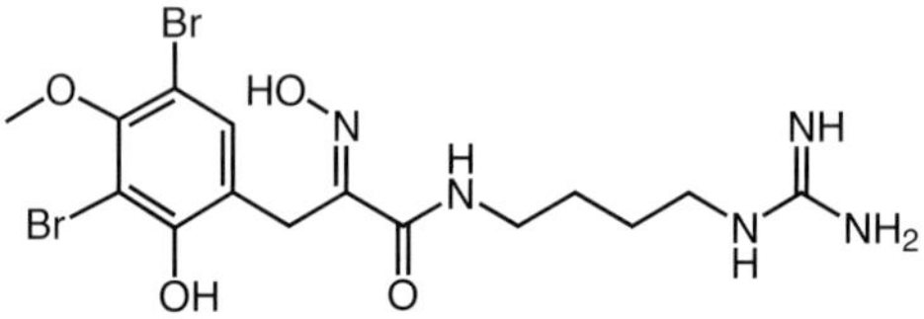

2006 (purealidin O)

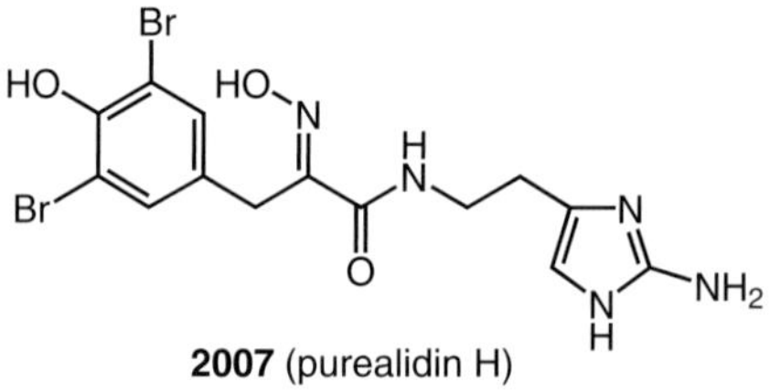

2007 (purealidin H)

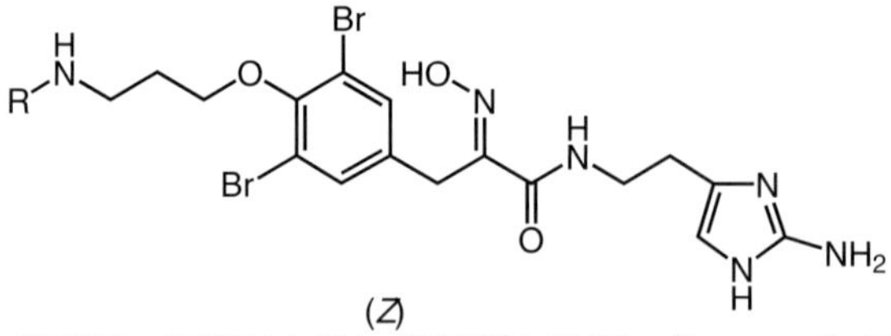

(Z)
2008 R = CO(CH₂)₇CH=CH(CH₂)₄CHMe₂ (lipopurealin D)
2009 R = CO(CH₂)₁₇CH₃ (lipopurealin E)

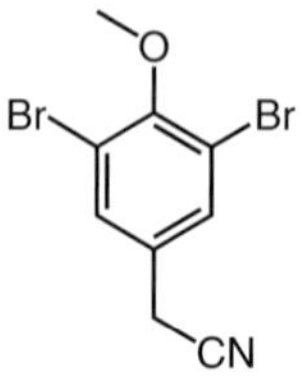

2010

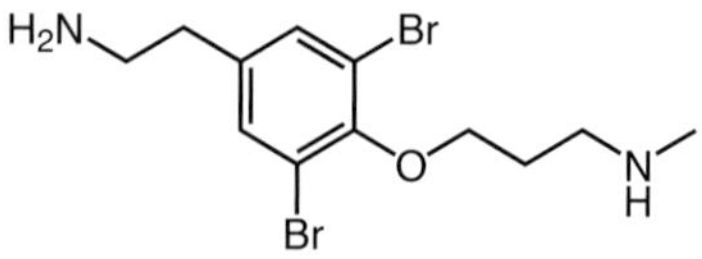

2012

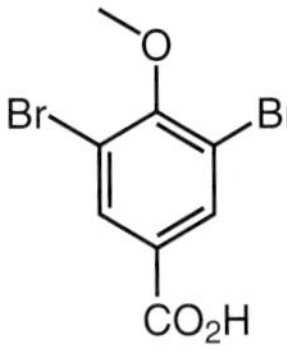

3.22 Complex Phenols

2011

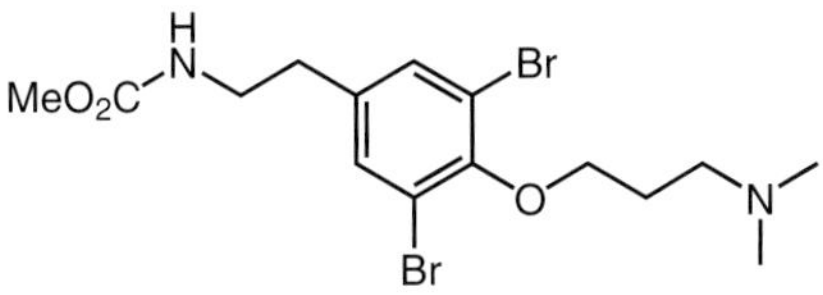

2013
2014 HX salt

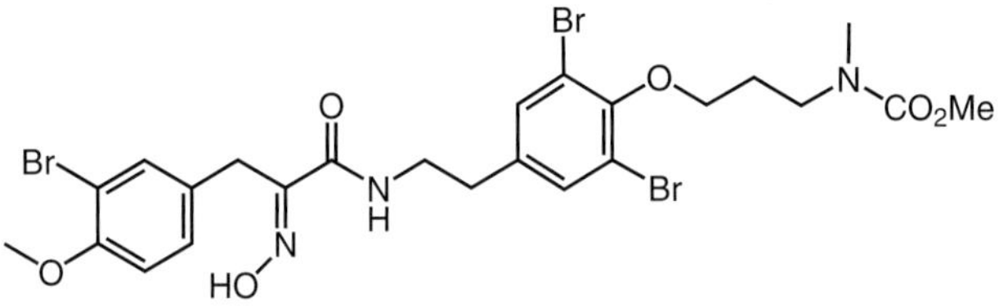

2015 (purpuramine K)

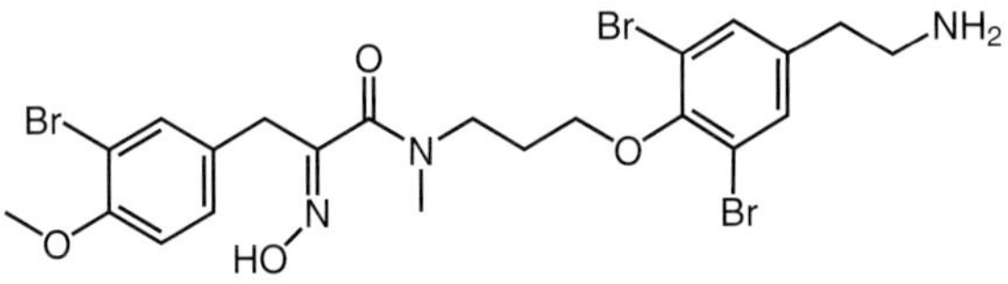

2016 (purpuramine L)

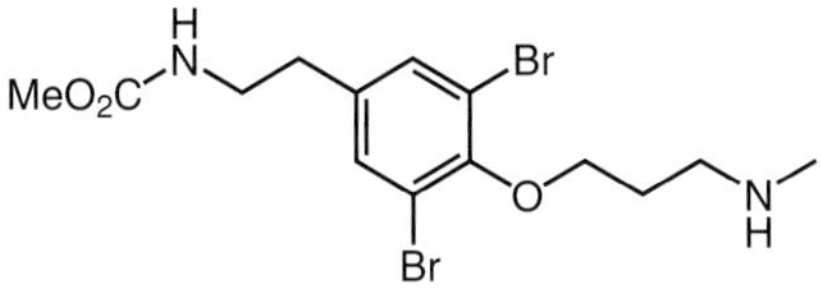

2017

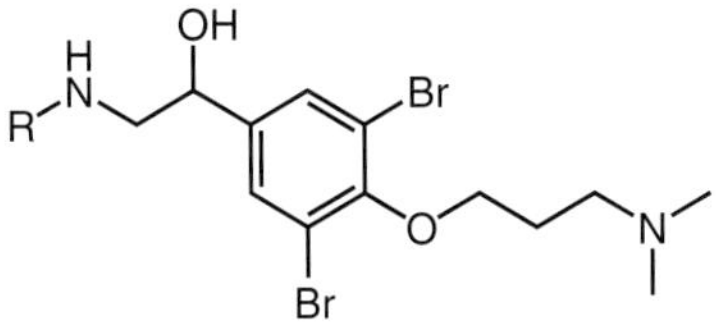

2018 R = H (purpurealidin F)
2019 R = CO₂Et (purpurealidin G)

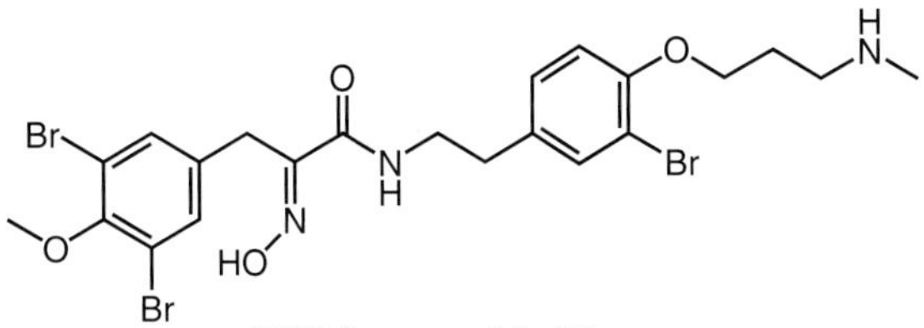

2020 (purpurealidin H)

An undescribed Verongid sponge from Molokai, Hawaii, has yielded *N*-methylceratinamine (**2021**) and the moloka'iamine derivatives, wai'anaeamines A (**2022**) and B (**2023**) (*1843*). This sponge is most likely of the genus *Psammaplysilla* or *Pseudoceratina*. Another collection of a Verongid sponge from Molokai has furnished a series of mololipids, **2024–2036**, which display anti-HIV activity. These amides are derivatives of the previously known moloka'iamine, also present in this sponge (*1844*). A collection of *Psammaplysilla* sp. from the Indian Ocean has provided the new psammaplysenes A (**2037**) and B (**2038**), which are inhibitors of the FOXO1a nuclear export (*1845*).

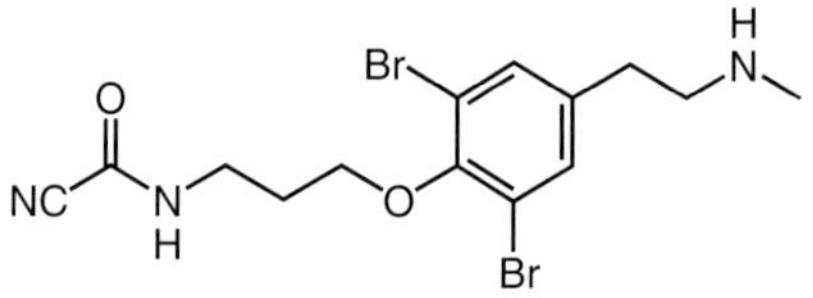

2021 (*N*-methylceratinamine)

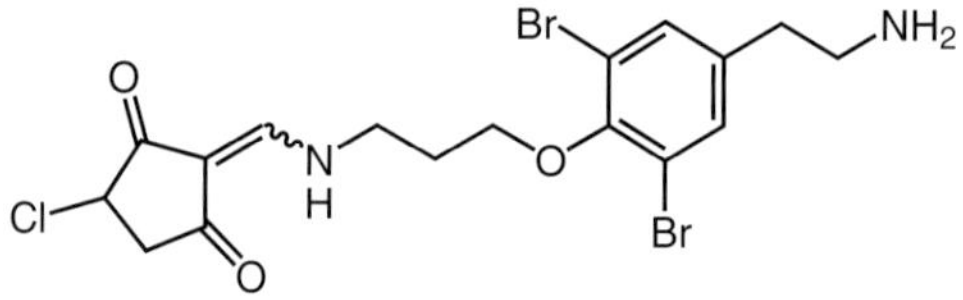

2022 (wai'anaeamine A)

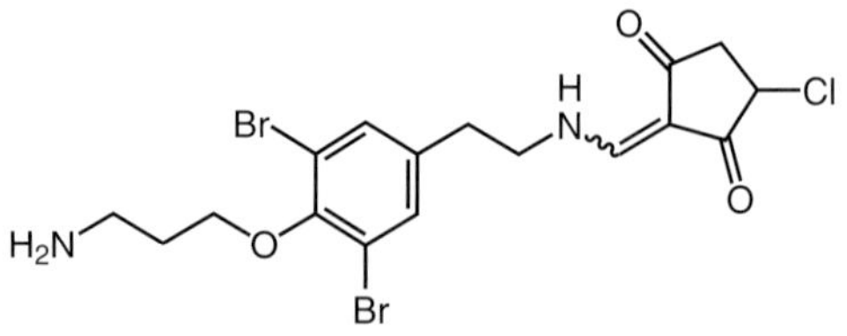

2023 (wai'anaeamine B)

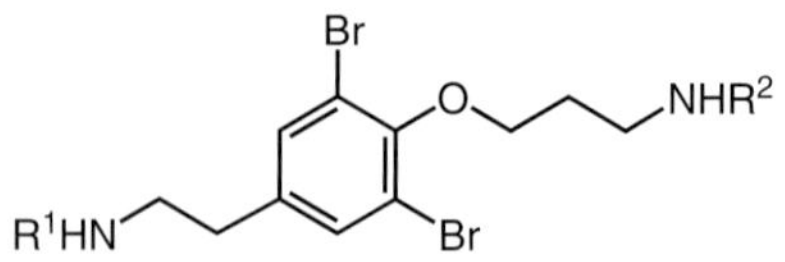

2024 - 2036 (mololipids)

R^1, R^2 = myristic, palmitic, margaric, oleic, stearic, arachidic, pentadecanoic, hexadecanoic (2), heptadecanoic, nondecanoic, unknown (2)

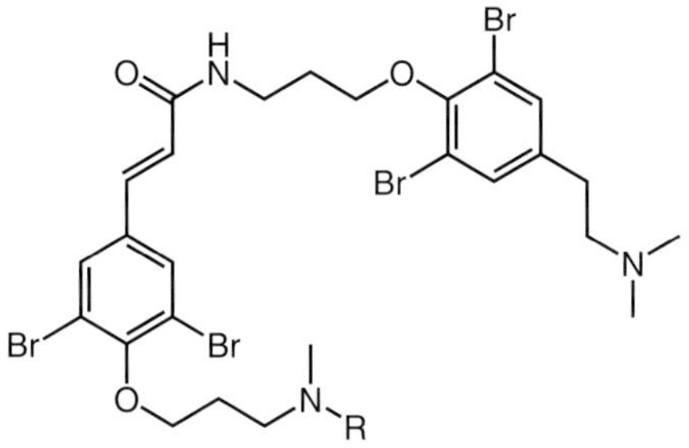

3.22 Complex Phenols

2037 R = Me (psammaplysene A)
2038 R = H (psammaplysene B)

Sponges of the genus *Aplysina* are abundant in the Caribbean and Mediterranean Seas and have yielded a variety of bromotyrosine metabolites (*1*). The Caribbean *Aplysina insularis* affords **2039** (*1846*), and an Indo-Pacific *Aplysina* sp. sponge has yielded aplyzanzine A (**2040**) (*1847*). A study of the Brazilian sponges *Aplysina cauliformis* and *Pachychalina* sp. has led to compounds **2041** and **2042**, respectively (*1848*). An *Aplysinella* sp. sponge from Micronesia contains 7-hydroxyceratinamine (**2043**) and dibromotyramine **2044** (*1849*). An Australian collection of *Aplysinella rhax* has furnished psammaplin A 11′-sulfate (**2045**) and bisaprasin 11′-sulfate (**2046**) (*1850*). An independent study of this sponge, which was collected in Guam, Palau, and Pohnpei, identified **2045** (as the *N*,*N*-dimethylguanidium salt) along with the new psammaplin A$_2$ (**2047**), aplysinellins A (**2048**) and B (**2049**) (*1851*). A Fijian version of *Aplysinella rhax* has yielded the new psammaplins K (**2050**) and L (**2051**) (*1852*).

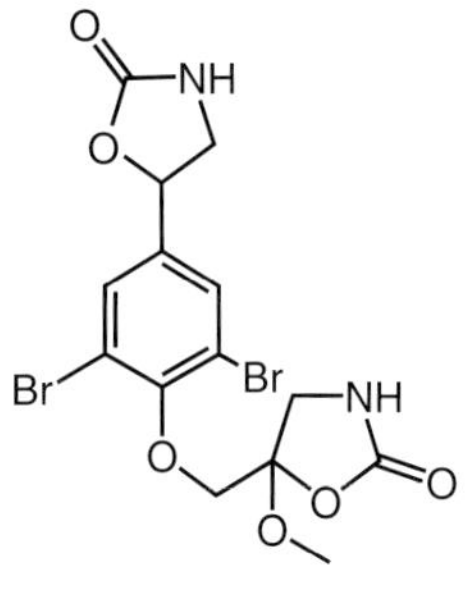

2039

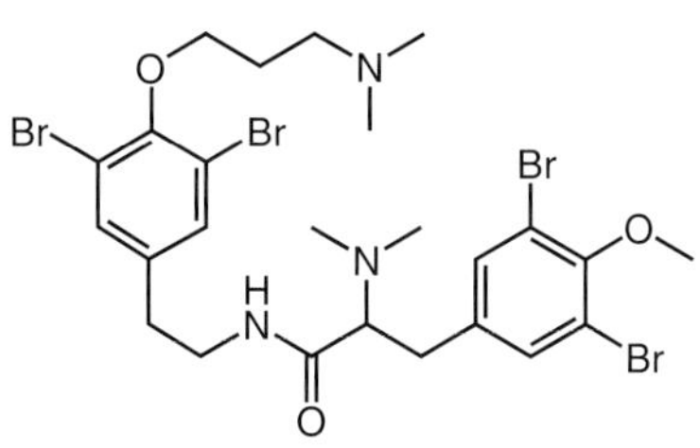

2040 (aplyzanzine A)

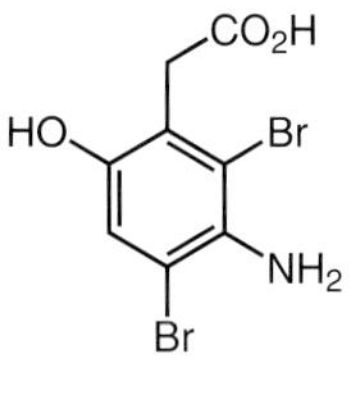

2041

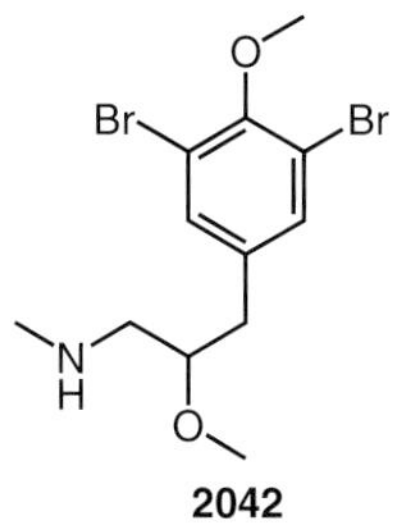

2042

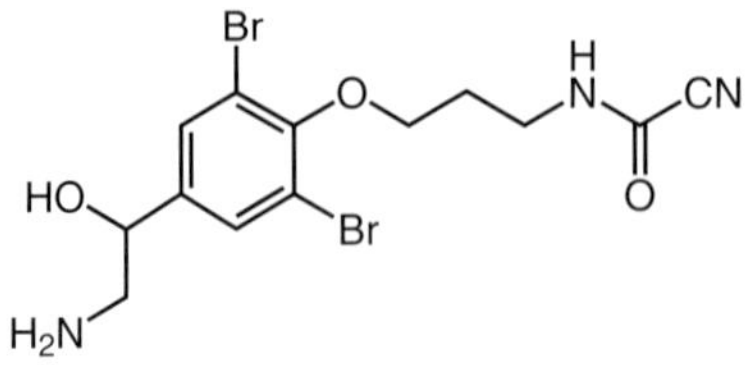

2043 (7-hydroxyceratinamine)

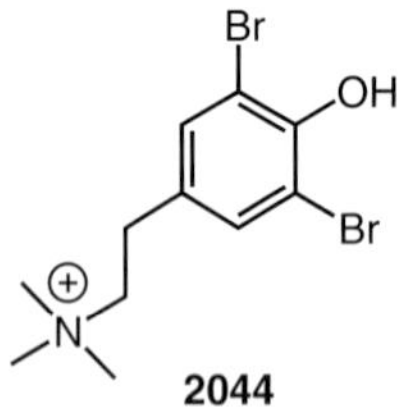

2044

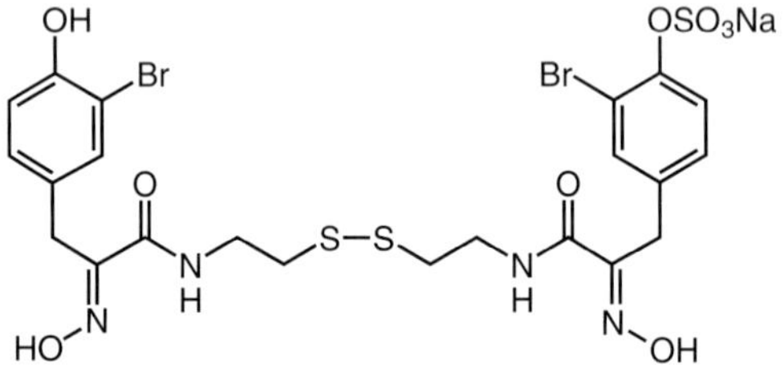

2045 (psammaplin A 11'-sulfate)

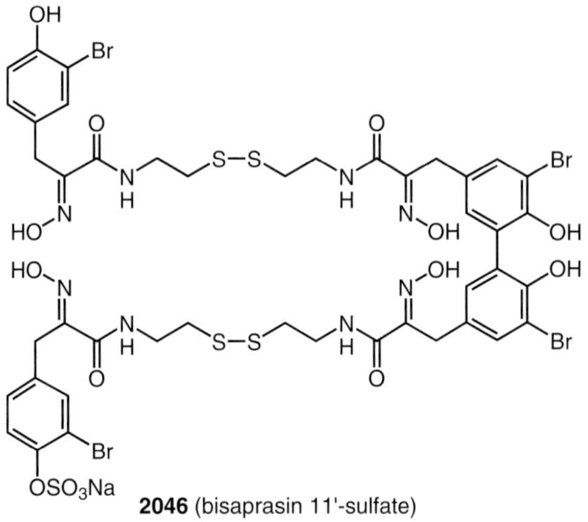

2046 (bisaprasin 11'-sulfate)

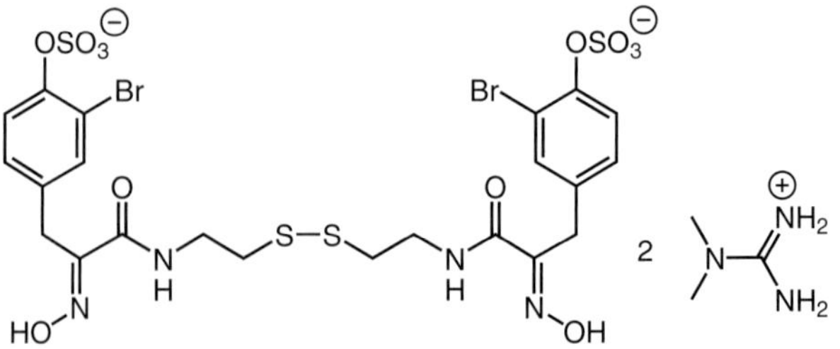

2047 (psammaplin A₂)

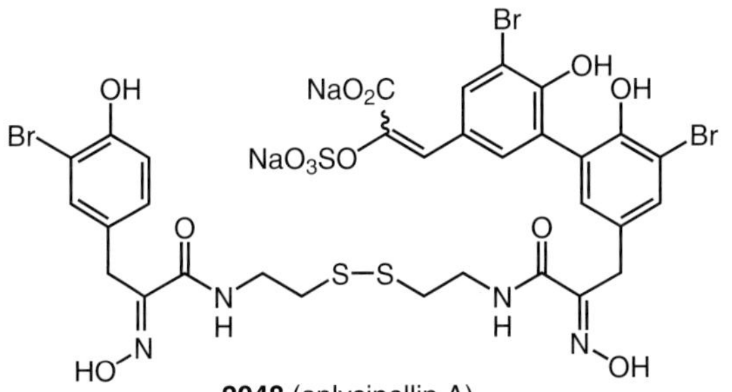

3.22 Complex Phenols

2048 (aplysinellin A)

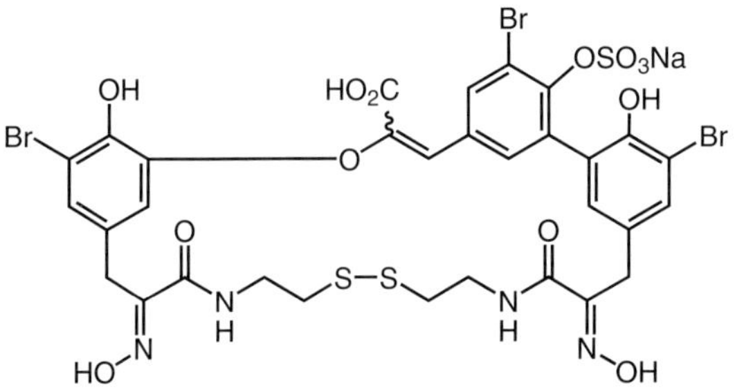

2049 (aplysinellin B)

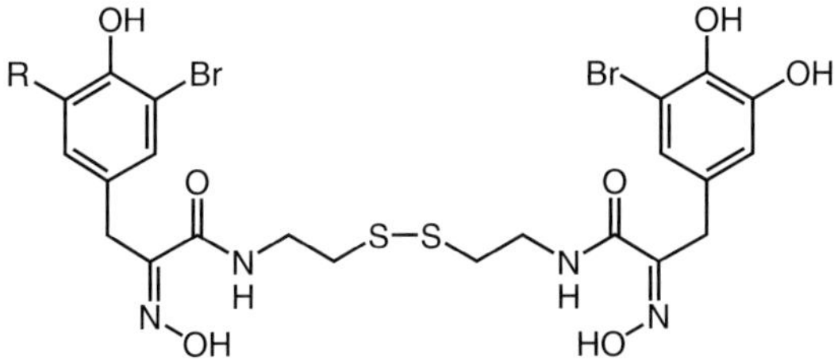

2050 R = H (psammaplin K)
2051 R = OH (psammaplin L)

The novel iodinated tyrosine derivative dakaramine (**2052**) is present in the Senegalese sponge *Ptilocaulis spiculifer* (*1853*). A Papua New Guinea sponge *Ianthella basta* (Fig. 3.30) has furnished nine new bromotyrosine compounds, hemibastadins 3 (**2053**), **2054**–**2056**, and hemibastadinols 1 (**2057**), 2 (**2058**), and 3 (**2059**) (*1854*). The previously known hemibastadins 1 and 2 were also isolated. A Guamanian collection of this sponge has afforded 1-*O*-sulfatohemibastadins-1 (**2060**) and -2 (**2061**) (*1855*). The Caribbean sponge *Verongula gigantea* ("Netted Barrel Sponge") contains the novel bromotyrosine metabolite **2062** (*1856*). An unidentified Okinawan sponge of order Verongid has afforded nakirodin A (**2063**) (*1857*).

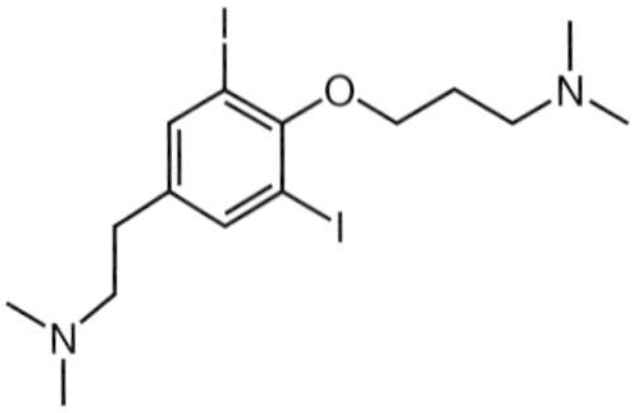

2052 (dakaramine)

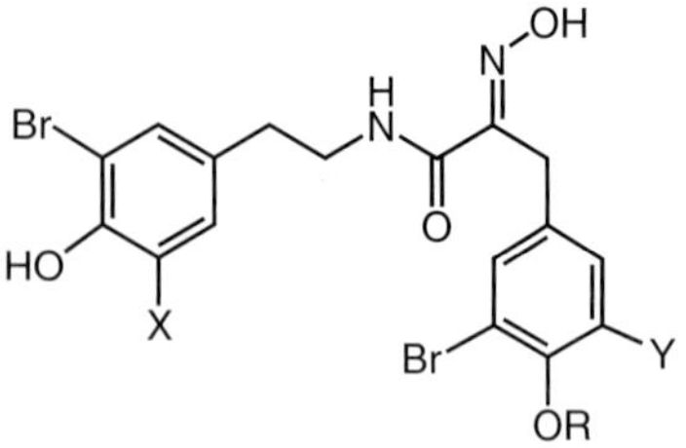

2053 Y = R = H, X = Br (hemibastadin 3)
2054 X = Y = H, R = Me
2055 X = H, R = Me, Y = Br

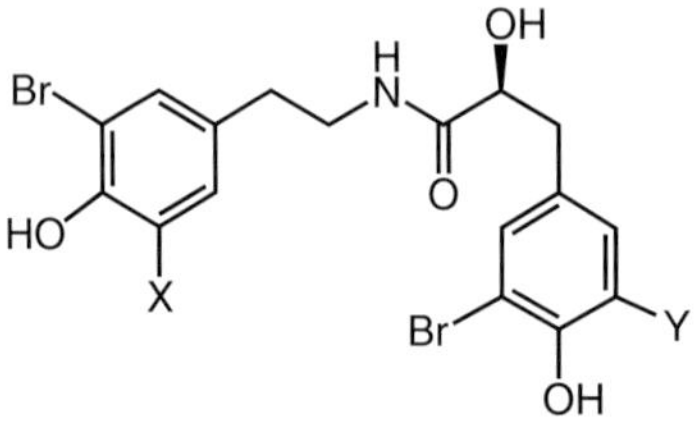

2057 X = Y = H (hemibastadinol 1)
2058 X = H, Y = Br (hemibastadinol 2)
2059 Y = H, X = Br (hemibastadinol 3)

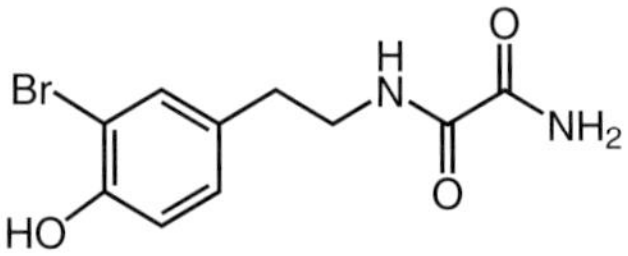

2056

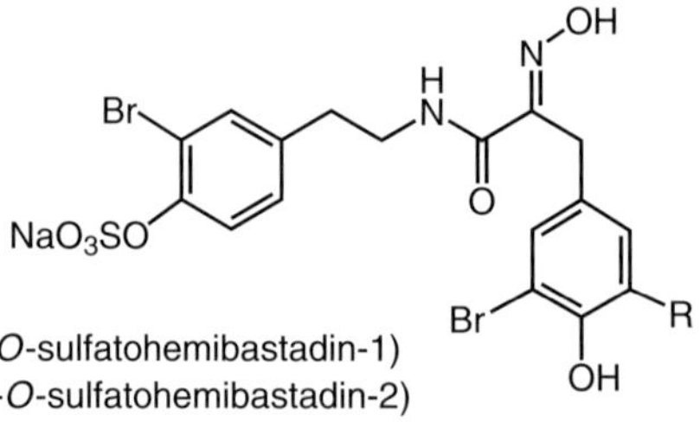

2060 R = H (1-*O*-sulfatohemibastadin-1)
2061 R = Br (1-*O*-sulfatohemibastadin-2)

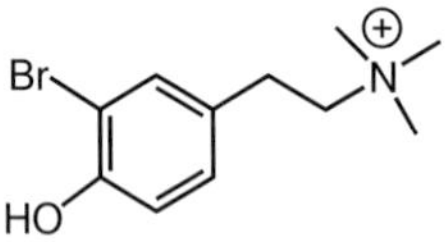

2062

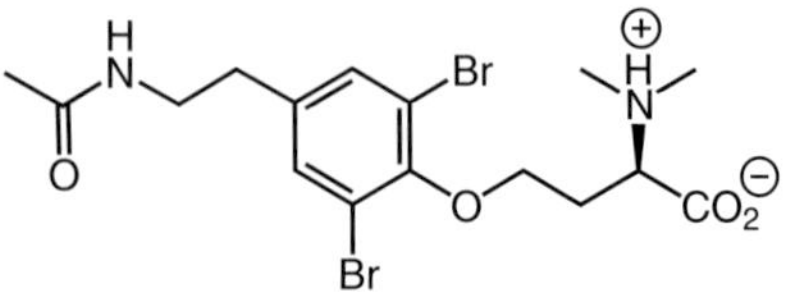

2063 (nakirodin A)

The novel pyrazinone bromotyrosines ma'edamines A (**2064**) and B (**2065**) were characterized from an Okinawan sponge *Suberea* sp. (*1858*). It is proposed that the pyrazinone ring may be derived from a dehydro form of the known aplysamine-2 or purpuramine H, which are also present in this sponge. A separate study of this sponge revealed the presence of the new suberedamines A (**2066**) and B (**2067**) (*1859*). An Australian non-Verongid sponge, *Oceanapia*

3.22 Complex Phenols

Fig. 3.30 *Ianthella basta*, a sponge rich in the bastadins **2053–2061** (Photo: F. J. Schmitz)

sp., has yielded **2068**, which is the first example of an inhibitor of the mycobacterial enzyme mycothiol *S*-conjugate amidase, found in *Mycobacterium* sp. (*1860*). The Fijian sponge *Druinella* sp. has afforded purpuramine J (**2069**), the first bromotyrosine *N*-oxide alkaloid to be discovered (*1861*). This metabolite is the *N*-oxide of aplysamine-2. The novel trisulfide **2070** and the two disulfides (*E*, *E*)-bromopsammaplin A (**2071**) and bispsammaplin A (**2072**) were found in a combined extract of the sponges *Jaspis wondoensis* and *Poecillastra wondoensis* (*1862*).

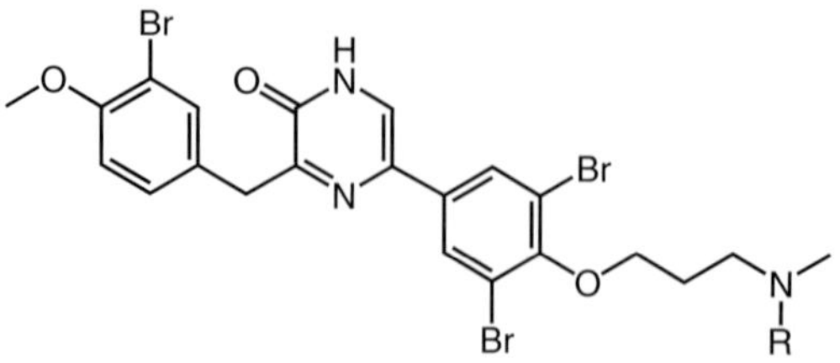

2064 R = Me (ma'edamine A)
2065 R = H (ma'edamine B)

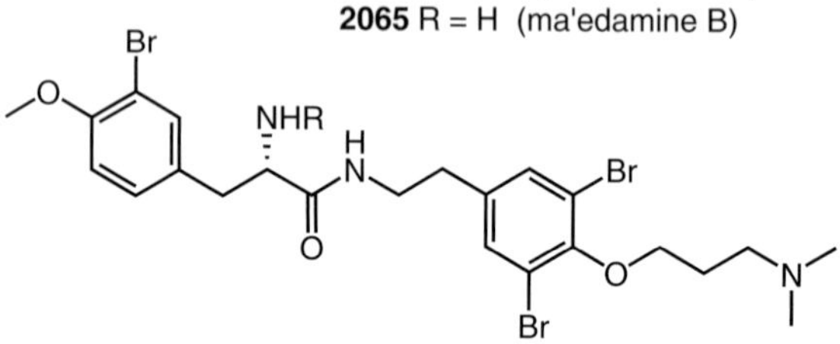

2066 R = H (suberedamine A)
2067 R = Me (suberedamine B)

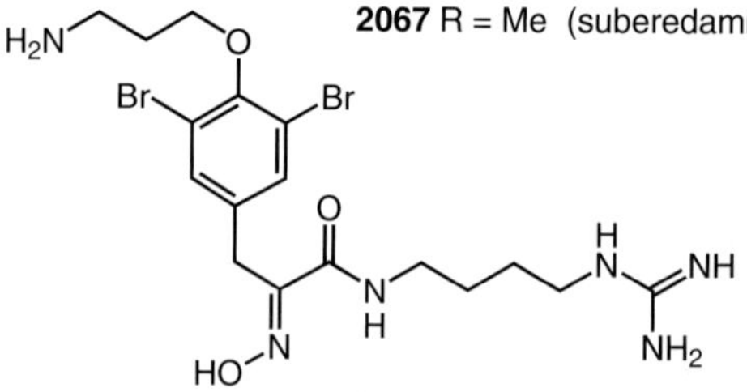

2068

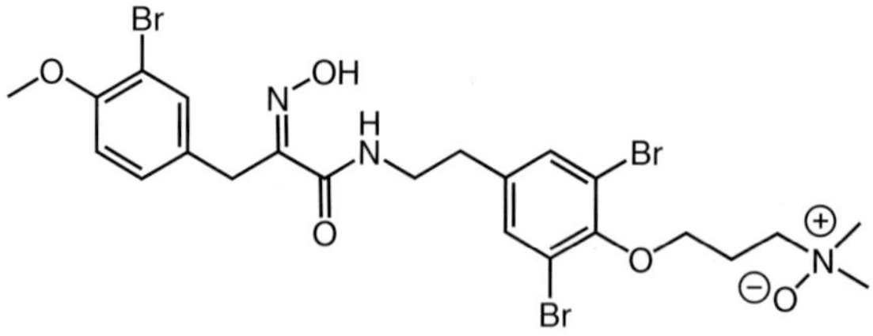

2069 (purpuramine J)

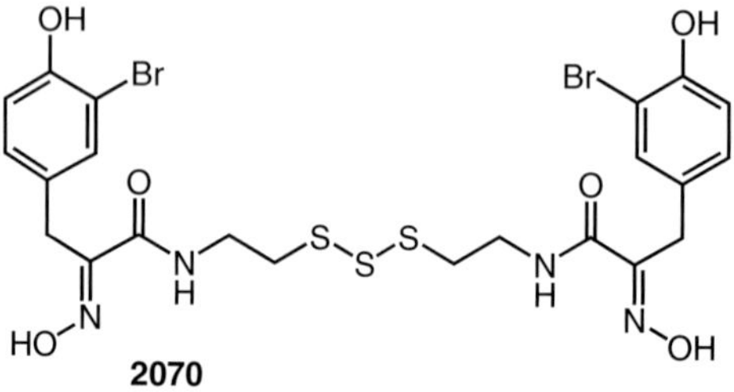

2070

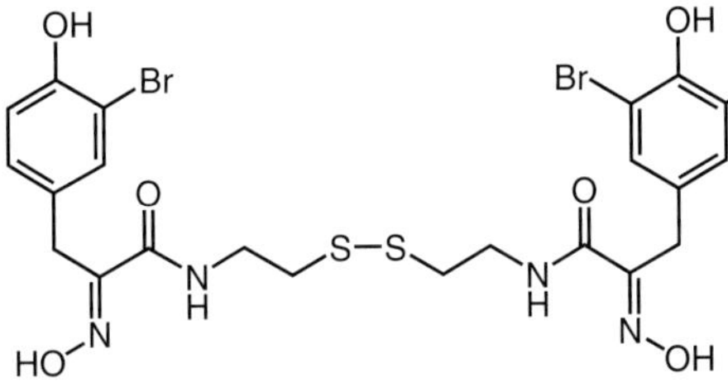

3.22 Complex Phenols

2071 ((E,E)-bromopsammaplin A)

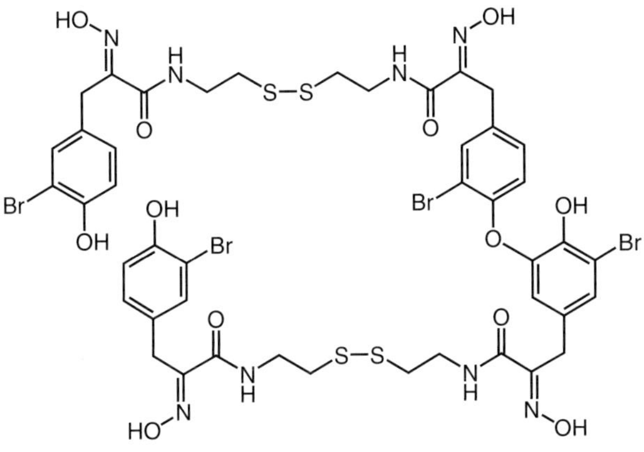

2072 (bispsammaplin A)

A southern Japan *Hexadella* sp. sponge has furnished the new moloka'iamines **2073** and **2074** and kuchinoenamine (**2075**), the latter having a unique tricyclo [$5.2.1.0^{2.6}$]decane skeleton (*1863*). These metabolites display antibacterial activity against the fish pathogenic bacterium *Aeromonas hydrophila*. A Madagascan sponge *Iotrochota purpurea* contains itampolins A (**2076**) and B (**2077**), which are comprised of three separate units including D-bromotyrosine (*1864*).

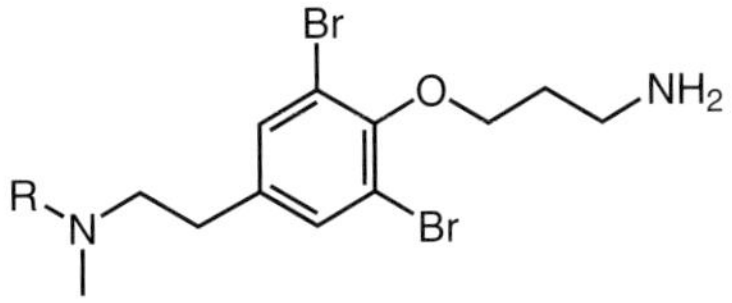

2073 R = H
2074 R = CN

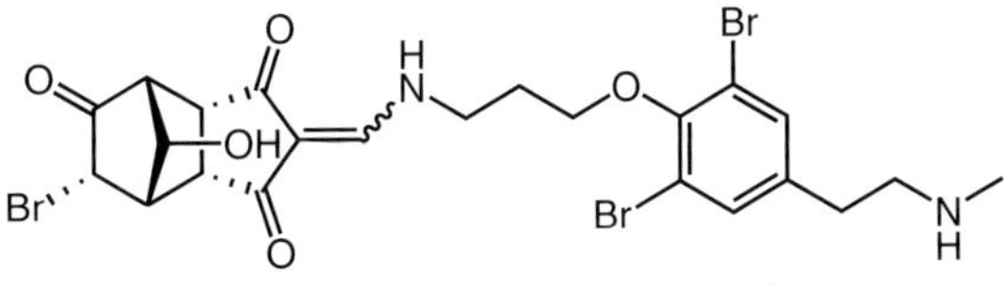

2075 (kuchinoenamine)

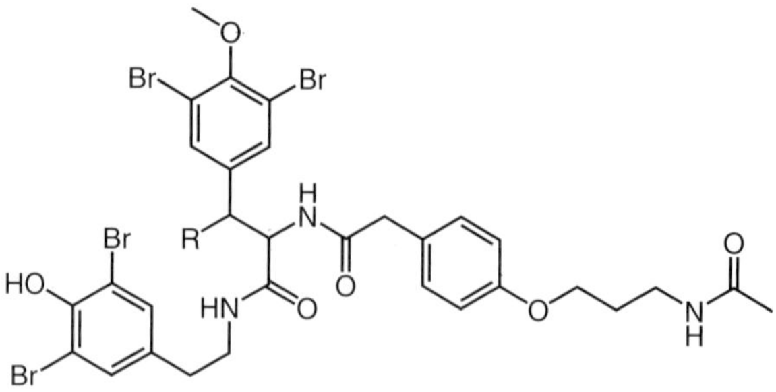

2076 R = H (itampolin A)
2077 R = OH (itampolin B)

While there exist too many syntheses of bromotyrosine alkaloids to delineate here, two illustrative examples are those of moloka'iamine (*1865*) and the mycothiol-*S*-conjugate amidase inhibitor **2068** (*1866*).

3.22.3.3 Transformed Multiple Tyrosines

As presented in the first survey (*1*), a large number of brominated tyrosines that contain a spirocyclohexadienyl isoxazoline ring are known, and 34 examples were described in the first survey (*1*). The first two such metabolites to be identified, aerothionin and homoaerothionin (*1*), are localized in the spherulous cells of the sponge *Aplysina fistularis*, which may suggest their release into the ectosome matrix and surrounding seawater as antifouling agents (*1867*).

The Okinawan sponge *Psammaplysilla purea* that contains purealidins M–O (**2004**–**2006**) also yields purealidins J (**2078**), K (**2079**), L (**2080**), P (**2081**), Q (**2082**), and R (**2083**) (*1835*). Purealidin J (**2078**) is the antipode of pseudoceratinine A (**2089**). The Indian sponge *Psammaplysilla purpurea*, which is the source of purpurealidins F–H (**2018**–**2020**) and other bromotyrosines (vide supra), also contains purpurealidins A (**2084**), B (**2085**), C (**2086**), and D (**2087**) (*1842*). A Caribbean *Pseudoceratina* sponge has afforded the simple carboxylic acid **2088** (*1868*). The New Caledonian sponge *Pseudoceratina verrucosa*, which is the source of pseudoceratinine B (**1990**), also contains pseudoceratinines A (**2089**) and C (**2090**), the absolute configurations of which are shown (*1829*).

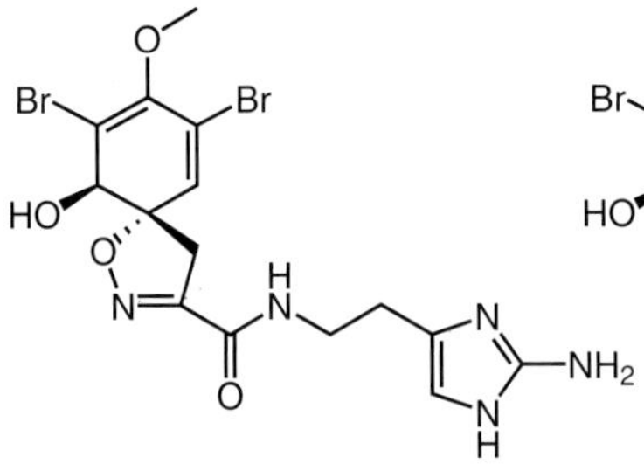

3.22 Complex Phenols

2078 (purealidin J)

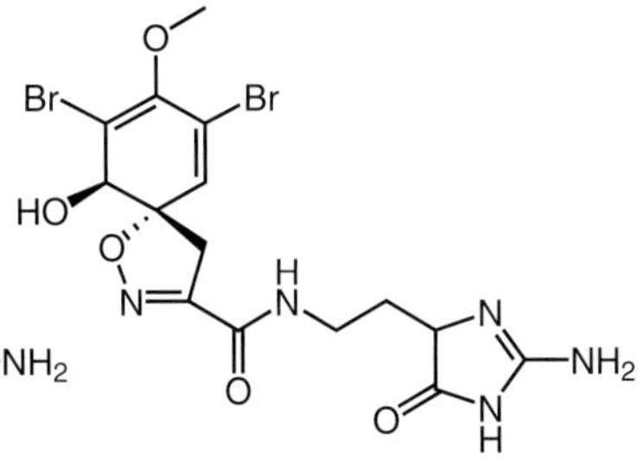

2079 (purealidin K)

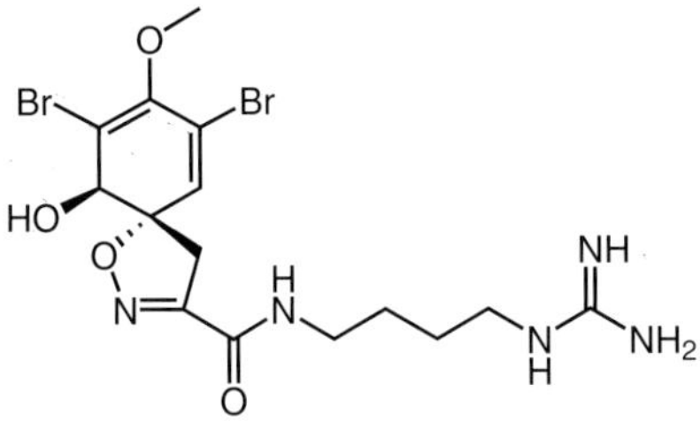

2080 (purealidin L)

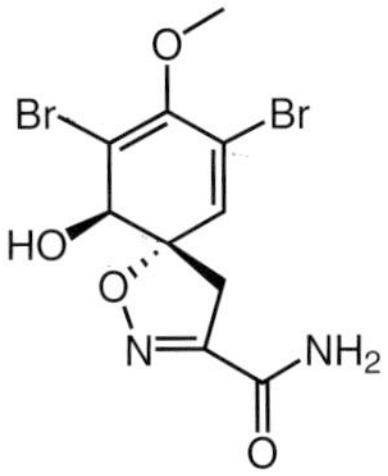

2083 (purealidin R)

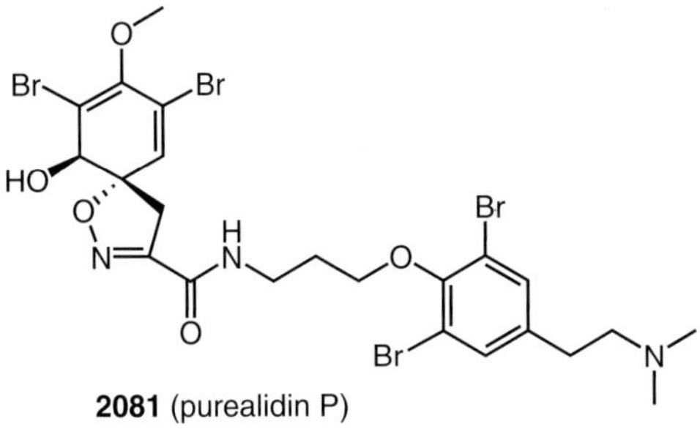

2081 (purealidin P)

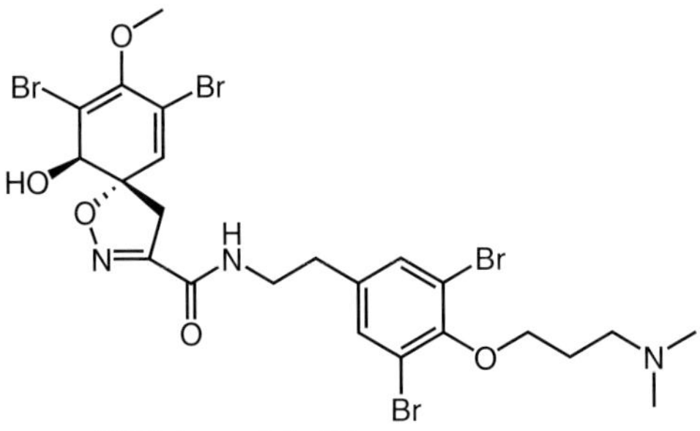

2082 (purealidin Q)

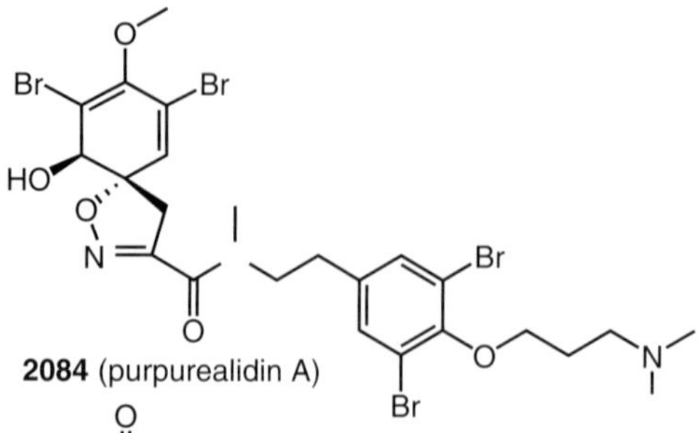

2084 (purpurealidin A)

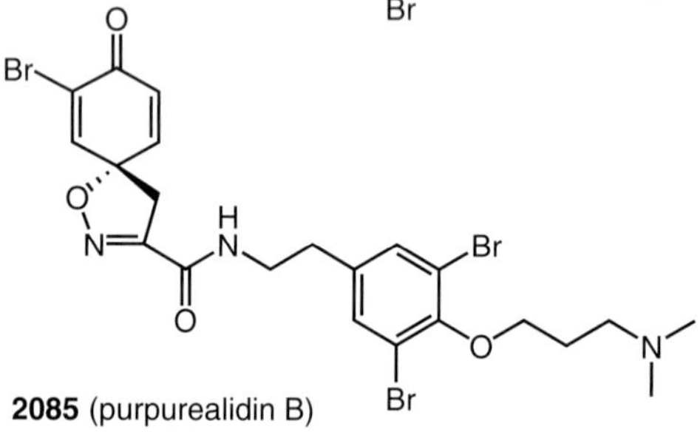

2085 (purpurealidin B)

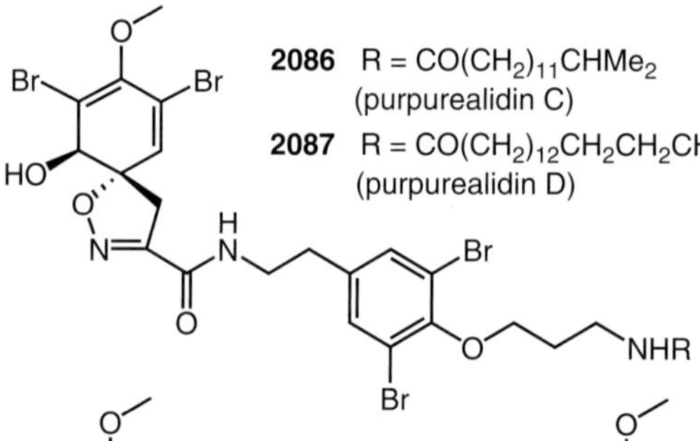

2086 R = $CO(CH_2)_{11}CHMe_2$
(purpurealidin C)

2087 R = $CO(CH_2)_{12}CH_2CH_2CH_3$
(purpurealidin D)

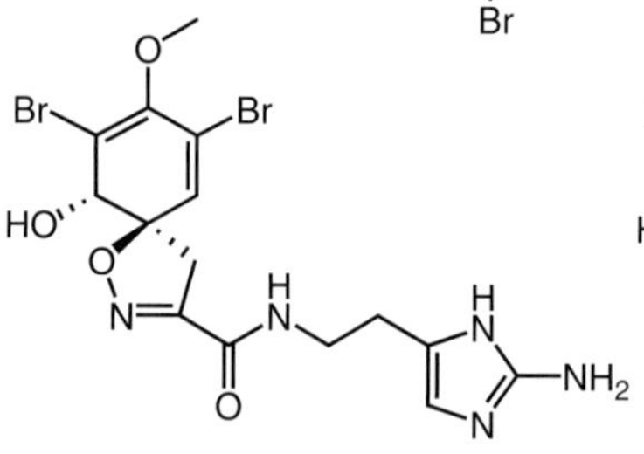

2089 (pseudoceratinine A)

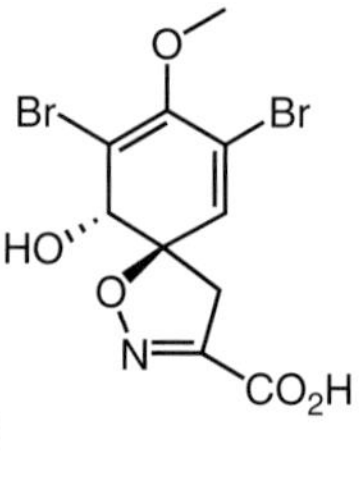

2088

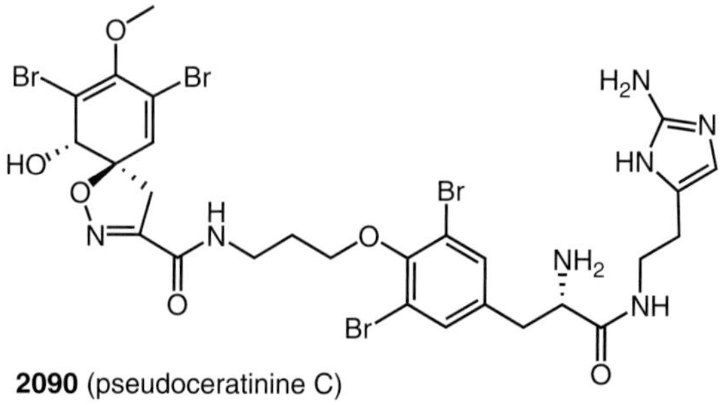

2090 (pseudoceratinine C)

3.22 Complex Phenols

A Japanese collection of *Pseudoceratina purpurea* has uncovered the presence of ceratinamides A (**2091**) and B (**2092**) in this sponge (*1869*). These new compounds are acyl derivatives of psammaplysin A (*1*), which is also present in this sponge. The closely related psammaplysin F (**2093**) was identified in an *Aplysinella* sp. from Chuuk (*1870*). A Gulf of Thailand sponge, *Pseudoceratina purpurea*, has yielded purpuroceratic acids A (**2094**) and B (**2095**) (*1871*). In contrast to the aforementioned simple bromotyrosines, the complex zamamistatin (**2096**) was isolated from an Okinawan collection of *Pseudoceratina purpurea* (*1872*) (revised in *1873*). This novel zamamistatin exhibits significant antibacterial activity against the marine bacterium *Rhodospirillum salexigens*, which has adhering properties (*1872*).

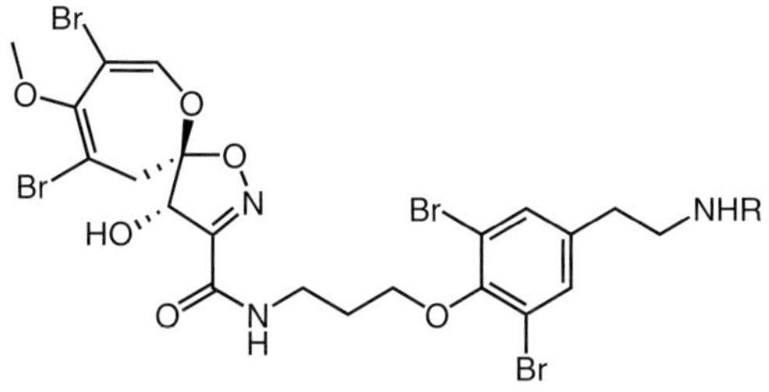

2091 R = CHO (ceratinamide A)
2092 R = CO(CH$_2$)$_{11}$CHMe$_2$ (ceratinamide B)
2093 R = Me (psammaplysin F)

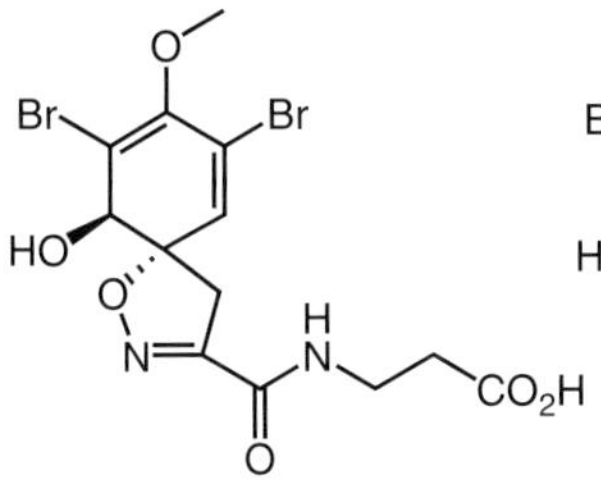

2094 (purpuroceratic acid A) 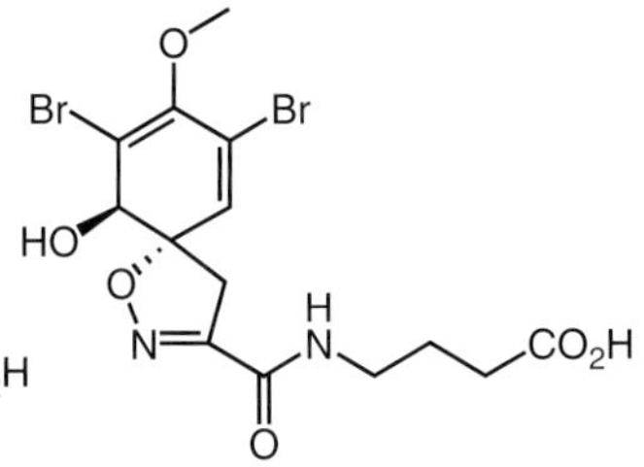   **2095** (purpuroceratic acid B)

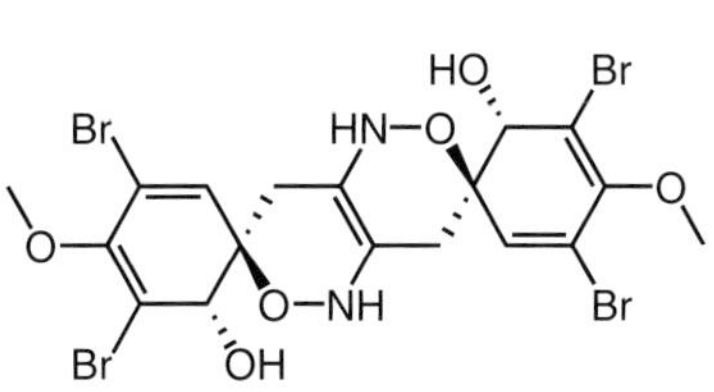

2096 (zamamistatin)

The Caribbean sponge *Aplysina insularis* has furnished **2097** (*1846*), 11-deoxyfistularin-3 (**2098**) (*1874*), and 14-oxoaerophobin-2 (**2099**) (*1875*), along with numerous previously known compounds. Similarly, the Verongida sponge *Aplysina archeri* (Fig. 3.31) contains a number of known bromotyrosine alkaloids in addition to the novel **2100** (*1876*). This Caribbean sponge has also afforded archerine (**2101**), a new metabolite that displays significant antihistamine activity (*1877*). The Mediterranean sponge *Aplysina cavernicola* has provided the new oxohomoaerothionin (**2102**) and 11-hydroxyfistularin-3 (**2103**) (*1878*). *Aplysina cauliformis*, from the Caribbean, has yielded the isomeric carbamates **2104** and **2105**, the latter of which inhibits mammalian protein synthesis and cell proliferation (*1879*).

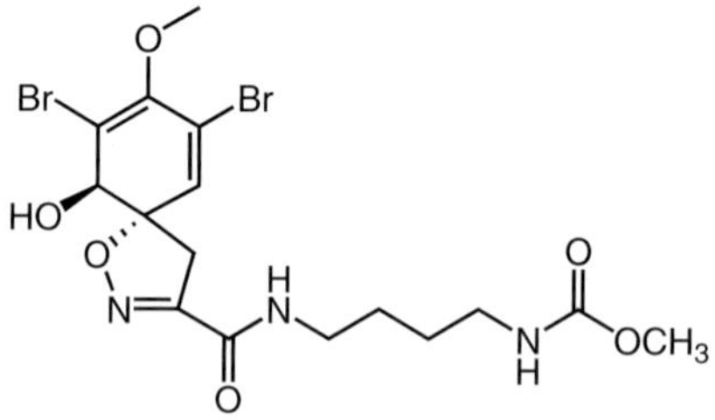

2097

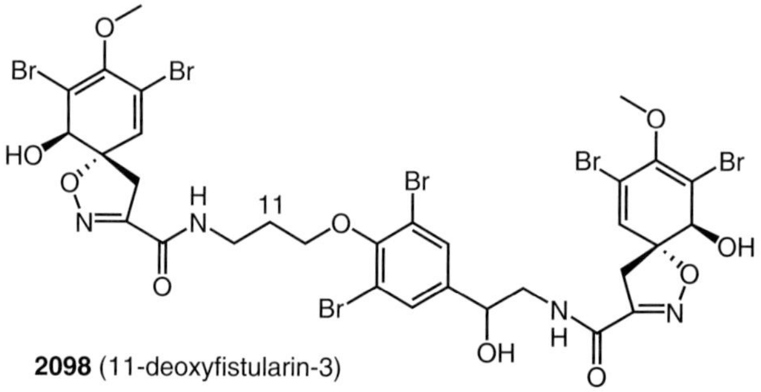

2098 (11-deoxyfistularin-3)

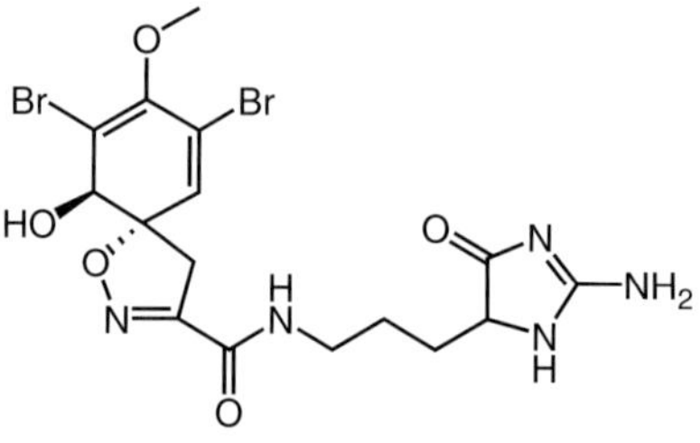

2099 (14-oxoaerophobin-2)

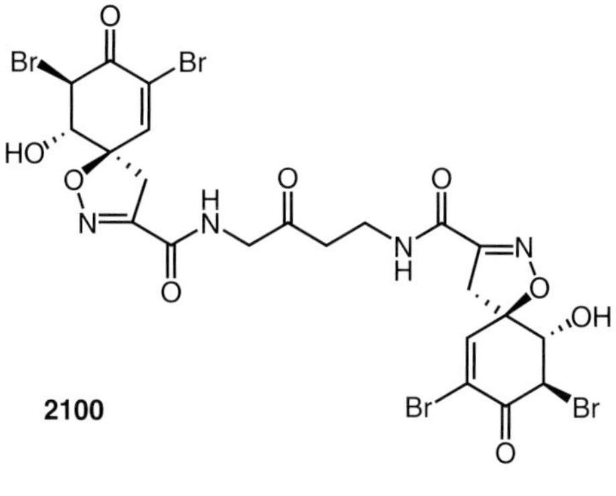

3.22 Complex Phenols

2100

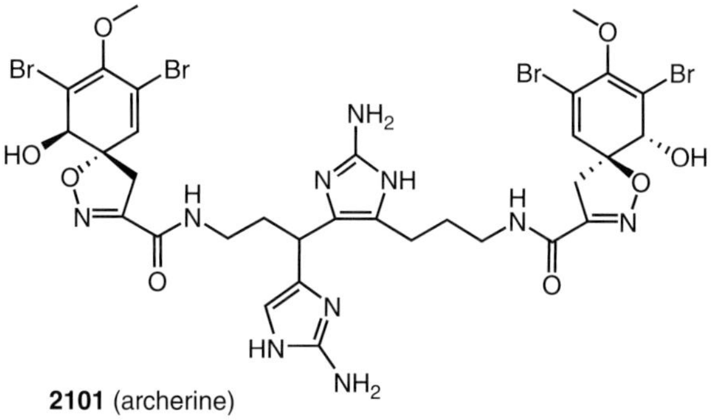

2101 (archerine)

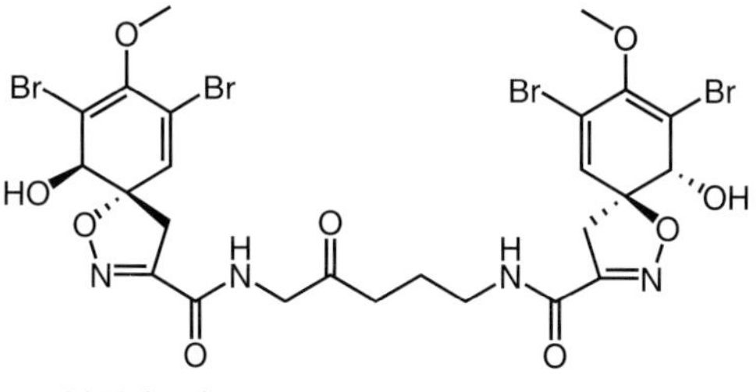

2102 (oxohomoaerothionin)

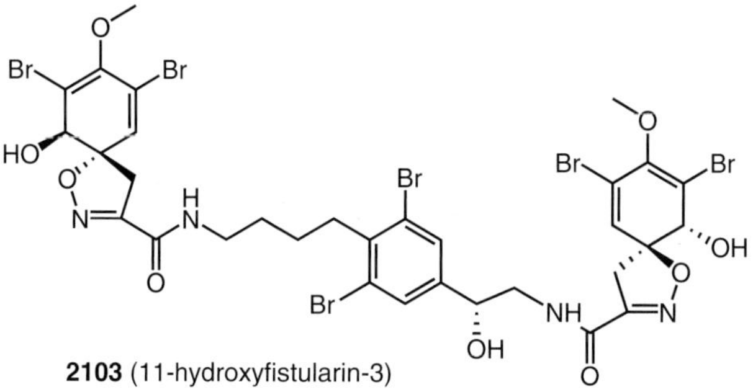

2103 (11-hydroxyfistularin-3)

2104

2105

Fig. 3.31 *Aplysina archeri*, a Caribbean sponge containing several bromotyrosines, including **2100** and archerine (**2101**) (Photo: J. R. Pawlik)

3.22 Complex Phenols

The Mexican sponge *Aplysina gerardogreeni* contains calafianin (**2106**) (*1880*) (structure revised and confirmed by total synthesis, (*1881–1883*)), the known aerothionin, and the new phenylacetic acid **2107** (*1880*). These studies confirm that calafianin (**2106**) and aerothionin have the same absolute configuration (*1883*). Whereas aerothionin displays antibacterial activity against *Mycobacterium tuberculosis*, calafianin does not (*1884*). The Brazilian sponge *Aplysina caissara* contains the new caissarines A (**2108**) and B (**2109**) (*1885*).

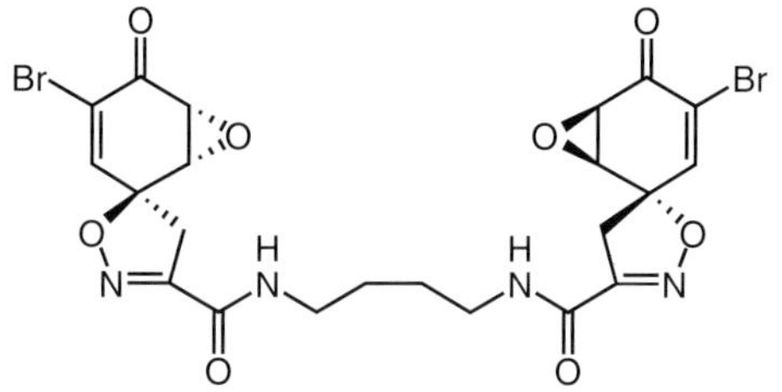

2106 (calafianin)

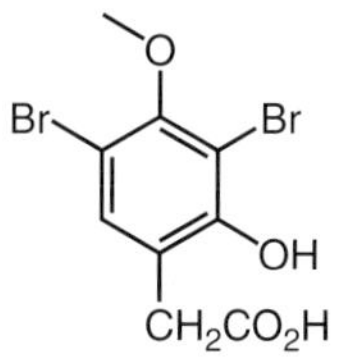

2107

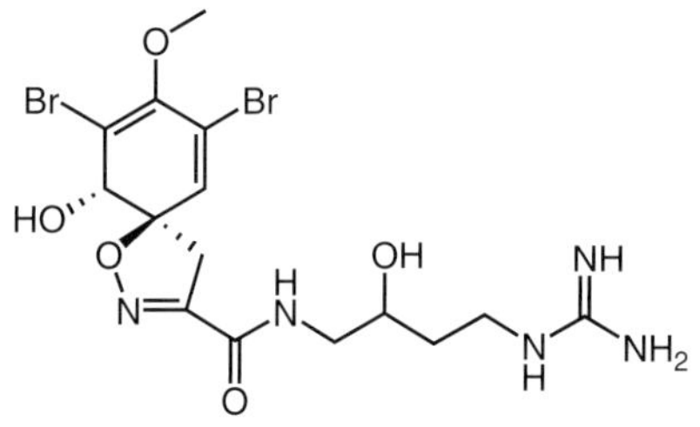

2108 (caissarine A)

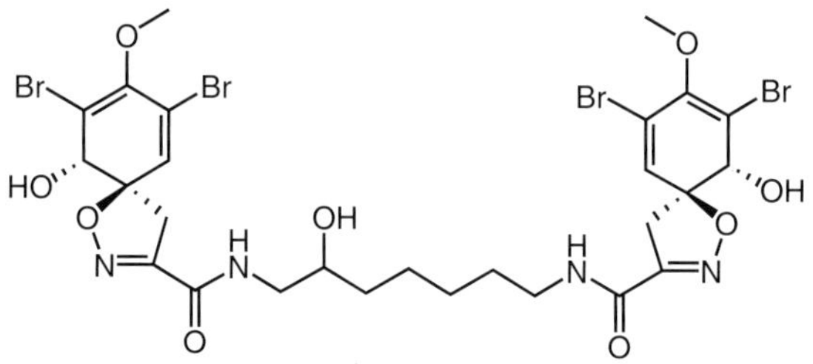

2109 (caissarine B)

A specimen of the Caribbean sponge *Aiolochroia crassa* has yielded the new *N*-methylaerophobin-2 (**2110**) (*1886*). Another collection of this sponge from Belize has afforded araplysillin III (**2111**) and hexadellin C (**2112**) (*1887*). This study established their absolute configurations as shown. A new stereoisomer, **FFF**, of fistularin-3 was reported from an Aegean Sea sample of the sponge *Verongia aerophoba* (*1888*). However, a determination of the absolute configuration of (+)-fistularin-3 and (+)-11-*epi*-fistularin-3 also reveals that **FFF** is, in fact, identical to the previously known 11-*epi*-fistularin-3 (*1892*). A Micronesian specimen of *Aplysinella* sp. has furnished (+)-aplysinillin (**2113**), which showed growth inhibition against the MCF-7 breast cancer cell line (*2649*).

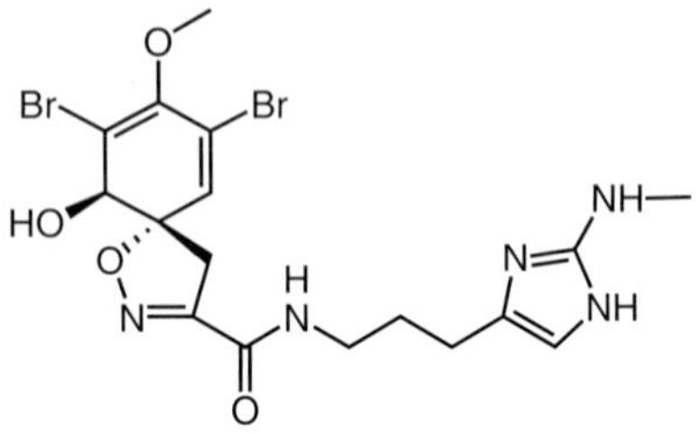

2110 (*N*-methylaerophobin-2)

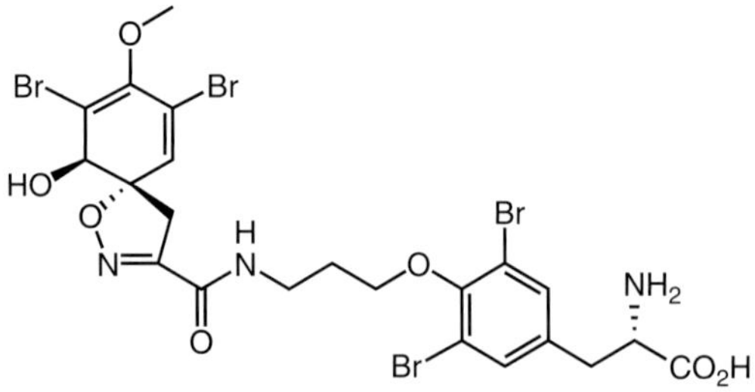

2111 (araplysillin III)

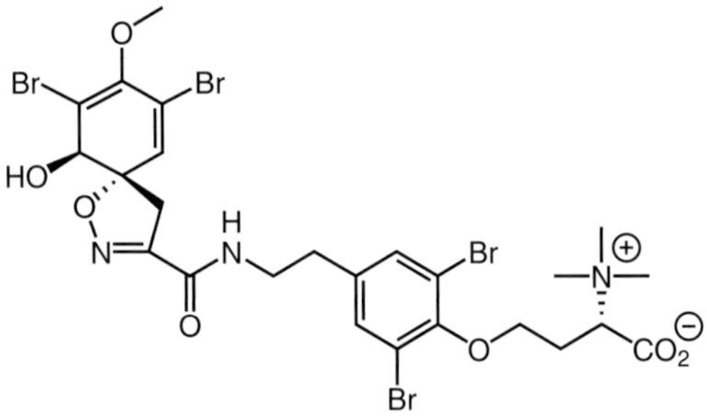

2112 (hexadellin C)

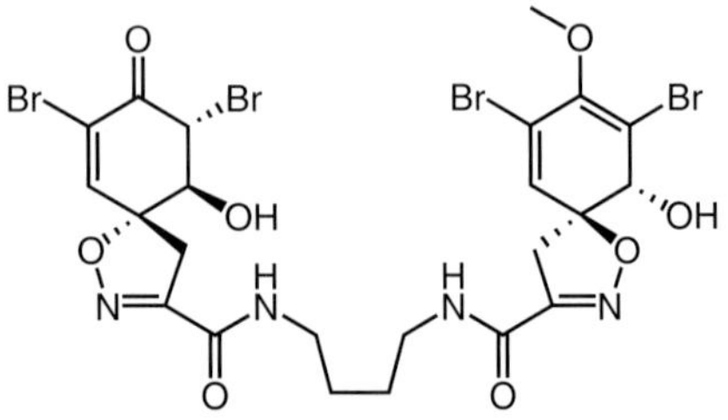

2113 (aplysinillin)

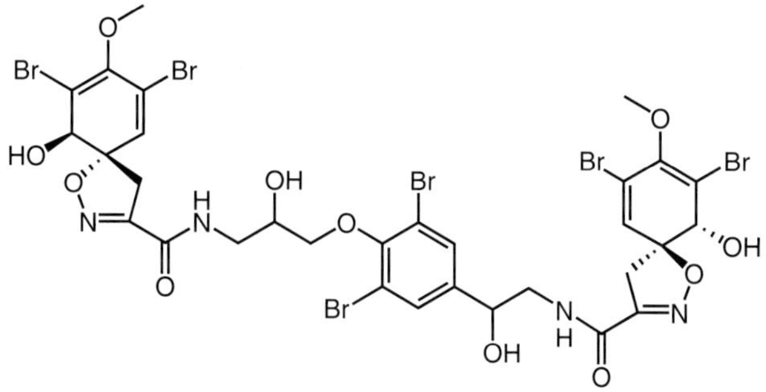

3.22 Complex Phenols

FFF (an epimer of fistularin-3)

A Great Barrier Reef *Ianthella* sp. sponge has yielded ianthesines A–D (**2114–2117**) (*1889*). Ianthesines B–D display Na,K-ATPase inhibitory activity in the range 50–440 μM. The isolation of **2118** from an Australian *Oceanapia* sp. sponge has been described, including determination of its absolute configuration (*1860*).

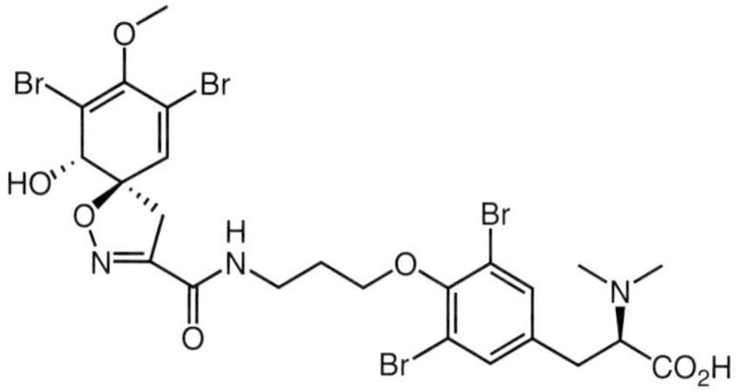

2114 (ianthesine A)

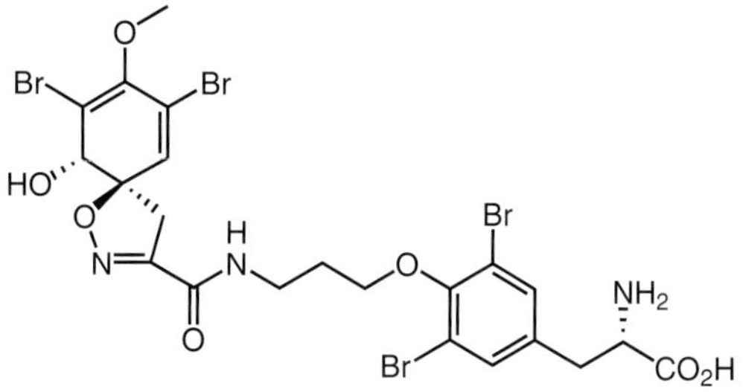

2115 (ianthesine B)

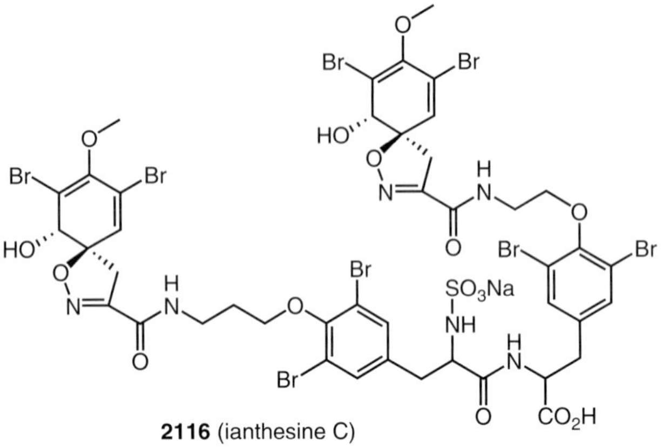

2116 (ianthesine C)

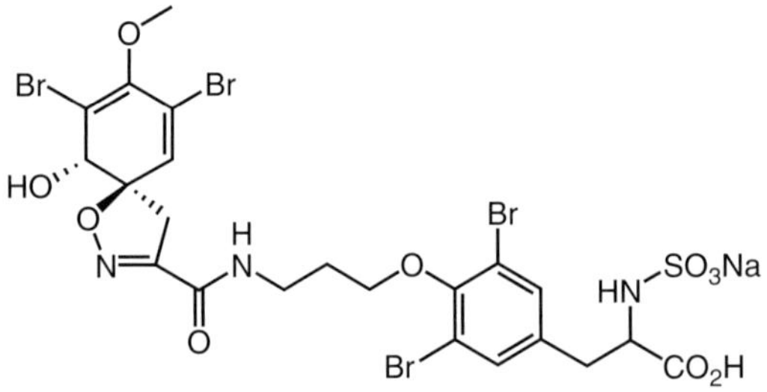

2117 (ianthesine D)

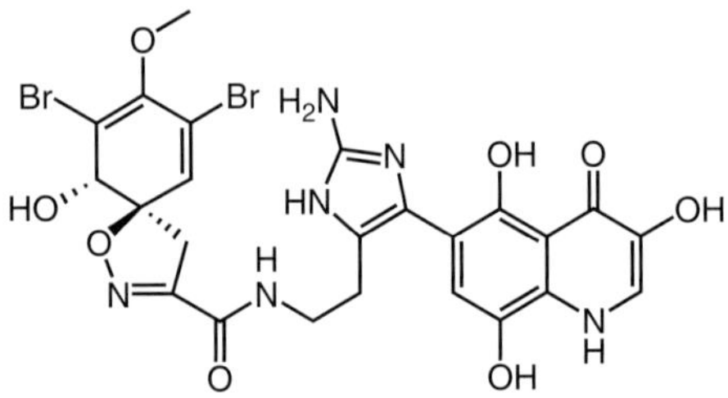

2118

A collection of the sponge *Suberea* aff. *praetensa* from the Gulf of Thailand contains 11,17-dideoxyagelorins A (**2119**) and B (**2120**) (*1890*), and the Fijian sponge *Druinella* sp. has afforded purealidin S (**2121**) (*1861*). The Malaysian crinoid *Himerometra magnipinna* has furnished (+)-12-hydroxyhomoaerothionin (**2122**) (*1891*).

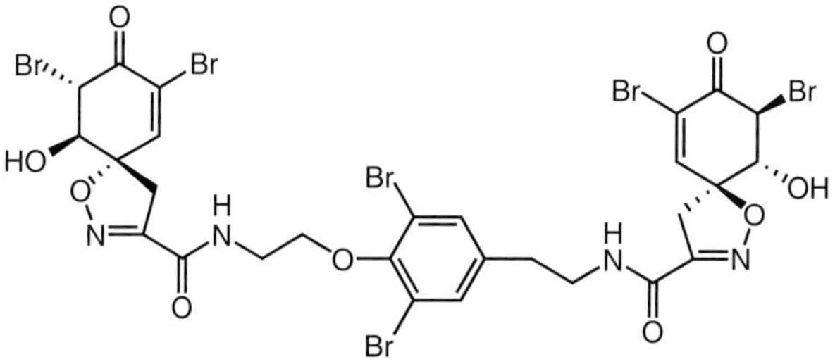

3.22 Complex Phenols

2119 (11,17-dideoxyagelorin A)

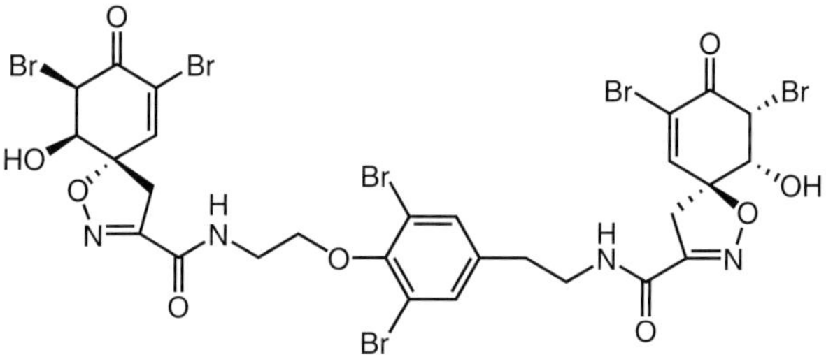

2120 (11,17-dideoxyagelorin B)

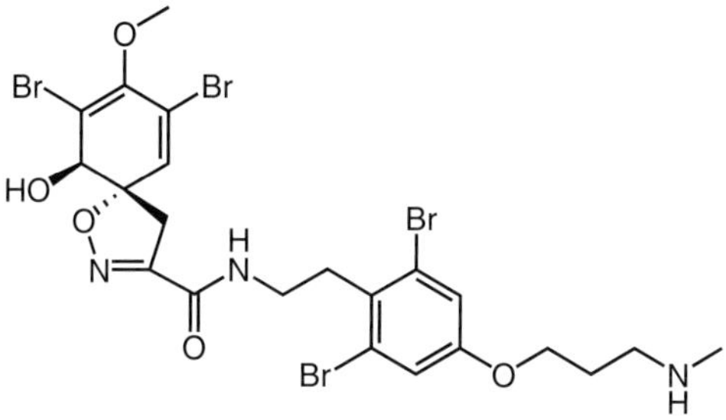

2121 (purealidin S)

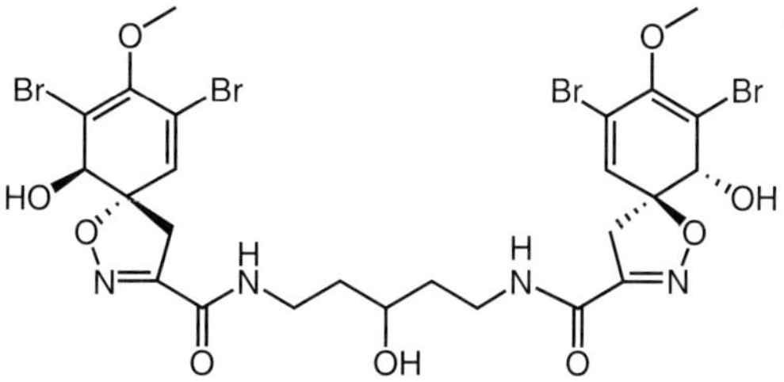

2122 (12-hydroxyhomoaerothionin)

Several syntheses of the bromotyrosine spiroisoxazolines have been described (*1893–1895*), but a full survey of these synthetic efforts cannot be covered here.

3.22.3.4 Bastadins

A study of the known bastadins-8, -10, and -12 from the Papua New Guinea sponge *Ianthella basta* has established the absolute configuration of these metabolites (*1896*). Bastadins-10 and -12 significantly inhibit the growth of several human cancer cell lines, and all three of these bastadins inhibit growth of *Staphylococcus aureus* and *Enterococcus faecalis*. Several new bastadins have been described since the first survey (*1*). A Western Australian *Ianthella basta* contains bastadin-20 (**2123**), 15,34-*O*-disulfatobastadin-7 (**2124**), and 10-*O*-sulfatobastadin-3 (**2125**) (*1897*). The Great Barrier Reef *Ianthella quadrangulata* has afforded bastadin-21 (**2126**) (*1898*).

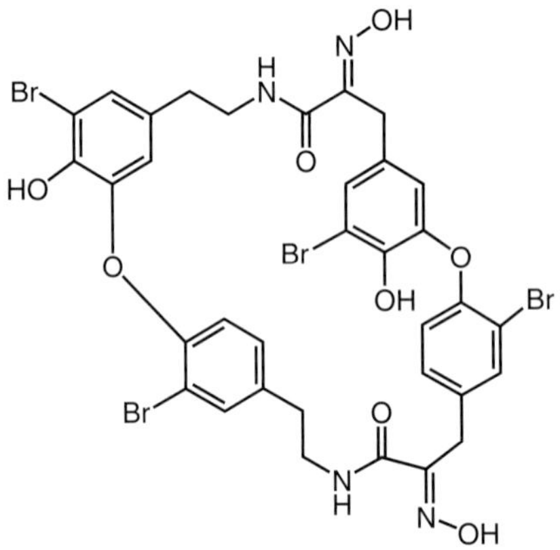

2123 (bastadin-20)

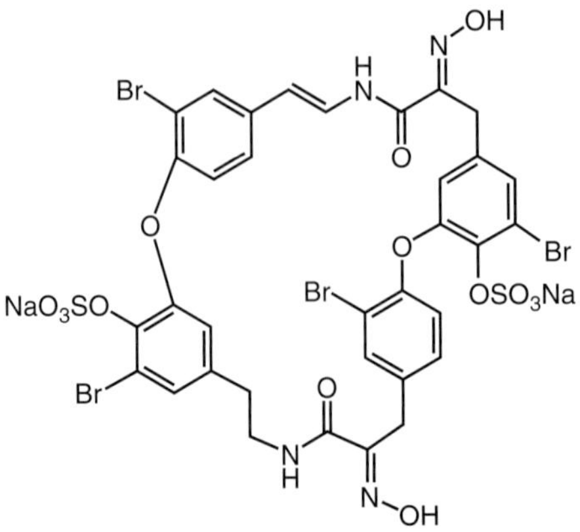

2124 (15,34-*O*-disulfatobastadin-7)

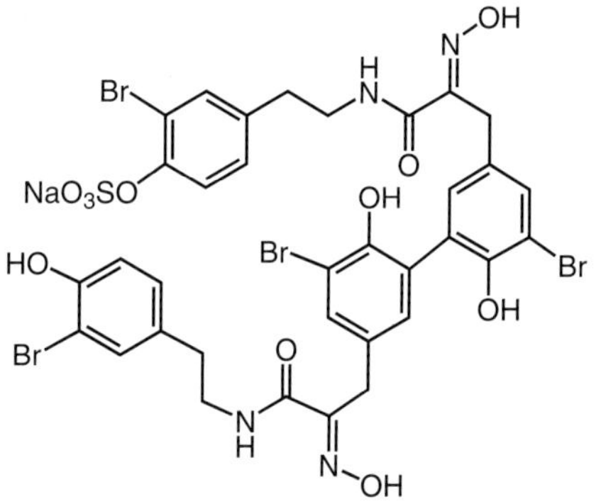

3.22 Complex Phenols

2125 (10-O-sulfatobastadin-3)

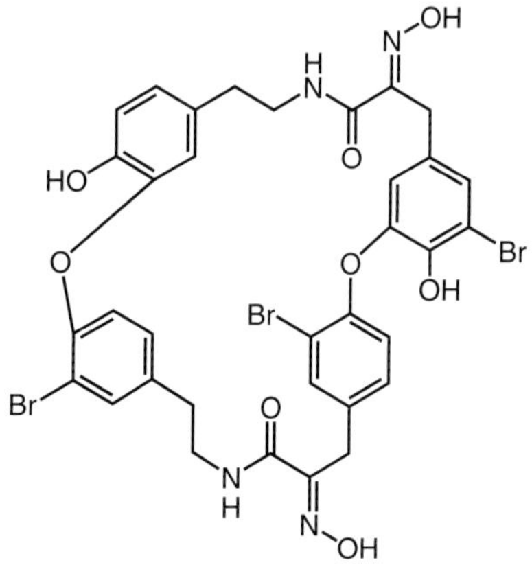

2126 (bastadin-21)

A Guam specimen of *Ianthella basta* has afforded the novel 34-*O*-sulfatobastadin-9 (**2127**) (*1855*), and the sponge *Dendrilla cactos* from India has yielded bastadins-22 (**2128**) and -23 (**2129**) (*1899*).

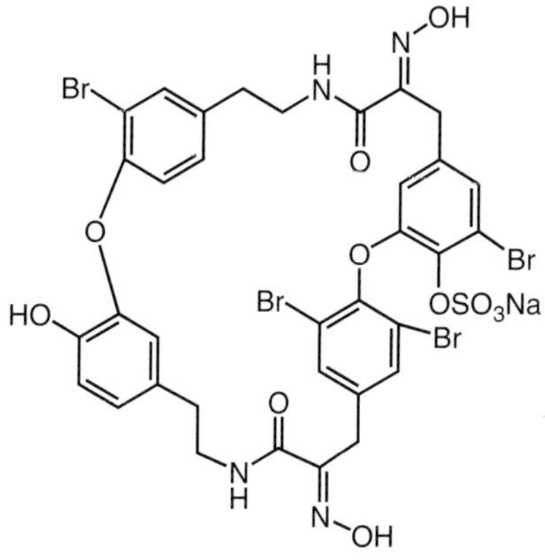

2127 (34-*O*-sulfatobastadin-9)

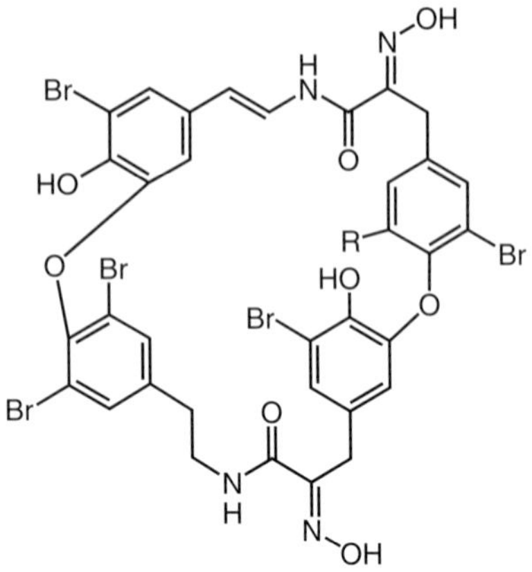

2128 R = Br (bastadin-22)
2129 R = H (bastadin-23)

These novel tetrameric bromotyrosine metabolites display a range of biological activities, including effects on calcium channels (*1900*), lipoxygenase inhibition (*1772*), tumor angiogenesis inhibition (*1901*), and endothelial cell anti-proliferation (*1902*). Syntheses of several bastadins have been accomplished (*1903, 1904*).

3.22.4 Depsides

The polyketide-derived depsides are ubiquitous lichen metabolites and some 15 chlorinated examples were cited in the first survey (*1*). A collection of the lichen *Lecanora jamesii* from England has yielded the new 2-*O*-methylsulphurellin (3,5-dichloro-4-*O*-demethylplanaic acid) (**2130**) (*1905, 1906*), while *Lecanora lividocinerea* from Spain has afforded 3,5-dichloro-2'-*O*-methylnorstenosporic acid (**2131**), 5-chloro-2'-*O*-methylanziaic acid (**2132**), and 3,5-dichloro-2'-*O*-methylnorhyperlatolic acid (**2133**) (*1906*). This latter study also confirmed the structure of **2130** by total synthesis. A Mexican sample of the lichen *Dimelaena* cf. *radiata* has yielded the new 5-chlorodivaricatic acid (**2134**) (*1907*). The wood-decaying fungus *Hypholoma fasciculare* contains **2135** (*1908*), and a marine fungus, *Emericella unguis*, which was collected from a Venezuelan mollusc (unidentified) and a medusa (*Stomolopus meliagris*; "Cannonball Jelly"), has afforded guisinol (**2136**) (*1909*). Another marine fungus, *Pestalotia* sp., found on the surface of the brown alga *Rosenvingea* sp. in the Bahamas, produces the novel antibiotic pestalone (**2137**), which displays potent antibacterial activity against both methicillin-resistant *Staphylococcus aureus* and vancomycin-resistant *Enterococcus faecium* (*1910*).

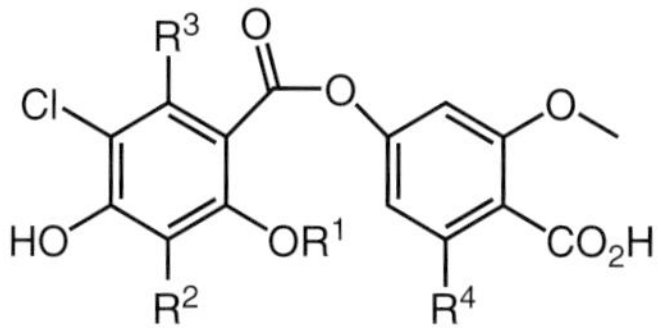

2130 $R^1 = CH_3, R^2 = Cl, R^3 = n-C_5H_{11}, R^4 = n-C_5H_{11}$
2131 $R^1 = H, \quad R^2 = Cl, R^3 = n-C_3H_7, \quad R^4 = n-C_5H_{11}$
2132 $R^1 = H, \quad R^2 = H, \: R^3 = n-C_5H_{11}, R^4 = n-C_5H_{11}$
2133 $R^1 = H, \quad R^2 = Cl, R^3 = n-C_5H_{11}, R^4 = n-C_7H_{15}$

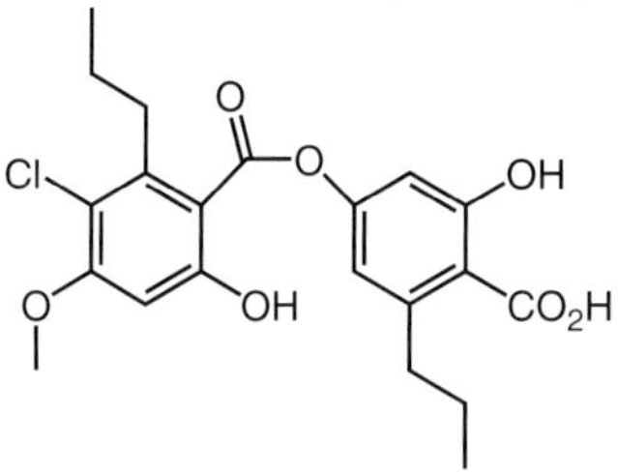

2134 (5-chlorodivaricatic acid)

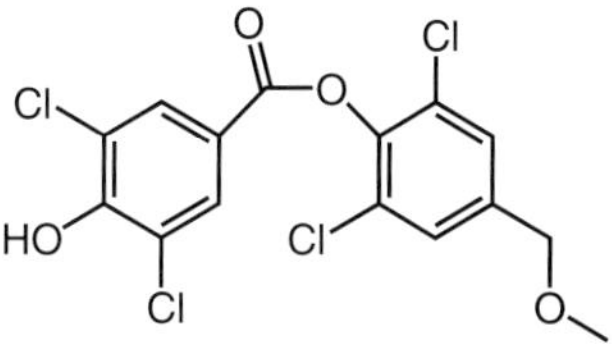

2135

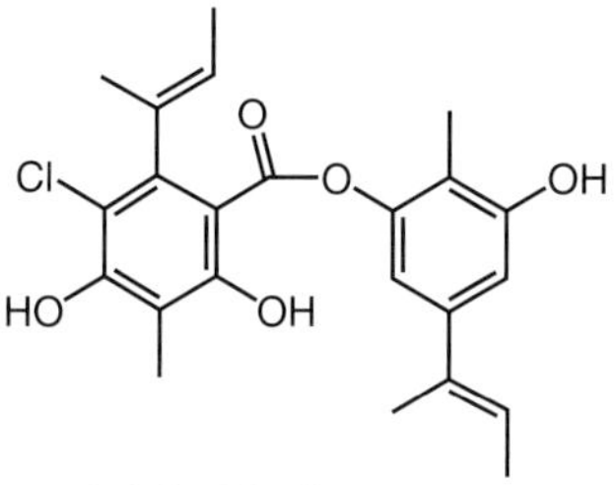

2136 (guisinol)

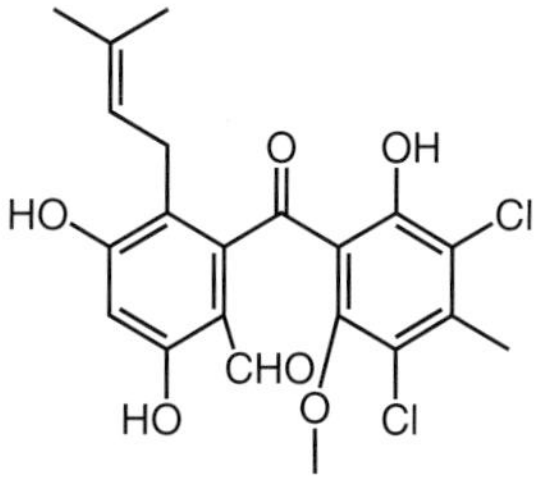

2137 (pestalone)

3.22.5 Depsidones

Depsidones are cyclized depsides that also seem to be confined to the world of lichens. Nearly 50 chlorinated depsidones were identified in the first survey (*1*). A study of *Lecanora chlarotera*, a lichen collected in southeast Scotland, contains the new norgangaleoidin (**2138**) (*1911*). Several collections of *Fulgensia fulgida* (France, Spain, and Israel) yield fulgoicin (**2139**) (*1912*), the structure of which is confirmed by total synthesis (*1913*). The related fulgidin was described earlier from this lichen (*1, 1912*), although a subsequent investigation showed, by synthesis, that fulgidin has the revised structure **GGG** (*1914, 1915*). Ironically, the incorrectly proposed structure of fulgidin, now named "isofulgidin" (**2140**), is a depsidone found in the lichens *Rinodina dissa*, *Hafellia parastata*, and *Fulgensia canariensis* (*1914*).

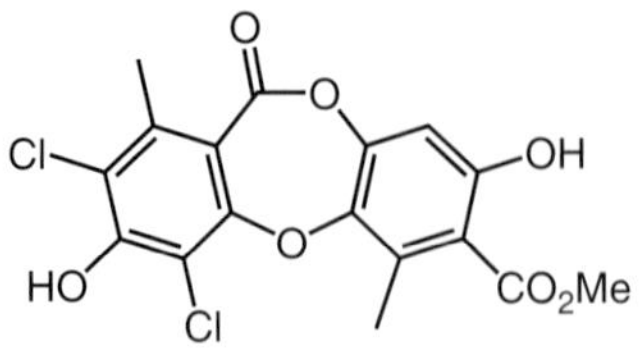

2138 (norgangaleoidin)

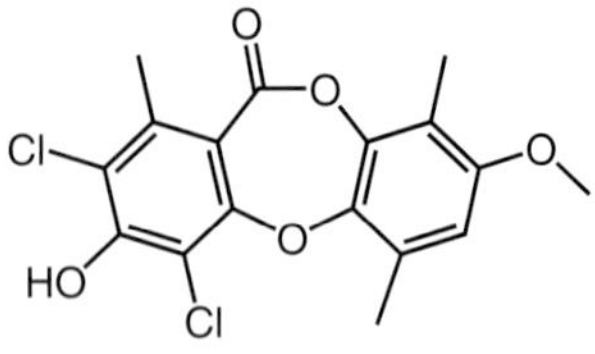

2139 (fulgoicin)

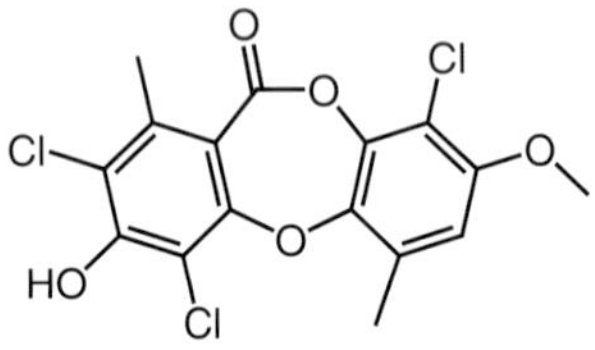

GGG (fulgidin)

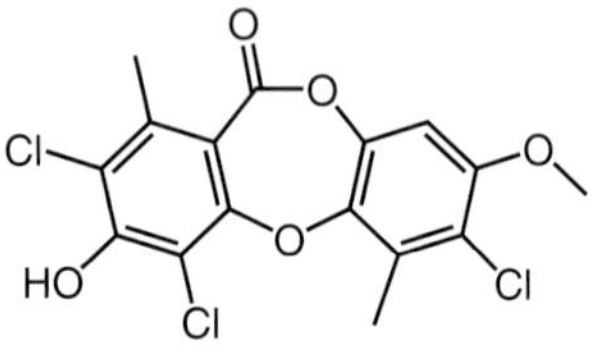

2140 (isofulgidin)

The new chlorolecideoidin (**2141**) is a minor depsidone from the lichens *Lecanora leprosa* and *Lecanora sulphurescens* (*1916*), and the novel 4-dechlorogangaleoidin (**2142**) has been identified in *Lecanora argentata* and *Lecanora californica* (*1917*). The Fijian lichen *Catarraphia dictyoplaca* has yielded cyclographin (**2143**) (*1918*). Cultures of the ascomycete *Coniochaeta tetraspora* have furnished CT-1 (**2144**) (*1919*). An unidentified *Xylaria* fungus contains maldoxone (**2145**) (*1795*), and the two brominated depsidones, acarogobiens A (**2146**) and B (**2147**), were characterized from the Central Asian lichen *Acarospora gobiensis* (Fig. 3.32) (*1920*). These compounds are the first brominated lichen metabolites to be discovered.

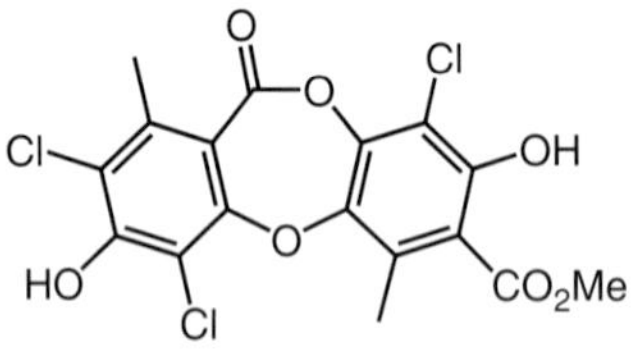

2141 (chlorolecideoidin)

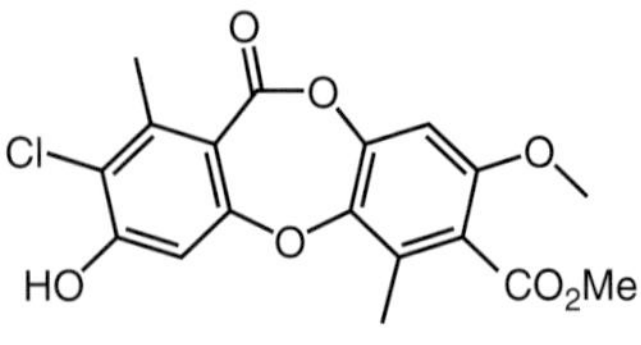

2142 (4-dechlorogangaleoidin)

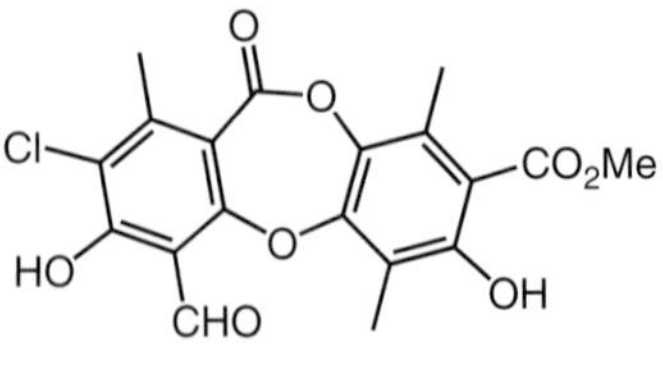

2143 (cyclographin)

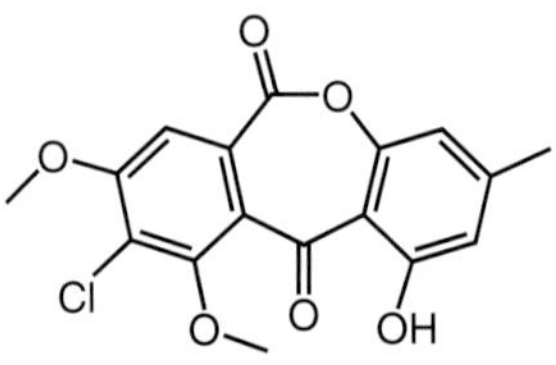

2144 (CT-1)

3.22 Complex Phenols

2145 (maldoxone)

2146 (acarogobien A)

2147 (acarogobien B)

Fig. 3.32 *Acarospora gobiensis*, a Central Asian lichen that contains the novel brominated depsidones, acarogobiens A and B (**2146** and **2147**) (Photo: T. Rezanka)

3.22.6 Xanthones

Like depsides and depsidones, most chlorinated xanthones are found in lichens, and more than 50 such compounds were described in the first survey (*1*). The lichen

Lecanora broccha contains the new 5,7-dichloro-3-*O*-methylnorlichexanthone (**2148**) (*1921*). The synthesis of 17 chlorinated xanthones, which were listed in the first survey (*1*), has been reported (*1922*). Demethylchodatin (**2149**) occurs in the lichen *Lecanora pachysoma* (*1923*), and *Byssoloma subdiscordans* has furnished the new 5,7-dichloro-6-*O*-methylnorlichexanthone (**2150**) (*1924*). *Sporopodium citrinum* contains 4-chlorolichexanthone (**2151**) and 4-chloro-3-*O*-methylnorlichexanthone (**2152**) (*1924*). The previously known vinetorin (*1*) is the first chloroxanthone to be isolated from a higher plant, *Hypericum ascyron* (*1925*). The new xanthone **2153** has been characterized from the Italian plant *Polygala vulgaris* (*1926*).

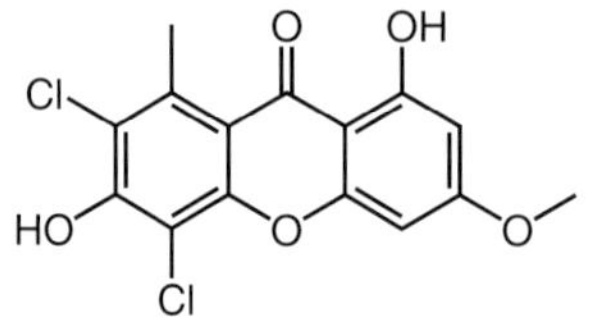

2148 (5,7-dichloro-3-*O*-methylnorlichexanthone)

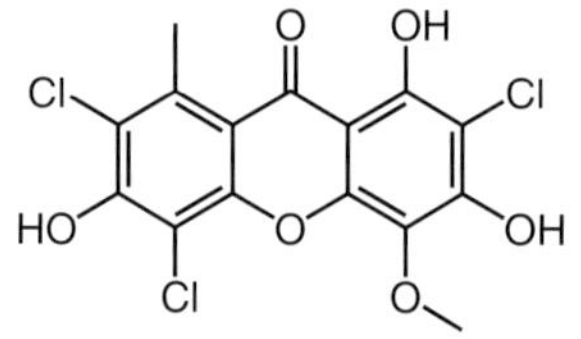

2149 (demethylchodatin)

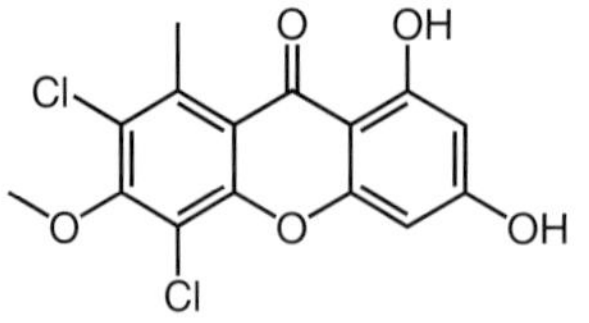

2150 (5,7-dichloro-6-*O*-methylnorlichexanthone)

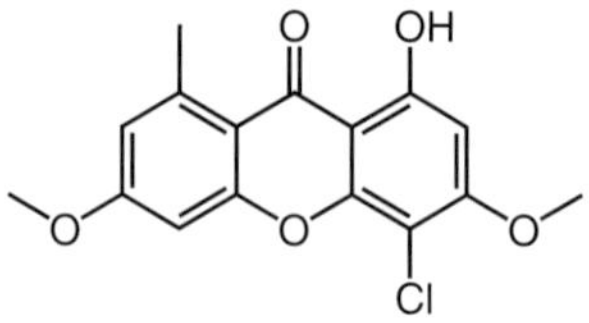

2151 (4-chlorolichexanthone)

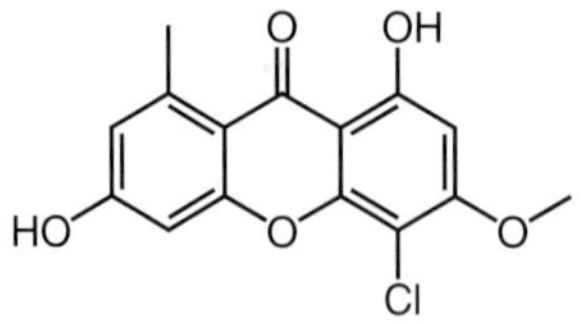

2152 (4-chloro-3-*O*-methylnorlichexanthone)

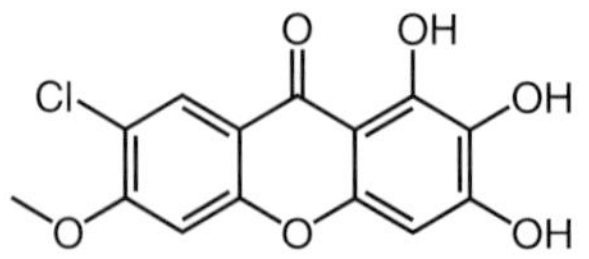

2153

The structures of the previously isolated beticolins 2 and 4 (*1*) have now been confirmed by X-ray crystallography (*1927*). A new isolate from the fungus *Cercospora beticola*, which is a highly destructive disease of sugar beets, is beticolin 0 (**2154**) (*1928*). The polycyclic xanthone Sch 54445 (**2155**) is produced by an *Actinoplanes* species, and is a very active antifungal agent (MIC, 0.00038 µg mL^{-1}) (*1929*). Xantholipin (**2156**) is a related substance from a *Streptomyces* sp. (*1930*), and is structurally similar to the previously known lysolipins from *Streptomyces violaceoniger* (*1*).

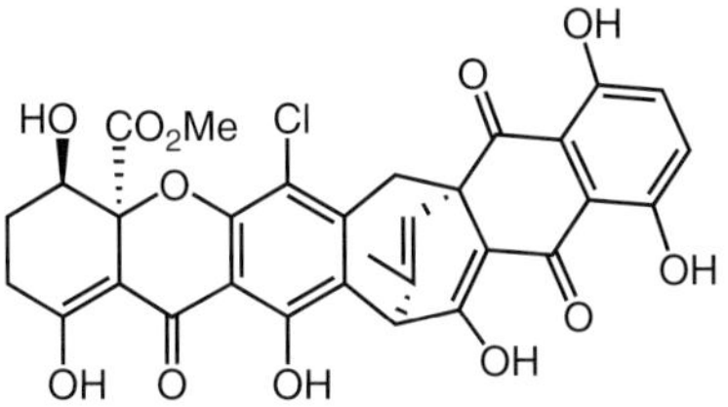

3.22 Complex Phenols

2154 (beticolin 0)

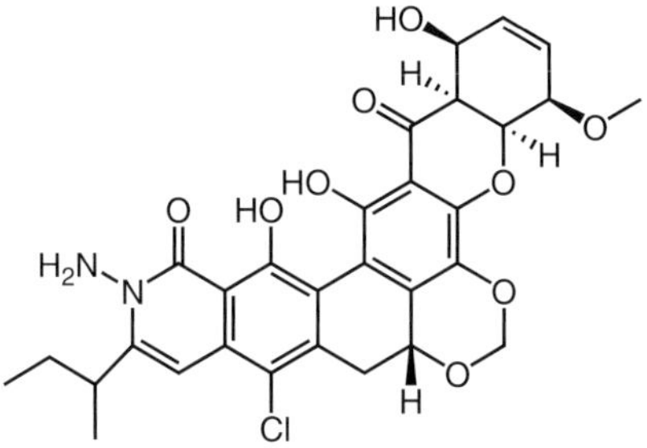

2155 (Sch 54445)

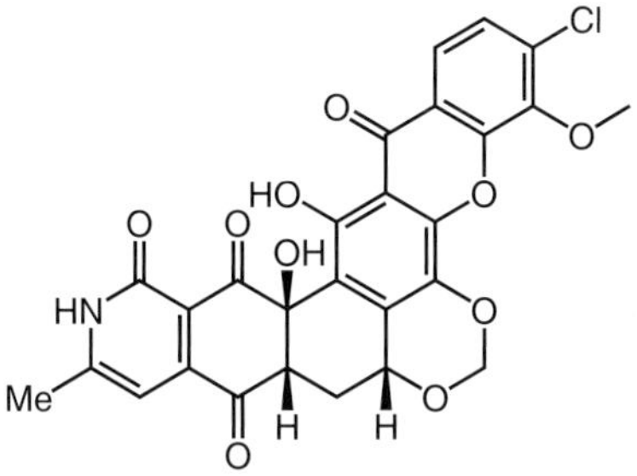

2156 (xantholipin)

3.22.7 *Anthraquinones and Related Compounds*

Most of the previously identified 25 chlorinated anthraquinones are found in lichen and fungi (*1*). The newly discovered examples have a wider range of sources. Studies of the lichen *Nephroma laevigatum* from the British Columbia coast have identified the new anthraquinone, 7-chloro-1-*O*-methyl-ω-hydroxyemodin (**2157**), and the two novel hypericins, 7,7′-dichlorohypericin (**2158**) and 2,2′,7,7′-tetrachlorohypericin (**2159**) (*1931*), as well as 5-chloroemodin (**2160**), 5-chloro-1-*O*-methyl-ω-hydroxyemodin (**2161**), and 5-chloro-ω-hydroxyemodin (**2162**) (*1932*). In addition to containing several known chlorinated anthraquinones, the Scandinavian fungus *Dermocybe sanguinea* has afforded the new 5,7-dichloroendocrocin (**2163**) (*1933*). The novel tetracyclic anthraquinones topyrones A (**2164**) and B (**2165**), and two non-chlorinated analogs, were isolated from cultures of the fungi *Phoma* sp. and *Penicillium* sp. (*1934, 1935*). These compounds are topoisomerase I inhibitors and topopyrone B has activity comparable to that of camptothecin. Topopyrone B is also potent against herpes virus VZV, and is 24 times more active than acyclovir. Syntheses of topopyrones have been described (*1936, 2650*).

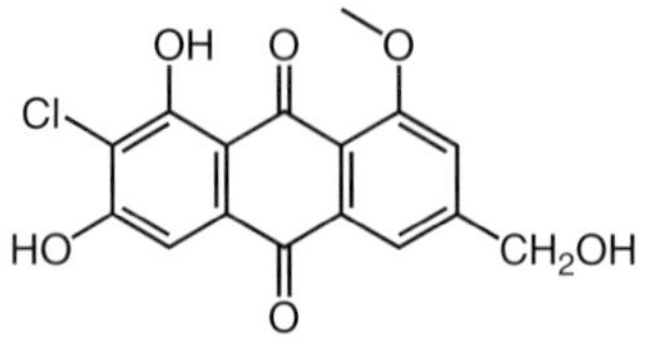

2157

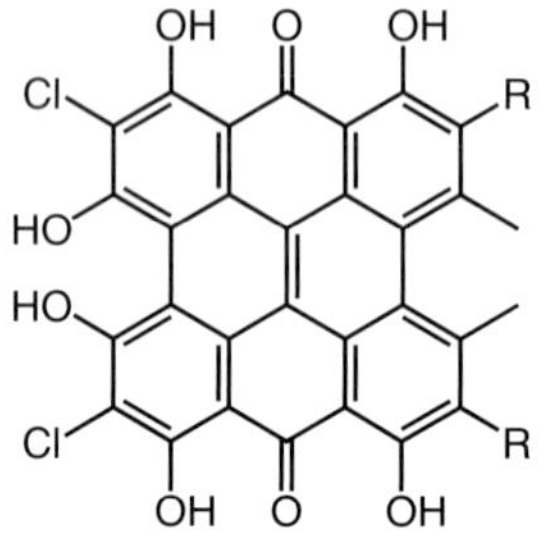

2158 R = H (7,7'-dichlorohypericin)
2159 R = Cl (2,2',7,7'-tetrachlorohypericin)

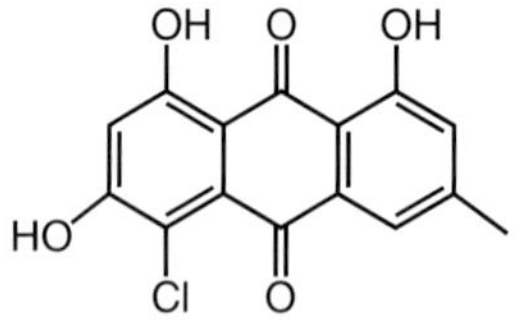

2160

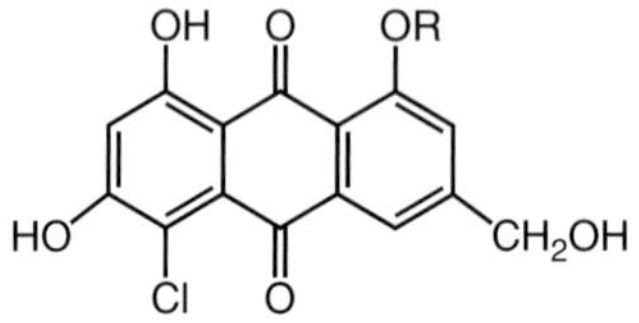

2161 (R = H)
2162 (R = Me)

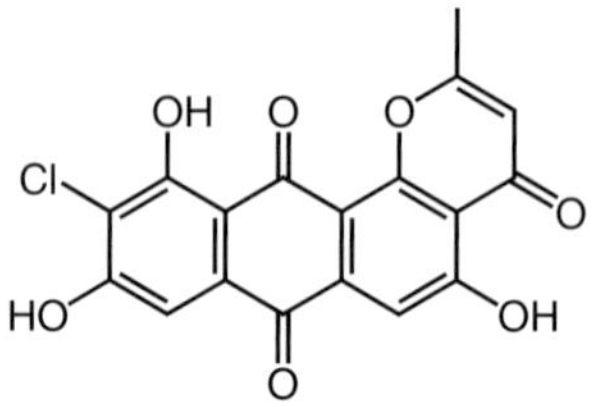

2164 (topopyrone A)

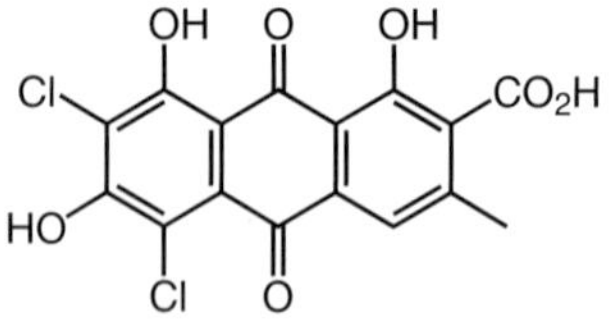

2163 (5,7-dichloroendocrocin)

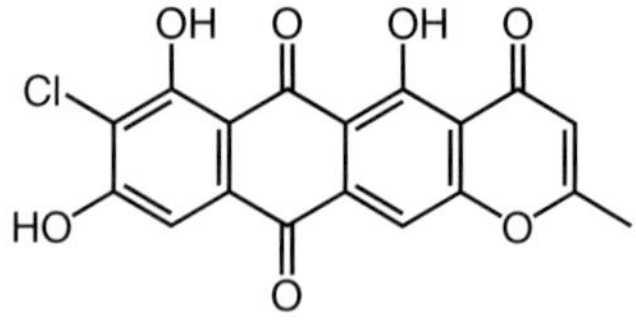

2165 (topopyrone B)

The *Streptomyces* strain that produces celastramycin A (**1212**) has also yielded celastramycin B (**2166**) (*1225*). Another *Streptomyces* sp. has afforded bischloroanthrabenzoxocinone ((−)-BABX) (**2167**), which has antibacterial activity and inhibits ligand-binding activity of liver X receptors (*1937*). An example of a rare chlorinated anthraquinone is anthrasesamone C (**2168**), which was characterized in the Japanese plant *Sesamum indicum* (*1938*). The angucycline-type marmycin B

3.22 Complex Phenols

(**2169**) was isolated from cultures of a *Streptomyces* strain, along with the dechloro marmycin A, which was more cytotoxic against several human cancer cell lines than marmycin B (*1941*). A Gram-positive strain of a *Bacillus* bacterium from Californian soil has yielded the novel fluorescent pyrene, chlorxanthomycin (**2170**), which has selective antibiotic activity (*1942*). The antitumor antibiotic BE-19412A (**2171**) is produced by a Streptomycete (*1943*).

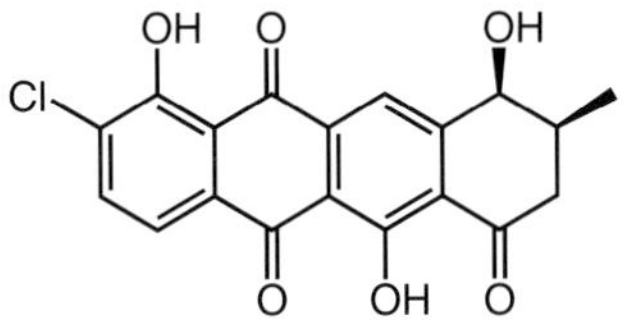

2166 (celastramycin B)

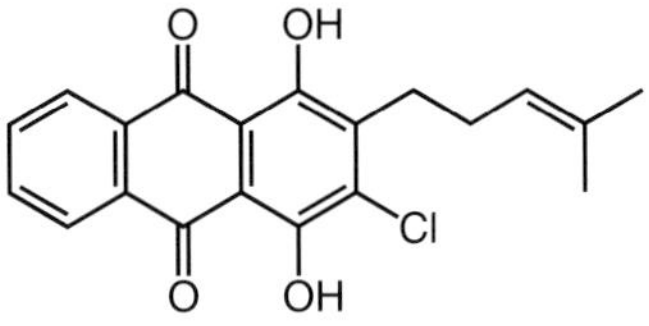

2168 (anthrasesamone C)

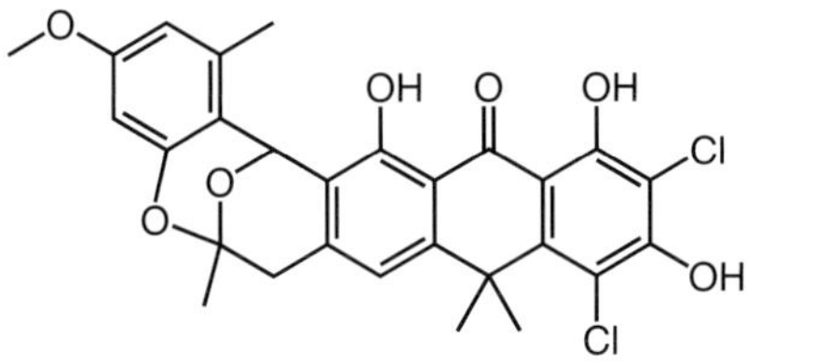

2167 (bischloroanthrabenzoxocinone; (–)-BABX)

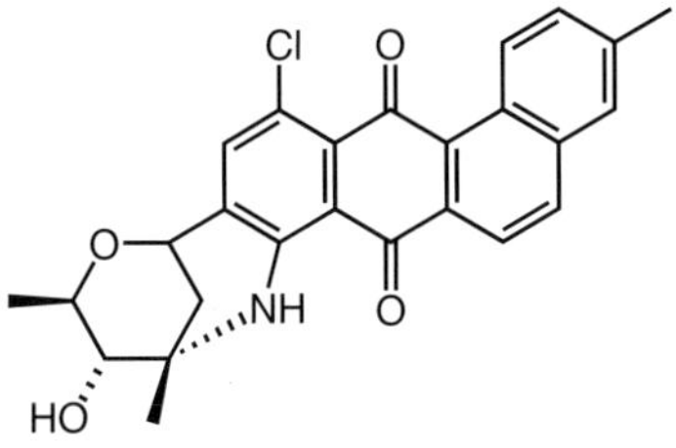

2169 (marmycin B)

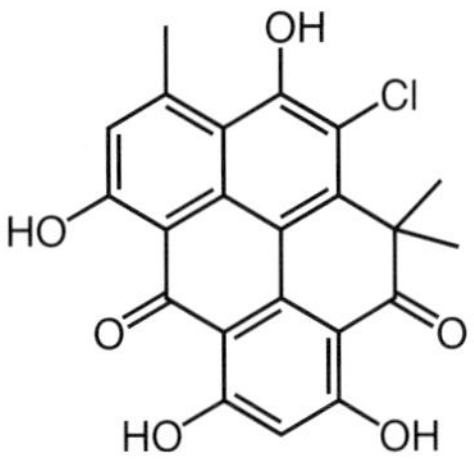

2170 (chlorxanthomycin)

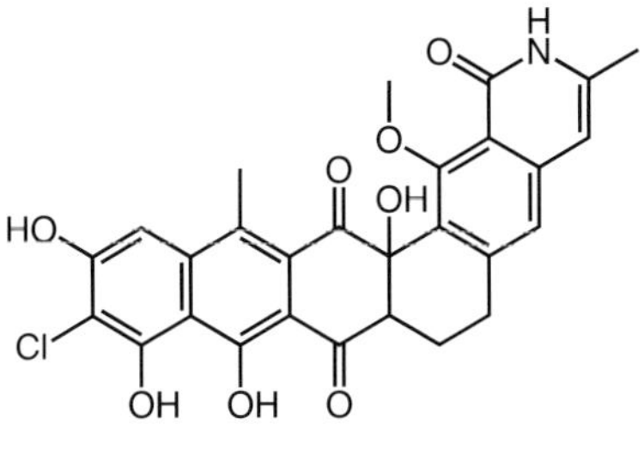

2171 (BE-19412A)

3.22.8 Griseofulvin and Related Compounds

One of the earliest recognized naturally occurring organohalogen compounds is griseofulvin (*1*), and this fungal metabolite is still used clinically to treat tinea pedis (athlete's foot) and, more recently, may have anticancer activity (*1944, 1945*). The new spirocyclohexadienone, maldoxin (**2172**), was isolated from a member of the fungus genus *Xylaria* (*1795*). A fermentation broth of *Aspergillus* sp. has afforded Sch 202596 (**2173**), which displays inhibitory activity in the galanin receptor GALR1 assay (*1946*). This fungus was isolated from the tailing piles of an abandoned uranium mine in California.

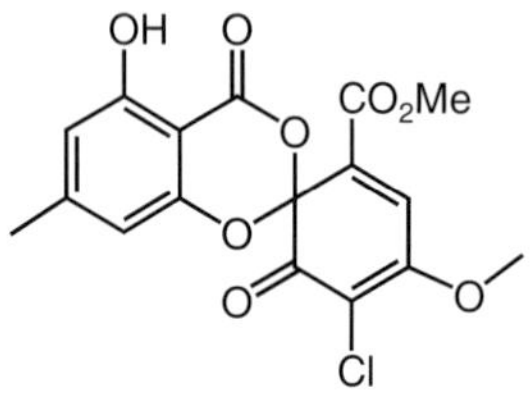

2172 (maldoxin) 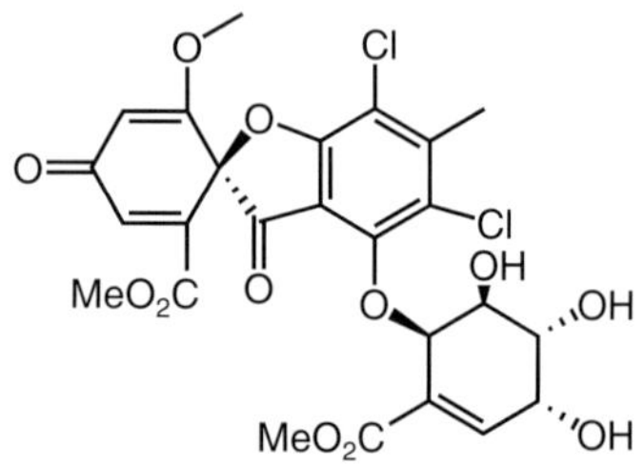 **2173** (Sch 202596)

3.22.9 Miscellaneous Fungal Metabolites and Other Complex Phenols

A large number of natural organohalogen compounds, mainly found in fungi, do not fit into the structural categories defined earlier and are presented here. Some new analogs of the well-known prenyl-phenol antibiotic ascochlorin have been reported. The literature on this class of fungal metabolites is confusing since several of the same compounds have been named differently in separate investigations. Thus, ascochlorin is also known as LL-Z1272γ and ilicicolin D, and the known cylindrochlorin (= ilicicolin E) was isolated more than 20 years after its initial discovery and named as 8′,9′-dehydroascochlorin from a *Verticillium* sp. (*1949*). Ilicicolin E is also found in the canker disease phytopathogenic fungus *Nectria galligena* (*1950*). Cylindrol A$_4$ (**2174**) was isolated from *Cylindrocarpon lucidum* (*1951*) and is related to the known corresponding acetate, chloronectrin (*1*). The insect pathogenic fungus *Verticillium hemipterigenum* from Thailand has yielded vertihemipterin A (**2175**), a glucoside of the previously known aglycone, along with 8′-hydroxyascochlorin (**2176**) (*1952*). Ascochlorin derivatives display significant biological activity such as antidiabetes (*1953, 1954*). The first synthesis of (−)-ascochlorin has been reported (*1955*).

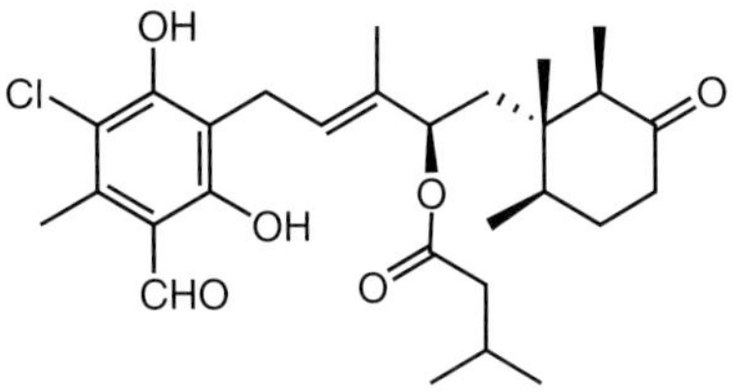

2174 (cylindrol A$_4$)

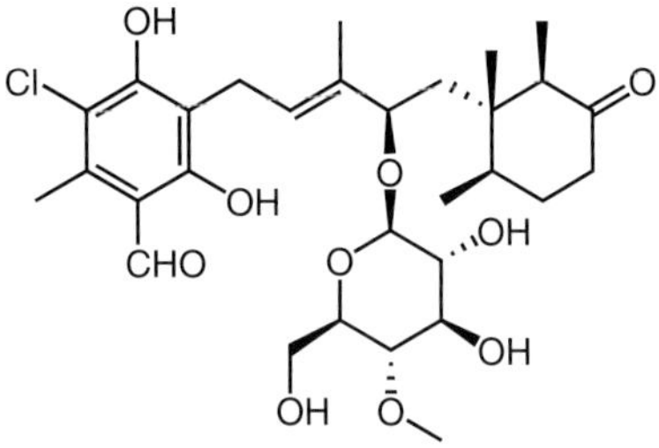

2175 (vertihemipterin A)

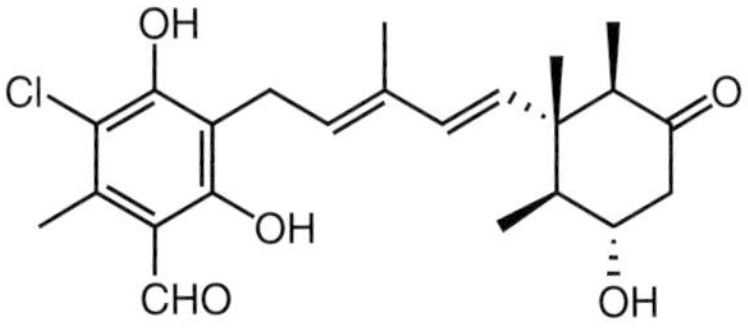

2176 (8'-hydroxyascochlorin)

The previously described antifungal strobilurin B, which has been synthesized (*1956*), is joined in kind by the discovery of oudemansin B (**2177**) from *Xerula longipes* and *Xerula melantricha* (*1957*). This class of substances holds promise for the development of new fungicides (*1958*). Like ascochlorin, the previously known aspirochlorine (*1*) has been frequently isolated (= A30641 = oryzachlorine), and displays potent antifungal activity (*1959*). The new analog tetrathioaspirochlorine (**2178**) and possibly the trisulfide derivative (not counted here) are found in extracts of *Aspergillus flavus* along with aspirochlorine (*1960*).

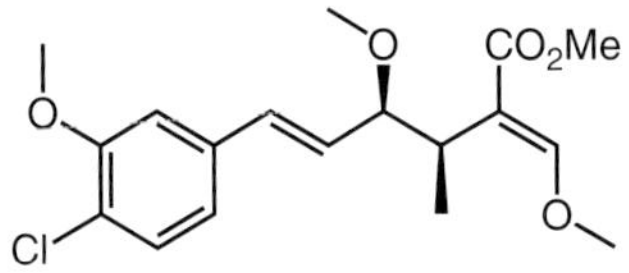

2177 (oudemansin B) 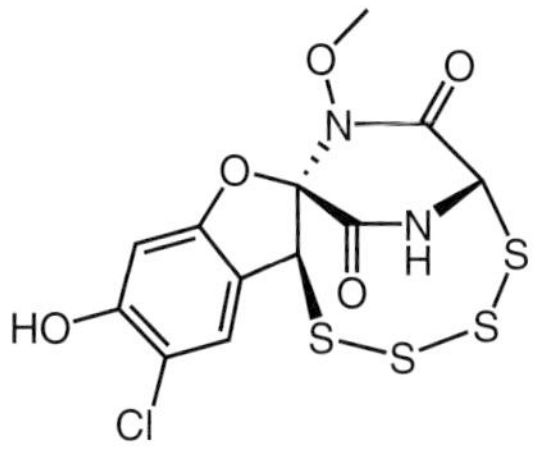   **2178** (tetrathioaspirochlorine)

As reported in the first survey, several fungi produce novel cyclobutane-containing metabolites, such as armillaridin, melleolides, and melledonals (*1*). Newly isolated members of this class include arnamiol (**2179**) from *Armillaria mellea* (*1961*), *Armillaria ostoyae* (*1962*), *Armillaria tabescens*, *Armillaria monadelpha*, *Armillaria gallica*, and *Armillaria cepestipes* (*1963*), armellide B (**2180**), melleolides I (**2181**) and J (**2182**) from *Armillaria novae-zelandiae* (*1964*), and melledonal D (**2183**) from *Clitocybe elegans* (*1965*). Melleolide J (**2182**) may be identical to armillarikin isolated from *Armillaria mellea* (*1966*). The pathogenic fungus *Armillaria novae-zelandiae* has also afforded 6′-chloro-10α-hydroxymelleolide (**2184**) (*1967*). Melleolides K (**2185**), L, and M (**2186**) were isolated from *Armillariella mellea* (*1968*), but melleolide L appears to be the same as **2184**.

2179 (arnamiol)

2181 (melleolide I)

2180 (armellide B)

2182 (melleolide J)

2183 (melledonal D)

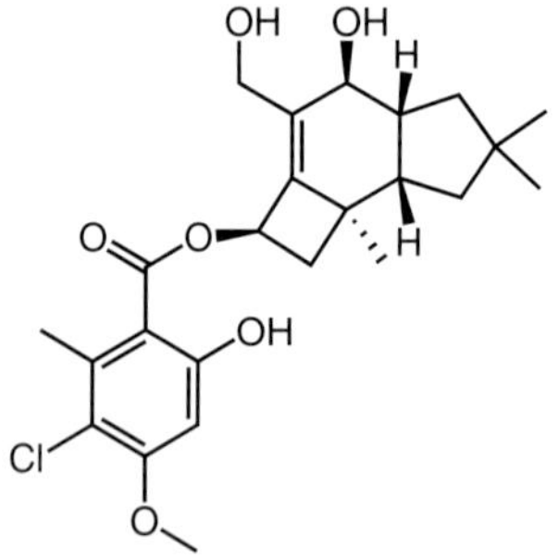

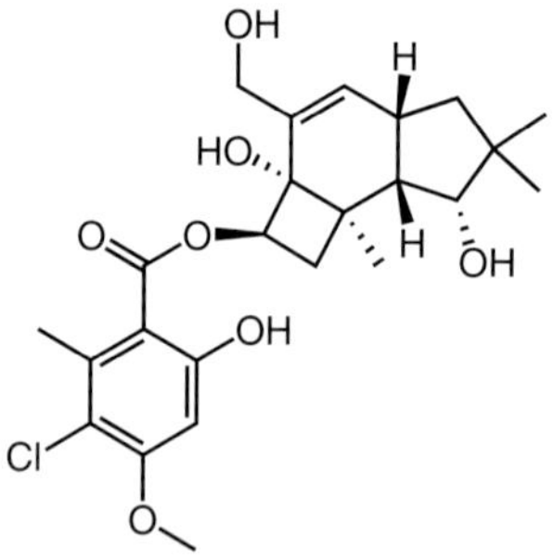

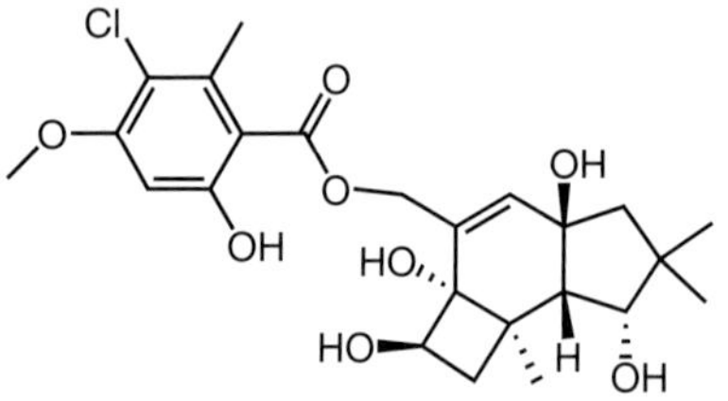

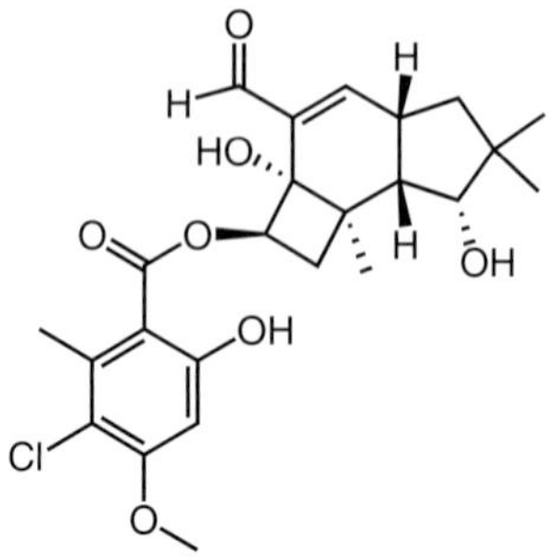

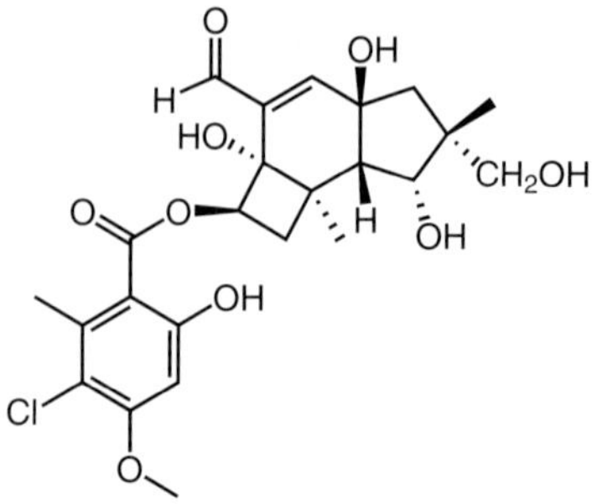

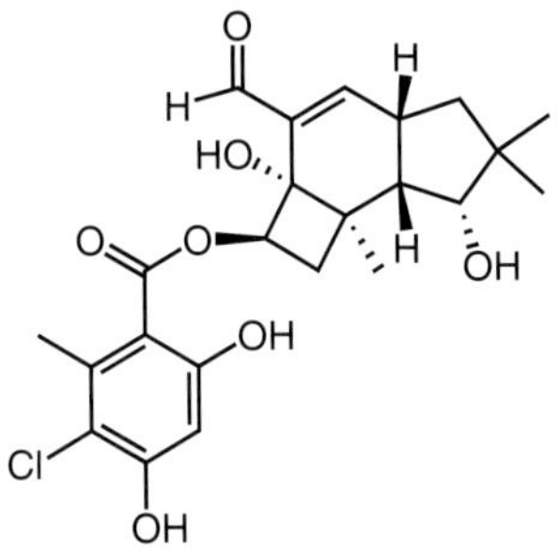

3.22 Complex Phenols

2184

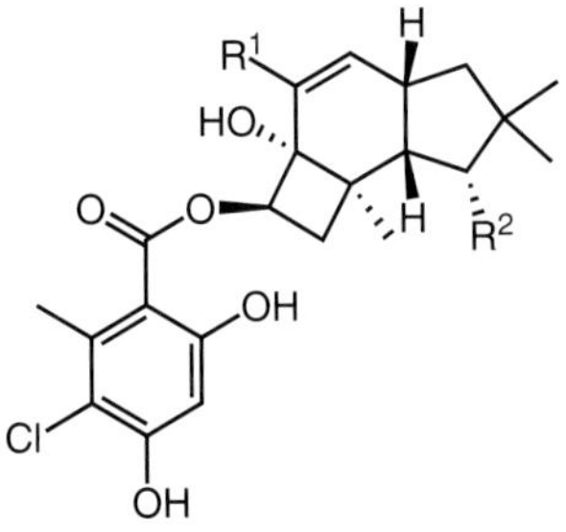

2185 R¹ = CHO, R² = H (melleolide K)
2186 R¹ = CH₂OH, R² = OH (melleolide M)

The fungus *Emericella falconensis* is the source of several azaphilones, the falconensins (*1*), and the new falconensin E (**2187**) has been identified in a Venezuelan soil sample containing this fungus (*1969*). The absolute configuration of the falconensins was established in this study. This fungus and *Emericella fruticulosa* have furnished the new falconensins K (**2188**), L (**2190**), M (**2189**), and N (**2191**) (*1970*). The culture broth of an *Amycolatopsis* strain produces the chlorine-containing epoxyquinomicins A and D (**2192**, **2193**) (*1971–1975*). The non-chlorinated epoxyquinomicins B and C are more active than A and D in inhibiting rat embryo histidine decarboxylase (*1975*). An unidentified *Coniothyrium* fungus has furnished coniothyriomycin (**2194**), which shows fungicidal and herbicidal activity (*1976, 1977*). This metabolite is related to **1776** from a *Xylaria* fungus. Cultures of *Actinoplanes* sp. yield BE-40665D (**2195**), a novel brominated antibacterial antibiotic (*1978*).

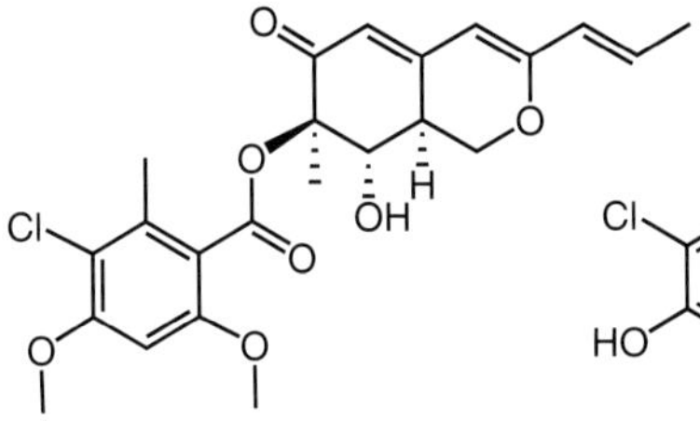

2187 (falconensin E)

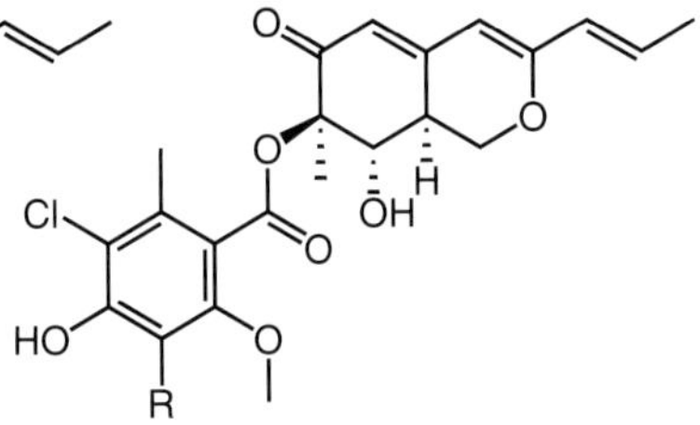

2188 R = H (falconensin K)
2189 R = Cl (falconensin M)

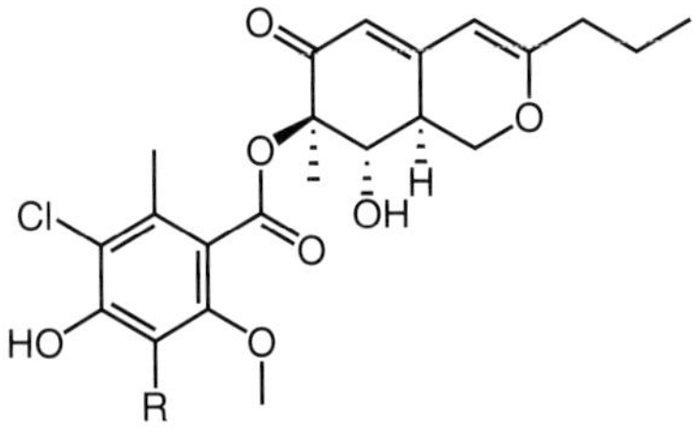

2190 R = H (falconensin L)
2191 R = Cl (falconensin N)

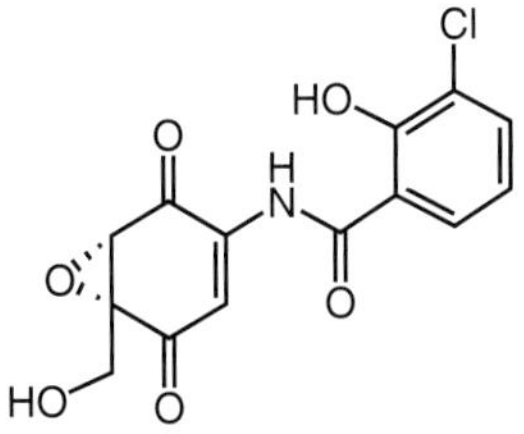

2192 (epoxyquinomicin A)

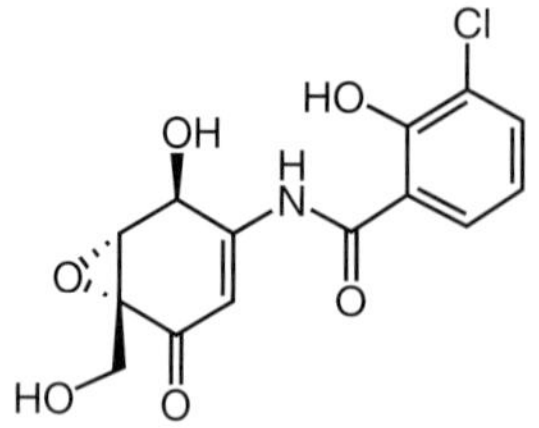

2193 (epoxyquinomicin D)

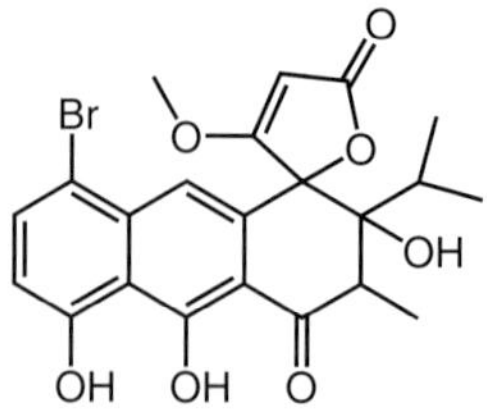

2195 (BE-40665D)

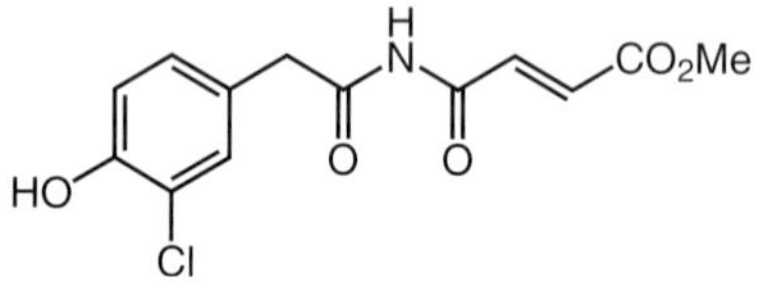

2194 (coniothyriomycin)

The tropical fungus *Scleroderma sinnamariense* has afforded methyl 2′,5′-dichloro-4,4′-di-*O*-methylatromentate (**2196**) (*1979*), and the related pulvinic acid derivative methyl 3′,5′-dichloro-4,4′-di-*O*-methylatromentate (**2197**) was isolated from the fruiting body of a *Scleroderma* sp. ("poison puff ball") (*1980*). A New Zealand *Chamonixia pachydermis* has yielded pachydermin (**2198**) (*1981*).

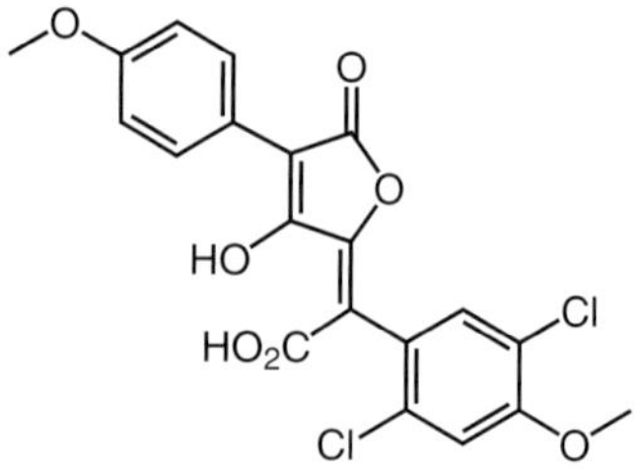

2196

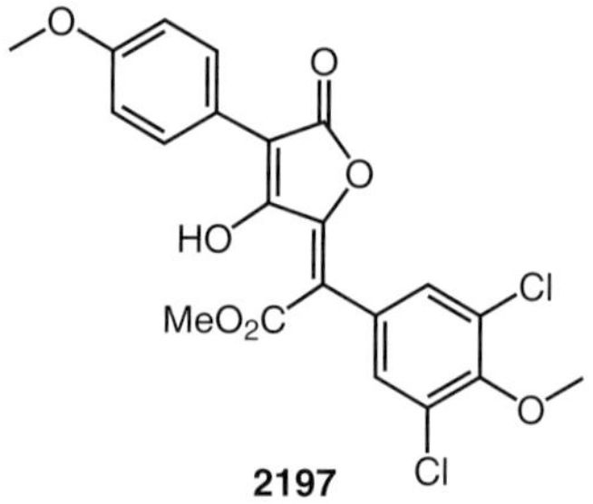

2197

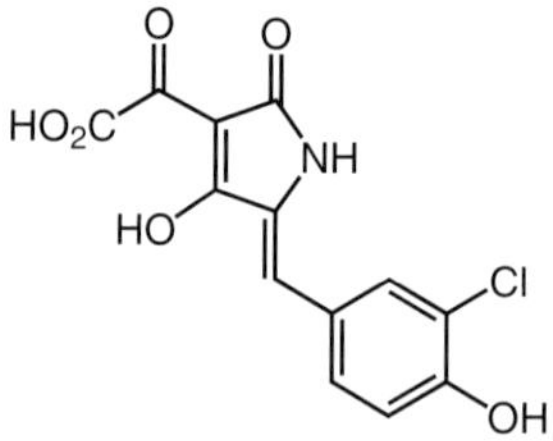

2198 (pachydermin)

The fungus *Chloridium* sp. produces CJ-21,164 (**2199**), a novel D-glucose-6-phosphate phosphohydrolase inhibitor (*1982*). The edible mushroom *Agaricus macrosporus* has yielded agaricoglycerides A (**2200**), B (**2201**), C (**2202**), D (**2203**), agaricic ester (**2204**), and monoacetylagaricoglycerides A (**2205**, **2206**) (*1983*).

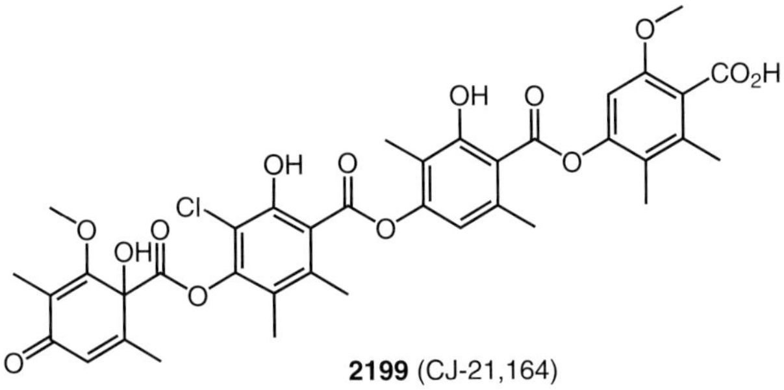

3.22 Complex Phenols

2199 (CJ-21,164)

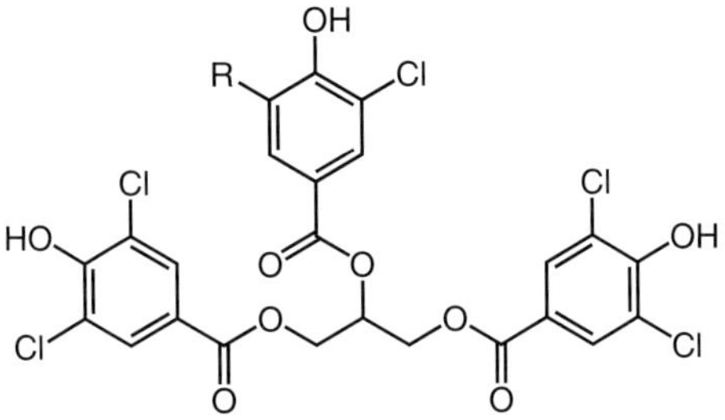

2200 R = Cl (agaricoglyceride A)
2201 R = H   (agaricoglyceride B)

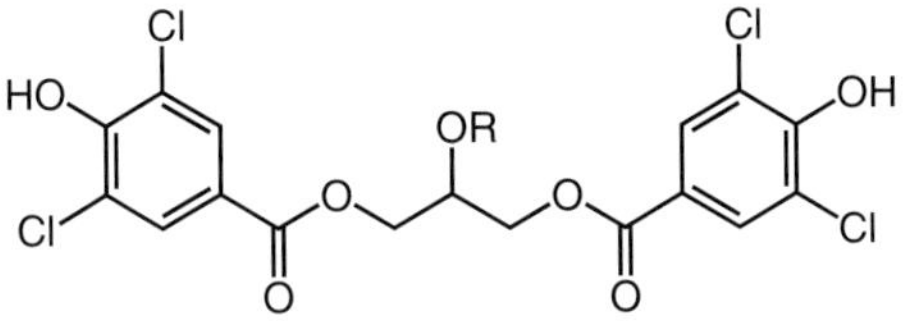

2202 R = H   (agaricoglyceride C)
2203 R = Ac (agaricoglyceride D)

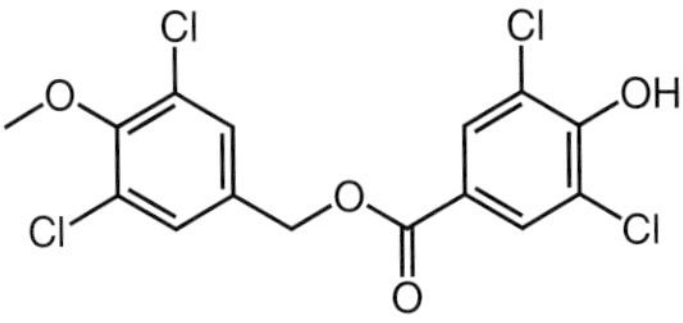

2204 (agaricic ester)

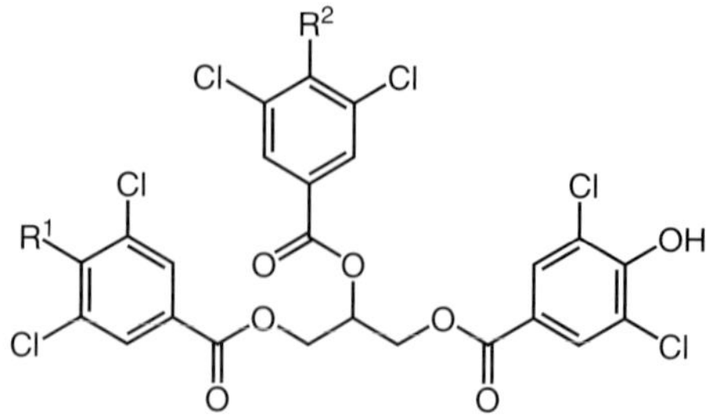

2205 R^1 = OAc, R^2 = OH (monoacetylagaricoglycerides A)
2206 R^1 = OH, R^2 = OAc

3.23 Glycopeptides

Probably no class of natural products with therapeutic potential has received more attention than the vancomycin glycopeptide antibiotics. In clinical use for more than 40 years, vancomycin – the antibiotic of "last resort" – has been extensively investigated regarding its mechanism of action (*1991–1997*), analog development to combat resistant bacteria (*1993, 1994, 1996–2001*), biosynthesis (*2002*), and total synthesis (*1993, 1997, 2003*). The enormity of the vancomycin and related glycopeptide literature renders full coverage not feasible here, but excellent general reviews are available (*1993, 1996, 1997, 1998, 2004, 2005*). A crystal structure of vancomycin was only relatively recently obtained (*2006*). Some 75 naturally occurring chlorinated glycopeptides were documented in the previous review (*1*), and several new examples have been described subsequently. The new A-40926-PA (**2207**) and A-40926-PB (**2208**), acetates of two previously known glycopeptides A-40926-A and -B (*1*), are produced by an *Actinomadura* strain (*2007*). All four of these metabolites are active against *Neisseria gonorrhoeae* and may offer a treatment for gonorrhea.

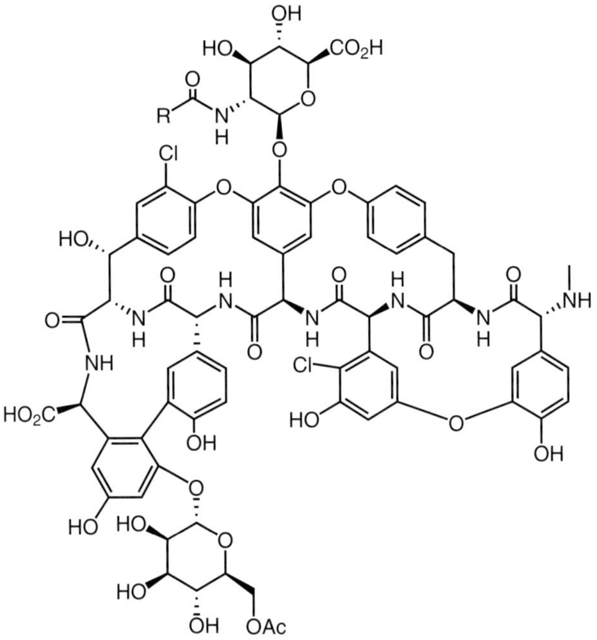

3.23 Glycopeptides

2207 R = n-C$_{10}$H$_{21}$ (A-40926-PA)
2208 R = (CH$_2$)$_8$CH(CH$_3$)$_2$ (A-40926-PB)

Cultures of *Amycolatopsis* sp. have yielded six new 4-oxovancosamine-containing glycopeptides, ureido-balhimycin (**2209**), rhamnosyl-balhimycin (**2210**), methylbalhimycin (**2211**), demethylbalhimycin (**2212**), balhimycin V (**2213**), devancosamine-vancomycin (**2214**), M43C (**2215**), and degluco-balhimycin (**2216**), along with the known balhimycin (*2008*). A crystal structure of ureido-balhimycin has been reported (*2009*).

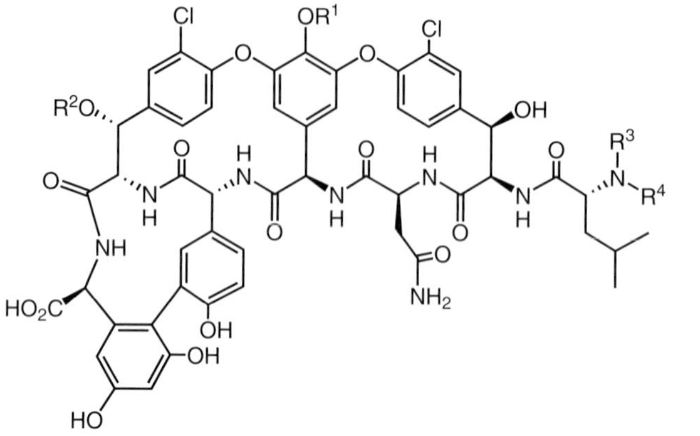

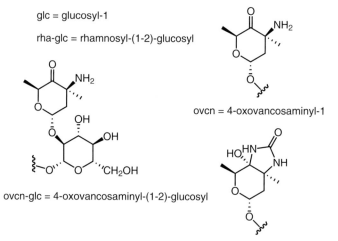

2209 R^1 = glc, R^2 = urvcn, R^3 = H, R^4 = Me (ureido-balhimycin)
2210 R^1 = rha-glc, R^2 = urvcn, R^3 = H, R^4 = Me (rhamnosyl-balhimycin)
2211 R^1 = glc, R^2 = ovcn, R^3 = R^4 = Me (methylbalhimycin)
2212 R^1 = glc, R^2 = ovcn, R^3 = R^4 = H (demethylbalhimycin)
2213 R^1 = ovcn-glc, R^2 = ovcn, R^3 = H, R^4 = Me (balhimycin V)
2214 R^1 = glc, R^2 = H, R^3 = H, R^4 = Me (devancosamine-vancomycin)
2215 R^1 = R^2 = H, R^3 = R^4 = Me (M43C)
2216 R^1 = R^3 = H, R^2 = ovcn, R^4 = Me (degluco-balhimycin)

glc = glucosyl-1

rha-glc = rhamnosyl-(1-2)-glucosyl

ovcn = 4-oxovancosaminyl-1

ovcn-glc = 4-oxovancosaminyl-(1-2)-glucosyl

urvcn = ureido-4-oxovancosaminyl-1

The previously known chloropeptin 1 and complestatin (= chloropeptin II) have been the object of synthetic and stereochemical studies and structural revisions (*2010–2012, 2653*). The new complestatins A (**2217**) and B (**2218**) were characterized from a *Streptomyces* sp. MA7-234 (*2013*), and these two compounds would appear to be the same as neuroprotectins A and B isolated from *Streptomyces* sp. Q27107 (*2014, 2015*). Another *Streptomyces* sp. has

3.23 Glycopeptides

furnished SCH 212394 (**2219**), which incorporates a 6-chloroindole unit (*2016*), and SCH 204698 (**2220**), which is a formal acetone addition product of chloropeptin I (*2017*).

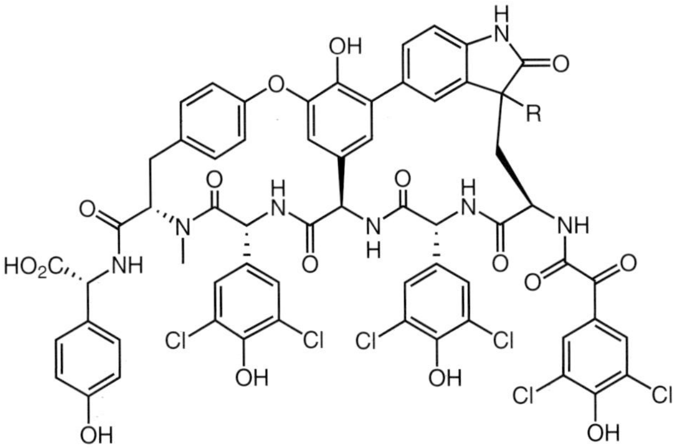

2217 R = H (complestatin A = neuroprotectin A)
2218 R = OH (complestatin B = neuroprotectin B)

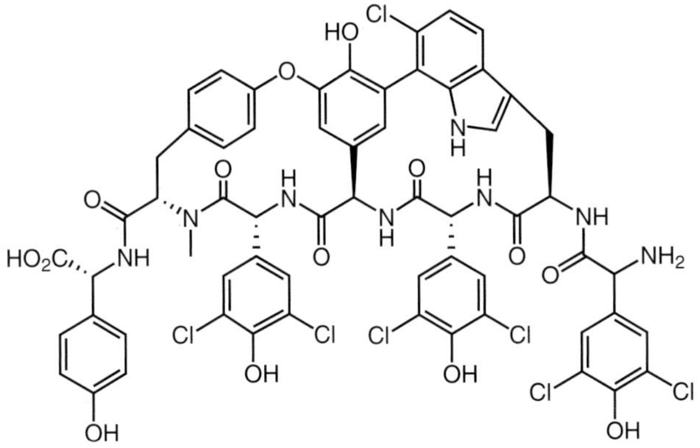

2219 (SCH 212394)

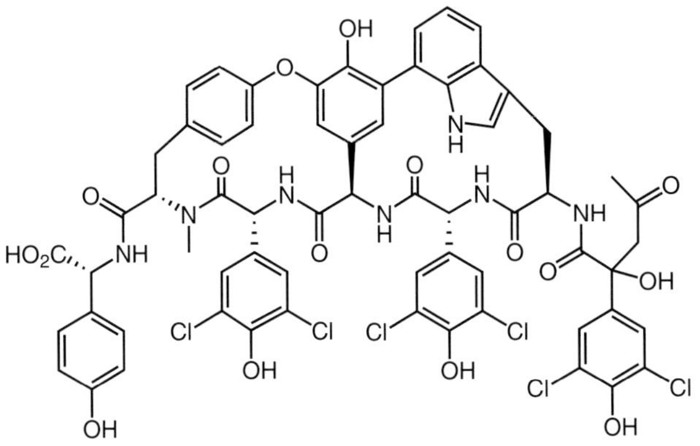

2220 (SCH 204698)

Two biosynthetic intermediates of the vancomycin glycopeptides, SP-969 (**2221**) and SP-1134 (**2222**), are found in cultures of *Amycolatopsis mediterranei* (*2018*). This is the first reported isolation of linear biosynthetic intermediates of the vancomycin family. Monodechlorovancomycin **2223** is found for the first time in fermentation broths of *Amycolatopsis orientalis* (*2019*). The other monodechlorovancomycin was synthesized for comparison with **2223**.

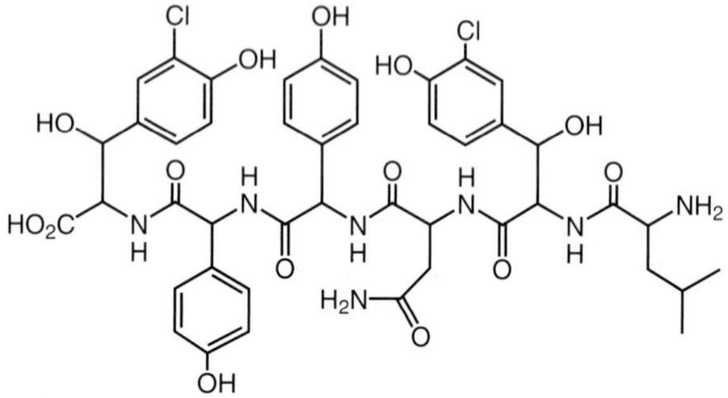

2221 (SP-969)

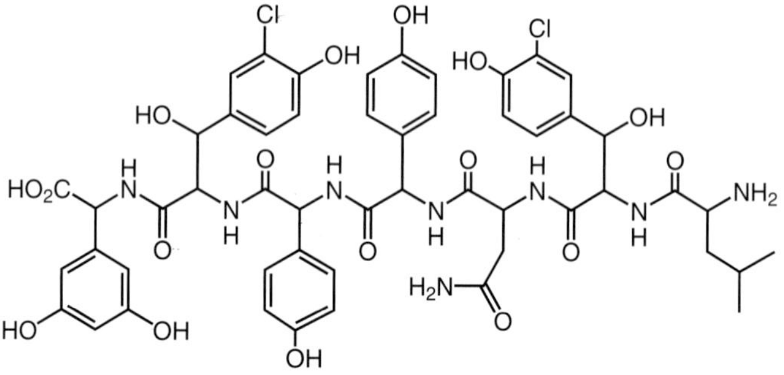

2222 (SP-1134)

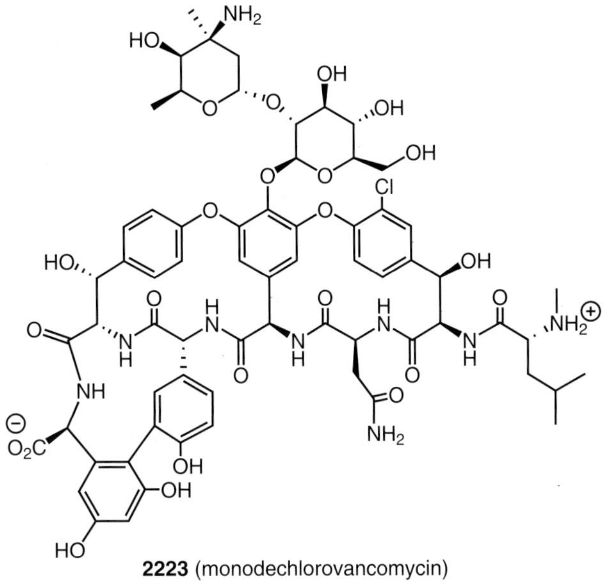

2223 (monodechlorovancomycin)

3.24 Orthosomycins

The small number of novel chlorophenol-oligosaccharide antibiotics (orthosomycins) presented in the earlier survey (*1*) has been expanded to include a few new examples. However, the highlight in this area is the total synthesis of everninomicin 13,384-1 (ziracin; Sch 27899) (**2224**) (*2020–2022*), which is found in cultures of *Micromonospora carbonacea var. africana* (*2023–2025*). This organism has also furnished the related everninomicins **2225**, **2226**, 13,384-5 (Sch 27900) (**2227**), Sch 49088 (**2228**) (*2023–2026*), and Sch 58761 (**2229**) (*2027*).

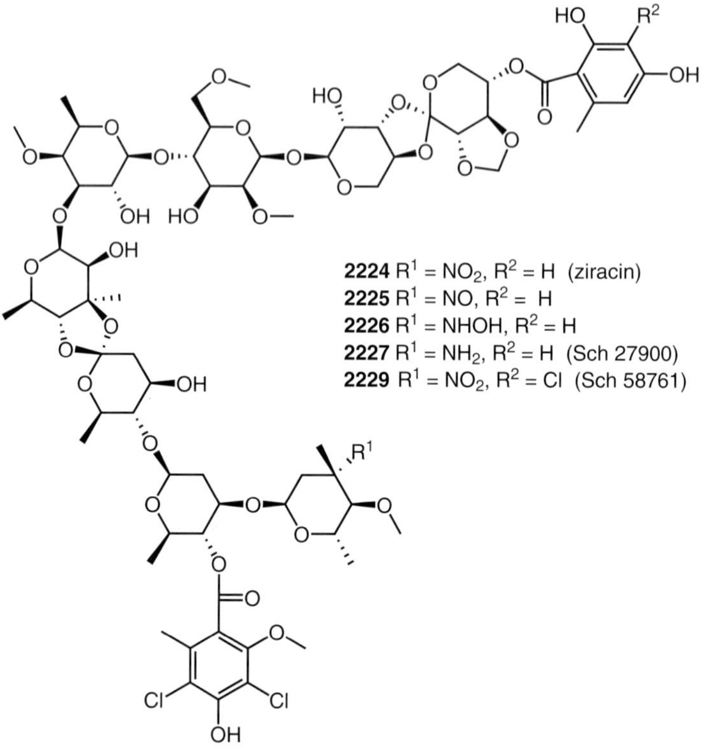

2224 $R^1 = NO_2$, $R^2 = H$ (ziracin)
2225 $R^1 = NO$, $R^2 = H$
2226 $R^1 = NHOH$, $R^2 = H$
2227 $R^1 = NH_2$, $R^2 = H$ (Sch 27900)
2229 $R^1 = NO_2$, $R^2 = Cl$ (Sch 58761)

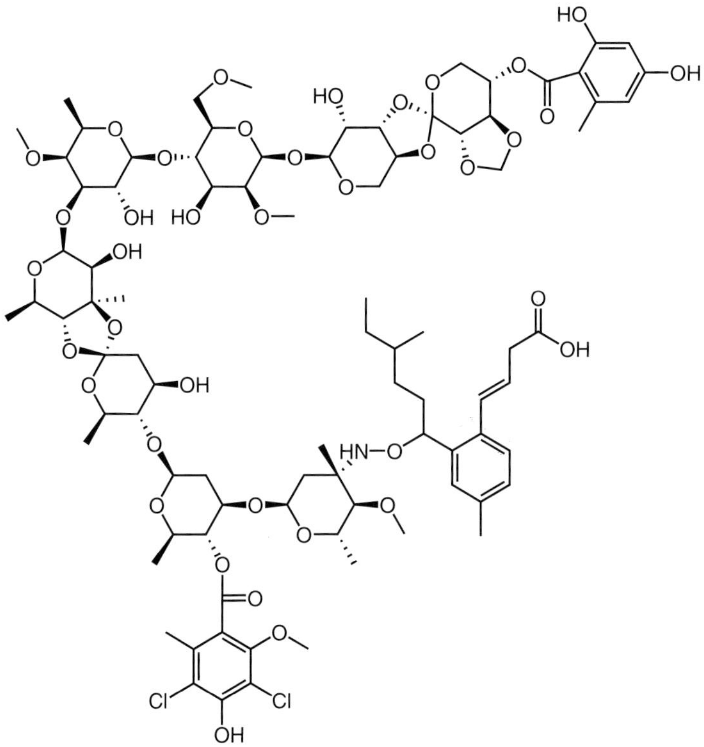

3.24 Orthosomycins

2228 (Sch 49088)

Additional studies of *Micromonospora carbonacea* have revealed the presence of everninomicin-6 (**2230**) (*2028*), and Sch 58769 (**2231**), Sch 58771 (**2232**), Sch 58773 (**2233**), and Sch 58775 (**2234**) (*2029*).

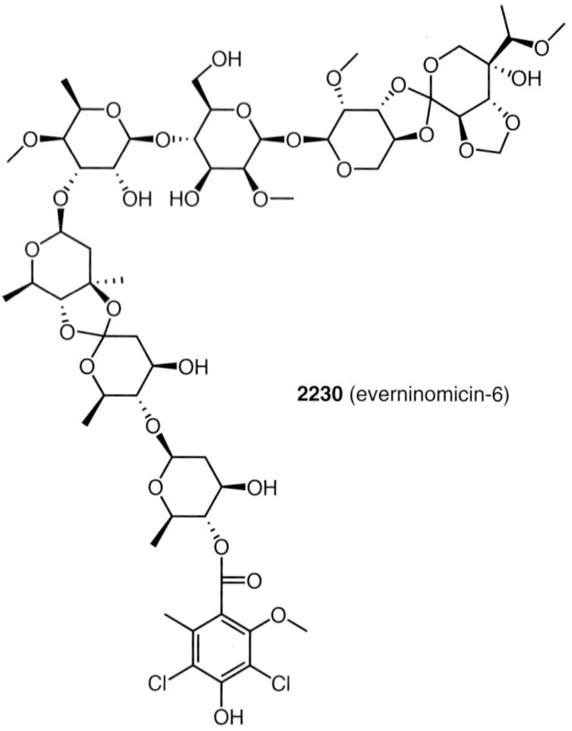

2230 (everninomicin-6)

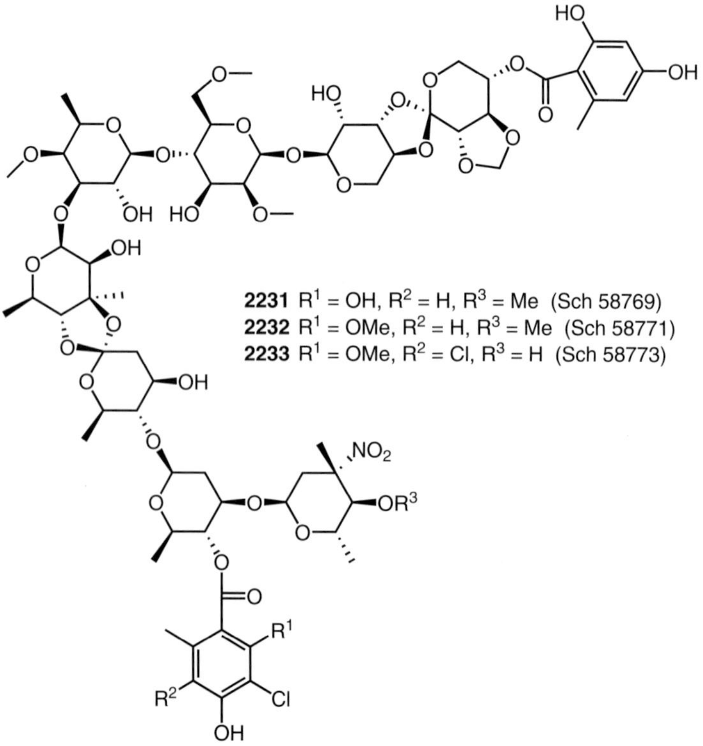

2231 R¹ = OH, R² = H, R³ = Me (Sch 58769)
2232 R¹ = OMe, R² = H, R³ = Me (Sch 58771)
2233 R¹ = OMe, R² = Cl, R³ = H (Sch 58773)

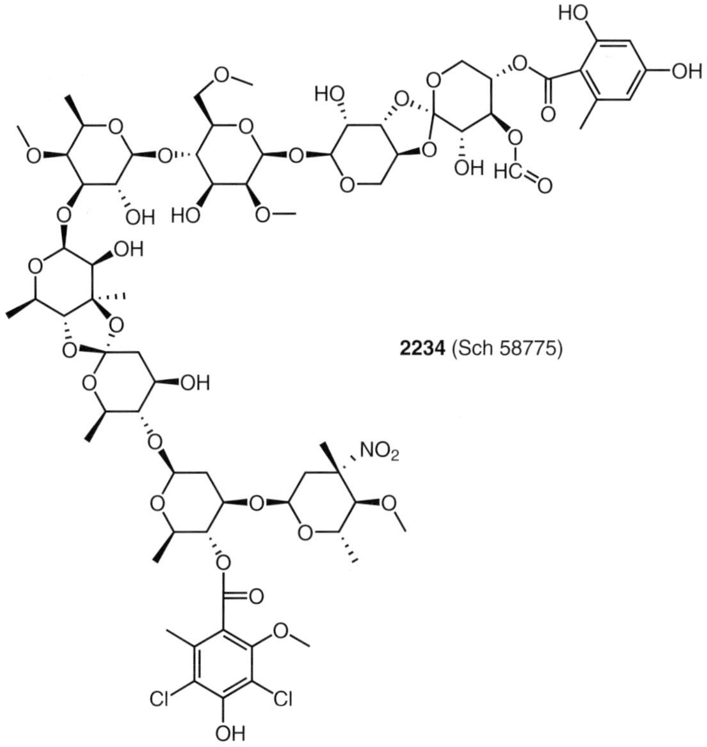

2234 (Sch 58775)

3.25 Dioxins and Dibenzofurans

For more than 30 years, the class of halogenated dibenzo-*p*-dioxins – "dioxin" – and the related dibenzofurans have probably received more attention by the lay press, the public, politicians, policy regulators, and environmental scientists than all other halogenated chemicals combined. The anthropogenic origins and biological effects of dioxins are summarized in the earlier survey (*1*). The intervening years since 1996 have clearly identified several new natural sources of both halogenated dioxins and dibenzofurans, both biogenic and abiotic, and confirmed previously discovered sources.

Given the huge number of polybrominated diphenyl ethers in marine sponges (vide supra, Sect. 3.22.2 (Diphenyl Ethers)) and the ubiquity of bromoperoxidase in these animals, it is not surprising that several polybrominated dibenzo-*p*-dioxins are found in sponges. Two examples were cited earlier (*1*). The Australian sponge *Dysidea dendyi* (Fig. 3.33) has yielded the new brominated dioxins spongiadioxin A (**2235**), the previously reported (*1*) spongiadioxin B (**2235a**) (*2030*),

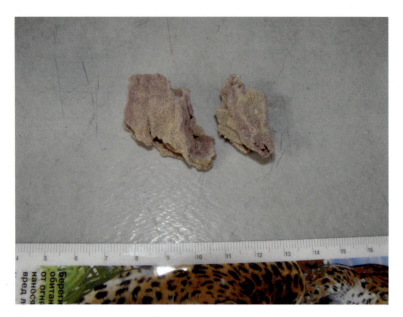

Fig. 3.33 *Dysidea dendyi*, an Australian sponge that contains the brominated dioxins, spongiadioxins A–C and related methyl ethers (**2235**–**2239**) (Photo: N. Utkina)

spongiadioxin C (**2236**), and methyl ethers **2237**–**2239** (*2031*). These five brominated dioxins inhibit the cell division of fertilized sea urchin eggs. Methyl ethers **2237**–**2239** are less active (IC_{50} 166, 141, and 94 µM, respectively) than the hydroxy-containing spongiadioxins A, B (**2235**), and C (**2236**) (IC_{50} 5.7, 4.8, and 1.1 µM, respectively). The highest activity of spongiadioxin B is consistent with the lateral arrangement of halogens on the dioxin framework, which is known to impart high biological activity (toxicity) to dioxins (*2032*). A study of three *Dysidea* sponge collections from Indonesia has also uncovered the presence of spongiadioxins A (**2235**) and C (**2236**) (*2033*). The first examples of non-hydroxylated dioxins, 1,3,7- (**2240**) and 1,3,8-tribromodibenzo-*p*-dioxin (**2241**) were characterized from blue mussels (*Mytilus edulis*) from the Baltic Sea (*2034*). A natural source is assumed for these two dioxins and five other brominated dioxins and one brominated dibenzofuran that are only tentatively identified. All of these polybrominated compounds are present in high levels in blue mussels and fish and are widely distributed in the Baltic environment (*2034, 2035*). A biosynthesis of these polybrominated dioxins from bromophenols has been advanced (*2035*). The Yellow Sea brown alga *Leathesia nana* contains the novel **2242** (*1738*). Interestingly, several novel phlorotannins, e.g., eckol (**HHH**), 2-phloroeckol (**III**), and dieckol, which are nonhalogenated dioxins, are found in the brown alga *Ecklonia kurome* Okamura (*2036*). This again illustrates that biohalogenation of electron-rich aromatic rings in the marine environment is not *fait accompli*.

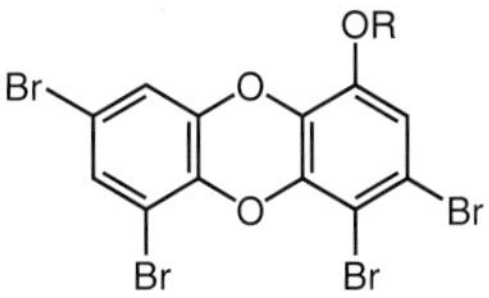

3.25 Dioxins and Dibenzofurans

2235 R = H (spongiadioxin A)
2237 R = Me

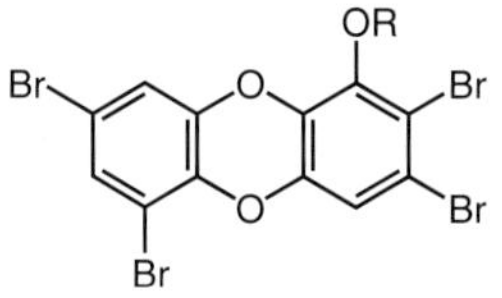

2235a R = H spongiadioxin B
2239 R = Me

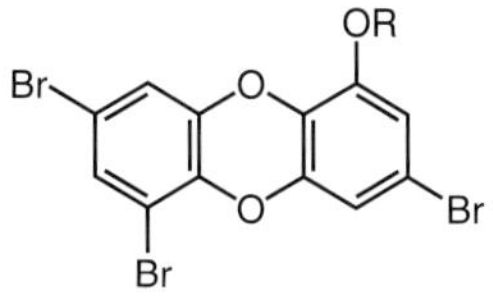

2236 R = H (spongiadioxin C)
2238 R = Me

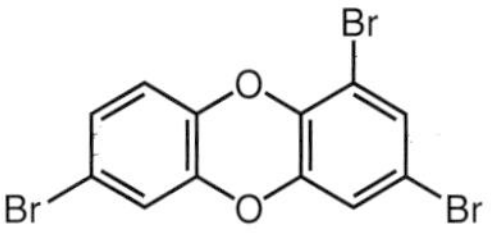

2240

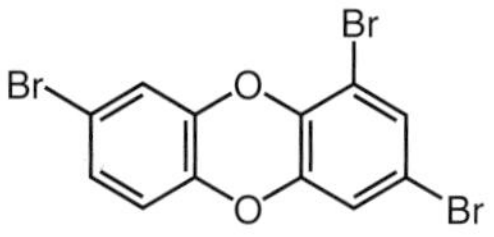

2241

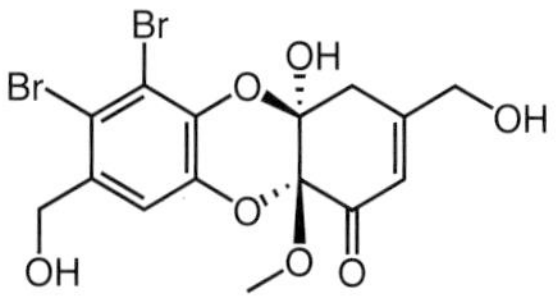

2242

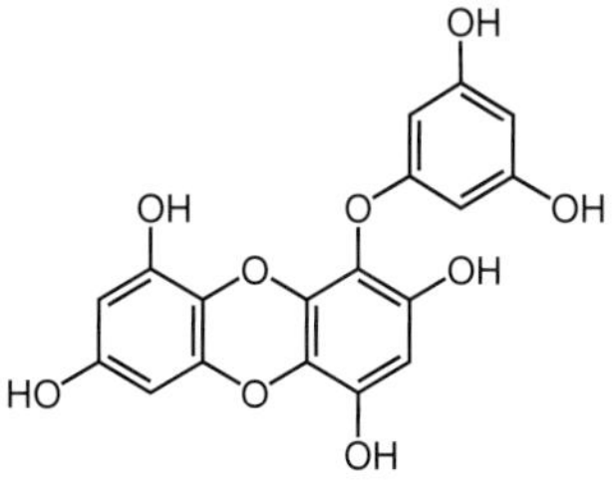

HHH (eckol)

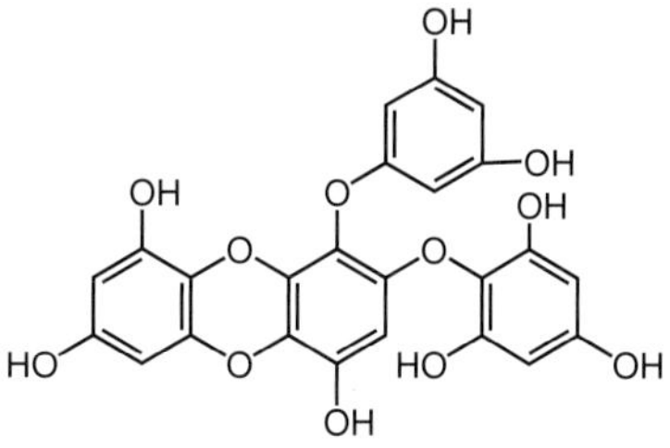

III (2-phloroeckol)

Whereas the earlier survey mentioned that polyhalogenated dibenzo-*p*-dioxins have myriad industrial and combustion sources (*1*), more recent studies have confirmed and extended the fact that natural sources of dioxins and the related polychlorinated dibenzofurans do exist. Thus, Canadian peat bogs are shown to produce several dioxins and dibenzofurans, including 1,3,6,8-tetrachlorodibenzo-*p*-dioxin (**2243**), 1,3,7,9-tetrachlorodibenzo-*p*-dioxin (**2244**), and 2,4,6,8-tetrachlorodibenzofuran (**2245**), along with several minor analogues (Fig. 3.34) (*2037*). Labeling studies with $^{36}Cl^-$-chloride demonstrated incorporation into the dioxins and dibenzofurans via 2,4-dichlorophenol, which was also identified in the peat, along with chloroform, a chlorocresol, chloromethoxybenzoic acids, and chlorocinnamic acids. Other dioxins and furans that are minor components in the peat samples are mono- through octachlorinated dioxins and furans. In one sample of the Richibucto, New Brunswick, bog there were identified four monochlorinated furans, nine dichlorinated furans, and six trichlorinated furans. These dioxin and furan isomer patterns are unique to these peat systems and differ from the patterns

Fig. 3.34 New Brunswick peat that produces the dioxins **2243–2245** and several chlorophenols, chloroform, and other organochlorines (Photo: P. Silk)

observed from atmospheric deposition associated with anthropogenic sources of dioxins and furans. The same peat dioxin and furan pattern is duplicated when 2,4-dichlorophenol is allowed to react with the fungal enzyme chloroperoxidase (*2037*).

2243   **2244**   **2245**

A study of the soil of a *Douglas* fir forest in The Netherlands spiked with $^{37}Cl^-$-chloride demonstrated that chlorinated phenols, dibenzo-*p*-dioxins, and dibenzofurans are produced naturally in the humic soil layer probably via chloroperoxidase chemistry (Scheme 3.5) (*1712*). Twenty polychlorinated dioxins and furans were found to be produced naturally in this study, including the highly toxic 2,3,7,8-tetra- (**2246**), 1,2,3,7,8-penta- (**2247**), and 1,2,3,7,8,9-hexachlorodibenzo-*p*-dioxin (**2248**). The major congeners found are 4-chloro- (**2249**), 1,7-dichloro-

3.25 Dioxins and Dibenzofurans

(**2250**), 1,2,3,4,6,8,9-heptachlorodibenzofuran (**2251**), and 1-chloro- (**2252**), 1,2,3,4,6-pentachloro- (**2253**), and 1,2,4,7-/1,2,4,8-/1,3,6,9-tetrachlorodibenzo-*p*-dioxin (**2254**) (isomers not distinguished) (*1712*).

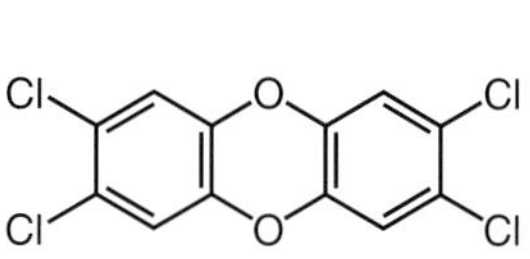

2246

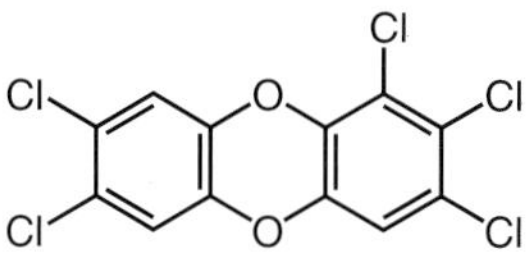

2247

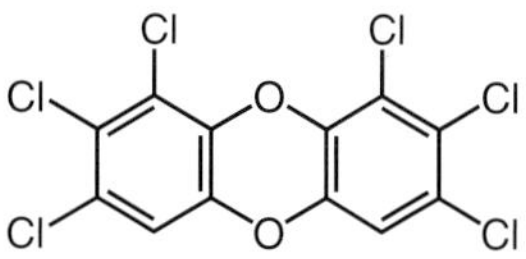

2248

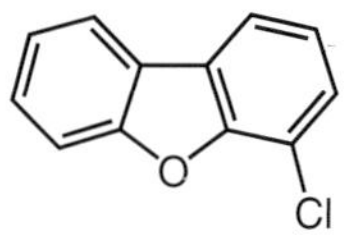

2249

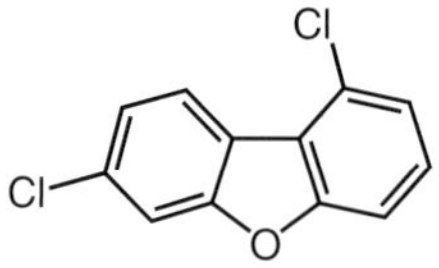

2250

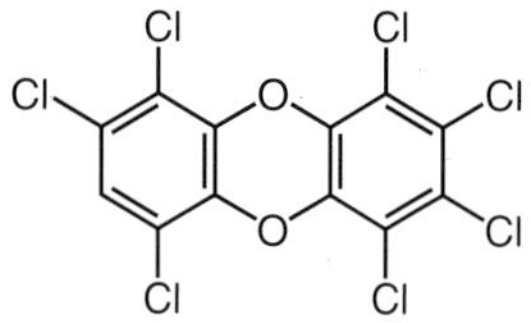

2251

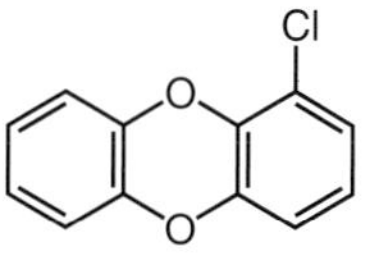

2252

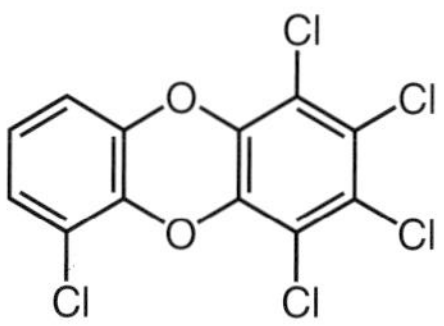

2253

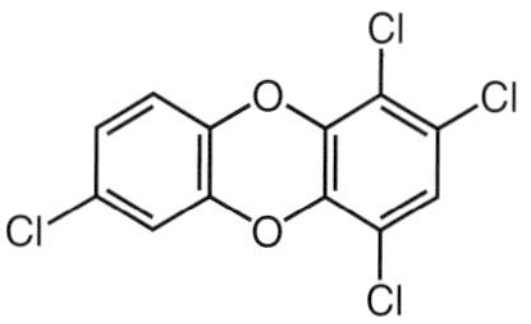

2254a

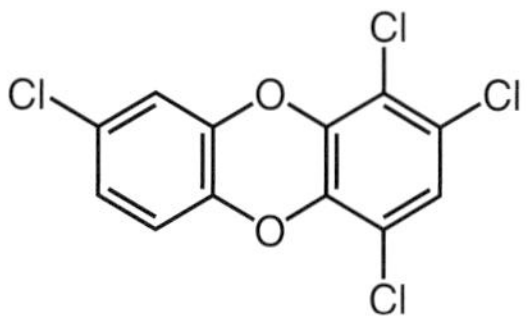

2254b

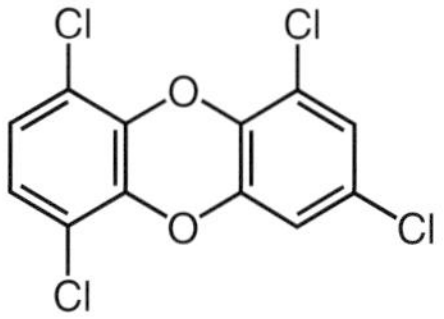

2254c

The cellular slime mold *Dictyostelium purpureum* K1001 contains AB0022A (**2255**), a novel antibacterial dibenzofuran, the structure of which was confirmed by total synthesis (*2038*). The lichen *Lecanora cinereocarnea* has yielded several new dibenzofurans, including three chlorinated analogues (**2256–2258**) (*2039*), and *Lecanora iseana* contains **2259** and **2260** (*2040*). The first naturally occurring polybrominated dibenzofuran, corallinafuran (**2261**), is present in a crustose coralline red alga that also contains corallinaether (**1913**) cited earlier (*1769*).

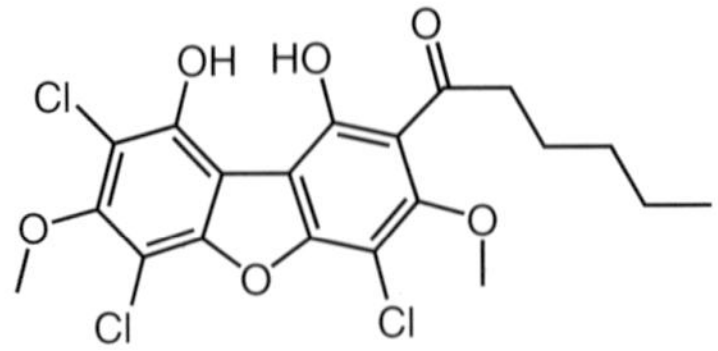

2255 (AB0022A)

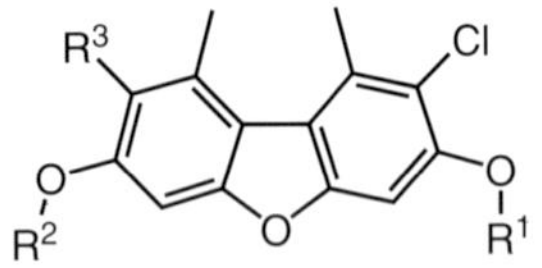

2256 R^1 = R^2 = R^3 = H
2257 R^1 = R^2 = H, R^3 = Cl
2258 R^1 = Me, R^2 = R^3 = H

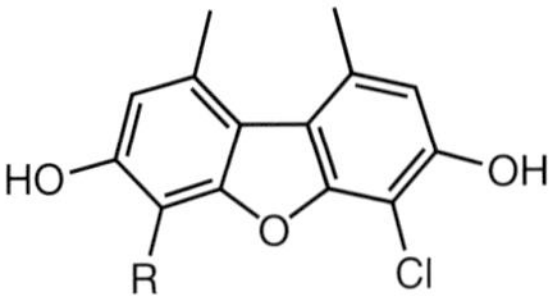

2259 R = H
2260 R = Cl

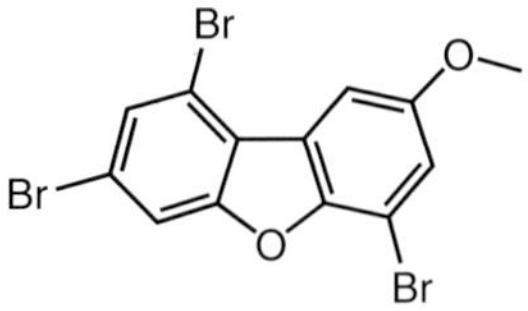

2261 (corallinafuran)

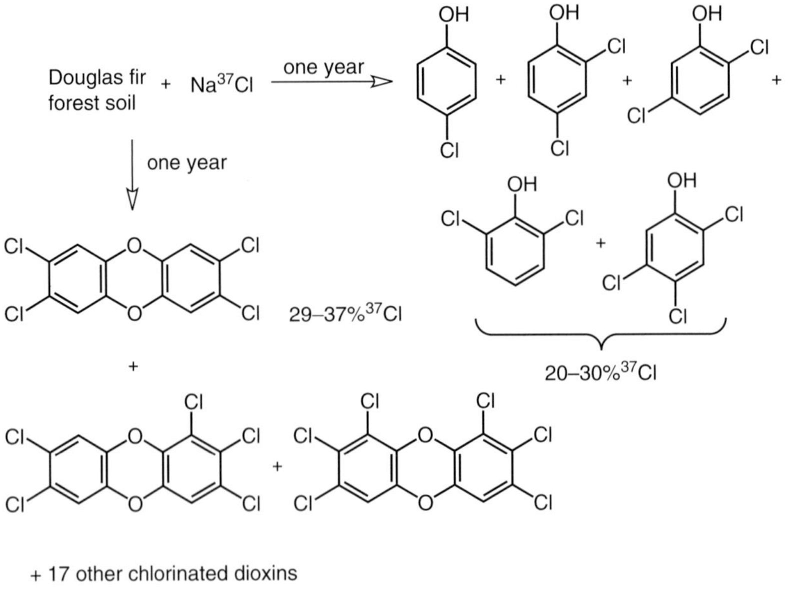

Douglas fir forest soil + Na^{37}Cl $\xrightarrow{\text{one year}}$ [chlorophenols] 29–37% ^{37}Cl

↓ one year

[chlorinated dioxins]

20–30% ^{37}Cl

+ 17 other chlorinated dioxins

Formation of dioxins and chlorophenols in soil (*1712*)

Scheme 3.5

3.25 Dioxins and Dibenzofurans

As discussed in the earlier survey (*1*), a biogenic source of polychlorinated dibenzo-*p*-dioxins and dibenzofurans is peroxidase-catalyzed transformation of chlorophenols as first reported by *Öberg* and *Rappe* (*2041–2044*). More recent studies confirm these observations (*2045–2048*). In addition to lactoperoxidase and horseradish peroxidase, human leukocyte myeloperoxidase catalyzes in vitro formation of dioxins and dibenzofurans from chlorophenols (*2046, 2047*). Formation rates are in the µmol/mol range (Scheme 3.6) demonstrating that a human biosynthesis of dioxins and furans is not only possible but also likely. These observations are reinforced by the reported in vivo (rats) conversion of the pre-dioxin nonachloro-2-phenoxyphenol to octachlorodibenzo-*p*-dioxin (OCDD) (*2049*), and the production of hepta- and octachlorodibenzo-*p*-dioxin in the feces of cows fed pentachlorophenol-treated wood (Scheme 3.7) (*2050, 2051*).

Similarly, polychlorinated dioxins and furans form in both compost and sewage sludge (*1*), but the major congeners in both systems are heptachloro- and octachlorodibenzo-*p*-dioxins and their origin is not understood (*2052–2056*). Several studies have attempted to elucidate the importance of natural combustion events as a source of polychlorinated dioxins and furans (*1*), but recent reports indicate that forest fires may not be a significant source of these compounds (*227, 2057*) despite earlier suggestions to the contrary (*1, 2058, 2059*). Nevertheless, numerous studies (wood stoves, control burns, etc.) clearly demonstrate that the combustion of wood

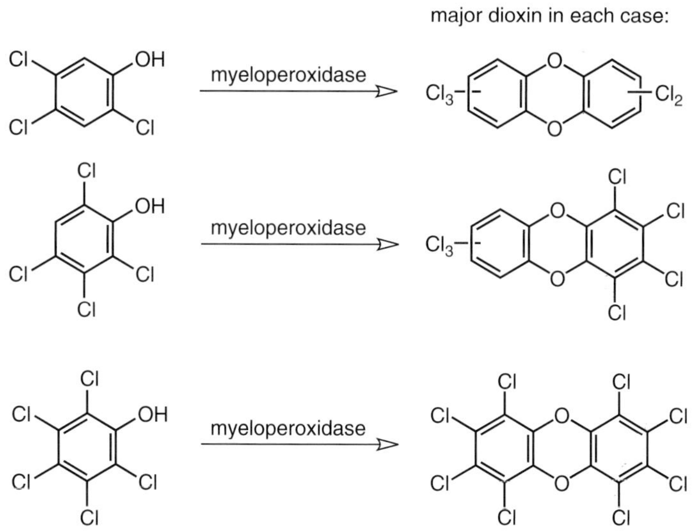

Myeloperoxidase-induced dioxin formation from chlorophenols (*2046, 2047*).

Scheme 3.6

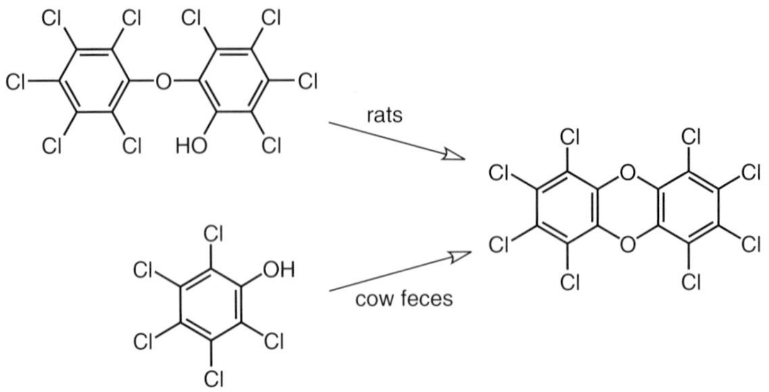

Octachlorodibenzo-p-dioxin formation in mammals (*2049-2051*)

Scheme 3.7

does lead to PCDDs and PCDFs (*1, 2060–2062*). Other reports reveal that the combustion of domesite lignite (*2063*), household waste (*2064*), and chemical waste (*2065*) produces PCDDs and PCDFs, and emissions from landfill fires (*2066*), bonfires and fireworks (*2067*), and crematories (*2068*) are sources of these chlorinated compounds. Quite astonishing is the observation that heating a mixture of methane, hydrogen chloride, and oxygen produces PCDDs and PCDFs containing up to three chlorine atoms (*232*).

Subsequent studies to those reported earlier (*1*) of preserved and ancient soil and sediment samples consistently reveal the presence of presumed naturally occurring PCDDs and (sometimes) PCDFs, but not PCBs. Thus, examination of ancient sediments (estimated at 1–10 million years old) from the Yellow Sea, the East China Sea, and the Pacific Ocean uncovered PCDDs but not PCDFs, the major compound being OCDD (*2069*). A study of Baltic Sea sediments detected both PCDDs and PCDFs "in small but significant levels during the period 1882–1962", including sediments from 1882, 1906, 1922, 1938, 1954, and 1962. Increased levels of these compounds were found in sediments from the period 1970–1985 as expected from anthropogenic contributions (*2070*). A natural origin is indicated for PCDDs and PCDFs found in sediments and clays in the southern United States, compounds that were also detected in catfish and chicken feed adulterated with these clays (*2071–2083*). Carbon and chlorine isotope studies suggest that these PCDDs form abiotically in situ in the sediments and clays (*2082, 2083*). The highest concentration of any congener in most samples is OCDD, and PCDFs are found in much lower amounts, if at all. Examination of ancient clays and sediments in Germany (*2084*), Queensland, Australia (*2085–2088*), and in ceramics and pottery produced from ball clay mined in the United States (*2089*) all reveal the presence of PCDDs, mainly OCDD with lesser amounts of 2,3,7,8-tetrachlorodibenzo-*p*-dioxin (TCDD) (**2246**), **2247**, 1,2,3,4,7,8-hexachlorodibenzo-*p*-dioxin (**2262**), 1,2,3,6,7,8-

hexachlorodibenzo-*p*-dioxin (**2263**), **2248**, 1,2,3,4,6,7,8-heptachlorodibenzo-*p*-dioxin (**2264**), and several other PCDDs (*2089*).

2262   **2263**   **2264**

Likewise, 8,000-year old sediments from a Finland lake show a similar PCDD profile to the clay samples (*2090*), as do sediments from Hong Kong (*2091*), and archived soil samples from the UK from the late 1800s and early 1900s (*2092–2094*). No PCBs were found in these latter preserved soil samples. A possible pre-industrial origin of these UK PCDDs is the burning of coastal peat, which is rich in chloride, over the millennia (*2095*). A sealed 1933 sample of municipal sewage sludge exhibits a suite of PCDDs, proposed to arise by in situ formation and condensation of chlorophenols (*2096*). Whether or not these myriad sources of PCDDs and PCDFs are formed biogenically or abiotically, the inescapable conclusion is that they have a natural origin. An excellent review of the occurrence of PCDDs and PCDFs in the environment is available (*22*).

3.26 Humic Acids

Numerous studies support the notion that organohalogen compounds originate on a massive scale via the natural in situ chlorination of humic and fulvic acids and their subsequent breakdown to chlorophenols, chloroacetic acids, chloroform, and other chlorinated and halogenated compounds (*1*). More recent investigations substantiate this ubiquitous route to natural organohalogens (Scheme 3.8) (*172, 2097–2109*). The electron-rich phenolic rings in humic acids (**2666**) are extremely susceptible to both biogenic and abiotic halogenation chemistry, and it is estimated that up to 10% of the aromatic rings in humic acids can be halogenated (*2097*). Evidence shows that the chlorination of humic and fulvic acids facilitates their further decomposition to nonaromatic compounds (*2107, 2109*), and that chloride (i.e., ^{36}Cl) is incorporated into humic acids (*2108*). Furthermore, presumed natural halogenation of humic material also occurs in Baltic Sea marine sediments leading to brominated and iodinated phenolic units in high molecular weight matter (*2110*). Several laboratory studies point to a chloro- or haloperoxidase-promoted halogenation of terrestrial humic and fulvic acids, e.g., (Scheme 3.9) (*315, 412, 2111, 2112*). Moreover, compelling evidence exists for the subsequent formation of chloroacetic acids and chloroform from chlorinated phenolic humic material (*278, 317, 324, 407, 410, 412, 2113, 2654*), including a novel abiotic pathway (Scheme 3.10) (*412*).

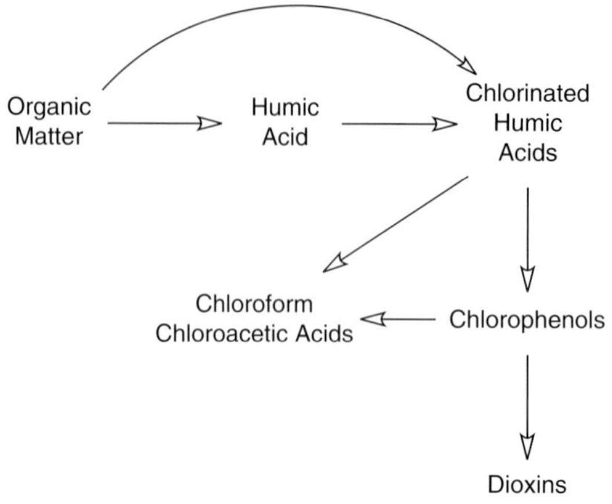

The possible natural formation of organochlorines
from humic acid substances.

Scheme 3.8

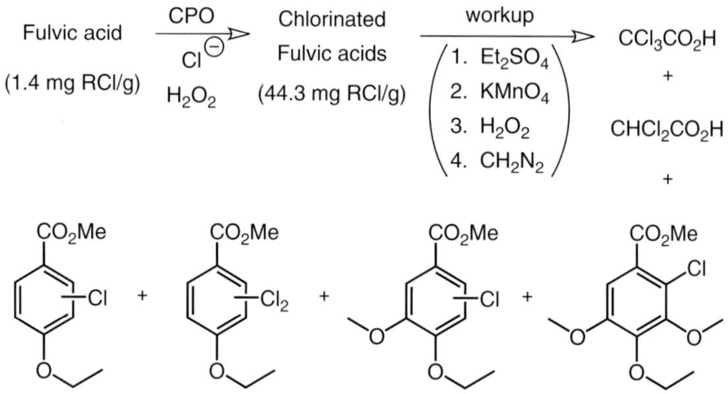

Chloroperoxidase-induced chlorination of fulvic acid (*412*).

Scheme 3.9

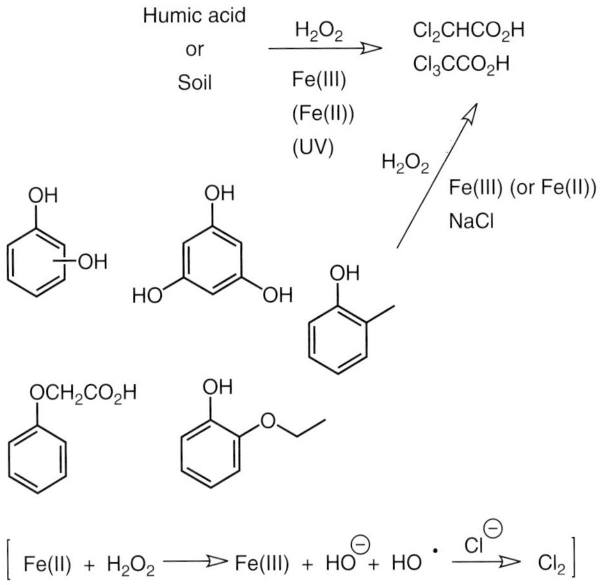

3.26 Humic Acids

Scheme 3.10

Abiotic formation of chloroacetic acids in soil (*412*).

More than 100 organochlorines have been identified and structurally characterized in the laboratory chlorination of terrestrial humic acid, although the major products are chloroform and trichloroacetic acid, followed by dichloroacetic acid and chlorinated C-4 dicarboxylic acids (*324*). In addition, other products that form in the chlorination of both humic acid and the model compound 3,4-dihydroxybenzoic acid are shown in Scheme 3.11. A more recently discovered source of natural organically bound chlorine is peat, reaching to 0.2% of the dry weight, and estimated to have accumulated globally to the extent of 280–1,000 million tons (*169*).

CCl$_3$CHO   CHCl=CClCO$_2$H   CCl$_3$CH$_2$CO$_2$H

CCl$_3$CHOHCO$_2$H   C$_3$H$_2$Cl$_2$CO$_2$H

HO$_2$CCCl=CClCO$_2$H   C$_3$HCl$_4$CO$_2$H

HO$_2$CC$_3$H$_2$Cl$_2$CO$_2$H   HO$_2$CCCl=C(CO$_2$H)$_2$

CCl$_3$CHOHCCl$_2$CHClCO$_2$H   CCl$_3$COCCl=C(CO$_2$H)$_2$

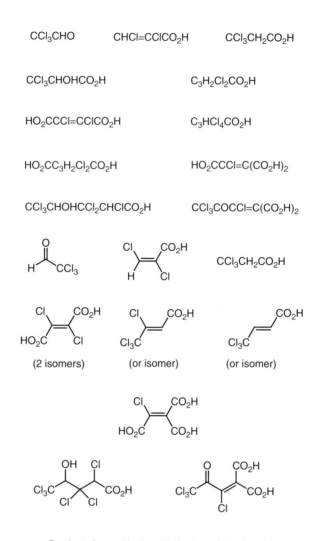

Products formed in the chlorination of humic acid and 3,5-dihydroxybenzoic acid (*324*).

Scheme 3.11

4 Biohalogenation

4.1 Introduction

While the question of how nature produces organohalogens lagged far behind their discovery, this situation has dramatically changed since the first review (*1*). Numerous excellent reviews of biohalogenation are available (*17, 59, 2114–2122, 2323*), and, as will be seen, several new halogen peroxidases, halogenases, and other enzymes capable of introducing halogen into organic compounds are known, including fluorine.

Specialized reviews on biohalogenation involving vanadium haloperoxidases (*2123–2126*), biochlorination (*2127*), biohalogenation by Basidiomycetes fungi (*2128*), haloperoxidases in organic synthesis (*2129–2131, 2327*), biohalogenation enzymatic mechanisms (*2132*), and halomethane biosynthesis (*2133*) are available. The role of hydrogen peroxide in defining the function of haloperoxidases and other plant enzymes has also been investigated (*2134–2137*).

4.2 Chloroperoxidase

The ubiquitous hemoprotein chloroperoxidase (CPO) (*1*) continues to be of great mechanistic and practical interest following its isolation more than 40 years ago from *Caldariomyces fumago* (*2138*). The CPO gene from this filamentous fungus has been isolated and sequenced (*2139*), an active recombinant CPO has been produced (*2140*), and the crystal structure of this CPO has been determined (*2141, 2142*). The fungus *Curvularia inaequalis* contains a vanadium CPO, which has been characterized (primary and X-ray structure) (Fig. 4.1) (*2143–2147*), as has the vanadium haloperoxidase from *Corallina officinalis* (*2324*). This enzyme has also been studied by density functional theory lending support to the proposed mechanism of action (Scheme 4.1) (*2325*). A related vanadium CPO, which shares 68% primary structural identity with the *Curvularia inaequalis* CPO, is produced

G.W. Gribble, *Naturally Occurring Organohalogen Compounds – A Comprehensive Update*, Progress in the Chemistry of Organic Natural Products, Vol. 91,
DOI 10.1007/978-3-211-99323-1_4, © Springer-Verlag/Wien 2010

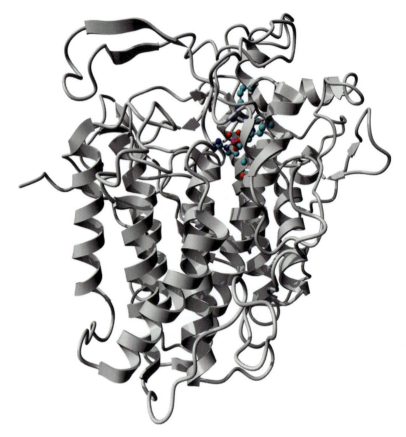

Fig. 4.1 A ribbon diagram of vanadium chloroperoxidase from the fungus *Curvularia inaequalis* (Photo: T. van Herk)

by the fungus *Embellisia didymospora* (*2148*). Some 10 *Caldariomyces* cultures produce CPOs with variable carbohydrate content but identical enzymatic activity (*2149*). CPO enzymes are found in bryophytes (liverworts) (*1710*), the marine worm *Notomastus lobatus* (*2150–2152*), and the bacteria *Streptomyces lividans* (*2153*) and *Serratia marcescens* (*2154*). The latter two CPOs do not contain a metal ion and the *Notomastus lobatus* CPO is the smallest hemoprotein known. Immobilized silica-supported heme-CPO from *Caldariomyces fumago* retains biological activity (*2155, 2156*), and CPO from this fungus also serves as a dehaloperoxidase in the dehalogenation of halophenols (*2157, 2158*).

The mechanism of CPO-induced halogenation has been of interest since the discovery of this extraordinary set of heme proteins, which exhibit catalase, peroxidase, and cytochrome P450 activities in addition to biohalogenation (*2159–2171*). A general consensus mechanism has been proposed that does not involve free

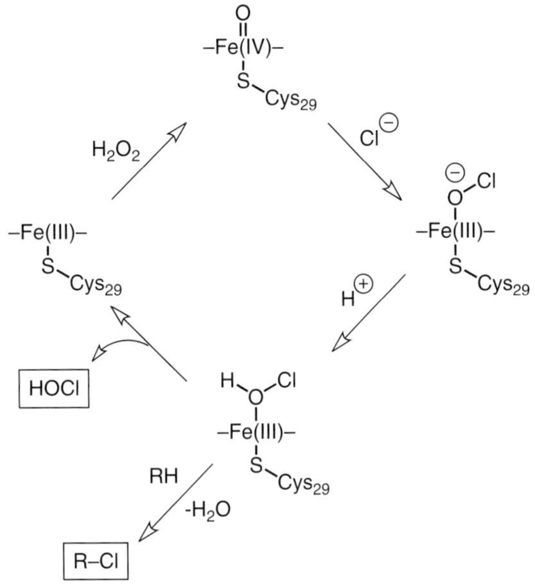

4.2 Chloroperoxidase

Scheme 4.1 Abbreviated proposed mechanism of heme-chloroperoxidase-catalyzed chlorination (*2163, 2165, 2168, 2202, 2303*).

chlorine or hypochlorite but rather an Fe(III)–OCl species that transfers chlorine to the organic substrate (Scheme 4.1).

Vanadium CPO from *Curvularia inaequalis* has also been the object of both experimental (*2146, 2172–2175, 2329*) (for an X-ray structure see Fig. 4.1) and theoretical studies (*2176*) to understand the biohalogenation operation of this enzyme. A reasonable mechanism has emerged from these data (Scheme 4.2) (active site amino acids and H-bonds are deleted for clarity). The nature of the halide-vanadium intermediate is unknown.

The CPO from *Caldariomyces fumago* has seen extensive use as a synthetic reagent *par excellence* (*1, 2129–2131, 2177, 2178*) and new applications are known. For example, the enantioselective CPO oxidation of sulfides to (*R*)-sulfoxides has been intensely pursued (*2179–2193*) in some cases displaying 100% enantiomeric excess and quantitative yields (Scheme 4.3). Another important and versatile reaction with CPO involving oxygen transfer is epoxidation (*2194–2207*) and some examples are shown in Scheme 4.3. The mechanism of these CPO-catalyzed oxygen insertion reactions has been examined (*2208*). Oxidation reactions that are catalyzed by heme CPO are benzylic hydroxylation (*2209*), propargylic oxidation (*2210, 2211*), benzylic alcohol oxidation (*2212*), cyclopropylmethanol oxidation (*2213*), 5-hydroxymethylfurfural oxidation (*2214*), the enantioselective oxidation of

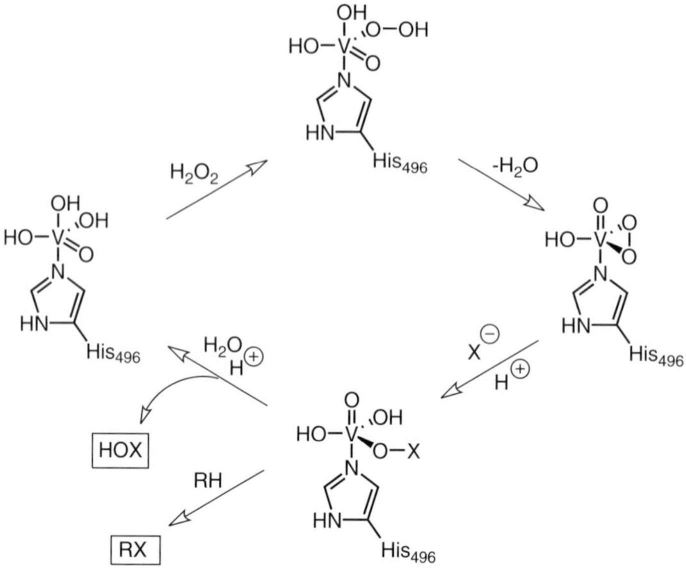

Abbreviated proposed mechanism of vanadium chloroperoxidase-catalyzed halogenation (*2123, 2124, 2146, 2172, 2173, 2175*).

Scheme 4.2

epoxyalcohols (*2215*), and phenol oxidation (*2216*). Indoles are oxidized to oxindoles in excellent yield (*2184, 2189, 2217–2219*), and both benzofurans and benzothiophenes are oxidized to various products (*2219*). CPO converts oximes to halonitro compounds (*2220*) and phosphorothioate pesticides to phosphates (*2221*), chlorinates aromatic hydrocarbons (*2222, 2223*), and effects polymerization of polychlorinated phenols (*2224*). A selection of these reactions is presented (Scheme 4.3).

Although less studied as a synthesis reagent, vanadium-CPO effects similar oxidation reactions to those of heme-CPO (*2225, 2226, 2326*). The CPO from *Streptomyces aureofaciens* can brominate pyrroles in the presence of bromide (*2227*). The synthesis performance of CPO has been improved by controlling the hydrogen peroxide delivery rate (*2228*), engineering CPO mutants resistant to deactivation (*2229–2231*), designing active site analogues (*2232*), and optimizing the role of organic solvents in these reactions (*2233*).

Despite the enormous versatility and efficiency of CPO in organic synthesis, the natural functions of this enzyme are no less important. In addition to its role in the biosynthesis of caldariomycin and other metabolites (*1*), CPO is involved in the degradative recycling of humic and fulvic acids (*315, 412, 2100, 2108, 2111–2113, 2234, 2235*). Both *Caldariomyces fumago* and *Curvularia inaequalis* CPO, which

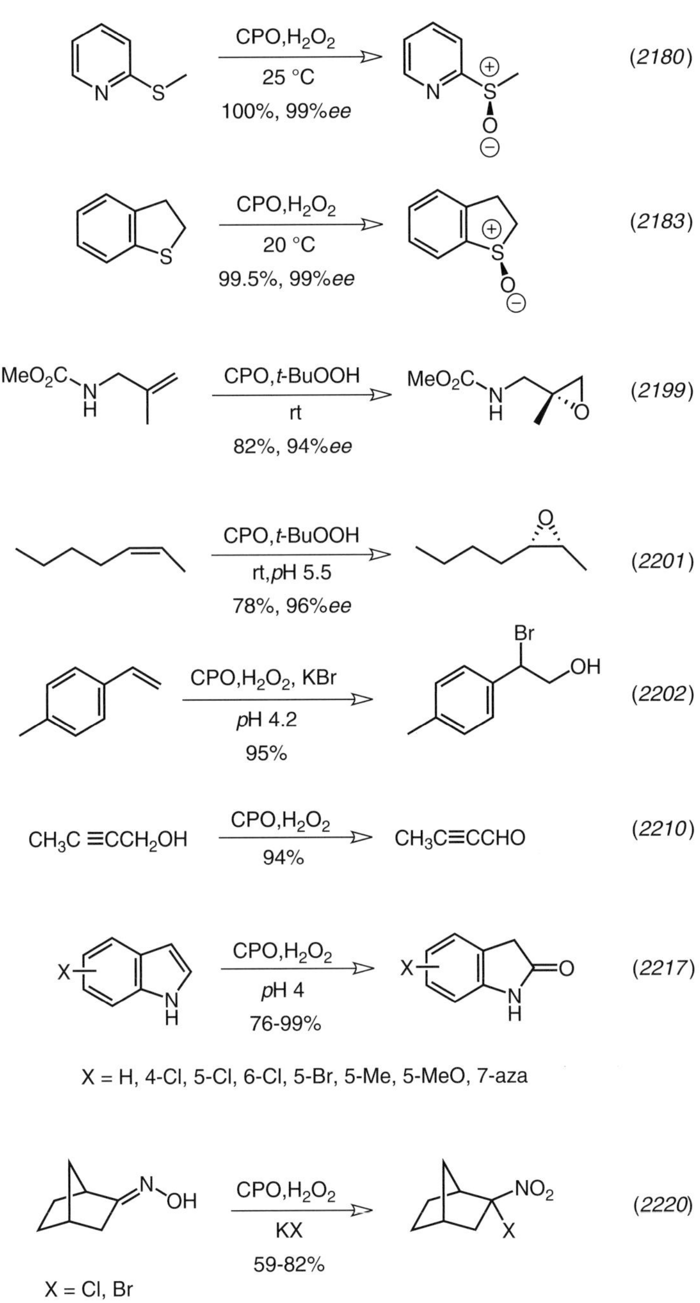

4.2 Chloroperoxidase

Oxidations with chloroperoxidase.

Scheme 4.3

occur in soils, chlorinate and cleave lignin structures (*2234, 2235*), results augmented by the specific incorporation of $^{36}Cl^-$ into humic acid (*2108*). A fern (*Athyrium filix-femina*) and a moss (*Polytrichum commune*) take in $^{36}Cl^-$ that is released as radiolabelled $CHCl_3$, CCl_4, and CH_3CCl_3, suggesting CPO activity in these forest plants (*2236*). Earlier studies also support the CPO production of $CHCl_3$ and trichloroacetic acid in soil and fungi (*317, 326, 410*), and chlorophenols and dioxins in peat (*2037*). It is estimated that global peatlands contain 280–1,000 million tons of peat-bound organochlorines, perhaps formed via humification by CPO (*169*). Likewise, CPO could play a role in the production of organochlorines in

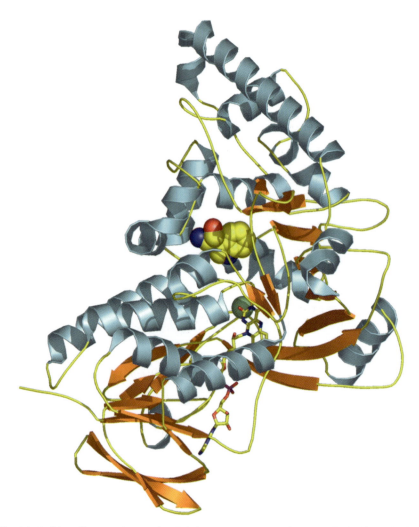

Fig. 4.2 A ribbon diagram of tryptophan 7-halogenase, an enzyme important in the biosynthesis of pyrrolnitrin and rebeccamycin. (Photo: K.-H. van Pée and J. H. Naismith)

weathering plant material (*172, 173*), and CPO-induced chlorination of surface water in a Swedish peat bog (Fig. 4.2) affords organochlorines such as 2,4,6-trichlorophenol (*2237*). CPO activity in *Laminaria digitata* seems to account for the formation of $CHCl_3$ in this macroalga (*306*), for the biosynthesis of chlorinated orcinols (**1783–1789**) in the Japanese lily *Lilium maximowiczii* (*1690*), and for chlorinated anthraquinones in the lichen *Nephroma laevigatum* (*1932, 2238*). The flavanones naringenin and hesperetin are chlorinated (and brominated) by CPO, although the resulting products are unnatural (*2239*).

4.3 Bromoperoxidase

As noted previously, bromoperoxidase (BPO) is a ubiquitous enzyme that brominates a wide variety of organic substrates (*1*). Both heme and vanadium BPOs are known and these enzymes are probably the main actor in the biosynthesis of the myriad marine organobromine metabolites (*2240–2242, 2329*).

In addition to the organisms cited earlier that contain BPO (*1*), new discoveries of BPO or BPO activity include the green algae *Ulva lactuca* (*2243*) and *Ulvella lens* (*366*), the red algae *Kappaphycus alvarezii* and *Eucheuma serra* (*2244*) and *Ochtodes secundiramea* (*2245*), and the Arctic brown algae *Laminaria saccharina* (*2246, 2247*) and *Laminaria digitata* (*2247, 2248*). A BPO has been isolated from the marine snail *Murex trunculus* (*2249*), and the nonheme BPO found in the bacterium *Pseudomonas putida* has been purified and characterized (*2250*). BPO genes have been cloned and expressed from *Streptomyces aureofaciens* (*2251–2253*), *Streptomyces venezuelae* (*2254*), *Corallina pilulifera* (*2255, 2256*), and *Corallina officinalis* (*2257*). X-ray crystal structure determinations have been reported for BPOs from *Streptomyces aureofaciens* (*2258*), *Corallina officinalis* (*2259, 2260*), and *Ascophyllum nodosum* (*2261–2263*). Based on these crystal structures and extensive model studies (*2124, 2264–2269*), a plausible mechanism for the *Ascophyllum nodosum* vanadium BPO bromination chemistry can be formulated (Scheme 4.4) (active site amino acids and H-bonds are deleted for clarity) (*2124, 2261, 2263, 2264, 2269*).

The synthetic utility of BPO is immature relative to that of CPO, but is showing promise as both a bromination reagent and a source of oxygen (*2131, 2326*). A vanadium-containing BPO from *Corallina officinalis* oxidizes sulfides to sulfoxides with the *S*-configuration, opposite to that observed with CPO (*2270–2273*), and forms bromohydrins from alkenes (*2194*). The indole ergot alkaloid agroclavine is oxidized to the corresponding oxindole and other products with BPO (*2274, 2275*). Several biomimetic studies with BPO have demonstrated the conversion of laurediols and related precursors to marine natural products (*2276–2278*), the cyclization of terpenes to brominated marine metabolites (*2279, 2280*), and the bromination of bromophenols (*1724, 2281*), examples of which are shown in Scheme 4.5. Interestingly, BPO from *Ascophyllum nodosum* contains brominated tyrosines at the surface of this enzyme (*2328*).

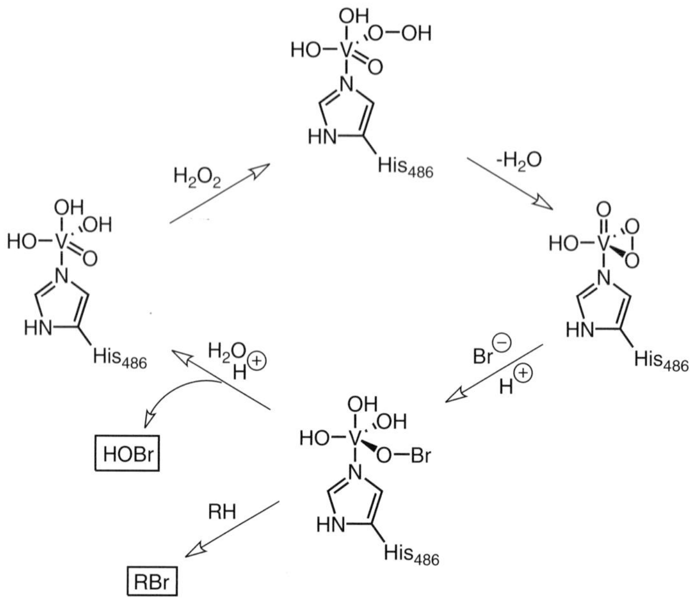

Scheme 4.4

Abbreviated proposed mechanism of vanadium bromoperoxidase-catalyzed bromination (*2123, 2124, 2175, 2261, 2263, 2269, 2280, 2328*).

In marine organisms, notably algae, the normal function of BPO is the production of brominated alkanes, such as $CHBr_3$, CH_2Br_2, $CHClBr_2$, and other bromoalkanes as has been established in studies of marine phytoplankton (*Nitzschia arctica, Porosira glacialis, Navicula* sp.) (*339*), *Corallina pilulifera* (*354, 2282*), and the red alga *Asparagopsis* sp. (*370*).

4.4 Halogenases, Other Haloperoxidases and Peroxidases

Several new enzymes capable of biohalogenation have been identified since the first review (*1*). Thus, given the significant number of naturally occurring organoiodine compounds, it is not surprising that iodoperoxidases (IPO) are known. For example, one species of *Navicula* marine phytoplankton produces CH_2I_2 and $ClCH_2I$ via an iodoperoxidase, an enzyme capable of oxidizing iodide but not bromide or chloride (*339*). A vanadium-dependent IPO has been purified and characterized from the brown alga *Saccorhiza polyschides* (*2283*), and also isolated from the brown alga *Phyllariopsis brevipes* (*2284*), and *Laminaria saccharina, Laminaria hyperborea, A. n. lusitanica, Pelvetia canaliculata,* and *Laminaria ochroleuca* (*2285, 2286,*

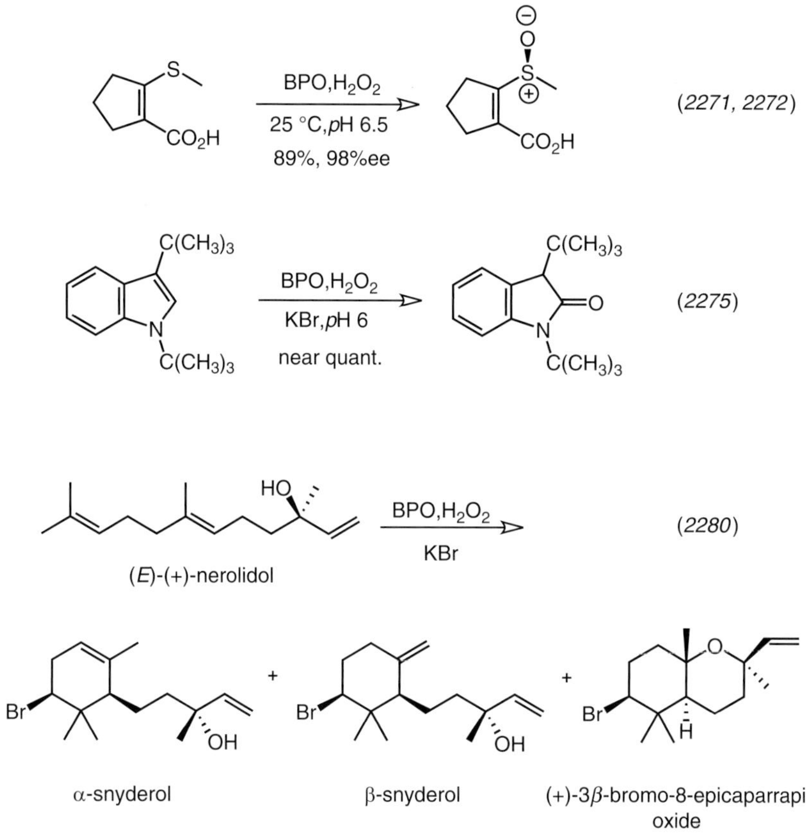

4.4 Halogenases, Other Haloperoxidases and Peroxidases

Scheme 4.5. Oxidations with bromoperoxidase.

2302). Other studies of *Laminaria digitata* and *Laminaria saccharina* indicate the presence of IPO (2246, 2247, 2287), as do studies of the marine microalga *Porphyridium purpureum* (2288) and the alga *Ascophyllum nodosum* (2289). The Arctic green algae *Acrosiphonia sonderi* and *Enteromorpha compressa* have high IPO activity (2247). Two peroxidase enzymes (2290) that catalyze the iodination of tyrosine are horseradish peroxidase (HRP) and lactoperoxidase (LPO) (2291). The latter enzyme is dominant for the iodination of tyrosine in mammals. The heme-containing HRP, which has been studied for more than one hundred years (2292), can also effect the oxidation of pentachlorophenol (2293). Similarly, a lignin peroxidase (LP) from *Phanerochaete chrysosporium* that is capable of oxidatively degrading lignin (2294, 2295) exhibits haloperoxidase activity (2296). This is the first report of biohalogenation in a white rot fungus, and this fungal LP and a related manganese peroxidase (MP) oxidize both bromide and iodide, thus functioning as a

BPO and IPO (*2296, 2297*). The basidiomycetous fungus *Agrocybe aegerita* also contains a haloperoxidase (*2298*), and a model oxomanganese (V) porphyrin is a haloperoxidase mimic (*2299*). The fresh water alga *Cladophora glomerata* contains a heme-haloperoxidase that oxidizes iodide to iodine and iodinates tyrosine and other phenols (*2300*). The actinomycete *Rhodococcus erythropolis* NI 86/21 produces a nonheme haloperoxidase that degrades thiocarbamate herbicides, and is the first such enzyme to be identified in a nocardioform actinomycete (*2301*). The importance of the occurrence and properties of heme peroxidases and their potential as biocatalysts with both biological and environmental applications has been succinctly summarized (*2303*).

A recent development is that of flavin-dependent halogenases discovered during studies of pyrrolnitrin biosynthesis from tryptophan (i.e., tryptophan 7-halogenase) (Fig. 4.2) (*1189–1191, 2122, 2304–2308, 2320*), which is discussed in Sect. 4.8 (Biosynthesis). Tryptophan 7-halogenase requires $FADH_2$ for halogenation and is the first member of this new type of halogenating enzyme. A similarly regioselective tryptophan 5-halogenase is present in *Streptomyces rugosporus* that produces pyrroindomycin B (**1468**) (*2309*), and a tryptophan 6-halogenase was found in the thienodolin (**2265**) producer *Streptomyces albogriseolus* (*2310, 2330, 2331*). A theoretical evaluation of flavin-dependent halogenase biohalogenation with oxidants such as O_2 or N_2O show that this reaction is thermodynamically feasible even without NADH (*2311*). Another halogenase has been isolated from the actinomycete *Actinoplanes* sp. ATCC 33002, a producer of pentachloropseudilin (**1155**) (*2312*), and the role of tryptophan 7-halogenase in the biosynthesis of rebeccamycin has been demonstrated (*1440*). The production of syringomycin E by *Pseudomonas syringae* pv. *syringae* B302D involves chlorination of a threonine (unactivated) methyl group by a novel halogenase, SyrB2, that is a nonheme Fe(II) protein utilizing α-ketoglutarate, O_2, and Cl^- to effect chlorination (*1145, 1146, 2313, 2323*) (Scheme 4.6). A similar enzyme chlorinates an unactivated methyl group of a L-*allo*-isoleucine residue en route to the biosynthesis of coronatine (*2314, 2315, 2323*).

Other enzymes capable of halogenation processes include a bacterial esterase from *Pseudomonas fluorescens* (*2316*), acid phosphatases from the bacteria *Shigella flexneri* and *Salmonella enterica* ser. *typhimurium* (*2317*), a lactonohydrolase from *Acinetobacter calcoaceticus* F46 (*2318*), and hydroperoxide halolyse from the marine diatom *Stephanopyxis turris* (*2319*). The biosynthesis of the ubiquitous methyl halides seems to involve methyl transferase enzymes, which have been isolated and purified in the plant *Brassica oleracea* (*S*-adenosyl-L-methionine:

2265 (thienodolin)

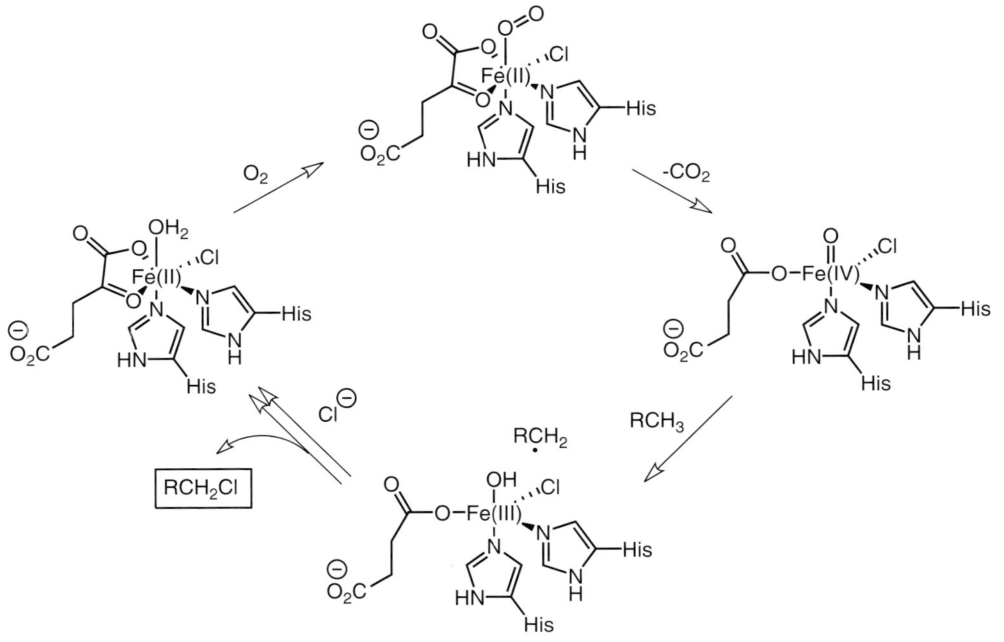

4.4 Halogenases, Other Haloperoxidases and Peroxidases

Scheme 4.6 Abbreviated proposed mechanism of non-heme iron halogenase chlorination (*1145, 1146, 2127, 2313, 2323*).

halide/bisulfide methyltransferase) (*2321, 2322*), and in the halophytic plant *Batis maritima*, a robust generator of CH_3Cl (*291, 292*).

4.5 Myeloperoxidase

The biohalogenating mammalian enzyme myeloperoxidase was extensively reviewed in Chap. 3 (*1*). Subsequent studies confirm and extend the importance of this white blood cell (neutrophil) enzyme and the related eosinophil peroxidase in the infection fighting process via, respectively, the generation of hypochlorous (HOCl) and hypobromous (HOBr) acid. Excellent reviews are available (*763, 764, 2332–2336*). The antimicrobial activity of HOCl is revealed by the fact that it is at least 10^3 times more effective than H_2O_2 and hydroxyl radical in killing *Escherichia coli* (*764*). The heme-containing myeloperoxidase, which is the most abundant protein in neutrophils amounting up to 5% of the dry weight (*2337*), has been characterized by X-ray crystallography (*2338*). This protein is the only human enzyme known to produce HOCl at physiological chloride concentrations (100 mM in plasma) (*2339*). Further reaction with cellular amino constituents such as taurine leads to taurine chloramine and other *N*-chloramines that are longer lived and less reactive (and potentially less destructive) chlorinating agents than HOCl (*2340–2343*). This early-demonstrated property of myeloperoxidase has been supported and amplified by many recent studies (*2344–2352*). Free chlorine gas is also implicated in some biochlorination reactions involving myeloperoxidase (*768, 1806, 2353*).

The active chlorinating species produced by the $MPO–Cl^-–H_2O_2$ system react with a myriad of biological targets including cholesterol (*766–769*), plasmalogens (*2354*), phospholipids (*2355, 2356*), amino acids (*2357–2362*), nucleosides (*2363*), and DNA (*1479, 2364*), accompanied by an array of chlorinated and oxidized by products (3-chlorotyrosine, 5-chlorouracil (**1555**), and others). Of great interest has been the role of MPO in the oxidation of both low- and high-density lipoprotein and implications in atherogenesis (*1807, 1810, 2365–2372*), although the function of MPO in atherosclerosis remains controversial (*2373*). Interestingly, vitamin C is reported to protect and reverse the HOCl- and chloramine-induced oxidation of low-density lipoprotein that may be involved in atherosclerosis (*2374*). In any event, it is clear that a deficiency of MPO can lead to severe fungal infection such as that from *Candida albicans* (*2375*).

In contrast to MPO, eosinophil peroxidase (EPO) prefers to oxidize plasma level bromide (20–100 μM) to hypobromous acid (HOBr) and several biological targets are implicated, including nucleic acids and nucleosides (*1480, 1482, 2376*), proteins (*1812, 1813, 2377, 2378*), unsaturated fatty acids (*2379*), and low-density lipoprotein (*2380, 2381*). This EPO-dependent bromination is suggested to be involved in the pathogenesis of asthma (*2382*). Accordingly, both 3-bromotyrosine and 3,5-dibromotyrosine (*1812, 1813*) are produced by EPO-induced bromination of tyrosine residues in lung tissue (*1813, 2382*).

4.6 Abiotic Processes

A major development since the previous review is the discovery that some organohalogen compounds can form in soils by a purely abiotic mechanism involving a Fenton oxidation pathway and the concomitant reduction of Fe(III) to Fe(II) (*2384–2386*). The formation of alkyl halides by this mechanism is shown in Scheme 4.6 (*2387*). The rates of production from soils decreased in this order: $CH_3X > CH_3CH_2X > CH_3CH_2CH_2X \gg CH_3CH_2CH_2CH_2X$, where X = Cl, Br, I. Subsequent studies show that iodoalkanes of 1–4 carbons (*2388*) and chloroacetic acids form abiotically in soil (*413*), in addition to their well-known biogenic enzymatic formation. An abiotic source of CH_3Br is suggested for the emission of this gas from ash (*Fraxinus excelsior*) and saltwort (*Batis maritima*), plants having known bromine content (*2389*). The emissions are a function of both temperature increase and bromine concentration. The natural formation of chloroethyne (**58**) in soil is also proposed to involve a Fenton reaction (*382*), as is the production of dichloroacetic and trichloroacetic acids from phenols and soil humic acid (*413*).

4.7 Biofluorination

Although few in number, fluoroacetic acid and the other naturally occurring ω-fluoro fatty acids (*34, 66, 2390, 2391*) are unsurpassed for their biogenetic intricacy, which has inspired enormous scientific interest, most notably from O'Hagan and his colleagues (*898–909, 911–913, 916, 2392–2394*). As noted earlier

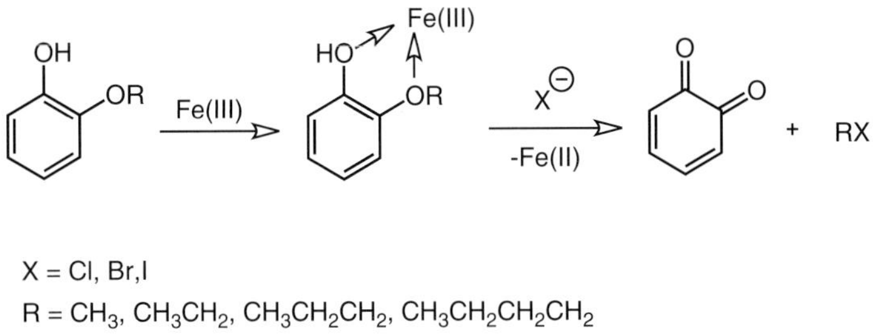

X = Cl, Br, I
R = CH_3, CH_3CH_2, $CH_3CH_2CH_2$, $CH_3CH_2CH_2CH_2$

Proposed abiotic formation of alkyl halides (*2386, 2388*).

Scheme 4.7

(Sect. 3.1.13 (Simple Organofluorines)), several reviews are available (*895, 914, 915, 2390, 2391, 2395*). Based on the available evidence, a proposed biosynthetic pathway for the formation of fluoroacetic acid and 4-fluorothreonine is shown in Scheme 4.7. Fluorinase has also served as a catalyst for the incorporation of [^{18}F]-fluoride into nucleosides (*2396, 2397*).

4.8 Biosynthesis

The biosynthesis of organohalogens has seen enormous interest since the first survey, and several examples are mentioned earlier in the present review. Space does not allow for full coverage of this topic, but some additional examples are presented here. The reader is also directed to general reviews on the biosynthesis of marine natural products, many of which contain halogen (*2398–2401*), terrestrial fungal (basidiomycetes) metabolites (*2402*), and halogenated alkaloids (*2403*).

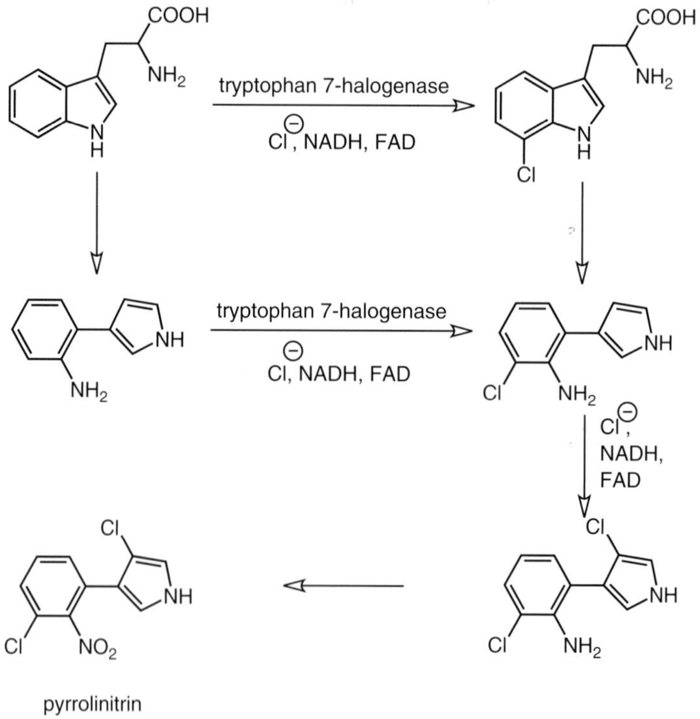

Proposed biosynthesis of pyrrolnitrin (*1189, 2310*).

Scheme 4.8

4.8 Biosynthesis

The biosynthesis of halogenated pyrroles has been of particular interest (*1189, 2404*). Extensive labeling experiments (^{13}C and ^{15}N) support acetate, propionate, proline, glucose, and methionine as the precursors of pyralomicin 1a (**1158**) (*2405, 2406*). A proline and polyketide origin is also established for the biosynthesis of streptopyrrole (**1165**) (*2407*). In both cases the timing of the chlorination step is unknown. Carbon-13 labelling studies show that the benzene ring in pentabromopseudilin evolves via 4-hydroxybenzoic acid and the shikimate pathway (*2408*), while the pyrrole ring is derived from proline (*2409*). Histidine, ornithine, and proline are incorporated into the brominated oroidin sponge alkaloid stevensine (*2410*). The biogenesis of the numerous pyrrole-imidazole alkaloids has received special attention (*2411, 2412*). Dioxapyrrolomycin also features a proline-polyketide pathway (*2413*), and other studies have explored the formation of the pyrrole ring in related metabolites (*2414, 2415*).

The biosynthesis of pyrrolnitrin and related phenylpyrroles has been extensively studied by van Pée (*1189, 2304, 2306, 2307, 2310*), and a proposed biosynthesis from tryptophan is illustrated in Scheme 4.8 (*1189, 2310*).

The anticancer indolocarbazole alkaloid rebeccamycin has been the subject of several biosynthetic studies (*1439–1441, 2416*), which is also proposed to involve the chlorination of tryptophan (Scheme 4.9).

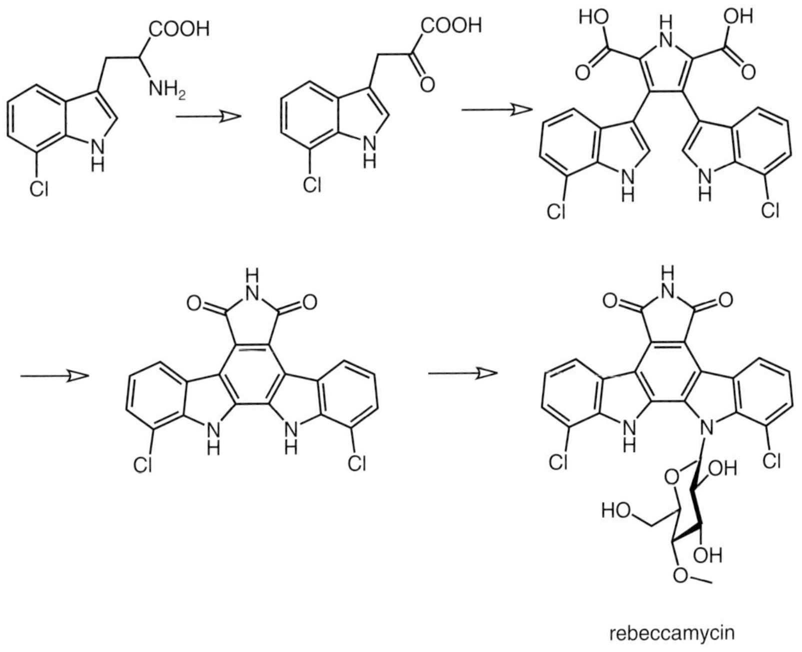

Proposed biosynthesis of rebeccamycin (*1439, 1440, 2416*).

Scheme 4.9

The biogenesis of vancomycin, the vanguard antibiotic of more than 200 naturally occurring glycopeptides, has been exhaustively studied (*2002, 2417*), as have been the venerable antitumor antibiotic maytansinoids (e.g., **1603–1605**), such as ansamitocin (*2418–2422*). In addition to a polyketide sequence, 3-amino-5-hydroxylbenzoic acid is the precursor to the chlorine-containing benzene ring of ansamitocin (*2421*). Similarly, the chlorine-containing benzene ring in the enediyne antitumor antibiotic C-1027 arises from (*S*)-3-chloro-4,5-dihydroxy-β-phenylalanine (*1552, 2423*). The biosynthesis of the mixed polypeptide-polyketide barbamide (**917**) was mentioned in Sect. 3.12 (Amino Acids and Peptides). The trichloromethyl group originates by chlorination of the pro-*R* methyl group of L-leucine, and subsequent conversion to trichloroisovaleric acid (*2424*). Biochlorination of a tyrosine derivative leads to the chlorine-containing coumarin ring of chlorobiocin (*1503*), and tyrosine is the precursor to several brominated tyrosines (*2425*). Likewise, tyrosine and several subsequent intermediates have been identified in the biosynthesis of thyroxine (*2426, 2427*). A polyketide pathway is implicated in the formation of numerous lichen chlorinated anthraquinones (e.g., **2157–2165**) and this lichen (*Nephroma laevigatum*) is able to chlorinate preexisting anthraquinones (*1932*). The origin of the chlorinated cyclopentene ring

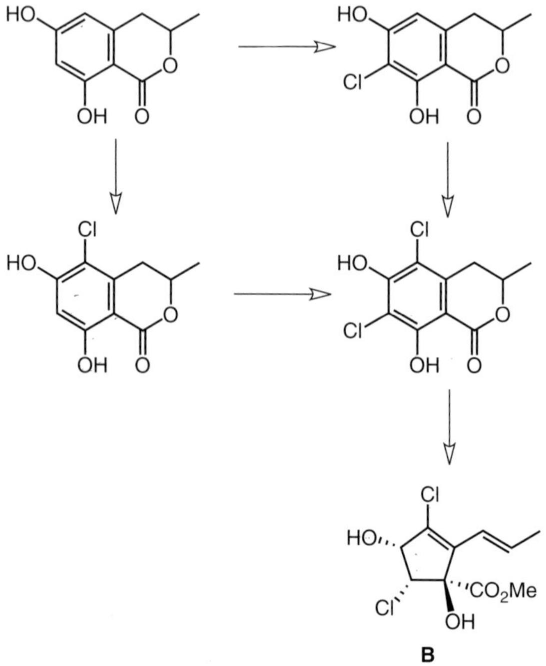

Proposed abbreviated biosynthesis of cryptosporiopsin
(**B**) (*415, 416*).

Scheme 4.10

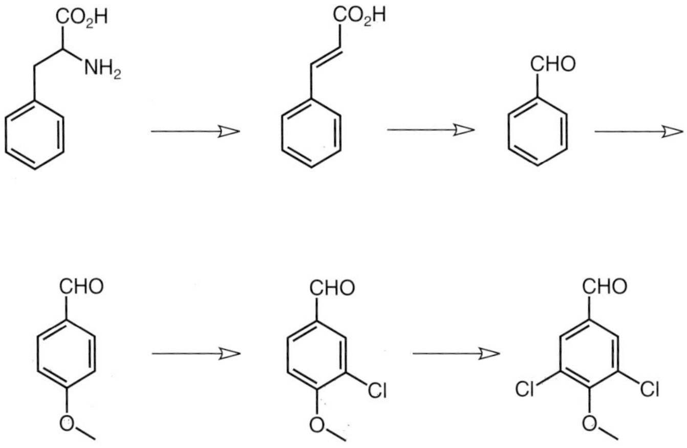

4.8 Biosynthesis

Proposed biosynthesis of 3-chloro- and 3,5-dichloro-4-methoxybenzaldehyde (*2430, 2431*).

Scheme 4.11

of the *Periconia macrospinosa* metabolites seems to involve an isocoumarin (Scheme 4.10).

The biosynthesis of the white rot fungus *Bjerkandera adusta* chlorinated aryl metabolites has been extensively studied by Silk and others (*1687, 2428–2431*), an abbreviated version of which is shown in Scheme 4.11 for 3-chloro- and 3,5-dichloro-4-methoxybenzaldehyde.

5 Biodegradation

An essential component of the biogenic halogen cycle is the degradation of organohalogens into their constitutive elements when an organism dies. Following the style of the earlier review (*1*), coverage here will be brief since biodegradation of organohalogens, both natural and anthropogenic, is an enormous topic and excellent reviews are available (*2119, 2432–2442*).

Several organisms are capable of degrading methyl halides, including *Methylobacterium chloromethanicum* (*2443, 2444*), *Hyphomicrobium chloromethanicum* (*2444*), *Aminobacter* spp. (*2444*), and others (*2445*), including marine bacteria (*2446, 2447*). The biodegradation of 1,2-dichloroethane has received particular interest, and the haloalkane dehalogenase from *Xanthobacter autotrophicus* has been extensively investigated (*2448–2452*). The key step in this degradation is an S_N2 displacement of chloride that is supported by chlorine isotope effect studies (*2453*). Other bacterial enzymes also biodegrade 1,2-dichloroethane and related haloalkanes and haloacetic acids (*278, 2454–2458*). The dry cleaning agents, trichloroethylene (TCE) and perchloroethylene (PERC), which also have a natural source, are degraded by bacterial enzymes (*2459–2462*), as are dichloroethylenes (*2462, 2464*) and vinyl chloride (*2464*). The white-rot fungus *Trametes versicolor* also mineralizes TCE (*2465*). Transgenic plants degrade TCE, 1,2-dibromoethane, and other organohalogens (*2466*).

Another large group of naturally occurring organohalogen compounds are the halogenated phenols. Both terrestrial and marine versions are readily degraded enzymatically. Marine worms (*Amphitrite ornata* and *Notomastus lobatus*) employ a haloperoxidase to degrade halophenols including fluorophenols (*2151, 2152, 2467–2469*). The enzyme from *Amphitrite ornata* has been purified and crystallized for an X-ray determination (*2469*). Bacteria associated with the marine sponge *Aplysina aerophoba* (*2470*), and the venerable *Caldariomyces fumago* fungal enzyme chloroperoxidase effect dehalogenation of halophenols (*2157, 2158*), as does a marine anaerobic *Desulfovibrio* bacterium strain (*2471*). The ubiquitous chlorophenols are degraded by a wide range of microorganisms, including fungi (*Paxillus involutus*, *Suillus variegatus*) (*2472*), (*Pycnoporus cinnabarinus*) (*2473*), anaerobic bacteria (*Desulfitobacterium* sp.) (*2474–2477*), horseradish peroxidase

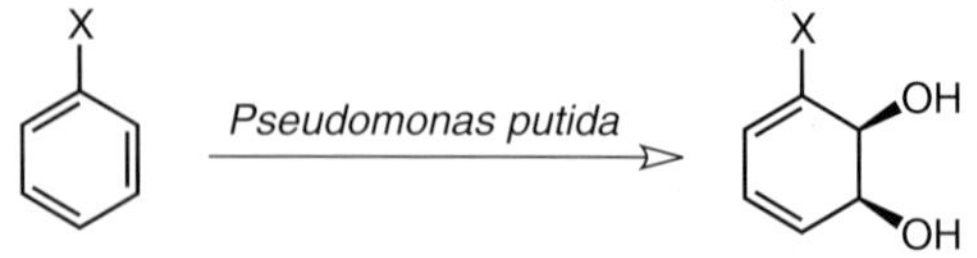

Scheme 5.1 Microbial oxidation of halobenzenes (*2482-2485*).

(*2293, 2478*), and the marine microalga *Tetraselmis marina* (*2479*). The latter organism converts 2,4-dichlorophenol to 2,4-dichlorophenyl-β-D-glucopyranoside for detoxification. Polychlorinated dibenzo-*p*-dioxins and dibenzofurans are biodegraded by an array of organisms, and this important area has been summarized (*2480, 2481*). Both aerobic bacteria (*Sphingomonas, Pseudomonas, Burkholderia*) and anaerobic sediments are capable of these biotransformations.

The microbial oxidation of halobenzenes to the corresponding *cis*-1,2-dihydrocatechols (Scheme 5.1) has proven to be a treasure trove for organic synthesis (*2482–2485*), most notably utilized by Hudlicky (*2483*). Thus, a myriad of natural products and analogs have been synthesized starting from the biooxidation of halobenzenes: vitamin C (*2486, 2487*), (+)-pericosine B (*2488*), *N*-acetylneuraminic acid (*2489*), combretastatins (*2490*), pancratistatin (*2491*), (−)-patchoulenone (*2492*), (−)-hirsutene (*2492*), (−)-cladospolide B (*2493*), (−)-cladospolide C (*2494*), shikimic acid analogs (*2495*), (−)-conduritol E (*2496*), (−)-conduramine C-4 (*2497*), phenylthioconduritol F (*2498*), (+)-codeine (*2499*), (+)-nangustine (*2500*), the anti-influenza agents Tamiflu and Tamiphosphor (*2501*), inositols (*2502, 2503*), and several other oxygenated benzene compounds (*2504–2513*). The power and versatility of the *Pseudomonas putida* oxidation of halobenzenes is beyond dispute.

6 Natural Function

The question that can be asked about all natural products, including naturally occurring halogenated compounds, is "Why do organisms produce organohalogens?" The first review provided evidence that seems to answer this question for several halogenated metabolites (*1*). For example, in the case of sessile marine organisms, a chemical defense function for these compounds seems paramount, and excellent reviews on this topic are available (*39, 2514–2520*). Nevertheless, a clear function for most of the identified biogenic organohalogens is presently unknown.

While natural chloromethane may have several functions (*42*), in Basidiomycetes wood-rot fungi (*Phellinus, Inonotus, Fomitosporia, Hymenochaete, Phaeolus,* and *Fomitopsis*) (*59*) chloromethane is a methyl donor in the biosynthesis of veratryl alcohol, the first step of which is methylation of 4-hydroxybenzoic acid. A second methylation of isovanillic acid affords ultimately veratryl alcohol, the function of which in these fungi is to stabilize lignin peroxidase thus promoting lignin degradation (*2521*).

A spectacular example of host defense involves the Japanese lily *Lilium maximowiczii* (Fig. 3.26) that produces seven chlorinated fungicides (**1783–1789**) in response to attack by the pathogenic fungus *Fusarium oxysporum* (*1690*).

It seems without argument that sponges, tunicates, algae, and other sessile marine organisms produce metabolites, halogenated and not, to prevent bacterial and barnacle overgrowth – "biofouling" – lest these animals be fatally smothered (*2522*), a plague of the shipping industry (*2523*). This antifouling activity is clearly expressed in sponges, such as *Acanthella cavernosa* (and/or their associated bacteria (*2524*)), against larvae of the barnacle *Balanus amphitrite*, most especially due to the action of chlorinated isocyanoterpenoids (*600, 2525*). This bacterial "cleansing" is reported for other organohalogens from many sponges (*2526*), such as the bromine-containing ianthellin (*2527*). The associated marine bacteria genus *Pseudoalteromonas* produces antifouling compounds (*2528*), and the sponge *Geodia barretti* displays antifouling properties from the secretion of barettin (**1362**), 8,9-dihydrobarettin (**1363**), and the new cyclopeptide **2266**, which inhibit larvae of the barnacle *Balanus improvisus* (*2529, 2530*). Coatings of these compounds on artificial surfaces inhibit fouling by this barnacle and by the blue mussel *Mytilus* *edulis*. Zebra mussel antifouling is also inhibited by moloka'iamine and other dibromotyramines on sponges of the order Verongida (*2531*).

2266

Sponges and their associated bacteria possess other chemical defense mechanisms that have evolved over 600 million years involving antibacterial, antiviral, and cytostatic compounds, many of which contain halogen (*118*). When *Aplysina* (Fig. 6.1) sponges are wounded or otherwise mechanically damaged, brominated isoxazoline alkaloids within the sponge are transformed to a potent fish-deterrent dienone compound (Scheme 6.1) (*2532–2535*). It is suggested that this biotransformation protects the damaged sponge from invasion by foreign bacteria, although contrary results have been claimed (*2536*).

Purealin, another brominated tyrosine metabolite of the sponge *Psammaplysilla purea*, blocks the sliding movement of sea urchin *Anthocidaris crassispina* sperm flagella (*2537*). Numerous sponge metabolites are feeding deterrents to predatory

Fig. 6.1 *Aplysina fistularis*, a sponge rich in bromophenols and bromotyrosines (Photo: F. J. Schmitz)

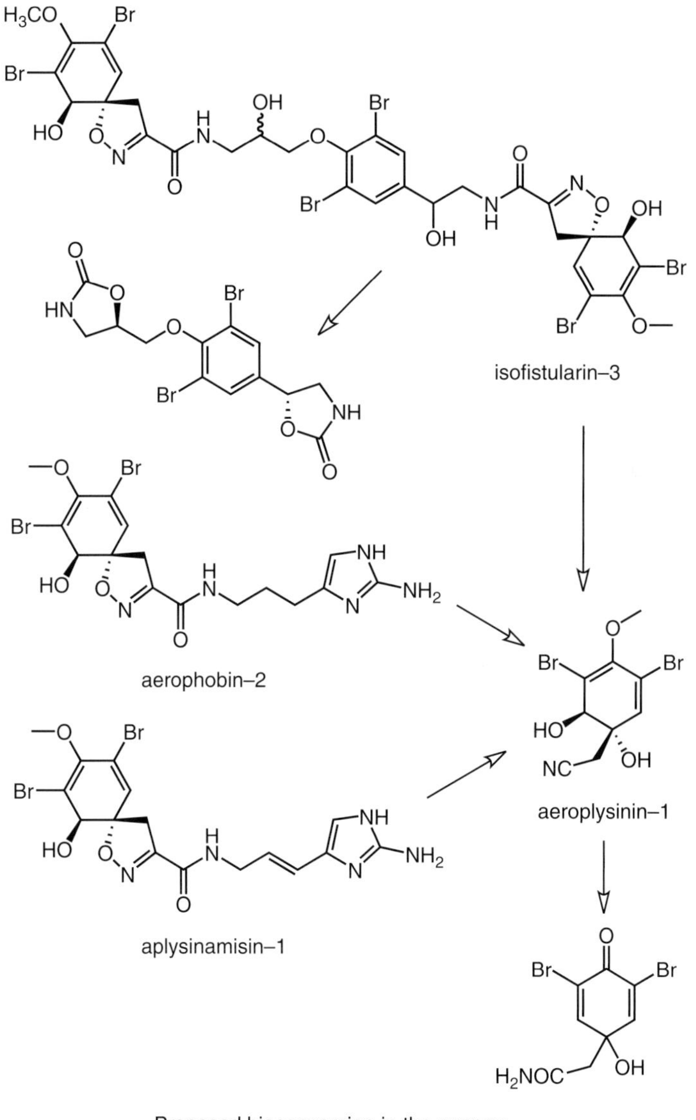

6 Natural Function

Scheme 6.1

Proposed bioconversion in the sponge
Aplysina aerophoba (*2535*).

reef fishes, including the brominated pyrrole stevensine from *Axinella corrugata* (Fig. 6.2) (*2538*), and several other brominated pyrrole alkaloids (i.e., dispacamide (**1239**), keramadine, oroidin, midpacamide, 4,5-dibromopyrrole-2-carboxylic acid, 4,5-dibromopyrrole-2-carboxamide) present in *Agelas* spp. along with some

Fig. 6.2 *Axinella corrugata,* a sponge containing the previously known antifeedant bromopyrrole alkaloid stevensine (Photo: J. R. Pawlik)

synthetic analogs (*2539, 2540*). A pyrrole ring is required for activity and bromine increases it further. In contrast to other *Agelas* spp., Caribbean *Agelas conifera* contains a mixture of antifeedant dimeric bromopyrrole alkaloids, sceptrin, dibromosceptrin, bromoageliferin, dibromoageliferin, ageliferin, bromosceptrin (**1296**), but dominated by sceptrin (*2540*). In addition to providing a purely defensive role for the sponge, brominated metabolites from *Aplysina fistularis* (Fig. 6.1) such as aerothionin and homoaerothionin may act to clump bacteria together for retention as a food source (*2541*).

Bromophenols represent an enormous class of marine natural products, particularly from acorn worms of families Polychaete and Hemichordata. These sediment dwelling animals can live anywhere from the intertidal zone to a depth of 1,400 m (*2542, 2543*). Thus, one function of 2,4,6-tribromophenol produced by the deep-sea acorn worm *Stereobalanus canadensis* living in the Norwegian Sea is to prevent encroachment by other organisms and to inhibit bacterial growth in the worm's burrow wall (*2543*). Bromophenols from worms of the genus the *Thelepus* may protect the mucous cocoon formed by the tentacles during reproduction and be an antiseptic in wound healing in those protruding and exposed body parts (*1793*). The worm *Notomastus lobatus* lives head down in marine sediments and has the highest concentration of bromophenols in the tail, the animal part first encountered by potential predators (*2150*). Interestingly, 4-bromophenol is also present in this worm, but is not antibacterial against marine sediment bacteria (*2544, 2545*). The worm *Saccoglossus kowalevskii* produces 2,3,4-tribromopyrrole (0.2% of worm dry weight), and this worm is highly unpalatable to predatory fishes (*2546*). Similarly, the burrow tubes of *Sabella pavonia* and *Spirographis spallanzanii* are thought to be strengthened by a halogenation tanning process involving iodination of tyrosine. Iodine can comprise 0.8% of the dry weight of these tubes (*2547*).

The gastropterids *Sagaminopteron nigropunctatum* and *S. psychedelicum*, which contain the diphenyl ether 3,5-dibromo-3-(2′,4′-dibromophenoxy)phenol, acquired through feeding on the sponge *Dysidea granulosa*, deters feeding by the sharpnose pufferfish (*Canthigaster solandri*). This metabolite is transferred to the egg masses of *S. nigropunctatum* where it may offer protection from bacteria (*2548*). The highly toxic cone snail toxin σ-conotoxin GVIIIA, a 41-amino acid peptide, is a highly selective inhibitor of the 5-HT_3 serotonin receptor. It is suggested that a 6-bromotryptophan residue is an important determinant of the pharmacological specificity of this peptide since the endogenous ligand for the 5-HT_3 receptor is 5-hydroxytryptamine, and the 6-bromotryptophan is perhaps situated within a constrained loop of the peptide and assumes a conformation favoring interaction with the 5-HT_3 binding site leading to inactivation of this receptor (*2549*).

A field study of the marine algae organohalogen terpenoids elatol, isolaurinterol, and cymopol, when coated on the palatable seagrass *Thalassia testudinium*, showed significant antifeeding activity towards the herbivorous sea urchin *Diadema antillarum* and reef fishes (*2550*). The brominated furanones present in the red alga *Delisea pulchra*, and acquired by the sea hare *Aplysia parvula* through feeding, not only seem to function as predator deterrents but also may serve as chemical camouflage, since the color of this sea hare closely mimics that of the alga (*2551*). This unique defensive strategy may be more common than commonly thought (*2552*). These furanones are strong inhibitors of both the acyl homoserine lactone and the AI-2 bacterial quorum sensing systems, thus preventing bacterial fouling on *Delisea pulchra* (*2553–2557, 2660*). The sea hare *Aplysia parvula* feeds on this red alga so as to acquire these furanones for apparent chemical defense (*2558*). The deactivation of these bacterial quorum sensing systems by natural haloperoxidases and oxidized halogen species has been reported (*2559*).

The red alga *Plocamium hamatum* produces a metabolite, chloromertensene, that exhibits allelopathy towards the octocoral *Sinularia cruciata*, causing tissue necrosis upon direct contact and dissuading overgrowth by local soft corals (*2560*). Other marine plants seem to generate chemical defenses in response to challenge by aggressors (*2561*), and this induced chemical defense is seen in other macroalgae

(*2562*). Whereas biofouling in the cultivation of *Gracilaria* spp. is a major, global problem (*2563*), the production of bromoform and dibromoacetaldehyde by the red algae *Corallina pilulifera* (*2282*) and *Kappaphycus alvarezii* (*2244*), respectively, seem to play an important role in preventing overgrowth by microalgae, at least with *Corallina pilulifera*. Antifeeding organohalogens have been identified in *Laurencia saitoi* (against young abalone and young sea urchin) (*2564*), and in *Laurencia obtusa* (against crab and sea urchin) (*2565*). Several halogenated monoterpenes, including furoplocamioid C (**229**), are efficient aphid repellents (*2566*), and four hapalindoles, two of which are chlorinated (hapalindole L and 12-*epi*-hapalindole E isonitrile), from the freshwater cyanobacterium *Fischerella* ATCC 43239 are potent insecticides against a dipteran (*2567*).

A role of natural haloalkanes is to cycle halogen/halide between the ocean, atmosphere, and land. This massive global halogen cycle is well established, and excellent reviews are available covering chlorine (*37, 85, 298, 299, 2568*), bromine (*96, 99*), and iodine (*104*). Chloromethane has a major impact on atmospheric ozone (*2569, 2576*) and recent studies suggest that abiotic (and biogenic) methylation of chloride in plants and soil produces the majority of atmospheric chloromethane (*2236, 2569–2571*). Thus, laboratory studies of ferns, a moss, and halophilous plants emit significant amounts of chloromethane (*2236, 2571*), and exhibit uptake of ^{36}Cl-chloride and release of ^{36}Cl-chloromethane (*2236*). It has been suggested that the biosynthesis of halomethanes is the result of "accidents of metabolism" and that the main function of haloperoxidases is to remove hydrogen peroxide (*2133*). Sea salt spray is known to be a source of atmospheric chlorine (Sect. 2.1 (Marine Environment)). A new mechanism for the oxidation of sea salt chloride to chlorine involves reaction with dinitrogen pentoxide (N_2O_5) to nitryl chloride (NO_2Cl) and then to chlorine (*2572*). A newly proposed role of sea spray is as a "cleansing agent" for air pollution over the ocean (*2573*). The relatively new role of natural bromine and iodine, particularly bromine oxide and iodine oxide, in atmospheric processes was mentioned in Sect. 2.1 (Marine Environment) (*90, 97, 98, 100, 103, 2574, 2575*). The highest concentration of iodine oxide (20 ppt) was recorded off the Antarctic coast, and is likely a photolysis-oxidation product of the marine algae metabolites CH_2I_2 and CH_2IBr (*2575*).

7 Significance

Combined with those presented in the first review (*1*), the number of naturally occurring organohalogens – biogenic and abiotic – is more than 4,700. Despite this staggering figure, the quantities of individual organohalogens present in the environment at any given time are largely unknown. A few examples were cited earlier (*1*).

Like other natural products, naturally occurring organohalogens can display a plethora of biological activities (*12, 2577*). In particular, marine natural products – 15–20% of which contain halogen – are of great interest and show enormous potential for the treatment of human disease (*2578–2587, 2667*), against cancer (*2588*), inflammation (*2589*), malaria (*2590*), tuberculosis (*2591*), and others (*1187, 2586, 2587, 2592, 2604*). Marine algae are a treasure trove of biologically active natural products (*2593–2595*), as are symbiotic bacteria in sponges (*121, 2596*). Marine organisms have also furnished numerous insecticidal agents, several examples of which are shown earlier (**682, 685, 1466**) (*2597, 2598*). Terpene isonitriles from sponges exhibit antimalarial activity (*2599*), and sponge-derived terpenoids are potent and selective lipoxygenase inhibitors, such as chloropuupehenone (*2600*). The organohalogen-rich sponges *Aplysina aerophoba* and *Aplysina cavernicola* possess antimicrobial activity (*2601*), and *Verongia aerophoba* displays both antibiotic and cytotoxic activity from aeroplysinin-1 and the dienone metabolite (Scheme 6.1) (*2602*). Hymenialdisine, a bromopyrrole from several sponges, is a novel cyclin-dependent kinase inhibitor (GSK-3β and CK1) (*2603*). The bryozoan *Flustra foliacea* metabolite deformylflustrabromine (**1346**) potentiates the human $\alpha 4\beta 2$ neuronal nicotinic receptor (*2605*). Certain natural organohalogens are calmodulin inhibitors, including eudistomidines A and B, malbrancheamide, konbamide, and several KS-504 compounds (*2606*). Maytansinoids have been conjugated with various agents for specific cell targeting and improved antitumor activity (*2607, 2608*). Several marine sponge metabolites, including the organohalogens bromotopsentin, bastadins 4, 8, and 9, and hymenialdisine display anti-inflammatory activity (*2589*). The chlorine-containing radiciol is a promising lead compound for new anticancer agents (*2609*), and dichloroacetate is in clinical use for the treatment of lactic acidosis (*2610*). Bromophenols from the red alga *Odonthalia corymbifera* are highly fungicidal against the rice pathogen *Magnaporthe grisea*

(*2611*). New vancomycin-type glycopeptide antibiotics are in clinical trials to combat methicillin-resistant bacteria (*2612, 2613*) such as the deadly *Staphylococcus aureus* (*2614*). For the first time, extracts of marine algae indigenous to Japan have shown activity against methicillin-resistant *Staphylococcus aureus* bacteria (*2615, 2661*). The therapeutic use of iodine has been rejuvenated (*2616*). For example, in the treatment of cyclic mastalgia (*2617*).

Cyanobacteria – the Jekyll and Hyde of marine organisms – are a novel source of potential new pharmaceutical compounds (*2618–2620, 2662*). On the other hand, toxic cyanobacterial blooms in lakes, rivers, and water storage reservoirs have occurred worldwide (*2621, 2663, 2664*). For example, 60 patients in a Brazil hemodialysis unit died after drinking water from a lake contaminated with cyanobacterial microcystins (*2622*), not unlike the toxicity of "red tides" (*2623*). Cyanobacteria also produce the highly toxic neurotoxin, β-*N*-methylamino-L-alanine, which may be produced by all cyanobacteria (*2624, 2665*).

There are many examples of the positive, beneficial effects of halogen substitution on organic compounds (*1*), and excellent reviews on this topic are available (*19, 2625*). A chlorinated imidazobenzodiazepinone is 20 times more active than the nonchlorinated analog and three times more active than AZT towards HIV-I (*2626*). Likewise, halogenated (chlorine-, bromide-, and iodine-substituted) gomisin J derivatives are more effective than the natural product itself as HIV-1 reverse transcriptase inhibitors (*2627*). Halogen substitution on aromatic rings greatly stabilizes cross-strand aromatic rings in model β-hairpin peptides (*2628*), which is similar to the iodine-aromatic ring interaction between the thyroid hormone triiodothyroxine, T_3, and the thyroid hormone receptor.

8 Outlook

The dozen years since the publication of the first survey of naturally occurring organohalogens (*1*) has seen an approximate doubling of these new natural compounds, from 2,448 to 4,715, with no sign of abatement. This increase parallels the revitalization of natural products research in general, and is a consequence of improved collection, isolation, and identification techniques. An awareness of ethnobotany and folk medicine leads natural products scientists to potentially biologically active organisms. Multidimensional nuclear magnetic resonance spectroscopy, and improved X-ray crystallography and high-resolution mass spectrometry methods allow for the characterization of minute quantities of compounds. Cultivation techniques like marine bioprocessing (*116, 117, 2629*) permit the harvesting of target marine organisms without plundering the ocean. Remote submersibles can access otherwise inaccessible ocean depths for new marine organisms, such as a new Woods Hole Oceanographic Institution vessel capable of diving to 6,500 m (*2630*). This will allow for the sampling of marine bacteria and other organisms on the ocean floor; for example, the iron-oxidizing bacteria living around deep-sea thermal vents (*2631*) and other deep-sea organisms (*2632, 2633*). Moreover, marine bacteria, in general, are a new field of natural products exploration with enormous possibilities for the discovery of new natural halogenated compounds (e.g., salinosporamide A (**1124**)) (*2634–2636*), especially considering that seawater contains as many as 10^6 bacteria cm^{-3} (*2634*). Marine and terrestrial fungi are also a relatively untapped source of natural products. Indeed, of the 1.5 million estimated terrestrial fungal species on Earth, only 70,000 have been described, let alone examined for their chemical content (*181, 186*). Similarly, marine fungi and terrestrial mosses (bryophytes) are virtually unexplored for their metabolites (*188*). New species of sponges continue to be discovered (*115, 2637*), and it has been suggested that in the oceans sponges can undergo comparatively rapid evolution leading to new species with novel metabolites (*2638*). Furthermore, with bacterial densities as high as 10^{10} bacteria g^{-1} of sponge wet weight, sponges are "microbial fermenters" (*2639*), and only a fraction of the 12,000 extant sponges have been studied for their chemical composition (*112*). The ocean crust is also an abundant repository of microbes (*2640*) and has been for 3.5 billion years (*2641*).

Abiotic sources of organohalogens will continue to be a major contributor to the environment. Global warming may be leading to more wildfires, adding 3.5×10^{15} g of carbon to the atmospheric carbon budget annually (40% of fossil fuel carbon emissions) (*225*). The world's volcanoes continue to be active, producing massive quantities of HCl and HF as they have done for eons (*2642*). Likewise, the interaction of lava with seawater produces significant quantities of HCl (*221*). Ancient sediments continue to reveal the presence of organohalogens presumably of natural origin (*e.g., 157, 2069–2091*), and the abiotic formation of simple organohalogens in soil is a rapidly developing area of research (*175, 177*). Of academic interest are the observations of HCl and HF on Venus (*2643*) and chloride salts on Mars (*2644*). A potentially highly significant newly discovered source of organohalogens is the abiotic (or biogenic) decomposition of leaf litter leading to as yet unidentified organohalogen compounds (*172–174*). Human breath contains a number of aliphatic and aromatic chlorine compounds that are suggested to be exogenous in origin (*2645*).

A major technical advance in the study of organohalogen compounds is the use of ^{14}C radiocarbon analysis to distinguish natural (high ^{14}C content) from anthropogenic (low or no ^{14}C content) organohalogens (*1223, 1746, 1789, 2646, 2647*). Given sufficient material for analysis, this technique would unequivocally identify the origin of chloroform, chlorophenols, bromophenols, dioxins, brominated diphenyl ethers, and several other compounds that have both natural and anthropogenic sources.

References

1. Gribble GW (1996) Naturally Occurring Organohalogen Compounds – A Comprehensive Survey. Prog Chem Org Nat Prod **68**: 1
2. Blunt JW, Copp BR, Hu W-P, Munro MHG, Northcote PT, Prinsep MR (2007) Marine Natural Products. Nat Prod Rep **24**: 31
3. Blunt JW, Copp BR, Munro MHG, Northcote PT, Prinsep MR (2006) Marine Natural Products. Nat Prod Rep **23**: 26
4. Vetter W (2006) Marine Halogenated Natural Products of Environmental Relevance. Rev Environ Contam Toxicol **188**: 1
5. Blunt JW, Copp BR, Munro MHG, Northcote PT, Prinsep MR (2005) Marine Natural Products. Nat Prod Rep **22**: 15
6. Gribble GW (2005) Organohalogens in Nature? Naturally! The Nucleus **83**(2): 8
7. Blunt JW, Copp BR, Munro MHG, Northcote PT, Prinsep MR (2004) Marine Natural Products. Nat Prod Rep **21**: 1
8. Faulkner DJ, Newman DJ, Cragg GM (2004) Investigations of the Marine Flora and Fauna of the Islands of Palau. Nat Prod Rep **21**: 50
9. Kladi M, Vagias C, Roussis V (2004) Volatile Halogenated Metabolites from Red Algae. Phytochem Rev **3**: 337
10. Gribble GW (2004) Amazing Organohalogens. Am Sci **92**: 342
11. Gribble GW (2004) Natural Organohalogens. Euro Chlor Science Dossier. Brussels, p 1
12. Gribble GW (2004) Natural Organohalogens: A New Frontier for Medicinal Agents? J Chem Ed **81**: 1441
13. Gribble GW (2003) Naturally Occurring Halogenated Pyrroles and Indoles. Prog Heterocycl Chem **15**: 58
14. Gribble GW (2003) Organohalogens, Naturally. Educ Chem **40**(3): 40
15. Blunt JW, Copp BR, Munro MHG, Northcote PT, Prinsep MR (2003) Marine Natural Products. Nat Prod Rep **20**: 1
16. Gribble GW (2003) Commercial Potential of Naturally Occurring Halo-Organics. Special Chem Mag **23**(6): 22
17. Murphy CD (2003) New Frontiers in Biological Halogenation. J Appl Microbiol **94**: 539
18. Gribble GW (2003) The Diversity of Naturally Produced Organohalogens. Chemosphere **52**: 289
19. Naumann K (2003) How Chlorine in Molecules Affects Biological Activity. Euro Chlor Science Dossier. Brussels, p 1
20. Neilson AH (2003) Biological Effects and Biosynthesis of Brominated Metabolites. In: Hutziner O (ed) The Handbook of Environmental Chemistry, vol 3, part R. Springer, Berlin, p 75

21. Häggblom MM, Bossert ID (2003) Halogenated Organic Compounds – A Global Perspective. In: Häggblom MM, Bossert ID (eds) Dehalogenation: Microbial Processes and Environmental Applications. Kluwer, Dordrecht, p 3
22. Jones KC (2003) Dioxins and Furans in the Environment. Euro Chlor Science Dossier. Brussels, p 1
23. Gribble GW (2003) The Diversity of Naturally Produced Organohalogens. In: Gribble GW (ed) Natural Production of Organohalogen Compounds, The Handbook of Environmental Chemistry, vol 3, part P. Springer, Berlin, p 1
24. Harper DB, Hamilton JTG (2003) The Global Cycles of the Naturally-Occurring Monohalomethanes. In: Gribble GW (ed) Natural Production of Organohalogen Compounds, The Handbook of Environmental Chemistry, vol 3, part P. Springer, Berlin, p 17
25. Öberg GM (2003) The Biogeochemistry of Chlorine in Soil. In: Gribble GW (ed) Natural Production of Organohalogen Compounds, The Handbook of Environmental Chemistry, vol 3, part P. Springer, Berlin, p 43
26. Schöler HF, Keppler F (2003) Abiotic Formation of Organohalogens During Early Diagenetic Processes. In: Gribble GW (ed) Natural Production of Organohalogen Compounds, The Handbook of Environmental Chemistry, vol 3, part P. Springer, Berlin, p 63
27. Moore RM (2003) Marine Sources of Volatile Organohalogens. In: Gribble GW (ed) Natural Production of Organohalogen Compounds, The Handbook of Environmental Chemistry, vol 3, part P. Springer, Berlin, p 85
28. Field JA, Wijnberg JBPA (2003) An Update on Organohalogen Metabolites Produced by Basidiomycetes. In: Gribble GW (ed) Natural Production of Organohalogen Compounds, The Handbook of Environmental Chemistry, vol 3, part P. Springer, Berlin, p 103
29. Jordan A (2003) Volcanic Formation of Halogenated Organic Compounds. In: Gribble GW (ed) Natural Production of Organohalogen Compounds, The Handbook of Environmental Chemistry, vol 3, part P. Springer, Berlin, p 121
30. Harper DB, O'Hagan D, Murphy CD (2003) Fluorinated Natural Products: Occurrence and Biosynthesis. In: Gribble GW (ed) Natural Production of Organohalogen Compounds, The Handbook of Environmental Chemistry, vol 3, part P. Springer, Berlin, p 141
31. van Pée K-H, Zehner S (2003) Enzymology and Molecular Genetics of Biological Halogenation. In: Gribble GW (ed) Natural Production of Organohalogen Compounds, The Handbook of Environmental Chemistry, vol 3, part P. Springer, Berlin, p 171
32. Henderson JP, Heinecke JW (2003) Myeloperoxidase and Eosinophil Peroxidase: Phagocyte Enzymes for Halogenation in Humans. In: Gribble GW (ed) Natural Production of Organohalogen Compounds, The Handbook of Environmental Chemistry, vol 3, part P. Springer, Berlin, p 201
33. Faulkner DJ (2002) Marine Natural Products. Nat Prod Rep **19**: 1
34. Gribble GW (2002) Naturally Occurring Organofluorines. In: Neilson AH (ed) The Handbook of Environmental Chemistry, vol 3, part N. Organofluorines. Springer, Berlin, p 121
35. Vetter W (2002) Natural Halogenated Organohalogens – A "New" Series of POPs. Organohalogen Compds **59**: 271
36. Dembitsky VM, Srebnik M (2002) Natural Halogenated Fatty Acids: Their Analogues and Derivatives. Prog Lipid Res **41**: 315
37. Öberg G (2002) The Natural Chlorine Cycle – Fitting the Scattered Pieces. Appl Microbiol Biotechnol **58**: 565
38. Faulkner DJ (2001) Marine Natural Products. Nat Prod Rep **18**: 1
39. Laus G (2001) Biological Activities of Natural Halogen Compounds. Stud Nat Prod Chem **25**: 757
40. Faulkner DJ (2000) Marine Natural Products. Nat Prod Rep **17**: 7
41. Gribble GW (2000) The Natural Production of Organobromine Compounds. Environ Sci Pollut Res **7**: 37

42. Harper DB (2000) The Global Chloromethane Cycle: Biosynthesis, Biodegradation and Metabolic Role. Nat Prod Rep **17**: 337
43. Winterton N (2000) Chlorine: The Only Green Element – Towards a Wider Acceptance of Its Role in Natural Cycles. Green Chem **2**: 173
44. Faulkner DJ (1999) Marine Natural Products. Nat Prod Rep **16**: 155
45. Gribble GW (1999) The Diversity of Naturally Occurring Organobromine Compounds. Chem Soc Rev **28**: 335
46. Naumann K (1999) Influence of Chlorine Substituents on Biological Activity of Chemicals. J Prakt Chem **341**: 417
47. Gribble GW (1999) Chlorine – Element from Hell or Gift from God? The Scientific Side of the Chlorine Story. Technology **6**: 193
48. Faulkner DJ (1998) Marine Natural Products. Nat Prod Rep **15**: 113
49. Gribble GW (1998) Naturally Occurring Organohalogen Compounds. Acc Chem Res **31**: 141
50. Gribble GW (1998) Chlorinated Compounds in the Biosphere, Natural Production. In: Meyers RA (ed) Encyclopedia of Environmental Analysis and Remediation. Wiley, New York, p 972
51. Mastalerz P (1998) Biogenic Organohalogen Compounds. Wiad Chem **52**: 643
52. Öberg G (1998) Chloride and Organic Chlorine in Soil. Acta Hydrochim Hydrobiolobiol **26**: 137
53. Faulkner DJ (1997) Marine Natural Products. Nat Prod Rep **14**: 259
54. Faulkner DJ (1996) Marine Natural Products. Nat Prod Rep **13**: 75
55. Salietti Vinué JM (1996) Chlorine Chemistry: I. Today's Problems. Afinidad **53**: 155
56. Gribble GW (1996) The Diversity of Natural Organochlorines in Living Organisms. Pure Appl Chem **68**: 1699
57. Ashby J (1996) Naturally Occurring Organohalogen Compounds. Mut Res **356**: 297
58. Faulkner DJ (1995) Marine Natural Products. Nat Prod Rep **12**: 223
59. Field JA, Verhagen FJM, de Jong E (1995) Natural Organohalogen Production by Basidiomycetes. TIBTECH **13**: 451
60. Hoekstra EJ, De Leer EWB (1995) Organohalogens: The Natural Alternatives. Chem Brit **31**: 127
61. Geckeler KE, Eberhardt W (1995) Biogenic Organic Chlorinated Compounds – Occurrence, Function and Environmental Relevance. Naturwissenschaften **82**: 2
62. Grimvall A, de Leer EWB (1995) Naturally Produced Organohalogens. Kluwer, Dordrecht
63. Mastalerz P (1995) Organic Halogen Compounds in the Biosphere. Wiad Chem **49**: 117
64. Humanes MM, Matoso CM, Da Silva JAL, Fraústo da Silva JJR (1995) Volatile Halogenated Compounds of Natural Origin: Some Aspects of Their Interaction with Atmospheric Ozone. Química **58**: 16
65. Faulkner DJ (1994) Marine Natural Products. Nat Prod Rep **11**: 355
66. Harper DB, O'Hagan D (1994) The Fluorinated Natural Products. Nat Prod Rep **11**: 123
67. McCarty LS (1994) Chlorine and Organochlorines in the Environment: A Perspective. Cand Chem News March 22
68. Gribble GW (1994) Natural Chlorine? You Bet! Priorities **6**(2): 9
69. Gribble GW (1994) Natural Organohalogens – Many More Than You Think! J Chem Ed **71**: 907
70. Naumann K (1994) Chlorine Chemistry in Nature. Actualite Chim Novembre 11
71. Garson MJ (1993) The Biosynthesis of Marine Natural Products. Chem Rev **93**: 1699
72. Strunz GM (1984) Microbial Chlorine-Containing Metabolites. In: Laskin AI, Lechevalier HA (eds) CRC Handbook of Microbiology, 2nd edn, vol V. CRC Press, Boca Raton, FL, p 749
73. Goldberg ED (1963) The Oceans as a Chemical System. In: Hill MN (ed) The Sea, vol 2. Wiley, New York, p 3

74. Hylin JW, Spenger RE, Gunther FA (1969) Potential Interferences in Certain Pesticide Residue Analyses from Organochlorine Compounds Occurring Naturally in Plants. Residue Rev **26**: 127
75. Stijve T (1984) Inorganic Bromide in Higher Fungi. Z Naturforsch **39C**: 863
76. Harper DB (1985) Halomethane from Halide Ion – A Highly Efficient Fungal Conversion of Environmental Significance. Nature **315**: 55
77. Knauth LP (1998) Salinity History of the Earth's Early Ocean. Nature **395**: 554
78. Monks PS (2005) Gas-Phase Radical Chemistry in the Troposphere. Chem Soc Rev **34**: 376
79. Lary DJ (2005) Halogens and the Chemistry of the Free Troposphere. Atmos Chem Phys **5**: 227
80. von Glasow R, von Kuhlmann R, Lawrence MG, Platt U, Crutzen PJ (2004) Impact of Reactive Bromine Chemistry in the Troposphere. Atmos Chem Phys **4**: 2481
81. Bedjanian Y, Poulet G (2003) Kinetics of Halogen Oxide Radicals in the Stratosphere. Chem Rev **103**: 4639
82. Platt U, Hönninger G (2003) The Role of Halogen Species in the Troposphere. Chemosphere **52**: 325
83. Foster KL, Plastridge RA, Bottenheim JW, Shepson PB, Finlayson-Pitts BJ, Spicer CW (2001) The Role of Br_2 and BrCl in Surface Ozone Destruction at Polar Sunrise. Science **291**: 471
84. Graedel TE, Keene WC (1996) The Budget and Cycle of Earth's Natural Chlorine. Pure Appl Chem **68**: 1689
85. Keene WC, Pszenny AAP, Jacob DJ, Duce RA, Galloway JN, Schultz-Tokos JJ, Sievering H, Boatman JF (1990) The Geochemical Cycling of Reactive Chlorine Through the Marine Troposphere. Global Biogeochem Cycles **4**: 407
86. Mozurkewich M (1995) Mechanisms for the Release of Halogens from Sea-Salt Particles by Free Radical Reactions. J Geophys Res **100**: 14199
87. Vogt R, Crutzen PJ, Sander R (1996) A Mechanism for Halogen Release from Sea-Salt Aerosol in the Remote Marine Boundary Layer. Nature **383**: 327
88. Oum KW, Lakin MJ, DeHaan DO, Brauers T, Finlayson-Pitts BJ (1998) Formation of Molecular Chlorine from the Photolysis of Ozone and Aqueous Sea-Salt Particles. Science **279**: 74
89. Spicer CW, Chapman EG, Finlayson-Pitts BJ, Plastridge RA, Hubbe JM, Fast JD, Berkowitz CM (1998) Unexpectedly High Concentrations of Molecular Chlorine in Coastal Air. Nature **394**: 353
90. Behnke W, Elend M, Krüger U, Zetzsch C (1999) The Influence of NaBr/NaCl Ratio on the Br^--Catalysed Production of Halogenated Radicals. J Atmos Chem **34**: 87
91. Moldanová J, Ljungström E (2001) Sea-Salt Aerosol Chemistry in Coastal Areas: A Model Study. J Geophys Res **106**: 1271
92. Herrmann H, Majdik Z, Ervens B, Weise D (2003) Halogen Production from Aqueous Tropospheric Particles. Chemosphere **52**: 485
93. Finlayson-Pitts BJ (2003) The Tropospheric Chemistry of Sea Salt: A Molecular-Level View of the Chemistry of NaCl and NaBr. Chem Rev **103**: 4801
94. Hunt SW, Roeselová M, Wang W, Wingen LM, Knipping EM, Tobias DJ, Dabdub D, Finlayson-Pitts BJ (2004) Formation of Molecular Bromine from the Reaction of Ozone with Deliquesced NaBr Aerosol: Evidence for Interface Chemistry. J Phys Chem A **108**: 11559
95. Tang T, McConnell JC (1996) Autocatalytic Release of Bromine from Arctic Snow Pack During Polar Sunrise. Geophys Res Lett **23**: 2633
96. Daniel JS, Solomon S, Portmann RW, Garcia RR (1999) Stratospheric Ozone Destruction: The Importance of Bromine Relative to Chlorine. J Geophys Res **104**: 23871
97. McElroy CT, McLinden CA, McConnell JC (1999) Evidence for Bromine Monoxide in the Free Troposphere during the Arctic Polar Sunrise. Nature **387**: 338

98. Kaleschke L, Richter A, Burrows J, Afe O, Heygster G, Notholt J, Rankin AM, Roscoe HK, Hollwedel J, Wagner T, Jacobi H-W (2004) Frost Flowers on Sea Ice as a Source of Sea Salt and Their Influence on Tropospheric Halogen Chemistry. Geophys Res Lett **31**: L16114
99. Yang X, Cox RA, Warwick NJ, Pyle JA, Carver GD, O'Connor FM, Savage NH (2005) Tropospheric Bromine Chemistry and Its Impact on Ozone: A Model Study. J Geophys Res **110**: D23311
100. Salawitch RJ, Weisenstein DK, Kovalenko LJ, Sioris CE, Wennberg PO, Chance K, Ko MKW, McLinden CA (2005) Sensitivity of Ozone to Bromine in the Lower Stratosphere. Geophys Res Lett **32**: L05811
101. Matveev V, Peleg M, Rosen D, Tov-Alper DS, Hebestreit K, Stuta J, Platt U, Blake D, Luria M (2001) Bromine Oxide–Ozone Interaction over the Dead Sea. J Geophys Res **106**: 10375
102. Stutz J, Ackermann R, Fast JD, Barrie L (2002) Atmospheric Reactive Chlorine and Bromine at the Great Salt Lake, Utah. Geophys Res Lett **29**: 1380
103. O'Dowd CD, Jimenez JL, Bahreini R, Flagan RC, Seinfeld JH, Hämeri K, Pirjola L, Kulmala M, Jennings SG, Hoffmann T (2002) Marine Aerosol Formation from Biogenic Iodine Emissions. Nature **417**: 632
104. Carpenter LJ (2003) Iodine in the Marine Boundary Layer. Chem Rev **103**: 4953
105. O'Dowd CD, Facchini MC, Cavalli F, Ceburnis D, Mircea M, Decesari S, Fuzzi S, Yoon YJ, Putaud J-P (2004) Biogenically Driven Organic Contribution to Marine Aerosol. Nature **431**: 676
106. Bhakuni DS, Rawat DS (2005) Bioactive Marine Natural Products. Springer, New York
107. Gribble GW, unpublished compilation from the data presented in references *2*, *3*, *5*, *7*, *15*, *33*, *38*, *40*.
108. Fenical W (1993) Chemical Studies of Marine Bacteria: Developing a New Resource. Chem Rev **93**: 1673
109. Barbier M (1981) Marine Natural Products. In: Scheuer PJ (ed) Marine Natural Products, vol IV. Academic, New York, p 147
110. Scheuer PJ (1971) Toxins from Marine Invertebrates. Naturwissenschaften **58**: 549
111. Schulte GR, Scheuer PJ (1982) Defense Allomones of Some Marine Mollusks. Tetrahedron **38**: 1857
112. Hooper JNA, van Soest RWM (2002) Systema Porifera: A Guide to the Classification of Sponges. Kluwer, New York
113. Christophersen C (1985) Marine Alkaloids. In: Brossi A (ed) The Alkaloids, vol 24. Academic, New York, p 25
114. Haseley SR, Vermeer HJ, Kamerling JP, Vliegenthart JFG (2001) Carbohydrate Self-Recognition Mediates Marine Sponge Cellular Adhesion. Proc Natl Acad Sci U S A **98**: 9419
115. Lehnert H, van Soest RWM (1996) North Jamaican Deep Fore-Reef Sponges. Beaufortia **46**: 53
116. Duckworth AR, Battershill CN (2003) Developing Farming Structures for Production of Biologically Active Sponge Metabolites. Aquaculture **217**: 139
117. Duckworth A, Battershill C (2003) Sponge Aquaculture for the Production of Biologically Active Metabolites: The Influence of Farming Protocols and Environment. Aquaculture **221**: 311
118. Müller WEG, Grebenjuk VA, Le Pennec G, Schröder H-C, Brümmer F, Hentschel U, Müller IM, Breter H-J (2004) Sustainable Production of Bioactive Compounds by Sponges – Cell Culture and Gene Cluster Approach: A Review. Mar Biotechnol **6**: 105
119. Vacelet J (1975) Electron-Microscope Study of Association Between Bacteria and Sponges of Genus *Verongia*. J Microscopic Biol Cell **23**: 271
120. Bewley CA, Faulkner DJ (1998) Lithistid Sponges: Star Performers or Hosts to the Stars. Angew Chem Int Ed **37**: 2162
121. Faulkner DJ, Harper MK, Haywood MG, Salomon CE, Schmidt EW (2000) Symbiotic Bacteria in Sponges: Sources of Bioactive Substances. In: Fusetani N (ed) Drugs from the Sea. Karger, Basel, p 107

122. Hentschel U, Fieseler L, Wehrl M, Gernert C, Steinert M, Hacker J, Horn M (2003) Microbial Diversity of Marine Sponges. In: Müller WEG (ed) Sponges (Porifera). Springer, Berlin, p 59
123. Piel J (2004) Metabolites from Symbiotic Bacteria. Nat Prod Rep **21**: 519
124. Friedrich AB, Merkert H, Fendert T, Hacker J, Proksch P, Hentschel U (1999) Microbial Diversity in the Marine Sponge *Aplysina cavernicola* (formerly *Verongia cavernicola*) Analyzed by Fluorescence in Situ Hybridization (FISH). Mar Biol **134**: 461
125. Thoms C, Horn M, Wagner M, Hentschel U, Proksch P (2003) Monitoring Microbial Diversity and Natural Product Profiles of the Sponge *Aplysina cavernicola* Following Transplantation. Mar Biol **142**: 685
126. Hentschel U, Schmid M, Wagner M, Fieseler L, Gernert C, Hacker J (2001) Isolation and Phylogenetic Analysis of Bacteria with Antimicrobial Activities from the Mediterranean Sponges *Aplysina aerophoba* and *Aplysina cavernicola*. FEMS Microbiol Ecol **35**: 305
127. Friedrich AB, Fischer I, Proksch P, Hacker J, Hentschel U (2001) Temporal Variation of the Microbial Community Associated with the Mediterranean Sponge *Aplysina aerophoba*. FEMS Microbiol Ecol **38**: 105
128. Hentschel U, Hopke J, Horn M, Friedrich AB, Wagner M, Hacker J, Moore BS (2002) Molecular Evidence for a Uniform Microbial Community in Sponges from Different Oceans. Appl Environ Microbiol **68**: 4431
129. Webster NS, Wilson KJ, Blackall LL, Hill RT (2001) Phylogenetic Diversity of Bacteria Associated with the Marine Sponge *Rhopaloeides odorabile*. Appl Environ Microbiol **67**: 434
130. Webster NS, Hill RT (2001) The Culturable Microbial Community of the Great Barrier Reef Sponge *Rhopaloeides odorabile* is Dominated by an α-Proteobacterium. Mar Biol **138**: 843
131. Montalvo NF, Mohamed NM, Enticknap JJ, Hill RT (2005) Novel Actinobacteria from Marine Sponges. Ant Leeuwenhoek **87**: 29
132. Bartram J, Carmichael WW, Chorus I, Jones G, Skulberg OV (1999) Introduction. In: Bartram J, Chorus I (eds) Toxic Cyanobacteria in Water. E&FN Spon, London, p 1
133. Namikoshi M, Rinehart KL (1996) Bioactive Compounds Produced by Cyanobacteria. J Ind Microbiol **17**: 373
134. Burja AM, Banaigs B, Abou-Mansour E, Burgess JG, Wright PC (2001) Marine Cyanobacteria – A Prolific Source of Natural Products. Tetrahedron **57**: 9347
135. Gerwick WH, Tan LT, Sitachitta N (2001) Nitrogen-Containing Metabolites from Marine Cyanobacteria. In: Cordell GA (ed) The Alkaloids, vol 57. Academic, San Diego, p 75
136. Carmichael WW (1994) The Toxins of Cyanobacteria. Sci Am **270**: 78
137. Hunter PR (1998) Cyanobacterial Toxins and Human Health. J Appl Microbiol **84**: 355
138. Osborne NJT, Webb PM, Shaw GR (2001) The Toxins of *Lyngbya majuscula* and Their Human and Ecological Health Effects. Environ Int **27**: 381
139. Berthold RJ, Borowitzka MA, Mackay MA (1982) The Ultrastructure of *Oscillatoria spongeliae*, the Blue-Green Algal Endosymbiont of the Sponge *Dysidea herbacea*. Phycologia **21**: 327
140. Faulkner DJ, He H, Unson MD, Bewley CA (1993) New Metabolites from Marine Sponges: Are Symbionts Important? Gazz Chim Ital **123**: 301
141. Hinde R, Pironet F, Borowitzka MA (1994) Isolation of *Oscillatoria spongeliae*, the Filamentous Cyanobacterial Symbiont of the Marine Sponge *Dysidea herbacea*. Mar Biol **119**: 99
142. Faulkner DJ, Unson MD, Bewley CA (1994) The Chemistry of Some Sponges and Their Symbionts. Pure Appl Chem **86**: 1983
143. Thacker RW, Starnes S (2003) Host Specificity of the Symbiotic Cyanobacterium *Oscillatoria spongeliae* in Marine Sponges, *Dysidea* spp. Mar Biol **142**: 643
144. Becerro MA, Paul VJ (2004) Effects of Depth and Light on Secondary Metabolites and Cyanobacterial Symbionts of the Sponge *Dysidea granulosa*. Mar Ecol Prog Ser **280**: 115

145. Cimino G, Fontana A, Gavagnin M (1999) Marine Opisthobranch Molluscs: Chemistry and Ecology in Sacoglossans and Dorids. Curr Org Chem **3**: 327
146. Yamada K, Kigoshi H (1997) Bioactive Compounds from the Sea Hares of Two Genera: *Aplysia* and *Dolabella*. Bull Chem Soc Jpn **70**: 1479
147. Ciminiello P, Fattorusso E (2004) Shellfish Toxins – Chemical Studies on Northern Adriatic Mussels. Eur J Org Chem 2533
148. Anthoni U, Nielsen PH, Pereira M, Christophersen C (1990) Bryozoan Secondary Metabolites: A Chemotaxonomical Challenge. Comp Biochem Physiol **96B**: 431
149. Coll JC (1992) The Chemistry and Chemical Ecology of Octocorals (Coelenterata, Anthozoa, Octocorallia). Chem Rev **92**: 813
150. Moore RM, Tokarczyk R, Tait VK, Poulin M, Geen C (1995) Marine Phytoplankton as a Natural Source of Volatile Organohalogens. In: Grimvall A, de Leer EWB (eds) Naturally-Produced Organohalogens. Kluwer, Dordrecht, p 283
151. Klöser H, Quartino ML, Wiencke C (1996) Distribution of Macroalgae and Macroalgal Communities in Gradients of Physical Conditions in Potter Cove, King George Island, Antarctica. Hydrobiologia **333**: 1
152. Laturnus F (2001) Marine Macroalgae in Polar Regions as Natural Sources for Volatile Organohalogens. Environ Sci Pollut Res **8**: 103
153. Jensen PR, Gontang E, Mafnas C, Mincer TJ, Fenical W (2005) Culturable Marine Actinomycete Diversity from Tropical Pacific Ocean Sediments. Environ Microbiol **7**: 1039
154. Maldonado LA, Stach JEM, Pathom-aree W, Ward AC, Bull AT, Goodfellow M (2005) Diversity of Cultivable Actinobacteria in Geographically Widespread Marine Sediments. Antonie van Leeuwenhoek **87**: 11
155. Schippers A, Neretin LN, Kallmeyer J, Ferdelman TG, Cragg BA, Parkes RJ, Jørgensen BB (2005) Prokaryotic Cells of the Deep Sub-Seafloor Biosphere Identified as Living Bacteria. Nature **433**: 861
156. Rouse GW, Goffredi SK, Vrijenhoek RC (2004) *Osedax*: Bone-Eating Marine Worms with Dwarf Males. Science **305**: 668
157. Müller G, Schmitz W (1985) Halogen-Containing Compounds in Aquatic Sediments. Anthropogens and Biogens. Chem-Zeit **109**: 415
158. Asplund G, Grimvall A, Jonsson S (1994) Determination of the Total and Leachable Amounts of Organohalogens in Soil. Chemosphere **28**: 1467
159. Asplund G (1995) Origin and Occurrence of Halogenated Organic Matter in Soil. In: Grimvall A, de Leer EWB (eds) Naturally-Produced Organohalogens. Kluwer, Dordrecht, p 35
160. Öberg G, Börjesson I, Samuelsson B (1996) Net Change in Organically Bound Halogens in Relation to Soil pH. Water, Air, Soil Pollut **89**: 351
161. Müller G, Nkusi G, Schöler HF (1996) Natural Organohalogens in Sediments. J Prakt Chem **338**: 23
162. Kankaanpää HT, Laurén MA, Saares RJ, Heitto LV, Suursaar UK (1997) Distribution of Halogenated Organic Material in Sediments from Anthropogenic and Natural Sources in the Gulf of Finland Catchment Area. Environ Sci Technol **31**: 96
163. Öberg G (1998) Chloride and Organic Chlorine in Soil. Acta Hydrochim Hydrobiol **26**: 137
164. Öberg G, Grøn C (1998) Sources of Organic Halogens in Spruce Forest Soil. Environ Sci Technol **32**: 1573
165. Johansson E, Ebenå G, Sandén P, Svensson T, Öberg G (2001) Organic and Inorganic Chlorine in Swedish Spruce Forest Soil: Influence of Nitrogen. Geoderma **101**: 1
166. Müller G (2003) Sense or *No*-Sense of the Sum Parameter for Water Soluble "Adsorbable Organic Halogens" (AOX) and "Adsorbed Organic Halogens" (AOX-S18) for the Assessment of Organohalogens in Sludges and Sediments. Chemosphere **52**: 371
167. Rodstedth M, Ståhlberg C, Sandén P, Öberg G (2003) Chloride Imbalances in Soil Lysimeters. Chemosphere **52**: 381

168. Suominen KP, Liukkonen M, Salkinoja-Salonen M (2001) Origin of Organic Halogen in Boreal Lakes. J Soils Sediments **1**: 2
169. Keppler F, Biester H (2003) Peatlands: A Major Sink of Naturally Formed Organic Chlorine. Chemosphere **52**: 451
170. Biester H, Keppler F, Putschew A, Martinez-Cortizas A, Petri M (2004) Halogen Retention, Organohalogens, and the Role of Organic Matter Decomposition on Halogen Enrichment in Two Chilean Peat Bogs. Environ Sci Technol **38**: 1984
171. Keppler F, Biester H, Putschew A, Silk PJ, Schöler HF, Müller G (2004) Organoiodine Formation during Humification in Peatlands. Environ Chem Lett **1**: 219
172. Myneni SCB (2002) Formation of Stable Chlorinated Hydrocarbons in Weathering Plant Material. Science **295**: 1039
173. Reina RG, Leri AC, Myneni SCB (2004) Cl K-edge X-ray Spectroscopic Investigation of Enzymatic Formation of Organochlorines in Weathering Plant Material. Environ Sci Technol **38**: 783
174. Leri AC, Hay MB, Lanzirotti A, Rao W, Myneni SCB (2006) Quantitative Determination of Absolute Organohalogen Concentrations in Environmental Samples by X-ray Absorption Spectroscopy. Anal Chem **78**: 5711
175. Keppler F, Eiden R, Niedan V, Pracht J, Schöler HF (2000) Halocarbons Produced by Natural Oxidation Processes During Degradation of Organic Matter. Nature **403**: 298
176. Pracht J, Boenigk J, Isenbeck-Schröter M, Keppler F, Schöler HF (2001) Abiotic Fe(III) Induced Mineralization of Phenolic Substances. Chemopshere **44**: 613
177. Schöler HF, Keppler F (2003) Abiotic Formation of Organohalogens in the Terrestrial Environment. Chimia **57**: 33
178. Bugni TS, Ireland CM (2004) Marine-Derived Fungi: A Chemically and Biologically Diverse Group of Microorganisms. Nat Prod Rep **21**: 143
179. Hunek S, Yoshimura I (1996) Identification of Lichen Substances. Springer, Berlin
180. Heckman DS, Geiser DM, Eidell BR, Stauffer RL, Kardos NL, Hedges SB (2001) Molecular Evidence for the Early Colonization of Land by Fungi and Plants. Science **293**: 1129
181. Hawksworth DL, Rossman AY (1997) Where Are All the Undescribed Fungi? Phytopathology **87**: 888
182. Verhagen FJM, Swarts HJ, Kuyper TW, Wijnberg JBPA, Field JA (1996) The Ubiquity of Natural Adsorbable Organic Halogen Production Among Basidiomycetes. Appl Microbiol Biotechnol **45**: 710
183. Partida-Martinez LP, Hertweck C (2005) Pathogenic Fungus Harbours Endosymbiotic Bacteria for Toxin Production. Nature **437**: 884
184. Yuan X, Xiao S, Taylor TN (2005) Lichen-Like Symbiosis 600 Million Years Ago. Science **308**: 1017
185. Schadt CW, Martin AP, Lipson DA, Schmidt SK (2003) Seasonal Dynamics of Previously Unknown Fungal Lineages in Tundra Soils. Science **301**: 1359
186. Isaka M, Kittakoop P, Kirtikara K, Hywel-Jones NL, Thebtaranonth Y (2005) Bioactive Substances from Insect Pathogenic Fungi. Acc Chem Res **38**: 813
187. Dembitsky VM, Rezanka T, Spízek J, Hanus LO (2005) Secondary Metabolites of Slime Molds (Myxomycetes). Phytochemistry **66**: 747
188. Zinsmeister HD, Becker H, Eicher T (1991) Bryophytes, a Source of Biologically Active, Naturally Occurring Material? Angew Chem Int Ed Engl **30**: 130
189. Asakawa Y (2004) Chemosystematics of the Hepaticae. Phytochemistry **65**: 623
190. Grimaldi FS, Ingram B, Cuttitta F (1955) Determination of Small and Large Amounts of Fluorine in Rocks. Anal Chem **27**: 918
191. Johns WD, Huang WH (1967) Distribution of Chlorine in Terrestrial Rocks. Geochim Cosmochim Acta **31**: 35
192. Fuge R, Power GM (1969) Fluorine and Chlorine in Some Granitic Rocks. Proc Ussher Soc **2**: 43

193. Fuge R, Power GM (1969) Chlorine and Fluorine in Granitic Rocks from S.W. England. Geochim Cosmochim Acta **33**: 888
194. Fuge R, Power GM (1969) Chlorine in Tourmalines from S.W. England. Mineralog Mag **37**: 293
195. Yoshida M, Takahashi K, Yonehara N, Ozawa T, Iwasaki I (1971) The Fluorine, Chlorine, Bromine, and Iodine Contents of Volcanic Rocks in Japan. Bull Chem Soc Jpn **44**: 1844
196. Sigvaldason GE, Oskarsson N (1976) Chlorine in Basalts from Iceland. Geochim Cosmochim Acta **40**: 777
197. Nedachi M (1980) Chlorine and Fluorine Contents of Rock-Forming Minerals of the Neogene Granitic Rocks in Kyushu, Japan. Mining Society Special Issue No. 8 39
198. Webster JD (1992) Fluid-Melt Interactions Involving Cl-Rich Granites: Experimental Study from 2 to 8 kbar. Geochim Cosmochim Acta **56**: 659
199. Harnisch J, Eisenhauer A (1998) Natural CF_4 and SF_6 on Earth. Geophys Res Lett **25**: 2401
200. Schilling J-G, Bergeron MB, Evans R (1980) Halogens in the Mantle Beneath the North Atlantic. Phil Trans R Soc London A **297**: 147
201. Michael PJ, Schilling J-G (1989) Chlorine in Mid-Ocean Ridge Magmas: Evidence for Assimilation of Seawater-Influenced Components. Geochim Cosmochim Acta **53**: 3131
202. Pennisi M, LeCloarec M-F (1998) Variations of Cl, F, and S in Mount Etna's Plume, Italy, Between 1992 and 1995. J Geophys Res **103**: 5061
203. Francis P, Burton MR, Oppenheimer C (1998) Remote Measurements of Volcanic Gas Compositions by Solar Occulation Spectroscopy. Nature **396**: 567
204. Aiuppa A, Federico C, Paonita A, Pecoraino G, Valenza M (2002) S, Cl and F Degassing as an Indicator of Volcanic Dynamics: The 2001 Eruption of Mount Etna. Geophys Res Lett **29**: 1559
205. Aiuppa A, Federico C, Giudice G, Gurrieri S, Paonita A, Valenza M (2004) Plume Chemistry Provides Insights into Mechanisms of Sulfur and Halogen Degassing in Basaltic Volcanoes. Earth Planet Sci Lett **222**: 469
206. Bureau H, Keppler H, Métrich N (2000) Volcanic Degassing of Bromine and Iodine: Experimental Fluid/Melt Partitioning Data and Applications to Stratospheric Chemistry. Earth Planet Sci Lett **183**: 51
207. Oppenheimer C, Francis P, Maciejewski AJH (1998) Spectroscopic Observation of HCl Degassing from Soufriere Hills Volcano, Montserrat. Geophys Res Lett **25**: 3689
208. Bobrowski N, Hönninger G, Galle B, Platt U (2003) Detection of Bromine Monoxide in a Volcanic Plume. Nature **423**: 273
209. Love SP, Goff F, Counce D, Siebe C, Delgado H (1998) Passive Infrared Spectroscopy of the Eruption Plume at Popocatepetl Volcano, Mexico. Nature **396**: 563
210. Shinohara H, Witter JB (2005) Volcanic Gases Emitted During Mild Strombolian Activity of Villarrica Volcano, Chile. Geophys Res Lett **32**: L20308
211. Snyder GT, Fehn U, Goff F (2002) Iodine Isotope Ratios and Halide Concentrations in Fluids of the Satsuma-Iwojima Volcano, Japan. Earth Planet Space **54**: 265
212. Lee C, Kim YJ, Tanimoto H, Bobrowski N, Platt U, Mori T, Yamamoto K, Hong CS (2005) High ClO and Ozone Depletion Observed in the Plume of Sakurajima Volcano, Japan. Geophys Res Lett **32**: L21809
213. Thordarson Th, Self S, Óskarsson N, Hulsebosch T (1996) Sulfur, Chlorine, and Fluorine Degassing and Atmospheric Loading by the 1783–1784 AD Laki (Skaftár Fires) Eruption in Iceland. Bull Volcanol **58**: 205
214. Oppenheimer C, Francis P, Burton M, Maciejewski AJH, Boardman L (1998) Remote Measurement of Volcanic Gases by Fourier Transform Infrared Spectroscopy. Appl Phys B **67**: 505
215. Gerlach TM (2004) Volcanic Sources of Tropospheric Ozone-Depleting Trace Gases. Geochem Geophys Geosyst **5**: 1

216. Schwandner FM, Seward TM, Gize AP, Hall PA, Dietrich VJ (2004) Diffuse Emission of Organic Trace Gases from the Flank and Crater of a Quiescent Active Volcano (Vulcano, Aeolian Islands, Italy). J Geophys Res **109**: D04301
217. Jordan A, Harnisch J, Borchers R, Le Guern F, Shinohara H (2000) Volcanogenic Halocarbons. Environ Sci Technol **34**: 1122
218. Isidorov VA (1990) Organic Chemistry of the Earth's Atmosphere. Springer, Berlin, p 72
219. Capaccioni B, Martini M, Mangani F (1995) Light Hydrocarbons in Hydrothermal and Magmatic Fumaroles: Hints of Catalytic and Thermal Reactions. Bull Volcanol **56**: 593
220. Hoffmann R (2001) Thermophiles in Kamchatka. Am Sci **89**: 20
221. Edmonds M, Gerlach TM (2006) The Airborne Lava–Seawater Interaction Plume at Kilauea Volcano, Hawai'i. Earth Planet Sci Lett **244**: 83
222. Goren-Inbar N, Alperson N, Kislev ME, Simchoni O, Melamed Y, Ben-Nun A, Werker E (2004) Evidence of Hominin Control of Fire at Gesher Benot Ya'aqov, Israel. Science **304**: 725
223. Levine JS, Cofer III WR, Cahoon Jr. DR, Winstead EL (1995) Biomass Burning. A Driver for Global Change. Environ Sci Technol **29**: 120A
224. Cochrane MA (2003) Fire Science for Rainforests. Nature **421**: 913
225. Running SW (2006) Is Global Warming Causing More, Larger Wildfires? Science **313**: 927
226. Page SE, Siegert F, Rieley JO, Boehm H-DV, Jaya A, Limin S (2002) The Amount of Carbon Released from Peatland Forest Fires in Indonesia During 1997. Nature **420**: 61
227. Gabos S, Ikonomou MG, Schopflocher D, Fowler BR, White J, Prepas E, Prince D, Chen W (2001) Characteristics of PAHs, PCDD/Fs and PCBs in Sediment Following Forest Fires in Northern Alberta. Chemosphere **43**: 709
228. Pfister G, Hess PG, Emmons LK, Lamarque J-F, Wiedinmyer C, Edwards DP, Pétron G, Gille JC, Sachse GW (2005) Quantifying CO Emissions from the 2004 Alaskan Wildfires using MOPITT CO Data. Geophys Res Lett **32**: L11809
229. Mollicone D, Eva HD, Achard F (2006) Human Role in Russian Wild Fires. Nature **440**: 436
230. Blake DR, Smith Jr. TW, Chen T-Y, Whipple WJ, Rowland FS (1994) Effects of Biomass Burning on Summertime Nonmethane Hydrocarbon Concentrations in the Canadian Wetlands. J Geophys Res **99**: 1699
231. Westerling AL, Hidalgo HG, Cayan DR, Swetnam TW (2006) Warming and Earlier Spring Increase Western US Forest Wildfire Activity. Science **313**: 940
232. Eklund G, Pedersen JR, Strömberg B (1988) Methane, Hydrogen Chloride and Oxygen Form a Wide Range of Chlorinated Organic Species in the Temperature Range 400°C–950° C. Chemosphere **17**: 575
233. Müller G (1995) Cadmium Concentration in Tobacco of Popular Smoked Cigarettes in Germany 1978, 1985, and 1994: A Comparison. Naturwissenschaften **82**: 135
234. Blake GA, Keene J, Phillips TG (1985) Chlorine in Dense Interstellar Clouds: The Abundance of HCl in OMC-1. Astrophys J **295**: 501
235. Neufeld DA, Zmuidzinas J, Schilke P, Phillips TG (1997) Discovery of Interstellar Hydrogen Fluoride. Astrophys J **488**: L141
236. Schöler HF, Nkusi G, Niedan VW, Müller G, Spitthoff B (2005) Screening of Organic Halogens and Identification of Chlorinated Benzoic Acids in Carbonaceous Meteorites. Chemosphere **60**: 1505
237. Mueller G (1953) The Properties and Theory of Genesis of the Carbonaceous Complex within the Cold Bokevelt Meteorite. Geochim Cosmochim Acta **4**: 1
238. Hayes JM (1967) Organic Constituents of Meteorites – A Review. Geochim Cosmochim Acta **31**: 1395
239. Studier MH, Hayatsu R, Anders E (1968) Origin of Organic Matter in Early Solar System – I. Hydrocarbons. Geochim Cosmochim Acta **32**: 151
240. Hayes JM, Biemann K (1968) High Resolution Mass Spectrometric Investigations of the Organic Constituents of the Murray and Holbrook Chondrites. Geochim Cosmochim Acta **32**: 239

241. Küppers M, Schneider NM (2000) Discovery of Chlorine in the Io Torus. Geophys Res Lett **27**: 513
242. Fegley Jr. B, Zolotov MYu (2000) Chemistry of Sodium, Potassium, and Chlorine in Volcanic Gases on Io. Icarus **148**: 193
243. Feldman PD, Ake TB, Berman AF, Moos HW, Sahnow DJ, Strobel DF, Weaver HA, Young PR (2001) Detection of Chlorine Ions in the Far Ultraviolet Spectroscopic Explorer Spectrum of the Io Plasma Torus. Astrophys J **554**: L123
244. Moses JI, Zolotov MYu, Fegley Jr. B (2002) Alkali and Chlorine Photochemistry in a Volcanically Driven Atmosphere on Io. Icarus **156**: 107
245. Lellouch E, Paubert G, Moses JI, Schneider NM, Strobel DF (2003) Volcanically Emitted Sodium Chloride as a Source for Io's Neutral Clouds and Plasma Torus. Nature **421**: 45
246. Khalil MAK, Rasmussen RA (1999) Atmospheric Methyl Chloride. Atmos Environ **33**: 1305
247. Harper DB (1995) The Contribution of Natural Halogenation Processes to the Atmospheric Halomethane Burden. In: Grimvall A, de Leer EWB (eds) Naturally-Produced Organohalogens. Kluwer, Dordrecht, p 21
248. Scarratt MG, Moore RM (1996) Production of Methyl Chloride and Methyl Bromide in Laboratory Cultures of Marine Phytoplankton. Mar Chem **54**: 263
249. Scarratt MG, Moore RM (1998) Production of Methyl Bromide and Methyl Chloride in Laboratory Cultures of Marine Phytoplankton II. Mar Chem **59**: 311
250. Baker JM, Sturges WT, Sugier J, Sunnenberg G, Lovett AA, Reeves CE, Nightingale PD, Penkett SA (2001) Emissions of CH_3Br, Organochlorines, and Organoiodines from Temperate Macroalgae. Chemosphere – Glob Change Sci **3**: 93
251. Itoh N, Tsujita M, Ando T, Hisatomi G, Higashi T (1997) Formation and Emission of Monohalomethanes from Marine Algae. Phytochemistry **45**: 67
252. Rhew RC, Miller BR, Weiss RF (2000) Natural Methyl Bromide and Methyl Chloride Emissions from Coastal Salt Marshes. Nature **403**: 292
253. Khalil MAK, Moore RM, Harper DB, Lobert JM, Erickson DJ, Koropalov V, Sturges WT, Keene WC (1999) Natural Emissions of Chlorine-Containing Gases: Reactive Chloride Emissions Inventory. J Geophys Res **104**: 8333
254. Butler JH, Battle M, Bender ML, Montzka SA, Clarke AD, Saltzman ES, Sucher CM, Severinghaus JP, Elkins JW (1999) A Record of Atmospheric Halocarbons During the Twentieth Century from Polar Firn Air. Nature **399**: 749
255. Li H-J, Yokouchi Y, Akimoto H (1999) Measurement of Methyl Halides in the Marine Atmosphere. Atmos Environ **33**: 1881
256. Cox ML, Sturrock GA, Fraser PJ, Siems ST, Krummel PD, O'Doherty S (2003) Regional Sources of Methyl Chloride, Chloroform and Dichloromethane Identified from AGAGE Observations at Cape Grim, Tasmania, 1998–2000. J Atmos Chem **45**: 79
257. Kaspars KA, van de Wal RSW, de Gouw JA, Hofstede CM, van den Broeke MR, van den Veen C, Neubert REM, Meijer HAJ, Brenninkmeijer CAM, Karlöf L, Winther J-G (2004) Analysis of Firn Gas Samples from Dronning Maud Land, Antarctica: Study of Nonmethane Hydrocarbons and Methyl Chloride. J Geophys Res **109**: D02307
258. Aydin M, Saltzman ES, De Bruyn WJ, Montzka SA, Butler JH, Battle M (2004) Atmospheric Variability of Methyl Chloride During the Last 300 Years from an Antarctic Ice Core and Firn Air. Geophys Res Lett **31**: L02109
259. Trudinger CM, Etheridge DM, Sturrock GA, Fraser PJ, Krummel PB, McCulloch A (2004) Atmospheric Histories of Halocarbons from Analysis of Antarctic Firn Air: Methyl Bromide, Methyl Chloride, Chloroform, and Dichloromethane. J Geophys Res **109**: D22310
260. Saito T, Yokouchi Y, Aoki S, Nakazawa T, Fujii Y, Watanabe O (2006) A Method for Determination of Methyl Chloride Concentration in Air Trapped in Ice Cores. Chemosphere **63**: 1209
261. Yokouchi Y, Noijiri Y, Barrie LA, Toom-Sauntry D, Machida T, Inuzuka Y, Akimoto H, Li H-J, Fujinuma Y, Aoki S (2000) A Strong Source of Methyl Chloride to the Atmosphere from Tropical Coastal Land. Nature **403**: 295

262. Redeker KR, Wang N-Y, Low JC, McMillan A, Tyler SC, Cicerone RJ (2000) Emissions of Methyl Halides and Methane from Rice Paddies. Science **290**: 966
263. Redeker KR, Andrews J, Fisher F, Sass R, Cicerone RJ (2002) Interfield and Intrafield Variability of Methyl Halide Emissions from Rice Paddies. Global Biogeochem Cycles **16**: 1125
264. Redeker KR, Manley SL, Walser M, Cicerone RJ (2004) Physiological and Biochemical Controls over Methyl Halide Emissions from Rice Plants. Global Biogeochem Cycles **18**: GB1007
265. Redeker KR, Cicerone RJ (2004) Environmental Controls over Methyl Halide Emissions from Rice Paddies. Global Biogeochem Cycles **18**: GB1027
266. Lee-Taylor J, Redeker KR (2005) Reevaluation of Global Emissions from Rice Paddies of Methyl Iodide and Other Species. Geophys Res Lett **32**: L15801
267. Dimmer CH, Simmonds PG, Nickless G, Bassford MR (2001) Biogenic Fluxes of Halomethanes from Irish Peatland Ecosystems. Atmos Environ **35**: 321
268. Yokouchi Y, Ikeda M, Inuzuka Y, Yukawa T (2002) Strong Emission of Methyl Chloride from Tropical Plants. Nature **416**: 163
269. Rhew RC, Miller BR, Vollmer MK, Weiss RF (2001) Shrubland Fluxes of Methyl Bromide and Methyl Chloride. J Geophys Res **106**: 20875
270. Varner RK, Crill PM, Talbot RW (1999) Wetlands: A Potentially Significant Source of Atmospheric Methyl Bromide and Methyl Chloride. Geophys Res Lett **26**: 2433
271. Watling R, Harper DB (1998) Chloromethane Production by Wood-Rotting Fungi and an Estimate of the Global Flux to the Atmosphere. Mycol Res **102**: 769
272. Moore RM, Gut A, Andreae MO (2005) A Pilot Study of Methyl Chloride Emissions from Tropical Woodrot Fungi. Chemosphere **58**: 221
273. Redeker KR, Treseder KK, Allen MF (2004) Ectomycorrhizal Fungi: A New Source of Atmospheric Methyl Halides? Global Change Biol **10**: 1009
274. Isidorov V, Jdanova M (2002) Volatile Organic Compounds from Leaves Litter. Chemosphere **48**: 975
275. Cox ML, Fraser PJ, Sturrock GA, Siems ST, Porter LW (2004) Terrestrial Sources and Sinks of Halomethanes Near Cape Grim, Tasmania. Atmos Environ **38**: 3839
276. Harper DB, Harvey BMR, Jeffers MR, Kennedy JT (1999) Emissions, Biogenesis and Metabolic Utilization of Chloromethane by Tubers of the Potato (*Solanum tuberosum*). New Phytol **142**: 5
277. Harper DB, Kalin RM, Hamilton JTG, Lamb C (2001) Carbon Isotope Ratios for Chloromethane of Biological Origin: Potential Tool in Determining Biological Emissions. Environ Sci Technol **35**: 3616
278. Laturnus F, Fahimi I, Gryndler M, Hartmann A, Heal MR, Matucha M, Schöler HF, Schroll R, Svensson T (2005) Natural Formation and Degradation of Chloroacetic Acids and Volatile Organochlorines in Soil. Environ Sci Pollut Res **12**: 233
279. Harper DB, Hamilton JTG, Ducrocq V, Kennedy JT, Downey A, Kalin RM (2003) The Distinctive Isotopic Signature of Plant-Derived Chloromethane: Possible Application in Constraining the Atmospheric Chloromethane Budget. Chemosphere **52**: 433
280. Keppler F, Harper DB, Röckmann T, Moore RM, Hamilton JTG (2005) New Insight into the Atmospheric Chloromethane Budget Gained using Stable Carbon Isotope Ratios. Atmos Chem Phys **5**: 2403
281. Hamilton JTG, McRoberts WC, Keppler F, Kalin RM, Harper DB (2003) Chloride Methylation by Plant Pectin: An Efficient Environmentally Significant Process. Science **301**: 206
282. Reinhardt TE, Ward DE (1995) Factors Affecting Methyl Chloride Emissions from Forest Biomass Combustion. Environ Sci Technol **29**: 825
283. Rudolph J, Koppmann AKR, Bonsang B (1995) Field Study of the Emissions of Methyl Chloride and Other Halocarbons from Biomass Burning in Western Africa. J Atmos Chem **22**: 67

284. Blake NJ, Blake DR, Sive BC, Chen T-Y, Rowland FS, Collins Jr. JE, Sachse GW, Anderson BE (1996) Biomass Burning Emissions and Vertical Distribution of Atmospheric Methyl Halides and Other Reduced Carbon Gases in the South Atlantic Region. J Geophys Res **101**: 24151
285. Andreae MO, Atlas E, Harris GW, Helas G, de Kock A, Koppmann R, Maenhaut W, Manø S, Pollock WH, Rudolph J, Scharffe D, Schebeske G, Welling M (1996) Methyl Halide Emissions from Savanna Fires in Southern Africa. J Geophys Res **101**: 23603
286. Lobert JM, Keene WC, Logan JA, Yevich R (1999) Global Chlorine Emissions from Biomass Burning: Reactive Chlorine Emissions Inventory. J Geophys Res **104**: 8373
287. Keene WC, Khalil MAK, Erickson III DJ, McCulloch A, Graedel TE, Lobert JM, Aucott ML, Gong SL, Harper DB, Kleiman G, Midgley P, Moore RM, Seuzaret C, Sturges WT, Benkovitz CM, Koropalov V, Barrie LA, Li YF (1999) Composite Global Emissions of Reactive Chlorine from Anthropogenic and Natural Sources: Reactive Chlorine Emissions Inventory. J Geophys Res **104**: 8429
288. Andreae MO, Merlet P (2001) Emission of Trace Gases and Aerosols from Biomass Burning. Global Biogeochem Cycles **15**: 955
289. Harper DB (1995) Biosynthesis and Metabolic Role of Chloromethane in Fungi. In: Grimvall A, de Leer EWB (eds) Naturally-Produced Organohalogens. Kluwer, Dordrecht, p 235
290. Urhahn T, Ballschmiter K (1998) Chemistry of the Biosynthesis of Halogenated Methanes: C1-Organohalogens as Pre-Industrial Chemical Stressors in the Environment? Chemosphere **37**: 1017
291. Ni X, Hager LP (1998) cDNA Cloning of *Batis maritima* Methyl Chloride Transferase and Purification of the Enzyme. Proc Natl Acad Sci U S A **95**: 12866
292. Ni X, Hager LP (1999) Expression of *Batis maritima* Methyl Chloride Transferase in *Escherichia coli*. Proc Natl Acad Sci U S A **96**: 3611
293. Harper DB, Kalin RM, Larkin MJ, Hamilton JTG, Coulter C (2000) Microbial Transhalogenation: A Complicating Factor in Determination of Atmospheric Chloro- and Bromomethane Budgets. Environ Sci Technol **34**: 2525
294. Streitwieser Jr. A (1962) Solvolytic Displacement Reactions. McGraw-Hill, New York, p 167
295. Yoshida Y, Wang Y, Zeng T, Yantosca R (2004) A Three-Dimensional Global Model Study of Atmospheric Methyl Chloride Budget and Distributions. J Geophys Res **109**: D24309
296. McCulloch A, Aucott ML, Benkovitz CM, Graedel TE, Kleiman G, Midgley PM, Li YF (1999) Global Emissions of Hydrogen Chloride and Chloromethane from Coal Combustion, Incineration, and Industrial Activities: Reactive Chlorine Emissions Inventory. J Geophys Res **104**: 8391
297. Lee-Taylor JM, Brasseur GP, Yokouchi Y (2001) A Preliminary Three-Dimensional Global Model Study of Atmospheric Methyl Chloride Distribution. J Geophys Res **106**: 34221
298. Öberg G, Sandén P (2005) Retention of Chloride in Soil and Cycling of Organic Matter-Bound Chlorine. Hydrol Process **19**: 2123
299. Öberg G, Holm M, Sandén P, Svensson T, Parikka M (2005) The Role of Organic-Matter-Bound Chlorine in the Chlorine Cycle: A Case Study of the Stubbetorp Catchment, Sweden. Biogeochem **75**: 241
300. Moore RM, Groszko W, Niven JS (1996) Ocean-Atmosphere Exchange of Methyl Chloride: Results from NW Atlantic and Pacific Ocean Studies. J Geophys Res **101**: 28529
301. McCulloch A, Midgley PM (1996) The Production and Global Distribution of Emissions of Trichloroethene, Tetrachloroethene and Dichloromethane over the Period 1988–1992. Atmos Environ **30**: 601
302. Laturnus F, Svensson T, Wiencke C, Öberg G (2004) Ultraviolet Radiation Affects Emission of Ozone-Depleting Substances by Marine Macroalgae: Results from a Laboratory Incubation Study. Environ Sci Technol **38**: 6605

303. Laturnus F (2002) Chloroform in the Environment – Occurrence, Sources, Sinks and Effects, Euro Chlor, Brussels, p 1
304. Laturnus F, Haselmann KF, Borch T, Grøn C (2002) Terrestrial Natural Sources of Trichloromethane (Chloroform, $CHCl_3$) – An Overview. Biogeochem **60**: 121
305. McCulloch A (2003) Chloroform in the Environment: Occurrence, Sources, Sinks and Effects. Chemosphere **50**: 1291
306. Nightingale PB, Malin G, Liss PS (1995) Production of Chloroform and Other Low-Molecular-Weight Halocarbons by Some Species of Macroalgae. Limnol Oceanogr **40**: 680
307. Mtolera MSP, Collén J, Pedersén M, Ekdahl A, Abrahamsson K, Semesi AK (1996) Stress-Induced Production of Volatile Halogenated Organic Compounds in *Eucheuma denticulatum* (Rhodophyta) Caused by Elevated pH and High Light Intensities. Eur J Phycol **31**: 89
308. Ekdahl A, Pedersén M, Abrahamsson K (1998) A Study of the Diurnal Variation of Biogenic Volatile Halocarbons. Mar Chem **63**: 1
309. Abrahamsson K, Choo K-S, Pedersén M, Johansson G, Snoeijs P (2003) Effects of Temperature on the Production of Hydrogen Peroxide and Volatile Hydrocarbons by Brackish-Water Algae. Phytochemistry **64**: 725
310. Scarratt MG, Moore RM (1999) Production of Chlorinated Hydrocarbons and Methyl Iodide by the Red Microalga *Porphyridium purpureum*. Limnol Oceanogr **44**: 703
311. Murphy CD, Moore RM, White RM (2000) An Isotope Labeling Method for Determining Production of Volatile Organohalogens by Marine Microalgae. Limnol Oceanogr **45**: 1868
312. Weissflog L, Lange CA, Pfenningsdorff A, Kotte K, Elansky N, Lisitzyna L, Putz E, Krueger G (2005) Sediments of Salt Lakes as a New Source of Volatile Highly Chlorinated C_1/C_2 Hydrocarbons. Geophys Res Lett **32**: L01401
313. Asplund G, Grimval A (1991) Organohalogens in Nature. Environ Sci Technol **25**: 1346
314. Hoekstra EJ, de Leer EWB (1995) Natural Production of Chlorinated Organic Compounds in Soil. In: Arendt F, Annokkée GJ, Bosman R, van den Brink (eds) Contaminated Soil '93. Kluwer, Dordrecht, p 215
315. Laturnus F, Mehrtens G, Grøn C (1995) Haloperoxidase-Like Activity in Spruce Forest Soil – A Source of Volatile Halogenated Organic Compounds? Chemosphere **31**: 3709
316. Hoekstra EJ, de Leer EWB, Brinkman UATh (1998) Natural Formation of Chloroform and Brominated Trihalomethanes in Soil. Environ Sci Technol **32**: 3724
317. Hoekstra EJ, de Leer EWB, Brinkman UATh (1999) Findings Supporting the Natural Formation of Trichloroacetic Acid in Soil. Chemosphere **38**: 2875
318. Haselmann KF, Ketola RA, Laturnus F, Lauritsen FR, Grøn C (2000) Occurrence and Formation of Chloroform at Danish Forest Sites. Atmos Environ **34**: 187
319. Haselmann KF, Laturnus F, Svensmark B, Grøn C (2000) Formation of Chloroform in Spruce Forest Soil – Results from Laboratory Incubation Studies. Chemosphere **41**: 1769
320. Laturnus F, Lauritsen FR, Grøn C (2000) Chloroform in a Pristine Aquifer System: Toward an Evidence of Biogenic Origin. Wat Resour Res **36**: 2999
321. Hoekstra EJ, Duyzer JH, de Leer EWB, Brinkman UATh (2001) Chloroform – Concentration Gradients in Soil Air and Atmosphere Air, and Emission Fluxes from Soil. Atmos Environ **35**: 61
322. Haselmann KF, Laturnus F, Grøn C (2002) Formation of Chloroform in Soil. A Year-Round Study at a Danish Spruce Forest Site. Water Air Soil Pollut **139**: 35
323. Svensson T, Laturnus F, Sandén P, Öberg G (2007) Chloroform in Runoff Water – A Two-Year Study in a Small Catchment in Southeast Sweden. Biogeochem **82**: 139
324. de Leer EWB, Damsté JSS, Erkelens C, de Galan L (1985) Identification of Intermediates Leading to Chloroform and C-4 Diacids in the Chlorination of Humic Acid. Environ Sci Technol **19**: 512
325. Olson TM, Gonzalez AC, Vasquez VR (2001) Gas Chromatography Analyses for Trihalomethanes: An Experiment Illustrating Important Sources of Disinfection By-Products in Water Treatment. J Chem Ed **78**: 1231

326. Hoekstra EJ, Verhagen FJM, Field JA, de Leer EWB, Brinkman UATh (1998) Natural Production of Chloroform by Fungi. Phytochemistry **49**: 91
327. Khalil MAK, Rasmussen RA, French JRJ, Holt JA (1990) The Influence of Termites on Atmospheric Trace Gases: CH_4, CO, $CHCl_3$, N_2O, CO, H_2, and Light Hydrocarbons. J Geophys Res **95**: 3619
328. Isidorov VA, Zenkevich IG, Karpov GA (1991) Volatile Organic Compounds in Steam-Gas Discharges from Some Volcanoes and Hydrothermal Systems of Kamchatka. Vulkanol Seismol 19
329. Isidorov VA, Prilepskii EB, Fedorov YuN (1991) Organic Components of Thermal Source Gases in Seismically Active Regions. Dokl Akad Nauk SSSR **319**: 1106
330. Buslaeva EYu (1994) Halogen-Substituted Hydrocarbons in Carbonaceous Black Shales. Geokhimiya 1130
331. Duniway JM (2002) Status of Chemical Alternatives to Methyl Bromide for Pre-Plant Fumigation of Soil. Phytopathology **92**: 1337
332. Yates SR, Gan J, Papiernik SK (2003) Environmental Fate of Methyl Bromide as a Soil Fumigant. Rev Environ Contam Toxicol **177**: 45
333. Gan J, Yates SR, Ohr HD, Sims JJ (1997) Volatilization and Distribution of Methyl Iodide and Methyl Bromide after Subsoil Application. J Environ Qual **26**: 1107
334. McCook A (2006) The Banned Pesticide in Our Soil. The Scientist January 40
335. Yang RSH, Witt KL, Alden CJ, Cockerham LG (1995) Toxicology of Methyl Bromide. Rev Environ Contam Toxicol **142**: 65
336. Laturnus F (1995) Release of Volatile Halogenated Organic Compounds by Unialgal Cultures of Polar Macroalgae. Chemosphere **31**: 3387
337. Laturnus F, Wiencke C, Adams FC (1998) Influence of Light Conditions on the Release of Volatile Halocarbons by Antarctic Macroalgae. Mar Environ Res **45**: 285
338. Laturnus F, Adams FC, Wiencke C (1998) Methyl Halides from Antarctic Macroalgae. Geophys Res Lett **25**: 773
339. Moore RM, Webb M, Tokarczyk R, Wever R (1996) Bromoperoxidase and Iodoperoxidase Enzymes and Production of Halogenated Methanes in Marine Diatom Cultures. J Geophys Res **101**: 20899
340. Saemundsdóttir S, Matrai PA (1998) Biological Production of Methyl Bromide by Cultures of Marine Phytoplankton. Limnol Oceanogr **43**: 81
341. Pilinis C, King DB, Saltzmann ES (1996) The Oceans: A Source or a Sink of Methyl Bromide? Geophys Res Lett **23**: 817
342. Schall C, Heumann KG, Kirst GO (1997) Biogenic Volatile Organoiodine and Organobromine Hydrocarbons in the Atlantic Ocean from 42°N to 72°S. Fresenius J Anal Chem **359**: 298
343. Baker JM, Reeves CE, Nightingale PD, Penkett SA, Gibb SW, Hatton AD (1999) Biological Production of Methyl Bromide in the Coastal Waters of the North Sea and Open Ocean of the Northeast Atlantic. Mar Chem **64**: 267
344. Carpenter LJ, Sturges WT, Penkett SA, Liss PS, Alicke B, Hebestreit K, Platt U (1999) Short-Lived Alkyl Iodides and Bromides at Mace Head, Ireland: Links to Biogenic Sources and Halogen Oxide Production. J Geophys Res **104**: 1679
345. Carpenter LJ, Malin G, Liss PS, Küpper FC (2000) Novel Biogenic Iodine-Containing Trihalomethanes and Other Short-Lived Halocarbons in the Coastal East Atlantic. Global Biogeochem Cycl **14**: 1191
346. Sturges WT, McIntyre HP, Penkett SA, Chappellaz J, Barnola J-M, Mulvaney R, Atlas E, Stroud V (2001) Methyl Bromide, Other Brominated Methanes, and Methyl Iodide in Polar Firn Air. J Geophys Res **106**: 1595
347. Gan J, Yates SR, Ohr HD, Sims JJ (1998) Production of Methyl Bromide by Terrestrial Higher Plants. Geophys Res Lett **25**: 3595
348. Jeffers PM, Wolfe NL, Nzengung V (1998) Green Plants: A Terrestrial Sink for Atmospheric CH_3Br. Geophys Res Lett **25**: 43

349. Xu S, Leri AC, Myneni SCB, Jeffe PR (2004) Uptake of Bromide by Two Wetland Plants (*Typha latifolia* L. and *Phragmites australis* (Cav.) Trin. ex Steud). Environ Sci Technol **38**: 5642
350. Lee-Taylor JM, Holland EA (2000) Litter Decomposition as a Potential Natural Source of Methyl Bromide. J Geophys Res **105**: 8857
351. Butler JH, Rodriguez JM (1996) Methyl Bromide in the Atmosphere. In: Bell CH, Price N, Chakrabarti B (eds) The Methyl Bromide Issue. Wiley, Chichester, UK, p 27
352. Baker JM, Reeves CE, Penkett SA, Cardenas LM, Nightingale PD (1998) An Estimate of the Global Emissions of Methyl Bromide from Automobile Exhausts. Geophys Res Lett **25**: 2405
353. Reeves CE (2003) Atmospheric Budget Implications of the Temporal and Spatial Trends in Methyl Bromide Concentration. J Geophys Res **108**: 4343
354. Itoh N, Shinya M (1994) Seasonal Evolution of Bromomethanes from Coralline Algae (Corallinaceae) and Its Effect on Atmospheric Ozone. Mar Chem **45**: 95
355. Tokarczyk R, Moore RM (1994) Production of Volatile Organohalogens by Phytoplankton Cultures. Geophys Res Lett **21**: 285
356. Schall C, Laturnus F, Heumann KG (1994) Biogenic Volatile Organoiodine and Organobromine Compounds Released from Polar Macroalgae. Chemosphere **28**: 1315
357. Fogelqvist E, Tanhua T (1995) Iodinated C_1–C_4 Hydrocarbons Released from Ice Algae in Antarctica. In: Grimvall A, de Leer EWB (eds) Naturally-Produced Organohalogens. Kluwer, Dordrecht, p 295
358. Sturges WT, Cota GF (1995) Biogenic Emission of Organobromine Compounds to the Arctic Ocean and Atmosphere. In: Grimvall A, de Leer EWB (eds) Naturally-Produced Organohalogens. Kluwer, Dordrecht, p 385
359. Schall C, Heumann KG, de Mora S, Lee PE (1996) Biogenic Brominated and Iodinated Organic Compounds in Ponds on the McMurdo Ice Shelf, Antarctica. Antarctic Sci **8**: 45
360. Laturnus F (1996) Volatile Halocarbons Released from Arctic Macroalgae. Mar Chem **55**: 359
361. Pedersén M, Collén J, Abrahamsson K, Ekdahl A (1996) Production of Halocarbons from Seaweeds: An Oxidative Stress Reaction? Sci Mar **60** (Suppl 1): 257
362. Laturnus F, Wiencke C, Klöser H (1996) Antarctic Macroalgae – Sources of Volatile Halogenated Organic Compounds. Mar Environ Res **41**: 159
363. Goodwin KD, North WJ, Lidstrom ME (1997) Production of Bromoform and Dibromomethane by Giant Kelp: Factors Affecting Release and Comparison to Anthropogenic Bromine Sources. Limnol Oceanogr **42**: 1725
364. Marshall RA, Harper DB, McRoberts WC, Dring MJ (1999) Volatile Bromocarbons Produced by *Falkenbergia* Stages of *Asparagopsis* spp. (Rhodophyta). Limnol Oceanogr **44**: 1348
365. Giese B, Laturnus F, Adams FC, Wiencke C (1999) Release of Volatile Iodinated C_1–C_4 Hydrocarbons by Marine Macroalgae from Various Climate Zones. Environ Sci Technol **33**: 2432
366. Ohshiro T, Nakano S, Takahashi Y, Suzuki M, Izumi Y (1999) Occurrence of Bromoperoxidase in the Marine Green Macro-Alga, *Ulvella lens*, and Emission of Volatile Brominated Methane by the Enzyme. Phytochemistry **52**: 1211
367. Laturnus F, Giese B, Wiencke C, Adams FC (2000) Low-Molecular-Weight Organoiodine and Organobromine Compounds Released by Polar Macroalgae – The Influence of Abiotic Factors. Fresenius J Anal Chem **368**: 297
368. Ohsawa N, Ogata Y, Okada N, Itoh N (2001) Physiological Function of Bromoperoxidase in the Red Marine Alga, *Corallina pitulifera*: Production of Bromoform as an Allelochemical and the Simultaneous Elimination of Hydrogen Peroxide. Phytochemistry **58**: 683
369. Yamamoto H, Yokouchi Y, Otsuki A, Itoh H (2001) Depth Profiles of Volatile Halogenated Hydrocarbons in Seawater in the Bay of Bengal. Chemosphere **45**: 371

370. Marshall RA, Hamilton JTG, Dring MJ, Harper DB (2003) Do Vesicle Cells of the Red Alga *Asparagopsis* (*Falkenbergia* stage) Play a Role in Bromocarbon Production? Chemosphere **52**: 471
371. Quack B, Atlas E, Petrick G, Stroud V, Schauffler S, Wallace DWR (2004) Oceanic Bromoform Sources for the Tropical Atmosphere. Geophys Res Lett **31**: L23S05
372. Quack B, Wallace DWR (2003) Air–Sea Flux of Bromoform: Controls, Rates, and Implications. Global Biogeochem Cycles **17**: 1023
373. Carpenter LJ, Liss PS (2000) On Temperate Sources of Bromoform and Other Reactive Organic Bromine Gases. J Geophys Res **105**: 20539
374. Muramatsu Y, Yoshida S (1995) Volatilization of Methyl Iodide from the Soil–Plant System. Atmos Environ **29**: 21
375. Moore RM, Groszko W (1999) Methyl Iodide Distribution in the Ocean and Fluxes to the Atmosphere. J Geophys Res **104**: 11163
376. Abrahamsson K, Ekdahl A, Collén J, Pedersén M (1995) Marine Algae – A Source of Trichloroethylene and Perchloroethylene. Limnol Oceanogr **40**: 1321
377. Abrahamsson K, Pedersén M (2000) Evidence of the Natural Production of Trichloroethylene (Reply to the Comment by Marshall *et al.*). Limnol Oceanogr **45**: 520
378. Marshall RA, Hamilton JTG, Dring MJ, Harper DB (2000) The Red Alga *Asparagopsis taxiformis/Falkenbergia hillebrandii* – A Possible Source of Trichloroethylene and Perchloroethylene? Limnol Oceanogr **45**: 516
379. Dimmer CH, McCulloch A, Simmonds PG, Nickless G, Bassford MR, Smythe-Wright D (2001) Tropospheric Concentrations of the Chlorinated Solvents, Tetrachloroethene and Trichloroethene, Measured in the Remote Northern Hemisphere. Atmos Environ **35**: 1171
380. Wood WF, Walsh A, Seyjagat J, Weldon PJ (2005) Volatile Components in Shoulder Gland Secretions of Male Flying Foxes, Genus *Pteropus* (Pteropodidae, Chiroptera). Z Naturforsch **60c**: 779
381. Keppler F, Borchers R, Pracht J, Rheinberger S, Schöler HF (2002) Natural Production of Vinyl Chloride in the Terrestrial Environment. Environ Sci Technol **36**: 2479
382. Keppler F, Borchers R, Hamilton JTG, Kilian G, Pracht J, Schöler HF (2006) De Novo Formation of Chloroethyne in Soil. Environ Sci Technol **40**: 130
383. Khalil MAK, Rasmussen RA, Culbertson JA, Prins JM, Grimsrud EP, Shearer MJ (2003) Atmospheric Perfluorocarbons. Environ Sci Technol **37**: 4358
384. Harnisch J, Eisenhauer A (1998) Natural CF_4 and SF_6 on Earth. Geophys Res Lett **25**: 2401
385. Harnisch J, Borchers R, Fabian P, Gaggeler HW, Schotterer U (1996) Effect of Natural Tetrafluoromethane. Nature **384**: 32
386. Harnisch J, Frische M, Borchers R, Eisenhauer A, Jordan A (2000) Natural Fluorinated Organics in Rocks. Geophys Res Lett **27**: 1883
387. Rudolph J, von Czapiewski K, Koppmann R (2000) Emissions of Methyl Chloroform (CH_3CCl_3) from Biomass Burning and the Tropospheric Methyl Chloroform Budget. Geophys Res Lett **27**: 1887
388. Studier MH, Hayatsu R, Anders E (1965) Organic Compounds in Carbonaceous Chondrites. Science **149**: 1455
389. Grossi V, Raphel D (2003) Long-Chain (C_{19}–C_{29}) 1-Chloro-*n*-alkanes in Leaf Waxes of Halophytes of the Chenopodiaceae. Phytochemistry **63**: 693
390. Fielman KT, Woodin SA, Walla MD, Lincoln DE (1999) Widespread Occurrence of Natural Halogenated Organics Among Temperate Marine Infauna. Mar Ecol Prog Ser **181**: 1
391. Woolard FX, Moore RE, Roller PP (1979) Halogenated Acetic and Acrylic Acids from the Red Alga *Asparagopsis taxiformis*. Phytochemistry **18**: 617
392. Sugano M, Sato A, Nagaki H, Yoshioka S, Shiraki T, Horikoshi H (1990) Aldose Reductase Inhibitors from the Red Alga, *Asparagopsis taxiformis*. Tetrahedron Lett **31**: 7015
393. Cueto M, Darias J, San-Martín A, Rovirosa J (1997) New Acetyl Derivatives from Antarctic *Delisea fimbriata*. J Nat Prod **60**: 279

394. Vetter W, Janussen D (2005) Halogenated Natural Products in Five Species of Antarctic Sponges: Compounds with POP-Like Properties? Environ Sci Technol **39**: 3889
395. Kigoshi H, Ichino T, Takada N, Suenaga K, Yamada A, Yamada K, Uemura D (2004) Isolation of Tribromoacetamide from an Okinawan Alga and Biological Activities of Its Analogs. Chem Lett **33**: 98
396. Li X, Kim S-K, Kang JS, Choi HD, Son BW (2004) Polyketide and Sesquiterpenediol Metabolites from a Marine-Derived Fungus. Bull Korean Chem Soc **25**: 607
397. Becker U, Anke T, Sterner O (1994) A Novel Halogenated Compound Possessing Antibiotic and Cytotoxic Activities Isolated from the Fungus *Resinicium pinicola* (J. Erikss.) Erikss. & Hjortst. Z Naturforsch **49c**: 772
398. de Jong E, Field JA (1997) Sulfur Tuft and Turkey Tail: Biosynthesis and Biodegradation of Organohalogens by Basidiomycetes. Ann Rev Microbiol **51**: 375
399. Blitzke T, Porzel A, Masaoud M, Schmidt J (2000) A Chlorinated Amide and Piperidine Alkaloids from *Aloe sabaea*. Phytochemistry **55**: 979
400. Orsini MA, Pannell LK, Erickson KL (2001) Polychlorinated Acetamides from the Cyanobacterium *Microcoleus lyngbyaceus*. J Nat Prod **64**: 572
401. Takeuchi M, Nakajima M, Ogita T, Inukai M, Kodama K. Furuya K, Nagaki H, Haneishi T (1989) Fosfonochlorin, a New Antibiotic with Spheroplast Forming Activity. J Antibiot **42**: 198
402. Yanagisawa I, Yoshikawa H (1973) A Bromine Compound Isolated from Human Cerebrospinal Fluid. Biochim Biophys Acta **329**: 283
403. Patricelli MP, Patterson JE, Boger DL, Cravatt BF (1998) An Endogenous Sleep-Inducing Compound is a Novel Competitive Inhibitor of Fatty Acid Amide Hydrolysis. Bioorg Med Chem Lett **8**: 613
404. Yanagisawa I, Torii S (2002) A Bromine Compound Existing in Blood. Tohoku J Exp Med **196**: 111
405. McCulloch A (2002) Trichloroacetic Acid in the Environment. Chemosphere **47**: 667
406. Schöler HF, Keppler F, Fahimi IJ, Niedan VW (2003) Fluxes of Trichloroacetic Acid between Atmosphere, Biota, Soil, and Groundwater. Chemosphere **52**: 339
407. Hoekstra E (2003) Review of Concentrations and Chemistry of Trichloroacetate in the Environment. Chemosphere **52**: 355
408. McCulloch A (2005) Trichloroacetic Acid in the Environment. Euro Chlor, Brussels, p 1
409. Christman RF, Norwood DL, Millington DS, Johnson JD, Stevens AA (1983) Identity and Yields of Major Halogenated Products of Aquatic Fulvic Acid Chlorination. Environ Sci Technol **17**: 625
410. Haiber G, Jacob G, Niedan V, Nkusi G, Schöler HF (1996) The Occurrence of Trichloroacetic Acid (TCAA) – Indications of a Natural Production? Chemosphere **33**: 839
411. Niedan V, Pavasars I, Öberg G (2000) Chloroperoxidase-Mediated Chlorination of Aromatic Groups in Fulvic Acid. Chemosphere **41**: 779
412. Fahimi IJ, Keppler F, Schöler HF (2003) Formation of Chloroacetic Acids from Soil, Humic Acid and Phenolic Moieties. Chemosphere **52**: 513
413. Schöler, HF (2004) Abiotic Formation of Organohalogens in the Terrestrial Environment. Euro Chlor Science Dossier. Brussels, p 13
414. Höller U, König GM, Wright AD (1999) Three New Metabolites from Marine-Derived Fungi of the Genera *Coniothyrium* and Microsphaeropsis. J Nat Prod **62**: 114
415. Holder JSE, Young K (1975) Biosynthesis of Metabolites of *Periconia macrospinosa* from [1-^{13}C]-, [2-^{13}C]-, and [1,2-^{13}C]-Acetate. J Chem Soc Chem Commun 525
416. Henderson GB, Hill RA (1982) The Biosynthesis of Chlorine-Containing Metabolites of *Periconia macrospinosa*. J Chem Soc, Perkin Trans 1: 3037
417. McGee DI, Mallais T, Strunz GM (2004) Asymmetric Formal Synthesis of (–)-Cryptosporiopsin, a Metabolite of *Phialophora asterris*, and its 3-Deschloro Congener. Can J Chem **82**: 1686

418. Schulz B, Sucker J, Aust HJ, Krohn K, Ludewig K, Jones, PG, Döring D (1995) Biologically Active Secondary Metabolites of Endophytic *Pezicula* Species. Mycol Res **99**: 1007
419. Lee CH, Chung MC, Lee HJ, Bae KS, Kho YH (1997) MR566A and MR566B, New Melanin Synthesis Inhibitors Produced by *Trichoderma harzianum* I. Taxonomy, Fermentation, Isolation and Biological Activities. J Antibiot **50**: 469
420. Lee CH, Koshino H, Chung MC, Lee HJ, Hong JK, Yoo JS, Kho YH (1997) MR566A and MR566B, New Melanin Synthesis Inhibitors Produced by *Trichoderma harzianum* II. Physio-Chemical Properties and Structural Elucidation. J Antibiot **50**: 474
421. Amagata T, Usami Y, Minoura K, Ito T, Numata A (1998) Cytotoxic Substances Produced by a Fungal Strain from a Sponge: Physico-Chemical Properties and Structures. J Antibiot **51**: 33
422. Usami Y, Ikura T, Amagata T, Numata A (2000) First Total Syntheses and Configurational Assignments of Cytotoxic Trichodenones A–C. Tetrahedron: Asym **11**: 3711
423. Weidler M, Rether J, Anke T, Erkel G, Sterner O (2001) New Bioactive Cyclopentenone Derivatives as Inhibitors of the IL-6 Dependent Signal Transduction. J Antibiot **54**: 679
424. Mierau V, Sterner O, Anke T (2004) Two New Biologically Active Cyclopentenones from *Dasyscyphus* sp. A47-98. J Antibiot **57**: 311
425. Rether J, Erkel G, Sterner O, Anke T (2005) Inhibition of TNF-α Promoter Activity and Synthesis by A11-99-1, a New Cyclopentenone from the Ascomycete *Mollisia melaleuca*. Z Naturforsch **60c**: 478
426. Coombes PH, Naidoo D, Mulholland DA, Randrianarivelojosia M (2005) Quassinoids from the Leaves of the Madagascan Simaroubaceae *Samadera madagascariensis*. Phytochemistry **66**: 2734
427. Uchida R, Tomoda H, Arai M, Omura S (2001) Chlorogentisylquinone, a New Neutral Sphingomyelinase Inhibitor, Produced by a Marine Fungus. J Antibiot **54**: 882
428. Numata A, Iritani M, Yamada T, Minoura K, Matsumura E, Yamori T, Tsuruo, T (1997) Novel Antitumour Metabolites Produced by a Fungal Strain from a Sea Hare. Tetrahedron Lett **38**: 8215
429. Usami Y, Ueda Y (2005) Synthetic Study Toward Antitumour Natural Product Pericosine A. Chem Lett **34**: 1062
430. Usami Y, Horibe Y, Takaoka I, Ichikawa H, Arimoto M (2006) First Total Synthesis of (–)-Pericosine A from (–)-Shikimic Acid: Structure Revision and Determination of the Absolute Configuration of Antitumour Natural Product Pericosine A. Synlett 1598
431. Block O, Klein G, Altenbach H-J, Brauer DJ (2000) New Stereoselective Route to the Epoxyquinol Core of Manumycin-Type Natural Products. Synthesis of Enantiopure (+)-Bromoxone, (–)-LL-Cl0037α, and (+)-KT 8110. J Org Chem **65**: 716
432. Milic DR, Kop T, Juranic Z, Gasic MJ, Solaja BA (2001) Synthesis and Antiproliferative Activity of Epoxy and Bromo Compounds Derived from Estrone. Bioorg Med Chem Lett **11**: 2197
433. Barros MR, Matias PM, Maycock CD, Ventura MR (2003) Aziridines as a Protecting and Directing Group. Stereoselective Synthesis of (+)-Bromoxone. Org Lett **5**: 4321
434. Miyashita K, Imanishi T (2005) Syntheses of Natural Products Having an Epoxyquinone Structure. Chem Rev **105**: 4515
435. Stadler M, Anke H, Sterner O (1995) Metabolites with Nematicidal and Antimicrobial Activities from the Ascomycete *Lachnum papyraceum* (Karst.) Karst V. Production, Isolation and Biological Activities of Bromine-Containing Mycorrhizin and Lachnumon Derivatives and Four Additional New Bioactive Metabolites. J Antibiot **48**: 149
436. Stadler M, Anke H, Sterner O (1995) New Metabolites with Nematicidal and Antimicrobial Activities from the Ascomycete *Lachnum papyraceum* (Karst.) Karst VII. Structure Determination of Brominated Lachnumon and Mycorrhizin A Derivatives. J Antibiot **48**: 158
437. Hayashi M, Kim Y-P, Takamatsu S, Preeprame S, Komiya T, Masuma R, Tanaka H, Komiyama K, Omura S (1996) Chlovalicin, a New Cytocidal Antibiotic Produced by Sporothrix sp. FO-4649 I. Taxonomy, Fermentation, Isolation and Biological Activities. J Antibiot **49**: 631

438. Takamatsu S, Kim Y-P, Komiya T, Sunazuka T, Hayashi M, Tanaka H, Komiyama K, Omura S (1996) Chlovalicin, a New Cytocidal Antibiotic Produced by Sporothrix sp FO-4649 II. Physiochemical Properties and Structure Elucidation. J Antibiot **49**: 635
439. Stevens-Miles S, Goetz MA, Bills GF, Giacobbe RA, Tkacz JS, Chang RSL, Mojena M, Martin I, Diez MT, Pelaez F, Hensens OD, Jones T, Burg RW, Kong YL, Huang L (1996) Discovery of an Angiotension II Binding Inhibitor from a *Cytospora* sp. Using Semi-Automated Screening Procedures. J Antibiot **49**: 119
440. Satoh Y, Yamazaki M (1989) Studies on the Monoamine Oxidase (MAO) Inhibitory Potency of TL-1, Isolated from a Fungus *Talaromyces luteus*. Chem Pharm Bull **37**: 206
441. Fujimoto H, Matsudo T, Yamaguchi A, Yamazaki M (1990) Two New Fungal Azaphilones from *Talaromyces luteus*, with Monoamine Oxidase Inhibitory Effect. Heterocycles **30**: 607
442. Yoshida E, Fujimoto H, Yamazaki M (1996) Isolation of Three New Azaphilones, Luteusins C, D, and E, from an Ascomycete, *Talaromyces luteus*. Chem Pharm Bull **44**: 284
443. Yoshida E, Fujimoto H, Yamazaki M (1996) Revised Stereostructures of Luteusins C and D. Chem Pharm Bull **44**: 1775
444. Yoshida E, Fujimoto H, Baba M, Yamasaki M (1995) Four New Chlorinated Azaphilones, Helicusins A-D, Closely Related to 7-*epi*-Sclerotiorin, from an Ascomycetous Fungus, *Talaromyces helicus*. Chem Pharm Bull **43**: 1307
445. Matsuzaki K, Ikeda H, Masuma R, Tanaka H, Omura S (1995) Isochromophilones I and II, Novel Inhibitors against gp120-CD4 Binding Produced by *Penicillium multicolor* FO-2338 I. Screening, Taxonomy, Ferrmentation, Isolation and Biological Activity. J Antibiot **48**: 703
446. Matsuzaki K, Tanaka H, Omura S (1995) Isochromophilones I and II, Novel Inhibitors against gp120-CD4 Binding Produced by *Penicillium multicolor* FO-2338 II. Structure Elucidation. J Antibiot **48**: 708
447. Arai N, Shiomi K, Tomoda H, Tabata N, Yang DJ, Masuma R, Kawakubo T, Omura S (1995) Isochromophilones III~VI, Inhibitors of Acyl-CoA: Cholesterol Acyltransferase Produced by *Penicillium multicolor* FO-3216. J Antibiot **48**: 696
448. Yang D-J, Tomoda H, Tabata N, Masuma R, Omura S (1996) New Isochromophilones VII and VIII Produced by *Penicillium* sp. FO-4164. J Antibiot **49**: 223
449. Tomoda H, Matsushima C, Tabata N, Namatame I, Tanaka H, Bamberger MJ, Arai H, Fukazawa M, Inoue K, Omura S (1999) Structure-Specific Inhibition of Cholesteryl Ester Transfer Protein by Azaphilones. J Antibiot **52**: 160
450. Pairet L, Wrigley SK, Chetland I, Reynolds EE, Hayes MA, Holloway J, Ainsworth AM, Katzer W, Cheng X-M, Hupe DJ, Charlton P, Doherty AM (1995) Azaphilones with Endothelin Receptor Binding Activity Produced by *Penicillium sclerotiorum*: Taxonomy, Fermentation, Isolation, Structure Elucidation and Biological Activity. J Antibiot **48**: 913
451. Matsuzaki K, Tahara H, Inokoshi J, Tanaka H, Masuma R, Omura S (1998) New Brominated and Halogen-Less Derivatives and Structure–Activity Relationships of Azaphilones Inhibiting gp120-CD4 Binding. J Antibiot **51**: 1004
452. Michael AP, Grace EJ, Kotiw M, Barrow RA (2003) Isochromophilone IX, a Novel GABA-Containing Metabolite Isolated from a Cultured Fungus, *Penicillium* sp. Aust J Chem **56**: 13
453. Toki S, Tanaka T, Uosaki Y, Yoshida M, Suzuki Y, Kita K, Mihara A, Ando K, Lokker NA, Giese NA, Matsuda Y (1999) RP-1551s, a Family of Azaphilones Produced by *Penicillium* sp., Inhibit the Binding of PDGF to the Extracellular Domain of Its Receptor. J Antibiot **52**: 235
454. Nam J-Y, Kim H-K, Kwon J-Y, Han MY, Son K-H, Lee UC, Choi J-D, Kwon B-M (2000) 8-*O*-Methylsclerotiorinamine, Antagonist of the Grb2-5H2 Domain, Isolated from *Penicillium multicolor*. J Nat Prod **63**: 1303
455. Duncan SJ, Grüschow S, Williams DH, McNicholas C, Purewal R, Hajek M, Gerlitz M, Martin S, Wrigley SK, Moore M (2001) Isolation and Structure Elucidation of a Chlorofusin, a Novel p53-MDM2 Antagonist from a *Fusarium* sp. J Am Chem Soc **123**: 554

456. Duncan SJ, Williams DH, Ainsworth M, Martin S, Ford R, Wrigley SK (2002) On the Biosynthesis of an Inhibitor of the p53/MDM2 Interaction. Tetrahedron Lett **43**: 1075
457. Schneider K, Nicholson G, Ströbele M, Baur S, Niehaus J, Fiedler H-P, Süssmuth RD (2006) The Structures of Fluostatins C, D and E, Novel Members of the Fluostatin Family. J Antibiot **59**: 105
458. Baur S, Niehaus J. Karagouni AD, Katsifas EA, Chalkou K, Meintanis C, Jones AL, Goodfellow M, Ward AC, Beil W, Schneider K, Süssmuth RD, Fiedler H-P (2006) Fluostatins C–E, Novel Members of the Fluostatin Family Produced by *Streptomyces* Strain Acta 1383. J Antibiot **59**: 293
459. Dembitsky VM, Tolstikov (1999) Natural Halogenated Monoterpenoids. Chemistry for the Interests of Stable Development **7**: 601 (in Russian)
460. Naylor S, Manes LV, Crews P (1985) C-13 Substituent Effects in Multifunctional Marine Natural Products. J Nat Prod **48**: 72
461. Rovirosa J, Sanchez I, Palacios Y, Darias J, San-Martin A (1990) Antimicrobial Activity of a New Monoterpene from *Plocamium cartilagineum* from Antarctic Peninsula. Bol Soc Chil Quím **35**: 131
462. Fuller RW, Cardellina II JH, Kato Y, Brinen LS, Clardy J, Snader KM, Boyd MR (1992) A Pentahalogenated Monoterpene from the Red Alga *Portieria hornemannii* Produces a Novel Cytotoxicity Profile Against a Diverse Panel of Human Tumor Cell Lines. J Med Chem **35**: 3007
463. Sotokawa T, Noda T, Pi S, Hirama M (2000) A Three-Step Synthesis of Halomon. Angew Chem Int Ed **39**: 3430
464. Fuller RW, Cardellina II JH, Jurek J, Scheuer PJ, Alvarado-Lindner B, McGuire M, Gray GN, Steiner JR, Clardy J, Menez E, Shoemaker RH, Neuman DJ, Snader KM, Boyd MR (1994) Isolation and Structure/Activity Features of Halomon-Related Antitumor Monoterpenes from the Red Alga *Portieria hornemannii*. J Med Chem **37**: 4407
465. Abreu PM, Galindro JM (1996) Polyhalogenated Monoterpenes from *Plocamium cartilagineum* from the Portuguese Coast. J Nat Prod **59**: 1159
466. Ortega MJ, Zubía E, Salvá J (1997) New Polyhalogenated Monoterpenes from the Sea Hare *Aplysia punctata*. J Nat Prod **60**: 482
467. Cueto M, Darias J, Rovirosa J, San-Martin A (1998) Pantoneurotriols: Possible Biogenetic Precursors of Oxygenated Monoterpenes from Antarctic *Pantoneura plocamioides*. Tetrahedron **54**: 3575
468. König GM, Wright AD, de Nys R (1999) Halogenated Monoterpenes from *Plocamium costatum* and Their Biological Activity. J Nat Prod **62**: 383
469. König GM, Wright AD, Linden A (1999) *Plocamium hamatum* and Its Monoterpenes: Chemical and Biological Investigations of the Tropical Marine Red Alga. Phytochemistry **52**: 1047
470. Jongaramruong J, Blackman AJ (2000) Polyhalogenated Monoterpenes from a Tasmanian Collection of the Red Seaweed *Plocamium cartilagineum*. J Nat Prod **63**: 272
471. Díaz-Marrero AR, Rovirosa J, Darias J, San-Martín, Cueto M (2002) Plocamenols A-C, Novel Linear Polyhalohydroxylated Monoterpenes from *Plocamium cartilagineum*. J Nat Prod **65**: 585
472. Díaz-Marrero AR, Cueto M, Dorta E, Rovirosa J, San-Martín A, Darias J (2002) Geometry and Halogen Regiochemistry Determination of Vicinal Vinyl Dihalides by ^{1}H and ^{13}C NMR. Application to the Structure Elucidation of Prefuroplocamioid, an Unusual Marine Monoterpene. Org Lett **4**: 2949
473. Díaz-Marrero AR, Cueto M, Dorta E, Rovirosa J, San-Martín A, Darias J (2002) New Halogenated Monoterpenes from the Red Alga *Plocamium cartilagineum*. Tetrahedron **58**: 8539
474. Díaz-Marrero AR, Dorta E, Cueto M, Rovirosa J, San-Martín A, Darias J (2004) Supporting the NMR-Based Empirical Rules to Determine the Stereochemistry and Halogen Regiochemistry of Vicinal Vinyl Dihalides. Naturally Occurring Monoterpenes as Chemical Models. Tetrahedron **60**: 5049

475. Knott MG, Mkivananzi H, Arendse CE, Hendricks DT, Bolton JJ, Beukes DR (2005) Plocoralides A–C, Polyhalogenated Monoterpenes from the Marine Alga *Plocamium corallorhiza*. Photochemistry **66**: 1108
476. Ankisetty S, Nandiraju S, Win H, Park YC, Amsler CD, McClintock JB, Baker JA, Diyabalanage TK, Pasaribu A, Singh MP, Maiese WM, Walsh RD, Zaworotko MJ, Baker BJ (2004) Chemical Investigation of Predator-Derived Macroalgae from the Antarctic Peninsula. J Nat Prod **67**: 1295
477. Andrianasolo EH, France D, Cornell-Kennon S, Gerwick WH (2006) DNA Methyl Transferase Inhibiting Halogenated Monoterpenes from the Madagascar Red Marine Alga *Portieria hornemannii*. J Nat Prod **69**: 576
478. Vetter W, Hiebl J, Oldham NJ (2001) Determination and Mass Spectrometric Investigation of a New Mixed Halogenated Persistent Component in Fish and Seal. Environ Sci Technol **35**: 4157
479. Herzke D, Berger U, Kallenborn R, Nygård T, Vetter W (2005) Brominated Flame Retardants and Other Organobromines in Norwegian Predatory Bird Eggs. Chemosphere **61**: 441
480. Argandoña VH, Rovirosa J, San-Martín A, Riquelme A, Díaz-Marrero AR, Cueto M, Darias J, Santana O, Guardaño A, González-Coloma A (2002) Antifeedant Effects of Marine Halogenated Monoterpenes. J Agric Food Chem **50**: 7029
481. Maliakal S, Cheney DP, Rorrer GL (2001) Halogenated Monoterpene Production in Regenerated Plantlet Cultures of *Ochtodes secundiramea* (Rhodophyta, Cryptonemiales). J Phycol **37**: 1010
482. Wise ML, Rorrer GL, Polzin JJ, Croteau R (2002) Biosynthesis of Marine Natural Products: Isolation and Characterization of a Myrcene Synthase from Cultured Tissues of the Marine Red Alga *Ochtodes secundiramea*. Arch Biochem Biophys **400**: 125
483. Polzin JJ, Rorrer GL, Cheney DP (2003) Metabolic Flux Analysis of Halogenated Monoterpene Biosynthesis in Microplantlets of the Macrophytic Red Alga *Ochtodes secundiramea*. Biomol Eng **20**: 205
484. Abreu PM, Galindro JM (1998) A New Polyhalogenated Epoxymonoterpene from *Plocamium cartilagineum*. Ind J Chem **37B**: 610
485. Gunatilaka AAL, Paul VJ, Park PU, Puglisi MP, Gitler AD, Eggleston DS, Haltiwanger RC, Kingston DGI (1999) Apakaochtodenes A and B: Two Tetrahalogenated Monoterpenes from the Red Marine Alga *Portieria hornemannii*. J Nat Prod **62**: 1376
486. Matlock DB, Ginsburg DW, Paul VJ (1999) Spatial Variability in Secondary Metabolite Production by the Tropical Red Alga *Portieria hornemannii*. Hydrobiologia **398/399**: 267
487. Kimura J, Tobita Y, Motoyama T, Ataka Y, Takada Y (2005) Ochtodene Derivatives from the Red Alga *Carpopeltis crispata*. J Nat Prod **68**: 585
488. Quiñoa E, Castedo L, Riguera R (1989) The Halogenated Monoterpenes of *Aplysia punctata*. A Comparative Study. Comp Biochem Physiol **92B**: 99
489. Hiebl J, Melcher J, Gundersen H, Schlabach M, Vetter W (2006) Identification and Quantification of Polybrominated Hexahydroxanthene Derivatives and Other Halogenated Natural Products in Commercial Fish and Other Marine Samples. J Agric Food Chem **54**: 2652
490. Argandoña V, Del Pozo T, San-Martín A, Rovirosa J (2000) Insecticidal Activity of *Plocamium cartilagineum* Monoterpenes. Bol Soc Chil Quím **45**: 371
491. Kuniyoshi M, Oshiro N, Miono T, Higa T (2003) Halogenated Monoterpenes Having a Cyclohexadienone from the Red Alga *Portieria hornemannii*. J Chin Chem Soc **50**: 167
492. Todd JS, Gerwick WH (1995) Isolation of a Cyclic Carbonate, a γ-Butyrolactone, and a New Indole Derivative from the Marine Cyanobacterium *Lyngbya majuscula*. J Nat Prod **58**: 586
493. Cueto M, Darias J (1996) Uncommon Tetrahydrofuran Monoterpenes from Antarctic *Pantoneura plocamioides*. Tetrahedron **52**: 5899
494. Cueto M, Darias J, Rovirosa J, San Martin A (1998) Unusual Polyoxygenated Monoterpenes from the Antarctic Alga *Pantoneura plocamioides*. J Nat Prod **61**: 17

495. Darias J, Rovirosa J, San Martin A, Díaz A-R, Dorta E, Cueto M (2001) Furoplocamioids A–C, Novel Polyhalogenated Furanoid Monoterpenes from *Plocamium cartilagineum*. J Nat Prod **64**: 1383
496. Cueto M, Darias J, Rovirosa J, San Martin A (1998) Tetrahydropyran Monoterpenes from *Plocamium cartilagineum* and *Pantoneura plocamioides*. J Nat Prod **61**: 1466
497. Ali MS, Saleem M, Ahmad W, Parvez M, Yamdagni R (2002) A Chlorinated Monoterpene Ketone, Acylated β-Sitosterol Glycosides and a Flavanone Glycoside from *Mentha longifolia* (Lamiaceae). Phytochemistry **59**: 889
498. Fraga BM (2006) Natural Sesquiterpenoids. Nat Prod Rep **23**: 943
499. Osawa T, Suzuki A, Tamura S, Ohashi Y, Sasada Y (1973) Structure of Chlorochrymorin, a Novel Sesquiterpene Lactone from *Chrysanthemum morifolium*. Tetrahedron Lett **14**: 5135
500. Herz W, Poplawski J, Sharma RP (1975) New Guaianolides from *Liatris* Species. J Org Chem **40**: 199
501. Aizawa S, Akutsu H, Satomi T, Kawabata S, Sasaki K (1978) AA-57, a New Antibiotic Related to Pentalenolactone. J Antibiot **31**: 729
502. Ito K, Sakakibara Y, Haruna M (1979) New Sesquiterpene Lactones from *Eupatorium chinese* var. *simplicifolium* (Makino) Kitam. Chem Lett 1473
503. Aguilar JM, Collado IG, Macías FA, Massanet GM, Rodríguez-Luis F, Fronczek FR, Waikins SF (1988) Sesquiterpene Lactones from *Artemisia lanata*. Phytochemistry **27**: 2229
504. González AG, Darias V, Alonso G, Estévez E (1980) The Cytostatic Activity of the Chlorohyssopifolins, Chlorinated Sesquiterpene Lactones from *Centaurea*. Planta Med **40**: 179
505. Hamburger M, Wolfender J-L, Hostettmann K (1993) Search for Chlorinated Sesquiterpene Lactones in the Neurotoxic Thistle *Centaurea solstitialis* by Liquid Chromatographic-Mass Spectrometry, and Model Studies on Their Possible Artifactual Formation. Nat Tox **1**: 315
506. Öksüz S, Topcu G (1994) Guaianolides from *Centaurea glastifolia*. Phytochemistry **37**: 487
507. Fernández I, Pedro JR, Polo E (1995) Sesquiterpene Lactones from *Centaurea alba* and *C. conifera*. Phytochemistry **38**: 655
508. de Hernández ZNJ, Hernández LR, Catalán CAN, Gedris TE, Herz W (1997) Guaianolides and Germacradienolides from *Stevia sanguinea*. Phytochemistry **46**: 721
509. Youssef DTA (1998) Sesquiterpene Lactones of *Centaurea scoparia*. Phytochemistry **49**: 1733
510. Krishnaswamy NR, Ramji N (1995) Sesquiterpene Lactones from *Enhydra fluctuans*. Phytochemistry **38**: 433
511. Morgenstern T, King RM, Jakupovic J (1996) Sesquiterpene Lactones from Two *Bejaranoa* Species. Phytochemistry **41**: 1543
512. Todorova MN, Tsankova ET, Taskova RM, Peev DR (1999) Terpenoids from *Achillea clusiana*. Z Naturforsch **54c**: 1011
513. Mahmoud AA, Ahmed AA, Bassuony AAEl (1999) A New Chlorosesquiterpene Lactone from *Ambrosia maritima*. Fitoterapia **70**: 575
514. Tori M, Takeichi Y, Kuga H, Nakashima K, Sono M (2000) Two New Chlorine-Containing Germacranolides, Eupaglehnins E and F from *Eupatorium glehni*. Heterocycles **52**: 1075
515. Tori M, Takeichi Y, Kuga H, Nakashima K, Sono M (2002) Seven Germacranolides, Eupaglehnins A, B, C, D, E, and F, and 2α-Acetoxyepitulipinolide from *Eupatorium glehni*. Chem Pharm Bull **50**: 1250
516. Ahmed AA, Gáti T, Hussein TA, Ali AT, Tzakou OA, Couladis MA, Mabry TJ, Tóth G (2003) Ligustolide A and B, Two Novel Sesquiterpenes with Rare Skeletons and Three 1,10-*seco*-Guaianolide Derivatives from *Achillea ligustica*. Tetrahedron **59**: 3729
517. Zidorn C, Ellmerer E-P, Konivalinka G, Schwaiger N, Stuppner H (2004) 13-Chloro-3-*O*-β-D-glucopyranosylsolstitialin from *Leontodon palisae*: The First Genuine Chlorinated Sesquiterpene Lactone Glucoside. Tetrahedron Lett **45**: 3433

518. Yang S-P, Huo J, Wang Y, Lou L-G, Yue J-M (2004) Cytotoxic Sesquiterpenoids from *Eupatorium chinense*. J Nat Prod **67**: 638
519. Huo J, Yang S-P, Ding J, Yue J-M (2004) Cytotoxic Sesquiterpene Lactones from *Eupatorium lindleyanum*. J Nat Prod **67**: 1470
520. Ahmed AA, El-Moghazy SA, El-Shanawany MA, Abdel-Ghani HF, Karchesy J, Sturtz G, Dalley K, Paré PW (2004) Polyol Monoterpenes and Sesquiterpene Lactones from the Pacific Northwest Plant *Artemisia suksdorfii*. J Nat Prod **67**: 1705
521. Trifunovic S, Aljancic I, Vajs V, Macura S, Milosavljevic S (2005) Sesquiterpene Lactones and Flavonoids of *Achillea depressa*. Biochem Syst Ecol **33**: 317
522. Bentamene A, Benayache S, Creche J, Petit G, Bermejo-Barrera J, Leon F, Benayache F (2005). A New Guaianolide and Other Sesquiterpene Lactones from *Centaurea acaulis* L. (Asteraceae). Biochem Syst Ecol **33**: 1061
523. Li X, Qian P, Liu Z, Zhao Y, Xu G, Tao D, Zhao Q, Sun H (2005) Sesquiterpenoids from *Cynara scolymus*. Heterocycles **65**: 287
524. Trifunovic S, Vajs V, Juranic Z, Zizak Z, Tesevic V, Macura S, Milosavljevic S (2006) Cytotoxic Constituents of *Achillea clavennae* from Montenegro. Phytochemistry **67**: 887
525. Chen X, Zhan Z-J, Zhang X-W, Ding J, Yu J-M (2005) Sesquiterpene Lactones with Potent Cytotoxic Activities from *Vernonia chinensis*. Planta Med **71**: 949
526. Chea A, Hout S, Long C, Marcourt L, Faure R, Azas N, Elias R (2006) Antimalarial Activity of Sesquiterpene Lactones from *Vernonia cinera*. Chem Pharm Bull **54**: 1437
527. Yamada K, Ojika M, Kigoshi H (1998) Isolation, Chemistry, and Biochemistry of Ptaquiloside, a Bracken Carcinogen. Angew Chem Int Ed **37**: 1818
528. Fukuyama Y, Shida N, Hata Y, Kodama M (1994) Prenylated C_6–C_3 Compounds from *Illicium tashiroi*. Phytochemistry **36**: 1497
529. Fukuyama Y, Okamoto K, Kubo Y, Shida N, Kodama M (1994) New Chlorine-Containing Prenylated C_6–C_3 Compounds Increasing Choline Acetyltransferase (ChAT) Activity in Culture of Postnatal Rat Septal Neurons from *Illicium tashiroi*. Chem Pharm Bull **42**: 2199
530. Ahmad VU, Khatvon R (1995) Two New Sesquiterpenes from *Pluchea arguta*. Sci Pharm **63**: 231
531. Ahmed AA, El-Seedi HR, Mahmoud AA, El-Aziz A El-Douski A, Zeid IF, Bohlin L (1998) Eudesmane Derivatives from *Laggera crispata* and *Pluchea carolonesis*. Phytochemistry **49**: 2421
532. Zhu Y, Yang L, Jia Z-J (1999) Novel Highly Oxygenated Bisabolane Sesquiterpenes from *Cremanthodium discoideum*. J Nat Prod **62**: 1479
533. Liu C-M, Fei D-Q, Wu Q-H, Gao K (2006) Bisabolane Sesquiterpenes from the Roots of *Ligularia cymbulifera*. J Nat Prod **69**: 695
534. Dai J, Krohn K, Flörke U, Gehle D, Aust H-J, Draeger S, Schulz B, Rheinheimer J (2005) Novel Highly Substituted Biraryl Ethers, Phomosines D–G, Isolated from the Endophytic Fungus *Phomopsis* sp. from *Adenocarpus foliolosus*. Eur J Org Chem 5100
535. Masuda M, Abe T, Suzuki T, Suzuki M (1996) Morphological and Chemotaxonomic Studies on *Laurencia composita* and *L. okamurae* (Ceramiales, Rhodophyta). Phycologia **35**: 650
536. Masuda M, Abe T, Sato S, Suzuki T, Suzuki M (1997) Diversity of Halogenated Secondary Metabolites in the Red Alga *Laurencia nipponica* (Rhodomelaceae, Ceramiales). J Phycol **33**: 196
537. Masuda M, Takahashi Y, Okamoto K, Matsuo Y, Suzuki M (1997) Morphology and Halogenated Secondary Metabolites of *Laurencia snackeyi* (Weber-van Bosse) Stat. Nov. (Ceramiales, Rhodophyta). Eur J Phycol **32**: 293
538. Masuda M, Kogame K, Arisawa S, Suzuki M (1998) Morphology and Halogenated Secondary Metabolites of Three Gran Canarian Species of *Laurencia* (Ceramiales, Rhodophyta). Bot Mar **41**: 265
539. de Nys R, Coll JC, Bowden BF (1993) Tropical Marine Algae. IX. A New Sesquiterpenoid Metabolite from the Red Alga *Laurencia marianensis*. Aust J Chem **46**: 933

540. Erickson KL, Beutler JA, Gray GN, Cardellina II JH, Boyd MR (1995) Majapolene A, a Cytotoxic Peroxide, and Related Sesquiterpenes from the Red Alga *Laurencia majuscula*. J Nat Prod **58**: 1848
541. Xu X-H, Zeng L-M, Su J-Y (1997) Tricyclic Sesquiterpene from *Laurencia majuscula*. Chem Res Chin Univer **13**: 176
542. Simpson JS, Raniga P, Garson MJ (1997) Biosynthesis of Dichloroimines in the Tropical Marine Sponge *Stylotella aurantium*. Tetrahedron Lett **38**: 7947
543. Hirota H, Okino T, Yoshimura E, Fusetani N (1998) Five New Antifouling Sesquiterpenes from Two Marine Sponges of the Genus *Axinyssa* and the Nudibranch *Phyllidia pustulosa*. Tetrahedron **54**: 13971
544. Kehraus S, König GM, Wright AD (2001) New Carbonimidic Dichlorides from the Australian Sponge *Ulosa spongia* and Their Possible Taxonomic Significance. J Nat Prod **64**: 939
545. Musman M, Tanaka J, Higa T (2001) New Sesquiterpene Carbonimidic Dichlorides and Related Compounds from the Sponge *Stylotella aurantium*. J Nat Prod **64**: 111
546. Brust A, Garson MJ (2004) Dereplication of Complex Natural Product Mixtures by 2D NMR: Isolation of a New Carbonimidic Dichloride of Biosynthetic Interest from the Tropical Marine Sponge *Stylotella aurantium*. ACGC Chem Res Commun **17**: 33
547. Brust A, Garson MJ (2003) Advanced Precursors in Marine Biosynthetic Study. Part 3: The Biosynthesis of Dichloroimines in the Tropical Marine Sponge *Stylotella aurantium*. Tetrahedron Lett **44**: 327
548. Simpson JS, Brust A, Garson MJ (2004) Biosynthetic Pathways to Dichloroimines; Precursor Incorporation Studies on Terpene Metabolites in the Tropical Marine Sponge *Stylotella aurantium*. Org Biomol Chem **2**: 949
549. Kuniyoshi M, Marma MS, Higa T, Bernardinelli G, Jefford CW (2001) New Bromoterpenes from the Red Alga *Laurencia luzonensis*. J Nat Prod **64**: 696
550. Topcu G, Aydogmus Z, Imre S, Gören AC, Pezzuto JM, Clement JA, Kingston DGI (2003) Brominated Sesquiterpenes from the Red Alga *Laurencia obtusa*. J Nat Prod **66**: 1505
551. Davyt D, Fernandez R, Suescun L, Mombrú AW, Saldaña J, Domínguez L, Fujii MT, Manta E (2006) Bisabolanes from the Red Alga *Laurenca scoparia*. J Nat Prod **69**: 1113
552. Brito I, Cueto M, Dorta E, Darias J (2002) Bromocyclococanol, a Halogenated Sesquiterpene with a Novel Carbon Skeleton from the Red Alga *Laurencia obtusa*. Tetrahedron Lett **43**: 2551
553. Guella G, Skropeta D, Breuils S, Mancini I, Pietra F (2001) Calenzanol, the First Member of a New Class of Sesquiterpene with a Novel Skeleton Isolated from the Red Seaweed *Laurencia microcladia* from the Bay of Calenzana, Elba Island. Tetrahedron Lett **42**: 723
554. Wessels M, König GM, Wright AD (2000) New Natural Product Isolation and Comparison of the Secondary Metabolite Content of Three Distinct Samples of the Sea Hare *Aplysia dactylomela* from Tenerife. J Nat Prod **63**: 920
555. McPhail KL, Davies-Coleman MT, Copley RCB, Eggleston DS (1999) New Halogenated Sesquiterpenes from South African Specimens of the Circumtropical Sea Hare *Aplysia dactylomela*. J Nat Prod **62**: 1618
556. Copley RCB, Davies-Coleman MT, Edmonds DR, Faulkner DJ, McPhail KL (2002) Absolute Stereochemistry of Ibhayinol from a South African Sea Hare. J Nat Prod **65**: 580
557. Brito I, Dias T, Díaz-Marrero AR, Darias J, Cueto M (2006) Aplysiadiol from *Aplysia dactylomela* Suggested a Key Intermediate for a Unified Biogenesis of Regular and Irregular Marine Algal Bisabolene-Type Metabolites. Tetrahedron **62**: 9755
558. Kuniyoshi M, Wahome PG, Miono T, Hashimoto T, Yokoyama M, Shrestha KL, Higa T (2005) Terpenoids from *Laurencia luzonensis*. J Nat Prod **68**: 1314
559. de Carvalho LR, Fujii MT, Roque NF, Kato MJ, Lago JHG (2003) Aldingenin A, New Brominated Sesquiterpene from Red Algae *Laurencia aldingensis*. Tetrahedron Lett **44**: 2637
560. de Carvalho LR, Fujii MT, Roque NF, Lago JHG (2006) Aldingenin Derivatives from the Red Alga *Laurencia aldingensis*. Phytochemistry **67**: 1331

561. Aydogmus Z, Imre S, Ersoy L, Wray V (2004) Halogenated Secondary Metabolites from *Laurencia obtusa*. Nat Prod Res **18**: 43
562. Fedorov SN, Reshetnyak MV, Stchedrin AP, Ilyin SG, Struchkov YuT, Stonik VA, Elyakov GB (1989) New Halogenated Chamigrane Sesquiterpenoid from the Mollusk *Aplysia* sp. Structure and Absolute Configuration. Dokl Akad Nauk SSSR **305**: 877
563. Juagdan EG, Kalidindi R, Scheuer P (1997) Two New Chamigranes from an Hawaiian Red Alga, *Laurencia cartilaginea*. Tetrahedron **53**: 521
564. Guella G, Öztunc A, Mancini I, Pietra F (1997) Stereochemical Features of Sesquiterpene Metabolites as a Distinctive Trait of Red Seaweeds in the Genus *Laurencia*. Tetrahedron Lett **38**: 8261
565. Francisco MEY, Turnbull MM, Erickson KL (1998) Cartilagineol, the Fourth Lineage of *Laurencia*-derived Polyhalogenated Chamigrene. Tetrahedron Lett **39**: 5289
566. König GM, Wright AD (1997) *Laurencia rigida*: Chemical Investigation of Its Antifouling Dichloromethane Extract. J Nat Prod **60**: 967
567. König GM, Wright AD (1997) Pulsed Field Gradient Spectroscopy (PFGS): Application to the Structure Elucidation of (+)-(10S)-10-Bromo-β-chamigrene. Phytochem Anal **8**: 167
568. Kaiser CR, Pitombo LF, Pinto AC (1998) C-13 and H-1 NMR Assignments of the Chamigrenes Prepacifenol and Dehydroxyprepacifenol Epoxides. Spectroscopy Lett **31**: 573
569. Kaiser CR, Pitombo LF, Pinto AC (2000) NMR Analysis of a Complex Spin System from a *Spiro*-Chamigrene. Spectroscopy Lett **33**: 457
570. Kaiser CR, Pitombo LF, Pinto AC (2001) Complete ^{1}H and ^{13}C NMR Assignments of Chamigrenes from *Aplysia dactilomela*. Mag Reson Chem **39**: 147
571. Guella G, Chiasera G, Pietra F (1992) Conformational Studies of Marine Polyhalogenated α-Chamigrenes Using Temperature-Dependent NMR Spectra. Cyclohexene-Ring Flipping and Rigid-Chair Cyclohexane Ring in the Presence of Equatorial Halogen Atoms at C(8) and C(9). Helv Chim Acta **75**: 2012
572. Guella G, Chiasera G, Pietra F (1992) Conformational Studies of Marine Polyhalogenated α-Chamigrenes Using Temperature-Dependent NMR Spectra. Inverted Chair and Twist-Boat Cyclohexane Moieties in the Presence of an Axial Halogen Atom at C(8). Helv Chim Acta **75**: 2026
573. Rovirosa J, Soto H, Cueto M, Dárias J, Herrera J, San-Martín A (1999) Sesquiterpenes from *Laurencia claviformis*. Phytochemistry **50**: 745
574. Kimura J, Kamada N, Tsujimoto Y (1999) Fourteen Chamigrane Derivatives from a Red Alga, *Laurencia nidifica*. Bull Chem Soc Jpn **72**: 289
575. Francisco MEY, Erickson KL (2001) Ma'iliohydrin, a Cytotoxic Chamigrene Dibromohydrin from a Philippine *Laurencia* Species. J Nat Prod **64**: 790
576. Suescun L, Mombrú AW, Mariezcurrena RA, Davyt D, Fernández R, Manta E (2001) Two Natural Products from the Algae *Laurencia scoparia*. Acta Cryst **C57**: 286
577. Vairappan CS, Suzuki M, Abe T, Masuda M (2001) Halogenated Metabolites with Antibacterial Activity from the Okinawan *Laurencia* Species. Phytochemistry **58**: 517
578. Suzuki M, Daitoh M, Vairappan CS, Abe T, Masuda M (2001) Novel Halogenated Metabolites from the Malaysian *Laurencia pannosa*. J Nat Prod **64**: 597
579. Jongaramruong J, Blackman AJ, Skelton BW, White AH (2002) Chemical Relationships between the Sea Hare *Aplysia parvula* and the Red Seaweed *Laurencia filiformis* from Tasmania. Aust J Chem **55**: 275
580. Davyt D, Fernandez R, Suescun L, Mombrú AW, Saldaña J, Dominguez L, Coll J, Fujii MT, Manta E (2001) New Sesquiterpene Derivatives from the Red Alga *Laurencia scoparia*. Isolation, Structure Determination, and Anthelmintic Activity. J Nat Prod **64**: 1552
581. Xu X-H, Lu J-H, Yao G-M, Li Y-M, Su J-Y, Zeng L-M (2001) Studies on the Chemical Constituent of the Alga *Laurencia majuscula*. Nat Prod Res Dev **13**: 5
582. Brito I, Cueto M, Díaz-Marrero AR, Darias J, San Martín A (2002) Oxachamigrenes, New Halogenated Sesquiterpenes from *Laurencia obtusa*. J Nat Prod **65**: 946

583. Fedorov SN, Radchenko OS, Shubina LK, Kalinovsky AI, Gerasimenko AV, Popov DY, Stonik VA (2001) Aplydactone, a New Sesquiterpenoid with an Unprecedented Carbon Skeleton from the Sea Hare *Aplysia dactylomela*, and Its Cargill-Like Rearrangement. J Am Chem Soc **123**: 504
584. Dias T, Brito I, Moujir L, Paiz N, Darias J, Cueto M (2005) Cytotoxic Sesquiterpenes from *Aplysia dactylomela*. J Nat Prod **68**: 1677
585. Rashid MA, Gustafson KR, Cardellina II JH, Boyd MR (1995) Brominated Chamigrane Sesquiterpenes Produce a Novel Profile of Differential Cytotoxicity in the NCI *in Vitro* Screen. Nat Prod Lett **6**: 255
586. Vairappan CS (2003) Potent Antibacterial Activity of Halogenated Metabolites from Malaysian Red Algae, *Laurencia majuscula* (Rhodomelaceae, Ceramiales). Biomol Engineer **20**: 255
587. Masuda M, Kawaguchi S, Abe T, Kawamoto T, Suzuki M (2002) Additional Analysis of Chemical Diversity of the Red Algal Genus *Laurencia* (Rhodomelaceae) from Japan. Phycol Res **50**: 135
588. Wu Q-X, Shi Y-P, Jia Z-J (2006) Eudesmane Sesquiterpenoids from the Asteraceae Family. Nat Prod Rep **23**: 699
589. Baker B, Ratnapala L, Mahindaratne MPD, de Silva ED, Tillekeratne LMV, Jeong JH, Scheuer PJ, Seff K (1988) Lankalapuol A and B: Two cis-Eudesmanes from the Sea Hare *Aplysia dactylomela*. Tetrahedron **44**: 4695
590. Paul VJ, Cronan Jr. JM, Cardellina II JH (1993) Isolation of New Brominated Sesquiterpene Feeding Deterrents from Tropical Green Alga *Neomeris annulata* (Dasycladaceae: Chlorophyta). J Chem Ecol **19**: 1847
591. Suzuki M, Takahashi Y, Mitome Y, Itoh T, Abe T, Masuda M (2002) Brominated Metabolites from an Okinawan *Laurencia intricata*. Phytochemistry **60**: 861
592. Guella G, Skropeta D, Mancini I, Petra F (2003) Calenzanane Sesquiterpenes from the Red Seaweed *Laurencia microcladia* from the Bay of Calenzana, Elba Island: Acid-Catalyzed Stereospecific Conversion of Calenzanol into Indene- and Guaiazulene-Type Sesquiterpenes. Chem Eur J **9**: 5770
593. Guella G, Skropeta D, Mancini I, Pietra F (2002) The First 6,8-Cycloeudesmane Sesquiterpene from a Marine Organism: The Red Seaweed *Laurencia microcladia* from the Baia di Calenzana, Elba Island. Z Naturforsch **57b**: 1147
594. Iliopoulou D, Roussis V, Pannecouque C, De Clercq E, Vagias C (2002) Halogenated Sesquiterpenes from the Red Alga *Laurencia obtusa*. Tetrahedron **58**: 6749
595. Kladi M, Xenaki H, Vagias C, Papazafiri P, Roussis V (2006) New Cytotoxic Sesquiterpenes from the Red Alga *Laurencia obtusa* and *Laurencia microcladia*. Tetrahedron **62**: 182
596. Findlay JA, Li G (2002) Novel Terpenoids from the Sea Hare *Aplysia punctata*. Can J Chem **80**: 1697
597. Iliopoulou D, Vagias C, Galanakis D, Argyropoulos D, Roussis V (2002) Brasilane-Type Sesquiterpenoids from *Laurencia obtusa*. Org Lett **4**: 3263
598. Huang H-C, Chao C-H, Liao J-H, Chiang MY, Dai C-F, Wu Y-C, Sheu J-H (2005) A Novel Chlorinated Norsesquiterpenoid and Two Related New Metabolites from the Soft Coral *Paralemnalia thyrsoides*. Tetrahedron Lett **46**: 7711
599. Urban S, Capon RJ (1996) Deoxyspongiaquinones: New Sesquiterpene Quinones and Hydroquinones from a Southern Australian Marine Sponge *Euryspongia* sp. Aust J Chem **49**: 611
600. Garson MJ, Simpson JS (2004) Marine Isocyanides and Related Natural Products – Structure, Biosynthesis and Ecology. Nat Prod Rep **21**: 164
601. Tanaka J, Higa T (1999) Two New Cytotoxic Carbonimidic Dichlorides from the Nudibranch *Reticulida fungia*. J Nat Prod **62**: 1339
602. Sun J, Han LJ, Shi DY, Fan X, Wang SJ, Li S, Yang YC, Shi JG (2005) Sesquiterpenes from Red Alga *Laurencia fristicha*. Chin Chem Lett **16**: 1611

603. Sun J, Shi D, Ma M, Li S, Wang S, Han L, Yang Y, Fan X, Shi J, He L (2005) Sesquiterpenes from the Red Alga *Laurencia tristicha*. J Nat Prod **68**: 915
604. Kladi M, Vagias C, Furnari G, Moreau D, Roussakis C, Rousis V (2005) Cytotoxic Cuparene Sesquiterpenes from *Laurencia microcladia*. Tetrahedron Lett **46**: 5723
605. Mao S-C, Guo Y-W (2005) Cuparene-Derived Sesquiterpenes from the Chinese Red Alga *Laurencia okamurai* Yamada. Helv Chim Acta **88**: 1034
606. Mao S-C, Guo Y-W (2006) A Laurane Sesquiterpene and Rearranged Derivatives from the Chinese Red Alga *Laurencia okamurai* Yamada. J Nat Prod **69**: 1209
607. Tsukamoto S, Yamashita Y, Ohta T (2005) New Cytotoxic and Antibacterial Compounds Isolated from the Sea Hare, *Aplysia kurodai*. Mar Drugs **3**: 22
608. Yoo S, Suh JH, Yi KY (1998) Total Synthesis of (±)-Filiforminol and (±)-Bromoether A. Synthesis 771
609. Harrowven DC, Lucas MC, Howes PD (2001) The Synthesis of a Natural Product Family: From Debromoisolaurinterol to the Aplysins. Tetrahedron **57**: 791
610. Ryu G, Park SH, Choi BW, Lee NH, Hwang HJ, Ryu, Lee BH (2002) Cytotoxic Activities of Brominated Sesquiterpenes from the Red Alga *Laurencia okamurae*. Nat Prod Sci **8**: 103
611. Dorta E, Darias J, San Martín A, Cueto M (2002) New Prenylated Bromoquinols from the Green Alga *Cymopolia barbata*. J Nat Prod **65**: 329
612. Takamatsu S, Hodges TW, Rajbhandari I, Gerwick WH, Hamann MT, Nagle DG (2003) Marine Natural Products as Novel Antioxidant Prototypes. J Nat Prod **66**: 605
613. Estrada DM, Martín JD, Pérez C (1987) A New Brominated Monoterpenoid Quinol from *Cymopolia barbata*. J Nat Prod **50**: 535
614. Martín MJ, Berrué F, Amade P, Fernández R, Francesch A, Reyes F, Cuevas C (2005) Halogenated Helianane Derivatives from the Sponge *Spirastrella hartmani*. J Nat Prod **68**: 1554
615. Wellington KD, Cambie RC, Rutledge PS, Bergquist PF (2000) Chemistry of Sponges. 19. Novel Bioactive Metabolites from *Hamigera tarangaensis*. J Nat Prod **63**: 79
616. Cambie RC, Lal AR, Kernan MR, Bergquist PR (1995) Chemistry of Sponges, 17. A Novel Brominated Benzocyclooctane Derivative from *Hamigera taragensis*. J Nat Prod **58**: 940
617. Nicolaou KC, Gray D, Tae J (2001) Total Synthesis of Hamigerans: Part 1. Development of Synthetic Technology for the Construction of Benzannulated Polycyclic Systems by the Intramolecular Trapping of Photogenerated Hydroxy-*o*-quinodimethanes and Synthesis of Key Building Blocks. Angew Chem Int Ed **40**: 3675
618. Nicolaou KC, Gray D, Tae J (2001) Total Synthesis of Hamigerans: Part 2. Implementation of the Intramolecular Diels-Alder Trapping of Photochemically Generated Hydroxy-*o*-quinodimethanes; Strategy and Completion of the Synthesis. Angew Chem Int Ed **40**: 3679
619. Nasu SS, Yeung BKS, Hamann MT, Scheuer PJ, Kelly-Borges M, Goins K (1995) Puupehenone-Related Metabolites from Two Hawaiian Sponges, *Hyrtios* spp. J Org Chem **60**: 7290
620. Urban S, Capon RJ (1996) Absolute Stereochemistry of Puupehenone and Related Metabolites. J Nat Prod **59**: 900
621. Palermo JA, Brasco MFR, Spagnuolo C, Seldes AM (2000) Illudalane Sesquiterpenoids from the Soft Coral *Alcyonium paessleri*: The First Natural Nitrate Esters. J Org Chem **65**: 4482
622. Hanson JR (2006) Diterpenoids. Nat Prod Rep **23**: 875
623. Pinto AC, Queiroz PPS, Garcez WS (1991) Diterpenes from *Vellozia bicolor*. J Braz Chem Soc **2**: 25
624. Bruno M, Fazio C, Piozzi F, Savona G, Rodríguez B, de la Torre MC (1995) *Neo*-Clerodane Diterpenoids from *Teucrium racemosum*. Phytochemistry **40**: 505
625. Beauchamp PS, Bottini AT, Caselles MC, Dev V, Hope H, Larter M, Lee G, Mathela CS, Melkani AB, Millar PD, Miyatake M, Pant AK, Raffel RJ, Sharma VK, Wyatt D (1996) *Neo*-Clerodane Diterpenoids from *Ajuga parviflora*. Phytochemistry **43**: 827

626. Konishi T, Konoshima T, Fujiwara Y, Kiyosawa S, Miyahara K, Nishi M (1999) Stereostructures of New Labdane-Type Diterpenes, Excoecarins F, G1, and G2 from the Wood of *Excoecaria agallocha*. Chem Pharm Bull **47**: 456
627. Anjuneyulu ASR, Rao VL (2000) Five Diterpenoids (Agallochins A–E) from the Mangrove Plant *Excoecaria agallocha* Linn. Phytochemistry **55**: 891
628. Han L, Huang X, Sattler I, Dahse H-M, Fu H, Lin W, Grabley S (2004) New Diterpenoids from the Marine Mangrove *Bruguiera gymnorrhiza*. J Nat Prod **67**: 1620
629. Ang HH, Hitotsuyanagi Y, Takeya K (2000) Eurycolactones A–C, Novel Quassinoids from *Eurycoma longifolia*. Tetrahedron Lett **41**: 6849
630. Park H-S, Yoda N, Fukaya H, Aoyagi Y, Takeya K (2004) Rakanmakilactones A–F, New Cytotoxic Sulfur-Containing Norterpene Dilactones from Leaves of *Podocarpus macrophyllus* var. *maki*. Tetrahedron **60**: 171
631. Park H-S, Kai N, Fukaya H, Aoyagi Y, Takeya K (2004) New Cytotoxic Norditerpene Dilactones from Leaves of *Podocarpus macrophyllus* var. *maki*. Heterocycles **63**: 347
632. Gavagnin M, Fontana A (2000) Diterpenes from Marine Opisthobranch Molluscs. Curr Org Chem **4**: 1201
633. Rochfort SJ, Capon RJ (1996) Parguerenes Revisited: New Brominated Diterpenes from the Southern Australian Marine Red Alga *Laurencia filiformis*. Aust J Chem **49**: 19
634. Lyakhova EG, Kalinovsky AI, Kolesnikova SA, Vaskovsky VE, Stonik VA (2004) Halogenated Diterpenoids from the Red Alga *Laurencia nipponica*. Phytochemistry **65**: 2527
635. Takahashi Y, Suzuki M, Abe T, Masuda M (1998) Anhydroaplysiadiol from *Laurencia japonensis*. Phytochemistry **48**: 987
636. Briand A, Kornprobst J-M, Al-Easa HS, Rizk AFM, Toupet L (1997) (–)-Paniculatol, a New *ent*-Labdane Bromoditerpene from *Laurencia paniculata*. Tetrahedron Lett **38**: 3399
637. Suzuki M, Nakano S, Takahashi Y, Abe T, Masuda M, Takahashi H, Kobayashi K (2002) Brominated Labdane-Type Diterpenoids from an Okinawan *Laurencia* sp. J Nat Prod **65**: 801
638. Suzuki M, Kawamoto T, Vairappan CS, Ishii T, Abe T, Masuda M (2005) Halogenated Metabolites from Japanese *Laurencia* spp. Phytochemistry **66**: 2787
639. Kuniyoshi M, Marma MS, Higa T, Bernardinelli G, Jefford CW (2000) 3-Bromobarekoxide, an Unusual Diterpene from *Laurencia luzonensis*. Chem Commun 1155
640. Iliopoulou D, Mihopoulos N, Roussis V, Vagías C (2003) New Brominated Labdane Diterpenes from the Red Alga *Laurencia obtusa*. J Nat Prod **66**: 1225
641. Mihopoulos N, Vagias C, Mikros E, Scoullos M, Roussis V (2001) Prevezols A and B: New Brominated Diterpenes from the Red Alga *Laurencia obtusa*. Tetrahedron Lett **42**: 3749
642. Iliopoulou D, Mihopoulos N, Vagias C, Papazafiri P, Roussis V (2003) Novel Cytotoxic Brominated Diterpenes from the Red Alga *Laurencia obtusa*. J Org Chem **68**: 7667
643. Takahashi Y, Daitoh M, Suzuki M, Abe T, Masuda M (2002) Halogenated Metabolites from the New Okinawan Red Alga *Laurencia yonaguniensis*. J Nat Prod **65**: 395
644. Guella G, Marchetti F, Pietra F (1997) Rogioldiol A, a New Obtusane Diterpene, and Rogiolol, a Degraded Derivative, of the Red Seaweed *Laurencia microcladia* from Il Rogiolio Along the Coast of Tuscany: a Synergism in Structural Elucidation. Helv Chim Acta **80**: 684
645. Guella G, Pietra F (1998) Antipodal Pathways to Secondary Metabolites in the Same Eukaryotic Organism. Chem Eur J **4**: 1692
646. Guella G, Pietra F (2000) A New-Skeleton Diterpenoid, New Prenylbisbolanes, and Their Putative Biogenetic Precursor, from the Red Seaweed *Laurencia microcladia* from Il Rogiolo: Assigning the Absolute Configuration When Two Chiral Halves are Connected By Single Bonds. Helv Chim Acta **83**: 2946
647. Bavoso A, Cafieri F, De Napoli L, Di Blasio B, Fattorusso E, Pavone V, Santacroce C (1987) Isolation and Structure Determination of Norsphaerol, a Bis-*nor*-Diterpene from the Red Alga *Sphaerococcus coronopifolius*. Gazz Chim Ital **117**: 87
648. Etahiri S, Bultel-Poncé V, Caux C, Guyot M (2001) New Bromoditerpenes from the Red Alga *Sphaerococcus coronopifolius*. J Nat Prod **64**: 1024

649. Williams PG, Yoshida WY, Moore RE, Paul VJ (2003) Novel Iodinated Diterpenes from a Marine Cyanobacterium and Red Alga Assemblage. Org Lett **5**: 4167
650. Kubanek J, Prusak AC, Snell TW, Giese RA, Hardcastle KI, Fairchild CR, Aalbersberg W, Raventos-Suarez C, Hay ME (2005) Antineoplastic Diterpene-Benzoate Macrolides from the Fijian Red Alga *Callophycus serratus*. Org Lett **7**: 5261
651. Kubanek J, Prusak AC, Snell TW, Giese RA, Fairchild CR, Aalbersberg W, Hay ME (2006) Bromophycolides C–I from the Fijian Red Alga *Callophycus serratus*. J Nat Prod **69**: 731
652. Chang CWJ, Scheuer PJ (1993) Marine Isocyano Compounds. Top Curr Chem **167**: 33
653. Chang CWJ, Patra A, Roll DM, Scheuer PJ, Matsumoto GK, Clardy J (1984) Kalihinol-A, a Highly Functionalized Diisocyano Diterpenoid Antibiotic from a Sponge. J Am Chem Soc **106**: 4644
654. Patra A, Chang CWJ, Scheuer PJ, Van Duyne GD, Matsumoto GK, Clardy J (1984) An Unprecedented Triisocyano Diterpenoid Antibiotic from a Sponge. J Am Chem Soc **106**: 7981
655. Shimomura M, Miyaoka H, Yamada Y (1999) Absolute Configuration of Marine Diterpenoid Kalihinol A. Tetrahedron Lett **40**: 8015
656. Okino T, Yoshimura E, Hirota H, Fusetani N (1995) Antifouling Kalihinenes from the Marine Sponge *Acanthella cavernosa*. Tetrahedron Lett **36**: 8637
657. Miyaoka H, Shida H, Yamada N, Mitome H, Yamada Y (2002) Total Synthesis of Marine Diterpenoid Kalihinene X. Tetrahedron Lett **43**: 2227
658. Hirota H, Tomono Y, Fusetani N (1996) Terpenoids with Antifouling Activity Against Barnacle Larvae from the Marine Sponge *Acanthella cavernosa*. Tetrahedron **52**: 2359
659. Okino T, Yoshimura E, Hirota H, Fusetani N (1996) New Antifouling Kalihipyrans from the Marine Sponge *Acanthella cavernosa*. J Nat Prod **59**: 1081
660. Miyaoka H, Shimomura M, Kimura H, Yamada Y (1998) Antimalarial Activity of Kalihinol A and New Relative Diterpenoids from the Okinawan Sponge, *Acanthella* sp. Tetrahedron **54**: 13467
661. Wolf D, Schmitz FJ (1998) New Diterpene Isonitriles from the Sponge *Phakellia pulcherrima*. J Nat Prod **61**: 1524
662. Manzo E, Ciavatta ML, Gavagnin M, Mollo E, Guo Y-W, Cimino G (2004) Isocyanide Terpene Metabolites of *Phyllidiella pustulosa*, a Nudibranch from the South China Sea. J Nat Prod **67**: 1701
663. Pika J, Faulkner DJ (1995) Unusual Chlorinated Homo-Diterpenes from the South African Nudibranch *Chromodoris hamiltoni*. Tetrahedron **51**: 8189
664. Gavagnin M, Ungur N, Castelluccio F, Cimino G (1997) Novel Verrucosins from the Skin of the Mediterranean Nudibranch *Doris verrucosa*. Tetrahedron **53**: 1491
665. Sung P-J, Sheu J-H, Xu J-P (2002) Survey of Briarane-Type Diterpenoids of Marine Origin. Heterocycles **56**: 535
666. Sung P-J, Chang P-C, Fang L-S, Sheu J-H, Chen W-C, Chen Y-P, Lin M-R (2005) Survey of Briarane-Related Diterpenoids – Part II. Heterocycles **65**: 195
667. Sung P-J, Gwo H-H, Fan T-Y, Li J-J, Dong J, Han C-C, Wu S-L, Fang L-S (2004) Natural Product Chemistry of Gorgonian Corals of the Genus *Junceella*. Biochem Syst Ecol **32**: 185
668. Fenical W, Pawlik JR (1991) Defensive Properties of Secondary Metabolites from the Caribbean Gorgonian Coral *Erythropodium caribaeorum*. Mar Ecol Prog Ser **75**: 1
669. Harvell CD, Fenical W, Roussis V, Ruesink JL, Griggs CC, Greene CH (1993) Local and Geographic Variation in the Defensive Chemistry of a West Indian Gorgonian Coral (*Briareum asbestinum*). Mar Ecol Prog Ser **93**: 165
670. Subrahmanyan C, Kulatheeswaran R, Ward RS (1998) Briarane Diterpenes from the Indian Ocean Gorgonian *Gorgonella umbraculum*. J Nat Prod **61**: 1120
671. Sung P-J, Fan T-F, Chen M-C, Fang L-S, Lin M-R, Chang P-C (2004) Junceellin and Praelolide, Two Briaranes from the Gorgonian Corals *Junceella fragilis* and *Junceella juncea* (Ellisellidae). Biochem Syst Ecol **32**: 111
672. Cheng J-F, Yamamura S, Terada Y (1992) Stereochemistry of the Brianolide Acetate (= Solenolide D) by the Molecular Mechanics Calculations. Tetrahedron Lett **33**: 101

673. Rodríguez AD, Ramíerz C, Cóbar OM (1996) Briareins C–L, Ten New Briarane Diterpenoids from the Common Caribbean Gorgonian *Briareum asbestinum*. J Nat Prod **59**: 15
674. Stierle DB, Carté B, Faulkner DJ, Tagle B, Clardy J (1980) The Asbestinins, a Novel Class of Diterpenes from the Gorgonian *Briareum asbestinum*. J Am Chem Soc **102**: 5088
675. Selover SJ, Crews P, Tagle B, Clardy J (1981) New Diterpenes from the Common Caribbean Gorgonian *Briareum asbestinum* (Pallus). J Org Chem **46**: 964
676. González N, Rodríguez J, Kerr RG, Jiménez C (2002) Cyclobutenbriarein A, the First Diterpene with a Tricyclo[8.4.0.0.^{3,6}]tetradec-4-ene Ring System Isolated from the Gorgonian *Briareum asbestinum*. J. Org. Chem **67**: 5117
677. Sheu J-H, Sung P-J, Cheng M-C, Liu H-Y, Fang L-S, Duh C-Y, Chiang MY (1998) Novel Cytotoxic Diterpenes, Excavatolides A–E, Isolated from the Formosan Gordonian *Briareum excavatum*. J Nat Prod **61**: 602
678. Sheu J-H, Sung P-J, Su J-H, Liu H-Y, Duh C-Y, Chiang MY (1999) Briaexcavatolides A–J, New Diterpenes from the Gorgonian *Briareum excavatum*. Tetrahedron **55**: 14555
679. Sung P-J, Su J-H, Duh C-Y, Chiang MY, Sheu J-H (2001) Briaexcavatolides K-N, New Briarane Diterpenes from the Gorgonian *Briareum excavatum*. J Nat Prod **64**: 318
680. Kwak JH, Schmitz FJ, Williams GC (2001) Milolides, New Briarane Diterpenoids from the Western Pacific Octocoral *Briareum stechei*. J Nat Prod **64**: 754
681. Kwak JH, Schmitz, JF, Williams GC (2002) Milolides G-N, New Briarane Diterpenoids from the Western Pacific Octocoral *Briareum stechei*. J Nat Prod **65**: 704
682. Iwagawa T, Babazono K, Okamura H, Nakatani M, Doe M, Morimoto Y, Shiro M, Takemura, K (2005) Briviolides, New Briarane Diterpenes from a Gorgonian *Briareum* sp. Heterocycles **65**: 2083
683. Hamann MT, Harrison KN, Carroll AR, Scheuer PJ (1996) Briarane Diterpenes from Micronesian Gorgonians. Heterocycles **42**: 325
684. Anjaneyulu ASR, Rav NSK (1997) Juncins G and H: New Briarane Diterpenoids of the Indian Ocean Gorgonian *Junceella juncea* Pallas. J Chem Soc, Perkin Trans 1: 959
685. Anjaneyulu ASR, Rao VL, Sastry VG, Venugopal MJRV, Schmitz FJ (2003) Juncins I–M, Five New Briarane Diterpenoids from the Indian Ocean Gorgonian *Juncella juncea* Pallas. J Nat Prod **66**: 507
686. Sung P-J, Fan T-Y, Fang L-S, Sheu J-H, Wu S-L, Wang G-H, Lin M-R (2003) Juncin N, a New Briarane-Type Diterpenoid from the Gorgonian Coral *Junceella juncea*. Heterocycles **61**: 587
687. Qi S-H, Zhang S, Huang H, Xiao Z-H, Huang J-S, Li Q-X (2004) New Briaranes from the South China Sea Gorgonian *Junceella juncea*. J Nat Prod **67**: 1907
688. Qi S-H, Zhang S, Qian P-Y, Xiao Z-H, Li M-Y (2006) Ten New Antifouling Briarane Diterpenoids from the South China Sea Gorgonian *Junceella juncea*. Tetrahedron **62**: 9123
689. Shen Y-C, Lin Y-C, Chiang MY (2002) Juncenolide A, a New Briarane from the Taiwanese Gorgonian *Junceella juncea*. J Nat Prod **65**: 54
690. Lin Y-C, Huang Y-L, Khalil AT, Chen M-H, Shen Y-C (2005) Juncenolides F and G, Two New Briarane Diterpenoids from Taiwanese Gorgonian *Junceella juncea*. Chem Pharm Bull **53**: 128
691. García M, Rodríguez J, Jiménez C (1999) Absolute Structures of New Briarane Diterpenoids from *Junceella fragilis*. J Nat Prod **62**: 257
692. Zhang W, Guo Y-W, Mollo E, Cimino G (2004) Junceellonoids A and B, Two New Briarane Diterpenoids from the Chinese Gorgonian *Junceella fragilis* Ridley. Helv Chim Acta **87**: 2341
693. Qi S-H, Zhang S, Wen Y-M, Xiao Z-H, Li Q-X (2005) New Briaranes from the South China Sea Gorgonian *Junceella fragilis*. Helv Chim Acta **88**: 2349
694. Kubota NK, Kobayashi Y, Iwamoto H, Fukazawa Y, Uchio Y (2006) Two New Halogenated Briarane Diterpenes from the Papuan Gorgonian Coral *Junceella fragilis*. Bull Chem Soc Jpn **79**: 634

695. Banjoo D, Maxwell AR, Mootoo BS, Lough AJ, McLean S, Reynolds WF (1998) An Unusual Erythrolide Containing a Bicyclo[9.2.1]tetradecane Skeleton. Tetrahedron Lett **39**: 1469
696. Maharaj D, Pascoe KO, Tinto WF (1999) Briarane Diterpenes from the Gorgonian Octocoral *Erythropodium caribaeorum* from the Northern Caribbean. J Nat Prod **62**: 313
697. Banjoo D, Mootoo BS, Ramsewak RS, Sharma R, Lough AJ, McLean S, Reynolds WJ (2002) New Erythrolides from the Caribbean Gorgonian Octocoral *Erythropodium caribaeorum*. J Nat Prod **65**: 314
698. Taglialatela-Scafati O, Craig KS, Rebérioux D, Roberge M, Andersen RJ (2003) Briarane, Erythrane, and Aquariane Diterpenoids from the Caribbean Gorgonian *Erythropodium caribaeorum*. Eur J Org Chem 3515
699. Taglialatela-Scafati O, Deo-Jangra U, Campbell M, Roberge M, Andersen RJ (2002) Diterpenoids from Cultured *Erythropodium caribaeorum*. Org Lett **4**: 4085
700. Subrahmanyam C, Ratnakumar S, Ward RS (2000) Umbraculolides B–D, Further Briarane Diterpenes from the Gorgonian *Gorgonella umbraculum*. Tetrahedron **56**: 4585
701. Tanaka C, Yamamoto Y, Otsuka M, Tanaka J, Ichiba T, Marriott G, Rachmat R, Higa T (2004) Briarane Diterpenes from Two Species of Octocorals, *Ellisella* sp. and *Pteroeides* sp. J Nat Prod **67**: 1368
702. Barsby T, Kubanek J (2005) Isolation and Structure Elucidation of Feeding Deterrent Diterpenoids from the Sea Pansy, *Renilla reniformis*. J Nat Prod **68**: 511
703. Iwasaki J, Ito H, Aoyagi M, Sato Y, Iguchi K (2006) Briarane-Type Diterpenoids from the Okinawan Soft Coral *Pachyclavularia violacea*. J Nat Prod **69**: 2
704. Yamada A, Kitamura H, Yamaguchi K, Fukuzawa S, Kamijima C, Yazawa K, Kuramoto M, Wang G-Y-S, Fujitani Y, Uemura D (1997) Development of Chemical Substances Regulating Biofilm Formation. Bull Chem Soc Jpn **70**: 3061
705. Lievens SC, Hope H, Molinski TF (2004) New 3-Oxo-chol-en-24-oic Acids from the Marine Soft Coral *Eleutherobia* sp. J Nat Prod **67**: 2130
706. Shen Y-C, Pan Y-L, Ko C-L, Kuo Y-H, Chen C-Y (2003) New Dolabellanes from the Taiwanese Soft Coral *Clavularia inflata*. J Chin Chem Soc **50**: 471
707. Wei X, Rodríguez AD, Baran P, Raptis RG, Sánchez JA, Ortega-Barria E, González J (2004) Antiplasmodial Cembradiene Diterpenoids from a Southwestern Caribbean Gorgonian Octocoral of the Genus *Eunicea*. Tetrahedron **60**: 11813
708. Rudi A, Shmul G, Benayahu Y, Kashman Y (2006) Sinularectin, a New Diterpenoid from the Soft Coral *Sinularia erecta*. Tetrahedron Lett **47**: 2937
709. Biard J-F, Malochet-Grivois C, Roussakis C, Cotelle P, Hénichart J-P, Débitus C, Verbist J-F (1994) Lissoclimides, Cytotoxic Diterpenes from *Lissoclinum voeltzkowi* Michaelsen. Nat Prod Lett **4**: 43
710. Toupet L, Biard J-F, Verbist J-F (1996) Dichlorolissoclimide from *Lissoclinum voeltzkowi* Michaelson (Urochordata): Crystal Structure and Absolute Stereochemistry. J Nat Prod **59**: 1203
711. Malochet-Grivois C, Roussakis C, Robillard N, Biard JF, Riou D, Débitus C, Verbist JF (1992) Effects *in Vitro* of Two Marine Substances, Chlorolissoclimide and Dichlorolissoclimide, on a Non-Small-Cell Bronchopulmonary Carcinoma Line (NSCLC-N6). Anticancer Drug Design **7**: 493
712. Uddin MJ, Kokubo S, Suenaga K, Ueda K, Uemura D (2001) Haterumaimides A-E, Five New Dichlorolissoclimide-Type Diterpenoids from an Ascidian, *Lissoclinum* sp. Heterocycles **54**: 1039
713. Uddin MJ, Kokubo S, Ueda K, Suenaga K, Uemura D (2001) Haterumaimides F-I, Four New Cytotoxic Diterpene Alkaloids from an Ascidian *Lissoclinum* Species. J Nat Prod **64**: 1169
714. Uddin MJ, Kokubo S, Ueda K, Suenaga K, Uemura D (2002) Haterumaimides J and K, Potent Cytotoxic Diterpene Alkaloids from the Ascidian *Lissoclinum* Species. Chem Lett 1028

715. Uddin J, Ueda K, Siwu ERO, Kita M, Uemura D (2006) Cytotoxic Labdane Alkaloids from an Ascidian *Lissoclinum* sp.: Isolation, Structure Elucidation, and Structure–Activity Relationship. Bioorg Med Chem **14**: 6954
716. Fu X, Palomar AJ, Hong EP, Schmitz FJ, Valeriote FA (2004) Cytotoxic Lissoclimide-Type Diterpenes from the Molluscs *Pleurobranchus albiguttatus* and *Pleurobranchus forskalii*. J Nat Prod **67**: 1415
717. Robert F, Gao HQ, Donia M, Merrick WC, Hamann MT, Pelletier J (2006) Chlorolissoclimides: New Inhibitors of Eukaryotic Protein Synthesis. RNA **12**: 717
718. Fernández JJ, Souto ML, Norte M (2000) Marine Polyether Triterpenes. Nat Prod Rep **17**: 235
719. Liu Y, Zhang S, Abreu PJM (2006) Heterocyclic Terpenes: Linear Furano- and Pyrroloterpenoids. Nat Prod Rep **23**: 630
720. Matsuo Y, Suzuki M, Masuda M (1995) Enshuol, a Novel Squalene-Derived Pentacyclic Triterpene Alcohol from a New Species of the Red Algal Genus *Laurencia*. Chem Lett 1043
721. Suzuki M, Matsuo Y, Takahashi Y, Masuda M (1995) Callicladol, a Novel Cytotoxic Bromotriterpene Polyether from a Vietnamese Species of the Red Algal Genus *Laurencia*. Chem Lett 1045
722. Norte M, Fernández JJ, Souto ML, Gavín JA, García-Grávalos MD (1997) Thyrsenols A and B, Two Unusual Polyether Squalene Derivatives. Tetrahedron **53**: 3173
723. Norte M, Fernández JJ, Souto ML, García-Grávalos MD (1996) Two New Antitumoral Polyether Squalene Derivatives. Tetrahedron Lett **37**: 2671
724. Norte M, Fernández JJ, Souto ML (1997) New Polyether Squalene Derivatives from *Laurencia*. Tetrahedron **53**: 4649
725. Gonzalez AG, Arteaga JM, Fernandez JJ, Martin JD, Norte M, Ruano JZ (1984) Terpenoids of the Red Alga *Laurencia pinnatifida*. Tetrahedron **40**: 2751
726. Manríquez CP, Souto ML, Gavín JA, Norte M, Fernández JJ (2001) Several New Squalene-Derived Triterpenes from *Laurencia*. Tetrahedron **57**: 3117
727. Souto ML, Manríquez CP, Norte M, Fernández JJ (2002) Novel Marine Polyethers. Tetrahedron **58**: 8119
728. Suenaga K, Shibata T, Takada N, Kigoshi H, Yamada K (1998) Aurilol, a Cytotoxic Bromotriterpene Isolated from the Sea Hare *Dolabella auricularia*. J Nat Prod **61**: 515
729. Morimoto Y, Nishikawa Y, Takaishi M (2005) Total Synthesis and Complete Assignment of the Stereostructure of a Cytotoxic Bromotriterpene Polyether (+)-Aurilol. J Am Chem Soc **127**: 5806
730. Ciavatta ML, Wahidulla S, D'Souza L, Scognamiglio G, Cimino G (2001) New Bromotriterpene Polyethers from the Indian Alga *Chondria armata*. Tetrahedron **57**: 617
731. González IC, Forsyth CJ (2000) Total Synthesis of Thyrsiferyl 23-Acetate, a Specific Inhibitor of Protein Phosphatase 2A and an Anti-Leukemic Inducer of Apoptosis. J Am Chem Soc **122**: 9099
732. Nishiguchi GA, Graham J, Bouraoui A, Jacobs RS, Little RD (2006) 7,11-*epi*-Thyrsiferol: Completion of Its Synthesis, Evaluation of Its Antimitotic Properties, and the Further Development of an SAR Model. J Org Chem **71**: 5936
733. Morimoto Y, Takaishi M, Adachi N, Okita T, Yata H (2006) Two-Directional Synthesis and Stereochemical Assignment Toward a C_2 Symmetric Oxasqualenoid (+)-Intricatetraol. Org Biomol Chem **4**: 3220
734. Matsuzawa S, Suzuki T, Suzuki M, Matsuda A, Kawamura T, Mizuno Y, Kikuchi K (1994) Thyrsiferyl 23-Acetate is a Novel Specific Inhibitor of Protein Phosphatase PP2A. FEBS Lett **356**: 272
735. Fernández JJ, Souto ML, Norte M (1998) Evaluation of the Cytotoxic Activity of Polyethers Isolated from *Laurencia*. Bioorg Med Chem **6**: 2237
736. Kikuchi K, Shima H, Mitsuhashi S, Suzuki M, Oikawa H (1999) The Apoptosis-Inducing Activity of the Two Protein Phosphatase Inhibitors, Tautomycin and Thyrsiferyl 23-Acetate, is not due to the Inhibition of Protein Phosphatases PP1 and PP2A (Review). Int J Mol Med **4**: 395

737. Matsuzawa S, Kawamura T, Mitsuhashi S, Suzuki T, Matsuo Y, Suzuki M, Mizuno Y, Kikuchi K (1999) Thyrsiferyl 23-Acetate and its Derivatives Induce Apoptosis in Various T- and B-Leukemia Cells. Bioorg Med Chem **7**: 381
738. Pec MK, Moser-Thier K, Fernández JJ, Souto ML, Kubista E (1999) Growth Inhibition by Dehydrothyrsiferol – A Non-Pgp Modulator, Derived from a Marine Red Alga – in Human Breast Cancer Cell Lines. Int J Oncol **14**: 739
739. Pec MK, Aguirre A, Moser-Thier K, Fernández JJ, Souto ML, Dorta J, Díaz-González F, Villar J (2003) Induction of Apoptosis in Estrogen Dependent and Independent Breast Cancer Cells by the Marine Terpenoid Dehydrothyrsiferol. Biochem Pharmacol **65**: 1451
740. Renner MK, Jensen PR, Fenical W (1998) Neomangicols: Structures and Absolute Stereochemistries of Unprecedented Halogenated Sesterterpenes from a Marine Fungus of the Genus *Fusarium*. J Org Chem **63**: 8346
741. Renner MK, Jensen PR, Fenical W (2000) Mangicols: Structures and Biosynthesis of a New Class of Sesterterpene Polyols from a Marine Fungus of the Genus *Fusarium*. J Org Chem **65**: 4843
742. Wang X-N, Yin S, Fan C-Q, Wang F-D, Lin L-P, Ding J, Yue J-M (2006) Turrapubesins A and B, First Examples of Halogenated and Maleimide-Bearing Limonoids in Nature from *Turraea pubescens*. Org Lett **8**: 3845
743. Issa HH, Tanaka J, Higa T (2003) New Cytotoxic Furanosesterterpenes from an Okinawan Marine Sponge, *Ircinia* sp. J Nat Prod **66**: 251
744. Kumagi H, Someno T, Dobashi K, Isshiki K, Ishizuka M, Ikeda D (2004) ICM0301s, New Angiogenesis Inhibitors from *Aspergillus* sp. F-1491 I. Taxonomy, Fermentation, Isolation and Biological Activities. J Antibiot **57**: 97
745. Someno T, Kumagai H, Ohba S, Amemiya M, Naganawa H, Ishizuka M, Ikeda D (2004) ICM0301s, New Angiogenesis Inhibitors from *Aspergillus* sp. F-1491 II. Physico-Chemical Properties and Structure Elucidation. J Antibiot **57**: 104
746. Luo X-D, Wu S-H, Ma Y-B, Wu D-G (2000) Dammarane Triterpenoids from *Amoora yunnanensis*. Heterocycles **53**: 2795
747. Monte FJQ, Kintzinger JP, Braz-Filho R (1998) Total Assignment of ^{1}H and ^{13}C Spectra of the Chlorinated Triterpenoid (Methyl 2α,3β,24-Tri-*O*-acetylolean-12α-chloro-28,13β-olide) by NMR Spectroscopy. Mag Reson Chem **36**: 381
748. Barton DHR, Holness NJ (1952) Triterpenoids. V. Some Relative Configurations in Rings C, D, and E of the β-Amyrin and the Lupeol Group of Triterpenoids. J Chem Soc 78
749. Makino B, Kawai M, Ogura T, Nakanishi M, Yamamura H, Butsugan Y (1995) Structural Revision of Physalin H Isolated from *Physalis angulata*. J Nat Prod **58**: 1668
750. Bonetto GM, Gil RR, Oberti JC, Veleiro AS, Burton G (1995) Novel Withanolides from *Jaborosa sativa*. J Nat Prod **58**: 705
751. Cirigliano AM, Veleiro AS, Bonetto GM, Oberti JC, Burton G (1996) Spiranoid Withanolides from *Jaborosa runcinata* and *Jaborosa araucana*. J Nat Prod **59**: 717
752. Cirigliano AM, Veleiro AS, Oberti JC, Burton G (2002) Spiranoid Withanolides from *Jaborosa odonelliana*. J Nat Prod **65**: 1049
753. Nicotra VE, Gil RR, Vaccarini C, Oberti JC, Burton G (2003) 15,21-Cyclowithanolides from *Jaborosa bergii*. J Nat Prod **66**: 1471
754. Misico RI, Song LL, Veleiro AS, Cirigliano AM, Tettamanzi MC, Burton G, Bonetto GM, Nicotra VE, Silva GL, Gil RR, Oberti JC, Kinghorn AD, Pezzuto JM (2002) Induction of Quinone Reductase by Withanolides. J Nat Prod **65**: 677
755. Jayaprakasam B, Zhang Y, Seeram NP, Nair MG (2003) Growth Inhibition of Human Tumor Cell Lines by Withanolides from *Withania somnifera* leaves. Life Sci **74**: 125
756. Choudhary MI, Yousuf S, Samreen, Shah SAA, Ahmed S, Atta-ur-Rahman (2006) Biotransformation of Physalin H and Leishmanicidal Activity of Its Transformed Products. Chem Pharm Bull **54**: 927
757. Kobayashi M, Chen Y-J, Higuchi K, Aoki S, Kitagawa I (1996) Marine Natural Products. XXXVII. Aragusteroketals A and C, Two Novel Cytotoxic Steroids from a Marine Sponge of *Xestospongia* sp. Chem Pharm Bull **44**: 1840

758. Iwashima M, Nara K, Nakamichi Y, Iguchi K (2001) Three New Chlorinated Marine Steroids, Yonarasterols G, H and I, Isolated from the Okinawan Soft Coral, *Clavularia viridis*. Steroids **66**: 25
759. Teruya T, Nakagawa S, Koyama T, Suenaga K, Kita M, Uemura D (2003) Nakiterpiosin, a Novel Cytotoxic C-nor-D-Homosteroid from the Okinawan Sponge *Terpios hoshinota*. Tetrahedron Lett **44**: 5171
760. Teruya T, Nakagawa S, Koyama T, Arimoto H, Kita M, Uemura D (2004) Nakiterpiosin and Nakiterpiosinone, Novel Cytotoxic C-nor-D-Homosteroids from the Okinawan Sponge *Terpios hoshinota*. Tetrahedron **60**: 6989
761. Fattorusso E, Taglialatela-Scafati O, Petrucci F, Bavestrello G, Calcinai B, Cerrano C, Di Meglio P, Ianaro A (2004) Polychlorinated Androstanes from the Burrowing Sponge *Cliona nigricans*. Org Lett **6**: 1633
762. Dorta E, Díaz-Marrero AR, Cueto M, D'Croz L, Maté JL, San-Martín A, Darias J (2004) Unusual Chlorinated Pregnanes from the Eastern Pacific Octocoral *Carijoa multiflora*. Tetrahedron Lett **45**: 915
763. Kettle AJ, Winterbourn CC (1997) Myeloperoxidase: a Key Regulator of Neutrophil Oxidant Production. Redox Rep **3**: 3
764. Hurst JK, Lymar SV (1999) Cellularly Generated Inorganic Oxidants as Natural Microbicidal Agents. Acc Chem Res **32**: 520
765. van den Berg JJM, Winterbourn CC, Kuypers FA (1993) Hypochlorous Acid-Mediated Oxidation of Cholesterol and Phospholipid: Analysis of Reaction Products by Gas Chromatography-Mass Spectrometry. J Lipid Res **34**: 2005
766. Heinecke JW, Li W, Mueller DM, Bohrer A, Turk J (1994) Cholesterol Chlorohydrin Synthesis by the Myeloperoxidase-Hydrogen Peroxide-Chloride System: Potential Markers for Lipoproteins Oxidatively Damaged by Phagocytes. Biochemistry **33**: 10127
767. Carr AC, van den Berg JJM, Winterbourn CC (1996) Chlorination of Cholesterol in Cell Membranes by Hypochlorous Acid. Arch Biochem Biophys **332**: 63
768. Hazen SL, Hsu FF, Duffin K, Heinecke JW (1996) Molecular Chlorine Generated by the Myeloperoxidase-Hydrogen Peroxide-Chloride System of Phagocytes Converts Low Density Lipoprotein Cholesterol into a Family of Chlorinated Steroids. J Biol Chem **271**: 23080
769. Carr AC, Winterbourn CC, Blunt JW, Phillips AJ, Abell AD (1997) Nuclear Magnetic Resonance Characterization of 6α-Chloro-5β-cholestane-3β,5-diol Formed from the Reaction of Hypochlorous Acid with Cholesterol. Lipids **32**: 363
770. Nakamura H, Kuruto-Niwa R, Uchida M, Terao Y (2007) Formation of Chlorinated Estrones via Hypochlorous Disinfection of Wastewater Effluent Containing Estrone. Chemosphere **66**: 1441
771. Hoffmann-Röder A, Krause N (2004) Synthesis and Properties of Allenic Natural Products and Pharmaceuticals. Angew Chem Int Ed **43**: 1196
772. Hall JG, Reiss JA (1986) Elatenyne – a Pyrano[3,2-*b*]pyranyl Vinyl Acetylene from the Red Alga *Laurencia elata*. Aust J Chem **39**: 1401
773. Takahashi Y, Suzuki M, Abe M, Masuda M (1999) Japonenynes, Halogenated C_{15} Acetogenins from *Laurencia japonensis*. Phytochemistry **50**: 799
774. Okamoto Y, Nitanda N, Ojika M, Sakagami Y (2001) Aplysiallene, a New Bromoallene as an Na, K-ATPase Inhibitor from the Sea Hare, *Aplysia kurodai*. Biosci Biotechnol Biochem **65**: 474
775. Okamoto Y (2003) Aplysiallene, a New Bromoallene as an Na, K-ATPase Inhibitor from the Sea Hare, *Aplysia kurodai* [Correction to the Original 2001 Article]. Biosci Biotechnol Biochem **67**: 460
776. Suzuki M, Takahashi Y, Matsuo Y, Masuda M (1996) Pannosallene, a Brominated C_{15} Nonterpenoid from *Laurencia pannosa*. Phytochemistry **41**: 1101
777. Lyakhova EG, Kalinovsky AI, Dmitrenok AS, Kolesnikova SA, Fedorov SN, Vaskovsky VE, Stonik VA (2006) Structures and Absolute Stereochemistry of Nipponallene and Neonipponallene, New Brominated Allenes from the Red Alga *Laurencia nipponica*. Tetrahedron Lett **47**: 6549

778. Miyamoto T, Ebisawa Y, Higuchi R (1995) Aplyparvunin, a Bioactive Acetogenin from the Sea Hare *Aplysia parvula*. Tetrahedron Lett **36**: 6073
779. Norte M, Fernandez JJ, Cataldo F, Gonzalez AG (1989) *E*-Dihydrorhodophytin, a C_{15} Acetogenin from the Red Alga *Laurencia pinnatifida*. Phytochemistry **28**: 647
780. San-Martín A, Darias J, Soto H, Contreras C, Herrera JS, Rovirosa J (1997) A New C_{15} Acetogenin from the Marine Alga *Laurencia claviformis*. Nat Prod Lett **10**: 303
781. Imre S, Aydogmus Z (1997) Secondary Metabolites from the Red Alga *Laurencia obtusa*. Pharmazie **52**: 883
782. Mihopoulos N, Vagias C, Scoullos M, Roussis V (1999) Laurencienyne B, a New Acetylenic Cyclic Ether from the Red Alga *Laurencia obtusa*. Nat Prod Lett **13**: 151
783. Aydogmus Z, Imre S (1999) a New Halogenated C_{15} Non-Terpenoid Compound from the Marine Red Alga, *Laurencia obtusa*. Acta Pharm Turc **41**: 93
784. Iliopoulou D, Vagias, C, Harvala C, Roussis V (2002) C_{15} Acetogenins from the Red Alga *Laurencia obtusa*. Phytochemistry **59**: 111
785. Manzo E, Ciavatta ML, Gavagnin M, Puliti R, Mollo E, Guo Y-W, Mattia CA, Mazzarella L, Cimino G (2005) Structure and Absolute Stereochemistry of Novel C_{15}-Halogenated Acetogenins from the Anaspidean Mollusc *Aplysia dactylomela*. Tetrahedron **61**: 7456
786. Suzuki M, Takahashi Y, Matsuo Y, Guiry MD, Masuda M (1997) Scanlonenyne, a Novel Halogenated C_{15} Acetogenin from the Red Alga *Laurencia obtusa* in Irish Waters. Tetrahedron **53**: 4271
787. Suzuki M, Nakano S, Takahashi Y, Abe T, Masuda M (1999) Bisezakyne-A and -B, Halogenated C_{15} Acetogenins from a Japanese *Laurencia* Species. Phytochemistry **51**: 657
788. Lorenzo M, Cueto M, San-Martín A, Fajardo V, Darias J (2005) Pyranosylmagellanicus a Novel Structural Class of Polyhalogenated Acetogenins from *Ptilonia magellanica*. Tetrahedron **61**: 9550
789. de Nys R, Coll JC, Carroll AR, Bowden BJ (1993) Tropical Marine Algae. X. Isolaurefucin Methyl Ether, a New Lauroxocane Derivative from the Red Alga *Dasyphila plumariodes*. Aust J Chem **46**: 1073
790. Suzuki M, Mizuno Y, Matsuo Y, Masuda M (1996) Neoisoprelaurefucin, a Halogenated C_{15} Non-Terpenoid Compound from *Laurencia nipponica*. Phytochemistry **43**: 121
791. Lee H, Kim H, Baek S, Kim S, Kim D (2003) Total Synthesis and Determination of the Absolute Configuration of (+)-Neoisoprelaurefucin. Tetrahedron Lett **44**: 6609
792. McPhail KL, Davies-Coleman MT (2005) (3Z)-Bromofucin from a South African Sea Hare. Nat Prod Res **19**: 449
793. Ciavatta ML, Gavagnin M, Puliti R, Cimino G, Martínez E, Ortea J, Mattia CA (1997) Dactylallene: A Novel Dietary C_{15} Bromoallene from the Atlantic Anaspidean Mollusc *Aplysia dactylomela*. Tetrahedron **53**: 17343
794. Guella G, Chiasera G, Mancini I, Öztunc A, Pietra F (1997) Twelve-Membered *O*-Bridged Cyclic Ethers of Red Seaweeds in the Genus *Laurencia* Exist in Solution as Slowly Interconverting Conformers. Chem Eur J **3**: 1223
795. Guella G, Mancini I, Öztunc A, Pietra F (2000) Conformational Bias in Macrocyclic Ethers and Observation of High Solvolytic Reactivity at a Masked Furfuryl (=2-Furylmethyl) C-Atom. Helv Chim Acta **83**: 336
796. Vairappan CS, Daitoh M, Suzuki M, Abe T, Masuda M (2001) Antibacterial Halogenated Metabolites from the Malaysian *Laurencia* Species. Phytochemistry **58**: 291
797. Berger D, Overman LE, Renhowe PA (1997) Total Synthesis of (+)-Isolaurepinnacin. Use of Acetal-Alkene Cyclizations to Prepare Highly Functionalized Seven-Membered Cyclic Ethers. J Am Chem Soc **119**: 2446
798. Boukouvalas J, Fortier G, Radu I-I (1998) Efficient Synthesis of (−)-*trans*-Kumausyne via Tandem Intramolecular Alkoxycarbonylation-Lactonization. J Org Chem **63**: 916
799. Evans PA, Murthy VS, Roseman JB, Rheingold AL (1999) Enantioselective Total Synthesis of the Nonisoprenoid Sesquiterpene (−)-Kumausallene. Angew Chem Int Ed **38**: 3175
800. Uemura T, Suzuki T, Onodera N, Hagiwara H, Hoshi T (2007) Total Synthesis of (+)-Obtusenyne. Tetrahedron Lett **48**: 715

801. Awakura D, Fujiwara K, Murai A (1999) Determination of the Absolute Configurations of Norte's Obtusenynes by Total Syntheses of (12R,13R)-(−)- and (12S,13R)-(+)-Obtusenynes. Chem Lett 461
802. Crimmins MT, Emmitte KA (1999) Total Synthesis of (+)-Laurencin: An Asymmetric Alkylation-Ring-Closing Metathesis Approach to Medium Ring Ethers. Org Lett **1**: 2029
803. Sugimoto M, Suzuki T, Hagiwara H, Hoshi T (2007) The First Total Synthesis of (+)-(Z)-Laureatin. Tetrahedron Lett **48**: 1109
804. Matsumura R, Suzuki T, Hagiwara H, Hoshi T, Ando M (2001) The First Total Synthesis of (+)-Rogioloxepane A. Tetrahedron Lett **42**: 1543
805. Crimmins MT, Emmitte KA (2001) Asymmetric Total Synthesis of (−)-Isolaurallene. J Am Chem Soc **123**: 1533
806. Boeckman Jr. RK, Zhang J, Reeder MR (2002) Synthetic and Mechanistic Studies of the Retro-Claisen Rearrangement 4. An Application to the Total Synthesis of (+)-Laurenyne. Org Lett **4**: 3891
807. Crimmins MT, DeBaillie AC (2003) Enantioselective Total Synthesis of (+)-Rogioloxepane A. Org Lett **5**: 3009
808. Kim H, Choi WJ, Jung J, Kim S, Kim D (2003) Construction of Eight-Membered Ether Rings by Olefin Geometry-Dependent Internal Alkylation: First Asymmetric Total Syntheses of (+)-3-(E)- and (+)-3-(Z)-Pinnatifidenyne. J Am Chem Soc **125**: 10238
809. Boukouvalas J, Pouliot M, Robichaud J, MacNeil S, Snieckus V (2006) Asymmetric Total Synthesis of (−)-Panacene and Correction of Its Relative Configuration. Org Lett **8**: 3597
810. Sheldrake HM, Jamieson C, Burton JW (2006) The Changing Faces of Halogenated Marine Natural Products. Total Synthesis of the Reported Structures of Elatenyne and an Enyne from *Laurencia majuscula*. Angew Chem Int Ed **45**: 7199
811. Dinda B, Delnath S, Harigara Y (2007) Naturally Occurring Iridoids. A Review Part 1. Chem Pharm Bull **55**: 159
812. Kasai R, Katagiri M, Ohtani K, Yamasaki K, Yang C-R, Tanaka O (1994) Iridoid Glycosides from *Phlomis younghusbandii* Roots. Phytochemistry **36**: 967
813. Catalano S, Flamini G, Bilia AR, Morelli I, Nicoletti M (1995) Iridoids from *Mentzella cordifolia*. Phytochemistry **38**: 895
814. Kitagawa I, Fukuda Y, Taniyama T, Yoshikawa M (1995) Chemical Studies on Crude Drug Processing. VIII. On the Constituents of Rehmanniae Radix. (2): Absolute Stereostructures of Rehmaglutin C and Glutinoside Isolated from Chinese Rehmanniae Radix, the Dried Root of *Rehmannia glutinosa* Libosch. Chem Pharm Bull **43**: 1096
815. Nass R, Rimpler H (1996) Distribution of Iridoids in Different Populations of *Physostegia virginiana* and Some Remarks on Iridoids from *Avicennia officinalis* and *Scrophularia ningpoensis*. Phytochemistry **41**: 489
816. Sudo H, Ide T, Otsuka H, Hirata E, Takushi A, Takeda Y (1997) 10-O-Acylated Iridoid Glucosides from Leaves of *Premna subscandens*. Phytochemistry **46**: 1231
817. Machida K, Ogawa M, Kikuchi M (1998) Studies on the Constituents of *Catalpa* Species. II. Iridoids from Calalpae Fructus. Chem Pharm Bull **46**: 1056
818. Harput US, Nagatsu A, Ogihara Y, Saracoglu I (2003) Iridoid Glucosides from *Veronica pectinata* var. *glandulosa*. Z Naturforsch **58c**: 481
819. Wang H, Wu F-H, Xiong F, Wu J-J, Zhang L-Y, Ye W-C, Li P, Zhao S-X (2006) Iridoids from *Neopicrorhiza scrophulariiflora* and Their Hepatoprotective Activities *in Vitro*. Chem Pharm Bull **54**: 1144
820. Es-Safi N-E, Khlifi S, Kollmann A, Kerhoas L, El Abbouyi A, Ducrot P-H (2006) Iridoid Glucosides from the Aerial Parts of *Globularia alypum* L. (Globulariaceae). Chem Pharm Bull **54**: 85
821. Xie H, Morikawa T, Matsuda H, Nakamura S, Muraoka O, Yoshikawa M (2006) Monoterpene Constituents from *Cistanche tubulosa* – Chemical Structures of Kankanosides A-E and Kankanol. Chem Pharm Bull **54**: 669

822. Bianco A, Caciola P, Guiso M, Iavarone C, Trogolo C (1981) Iridoids. XXXI. Carbon-13 Nuclear Magnetic Resonance Spectroscopy of Free Iridoid Glucosides in D_2O Solution. Gazz Chim Ital **111**: 201
823. Ciminiello P, Fattorusso E, Forino M, Di Rosa M, Ianaro A, Poletti R (2001) Structural Elucidation of a New Cytotoxin Isolated from Mussels of the Adriatic Sea. J Org Chem **66**: 578
824. Ciminiello P, Dell'Aversano C, Fattorusso E, Forino M, Magno S, Di Rosa M, Ianaro A, Poletti R (2002) Structure and Stereochemistry of a New Cytotoxic Polychlorinated Sulfolipid from Adriatic Shellfish. J Am Chem Soc **124**: 13114
825. Ciminiello P, Dell'Aversano C, Fattorusso E, Forino M, Magno S, Di Meglio P, Ianaro A, Poletti R (2004) A New Cytotoxic Polychlorinated Sulfolipid from Contaminated Adriatic Mussels. Tetrahedron **60**: 7093
826. Mu H, Wesén C, Sundin P (1997) Halogenated Fatty Acids I. Formation and Occurrence in Lipids. Trends Anal Chem **16**: 266
827. Mu H, Sundin P, Wesén C (1997) Halogenated Fatty Acids II. Methods of Determination in Lipids. Trends Anal Chem **16**: 274
828. Ewald G (1998) Chlorinated Fatty Acids – Environmental Pollutants with Intriguing Properties. Chemosphere **37**: 2833
829. Ewald G (1999) Ecotoxicological Aspects of Chlorinated Fatty Acids. Aquat Ecosys Health Manage **2**: 71
830. Dembitsky VM, Srebnik M (2002) Natural Halogenated Fatty Acids: Their Analogues and Derivatives. Prog Lipid Res **41**: 315
831. Ichiba T, Scheuer PJ, Kelly-Borges M (1993) Sponge-Derived Polyunsaturated C_{16} Di- and Tribromocarboxylic Acids. Helv Chim Acta **76**: 2814
832. Garson MJ, Zimmermann MP, Hoberg M, Larsen RM, Battershill CN, Murphy PT (1993) Isolation of Brominated Long-Chain Fatty Acids from the Phospholipids of the Tropical Marine Sponge *Amphimedon terpenensis*. Lipids **28**: 1011
833. Garson MJ, Zimmermann MP, Battershill CN, Holden JL, Murphy PT (1994) The Distribution of Brominated Long-Chain Fatty Acids in Sponge and Symbiont Cell Types from the Tropical Marine Sponge *Amphimedon terpenensis*. Lipids **29**: 509
834. Carballeira NM, Emiliano A (1993) Novel Brominated Phospholipid Fatty Acids in the Caribbean Sponge *Agelas* sp. Lipids **28**: 763
835. Carballeira NM, Medina JR (1994) New $\Delta^{5,9}$ Fatty Acids in the Phospholipids of the Sea Anemone *Stoichactis helianthus*. J Nat Prod **57**: 1688
836. Carballeira NM, Reyes M (1995) Identification of a New 6-Bromo-5,9-eicosadienoic Acid from the Anemone *Condylactis gigantea* and the Zoanthid *Palythoa caribaeorum*. J Nat Prod **58**: 1689
837. Li Y, Ishibashi M, Sasaki T, Kobayashi J (1995) New Bromine-Containing Unsaturated Fatty Acid Derivatives from the Okinawan Marine Sponge *Xestospongia* sp. J Chem Res (S) 126
838. Brantley SE, Molinski TF, Preston CM, DeLong EF (1995) Brominated Acetylenic Fatty Acids from *Xestospongia* sp., a Marine Sponge-Bacteria Association. Tetrahedron **51**: 7667
839. Pham NB, Butler MS, Hooper JNA, Moni RW, Quinn RJ (1992) Isolation of Xestosterol Esters of Brominated Acetylenic Fatty Acids from the Marine Sponge *Xestospongia testudinaria*. J Nat Prod **62**: 1439
840. Rezanka T, Dembitsky V (1999) Brominated Fatty Acids from Lichen *Acorospora gobiensis*. Phytochemistry **50**: 97
841. Rezanka T, Dembitsky V (1999) Novel Brominated Lipidic Compounds from Lichens of Central Asia. Phytochemistry **51**: 963
842. Rezanka T, Dembitsky VM (2001) Bromoallenic Lipid Compounds from Lichens of Central Asia. Phytochemistry **56**: 869
843. Rezanka T, Dembitsky VM (2001) Polyhalogenated Homosesquiterpenic Fatty Acids from *Plocamium cartilagineum*. Phytochemistry **57**: 607

844. Salomon CE, Williams DH, Faulkner DJ (1995) New Azacyclopropene Derivatives from *Dysidea fragilis* Collected in Pohnpei. J Nat Prod **58**: 1463
845. MacMillan JB, Molinski TF (2005) Majusculoic Acid, a Brominated Cyclopropyl Fatty Acid from a Marine Cyanobacterial Mat Assemblage. J Nat Prod **68**: 604
846. Sitachitta N, Gerwick WH (1998) Grenadadiene and Grenadamide, Cyclopropyl-Containing Fatty Acid Metabolites from the Marine Cyanobacterium *Lyngbya majuscula*. J Nat Prod **61**: 681
847. Izumi AK, Moore RE (1987) Seaweed (*Lyngbya majuscula*) Dermatitis. Clin Dermatol **5**: 92
848. Anderson BS, Sims JK, Liang AP, Minette HP (1988) Outbreak of Eye and Respiratory Irritation in Lahaina, Maui, Possibly Associated with *Microcoleus lyngbyaceus*. J Environ Health **50**: 205
849. Dodds WK, Gudder DA, Mollenhauer D (1995) The Ecology of *Nostoc*. J Phycol **31**: 2
850. Todd JS, Gerwick WH (1995) Malyngamide I from the Tropical Marine Cyanobacterium *Lyngbya majuscula* and the Probable Structure Revision of Stylocheilamide. Tetrahedron Lett **36**: 7837
851. Wu M, Milligan KE, Gerwick WH (1997) Three New Malyngamides from the Marine Cyanobacterium *Lyngbya majuscula*. Tetrahedron **53**: 15983
852. Milligan KE, Márquez B, Williamson RT, Davies-Coleman M, Gerwick WH (2000) Two New Malyngamides from a Madagascan *Lyngbya majuscula*. J Nat Prod **63**: 965
853. Nogle LM, Gerwick WH (2003) Diverse Secondary Metabolites from a Puerto Rican Collection of *Lyngbya majuscula*. J Nat Prod **66**: 217
854. Gallimore WA, Scheuer PJ (2000) Malyngamides O and P from the Sea Hare *Stylocheilus longicauda*. J Nat Prod **63**: 1422
855. Appleton DR, Sewell MA, Berridge MV, Copp BR (2002) A New Biologically Active Malyngamide from a New Zealand Collection of the Sea Hare *Bursatella leachii*. J Nat Prod **65**: 630
856. Kan Y, Fujita T, Nagai H, Sakamoto B, Hokama Y (1998) Malyngamides M and N from the Hawaiian Red Alga *Gracilaria coronopifolia*. J Nat Prod **61**: 152
857. Kan Y, Sakamoto B, Fujita T, Nagai H (2000) New Malyngamides from the Hawaiian Cyanobacterium *Lyngbya majuscula*. J Nat Prod **63**: 1599
858. Moore RE (1981) Constituents of Blue-Green Algae. In: Scheuer PJ (ed) Marine Natural Products, vol. 4. Academic Press, London, p 1
859. Nagle DG, Park PU, Paul VJ (1997) Pitiamide A, a New Chlorinated Lipid from a Mixed Marine Cyanobacterial Assemblage. Tetrahedron Lett **38**: 6969
860. Ribe S, Kondru RK, Beratan DN, Wipf P (2000) Optical Rotation Computation, Total Synthesis, and Stereochemistry Assignment of the Marine Natural Product Pitiamide A. J Am Chem Soc **122**: 4608
861. Edwards DJ, Marquez BL, Nogle LM, McPhail K, Goeger DE, Roberts MA, Gerwick WH (2004) Structure and Biosynthesis of the Jamaicamides, New Mixed Polyketide-Peptide Neurotoxins from the Marine Cyanobacterium *Lyngbya majuscula*. Chem Biol **11**: 817
862. Williamson RT, Marquez BL, Gerwick WH, Koehn FE (2001) ACCORD-ADEQUATE: An Improved Technique for the Acquisition of Inverse-Detected INADEQUATE Data. Mag Reson Chem **39**: 544
863. Gallimore WA, Galario DL, Lacy C, Zhu Y, Scheuer PJ (2000) Two Complex Proline Esters from the Sea Hare *Stylocheilus longicauda*. J Nat Prod **63**: 1022
864. Shoeb M, Jaspars M (2003) Chlorinated C12 Fatty Acid Metabolites from the Red Alga *Gracilaria verrucosa*. J Nat Prod **66**: 1509
865. Herrmann M, Böhlendorf B, Irschik H, Reichenbach H, Höfle G (1998) Maracin and Maracen: New Types of Ethynyl Vinyl Ether and α-Chloro Divinyl Ether Antibiotics from *Sorangium cellulosum* with Specific Activity Against Mycobacteria. Angew Chem Int Ed **37**: 1253
866. Numata A, Amagata T, Minoura K, Ito T (1997) Gymnastatins, Novel Cytotoxic Metabolites Produced by a Fungal Strain from a Sponge. Tetrahedron Lett **38**: 5675

867. Amagata T, Doi M, Ohta T, Minoura K, Numata A (1998) Absolute Stereostructures of Novel Cytotoxic Metabolites, Gymnastatins A–E, from a *Gymnascella* Species Separated from a *Halichondria* Sponge. J Chem Soc Perkin Trans 1: 3585
868. Amagata T, Minoura K, Numata A (2006) Gymnastatins F-H, Cytostatic Metabolites from the Sponge-Derived Fungus *Gymnascella dankaliensis*. J Nat Prod **69**: 1384
869. Gurjar MK, Bhaket P (2000) Total Synthesis of a Novel Cytotoxic Metabolite Gymnastatin A. Heterocycles **53**: 143
870. Mukhopadhyay T, Bhat RG, Roy K, Vijayakumar EKS, Ganguli BN (1998) Aranochlor A and Aranochlor B, Two New Metabolites from *Pseudoarachniotus roseus*: Production, Isolation, Structure Elucidation and Biological Activities. J Antibiot **51**: 439
871. Wolf D, Schmitz FJ, Qiu F, Kelly-Borges M (1999) Aurantoside C, a New Tetramic Acid Glycoside from the Sponge *Homophymia conferta*. J Nat Prod **62**: 170
872. Sata NU, Matsunaga S, Fusetani N, van Soest RWM (1999) Aurantosides D, E, and F: New Antifungal Tetramic Acid Glycosides from the Marine Sponge *Siliquariaspongia japonica*. J Nat Prod **62**: 969
873. Ratnayake AS, Davis RA, Harper MK, Veltri CA, Andjelic CD, Barrows LR, Ireland CM (2005) Aurantosides G, H, and I: Three New Tetramic Acid Glycosides from a Papua New Guinea *Theonella swinhoei*. J Nat Prod **68**: 104
874. Sata NU, Wada S, Matsunaga S, Watabe S, van Soest RWM, Fusetani N (1999) Rubrosides A–H, New Bioactive Tetramic Acid Glycosides from the Marine Sponge *Siliquariaspongia japonica*. J Org Chem **64**: 2331
875. Sirirath S, Tanaka J, Ohtani II, Ichiba T, Rachmat R, Ueda K, Usui T, Osada H, Higa T (2002) Bitungolides A–F, New Polyketides from the Indonesian Sponge *Theonella* cf. *swinhoei*. J Nat Prod **65**: 1820
876. Campagnuolo C, Fattorusso E, Taglialatela-Scafati O, Ianaro A, Pisano B (2002) Plakortethers A–G: A New Class of Cytotoxic Plakortin-Derived Metabolites. Eur J Org Chem 61
877. Ojika M, Itou Y, Sakagami Y (2003) Structural Studies and Antifungal Activity of Unique Polyene Amides, Clathrynamide A and Three New Derivatives from a Marine Sponge, *Psammoclemma* sp. Biosci Biotechnol Biochem **67**: 1568
878. Yoshikawa K, Adachi K, Nishida F, Mochida K (2003) Planar Structure and Antibacterial Activity of Korormicin Derivatives Isolated from *Pseudoalteromonas* sp. F-420. J Antibiot **56**: 866
879. Jansen R, Kunze B, Reichenbach H, Höfle G (2003) Chondrochloren A and B, New β-Amino Styrenes from *Chondromyces crocatus* (Myxobacteria). Eur J Org Chem 2684
880. Li F, Maskey RP, Qin S, Sattler I, Fiebig HH, Maier A, Zeeck A, Laatsch H (2005) Chinikomycins A and B: Isolation, Structure Elucidation, and Biological Activity of Novel Antibiotics from a Marine *Streptomyces* sp. Isolate MO45. J Nat Prod **68**: 349
881. Takeuchi R, Kiyota H, Yaosaka M, Watanabe T, Enari K, Sugiyama T, Oritami T (2001) Stereochemistry of Enacyloxins. Part 3. (12'S,17'R,18'S,19'R)-Absolute Configuration of Enacyloxins, a Series of Antibiotics from *Frateuria* sp. W-315. J Chem Soc Perkin Trans 1: 2676
882. Watanabe T, Sugiyama T, Takahashi M, Shima J, Yamashita K, Izaki K, Furihata K, Seto H (1992) New Polyenic Antibiotics Active Against Gram-Positive and Gram-Negative Bacteria IV. Structural Elucidation of Enacyloxin IIa. J Antibiot **45**: 470
883. Watanabe T, Shima J, Izaki K, Sugiyama T (1992) New Polyenic Antibiotics Active Against Gram-Positive and Gram-Negative Bacteria VII. Isolation and Structure of Enacyloxin IVa, a Possible Biosynthetic Intermediate of Enacyloxin IIa. J Antibiot **45**: 575
884. Watanabe T, Suzuki T, Izaki K (1991) New Polyenic Antibiotics Active Against Gram-Positive and Gram-Negative Bacteria V. Mode of Action of Enacyloxin IIa. J Antibiot **44**: 1457
885. Cetin R, Krab IM, Anborgh PH, Cool RH, Watanabe T, Sugiyama T, Izaki K, Parmeggiani A (1996) Enacyloxin IIa, an Inhibitor of Protein Biosynthesis That Acts on Elongation Factor Tu and the Ribosome. EMBO J **15**: 2604

886. Clough B, Rangachari K, Strath M, Preiser PR, Wilson RJM (1999) Antibiotic Inhibitors of Organellar Protein Synthesis of *Plasmodium falciparum*. Protist **150**: 189
887. Zuurmond A-M, Olsthoorn-Tieleman LN, de Graaf JM, Parmeggiani A, Kraal B (1999) Mutant EF-Tu Species Reveal Novel Features of the Enacyloxin IIa Inhibition Mechanism on the Ribosome. J Mol Biol **294**: 627
888. Pereira A, Braekman J-C, Dumont JE, Boeynaems J-M (1990) Identification of a Major Iodolipid from the Horse Thyroid Gland as 2-Iodohexadecanal. J Biol Chem **265**: 17018
889. Panneels V, Van Sande J, Van den Bergen H, Jacoby C, Braekman JC, Dumont JE, Boeynaems JM (1994) Inhibition of Human Thyroid Adenylyl Cyclase by 2-Iodoaldehydes. Mol Cell Endocrinol **106**: 41
890. Panneels V, Van den Bergen H, Jacoby C, Braekman JC, Van Sande J, Dumont JE, Boeynaems JM (1994) Inhibition of H_2O_2 Production by Iodoaldehydes in Cultured Dog Thyroid Cells. Mol Cell Endocrinol **102**: 167
891. Gärtner R, Dugrillon A, Bechtner G (1996) Evidence that Iodolactones are the Mediators of Growth Inhibition by Iodine on the Thyroid. Acta Med Austriaca **23**: 47
892. Gärtner R, Schopohl D, Schaefer S, Dugrillon A, Erdmann A, Toda S, Bechtner G (1997) Regulation of Transforming Growth Factor β_1 Messenger Ribonucleic Acid Expression in Porcine Thyroid Follicles *In vitro* by Growth Factors, Iodine, or δ-Iodolactone. Thyroid **7**: 633
893. Reid KA, Hamilton JT, Bowden RD, O'Hagan D, Dasaradhi L, Amin MR, Harper DB (1995) Biosynthesis of Fluorinated Secondary Metabolites by *Streptomyces cattleya*. Microbiology **141**: 1385
894. Amin MR, Harper DB, Moloney JM, Murphy CD, Howard JAK, O'Hagan D (1997) A Short Highly Stereoselective Synthesis of the Fluorinated Natural Product (2S,3S)-4-Fluorothreonine. Chem Commun 1471
895. Hamilton JTG, Harper DB (1997) Fluoro Fatty Acids in Seed Oil of *Dichapetalum toxicarium*. Phytochemistry **44**: 1129
896. Christie WW, Hamilton JTG, Harper DB (1998) Mass Spectrometry of Fluorinated Fatty Acids in the Seed Oil of *Dichapetalum toxicarium*. Chem Phys Lipids **97**: 41
897. Tamura T, Wada M, Esaki N, Soda K (1995) Synthesis of Fluoroacetate from Fluoride, Glycerol, and β-Hydroxypyruvate by *Streptomyces cattleya*. J Bacteriol **177**: 2265
898. Hamilton JTG, Amin MR, Harper DB, O'Hagan D (1997) Biosynthesis of Fluoroacetate and 4-Fluorothreonine by *Streptomyces cattleya*. Glycine and Pyruvate as Precursors. Chem Commun 797
899. Nieschalk J, Hamilton JTG, Murphy CD, Harper DB, O'Hagan D (1997) Biosynthesis of Fluoroacetate and 4-Fluorothreonine by *Streptomyces cattleya*. The Stereochemical Processing of Glycerol. Chem Commun 799
900. Hamilton JTG, Murphy CD, Amin MR, O'Hagan D, Harper DB (1998) Exploring the Biosynthetic Origin of Fluoroacetate and 4-Fluorothreonine in *Streptomyces cattleya*. J Chem Soc Perkin Trans 1: 759
901. Moss SJ, Murphy CD, Hamilton JTG, McRoberts WC, O'Hagan D, Schaffrath C, Harper DB (2000) Fluoroacetaldehyde: A Precursor of Both Fluoroacetate and 4-Fluorothreonine in *Streptomyces cattleya*. Chem Commun 2281
902. Murphy CD, Moss SJ, O'Hagan D (2001) Isolation of an Aldehyde Dehydrogenase Involved in the Oxidation of Fluoroacetaldehyde to Fluoroacetate in *Streptomyces cattleya*. Appl Environ Microbiol **67**: 4919
903. Murphy CD, O'Hagan D, Schaffrath C (2001) Identification of a PLP-Dependent Threonine Transaldolase: A Novel Enzyme Involved in 4-Fluorothreonine Biosynthesis in *Streptomyces cattleya*. Angew Chem Int Ed **40**: 4479
904. Schaffrath C, Murphy CD, Hamilton JTG, O'Hagan D (2001) Biosynthesis of Fluoroacetate and 4-Fluorothreonine in *Streptomyces cattleya*. Incorporation of Oxygen-18 from [2-^{2}H,2-^{18}O]-Glycerol and the Role of Serine Metabolites in Fluoroacetaldehyde Biosynthesis. J Chem Soc Perkin Trans 1: 3100

905. Schaffrath C, Cobb SL, O'Hagan D (2002) Cell-Free Biosynthesis of Fluoroacetate and 4-Fluorothreonine in *Streptomyces cattleya*. Angew Chem Int Ed **41**: 3913
906. O'Hagan D, Schaffrath C, Cobb SL, Hamilton JTG, Murphy CD (2002) Biosynthesis of an Organofluorine Molecule. Nature **416**: 279
907. O'Hagan D, Goss RJM, Meddour A, Courtieu J (2003) Assay for the Enantiomeric Analysis of [^{2}H$_1$]-Fluoroacetic Acid: Insight in the Stereochemical Course of Fluorination During Fluorometabolite Biosynthesis in *Streptomyces cattleya*. J Am Chem Soc **125**: 379
908. Cobb SL, Deng H, Hamilton JTG, McGlinchey RP, O'Hagan D (2004) Identification of 5-Fluoro-5-deoxy-D-ribose-1-phosphate as an Intermediate in Fluorometabolite Biosynthesis in *Streptomyces cattleya*. Chem Commun 592
909. Cadicamo CD, Courtieu J, Deng H, Meddour A, O'Hagan D (2004) Enzymatic Fluorination in *Streptomyces cattleya* Takes Place with an Inversion of Configuration Consistent with an S_N2 Reaction Mechanism. ChemBioChem **5**: 685
910. Senn HM, O'Hagan D, Thief W (2005) Insight into Enzymatic C–F Bond Formation from QM and QM/MM Calculations. J Am Chem Soc **127**: 13643
911. Huang F, Haydock SF, Spiteller D, Mironenko T, Li T-L, O'Hagan D, Leadlay PF, Spencer JB (2006) The Gene Cluster for Fluorometabolite Biosynthesis in *Streptomyces cattleya*: A Thioesterase Confers Resistance to Fluoroacetyl-Coenzyme A. Chem Biol **13**: 475
912. Deng H, Cobb SL, McEwan AR, McGlinchey RP, Naismith JH, O'Hagan D, Robinson DA, Spencer JB (2006) The Fluorinase from *Streptomyces cattleya* Is Also a Chlorinase. Angew Chem Int Ed **45**: 759
913. Cobb SL, Deng H, McEwan AR, Naismith JH, O'Hagan D, Robinson DA (2006) Substrate Specificity in Enzymatic Fluorination. The Fluorinase from *Streptomyces cattleya* Accepts 2'-Deoxyadenosine Substrates. Org Biomol Chem **4**: 1458
914. Murphy CD, Schaffrath C, O'Hagan D (2003) Fluorinated Natural Products: The Biosynthesis of Fluoroacetate and 4-Fluorothreonine in *Streptomyces cattleya*. Chemosphere **52**: 455
915. Deng H, O'Hagan D, Schaffrath C (2004) Fluorometabolite Biosynthesis and the Fluorinase from *Streptomyces cattleya*. Nat Prod Rep **21**: 773
916. Dong CJ, Huang FL, Deng H, Schaffrath C, Spencer JB, O'Hagan D, Naismith JH (2004) Crystal Structure and Mechanism of a Bacterial Fluorinating Enzyme. Nature **427**: 561
917. Zurita JL, Jos A, Cameán AM, Salguero M, Lopez-Artíguez M, Repetto G (2007) Ecotoxicological Evaluation of Sodium Fluoroacetate on Aquatic Organisms and Investigation of the Effects on Two Fish Cell Lines. Chemosphere **67**: 1
918. Gregg K, Hamdorf B, Henderson K, Kopecny J, Wong C (1998) Genetically Modified Ruminal Bacteria Protect Sheep from Fluoroacetate Poisoning. Appl Environ Microbiol **64**: 3496
919. Lauble H, Kennedy MC, Emptage MH, Beinert H, Stout CD (1996) The Reaction of Fluorocitrate with Aconitase and the Crystal Structure of the Enzyme-Inhibitor Complex. Proc Natl Acad Sci U S A **93**: 13699
920. Frank H, Christoph EH, Holm-Hansen O, Bullister JH (2002) Trifluoroacetate in Ocean Waters. Environ Sci Technol **36**: 12
921. Grechkin AN (1995) Clavulones and Related *tert*-Hydroxycyclopentenone Fatty Acids: Occurrence, Physiological Activity and Problem of Biogenetic Origin. J Lipid Mediators Cell Signal **11**: 205
922. Ciufolini MA, Zhu S (1998) Practical Synthesis of (±)-Chlorovulone II. J Org Chem **63**: 1668
923. Watanabe K, Sekine M, Takahashi H, Iguchi K (2001) New Halogenated Marine Prostaglandins with Cytotoxic Activity from the Okinawan Soft Coral *Clavularia viridis*. J Nat Prod **64**: 1421
924. Watanabe K, Sekine M, Iguchi K (2003) Isolation and Structures of New Halogenated Prostanoids from the Okinawan Soft Coral *Clavularia viridis*. J Nat Prod **66**: 1434
925. Shen Y-C, Cheng Y-B, Lin Y-C, Guh J-H, Teng C-M, Ko C-L (2004) New Prostanoids with Cytotoxic Activity from Taiwanese Octocoral *Clavularia viridis*. J Nat Prod **67**: 542

926. Watanabe K, Sekine M, Iguchi K (2003) Isolation of Three Marine Prostanoids, Possible Biosynthetic Intermediates for Clavulones, from the Okinawan Soft Coral *Clavularia viridis*. Chem Pharm Bull **51**: 909
927. Rezanka T, Dembitsky VM (2003) Brominated Oxylipins and Oxylipin Glycosides from Red Sea Corals. Eur J Org Chem 309
928. Kousaka K, Ogi N, Akazawa Y, Fujieda M, Yamamoto Y, Takada Y, Kimura J (2003) Novel Oxylipin Metabolites from the Brown Alga *Eisenia bicyclis*. J Nat Prod **66**: 1318
929. Dorta E, Díaz-Marrero A-R, Cueto M, D'Croz L, Maté JL, Darias J (2004) Carijenone, a Novel Class of Bicyclic Prostanoid from the Eastern Pacific Octocoral *Carijoa multiflora*. Org Lett **6**: 2229
930. Verbitski SM, Mullally JE, Fitzpatrick FA, Ireland CM (2004) Punaglandins, Chlorinated Prostaglandins, Function as Potent Michael Receptors to Inhibit Ubiquitin Isopeptidase Activity. J Med Chem **47**: 2062
931. Ratnayake AS, Bugni TS, Veltri CA, Skalicky JJ, Ireland CM (2006) Chemical Transformation of Prostaglandin-A_2: A Novel Series of C-10 Halogenated, C-12 Hydroxylated Prostaglandin-A_2 Analogues. Org Lett **8**: 2171
932. König GM, Wright AD, Bernardinelli G (1995) Determination of the Absolute Configuration of a Series of Halogenated Furanones from the Marine Alga *Delisea pulchra*. Helv Chim Acta **78**: 758
933. Steinberg PD, de Nys R, Kjelleberg S (1998) Chemical Inhibition of Epibiota by Australian Seaweeds. Biofouling **12**: 227
934. Maximilien R, de Nys R, Holmström C, Gram L, Givskov M, Crass K, Kjelleberg S, Steinberg PD (1998) Chemical Mediation of Bacterial Surface Colonisation by Secondary Metabolites from the Red Alga *Delisea pulchra*. Aquat Microb Ecol **15**: 233
935. Manefield M, de Nys R, Kumar N, Read R, Givskov M, Steinberg P, Kjelleberg S (1999) Evidence that Halogenated Furanones from *Delisea pulchra* Inhibit Acylated Homoserine Lactone (AHL)-Mediated Gene Expression by Displacing the AHL Signal from Its Receptor Protein. Microbiology **145**: 283
936. Rasmussen TB, Manefield M, Andersen JB, Eberl L, Anthoni U, Christophersen C, Steinberg P, Kjelleberg S, Givskov M (2000) How *Delisea pulchra* Furanones Affect Quorum Sensing and Swarming Motility in *Serratia liquefaciens* MG1. Microbiology **146**: 3237
937. Manefield M, Welch M, Givskov M, Salmond GPC, Kjelleberg S (2001) Halogenated Furanones from the Red Alga, *Delisea pulchra*, Inhibit Carbapenem Antibiotic Synthesis and Exoenzyme Virulence Factor Production in the Phytopathogen *Erivinia carotovora*. FEMS Microbiol Lett **205**: 131
938. Hentzer M, Riedel K, Rasmussen TB, Heydorn A, Andersen JB, Parsek MR, Rice SA, Eberl L, Molin S, Høiby N, Kjelleberg S, Givskov M (2002) Inhibition of Quorum Sensing in *Pseudomonas aeruginosa* Biofilm Bacteria by a Halogenated Furanone Compound. Microbiology **148**: 87
939. Manefield M, Rasumssen TB, Henzter M, Andersen JB, Steinberg P, Kjelleberg S, Givskov M (2002) Halogenated Furanones Inhibit Quorum Sensing Through Accelerated LuxR Turnover. Microbiology **148**: 1119
940. Suga H, Smith KM (2003) Molecular Mechanisms of Bacterial Quorum Sensing as a New Drug Target. Curr Opin Chem Biol **7**: 586
941. Hentzer M, Wu H, Andersen JB, Riedel K, Rasmussen TB, Bagge N, Kumar N, Schembri MA, Song Z, Kristoffersden P, Manefield M, Costerton JW, Molin S, Eberl L, Steinberg P, Kjelleberg S, Høiby N, Givskov M (2003) Attenuation of *Pseudomonas aeruginosa* Virulence by Quorum Sensing Inhibitors. EMBO J **22**: 3803
942. Ren D, Bedzyk LA, Ye RW, Thomas SM, Wood TK (2004) Differential Gene Expression Shows Natural Brominated Furanones Interfere with the Autoinducer-2 Bacterial Signaling System of *Escherichia coli*. Biotechnol Bioengin **88**: 630
943. Wright JT, de Nys R, Poore AGB, Steinberg PD (2004) Chemical Defense in a Marine Alga: Heritability and the Potential for Selection by Herbivores. Ecology **85**: 2946

944. Rice SA, McDougald D, Kumar N, Kjelleberg S (2005) The Use of Quorum-Sensing Blockers as Therapeutic Agents for the Control of Biofilm-Associated Infections. Curr Opin Invest Drugs **6**: 178
945. Wright AD, de Nys R, Angerhofer CK, Pezzuto JM, Gurrath M (2006) Biological Activities and 3D QSAR Studies of a Series of *Delisea pulchra* (cf. *fimbriata*) Derived Natural Products. J Nat Prod **69**: 1180
946. de Nys R, Steinberg PD, Rogers CN, Charlton TS, Duncan MW (1996) Quantitative Variation of Secondary Metabolites in the Sea Hare *Aplysia parvula* and Its Host Plant, *Delisea pulchra*. Mar Ecol Prog Ser **130**: 135
947. Ortega MJ, Zubía E, Ocaña JM, Naranjo S, Salvá J (2000) New Rubrolides from the Ascidian *Synoicum blochmanni*. Tetrahedron **56**: 3963
948. Pearce AN, Chia EW, Berridge, MV, Maas EW, Page MJ, Webb VL, Harper JL, Copp BR (2007) *E/Z*-Rubrolide O, an Anti-inflammatory Halogenated Furanone from the New Zealand Ascidian *Synoicum* n. sp. J Nat Prod **70**: 111
949. Boukouvalas J, Lachance N, Ouellet M, Trudeau M (1998) Facile Access to 4-Aryl-2(5H)-furanones by Suzuki Cross Coupling: Efficient Synthesis of Rubrolides C and E. Tetrahedron Lett **39**: 7665
950. Smith CJ, Hettich RL, Jompa J, Tahir A, Buchanan MV, Ireland CM (1998) Cadiolides A and B, New Metabolites from an Ascidian of the Genus *Botryllus*. J Org Chem **63**: 4147
951. Carroll AR, Healy PC, Quinn RJ, Tranter CJ (1999) Prunolides A, B, and C: Novel Tetraphenolic Bis-Spiroketals from the Australian Ascidian *Synoicum prunum*. J Org Chem **64**: 2680
952. Nair MSR, Hervey A (1979) Structure of Lepiochlorin, an Antibiotic Metabolite of a Fungus Cultivated by Ants. Phytochemistry **18**: 326
953. Wood WF, Clark TJ, Bradshaw DE, Foy BD, Largent DL, Thompson BL (2004) Clitolactone: A Banana Slug Antifeedant from *Clitocybe flaccida*. Mycologia **96**: 23
954. Hu JY, Qiao W, Takaishi Y, Duan HQ (2006) A New Hemiterpene Derivative from *Prinsepia utilis*. Chin Chem Lett **17**: 198
955. Abrell LM, Borgeson B, Crews P (1996) Chloro Polyketides from the Cultured Fungus (*Aspergillus*) Separated from a Marine Sponge. Tetrahedron Lett **37**: 2331
956. Namikoshi M, Negishi R, Nagai H, Dmitrenok A, Kobayashi H (2003) Three New Chlorine-Containing Antibiotics from a Marine-Derived Fungus *Aspergillus ositanus* Collected in Pohnpei. J Antibiot **56**: 755
957. Gross F, Lewis EA, Piraee M, van Pée K-H, Vining LC, White RL (2002) Isolation of 3'-*O*-Acetylchloramphenicol: A Possible Intermediate in Chloramphenicol Biosynthesis. Bioorg Med Chem Lett **12**: 283
958. Buarque de Gusmao N, Kaouadji M, Seigle-Murandi F, Steinman R, Thomasson F (1993) Acrodontiolamide, a Chlorinated Fungal Metabolite from *Acrodontium salmoneum*. Spectrosc Lett **26**: 1373
959. Steinman R, Benoit-Guyod JL, Guiraud P, Seigle-Murandi F (1995) Evaluation of Acrodontiolamide, a Chlorinated Compound Produced by *Acrodontium salmoneum* de Hoog for Cytotoxicity and Antimicrobial Activity. Pharmazie **50**: 693
960. George S, Narina SV, Sudalai A (2006) A Short Enantioselective Synthesis of (−)-Chloramphenicol and (+)-Thiamphenicol Using Tethered Aminohydroxylation. Tetrahedron **62**: 10202
961. He J, Magarvey N, Piraee M, Vining LC (2001) The Gene Cluster for Chloramphenicol Biosynthesis in *Streptomyces venezuelae* ISP5230 Includes Novel Shikimate Pathway Homologues and a Monomodular Nonribosomal Peptide Synthetase Gene. Microbiology **147**: 2817
962. Piraee M, White RL, Vining LC (2004) Biosynthesis of the Dichloroacetyl Component of Chloramphenicol in *Streptomyces venezuelae* ISP5230: Genes Required for Halogenation. Microbiology **150**: 85
963. Adachi H, Nishimura Y, Takeuchi T (2002) Synthesis and Activities of Bactobolin Derivatives Having New Functionality at C-3. J Antibiot **55**: 92

964. Ohta T, Matsuda M, Takahashi T, Nakajima S, Nozoe S (1995) (*S*)-cis-2-Amino-5-chloro-4-pentenoic Acid from the Fungus *Amanita vergineoides*. Chem Pharm Bull **43**: 899
965. Hatanaka S, Furukawa J, Aoki T, Akatsuka H, Nagasawa E (1994) (2*S*)-2-Amino-5-chloro-4-hydroxy-5-hexenoic Acid, a New Chloroamino Acid, and Related Compounds from *Amanita gymnopus*. Mycoscience **35**: 391
966. Hatanaka S, Okada K, Nagasawa E (1995) Isolation and Identification of (2*S*)-2-Amino-5-chloro-4-hydroxy-5-hexenoic Acid from an *Amanita* of the Section *Roanokenses* (Amanitaceae). Mycoscience **36**: 395
967. Hatanaka S, Niimura Y, Takishima K, Sugiyama J (1998) (2*R*)-2-Amino-6-hydroxy-4-hexynoic Acid, and Related Amino Acids in the Fruiting Bodies of *Amanita miculifera*. Phytochemistry **49**: 573
968. Yoshimura H, Takegami K, Doe M, Yamashita T, Shibata K, Wakabayashi K, Soga K, Kamisaka S (1999) α-Amino Acids from a Mushroom, *Amanita castanopsidis* Hongo, with Growth-Inhibiting Activity. Phytochemistry **52**: 25
969. Liu K, White RL, He J-Y, Vining LC (1995) Biosynthesis of Armentomycin: A Chlorinated Nonprotein Amino Acid. J Antibiot **48**: 347
970. Carmichael WW (1994) The Toxins of Cyanobacteria. Sci Am **270**: 78
971. Kuiper-Goodman T, Falconer I, Fitzgerald J (1999) Human Health Aspects. In: Chorus I, Bartram J (eds) Toxic Cyanobacteria in Water. E&FN Spon, London, p 113
972. Falconer IR (2004) Cyanobacterial Toxins of Drinking Water Supplies. CRC Press, Boca Raton, FL
973. Sano T, Kaya K (1997) A 3-Amino-10-chloro-2-hydroxydecanoic Acid-Containing Tetrapeptide from *Oscillatoria agardhii*. Phytochemistry **44**: 1503
974. Orjala J, Gerwick WH (1996) Barbamide, a Chlorinated Metabolite with Molluscicidal Activity from the Caribbean Cyanobacterium *Lyngbya majuscula*. J Nat Prod **59**: 427
975. Sitachitta N, Márquez BL, Williamson RT, Rossi J, Roberts MA, Gerwick WH, Nguyen V-A, Willis CL (2000) Biosynthetic Pathway and Origin of the Chlorinated Methyl Group in Barbamide and Dechlorobarbamide, Metabolites from the Marine Cyanobacterium *Lyngbya majuscula*. Tetrahedron **56**: 9103
976. Sitachitta N, Rossi J, Roberts MA, Gerwick WH, Fletcher MD, Willis CL (1998) Biosynthesis of the Marine Cyanobacterial Metabolite Barbamide. 1. Origin of the Trichloromethyl Group. J Am Chem Soc **120**: 7131
977. Williamson RT, Sitachitta N, Gerwick WH (1999) Biosynthesis of the Marine Cyanobacterial Metabolite Barbamide. 2: Elucidation of the Origin of the Thiazole Ring by Application of a New GHNMBC Experiment. Tetrahedron Lett **40**: 5175
978. Chang Z, Flatt P, Gerwick WH, Nguyen V-A, Willis CL, Sherman DH (2002) The Barbamide Biosynthetic Gene Cluster: A Novel Marine Cyanobacterial System of Mixed Polyketide Synthase (PKS)-non-ribosomal Peptide Synthetase (NRPS) Origin Involving an Unusual Trichloroleucyl Starter Unit. Gene **296**: 235
979. Gerwick WH, Leslie P, Long GC, Marquez BL, Willis CL (2003) [6-^{13}C]-(2*S*,4*S*)-5-Chloroleucine: Synthesis and Incubation Studies with Cultures of the Cyanobacterium, *Lyngbya majuscula*. Tetrahedron Lett **44**: 285
980. Galonić DP, Vaillancourt FH, Walsh CT (2006) Halogenation of Unactivated Carbon Centers in Natural Product Biosynthesis: Trichlorination of Leucine During Barbamide Biosynthesis. J Am Chem Soc **128**: 3900
981. Flatt PM, O'Connell SJ, McPhail KL, Zeller G, Willis CL, Sherman DH, Gerwick WH (2006) Characterization of the Initial Enzymatic Steps of Barbamide Biosynthesis. J Nat Prod **69**: 938
982. Hartung J (1999) The Biosynthesis of Barbamide – A Radical Pathway for "Biohalogenation". Angew Chem Int Ed **38**: 1209
983. Schnarr NA, Khosla C (2005) Just Add Chlorine. Nature **436**: 1094
984. Clark WD, Crews P (1995) A Novel Chlorinated Ketide Amino Acid, Herbamide A, from the Marine Sponge *Dysidea herbacea*. Tetrahedron Lett **36**: 1185

985. Dumdei EJ, Simpson JS, Garson MJ, Byriel KA, Kennard CHL (1997) New Chlorinated Metabolites from the Tropical Marine Sponge *Dysidea herbacea*. Aust J Chem **50**: 139
986. Fu X, Zeng L-M, Su J-Y, Pais M (1997) A New Diketopiperazine Derivative from the South China Sea Sponge *Dysidea fragilis*. J Nat Prod **60**: 695
987. Fu X, Ferreira MLG, Schmitz FJ, Kelly-Borges M (1998) New Diketopiperazines from the Sponge *Dysidea chlorea*. J Nat Prod **61**: 1226
988. Su JY, Zeng LM, Zhong YL, Fu X (1995) Biologically-Active Metabolites from Marine Organisms. J Chin Chem Soc **42**: 735
989. Fu X, Su J-Y, Zeng L-M (2000) Dysamide U, a New Trichlorinated Diketopiperazine from the Sponge *Dysidea* sp. Chin J Chem **18**: 882
990. MacMillan JB, Molinski TF (2000) Herbacic Acid, a Simple Prototype of 5,5,5-Trichloroleucine Metabolites from the Sponge *Dysidea herbacea*. J Nat Prod **63**: 155
991. Stapleton BL, Cameron GM, Garson MJ (2001) New Chlorinated Peptides from the Tropical Marine Sponge *Dysidea* sp. Tetrahedron **57**: 4603
992. MacMillan JB, Trousdale EK, Molinski TF (2000) Structure of (−)-Neodysidenin from *Dysidea herbacea*. Implications for Biosynthesis of 5,5,5-Trichloroleucine Peptides. Org Lett **2**: 2721
993. Dumrongchai N, Ponglimanont C, Stapleton BL, Garson MJ (2001) Chemical Diversity in the Tropical Marine Sponge *Dysidea herbacea*. ACGC Chem Commun **13**: 17
994. Jiménez JI, Scheuer PJ (2001) New Lipopeptides from the Caribbean Cyanobacterium *Lyngbya majuscula*. J Nat Prod **64**: 200
995. Harrigan GG, Goetz GH, Luesch H, Yang S, Likos J (2001) Dysideaprolines A-F and Barbaleucamides A-B, Novel Polychlorinated Compounds from a *Dysidea* Species. J Nat Prod **64**: 1133
996. Fahey SJ, Garson MJ (2002) Geographic Variation of Natural Products of Tropical Nudibranch *Asteronotus cespitosus*. J Chem Ecol **28**: 1773
997. Ardá A, Rodríguez J, Nieto RM, Bassarello C, Gomez-Paloma L, Bifulco G, Jiménez C (2005) NMR *J*-Based Analysis of Nitrogen-Containing Moieties and Application to Dysithiazolamide, a New Polychlorinated Dipeptide from *Dysidea* sp. Tetrahedron **61**: 10093
998. Sauleau P, Retailleau P, Vacelet J, Bourquet-Kondracki M-L (2005) New Polychlorinated Pyrrolidinones from the Red Sea Marine Sponge *Lamellodysidea herbacea*. Tetrahedron **61**: 955
999. Trousdale EK, Taylor SW, Parkin S, Hope H, Mokinski TF (1998) Reductive Dechlorination of Dysidenin from *Dysidea herbacea*. Structure of a Novel Binuclear Zinc Metallocycle. Nat Prod Lett **12**: 61
1000. Van Sande J, Deneubourg F, Beauwens R, Braekman JC, Daloze D, Dumont JE (1990) Inhibition of Iodide Transport in Thyroid Cells by Dysidenin, a Marine Toxin, and Some of Its Analogs. Mol Pharmacol **37**: 583
1001. Vroye L, Beauwens R, Van Sande J, Daloze D, Braekmann JC, Golstein PE (1998) The Na^+–I Cotransporter of the Thyroid: Characterisation of New Inhibitors. Eur J Physiol **435**: 259
1002. Flowers AE, Garson MJ, Webb RI, Dumdei EJ, Charan RD (1998) Cellular Origin of Chlorinated Diketopiperazines in the Dictyoceratid Sponge *Dysidea herbacea* (Keller). Cell Tissue Res **292**: 597
1003. Ridley CP, Berquist PR, Harper MK, Faulkner DJ, Hooper JNA, Haygood MG (2005) Speciation and Biosynthetic Variation in Four Dictyoceratid Sponges and Their Cyanobacterial Symbiont, *Oscillatoria spongeliae*. Chem Biol **12**: 397
1004. Sone H, Kondo T, Kiryu M, Ishiwata H, Ojika M, Yamada K (1995) Dolabellin, a Cytotoxic Bisthiazole Metabolite from the Sea Hare *Dolabella auricularia*: Structural Determination and Synthesis. J Org Chem **60**: 4774
1005. Luesch H, Yoshida WY, Moore RE, Paul VJ, Mooberry SL (2000) Isolation, Structure Determination, and Biological Activity of Lyngbyabellin A from the Marine Cyanobacterium *Lyngbya majuscula*. J Nat Prod **63**: 611

References

1006. Luesch H, Yosida WY, Moore RE, Paul VJ (2000) Isolation and Structure of the Cytotoxin Lyngbyabellin B and Absolute Configuration of Lyngbyapeptin A from the Marine Cyanobacterium *Lyngbya majuscula.* J Nat Prod **63**: 1437

1007. Milligan KE, Marquez BL, Williamson RT, Gerwick WH (2000) Lyngbyabellin B, a Toxic and Antifungal Secondary Metabolite from the Marine Cyanobacterium *Lyngbya majuscula.* J Nat Prod **63**: 1440

1008. Yokokawa F, Sameshima H, Shiori T (2001) Total Synthesis of Lyngbyabellin A, a Potent Cytotoxic Metabolite from the Marine Cyanobacterium *Lyngbya majuscula.* Tetrahedron Lett **42**: 4171

1009. Luesch H, Yoshida WY, Moore RE, Paul VJ (2002) Structurally Diverse New Alkaloids from Palauan Collections of the Apratoxin-Producing Marine Cyanobacterium *Lyngbya* sp. Tetrahedron **58**: 7959

1010. Marquez BL, Watts KS, Yokochi A, Roberts MA, Verdier-Pinard P, Jimenez JI, Hamel E, Scheuer PJ, Gerwick WH (2002) Structure and Absolute Stereochemistry of Hectochlorin, a Potent Stimulator of Actin Assembly. J Nat Prod **65**: 866

1011. Cetusic JRP, Green III FR, Graupner PR, Oliver MP (2002) Total Synthesis of Hectochlorin. Org Lett **4**: 1307

1012. Suntornchashwej S, Chaichit N, Isobe M, Suwanborirux K (2005) Hectochlorin and Morpholine Derivatives from the Thai Sea Hare, *Bursatella leachii*. J Nat Prod **68**: 951

1013. Williams PG, Luesch H, Yoshida WY, Moore RE, Paul VJ (2003) Continuing Studies on the Cyanobacterium *Lyngbya* sp.: Isolation and Structure Determination of 15-Norlyngbyapeptin A and Lyngbyabellin D. J Nat Prod **66**: 595

1014. Han B, McPhail KL, Gross H, Goeger DE, Mooberry SL, Gerwick WH (2005) Isolation and Structure of Five Lyngbyabellin Derivatives from a Papua New Guinea Collection of the Marine Cyanobacterium *Lyngbya majuscula*. Tetrahedron **61**: 11723

1015. Igarashi M, Kinoshita N, Ikeda T, Kameda M, Hamada M, Takeuchi T (1997) Resormycin, a Novel Herbicidal and Antifungal Antibiotic Produced by a Strain of *Streptomyces platensis* I. Taxonomy, Production, Isolation and Biological Properties. J Antibiot **50**: 1020

1016. Igarashi M, Nakamura H, Naganawa H, Takeuchi T (1997) Resormycin, a Novel Herbicidal and Antifungal Antibiotic Produced by a Strain of *Streptomyces platensis* II. Structure Elucidation of Resormycin. J Antibiot **50**: 1026

1017. Shin-ya K, Kim J-S, Furihata K, Hayakawa Y, Seto H (1997) Structure of Kaitocephalin, a Novel Glutamate Receptor Antagonist Produced by *Eupenicillium shearii*. Tetrahedron Lett **38**: 7079

1018. Kobayashi H, Shin-ya K, Furikata K, Hayakawa Y, Seto H (2001) Absolute Configuration of a Novel Glutamate Receptor Antagonist Kaitocephalin. Tetrahedron Lett **42**: 4021

1019. Okue M, Kobayashi H, Shin-ya K, Furihata K, Hayakawa Y, Seto H, Watanabe H, Kitahara T (2002) Synthesis of the Proposed Structure and Revision of Stereochemistry of Kaitocephalin. Tetrahedron Lett **43**: 857

1020. Watanabe H, Okue M, Kobayashi H, Kitahara T (2002) The First Synthesis of Kaitocephalin Based on the Structure Revision. Tetrahedron Lett **43**: 861

1021. Ma D, Yang J (2001) Total Synthesis of Kaitocephalin, the First Naturally Occurring AMPA/KA Receptor Antagonist. J Am Chem Soc **123**: 9706

1022. Kawasaki M, Shinada T, Hamada M, Ohfune Y (2005) Total Synthesis of (−)-Kaitocephalin. Org Lett **7**: 4165

1023. Hoshino Y, Mukai A, Yazawa K, Uno J, Ishikawa J, Ando A, Fukai T, Mikami Y (2004) Transvalencin A, a Thiazolidine Zinc Complex Antibiotic Produced by a Clinical Isolate of *Nocardia transvalensis* I. Taxonomy, Fermentation, Isolation and Biological Activities. J Antibiot **57**: 797

1024. Hoshino Y, Mukai A, Yazawa K, Uno J, Ando A, Mikami Y, Fukai T, Ishikawa J, Yamaguchi K (2004) Transvalencin A, a Thiazolidine Zinc Complex Antibiotic Produced by a Clinical Isolate of *Nocardia transvalensis* II. Structure Elucidation. J Antibiot **57**: 803

1025. Ghosh AC, Ramgopal M (1980) Cyclic Peptides from *Penicillium islandicum*. A Review and a Reevaluation of the Structure of Islanditoxin. J Heterocycl Chem **17**: 1809
1026. Morita H, Nagashima S, Takeya K, Itokawa H (1994) A Novel Cyclic Pentapeptide with a β-Hydroxy-γ-chloroproline from *Aster tataricus*. Chem Lett 2009
1027. Morita H, Nagashima S, Takeya K, Itokawa H, Iitaka Y (1995) Structures and Conformation of Antitumour Cyclic Pentapeptides, Astins A, B and C, from *Aster tataricus*. Tetrahedron **51**: 1121
1028. Morita H, Nagashima S, Takeya K, Itokawa H (1995) Solution Forms of Antitumor Cyclic Pentapeptides with 3,4-Dichlorinated Proline Residues, Astins A and C, from *Aster tataricus*. Chem Pharm Bull **43**: 1395
1029. Morita H, Nagashima S, Uchiumi Y, Kuroki O, Takeya K, Itokawa H (1996) Cyclic Peptides from Higher Plants. XXVIII. Antitumor Activity and Hepatic Microsomal Biotransformation of Cyclic Pentapeptides, Astins, from *Aster tataricus*. Chem Pharm Bull **44**: 1026
1030. Schumacher KK, Hauze DB, Jiang J, Szewczyk J, Reddy RE, Davis FA, Joullié MM (1999) First Total Synthesis of Astin G. Tetrahedron Lett **40**: 455
1031. Cai P, Smith D, Katz B, Pearce C, Venables D, Houck D (1998) Destruxin-A4 Chlorohydrin, a Novel Destruxin from the Fungus OS-F68576: Isolation, Structure Determination, and Biological Activity as an Inducer of Erythropoietin. J Nat Prod **61**: 290
1032. Lira SP, Vita-Marques A, Seleghim MHR, Bugni TS, LaBarbera DV, Sette LD, Sponchiado SRP, Ireland CM, Berlinck RGS (2006) New Destruxins from the Marine-Derived Fungus *Beauveria felina*. J Antibiot **59**: 553
1033. Murakami M, Ishida K, Okino T, Okita Y, Matsuda H, Yamaguchi K (1995) Aeruginosins 98-A and B, Trypsin Inhibitors from the Blue-Green Alga *Microcystis aeruginosa* (NIES-98). Tetrahedron Lett **36**: 2785
1034. Shin HJ, Matsuda H, Murakami M, Yamaguchi K (1997) Aeruginosins 205A and -B, Serine Protease Inhibitory Glycopeptides from the Cyanobacterium *Oscillatoria agardhii* (NIES-205). J Org Chem **62**: 1810
1035. Ishida K, Matsuda H, Murakami M, Yamaguchi K (1997) Microginins 299-A and -B, Leucine Aminopeptidase Inhibitors from the Cyanobacterium *Microcystis aeruginosa* (NIES-299). Tetrahedron **53**: 10281
1036. Ishida K, Matsuda H, Murakami M, Yamaguchi K (1997) Micropeptins 478-A and -B, Plasmin Inhibitors from the Cyanobacterium *Microcystis aeruginosa*. J Nat Prod **60**: 184
1037. Ishida K, Matsuda H, Murakami M (1998) Four New Microginins, Linear Peptides from the Cyanobacterium *Microcystis aeruginosa*. Tetrahedron **54**: 13475
1038. Ishida K, Okita Y, Matsuda H, Okino T, Murakami M (1999) Aeruginosins, Protease Inhibitors from the Cyanobacterium *Microcystis aeruginosa*. Tetrahedron **55**: 10971
1039. Ishida K, Kato T, Murakami M, Watanabe M, Watanabe MF (2000) Microginins, Zinc Metalloproteases Inhibitors from the Cyanobacterium *Microcystis aeruginosa*. Tetrahedron **56**: 8643
1040. Valls N, López-Canet M, Vallribera M, Bonjoch J (2000) Total Synthesis and Reassignment of Configuration of Aeruginosin 298-A. J Am Chem Soc **122**: 11248
1041. von Elert E, Oberer L, Merkel P, Huhn T, Blom JF (2005) Cyanopeptolin 954, a Chlorine-Containing Chymotrypsin Inhibitor of *Microcystis aeruginosa* IVIVA Cya 43. J Nat Prod **68**: 1324
1042. Hanessian S, Del Valle JR, Blomberg N (2006) Total Synthesis and Structural Confirmation of Chlorodysinosin A. J Am Chem Soc **128**: 10491
1043. Macko V, Wolpert TJ, Acklin W, Jaun B, Seibl J, Meili J, Arigoni D (1985) Characterization of Victorin C, the Major Host-Selective Toxin from *Cochliobolus victoriae*: Structure of Degradation Products. Experientia **41**: 1366
1044. Gloer JB, Meinwald J, Walton JD, Earle ED (1985) Studies on the Fungal Phytotoxin Victorin: Structures of Three Novel Amino Acids from the Acid Hydrolyzate. Experientia **41**: 1370

1045. Wolpert TJ, Macko V, Acklin W, Jaun B, Seibl J, Meili J, Arigoni D (1985) Structure of Victorin C, the Major Host-Selective Toxin from *Cochliobolus victoriae*. Experientia **41**: 1524
1046. Wolpert TJ, Macko V, Acklin W, Jaun B, Arigoni D (1986) Structure of Minor Host-Selective Toxins from *Cochliobolus victoriae*. Experientia **42**: 1296
1047. Wolpert TJ, Macko V (1989) Specific Binding of Victorin to a 100-kDa Protein from Oats. Proc Natl Acad Sci U S A **86**: 4092
1048. Wolpert TJ, Navarre DA, Moore DL, Macko V (1994) Identification of the 100-kD Victorin Binding Protein from Oats. The Plant Cell **6**: 1145
1049. Macko V, Stimmel MB, Wolpert TJ, Dunkle LD, Acklin W, Bänteli R, Jaun B, Arigoni D (1992) Structure of the Host-Specific Toxins Produced by the Fungal Pathogen *Periconia circinata*. Proc Natl Acad Sci U S A **89**: 9574
1050. Bevan K, Davies JS, Hassall CH, Morton RB, Phillips DAS (1971) Amino-Acids and Peptides. Part X. Characterisation of the Monamycins, Members of a New Family of Cyclodepsipeptide Antibiotics. J Chem Soc (C) 514
1051. Hassall CH, Ogihara Y, Thomas WA (1971) Amino-Acids and Peptides. Part XI. (3R,5S)-5-Chloropiperazic Acid and (3S,5S)-5-Hydroxypiperazic Acid, Products of Hydrolysis of Monamycin. J Chem Soc (C) 522
1052. Hassall CH, Morton RB, Ogihara Y, Phillips DAS (1971) Amino-Acids and Peptides. Part XII. The Molecular Structures of the Monamycins, Cyclodepsipeptide Antibiotics. J Chem Soc (C) 526
1053. Kunze B, Jansen R, Sasse F, Höfle G, Reichenbach (1995) Chondramides A-D, New Antifungal and Cytostatic Depsipeptides from *Chondromyces crocatus* (Myxobacteria). Production, Physico-Chemical and Biological Properties. J Antibiot **48**: 1262
1054. Jansen R, Kunze B, Reichenbach H, Höfle G (1996) Chondramides A-D, New Antifungal and Cytostatic Depsipeptides from *Chondromyces crocatus* (Myxobacteria). Isolation and Structure Elucidation. Liebigs Ann 285
1055. Sasse F, Kunze B, Gronewold TMA, Reichenbach H (1998) The Chondramides: Cytostatic Agents from Myxobacteria Acting on the Actin Cytoskeleton. J Natl Cancer Inst **90**: 1559
1056. Zampella A, Giannini C, Debitus C, Roussakis C, D'Auria MV (1999) New Jaspamide Derivatives from the Marine Sponge *Jaspis splendans* Collected in Vanuatu. J Nat Prod **62**: 332
1057. Bubb MR, Senderowicz AMJ, Sausville EA, Duncan KLK, Korn ED (1994) Jasplakinolide, a Cytotoxic Natural Product, Induces Actin Polymerization and Competitively Inhibits the Binding of Phalloidin to F-Actin. J Biol Chem **269**: 14869
1058. Senderowicz AMJ, Kaur G, Sainz E, Laing C, Inman WD, Rodríguez J, Crews P, Malspeis L, Grever MR, Sausville EA, Duncan KLK (1995) Jasplakinolide's Inhibition of the Growth of Prostate Carcinoma Cells *in Vitro* with Disruption of the Actin Cytoskeleton. J Natl Cancer Inst **87**: 46
1059. Tabudravu JN, Morris LA, Milne BF, Jaspars M (2005) Conformational Studies of Free and Li$^+$ Complexed Jasplakinolide, a Cyclic Depsipeptide from the Fijian Marine Sponge *Jaspis splendans*. Org Biomol Chem **3**: 745
1060. Terracciano S, Bruno I, Bifulco G, Copper JE, Smith CD, Gomez-Paloma L, Riccio R (2004) Synthesis, Conformational Analysis, and Cytotoxicity of New Analogues of the Natural Cyclodepsipeptide Jaspamide. J Nat Prod **67**: 1325
1061. Terracciano S, Bruno I, Bifulco G, Avallone E, Smith CD, Gomez-Paloma L, Riccio R (2005) Synthesis, Solution Structure, and Bioactivity of Six New Simplified Analogues of the Natural Cyclodepsipeptide Jaspamide. Bioorg Med Chem **13**: 5225
1062. Palermo JA, Brasco MFR, Cabezas E, Balzaretti V, Seldes AM (1998) Celenamide E, a Tripeptide Alkaloid from the Patagonian Sponge *Cliona chilensis*. J Nat Prod **61**: 488
1063. Clark WD, Corbett T, Valeriote F, Crews P (1997) Cyclocinamide A. An Unusual Cytotoxic Halogenated Hexapeptide from the Marine Sponge *Psammocinia*. J Am Chem Soc **119**: 9285

1064. Schmidt EW, Faulkner DJ (1998) Microsclerodermins C–E, Antifungal Cyclic Peptides from the Lithisid Marine Sponges *Theonella* sp. and *Microscleroderma* sp. Tetrahedron **54**: 3043
1065. Erdogan I, Tanaka J, Higa T (2000) Two Cyclic Hexapeptides from the Marine Sponge *Theonella cupola*. FABAD J Pharm Sci **25**: 7
1066. Ishibashi M, Li Y, Sato M, Kobayashi J (1994) Stereochemistry of the Modified Tryptophan Residues Contained in Konbamide and Keramamide A Isolated from Marine Sponges of the Genus *Theonella*. Nat Prod Lett **4**: 293
1067. Schmidt U, Weinbrenner S (1996) What is the Structure of the Calmodulin Antagonist Konbamide from *Theonella* sp.? Synthesis of Two Isomers by Direct Biomimetic Introduction of Bromine in Hydroxytryptophan-Containing Cyclic Peptides. Angew Chem Int Ed Engl **35**: 1336
1068. Kobayashi J, Itagaki F, Shigemori H, Takao T, Shimonishi Y (1995) Keramamides E, G, H, and J, New Cyclic Peptides Containing an Oxazole or a Thiazole Ring from a *Theonella* sponge. Tetrahedron **51**: 2525
1069. Uemoto H, Yahiro Y, Shigemori H, Tsuda M, Takao T, Shimonishi Y, Kobayashi J (1998) Keramamides K and L, New Cyclic Peptides Containing Unusual Tryptophan Residue from *Theonella* Sponge. Tetrahedron **54**: 6719
1070. Tsuda M, Ishiyama H, Masuko K, Takao T, Shimonishi Y, Kobayashi J (1999) Keramamides M and N, Two New Cyclic Peptides with a Sulfate Ester from *Theonella* Sponge. Tetrahedron **55**: 12543
1071. Broberg A, Menkis A, Vasiliauskas R (2006) Kutznerides 1-4, Depsipeptides from the Actinomycete *Kutzneria* sp. 744 Inhabiting Mycorrhizal Roots of *Picea abies* Seedlings. J Nat Prod **69**: 97
1072. Pohanka A, Menkis A, Levenfors J, Broberg A (2006) Low-Abundance Kutznerides from *Kutzneria* sp. 744. J Nat Prod **69**: 1776
1073. Umezawa K, Ikeda Y, Uchihata Y, Naganawa H, Kondo S (2000) Chloptosin, an Apoptosis-Inducing Dimeric Cyclohexapeptide Produced by *Streptomyces*. J Org Chem **65**: 459
1074. Myers RA, Cruz LJ, Rivier JE, Olivera BM (1993) *Conus* Peptides as Chemical Probes for Receptors and Ion Channels. Chem Rev **93**: 1923
1075. Olivera BM (2000) ω-Conotoxin MVIIA: From Marine Snail Venom to Analgesic Drug. In: Fusetani N (ed) Drugs from the Sea, Karger, Basel, p 74
1076. Alonso D, Khalil Z, Satkunanthan N, Livett BG (2003) Drugs from the Sea: Conotoxins as Drug Leads for Neuropathic Path and Other Neurological Conditions. Mini Rev Med Chem **3**: 785
1077. Alewood P, Hopping G, Armishaw C (2003) Marine Toxins as Sources of Drug Leads. Aust J Chem **56**: 769
1078. Livett BG, Gayler KR, Khalil Z (2004) Drugs from the Sea: Conopeptides as Potential Therapeutics. Curr Med Chem **11**: 1715
1079. Staats PS, Yearwood T, Charapata SG, Presley RW, Wallace MS, Byas-Smith M, Fisher R, Bryce DA, Mangieri EA, Luther RR, Mayo M, McGuire D, Ellis D (2004) Intrathecal Ziconotide in the Treatment of Refractory Pain in Patients with Cancer or AIDS. J Am Med Assoc **291**: 63
1080. Nelson L (2004) One Slip, and You're Dead... Nature **429**: 798
1081. Craig AG, Jimenez EC, Dykert J, Nielsen DB, Gulyas J, Abogadie FC, Porter J, Rivier JE, Cruz LJ, Olivera BM, McIntosh JM (1997) A Novel Post-translational Modification Involving Bromination of Tryptophan. J Biol Chem **272**: 4689
1082. Jimenez EC, Craig AG, Watkins M, Hillyard DR, Gray WR, Gulyas J, Rivier JE, Cruz LJ, Olivera BM (1997) Bromocontryphan: Post-Translational Bromination of Tryptophan. Biochemistry **36**: 989
1083. Lirazan MB, Craig AG, Shetty R, Walker CS, Olivera BM, Cruz LJ (1999) Multiple Bromotryptophan and γ-Carboxyglutamate Residues in a *Conus* Peptide. Philippine J Sci **128**: 239

1084. Walker CS, Steel D, Jacobsen RB, Lirazan MB, Cruz LJ, Hooper D, Shetty R, DelaCruz RC, Nielsen JS, Zhou LM, Bandyopadhyay P, Craig AG, Olivera BM (1999) The T-Superfamily of Conotoxins. J Biol Chem **274**: 30664
1085. Taylor SW, Kammerer B, Nicholson GJ, Pusecker K, Walk T, Bayer E, Scippa S, de Vincentüs M (1997) Morulin Pm: A Modified Polypeptide Containing TOPA and 6-Bromotryptophan from the Morula Cells of the Ascidian, *Phallusia mammillata*. Arch Biochem Biophys **348**: 278
1086. Taylor SW, Craig AG, Fischer WH, Park M, Lehrer RI (2000) Styelin D, an Extensively Modified Antimicrobial Peptide from Ascidian Hemocytes. J Biol Chem **275**: 38417
1087. Lehrer RI, Tincu JA, Taylor SW, Menzel LP, Waring AJ (2003) Natural Peptide Antibiotics from Tunicates: Structures, Functions and Potential Uses. Integr Comp Biol **43**: 313
1088. Shinnar AE, Butler KL, Park HJ (2003) Cathelicidin Family of Antimicrobial Peptides: Proteolytic Processing and Protease Resistance. Bioorg Chem **31**: 425
1089. Li J, Jeong S, Esser L, Harran PG (2001) Total Synthesis of Nominal Diazonamides – Part 1: Convergent Preparation of the Structure Proposed for (–)-Diazonamide A. Angew Chem Int Ed **40**: 4765
1090. Li J, Burgett AWG, Esser L, Amezcua C, Harran PG (2001) Total Synthesis of Nominal Diazonamides – Part 2: On the True Structure and Origin of Natural Isolates. Angew Chem Int Ed **40**: 4770
1091. Ritter T, Carreira EM (2002) The Diazonamides: The Plot Thickens. Angew Chem Int Ed **41**: 2489
1092. Nicolaou KC, Chen D Y-K, Huang X, Ling T, Bella M, Snyder SA (2004) Chemistry and Biology of Diazonamide A: First Total Synthesis and Confirmation of the True Structure. J Am Chem Soc **126**: 12888
1093. Nicolaou KC, Hao J, Reddy MV, Rao PB, Rassias G, Snyder SA, Huang X, Chen D Y-K, Brenzovich WE, Giuseppone N, Giannakakou P, O'Brate A (2004) Chemistry and Biology of Diazonamide A: Second Total Synthesis and Biological Investigations. J Am Chem Soc **126**: 12897
1094. Burgett AWG, Li Q, Wei Q, Harran PG (2003) A Concise and Flexible Total Synthesis of (–)-Diazonamide A. Angew Chem Int Ed **42**: 4961
1095. Cruz-Monserrate Z, Vervoort HC, Bai R, Newman DJ, Howell SB, Los G, Mullaney JT, Williams MD, Pettit GR, Fenical W, Hamel E (2003) Diazonamide A and a Synthetic Structural Analog: Disruptive Effects on Mitosis and Cellular Microtubules and Analysis of Their Interactions with Tubulin. Mol Pharmacol **63**: 1273
1096. Coleman JE, de Silva ED, Kong F, Andersen RJ, Allen TM (1995) Cytotoxic Peptides from the Marine Sponge *Cymbastela* sp. Tetrahedron **51**: 10653
1097. Tinto WF, Lough AJ, McLean S, Reynolds WF, Yu M, Chan WR (1998) Geodiamolides H and I, Further Cyclodepsipeptides from the Marine Sponge *Geodia* sp. Tetrahedron **54**: 4451
1098. Coleman JE, Van Soest R, Andersen RJ (1999) New Geodiamolides from the Sponge *Cymbastela* sp. Collected in Papua New Guinea. J Nat Prod **62**: 1137
1099. Talpir R, Benayahu Y, Kashman Y, Pannell L, Schleyer M (1994) Hemiasterlin and Geodiamolide TA; Two New Cytotoxic Peptides from the Marine Sponge *Hemiasterella minor* (Kirkpatrick). Tetrahedron Lett **35**: 4453
1100. D'Auria MV, Paloma LG, Minale L, Zampella A, Debitus C, Perez J (1995) Neosiphoniamolide A, a Novel Cyclodepsipeptide, with Antifungal Activity from the Marine Sponge *Neosiphonia superstes*. J Nat Prod **58**: 121
1101. Tanaka C, Tanaka J, Bolland RF, Marriott G, Higa T (2006) Seragamides A-F, New Actin-Targeting Depsipeptides from the Sponge *Suberites japonicus* Thiele. Tetrahedron **62**: 3536
1102. Iizuka T, Fudou R, Jojima Y, Ogawa S, Yamanaka S, Inukai Y, Ojika M (2006) Miuraenamides A and B, Novel Antimicrobial Cyclic Depsipeptides from a New Slightly Halophilic Myxobacterium: Taxonomy, Production, and Biological Properties. J Antibiot **59**: 385

1103. Hinterding K, Hagenbuch P, Rétey J, Waldmann H (1998) Synthesis and *In Vitro* Evaluation of the Ras Farnesyltransferase Inhibitor Pepticinnamin E. Angew Chem Int Ed **37**: 1236
1104. Thutewohl M, Kissau L, Popkirova B, Karaguni I-M, Nowak T, Bate M, Kuhlmann J, Müller O, Waldmann H (2002) Solid-Phase Synthesis and Biological Evaluation of a Pepticinnamin E Library. Angew Chem Int Ed **41**: 3616
1105. Thutewohl M, Waldmann H (2003) Solid-Phase Synthesis of a Pepticinnamin E Library. Bioorg Med Chem **11**: 2591
1106. Thutewohl M, Kissau L, Popkirova B, Karaguni I-M, Nowak T, Bate M, Kuhlmann J, Müller O, Waldmann H (2003) Identification of Mono- and Bisubstrate Inhibitors of Protein Farnesyltransferase and Inducers of Apoptosis from a Pepticinnamin E Library. Bioorg Med Chem **11**: 2617
1107. Eggen M, Georg GI (2002) The Cryptophycins: Their Synthesis and Anticancer Activity. Med Res Rev **22**: 85
1108. Li T, Shih C (2002) Structure Activity Relationships of Cryptophycins, A Novel Class of Antitumor Antimitotic Agents. Front Biotechnol Pharmaceut **3**: 1972
1109. Tius, MA (2003) Cryptophycin Synthesis. In: Gribble GW (ed) Handbook of Environmental Chemistry. Springer, Berlin, p 265
1110. Barrow RA, Hemscheidt T, Liang J, Paik S, Moore RE, Tius MA (1995) Total Synthesis of Cryptophycins. Revision of the Structures of Cryptophycins A and C. J Am Chem Soc **117**: 2479
1111. Golakoti T, Ogino J, Heltzel CE, Husebo TL, Jensen CM, Larsen LK, Patterson GML, Moore RE, Mooberry SL, Corbett TH, Valeriote FA (1995) Structure Determination, Conformational Analysis, Chemical Stability Studies, and Antitumor Evaluation of the Cryptophycins. Isolation of 18 New Analogs from *Nostoc* sp. Strain GSV 224. J Am Chem Soc **117**: 12030
1112. Subbaraju GV, Golakoti T, Patterson GML, Moore RE (1997) Three New Cryptophycins from *Nostoc* sp. GSV 224. J Nat Prod **60**: 302
1113. Chaganty S, Golakoti T, Heltzel C, Moore RE, Yoshida WY (2004) Isolation and Structure Determination of Cryptophycins 38, 326, and 327 from the Terrestrial Cyanobacterium *Nostoc* sp. GSV 224. J Nat Prod **67**: 1403
1114. Eisser S, Stoncius A, Nahrwold M, Sewald N (2006) The Synthesis of Cryptophycins. Synthesis 3747
1115. McCubbin JA, Maddess ML, Lautens M (2006) Total Synthesis of Cryptophycin Analogues via a Scaffold Approach. Org Lett **8**: 2993
1116. Wagner MM, Paul DC, Shih C, Jordan MA, Wilson L, Williams DC (1999) *In Vitro* Pharmacology of Cryptophycin 52 (LY355703) in Human Tumor Cell Lines. Cancer Chemother Pharmacol **43**: 115
1117. Menon K, Alvarez E, Forler P, Phares V, Amsrud T, Shih C, Al-Awar R, Teicher BA (2000) Antitumor Activity of Cryptophycins: Effect of Infusion Time and Combination Studies. Cancer Chemother Pharmacol **46**: 142
1118. Stevenson JP, Sun W, Gallagher M, Johnson R, Vaughn D, Schuchter L, Algazy K, Hahn S, Enas N, Ellis D, Thornton D, O'Dwyer PJ (2002) Phase I Trial of the Cryptophycin Analogue LY355703 Administered as an Intravenous Infusion on a Day 1 and 8 Schedule Every 21 Days. Clin Cancer Res **8**: 2524
1119. Edelman MJ, Gandara DR, Hausner P, Israel V, Thornton D, DeSanto J, Doyle LA (2003) Phase 2 Study of Cryptophycin 52 (LY355703) in Patients Previously Treated with Platinum Based Chemotherapy for Advanced Non-Small Cell Lung Cancer. Lung Cancer **39**: 197
1120. Davis RA (2005) Isolation and Structure Elucidation of the New Fungal Metabolite (–)-Xylariamide A. J Nat Prod **68**: 769
1121. Davis RA, Kotiw M (2005) Synthesis of the Fungal Natural Product (–)-Xylariamide A. Tetrahedron Lett **46**: 5199

1122. Gulavita NK, Pomponi SA, Wright AF, Yarwood D, Sills MA (1994) Isolation and Structure Elucidation of Perthamide B, a Novel Peptide from the Sponge *Theonella* sp. Tetrahedron Lett **35**: 6815
1123. Li H, Matsunaga S, Fusetani N (1995) Halicylindramides A–C, Antifungal and Cytotoxic Depsipeptides from the Marine Sponge *Halichondria cylindrata*. J Med Chem **38**: 338
1124. Li H, Matsunaga S, Fusetani N (1996) Halicylindramides D and E, Antifungal Peptides from the Marine Sponge *Halichondria cylindrata*. J Nat Prod **59**: 163
1125. Matsunaga S, Fusetani N (1995) Theonellamides A–E, Cytotoxic Bicyclic Peptides, from a Marine Sponge *Theonella* sp. J Org Chem **60**: 1177
1126. Bewley CA, Holland ND, Faulkner DJ (1996) Two Classes of Metabolites from *Theonella swinhoei* are Localized in Distinct Populations of Bacterial Symbionts. Experientia **52**: 716
1127. Schmidt EW, Bewley CA, Faulkner DJ (1998) Theopalauamide, a Bicyclic Glycopeptide from Filamentous Bacterial Symbionts of the Lithistid Sponge *Theonella swinhoei* from Palau and Mozambique. J Org Chem **63**: 1254
1128. Bewley CA, Faulkner DJ (1995) Theonegramide, an Antifungal Glycopeptide from the Philippine Lithistid Sponge *Theonella swinhoei*. J Org Chem **60**: 2644
1129. Clark DP, Carroll J, Naylor S, Crews P (1998) An Antifungal Cyclodepsipeptide, Cyclolithistide A, from the Sponge *Theonella swinhoei*. J Org Chem **63**: 8757
1130. Rouhiainen L, Paulin L, Suomalainen S, Hyytiäinen H, Buikema W, Haselkorn R, Sivonen K (2000) Genes Encoding Synthetases of Cyclic Depsipeptides, Anabaenopeptilides, in *Anabaena* Strain 90. Mol Microbiol **37**: 156
1131. Rashid MA, Gustafson KR, Cartner LK, Shigematsu N, Pannell LK, Boyd MR (2001) Microspinosamide, a New HIV-Inhibitory Cyclic Depsipeptide from the Marine Sponge *Sidonops microspinosa*. J Nat Prod **64**: 117
1132. Capon RJ, Ford J, Lacey E, Gill JH, Heiland K, Friedel T (2002) Phoriospongin A and B: Two New Nematocidal Depsipeptides from the Australian Marine Sponge *Phoriospongia* sp. and *Callyspongia bilamellata*. J Nat Prod **65**: 358
1133. Ishida K, Matsuda H, Murakami M, Yamaguchi K (1997) Micropeptins 478-A and -B, Plasmin Inhibitors from the Cyanobacterium *Microcystis aeruginosa*. J Nat Prod **60**: 184
1134. Matern U, Oberer L, Falchetto RA, Erhard M, König WA, Herdman M, Weckesser J (2001) Scyptolin A and B, Cyclic Depsipeptides from Axenic Cultures of *Scytonema hofmanni* PCC 7110. Phytochemistry **58**: 1087
1135. Matern U, Schleberger C, Jelakovic S, Weckesser J, Schulz GE (2003) Binding Structure of Elastase Inhibitor Scyptolin A. Chem Biol **10**: 997
1136. Omura S, Mamada H, Wang N-J, Imamura N, Oiwa R, Iwai Y, Muto, N (1984) Takaokamycin, a New Peptide Antibiotic Produced by *Streptomyces* sp. J Antibiot **37**: 700
1137. Otoguro K, Ui H, Ishiyama A, Arai N, Kobayashi M, Takahashi Y, Masuma R, Shiomi K, Yamada H, Omura S (2003) *In Vitro* Antimalarial Activities of the Microbial Metabolites. J Antibiot **56**: 322
1138. Andres N, Wolf H, Zähner H, Rössner E, Zeeck A, König WA, Sinnwell V (1989) Hormaomycin, a Novel Peptide Lactone with Morphogenetic Actvity on *Streptomyces*. Helv Chim Acta **72**: 426
1139. Rössner E, Zeeck A, König WA (1990) Elucidation of the Structure of Hormaomycin. Angew Chem Int Ed Engl **29**: 64
1140. Zlatopolskiy BD, Loscha K, Alvermann P, Kozhushkov SI, Nikolaev SV, Zeeck A, de Meijere A (2004) Final Elucidation of the Absolute Configuration of the Signal Metabolite Hormaomycin. Chem Eur J **10**: 4708
1141. Zlatopolskiy BD, de Meijere A (2004) Total Synthesis of Hormaomycin, a Naturally Occurring Depsipeptide with Interesting Biological Activities. Chem Eur J **10**: 4718
1142. Lackner H, Bahner I, Shigematsu N, Pannell LK, Mauger AB (2000) Structures of Five Components of the Actinomycin Z Complex from *Streptomyces fradiae*, Two of Which Contain 4-Chlorothreonine. J Nat Prod **63**: 352
1143. Bitzer J, Gesheva V, Zeeck A (2006) Actinomycins with Altered Threonine Units in the β-Peptidolactone. J Nat Prod **69**: 1153

1144. Grgurina I, Barca A, Cervigni S, Gallo M, Scaloni A, Pucci P (1993) Relevance of Chlorine-Substituent for the Antifungal Activity of Syringomycin and Syringotoxin, Metabolites of the Phytopathogenic Bacterium *Pseudomonas syringae* pv. *syringae*. Experientia **50**: 130

1145. Vaillancourt FH, Yin J, Walsh CT (2005) SyrB2 in Syringomycin E Biosynthesis is a Nonheme Fe^{II} α-Ketoglutarate- and O$_2$-Dependent Halogenase. Proc Natl Acad Sci U S A **102**: 10111

1146. Blasiak LC, Vaillancourt FH, Walsh CT, Drennan CL (2006) Crystal Structure of the Non-Haem Iron Halogenase SyrB2 in Syringomycin Biosynthesis. Nature **440**: 368

1147. Feling RH, Buchanan GO, Mincer TJ, Kauffman CA, Jensen PR, Fenical W (2003) Salinosporamide A: A Highly Cytotoxic Proteasome Inhibitor from a Novel Microbial Source, a Marine Bacterium of the New Genus *Salinospora*. Angew Chem Int Ed **42**: 355

1148. Chauhan D, Catley L, Li G, Podar K, Hideshima T, Velankar M, Mitsiades C, Mitsiades N, Yasui H, Letai A, Ovaa H, Berkers C, Nicholson B, Chao T-H, Neuteboom STC, Richardson P, Palladino MA, Anderson KC (2005) A Novel Orally Active Proteasome Inhibitor Induces Apoptosis in Multiple Mycloma Cells with Mechanisms Distinct from Bortezomib. Cancer Cell **8**: 407

1149. Williams PG, Buchanan GO, Fehling RH, Kauffman CA, Jensen PR, Fenical W (2005) New Cytotoxic Salinosporamides from the Marine Actinomycete *Salinispora tropica*. J Org Chem **70**: 6196

1150. Reed KA, Manam RR, Mitchell SS, Xu J, Teisan S, Chao T-H, Deyanat-Yazdi G, Neuteboom STC, Lam KS, Potts BCM (2007) Salinosporamides D–J from the Marine Actinomycete *Salinispora tropica*, Bromosalinosporamide, and Thioester Derivatives Are Potent Inhibitors of the 20S Proteasome. J Nat Prod **70**: 269

1151. Macherla VR, Mitchell SS, Manam RR, Reed KA, Chao T-H, Nichoson B, Deyanat-Yazdi G, Mai B, Jensen PR, Fenical WF, Neuteboom STC, Lam KS, Palladino MA, Potts BCM (2005) Structure–Activity Relationship Studies of Salinosporamide A (NPI-0052), a Novel Marine Derived Proteasome Inhibitor. J Med Chem **48**: 3684

1152. Groll M, Huber R, Potts BCM (2006) Crystal Structures of Salinosporamide A (NPI-0052) and B (NPI-0047) in Complex with the 20S Proteasome Reveal Important Consequences of β-Lactone Ring Opening and a Mechanism for Irreversible Binding. J Am Chem Soc **128**: 5136

1153. Shibasaki M, Kanai M, Fukuda N (2007) Total Synthesis of Lactacystin and Salinosporamide A. Chem Asian J **2**: 20

1154. Garo E, Starks CM, Jensen PR, Fenical W, Lobkovsky E, Clardy J (2003) Trichodermamides A and B, Cytotoxic Modified Dipeptides from the Marine-Derived Fungus *Trichoderma virens*. J Nat Prod **66**: 423

1155. Ohtsu Y, Sasamura H, Tsurumi Y, Yoshimura S, Takase S, Hashimoto M, Shibata T, Hino M, Fujii T (2003) The Novel Gluconeogenesis Inhibitors FR225659 and Related Compounds that Originate from *Helicomyces* sp. No. 19353 I. Taxonomy, Fermentation, Isolation and Physico-chemical Properties. J Antibiot **56**: 682

1156. Ohtsu Y, Sasamura H, Shibata T, Nakajima H, Hino M, Fujii T (2003) The Novel Gluconeogenesis Inhibitors FR225659 and Related Compounds that Originate from *Helicomyces* sp. No. 19353 II. Biological Profiles. J Antibiot **56**: 689

1157. Zenkoh T, Ohtsu Y, Yoshimura S, Shigematsu N, Takase S, Hino M (2003) The Novel Gluconeogenesis Inhibitors FR225659 and Related Compounds that Originate from *Helicomyces* sp. No. 19353 III. Structure Determination. J Antibiot **56**: 694

1158. Hatori H, Zenkoh T, Kobayashi M, Ohtsu Y, Shigematsu N, Setoi H, Hino M, Handa H (2004) FR225659-Binding Proteins: Identification as Serine/Threonine Protein Phosphatase PP1 and PP2A Using High-Performance Affinity Beads. J Antibiot **57**: 456

1159. Carroll FI (2004) Epibatidine Structure Activity Relationships. Bioorg Med Chem Lett **14**: 1889

1160. White R, Malpass JR, Handa S, Baker SR, Broad LM, Folly L, Mogg A (2006) Epibatidine Isomers and Analogues: Structure–Activity Relationships. Bioorg Med Chem Lett **16**: 5493

1161. Olivo HF, Hemenway MS (2002) Recent Syntheses of Epibatidine. A Review. Org Prep Proc Int **34**: 1
1162. Kesingland AC, Gentry CT, Panesar MS, Bowes MA, Vernier J-M, Cube R, Walker K, Urban L (2000) Analgesic Profile of the Nicotinic Acetylcholine Receptor Agonists, (+)-Epibatidine and ABT-594 in Models of Persistent Inflammatory and Neuropathic Pain. Pain **86**: 113
1163. Berthou S, Leboeuf M, Cavé A, Mahuteau J, David B, Guinaudeau H (1989) Limalongine, a Modified Hasubanan Type Alkaloid, and Clolimalongine, Its Chlorinated Derivative. J Org Chem **54**: 3491
1164. Sugimoto Y, Babiker HAA, Saisho T, Furumoto T, Inanaga S, Kato M (2001) Chlorinated Alkaloids in *Menispermum dauricum* DC. Root Culture. J Org Chem **66**: 3299
1165. Sugimoto Y, Uchida S, Inanaga S, Kimura Y, Hashimoto M, Isogai A (1996) Early Steps of Dauricine Biosynthesis in Cultured Roots of *Menispermum dauricum*. Biosci Biotech Biochem **60**: 503
1166. Krebs HC, Carl T, Habermehl GG (1996) Pyrrolizidine Alkaloid Composition in Six Brazilian *Senecio* Species. Phytochemistry **43**: 1227
1167. Carroll AR, Ngo A, Quinn RJ, Redburn J, Hooper JNA (2005) Petrosamine B, an Inhibitor of the *Helicobacter pylori* Enzyme Aspartyl Semialdehyde Dehydrogenase from the Australian Sponge *Oceanapia* sp. J Nat Prod **68**: 804
1168. Nakahara S, Matsui J, Kubo A (1998) Synthesis of Pantherinine, a Cytotoxic Fused Tetracyclic Aromatic Alkaloid. Tetrahedron Lett **39**: 5521
1169. Chang F-R, Chen C-Y, Wu P-H, Kuo R-Y, Chang Y-C, Wu Y-C (2000) New Alkaloids from *Annona purpurea*. J Nat Prod **63**: 746
1170. Kuo R-Y, Chang F-R, Chen C-Y, Teng C-M, Yen H-F, Wu Y-C (2001) Antiplatelet Activity of N-Methoxycarbonyl Aporphines from *Rollinia mucosa*. Phytochemistry **57**: 421
1171. Sobarzo-Sánchez EM, Arbaoui J, Protais P, Cassels BK (2000) Halogenated Boldine Derivatives with Enhanced Monoamine Receptor Selectivity. J Nat Prod **63**: 480
1172. Iwasa K, Moriyasu M, Yamori T, Turuo T, Lee D-U, Wiegrebe W (2001) *In Vitro* Cytotoxicity of the Protoberberine-Type Alkaloids. J Nat Prod **64**: 896
1173. Manzo E, van Soest R, Matainaho L, Roberge M, Andersen RJ (2003) Ceratamines A and B, Antimitotic Heterocyclic Alkaloids Isolated from the Marine Sponge *Pseudoceratina* sp. Collected in Papua New Guinea. Org Lett **5**: 4591
1174. Didier C, Critcher DJ, Walshe ND, Kojima Y, Yamauchi Y, Barrett AGM (2004) Full Stereochemical Assignment and Synthesis of the Potent Anthelmintic Pyrrolobenzoxazine Natural Product CJ-12662. J Org Chem **69**: 7875
1175. Bugni TS, Woolery M, Kauffman CA, Jensen PR, Fenical W (2006) Bohemamines from a Marine-Derived *Streptomyces* sp. J Nat Prod **69**: 1626
1176. Wang H-B, Tan C-H, Tan J-J, Gao M-Y, Li Y-M, Jiang S-H, Zhu D-Y (2007) Lycopodium Alkaloids from *Huperzia serrata*. Helv Chim Acta **90**: 153
1177. Kuramoto M, Tong C, Yamada K, Chiba T, Hayashi Y, Uemura D (1996) Halichlorine, an Inhibitor of VCAM-1 Induction from the Marine Sponge *Halichondria okadai* Kadota. Tetrahedron Lett **37**: 3867
1178. Arimoto H, Hayakawa I, Kuramoto M, Uemura D (1998) Absolute Stereochemistry of Halichlorine; a Potent Inhibitor of VCAM-1 Induction. Tetrahedron Lett **39**: 861
1179. Chou T, Kuramoto M, Otani Y, Shikano M, Yazawa K, Uemura D (1996) Pinnaic Acid and Tauropinnaic Acid: Two Novel Fatty Acids Composing a 6-Azaspiro[4.5]decane Unit from the Okinawan Bivalve *Pinna muricata*. Tetrahedron Lett **37**: 3871
1180. Clive DLJ, Yu M, Wang J, Yeh VSC, Kang S (2005) Synthetic Chemistry of Halichlorine and the Pinnaic Acids. Chem Rev **105**: 4483
1181. Trauner D, Schwarz JB, Danishefsky SJ (1999) Total Synthesis of (+)-Halichlorine: An Inhibitor of VCAM-1 Expression. Angew Chem Int Ed **38**: 3542
1182. Carson MW, Kim G, Danishefsky SJ (2001) Total Synthesis and Proof of Stereochemistry of Natural and Unnatural Pinnaic Acids: A Remarkable Long-Range Stereochemical Effect in the Reduction of 17-Oxo Precursors of the Pinnaic Acids. Angew Chem Int Ed **40**: 4453

1183. Weinreb SM (2006) Studies on Total Synthesis of the Cylindricine/Fasicularin/Lepadiformine Family of Tricyclic Marine Alkaloids. Chem Rev **106**: 2531
1184. Cherin L, Brandis A, Ismailov Z, Chet I (1996) Pyrrolnitrin Production by an *Enterobacter agglomerans* Strain with a Broad Spectrum of Antagonistic Activity Towards Fungal and Bacterial Phytopathogens. Curr Microbiol **32**: 208
1185. Santo RD, Costi R, Artico M, Massa S, Lampis G, Deidda D, Pompei R (1998) Pyrrolnitrin and Related Pyrroles Endowed with Antibacterial Activities Against *Mycobacterium tuberculosis*. Bioorg Med Chem Lett **8**: 2931
1186. El-Banna N, Winkelmann G (1998) Pyrrolnitrin from *Burkholderia cepacia*: Antibiotic Activity Against Fungi and Novel Activities Against Streptomycetes. J Appl Microbiol **85**: 69
1187. Ligon JM, Hill DS, Hammer PE, Torkewitz NR, Hofmann D, Kempf H-J, van Pée K-H (2000) Natural Products with Antifungal Activity from *Pseudomonas* Biocontrol Bacteria. Pest Manag Sci **56**: 688
1188. Sako M, Kihara T, Tanisaki M, Maki Y, Miyamae, Azuma T, Kohda S, Masugi T (2002) Novel Photodegradation of the Antifungal Antibiotic Pyrrolnitrin in Anhydrous and Aqueous Aprotic Solvents. J Org Chem **67**: 668
1189. van Pée K-H, Ligon JM (2000) Biosynthesis of Pyrrolnitrin and Other Phenylpyrrole Derivatives by Bacteria. Nat Prod Rep **17**: 157
1190. Hölzer M, Burd W, Reissig H-U, van Pée K-H (2001) Substrate Specificity and Regioselectivity of Tryptophan 7-Halogenase from *Pseudomonas fluorescens* BL915. Adv Synth Catal **343**: 591
1191. Dong C, Kotzsch A, Dorward M, van Pée K-H, Naismith JH (2004) Crystallization and X-ray Diffraction of a Halogenating Enzyme, Tryptophan 7-Halogenase, from *Pseudomonas fluorescens*. Acta Cryst D**60**: 1438
1192. Charan RD, Schlingmann G, Berman VS, Feng X, Carter GT (2005) Additional Pyrrolomycins from Cultures of *Streptomyces fumanus*. J Nat Prod **68**: 277
1193. Cavalleri B, Volpe G, Tuan G, Berti M, Parenti F (1978) A Chlorinated Phenylpyrrole Antibiotic from *Actinoplanes*. Curr Microbiol **1**: 319
1194. Funabashi Y, Takizawa M, Tsubotani S, Tanida S, Harada S, Kamiya K (1992) Chemistry and Biological Activities of New Pyrrole Antibiotics, TAN-876A and B. J Takeda Res Lab **51**: 73
1195. Kawamura N, Sawa R, Takahashi Y, Isski K, Sawa T, Kinoshita N, Naganawa H, Hamada M, Takeuchi T (1995) Pyralomicins, New Antibiotics from *Antinomadura spiralis*. J Antibiot **48**: 435
1196. Kawamura N, Sawa R, Takahashi Y, Isshiki K, Sawa T, Naganawa H, Takeuchi T (1996) Pyralomicins, Novel Antibiotics from *Microtetraspora spiralis* II. Structure Determination. J Antibiot **49**: 651
1197. Kawamura N, Kinoshita N, Sawa R, Takahashi Y, Sawa T, Naganawa H, Hamada M, Takeuchi T (1996) Pyralomicins, Novel Antibiotics from *Microtetraspora spiralis* I. Taxonomy and Production. J Antibiot **49**: 706
1198. Kawamura N, Nakamura H, Sawa R, Takahashi Y, Sawa T, Naganawa H, Takeuchi T (1997) Pyralomicins, Novel Antibiotics from *Microtetraspora spiralis* IV. Absolute Configuration. J Antibiot **50**: 147
1199. Breinholt J, Gürtler H, Kjaer A, Nielsen SE, Olsen CE (1998) Streptopyrrole: An Antimicrobial Metabolite from *Streptomyces armeniacus*. Acta Chem Scand **52**: 1040
1200. Trew SJ, Wrigley SK, Pairet L, Sohal J, Shanu-Wilson P, Hayes MA, Martin SM, Manohar RN, Chicarelli-Robinson MI, Kau DA, Byrne CV, Wellington EMH, Moloney JM, Howard J, Hupe D, Olson ER (2000) Novel Streptopyrroles from *Streptomyces rimosus* with Bacterial Protein Histidine Kinase Inhibitory and Antimicrobial Activities. J Antibiot **53**: 1
1201. Fielman KT, Targett NM (1995) Variation of 2,3,4-Tribromopyrrole and its Sodium Sulfamate Salt in the Hemichordate *Saccoglossus kowalevskii*. Mar Ecol Prog Ser **116**: 125

1202. Blackman AJ, Li C (1994) New Tambjamine Alkaloids from the Marine Bryozoan *Bugula dentata*. Aust J Chem **47**: 1625
1203. Tittlemier S, Simon M, Jarman WM, Elliott JE, Norstrom RJ (1999) Identification of a Novel $C_{10}H_6N_2Br_4Cl_2$ Heterocyclic Compound in Seabird Eggs. A Bioaccumulating Marine Natural Product? Environ Sci Technol **33**: 26
1204. Gribble GW, Blank DH, Jasinski JP (1999) Synthesis and Identification of Two Halogenated Bipyrroles Present in Seabird Eggs. Chem Commun 2195
1205. Blank DH, Gribble GW, Schneekloth Jr. JS, Jasinski JP (2002) A Novel Class of Naturally Occurring Halogenated Pyrroles, 1,1'-Dimethyl-3,3',4,4',5,5'-hexabromo-2,2'-bipyrrole, 5,5'-Dichloro-1,1'-dimethyl-3,3',4,4'-tetrabromo-2,2'-bipyrrole, and 1,1'-Dimethyl-3,3',4,4',5,5'-hexachloro-2,2'-bipyrrole. J Chem Crystallogr **32**: 541
1206. Tittlemier SA, Blank DH, Gribble GW, Norstrom RJ (2002) Structure Elucidation of Four Possible Biogenic Organohalogens Using Isotope Exchange Mass Spectrometry. Chemosphere **46**: 511
1207. Tittlemier SA, Fisk AT, Hobson KA, Norstrom RJ (2002) Examination of the Bioaccumulation of Halogenated Dimethyl Bipyrroles in an Arctic Marine Food Web Using Stable Nitrogen Isotope Analysis. Environ Pollut **116**: 85
1208. Tittlemier S, Borrell A, Duffe J, Duignan PJ, Fair P, Hall A, Hoekstra P, Kovacs KM, Krahn MM, Lebeuf M, Lydersen C, Muir D, O'Hara T, Olsson M, Pranschke J, Ross P, Siebert U, Stern G, Tanabe S, Norstrom R (2002) Global Distribution of Halogenated Dimethyl Bipyrroles in Marine Mammal Blubber. Arch Environ Contam Toxicol **43**: 244
1209. Haraguchi K, Hisamichi Y, Endo T (2006) Bioaccumulation of Naturally Occurring Mixed Halogenated Dimethylbipyrroles in Whale and Dolphin Products on the Japanese Market. Arch Environ Contam Toxicol **51**: 135
1210. Weichbrodt M, Vetter W, Scholz E, Luckas B, Reinhardt K (1999) Determination of Organochlorine Levels in Antarctic Skua and Penguin Eggs by Application of Combined Focused Open-Vessel Microwave Assisted Extraction, Gel-Permeation Chromatography, Adsorption Chromatography, and GC/ECD. Inter J Environ Anal Chem **73**: 309
1211. Vetter W, Alder L, Palavinskas R (1999) Mass Spectrometric Characterization of Q1, a $C_9H_3Cl_7N_2$ Contaminant in Environmental Samples. Rapid Commun Mass Spectrom **13**: 2118
1212. Vetter W, Weichbrodt M, Scholz E, Luckas B, Oelschläger H (1999) Levels of Organochlorines (DDT, PCBs, Toxaphene, Chlordane, Dieldrin, and HCHs) in Blubber of South African Fur Seals (*Arctocephalus pusillus pusillus*) from Cape Cross/Namibia. Mar Pollut Bull **38**: 830
1213. Vetter W, Alder L, Kallenborn R, Schlabach M (2000) Determination of Q1, an Unknown Organochlorine Contaminant, in Human Milk, Antarctic Air, and Further Environmental Samples. Environ Pollut **110**: 401
1214. Wu J, Vetter W, Gribble GW, Schneekloth Jr. JS, Blank DH, Görls H (2002) Structure and Synthesis of the Natural Heptachloro-1'-methyl-1,2'-bipyrrole (Q1). Angew Chem Int Ed **41**: 1740
1215. Vetter W, Scholz E, Gaus C, Müller JF, Haynes D (2001) Anthropogenic and Natural Organohalogen Compounds in Blubber of Dolphins and Dugongs (*Dugong dugon*) from Northeastern Australia. Arch Environ Contam Toxicol **41**: 221
1216. Vetter W (2002) Environment Occurrence of Q1, a $C_9H_3Cl_7N_2$ Compound, That Has Been Identified as a Natural Bioaccumulative Organochlorine. Chemosphere **46**: 1477
1217. Vetter W, Jun W, Althoff G (2003) Non-Polar Halogenated Natural Products Bioaccumulated in Marine Samples. I. 2,3,3',4,4',5,5'-Heptachloro-1'-methyl-1,2'-bipyrrole (Q1). Chemosphere **52**: 415
1218. Vetter W, Hahn ME, Tomy G, Ruppe S, Vatter S, Chalbane N, Lenoir D, Schramm K-W, Scherer G (2004) Biological Activity and Physiochemical Parameters of Marine Halogenated Natural Products 2,3,3',4,4',5,5'-Heptachloro-1'-methyl-1,2'-bipyrrole and 2,4,6-Tribromoanisole. Arch Environ Contam Toxicol **48**: 1

1219. Teuten EL, Pedler BE, Hangsterfer AN, Reddy CM (2006) Identification of Highly Brominated Analogues of Q1 in Marine Mammals. Environ Pollut **144**: 336
1220. Teuten EL, Saint-Louis R, Pedler BE, Xu L, Pelletier E, Reddy CM (2006) Expanding the Range of Brominated Q1 Analogues. Mar Pollut Bull **52**: 572
1221. Vetter W, Gaul S, Olbrich D, Gaus C (2007) Monobromo and Higher Brominated Congeners of the Marine Halogenated Natural Product 2,3,3',4,4',5,5'-Heptachloro-1'-methyl-1,2'-bipyrrole (Q1). Chemosphere **66**: 2011
1222. Teuten EL, Reddy CM (2007) Halogenated Organic Compounds in Archived Whale Oil: A Pre-Industrial Record. Environ Pollut **145**: 668
1223. Reddy CM, Xu L, O'Neil GW, Nelson RK, Eglinton TI, Faulkner DJ, Norstrom R, Ross PS, Tittlemier SA (2004) Radiocarbon Evidence for a Naturally Produced, Bioaccumulating Halogenated Organic Compound. Environ Sci Technol **38**: 1992
1224. Saint-Louis R, Pelletier E (2005) Unsuspected Organic Pollutants in Marine Mammals: Halogenated Naphthols. Mar Pollut Bull **50**: 889
1225. Pullen C, Schmitz P, Meurer K, Bamberg DD, Lohmann S, De Castro Franca S, Groth I, Schlegel B, Möllmann U, Gollmick F, Gräfe U, Leistner E (2002) New and Bioactive Compounds from *Streptomyces* Strains Residing in the Wood of Celastraceae. Planta **216**: 162
1226. Clark BR, Capon RJ, Lacey E, Tennant S, Gill JH (2006) Polyenylpyrroles and Polyenylfurans from an Australian Isolate of the Soil Ascomycete *Gymnoascus reessii*. Org Lett **8**: 701
1227. Hosoe T, Fukushima K, Takizawa K, Miyaji M, Kawai K (1999) Three Pyrrolyloctatetraenyl-α-pyrones from *Auxarthron conjugatum*. Phytochemistry **52**: 459
1228. Lang G, Cole ALJ, Blunt JW, Munro MHG (2006) An Unusual Oxalylated Tetramic Acid from the New Zealand Basidiomycete *Chamonixia pachydermis*. J Nat Prod **69**: 151
1229. Momose I, Iinuma H, Kinoshita N, Momose Y, Kunimoto S, Hamada M, Takeuchi T (1999) Decatromicins A and B, New Antibiotics Produced by *Actinomadura* sp. MK73-NF4 I. Taxonomy, Isolation, Physico-Chemical Properties and Biological Activities. J Antibiot **52**: 781
1230. Momose I, Hirosawa S, Nakamura H, Naganawa H, Iinuma H, Ikeda D, Takeuchi T (1999) Decatromicins A and B, New Antibiotics Produced by *Actinomadura* sp. MK73-NF4 II. Structure Determination. J Antibiot **52**: 787
1231. Lam KS, Hesler GA, Gustavson DR, Berry RL, Tomita K, MacBeth JL, Ross J, Miller D, Forenza S (1996) Pyrrolosporin A, a New Antitumor Antibiotic from *Micromonospora* sp. C39217-R109-7 I. Taxonomy of Producing Organism, Fermentation and Biological Activity. J Antibiot **49**: 860
1232. Schroeder DR, Colson KL, Klohr SE, Lee MS, Matson JA, Brinen LS, Clardy J (1996) Pyrrolosporin A, a New Antitumor Antibiotic from *Micromonospora* sp. C39217-R109-7 II. Isolation, Physico-chemical Properties, Spectroscopic Study and X-ray Analysis. J Antibiot **49**: 865
1233. Rudi A, Evan T, Aknin M, Kashman Y (2000) Polycitone B and Prepolycitrin A: Two Novel Alkaloids from the Marine Ascidian *Polycitor africanus*. J Nat Prod **63**: 832
1234. Loya S, Rudi A, Kashman Y, Hizi A (1999) Polycitone A, a Novel and Potent General Inhibitor of Retroviral Reverse Transcriptases and Cellular DNA Polymerases. Biochem J **344**: 85
1235. Beccalli EM, Clerici F, Marchesini A (2000) First Total Synthesis of the Alkaloid Polycitrin B. Tetrahedron **56**: 2699
1236. Hoffmann H, Lindel T (2003) Synthesis of the Pyrrole-Imidazole Alkaloids. Synthesis 1753
1237. Jacquot DEN, Lindel T (2005) Challenge Palau'amine: Current Standings. Curr Org Chem **9**: 1551
1238. Liu J-F, Guo S-P, Jiang B (2005) Progress in the Study of Marine Bromopyrrole Alkaloids. Chin J Org Chem **25**: 788
1239. Iwagawa T, Kaneko M, Okamura H, Nakatani M, van Soest RWM (1998) New Alkaloids from the Papua New Guinean Sponge *Agelas nakamurai*. J Nat Prod **61**: 1310

1240. Umeyama A, Ito S, Yuasa E, Arihara S, Yamada T (1998) A New Bromopyrrole Alkaloid and the Optical Resolution of the Racemate from the Marine Sponge *Homaxinella* sp. J Nat Prod **61**: 1433
1241. Reddy NS, Venkateswarlu Y (2000) *S*-(+)-Methyl Ester of Hanishin from the Marine Sponge *Agelas ceylonica*. Biochem System Ecol **28**: 1035
1242. Cafieri F, Fattorusso E, Mangoni A, Taglialatela-Scafati O (1995) Longamide and 3,7-Dimethylisoguanine, Two Novel Alkaloids from the Marine Sponge *Agelas longissima*. Tetrahedron Lett **36**: 7893
1243. Mancini I, Guella G, Amade P, Roussakis C, Pietra F (1997) Hanishin, a Semiracemic, Bioactive C_9 Alkaloid of the Axinellid Sponge *Acanthella carteri* from the Hanish Islands. A Shunt Metabolite? Tetrahedron Lett **38**: 6271
1244. Reddy NS, Venkateswarlu Y (2000) A New Bromopyrrole Alkaloid from the Sponge *Axinella tenuidigitata*. Ind J Chem **39B**: 971
1245. Cafieri F, Fattorusso E, Taglialatela-Scafati O (1998) Novel Bromopyrrole Alkaloids from the Sponge *Agelas dispar*. J Nat Prod **61**: 122
1246. Li C-J, Schmitz FJ, Kelly-Borges M (1998) A New Lysine Derivative and New 3-Bromopyrrole Carboxylic Acid Derivatives from Two Marine Sponges. J Nat Prod **61**: 387
1247. Uemoto H, Tsuda M, Kobayashi J (1999) Mukanadins A-C, New Bromopyrrole Alkaloids from Marine Sponge *Agelas nakamurai*. J Nat Prod **62**: 1581
1248. Cafieri F, Fattorusso E, Mangoni A, Taglialatela-Scafati (1996) Clathramides, Unique Bromopyrrole Alkaloids from the Caribbean Sponge *Agelas clathrodes*. Tetrahedron **52**: 13713
1249. Cafieri F, Fattorusso E, Mangoni A, Taglialatela-Scafati O (1995) A Novel Bromopyrrole Alkaloid from the Sponge *Agelas longissima* with Antiserotonergic Activity. Bioorg Med Chem Lett **5**: 799
1250. Cafieri F, Fattorusso E, Mangoni A, Taglialatela-Scafati O (1996) Dispacamides, Anti-Histamine Alkaloids from Caribbean *Agelas* Sponges. Tetrahedron Lett **37**: 3587
1251. Cafieri F, Carnuccio R, Fattorusso E, Taglialatela-Scafati O, Vallefuoco T (1997) Anti-Histaminic Activity of Bromopyrrole Alkaloids Isolated from Caribbean *Agelas* Sponges. Bioorg Med Chem Lett **7**: 2283
1252. Hu J-F, Peng J, Kazi AB, Kelly M, Hamann MT (2005) Bromopyrrole Alkaloids from the Jamaican Sponge *Didiscus oxeata*. J Chem Res 427
1253. Tsukamoto S, Kato H, Hirota H, Fusetani N (1996) Pseudoceratidine: A New Antifouling Spermidine Derivative from the Marine Sponge *Pseudoceratina purpurea*. Tetrahedron Lett **37**: 1439
1254. Kobayashi J, Inaba K, Tsuda M (1997) Tauroacidins A and B, Two Bromopyrrole Alkaloids Possessing a Taurine Residue *Hymeniacidon* Sponge. Tetrahedron **53**: 16679
1255. Fattorusso E, Taglialatela-Scafati (2000) Two Novel Pyrrole-Imidazole Alkaloids from the Mediterranean Sponge *Agelas oroides*. Tetrahedron Lett **41**: 9917
1256. Assmann M, Lichte E, van Soest RWM, Köck M (1999) New Bromopyrrole Alkaloid from the Marine Sponge *Agelas wiedenmayeri*. Org Lett **1**: 455
1257. Williams DE, Patrick BO, Behrisch HW, van Soest R, Roberge M, Andersen RJ (2005) Dominicin, a Cyclic Octapeptide, and Laughine, a Bromopyrrole Alkaloid, Isolated from the Caribbean Marine Sponge *Eurypon laughlini*. J Nat Prod **68**: 327
1258. Grube A, Lichte E, Köck M (2006) Isolation and Synthesis of 4-Bromopyrrole-2-carboxy-arginine and 4-Bromopyrrole-2-carboxy-$N(\varepsilon)$-lysine from the Marine Sponge *Stylissa caribica*. J Nat Prod **69**: 125
1259. Lindel T, Hochgürtel M, Assmann M, Köck M (2000) Synthesis of the Marine Natural Product $N\alpha$-(4-Bromopyrrolyl-2-carbonyl)-L-homoarginine, a Putative Biogenetic Precursor of the Pyrrole-Imidazole Alkaloids. J Nat Prod **63**: 1566
1260. Assmann M, Zea S, Köck M (2001) Sventrin, a New Bromopyrrole Alkaloid from the Caribbean Sponge *Agelas sventres*. J Nat Prod **64**: 1593

1261. Aiello A, D'Esposito M, Fattorusso E, Menna M, Müller WEG, Perović-Ottstadt S, Schröder HC (2006) Novel Bioactive Bromopyrrole Alkaloids from the Mediterranean Sponge *Axinella verrucosa*. Bioorg Med Chem **14**: 17
1262. Tsuda M, Uemoto H, Kobayashi J (1999) Slagenins A~C, Novel Bromopyrrole Alkaloids from Marine Sponge *Agelas nakamurai*. Tetrahedron Lett **40**: 5709
1263. Jiang B, Liu J-F, Zhao S-Y (2002) Enantioselective Synthesis of Slagenins A-C. Org Lett **4**: 3951
1264. Tsukamoto S, Kato H, Hirota H, Fusetani N (1996) Mauritiamine, a New Antifouling Oroidin Dimer from the Marine Sponge *Agelas mauritiana*. J Nat Prod **59**: 501
1265. Goetz GH, Harrigan GG, Likos J (2001) Ugibohlin: A New Dibromo-seco-isophakellin from *Axinella carteri*. J Nat Prod **64**: 1581
1266. Patel J, Pelloux-Léon N, Minassian F, Vallée Y (2006) Total Synthesis of (*S*)-(−)-Cyclooroidin. Tetrahedron Lett **47**: 5561
1267. Patel J, Pelloux-Léon N, Minassian F, Vallée Y (2005) Syntheses of *S*-Enantiomers of Hanishin, Longamide B, and Longamide B Methyl Ester from L-Cispartic Acid β-Methyl Ester: Establishment of Absolute Stereochemistry. J Org Chem **70**: 9081
1268. Fujita M, Nakao Y, Matsunaga S, Seiki M, Itoh Y, Yamashita J, van Soest RWM, Fusetani N (2003) Ageladine A: An Antiangiogenic Matrixmetalloproteinase Inhibitor from the Marine Sponge *Agelas nakamurai*. J Am Chem Soc **125**: 15700
1269. Meketa ML, Weinreb SM (2006) Total Synthesis of Ageladine A, an Angiogenesis Inhibitor from the Marine Sponge *Agelas nakamurai*. Org Lett **8**: 1443
1270. Shengule SR, Karuso P (2006) Concise Total Synthesis of the Marine Natural Product Ageladine A. Org Lett **8**: 4083
1271. Tsuda M, Yasuda T, Fukushi E, Kawabata J, Sekiguchi M, Fromont J, Kobayashi J (2006) Agesamides A and B, Bromopyrrole Alkaloids from Sponge *Agelas* Species: Application of DOSY for Chemical Screening of New Metabolites. Org Lett **8**: 4235
1272. Grube A, Köck M (2006) Oxocyclostylidol, an Intramolecular Cyclized Oroidin Derivative from the Marine Sponge *Stylissa caribica*. J Nat Prod **69**: 1212
1273. Supriyono A, Schwarz B, Wray V, Witte L, Müller WEG, van Soest R, Sumaryono W, Proksch P (1995) Bioactive Alkaloids from the Tropical Marine Sponge *Axinella carteri*. Z Naturforsch **50c**: 669
1274. Williams DH, Faulkner DJ (1996) Isomers and Tautomers of Hymenialdisine and Debromohymenialdisine. Nat Prod Lett **9**: 57
1275. Patil AD, Freyer AJ, Killmer L, Hofmann G, Johnson RK (1997) Z-Axinohydantoin and Debromo-Z-axinohydantoin from the Sponge *Stylotella aurantium*: Inhibitors of Protein Kinase C. Nat Prod Lett **9**: 201
1276. Inaba K, Sato H, Tsuda M, Kobayashi J (1998) Spongiacidins A-D, New Bromopyrrole Alkaloids from *Hymeniacidon* Sponge. J Nat Prod **61**: 693
1277. Eder C, Proksch P, Wray V, Steube K, Bringmann G, van Soest RWM, Sudarsono, Ferdinandus E, Pattisina LA, Wiryowidagdo S, Moka W (1999) New Alkaloids from the Indopacific Sponge *Stylissa carteri*. J Nat Prod **62**: 184
1278. Sosa ACB, Yakushijin K, Horne DA (2002) Synthesis of Axinohydantoins. J Org Chem **67**: 4498
1279. Papeo G, Posteri H, Borghi D, Varasi M (2005) A New Glycociamidine Ring Precursor: Syntheses of (*Z*)-Hymenialdisine, (*Z*)-2-Debromohymenialdisine, and (±)-*endo*-2-Debromohymenialdisine. Org Lett **7**: 5641
1280. Linington RG, Williams DE, Tahir A, van Soest R, Andersen RJ (2003) Latonduines A and B, New Alkaloids Isolated from the Marine Sponge *Stylissa carteri*: Structure Elucidation, Synthesis, and Biogenetic Implications. Org Lett **5**: 2735
1281. Pettit GR, McNulty J, Herald DL, Doubek DL, Chapuis J-C, Schmidt JM, Tackett LP, Boyd MR (1997) Antineoplastic Agents. 362. Isolation and X-ray Crystal Structure of Dibromophakellstatin from the Indian Ocean Sponge *Phakellia mauritiana*. J Nat Prod **60**: 180

1282. Assmann M, van Soest RWM, Köck M (2001) New Antifeedant Bromopyrrole Alkaloid from the Caribbean Sponge *Stylissa caribica*. J Nat Prod **64**: 1345
1283. Assmann M, Köck M (2002) Monobromoisophakellin, a New Bromopyrrole Alkaloid from the Caribbean Sponge *Agelas* sp. Z Naturforsch **57c**: 153
1284. Wiese KJ, Yakushijin K, Horne DA (2002) Synthesis of Dibromophakellstatin and Dibromoisophakellin. Tetrahedron Lett **43**: 5135
1285. Feldman KS, Skoumbourdis AP (2005) Extending Pummerer Reaction Chemistry. Synthesis of (±)-Dibromophakellstatin by Oxidative Cyclization of an Imidazole Derivative. Org Lett **7**: 929
1286. Zöllinger M, Mayer P, Lindel T (2006) Total Synthesis of the Cytostatic Marine Natural Product Dibromophakellstatin via Three-Component Imidazolidinone Anellation. J Org Chem **71**: 9431
1287. Tsukamoto S, Tane K, Ohta T, Matsunaga S, Fusetani N, van Soest RWM (2001) Four New Bioactive Pyrrole-Derived Alkaloids from the Marine Sponge *Axinella brevistyla*. J Nat Prod **64**: 1576
1288. Gautschi JT, Whitman S, Holman TR, Crews P (2004) An Analysis of Phakellin and Oroidin Structures Stimulated by Further Study of an *Agelas* Sponge. J Nat Prod **67**: 1256
1289. Williams DH, Faulkner DJ (1996) *N*-Methylated Ageliferins from the Sponge *Astrosclera willeyana* from Pohnpei. Tetrahedron **52**: 5381
1290. Eder C, Proksch P, Wray V, van Soest RWM, Ferdinandus E, Pattisina LA, Sudarsono (1999) New Bromopyrrole Alkaloids from the Indopacific Sponge *Agelas nakamurai*. J Nat Prod **62**: 1295
1291. Assmann M, Köck M (2002) Bromosceptrin, an Alkaloid from the Marine Sponge *Agelas conifera*. Z Naturforsch **57c**: 157
1292. Baran PS, Li K, O'Malley DP, Mitsos C (2006) Short, Enantioselective Total Synthesis of Sceptrin and Ageliferin by Programmed Oxaquadricyclane Fragmentation. Angew Chem Int Ed **45**: 249
1293. Kawasaki I, Sakaguchi N, Khadeer A, Yamashita M, Ohta S (2006) Homonuclear Diels-Alder Dimerization of 5-Ethenyl-2-phenylsulfanyl-1*H*-imidazoles and Its Application to Synthesis of 12,12'-Dimethylageliferin. Tetrahedron **62**: 10182
1294. Hong TW, Jímenez DR, Molinski TF (1998) Agelastatins C and D, New Pentacyclic Bromopyrroles from the Sponge *Cymbastela* sp., and Potent Arthropod Toxicity of (−)-Agelastatin A. J Nat Prod **61**: 158
1295. Stien D, Anderson GT, Chase CE, Koh Y, Weinreb SM (1999) Total Synthesis of the Antitumor Marine Sponge Alkaloid Agelastatin A. J Am Chem Soc **121**: 9574
1296. Feldman KS, Saunders JC, Wrobleski ML (2002) Alkynyliodonium Salts in Organic Synthesis. Development of a Unified Strategy for the Syntheses of (−)-Agelastatin A and (−)-Agelastatin B. J Org Chem **67**: 7096
1297. Domostoj MM, Irving E, Scheinmann F, Hale KJ (2004) New Total Syntheses of the Marine Antitumor Alkaloid (−)-Agelastatin A. Org Lett **6**: 2615
1298. Davis FA, Deng J (2005) Asymmetric Total Syntheses of (−)-Agelastatin A Using Sulfinimine (*N*-Sulfinyl Imine) Derived Methodologies. Org Lett **7**: 621
1299. Endo T, Tsuda M, Okada T, Mitsuhashi S, Shima H, Kikuchi K, Mikami Y, Fromont J, Kobayashi J (2004) Nagelamides A-H, New Dimeric Bromopyrrole Alkaloids from Marine Sponge *Agelas* Species. J Nat Prod **67**: 1262
1300. Grube A, Köck M (2006) Stylissadines A and B: The First Tetrameric Pyrrole-Imidazole Alkaloids. Org Lett **8**: 4675
1301. Urban S, de Almeida Leone P, Carroll AR, Fechner GA, Smith J, Hooper JNA, Quinn RJ (1999) Axinellamines A-D, Novel Imidazo-Azolo-Imidazole Alkaloids from the Australian Marine Sponge *Axinella* sp. J Org Chem **64**: 731
1302. Nishimura S, Matsunaga S, Shibazaki M, Suzuki K, Furihata K, van Soest RWM, Fusetani N (2003) Massadine, a Novel Geranylgeranyltransferase Type I Inhibitor from the Marine Sponge *Stylissa* aff. *massa*. Org Lett **5**: 2255

1303. Kinnel RB, Gehrken H-P, Scheuer PJ (1993) Palau'amine: A Cytotoxic and Immunosuppressive Hexacyclic Bisguanidine Antibiotic from the Sponge *Stylotella agminata*. J Am Chem Soc **115**: 3376
1304. Kinnel RB, Gehrken H-P, Swali R, Skoropowski G, Scheuer PJ (1998) Palau'amine and Its Congeners: A Family of Bioactive Bisguanidines from the Marine Sponge *Stylotella aurantium*. J Org Chem **63**: 3281
1305. Kato T, Shizuri Y, Izumida H, Yokoyama A, Endo M (1995) Styloguanidines, New Chitinase Inhibitors from the Marine Sponge *Stylotella aurantium*. Tetrahedron Lett **36**: 2133
1306. Kobayashi J, Suzuki M, Tsuda M (1997) Konbu'acidin A, a New Bromopyrrole Alkaloid with cdk4 Inhibitory Activity from *Hymeniacidon* Sponge. Tetrahedron **53**: 15681
1307. Buchanan MS, Carroll AR, Addepalli R, Avery VM, Hooper JNA, Quinn RJ (2007) Natural Products, Stylissadines A and B, Specific Antagonists of the $P2X_7$ Receptor, an Important Inflammatory Target. J Org Chem **72**: 2309
1308. Grube A, Köck M (2007) Structural Assignment of the Tetrabromostyloguanidine: Does the Related Configuration of the Palau'amines Need Revision? Angew Chem Int Ed **46**: 2320
1309. Kobayashi H, Kitamura K, Nagai K, Nakao Y, Fusetani N, van Soest RWM, Matsunaga S (2007) Carteramine A, an Inhibitor of Neutrophil Chemotaxis, from the Marine Sponge *Stylissa carteri*. Tetrahedron Lett **48**: 2127
1310. Qureshi A, Faulkner DJ (1999) 3,6-Dibromoindole, a New Indole from the Palauan Ascidian *Distaplia regina*. Nat Prod Lett **13**: 59
1311. Sridevi KV, Venkatesham U, Reddy AV, Venkateswarlu Y (2003) Chemical Constituents of the Red Alga *Nitophyllum marginata*. Biochem System Ecol **31**: 335.
1312. Wood WF, Smith S, Wayman K, Largent DL (2003) Indole and 3-Chloroindole: The Source of the Disagreeable Odor of *Hygrophorus paupertinus*. Mycologia **95**: 807
1313. Maruya KA (2003) Di- and Tribromoindoles in the Common Oyster (*Crassostrea virginica*). Chemosphere **52**: 409
1314. Reineke N, Biselli S, Franke S, Franke W, Heinzel N, Hühnerfuss H, Iznaguen H, Kammann U, Theobald N, Vobach M, Wosniok W (2006) Brominated Indoles and Phenols in Marine Sediments and Water Extracts from the North and Baltic Seas – Concentrations and Effects. Arch Environ Contam Toxicol **51**: 186
1315. Ji N-Y, Li X-M, Ding L-P, Wang B-G (2007) Aristolane Sesquiterpenes and Highly Brominated Indoles from the Marine Red Alga *Laurencia similis* (Rhodomelaceae). Helv Chim Acta **90**: 385
1316. Ochi M, Kataoka K, Ariki S, Iwatsuki C, Kodama M, Fukuyama Y (1998) Antioxidative Bromoindole Derivatives from the Mid-Intestinal Gland of the Muricid Gastropod *Drupella fragum*. J Nat Prod **61**: 1043
1317. Meragelman KM, West LM, Northcote PT, Pannell LK, McKee TC, Boyd MR (2002) Unusual Sulfamate Indoles and a Novel Indolo[3,2-*a*]carbazole from *Ancorina* sp. J Org Chem **67**: 6671
1318. El-Gamal AA, Wang W-L, Duh C-Y (2005) Sulfur-Containing Polybromoindoles from the Formosan Red Alga *Laurencia brongniartii*. J Nat Prod **68**: 815
1319. Narkowicz CK, Blackman AJ, Lacey E, Gill JH, Heiland K (2002) Convolutindole A and Convolutamine H, New Nematocidal Brominated Alkaloids from the Marine Bryozoan *Amathia convoluta*. J Nat Prod **65**: 938
1320. Peters L, Wright AD, Kehraus S, Gündisch D, Tilotta MC, König GM (2004) Prenylated Indole Alkaloids from *Flustra foliacea* with Subtype Specific Binding on NAChRs. Planta Med **70**: 883
1321. Peters L, König GM, Terlau H, Wright AD (2002) Four New Bromotryptamine Derivatives from the Marine Bryozoan *Flustra foliacea*. J Nat Prod **65**: 1633
1322. Tasdemir D, Bugni TS, Mangalindan GC, Concepción GP, Harper MK, Ireland CM (2002) Cytotoxic Bromoindole Derivatives and Terpenes from the Philippine Marine Sponge *Smenospongia* sp. Z Naturforsch **57c**: 914

1323. Segraves NL, Crews P (2005) Investigation of Brominated Tryptophan Alkaloids from Two Thorectidae Sponges: *Thorectandra* and *Smenospongia*. J Nat Prod **68**: 1484
1324. Campagnuolo C, Fattorusso E, Taglialatela-Scafati O (2003) Plakohypaphorines A-C, Iodine-Containing Alkaloids from the Caribbean Sponge *Plakortis simplex*. Eur J Org Chem 284
1325. Borrelli F, Campagnuolo C, Capasso R, Fattorusso E, Taglialatela-Scafati O (2004) Iodinated Indole Alkaloids from *Plakortis simplex* – New Plakohypaphorines and an Evaluation of their Antihistamine Activity. Eur J Org Chem 3227
1326. Kelley WP, Wolters AM, Sacks JT, Jockusch RA, Jurchen JC, Williams ER, Sweedler JV, Gilly WF (2003) Characterization of a Novel Gastropod Toxin (6-Bromo-2-mercaptotryptamine) That Inhibits Shaker K Channel Activity. J Biol Chem **278**: 34934
1327. Lidgren G, Bohlin L, Bergman J (1986) Studies of Swedish Marine Organisms VII. A Novel Biologically Active Indole Alkaloid from the Sponge *Geodia baretti*. Tetrahedron Lett **27**: 3283
1328. Sölter S, Dieckmann R, Blumenberg M, Francke W (2002) Barettin, Revisited? Tetrahedron Lett **43**: 3385
1329. Lieberknecht A, Griesser H (1987) What is the Structure of Barettin? Novel Synthesis of Unsaturated Diketopiperazines. Tetrahedron Lett **28**: 4275
1330. Hedner E, Sjögren M, Frändberg P-A, Johansson T, Göransson U, Dahlström M, Jonsson P, Nyberg F, Bohlin L (2006) Brominated Cyclodipeptides from the Marine Sponge *Geodia barretti* as Selective 5-HT Ligands. J Nat Prod **69**: 1421
1331. García A, Vázquez MG, Quiñoá E, Riguera R, Debitus C (1996) New Amino Acid Derivatives from the Marine Ascidian *Leptoclinides dubius*. J Nat Prod **59**: 782
1332. Lee N-K, Fenical W, Lindquist N (1997) Alternatamides A-D: New Bromotryptamine Peptide Antibiotics from the Atlantic Marine Bryozoan *Amathia alternata*. J Nat Prod **60**: 697
1333. Lawrence NJ, Bushell SM (2001) The Asymmetric Synthesis and Stereochemical Assignment of Chelonin B. Tetrahedron Lett **42**: 7671
1334. Aoki S, Ye Y, Higuchi K, Takashima A, Tanaka Y, Kitagawa I, Kobayashi M (2001) Novel Neuronal Nitric Oxide Synthase (nNOS) Selective Inhibitor, Aplysinopsin-Type Indole Alkaloid, from Marine Sponge *Hyrtios erecta*. Chem Pharm Bull **49**: 1372
1335. Appleton DR, Page MJ, Lambert G, Berridge MV, Copp BR (2002) Kottamides A-D: Novel Bioactive Imidazolone-Containing Alkaloids from the New Zealand Ascidian *Pycnoclavella kottae*. J Org Chem **67**: 5402
1336. Appleton DR, Copp BR (2003) Kottamide E, the First Example of a Natural Product Bearing the Amino Acid 4-Amino-1,2-dithiolane-4-carboxylic Acid (Adt). Tetrahedron Lett **44**: 8963
1337. Butler MS, Capon RJ, Lu CC (1992) Psammopemmins (A-C), Novel Brominated 4-Hydroxyindole Alkaloids from an Antarctic Sponge, *Psammopemma* sp. Aust J Chem **45**: 1871
1338. Hernández Franco L, Bal de Kier Joffé E, Puricelli L, Tatian M, Seldes AM, Palermo JA (1998) Indole Alkaloids from the Tunicate *Aplidium meridianum*. J Nat Prod **61**: 1130
1339. Gompel M, Leost M, Bal de Kier Joffe E, Puricelli L, Hernandez Franco L, Palermo J, Meijer L (2004) Meridianins, a New Family of Protein Kinase Inhibitors Isolated from the Ascidian *Aplidium meridianum*. Bioorg Med Chem Lett **14**: 1703
1340. Karpov AS, Merkul E, Rominger F, Müller TJJ (2005) Concise Syntheses of Meridianins by Carbonylative Alkynylation and a Four-Component Pyrimidine Synthesis. Angew Chem Int Ed **44**: 6951
1341. Cohen J, Paul GK, Gunasekera SP, Longley RE, Pomponi SA (2004) 6-Hydroxydiscodermindole, a New Discodermindole from the Marine Sponge *Discodermia polydiscus*. Pharm Biol **42**: 59
1342. Murray LM, Lim TK, Hooper JNA, Capon RJ (1995) Isobromotopsentin: A New Bis(Indole) Alkaloid from a Deep-Water Marine Sponge *Spongosorites* sp. Aust J Chem **48**: 2053

1343. Vervoort HC, Richards-Gross SE, Fenical W, Lee AY, Clardy J (1997) Didemnimides A–D: Novel, Predator-Deterrent Alkaloids from the Caribbean Mangrove Ascidian *Didemnum conchyliatum*. J Org Chem **62**: 1486
1344. Hughes TV, Cava MP (1998) Total Synthesis of Didemnimide A and B. Tetrahedron Lett **39**: 9629
1345. Shin J, Seo Y, Cho KW, Rho J-R, Sim CJ (1999) New Bis(Indole) Alkaloids of the Topsentin Class from the Sponge *Spongosorites genitrix*. J Nat Prod **62**: 647
1346. Sato H, Tsuda M, Watanabe K, Kobayashi J (1998) Rhopaladins A–D, New Indole Alkaloids from Marine Tunicate *Rhopalaea* sp. Tetrahedron **54**: 8687
1347. Janosik T, Johnson A-L, Bergman J (2002) Synthesis of the Marine Alkaloids Rhopaladins A, B, C and D. Tetrahedron **58**: 2813
1348. Casapullo A, Bifulco G, Bruno I, Ricco R (2000) New Bisindole Alkaloids of the Topsentin and Hamacanthin Classes from the Mediterranean Marine Sponge *Rhaphisia lacazei*. J Nat Prod **63**: 447
1349. Bao B, Sun Q, Yao X, Hong J, Lee C-O, Sim CJ, Im KS, Jung JH (2005) Cytotoxic Bisindole Alkaloids from a Marine Sponge *Spongosorites* sp. J Nat Prod **68**: 711
1350. Oh K-B, Mar W, Kim S, Kim J-Y, Oh M-N, Kim J-G, Shin D, Sim CJ, Shin J (2005) Bis (Indole) Alkaloids as Sortase A Inhibitors from the Sponge *Spongosorites* sp. Bioorg Med Chem Lett **15**: 4927
1351. Bao B, Sun Q, Yao X, Hong J, Lee C-O, Cho HY, Jung JH (2007) Bisindole Alkaloids of the Topsentin and Hamacanthin Classes from a Marine Sponge *Spongosorites* sp. J Nat Prod **70**: 2
1352. Oh K-B, Mar W, Kim S, Kim J-Y, Lee T-H, Kim J-G, Shin D, Sim CJ, Shin J (2006) Antimicrobial Activity and Cytotoxicity of Bis(indole) Alkaloids from the Sponge *Spongosorites* sp. Biol Pharm Bull **29**: 570
1353. Miyake FY, Yakushijin K, Horne DA (2002) Synthesis of Marine Sponge Bisindole Alkaloids Dihydrohamacanthins. Org Lett **4**: 941
1354. Kouko T, Matsumura K, Kawasaki T (2005) Total Synthesis of Marine Bisindole Alkaloids, (+)-Hamacanthins A, B and (–)-Antipode of *cis*-Dihydrohamacanthin B. Tetrahedron **61**: 2309
1355. Mancini I, Guella G, Debitus C, Waikedre J, Pietra F (1996) From Inactive Nortopsentin D, a Novel Bis(indole) Alkaloid Isolated from the Axinellid Sponge *Dragmacidon* sp. from Deep Waters South of New Caledonia, to a Strongly Cytotoxic Derivative. Helv Chim Acta **79**: 2075
1356. Capon RJ, Rooney F, Murray LM, Collins E, Sim ATR, Rostas JAP, Butler MS, Carroll AR (1998) Dragmacidins: New Protein Phosphatase Inhibitors from a Southern Australian Deep-Water Marine Sponge, *Spongosorites* sp. J Nat Prod **61**: 660
1357. Cutignano A, Bifulco G, Bruno I, Casapullo A, Gomez-Paloma L, Riccio R (2000) Dragmacidin F: A New Antiviral Bromoindole Alkaloid from the Mediterranean Sponge *Halicortex* sp. Tetrahedron **56**: 3743
1358. Garg NK, Stoltz BM (2006) A Unified Synthetic Approach to the Pyrazinone Dragmacidins. Chem Commun 3769
1359. Iwagawa T, Miyazaki M, Okamura H, Nakatani M, Doe M, Takemura K (2003) Three Novel Bis(indole) Alkaloids from a Stony Coral, *Tubastracea* sp. Tetrahedron Lett **44**: 2533
1360. Bifulco G, Bruno I, Riccio R, Lavayre J, Bourdy G (1995) Further Brominated Bis- and Tris-Indole Alkaloids from the Deep-Water New Caledonian Marine Sponge *Orina* sp. J Nat Prod **58**: 1254
1361. Ovenden SPB, Capon RJ (1999) Echinosulfonic Acids A–C and Echinosulfone A: Novel Bromoindole Sulfonic Acids and a Sulfone from a Southern Australian Marine Sponge, *Echinodictyum*. J Nat Prod **62**: 1246
1362. Rubnov S, Chevallier C, Thoison O, Debitus C, Laprevote O, Guénard D, Sévenet T (2005) Echinosulfonic Acid D: An ESI MS^n Evaluation of a New Cytotoxic Alkaloid from the New-Caledonian Sponge *Psammoclemma* sp. Nat Prod Res **19**: 75

1363. Bokesch HR, Pannell LK, McKee TC, Boyd MR (2000) Coscinamides A, B and C, Three New Bis Indole Alkaloids from the Marine Sponge *Coscinoderma* sp. Tetrahedron Lett **41**: 6305
1364. Kubota NK, Iwamoto H, Fukazawa Y, Uchio Y (2005) Five New Sulfur-Containing Polybrominated Bisindoles from the Red Alga *Laurencia brongniartii*. Heterocycles **65**: 2675
1365. Shi DY, Han LJ, Sun J, Li S, Wang SJ, Yang YC, Fan X, Shi JG (2005) A New Halogenated Biindole and a New *Apo*-Carotenone from the Green Alga *Chaetomorpha basiretorsa* Setchell. Chin Chem Lett **16**: 777
1366. Tsuda M, Takahasi Y, Fromont J, Mikami Y, Kobayashi J (2005) Dendridine A, a Bisindole Alkaloid from a Marine Sponge *Dictyodendrilla* Species. J Nat Prod **68**: 1277
1367. Verbitski SM, Mayne CL, Davis RA, Concepcion GP, Ireland CM (2002) Isolation, Structure Determination, and Biological Activity of a Novel Alkaloid, Perophoramidine, from the Philippine Ascidian *Perophora namei*. J Org Chem **67**: 7124
1368. Reyes F, Fernández R, Rodríguez A, Francesch A, Taboada S, Ávila C, Cuevas C (2008) Aplicyanins A-F, New Cytotoxic Bromoindole Derivatives from the Marine Tunicate *Aplidium cyaneum*. Tetrahedron **64**: 5119
1369. Edmonds J (2000) The Mystery of Imperial Purple Dye. Historic Dyes Series No. 7. Little Chalfont, Bucks, England
1370. Schatz PF (2001) Indigo and Tyrian Purple – In Nature and in the Lab. J Chem Ed **78**: 1442
1371. Steinhart CE (2001) Biology of the Blues: The Snail Behind the Ancient Dyes. J Chem Ed **78**: 1444
1372. Cooksey CJ (2001) Tyrian Purple: 6,6'-Dibromoindigo and Related Compounds. Molecules **6**: 736
1373. Ferreira ESB, Hulme AN, McNab H, Ouye A (2004) The Natural Constituents of Historical Textile Dyes. Chem Soc Rev **33**: 329
1374. Gore R (2004) Am a Phoenician. Nat Geog October 34
1375. Clark RJH, Cooksey CJ (1997) Bromoindirubins: The Synthesis and Properties of Minor Components of Tyrian Purple and the Composition of the Colorant from *Nucella lapillus*. J Soc Dyers Colour **113**: 316
1376. Clark RJH, Cooksey CJ (1999) Monobromoindigos: A New General Synthesis, the Characterization of All Four Isomers and an Investigation into the Purple Colour of 6,6'-Dibromoindigo. New J Chem 323
1377. Benkendorff K, Bremner JB, Davis AR (2000) Tyrian Purple Precursors in the Egg Masses of the Australian Muricid, *Dicathais orbita*: A Possible Defensive Role. J Chem Ecol **26**: 1037
1378. Benkendorff K, Bremner JB, Davis AR (2001) Indole Derivatives from the Egg Masses of Muricid Molluscs. Molecules **6**: 70
1379. Meijer L, Skaltsounis A-L, Magiatis P, Polychronopoulos P, Knockaert M, Leost M, Ryan XP, Vonica CA, Brivanlou A, Dajani R, Crovace C, Tarricone C, Musacchio A, Roe SM, Pearl L, Greengard P (2003) GSK-3-Selective Inhibitors Derived from Tyrian Purple Indirubins. Chem Biol **10**: 1235
1380. Andreotti A, Bonaduce I, Colombini MP, Ribechini E (2004) Characterisation of Natural Indigo and Shellfish Purple by Mass Spectrometric Techniques. Rapid Commun Mass Spectrom **18**: 1213
1381. Maskey RP, Grün-Wollny I, Fiebig HH, Laatsch H (2002) Akashins A, B, and C: Novel Chlorinated Indigoglycosides from *Streptomyces* sp. GW 48/1497. Angew Chem Int Ed **41**: 597
1382. Rahbaek L, Anthoni U, Christophersen C, Nielsen PH, Petersen BO (1996) Marine Alkaloids. 18. Securamines and Securines, Halogenated Indole-Imidazole Alkaloids from the Marine Bryozoan *Securiflustra securifrons*. J Org Chem **61**: 887
1383. Rahbaek L, Christophersen C (1997) Marine Alkaloids. 19. Three New Alkaloids, Securamines E-G, from the Marine Bryozoan *Securiflustra securifrons*. J Org Chem **60**: 175

1384. Fuchs JR, Funk RL (2005) Indol-2-one Intermediates: Mechanistic Evidence and Synthetic Utility. Total Syntheses of (±)-Flustramines A and C. Org Lett **7**: 677
1385. Lindel T, Bräuchle L, Golz G, Böhrer P (2007) Total Synthesis of Flustramine C via Dimethylallyl Rearrangement. Org Lett **9**: 283
1386. Kawasaki T, Shinada M, Kamimura D, Ohzono M, Ogawa A (2006) Enantioselective Total Synthesis of (−)-Flustramines A, B and (−)-Flustramides A, B via Domino Olefination/Isomerization/Claisen Rearrangement Sequence. Chem Commun 420
1387. Baran PS, Shenvi RA, Mitsos CA (2005) A Remarkable Ring Contraction En Route to the Chartelline Alkaloids. Angew Chem Int Ed **44**: 3714
1388. Baran PS, Shenvi RA (2006) Total Synthesis of (±)-Chartelline C. J Am Chem Soc **128**: 14028
1389. Klein D, Daloze D, Braekman JC, Hoffmann L, Demoulin V (1995) New Hapalindoles from the Cyanophyte *Hapalosiphon laingii*. J Nat Prod **58**: 1781
1390. Huber U, Moore RE, Patterson GML (1998) Isolation of a Nitrile-Containing Indole Alkaloid from the Terrestrial Blue-Green Alga *Hapalosiphon delicatulus*. J Nat Prod **61**: 1304
1391. Jimenez JI, Huber U, Moore RE, Patterson GML (1999) Oxidized Welwitindolinones from Terrestrial *Fischerella* ssp. J Nat Prod **62**: 569
1392. Smith CD, Zilfou JT, Stratmann K, Patterson GML, Moore RE (1995) Welwitindolinone Analogues that Reverse P-Glycoprotein Mediated Multiple Drug Resistance. Mol Pharmacol **47**: 241
1393. Zhang X, Smith CD (1996) Microtubule Effects of Welwistatin, a Cyanobacterial Indolinone that Circumvents Multiple Drug Resistance. Mol Pharmacol **49**: 288
1394. Baran PS, Richter JM (2005) Enantioselective Total Syntheses of Welwitindolinone A and Fischerindoles I and G. J Am Chem Soc **127**: 15394
1395. Makarieva TN, Ilyin SG, Stonik VA, Lyssenko KA, Denisenko VA (1999) Pibocin, the First Ergoline Marine Alkaloid from the Far-Eastern Ascidian *Eudistoma* sp. Tetrahedron Lett **40**: 1591
1396. Makarieva TN, Dmitrenok AS, Dmitrenok PS, Grebnev BB, Stonik VA (2001) Pibocin B, the First *N-O*-Methylindole Marine Alkaloid, a Metabolite from the Far-Eastern Ascidian *Eudistoma* Species. J Nat Prod **64**: 1559
1397. González MC, Lull C, Moya P, Ayala I, Primo J, Yúfera EP (2003) Insecticidal Activity of Penitrems, Including Penitrem G, a New Member of the Family Isolated from *Penicillium crustosum*. J Agric Food Chem **51**: 2156
1398. Rundberget T, Wilkins AL (2002) Thomitrems A and E, Two Indole-Alkaloid Isoprenoids from *Penicillium crustosum* Thom. Phytochemistry **61**: 979
1399. Martínez-Luis S, Rodríguez R, Acevedo L, González MC, Lira-Rocha A, Mata R (2006) Malbrancheamide, a New Calmodulin Inhibitor from the Fungus *Malbranchea aurantiaca*. Tetrahedron **62**: 1817
1400. Ding W, Williams DR, Northcote P, Siegel MM, Tsao R, Ashcroft J, Morton GO, Alluri M, Abbanat D, Maiese WM, Ellestad GA (1994) Pyrroindomycins, Novel Antibiotics Produced by *Streptomyces rugosporus* sp. LL-42D005 I. Isolation and Structure Determination. J Antibiot **47**: 1250
1401. Buchanan MS, Carroll AR, Pass D, Quinn RJ (2007) NMR Spectral Assignments of a New Chlorotryptamine Alkaloid and Its Analogues from *Acacia confusa*. Magn Reson Chem **45**: 359
1402. Barrows LR, Radisky DC, Copp BR, Swaffar DS, Kramer RA, Warters RL, Ireland CM (1993) Makaluvamines, Marine Natural Products, are Active Anti-Cancer Agents and DNA Topo II Inhibitors. Anti-Cancer Drug Design **8**: 333
1403. Harayama Y, Kita Y (2005) Pyrroloiminoquinone Alkaloids: Discorhabdins and Makaluvamines. Curr Org Chem **9**: 1567
1404. Antunes EM, Copp BR, Davies-Coleman MT, Samaai T (2005) Pyrroloiminoquinone and Related Metabolites from Marine Sponges. Nat Prod Rep **22**: 62

1405. Yang A, Baker BJ, Grimwade J, Leonard A, McClintock JB (1995) Discorhabdin Alkaloids from the Antarctic Sponge *Latrunculia apicalis*. J Nat Prod **58**: 1596
1406. Furrow FB, Amsler CD, McClintock JB, Baker BJ (2003) Surface Sequestration of Chemical Feeding Deterrents in the Antarctic Sponge *Latrunculia apicalis* as an Optimal Defense Against Sea Star Spongivory. Mar Biol **143**: 443
1407. Hooper GJ, Davies-Coleman MT, Kelly-Borges M, Coetzee PS (1996) New Alkaloids from a South African Latrunculid Sponge. Tetrahedron Lett **37**: 7135
1408. Gunasekera SP, McCarthy PJ, Longley RE, Pomponi SA, Wright AE, Lobkovsky E, Clardy J (1999) Discorhabdin P, a New Enzyme Inhibitor from a Deep-Water Caribbean Sponge of the Genus *Batzella*. J Nat Prod **62**: 173
1409. Dijoux M-G, Gamble WR, Hallock YF, Cardellina II JH, van Soest R, Boyd MR (1999) A New Discorhabdin from Two Sponge Genera. J Nat Prod **62**: 636
1410. Gunasekera SP, Zuleta IA, Longley RE, Wright AE, Pomponi SA (2003) Discorhabdins S, T, and U, New Cytotoxic Pyrroloiminoquinones from a Deep-Water Caribbean Sponge of the Genus *Batzella*. J Nat Prod **66**: 1615
1411. Antunes EM, Beukes DR, Kelly M, Samaai T, Barrows LR, Marshall KM, Sincich C, Davies-Coleman MT (2004) Cytotoxic Pyrroloiminoquinones from Four New Species of South African Latrunculid Sponges. J Nat Prod **67**: 1268
1412. Lang G, Pinkert A, Blunt JW, Munro MHG (2005) Discorhabdin W, the First Dimeric Discorhabdin. J Nat Prod **68**: 1796
1413. Tohma H, Harayama Y, Hashizume M, Iwata M, Egi M, Kita Y (2002) Synthetic Studies on the Sulfur-Cross-Linked Core of Antitumor Marine Alkaloid, Discorhabdins: Total Synthesis of Discorhabdin A. Angew Chem Int Ed **41**: 348
1414. Harayama Y, Yoshida M, Kamimura D, Wada Y, Kita Y (2006) The Efficient Direct Synthesis of N,O-Acetal Compounds as Key Intermediates of Discorhabdin A: Oxidative Fragmentation Reaction of α-Amino Acids or β-Amino Alcohols by Using Hypervalent Iodine(III) Reagents. Chem Eur J **12**: 4893
1415. Grkovic T, Kaur B, Webb VL, Copp BR (2006) Semi-Synthetic Preparation of the Rare, Cytotoxic, Deep-Sea Sourced Sponge Metabolites Discorhabdins P and U. Bioorg Med Chem Lett **16**: 1944
1416. D'Ambrosio M, Guerriero A, Chiasera G, Pietra F, Tato M (1996) Epinardins A-D, New Pyrroloiminoquinone Alkaloids of Undetermined Deep-Water Green Demosponges from Pre-Antarctic Indian Ocean. Tetrahedron **52**: 8899
1417. Venables DA, Concepción GP, Matsumoto SS, Barrows LR, Ireland CM (1997) Makaluvamine N: A New Pyrroloiminoquinone from *Zyzzya fuliginosa*. J Nat Prod **60**: 408
1418. Chang LC, Otero-Quintero S, Hooper JNA, Bewley CA (2002) Batzelline D and Isobatzelline E from the Indopacific Sponge *Zyzzya fuliginosa*. J Nat Prod **65**: 776
1419. Dijoux M-G, Schnabel PC, Hallock YF, Boswell JL, Johnson TR, Wilson JA, Ireland CM, van Soest R, Boyd MR, Barrows LR, Cardellina II JH (2005) Antitumor Activity and Distribution of Pyrroloiminoquinones in the Sponge Genus *Zyzzya*. Bioorg Med Chem **13**: 6035
1420. Gunasekera SP, McCarthy PJ, Longley RE, Pomponi SA, Wright AE (1999) Secobatzellines A and B, Two New Enzyme Inhibitors from a Deep-Water Caribbean Sponge of the Genus *Batzella*. J Nat Prod **62**: 1208
1421. Hu J-F, Schetz JA, Kelly M, Peng J-N, Ang KKH, Flotow H, Leong CY, Ng SB, Buss AD, Wilkins SP, Hamann MT (2002) New Antiinffective and Human 5-HT2 Receptor Binding Natural and Semisynthetic Compounds from the Jamaican Sponge *Smenospongia aurea*. J Nat Prod **65**: 476
1422. Isshiki K, Takahashi Y, Okada M, Sawa T, Hamada M, Naganawa H, Takita T, Takeuchi T, Umezawa H (1987) A New Antibiotic Indisocin and *N*-Methylindisocin. J Antibiot **40**: 1195
1423. Kamano Y, Zhang H, Ichihara Y, Kizu H, Komiyama K, Pettit GR (1995) Convolutamydine A, a Novel Bioactive Hydroxyoxindole Alkaloid from Marine Bryozoan *Amathia convoluta*. Tetrahedron Lett **36**: 2783

1424. Zhang H, Kamano Y, Ichihara Y, Kizu H, Komiyama K, Itokawa H, Pettit GR (1995) Isolation and Structure of Convolutamydines B~D from Marine Bryozoan *Amathia convoluta*. Tetrahedron **51**: 5523
1425. Kamano Y, Kotake A, Hashima H, Hayakawa I, Hiraide H, Zhang H-P, Kizu H, Komiyama K, Hayashi M, Pettit GR (1999) Three New Alkaloids, Convolutamines F and G, and Convolutamydine E, from the Floridian Marine Bryozoan *Amathia convoluta*. Collect Czech Chem Commun **64**: 1147
1426. Nakamura T, Shirokawa S, Hosokawa S, Nakazaki A, Kobayashi S (2006) Enantioselective Total Synthesis of Convolutamydines B and E. Org Lett **8**: 677
1427. Luppi G, Monari M, Correa RJ, Violante F de A, Pinto AC, Kaptein B, Broxterman QB, Garden SJ, Tomasini C (2006) The First Total Synthesis of (*R*)-Convolutamydine A. Tetrahedron **62**: 12017
1428. Cravotto G, Giovenzana GB, Palmisano G, Penoni A, Pilati T, Sisti M, Stazi F (2006) Convolutamydine A: The First Authenticated Absolute Configuration and Enantioselective Synthesis. Tetrahedron: Asym **17**: 3070
1429. Carletti I, Banaigs B, Amade P (2000) Matemone, a New Bioactive Bromine-Containing Oxindole Alkaloid from the Indian Ocean Sponge *Iotrochota purpurea*. J Nat Prod **63**: 981
1430. Abourriche A, Abboud Y, Maoufoud S, Mohou H, Seffaj T, Charrouf M, Chaib N, Bennamara A, Bontemps N, Francisco C (2003) Cynthichlorine: A Bioactive Alkaloid from the Tunicate *Cynthia savignyi*. Il Farm **58**: 1351
1431. Prudhomme M (2000) Recent Developments of Rebeccamycin Analogues as Topoisomerase I Inhibitors and Antitumor Agents. Curr Med Chem **7**: 1189
1432. Long BH, Rose WC, Vyas DM, Matson JA, Forenza S (2002) Discovery of Antitumor Indolocarbazoles: Rebeccamycin, NSC 655649, and Fluoroindolocarbazoles. Curr Med Chem-Anti-Cancer Agents **2**: 255
1433. Lam KS, Schroeder DR, Veitch JM, Colson KL, Matson JA, Rose WC, Doyle TW, Forenza S (2001) Production, Isolation and Structure Determination of Novel Fluoroindolocarbazoles from *Saccharothrix aerocolonigenes* ATCC 39243. J Antibiot **54**: 1
1434. Anizon F, Moreau P, Sancelme M, Laine W, Bailly C, Prudhomme M (2003) Rebeccamycin Analogues Bearing Amine Substituents or Other Groups on the Sugar Moiety. Bioorg Med Chem **11**: 3709
1435. Moreau P, Gaillard N, Marminon C, Anizon F, Dias N, Baldeyrou B, Bailly C, Pierré A, Hickman J, Pfeiffer B, Renard P, Prudhomme M (2003) Semi-Synthesis, Topoisomerase I and Kinase Inhibitory Properties, and Antiproliferative Activities of New Rebeccamycin Derivatives. Bioorg Med Chem **11**: 4871
1436. Voldoire A, Moreau P, Sancelme M, Matulova M, Léonce S, Pierré A, Hickman J, Pfeiffer B, Renard P, Dias N, Bailly C, Prudhomme M (2004) Analogues of Antifungal Tjipanazoles from Rebeccamycin. Bioorg Med Chem **12**: 1955
1437. Sánchez C, Zhu L, Braña AF, Salas AP, Rohr J, Méndez C, Salas JA (2005) Combinatorial Biosynthesis of Antitumor Indolocarbazole Compounds. Proc Natl Acad Sci U S A **102**: 461
1438. Hussain M, Vaishampayan U, Heilbrun LK, Jain V, LoRusso PM, Ivy P, Flaherty L (2003) A Phase II Study of Rebeccamycin Analog (NSC-655649) in Metastatic Renal Cell Cancer. Invest New Drugs **21**: 465
1439. Onaka H, Taniguchi S, Igarashi Y, Furumai T (2003) Characterization of the Biosynthetic Gene cluster of Rebeccamycin from *Lechevalieria aerocolonigenes* ATCC 39243. Biosci Biotechnol Biochem **67**: 127
1440. Yeh E, Garneau S, Walsh CT (2005) Robust *in vitro* Activity of RebF and RebH, a Two-Component Reductase/Halogenase, Generating 7-Chlorotryptophan During Rebeccamycin Biosynthesis. Proc Natl Acad Sci U S A **102**: 3960
1441. Howard-Jones AR, Walsh CT (2006) Staurosporine and Rebeccamycin Aglycones Are Assembled by the Oxidative Action of StaP, StaC, and RebC on Chromopyrrolic Acid. J Am Chem Soc **128**: 12289

1442. Schumacher RW, Davidson BS (1995) Didemnolines A–D, New N9-Substituted β-Carbolines from the Marine Ascidian *Didemnum* sp. Tetrahedron **51**: 10125
1443. Kang H, Fenical W (1996) New Isoeudistomin Class Dihydro-β-carbolines from an Undescribed Ascidian of the Genus *Eudistoma*. Nat Prod Lett **9**: 7
1444. Davis RA, Carroll AR, Quinn RJ (1998) Eudistomin V, a New β-Carboline from the Australian Ascidian *Pseudodistoma aureum*. J Nat Prod **61**: 959
1445. Rashid MA, Gustafson KR, Boyd MR (2001) New Cytotoxic *N*-Methylated β-Carboline Alkaloids from the Marine Ascidian *Eudistoma gilboverde*. J Nat Prod **64**: 1454
1446. Sandler JS, Colin PL, Hooper JNA, Faulkner DJ (2002) Cytotoxic β-Carbolines and Cyclic Peroxides from the Palauan Sponge *Plakortis nigra*. J Nat Prod **65**: 1258
1447. Liu J-J, Hino T, Tsuruoka A, Harada N, Nakagawa M (2000) Total Synthesis of (–)-Eudistomins with an Oxathiazepine Ring. Part 2. Synthesis of (–)-Eudistomins C, E, F, K, and L. J Chem Soc Perkin Trans 1: 3487
1448. Yamashita T, Kawai N, Tokuyama H, Fukuyama T (2005) Stereocontrolled Total Synthesis of (–)-Eudistomin C. J Am Chem Soc **127**: 15038
1449. Segraves NL, Lopez S, Johnson TA, Said SA, Fu X, Schmitz FJ, Pietraszkiewicz H, Valeriote FA, Crews P (2003) Structures and Cytotoxicities of Fascaplysin and Related Alkaloids from Two Marine Phyla – *Fascaplysinopsis* Sponges and *Didemnum* Tunicates. Tetrahedron Lett **44**: 3471
1450. Segraves NL, Robinson SJ, Garcia D, Said SA, Fu X, Schmitz FJ, Pietraszkiewicz H, Valeriote FA, Crews P (2004) Comparison of Fascaplysin and Related Alkaloids: A Study of Structures, Cytotoxicities, and Sources. J Nat Prod **67**: 783
1451. Becher PG, Beuchat J, Gademann K, Jüttner F (2005) Nostocarboline: Isolation and Synthesis of a New Cholinesterase Inhibitor from *Nostoc* 78-12A. J Nat Prod **68**: 1793
1452. Blom JF, Brütsch T, Barbaras D, Bethuel Y, Locher HH, Hubschwerlen C, Gademann K (2006) Potent Algicides Based on the Cyanobacterial Alkaloid Nostocarboline. Org Lett **8**: 737
1453. Kanchanapoom T, Kamel MS, Kasai R, Picheansoonthon C, Hiraga Y, Yamasaki K (2001) Benzoxazinoid Glucosides from *Acanthus ilicifolius*. Phytochemistry **58**: 637
1454. Kanchanapoom T, Kasai R, Picheansoonthon C, Yamasaki K (2001) Megastigmane, Aliphatic Alcohol and Benzoxazinoid Glycosides from *Acanthus ebracteatus*. Phytochemistry **58**: 811
1455. Ma M, Zhao S, Li S, Yang Y, Shi J, Fan X, He L (2007) Bromophenols Coupled with Nucleoside Bases and Brominated Tetrahydroisoquinolines from the Red Alga *Rhodomela confervoides*. J Nat Prod **70**: 337
1456. Milanowski DJ, Gustafson KR, Kelley JA, McMahon JB (2004) Caulibugulones A-F, Novel Cytotoxic Isoquinoline Quinones and Iminoquinones from the Marine Bryozoan *Caulibugula intermis*. J Nat Prod **67**: 70
1457. Wipf P, Joo B, Nguyen T, Lazo JS (2004) Synthesis and Biological Evaluation of Caulibugulones A-E. Org Biomol Chem **2**: 2173
1458. Kim W-G, Kim J-P, Kim C-J, Lee K-H, Yoo I-D (1996) Benzastatins A, B, C, and D: New Free Radical Scavengers from *Streptomyces nitrosporeus* 30643 I. Taxonomy, Fermentation, Isolation, Physico-Chemical Properties and Biological Activities. J Antibiot **49**: 20
1459. Kim W-G, Kim J-P, Yoo I-D (1996) Benzastatins A, B, C, and D: New Free Radical Scavengers from *Streptomyces nitrosporeus* 30643 II. Structure Determination. J Antibiot **49**: 26
1460. Morimoto Y, Matsuda F, Shirahama H (1996) Synthetic Studies on Virantmycin. 1. Total Synthesis of (±)-Virantmycin and Determination of Its Relative Stereochemistry. Tetrahedron **52**: 10609
1461. Morimoto Y, Shirahama H (1996) Synthetic Studies on Virantmycin. 2. Total Synthesis of Unnatural (+)-Virantmycin and Determination of Its Absolute Stereochemistry. Tetrahedron **52**: 10631
1462. Morris BD, Prinsep MR (1998) Euthyroideones, Novel Brominated Quinone Methides from the Bryozoan *Euthyroides episcopalis*. J Org Chem **63**: 9545

1463. Al-Rehaily AJ, Ahmad MS, Muhammad I, Al-Thukair AA, Perzanowski HP (2003) Furoquinoline Alkaloids from *Teclea nobilis*. Phytochemistry **64**: 1405
1464. Baumert A, Gröger D, Kuzovkina IN, Reisch J (1992) Secondary Metabolites Produced by Callus Cultures of Various *Ruta* Species. Plant Cell Tissue Organ Cult **28**: 159
1465. Anai T, Aizawa H, Ohtake N, Kosemura S, Yamamura S, Hasegawa K (1996) A New Auxin-Inhibiting Substance, 4-Cl-6,7-dimethoxy-2-benzoxazolinone, From Light-Grown Maize Shoots. Phytochemistry **42**: 273
1466. Kato-Noguchi H, Kosemura S, Yamamura S (1998) Allelopathic Potential of 5-Chloro-6-methoxy-2-benzoxazolinone. Phytochemistry **48**: 433
1467. Benkendorff K, Pillai R, Bremner JB (2004) 2,4,5-Tribromo-1H-imidazole in the Egg Masses of Three Muricid Molluscs. Nat Prod Res **18**: 427
1468. Das B, Reddy MR, Ravindranath N, Kishore KH, Murthy USN (2005) A Substituted Imidazole Derivative from *Jatropha curcas*. Ind J Chem **44B**: 1119
1469. Arison BH, Beck JL (1973) The Structure of Compound 593A. A New Anti-tumor Agent. Tetrahedron **29**: 2743
1470. Cen Y-Z, Su J-Y, Zeng L-M (1997) Studies on the Chemical Compositions of the Marine Sponge *Rhaphisia pallida* (I) – The Structure Elucidation of Two Novel Ten-membered Heterocyclic Compounds. Chem J Chin Univer **18**: 1057
1471. Kumagai H, Nishida H, Imamura N, Tomoda H, Ōmura S, Bordner J (1990) The Structures of Atpenins A4, A5 and B, New Antifungal Antibiotics Produced by *Penicillium* sp. J Antibiotics **43**: 1553
1472. Lis AW, McLaughlin RK, McLaughlin DI, Daves Jr. GG, Anderson Jr. WR (1973) 5-Chlorocytosine. Occurrence in Salmon Sperm Deoxyribonucleic Acid. J Am Chem Soc **95**: 5789
1473. Ichiba T, Nakao Y, Scheuer PJ, Sata NU, Kelly-Borges M (1995) Kumusine, a Chloroadenine Riboside from a Sponge, *Theonella* sp. Tetrahedron Lett **36**: 3977
1474. Searle PA, Molinski TF (1995) Trachycladines A and B: 2'-C-Methyl-5'-deoxyribofuranosyl Nucleosides from the Marine Sponge *Trachycladus laevispirulifer*. J Org Chem **60**: 4296
1475. Erdogan I, Higa T (2000) A Chloroadenine Riboside-Type Nucleoside from the Marine Sponge *Theonella cupola*. J Fac Pharm Ankara **29**: 1
1476. Mitchell SS, Pomerantz SC, Concepción GP, Ireland CM (1996) Tubercidin Analogs from the Ascidian *Didemnum voeltzkowi*. J Nat Prod **59**: 1000
1477. Xu X-H, Yao G-M, Li Y-M, Lu J-H, Lin C-J, Wang X, Kong C-H (2003) 5-Fluorouracil Derivatives from the Sponge *Phakellia fusca*. J Nat Prod **66**: 285
1478. Henderson JP, Byun J, Takeshita J, Heinecke JW (2003) Phagocytes Produce 5-Chlorouracil and 5-Bromouracil, Two Mutagenic Products of Myeloperoxidase, in Human Inflammatory Tissue. J Biol Chem **278**: 23522
1479. Takeshita J, Byun J, Nhan TQ, Pritchard DK, Pennathur S, Schwartz SM, Chait A, Heinecke JW (2006) Myeloperoxidase Generates 5-Chlorouracil in Human Atherosclerotic Tissue. J Biol Chem **281**: 3096
1480. Henderson JP, Byun J, Mueller DM, Heinecke JW (2001) The Eosinophil Peroxidase-Hydrogen Peroxide-Bromide System of Human Eosinophils Generates 5-Bromouracil, a Mutagenic Thymine Analogue. Biochemistry **40**: 2052
1481. Whiteman M, Jenner A, Halliwell B (1999) 8-Chloroadenine: A Novel Product Formed from Hypochlorous Acid-Induced Damage to Calf Thymus DNA. Biomarkers **4**: 303
1482. Henderson JP, Byun J, Williams MV, Mueller DM, McCormick ML, Heinecke JW (2001) Production of Brominating Intermediates by Myeloperoxidase. J Biol Chem **276**: 7867
1483. Valinluck V, Wu W, Liu P, Neidigh JW, Sowers LC (2006) Impact of Cytosine 5-Halogens on the Interaction of DNA with Restriction Endonucleases and Methyltransferase. Chem Res Toxicol **19**: 556
1484. Sand T, Kavvadias D, Feineis D, Riederer P, Schreier P, Kleinschnitz M, Czygan F-C, Abou-Mandour A, Bringmann G, Beckmann H (2000) Naturally Occurring Benzodiazepines: Current Status of Research and Clinical Implications. Eur Arch Psychiatry Clin Neurosci **250**: 194

1485. Kavvadias D, Abou-Mandour AA, Czygan F-C, Beckmann H, Sand P, Riederer P, Schreier P (2000) Identification of Benzodiazepines in *Artemisia dracunculus* and *Solanum tuberosum* Rationalizing Their Endogenous Formation in Plant Tissue. Biochem Biophys Res Commun **269**: 290
1486. Engler M, Anke T, Sterner O, Brandt U (1997) Pterulinic Acid and Pterulone, Two Novel Inhibitors of NADH: Ubiquinone Oxidoreductase (Complex I) Produced by a *Pterula* Species I. Production, Isolation and Biological Activities. J Antibiot **50**: 325
1487. Engler M, Anke T, Sterner O (1997) Pterulinic Acid and Pterulone, Two Novel Inhibitors of NADH: Ubiquinone Oxidoreductase (Complex I) Produced by a Pterula Species II. Physico-chemical Properties and Structure Elucidation. J Antibiot **50**: 330
1488. Engler M, Anke T, Sterner O (1998) Production of Antibiotics by *Collybia nivalis*, *Omphalotus olearius*, a *Favolaschia* and a *Pterula* Species on Natural Substrates. Z Naturforsch **53c**: 318
1489. Wijnberg JBPA, van Veldhuizen A, Swarts HJ, Frankland JC, Field JA (1999) Novel Monochlorinated Metabolites with a 1-Benzoxepin Skeleton from *Mycena galopus*. Tetrahedron Lett **40**: 5767
1490. Gruijters BWT, van Veldhuizen A, Weijers CAGM, Wijnberg JBPA (2002) Total Synthesis and Bioactivity of Some Naturally Occurring Pterulones. J Nat Prod **65**: 558
1491. Okamoto Y, Ojika M, Sakagami Y (1999) Iantheran A, a Dimeric Polybrominated Benzofuran as a Na,K-ATPase Inhibitor from a Marine Sponge, *Ianthella* sp. Tetrahedron Lett **40**: 507
1492. Okamoto Y, Ojika M, Suzuki S, Murakami M, Sakagami Y (2001) Iantherans A and B, Unique Dimeric Polybrominated Benzofurans as Na,K-ATPase Inhibitors from a Marine Sponge, *Ianthella* sp. Bioorg Med Chem **9**: 179
1493. Atta-ur-Rahman, Choudhard MI, Hayat S, Khan AM, Ahmed A (2001) Two New Aurones from Marine Brown Alga *Spatoglossum variabile*. Chem Pharm Bull **49**: 105
1494. Lan Y-H, Chang F-R, Yu J-H, Yang Y-L, Chang Y-L, Lee S-J, Wu Y-C (2003) Cytotoxic Styrylpyrones from *Goniothalamus amuyon*. J Nat Prod **66**: 487
1495. Yagura T, Ito M, Kiuchi F, Honda G, Shimada Y (2003) Four New 2-(2-Phenylethyl) chromone Derivatives from Withered Wood of *Aquilaria sinensis*. Chem Pharm Bull **51**: 560
1496. Kim S-U, Lee S-Y, Kim S-K, Son K-H, Kim Y-K, Moon S-S, Bok S-H (1996) Isolation and Characterization of Antifungal Compound Produced by *Aspergillus candidus* F1484. Kor J Appl Microbiol Biotechnol **24**: 574
1497. Quaghebeur K, Coosemans J, Toppet S, Compernolle F (1994) Cannabiorci- and 8-Chlorocannabiorcichromenic Acid as Fungal Antagonists from *Cylindrocarpon olidum*. Phytochemistry **37**: 159
1498. Ohashi K, Watanabe H, Ohi K, Arimoto H, Okumura Y (1995) Two New 7-Geranyloxycoumarins from the Bark of *Aegle marmelos*, an Indonesian Medicinal Plant. Chem Lett 881
1499. Schimana J, Fiedler H-P, Groth I, Süssmuth R, Beil W, Walker M, Zeeck A (2000) Simocyclinones, Novel Cytostatic Angucyclinone Antibiotics Produced by *Streptomyces antibioticus* Tü6040 I. Taxonomy, Fermentation, Isolation and Biological Activities. J Antibiot **53**: 779
1500. Holzenkämpfer M, Walker M, Zeeck A, Schimana J, Fiedler H-P (2002) Simocyclinones, Novel Cytostatic Angucyclinone Antibiotics Produced by *Streptomyces antibioticus* Tü6040 II. Structure Elucidation and Biosynthesis. J Antibiot **55**: 301
1501. Holzenkämpfer M, Zeeck A (2002) Biosynthesis of Simocyclinone D8 in an $^{18}O_2$-rich Atmosphere. J Antibiot **55**: 341
1502. Li S-M, Heide L (2005) New Aminocoumarin Antibiotics from Genetically Engineered *Streptomyces* Strains. Curr Med Chem **12**: 419
1503. Eustáquio AS, Gust B, Luft T, Li S-M, Chater KF, Heide L (2003) Clorobiocin Biosynthesis in *Streptomyces*: Identification of the Halogenase and Generation of Structural Analogs. Chem Biol **10**: 279

1504. Guz NR, Lorenz P, Stermitz FR (2001) New Coumarins from *Harbouria trachypleura*: Isolation and Synthesis. Tetrahedron Lett **42**: 6491
1505. Duan H, Takaishi Y, Fujimoto Y, Garzon C, Osorio C, Duque C (2002) Chemical Constituents from the Colombian Medicinal Plant *Niphogeton ternata*. Chem Pharm Bull **50**: 115
1506. Patnam R, Kadali SS, Koumaglo KH, Roy R (2005) A Chlorinated Coumarinolignan from the African Medicinal Plant, *Mondia whitei*. Phytochemistry **66**: 683
1507. Hill RA (1986) Naturally Occurring Isocoumarins. Prog Chem Org Nat Prod **49**: 1
1508. Steyn PS, Holzapfel CW (1967) The Isolation of the Methyl and Ethyl Esters of Ochratoxins A and B, Metabolites of *Aspergillus ochraceus* Wilh. J South African Chem Inst **20**: 186
1509. Sonnenbichler J, Bliestle IM, Peipp H, Holdenreider O (1989) Secondary Fungal Metabolites and Their Biological Activities, I. Isolation of Antibiotic Compounds from Cultures of *Heterobasidion annosum* Synthesized in the Presence of Antagonistic Fungi or Host Plant Cells. Biol Chem Hoppe-Seyler **370**: 1295
1510. Stadler M, Anke H, Sterner O (1995) New Metabolites with Nematicidal and Antimicrobial Activities from the Ascomycete *Lachnum papyraceum* (Karst.) Karst IV. Structural Elucidation of Novel Isocoumarin Derivatives. J Antibiot **48**: 267
1511. Krohn K, Bahramsari R, Flörke U, Ludewig K, Kliche-Spory C, Michel A, Aust H-J, Draeger S, Schulz B, Antus S (1997) Dihydroisocoumarins from Fungi: Isolation, Structure Elucidation, Circular Dichroism and Biological Activity. Phytochemistry **45**: 313
1512. Uchida K, Watanabe H, Kitahara T (1998) Synthesis of Both Enantiomers of Hiburipyranone. Tetrahedron **54**: 8975
1513. Larsen TO, Breinholt J (1999) Dichlorodiaportin, Diaportinol, and Diaportinic Acid: Three Novel Isocoumarins from *Penicillium nalgiovense*. J Nat Prod **62**: 1182
1514. Tanahashi T, Takenaka Y, Nagakura N, Hamada N, Miyawaki H (2000) Two Isocoumarins from the Cultured Lichen Mycobiont of *Graphis* sp. Heterocycles **53**: 723
1515. Wan J, Lin Y-C, Wu X-Y, Zhou S-N, Vrijmoed LLP (2001) Avicennin A, a New Isocoumarin from Mangrove Entophytic Fungus No. 2533 from the South China Sea. Zhongshan Daxue Xuebao, Ziran Kexueban **40**: 127
1516. Wijeratne EMK, Paranagama PA, Gunatilaka AAL (2006) Five New Isocoumarins from Sonoran Desert Plant-Associated Fungal Strains *Paraphaeosphaeria quadriseptata* and *Chaetomium chiversii*. Tetrahedron **62**: 8439
1517. Sakurai M, Nishio M, Yamamoto K, Okuda T, Kawano K, Ohnuki T (2003) TMS-264, a Novel Antiallergic Heptaketide Produced by the Fungus *Phoma* sp. TC 1674. Org Lett **5**: 1083
1518. Sakurai M, Nishio M, Yamamoto K, Okuda T, Kawano K, Ohnuki T (2003) TMS-264, a Novel Inhibitor of STAT6 Activation Produced by *Phoma* sp. TC 1674. J Antibiot **56**: 513
1519. Zhang H-W, Huang W-Y, Song Y-C, Chen J-R, Tan R-X (2005) Four 6*H*-Dibenzo[*b*,*d*]pyran-6-one Derivatives Produced by the Endophyte *Cephalosporium acremonium* IFB-E007. Helv Chim Acta **88**: 2861
1520. Pittet A (1998) Natural Occurrence of Mycotoxins in Foods and Feeds – An Updated Review. Revue Méd Vét **149**: 479
1521. Abarca ML, Accensi F, Bragulat MR, Cabañes FJ (2001) Current Importance of Ochratoxin A-Producing *Aspergillus* spp. J Food Prot **64**: 903
1522. Creppy EE (1999) Human Ochratoxicosis. J Toxicol Toxin Rev **18**: 277
1523. Petzinger E, Ziegler K (2000) Ochratoxin A from a Toxicological Perspective. J Vet Pharmacol Therap **23**: 91
1524. Schwartz GG (2002) *Hypothesis*: Does Ochratoxin A Cause Testicular Cancer? Cancer Causes Control **13**: 91
1525. Zimmerli B, Dick R (1996) Ochratoxin A in Table Wine and Grape-Juice: Occurrence and Risk Assessment. Food Addit Contam **13**: 655
1526. Otteneder H, Majerus P (2000) Occurrence of Ochratoxin A (OTA) in Wines. Influence of the Type of Wine and Its Geographical Origin. Food Addit Contam **17**: 793

1527. Cabañes FJ, Accensi F, Bragulat MR, Abarca ML, Castellá G, Minguez S, Pons A (2002) What is the Source of Ochratoxin A in Wine? Int J Food Microbiol **79**: 213
1528. Sage L, Krivobok S, Delbos E, Seigle-Murandi F, Creppy EE (2002) Fungal Flora and Ochratoxin A Production in Grapes and Musts from France. J Agric Food Chem **50**: 1306
1529. Shephard GS, Fabiani A, Stockenström S, Mshicileli N, Sewram V (2003) Quantitation of Ochratoxin A in South African Wines. J Agric Food Chem **51**: 1102
1530. Marín S, Bellí N, Lasram S, Chebil S, Ramos AJ, Ghorbel A, Sanchis V (2006) Kinetics of Ochratoxin A Production and Accumulation by *Aspergillus carbonarius* on Synthetic Grape Medium at Different Temperature Levels. J Food Sci **71**: M196
1531. Khoury AEl, Rizk T, Lteif R, Azouri H, Delia M-L, Lebrihi A (2006) Occurrence of Ochratoxin A- and Aflatoxin B1-Producing Fungi in Lebanese Grapes and Ochratoxin A Content in Musts and Finished Wines during 2004. J Agric Food Chem **54**: 8977
1532. Fernandes A, Ratola N, Cerdeira A, Alves A, Venãncio A (2007) Changes in Ochratoxin A Concentration during Winemaking. Am J Enol Vitic **58**: 92
1533. Savino M, Limosani P, Garcia-Moruno E (2007) Reduction of Ochratoxin A Contamination in Red Wines by Oak Wood Fragments. Am J Enol Vitic **58**: 97
1534. Vrabcheva T, Usleber E, Dietrich R, Märtlbauer E (2000) Co-occurrence of Ochratoxin A and Citrinin in Cereals from Bulgarian Villages with a History of Balkan Endemic Nephropathy. J Agric Food Chem **48**: 2483
1535. Visconti A, Pascale M, Centonze G (2000) Determination of Ochratoxin A in Domestic and Imported Beers in Italy by Immunoaffinity Clean-Up and Liquid Chromatography. J Chromatogr A **888**: 321
1536. Aresta A, Palmisano F, Vatinno R, Zambonin CG (2006) Ochratoxin A Determination in Beer by Solid-Phase Microextraction Coupled to Liquid Chromatography with Florescence Detection: A Fast and Sensitive Method for Assessment of Noncompliance to Legal Limits. J Agric Food Chem **54**: 1594
1537. MacDonald S, Wilson P, Barnes K, Damant A, Massey R, Mortby E, Shepherd MJ (1999) Ochratoxin A in Dried Vine Fruit: Method Development and Survey. Food Addit Contam **16**: 253
1538. Sibanda L, De Saeger S, Barna-Vetro I, Van Peteghem C (2002) Development of a Solid-Phase Cleanup and Portable Rapid Flow-Through Enzyme Immunoassay for the Detection of Ochratoxin A in Roasted Coffee. J Agric Food Chem **50**: 6964
1539. Bonvehí JS (2004) Occurrence of Ochratoxin A in Cocoa Products and Chocolate. J Agric Food Chem **52**: 6347
1540. Watanabe S, Hirai H, Kato Y, Nishida H, Saito T, Yoshikawa N, Parkinson T, Kojima Y (2001) CJ-19,784, a New Antifungal Agent from a Fungus, *Acanthostigmella* sp. J Antibiot **54**: 1031
1541. Demirezer LÖ, Kuruüzüm-Uz A, Bergere I, Schiewe H-J, Zeeck A (2001) The Structures of Antioxidant and Cytotoxic Agents from Natural Source: Anthraquinones and Tannins from Roots of *Rumex patientia*. Phytochemistry **58**: 1213
1542. Nakajima H, Sato B, Fujita T, Takase S, Terano H, Okuhara M (1996) New Antitumor Substances, FR901463, FR901464 and FR901465 I. Taxonomy, Fermentation, Isolation, Physico-Chemical Properties and Biological Activities. J Antibiot **49**: 1196
1543. Nakajima H, Hori Y, Terano H, Okuhara M, Manda T, Matsumoto S, Shimomura K (1996) New Antitumor Substances, FR901463, FR901464 and FR901465 II. Activities Against Experimental Tumors in Mice and Mechanism of Action. J Antibiot **49**: 1204
1544. Nakajima H, Takase S, Terano H, Tanaka H (1997) New Antitumor Substances, FR901463, FR901464 and FR901465 III. Structures of FR901463, FR901464 and FR901465. J Antibiot **50**: 96
1545. Nicolaou KC, Dai W-M (1991) Chemistry and Biology of the Enediyne Anticancer Antibiotics. Angew Chem Int Ed **30**: 1387
1546. Jones GB, Fouad FS (2002) Designed Enediyne Antitumor Agents. Curr Pharm Design **8**: 2415

1547. Shen B, Liu W, Nonaka K (2003) Enediyne Natural Products: Biosynthesis and Prospect Towards Engineering Novel Antitumor Agents. Curr Med Chem **10**: 2317
1548. Galm U, Hager MH, Van Lanen SG, Ju J, Thorson JS, Shen B (2005) Antitumor Antibiotics: Bleomycin, Enediynes, and Mitomycin. Chem Rev **105**: 739
1549. Watanabe CMH, Supekova L, Schultz PG (2002) Transcriptional Effects of the Potent Enediyne Anti-Cancer Agent Calicheamicin γ_1'. Chem Biol **9**: 245
1550. Usuki T, Nakanishi K, Ellestad GA (2006) Spin-Trapping of the *p*-Benzyne Intermediates from Ten-Membered Enediyne Calicheamicin γ_1'. Org Lett **8**: 5461
1551. Inoue M (2006) Exploring the Chemistry and Biology of Antitumor Enediyne Chromoprotein C-1027. Bull Chem Soc Jpn **79**: 501
1552. Van Lanen SG, Dorrestein PC, Christenson SD, Liu W, Ju J, Kelleher NL, Shen B (2005) Biosynthesis of the β-Amino Acid Moiety of the Enediyne Antitumor Antibiotic C-1027 Featuring β-Amino Acyl-*S*-carrier Protein Intermediates. J Am Chem Soc **127**: 11594
1553. Inoue M, Sasaki T, Hatano S, Hirama M (2004) Synthesis of the C-1027 Chromophore Framework through Atropselective Macrolactonization. Angew Chem Int Ed **43**: 6500
1554. Usuki T, Inoue M, Hirama M, Tanaka T (2004) Rational Design of a Supra C-1027: Kinetically Stabilized Analogue of the Antitumor Enediyne Chromoprotein. J Am Chem Soc **126**: 3022
1555. Usuki T, Inoue M, Akiyama K, Hirama M (2005) ESR Studies on DNA Cleavage Induced by Enediyne C-1027 Chromophore. Bioorg Med Chem **13**: 5218
1556. McHugh MM, Gawron LS, Matsui S-I, Beerman TA (2005) The Antitumor Enediyne C-1027 Alters Cell Cycle Progression and Induces Chromosomal Aberrations and Telomere Dysfunction. Cancer Res **65**: 5344
1557. Iida K, Fukuda S, Tanaka T, Hirama M, Imajo S, Ishiguro M, Yoshida K, Otani T (1996) Absolute Configuration of C-1027 Chromophore. Tetrahedron Lett **37**: 4997
1558. Otani T, Yoshida K, Sasaki T, Minami Y (1999) C-1027 Enediyne Chromophore: Presence of Another Active Form and Its Chemical Structure. J Antibiot **52**: 415
1559. Kawata S, Ashizawa S, Hirama M (1997) Synthetic Study of Kedarcidin Chromophore: Revised Structure. J Am Chem Soc **119**: 12012
1560. Myers AG, Hurd AR, Hogan PC (2002) Evidence for Facile Atropisomerism and Simple (Non-Nucleophilic) Biradical-Forming Cycloaromatization within Kedarcidin Chromophore Aglycon. J Am Chem Soc **124**: 4583
1561. Hanada M, Ohkuma H, Yonemoto T, Tomita K, Ohbayashi M, Kamei H, Miyaki T, Konishi M, Kawaguchi H, Forenza S (1991) Maduropeptin, a Complex of New Macromolecular Antitumor Antibiotics. J Antibiot **44**: 403
1562. Schroeder DR, Colson KL, Klohr SE, Zein N, Langley DR, Lee MS, Matson JA, Doyle TW (1994) Isolation, Structure Determination, and Proposed Mechanism of Action for Artifacts of Maduropeptin Chromophore. J Am Chem Soc **116**: 9351
1563. Suffert J, Toussaint D (1997) Synthesis of a Cyclic Dienediyne Related to the Maduropeptin Chromophore. Tetrahedron Lett **38**: 5507
1564. Ando T, Ishii M, Kajiura T, Kameyama T, Miwa K, Sugiura Y (1998) A New Non-Protein Enediyne Antibiotic N1999A2: Unique Enediyne Chromophore Similar to Neocarzinostatin and DNA Cleavage Feature. Tetrahedron Lett **39**: 6495
1565. Kobayashi S, Reddy RS, Sugiura Y, Sasaki D, Miyagawa N, Hirama M (2001) Investigation of the Total Synthesis of N1999-A2: Implication of Stereochemistry. J Am Chem Soc **123**: 2887
1566. Kobayashi S, Ashizawa S, Takahashi Y, Sugiura Y, Nagaoka M, Lear MJ, Hirama M (2001) The First Total Synthesis of N1999-A2: Absolute Stereochemistry and Stereochemical Implications into DNA Cleavage. J Am Chem Soc **123**: 11294
1567. Cassady JM, Chan KK, Floss HG, Leistner E (2004) Recent Developments in the Maytansinoid Antitumor Agents. Chem Pharm Bull **52**: 1
1568. Widdison WC, Wilhelm SD, Cavanagh EE, Whiteman KR, Leece BA, Kovtun Y, Goldmacher VS, Xie H, Steeves RM, Lutz RJ, Zhao R, Wang L, Blättler WA, Chari RVJ

References

(2006) Semisynthetic Maytansine Analogues for the Targeted Treatment of Cancer. J Med Chem **49**: 4392
1569. Larson GM, Schaneberg BT, Sneden AT (1999) Two New Maytansinoids from *Maytenus buchananii*. J Nat Prod **62**: 361
1570. Lu C, Bai L, Shen Y (2004) A Novel Amide *N*-Glycoside of Ansamitocins from *Actinosynnema pretiosum*. J Antibiot **57**: 348
1571. Sharma SV, Agatsuma T, Nakano H (1998) Targeting of the Protein Chaperone, HSP90, by the Transformation Suppressing Agent, Radicicol. Oncogen **16**: 2639
1572. Roe SM, Prodromou C, O'Brien R, Ladbury JE, Piper PW, Pearl LH (1999) Structural Basis for Inhibition of the Hsp90 Molecular Chaperone by the Antitumor Antibiotics Radicicol and Geldanamycin. J Med Chem **42**: 260
1573. Tanaka Y, Kamei K, Otoguro K, Omura S (1999) Heme-Dependent Radical Generation: Possible Involvement in Antimalarial Action of Non-Peroxide Microbial Metabolites, Nanaomycin A and Radicicol. J Antibiot **52**: 880
1574. Yang Z-Q, Geng X, Solit D, Pratilas CA, Rosen N, Danishefsky SJ (2004) New Efficient Synthesis of Resorcinylic Macrolides via Ynolides: Establishment of Cyclopropadicicol as Synthetically Feasible Preclinical Anticancer Agent Based on Hsp90 as the Target. J Am Chem Soc **126**: 7881
1575. Moulin E, Zoete V, Barluenga S, Karplus M, Winssinger N (2005) Design, Synthesis, and Biological Evaluation of HSP90 Inhibitors Based on Conformational Analysis of Radicicol and Its Analogues. J Am Chem Soc **127**: 6999
1576. Shen G, Wang M, Welch TR, Blagg BSJ (2006) Design, Synthesis, and Structure–Activity Relationships for Chimeric Inhibitors of Hsp90. J Am Chem Soc **71**: 7618
1577. Wicklow DT, Joshi BK, Gamble WR, Gloer JB, Dowd PF (1998) Antifungal Metabolites (Monorden, Monocillin IV, and Cerebrosides) from *Humicola fuscoatra* Traaen NRRL 22980, Mycoparasite of *Aspergillus flavus* Sclerotia. Appl Environ Microbiol **64**: 4482
1578. Hellwig V, Mayer-Bartschmid A, Müller H, Greif G, Kleymann G, Zitzmann W, Tichy H-V, Stadler M (2003) Pochonins A-F, New Antiviral and Antiparasitic Resorcylic Acid Lactones from *Pochonia chlamydosporia* var. *catenulata*. J Nat Prod **66**: 829
1579. Arai M, Yamamoto K, Namatame I, Tomoda H, Ōmura S (2003) New Monordens Produced by Amidepsine-Producing Fungus *Humicola* sp. FO-2942. J Antibiot **56**: 526
1580. Yamamoto K, Hatano H, Arai M, Shiomi K, Tomoda H, Ōmura S (2003) Structure Elucidation of New Monordens Produced by *Humicola* sp. FO-2942. J Antibiot **56**: 533
1581. Moulin E, Barluenga S, Totzke F, Winssinger N (2006) Diversity-Oriented Synthesis of Pochonins and Biological Evaluation Against a Panel of Kinases. Chem Eur J **12**: 8819
1582. Knust H, Hoffmann RW (2003) Synthesis and Conformational Analysis of Macrocyclic Dilactones Mimicking the Pharmacophore of Aplysiatoxin. Helv Chim Acta **86**: 1871
1583. Nagai H, Yasumoto T, Hokama Y (1997) Manauealides, Some of the Causative Agents of a Red Alga *Gracilaria coronopifolia* Poisoning in Hawaii. J Nat Prod **60**: 925
1584. Issa HH, Tanaka J, Rachmat R, Higa T (2003) Floresolides, New Metacyclophane Hydroquinone Lactones from an Ascidian, *Aplidium* sp. Tetrahedron Lett **44**: 1243
1585. Buchanan GO, Williams PG, Feling RH, Kauffman CA, Jensen PR, Fenical W (2005) Sporolides A and B: Structurally Unprecedented Halogenated Macrolides from the Marine Actinomycete *Salinispora tropica*. Org Lett **7**: 2731
1586. Yeung K-S, Paterson I (2005) Advances in the Total Synthesis of Biologically Important Marine Macrolides. Chem Rev **105**: 4237
1587. Pietruszka J (1998) Spongistatins, Cynachyrolides, or Altohyrtins? Marine Macrolides in Cancer Therapy. Angew Chem Int Ed **37**: 2629
1588. Ball M, Gaunt MJ, Hook DF, Jessiman AS, Kawahara S, Orsini P, Scolaro A, Talbot AC, Tanner HR, Yamanoi S, Ley SV (2005) Total Synthesis of Spongistatin 1: A Synthetic Strategy Exploiting Its Latent Pseudo-Symmetry. Angew Chem Int Ed **44**: 5433
1589. Arnone A, Nasini G, Cavalleri B (1987) Structure Elucidation of the Macrocyclic Antibiotic Lipiarmycin. J Chem Soc Perkin Trans 1: 1353

1590. Cavalleri B, Arnone A, Di Modugno E, Nasini G, Goldstein BP (1988) Structure and Biological Activity of Lipiarmycin B. J Antibiot **41**: 308
1591. Omura S, Imamura N, Oiwa R, Kuga H, Iwata R, Masuma R, Iwai Y (1986) Clostomicins, New Antibiotics Produced by *Micromonospora echinospora* subsp. *armeniaca* subsp. nov. I. Production, Isolation, and Physico-Chemical and Biological Properties. J Antibiot **39**: 1407
1592. Takahashi Y, Iwai Y, Omura S (1986) Clostomicins, New Antibiotics Produced by *Micromonospora echinospora* subsp. *armeniaca* subsp. nov. II. Taxonomic Study of the Producing Organism. J Antibiot **39**: 1413
1593. Theriault RJ, Karwowski JP, Jackson M, Girolami RL, Sunga GN, Vojtko CM, Coen LJ (1987) Tiacumicins, A Novel Complex of 18-Membered Macrolide Antibiotics I. Taxonomy, Fermentation and Antibacterial Activity. J Antibiot **40**: 567
1594. Hochlowski JE, Swanson SJ, Ranfranz LM, Whitten DN, Buko AM, McAlpine JB (1987) Tiacumicins, A Novel Complex of 18-Membered Macrolide Antibiotics II. Isolation and Structure Determination. J Antibiot **40**: 575
1595. Kong F, Liu DQ, Nietsche J, Tischler M, Carter GT (1999) Colubricidin A, a Novel Macrolide Antibiotic from a *Streptomyces* sp. Tetrahedron Lett **40**: 9219
1596. Williams DE, Roberge M, Van Soest R, Andersen RJ (2003) Spirastrellolide A, an Antimitotic Macrolide Isolated from the Caribbean Marine Sponge *Spirastrella coccinea*. J Am Chem Soc **125**: 5296
1597. Williams DE, Lapawa M, Feng X, Tarling T, Roberge M, Andersen RJ (2004) Spirastrellolide A: Revised Structure, Progress Toward the Relative Configuration, and Inhibition of Protein Phosphatase 2A. Org Lett **6**: 2607
1598. Sone H, Kigoshi H, Yamada K (1996) Aurisides A and B. Cytotoxic Macrolide Glycosides from the Japanese Sea Hare *Dolabella auricularia*. J Org Chem **61**: 8956
1599. Paterson I, Florence GJ, Heimann AC, Mackay AC (2005) Stereocontrolled Total Synthesis of (–)-Aurisides A and B. Angew Chem Int Ed **44**: 1130
1600. Zampella A, D'Auria MV, Minale L, Debitus C, Roussakis C (1996) Callipeltoside A: A Cytotoxic Aminodeoxy Sugar-Containing Macrolide of a New Type from the Marine Lithistida Sponge *Callipelta* sp. J Am Chem Soc **118**: 11085
1601. Zampella A, D'Auria MV, Minale L, Debitus C (1997) Callipeltosides B and C, Two Novel Cytotoxic Glycoside Macrolides from a Marine Lithistida Sponge *Callipelta* sp. Tetrahedron **53**: 3243
1602. Skepper CK, MacMillan JB, Zhou G-X, Masuno MN, Molinski TF (2007) Chlorocyclopropane Macrolides from the Marine Sponge *Phorbas* sp. Assignment of the Configurations of Phorbasides A and B by Quantitative CD. J Am Chem Soc **129**: 4150
1603. Klein D, Braekman JC, Daloze D, Hoffmann L, Demoulin V (1997) Lyngbyaloside, a Novel 2,3,4-Tri-*O*-methyl-6-deoxy-α-mannopyranoside Macrolide from *Lyngbya bouillonii* (Cyanobacteria). J Nat Prod **60**: 1057
1604. Luesch H, Yoshida WY, Harrigan GG, Doom JP, Moore RE, Paul VJ (2002) Lyngbyaloside B, a New Glycoside Macrolide from a Palauan Marine Cyanobacterium, *Lyngbya* sp. J Nat Prod **65**: 1945
1605. Searle PA, Molinski TF (1995) Phorboxazoles A and B: Potent Cytostatic Macrolides from Marine Sponge *Phorbas* Sp. J Am Chem Soc **117**: 8126
1606. Searle PA, Molinski TF, Brzezinski LJ, Leahy JW (1996) Absolute Configuration of Phorboxazoles A and B from the Marine Sponge *Phorbas* sp. 1. Macrolide and Hemiketal Rings. J Am Chem Soc **118**: 9422
1607. Molinski TF (1996) Absolute Configuration of Phorboxazoles A and B from the Marine Sponge, *Phorbas* sp. 2. C43 and Complete Stereochemistry. Tetrahedron Lett **37**: 7879
1608. Molinski TF, Brzezinski LJ, Leahy JW (2002) Absolute Configuration of Phorboxazole A C32-C43 Analogs by CD Exciton-Coupling of Allylic 2-Naphthoate Esters. Tetrahedron: Asym **13**: 1013
1609. Haustedt LO, Hartung IV, Hoffmann HMR (2003) The Total Synthesis of Phorboxazoles – New Classics in Natural Product Synthesis. Angew Chem Int Ed **42**: 2711

References

1610. Li D-R, Zhang D-H, Sun C-Y, Zhang J-W, Yang L, Chen J, Liu B, Su C, Zhou W-S, Lin G-Q (2006) Total Synthesis of Phorboxazole B. Chem Eur J **12**: 1185
1611. White JD, Kuntiyong P, Lee TH (2006) Total Synthesis of Phorboxazole A. 1. Preparation of Four Subunits. Org Lett **8**: 6039
1612. White JD, Lee TH, Kuntiyong P (2006) Total Synthesis of Phorboxazole A. 2. Assembly of Subunits and Completion of the Synthesis. Org Lett **8**: 6043
1613. Pettit GR, Xu J-P, Doubek DL, Chapuis J-C, Schmidt JM (2004) Antineoplastic Agents 510. Isolation and Structure of Dolastatin 19 from the Gulf of California Sea Hare *Dolabella auricularia*. J Nat Prod **67**: 1252
1614. Paterson I, Findlay AD, Florence GJ (2006) Total Synthesis and Stereochemical Reassignment of (+)-Dolastatin 19. Org Lett **8**: 2131
1615. Sandler JS, Colin PL, Kelly M, Fenical W (2006) Cytotoxic Macrolides from a New Species of the Deep-Water Marine Sponge *Leiodermatium*. J Org Chem **71**: 7245
1616. Lu C, Shen Y (2007) A Novel Ansamycin, Naphthomycin K from *Streptomyces* sp. J Antibiot **60**: 649
1617. Snipes CE, Duebelbeis DO, Olson M, Hahn DR, Dent III WH, Gilbert JR, Werk TL, Davis GE, Lee-Lu R, Graupner PR (2007) The Ansacarbamitocins: Polar Ansamitocin Derivatives. J Nat Prod **70**: 1578
1618. Williamson RT, Boulanger A, Vulpanovici A, Roberts MA, Gerwick WH (2002) Structure and Absolute Stereochemistry of Phormiodolide, a New Toxic Metabolite from the Marine Cyanobacterium *Phormidium* sp. J Org Chem **67**: 7927
1619. Rezanka T, Hanus LO, Dembitsky VM (2004) Lytophilippines A–C: Novel Macrolactones from the Red Sea Hydroid *Lytocarpus philippinus*. Tetrahedron **60**: 12191
1620. Rezanka T, Hanus L, Dembitsky VM (2003) Chagosensine, a New Chlorinated Macrolide from the Red Sea Sponge *Leucetta chagosensis*. Eur J Org Chem 4073
1621. Ueda K, Hu Y (1999) Haterumalide B: A New Cytotoxic Macrolide from an Okinawan Ascidian *Lissoclinum* sp. Tetrahedron Lett **40**: 6305
1622. Takada N, Sato H, Suenaga K, Arimoto H, Yamada K, Ueda K, Uemura D (1999) Isolation and Structures of Haterumalides NA, NB, NC, ND, and NE, Novel Macrolides from an Okinawan Sponge *Ircinia* sp. Tetrahedron Lett **40**: 6309
1623. Kigoshi H, Kita M, Ogawa S, Itoh M, Uemura D (2003) Enantioselective Synthesis of 15-*epi*-Haterumalide NA Methyl Ester and Revised Structure of Haterumalide NA. Org Lett **5**: 957
1624. Gu Y, Snider BB (2003) Synthesis of *ent*-Haterumalide NA (*ent*-Oocydin A) Methyl Ester. Org Lett **5**: 4385
1625. Strobel G, Li J-Y, Sugawara F, Koshino H, Harper J, Hess WM (1999) Oocydin A, a Chlorinated Macrocyclic Lactone with Potent Anti-Oomycete Activity from *Serratia marcescens*. Microbiology **145**: 3557
1626. Thaning C, Welch CJ, Borowicz JJ, Hedman R, Gerhardson B (2001) Suppression of *Sclerotinia sclerotiorum* Apothecial Formation by the Soil Bacterium *Serratia plymuthica*: Identification of a Chlorinated Macrolide as One of the Causal Agents. Soil Biol Biochem **33**: 1817
1627. Teruya T, Shimogawa H, Suenaga K, Kigoshi H (2004) Biselides A and B, Novel Macrolides from an Okinawan Ascidian Didemnidae sp. Chem Lett **33**: 1184
1628. Teruya T, Suenaga K, Maruyama S, Kurotaki M, Kigoshi H (2005) Biselides A–E: Novel Polyketides from an Okinawan Ascidian Didemnidae sp. Tetrahedron **61**: 6561
1629. Sato B, Nakajima H, Fujita T, Takase S, Yoshimura S, Kinoshita T, Terano H (2005) FR177391, a New Anti-Hyperlipidemic Agent from *Serratia* 1. Taxonomy, Fermentation, Isolation, Physico-Chemical Properties, Structure Elucidation and Biological Activities. J Antibiot **58**: 634
1630. Rezanka T, Dembitsky VM (2003) Ten-Membered Substituted Cyclic 2-Oxecanone (Decalactone) Derivatives from *Latrunculia corticata*, a Red Sea Sponge. Eur J Org Chem 2144
1631. Matsunaga S, Wakimoto T, Fusetani N (1997) Isolation of Four New Calyculins from the Marine Sponge *Discodermia calyx*. J Org Chem **62**: 2640

1632. Igarashi T, Satake M, Yasumoto T (1996) Prymnesin-2: A Potent Ichthyotoxic and Hemolytic Glycoside Isolated from the Red Tide Alga *Prymnesium parvum*. J Am Chem Soc **118**: 479
1633. Igarashi T, Satake M, Yasumoto T (1999) Structures and Partial Stereochemical Assignments for Prymnesin-1 and Prymnesin-2: Potent Hemolytic and Ichthyotoxic Glycosides Isolated from the Red Tide Alga *Prymnesium parvum*. J Am Chem Soc **121**: 8499
1634. Igarashi T, Aritake S, Yasumoto T (1998) Biological Activities of Prymnesin-2 Isolated from a Red Tide Alga *Prymnesium parvum*. Nat Toxins **6**: 35
1635. Fukuda DS, Mynderse JS, Baker PJ, Berry DM, Boeck LD, Yao RC, Mertz FP, Nakatsukasa WM, Mabe J, Ott J, Counter FT, Ensminger PW, Allen NE, Alborn Jr. WE, Hobbs Jr. JN (1990) A80915, a New Antibiotic Complex Produced by *Streptomyces aculeolatus*. Discovery, Taxonomy, Fermentation, Isolation, Characterization, and Antibacterial Evaluation. J Antibiot **43**: 623
1636. Soria-Mercado IE, Prieto-Davo A, Jensen PR, Fenical W (2005) Antibiotic Terpenoid Chloro-Dihydroquinones from a New Marine Actinomycete. J Nat Prod **68**: 904
1637. Soria-Mercado IE, Jensen PR, Fenical W, Kassel S, Golen J (2004) 3,4a-Dichloro-10a-(3-chloro-6-hydroxy-2,6-trimethylcyclohexylmethyl)-6,8-dihydroxy-2,2,7-trimethyl-3,4,4a,10a-tetrahydro-2*H*-benzo[*g*]chromene-5,10-dione. Acta Cryst E**60**: o1627
1638. Hardt IH, Jensen PR, Fenical W (2000) Neomarinone, and New Cytotoxic Marinone Derivatives, Produced by a Marine Filamentous Bacterium (Actinomycetales). Tetrahedron Lett **41**: 2073
1639. Cho JY, Kwon HC, Williams PG, Jensen PR, Fenical W (2006) Azamerone, a Terpenoid Phthalazinone from a Marine-Derived Bacterium Related to the Genus *Streptomyces* (Actinomycetales). Org Lett **8**: 2471
1640. Saxena G, Farmer SW, Hancock REW, Towers GHN (1996) Chlorochimaphilin: A New Antibiotic from *Moneses uniflora*. J Nat Prod **59**: 62
1641. Hasan AFMF, Begum S, Furumoto T, Fukui H (2000) A New Chlorinated Red Naphthoquinone from Roots of *Sesamum indicum*. Biosci Biotechnol Biochem **64**: 873
1642. He H, Yang HY, Luckman SW, Roll DM, Carter GT (2002) Chloroquinocin, a Novel Chlorinated Naphthoquinone Antibiotic from *Streptomyces* sp., *LL*-A9227. J Antibiot **55**: 1072
1643. Concepción GP, Foderaro TA, Eldredge GS, Lobkovsky E, Clardy J, Barrows LR, Ireland CM (1995) Topoisomerase II–Mediated DNA Cleavage by Adocia- and Xestoquinones from the Philippine Sponge *Xestospongia* sp. J Med Chem **38**: 4503
1644. Dey S, Mal D (2005) Total Synthesis of BE-23254, a Chlorinated Angucycline Antibiotic. Tetrahedron Lett **46**: 5483
1645. Ratnayake R, Lacey E, Tennant S, Gill JH, Capon RJ (2007) Kibdelones: Novel Anticancer Polyketides from a Rare Australian Actinomycete. Chem Eur J **13**: 1610
1646. Ratnayake R, Lacey E, Tennant S, Gill JH, Capon RJ (2006) Isokibdelones: Novel Heterocyclic Polyketides from a *Kibdelosporangium* sp. Org Lett **8**: 5267
1647. Krohn K, Michel A, Flörke U, Aust H-J, Draeger S, Schulz B (1994) Palmarumycins C_1-C_{16} from *Coniothyrium* sp.: Isolation, Structure Elucidation, and Biological Activity. Liebigs Ann Chem 1099
1648. McDonald LA, Abbanat DR, Barbieri LR, Bernan VS, Discafani CM, Greenstein M, Janota K, Korshalla JD, Lassota P, Tischler M, Carter GT (1999) Spiroxins, DNA Cleaving Antitumor Antibiotics from a Marine-Derived Fungus. Tetrahedron Lett **40**: 2489
1649. Wang T, Shirota O, Nakanishi K, Berova N, McDonald LA, Barbieri LR, Carter GT (2001) Absolute Stereochemistry of the Spiroxins. Can J Chem **79**: 1786
1650. Miyashita K, Sakai T, Imanishi T (2003) Total Synthesis of (±)-Spiroxin C. Org Lett **5**: 2683
1651. Dairi T, Nakano T, Aisaka K, Katsumata R, Hasegawa M (1995) Cloning and Nucleotide Sequence of the Gene Responsible for Chlorination of Tetracycline. Biosci Biotechnol Biochem **59**: 1099
1652. Schöler HF, Nkusi G, Niedan VW, Müller G, Spitthoff B (2005) Screening of Organic Halogens and Identification of Chlorinated Benzoic Acids in Carbonaceous Meteorites. Chemosphere **60**: 1505

1653. Niedan V, Schöler HF (1997) Natural Formation of Chlorobenzoic Acids (CBA) and Distinction Between PCB-Degraded CBA. Chemosphere **35**: 1233
1654. Wright AD, Papendorf O, König GM (2005) Ambigol C and 2,4-Dichlorobenzoic Acid, Natural Products Produced by the Terrestrial Cyanobacterium *Fischerella ambigua*. J Nat Prod **68**: 459
1655. Repeta DJ, Hartman NT, John S, Jones AD, Goericke R (2004) Structure Elucidation and Characterization of Polychlorinated Biphenyl Carboxylic Acids as Major Constituents of Chromophoric Dissolved Organic Matter in Seawater. Environ Sci Tech **38**: 5373
1656. Niedan VW, Keppler F, Ahlsdorf B, Schöler HF (2003) *De novo* Formation of Organochlorines in a Sewage Treatment Plant. Biogeochemistry **62**: 277
1657. Nagaoka T, Umezu K, Kouno K, Yoshida S, Ishiguro Y, Ando T (1996) Selective Inhibitors of Legume Seeds in Activated Sludge Compost. Plant Growth Regul **20**: 295
1658. Oh D-C, Williams PG, Kauffmann CA, Jensen PR, Fenical W (2006) Cyanosporasides A and B, Chloro- and Cyano-cyclopenta[*a*]indene Glycosides from the Marine Actinomycete *"Salinispora pacifica"*. Org Lett **8**: 1021
1659. Bui HTN, Jensen R, Pham HTL, Mundt S (2007) Carbamidocyclophanes A-E, Chlorinated Paracyclophanes with Cytotoxic and Antibiotic Activity from the Vietnamese Cyanobacterium *Nostoc* sp. J Nat Prod **70**: 499
1660. Johnsen S, Gribbestad IS, Johansen S (1989) Formation of Chlorinated PAH – A Possible Health Hazard from Water Chlorination. Sci Total Environ **81/82**: 231
1661. Meehan T, Wolfe AR, Negrete GR, Song Q (1997) Benzo[*a*]pyrene Diol Epoxide-DNA cis Adduct Formation Through a trans Chlorohydrin Intermediate. Proc Natl Acad Sci U S A **94**: 1749
1662. Lusby WR, Sonenshine DE, Yunker CE, Norval RA, Burridge MJ (1991) Comparison of Known and Suspected Pheromonal Constituents in Males of the African Ticks, *Amblyomma hebraeum* Koch and *Amblyomma variegatum* (Fabricius). Exp Appl Acarol **13**: 143
1663. Yunker CE, Peter T, Norval RAI, Sonenshine DE, Burridge MJ, Butler JF (1992) Olfactory Responses of Adult *Amblyomma hebraeum* and *A. variegatum* (Acari: Ixodidae) to Attractant Chemicals in Laboratory Tests. Exp Appl Acarol **13**: 295
1664. Norval RAI, Sonenshine DE, Allan SA, Burridge MJ (1996) Efficacy of Pheromone-Acaricide-Impregnated Tail-Tag Decoys for Controlling the Bont Tick, *Amblyomma hebraeum* (Acari: Ixodidae), on Cattle in Zimbabwe. Exp Appl Acarol **20**: 31
1665. Sonenshine DE, Taylor D, Corrigan G (1985) Studies to Evaluate the Effectiveness of Sex Pheromone-Impregnated Formulations for Control of Populations of the American Dog Tick, *Dermacentor variabilis* (Say) (Acari: Ixodidae). Exp Appl Acarol **1**: 23
1666. Haggart DA, David EE (1981) Neurons Sensitive to 2,6-Dichlorophenol on the Tarsi of the Tick *Amblyomma americanum* (Acari: Ixodidae). J Med Entomol **18**: 187
1667. Gavin J, Tabacchi R (1975) Isolation and Identification of Phenolic Compounds and Monoterpenes from Oakmoss (*Evernia prunastri* (L.) Ach.). Helv Chim Acta **58**: 190
1668. Qian FG, Xu GY, Du SJ, Li MH (1990) Isolation and Identification of Two New Pyrone Compounds from the Culture of *Hericium erinaceus*. Acta Pharm Sinica **25**: 522
1669. Okamoto K, Shimada A, Shirai R, Sakamoto H, Yoshida S, Ojima F, Ishiguro Y, Sakai T, Kawagishi H (1993) Antimicrobial Chlorinated Oricinol Derivatives from Mycelia of *Hericium erinaceum*. Phytochemistry **34**: 1445
1670. Pfefferle W, Anke H, Bross M, Steglich W (1990) Inhibition of Solubilized Chitin Synthase by Chlorinated Aromatic Compounds Isolated from Mushroom Cultures. Agric Biol Chem **54**: 1381
1671. Swarts HJ, Verhagen FJM, Field JA, Wijnberg JBPA (1998) Trichlorinated Phenols from *Hypholoma elongatum*. Phytochemistry **49**: 203
1672. Takahashi A, Agatsuma T, Ohta T, Nunozawa T, Endo T, Nozoe S (1993) Russuphelins B, C, D, E and F, New Cytotoxic Substances from the Mushroom *Russula subnigricans* HONGO. Chem Pharm Bull **41**: 1726
1673. Abraham BG, Berger RG (1994) Higher Fungi for Generating Aroma Components through Novel Biotechnologies. J Agric Food Chem **42**: 2344

1674. Huff T, Kuball HG, Anke T (1994) 7-Chloro-4,6-dimethoxy-1(3H)-isobenzofurane and Basidalin: Antibiotic Secondary Metabolites from *Leucoagaricus carneifolia* Gillet (Basidiomycetes). Z Naturforsch C **49**: 407
1675. Becker U, Anke T, Sterner O (1994) A Novel Bioactive Illudalane Sesquiterpene from the Fungus *Pholiota destruens*. Nat Prod Lett **5**: 171
1676. Wood WF, Lefevre CK (2007) Changing Volatile Compounds from Mycelium and Sporocarp of American Matsutake Mushroom, *Tricholoma magnivelare*. Biochem Syst Ecol **35**: 634
1677. Lauritsen FR, Lunding A (1998) A Study of the Bioconversion Potential of the Fungus *Bjerkandera adusta* with Respect to a Production of Chlorinated Aromatic Compounds. Enzyme Microbial Technol **22**: 459
1678. Swarts HJ, Verhagen FJM, Field JA, Wijnberg JBPA (1996) Novel Chlorometabolites Produced by *Bjerkandera* Species. Phytochemistry **42**: 1699
1679. Hjelm O, Borén H, Öberg G (1996) Analysis of Halogenated Organic Compounds in Coniferous Forest Soil from a *Lepista nuda* (Wood Blewitt) Fairy Ring. Chemosphere **32**: 1719
1680. Daferner M, Anke T, Hellwig V, Steglich W, Sterner O (1998) Strobilurin M, Tetrachloropyrocatechol and Tetrachloropyrocatechol Methyl Ether: New Antibiotics from a *Mycena* Species. J Antibiot **51**: 816
1681. Peters S, Spiteller P (2006) Chloro- and Bromophenols from Cultures of *Mycena alcalina*. J Nat Prod **69**: 1809
1682. Swarts HJ, Mester T, Verhagen FJM, Field JA, Wijnberg JBPA (1997) The Formation of Veratryl Chloride by *Bjerkandera* sp. Strain BOS55. Phytochemistry **46**: 1011
1683. Swarts HJ, Verhagen FJM, Field JA, Wijnberg JBPA (1998) Identification and Synthesis of Novel Chlorinated *p*-Anisylpropanoid Metabolites from *Bjerkandera* Species. J Nat Prod **61**: 1110
1684. Brambilla U, Nasini G, de Pava OV (1995) Secondary Mold Metabolites, Part 49. Isolation, Structural Elucidation, and Biomimetic Synthesis of Trametol, a New 1-Arylpropane-1,2-diol Produced by the Fungus *Trametes* sp. J Nat Prod **58**: 1251
1685. Davis RA, Watters D, Healy PC (2005) The Isolation and Synthesis of 3-Chloro-4-hydroxyphenylacetamide Produced by a Plant-Associated Microfungus of the Genus *Xylaria*. Tetrahedron Lett **46**: 919
1686. Swarts HJ, Teunissen PJM, Verhagen FJM, Field JA, Wijnberg JBPA (1997) Chlorinated Anisyl Metabolites Produced by Basidiomycetes. Mycol Res **101**: 372
1687. Silk PJ, Aubry C, Lonergan GC, Macaulay JB (2001) Chlorometabolite Production by the Ecologically Important White Rot Fungus *Bjerkandera adusta*. Chemosphere **44**: 1603
1688. Lang M, Spiteller P, Hellwig V, Steglich W (2001) Stephanosporin, a "Traceless" Precursor of 2-Chloro-4-nitrophenol in the Gasteromycete *Stephanospora caroticolor*. Angew Chem Int Ed **40**: 1704
1689. Baek S-H, Phipps RK, Perry NB (2004) Antimicrobial Chlorinated Bibenzyls from the Liverwort *Riccardia marginata*. J Nat Prod **67**: 718
1690. Monde K, Satoh H, Nakamura M, Tamura M, Takasugi M (1998) Organochlorine Compounds from a Terrestrial Higher Plant: Structures and Origin of Chlorinated Orcinol Derivatives from Diseased Bulbs of *Lilium maximowiczii*. J Nat Prod **61**: 913
1691. Choudhary MI, Azizuddin, Jalil S, Atta-ur-Rahman (2005) Bioactive Phenolic Compounds from a Medicinal Lichen, *Usnea longissima*. Phytochemistry **66**: 2346
1692. Marante FJT, Castellano AG, Rosas FE, Aguiar JQ, Barrera JB (2003) Identification and Quantitation of Allelochemicals from the Lichen *Lethariella canariensis*: Phytotoxicity and Antioxidative Activity. J Chem Ecol **29**: 2049
1693. Wunder A, Anke T, Klostermeyer D, Steglich W (1996) Lactarane Type Sesquiterpenoids as Inhibitors of Leukotriene Biosynthesis and Other, New Metabolites from Submerged Cultures of *Lentinellus cochleatus* (Pers. ex Fr.) Karst. Z Naturforsch **51c**: 493
1694. Passreiter CM, Matthiesen U, Willuhn G (1998) 10-Acetoxy-9-chloro-8,9-dehydrothymol and Further Thymol Derivatives from *Arnica sachalinensis*. Phytochemistry **49**: 777

1695. Takaichi S, Maoka T, Akimoto N, Sorokin D Yu, Banciu H, Kuenen JG (2004) Two Novel Yellow Pigments Natronochrome and Chloronatronochrome from the Natrono(alkali)philic Sulfur-Oxidizing Bacterium *Thialkalivibrio versutus* Strain ALJ 15. Tetrahedron Lett **45**: 8303
1696. Kuruüzüm A, Demirezer LÖ, Bergere I, Zeeck A (2001) Two New Chlorinated Naphthalene Glycosides from *Rumex patientia*. J Nat Prod **64**: 688
1697. Akdemir ZS, Tatli II, Saracoglu I, Ismailoglu UB, Sahin-Erdemli I, Calis I (2001) Polyphenolic Compounds from *Geranium pratense* and Their Free Radical Scavenging Activities. Phytochemistry **56**: 189
1698. Anjaneyulu ASR, Rao AM, Rao VK, Row LR, Pelter A, Ward RS (1975) The Isolation and Structure of 6"-Bromo-isoarboreol – The First Bromine Containing Lignan. Tetrahedron Lett **16**: 4697
1699. Keseru GM, Nogradi M (1995) The Chemistry of Macrocyclic Bis(bibenzyls). Nat Prod Rep **12**: 69
1700. Anton H, Kraut L, Mues R, Morales Z, MI (1997) Phenanthrenes and Bibenzyls from a *Plagiochila* Species. Phytochemistry **46**: 1069
1701. Martini U, Zapp J, Becker H (1998) Chlorinated Macrocyclic *Bis*bibenzyls from the Liverwort *Bazzania trilobata*. Phytochemistry **47**: 89
1702. Hashimoto T, Toyota M, Irita H, Asakawa Y (2000) Chemical Constituents of the Liverworts *Herbertus sakuraii* and *Herbertus aduncus*. J Hattori Bot Lab No 89 267
1703. Hashimoto T, Irita H, Takaoka S, Tanaka M, Asakawa Y (2000) New Chlorinated Cyclic Bis(bibenzyls) from the Liverworts *Herbertus sakuraii* and *Mastigophora diclados*. Tetrahedron **56**: 3153
1704. Wu, C-L, Liou C-S, Ean U-J (2001) Fusicoccane Diterpenoids and Bisbibenzyls from the Liverwort *Plagiochila peculiaris*. J Chin Chem Soc **48**: 1197
1705. Scher JM, Zapp J, Schmidt A, Becker H (2003) Bazzanins L-R, Chlorinated Macrocyclic Bisbibenzyls from the Liverwort *Lepidozia incurvata*. Phytochemistry **64**: 791
1706. Scher JM, Zapp J, Becker H, Kather N, Kolz J, Speicher A, Dreyer M, Maksimenka K, Bringmann G (2004) Optically Active Bisbibenzyls from *Bazzania trilobata*: Isolation and Stereochemical Analysis by Chromatographic, Chiroptical, and Computational Methods. Tetrahedron **60**: 9877
1707. Hertewich UM, Zapp J, Becker H (2003) Secondary Metabolites from the Liverwort *Jamesoniella colorata*. Phytochemistry **63**: 227
1708. Speicher A, Hollemeyer K, Heinzle E (2001) Rapid Detection of Chlorinated Bisbibenzyls in *Bazzania trilobata* using MALD1-TOF Mass Spectrometry. Phytochemistry **57**: 303
1709. Speicher A, Hollemeyer K, Heinzle E (2001) Rapid Detection of Multiple Chlorinated Bis (bibenzyls) in Bryophyte Crude Extracts Using Laser Desorption/Ionization Time-of-Flight Mass Spectrometry. Rapid Commun Mass Spectrom **15**: 124
1710. Speicher A, Heisel R, Kolz J (2003) First Detection of a Chloroperoxidase in Bryophytes. Phytochemistry **62**: 679
1711. Eklind Y, Hjelm O, Kothéus M, Kirchmann H (2004) Formation of Chloromethoxybenzaldehyde During Composting of Organic Household Waste. Chemosphere **56**: 475
1712. Hoekstra EJ, De Weerd H, De Leer EWB, Brinkman UATh (1999) Natural Formation of Chlorinated Phenols, Dibenzo-*p*-dioxins, and Dibenzofurans in Soil of a Douglas Fir Forest. Environ Sci Technol **33**: 2543
1713. Hjelm O, Johansson E, Öberg G (1999) Production of Organically Bound Halogens by the Litter-Degrading Fungus *Lepista nuda*. Soil Biol Biochem **31**: 1509
1714. Glombitza K-W, Sukopp I, Wiedenfeld H (1985) Antibiotics from Algae XXXVII. Rhodomelol and Methylrhodomelol from *Polysiphonia lanosa*. Planta Med 437
1715. Aknin M, Samb A, Mirailles J, Costantino V, Fattorusso E, Mangoni A (1992) Polysiphenol, a New Brominated 9,10-Dihydrophenanthrene from the Senegalese Red Alga *Polysiphonia ferulacea*. Tetrahedron Lett **33**: 555
1716. Woodin SA, Walla MC, Lincoln DE (1987) Occurrence of Brominated Compounds in Soft-Bottom Benthic Organisms. J Exp Mar Biol Ecol **107**: 209

1717. Stewart CC, Pinckney J, Piceno Y, Lovell CR (1992) Bacterial Numbers and Activity, Microalgal Biomass and Productivity, and Meiofaunal Distribution in Sediments Naturally Contaminated with Biogenic Bromophenols. Mar Ecol Prog Ser **90**: 61
1718. Fu X, Schmitz FJ, Govindan M, Abbas SA, Hanson KM, Horton PA, Crews P, Laney M, Schatzman RC (1995) Enzyme Inhibitors: New and Known Polybrominated Phenols and Diphenyl Ethers from Four Indo-Pacific *Dysidea* Sponges. J Nat Prod **58**: 1384
1719. Gulavita NK, Pomponi SA, Wright AE, Garay M, Sills MA (1995) Aplysillin A, a Thombin Receptor Antagonist from the Marine Sponge *Aplysina fistularis fulva*. J Nat Prod **58**: 954
1720. Utkina NK, Fedoreyev SA, Ilyin SG, Antipin MYu (1998) 3,5-Dibromo-2-methoxybenzoic Acid from Sea Sponge *Didiscus* sp. Russ Chem Bull **47**: 2292
1721. Kang H, Fenical W (1997) Aplidiamine, a Unique Zwitterionic Benzyl Hydroxyadenine from the Western Australian Marine Ascidian *Aplidiopsis* sp. Tetrahedron Lett **38**: 941
1722. Itaya T, Hozumi Y, Kanai T, Ohta T (1998) Syntheses of the Marine Ascidian Purine Aplidiamine and Its 9-β-D-Ribofuranoside. Tetrahedron Lett **39**: 4695
1723. Itaya T, Hozumi Y, Kanai T, Ohta T (1999) Synthesis and Structure of the Marine Ascidian 8-Oxoadenine Aplidiamine. Chem Pharm Bull **47**: 1297
1724. Flodin C, Whitfield FB (1999) 4-Hydroxybenzoic Acid: A Likely Precursor of 2,4,6-Tribromophenol in *Ulva lactuca*. Phytochemistry **51**: 249
1725. Venkateswarlu Y, Chavakula R (1995) Brominated Benzeneacetonitriles, the Dibromotyrosine Metabolites from the Sponge *Psammaplysilla purpurea*. J Nat Prod **58**: 1087
1726. Jiang Z, Boyd KG, Mearns-Spragg A, Adams DR, Wright PC, Burgess JG (2000) Two Diketopiperazines and One Halogenated Phenol from Cultures of the Marine Bacterium, *Pseudoalteromonas luteoviolacea*. Nat Prod Lett **14**: 435
1727. Flodin C, Whitfield FB (2000) Brominated Anisoles and Cresols in the Red Alga *Polysiphonia sphaerocarpa*. Phytochemistry **53**: 77
1728. Führer U, Ballschmiter K (1998) Bromochloromethoxybenzenes in the Marine Troposphere of the Atlantic Ocean: A Group of Organohalogens with Mixed Biogenic and Anthropogenic Origin. Environ Sci Technol **32**: 2208
1729. Chatonnet P, Bonnet S, Boutou S, Labadie M-D (2004) Identification and Responsibility of 2,4,6-Tribromoanisole in Musty, Corked Odors in Wine. J Agric Food Chem **52**: 1255
1730. Boyle JL, Lindsay RC, Stuiber DA (1993) Occurrence and Properties of Flavor-Related Bromophenols Found in the Marine Environment: A Review. J Aquat Food Prod Technol **2**: 75
1731. Choi JS, Park HJ, Jung HA, Chung HY, Jung JH, Choi WC (2000) A Cyclohexanonyl Bromophenol from the Red Alga *Symphyocladia latiuscula*. J Nat Prod **63**: 1705
1732. Son BW, Choi JS, Kim JC, Nam KW, Kim D-S, Chung HY, Kang JS, Choi HD (2002) Parasitenone, a New Epoxycyclohexenone Related to Gabosine from the Marine-Derived Fungus *Aspergillus parasiticus*. J Nat Prod **65**: 794
1733. Fan X, Xu N-J, Shi J-G (2003) Bromophenols from the Red Alga *Rhodomela confervoides*. J Nat Prod **66**: 455
1734. Fan X, Xu NJ, Shi JG (2003) A New Brominated Phenylpropylaldehyde and Its Dimethyl Acetal from Red Alga *Rhodomela confervoides*. Chin Chem Lett **14**: 1045
1735. Zhao J, Fan X, Wang S, Li S, Shang S, Yang Y, Xu N, Lü Y, Shi J (2004) Bromophenol Derivatives from the Red Alga *Rhodomela confervoides*. J Nat Prod **67**: 1032
1736. Xu XL, Fan X, Song FH, Zhao JL, Han LJ, Shi JG (2004) A New Bromophenol from the Brown Alga *Leathesia nana*. Chin Chem Lett **15**: 661
1737. Xu X-L, Fan X, Song F-H, Zhao J-L, Han L-J, Yang Y-C, Shi J-G (2004) Bromophenols from the Brown Alga *Leathesia nana*. J Asian Nat Prod Res **6**: 217
1738. Xu X, Song F, Wang S, Li S, Xiao F, Zhao J, Yang Y, Shang S, Yang L, Shi J (2004) Dibenzyl Bromophenols with Diverse Dimerization Patterns from the Brown Alga *Leathesia nana*. J Nat Prod **67**: 1661
1739. Ma M, Zhao J, Wang S, Li S, Yang Y, Shi J, Fan X, He L (2006) Bromophenols Coupled with Methyl γ-Ureidobutyrate and Bromophenol Sulfates from the Red Alga *Rhodomela confervoides*. J Nat Prod **69**: 206

1740. De Carvalho LR, Guimarâes SMP De B, Roque NF (2006) Sulfated Bromophenols from *Osmundaria obtusiloba* (C. Agardh) R. E. Norris (Rhodophyta, Ceramiales). Rev Brasil Bot **29**: 453
1741. Kanakubo A, Isobe M (2005) Isolation of Brominated Quinones Showing Chemiluminescence Activity from Luminous Acorn Worm, *Ptychodera flava*. Bioorg Med Chem **13**: 2741
1742. Wang W, Okada Y, Shi H, Wang Y, Okuyama T (2005) Structures and Aldose Reductase Inhibitory Effects of Bromophenols from the Red Alga *Symphyocladia latiuscula*. J Nat Prod **68**: 620
1743. Liu Q-W, Tan C-H, Zhang T, Zhang S-J, Han L-J, Fan X, Zhu D-Y (2006) Urceolatol, a Tetracyclic Bromobenzaldehyde Dimer from *Polysiphonia urceolata*. J Asian Nat Prod Res **8**: 379
1744. Zhao J, Ma M, Wang S, Li S, Cao P, Yang Y, Lii Y, Shi J, Xu N, Fan X, He L (2005) Bromophenols Coupled with Derivatives of Amino Acids and Nucleosides from the Red Alga *Rhodomela confervoides*. J Nat Prod **68**: 691
1745. Sata N, Galario MA, Sitachitta N, Scheuer PJ, Kelly M (2005) Poipuol, a New Metabolite from a Hawaiian Sponge of Genus *Hyrtios*. J Nat Prod **68**: 262
1746. Teuten EL, King GM, Reddy CM (2006) Natural ^{14}C in *Saccoglossus bromophenolosus* Compared to ^{14}C in Surrounding Sediments. Mar Ecol Prog Ser **324**: 167
1747. Boyle JL, Lindsay RC, Stuiber DA (1992) Bromophenol Distribution in Salmon and Selected Seafoods of Fresh- and Saltwater Origin. J Food Sci **57**: 918
1748. Whitfield FB, Helidoniotis F, Shaw KJ, Svoronos D (1997) Distribution of Bromophenols in Australian Wild-Harvested and Cultivated Prawns (Shrimp). J Agric Food Chem **45**: 4398
1749. Whitfield FB, Helidoniotis F, Shaw KJ, Svoronos D (1998) Distribution of Bromophenols in Species of Ocean Fish from Eastern Australia. J Agric Food Chem **46**: 3750
1750. Whitfield FB, Drew M, Helidoniotis F, Svoronos D (1999) Distribution of Bromophenols in Species of Marine Polychaetes and Bryozoans from Eastern Australia and the Role of Such Animals in the Flavor of Edible Ocean Fish and Prawns (Shrimp). J Agric Food Chem **47**: 4756
1751. Chung HY, Ma WCJ, Kim J-S (2003) Seasonal Distribution of Bromophenols in Selected Hong Kong Seafood. J Agric Food Chem **51**: 6752
1752. Whitfield FB, Helidoniotis F, Shaw KJ, Svoronos D (1999) Distribution of Bromophenols in Species of Marine Algae from Eastern Australia. J Agric Food Chem **47**: 2367
1753. Chung HY, Ma WCJ, Ang Jr. PO, Kim J-S, Chen F (2003) Seasonal Variations of Bromophenols in Brown Algae (*Padina arborescens*, *Sargassum siliquastrum* and *Lobophora variegata*) Collected in Hong Kong. J Agric Food Chem **51**: 2619
1754. Gribble GW, Leese RM, Evans BE (1977) Reactions of Sodium Borohydride in Acidic Media; IV. Reduction of Diarylmethanols and Triarylmethanols in Trifluoroacetic Acid. Synthesis 172
1755. Kurata K, Taniguchii K, Takashima K, Hayashi I, Suzuki M (1997) Feeding-Deterrent Bromophenols from *Odonthalia corymbifera*. Phytochemistry **45**: 485
1756. Isnansetyo A, Kamei Y (2003) MC21-A, a Bactericidal Antibiotic Produced by a New Marine Bacterium, *Pseudoalteromonas phenolica* sp. nov. O-BC30^T, Against Methicillin-Resistant *Staphylococcus aureus*. Antimicrob Agents Chemother **47**: 480
1757. Marsh G, Athanasiadou M, Bergman Å, Athanassiadis I, Endo T, Haraguchi K (2004) Identification of a Novel Dimethoxylated Polybrominated Biphenyl Bioaccumulating in Marine Animals. Organohalogen Cpds **66**: 3823
1758. Marsh G, Athanasiadou M, Athanassiadis I, Bergman Å, Endo T, Haraguchi K (2005) Identification, Quantification, and Synthesis of a Novel Dimethoxylated Polybrominated Biphenyl in Marine Mammals Caught off the Coast of Japan. Environ Sci Technol **39**: 8684
1759. Laurenti E, Ghibaudi E, Ardissone S, Ferrari RP (2003) Oxidation of 2,4-Dichlorophenol Catalyzed by Horseradish Peroxidase: Characterization of the Reaction Mechanism by UV-Visible Spectroscopy and Mass Spectrometry. J Inorg Biochem **95**: 171
1760. Endo M, Nakagawa M, Hamamoto Y, Ishihama M (1986) Pharmacologically Active Substances from Southern Pacific Marine Invertebrates. Pure Appl Chem **58**: 387

1761. Klein DM (1986) Secondary Metabolites of Several Marine Invertebrates. PhD Thesis, University of Hawaii, p 165
1762. Fu X, Schmitz FJ (1996) New Brominated Diphenyl Ether from an Unidentified Species of *Dysidea* Sponge. ^{13}C NMR Data for Some Brominated Diphenyl Ethers. J Nat Prod **59**: 1102
1763. Anjaneyulu V, Nageswara Rao K, Radhika P, Muralikrishna M, Connolly JD (1996) A New Tetrabromodiphenyl Ether from the Sponge *Dysidea herbacea* of the Indian Ocean. Indian J Chem **35B**: 89
1764. Handayani D, Edrada RA, Proksch P, Wray V, Witte L, Van Soest RWM, Kunzmann A, Soedarsono (1997) Four New Bioactive Polybrominated Diphenyl Ethers of the Sponge *Dysidea herbacea* from West Sumatra, Indonesia. J Nat Prod **60**: 1313
1765. Kurihara H, Mitani T, Kawabata J, Takahashi K (1999) Two New Bromophenols from the Red Alga *Odonthalia corymbifera*. J Nat Prod **62**: 882
1766. Bowden BF, Towerzey L, Junk PC (2000) A New Brominated Diphenyl Ether from the Marine Sponge *Dysidea herbacea*. Aust J Chem **53**: 299
1767. Hattori T, Konno A, Adachi K, Shizuri Y (2001) Four New Bioactive Bromophenols from the Palauan Sponge *Phyllospongia dendyi*. Fish Sci **67**: 899
1768. Liu H, Namikoshi M, Meguro S, Nagai H, Kobayashi H, Yao X (2004) Isolation and Characterization of Polybrominated Diphenyl Ethers as Inhibitors of Microtubule Assembly from the Marine Sponge *Phyllospongia dendyi* Collected at Palau. J Nat Prod **67**: 472
1769. Kitamura M, Koyama T, Nakano Y, Uemura D (2005) Corallinafuran and Corallinaether, Novel Toxic Compounds from Crustose Coralline Red Algae. Chem Lett **34**: 1272
1770. Agrawal MS, Bowden BF (2005) Marine Sponge *Dysidea herbacea* Revisited: Another Brominated Diphenyl Ether. Mar Drugs **3**: 9
1771. Hanif N, Tanaka J, Setiawan A, Trianto A, de Voogd NJ, Murni A, Tanaka C, Higa T (2007) Polybrominated Diphenyl Ethers from the Indonesian Sponge *Lamellodysidea herbacea*. J Nat Prod **70**: 432
1772. Segraves EN, Shah RR, Segraves NL, Johnson TA, Whitman S, Sui JK, Kenyon VA, Cichewicz RH, Crews P, Holman TR (2004) Probing the Activity Differences of Simple and Complex Brominated Aryl Compounds Against 15-Soybean, 15-Human, and 12-Human Lipoxygenase. J Med Chem **47**: 4060
1773. Malmvärn A, Marsh G, Kautsky L, Athanasiadou M, Bergman A, Asplund L (2005) Hydroxylated and Methoxylated Brominated Diphenyl Ethers in the Red Algae *Ceramium tenuicorne* and Blue Mussels from the Baltic Sea. Environ Sci Technol **39**: 2990
1774. de Wit CA (2002) An Overview of Brominated Flame Retardants in the Environment. Chemosphere **46**: 583
1775. Hites RA (2004) Polybrominated Diphenyl Ethers in the Environment and in People: A Meta-Analysis of Concentrations. Environ Sci Technol **38**: 945
1776. Zhu LY, Hites RA (2005) Brominated Flame Retardants in Sediment Cores from Lakes Michigan and Erie. Environ Sci Technol **39**: 3488
1777. Fu X, Hossian MB, Schmitz FJ, van der Helm D (1997) Longithorones, Unique Prenylated Para- and Metacyclophane Type Quinones from the Tunicate *Aplidium longithorax*. J Org Chem **62**: 3810
1778. Carté B, Kernan MR, Barrabee EB, Faulkner DJ, Matsumoto GK, Clardy J (1986) Metabolites of the Nudibranch *Chromodoris tunerea* and the Singlet Oxygen Oxidation Products of Furodysin and Furodysinin. J Org Chem **51**: 3528
1779. Asplund L, Athanasiadou M, Sjödin A, Bergman A, Börjeson H (1999) Organohalogen Substances in Muscle, Egg and Blood from Healthy Baltic Salmon (*Salmo salar*) and Baltic Salmon that Produced Offspring with the M74 Syndrome. Ambio **28**: 67
1780. Marsh G, Athanasiadou M, Bergman A, Asplund L (2004) Identification of Hydroxylated and Methoxylated Polybrominated Diphenyl Ethers in Baltic Sea Salmon (*Salmo salar*) Blood. Environ Sci Technol **38**: 10
1781. Kierkegaard A, Bignert A, Sellström U, Olsson M, Asplund L, Jansson B, de Wit CA (2004) Polybrominated Diphenyl Ethers (PBDEs) and Their Methoxylated Derivatives in Pike

from Swedish Waters with Emphasis on Temporal Trends, 1967–2000. Environ Pollut **130**: 187
1782. Tittlemier SA, Forsyth D, Breakell K, Verigin V, Ryan JJ, Hayward S (2004) Polybrominated Diphenyl Ethers in Retail Fish and Shellfish Samples Purchased from Canadian Markets. J Agric Food Chem **52**: 7740
1783. Sinkkonen S, Rantalainen A-L, Paasivirta J, Lahtiperä M (2004) Polybrominated Methoxy Diphenyl Ethers (MeO-PBDEs) in Fish and Guillemot of Baltic, Atlantic and Arctic Environments. Chemosphere **56**: 767
1784. Vetter W, Stoll E, Garson MJ, Fahey SJ, Gaus C, Müller JF (2002) Sponge Halogenated Natural Products Found at Parts-Per-Million Levels in Marine Mammals. Environ Toxicol Chem **21**: 2014
1785. Pettersson A, van Bavel B, Engwall M, Jimenez B (2004) Polybrominated Diphenylethers and Methoxylated Tetrabromodiphenylethers in Cetaceans from the Mediterranean Sea. Arch Environ Contam Toxicol **47**: 542
1786. Melcher J, Olbrich D, Marsh G, Nikiforov V, Gaus C, Gaul S, Vetter W (2005) Tetra- and Tribromophenoxyanisoles in Marine Samples from Oceania. Environ Sci Technol **39**: 7784
1787. Teuten EL, Johnson CG, Mandalakis M, Asplund L, Gustafsson Ö, Unger M, Marsh G, Reddy CM (2006) Spectral Characterization of Two Bioaccumulated Methoxylated Polybrominated Diphenyl Ethers. Chemosphere **62**: 197
1788. Vetter W, Jun W (2003) Non-Polar Halogenated Natural Products Bioaccumulated in Marine Samples. II. Brominated and Mixed Halogenated Compounds. Chemosphere **52**: 423
1789. Teuten EM, Xu L, Reddy CM (2005) Two Abundant Bioaccumulated Halogenated Compounds are Natural Products. Science **307**: 917
1790. Marsh G, Hu J, Jakobsson E, Rahm S, Bergman A (1999) Synthesis and Characterization of 32 Polybrominated Diphenyl Ethers. Environ Sci Technol **33**: 3033
1791. Marsh G, Stenutz R, Bergman A (2003) Synthesis of Hydroxylated and Methoxylated Polybrominated Diphenyl Ethers – Natural Products and Potential Polybrominated Diphenyl Ether Metabolites. Eur J Org Chem 2566
1792. Liu H, Bernhardsen M, Fiksdahl A (2006) Polybrominated Diphenyl Ethers (BDEs); Preparation of Reference Standards and Fluorinated Internal Analytical Standards. Tetrahedron **62**: 3564
1793. Goerke H, Emrich R, Weber K, Duchene J-C (1991) Concentrations and Localization of Brominated Metabolites in the Genus *Thelepus* (Polychaeta: Terebellidae). Comp Biochem Physiol **99B**: 203
1794. Ogawa T, Ando K, Aotani Y, Shinoda K, Tanaka T, Tsukuda E, Yoshida M, Matsuda Y (1995) RES-1214-1 and -2, Novel Non-Peptidic Endothelin Type A Receptor Antagonists Produced by *Pestalotiopsis* sp. J Antibiot **48**: 1401
1795. Adeboya MO, Edwards RL, Lassøe T, Maitland DJ, Shields L, Whalley AJS (1996) Metabolites of the Higher Fungi. Part 29. Maldoxin, Maldoxone, Dihydromaldoxin, Isodihydromaldoxin and Dechlorodihydromaldoxin. A Spirocyclohexadienone, a Depsidone and Three Diphenyl Ethers: Keys in the Depsidone Biosynthetic Pathway from a Member of the Fungus Genus *Xylaria*. J Chem Soc Perkin Trans 1: 1419
1796. Hargreaves J, Park J, Ghisalberti EL, Swasithamparam K, Skelton BW, White AH (2002) New Chlorinated Diphenyl Ethers from an *Aspergillus* Species. J Nat Prod **65**: 7
1797. Lee HJ, Lee JH, Hwang BY, Kim HS, Lee JJ (2002) Fungal Metabolites, Asterric Acid Derivatives Inhibit Vascular Endothelial Growth Factor (VEGF)-Induced Tube Formation of HUVECs. J Antibiot **55**: 552
1798. Katano T, Goto K, Murakami E, Yamazaki R, Venoyama T, Sugimoto T, Kawashima Y (1985) Production of Phosphodiesterase Inhibitors. Jpn Kokai Tokkyo Koho JP 60188084 Chem Abst **104**: 49846
1799. Rocha JHC, Cardoso MP, David JP, David JM (2006) A Novel Diphenyl Ether from *Byrsonima microphylla* (Malpighiaceae). Biosci Biotechnol Biochem **70**: 2759

1800. Takahashi A, Agatsuma T, Matsuda M, Ohta T, Nunozawa T, Endo T, Nozoe S (1992) Russuphelin A, a New Cytotoxic Substance from the Mushroom *Russula subnigricans* HONGO. Chem Pharm Bull **40**: 3185
1801. Ohta T, Takahashi A, Matsuda M, Kamo S, Agatsuma T, Endo T, Nozoe S (1995) Russuphelol, a Novel Optically Active Chlorohydroquinone Tetramer from the Mushroom *Russula subnigricans*. Tetrahedron Lett **36**: 5223
1802. Glombitza K-W, Keusgen M, Hauperich S (1997) Fucophlorethols from the Brown Algae *Sargassum spinuligerum* and *Cystophora torulosa*. Phytochemistry **46**: 1417
1803. Sailler B, Glombitza K-W (1999) Halogenated Phlorethols and Fucophlorethols from the Brown Alga *Cystophora retroflexa*. Nat Toxins **7**: 57
1804. Singh IP, Bharate SB (2006) Phloroglucinol Compounds of Natural Origin. Nat Prod Rep **23**: 558
1805. Kettle AJ (1996) Neutrophils Convert Tyrosyl Residues in Albumin to Chlorotyrosine. FEBS Letters **379**: 103
1806. Hazen SL, Hsu FF, Mueller DM, Crowley JR, Heinecke JW (1996) Human Neutrophils Employ Chlorine Gas as an Oxidant During Phagocytosis. J Clin Invest **98**: 1283
1807. Hazen SL, Heinecke JW (1997) 3-Chlorotyrosine, a Specific Marker of Myeloperoxidase-Catalyzed Oxidation, Is Markedly Elevated in Low Density Lipoprotein Isolated from Human Atherosclerotic Intima. J Clin Invest **99**: 2075
1808. Himmelfarb J, McMenamin ME, Loseto G, Heinecke JW (2001) Myeloperoxidase-Catalyzed 3-Chlorotyrosine Formation in Dialysis Patients. Free Radic Biol Med **31**: 1163
1809. van der Vliet A, Nguyen MN, Shigenaga MK, Eiserich JP, Marelich GP, Cross CE (2000) Myeloperoxidase and Protein Oxidation in Cystic Fibrosis. Am J Physiol Lung Cell Mol Physiol **279**: L537
1810. Brennan M-L, Anderson MM, Shih DM, Qu X-D, Wang X, Mehta AC, Lim LL, Shi W, Hazen SL, Jacob JS, Crowley JR, Heinecke JW, Lusis AJ (2001) Increased Atherosclerosis in Myeloperoxidase-Deficient Mice. J Clin Invest **107**: 419
1811. Buss IH, Senthilmohan R, Darlow BA, Mogridge N, Kettle AJ, Winterbourn CC (2003) 3-Chlorotyrosine as a Marker of Protein Damage by Myeloperoxidase in Tracheal Aspirates from Preterm Infants: Association with Adverse Respiratory Outcome. Pediatr Res **53**: 455
1812. Wu W, Chen Y, d'Avignon A, Hazen SL (1999) 3-Bromotyrosine and 3,5-Dibromotyrosine Are Major Products of Protein Oxidation by Eosinophil Peroxidase: Potential Markers for Eosinophil-Dependent Tissue Injury *in Vivo*. Biochemistry **38**: 3538
1813. Wu W, Samoszuk MK, Comhair SAA, Thomassen MJ, Farver CF, Dweik RA, Kavuru MS, Erzurum SC, Hazen SL (2000) Eosinophils Generate Brominating Oxidants in Allergen-Induced Asthma. J Clin Invest **105**: 1455
1814. Blackman AJ, Eldershaw TPD, Garland SM (1993) Alkaloids from Two Further *Amathia* Bryozoan Species. Aust J Chem **46**: 401
1815. Zhang H, Kamano Y, Kizu H, Itokawa H, Pettit GR, Herald CL (1994) Convolutamines A-E, Novel β-Phenylethylamine Alkaloids from Marine Bryozoan *Amathia convoluta*. Chem Lett 2271
1816. Montanari AM, Fenical W, Lindquist N, Lee AY, Clardy J (1996) Volutamides A-E, Halogenated Alkaloids with Antifeedant Properties from the Atlantic Bryozoan *Amathia convoluta*. Tetrahedron **52**: 5371
1817. Morris BD, Prinsep MR (1999) Amathaspiramides A-F, Novel Brominated Alkaloids from the Marine Bryozoan *Amathia wilsoni*. J Nat Prod **62**: 688
1818. Glombitza K-W, Schmidt A (1999) Nonhalogenated and Halogenated Phlorotannins from the Brown Alga *Carpophyllum angustifolium*. J Nat Prod **62**: 1238
1819. Carroll AR, Bowden BJ, Coll JC (1993) Studies of Australian Ascidians. II Novel Cytotoxic Iodotyrosine-Based Alkaloids from Colonial Ascidians, *Aplidium* sp. Aust J Chem **46**: 825

1820. McDonald LA, Swersey JC, Ireland CM, Carroll AR, Coll JC, Bowden BF, Fairchild CR, Cornell L (1995) Botryllamides A-D, Two Brominated Tyrosine Derivatives from Styelid Ascidians of the Genus *Botryllus*. Tetrahedron **51**: 5237
1821. Rao MR, Faulkner DJ (2004) Botryllamides E-H, Four New Tyrosine Derivatives from the Ascidian *Botrylloides tyreum*. J Nat Prod **67**: 1064
1822. Lindsay BS, Battershill CN, Copp BR (1998) Isolation of 2-(3'-Bromo-4'-hydroxyphenol) ethanamine from the New Zealand Ascidian *Cnemidocarpa bicornuta*. J Nat Prod **61**: 857
1823. Van Wagoner RM, Jompa J, Tahir A, Ireland CM (1999) Trypargine Alkaloids from a Previously Undescribed *Eudistoma* sp. Ascidian. J Nat Prod **62**: 794
1824. Kigoshi H, Kanematsu K, Uemura D (1999) Turbotoxins A and B, Novel Diiodotyramine Derivatives from the Japanese Gastropod *Turbo marmorata*. Tetrahedron Lett **40**: 5745
1825. Kigoshi H, Kanematsu K, Yokota K, Uemura D (2000) Turbotoxins A and B, Novel Diiodotyramine Derivatives from the Japanese Gastropod *Turbo marmorata*. Tetrahedron **56**: 9063
1826. Kotaki Y, Yasumoto T (1977) Toxicity Studies on Marine Snails. VIII. Identification of Iodomethyltrimethylammonium as a Minor Toxin of the Green Turban Shell. Bull Japan Soc Sci Fisheries **43**: 1467
1827. Chantraine J-M, Combaut G, Teste J (1973) Bromophenols of a Red Algae, *Halopytis incurvus*. Carboxylic Acids. Phytochemistry **12**: 1793
1828. Ciminiello P, Fattorusso E, Magno S, Pansini M (1995) Chemistry of Verongida Sponges, IV. Comparison of the Secondary Metabolite Composition of Several Specimens of *Pseudoceratina crassa*. J Nat Prod **58**: 689
1829. Benharref A, Païs M, Debitus C (1996) Bromotyrosine Alkaloids from the Sponge *Pseudoceratina verrucosa*. J Nat Prod **59**: 177
1830. Thirionet I, Daloze D, Braekman JC, Willemsen P (1998) 5-Bromoverongamine, A Novel Antifouling Tyrosine Alkaloid from the Sponge *Pseudoceratina* sp. Nat Prod Lett **12**: 209
1831. Tsukamoto S, Kato H, Hirota H, Fusetani N (1996) Ceratinamine: An Unprecedented Antifouling Cyanoformamide from the Marine Sponge *Pseudoceratina purpurea*. J Org Chem **61**: 2936
1832. Fusetani N, Masuda Y, Nakao Y, Matsunaga S, van Soest RWM (2001) Three New Bromotyrosine Derivatives Lethal to Crab from the Marine Sponge, *Pseudoceratina purpurea*. Tetrahedron **57**: 7507
1833. Piña IC, Gautschi JT, Wang G-Y-S, Sanders ML, Schmitz FJ, France D, Cornell-Kennon S, Sambucetti LC, Remiszewski SW, Perez LB, Bair KW, Crews P (2003) Psammaplins from the Sponge *Pseudoceratina purpurea*: Inhibition of Both Histone Deacetylase and DNA Methyltransferase. J Org Chem **68**: 3866
1834. Jang J-H, van Soest RWM, Fusetani N, Matsunaga S (2007) Pseudoceratins A and B, Antifungal Bicyclic Bromotyrosine-Derived Metabolites from the Marine Sponge *Pseudoceratina purpurea*. J Org Chem **72**: 1211
1835. Kobayashi J, Honma K, Sasaki T, Tsuda M (1995) Purealidins J-R, New Bromotyrosine Alkaloids from the Okinawan Marine Sponge *Psammaplysilla purea*. Chem Pharm Bull **43**: 403
1836. Kobayashi J, Honma K, Tsuda M, Kosaka T (1995) Lipopurealins D and E and Purealidin H, New Bromotyrosine Alkaloids from the Okinawan Marine Sponge *Psammaplysilla purea*. J Nat Prod **58**: 467
1837. Venkateswarlu Y, Chavakula R (1995) Brominated Benzeneacetonitriles, the Dibromotyrosine Metabolites from the Sponge *Psammaplysilla purpurea*. J Nat Prod **58**: 1087
1838. Venkateswarlu Y, Rama Rao M, Venkatesham U (1998) A New Dibromotyrosine-Derived Metabolite from the Sponge *Psammaplysilla purpurea*. J Nat Prod **61**: 1388
1839. Venkateswarlu Y, Venkatesham U, Ramo Rao M (1999) Novel Bromine-Containing Constituents of the Sponge *Psammaplysilla purpurea*. J Nat Prod **62**: 893
1840. Venkateshwar Goud T, Srinivasulu M, Niranjan Reddy VL, Vijender Reddy A, Prabhakar Rao T, Srujan Kumar D, Suryanaramyana Murty U, Venkateswarlu Y (2003) Two New

Bromotyrosine-Derived Metabolites from the Sponge *Psammaplysilla purpurea*. Chem Pharm Bull **51**: 990

1841. Ravinder K, Vijender Reddy A, Raju TV, Venkateswarlu Y (2005) A New Dibromotyrosine-Derived Metabolite from the Sponge *Psammaplysilla purpurea*. Arikov 51
1842. Tilvi S, Rodrigues C, Naik CG, Parameswaran PS, Wahidhulla S (2004) New Bromotyrosine Alkaloids from the Marine Sponge *Psammaplysilla purpurea*. Tetrahedron **60**: 10207
1843. Lacy C, Scheuer PJ (2000) New Moloka'iamine Derivatives from an Undescribed Verongid Sponge. J Nat Prod **63**: 119
1844. Ross SA, Weete JD, Schinazi RF, Wirtz SS, Tharnish P, Scheuer PJ, Hamann MT (2000) Mololipids, A New Series of Anti-HIV Bromotyramine-Derived Compounds from a Sponge of the Order Verongida. J Nat Prod **63**: 501
1845. Schroeder FC, Kau TR, Silver PA, Clardy J (2005) The Psammaplysenes, Specific Inhibitors of FOXO1a Nuclear Export. J Nat Prod **68**: 574
1846. Ciminiello P, Fattorusso E, Magno S, Pansini M (1996) Chemistry of Verongida Sponges VI. Comparison of the Secondary Metabolic Composition of *Aplysina insularis* and *Aplysina fulva*. Biochem Syst Ecol **24**: 105
1847. Evan T, Rudi A, Ilan M, Kashman Y (2001) Aplyzanzine A, a New Dibromotyrosine Derivative from a Verongida Sponge. J Nat Prod **64**: 226
1848. de Oliveira MF, de Oliveira JHHL, Galetti FCS, de Souza AO, Silva CL, Hajdu E, Peixinho S, Berlinck RGS (2006) Antimycobacterial Brominated Metabolites from Two Species of Marine Sponges. Planta Med **72**: 437
1849. Fu X, Schmitz FJ (1999) 7-Hydroxyceratinamine, a New Cyanoformamide-Containing Metabolite from a Sponge, *Aplysinella* sp. J Nat Prod **62**: 1072
1850. Pham NB, Butler MS, Quinn RJ (2000) Isolation of Psammaplin A 11'-Sulfate and Bisaprasin 11'-Sulfate from the Marine Sponge *Aplysinella rhax*. J Nat Prod **63**: 393
1851. Shin J, Lee H-S, Seo Y, Rho J-R, Cho KW, Paul VJ (2000) New Bromotyrosine Metabolites from the Sponge *Aplysinella rhax*. Tetrahedron **56**: 9071
1852. Tabudravu JN, Eijsink VGH, Gooday GW, Jaspars M, Komander D, Legg M, Synstad B, van Aalten DMF (2002) Psammaplin A, a Chitinase Inhibitor Isolated from the Fijian Marine Sponge *Aplysinella rhax*. Bioorg Med Chem **10**: 1123
1853. Diop M, Samb A, Costantino V, Fattorusso E, Mangoni A (1996) A New Iodinated Metabolite and a New Alkyl Sulfate from the Senegalese Sponge *Ptilocaulis spiculifer*. J Nat Prod **59**: 271
1854. Pettit GR, Butler MS, Williams MD, Filiatrault MJ, Pettit RK (1996) Isolation and Structure of Hemibastadinols 1-3 from the Papua New Guinea Marine Sponge *Ianthella basta*. J Nat Prod **59**: 927
1855. Masuno MN, Hoepker AC, Pessah IN, Molinski TF (2004) 1-*O*-Sulfatobastadins-1 and -2 from *Ianthella basta* (Pallas). Antagonists of the RyR1-FKBP12 Ca^{2+} Channel. Mar Drugs **2**: 176
1856. Ciminiello P, Dell'Aversano C, Fattorusso E, Magno S, Pansini M (2000) Chemistry of Verongida Sponges. 10. Secondary Metabolite Composition of the Caribbean Sponge *Verongula gigantea*. J Nat Prod **63**: 263
1857. Tsuda M, Endo T, Watanabe K, Fromont J, Kobayashi J (2002) Nakirodin A, a Bromotyrosine Alkaloid from a Verongid Sponge. J Nat Prod **65**: 1670
1858. Hirano K, Kubota T, Tsuda M, Watanabe K, Fromont J, Kobayashi J (2000) Ma'edamines A and B, Cytotoxic Bromotyrosine Alkaloids with a Unique 2(1H)Pyrazinone Ring from Sponge *Suberea* sp. Tetrahedron **56**: 8107
1859. Tsuda M, Sakuma Y, Kobayashi J (2001) Subserediamines A and B, New Bromotyrosine Alkaloids from a Sponge *Subera* Species. J Nat Prod **64**: 980
1860. Nicholas GM, Newton GL, Fahey RC, Bewley CA (2001) Novel Bromotyrosine Alkaloids: Inhibitors of Mycothiol *S*-Conjugate Amidase. Org Lett **3**: 1543
1861. Tabudravu JN, Jaspars M (2002) Purealidin S and Purpuramine J, Bromotyrosine Alkaloids from the Fijian Marine Sponge *Druinella* sp. J Nat Prod **65**: 1798

1862. Park Y, Liu Y, Hong J, Lee C-O, Cho H, Kim D-K, Im KS, Jung JH (2003) New Bromotyrosine Derivatives from an Association of Two Sponges, *Jaspis wondoensis* and *Poecillastra wondoensis*. J Nat Prod **66**: 1495
1863. Matsunaga S, Kobayashi H, van Soest RWM, Fusetani N (2005) Novel Bromotyrosine Derivatives That Inhibit Growth of the Fish Pathogenic Bacterium *Aeromonas hydrophila*, from a Marine Sponge *Hexadella* sp. J Org Chem **70**: 1893
1864. Sorek H, Rudi A, Aknin M, Gaydou E, Kashman Y (2006) Itampolins A and B, New Brominated Tyrosine Derivatives from the Sponge *Iotrochota purpurea*. Tetrahedron Lett **47**: 7237
1865. Schoenfeld RC, Conova S, Rittschof D, Ganem B (2002) Cytotoxic, Antifouling Bromotyramines: A Synthetic Study on Simple Marine Natural Products and Their Analogues. Bioorg Med Chem Lett **12**: 823
1866. Chanda BM, Sulake RS (2005) An Expeditious Convergent Synthesis of a Dibromotyrosine Alkaloid Inhibitor of Mycothiol-*S*-Conjugate Amidase. Tetrahedron Lett **46**: 6461
1867. Thompson JE, Barrow KD, Faulkner DJ (1983) Localization of Two Brominated Metabolites, Aerothionin and Homoaerothionin, in Spherulous Cells of the Marine Sponge *Aplysina fistularis* (=*Verongia thiona*). Acta Zoologica **64**: 199
1868. Aiello A, Fattorusso E, Menna M, Pansini M (1995) Chemistry of Verongida Sponges – V. Brominated Metabolites from the Caribbean Sponge *Pseudoceratina* sp. Biochem Syst Ecol **23**: 377
1869. Tsukamoto S, Kato H, Hirota H, Fusetani N (1996) Ceratinamides A and B: New Antifouling Dibromotyrosine Derivatives from the Marine Sponge *Pseudoceratina purpurea*. Tetrahedron **52**: 8181
1870. Liu S, Fu X, Schmitz FJ, Kelly-Borges M (1997) Psammaplysin F, a New Bromotyrosine Derivative from a Sponge, *Aplysinella* sp. J Nat Prod **60**: 614
1871. Kijjoa A, Bessa J, Wattanadilok R, Sawangwong P, Nascimento MSJ, Pedro M, Silva AMS, Eaton G, van Soest R, Herz W (2005) Dibromotyrosine Derivatives, a Maleimide, Aplysamine-2 and Other Constituents of the Marine Sponge *Pseudoceratina purpurea*. Z Naturforsch **60b**: 904
1872. Takada N, Watanabe R, Suenaga K, Yamada K, Ueda K, Kita M, Uemura D (2001) Zamamistatin, a Significant Antibacterial Bromotyrosine Derivative, from the Okinawan Sponge *Pseudoceratina purpurea*. Tetrahedron Lett **42**: 5265
1873. Hayakawa I, Teruya T, Kigoshi H (2006) Revised Structure of Zamamistatin. Tetrahedron Lett **47**: 155
1874. Compagnone RS, Avila R, Suárez AI, Abrams OV, Rangel HR, Arvelo F, Piña IC, Merentes E (1999) 11-Deoxyfistularin-3, a New Cytotoxic Metabolite from the Caribbean Sponge *Aplysina fistularis insularis*. J Nat Prod **62**: 1443
1875. Fendert T, Wray V, van Soest RWM, Proksch P (1999) Bromoisoxazoline Alkaloids from the Caribbean Sponge *Aplysina insularis*. Z Naturforsch **54c**: 246
1876. Ciminiello P, Dell'Aversano C, Fattorusso E, Magno S, Carrano L, Pansini M (1996) Chemistry of Verongida Sponges. VII Bromocompounds from the Caribbean Sponge *Aplysina archeri*. Tetrahedron **52**: 9863
1877. Ciminiello P, Dell'Aversano C, Fattorusso E, Magno S (2001) Archerine, a Novel Anti-Histaminic Bromotyrosine-Derived Compound from the Caribbean Marine Sponge *Aplysina archeri*. Eur J Org Chem 55
1878. Ciminiello P, Fattorusso E, Forino M, Magno S, Pansini M (1997) Chemistry of Verongida Sponges VIII. Bromocompounds from the Mediterranean Sponges *Aplysina aerophoba* and *Aplysina cavernicola*. Tetrahedron **53**: 6565
1879. Ciminiello P, Dell'Aversano C, Fattorusso E, Magno S, Pansini M (1999) Chemistry of Verongida Sponges. 9. Secondary Metabolite Composition of the Caribbean Sponge *Aplysina cauliformis*. J Nat Prod **62**: 590
1880. Encarnación RD, Sandoval E, Malmstrøm J, Christophersen C (2000) Calafianin, a Bromotyrosine Derivative from the Marine Sponge *Aplysina gerardogreeni*. J Nat Prod **63**: 874

1881. Ogamino T, Nishiyama S (2005) Synthesis and Structure Revision of Calafianin, a Member of the Spiroisoxazole Family Isolated from the Marine Sponge, *Aplysina gerardogreeni*. Tetrahedron Lett **46**: 1083
1882. Ogamino T, Obata R, Tomoda H, Nishiyama S (2006) Total Synthesis, Structure Revision, and Biological Evaluation of Calafianin, a Marine Spiroisoxazoline from the Sponge, *Aplysina gerardogreeni*. Bull Chem Soc Jpn **79**: 134
1883. Bardhan S, Schmitt DC, Porco Jr. JA (2006) Total Synthesis and Stereochemical Assignment of the Spiroisoxazoline Natural Product (+)-Calafianin. Org Lett **8**: 927
1884. Encarnación-Dimayuga R, Ramírez MR, Luna-Herrera J (2003) Aerothionin, a Bromotyrosine Derivative with Antimycobacterial Activity from the Marine Sponge *Aplysina gerardogreeni* (Demospongia). Pharm Biol **41**: 384
1885. Saeki BM, Granato AC, Berlinck RGS, Magalhâes A, Schefer AB, Ferreira AG, Pinheiro US, Hajdu E (2002) Two Unprecedented Dibromotyrosine-Derived Alkaloids from the Brazilian Endemic Marine Sponge *Aplysina caissara*. J Nat Prod **65**: 796
1886. Assmann M, Wray V, van Soest RWM, Proksch P (1998) A New Bromotyrosine Alkaloid from the Caribbean Sponge *Aiolochroia crassa*. Z Naturforsch **53c**: 398
1887. Gao H, Kelly M, Hamann MT (1999) Bromotyrosine-Derived Metabolites from the Sponge *Aiolochroia crassa*. Tetrahedron **55**: 9717
1888. Aydogmus Z, Ersoy N, Imre S (1999) Chemical Investigation of the Sponge *Verongia aerophoba*. Turk J Chem **23**: 339
1889. Okamoto Y, Ojika M, Kato S, Sakagami Y (2000) Ianthesines A-D, Four Novel Dibromotyrosine-Derived Metabolites from a Marine Sponge, *Ianthella* sp. Tetrahedron **56**: 5813
1890. Kijjoa A, Watanadilok R, Sonchaeng P, Silva AMS, Eaton G, Herz W (2001) 11,17-Dideoxyagelorin A and B, New Bromotyrosine Derivatives and Analogs from the Marine Sponge *Suberea* aff. *praetensa*. Z Naturforsch **56c**: 1116
1891. Shao N, Yao G, Chang LC (2007) Bioactive Constituents from the Marine Crinoid *Himerometra magnipinna*. J Nat Prod **70**: 869
1892. Rogers EW, de Oliveira MF, Berlinck RGS, König GM, Molinski TF (2005) Stereochemical Heterogeneity in Verongid Sponge Metabolites. Absolute Stereochemistry of (+)-Fistularin-3 and (+)-11-*epi*-Fistularin-3 by Microscale LCMS-Marfey's Analysis. J Nat Prod **68**: 891
1893. Boehlow TR, Harburn JJ, Spilling CD (2001) Approaches to the Synthesis of Some Tyrosine-Derived Marine Sponge Metabolites: Synthesis of Verongamine and Purealidin N. J Org Chem **66**: 3111
1894. Ogamino T, Ishikawa Y, Nishiyama S (2003) Electrochemical Synthesis of Spiroisoxazole Derivatives and Its Application to Natural Products. Heterocycles **61**: 73
1895. Ogamino T, Nishiyama S (2003) A New Ring-Opening Access to Aeroplysinin-1, a Secondary Metabolite of *Verongia aerophoba*. Tetrahedron **59**: 9419
1896. Pettit GR, Butler MS, Bass CG, Doubek DL, Williams MD, Schmidt JM, Pettit RK, Hooper JNA, Tackett LP, Filiatrault MJ (1995) Antineoplastic Agents, 326. The Stereochemistry of Bastadins 8, 10, and 12 from the Bismark Archipelago Marine Sponge *Ianthella basta*. J Nat Prod **58**: 680
1897. Franklin MA, Penn SG, Lebrilla CB, Lam TH, Pessah IN, Molinski TF (1996) Bastadin 20 and Bastadin *O*-Sulfate Esters from *Ianthella basta*: Novel Modulators of the Ry_1R FKBP12 Receptor Complex. J Nat Prod **59**: 1121
1898. Coll JC, Kearns PS, Rideout JA, Sandar V (2002) Bastadin 21, a Novel Isobastarane Metabolite from the Great Barrier Reef Marine Sponge *Ianthella quadrangulata*. J Nat Prod **65**: 753
1899. Reddy AV, Ravinder K, Narasimhulu M, Sridevi A, Satyanarayana N, Kondapi AK, Venkateswarlu Y (2006) New Anticancer Bastadin Alkaloids from the Sponge *Dendrilla cactos*. Bioorg Med Chem **14**: 4452
1900. Masuno MN, Pessah IN, Olmstead MM, Molinski TF (2006) Simplified Cyclic Analogues of Bastadin-5. Structure–Activity Relationships for Modulation of the RyR1/FKBP12 Ca^{+2} Channel Complex. J Med Chem **49**: 4497

1901. Aoki S, Cho S, Ono M, Kuwano T, Nakao S, Kuwano M, Nakagawa S, Gao J-Q, Mayumi T, Shibuya M, Kobayashi M (2006) Bastadin 6, a Spongean Brominated Tyrosine Derivative, Inhibits Tumor Angiogenesis by Inducing Selective Apoptosis to Endothelial Cells. Anti-Cancer Drugs **17**: 269
1902. Aoki S, Cho S, Hiramatsu A, Kotoku N, Kobayashi M (2006) Bastadins, Cyclic Tetramers of Brominated-Tyrosine Derivatives, Selectively Inhibit the Proliferation of Endothelial Cells. J Nat Med **60**: 231
1903. Couladouros EA, Pitsinos EN, Moutsos VI, Sarakinos G (2005) A General Method for the Synthesis of Bastaranes and Isobastaranes: First Total Synthesis of Bastadins 5, 10, 12, 16, 20, and 21. Chem Eur J **11**: 406
1904. Kotoku N, Tsujita H, Hiramatsu A, Mori C, Koizumi N, Kobayashi M (2005) Efficient Total Synthesis of Bastadin 6, an Anti-Angiogenic Brominated Tyrosine-Derived Metabolite from Marine Sponge. Tetrahedron **61**: 7211
1905. Brodo IM, Elix JA (1993) *Lecanora jamesii* and the Relationship Between *Lecanora* s. str. and *Straminella*. Bibl Lichenologica **53**: 19
1906. Elix JA, Barclay CE, Lumbsch HT, Wardlaw JH (1997) New Chloro Depsides from the Lichen *Lecanora lividocinera*. Aust J Chem **50**: 971
1907. Elix JA, Mayrhofer H, Wippel A (1995) 5-Chlorodivaricatic Acid, a New Depside from the Lichen Genus *Dimelaena*. Aust Lichen News **36**: 25
1908. Griffith GS, Rayner ADM, Wildman HG (1994) Extracellular Metabolites and Mycelial Morphogenesis of *Hypholoma fasciculare* and *Phlebia radiata* (Hymenomycetes). Nova Hedivigia **59**: 311
1909. Nielsen J, Nielsen PH, Frisvad JC (1999) Fungal Depside, Guisinol, from a Marine Derived Strain of *Emericella unguis*. Phytochemistry **50**: 263
1910. Cueto M, Jensen PR, Kauffman C, Fenical W, Lobkovsky E, Clardy J (2001) Pestalone, a New Antibiotic Produced by a Marine Fungus in Response to Bacterial Challenge. J Nat Prod **64**: 1444
1911. Mahandru MM, Gilbert OL (1979) Norgangaleoidin, a Dichlorodepsidone from *Lecanora chlarotera*. Bryologist **82**: 292
1912. Mahandru MM, Gilbert OL (1979) Chemical Studies in *Fulgensia*: Structures of Two New Chlorodepsidones. Bryologist **82**: 302
1913. Mahandru MM, Tajbakhsh A (1983) Fulgoicin, a New Depsidone from the Lichen *Fulgensia fulgida* (Nyl.) Szat. J Chem Soc Perkin Trans 1: 2249
1914. Birkbeck AA, Sargent MV, Elix JA (1990) The Structures of the Lichen Depsidones Fulgidin and Isofulgidin. Aust J Chem **43**: 419
1915. Sala T, Sargent MV (1981) Depsidone Synthesis. Part 16. Benzophenone – Grisa-3',5'-diene-2',3-dione – Depsidone Interconversion: A New Theory of Depsidone Biosynthesis. J Chem Soc Perkin Trans 1: 855
1916. Lumbsch HT, Elix JA (1993) Notes on the Circumscription of the Lichens *Lecanora leprosa* and *L. sulphurescens* (Lecanoraceae, lichenised Ascomycotina). Trop Bryol **7**: 71
1917. Elix JA, Venables DA, Lumbsch HT, Brako L (1994) Further New Metabolites from Lichens. Aust J Chem **47**: 1619
1918. Elix JA, Naidu R, Thor G (1995) Cyclographin, a New Depsidone from the Lichen *Catarraphia dictyoplaca*. Aust J Chem **48**: 635
1919. Fujimoto H, Inagaki M, Satoh Y, Yoshida E, Yamazaki M (1996) Monoamine Oxidase-Inhibitory Components from an Ascomycete, *Coniochaeta tetraspora*. Chem Pharm Bull **44**: 1090
1920. Rezakhana T, Guschina IA (1999) Brominated Depsidones from *Acarospora gobiensis*, a Lichen of Central Asia. J Nat Prod **62**: 1675
1921. Elix JA, Jiang H (1990) 5,7-Dichloro-3-*O*-methylnorlichexanthone, a New Xanthone from the Lichen *Lecanora broccha*. Aust J Chem **43**: 1591
1922. Elix JA, Crook CE, Hui J, Zhu ZN (1992) Synthesis of New Lichen Xanthones. Aust J Chem **45**: 845

1923. Elix JA, Robertson F, Wardlaw JH, Willis AC (1994) Isolation and Structure Determination of Demethylchodatin – a New Lichen Xanthone. Aust J Chem **47**: 2291
1924. Elix JA, Lumbsch HT, Lücking R (1995) The Chemistry of Foliicolous Lichens 2. Constituents of Some *Byssoloma* and *Sporopodium* Species. Bibl Lichenol **58**: 81
1925. Hu L-H, Yip S-C, Sim K-Y (1999) Xanthones from *Hypericum ascyron*. Phytochemistry **52**: 1371
1926. Dall'Acqua S, Innocenti G, Viola G, Piovan A, Caniato R, Cappelletti EM (2002) Cytotoxic Compounds from *Polygala vulgaris*. Chem Pharm Bull **50**: 1499
1927. Prangé T, Newman A, Milat M-L, Blein JP (1995) The Yellow Toxins Produced by *Cercospora beticola*. V. Structure of Beticolins 2 and 4. Acta Cryst **B51**: 308
1928. Ducrot P-H, Einhorn J, Kerhoas L, Lallemand J-Y, Milat M-L, Blein J-P, Newman A, Prangé T (1996) *Cercospora beticola* Toxins. Part XI: Isolation and Structure of Beticolin 0. Tetrahedron Lett **37**: 3121
1929. Chu M, Truumees I, Mierzwa R, Terracciano J, Patel M, Loebenberg D, Kaminski JJ, Das P, Puar MS (1997) Sch 54445: A New Polycyclic Xanthone with Highly Potent Antifungal Activity Produced by *Actinoplanes* sp. J Nat Prod **60**: 525
1930. Terui Y, Yiwen C, Jun-ying L, Ando T, Yamamoto H, Kawamura Y, Tomishima Y, Uchida S, Okazaki T, Munetomo E, Seki T, Yamamoto K, Murakami S, Kawashima A (2003) Xantholipin, a Novel Inhibitor of HSP47 Gene Expression Produced by *Streptomyces* sp. Tetrahedron Lett **44**: 5427
1931. Cohen PA, Towers GHN (1995) Anthraquinones and Phenanthroperylenequinones from *Nephroma laevigatum*. J Nat Prod **58**: 520
1932. Cohen PA, Towers GHN (1996) Biosynthetic Studies on Chlorinated Anthraquinones in the Lichen *Nephroma laevigtum*. Phytochemistry **42**: 1325
1933. Räisänen R, Björk H, Hynninen PH (2000) Two-Dimensional TLC Separation and Mass Spectrometric Identification of Anthraquinones Isolated from the Fungus *Dermocybe sanguinea*. Z Naturforsch **55c**: 195
1934. Kanai Y, Ishiyama D, Senda H, Iwatani W, Takahashi H, Konno H, Iwatani W, Takahashi H, Konno H, Tokumasu S, Kanazawa S (2000) Novel Human Topoisomerase I Inhibitors, Topopyrones A, B, C and D I. Producing Strain, Fermentation, Isolation, Physico-Chemical Properties and Biological Activity. J Antibiot **53**: 863
1935. Ishiyama D, Kanai Y, Senda H, Iwatani W, Takahashi H, Konno H, Kanazawa S (2000) Novel Human Topoisomerase I Inhibitors, Topopyrones A, B, C and D II. Structure Elucidation. J Antibiot **53**: 873
1936. Tan JS, Ciufolini MA (2006) Total Synthesis of Topopyrones B and D. Org Lett **8**: 4771
1937. Herath KB, Jayasuriya H, Guan Z, Schulman M, Ruby C, Sharma N, MacNaul K, Menke JG, Kodali S, Galgoci A, Wang J, Singh SB (2005) Anthrabenzoxocinones from *Streptomyces* sp. as Liver X Receptor Ligands and Antibacterial Agents. J Nat Prod **68**: 1437
1938. Furumoto T, Iwata M, Hasan AFMF, Fukui H (2003) Anthrasesamones from Roots of *Sesamum indicum*. Phytochemistry **64**: 863
1939. Okabe T, Ogino H, Suzuki H, Okuyama A, Suda H (1992) Anticancer BE-23254, its Manufacture, and BE-23254-Producing *Streptomyces*. Jpn Kokai Tokyo Koho 92, 316, 492; Chem Abst **118**: 167614v
1940. Mal D, Dey S (2006) Synthesis of Chlorine-Containing Angucycline BE-23254 and Its Analogs. Tetrahedron **62**: 9589
1941. Martin GDA, Tan LT, Jensen PR, Dimayuga RE, Fairchild CR, Raventos-Suarez C, Fenical W (2007) Marmycins A and B, Cytotoxic Pentacyclic C-Glycosides from a Marine Sediment-Derived Actinomycete Related to the Genus *Streptomyces*. J Nat Prod **70**: 1406
1942. Magyarosy A, Ho JZ, Rapoport H, Dawson S, Hancock J, Keasling JD (2002) Chlorxanthomycin, a Fluorescent, Chlorinated, Pentacyclic Pyrene from a *Bacillus* sp. Appl Environ Microbiol **68**: 4095
1943. Tsukamoto M, Nakajima S, Arakawa H, Sugiura Y, Suzuki H, Hirayama M, Kamiya S, Teshima Y, Kondo H, Kojiri K, Suda H (1998) A New Antitumor Antibiotic, BE-19412A, Produced by a Streptomycete. J Antibiot **51**: 908

1944. Oda T (2006) Effects of 2'-Demethoxy-2'-propoxygriseofulvin on Microtubule Distribution in Chinese Hamster V79 Cells. J Antibiot **59**: 114
1945. Panda D, Rathinasamy K, Santra MK, Wilson L (2005) Kinetic Suppression of Microtubule Dynamic Instability by Griseofulvin: Implications for its Possible Use in the Treatment of Cancer. Proc Natl Acad Sci U S A **102**: 9878
1946. Chu M, Mierzwa R, Truumees I, King A, Sapidou E, Barrabee E, Terracciano J, Patel MG, Gullo VP, Burrier R, Das PR, Mittelman S, Puar MS (1997) A New Fungal Metabolite, Sch 202596, with Inhibitory Activity in the Galanin Receptor GALR1 Assay. Tetrahedron Lett **38**: 6111
1947. Asakawa Y, Toyota M, Tori M, Hashimoto T (2000) Chemical Structures of Macrocyclic Bis(bibenzyls) Isolated from Liverworts (Hepaticae). Spectroscopy **14**: 149
1948. Bringmann G, Mühlbacher J, Reichert M, Dreyer M, Kolz J, Speicher A (2004) Stereochemistry of Isoplagiochin C, a Macrocyclic Bisbibenzyl from Liverworts. J Am Chem Soc **126**: 9283
1949. Takamatsu S, Rho M-C, Masuma R, Hayashi M, Komiyama K, Tanaka H, Ōmura S (1994) A Novel Testosterone 5α-Reductase Inhibitor, 8',9'-Dehydroascochlorin Produced by *Verticillium* sp. FO-2787. Chem Pharm Bull **42**: 953
1950. Gutiérrez M, Theoduloz C, Rodríguez J, Lolas M, Schmeda-Hirschmann G (2005) Bioactive Metabolites from the Fungus *Nectria galligena*, the Main Apple Canker Agent in Chile. J Agric Food Chem **53**: 7701
1951. Singh SB, Ball RG, Bills GF, Cascales C, Gibbs JB, Goetz MA, Hoogsteen K, Jenkins RG, Liesch JM, Lingham RB, Silverman KC, Zink DL (1996) Chemistry and Biology of Cylindrols: Novel Inhibitors of Ras Farnesyl-Protein Transferase from *Cylindrocarpon lucidum*. J Org Chem **61**: 7727
1952. Seephonkai P, Isaka M, Kittakoop P, Boonudomlap U, Thebtaranonth Y (2004) A Novel Ascochlorin Glycoside from the Insect Pathogenic Fungus *Verticillium hemipterigenum* BCC 2370. J Antibiot **57**: 10
1953. Togashi M, Masuda H, Kawada T, Tanaka M, Saida K, Ando K, Tamura G, Magae J (2002) PPARγ Activation and Adipocyte Differentiation Induced by AS-6, a Prenyl-phenol Antidiabetic Antibiotic. J Antibiot **55**: 417
1954. Togashi M, Ozawa S, Abe S, Nishimura T, Tsuruga M, Ando K, Tamura G, Kuwahara S, Ubukata M, Magae J (2003) Ascochlorin Derivatives as Ligands for Nuclear Hormone Receptors. J Med Chem **46**: 4113
1955. Aoki M, Kawashima K, Fujihara H, Shimizu I (2007) Facile Synthesis of (–)-Ascochlorin Using Palladium-Catalyzed Three Component Coupling Reaction. Chem Lett **36**: 654
1956. Coleman RS, Lu X (2006) Total Synthesis of Strobilurin B Using a Hetero-*bis*-Metallated Pentadiene Linchpin. Chem Commun 423
1957. Anke T, Besl H, Mocek U, Steglich W (1983) Antibiotics from Basidiomycetes. XIII Strobilurin C and Oudemansin B, Two New Antifungal Metabolites from *Xerula* Species (Agaricales). J Antibiot **36**: 661
1958. Sauter H, Steglich W, Anke T (1999) Strobilurins: Evolution of a New Class of Active Substances. Angew Chem Int Ed **38**: 1328
1959. Monti F, Ripamonti F, Hawser SP, Islam K (1999) Aspirochlorine: A Highly Selective and Potent Inhibitor of Fungal Protein Synthesis. J Antibiot **52**: 311
1960. Klausmeyer P, McCloud TG, Tucker KD, Cardellina II JH, Shoemaker RH (2005) Aspirochlorine Class Compounds from *Aspergillus flavus* Inhibit Azole-Resistant *Candida albicans*. J Nat Prod **68**: 1300
1961. Donnelly DMX, Coveney DJ, Fukuda N, Polonsky J (1986) New Sesquiterpene Aryl Esters from *Armillaria mellea*. J Nat Prod **49**: 111
1962. Peipp H, Sonnenbichler J (1992) Secondary Fungal Metabolites and Their Biological Activities, II. Occurrence of Antibiotic Compounds in Cultures of *Armillaria ostoyae* Growing in the Presence of an Antagonistic Fungus on Host Plant Cells. Biol Chem Hoppe-Seyler **373**: 675

1963. Cremin P, Donnelly DMX, Wolfender J-L, Hostettmann K (1995) Liquid Chromatographic-Thermospray Mass Spectrometric Analysis of Sesquiterpenes of *Armillaria* (Eumycota: Basidiomycotina) Species. J Chromatogr **710**: 273
1964. Arnone A, Cardillo R, Nasini B (1988) Secondary Mould Metabolites. XXIII. Isolation and Structure Elucidation of Melleolides I and J and Armellides A and B, Novel Sesquiterpenoid Aryl Esters from *Armillaria novae-zelandiae*. Gazz Chim Ital **118**: 523
1965. Arnone A, Cardillo R, Di Modugno V, Nasini G (1988) Secondary Mould Metabolites. XXII. Isolation and Structure Elucidation of Melledonals D and E and Melleolides E-H, Novel Sesquiterpenoid Aryl Esters from *Clitocybe elegans* and *Armillaria mellea*. Gazz Chim Ital **118**: 517
1966. Yang JS, Su YL, Wang YL, Feng XZ, Yu DQ, Cong PZ, Tamai M, Obuchi T, Kondoh H, Liang XT (1989) Isolation and Structures of Two New Sesquiterpenoid Aromatic Esters: Armillarigin and Armillarikin. Planta Med **55**: 479
1967. Cremin P, Guiry PJ, Wolfender J-L, Hostettmann K, Donnelly DMX (2000) A Liquid Chromatography-Thermospray Ionisation-Mass Spectrometry Guided Isolation of a New Sesquiterpene Aryl Ester from *Armillaria novae-zelandiae*. J Chem Soc Perkin Trans 1: 2325
1968. Momose I, Sekizawa R, Hosokawa N, Iinuma H, Matsui S, Nakamura H, Naganawa H, Hamada M, Takeuchi T (2000) Melleolides K, L and M, New Metabolites from *Armillariella mellea*. J Antibiot **53**: 137
1969. Itabashi T, Ogasawara N, Nozawa K, Kawai K (1996) Isolation and Structures of New Azaphilone Derivatives, Falconensins E-G, from *Emericella falconensis* and Absolute Configurations of Falconensins A-G. Chem Pharm Bull **44**: 2213
1970. Ogasawara N, Kawai K (1998) Hydrogenated Azaphilones from *Emericella falconensis* and *Emericella fruticulosa*. Phytochemistry **47**: 1131
1971. Tsuchida T, Umekita M, Kinoshita N, Iinuma H, Nakamura H, Nakamura KT, Naganawa H, Sawa T, Hamada M, Takeuchi T (1996) Epoxyquinomicins A and B, New Antibiotics from *Amycolatopsis*. J Antibiot **49**: 326
1972. Matsumoto N, Tsuchida T, Umekita M, Kinoshita N, Iinuma H, Sawa T, Hamada M, Takeuchi T (1997) Epoxyquinomicins A, B, C and D, New Antibiotics from *Amycolatopsis* I. Taxonomy, Fermentation, Isolation and Antimicrobial Activities. J Antibiot **50**: 900
1973. Matsumoto N, Iinuma H, Sawa T, Takeuchi T, Hirano S, Yoshioka T, Ishizuka M (1997) Epoxyquinomicins A, B, C and D, New Antibiotics from *Amycolatopsis* II. Effect on Type II Collagen-Induced Arthritis in Mice. J Antibiot **50**: 906
1974. Matsumoto N, Tsuchida T, Sawa R, Iinuma H, Nakamura H, Naganawa H, Sawa T, Takeuchi T (1997) Epoxyquinomicins A, B, C and D, New Antibiotics from *Amycolatopsis* III. Physico-Chemical Properties and Structure Determination. J Antibiot **50**: 912
1975. Matsumoto N, Agata N, Kuboki H, Iinuma H, Sawa T, Takeuchi T, Umezawa K (2000) Inhibition of Rat Embryo Histidine Decarboxylase by Epoxyquinomicins. J Antibiot **53**: 637
1976. Krohn K, Franke C, Jones PG, Aust H-J, Draeger S, Schulz B (1992) Isolation, Synthesis, and Biological Activity of Coniothyriomycin as well as Bioassay of Analogous Open-Chain Imides. Liebigs Ann Chem 789
1977. Krohn K, Elsässer B, Antus S, Konya K, Ammermann E (2003) Synthesis and Structure–Activity Relationship of Antifungal Coniothyriomycin Analogues. J Antibiot **56**: 296
1978. Tsukamoto M, Nakajima S, Murooka K, Suzuki H, Hirayama M, Egawa M, Kondo H, Kojiri K, Suda H (1999) BE-40665D, a New Antibacterial Antibiotic Produced by an *Actinoplanes* sp. J Antibiot **52**: 178
1979. Arnold N, Steglich W, Besl H (1996) Derivatives of Pulvinic Acid in the Genus *Scleroderma*. Z Mykol **62**: 69
1980. van der Sar SA, Blunt JW, Cole ALJ, Din LB, Munro MHG (2005) Dichlorinated Pulvinic Acid Derivative from a Malaysian *Scleroderma* sp. J Nat Prod **68**: 1799
1981. Lang G, Cole ALJ, Blunt JW, Munro MHG (2006) An Unusual Oxalylated Tetramic Acid from the New Zealand Basidiomycete *Chamonixia pachydermis*. J Nat Prod **69**: 151

1982. Kim Y-J, Nishida H, Pang C-H, Saito T, Sakemi S, Tonai-Kachi H, Yoshikawa N, Van Volkenberg MA, Parker JC, Kojima Y (2002) CJ-21,164, a New D-Glucose-6-phosphate Phosphohydrolase Inhibitor Produced by a Fungus *Chloridium* sp. J Antibiot **55**: 121
1983. Stadler M, Hellwig V, Mayer-Bartschmid, Denzer D, Wiese B, Burkhardt N (2005) Novel Analgesic Triglycerides from Cultures of *Agaricus macrosporus* and Other Basidiomycetes as Selective Inhibitors of Neurolysin. J Antibiot **58**: 775
1984. Rigby AC, Lucas-Meunier E, Kalume DE, Czerwiec E, Hambe B, Dahlqvist I, Fossier P, Baux G, Roepstorff P, Baleja JD, Furie BC, Furie B, Stenflo J (1999) A Conotoxin from *Conus textile* with Unusual Posttranslational Modifications Reduces Presynaptic Ca^{+2} Influx. Proc Natl Acad Sci U S A **96**: 5758
1985. Kalume DE, Stenflo J, Czerwiec E, Hambe B, Furie BC, Furie B, Roepstorff P (2000) Structure Determination of Two Conotoxins from *Conus textile* by a Combination of Matrix-Assisted Laser Desorption/Ionization Time-of-Flight and Electrospray Ionization Mass Spectrometry and Biochemical Methods. J Mass Spectrom **35**: 145
1986. Steen H, Mann M (2002) Analysis of Bromotryptophan and Hydroxyproline Modifications by High-Resolution, High-Accuracy Precursor Ion Scanning Utilizing Fragment Ions with Mass-Deficient Mass Tags. Anal Chem **74**: 6230
1987. Czerwiec E, Kalume DE, Roepstorff P, Hambe B, Furie B, Furie BC, Stenflo J (2006) Novel γ-Carboxyglutamic Acid-Containing Peptides from the Venom of *Conus textile*. FEBS J **273**: 2779
1988. Jimenez EC, Watkins M, Olivera BM (2004) Multiple 6-Bromotryptophan Residues in a Sleep-Inducing Peptide. Biochemistry **43**: 12343
1989. Aguilar MB, López-Vera E, Ortiz E, Becerril B, Possani LD, Olivera BM, Heimer de la Cotera EP (2005) A Novel Conotoxin from *Conus delessertii* with Posttranslationally Modified Lysine Residues. Biochemistry **44**: 11130
1990. Nair SS, Nilsson CL, Emmett MR, Schaub TM, Gowd KH, Thakur SS, Krishnan KS, Balaram P, Marshall AG (2006) De Novo Sequencing and Disulfide Mapping of a Bromotryptophan-Containing Conotoxin by Fourier Transform Ion Cyclotron Resonance Mass Spectrometry. Anal Chem **78**: 8082
1991. Williams DH (1996) The Glycopeptide Story – How to Kill the Deadly 'Superbugs'. Nat Prod Rep **13**: 469
1992. Williams DH, Bardsley B (1999) The Vancomycin Group of Antibiotics and the Fight Against Resistant Bacteria. Angew Chem Int Ed **38**: 1172
1993. Nicolaou KC, Boddy CNC, Bräse S, Winssinger N (1999) Chemistry, Biology, and Medicine of the Glycopeptide Antibiotics. Angew Chem Int Ed **38**: 2096
1994. Loll PJ, Axelsen PH (2000) The Structural Biology of Molecular Recognition by Vancomycin. Ann Rev Biophys Biomol Struct **29**: 265
1995. Allen NE, Nicas TI (2003) Mechanism of Action of Oritavancin and Related Glycopeptide Antibiotics. FEMS Microbiol Rev **26**: 511
1996. Kahne D, Leimkuhler C, Lu W, Walsh C (2005) Glycopeptide and Lipoglycopeptide Antibiotics. Chem Rev **105**: 425
1997. Walker S, Chen L, Hu Y, Rew Y, Shin D, Boger DL (2005) Chemistry and Biology of Ramoplanin: A Lipoglycodepsipeptide with Potent Antibiotic Activity. Chem Rev **105**: 449
1998. Malabarba A, Nicas TI, Thompson RC (1997) Structural Modifications of Glycopeptide Antibiotics. Med Res Rev **17**: 69
1999. Gao Y (2002) Glycopeptide Antibiotics and Development of Inhibitors to Overcome Vancomycin Resistance. Nat Prod Rep **19**: 100
2000. Pootoolal J, Neu J, Wright GD (2002) Glycopeptide Antibiotic Resistance. Annu Rev Pharmacol Toxicol **42**: 381
2001. Van Bambeke F, Van Laethem Y, Courvalin P, Tulkens PM (2004) Glycopeptide Antibiotics, from Conventional Molecules to New Derivatives. Drugs **64**: 913
2002. Hubbard BK, Walsh CT (2003) Vancomycin Assembly: Nature's Way. Angew Chem Int Ed **42**: 730

2003. Rao AVR, Gurjar MK, Reddy KL, Rao AS (1995) Studies Directed Towards the Synthesis of Vancomycin and Related Cyclic Peptides. Chem Rev **95**: 2135
2004. Williams DH, Westwell MS (1996) The Fight Against Antibiotic-Resistant Bacteria. Chemtech **26**(3): 17
2005. Nicolaou KC, Boddy CNC (2001) Behind Enemy Lines. Sci Am **284**(5): 54
2006. Schäfer M, Schneider TR, Sheldrick GM (1996) Crystal Structure of Vancomycin. Structure **4**: 1509
2007. Goldstein BP, Selva E, Gastaldo L, Berti M, Pallanza R, Ripamonti F, Ferrari P, Denaro M, Arioli V, Cassani G (1987) A-40926, a New Glycopeptide Antibiotic with Anti-*Neisseria* Activity. Antimicrob Agents Chemother **31**: 1961
2008. Vertesy L, Fehlhaber H-W, Kogler H, Limbert M (1996) New 4-Oxovancosamine-Containing Glycopeptide Antibiotics from *Amycolatopsis* sp. Y-86, 21022. J Antibiot **49**: 115
2009. Sheldrick GM, Paulus E, Vértesy L, Hahn F (1995) Structure of Ureido-Balhimycin. Acta Cryst **B51**: 89
2010. Gouda H, Matsuzaki K, Tanaka H, Hirono S, Ōmura S, McCauley JA, Sprengeler PA, Furst GT, Smith III AB (1996) Stereostructure of (–)-Chloropeptin I, a Novel Inhibitor of gp 120-CD4 Binding, via High-Temperature Molecular Dynamics, Monte Carlo Conformational Searching, and NMR Spectroscopy. J Am Chem Soc **118**: 13087
2011. Jayasuriya H, Salituro GM, Smith SK, Heck JV, Gould SJ, Singh SB, Homnick CF, Holloway MK, Pitzenberger SM, Patane MA (1998) Complestatin to Chloropeptin I via a Quantitative Acid Catalyzed Rearrangement. Absolute Stereochemical Determination of Complestatin. Tetrahedron Lett **39**: 2247
2012. Shinohara T, Deng H, Snapper ML, Hoveyda AH (2005) Isocomplestatin: Total Synthesis and Stereochemical Revision. J Am Chem Soc **127**: 7334
2013. Singh SB, Jayasuriya H, Salituro GM, Zink DL, Shafiee A, Heimbuch B, Silverman KC, Lingham RB, Genilloud O, Teran A, Vilella D, Felock P, Hazuda D (2001) The Complestatins as HIV-1 Integrase Inhibitors. Efficient Isolation, Structure Elucidation, and Inhibitory Activities of Isocomplestatin, Chloropeptin I, New Complestatins, A and B, and Acid-Hydrolysis Products of Chloropeptin I. J Nat Prod **64**: 874
2014. Kobayashi H, Shin-ya K, Nagai K, Suzuki K, Hayakawa Y, Seto H, Yun B-S, Ryoo I-J, Kim J-S, Kim C-J, Yoo I-D (2001) Neuroprotectins A and B, Bicyclohexapeptides Protecting Chick Telencephalic Neuronal Cells from Excitotoxicity I. Fermentation, Isolation, Physico-Chemical Properties and Biological Activities. J Antibiot **54**: 1013
2015. Kobayashi, H, Shin-ya K, Furihata K, Nagai K, Suzuki K, Hayakawa Y, Seto H, Yun B-S, Ryoo I-J, Kim J-S, Kim C-J, Yoo I-D (2001) Neuroprotectins A and B, Bicyclohexapeptides Protecting Chick Telencephalic Neuronal Cells from Excitotoxicity II. Structure Determination. J Antibiot **54**: 1019
2016. Hegde VR, Puar MS, Dai P, Patel M, Gullo VP, Pramanik B, Jenh C-H (2002) A Microbial Metabolite Inhibitor of CD28–CD80 Interactions. Tetrahedron Lett **43**: 5339
2017. Hegde VR, Puar MS, Dai P, Patel M, Gullo VP, Chan T-M, Silver J, Pramanik BN, Jenh C-H (2003) Condensed Aromatic Peptide Family of Microbial Metabolites, Inhibitors of CD28-CD80 Interactions. Bioorg Med Chem Lett **13**: 573
2018. Süssmuth RD, Pelzer S, Nicholson G, Walk T, Wohlleben W, Jung G (1999) New Advances in the Biosynthesis of Glycopeptide Antibiotics of the Vancomycin Type from *Amycolatopsis mediterranei*. Angew Chem Int Ed **38**: 1976
2019. Sun-L, Lindbeck AC, Nilius AM, Towne TB, Zhou CC, Paulus TJ (2004) Synthesis, Isolation, and Antibacterial Activities of Monodechlorovancomycins. J Antibiot **57**: 691
2020. Nicolaou KC, Mitchell HJ, Suzuki H, Rodríguez RM, Baudoin O, Fylaktakidou KC (1999) Total Synthesis of Everninomicin 13,384-1 – Part 1: Synthesis of the $A_1B(A)C$ Fragment. Angew Chem Int Ed **38**: 3334
2021. Nicolaou KC, Rodríguez RM, Fylaktakidou KC, Suzuki H, Mitchell JH (1999) Total Synthesis of Everninomicin 13,384-1 – Part 2: Synthesis of the $FGHA_2$ Fragment. Angew Chem Int Ed **38**: 3340

2022. Nicolaou KC, Mitchell HJ, Rodríguez RM, Fyllaktakidou KC, Suzuki H (1999) Total Synthesis of Everninomicin 13,384-1 – Part 3: Synthesis of the DE Fragment and Completion of the Total Synthesis. Angew Chem Int Ed **38**: 3345
2023. Ganguly AK, Pramanik B, Chan TM, Sarre O, Liu Y-T, Morton J, Girijavallabhan (1989) The Structure of New Oligosaccharide Antibiotics, 13-384 Components 1 and 5. Heterocycles **28**: 83
2024. Ganguly AK, McCormick JL, Chan T-M, Saksena AK, Das PR (1997) Determination of the Absolute Stereochemistry at the C16 Orthoester of Everninomicin Antibiotics; A Novel Acid-Catalyzed Isomerization of Orthoesters. Tetrahedron Lett **38**: 7989
2025. Chan T-M, Osterman RM, Morton JB, Ganguly AK (1997) Complete ^{1}H and ^{13}C NMR Assignments of the Oligosaccharide Antibiotic Sch 27899. Mag Res Chem **35**: 529
2026. Saksena AK, Jao E, Murphy B, Schumacher D, Chan T-M, Puar MS, Jenkins JK, Maloney D, Cordero M, Pramanik BN, Bartner P, Das PR, McPhail AT, Girijavallabhan VM, Ganguly AK (1998) Structure Elucidation of Sch 49088, A Novel Everninomicin Antibiotic Containing an Unusual Hydroxylaminose. Tetrahedron Lett **39**: 8441
2027. Chu M, Mierzwa R, Patel M, Jenkins J, Das P, Pramanik B, Chan T-M (2000) A Novel Everninomicin Antibiotic Active Against Multidrug-Resistant Bacteria. Tetrahedron Lett **41**: 6689
2028. Bartner P, Pramanik BN, Saksena AK, Liu Y-H, Das PR, Sarre O, Ganguly AK (1997) Structure Elucidation of Everninomicin-6, a New Oligosaccharide Antibiotic, by Chemical Degradation and FAB-MS Methods. J Am Chem Soc Mass Spectrom **8**: 1134
2029. Chu M, Mierzwa R, Jenkins J, Chan T-M, Das P, Pramanik B, Patel M, Gullo V (2002) Isolation and Characterization of Novel Oligosaccharides Related to Ziracin. J Nat Prod **65**: 1588
2030. Utkina NK, Denisenko VA, Scholokova OV, Virovaya MV, Gerasimenko AV, Popov DY, Krasokhin VB, Popov AM (2001) Spongiadioxins A and B, Two New Polybrominated Dibenzo-*p*-dioxins from an Australian Marine Sponge *Dysidea dendyi*. J Nat Prod **64**: 151
2031. Utkina NK, Denisenko VA, Virovaya MV, Scholokova OV, Prokofeva NG (2002) Two New Minor Polybrominated Dibenzo-*p*-dioxins from the Marine Sponge *Dysidea dendyi*. J Nat Prod **65**: 1213
2032. Mason G, Zacharewiski T, Denomme MA, Safe L, Safe S (1987) Polybrominated Dibenzo-*p*-dioxins and Related Compounds: Quantitative *In Vivo* and *In Vitro* Structure Activity Relationships. Toxicology **44**: 245
2033. Clark WD (1997) Investigations of Halogenated Constituents Isolated from Marine Sponges Associated with Cyanobacterial Symbionts. PhD Thesis, University of California, Santa Cruz
2034. Malmvärn A, Zebühr Y, Jensen S, Kautsky L, Greyerz E, Nakano T, Asplund L (2005) Identification of Polybrominated Dibenzo-*p*-dioxins in Blue Mussels (*Mytilus edulis*) from the Baltic Sea. Environ Sci Technol **39**: 8235
2035. Haglund P, Malmvärn A, Bergek S, Bignert A, Kautsky L, Nakano T, Wiberg K, Asplund L (2007) Brominated Dibenzo-*p*-Dioxins: A New Class of Marine Toxins? Environ Sci Technol **41**: 3069
2036. Fukuyama Y, Miura I, Kinzyo Z, Mori H, Kido M, Nakayama Y, Takahashi M, Ochi M (1985) Eckols, Novel Phlorotannins with a Dibenzo-*p*-Dioxin Skeleton Possessing Inhibitory Effects on α_2-Macroglobulin from the Brown Alga *Ecklonia kurome* Okamura. Chem Lett 739
2037. Silk PJ, Lonergan GC, Arsenault TL, Boyle CD (1997) Evidence of Natural Organochlorine Formation in Peat Bogs. Chemosphere **35**: 2865
2038. Sawada T, Aono M, Asakawa S, Ito A, Awano K (2000) Structure Determination and Total Synthesis of a Novel Antibacterial Substance, AB0022A, Produced by a Cellular Slime Mold. J Antibiotics **53**: 959
2039. Tanahashi T, Takenaka Y, Nagakura N, Hamada N (2001) Dibenzofurans from the Cultured Lichen Mycobionts of *Lecanora cinereocarnea*. Phytochemistry **58**: 1129

2040. Takenaka Y, Hamada N, Tanahashi T (2005) Monomeric and Dimeric Dibenzofurans from Cultured Mycobionts of *Lecanora iseana*. Phytochemistry **66**: 665
2041. Öberg LG, Paul KG (1985) The Transformation of Chlorophenols by Lactoperoxidase. Biochim Biophys Acta **842**: 30
2042. Svenson A, Kjeller L-O, Rappe C (1989) Enzyme-Mediated Formation of 2,8-Tetra-Substituted Chlorinated Dibenzodioxins and Dibenzofurans. Environ Sci Technol **23**: 900
2043. Öberg LG, Glas B, Swanson SE, Rappe C, Paul KG (1990) Peroxidase-Catalyzed Oxidation of Chlorophenols to Polychlorinated Dibenzo-*p*-dioxins and Dibenzofurans. Arch Environ Contam Toxicol **19**: 930
2044. Öberg LG, Rappe C (1992) Biochemical Formation of PCDD/Fs from Chlorophenols. Chemosphere **25**: 49
2045. Wagner H-C, Schramm K-W, Hutzinger O (1990) Biogenes polychloriertes Dioxin aus Trichlorophenol. UWSF - Z Umweltchem Ökotox **2**: 63
2046. Wittsiepe J, Kullmann Y, Schrey P, Selenka F, Wilhelm M (1999) Peroxidase-Catalyzed *in Vitro* Formation of Polychlorinated Dibenzo-*p*-dioxins and Dibenzofurans from Chlorophenols. Toxicol Lett **106**: 191
2047. Wittsiepe J, Kullmann Y, Schrey P, Selenka F, Wilhelm M (2000) Myeloperoxidase-Catalyzed Formation of PCDD/F from Chlorophenols. Chemosphere **40**: 963
2048. Morimoto K, Kenji T (1995) Effect of Humic Substances on the Enzymatic Formation of OCDD from PCP. Organohalogen Compds **23**: 387
2049. Huwe JK, Feil VJ, Zaylskie RG, Tiernan TO (2000) An Investigation of the *in Vivo* Formation of Octachlorodibenzo-*p*-dioxin. Chemosphere **40**: 957
2050. Fries GF, Dawson TE, Paustenbach DJ, Mathur DB, Luksemburg WJ (1997) Biosynthesis of Hepta- and Octa-chlorodioxins in Cattle and Evidence for Lack of Involvement by Rumen Microorganisms. Organohalogen Compds **33**: 296
2051. Fries G, Paustenbach D, Luksemburg W, Lorber M, Ferrario J (2000) The Formation of Hepta- and Octa-dioxins in Feces of Cows Fed Pentachlorophenol Treated Wood. Organohalogen Compds **46**: 1
2052. Malloy TA, Goldfarb TD, Surico MTJ (1993) PCDDs, PCDFs, PCBs, Chlorophenols (CPs) and Chlorobenzenes (CBzs) in Samples from Various Types of Composting Facilities in the United States. Chemosphere **27**: 325
2053. Krauß Th, Krauß P, Hagenmaier H (1994) Formation of PCDD/PCDF During Composting? Chemosphere **28**: 155
2054. Grossi G, Lichtig J, Krauß P (1998) PCDD/F, PCB and PAH Content of Brazilian Compost. Chemosphere **37**: 2153
2055. Klimm C, Schramm K-W, Henkelmann B, Martens D, Kettrup A (1998) Formation of Octa- and Heptachlorodibenzo-*p*-dioxins During Semi Anaerobic Digestion of Sewage Sludge. Chemosphere **37**: 2003
2056. Stevens J, Green NJL, Jones KC (2001) Survey of PCDD/Fs and Non-*ortho* PCBs in UK Sewage Sludges. Chemosphere **44**: 1455
2057. Martínez M, Díaz-Ferrero J, Martí R, Broto-Puig F, Comellas L, Rodríguez-Larena MC (2000) Analysis of Dioxin-like Compounds in Vegetation and Soil Samples Burned in Catalan Forest Fires. Comparison with the Corresponding Unburned Material. Chemosphere **41**: 1927
2058. Tashiro C, Clement RE, Stocks BJ, Radke L, Cofer WR, Ward P (1990) Preliminary Report: Dioxins and Furans in Prescribed Burns. Chemosphere **20**: 1533
2059. Bumb RR, Crummett WB, Cutie SS, Gledhill JR, Hummel RH, Kagel RO, Lamparski LL, Luoma EV, Miller DL, Nestrick TJ, Shadoff LA, Stehl RH, Woods JS (1980) Trace Chemistries of Fire: A Source of Chlorinated Dioxins. Science **210**: 385
2060. Schatowitz B, Brandt G, Gafner F, Schlumpf E, Bühler R, Hasler P, Nussbaumer T (1994) Dioxin Emissions from Wood Combustion. Chemosphere **29**: 2005
2061. Vikelsøe J, Madsen H, Hansen K (1994) Emission of Dioxins from Danish Wood-Stoves. Chemosphere **29**: 2019

2062. Wunderli S, Zennegg M, Doležal IS, Gujer E, Moser U, Wolfensberger M, Hasler P, Noger D, Studer C, Karlaganis G (2000) Determination of Polychlorinated Dibenzo-p-dioxins and Dibenzo-furans in Solid Residues from Wood Combustion by HRGC/HRMS. Chemosphere **40**: 641

2063. Thuß U, Popp P, Ehrlich Chr, Kalkoff W-D (1995) Domestic Lignite Combustion as Source of Polychlorodibenzodioxins and -Furans (PCDD/F) Chemosphere **31**: 2591

2064. Lemieux PM, Lutes CC, Abbott JA, Aldous KM (2000) Emission of Polychlorinated Dibenzo-p-dioxins and Polychlorinated Dibenzofurans from the Open Burning of Household Waste in Barrels. Environ Sci Technol **34**: 377

2065. Eduljee GH, Atkins DHF, Eggleton AE (1986) Observations and Assessment Relating to Incineration of Chlorinated Chemical Wastes. Chemosphere **15**: 1577

2066. Ruokojärvi P, Ettala M, Rahkonen P, Tarhanen J, Ruuskanen J (1995) Polychlorinated Dibenzo-p-dioxins and -Furans (PCDDs and PCDFs) in Municipal Waste Landfill Fires. Chemosphere **30**: 1697

2067. Dyke P, Coleman P (1995) Dioxins in Ambient Air, Bonfire Night 1994. Organohalogen Compds **24**: 213

2068. Takeda N, Takaoka M, Fujiwara T, Takeyama H, Eguchi S (2000) PCDDs/DFs Emissions from Crematories in Japan. Chemosphere **40**: 575

2069. Hashimoto S, Wakimoto T, Tatsukawa R (1995) Possible Natural Formation of Polychlorinated Dibenzo-p-Dioxins as Evidenced by Sediment Analysis from the Yellow Sea, the East China Sea and the Pacific Ocean. Mar Pollut Bull **30**: 341

2070. Kjeller L-O, Rappe C (1995) Time Trends in Levels, Patterns, and Profiles for Polychlorinated Dibenzo-p-dioxins, Dibenzofurans, and Biphenyls in a Sediment Core from the Baltic Proper. Environ Sci Technol **29**: 346

2071. Fiedler H, Lau C, Kjeller L-O, Rappe C (1996) Patterns and Sources of Polychlorinated Dibenzo-p-dioxins and Dibenzofurans Found in Soil and Sediment Samples in Southern Mississippi. Chemosphere **32**: 421

2072. Rappe C, Andersson R, Bonner M, Cooper K, Fiedler H, Howell F, Kulp SE, Lau C (1997) PCDDs and PCDFs in Soil and River Sediment Samples from a Rural Area in the United States of America. Chemosphere **34**: 1297

2073. Rappe C, Andersson R, Cooper K, Fiedler H, Lau C, Bonner M, Howell F (1997) PCDDs and PCDFs in Lake Sediment Cores from Southern Mississippi, USA. Organohalogen Compds **32**: 18

2074. Rappe C, Andersson R, Bonner M, Cooper K, Fiedler H, Lau C, Howell F (1997) PCDDs and PCDFs in Lake Sediments from a Rural Area in the USA. Organohalogen Compds **32**: 88

2075. Rappe C, Bergek S, Fiedler H, Cooper KR (1998) PCDD and PCDF Contamination in Catfish Feed from Arkansas, USA. Chemosphere **36**: 2705

2076. Hayward DG, Nortrup D, Gardner A, Clower Jr. M (1999) Elevated TCDD in Chicken Eggs and Farm-Raised Catfish Fed a Diet with Ball Clay from a Southern United States Mine. Environ Res Sec A **81**: 248

2077. Ferrario JB, Byrne CJ, Cleverly DH (2000) 2,8-Dibenzo-p-dioxins in Mined Clay Products from the United States: Evidence for Possible Natural Origin. Environ Sci Technol **34**: 4524

2078. Rappe C, Andersson R (2000) Concentrations of PCDDs in Ball Clay and Kaolin. Organohalogen Compds **46**: 9

2079. Rappe C, Andersson R, Cooper K, Bopp R, Fiedler H, Howell F, Bonner M (2000) PCDDs in Naturally-Formed and Man-Made Lake Sediment Cores from Southern Mississippi, USA. Organohalogen Compds **46**: 19

2080. Ferrario J, Byrne C, Cleverly D (2000) Summary of Evidence for the Possible Natural Formation of Dioxins in Mined Clay Products. Organohalogen Compds **46**: 23

2081. Gadomski D, Tysklind M, Irvine RL, Burns PC, Andersson R (2004) Investigations into the Vertical Distribution of PCDDs and Mineralogy in Three Ball Clay Cores from the United States Exhibiting the Natural Formation Pattern. Environ Sci Technol **38**: 4956

2082. Holmstrand H, Gadomski D, Mandalakis M, Tysklind M, Irvine R, Andersson P, Gustafsson Ö (2006) Origin of PCDDs in Ball Clay Assessed with Compound-Specific Chlorine Isotope Analysis and Radiocarbon Dating. Environ Sci Technol **40**: 3730
2083. Horii Y, van Bavel B, Kannan K, Petrick G, Nachtigall K, Yamashita N (2008) Novel Evidence for Natural Formation of Dioxins in Ball Clay. Chemosphere **70**: 1280
2084. Jobst H, Aldag R (2000) Dioxine in Lagerstätten-Tonen (Dioxins in Clay Deposits.) UWSF - Z Umweltchem Ökotox **12**: 2
2085. Müller JF, Haynes D, McLachlan M, Böhme F, Will S, Shaw GR, Mortimer M, Sadler R, Connell DW (1999) PCDDs, PCDFs, PCBs and HCB in Marine and Estuarine Sediments from Queensland, Australia. Chemosphere **39**: 1707
2086. Gaus C, Päpke O, Dennison N, Haynes D, Shaw GR, Connell DW, Müller JF (2001) Evidence for the Presence of a Widespread PCDD Source in Coastal Sediments and Soils from Queensland, Australia. Chemosphere **43**: 549
2087. Gaus C, Brunskill GJ, Weber R, Papke O, Müller JF (2001) Historical PCDD Inputs and Their Source Implications from Dated Sediment Cores in Queensland (Australia). Environ Sci Technol **35**: 4597
2088. Prange JA, Gaus C, Päpke O, Müller JF (2002) Investigations into the PCDD Contamination of Topsoil, River Sediments and Kaolinite Clay in Queensland, Australia. Chemosphere **46**: 1335
2089. Ferrario J, Byrne C (2002) Dibenzo-*p*-dioxins in the Environment from Ceramics and Pottery Produced from Ball Clay Mined in the United States. Chemosphere **46**: 1297
2090. Isosaari P, Pajunen H, Vartiainen T (2002) PCDD/F and PCB History in Dated Sediments of a Rural Lake. Chemosphere **47**: 575
2091. Müller JF, Gaus C, Prange JA, Paepke O, Poon KF, Lam MHW, Lam PKS (2002) Polychlorinated Dibenzo-*p*-dioxins and Polychlorinated Dibenzofurans in Sediments from Hong Kong. Mar Pollut Bull **45**: 372
2092. Alcock RE, McLachlan MS, Johnston AE, Jones KC (1998) Evidence for the Presence of PCDD/Fs in the Environment Prior to 1900 and Further Studies on Their Temporal Trends. Environ Sci Technol **32**: 1580
2093. Green NJL, Alcock RE, Johnston AE, Jones KC (2000) Are There Natural Dioxins? Evidence from Deep Soil Samples. Organohalogen Compds **46**: 12
2094. Green NJL, Jones JL, Johnston AE, Jones KC (2001) Further Evidence for the Existence of PCDD/Fs in the Environment Prior to 1900. Environ Sci Technol **35**: 1974
2095. Meharg AA, Killham K (2003) A Pre-industrial Source of Dioxins and Furans. Nature **421**: 909
2096. Lamparski LL, Nestrick TJ, Stenger VA (1984) Presence of Chlorodibenzo Dioxins in a Sealed 1933 Sample of Dried Municipal Sewage Sludge. Chemosphere **13**: 361
2097. Dahlman O, Mörck R, Ljungquist P, Relmann A, Johansson C, Borén H, Grimvall A (1993) Chlorinated Structural Elements in High Molecular Weight Organic Matter from Unpolluted Waters and Bleached-Kraft Mill Effluents. Environ Sci Technol **27**: 1616
2098. Dahlman O, Reimann A, Ljungquist P, Mörck R, Johansson C, Borén H, Grimvall A (1994) Characterization of Chlorinated Aromatic Structures in High Molecular Weight BKME-Materials and in Fulvic Acids from Industrially Unpolluted Waters. Wat Sci Tech **29**: 81
2099. Grimvall A, Laniewski K, Borén H, Jonsson S, Kaugare S (1994) Organohalogens of Natural or Unknown Origin in Surface Water and Precipitation. Toxicol Environ Chem **46**: 183
2100. Grimvall A (1995) Evidence of Naturally Produced and Man-Made Organohalogens in Water and Sediments. In: de Leer EWB (ed) Naturally-Produced Organohalogens. Kluwer, Dordrecht, p 3
2101. Hjelm O, Johansson M-B, Öberg-Asplund G (1995) Organically Bound Halogens in Coniferous Forest Soil. Chemosphere **30**: 2353
2102. Hjelm O, Asplund G (1995) Chemical Characterization of Organohalogens in a Coniferous Forest Soil. In: Grimvall A, de Leer EWB (eds) Naturally-Produced Organohalogens. Kluwer, Dordrecht, p 105

2103. Michaelis W, Richnow HH, Seifert R (1995) Chemically Bound Chlorinated Aromatics in Humic Substances. Naturwissenschaften **82**: 139
2104. Öberg G, Nordlund E, Berg B (1996) In situ Formation of Organically Bound Halogens During Decomposition of Norway Spruce Needles: Effects of Fertilization. Can J For Res **26**: 1040
2105. Flodin C, Johansson E, Borén H, Grimvall A, Dahlman O, Mörck R (1997) Chlorinated Structures in High Molecular Weight Organic Matter Isolated from Fresh and Decaying Plant Material and Soil. Environ Sci Technol **31**: 2464
2106. Flodin C, Ekelund M, Borén H, Grimvall A (1997) Pyrolysis-GC/AED and Pyrolysis-GC/MS Analysis of Chlorinated Structures in Aquatic Fulvic Acids and Chlorolignins. Chemosphere **34**: 2319
2107. Johansson E, Krantz-Rülcker C, Zhang BX, Öberg G (2000) Chlorination and Biodegradation of Lignin. Soil Biol Biochem **32**: 1029
2108. Lee RT, Shaw G, Wadey P, Wang X (2001) Specific Association of ^{36}Cl with Low Molecular Weight Humic Substances in Soils. Chemosphere **43**: 1063
2109. Hambsch B, Schmiedel U, Werner P, Frimmel FH (1993) Investigations on the Biodegradability of Chlorinated Fulvic Acids. Acta Hydrochim Hydrobiol **21**: 167
2110. Johansson C, Pavasars I, Borén H, Grimvall A, Dahlman O, Mörck R, Reimann A (1994) A Degradation Procedure for Determination of Halogenated Structural Elements in Organic Matter from Marine Sediments. Environ Int **20**: 103
2111. Carlsen L, Lassen P (1992) Enzymatically Mediated Formation of Chlorinated Humic Acids. Org Geochem **18**: 477
2112. Ortiz-Bermúdez P, Srebotnik E, Hammel KE (2003) Chlorination and Cleavage of Lignin Structures by Fungal Chloroperoxidases. Appl Environ Microbiol **69**: 5015
2113. Hoekstra EJ, Lassen P, van Leeuwen JGE, de Leer EWB, Carlsen L (1995) Formation of Organic Chlorine Compounds of Low Molecular Weight in the Chloroperoxidase-Mediated Reaction Between Chloride and Humic Material. In: Grimvall A, de Leer EWB (eds) Naturally-Produced Organohalogens. Kluwer, Dordrecht, p 149
2114. van Pée K-H (1996) Biosynthesis of Halogenated Metabolites by Bacteria. Annu Rev Microbiol **50**: 375
2115. Littlechild J (1999) Haloperoxidases and Their Role in Biotransformation Reactions. Curr Opin Chem Biol **3**: 28
2116. Dunford HB (1999) Heme Peroxidases. Wiley, New York
2117. van Pée K-H (2001) Microbial Biosynthesis of Halometabolites. Arch Microbiol **175**: 250
2118. van Pée K-H, Zehner S (2003) Enzymology and Molecular Genetics of Biological Halogenation. In: Gribble G (ed) The Handbook of Environmental Chemistry vol. 3, part P. Springer, Berlin, p 171
2119. van Pée K-H, Unversucht S (2003) Biological Dehalogenation and Halogenation Reactions. Chemosphere **52**: 299
2120. Ballschmiter K (2003) Pattern and Sources of Naturally Produced Organohalogens in the Marine Environment: Biogenic Formation of Organohalogens. Chemosphere **52**: 313
2121. Anderson JLR, Chapman SK (2006) Molecular Mechanisms of Enzyme-Catalysed Halogenation. Mol BioSyst **2**: 350
2122. van Pée K-H, Dong C, Flecks S, Naismith J, Patallo EP, Wage T (2006) Biological Halogenation has Moved Far Beyond Haloperoxidases. Adv Appl Microbiol **59**: 127
2123. Butler A (1998) Vanadium Haloperoxidases. Curr Opin Chem Biol **2**: 279
2124. Butler A (1999) Mechanistic Considerations of the Vanadium Haloperoxidases. Coord Chem Rev **187**: 17
2125. Rehder D (2003) Biological and Medicinal Aspects of Vanadium. Inorg Chem Commun **6**: 604
2126. Rehder D (2008) Is Vanadium a More Versatile Target in the Activity of Primordial Life Forms than Hitherto Anticipated? Org Biomol Chem **6**: 957
2127. Murphy CD (2006) Recent Developments in Enzymatic Chlorination. Nat Prod Rep **23**: 147

2128. de Jong E, Field JA (1997) Sulfur Tuft and Turkey Tail: Biosynthesis and Biodegradation of Organohalogens by Basidiomycetes. Annu Rev Microbiol **51**: 375
2129. Franssen MCR (1994) Halogenation and Oxidation Reactions with Haloperoxidases. Biocatalysis **10**: 87
2130. van de Velde F, van Rantwijk F, Sheldon RA (2001) Improving the Catalytic Performance of Peroxidases in Organic Synthesis. Trends Biotech **19**: 73
2131. Dembitsky VM (2003) Oxidation, Epoxidation and Sulfoxidation Reactions Catalysed by Haloperoxidases. Tetrahedron **59**: 4701
2132. Naismith JH (2006) Inferring the Chemical Mechanism from Structures of Enzymes. Chem Soc Rev **35**: 763
2133. Manley SL (2002) Phytogenesis of Halomethanes: A Product of Selection or a Metabolic Accident? Biogeochemistry **60**: 163
2134. Mtolera MSP, Collén J, Pedersén M, Semesi AK (1995) Destructive Hydrogen Peroxide Production in *Eucheuma denticulatum* (Rhodophyta) During Stress Caused by Elevated pH, High Light Intensities and Competition with Other Species. Eur J Phycol **30**: 289
2135. Liu L, Eriksson K-EL, Dean JFD (1999) Localization of Hydrogen Peroxide Production in *Zinnia elegans* L. Stems. Phytochemistry **52**: 545
2136. Orozco-Cardenas M, Ryan CA (1999) Hydrogen Peroxide is Generated Systematically in Plant Leaves by Wounding and Systemin Via the Octadecanoid Pathway. Proc Natl Acad Sci U S A **96**: 6553
2137. Petigara BR, Blough NV, Mignerey AC (2002) Mechanisms of Hydrogen Peroxide Decomposition in Soils. Environ Sci Technol **36**: 639
2138. Morris DR, Hager LP (1966) Chloroperoxidase I. Isolation and Properties of the Crystalline Glycoprotein. J Biol Chem **241**: 1763
2139. Nuell MJ, Fang G-H, Axley MJ, Kenigsberg P, Hager LP (1988) Isolation and Nucleotide Sequence of the Chloroperoxidase Gene from *Caldariomyces fumago*. J Bacteriol **170**: 1007
2140. Zong Q, Osmulski PA, Hagar LP (1995) High-Pressure-Assisted Reconstitution of Recombinant Chloroperoxidase. Biochemistry **34**: 12420
2141. Sundaramoorthy M, Mauro JM, Sullivan AM, Terner J, Poulos TL (1995) Preliminary Crystallographic Analysis of Chloroperoxidase from *Caldariomyces fumago*. Acta Cryst D**51**: 842
2142. Sundaramoorthy M, Terner J, Poulos TL (1995) The Crystal Structure of Chloroperoxidase: A Heme Peroxidase-Cytochrome P450 Functional Hybrid. Structure **3**: 1367
2143. Simons BH, Barnett P, Vollenbroek EGM, Dekker HL, Muijsers AO, Messerschmidt A, Wever R (1995) Primary Structure and Characterization of the Vanadium Chloroperoxidase from the Fungus *Curvularia inaequalis*. Eur J Biochem **229**: 566
2144. Messerschmidt A, Wever R (1996) X-ray Structure of a Vanadium-Containing Enzyme: Chloroperoxidase from the Fungus *Curvularia inaequalis*. Proc Natl Acad Sci U S A **93**: 392
2145. Hemrika W, Renirie R, Dekker HL, Barnett P, Wever R (1997) From Phosphatases to Vanadium Peroxidases: A Similar Architecture of the Active Site. Proc Natl Acad Sci U S A **94**: 2145
2146. Messerschmidt A, Prade L, Wever R (1997) Implications for the Catalytic Mechanism of the Vanadium-Containing Enzyme Chloroperoxidase from the Fungus *Curvularia inaequalis* by X-Ray Structures of the Native and Peroxide Form. Biol Chem **378**: 309
2147. Macedo-Ribeiro S, Hemrika W, Renirie R, Wever R, Messerschmidt A (1999) X-Ray Crystal Structures of Active Site Mutants of the Vanadium-Containing Chloroperoxidase from the Fungus *Curvularia inaequalis*. J Biol Inorg Chem **4**: 209
2148. Barnett P, Hemrika W, Dekker HL, Muijsers AO, Renirie R, Wever R (1998) Isolation, Characterization, and Primary Structure of the Vanadium Chloroperoxidase from the Fungus *Embellisia didymospora*. J Biol Chem **273**: 23381
2149. Hashimoto A, Pickard MA (1984) Chloroproxidases from *Caldariomyces* (= *Leptoxyphium*) Cultures: Glycoproteins with Variable Carbohydrate Content and Isoenzymic Forms. J Gen Microbiol **130**: 2051

2150. Yoon KS, Chen YP, Lovell CR, Lincoln DE, Knapp LW, Woodin SA (1994) Localization of the Chloroperoxidase of the Capitellid Polychaete *Notomastus lobatus*. Biol Bull **187**: 215
2151. Roach MP, Chen YP, Woodin SA, Lincoln DE, Lovell CR, Dawson JH (1997) *Notomastus lobatus* Chloroperoxidase and *Amphitrite ornata* Dehaloperoxidase Both Contain Histidine as Their Proximal Heme Iron Ligand. Biochemistry **36**: 2197
2152. Franzen S, Roach MP, Chen Y-P, Dyer RB, Woodruff WH, Dawson JH (1998) The Unusual Reactivities of *Amphitrite ornata* Dehaloperoxidase and *Notomastus lobatus* Chloroperoxidase Do Not Arise from a Histidine Imidazolate Proximal Heme Iron Ligand. J Am Chem Soc **120**: 4658
2153. Bantleon R, Altenbuchner J, van Pée K-H (1994) Chloroperoxidase from *Streptomyces lividans*: Isolation and Characterization of the Enzyme and the Corresponding Gene. J Bacteriol **176**: 2339
2154. Burd W, Yourkevich O, Voskoboev AJ, van Pée K-H (1995) Purification and Properties of a Non-Haem Chloroperoxidase from *Serratia marcescens*. FEMS Microbiol Lett **129**: 255
2155. Han Y-J, Watson JT, Stucky GD, Butler A (2002) Catalytic Activity of Mesoporous Silicate-Immobilized Chloroperoxidase. J Mol Catal B: Enzym **17**: 1
2156. Petri A, Gambicorti T, Salvadori P (2004) Covalent Immobilization of Chloroperoxidase on Silica Gel and Properties of the Immobilized Biocatalyst. J Mol Catal B: Enzym **27**: 103
2157. Osborne RL, Raner GM, Hager LP, Dawson JH (2006) *C. fumago* Chloroperoxidase is also a Dehaloperoxidase: Oxidative Dehalogenation of Halophenols. J Am Chem Soc **128**: 1036
2158. Osborne RL, Coggins MK, Terner J, Dawson JH (2007) *Caldariomyces fumago* Chloroperoxidase Catalyzes the Oxidative Dehalogenation of Chlorophenols by a Mechanism Involving Two One-Electron Steps. J Am Chem Soc **129**: 14838
2159. Dunford HB, Lambeir A-M, Kashem MA, Pickard M (1987) On the Mechanism of Chlorination by Chloroperoxidase. Arch Biochem Biophys **252**: 292
2160. Libby RD, Shedd AL, Phipps AK, Beachy TM, Gerstberger SM (1992) Defining the Involvement of HOCl or Cl_2 as Enzyme-Generated Intermediates in Chloroperoxidase-catalyzed Reactions. J Biol Chem **267**: 1769
2161. Sun W, Kadima TA, Pickard MA, Dunford HB (1994) Catalase Activity of Chloroperoxidase and Its Interaction with Peroxidase Activity. Biochem Cell Biol **72**: 321
2162. Yang ZP, Shelton KD, Howard JC, Woods AE (1995) Mechanism of the Chloroperoxidase-Catalyzed Bromination of Tyrosine. Comp Biochem Physiol **111B**: 417
2163. Libby RD, Beachy TM, Phipps AK (1996) Quantitating Direct Chlorine Transfer from Enzyme to Substrate in Chloroperoxidase-Catalyzed Reactions. J Biol Chem **271**: 21820
2164. Wagenknecht H-A, Woggon W-D (1997) Identification of Intermediates in the Catalytic Cycle of Chloroperoxidase. Chem Biol **4**: 367
2165. Sundaramoorthy M, Terner J, Poulos TL (1998) Stereochemistry of the Chloroperoxidase Active Site: Crystallographic and Molecular-Modeling Studies. Chem Biol **5**: 461
2166. Wagenknecht H-A, Claude C, Woggon W-D (1998) New Enzyme Models of Chloroperoxidase: Improved Stability and Catalytic Efficiency of Iron Porphyrinates Containing a Thiolato Ligand. Helv Chim Acta **81**: 1506
2167. Yi X, Mroczko M, Manoj KM, Wang X, Hager LP (1999) Replacement of the Proximal Heme Thiolate Ligand in Chloroperoxidase with a Histidine Residue. Proc Natl Acad Sci U S A **96**: 12412
2168. Reddy CM, Xu L, Drenzek NJ, Sturchio NC, Heraty LJ, Kimblin C, Butler A (2002) A Chlorine Isotope Effect for Enzyme-Catalyzed Chlorination. J Am Chem Soc **124**: 14526
2169. Yi X, Conesa A, Punt PJ, Hager LP (2003) Examining the Role of Glutamic Acid 183 in Chloroperoxidase Catalysis. J Biol Chem **278**: 13855
2170. Kim SH, Perera R, Hager LP, Dawson JH, Hoffman BM (2006) Rapid Freeze-Quench ENDOR Study of Chloroperoxidase Compound I: The Site of the Radical. J Am Chem Soc **128**: 5598

2171. Zhang R, Nagraj N, Lansakara-P DSP, Hager LP, Newcomb M (2006) Kinetics of Two-Electron Oxidations by the Compound I Derivative of Chloroperoxidase, A Model for Cytochrome P450 Oxidants. Org Lett **8**: 2731
2172. Hemrika W, Renirie R, Macedo-Ribeiro S, Messerschmidt A, Wever R (1999) Heterologous Expression of the Vanadium-Containing Chloroperoxidase from *Curvularia inaequalis* in *Saccharomyces cerevisiae* and Site-directed Mutagenesis of the Active Site Residues His^{496}, Lys^{353}, Arg^{360}, and Arg^{490}. J Biol Chem **274**: 23820
2173. Renirie R, Hemrika W, Wever R (2000) Peroxidase and Phosphatase Activity of Active-Site Mutants of Vanadium Chloroperoxidase from the Fungus *Curvularia inaequalis*. J Biol Chem **275**: 11650
2174. Tanaka N, Hasan Z, Wever R (2003) Kinetic Characterization of Active Site Mutants Ser402Ala and Phe397His of Vanadium Chloroperoxidase from the Fungus *Curvularia inaequalis*. Inorg Chim Acta **356**: 288
2175. Hasan Z, Renirie R, Kerkman R, Ruijssenaars HJ, Hartog AF, Wever R (2006) Laboratory-Evolved Vanadium Chloroperoxidase Exhibits 100-Fold Higher Halogenating Activity at Alkaline pH. Catalytic Effects from First and Second Coordination Sphere Mutations. J Biol Chem **281**: 9738
2176. Waller MP, Bühl M, Geethalakshmi KR, Wang D, Thiel W (2007) ^{51}V NMR Chemical Shifts Calculated from QM/MM Models of Vanadium Chloroperoxidase. Chem Eur J **13**: 4723
2177. Colonna S, Gaggero N, Richelmi C, Pasta P (1999) Recent Biotechnological Developments in the Use of Peroxidases. Trends Biotech **17**: 163
2178. van Rantwijk F, Sheldon RA (2000) Selective Oxygen Transfer Catalysed by Heme Peroxidases: Synthetic and Mechanistic Aspects. Curr Opin Biotech **11**: 554
2179. Colonna S, Gaggero N, Manfredi A, Casella L, Gullotti M, Carrea G, Pasta P (1990) Enantioselective Oxidations of Sulfides Catalyzed by Chloroperoxidase. Biochemistry **29**: 10465
2180. Colonna S, Gaggero N, Casella L, Carrea G, Pasta P (1992) Chloroperoxidase and Hydrogen Peroxide: An Efficient System for Enzymatic Enantioselective Sulfoxidations. Tetrahedron: Asym **3**: 95
2181. Fu H, Kondo H, Ichikawa Y, Look GC, Wong C-H (1992) Chloroperoxidase-Catalyzed Asymmetric Synthesis: Enantioselective Reactions of Chiral Hydroperoxides with Sulfides and Bromohydration of Glycals. J Org Chem **57**: 7265
2182. Pasta P, Carrea G, Colonna S, Gaggero N (1994) Effects of Chloride on the Kinetics and Stereochemistry of Chloroperoxidase Catalyzed Oxidation of Sulfides. Biochim Biophys Acta **1209**: 203
2183. Allenmark SG, Andersson MA (1996) Chloroperoxidase-Catalyzed Asymmetric Synthesis of a Series of Aromatic Cyclic Sulfoxides. Tetrahedron: Asym **7**: 1089
2184. van Deurzen MPJ, Remkes IJ, van Rantwijk F, Sheldon RA (1997) Chloroperoxidase Catalyzed Oxidations in *t*-Butyl Alcohol/Water Mixtures. J Mol Catal A: Chem **117**: 329
2185. Colonna S, Gaggero N, Carrea G, Pasta P (1997) A New Enzymatic Enantioselective Synthesis of Dialkyl Sulfoxides Catalysed by Monooxygenases. Chem Commun 439
2186. Allenmark SG, Andersson MA (1998) Chloroperoxidase-Induced Asymmetric Sulfoxidation of Some Conformationally Restricted Sulfides. Chirality **10**: 246
2187. Pasta P, Carrea G, Monzani E, Gaggero N, Colonna S (1999) Chloroperoxidase-Catalyzed Enantioselective Oxidation of Methyl Phenyl Sulfide with Dihydroxyfumaric Acid/Oxygen or Ascorbic Acid/Oxygen as Oxidants. Biotechnol Bioeng **62**: 489
2188. Vargas RR, Bechara EJH, Marzorati L, Wladislaw B (1999) Asymmetric Sulfoxidation of a β-Carbonyl Sulfide Series by Chloroperoxidase. Tetrahedron: Asym **10**: 3219
2189. van de Velde F, van Rantwijk F, Sheldon RA (1999) Selective Oxidations with Molecular Oxygen, Catalyzed by Chloroperoxidase in the Presence of a Reductant. J Mol Catal B: Enzym **6**: 453
2190. Holland HL (2001) Biotransformation of Organic Sulfides. Nat Prod Rep **18**: 171

2191. Colonna S, Del Sordo S, Gaggero N, Carrea G, Pasta P (2002) Enzyme-Mediated Catalytic Asymmetric Oxidations. Heteroatom Chem **13**: 467
2192. Trevisan V, Signoretto M, Colonna S, Pironti V, Strukul G (2004) Microencapsulated Chloroperoxidase as a Recyclable Catalyst for the Enantioselective Oxidation of Sulfides with Hydrogen Peroxide. Angew Chem Int Ed **43**: 4097
2193. Holland HL, Brown FM, Lozada D, Mayne B, Szerminski WR, van Vliet AJ (2002) Chloroperoxidase-Catalyzed Oxidation of Methionine Derivatives. Can J Chem **80**: 633
2194. Coughlin P, Roberts S, Rush C, Willetts A (1993) Biotransformation of Alkenes by Haloperoxidases: Regiospecific Bromohydrin Formation from Cinnamyl Substrates. Biotechnol Lett **15**: 907
2195. Colonna S, Gaggero N, Casella L, Carrea G, Pasta P (1993) Enantioselective Epoxidation of Styrene Derivatives by Chloroperoxidase Catalysis. Tetrahedron: Asym **4**: 1325
2196. Dexter AF, Lakner FJ, Campbell RA, Hager LP (1995) Highly Enantioselective Epoxidation of 1,1-Disubstituted Alkenes Catalyzed by Chloroperoxidase. J Am Chem Soc **117**: 6412
2197. Zaks A, Dodds DR (1995) Chloroperoxidase-Catalyzed Asymmetric Oxidations: Substrate Specificity and Mechanistic Study. J Am Chem Soc **117**: 10419
2198. Lakner FJ, Hager LP (1996) Chloroperoxidase as Enantioselective Epoxidation Catalyst: An Efficient Synthesis of (R)-(–)-Mevalonolactone. J Org Chem **61**: 3923
2199. Lakner FJ, Hager LP (1997) Chloroperoxidase-Mediated Asymmetric Epoxidation. Synthesis of (R)-Dimethyl 2-Methylaziridine-1,2-dicarboxylate – A Potential α-Methylamino Acid Synthon. Tetrahedron: Asym **8**: 3547
2200. Lakner FJ, Cain KP, Hager LP (1997) Enantioselective Epoxidation of ω-Bromo-2-methyl-1-alkenes Catalyzed by Chloroperoxidase. Effect of Chain Length on Selectivity and Efficiency. J Am Chem Soc **119**: 443
2201. Hager LP, Lakner FJ, Basavapathruni A (1998) Chiral Synthons via Chloroperoxidase Catalysis. J Mol Catal B: Enzym **5**: 95
2202. Aoun S, Baboulène M (1998) Regioselective Bromohydroxylation of Alkenes Catalyzed by Chloroperoxidase: Advantages of the Immobilization of Enzyme on Talc. J Mol Catal B: Enzym **4**: 101
2203. Hu S, Hager LP (1999) Asymmetric Epoxidation of Functionalized cis-Olefins Catalyzed by Chloroperoxidase. Tetrahedron Lett **40**: 1641
2204. Manoj KM, Lakner FJ, Hager LP (2000) Epoxidation of Indene by Chloroperoxidase. J Mol Catal B: Enzym **9**: 107
2205. Sanfilippo C, Patti A, Nicolosi G (2000) Asymmetric Oxidation of 1,3-Cyclohexadiene Catalysed by Chloroperoxidase from *Caldariomyces fumago*. Tetrahedron: Asym **11**: 3269
2206. Bougioukou DJ, Smonou I (2002) Chloroperoxidase-Catalyzed Oxidation of Conjugated Dienoic Esters. Tetrahedron Lett **43**: 339
2207. Bougioukou DJ, Smonou I (2002) Mixed Peroxides from the Chloroperoxidase-Catalyzed Oxidation of Conjugated Dienoic Esters with a Trisubstituted Terminal Double Bond. Tetrahedron Lett **43**: 4511
2208. Manoj KM, Hager LP (2001) Utilization of Peroxide and Its Relevance in Oxygen Insertion Reactions Catalyzed by Chloroperoxidase. Biochim Biophys Acta **1547**: 408
2209. Miller VP, Tschirret-Guth RA, Ortiz de Montellano PR (1995) Chloroperoxidase-Catalyzed Benzylic Hydroxylation. Arch Biochem Biophys **319**: 333
2210. Hu S, Hager LP (1998) Unusual Propargylic Oxidations Catalyzed by Chloroperoxidase. Biochem Biophys Res Commun **253**: 544
2211. Hu S, Hager LP (1999) Highly Enantioselective Propargylic Hydroxylations Catalyzed by Chloroperoxidase. J Am Chem Soc **121**: 872
2212. Baciocchi E, Fabbrini M, Lanzalunga O, Manduchi L, Pochetti G (2001) Prochiral Selectivity in H_2O_2-Promoted Oxidation of Arylalkanols Catalysed by Chloroperoxidase. Eur J Biochem **268**: 665

2213. Hu S, Dordick JS (2002) Highly Enantioselective Oxidation of *cis*-Cyclopropylmethanols to Corresponding Aldehydes Catalyzed by Chloroperoxidase. J Org Chem **67**: 314
2214. van Deurzen MPJ, van Rantwijk F, Sheldon RA (1997) Chloroperoxidase-Catalyzed Oxidation of 5-Hydroxymethylfurfural. J Carbohydrate Chem **16**: 299
2215. Kiljunen E, Kanerva LT (1999) Novel Applications of Chloroperoxidase: Enantioselective Oxidation of Racemic Epoxyalcohols. Tetrahedron: Asym **10**: 3529
2216. Casella L, Poli S, Gullotti M, Selvaggini C, Beringhelli T, Marchesini A (1994) The Chloroperoxidase-Catalyzed Oxidation of Phenols. Mechanism, Selectivity, and Characterization of Enzyme-Substrate Complexes. Biochemistry **33**: 6377
2217. van Deurzen MPJ, van Rantwijk F, Sheldon RA (1996) Synthesis of Substituted Oxindoles by Chloroperoxidase Catalyzed Oxidation of Indoles. J Mol Catal B: Enzym **2**: 33
2218. Seelbach K, van Deurzen MPJ, van Rantwijk F, Sheldon RA, Kragl U (1997) Improvement of the Total Turnover Number and Space-Time Yield for Chloroperoxidase Catalyzed Oxidation. Biotechnol Bioeng **55**: 283
2219. Alvarez RG, Hunter IS, Suckling CJ, Thomas M, Vitinius U (2001) A Novel Biotransformation of Benzofurans and Related Compounds Catalysed by a Chloroperoxidase. Tetrahedron **57**: 8581
2220. Zaks A, Yabannavar AV, Dodds DR, Evans CA, Das PR, Malchow R (1996) A Novel Application of Chloroperoxidase: Preparation of *gem*-Halonitro Compounds. J Org Chem **61**: 8692
2221. Hernandez J, Robledo NR, Velasco L, Quintero R, Pickard MA, Vazquez-Duhalt R (1998) Chloroperoxidase-Mediated Oxidation of Organophosphorus Pesticides. Pestic Biochem Physiol **61**: 87
2222. Marquez-Rocha FJ, Pica-Granados Y, Sandoval-Villasana AM, Vazquez-Duhalt R (1997) Determination of Genotoxicity Using a Chloroperoxidase-Mediated Model of PAH-DNA Adduct Formation. Bull Environ Contam Toxicol **59**: 788
2223. Vázquez-Duhalt F, Ayala M, Márquez-Rocha FJ (2001) Biocatalytic Chlorination of Aromatic Hydrocarbons by Chloroperoxidase of *Caldariomyces fumago*. Phytochemistry **58**: 929
2224. Longoria A, Tinoco R, Vázquez-Duhalt R (2008) Chloroperoxidase-Mediated Transformation of Highly Halogenated Monoaromatic Compounds. Chemosphere **72**: 485
2225. van de Velde F, Arends IWCE, Sheldon RA (2000) Biocatalytic and Biomimetic Oxidations with Vanadium. J Inorg Biochem **80**: 81
2226. ten Brink HB, Dekker HL, Schoemaker HE, Wever R (2000) Oxidation Reactions Catalyzed by Vanadium Chloroperoxidase from *Curvularia inaequalis*. J Inorg Biochem **80**: 91
2227. Laatsch H, Pudleiner H, Pelizaeus B, van Pée K-H (1994) Enzymatische Bromierung von Pseudilinen und verwandten Heteroarylphenolen mit der Chlorperoxidase aus *Streptomyces aureofaciens* Tü 24. Liebigs Ann Chem: 65
2228. van Deurzen MPJ, Seelbach K, van Rantwijk F, Kragl U, Sheldon RA (1997) Chloroperoxidase: Use of a Hydrogen Peroxide-Stat for Controlling Reactions and Improving Enzyme Performance. Biocatal Biotransform **15**: 1
2229. Rai GP, Zong Q, Hager LP (2000) Isolation of Directed Evolution Mutants of Chloroperoxidase Resistant to Suicide Inactivation by Primary Olefins. Israel J Chem **40**: 63
2230. Rai GP, Sakai S, Flórez AM, Mogollon L, Hager LP (2001) Directed Evolution of Chloroperoxidase for Improved Epoxidation and Chlorination Catalysis. Adv Synth Catal **343**: 638
2231. van de Velde F, Bakker M, van Rantwijk F, Rai GP, Hager LP, Sheldon RA (2001) Engineering Chloroperoxidase for Activity and Stability. J Mol Catal B: Enzym **11**: 765
2232. Wagenknecht H-A, Woggon W-D (1997) New Active-Site Analogues of Chloroperoxidase – Syntheses and Catalytic Reactions. Angew Chem Int Ed Engl **36**: 390
2233. Loughlin WA, Hawkes DB (2000) Effect of Organic Solvents on a Chloroperoxidase Biotransformation. Bioresource Technol **71**: 167

2234. Ortiz-Bermúdez P, Srebotnik E, Hammel KE (2003) Chlorination and Cleavage of Lignin Structures by Fungal Chloroperoxidases. Appl Environ Microbiol **69**: 5015
2235. Ortiz-Bermúdez P, Hirth KC, Srebotnik E, Hammel KE (2007) Chlorination of Lignin by Ubiquitous Fungi has a Likely Role in Global Organochlorine Production. Proc Natl Acad Sci U S A **104**: 3895
2236. Laturnus F, Matucha M (2008) Chloride – A Precursor in the Formation of Volatile Organochlorines by Forest Plants? J Environ Radioact **99**: 119
2237. Hodin F, Borén H, Grimvall A, Karlsson S (1991) Formation of Chlorophenols and Related Compounds in Natural and Technical Chlorination Processes. Wat Sci Tech **24**: 403
2238. Cohen PA, Towers GHN (1997) Chlorination of Anthraquinones by Lichen and Fungal Enzymes. Phytochemistry **44**: 271
2239. Yaipakdee P, Robertson LW (2001) Enzymatic Halogenation of Flavanones and Flavones. Phytochemistry **57**: 341
2240. Moore CA, Okuda RK (1996) Bromoperoxidase Activity in 94 Species of Marine Algae. J Nat Tox **5**: 295
2241. Moore CA, Okuda RK (1999) Marine Bromoperoxidases – Chemoenzymatic Applications. In: Cooper R, Snyder JK (eds) The Biology–Chemistry Interface. Marcel Dekker, New York, p 43
2242. Butler A, Carter-Franklin JN (2004) The Role of Vanadium Bromoperoxidase in the Biosynthesis of Halogenated Marine Natural Products. Nat Prod Rep **21**: 180
2243. Flodin C, Helidoniotis F, Whitfield FB (1999) Seasonal Variation in Bromophenol Content and Bromoperoxidase Activity in *Ulva lactuca*. Phytochemistry **51**: 135
2244. Kamenarska Z, Taniguchi T, Ohsawa N, Hiraoka M, Itoh N (2007) A Vanadium-Dependent Bromoperoxidase in the Marine Red Alga *Kappaphycus alvarezii* (Doty) Doty Displays Clear Substrate Specificity. Phytochemistry **68**: 1358
2245. Rorrer GL, Tucker MP, Cheney DP, Maliakal S (2001) Bromoperoxidase Activity in Microplantlet Suspension Cultures of the Macrophytic Red Alga *Ochtodes secundiramea*. Biotechnol Bioengin **74**: 389
2246. Mehrtens G, Laturnus F (1997) Halogenating Activity in an Arctic Population of Brown Macroalga *Laminaria saccharina* (L.) Lamour. Polar Res **16**: 19
2247. Mehrtens G (1994) Haloperoxidase Activities in Arctic Macroalgae. Polar Biol **14**: 351
2248. Colin C, Leblanc C, Wagner E, Delage L, Leize-Wagner E, Van Dorsselaer A, Kloareg B, Potin P (2003) The Brown Algal Kelp *Laminaria digitata* Features Distinct Bromoperoxidase and Iodoperoxidase Activities. J Biol Chem **278**: 23545
2249. Jannun R, Coe EL (1987) Bromoperoxidase from the Marine Snail, *Murex trunculus*. Comp Biochem Physiol **88B**: 917
2250. Itoh N, Morinaga N, Kouzai T (1994) Purification and Characterization of a Novel Metal-Containing Nonheme Bromoperoxidase from *Pseudomonas putida*. Biochim Biophys Acta **1207**: 208
2251. van Pée K-H (1988) Molecular Cloning and High-Level Expression of a Bromoperoxidase Gene from *Streptomyces aureofaciens* Tü 24. J Bacteriol **170**: 5890
2252. Pfeifer O, Pelletier I, Altenbuchner J, van Pée K-H (1992) Molecular Cloning and Sequencing of a Non-Haem Bromoperoxidase Gene from *Streptomyces aureofaciens* ATCC 10762. J Gen Microbiol **138**: 1123
2253. Pelletier I, Pfeifer O, Altenbuchner J, van Pée K-H (1994) Cloning of a Second Non-Haem Bromoperoxidase Gene from *Streptomyces aureofaciens* ATCC 10762: Sequence Analysis, Expression in *Streptomyces lividans* and Enzyme Purification. Microbiology **140**: 509
2254. Facey SJ, Groβ F, Vining LC, Yang K, van Pée K-H (1996) Cloning, Sequencing and Disruption of a Bromoperoxidase-Catalase Gene in *Streptomyces venezuelae*: Evidence That it Is Not Required for Chlorination in Chloramphenicol Biosynthesis. Microbiology **142**: 657
2255. Shimonishi M, Kuwamoto S, Inoue H, Wever R, Ohshiro T, Izumi Y, Tanabe T (1998) Cloning and Expression of the Gene for a Vanadium-Dependent Bromoperoxidase from a Marine Macro-alga, *Corallina pilulifera*. FEBS Lett **428**: 105

2256. Ohshiro T, Hemrika W, Aibara T, Wever R, Izumi Y (2002) Expression of the Vanadium-Dependent Bromoperoxidase Gene from a Marine Macro-alga *Corallina pilulifera* in *Saccharomyces cerevisiae* and Characterization of the Recombinant Enzyme. Phytochemistry **60**: 595
2257. Carter JN, Beatty KE, Simpson MT, Butler A (2002) Reactivity of Recombinant and Mutant Vanadium Bromoperoxidase from the Red Alga *Corallina officinalis*. J Inorg Biochem **91**: 59
2258. Hecht HJ, Sobek H, Haag T, Pfeifer O, van Pée K-H (1994) The Metal-Ion-Free Oxidoreductase from *Streptomyces aureofaciens* has an α/β Hydrolase Fold. Struct Biol **1**: 532
2259. Brindley AA, Dalby AR, Isupov MN, Littlechild JA (1998) Preliminary X-ray Analysis of a New Crystal Form of the Vanadium-Dependent Bromoperoxidase from *Corallina officinalis*. Acta Cryst D**54**: 454
2260. Isupov MN, Dalby AR, Brindley AA, Izumi Y, Tanabe T, Murshudov GN, Littlechild JA (2000) Crystal Structure of Dodecameric Vanadium-dependent Bromoperoxidase from the Red Algae *Corallina officinalis*. J Mol Biol **299**: 1035
2261. Weyand M, Hecht H-J, Kieß M, Liaud M-F, Vilter H, Schomburg D (1999) X-ray Structure Determination of a Vanadium-dependent Haloperoxidase from *Ascophyllum nodosum* at 2.0 Å Resolution. J Mol Biol **293**: 595
2262. Dau H, Dittmer J, Epple M, Hanss J, Kiss E, Rehder D, Schulzke C, Vilter H (1999) Bromine K-edge EXAFS Studies of Bromide Binding to Bromoperoxidase from *Ascophyllum nodosum*. FEBS Lett **457**: 237
2263. Christmann U, Dau H, Haumann M, Kiss E, Liebisch P, Rehder D, Santoni G, Schulzke C (2004) Substrate Binding to Vanadate-Dependent Bromoperoxidase from *Ascophyllum nodosum*: A Vanadium K-edge XAS Approach. Dalton Trans 2534
2264. Clague MJ, Keder NL, Butler A (1993) Biomimics of Vanadium Bromoperoxidase: Vanadium(V)-Schiff Base Catalyzed Oxidation of Bromide by Hydrogen Peroxide. Inorg Chem **32**: 4754
2265. Andersson M, Conte V, Di Furia F, Moro S (1995) Vanadium Bromoperoxidases Mimicking Systems: Bromohydrins Formation as Evidence of the Occurrence of a Hypobromite-Like Vanadium Complex. Tetrahedron Lett **36**: 2675
2266. Clague MJ, Butler A (1995) On the Mechanism of *cis*-Dioxovanadium(V)-Catalyzed Oxidation of Bromide by Hydrogen Peroxide: Evidence for a Reactive, Binuclear Vanadium(V) Peroxo Complex. J Am Chem Soc **117**: 3475
2267. Soedjak HS, Walker JV, Butler A (1995) Inhibition and Inactivation of Vanadium Bromoperoxidase by the Substrate Hydrogen Peroxide and Further Mechanistic Studies. Biochemistry **34**: 12689
2268. Rothenberg G, Clark JH (2000) On Oxyhalogenation, Acids, and Non-Mimics of Bromoperoxidase Enzymes. Green Chem **2**: 248
2269. Kimblin C, Bu X, Butler A (2002) Modeling the Catalytic Site of Vanadium Bromoperoxidase: Synthesis and Structural Characterization of Intramolecularly H-bonded Vanadium(V) Oxoperoxo Complexes, [VO(O$_2$)(^{NH}2pyg$_2$)]K and [VO(O$_2$)(^{BrNH}2pyg$_2$)]K. Inorg Chem **41**: 161
2270. Andersson M, Willetts A, Allenmark S (1997) Asymmetric Sulfoxidation Catalyzed by a Vanadium-Containing Bromoperoxidase. J Org Chem **62**: 8455
2271. Andersson MA, Allenmark SG (1998) Asymmetric Sulfoxidation Catalyzed by a Vanadium Bromoperoxidase: Substrate Requirements of the Catalyst. Tetrahedron **54**: 15293
2272. Andersson M, Sundell S, Allenmark S (1999) Absolute Configuration of Sulfoxides Obtained Via Haloperoxidase-catalyzed Reactions. Enantiomer **4**: 411
2273. Andersson M, Allenmark S (2000) The Potential of Vanadium Bromoperoxidase as a Catalyst in Preparative Asymmetric Sulfoxidation. Biocatal Biotransform **18**: 79
2274. Křen V, Kawuloková L, Sedmera P, Polášek M, Lindhorst TK, van Pée K-H (1997) Biotransformation of Ergot Alkaloids by Haloperoxidase from *Streptomyces aureofaciens*: Stereoselective Acetoxylation and Propionoxylation. Liebigs Ann/Recueil 2379

2275. Martinez JS, Carroll GL, Tschirret-Guth RA, Altenhoff G, Little RD, Butler A (2001) On the Regiospecificity of Vanadium Bromoperoxidase. J Am Chem Soc **123**: 3289
2276. Fukuzawa A, Aye M, Takasugi Y, Nakamura M, Tamura M, Murai A (1994) Enzymatic Bromo-ether Cyclization of Laurediols with Bromoperoxidase. Chem Lett 2307
2277. Ishihara J, Kanoh N, Murai A (1995) Enzymatic Reaction of (3E,6S,7S)-Laurediol and the Molecular Modeling Studies on the Cyclization of Laurediols. Tetrahedron Lett **36**: 737
2278. Ishihara J, Shimada Y, Kanoh N, Takasugi Y, Fukuzawa A, Murai A (1997) Conversion of Prelaureatin into Laurallene, a Bromo-Allene Compound, by Enzymatic and Chemical Bromo-Etherification Reactions. Tetrahedron **53**: 8371
2279. Carter-Franklin JN, Parrish JD, Tschirret-Guth RA, Little RD, Butler A (2003) Vanadium Haloperoxidase-Catalyzed Bromination and Cyclization of Terpenes. J Am Chem Soc **125**: 3688
2280. Carter-Franklin JN, Butler A (2004) Vanadium Bromoperoxidase-Catalyzed Biosynthesis of Halogenated Marine Natural Products. J Am Chem Soc **126**: 15060
2281. Flodin C, Whitfield FB (1999) Biosynthesis of Bromophenols in Marine Algae. Wat Sci Tech **40**: 53
2282. Ohsawa N, Ogata Y, Okada N, Itoh N (2001) Physiological Function of Bromoperoxidase in the Red Marine Alga, *Corallina pilulifera*: Production of Bromoform as an Allelochemical and the Simultaneous Elimination of Hydrogen Peroxide. Phytochemistry **58**: 683
2283. Almeida M, Humanes M, Melo R, Silva JA, Fraústo da Silva JJR, Vilter H, Wever R. (1998) *Saccorhiza polyschides* (Phaeophyceae; Phyllariaceae) a New Source for Vanadium-Dependent Haloperoxidases. Phytochemistry **48**: 229
2284. Almeda M, Humanes M, Melo R, Silva, JAL, Fraústo da Silva JJR (1996) *Phyllariopsis brevipes* – A Brown Alga with Vanadium Dependent Iodoperoxidase. In: Obringer C, Burner U, Ebermann R, Penel C, Greppin H (eds) Plant Peroxidases – Biochemistry and Physiology. University of Geneva, Geneva, p 146
2285. Almeida M, Almeida MG, Humanes M, Melo R, Silva JA, Fraústo da Silva JJR (1997) Novel Vanadium-Dependent Haloperoxidases in Brown Algae (Fucaceae and Laminariaceae) from Portugal. Phycologia **36**: 1
2286. Almeida MG, Humanes M, Melo R, Silva JA, Fraústo da Silva JJR, Wever R (2000) Purification and Characterisation of Vanadium Haloperoxidases from the Brown Alga *Pelvetia canaliculata*. Phytochemistry **54**: 5
2287. Küpper FC, Schweigert N, Ar Gall E, Legendre J-M, Vilter H, Kloareg B (1998) Iodine Uptake in Laminariales Involves Extracellular, Haloperoxidase-Mediated Oxidation of Iodide. Planta **207**: 163
2288. Murphy CD, Moore RM, White RL (2000) Peroxidases from Marine Microalgae. J Appl Phycol **12**: 507
2289. Vilter H (1983) Peroxidases from Phaeophyceae III: Catalysis of Halogenation by Peroxidases from *Ascophyllum nodosum* (L.) Le Jol. Bot Mar **XXVI**: 429
2290. van Deurzen MPJ, van Rantwijk F, Sheldon RA (1997) Selective Oxidations Catalyzed by Peroxidases. Tetrahedron **53**: 13183
2291. Sun W, Dunford HB (1993) Kinetics and Mechanism of the Peroxidase-Catalyzed Iodination of Tyrosine. Biochemistry **32**: 1324
2292. Veitch NC (2004) Horseradish Peroxidase: A Modern View of a Classic Enzyme. Phytochemistry **65**: 249
2293. Kazunga C, Aitken MD, Gold A (1999) Primary Product of the Horseradish Peroxidase-Catalyzed Oxidation of Pentachlorophenol. Environ Sci Technol **33**: 1408
2294. Srebotnik E, Jensen Jr. KA, Hammel KE (1994) Fungal Degradation of Recalcitrant Nonphenolic Lignin Structures without Lignin Peroxidase. Proc Natl Acad Sci U S A **91**: 12794
2295. Johjima T, Itoh N, Kabuto M, Tokimura F, Nakagawa T, Wariishi H, Tanaka H (1999) Direct Interaction of Lignin and Lignin Peroxidase from *Phanerochaete chrysosporium*. Proc Natl Acad Sci U S A **96**: 1989

2296. Renganathan V, Miki K, Gold MH (1987) Haloperoxidase Reactions Catalyzed by Lignin Peroxidase, an Extracellular Enzyme from the Basidiomycete *Phanerochaete chrysosporium*. Biochemistry **26**: 5127
2297. Sheng D, Gold MH (1997) Haloperoxidase Activity of Manganese Peroxidase from *Phanerochaete chrysosporium*. Arch Biochem Biophys **345**: 126
2298. Ullrich R, Nüske J, Scheibner K, Spantzel J, Hofrichter M (2004) Novel Haloperoxidase from the Agaric Basidiomycete *Agrocybe aegerita* Oxidizes Aryl Alcohols and Aldehydes. Appl Environ Microbiol **70**: 4575
2299. Jin N, Bourassa JL, Tizio SC, Groves JT (2000) Rapid, Reversible Oxygen Atom Transfer between an Oxomanganese(V) Porphyrin and Bromide: A Haloperoxidase Mimic with Enzymatic Rates. Angew Chem Int Ed **39**: 3849
2300. Verdel EF, Kline PC, Wani S, Woods AE (2000) Purification and Partial Characterization of Haloperoxidase from Fresh Water Algae *Cladophora glomerata*. Comp Biochem Physiol **125B**: 179
2301. de Schrijver A, Nagy I, Schoofs G, Proost P, Vanderleyden J, van Pée K-H, de Mot R (1997) Thiocarbamate Herbicide-Inducible Nonheme Haloperoxidase of *Rhodococcus erythropolis* NI86/21. Appl Environ Microbiol **63**: 1911
2302. Almeida M, Filipe S, Humanes M, Maia MF, Melo R, Severino N, da Silva JAL, Fraústo da Silva JJR, Wever R (2001) Vanadium Haloperoxidases from Brown Algae of the Laminariaceae Family. Phytochemistry **57**: 633
2303. Hofrichter M, Ullrich R (2006) Heme-Thiolate Haloperoxidases: Versatile Biocatalysts with Biotechnological and Environmental Significance. Appl Microbiol Biotechnol **71**: 276
2304. Hohaus K, Altmann A, Burd W, Fischer I, Hammer PE, Hill DS, Ligon JM, van Pée K-H (1997) NADH-Dependent Halogenases Are More Likely To Be Involved in Halometaolite Biosynthesis Than Haloperoxidases. Angew Chem Int Ed Engl **36**: 2012
2305. Keller S, Wage T, Hohaus K, Hölzer M, Eichhorn E, van Pée K-H (2000) Purification and Partial Characterization of Tryptophan 7-Halogenase (PrnA) from *Pseudomonas fluorescens*. Angew Chem Int Ed **39**: 2300
2306. Unversucht S, Hollmann F, Schmid A, van Pée K-H (2005) $FADH_2$-Dependence of Tryptophan 7-Halogenase. Adv Synth Catal **347**: 1163
2307. Dong C, Flecks S, Unversucht S, Haupt C, van Pée K-H, Naismith JH (2005) Tryptophan 7-Halogenase (PrnA) Structure Suggests a Mechanism for Regioselective Chlorination. Science **309**: 2216
2308. van Pée K-H, Patallo EP (2006) Flavin-Dependent Halogenases Involved in Secondary Metabolism in Bacteria. Appl Microbiol Biotechnol **70**: 631
2309. Zehner S, Kotzsch A, Bister B, Süssmuth RD, Méndez C, Salas JA, van Pée K-H (2005) A Regioselective Tryptophan 5-Halogenase Is Involved in Pyrroindomycin Biosynthesis in *Streptomyces rugosporus* LL-42D005. Chem Biol **12**: 445
2310. Seibold C, Schnerr H, Rumpf J, Kunzendorf A, Hatscher C, Wage T, Ernyei AJ, Dong C, Naismith JH, van Pée K-H (2006) A Flavin-Dependent Tryptophan 6-Halogenase and Its Use in Modification of Pyrrolnitrin Biosynthesis. Biocatal Biotransform **24**: 401
2311. Dolfing J (1998) Halogenation of Aromatic Compounds: Thermodynamic, Mechanistic and Ecological Aspects. FEMS Microbiol Lett **167**: 271
2312. Wynands I, van Pée K-H (2004) A Novel Halogenase Gene from the Pentachloropseudilin Producer *Actinoplanes* sp. ATCC 33002 and Detection of *In Vitro* Halogenase Activity. FEMS Microbiol Lett **237**: 363
2313. Yarnell A (2006) Nature's X-Factors. C&EN **84**: 12
2314. Vaillancourt FH, Yeh E, Vosburg DA, O'Connor SE, Walsh CT (2005) Cryptic Chlorination by a Non-Haem Iron Enzyme During Cyclopropyl Amino Acid Biosynthesis. Nature **436**: 1191
2315. Schnarr NA, Khosla C (2005) Just Add Chlorine. Nature **436**: 1094
2316. Pelletier I, Altenbuchner J (1995) A Bacterial Esterase is Homologous with Non-Haem Haloperoxidases and Displays Brominating Activity. Microbiology **141**: 459

2317. Tanaka N, Dumay V, Liao Q, Lange AJ, Wever R (2002) Bromoperoxidase Activity of Vanadate-Substituted Acid Phosphatases from *Shigella flexneri* and *Salmonella enterica* ser. *typhimurium*. Eur J Biochem **269**: 2162
2318. Kataoka M, Honda K, Shimizu S (2000) 3,4-Dihydrocoumarin Hydrolase with Haloperoxidase Activity from *Acinetobacter calcoaceticus* F46. Eur J Biochem **267**: 3
2319. Wichard T, Pohnert G (2006) Formation of Halogenated Medium Chain Hydrocarbons by a Lipoxygenase/Hydroperoxide Halolyase-Mediated Transformation in Planktonic Microalgae. J Am Chem Soc **128**: 7114
2320. van Pée K-H, Keller S, Wage T, Wynands I, Schnerr H, Zehner S (2000) Enzymatic Halogenation Catalyzed via a Catalytic Triad and by Oxidoreductases. Biol Chem **381**: 1
2321. Attieh JM, Hanson AD, Saini HS (1995) Purification and Characterization of a Novel Methyltransferase Responsible for Biosynthesis of Halomethanes and Methanethiol in *Brassica oleracea*. J Biol Chem **270**: 9250
2322. Saini HS, Attieh JM, Hanson AD (1995) Biosynthesis of Halomethanes and Methanethiol by Higher Plants via a Novel Methyltransferase Reaction. Plant Cell Environ **18**: 1027
2323. Vaillancourt FH, Yeh E, Vosburg DA, Garneau-Tsodikova S, Walsh CT (2006) Nature's Inventory of Halogenation Catalysts: Oxidative Strategies Predominate. Chem Rev **106**: 3364
2324. Rush C, Willetts A, Davies G, Dauter Z, Watson H, Littlechild J (1995) Purification, Crystallisation and Preliminary X-Ray Analysis of the Vanadium-Dependent Haloperoxidase from *Corallina officinalis*. FEBS Lett **359**: 244
2325. Zampella G, Fantucci P, Pecoraro VL, De Gioia L (2005) Reactivity of Peroxo Forms of the Vanadium Haloperoxidase Cofactor. A DFT Investigation. J Am Chem Soc **127**: 953
2326. ten Brink HB, Tuynman A, Dekker HL, Hemrika W, Izumi Y, Oshiro T, Schoemaker HE, Wever R (1998) Enantioselective Sulfoxidation Catalyzed by Vanadium Haloperoxidases. Inorg Chem **37**: 6780
2327. van Deurzen MPJ, van Rantwijk F, Sheldon RA (1997) Selective Oxidations Catalyzed by Peroxidases. Tetrahedron **53**: 13183
2328. Feiters MC, Leblanc C, Küpper FC, Meyer-Klaucke W, Michel G, Potin P (2005) Bromine is an Endogenous Component of a Vanadium Bromoperoxidase. J Am Chem Soc **127**: 15340
2329. Rehder D (2008) Bioinorganic Vanadium Chemistry. Wiley, Chichester, UK, p 105
2330. Kanbe K, Okamura M, Hattori S, Naganawa H, Hamada M, Okami Y, Takeuchi T (1993) Thienodolin, a New Plant Growth-Regulating Substance Produced by a Streptomycete Strain: I. Taxonomy and Fermentation of the Producing Strain, and the Isolation and Characterization of Thienodolin. Biosci Biotech Biochem **57**: 632
2331. Kanbe K, Naganawa H, Nakamura KT, Okami Y, Takeuchi T (1993) Thienodolin, a New Plant Growth-Regulating Substance Produced by a Streptomycete Strain: II. Structure of Thienodolin. Biosci Biotech Biochem **57**: 636
2332. Schaur JR, Jerlich A, Stelmaszyńska T (1998) Hypochlorous Acid as Reactive Oxygen Species. Curr Top Biophys **22** (Supplement B): 176
2333. Spickett CM, Jerlich A, Panasenko OM, Arnhold J, Pitt AR, Stelmaszyńska T, Schaur RJ (2000) The Reactions of Hypochlorous Acid, the Reactive Oxygen Species Produced by Myeloperoxidase, with Lipids. Acta Biochim Pol **47**: 889
2334. Henderson JP, Heinecke JW (2003) Myeloperoxidase and Eosinophil Peroxidase: Phagocyte Enzymes for Halogenation in Humans. In: Gribble GW (ed) Natural Production of Organohalogen Compounds, The Handbook of Environmental Chemistry, vol 3, part P. Springer, Berlin, p 201
2335. Hawkins CL, Pattison DI, Davies MJ (2003) Hypochlorite-Induced Oxidation of Amino Acids, Peptides and Proteins. Amino Acids **25**: 259
2336. Davies MJ, Hawkins CL, Pattison DI, Rees MD (2008) Mammalian Heme Peroxidases: From Molecular Mechanisms to Health Implications. Antioxid Redox Signaling **10**: 1199

2337. Suzuki T, Ohshima H (2002) Nicotine-Modulated Formation of Spiroiminodihydantoin Nucleoside via 8-Oxo-7,8-dihydro-2'-deoxyguanosine in 2'-Deoxyguanosine-Hypochlorous Acid Reaction. FEBS Lett **516**: 67
2338. Fiedler TJ, Davey CA, Fenna RE (2000) X-Ray Crystal Structure and Characterization of Halide-Binding Sties of Human Myeloperoxidase at 1.8Å Resolution. J Biol Chem **275**: 11964
2339. Gaut JP, Yeh GC, Tran HD, Byun J, Henderson JP, Richter GM, Brennan M-L, Lusis AJ, Belaaouaj A, Hotchkiss RS, Heinecke JW (2001) Neutrophils Employ the Myeloperoxidase System to Generate Antimicrobial Brominating and Chlorinating Oxidants During Sepsis. Proc Natl Acad Sci U S A **98**: 11961
2340. Weiss SJ, Klein R, Slivka A, Wei M (1982) Chlorination of Taurine by Human Neutrophils. J Clin Invest **70**: 598
2341. Weiss SJ, Lampert MB, Test ST (1983) Long-Lived Oxidants Generated by Human Neutrophils: Characterization and Bioactivity. Science **222**: 625
2342. Grisham MB, Jefferson MM, Melton DF, Thomas EL (1984) Chlorination of Endogenous Amines by Isolated Neutrophils. Ammonia-Dependent Bactericidal, Cytotoxic, and Cytolytic Activities of the Chloramines. J Biol Chem **259**: 10404
2343. Marquez LA, Dunford HB (1994) Chlorination of Taurine by Myeloperoxidase. Kinetic Evidence for an Enzyme-Bound Intermediate. J Biol Chem **269**: 7950
2344. Domigan NM, Charlton TS, Duncan MW, Winterbourn CC, Kettle AJ (1995) Chlorination of Tyrosyl Residues in Peptides by Myeloperoxidase and Human Neutrophils. J Biol Chem **270**: 16542
2345. Pero RW, Sheng Y, Olsson A, Bryngelsson C, Lund-Pero M (1996) Hypochlorous Acid/N-Chloramines are Naturally Produced DNA Repair Inhibitors. Carcinogenesis **17**: 13
2346. Marcinkiewicz J (1997) Neutrophil Chloramines: Missing Links Between Innate and Acquired Immunity. Immunol Today **18**: 577
2347. Hazell LJ, Davies MJ, Stocker R (1999) Secondary Radicals Derived from Chloramines of Apolipoprotein B-10 Contribute to HOCl-Induced Lipid Peroxidation of Low-Density Lipoproteins. Biochem J **339**: 489
2348. Hawkins CL, Davies MJ (2001) Hypochlorite-Induced Damage to Nucleosides: Formation of Chloramines and Nitrogen-Centered Radicals. Chem Res Toxicol **14**: 1071
2349. Hawkins CL, Davies MJ (2002) Hypochlorite-Induced Damage to DNA, RNA, and Polynucleotides: Formation of Chloramines and Nitrogen-Centered Radicals. Chem Res Toxicol **15**: 83
2350. Hawkins CL, Pattison DI, Davies MJ (2002) Reaction of Protein Chloramines with DNA and Nucleosides: Evidence for the Formation of Radicals, Protein-DNA Cross-Links and DNA Fragmentation. Biochem J **365**: 605
2351. Pattison DI, Hawkins CL, Davies MJ (2007) Hypochlorous Acid-Mediated Protein Oxidation: How Important Are Chloramine Transfer Reactions and Protein Tertiary Structure? Biochemistry **46**: 9853
2352. Wang L, Khosrovi B, Najafi R (2008) N-Chloro-2,2-dimethyltaurines: A New Class of Remarkably Stable N-Chlorotaurines. Tetrahedron Lett **49**: 2193
2353. Henderson JP, Byun J, Heinecke JW (1999) Molecular Chlorine Generated by the Myeloperoxidase-Hydrogen Peroxide-Chloride System of Phagocytes Produces 5-Chlorocytosine in Bacterial RNA. J Biol Chem **274**: 33440
2354. Albert CJ, Crowley JR, Hsu F-F, Thukkani AK, Ford DA (2001) Reactive Chlorinating Species Produced by Myeloperoxidase Target the Vinyl Ether Bond of Plasmalogens. J Biol Chem **276**: 23733
2355. Jerlich A, Pitt AR, Schaur RJ, Spickett CM (2000) Pathways of Phospholipid Oxidation by HOCl in Human LDL Detected by LC-MS. Free Radic Biol Med **28**: 673
2356. Arnhold J, Osipov AN, Spalteholz H, Panasenko OM, Schiller J (2001) Effects of Hypochlorous Acid on Unsaturated Phosphatidylcholines. Free Radic Biol Med **31**: 1111

2357. Anderson MM, Hazen SL, Hsu FF, Heinecke JW (1997) Human Neutrophils Employ the Myeloperoxidase-Hydrogen Peroxide-Chloride System to Convert Hydroxy-Amino Acids into Glycolaldehyde, 2-Hydroxypropanal, and Acrolein. J Clin Invest **99**: 424
2358. Hazen SL, d'Avignon A, Anderson MM, Hsu FF, Heinecke JW (1998) Human Neutrophils Employ the Myeloperoxidase-Hydrogen Peroxide-Chloride System to Oxidize α-Amino Acids to a Family of Reactive Aldehydes. J Biol Chem **273**: 4997
2359. Bergt C, Fu X, Huq NP, Kao J, Heinecke JW (2004) Lysine Residues Direct the Chlorination of Tyrosines in YXXK Motifs of Apolipoprotein A-I When Hypochlorous Acid Oxidizes High Density Lipoprotein. J Biol Chem **279**: 7856
2360. Zheng L, Settle M, Brubaker G, Schmitt D, Hazen SL, Smith JD, Kinter M (2005) Localization of Nitration and Chlorination Sites on Apolipoprotein A–I Catalyzed by Myeloperoxidase in Human Atheroma and Associated Oxidative Impairment in ABCA1-Dependent Cholesterol Efflux from Macrophages. J Biol Chem **280**: 38
2361. Shao B, Bergt C, Fu X, Green P, Voss JC, Oda MN, Oram JF, Heinecke JW (2005) Tyrosine 192 in Apolipoprotein A-I Is the Major Site of Nitration and Chlorination by Myeloperoxidase, but Only Chlorination Markedly Impairs ABCA1-Dependent Cholesterol Transport. J Biol Chem **280**: 5983
2362. Fu X, Wang Y, Kao J, Irwin A, d'Avignon A, Mecham RP, Parks WC, Heinecke JW (2006) Specific Sequence Motifs Direct the Oxygenation and Chlorination of Tryptophan by Myeloperoxidase. Biochemistry **45**: 3961
2363. Masuda M, Suzuki T, Friesen MD, Ravanat J-L, Cadet J, Pignatelli B, Nishino H, Ohshima H (2001) Chlorination of Guanosine and Other Nucleosides by Hypochlorous Acid and Myeloperoxidase of Activated Human Neutrophils. J Biol Chem **276**: 40486
2364. Jiang Q, Blount BC, Ames BN (2003) 5-Chlorouracil, a Marker of DNA Damage from Hypochlorous Acid During Inflammation. J Biol Chem **278**: 32834
2365. Daugherty A, Dunn JL, Rateri DL, Heinecke JW (1994) Myeloperoxidase, a Catalyst for Lipoprotein Oxidation, Is Expressed in Human Atherosclerotic Lesions. J Clin Invest **94**: 437
2366. Hazell LJ, Arnold L, Flowers D, Waeg G, Malle E, Stocker R (1996) Presence of Hypochlorite-Modified Proteins in Human Atherosclerotic Lesions. J Clin Invest **97**: 1535
2367. Heinecke JW (1997) Mechanisms of Oxidative Damage of Lipoprotein in Human Atherosclerosis. Curr Opin Lipidol **8**: 268
2368. Kulcharyk PA, Heinecke JW (2001) Hypochlorous Acid Produced by the Myeloperoxidase System of Human Phagocytes Induces Covalent Cross-Links Between DNA and Protein. Biochemistry **40**: 3648
2369. Zhang R, Brennan M-L, Fu X, Aviles RJ, Pearce GL, Penn MS, Topol EJ, Sprecher DL, Hazen SL (2001) Association Between Myeloperoxidase Levels and Risk of Coronary Artery Disease. JAMA **286**: 2136
2370. Fu X, Mueller DM, Heinecke JW (2002) Generation of Intramolecular and Intermolecular Sulfenamides, Sulfinamides, and Sulfonamides by Hypochlorous Acid: A Potential Pathway for Oxidative Cross-Linking of Low-Density Lipoprotein by Myeloperoxidase. Biochemistry **41**: 1293
2371. Pattison DI, Hawkins CL, Davies MJ (2003) Hypochlorous Acid-Mediated Oxidation of Lipid Components and Antioxidants Present in Low-Density Lipoproteins: Absolute Rate Constants, Product Analysis, and Computational Modeling. Chem Res Toxicol **16**: 439
2372. Shao B, Oda MN, Bergt C, Fu X, Green PS, Brot N, Oram JF, Heinecke JW (2006) Myeloperoxidase Impairs ABCA1-Dependent Cholesterol Efflux through Methionine Oxidation and Site-Specific Tyrosine Chlorination of Apolipoprotein A–I. J Biol Chem **281**: 9001
2373. Nauseef WM (2001) The Proper Study of Mankind. J Clin Invest **107**: 401
2374. Carr AC, Tijerina T, Frei B (2000) Vitamin C Protects Against and Reverses Specific Hypochlorous Acid- and Chloramine-Dependent Modifications of Low-Density Lipoprotein. Biochem J **346**: 491
2375. Aratani Y, Koyama H, Nyui S-I, Suzuki K, Kura F, Maeda N (1999) Severe Impairment in Early Host Defense Against *Candida albicans* in Mice Deficient in Myeloperoxidase. Infect Immun **67**: 1828

2376. Shen Z, Mitra SN, Wu W, Chen Y, Yang Y, Qin J, Hazen SL (2001) Eosinophil Peroxidase Catalyzes Bromination of Free Nucleosides and Double-Stranded DNA. Biochemistry **40**: 2041
2377. Hawkins CL, Davies MJ (2005) The Role of Reactive *N*-Bromo Species and Radical Intermediates in Hypobromous Acid-Induced Protein Oxidation. Free Radic Biol Med **39**: 900
2378. Hawkins CL, Davies MJ (2005) The Role of Aromatic Amino Acid Oxidation, Protein Unfolding, and Aggregation in the Hypobromous Acid-Induced Inactivation of Trypsin Inhibitor and Lysozyme. Chem Res Toxicol **18**: 1669
2379. Carr, AC, Winterbourn CC, van den Berg JJM (1996) Peroxidase-Mediated Bromination of Unsaturated Fatty Acids to Form Bromohydrins. Arch Biochem Biophys **327**: 227
2380. Carr AC, Decker EA, Park Y, Frei B (2001) Comparison of Low-Density Lipoprotein Modification by Myeloperoxidase-Derived Hypochlorous and Hypobromous Acids. Free Radic Biol Med **31**: 62
2381. Skaff O, Pattison DI, Davies MJ (2007) Kinetics of Hypobromous Acid-Mediated Oxidation of Lipid Components and Antioxidants. Chem Res Toxicol **20**: 1980
2382. Heinecke JW (2000) Eosinophil-Dependent Bromination in the Pathogenesis of Asthma. J Clin Invest **105**: 1331
2383. Winterton N (1997) Are Organochlorine Compounds Created in the Human Body? Mutation Res **373**: 293
2384. Schöler HF, Keppler F (2003) Abiotic Formation of Organohalogens During Early Diagenetic Processes. In: Gribble GW (ed) Natural Production of Organohalogen Compounds, The Handbook of Environmental Chemistry, vol 3, part P. Springer-Verlag, Berlin Heidelberg, p 63
2385. Kang N, Lee DS, Yoon J (2002) Kinetic Modeling of Fenton Oxidation of Phenol and Monochlorophenols. Chemosphere **47**: 915
2386. Duesterberg CK, Waite TD (2006) Process Optimization of Fenton Oxidation Using Kinetic Modeling. Environ Sci Technol **40**: 4189
2387. Keppler F, Eiden R, Niedan V, Pracht J, Schöler HF (2000) Halocarbons Produced by Natural Oxidation Processes During Degradation of Organic Matter. Nature **403**: 298
2388. Keppler F, Borchers R, Elsner P, Fahimi I, Pracht J, Schöler HF (2003) Formation of Volatile Iodinated Alkanes in Soil: Results from Laboratory Studies. Chemosphere **52**: 477
2389. Wishkerman A, Gebhardt S, McRoberts CW, Hamilton JTG, Williams J, Keppler F (2008) Abiotic Methyl Bromide Formation from Vegetation, and Its Strong Dependence on Temperature. Environ Sci Technol **42**: 6837
2390. O'Hagan D, Harper DB (1999) Fluorine-Containing Natural Products. J Fluorine Chem **100**: 127
2391. Harper DB, O'Hagan D, Murphy CD (2003) Fluorinated Natural Products: Occurrence and Biosynthesis. In: Gribble GW (ed) Natural Production of Organohalogen Compounds, The Handbook of Environmental Chemistry, vol 3, part P. Springer, Berlin, p 141
2392. Schaffrath C, Deng H, O'Hagan D (2003) Isolation and Characterisation of 5'-Fluorodeoxyadenosine Synthase, a Fluorination Enzyme from *Streptomyces cattleya*. FEBS Lett **547**: 111
2393. Dong C, Deng H, Dorward M, Schaffrath C, O'Hagan D, Naismith JH (2003) Crystallization and X-Ray Diffraction of 5'-Fluoro-5'-deoxyadenosine Synthase, a Fluorination Enzyme from *Streptomyces cattleya*. Acta Cryst D**59**: 2292
2394. Zhu X, Robinson DA, McEwan AR, O'Hagan D, Naismith JH (2007) Mechanism of Enzymatic Fluorination in *Streptomyces cattleya*. J Am Chem Soc **129**: 14597
2395. O'Hagan D (2006) Recent Developments on the Fluorinase from *Streptomyces cattleya*. J Fluorine Chem **127**: 1479
2396. Martarello L, Schaffrath C, Deng H, Gee AD, Lockhart A, O'Hagan D (2003) The First Enzymatic Method for C-^{18}F Bond Formation: The Synthesis of 5'-[^{18}F]-Fluoro-5'-deoxyadenosine for Imaging with PET. J Label Compd Radiopharm **46**: 1181

2397. Deng H, Cobb SL, Gee AD, Lockhart A, Martarello L, McGlinchey RP, O'Hagan D, Onega M (2006) Fluorinase Mediated C-^{18}F Bond Formation, an Enzymatic Tool for PET Labelling. Chem Commun 652
2398. Garson MJ (1989) Biosynthetic Studies on Marine Natural Products. Nat Prod Rep **6**: 143
2399. Garson MJ (1993) The Biosynthesis of Marine Natural Products. Chem Rev **93**: 1699
2400. Moore BS (2005) Biosynthesis of Marine Natural Products: Microorganisms (Part A). Nat Prod Rep **22**: 580
2401. Moore BS (2006) Biosynthesis of Marine Natural Products: Macroorganisms (Part B). Nat Prod Rep **23**: 615
2402. Field JA, Wijnberg JBPA (2003) An Update on Organohalogen Metabolites Produced by Basidiomycetes. In: Gribble GW (ed) Natural Production of Organohalogen Compounds, The Handbook of Environmental Chemistry, vol 3, part P. Springer, Berlin, p 103
2403. Gribble GW (2008) Structure and Biosynthesis of Halogenated Alkaloids. In: Fattorusso E, Taglialatela-Scafati O (eds) Modern Alkaloids: Structure, Isolation, Synthesis and Biology. Wiley, Weinheim, p 591
2404. Walsh CT, Garneau-Tsodikova S, Howard-Jones AR (2006) Biological Formation of Pyrroles: Nature's Logic and Enzymatic Machinery. Nat Prod Rep **23**: 517
2405. Kawamura N, Sawa R, Takahashi Y, Sawa T, Naganawa H, Takeuchi T (1996) Pyralomicins, Novel Antibiotics from *Microtetraspora spiralis*. III. Biosynthesis of Pyralomicin 1a. J Antibiot **49**: 657
2406. Naganawa H, Hashizume H, Kubota Y, Sawa R, Takahashi Y, Arakaws K, Bowers SG, Mahmud T (2002) Biosynthesis of the Cyclitol Moiety of Pyralomicin 1a in *Nonomuraea spiralis* MI178-34F18. J Antibiot **55**: 578
2407. Raggatt ME, Simpson TJ, Wrigley SK (1999) Biosynthesis of XR587 (Streptopyrrole) in *Streptomyces rimosus* Involves a Novel Carbon-to-Nitrogen Rearrangement of a Proline-Derived Unit. Chem Commun 1039
2408. Hanefeld U, Floss HG, Laatsch H (1994) Biosynthesis of the Marine Antibiotic Pentabromopseudilin. 1. The Benzene Ring. J Org Chem **59**: 3604
2409. Peschke JD, Hanefeld U, Laatsch H (2005) Biosynthesis of the Marine Antibiotic Pentabromopseudilin. 2. The Pyrrole Ring. Biosci Biotechnol Biochem **69**: 628
2410. Andrade P, Willoughby R, Pomponi SA, Kerr RG (1999) Biosynthetic Studies of the Alkaloid, Stevensine, in a Cell Culture of the Marine Sponge *Teichaxinella morchella*. Tetrahedron Lett **40**: 4775
2411. Mourabit AA, Potier P (2001) Sponge's Molecular Diversity Through the Ambivalent Reactivity of 2-Aminoimidazole: A Universal Chemical Pathway to the Oroidin-Based Pyrrole-Imidazole Alkaloids and Their Palau'amine Congeners. Eur J Org Chem 237
2412. Travert N, Al-Mourabit A (2004) A Likely Biogenetic Gateway Linking 2-Aminoimidazolinone Metabolites of Sponges to Proline: Spontaneous Oxidative Conversion of the Pyrrole-Proline-Guanidine Pseudo-Peptide to Dispacamide A. J Am Chem Soc **126**: 10252
2413. Charan RD, Schlingmann G, Bernan VS, Feng X, Carter GT (2006) Dioxapyrrolomycin Biosynthesis in *Streptomyces fumanus*. J Nat Prod **69**: 29
2414. Garneau S, Dorrestein PC, Kelleher NL, Walsh CT (2005) Characterization of the Formation of the Pyrrole Moiety During Clorobiocin and Coumermycin A$_1$ Biosynthesis. Biochemistry **44**: 2770
2415. Stanley AE, Walton LJ, Zerikly MK, Corre C, Challis GL (2006) Elucidation of the *Streptomyces coelicolor* Pathway to 4-Methoxy-2,2'-bipyrrole-5-carboxaldehyde, An Intermediate in Prodiginine Biosynthesis. Chem Commun 3981
2416. Sánchez C, Zhu L, Braña AF, Salas AP, Rohr J, Méndez C, Salas JA (2005) Combinatorial Biosynthesis of Antitumor Indolocarbazole Compounds. Proc Natl Acad Sci U S A **102**: 461
2417. Walsh C, Meyers CLF, Losey HC (2003) Antibiotic Glycosyltransferases: Antibiotic Maturation and Prospects for Reprogramming. J Med Chem **46**: 3425

2418. Kim C-G, Kirschning A, Bergon P, Zhou P, Su E, Sauerbrei B, Ning S, Ahn Y, Breuer M, Leistner E, Floss HG (1996) Biosynthesis of 3-Amino-5-hydroxybenzoic Acid, the Precursor of mC$_7$N Units in Ansamycin Antibiotics. J Am Chem Soc **118**: 7486

2419. Moss SJ, Bai L, Toelzer S, Carroll BJ, Mahmud T, Yu T-W, Floss HG (2002) Identification of Asm19 as an Acyltransferase Attaching the Biologically Essential Ester Side Chain of Ansamitocins Using *N*-Desmethyl-4,5-desepoxymaytansinol, Not Maytansinol, as Its Substrate. J Am Chem Soc **124**: 6544

2420. Yu T-W, Bai L, Clade D, Hoffmann D, Toelzer S, Trinh KQ, Xu J, Moss SJ, Leistner E, Floss HG (2002) The Biosynthetic Gene Cluster of the Maytansinoid Antitumor Agent Ansamitocin from *Actinosynnema pretiosum*. Proc Natl Acad Sci U S A **99**: 7968

2421. Spiteller P, Bai L, Shang G, Carroll BJ, Yu T-W, Floss HG (2003) The Post-Polyketide Synthase Modification Steps in the Biosynthesis of the Antitumor Agent Ansamitocin by *Actinosynnema pretiosum*. J Am Chem Soc **125**: 14236

2422. Ghisalba O (1985) Biosynthesis of Rifamycins (Ansamycins) and Microbial Production of Shikimate Pathway Precursors, Intermediates, and Metabolites. Chimia **39**: 79

2423. Lin S, Van Lanen SG, Shen B (2007) Regiospecific Chlorination of (*S*)-β-Tyrosyl-*S*-Carrier Protein Catalyzed by SgcC3 in the Biosynthesis of the Enediyne Antitumor Antibiotic C-1027. J Am Chem Soc **129**: 12432

2424. Flatt PM, O'Connell SJ, McPhail KL, Zeller G, Willis CL, Sherman DH, Gerwick WH (2006) Characterization of the Initial Enzymatic Steps of Barbamide Biosynthesis. J Nat Prod **69**: 938

2425. Carney JR, Rinehart KL (1995) Biosynthesis of Brominated Tyrosine Metabolites by *Aplysina fistularis*. J Nat Prod **58**: 971

2426. Oza VB, Salamonczyk GM, Guo Z, Sih CJ (1997) Model Reactions of Thyroxine Biosynthesis. Identification of the Key Intermediates in Thyroxine Formation from 3,5-Diiodo-L-tyrosine and 4-Hydroxy-3,5-diiodophenylpyruvic Acid. J Am Chem Soc **119**: 11315

2427. Ma Y-A, Sih CJ, Harms A (1999) Enzymatic Mechanism of Thyroxine Biosynthesis. Identification of the "Lost Three-Carbon Fragment". J Am Chem Soc **121**: 8967

2428. Silk PJ, Macaulay JB (2003) Stereoselective Biosynthesis of Chloroarylpropane Diols by the Basidiomycete *Bjerkandera adusta*: Exploring the Roles of Amino Acids, Pyruvate, Glycerol and Phenyl Acetyl Carbinol. FEMS Microbiol Lett **11243**: 1

2429. Silk PJ, Macaulay JB (2003) Stereoselective Biosynthesis of Chloroarylpropane Diols by the Basidiomycete *Bjerkandera adusta*. Chemosphere **52**: 503

2430. Beck HC (1997) Biosynthetic Pathway for Halogenated Methoxybenzaldehydes in the White Rot Fungus *Bjerkandera adusta*. FEMS Microbiol Lett **149**: 233

2431. Lapadatescu C, Giniès C, Le Quéré J-L, Bonnarme P (2000) Novel Scheme for Biosynthesis of Aryl Metabolites from L-Phenylalanine in the Fungus *Bjerkandera adusta*. Appl Environ Microbiol **66**: 1517

2432. Neilson AH (1990) The Biodegradation of Halogenated Organic Compounds. J Appl Bacteriol **69**: 445

2433. Mohn WW, Tiedje JM (1992) Microbial Reductive Dehalogenation. Microbiol Rev **56**: 482

2434. Neilson AH (1996) An Environmental Perspective on the Biodegradation of Organochlorine Xenobiotics. Int Biodeterior Biodegrad **37**: 3

2435. Fetzner S (1998) Bacterial Dehalogenation. Appl Microbiol Biotechnol **50**: 633

2436. Castro CE (1998) Environmental Dehalogenation: Chemistry and Mechanism. Rev Environ Contam Toxicol **155**: 1

2437. Lee MD, Odom JM, Buchanan RJ Jr. (1998) New Perspectives on Microbial Dehalogenation of Chlorinated Solvents: Insights from the Field. Annu Rev Microbiol **52**: 423

2438. Janssen DB, Oppentocht JE, Poelarends GJ (2001) Microbial Dehalogenation. Curr Opin Biotechnol **12**: 254

2439. Castro CE (2003) The Natural Destruction of Organohalogen Compounds. Environmental Dehalogenation. In: Gribble GW (ed) Natural Production of Organohalogen Compounds, The Handbook of Environmental Chemistry, vol 3, part P. Springer, Berlin, p 235

References

2440. Häggblom MM, Bossert ID (2003) Dehalogenation. Microbial Processes and Environmental Applications. Kluwer, Dordrecht
2441. Smidt H, de Vos WM (2004) Anaerobic Microbial Dehalogenation. Annu Rev Microbiol **58**: 43
2442. Solyanikova IP, Golovleva LA (2004) Bacterial Degradation of Chlorophenols: Pathways, Biochemica, and Genetic Aspects. J Environ Sci Health **B39**: 333
2443. Vannelli T, Studer A, Kertesz M, Leisinger T (1998) Chloromethane Metabolism by *Methylobacterium* sp. Strain CM4. Appl Environ Microbiol **64**: 1933
2444. McDonald IR, Warner KL, McAnulla C, Woodall CA, Oremland RS, Murrell JC (2002) A Review of Bacterial Methyl Halide Degradation: Biochemistry, Genetics and Molecular Ecology. Environ Microbiol **4**: 193
2445. Coulter C, Hamilton JTG, McRoberts WC, Kulakov L, Larkin MJ, Harper DB (1999) Halomethane:Bisulfide/Halide Ion Methyltransferase, an Unusual Corrinoid Enzyme of Environmental Significance Isolated from an Aerobic Methyltroph Using Chloromethane as the Sole Carbon Source. Appl Environ Microbiol **65**: 4301
2446. Goodwin KD, Lidstrom ME, Oremland RS (1997) Marine Bacterial Degradation of Brominated Methanes. Environ Sci Technol **31**: 3188
2447. Tokarczyk R, Saltzman ES, Moore RM, Yvon-Lewis SA (2003) Biological Degradation of Methyl Chloride in Coastal Seawater. Global Biogeochem Cycles **17**: 26/1
2448. Lightstone FC, Zheng Y-J, Bruice TC (1998) Molecular Dynamics Simulations of Ground and Transition States for the S_N2 Displacement of Cl$^-$ from 1,2-Dichloroethane at the Active Site of *Xanthobacter autotrophicus* Haloalkane Dehalogenase. J Am Chem Soc **120**: 5611
2449. Lewandowicz A, Rudziński J, Tronstad L, Widersten M, Ryberg P, Matsson O, Paneth P (2001) Chlorine Kinetic Isotope Effects on the Haloalkane Dehalogenase Reaction. J Am Chem Soc **123**: 4550
2450. Pieters RJ, Lutje Spelberg JH, Kellogg RM, Janssen DB (2001) The Enantioselectivity of Haloalkane Dehalogenases. Tetrahedron Lett **42**: 469
2451. Hur S, Kahn K, Bruice TC (2003) Comparison of Formation of Reactive Conformers for the S_N2 Displacements by CH_3CO_2 in Water and by Asp124-CO_2^- in a Haloalkane Dehalogenase. Proc Natl Acad Sci U S A **100**: 2215
2452. de Jong RM, Dijkstra BW (2003) Structure and Mechanism of Bacterial Dehalogenases: Different Ways to Cleave a Carbon-Halogen Bond. Curr Opin Struct Biol **13**: 722
2453. Paneth P (2003) Chlorine Kinetic Isotope Effects on Enzymatic Dehalogenations. Acc Chem Res **36**: 120
2454. Kurihara T, Esaki N, Soda K (2000) Bacterial 2-Haloacid Dehalogenases: Structures and Reaction Mechanisms. J Mol Catal B: Enzym **10**: 57
2455. van Hylckama Vlieg JET, Tang L, Lutje Spelberg JH, Smilda T, Poelarends GJ, Bosma T, van Merode AEJ, Fraaije MW, Janssen DB (2001) Halohydrin Dehalogenases Are Structurally and Mechanistically Related to Short-Chain Dehydrogenases/Reductases. J Bacteriol **183**: 5058
2456. Lewandowicz A, Sicinska D, Rudzinski J, Ichiyama S, Kurihara T, Esaki N, Paneth P (2001) Chlorine Kinetic Isotope Effect on the Fluoroacetate Dehalogenase Reaction. J Am Chem Soc **123**: 9192
2457. Olaniran AO, Pillay D, Pillay B (2004) Haloalkane and Haloacid Dehalogeases from Aerobic Bacterial Isolates Indigenous to Contaminated Sites in Africa Demonstrate Diverse Substrate Specificities. Chemosphere **55**: 27
2458. Moghaddam AP, Abbas R, Fisher JW, Stavrou S, Lipscomb JC (1996) Formation of Dichloroacetic Acid by Rat and Mouse Gut Microflora, an *in Vitro* Study. Biochem Biophys Res Commun **228**: 639
2459. Shang TQ, Doty SL, Wilson AM, Howald WN, Gordon MP (2001) Trichloroethylene Oxidative Metabolism in Plants: The Trichloroethanol Pathway. Phytochemistry **58**: 1055
2460. Kräutler B, Fieber W, Ostermann S, Fasching M, Ongania K-H, Gruber K, Kratky C, Miki C, Siebert A, Diekert G (2003) The Cofactor of Tetrachloroethene Reductive Dehalogenase

of *Dehalospirillum multivorans* Is Norpseudo-B_{12}, a New Type of a Natural Corrinoid. Helv Chim Acta **86**: 3698
2461. Friis AK, Edwards EA, Albrechtsen H-J, Udell KS, Duhamel M, Bjerg PL (2007) Dechlorination After Thermal Treatment of a TCE-Contaminated Aquifer: Laboratory Experiments. Chemosphere **67**: 816
2462. Nijenhuis I, Nikolausz M, Köth A, Felföldi T, Weiss H, Drangmeister J, Großmann J, Kästner M, Richnow H-H (2007) Assessment of the Natural Attenuation of Chlorinated Ethenes in an Anaerobic Contaminated Aquifer in the Bitterfeld/Wolfen Area Using Stable Isotope Techniques, Microcosm Studies and Molecular Biomarkers. Chemosphere **67**: 300
2463. Olaniran AO, Pillay D, Pillay B (2008) Aerobic Biodegradation of Dichloroethenes by Indigenous Bacteria Isolated from Contaminated Sites in Africa. Chemosphere **73**: 24
2464. He J, Ritalahti KM, Yang K-L, Koenigsberg SS, Löffler FE (2003) Detoxification of Vinyl Chloride to Ethene Coupled to Growth of an Anaerobic Bacterium. Nature **424**: 62
2465. Marco-Urrea E, Parella T, Gabarrell X, Caminal G, Vicent T, Reddy CA (2008) Mechanistics of Trichloroethylene Mineralization by the White-Rot Fungus *Trametes versicolor*. Chemosphere **70**: 404
2466. Doty SL, Shang TQ, Wilson AM, Tangen J, Westergreen AD, Newman LA, Strand SE, Gordon MP (2000) Enhanced Metabolism of Halogenated Hydrocarbons in Transgenic Plants Containing Mammalian Cytochrome P450 2E1. Proc Natl Acad Sci U S A **97**: 6287
2467. Chen YP, Woodin SA, Lincoln DE, Lovell CR (1996) An Unusual Dehalogenating Peroxidase from the Marine Terebellid Polychaete *Amphitrite ornata*. J Biol Chem **271**: 4609
2468. Steward CC, Lovell CR (1997) Respiration and Assimilation of 4-Bromophenol by Estuarine Sediment Bacteria. Microb Ecol **33**: 198
2469. LaCount MW, Zhang E, Chen YP, Han K, Whitton MM, Lincoln DE, Woodin SA, Lebioda L (2000) The Crystal Structure and Amino Acid Sequence of Dehaloperoxidase from *Amphitrite ornata* Indicate Common Ancestry with Globins. J Biol Chem **275**: 18712
2470. Ahn Y-B, Rhee S-K, Fennell DE, Kerkhof LJ, Hentschel U, Häggblom MM (2003) Reductive Dehalogenation of Brominated Phenolic Compounds by Microorganisms Associated with the Marine Sponge *Aplysina aerophoba*. Appl Environ Microbiol **69**: 4159
2471. Boyle AW, Phelps CD, Young LY (1999) Isolation from Estuarine Sediments of a *Desulfovibrio* Strain Which Can Grow on Lactate Coupled to the Reductive Dehalogenation of 2,4,6-Tribromophenol. Appl Environ Microbiol **65**: 1133
2472. Meharg AA, Cairney JWG, Maguire N (1997) Mineralization of 2,4-Dichlorophenol by Ectomycorrhizal Fungi in Axenic Culture and in Symbiosis with Pine. Chemosphere **34**: 2495
2473. Schultz A, Jonas U, Hammer E, Schauer F (2001) Dehalogenation of Chlorinated Hydroxybiphenyls by Fungal Laccase. Appl Environ Microbiol **67**: 4377
2474. Christiansen N, Ahring BK, Wohlfarth G, Diekert G (1998) Purification and Characterization of the 3-Chloro-4-hydroxy-phenylacetate Reductive Dehalogenase of *Desulfitobacterium hafniense*. FEBS Lett **436**: 159
2475. van de Pas BA, Smidt H, Hagen WR, van der Oost J, Schraa G, Stams AJM, de Vos WM (1999) Purification and Molecular Characterization of *ortho*-Chlorophenol Reductive Dehalogenase, a Key Enzyme of Halorespiration in *Desulfitobacterium dehalogenans*. J Biol Chem **274**: 20287
2476. Milliken CE, Meier GP, Watts JEM, Sowers KR, May HD (2004) Microbial Anaerobic Demethylation and Dechlorination of Chlorinated Hydroquinone Metabolites Synthesized by Basidiomycete Fungi. Appl Environ Microbiol **70**: 385
2477. Milliken CE, Meier GP, Sowers KR, May HD (2004) Chlorophenol Production by Anaerobic Microorganisms: Transformation of a Biogenic Chlorinated Hydroquinone Metabolite. Appl Environ Microbiol **70**: 2494
2478. Ferrari RP, Laurenti E, Trotta F (1999) Oxidative 4-Dechlorination of 2,4,6-Trichlorophenol Catalyzed by Horseradish Peroxidase. J Biol Inorg Chem **4**: 232

2479. Petroutsos D, Katapodis P, Samiotaki M, Panayotou G, Kekos D (2008) Detoxification of 2,4-Dichlorophenol by the Marine Microalga *Tetraselmis marina*. Phytochemistry **69**: 707
2480. Field JA, Sierra-Alvarez R (2008) Microbial Degradation of Chlorinated Dioxins. Chemosphere **71**: 1005
2481. van Pée K-H (2003) Dehalogenation of Polyhalogenated Dioxins. Angew Chem Int Ed **42**: 3718
2482. Boyd DR, Sheldrake GN (1998) The Dioxygenase-Catalysed Formation of Vicinal *cis*-Diols. Nat Prod Rep **15**: 309
2483. Hudlicky T, Gonzalez D, Gibson DT (1999) Enzymatic Dihydroxylation of Aromatics in Enantioselective Synthesis: Expanding Asymmetric Methodology. Aldrichimica Acta **32**: 35
2484. Johnson RA (2004) Microbial Arene Oxidations. Org React **63**: 117
2485. Boyd DR, Bugg TDH (2006) Arene *cis*-Dihydrodiol Formation: From Biology to Application. Org Biomol Chem **4**: 181
2486. Banwell M, Blakey S, Harfoot G, Longmore R (1998) First Synthesis of L-Ascorbic Acid (Vitamin C) from a Non-Carbohydrate Source. J Chem Soc Perkin Trans 1: 3141
2487. Banwell MG, Blakey S, Harfoot G, Longmore RW (1999) *cis*-1,2-Dihydrocatechols in Chemical Synthesis of L-Ascorbic Acid (Vitamin C) from a Non-Carbohydrate Source. Aust J Chem **52**: 137
2488. Donohoe TJ, Blades K, Helliwell M, Waring MJ, Newcombe NJ (1998) The Synthesis of (+)-Pericosine B. Tetrahedron Lett **39**: 8755
2489. Banwell M, De Savi C, Watson K (1998) Diastereoselective Synthesis of (–)-*N*-Acetylneuraminic Acid (Neu5Ac) from a Non-Carbohydrate Source. J Chem Soc Perkin Trans 1: 2251
2490. Bui VP, Hudlicky T, Hansen TV, Stenstrom Y (2002) Direct Biooxidation of Arenes to Corresponding Catechols with *E. coli* JM109 (pDTG602). Application to Synthesis of Combretastatins A-1 and B-1. Tetrahedron Lett **43**: 2839
2491. Rinner U, Hudlicky T, Gordon H, Pettit GR (2004) A β-Carboline-1-one Mimic of the Anticancer Amaryllidaceae Constituent Pancratistatin: Synthesis and Biological Evaluation. Angew Chem Int Ed **43**: 5342
2492. Banwell MG, Edwards AJ, Harfoot GJ, Jolliffe KA, McLeod MD, McRae KJ, Stewart SG, Vögtle M (2003) Chemoenzymatic Methods for the Enantioselective Preparation of Sesquiterpenoid Natural Products from Aromatic Precursors. Pure Appl Chem **75**: 223
2493. Austin KAB, Banwell MG, Loong DTJ, Rae AD, Willis AC (2005) A Chemoenzymatic Total Synthesis of the Undecenolide (–)-Cladospolide B *via* a Mid-Stage Ring-Closing Metathesis and a Late-Stage Photo-Rearrangement of the *E*-Isomer. Org Biomol Chem **3**: 1081
2494. Banwell MG, Loong DTJ, Willis AC (2005) A Chemoenzymatic Total Synthesis of the Undecenolide (–)-Cladospolide C. Aust J Chem **58**: 511
2495. Humphreys JL, Lowes DJ, Wesson KA, Whitehead RC (2006) Arene *cis*-Dihydrodiols – Useful Precursors for the Preparation of Antimetabolites of the Shikimic Acid Pathway: Application to the Synthesis of 6,6-Difluoroshikimic Acid and (6*S*)-6-Fluoroshikimic Acid. Tetrahedron **62**: 5099
2496. Finn KJ, Collins J, Hudlicky T (2006) Toluene Dioxygenase-Mediated Oxidation of Dibromobenzenes. Absolute Stereochemistry of New Metabolites and Synthesis of (–)-Conduritol E. Tetrahedron **62**: 7471
2497. Bellomo A, Giacomini C, Brena B, Seoane G, Gonzalez D (2007) Chemoenzymatic Synthesis and Biological Evaluation of (–)-Conduramine C-4. Synth Commun **37**: 3509
2498. Bellomo A, Gonzalez D (2006) Catalytic Thiolysis of Chemoenzymatically Derived Vinylepoxides. Efficient Synthesis of Homochiral Phenylthioconduritol F. Tetrahedron: Asym **17**: 474
2499. Omori AT, Finn KJ, Leisch H, Carroll RJ, Hudlicky T (2007) Chemoenzymatic Total Synthesis of (+)-Codeine by Sequential Intramolecular Heck Cyclizations via C-B-D Ring Construction. Synlett 2859

2500. Kokas OJ, Banwell MG, Willis AC (2008) Chemoenzymatic Approaches to the Montanine Alkaloids: A Total Synthesis of (+)-Nangustine. Tetrahedron **64**: 6444
2501. Shie J-J, Fang J-M, Wong C-H (2008) A Concise and Flexible Synthesis of the Potent Anti-Influenza Agents Tamiflu and Tamiphosphor. Angew Chem Int Ed **47**: 5788
2502. Oppong KA, Hudlicky T, Yan F, York C, Nguyen BV (1999) Chemoenzymatic Enantio-divergent Synthesis of 1,2-Dideoxy-2-amino-1-fluoro-*allo*-inositol. Tetrahedron **55**: 2875
2503. Paul BJ, Willis J, Martinot TA, Ghiviriga I, Abboud KA, Hudlicky T (2002) Synthesis, Structure, and Biological Evaluation of Novel *N*- and *O*-Linked Diinositols. J Am Chem Soc **124**: 10416
2504. Boyd DR, Sharma ND, O'Dowd CR, Hempenstall F (2000) Enantiopure Arene Dioxides: Chemoenzymatic Synthesis and Application in the Production of *trans*-3,4-Dihydrodiols. Chem Commun 2151
2505. Schapiro V, Cavalli G, Seoane GA, Faccio R, Mombrú AW (2002) Chemoenzymatic Synthesis of Chiral Enones from Aromatic Compounds. Tetrahedron: Asymmetry **13**: 2453
2506. Fonseca G, Seoane GA (2005) Chemoenzymatic Synthesis of Enantiopure α-Substituted Cyclohexanones from Aromatic Compounds. Tetrahedron: Asymmetry **16**: 1393
2507. Boyd DR, Sharma ND, Llamas NM, O'Dowd CR, Allen CCR (2006) Chemoenzymatic Synthesis of the *trans*-Dihydrodiol Isomers of Monosubstituted Benzenes via *anti*-Benzene Dioxides. Org Biomol Chem **4**: 2208
2508. Arthurs CL, Raftery J, Whitby HL, Whitehead RC, Wind NS, Stratford IJ (2007) Arene *cis*-Dihydrodiols: Useful Precursors for the Preparation of Analogues of the Anti-Tumour Agent, 2-Crotonyloxymethyl-(4*R*,5*R*,6*R*)-4,5,6-trihydroxycyclohex-2-enone (COTC). Bioorg Med Chem Lett **17**: 5974
2509. Boyd DR, Sharma ND, Coen GP, Gray PJ, Malone JF, Gawronski J (2007) Enzyme-Catalysed Synthesis and Absolute Configuration Assignments of *cis*-Dihydrodiol Metabolites from 1,4-Disubstituted Benzenes. Chem Eur J **13**: 5804
2510. Boyd DR, Sharma ND, Llamas NM, Coen GP, McGeehin PKM, Allen CCR (2007) Chemoenzymatic Synthesis of *trans*-Dihydrodiol Derivatives of Monosubstituted Benzenes from the Corresponding *cis*-Dihydrodiol Isomers. Org Biomol Chem **5**: 514
2511. Carrera I, Brovetto MC, Seoane G (2007) Selectivity in the Halohydroxylation of Cyclohexadienediols. Tetrahedron **63**: 4095
2512. Bellomo A, Gonzalez D (2007) Diasterodivergent Synthesis of Optically Pure Vinyl Episulfides and β-Hydroxy Thiocyanates from a Bacterial Metabolite. Tetrahedron Lett **48**: 3047
2513. Brovetto M, Seoane G (2008) Stereoselective Synthesis of 3-Oxygenated-*cis*-dialkyl-2,5-substituted Tetrahydrofurans from Cyclohexadienediols. J Org Chem **73**: 5776
2514. Pawlik JR (1993) Marine Invertebrate Chemical Defenses. Chem Rev **93**: 1911
2515. Hay ME (1996) Marine Chemical Ecology: What's Known and What's Next? J Exp Mar Biol Ecol **200**: 103
2516. McClintock JB, Baker BJ (1997) A Review of the Chemical Ecology of Antarctic Marine Invertebrates. Amer Zool **37**: 329
2517. Paul VJ, Puglisi MP (2004) Chemical Mediation of Interactions Among Marine Organisms. Nat Prod Rep **21**: 189
2518. Paul VJ, Puglisi MP, Ritson-Williams R (2006) Marine Chemical Ecology. Nat Prod Rep **23**: 153
2519. Paul VJ, Arthur KE, Ritson-Williams R, Ross C, Sharp K (2007) Chemical Defenses: From Compounds to Communities. Biol Bull **213**: 226
2520. Paul VJ, Ritson-Williams R (2008) Marine Chemical Ecology. Nat Prod Rep **25**: 662
2521. Harper DB, McRoberts WC, Kennedy JT (1996) Comparison of the Efficacies of Chloromethane, Methionine, and *S*-Adenosylmethionine as Methyl Precursors in the Biosynthesis of Veratryl Alcohol and Related Compounds in *Phanerochaete chrysosporium*. Appl Environ Microbiol **62**: 3366
2522. Fusetani N (2004) Biofouling and Antifouling. Nat Prod Rep **21**: 94

2523. Omae I (2003) General Aspects of Tin-Free Antifouling Paints. Chem Rev **103**: 3431
2524. König GM, Kehraus S, Seibert SF, Abdel-Lateff A, Müller D (2006) Natural Products from Marine Organisms and Their Associated Microbes. ChemBioChem **7**: 229
2525. Fusetani N, Hiroto H, Okino T, Tomono Y, Yoshimura E (1996) Antifouling Activity of Isocyanoterpenoids and Related Compounds Isolated from a Marine Sponge and Nudibranchs. J Nat Toxins **5**: 249
2526. Kelly SR, Jensen PR, Henkel TP, Fenical W, Pawlik JR (2003) Effects of Caribbean Sponge Extracts on Bacterial Attachment. Aquat Microb Ecol **31**: 175
2527. Kelly SR, Garo E, Jensen PR, Fenical W, Pawlik JR (2005) Effects of Caribbean Sponge Secondary Metabolites on Bacterial Surface Colonization. Aquat Microb Ecol **40**: 191
2528. Holmström C, Egan S, Franks A, McCloy S, Kjelleberg S (2002) Antifouling Activities Expressed by Marine Surface Associated *Pseudoalteromonas* Species. FEMS Microbiol Ecol **41**: 47
2529. Sjögren M, Göransson U, Johnson A-L, Dahlström M, Andersson R, Bergman J, Jonsson PR, Bohlin L (2004) Antifouling Activity of Brominated Cyclopeptides from the Marine Sponge *Geodia barretti*. J Nat Prod **67**: 368
2530. Hedner E, Sjögren M, Hodzic S, Andersson R, Göransson U, Jonsson PR, Bohlin L (2008) Antifouling Activity of a Dibrominated Cyclopeptide from the Marine Sponge *Geodia barretti*. J Nat Prod **71**: 330
2531. Diers JA, Pennaka HK, Peng J, Bowling JJ, Duke SO, Hamann MT (2004) Structural Activity Relationship Studies of Zebra Mussel Antifouling and Antimicrobial Agents from Verongid Sponges. J Nat Prod **67**: 2117
2532. Teeyapant R, Proksch P (1993) Biotransformation of Brominated Compounds in the Marine Sponge *Verongia aerophoba* – Evidence for an Induced Chemical Defense? Naturwissenschaften **80**: 369
2533. Weiss B, Ebel R, Elbrächter M, Kirchner M, Proksch P (1996) Defense Metabolites from the Marine Sponge *Verongia aerophoba*. Biochem Syst Ecol **24**: 1
2534. Ebel R, Brenzinger M, Kunze A, Gross HJ, Proksch P (1997) Wound Activation of Protoxins in Marine Sponge *Aplysina aerophoba*. J Chem Ecol **23**: 1451
2535. Thoms C, Ebel R, Proksch P (2006) Activated Chemical Defense in *Aplysina* Sponges Revisited. J Chem Ecol **32**: 97
2536. Puyana M, Fenical W, Pawlik JR (2003) Are There Activated Chemical Defenses in Sponges of the Genus *Aplysina* from the Caribbean? Mar Ecol Prog Ser **246**: 127
2537. Fang Y-I, Yokota E, Mabuchi I, Nakamura H, Ohizumi Y (1997) Purealin Blocks the Sliding Movement of Sea Urchin Flagellar Axonemes by Selective Inhibition of Half the ATPase Activity of Axonemal Dyneins. Biochemistry **36**: 15561
2538. Wilson DM, Puyana M, Fenical W, Pawlik JR (1999) Chemical Defense of the Caribbean Reef Sponge *Axinella corrugata* Against Predatory Fishes. J Chem Ecol **25**: 2811
2539. Lindel T, Hoffmann H, Hochgürtel M, Pawlik JR (2000) Structure–Activity Relationship of Inhibition of Fish Feeding by Sponge-Derived and Synthetic Pyrrole-Imidazole Alkaloids. J Chem Ecol **26**: 1477
2540. Assmann M, Lichte E, Pawlik JR, Köck M (2000) Chemical Defenses of the Caribbean Sponges *Agelas wiedenmayeri* and *Agelas conifera*. Mar Ecol Prog Ser **207**: 255
2541. Thompson JE, Barrow KD, Faulkner DJ (1983) Localization of Two Brominated Metabolites, Aerothionin and Homoaerothionin, in Spherulous Cells of the Marine Sponge *Aplysina fistularis* (= *Verongia thiona*). Acta Zool **64**: 199
2542. Shaffer PL (1983) Population Ecology of *Heteromastus filiformis* (Polychaeta: Capitellidae). Neth J Sea Res **17**: 106
2543. Jensen P, Emrich R, Weber K (1992) Brominated Metabolites and Reduced Numbers of Meiofauna Organisms in the Burrow Wall Lining of the Deep-Sea Enteropneust *Stereobalanus canadensis*. Deep-Sea Res **39**: 1247
2544. Lovell CR, Steward CC, Phillips T (1999) Activity of Marine Sediment Bacterial Communities Exposed to 4-Bromophenol, a Polychaete Secondary Metabolite. Mar Ecol Prog Ser **179**: 241

2545. Steward CC, Nold SC, Ringelberg DB, White DC, Lovell CR (1996) Microbial Biomass and Community Structures in the Burrows of Bromophenol Producing and Non-Producing Marine Worms and Surrounding Sediments. Mar Ecol Prog Ser **133**: 149
2546. Kicklighter CE, Kubanek J, Hay ME (2004) Do Brominated Natural Products Defend Marine Worms from Consumers? Some Do, Most Don't. Limnol Oceanogr **49**: 430
2547. Vovelle J, Rusaouen-Innocent M, Grasset M, Truchet M (1994) Halogenation and Quinone-Taning of the Organic Tube Components of Some Sabellidae (Annelida Polychaeta). Cah Biol Mar **35**: 441
2548. Becerro MA, Starmer JA, Paul VJ (2006) Chemical Defenses of Cryptic and Aposematic Gastropterid Molluscs Feeding on Their Host Sponge *Dysidea granulosa*. J Chem Ecol **32**: 1491
2549. England LJ, Imperial J, Jacobsen R, Craig AG, Gulyas J, Akhtar M, Rivier J, Julius D, Olivera BM (1998) Inactivation of a Serotonin-Gated Ion Channel by a Polypeptide Toxin from Marine Snails. Science **281**: 575
2550. Hay ME, Fenical W, Gustafson K (1987) Chemical Defense Against Diverse Coral-Reef Herbivores. Ecology **68**: 1581
2551. de Nys R, Steinberg PD, Rogers CN, Charlton TS, Duncan MW (1996) Quantitative Variation of Secondary Metabolites in the Sea Hare *Aplysia parvula* and Its Host Plant, *Delisea pulchra*. Mar Ecol Prog Ser **130**: 135
2552. Bandaranayake WM (2006) The Nature and Role of Pigments of Marine Invertebrates. Nat Prod Rep **23**: 223
2553. Lowery CA, Dickerson TJ, Janda KD (2008) Interspecies and Interkingdom Communication Mediated by Bacterial Quorum Sensing. Chem Soc Rev **37**: 1337
2554. de Nys R, Steinberg PD, Willemsen P, Dworjanyn SA, Gabelish CL, King RJ (1995) Broad Spectrum Effects of Secondary Metabolites from the Red Alga *Delisea pulchra* in Antifouling Assays. Biofouling **8**: 259
2555. Dworjanyn SA, Wright JT, Paul NA, de Nys R, Steinberg PD (2006) Cost of Chemical Defence in the Red Alga *Delisea pulchra*. OIKOS **113**: 13
2556. Hentzer M, Riedel K, Rasmussen TB, Heydor A, Andersen JB, Parsek MR, Rice SA, Eberl L, Molin S, Høiby N, Kjelleberg S, Givskov M (2002) Inhibition of Quorum Sensing in *Pseudomonas aeruginosa* Biofilm Bacteria by a Halogenated Furanone Compound. Microbiology **148**: 87
2557. Dworjanyn SA, de Nys R, Steinberg PD (2006) Chemically Mediated Antifouling in the Red Alga *Delisea pulchra*. Mar Ecol Prog Ser **318**: 153
2558. Rogers CN, Steinberg PD, de Nys R (1995) Factors Associated with Oligophagy in Two Species of Sea Hares (Mollusca: Anaspidea). J Exp Mar Biol Ecol **192**: 47
2559. Borchardt SA, Allain EJ, Michels JJ, Stearns GW, Kelly RF, McCoy WF (2001) Reaction of Acylated Homoserine Lactone Bacterial Signaling Molecules with Oxidized Halogen Antimicrobials. Appl Environ Microbiol **67**: 3174
2560. de Nys R, Coll JC, Price IR (1991) Chemically Mediated Interactions Between the Red Alga *Plocamium hamatum* (Rhodophyta) and the Octocoral *Sinularia cruciata* (Alcyonacea). Mar Biol **108**: 315
2561. Potin P, Bouarab K, Salaün J-P, Pohnert G, Kloareg B (2002) Biotic Interactions of Marine Algae. Curr Opin Plant Biol **5**: 308
2562. Amsler CD (2001) Induced Defenses in Macroalgae: The Herbivore Makes a Difference. J Phycol **37**: 353
2563. Fletcher RL (1995) Epiphytism and Fouling in *Gracilaria* Cultivation: An Overview. J Appl Phycol **7**: 325
2564. Kurata K, Taniguchi K, Agatsuma Y, Suzuki M (1998) Diterpenoid Feeding-Deterrents from *Laurencia saitoi*. Phytochemistry **47**: 363
2565. Pereira RC, Da Gama BAP, Teixeira VL, Yoneshigue-Valentin Y (2003) Ecological Roles of Natural Products of the Brazilian Red Seaweed *Laurencia obtusa*. Braz J Biol **63**: 665

2566. Argandoña VH, Rovirosa J, San-Martín A, Riquelme A, Díaz-Marrero AR, Cueto M, Darias J, Santana O, Guadaño A, González-Coloma A (2002) Antifeedant Effects of Marine Halogenated Monoterpenes. J Agric Food Chem **50**: 7029
2567. Becher PG, Keller S, Jung G, Süssmuth RD, Jüttner F (2007) Insecticidal Activity of 12-*epi*-Hapalindole J Isonitrile. Phytochemistry **68**: 2493
2568. Graedel TE, Keene WC (1996) The Budget and Cycle of Earth's Natural Chlorine. Pure Appl Chem **68**: 1689
2569. Butler JH (2000) Better Budgets for Methyl Halides? Nature **403**: 260
2570. Keppler F, Harper DB, Röckmann T, Moore RM, Hamilton JTG (2005) New Insight into the Atmospheric Chloromethane Budget Gained Using Stable Carbon Isotope Ratios. Atmos Chem Phys **5**: 2403
2571. Yokouchi Y, Saito T, Ishigaki C, Aramoto M (2007) Identification of Methyl Chloride-Emitting Plants and Atmospheric Measurements on a Subtropical Island. Chemosphere **69**: 549
2572. Roberts JM, Osthoff HD, Brown SS, Ravishankara AR (2008) N_2O_5 Oxidizes Chloride to Cl_2 in Acidic Atmospheric Aerosol. Science **321**: 1059
2573. Rosenfeld D, Lahav R, Khain A, Pinsky M (2002) The Role of Sea Spray in Cleansing Air Pollution Over Ocean via Cloud Processes. Science **297**: 1667
2574. Wagner T, Platt U (1998) Satellite Mapping of Enhanced BrO Concentrations in the Troposphere. Nature **395**: 486
2575. Saiz-Lopez A, Mahajan AS, Salmon RA, Bauguitte SJ-B, Jones AE, Roscoe HK, Plane JMC (2007) Boundary Layer Halogens in Coastal Antarctica. Science **317**: 348
2576. Read KA, Mahajan AS, Carpenter LJ, Evans MJ, Faria BVE, Heard DE, Hopkins JR, Lee JD, Moller SJ, Lewis AC, Mendes L, McQuaid JB, Oetjen H, Saiz-Lopez A, Pilling MJ, Plane JMC (2008) Extensive Halogen-Mediated Ozone Destruction Over the Tropical Atlantic Ocean. Nature **453**: 1232
2577. Gribble GW (2003) Commercial Potential of Naturally Occurring Halo-Organics. Specialty Chem Mag **23**(6): 22
2578. Bongiorni L, Pietra F (1996) Marine Natural Products for Industrial Applications. Chem Ind 54
2579. Carté BK (1996) Biomedical Potential of Marine Natural Products. BioScience **46**: 271
2580. Urban S, Hickford SJH, Blunt JW, Munro MHG (2000) Bioactive Marine Alkaloids. Curr Org Chem **4**: 765
2581. Capon RJ (2001) Marine Bioprospecting – Trawling for Treasure and Pleasure. Eur J Org Chem 633
2582. Krajick K (2004) Medicine from the Sea. Smithsonian **35**(5): 50
2583. Newman DJ, Cragg GM (2004) Marine Natural Products and Related Compounds in Clinical and Advanced Preclinical Trials. J Nat Prod **67**: 1216
2584. Newman DJ, Cragg GM (2004) Advanced Preclinical and Clinical Trials of Natural Products and Related Compounds from Marine Sources. Curr Med Chem **11**: 1693
2585. Jha RK, Xu Z (2004) Biomedical Compounds from Marine Organisms. Mar Drugs **2**: 123
2586. Donia M, Hamann MT (2003) Marine Natural Products and Their Potential Applications as Anti-Infective Agents. Lancet **3**: 338
2587. Mayer AMS, Hamann MT (2004) Marine Pharmacology in 2000: Marine Compounds with Antibacterial, Anticoagulant, Antifungal, Anti-Inflammatory, Antimalarial, Antiplatelet, Antituberculosis, and Antiviral Activities; Affecting the Cardiovascular, Immune, and Nervous Systems and Other Miscellaneous Mechanisms of Action. Mar Biotechnol **6**: 37
2588. Simons TL, Andrianasolo E, McPhail K, Flatt P, Gerwick WH (2005) Marine Natural Products as Anticancer Drugs. Mol Cancer Ther **4**: 333
2589. Keyzers RA, Davies-Coleman MT (2005) Anti-Inflammatory Metabolites from Marine Sponges. Chem Soc Rev **34**: 355

2590. El Sayed KA, Dunbar DC, Goins DK, Cordova CR, Perry TL, Wesson KJ, Sanders SC, Janus SA, Hamann MT (1996) The Marine Environment: A Resource for Prototype Antimalarial Agents. J Nat Tox **5**: 261
2591. König GM, Wright AD, Franzblau SG (2000) Assessment of Antimycobacterial Activity of a Series of Mainly Marine Derived Natural Products. Planta Med **66**: 337
2592. König GM, Wright AD (1997) Marine Organisms – Producers of Pharmacologically Active Secondary Metabolites. Pharm Unserer Zeit **26**(6): 281
2593. Cannell RJP (1993) Algae as a Source of Biologically Active Products. Pestic Sci **39**: 147
2594. Tringali C (1997) Bioactive Metabolites From Marine Algae: Recent Results. Curr Org Chem **1**: 375
2595. Smit AJ (2004) Medicinal and Pharmaceutical Uses of Seaweed Natural Products: A Review. J Appl Phycol **16**: 245
2596. Hill RT (2004) Microbes from Marine Sponges: A Treasure Trove of Biodiversity for Natural Products Discovery. In: Bull AT (ed) Microbial Diversity and Bioprospecting. AMS Press, Washington, p 177
2597. El Sayed KA, Dunbar DC, Perry TL, Wilkins SP, Hamann MT, Greenplate JT, Wideman MA (1997) Marine Natural Products as Prototype Insecticidal Agents. J Agric Food Chem **45**: 2735
2598. Peng J, Shen X, El Sayed KA, Dunbar DC, Perry TL, Wilkins SP, Hamann MT, Bobzin S, Huesing J, Camp R, Prinsen M, Krupa D, Wideman MA (2003) Marine Natural Products as Prototype Agrochemical Agents. J Agric Food Chem **51**: 2246
2599. Wright AD, Wang H, Gurrath M, König GM, Kocak G, Neumann G, Loria P, Foley M, Tilley L (2001) Inhibition of Heme Detoxification Processes Underlies the Antimalarial Activity of Terpene Isonitrile Compounds from Marine Sponges. J Med Chem **44**: 873
2600. Amagata T, Whitman S, Johnson TA, Stessman CC, Loo CP, Lobkovsky E, Clardy J, Crews P, Holman TR (2003) Exploring Sponge-Derived Terpenoids for Their Potency and Selectivity Against 12-Human, 15-Human, and 15-Soybean Lipoxygenases. J Nat Prod **66**: 230
2601. Hentschel U, Schmid M, Wagner M, Fieseler L, Gernert C, Hacker J (2001) Isolation and Phylogenetic Analysis of Bacteria with Antimicrobial Activities from the Mediterranean Sponges *Aplysina aerophoba* and *Aplysina cavernicola*. FEMS Microbiol Ecol **35**: 305
2602. Teeyapant R, Woerdenbag HJ, Kreis P, Hacker J, Wray V, Witte L, Proksch P (1993) Antibiotic and Cytotoxic Activity of Brominated Compounds from the Marine Sponge *Verongia aerophoba*. Z Naturforsch **48c**: 939
2603. Meijer L, Thunnissen A-MWH, White AW, Garnier M, Nikolic M, Tsai L-H, Walter J, Cleverley KE, Salinas PC, Wu Y-Z, Biernat J, Mandelkow E-M, Kim S-H, Pettit GR (2000) Inhibition of Cyclin-Dependent Kinases, GSK-3β and CK1 by Hymenialdisine, a Marine Sponge Constituent. Chem Biol **7**: 51
2604. Encarnación DR, Franzblau SG, Tapia CA, Cedillo-Rivera R (2000) Screening of Marine Organisms for Antimicrobial and Antiprotozoal Activity. Pharm Biol **38**: 379
2605. Sala F, Mulet J, Reddy KP, Bernal JA, Wikman P, Valor LM, Peters L, König GM, Criado M, Sala S (2005) Potentiation of Human α4β2 Neuronal Nicotinic Receptors by a *Flustra foliacea* Metabolite. Neurosci Lett **373**: 144
2606. Martínez-Luis S, Pérez-Vásquez A, Mata R (2007) Natural Products with Calmodulin Inhibitor Properties. Phytochemistry **68**: 1882
2607. Liu C, Tadayoni BM, Bourret LA, Mattocks KM, Derr SM, Widdison WC, Kedersha NL, Ariniello PD, Goldmacher VS, Lambert JM, Blättler WA, Chari RVJ (1996) Eradication of Large Colon Tumor Xenografts by Targeted Delivery of Maytansinoids. Proc Natl Acad Sci U S A **93**: 8618
2608. Ladino CA, Chari RVJ, Bourret LA, Kedersha NL, Goldmacher VS (1997) Folate-Maytansinoids: Target-Selective Drugs of Low Molecular Weight. Int J Cancer **73**: 859
2609. Geng X, Yang Z-Q, Danishefsky SJ (2004) Synthetic Development of Radicicol and Cycloproparadicicol: Highly Promising Anticancer Agents Targeting Hsp90. Synlett 1325

2610. Ammini CV, Stacpoole PW (2003) Biotransformation, Toxicology and Pharmacogenomics of Dichloroacetate. In: Gribble GW (ed) Natural Production of Organohalogen Compounds, The Handbook of Environmental Chemistry, vol 3, part P. Springer, Berlin, p 215
2611. Lee H-S, Lee T-H, Lee JH, Chae C-S, Chung S-C, Shin D-S, Shin J, Oh K-B (2007) Inhibition of the Pathogenicity of *Magnaporthe grisea* by Bromophenols, Isocitrate Lyase Inhibitors, from the Red Alga *Odonthalia corymbifera*. J Agric Food Chem **55**: 6923
2612. Bush K, Macielag M, Weidner-Wells M (2004) Taking Inventory: Antibacterial Agents Currently at or Beyond Phase 1. Curr Opin Microbiol **7**: 466
2613. Henry CM (2000) Antibiotic Resistance. Chem Eng News **78**(10): 41
2614. Lowy FD (2003) Antimicrobial Resistance: The Example of *Staphylococcus aureus*. J Clin Invest **111**: 1265
2615. Horikawa M, Noro T, Kamei Y (1999) *In Vitro* Anti-Methicillin-Resistant *Staphylococcus aureus* Activity Found in Extracts of Marine Algae Indigenous to the Coastline of Japan. J Antibiot **52**: 186
2616. Rosenfeld L (2000) Discovery and Early Uses of Iodine. J Chem Ed **77**: 984
2617. Kessler JH (2004) The Effect of Supraphysiologic Levels of Iodine on Patients with Cyclic Mastalgia. Breast J **10**: 328
2618. Jaspars M, Lawton LA (1988) Cyanobacteria – A Novel Source of Pharmaceuticals. Curr Opin Drug Discovery Dev **1**: 77
2619. Luesch H, Harrigan GG, Goetz G, Horgen FD (2002) The Cyanobacterial Origin of Potent Anticancer Agents Originally Isolated from Sea Hares. Curr Med Chem **9**: 1791
2620. Tan LT (2007) Bioactive Natural Products from Marine Cyanobacteria for Drug Discovery. Phytochemistry **68**: 954
2621. Haider S, Naithani V, Viswanathan PN, Kakkar P (2003) Cyanobacterial Toxins: A Growing Environmental Concern. Chemosphere **52**: 1
2622. Pouria S, de Andrade A, Barbosa J, Cavalcanti RL, Barreto VTS, Ward CJ, Preiser W, Poon GK, Neild GH, Codd GA (1998) Fatal Microcystin Intoxication in Haemodialysis Unit in Caruaru, Brazil. Lancet **352**: 21
2623. Anderson DM (1994) Red Tides. Sci Am **271**(8): 62
2624. Cox PA, Banack SA, Murch SJ, Rasmussen U, Tien G, Bidigare RR, Metcalf JS, Morrison LF, Codd GA, Bergman B (2005) Diverse Taxa of Cyanobacteria Produce β-*N*-Methyl-amino-L-alanine, a Neurotoxic Amino Acid. Proc Natl Acad Sci U S A **102**: 5074
2625. Naumann K (1999) Influence of Chlorine Substituents on Biological Activity of Chemicals. J Prakt Chem **341**: 417
2626. Pauwels R, Andries K, Desmyter J, Schols D, Kukla MJ, Breslin HJ, Raeymaeckers A, Van Gelder J, Woestenborghs R, Heykants J, Schellekens K, Janssen MAC, De Clercq E, Janssen PAJ (1990) Potent and Selective Inhibition of HIV-1 Replication *in Vitro* by a Novel Series of TIBO Derivatives. Nature **343**: 470
2627. Fujihashi T, Hara H, Sakata T, Mori K, Higuchi H, Tanaka A, Kaji H, Kaji A (1995) Anti-Human Immunodeficiency Virus (HIV) Activities of Halogenated Gomisin J Derivatives, New Nonnucleoside Inhibitors of HIV Type 1 Reverse Transcriptase. Antimicrob Agents Chemother **39**: 2000
2628. Tatko CD, Waters ML (2004) Effect of Halogenation on Edge-Face Aromatic Interactions in a β-Hairpin Peptide: Enhanced Affinity with Iodo-Substituents. Org Lett **6**: 3969
2629. Pomponi SA (1999) The Bioprocess–Technological Potential of the Sea. J Biotechnol **70**: 5
2630. Cooke R (2005) Back to the Bottom. Nature **437**: 612
2631. Reed C (2006) Boiling Points. Nature **439**: 905
2632. Holland ND, Clague DA, Gordon DP, Gebruk A, Pawson DL, Vecchione M (2005) 'Lophenteropneust' Hypothesis Refuted by Collection and Photos of New Deep-Sea Hemichordates. Nature **434**: 374
2633. van der Wielen PWJJ, Bolhuis H, Borin S, Daffonchio D, Corselli C, Giuliano L, D'Auria G, de Lange GJ, Huebner A, Varnavas SP, Thomson J, Tamburini C, Marty D, McGenity TJ, Timmis KN, BioDeep Scientific Party (2005) The Enigma of Prokaryotic Life in Deep Hypersaline Anoxic Basins. Science **307**: 121

2634. Jensen PR, Fenical W (1994) Strategies for the Discovery of Secondary Metabolites from Marine Bacteria: Ecological Perspectives. Annu Rev Microbiol **48**: 559
2635. Mincer TJ, Jensen PR, Kauffman CA, Fenical W (2002) Widespread and Persistent Populations of a Major New Marine Actinomycete Taxon in Ocean Sediments. Appl Environ Microbiol **68**: 5005
2636. Rouhi AM (2006) Deep-Sea Harvest. Chem Eng News **84**(15): 67
2637. Samaai T, Gibbons MJ, Kelly M, Davies-Coleman M (2003) South African Latrunculiidae (Porifera: Demospongiae: Poecilosclerida): Descriptions of New Species of *Latrunculia* du Bocage, *Strongylodesma* Lévi, and *Tsitsikamma* Samaai & Kelly. Zootaxa **371**: 1
2638. Müller WEG, Brümmer F, Batel R, Müller IM, Schröder HC (2003) Molecular Biodiversity. Case Study: Porifera (Sponges). Naturwissenschaften **90**: 103
2639. Hentschel U, Usher KM, Taylor MW (2006) Marine Sponges as Microbial Fermenters. FEMS Microbiol Ecol **55**: 167
2640. Santelli CM, Orcutt BN, Banning E, Bach W, Moyer CL, Sogin ML, Staudigel H, Edwards KJ (2008) Abundance and Diversity of Microbial Life in Ocean Crust. Nature **453**: 653
2641. Canfield DE (2006) Gas With an Ancient History. Nature **440**: 426
2642. Self S, Blake S, Sharma K, Widdowson M, Sephton S (2008) Sulfur and Chlorine in Late Cretaceous Deccan Magmas and Eruptive Gas Release. Science **319**: 1654
2643. Bertaux J-L, Vandaele A-C, Korablev O, Villard E, Fedorova A, Fussen D, Quémerais E, Belyaev D, Mahieux A, Montmessin F, Muller C, Neefs E, Nevejans D, Wilquet V, Dubois JP, Hauchecorne A, Stepanov A, Vinogradov I, Rodin A, and the SPICAV/SOIR Team (2007) A Warm Layer in Venus' Cryosphere and High-Altitude Measurements of HF, HCl, H_2O and HDO. Nature **450**: 646
2644. Osterloo MM, Hamilton VE, Bandfield JL, Glotch TD, Baldridge AM, Christensen PR, Tornabene LL, Anderson FS (2008) Chloride-Bearing Materials in the Southern Highlands of Mars. Science **319**: 1651
2645. Bernier UR, Kline DL, Barnard DR, Schreck CE, Yost RA (2000) Analysis of Human Skin Emanations by Gas Chromatography/Mass Spectrometry. 2. Identification of Volatile Compounds That Are Candidate Attractants for the Yellow Fever Mosquito (*Aedes aegypti*). Anal Chem **72**: 747
2646. Reddy CM, Xu L, Eglinton TI, Boon JP, Faulkner DJ (2002) Radiocarbon Content of Synthetic and Natural Semi-Volatile Halogenated Organic Compounds. Environ Pollut **120**: 163
2647. Blackman S (2005) "Industrial" Pollutants Reveal a Surprising Origin. The Scientist **19**(June 6): 24
2648. Brown LE, Konopelski JP (2008) Turning the Corner: Recent Advances in the Synthesis of the Welwitindolinones. Org Prep Proc Int **40**: 411
2649. Ankudey FJ, Kiprof P, Stromquist ER, Chang LC (2008) New Bioactive Bromotyrosine-Derived Alkaloid from a Marine Sponge *Aplysinella* sp. Planta Med **74**: 555
2650. Elban MA, Hecht SM (2008) Total Synthesis of the Topopyrones: A New Class of Topoisomerase I Poisons. J Org Chem **73**: 785
2651. Paterson I, Findlay AD, Florence GJ (2007) Total Synthesis and Stereochemical Reassignment of (+)-Dolastatin 19, a Cytotoxic Marine Macrolide Isolated from *Dolabella auricularia*. Tetrahedron **63**: 5806
2652. Carpenter J, Northrup AB, Chung dM, Wiener JJM, Kim S-G, MacMillan DWC (2008) Total Synthesis and Structural Revision of Callipeltoside C. Angew Chem Int Ed **47**: 3568
2653. Jia Y, Bois-Choussy M, Zhu J (2008) Synthesis of Diastereomers of Complestatin and Chloropeptin I: Substrate-Dependent Atropstereoselectivity of the Intramolecular Suzuki-Miyaura Reaction. Angew Chem Int Ed **47**: 4167
2654. Dickenson ERV, Summers RS, Croué J-P, Gallard H (2008) Haloacetic Acid and Trihalomethane Formation from the Chlorination and Bromination of Aliphatic β-Dicarbonyl Acid Model Compounds. Environ Sci Technol **42**: 3226

2655. Mulholland NP, Pattenden G, Walters IAS (2008) A Concise and Straightforward Total Synthesis of (±)-Salinosporamide A, Based on a Biosynthesis Model. Org Biomol Chem **6**: 2782
2656. Zhang D, Zuo P, Ye B-C (2008) Bead-Based Mesofluidic System for Residue Analysis of Chloramphenicol. J Agric Food Chem **56**: 9862
2657. Clark RC, Lee SY, Boger DL (2008) Total Synthesis of Chlorofusin, Its Seven Chromophore Diastereomers, and Key Partial Structures. J Am Chem Soc **130**: 12355
2658. Ji N-Y, Li X-M, Li K, Ding L-P, Gloer JB, Wang B-G (2007) Diterpenes, Sesquiterpenes, and a C_{15}-Acetogenin from the Marine Red Alga *Laurencia mariannensis*. J Nat Prod **70**: 1901
2659. Araki A, Tsuda M, Kubota T, Mikami Y, Fromont J, Kobayashi J (2007) Nagelamide J, a Novel Dimeric Bromopyrrole Alkaloid from a Sponge *Agelas* Species. Org Lett **9**: 2369
2660. de Nys R, Givskov M, Kumar N, Kjelleberg S, Steinberg PD (2006) Furanones. In: Fusetani N, Clare AS (eds) Antifouling Compounds. Progress in Molecular and Subcellular Biology Subseries Marine Molecular Biotechnology. Springer, Berlin, p 55
2661. Vairappan CS, Kawamoto T, Miwa H, Suzuki M (2004) Potent Antibacterial Activity of Halogenated Compounds Against Antibiotic-Resistant Bacteria. Planta Med **70**: 1087
2662. Moore RE, Corbett TH, Patterson GML, Valeriote FA (1996) The Search for New Antitumor Drugs from Blue-Green Algae. Curr Pharm Des **2**: 317
2663. Codd GA (1995) Cyanobacterial Toxins: Occurrence, Properties and Biological Significance. Wat Sci Tech **32**: 149
2664. Codd GA, Bell SG, Kaya K, Ward CJ, Beattie KA, Metcalf JS (1999) Cyanobacterial Toxins, Exposure Routes and Human Health. Eur J Phycol **34**: 405
2665. Cox PA, Banack SA, Murch SJ (2003) Biomagnification of Cyanobacterial Neurotoxins and Neurodegenerative Disease Among the Chamorro People of Guam. Proc Natl Acad Sci U S A **100**: 13380
2666. Davies G, Ghabbour EA, Steelink C (2001) Humic Acids: Marvelous Products of Soil Chemistry. J Chem Ed **78**: 1609
2667. Gross H, König GM (2006) Terpenoids From Marine Organisms: Unique Structures and Their Pharmacological Potential. Phytochem Rev **5**: 115
2668. Gossauer A (2003) Monopyrrolic Natural Compounds Including Tetramic Acid Derivatives. Prog Chem Org Nat Prod **86**: 1
2669. Krohn K (2003) Natural Products Derived from Naphthalenoid Precursors by Oxidative Dimerization. Prog Chem Org Nat Prod **85**: 1
2670. Murai K, Morishita M, Nakatani R, Kubo O, Fujioka H, Kita Y (2007) Concise Total Synthesis of (−)-Spongotine A. J Org Chem **72**: 8947
2671. McMurry J, Fay RC (1998) Chemistry, 2nd Ed. Prentice Hall, Upper Saddle River, New Jersey, p 739

Author Index

A

Aalbersberg, W., 408
Abarca, ML., 450, 451
Abbanat, D., 444
Abbanat, DR., 456
Abbas, R., 495
Abbas, SA., 460
Abbott, JA., 477
Abboud, KA., 498
Abboud, Y., 446
Abdel-Ghani, HF., 402
Abdel- Lateff, A., 499
Abe, M., 413
Abe, S., 471
Abe, T., 402, 404, 405, 407, 414
Abell, AD., 413
Abogadie, FC., 428
Abou-Mandour, A., 448
Abou-Mandour, AA., 449
Abou-Mansour, E., 384
Abourriche, A., 446
Abraham, BG., 457
Abrahamsson, K., 392, 394, 395
Abrams, OV., 467
Abrell, LM., 422
Abreu, PJM., 411
Abreu, PM., 399, 400
Accensi, F., 450, 451
Acevedo, L., 444
Achard, F., 388
Ackermann, R., 383
Acklin, W., 426, 427
Adachi, H., 422

Adachi, K., 418, 462
Adachi, N., 411
Adams, DR., 460
Adams, FC., 393, 394
Addepalli, R., 440
Adeboya, MO., 463
Afe, O., 383
Agata, N., 472
Agatsuma, T., 453, 457, 464
Agatsuma, Y., 500
Agrawal, MS., 462
Aguiar, JQ., 458
Aguilar, JM., 401
Aguilar, MB., 473
Aguirre, A., 412
Ahlsdorf, B., 457
Ahmad, MS., 448
Ahmad, VU., 402
Ahmad, W., 401
Ahmed, A., 449
Ahmed, AA., 401, 402
Ahmed, S., 412
Ahn, Y., 494
Ahn, Y-B., 496
Ahring, BK., 496
Aibara, T., 486
Aiello, A., 438, 467
Ainsworth, AM., 398
Ainsworth, M., 399
Aisaka, K., 456
Aitken, MD., 487
Aiuppa, A., 387
Aizawa, H., 448

Aizawa, S., 401
Akatsuka, H., 423
Akazawa, Y., 421
Akdemir, ZS., 459
Ake, TB., 389
Akhtar, M., 500
Akimoto, H., 389
Akimoto, N., 459
Akiyama, K., 452
Aknin, M., 436, 459, 467
Akutsu, H., 401
Al-Awar, R., 430
Albert, CJ., 490
Alborn, WE. Jr., 456
Albrechtsen, H-J., 496
Alcock, RE., 478
Aldag, R., 478
Alden, CJ., 393
Alder, L., 435
Aldous, KM., 477
Al-Easa, HS., 407
Alewood, P., 428
Algazy, K., 430
Ali, MS., 401
Alicke, B., 393
Aljancic, I., 402
Allain, EJ., 500
Allan, SA., 457
Allen, CCR., 498
Allen, MF., 390
Allen, NE., 456, 473
Allen, TM., 429
Allenmark, S., 486
Allenmark, SG., 482, 486
Alluri, M., 444
Almeida, M., 487, 488
Almeida, MG., 487
Al-Mourabit, A., 493
Alonso, G., 401
Alperson, N., 388
Al-Rehaily, AJ., 448
Altenbach, H-J., 397
Altenbuchner, J., 481, 485, 488
Altenhoff, G., 487
Althoff, G., 435
Al-Thukair, AA., 448
Altmann, A., 488
Alvarado-Lindner, B., 399

Alvarez, E., 430
Alvarez, RG., 484
Alvermann, P., 431
Alves, A., 451
Amade, P., 406, 437, 446
Amagata, T., 397, 417, 418, 502
Amemiya, M., 412
Ames, BN., 491
Amezcua, C., 429
Amin, MR., 419
Ammermann, E., 472
Ammini, CV., 503
Amsler, CD., 400, 445, 500
Amsrud, T., 430
Anai, T., 448
Anborgh, PH., 418
Anders, E., 388, 395
Andersen, JB., 421, 500
Andersen, RJ., 410, 429
Andersen, RJ., 433, 437, 438, 454
Anderson, BE., 391
Anderson, BS., 417
Anderson, DM., 503
Anderson, FS., 506
Anderson, GT., 439
Anderson, JLR., 479
Anderson, WR. Jr., 448
Anderson, KC., 432
Anderson, MM., 464, 491
Andersson, M., 486
Andersson, MA., 482, 486
Andersson, P., 478
Andersson, R., 477, 499
Andjelic, CD., 418
Ando, A., 425
Ando, K., 398, 463, 471
Ando, M., 415
Ando, T., 389, 452, 457, 470
Andrade, P., 493
Andreae, MO., 390, 391
Andreotti, A., 443
Andres, N., 431
Andrews, J., 390
Andrianasolo, E., 501
Andrianasolo, EH., 400
Andries, K., 503
Ang, HH., 407
Ang, PO. Jr., 461

Ang, KKH., 445
Angerhofer, CK., 422
Anizon, F., 446
Anjaneyulu, ASR., 409, 459
Anjaneyulu, V., 462
Anjuneyulu, ASR., 407
Anke, H., 397, 450, 457
Anke, T., 396, 397, 449, 458, 471
Ankisetty, S., 400
Ankudey, FJ., 506
Annokkée, GJ., 392
Anthoni, U., 385, 421, 443
Antipin Myu, 460
Anton, H., 459
Antunes, EM., 444, 445
Antus, S., 450, 472
Aoki, M., 471
Aoki, S., 389, 412, 441, 469
Aoki, T., 423
Aono, M., 475
Aotani, Y., 463
Aoun, S., 483
Aoyagi, M., 410
Aoyagi, Y., 407
Appleton, DR., 417, 441
Ar Gall, E., 487
Arai, H., 398
Arai, M., 397, 453
Arai, N., 398, 431
Arakawa, H., 470
Arakaws, K., 493
Araki, A., 507
Aramoto, M., 501
Aratani, Y., 491
Arbaoui, J., 433
Ardá, A., 424
Ardissone, S., 461
Arends, IWCE., 484
Arendse, CE., 400
Arendt, F., 392
Aresta, A., 451
Argandoña, VH., 400, 501
Argyropoulos, D., 405
Arigoni, D., 426, 427
Arihara, S., 437
Ariki, S., 440
Arimoto, H., 413, 433, 449, 455
Arimoto, M., 397

Ariniello, PD., 502
Arioli, V., 474
Arisawa, S., 402
Arison, BH., 448
Aritake, S., 456
Armishaw, C., 428
Arnhold, J., 489, 490
Arnold, L., 491
Arnold, N., 472
Arnone, A., 453, 454, 472
Arsenault, TL., 475
Arteaga, JM., 411
Arthur, KE., 498
Arthurs, CL., 498
Artico, M., 434
Arvelo, F., 467
Asakawa, S., 475
Asakawa, Y., 386, 459, 471
Ashby, J., 381
Ashcroft, J., 444
Ashizawa, S., 452
Asplund, G., 14, 385, 392, 478
Asplund, L., 462, 463, 475
Assmann, M., 437, 439, 468, 499
Ataka, Y., 400
Athanasiadou, M., 461, 462
Athanassiadis, I., 461
Atkins, DHF., 477
Atlas, E., 391, 393, 395
Atta-ur-Rahman, 412, 449, 458
Attieh, JM., 489
Aubry, C., 458
Aucott, ML., 391
Aust, HJ., 397
Aust, H-J., 402, 450, 456, 472
Austin, KAB., 497
Avallone, E., 427
Avery, VM., 440
Ávila, C., 443
Avila, R., 467
Aviles, RJ., 491
Awakura, D., 415
Awano, K., 475
Axelsen, PH., 473
Axley, MJ., 480
Ayala, I., 444
Ayala, M., 484
Aydin, M., 389

Aydogmus, Z., 403, 404, 414, 468
Aye, M., 487
Azas, N., 402
Azizuddin, 458
Azouri, H., 451
Azuma, T., 434

B

Baba, M., 398
Babazono, K., 409
Babiker, HAA., 433
Babouléne, M., 483
Bach, W., 506
Baciocchi, E., 483
Bae, KS., 397
Baek, S., 414
Baek, S-H., 458
Bagge, N., 421
Bahner, I., 431
Bahramsari, R., 450
Bahreini, R., 383
Bai, L., 453, 494
Bai, R., 429
Bailly, C., 446
Bair, KW., 465
Baker, B., 405
Baker, BJ., 445, 498
Baker, JA., 400
Baker, JM., 389, 393, 394
Baker, PJ., 456
Baker, SR., 432
Bakker, M., 484
Bal de Kier Joffé, E., 441
Balaram, P., 473
Baldeyrou, B., 446
Baldridge, AM., 506
Baleja, JD., 473
Ball, M., 453
Ball, RG., 471
Ballschmiter, K., 391, 460, 479
Balzaretti, V., 427
Bamberg, DD., 436
Bamberger, MJ., 398
Banack, SA., 503, 507
Banaigs, B., 384, 446
Banciu, H., 459
Bandaranayake, WM., 500

Bandfield, JL., 506
Bandyopadhyay, P., 429
Banjoo, D., 410
Banning, E., 506
Bänteli, R., 427
Bantleon, R., 481
Banwell, M., 497
Banwell, MG., 497, 498
Bao, B., 442
Baran, P., 410
Baran, PS., 439, 444
Barbaras, D., 447
Barbier, M., 383
Barbieri, LR., 456
Barbosa, J., 503
Barca, A., 432
Barclay, CE., 469
Bardhan, S., 468
Bardsley, B., 473
Barluenga, S., 453
Barnard, DR., 506
Barna-Vetro, I., 451
Barnes, K., 451
Barnett, P., 480
Barnola, J-M., 393
Barrabee, E., 471
Barrabee, EB., 462
Barrera, JB., 458
Barreto, VTS., 503
Barrett, AGM., 433
Barrie, L., 383
Barrie, LA., 389, 391
Barros, MR., 397
Barrow, KD., 467, 499
Barrow, RA., 398, 430
Barrows, LR., 418, 444, 445, 456
Barsby, T., 410
Bartner, P., 475
Barton, DHR., 412
Bartram, J., 384, 423
Basavapathruni, A., 483
Bass, CG., 468
Bassarello, C., 424
Bassford, MR., 390, 395
Bassuony, AAEl., 401
Bate, M., 430
Batel, R., 506
Battershill, CN., 383, 416, 465

Battle, M., 389
Baudoin, O., 474
Bauguitte, SJ-B., 501
Baumert, A., 448
Baur, S., 399
Baux, G., 473
Bavestrello, G., 413
Bavoso, A., 407
Bayer, E., 429
Beachy, TM., 481
Beattie, KA., 507
Beatty, KE., 486
Beauchamp, PS., 406
Beauwens, R., 424
Beccalli, EM., 436
Becerril, B., 473
Becerro, MA., 384, 500
Bechara, EJH., 482
Becher, PG., 447, 501
Bechtner, G., 419
Beck, HC., 494
Beck, JL., 448
Becker, H., 386, 459
Becker, U., 396, 458
Beckmann, H., 448, 449
Bedjanian, Y., 382
Bedzyk, LA., 421
Beerman, TA., 452
Begum, S., 456
Behnke, W., 382
Behrisch, HW., 437
Beil, W., 399, 449
Beinert, H., 420
Belaaouaj, A., 490
Bell, CH., 394
Bell, SG., 507
Bella, M., 429
Bellí, N., 451
Bellomo, A., 497, 498
Belyaev, D., 506
Benayache, F., 402
Benayache, S., 402
Benayahu, Y., 410, 429
Bender, ML., 389
Benharref, A., 465
Benkendorff, K., 443, 448
Benkovitz, CM., 391
Bennamara, A., 446

Ben-Nun, A., 388
Benoit-Guyod, JL., 422
Bentamene, A., 402
Beratan, DN., 417
Berg, B., 479
Bergek, S., 475, 477
Berger, D., 414
Berger, RG., 457
Berger, U., 400
Bergere, I., 459
Bergeron, MB., 387
Bergman, Å., 461
Bergman, A., 462, 463
Bergman, B., 503
Bergman, J., 441, 442, 499
Bergon, P., 494
Bergquist, PF., 406
Bergquist, PR., 406
Bergt, C., 491
Beringhelli, T., 484
Berkers, C., 432
Berkowitz, CM., 382
Berlinck, RGS., 426, 466, 468
Berman, AF., 389
Berman, VS., 434
Bermejo-Barrera, J., 402
Bernal, JA., 502
Bernan, VS., 456, 493
Bernardinelli, G., 403, 407, 421
Bernhardsen, M., 463
Bernier, UR., 506
Berova, N., 456
Berquist, PR., 424
Berridge, MV., 417, 422, 441
Berrué, F., 406
Berry, DM., 456
Berry, RL., 436
Bertaux, J-L., 506
Berthold, RJ., 384
Berthou, S., 433
Berti, M., 434, 474
Besl, H., 471, 472
Bessa, J., 467
Bethuel, Y., 447
Beuchat, J., 447
Beukes, DR., 400, 445
Beutler, JA., 403
Bevan, K., 427

Bewley, CA., 383, 384, 431, 445, 466
Bhaket, P., 418
Bhakuni, DS., 383
Bharate, SB., 464
Bhat, RG., 418
Bianco, A., 416
Biard, J-F., 410
Bidigare, RR., 503
Biemann, K., 388
Biernat, J., 502
Biester, H., 386
Bifulco, G., 424, 427, 442
Bignert, A., 462, 475
Bilia, AR., 415
Bills, GF., 398, 471
Birkbeck, AA., 469
Biselli, S., 440
Bister, B., 488
Bitzer, J., 431
Bjerg, PL., 496
Björk, H., 470
Blackall, LL., 384
Blackman, AJ., 399, 404, 435, 440, 464
Blackman, S., 506
Blades, K., 497
Blagg, BSJ., 453
Blake, D., 383
Blake, DR., 388, 391
Blake, GA., 388
Blake, NJ., 391
Blake, S., 506
Blakey, S., 497
Blank, DH., 435
Blasiak, LC., 432
Blättler, WA., 452, 502
Blein, JP., 470
Bliestle, IM., 450
Blitzke, T., 396
Block, O., 397
Blom, JF., 426, 447
Blomberg, N., 426
Blough, NV., 480
Blount, BC., 491
Blumenberg, M., 441
Blunt, JW., 379, 413, 436, 445, 472, 501
Boardman, L., 387
Boatman, JF., 382
Bobrowski, N., 387

Bobzin, S., 502
Boddy, CNC., 473, 474
Boeck, LD., 456
Boeckman, RK. Jr., 415
Boehlow, TR., 468
Boehm, H-DV., 388
Boenigk, J., 386
Boeynaems, JM., 419
Boger, DL., 396, 473, 507
Böhlendorf, B., 417
Bohlin, L., 402, 441, 499
Böhme, F., 478
Bohrer, A., 413
Böhrer, P., 444
Bois-Choussy, M., 506
Bok, S-H., 449
Bokesch, HR., 443
Bolhuis, H., 503
Bolland, RF., 429
Bolton, JJ., 400
Bonaduce, I., 443
Bonetto, GM., 412
Bongiorni, L., 501
Bonjoch, J., 426
Bonnarme, P., 494
Bonner, M., 477
Bonnet, S., 460
Bonsang, B., 390
Bontemps, N., 446
Bonvehí, JS., 451
Boon, JP., 506
Boonudomlap, U., 471
Bopp, R., 477
Borch, T., 392
Borchardt, SA., 500
Borchers, R., 388, 395, 492
Bordner, J., 448
Borén, H., 458, 478, 479, 485
Borgeson, B., 422
Borghi, D., 438
Borin, S., 503
Börjeson, H., 462
Börjesson, I., 385
Borowicz, JJ., 455
Borowitzka, MA., 384
Borrell, A., 435
Borrelli, F., 441
Bosma, T., 495

Bosman, R., 392
Bossert, ID., 380, 495
Boswell, JL., 445
Bottenheim, JW., 382
Bottini, AT., 406
Bouarab, K., 500
Bougioukou, DJ., 483
Boukouvalas, J., 414, 415, 422
Boulanger, A., 455
Bouraoui, A., 411
Bourassa, JL., 488
Bourdy, G., 442
Bourquet-Kondracki, M-L., 424
Bourret, LA., 502
Boutou, S., 460
Bowden, BF., 402, 414, 462, 465,
Bowden, BJ., 464
Bowden, RD., 419
Bowers, SG., 493
Bowes, MA., 433
Bowling, JJ., 499
Boyd, DR., 497, 498
Boyd, KG., 460
Boyd, MR., 399, 403, 405, 431, 438, 440, 443, 445, 447
Boyle, AW., 496
Boyle, CD., 475
Boyle, JL., 460, 461
Bradshaw, DE., 422
Braekman, JC., 419, 424, 444, 454, 465
Bragulat, MR., 450, 451
Brako, L., 469
Brambilla, U., 458
Braña, AF., 446, 493
Brandis, A., 434
Brandt, G., 449, 476
Brantley, SE., 416
Brasco, MFR., 406, 427
Bräse, S., 473
Brasseur, GP., 391
Bräuchle, L., 444
Brauer, DJ., 397
Brauers, T., 382
Braz-Filho, R., 412
Breakell, K., 463
Breinholt, J., 434, 450
Bremner, JB., 443, 448
Brena, B., 497

Brennan, M-L., 464, 490, 491
Brenninkmeijer, CAM., 389
Brenzinger, M., 499
Brenzovich, WE., 429
Breslin, HJ., 503
Breter, H-J., 383
Breuer, M., 494
Breuils, S., 403
Briand, A., 407
Brindley, AA., 486
Brinen, LS., 399, 436
Bringmann, G., 438, 448, 459, 471
Brinkman, UATh., 392, 393, 459
Brito, I., 403–405
Brivanlou, A., 443
Broad, LM., 432
Broberg, A., 428
Brodo, IM., 469
Bross, M., 457
Brossi, A., 383
Brot, N., 491
Broto-Puig, F., 476
Brovetto, M., 498
Brovetto, MC., 498
Brown, FM., 483
Brown, LE., 506
Brown, SS., 501
Broxterman, QB., 446
Brubaker, G., 491
Bruice, TC., 495
Brümmer, F., 383, 506
Bruno, I., 427, 442
Bruno, M., 406
Brunskill, GJ., 478
Brust, A., 403
Brütsch, T., 447
Bryce, DA., 428
Bryngelsson, C., 490
Brzezinski, LJ., 454
Bu, X., 486
Buarque de Gusmao, N., 422
Bubb, MR., 427
Buchanan, GO., 432, 453
Buchanan, RJ. Jr., 494
Buchanan, MS., 440, 444
Buchanan, MV., 422
Bugg, TDH., 497
Bugni, TS., 386, 421, 426, 433, 440

Bühl, M., 482
Bühler, R., 476
Bui, HTN., 457
Bui, VP., 497
Buikema, W., 431
Buko, AM., 454
Bull, AT., 385, 502
Bullister, JH., 420
Bultel-Poncé, V., 407
Bumb, RR., 476
Burd, W., 434, 481, 488
Bureau, H., 387
Burg, RW., 398
Burgess, JG., 384, 460
Burgett, AWG., 429
Burja, AM., 384
Burkhardt, N., 473
Burns, PC., 477
Burridge, MJ., 457
Burrier, R., 471
Burrows, J., 383
Burton, G., 412
Burton, JW., 415
Burton, MR., 387
Bush, K., 503
Bushell, SM., 441
Buslaeva Eyu, 393
Buss, AD., 445
Buss, IH., 464
Butler, A., 479, 481, 485–487
Butler, JF., 457
Butler, JH., 389, 394, 501
Butler, KL., 429
Butler, MS., 416, 441, 442, 466, 468
Butsugan, Y., 412
Byas-Smith, M., 428
Byriel, KA., 424
Byrne, C., 477, 478
Byrne, CV., 434
Byun, J., 448, 490

C

Cabañes, FJ., 450, 451
Cabezas, E., 427
Caciola, P., 416
Cadet, J., 491
Cadicamo, CD., 420
Cafieri, F., 407, 437
Cahoon, DR. Jr., 388
Cai, P., 426
Cain, KP., 483
Cairney, JWG., 496
Calcinai, B., 413
Calis, I., 459
Cambie, RC., 406
Cameán, AM., 420
Cameron, GM., 424
Caminal, G., 496
Camp, R., 502
Campagnuolo, C., 418, 441
Campbell, M., 410
Campbell, RA., 483
Canfield, DE., 506
Caniato, R., 470
Cannell, RJP., 502
Cao, P., 461
Capaccioni, B., 388
Capasso, R., 441
Capon, RJ., 405–407, 431, 436,
    441, 442, 456, 501
Cappelletti, EM., 470
Carballeira, NM., 416
Cardellina, II JH., 399, 403, 405, 445, 471
Cardenas, LM., 394
Cardillo, R., 472
Cardoso, MP., 463
Carl, T., 433
Carletti, I., 446
Carlsen, L., 479
Carmichael, WW., 384, 423
Carney, JR., 494
Carnuccio, R., 437
Carpenter, J., 506
Carpenter, LJ., 383, 393, 395, 501
Carr, AC., 413, 491, 492
Carrano, L., 467
Carrea, G., 482, 483
Carreira, EM., 429
Carrera, I., 498
Carroll, AR., 409, 414, 422, 433, 439,
    440, 442, 444, 447, 464, 465
Carroll, BJ., 494
Carroll, GL., 487
Carroll, J., 431
Carroll, RJ., 497

Carson, MW., 433
Carté, B., 409, 462
Carté, BK., 501
Carter, GT., 434, 454, 456, 493
Carter, JN., 486
Carter-Franklin, JN., 485, 487
Cartner, LK., 431
Carver, GD., 383
Casapullo, A., 442
Cascales, C., 471
Casella, L., 482–484
Caselles, MC., 406
Cassady, JM., 452
Cassani, G., 474
Cassels, BK., 433
Castedo, L., 400
Castellá, G., 451
Castellano, AG., 458
Castelluccio, F., 408
Castro, CE., 494
Catalán, CAN., 401
Catalano, S., 415
Cataldo, F., 414
Catley, L., 432
Caux, C., 407
Cava, MP., 442
Cavalcanti, RL., 503
Cavalleri, B., 434, 453, 454
Cavalli, F., 383
Cavalli, G., 498
Cavanagh, EE., 452
Cavé, A., 433
Cayan, DR., 388
Ceburnis, D., 383
Cedillo-Rivera, R., 502
Cen, Y-Z., 448
Centonze, G., 451
Cerdeira, A., 451
Cerrano, C., 95, 413
Cervigni, S., 432
Cetin, R., 418
Cetusic, JRP., 425
Chae, C-S., 503
Chaganty, S., 430
Chaib, N., 446
Chaichit, N., 425
Chait, A., 448
Chakrabarti, B., 394

Chalbane, N., 435
Chalkou, K., 399
Challis, GL., 493
Chan, KK., 452
Chan, T-M., 474, 475
Chan, WR., 429
Chance, K., 383
Chanda, BM., 467
Chang, CWJ., 408
Chang, F-R., 433, 449
Chang, LC., 445, 468, 506
Chang, P-C., 408
Chang, RSL., 398
Chang, Y-C., 433
Chang, Y-L., 449
Chang, Z., 423
Chantraine, J-M., 465
Chao, C-H., 405
Chao, T-H., 432
Chapman, EG., 382
Chapman, SK., 479
Chappellaz, J., 393
Chapuis, J-C., 438, 455
Charan, RD., 424, 434, 493
Charapata, SG., 428
Chari, RVJ., 452, 502
Charlton, P., 398
Charlton, TS., 422, 490, 500
Charrouf, M., 446
Chase, CE., 439
Chater, KF., 449
Chatonnet, P., 460
Chauhan, D., 432
Chavakula, R., 460, 465
Chea, A., 402
Chebil, S., 451
Chen, C-Y., 410, 433
Chen, D Y-K., 429
Chen, F., 461
Chen, J., 455
Chen, J-R., 450
Chen, L., 473
Chen, M-C., 408
Chen, M-H., 409
Chen, T-Y., 388, 391
Chen, W., 388
Chen, W-C., 408
Chen, X., 402

Chen, Y., 464, 492
Chen, Y-J., 412
Chen, YP., 408, 481, 496
Cheney, DP., 400, 485
Cheng, J-F., 408
Cheng, M-C., 409
Cheng, X-M., 398
Cheng, Y-B., 420
Cherin, L., 434
Chet, I., 434
Chetland, I., 398
Chevallier, C., 442
Chia, EW., 422
Chiag, MY., 409
Chiang, MY., 405, 409
Chiasera, G., 404, 414, 445
Chiba, T., 433
Chicarelli-Robinson, MI., 434
Cho, H., 467
Cho, HY., 442
Cho, JY., 456
Cho, KW., 442, 466
Cho, S., 469
Choi, BW., 406
Choi, HD., 396, 460
Choi, J-D., 398
Choi, JS., 460
Choi, WC., 460
Choi, WJ., 415
Choo, K-S., 392
Chorus, I., 384, 423
Chou, T., 433
Choudhard, MI., 412, 449, 458
Christensen, PR., 506
Christenson, SD., 452
Christiansen, N., 496
Christie, WW., 419
Christman, RF., 396
Christmann, U., 486
Christoph, EH., 420
Christophersen, C., 383, 385, 421, 443, 467
Chu, M., 470, 471, 475
Chung, d, M., 506
Chung, HY., 460, 461
Chung, MC., 397
Chung, S-C., 503
Ciavatta, ML., 408, 411, 414
Cicerone, RJ., 390

Cichewicz, RH., 462
Ciminiello, P., 385, 416, 465–467
Cimino, G., 385, 408, 409, 411, 414
Cirigliano, AM., 412
Ciufolini, MA., 420, 470
Clade, D., 494
Clague, DA., 503
Clague, MJ., 486
Clardy, J., 399, 408, 409, 432, 436, 442, 445, 456, 462, 464, 466, 469, 502
Clare, AS., 507
Clark, BR., 436
Clark, DP., 431
Clark, JH., 486
Clark, RC., 507
Clark, RJH., 443
Clark, TJ., 422
Clark, WD., 423, 427, 475
Clarke, AD., 389
Claude, C., 481
Clement, JA., 403
Clement, RE., 476
Clerici, F., 436
Cleverley, KE., 502
Cleverly, D., 477
Clive, DLJ., 433
Clough, B., 419
Clower, M. Jr., 477
Cóbar, OM., 409
Cobb, SL., 420, 493
Cochrane, MA., 388
Cockerham, LG., 393
Codd, GA., 503, 507
Coe, EL., 485
Coen, GP., 498
Coen, LJ., 454
Coetzee, PS., 445
Cofer, III WR., 388
Cofer, WR., 476
Coggins, MK., 481
Cohen, J., 441
Cohen, PA., 470, 485
Cole, ALJ., 436, 472
Coleman, JE., 429
Coleman, P., 477
Coleman, RS., 471
Colin, C., 485
Colin, PL., 447, 455

Coll, J., 404
Coll, JC., 402, 414, 464, 465, 468, 500
Collado, IG., 401
Collén, J., 392, 394, 395, 480
Collins, E., 442
Collins, J., 497
Collins, JE. Jr., 391
Colombini, MP., 443
Colonna, S., 482, 483
Colson, KL., 436, 446, 452
Combaut, G., 465
Comellas, L., 476
Comhair, SAA., 464
Compagnone, RS., 467
Compernolle, F., 449
Concepción, GP., 440, 443, 445, 448, 456
Conesa, A., 481
Cong, PZ., 472
Connell, DW., 478
Connolly, JD., 462
Conova, S., 467
Conte, V., 486
Contreras, C., 414
Cooke, R., 503
Cooksey, CJ., 443
Cool, RH., 418
Coombes, PH., 397
Cooper, K., 477
Cooper, R., 485
Coosemans, J., 449
Copley, RCB., 403
Copp, BR., 379, 417, 422, 441, 444, 445, 465
Copper, JE., 427
Corbett, T., 427
Corbett, TH., 430, 507
Cordell, GA., 384
Cordero, M., 475
Cordova, CR., 502
Cornell, L., 465
Cornell-Kennon, S., 400, 465
Corre, C., 493
Correa, RJ., 446
Corrigan, G., 457
Corselli, C., 503
Costantino, V., 459, 466
Costerton, JW., 421
Costi, R., 434
Cota, GF., 394

Cotelle, P., 410
Coughlin, P., 483
Couladis, MA., 401
Couladouros, EA., 469
Coulter, C., 391, 495
Counce, D., 387
Counter, FT., 456
Courtieu, J., 420
Courvalin, P., 473
Coveney, DJ., 471
Cox, ML., 389, 390
Cox, PA., 503, 507
Cox, RA., 383
Cragg, BA., 385
Cragg, GM., 379, 501
Craig, AG., 428, 429, 500
Craig, KS., 410
Crass, K., 421
Cravatt, BF., 396
Cravotto, G., 446
Creche, J., 402
Cremin, P., 472
Creppy, EE., 450, 451
Crews, P., 150, 151, 287, 399, 409, 422, 423, 427, 431, 439, 441, 447, 460, 462, 465, 502
Criado, M., 502
Crill, PM., 390
Crimmins, MT., 415
Critcher, DJ., 433
Cronan, JM. Jr., 405
Crook, CE., 469
Cross, CE., 464
Croteau, R., 400
Croué., J-P., 506
Crovace, C., 443
Crowley, JR., 464, 490
Crummett, WB., 476
Crutzen, PJ., 382
Cruz, LJ., 428, 429
Cruz-Monserrate, Z., 429
Cube, R., 433
Cueto, M., 395, 400, 401, 403–406, 413, 414, 421, 469, 501
Cuevas, C., 406, 443
Culbertson, JA., 395
Cutie, SS., 476
Cutignano, A., 442

Cuttitta, F., 386
Czerwiec, E., 473
Czygan, F-C., 448, 449

D

D'Ambrosio, M., 445
D'Auria, G., 503
D'Auria, MV., 427, 429, 454
D'Avignon, A., 464, 491
D'Croz, L., 413, 421
D'Esposito, M., 438
D'Souza, L., 411
Da Gama, BAP., 500
Da Silva, JAL., 381, 488
Dabdub, D., 382
Daferner, M., 458
Daffonchio, D., 503
Dahlman, O., 478, 479
Dahlqvist, I., 473
Dahlström, M., 441, 499
Dahse, H-M., 407
Dai, C-F., 405
Dai, J., 402
Dai, P., 474
Dai, W-M., 451
Dairi, T., 456
Daitoh, M., 404, 407, 414
Dajani, R., 443
Dalby, AR., 486
Dall'Acqua, S., 470
Dalley, K., 402
Daloze, D., 424, 444, 454, 465
Damant, A., 451
Damsté, JSS., 392
Daniel, JS., 382
Danishefsky, SJ., 433, 453, 502
Darias, J., 395, 399–401, 403–406, 413, 414, 421, 501
Darias, V., 401
Darlow, BA., 464
Das, B., 448
Das, P., 470
Das, PR., 471, 475, 484
Dasaradhi, L., 419
Dau, H., 486
Daugherty, A., 491
Dauter, Z., 489

Daves, GG. Jr., 448
Davey, CA., 490
David, B., 433
David, EE., 457
David, JM., 463
David, JP., 463
Davidson, BS., 447
Davies, G., 489, 507
Davies, JS., 427
Davies, MJ., 489–492
Davies-Coleman, M., 417, 506
Davies-Coleman, MT., 403, 414, 444, 445, 501
Davis, AR., 443
Davis, FA., 426, 439
Davis, GE., 455
Davis, RA., 418, 430, 443, 447, 458
Davyt, D., 403, 404
Dawson, JH., 481
Dawson, S., 470
Dawson, TE., 476
De Almeida Leone, P., 439
De Andrade, A., 503
De Bruyn, WJ., 389
De Carvalho, LR., 403, 461
De Castro Franca, S., 436
De Clercq, E., 405, 503
De Galan, L., 392
De Gioia, L., 489
De Gouw, JA., 389
De Graaf, JM., 419
De Haan, DO., 382
De Hernández, ZNJ., 401
De Jong, E., 381, 396, 480
De Jong, RM., 495
De Kock, A., 391
De la Torre, MC., 406
De Lange, GJ., 503
De Leer, EWB., 381, 385, 389, 391–394, 459, 478, 479
De Meijere, A., 431
De Mora, S., 394
De Mot, R., 488
De Napoli, L., 407
De Nys, R., 399, 402, 414, 421, 422, 500, 507
De Oliveira, JHHL., 466
De Oliveira, MF., 466, 468
De Pava, OV., 458

De Saeger, S., 451
De Savi, C., 497
De Schrijver, A., 488
De Silva, ED., 405, 429
De Souza, AO., 466
De Vincentüs, M., 429
De Voogd, NJ., 462
De Vos, WM., 495, 496
De Weerd, H., 459
De Wit, CA., 462
Dean, JFD., 480
DeBaillie, AC., 415
Debitus, C., 410, 427, 429, 441, 442, 454, 465
Decesari, S., 383
Decker, EA., 492
Deidda, D., 434
Dekker, HL., 480, 484, 489
Del Pozo, T., 400
Del Sordo, S., 483
Del Valle, JR., 426
Dela Cruz, RC., 429
Delage, L., 485
Delbos, E., 451
Delgado, H., 387
Delia, M-L., 451
Dell'Aversano, C., 416, 466, 467
Delnath, S., 415
DeLong, EF., 416
Dembitsky, VM., 380, 386, 399, 416, 421, 455, 480
Demirezer, LO., 451, 459
Demoulin, V., 444, 454
Denaro, M., 474
Deneubourg, F., 424
Deng, H., 420, 474, 492, 493
Deng, J., 439
Denisenko, VA., 444, 475
Dennison, N., 478
Denomme, MA., 475
Dent, III WH., 455
Denzer, D., 473
Deo-Jangra, U., 410
Derr, SM., 502
DeSanto, J., 430
Desmyter, J., 503
Dev, V., 406
Dexter, AF., 483

Dey, S., 456, 470
Deyanat-Yazdi, G., 432
Di Blasio, B., 407
Di Furia, F., 486
Di Meglio, P., 413, 416
Di Modugno, E., 454
Di Modugno, V., 472
Di Rosa, M., 416
Dias, N., 446
Dias, T., 403, 405
Díaz, A-R., 401
Díaz-Ferrero, J., 476
Díaz-González, F., 412
Díaz-Marrero, AR., 399, 400, 403, 404, 413, 421, 501
Dick, R., 450
Dickenson, ERV., 506
Dickerson, TJ., 500
Didier, C., 433
Dieckmann, R., 441
Diekert, G., 495, 496
Diers, JA., 499
Dietrich, R., 451
Dietrich, VJ., 388
Diez, MT., 398
Dijkstra, BW., 495
Dijoux, M-G., 445
Dimayuga, RE., 470
Dimmer, CH., 390, 395
Din, LB., 472
Dinda, B., 415
Ding, J., 402, 412
Ding, L-P., 440, 507
Ding, W., 444
Diop, M., 466
Discafani, CM., 456
Dittmer, J., 486
Dmitrenok, A., 422
Dmitrenok, AS., 413, 444
Dmitrenok, PS., 444
Dobashi, K., 412
Dodds, DR., 483, 484
Dodds, WK., 417
Doe, M., 409, 423, 442
Doherty, AM., 398
Doi, M., 418
Doležal, IS., 477
Dolfing, J., 488

Domigan, NM., 490
Domínguez, L., 403, 404
Domostoj, MM., 439
Dong, C., 434, 479, 488, 492
Dong, CJ., 420
Dong, J., 408
Donia, M., 411, 501
Donnelly, DMX., 471, 472
Donohoe, TJ., 497
Doom, JP., 454
Dordick, JS., 484
Döring, D., 397
Dorrestein, PC., 452, 493
Dorta, E., 399, 401, 403, 406, 413, 421
Dorta, J., 412
Dorward, M., 434, 492
Doty, SL., 495, 496
Doubek, DL., 438, 455, 468
Dowd, PF., 453
Downey, A., 390
Doyle, LA., 430
Doyle, TW., 446, 452
Draeger, S., 402, 450, 456, 472
Drangmeister, J., 496
Drennan, CL., 432
Drenzek, NJ., 481
Drew, M., 461
Dreyer, M., 459, 471
Dring, MJ., 394, 395
Du, SJ., 457
Duan, H., 450
Duan, HQ., 422
Dubois, JP., 506
Duce, RA., 382
Duchene, J-C., 463
Duckworth, AR., 383
Ducrocq, V., 390
Ducrot, P-H., 415, 470
Duebelbeis, DO., 455
Duesterberg, CK., 492
Duffe, J., 435
Duffin, K., 413
Dugrillon, A., 419
Duh, C-Y., 409, 440
Duhamel, M., 496
Duignan, PJ., 435
Duke, SO., 499
Dumay, V., 489

Dumdei, EJ., 424
Dumont, JE., 419
Dumrongchai, N., 424
Dunbar, DC., 502
Duncan, KLK., 427
Duncan, MW., 422, 490, 500
Duncan, SJ., 398, 399
Dunford, HB., 479, 481, 487, 490
Duniway, JM., 393
Dunkle, LD., 427
Dunn, JL., 491
Duque, C., 450
Duyzer, JH., 392
Dweik, RA., 464
Dworjanyn, SA., 500
Dyer, RB., 481
Dyke, P., 477
Dykert, J., 428

E
Ean, U-J., 459
Earle, ED., 426
Eaton, G., 467, 468
Ebel, R., 499
Ebenå, G., 385
Eberhardt, W., 381
Eberl, L., 421, 500
Ebisawa, Y., 414
Edelman, MJ., 430
Eder, C., 438, 439
Edmonds, DR., 403
Edmonds, J., 443
Edmonds, M., 388
Edrada, RA., 462
Eduljee, GH., 477
Edwards, AJ., 497
Edwards, DJ., 417
Edwards, DP., 388
Edwards, EA., 496
Edwards, KJ., 506
Edwards, RL., 463
Egan, S., 499
Egawa, M., 472
Eggen, M., 430
Eggleston, DS., 400, 403
Eggleton, AE., 477
Egi, M., 445

Eglinton, TI., 436, 506
Eguchi, S., 477
Ehrlich, Chr, 477
Eicher, T., 386
Eichhorn, E., 488
Eidell, BR., 386
Eiden, R., 386, 492
Eijsink, VGH., 466
Einhorn, J., 470
Eisenhauer, A., 387, 395
Eiserich, JP., 464
Eisser, S., 430
Ekdahl, A., 392, 394, 395
Ekelund, M., 479
Eklind, Y., 459
Eklund, G., 388
El Abbouyi, A., 415
El Sayed, KA., 502
Elansky, N., 392
El-Aziz, A., 402
Elban, MA., 506
El-Banna, N., 434
Elbrächter, M., 499
Eldershaw, TPD., 464
El-Douski, A., 402
Eldredge, GS., 456
Elend, M., 382
El-Gamal, AA., 440
Elias, R., 402
Elix, JA., 469, 470
Elkins, JW., 389
Ellestad, GA., 444, 452
Elliott, JE., 435
Ellis, D., 428, 430
Ellmerer, E-P., 401
El-Moghazy, SA., 402
Elsässer, B., 472
El-Seedi, HR., 402
El-Shanawany, MA., 402
Elsner, P., 492
Elyakov, GB., 404
Emiliano, A., 416
Emmett, MR., 473
Emmitte, KA., 415
Emmons, LK., 388
Emptage, MH., 420
Emrich, R., 463, 499
Enari, K., 418

Enas, N., 430
Encarnación, DR., 502
Encarnación, RD., 467
Encarnación-Dimayuga, R., 468
Endo, M., 440, 461
Endo, T., 435, 439, 457, 461, 464, 466
England, LJ., 500
Engler, M., 449
Engwall, M., 463
Ensminger, PW., 456
Enticknap, JJ., 384
Epple, M., 486
Erdmann, A., 419
Erdogan, I., 428, 448
Erhard, M., 431
Erickson, DJ., 389
Erickson, III DJ., 391
Erickson, KL., 396, 403, 404
Eriksson, K-EL., 480
Erkel, G., 397
Erkelens, C., 392
Ernyei, AJ., 488
Ersoy, L., 404
Ersoy, N., 468
Ervens, B., 382
Erzurum, SC., 464
Esaki, N., 419, 495
Es-Safi, N-E., 415
Esser, L., 429
Estévez, E., 401
Estrada, DM., 406
Etahiri, S., 407
Etheridge, DM., 389
Ettala, M., 477
Eustáquio, AS., 449
Eva, HD., 388
Evan, T., 436, 466
Evans, BE., 461
Evans, CA., 484
Evans, MJ., 501
Evans, PA., 414
Evans, R., 387
Ewald, G., 416

F
Fabbrini, M., 483
Fabian, P., 395

Fabiani, A., 451
Facchini, MC., 383
Faccio, R., 498
Facey, SJ., 485
Fahey, RC., 466
Fahey, SJ., 424, 463
Fahimi, I., 390, 492
Fahimi, IJ., 396
Fair, P., 435
Fairchild, CR., 408, 465, 470
Fajardo, V., 414
Falchetto, RA., 431
Falconer, I., 423
Falk, H., 11
Fan, C-Q., 412
Fan, T-F., 408
Fan, T-Y., 408, 409
Fan, X., 405, 406, 443, 447, 460, 461
Fang, G-H., 480
Fang, J-M., 498
Fang, L-S., 408, 409
Fang, Y-I., 499
Fantucci, P., 489
Faria, BVE., 501
Farmer, SW., 456
Farver, CF., 464
Fasching, M., 495
Fast, JD., 382
Fast, JD., 383
Fattorusso, E., 385, 407, 413, 416, 418, 437, 438, 441, 459, 465–467, 493
Faulkner, DJ., 379–381, 383, 384, 403, 408, 409, 417, 424, 428, 431, 436, 438–440, 447, 462, 465, 467, 499, 506
Faure, R., 402
Fay, RC., 507
Fazio, C., 406
Fechner, GA., 439
Federico, C., 387
Fedoreyev, SA., 460
Fedorov, SN., 404, 405, 413
Fedorov, YuN., 393
Fedorova, A., 506
Fegley, B. Jr., 389
Fehlhaber, H-W., 474
Fehn, U., 387
Fei, D-Q., 402
Feil, VJ., 476

Feineis, D., 448
Feiters, MC., 489
Feldman, KS., 439
Feldman, PD., 389
Felföldi, T., 496
Feling, RH., 432, 453
Felock, P., 474
Fendert, T., 384, 467
Feng, X., 434, 454, 493
Feng, XZ., 472
Fenical, W., 45, 72, 79, 383, 385, 408, 412, 429, 432, 433, 441, 442, 447, 453, 455–457, 460, 464, 469, 470, 499, 500, 506
Fenna, RE., 490
Fennell, DE., 496
Ferdelman, TG., 385
Ferdinandus, E., 439
Fernandes, A., 451
Fernández, I., 401
Fernández, JJ., 411, 412, 414
Fernández, R., 403, 404, 406, 443
Ferrari, P., 474
Ferrari, RP., 461, 496
Ferrario, J., 476–478
Ferrario, JB., 477
Ferreira, AG., 468
Ferreira, ESB., 443
Ferreira, MLG., 424
Fetzner, S., 494
Fieber, W., 495
Fiebig, HH., 418, 443
Fiedler, H., 477
Fiedler, H-P., 399, 449
Fiedler, TJ., 490
Field, JA., 380, 381, 386, 393, 396, 449, 457, 458, 480, 493, 497
Fielman, KT., 395, 434
Fieseler, L., 384, 502
Fiksdahl, A., 463
Filiatrault, MJ., 466, 468
Filipe, S., 488
Findlay, AD., 455, 506
Findlay, JA., 405
Finlayson-Pitts, BJ., 382
Finn, KJ., 497
Fischer, I., 384, 488
Fischer, WH., 429

Fisher, F., 390
Fisher, JW., 495
Fisher, R., 428
Fisk, AT., 435
Fitzgerald, J., 423
Fitzpatrick, FA., 421
Flagan, RC., 383
Flaherty, L., 446
Flamini, G., 415
Flatt, P., 423, 501
Flatt, PM., 423, 494
Flecks, S., 479, 488
Fletcher, MD., 423
Fletcher, RL., 500
Flodin, C., 460, 479, 485, 487
Florence, GJ., 454, 455, 506
Flórez, AM., 484
Flörke, U., 402, 450, 456
Floss, HG., 452, 493, 494
Flotow, H., 445
Flowers, A., 47
Flowers, AE., 424
Flowers, D., 491
Foderaro, TA., 456
Fogelqvist, E., 394
Foley, M., 502
Folly, L., 432
Fonseca, G., 498
Fontana, A., 385, 407
Ford, DA., 490
Ford, J., 431
Ford, R., 399
Forenza, S., 436, 446, 452
Forino, M., 416, 467
Forler, P., 430
Forsyth, CJ., 411
Forsyth, D., 463
Fortier, G., 414
Fossier, P., 473
Foster, KL., 382
Fouad, FS., 451
Fowler, BR., 388
Foy, BD., 422
Fraaije, MW., 495
Fraga, BM., 401
France, D., 400, 465
Francesch, A., 406, 443
Francis, P., 387

Francisco, C., 446
Francisco, MEY., 404
Francke, W., 441
Frändberg, P-A., 441
Frank, H., 420
Franke, C., 472
Franke, S., 440
Franke, W., 440
Frankland, JC., 449
Franklin, MA., 468
Franks, A., 499
Franssen, MCR., 480
Franzblau, SG., 502
Franzen, S., 481
Fraser, PJ., 389, 390
Fraústo da Silva, JJR., 381, 487, 488
Frei, B., 491, 492
French, JRJ., 393
Freyer, AJ., 438
Friedel, T., 431
Friedrich, AB., 384
Fries, G., 476
Fries, GF., 476
Friesen, MD., 491
Friis, AK., 496
Frimmel, FH., 479
Frische, M., 395
Frisvad, JC., 469
Fromont, J., 438, 439, 443, 466, 507
Fronczek, FR., 401
Fu, H., 407, 482
Fu, X., 411, 424, 447, 460, 462, 466, 467, 491
Fuchs, JR., 444
Fudou, R., 429
Fuge, R., 386, 387
Führer, U., 460
Fujieda, M., 421
Fujihara, H., 471
Fujihashi, T., 503
Fujii, MT., 403, 404
Fujii, T., 432
Fujii, Y., 389
Fujimoto, H., 398, 469
Fujimoto, Y., 450
Fujinuma, Y., 389
Fujioka, H., 507
Fujita, M., 438
Fujita, T., 417, 451, 455

Fujitani, Y., 410
Fujiwara, K., 415
Fujiwara, T., 477
Fujiwara, Y., 407
Fukai, T., 425
Fukaya, H., 407
Fukazawa, M., 398
Fukazawa, Y., 409, 443
Fukuda, DS., 456
Fukuda, N., 432, 471
Fukuda, S., 452
Fukuda, Y., 415
Fukui, H., 456, 470
Fukushi, E., 438
Fukushima, K., 436
Fukuyama, T., 447
Fukuyama, Y., 402, 440, 475
Fukuzawa, A., 487
Fukuzawa, S., 410
Fuller, RW., 399
Funk, RL., 444
Furie, B., 473
Furie, BC., 473
Furihata, K., 418, 425, 439, 474
Furnari, G., 406
Furrow, FB., 445
Furst, GT., 474
Furukawa, J., 423
Furumai, T., 446
Furumoto, T., 433, 456, 470
Furuya, K., 396
Fusetani, N., 383, 403, 408, 418, 428, 431, 437–440, 455, 465, 467, 498, 499, 507
Fussen, D., 506
Fuzzi, S., 383
Fylaktakidou, KC., 474, 475

G
Gabarrell, X., 496
Gabelish, CL., 500
Gabos, S., 388
Gademann, K., 447
Gadomski, D., 477, 478
Gafner, F., 476
Gaggeler, HW., 395
Gaggero, N., 482, 483
Gaillard, N., 446

Galanakis, D., 405
Galario, DL., 417
Galario, MA., 461
Galetti, FCS., 466
Galgoci, A., 470
Galindro, JM., 399, 400
Gallagher, M., 430
Gallard, H., 506
Galle, B., 387
Gallimore, WA., 417
Gallo, M., 432
Galloway, JN., 382
Galm, U., 452
Galonić, DP., 423
Gambicorti, T., 481
Gamble, WR., 445, 453
Gan, J., 393
Gandara, DR., 430
Ganem, B., 467
Ganguli, BN., 418
Ganguly, AK., 475
Gao, H., 468
Gao, HQ., 411
Gao, J-Q., 469
Gao, K., 402
Gao, M-Y., 433
Gao, Y., 473
Garay, M., 460
Garcez, WS., 406
García, A., 441
Garcia, D., 447
García, M., 409
Garcia, RR., 382
García-Grávalos, MD., 411
Garcia-Moruno, E., 451
Garden, SJ., 446
Gardner, A., 477
Garg, NK., 442
Garland, SM., 464
Garneau, S., 446, 493
Garneau-Tsodikova, S., 489, 493
Garnier, M., 502
Garo, E., 432, 499
Garson, MJ., 381, 403, 405, 416, 424, 463, 493
Gärtner, R., 419
Garzon, C., 450
Gasic, MJ., 397

Gastaldo, L., 474
Gáti, T., 401
Gaul, S., 436, 463
Gaunt, MJ., 453
Gaus, C., 435, 436, 463, 478
Gaut, JP., 490
Gautschi, JT., 439, 465
Gavagnin, M., 385, 407, 408, 414
Gavin, J., 457
Gavín, JA., 411
Gawron, LS., 452
Gawronski, J., 498
Gaydou, E., 467
Gayler, KR., 428
Gebhardt, S., 492
Gebruk, A., 503
Geckeler, KE., 381
Gedris, TE., 401
Gee, AD., 492, 493
Geen, C., 385
Geethalakshmi, KR., 482
Gehle, D., 402
Gehrken, H-P., 440
Geiser, DM., 386
Geng, X., 453, 502
Genilloud, O., 474
Gentry, CT., 433
Georg, GI., 430
George, S., 422
Gerasimenko, AV., 405, 475
Gerhardson, B., 455
Gerlach, TM., 387, 388
Gerlitz, M., 398
Gernert, C., 384, 502
Gerstberger, SM., 481
Gerwick, WH., 384, 400, 406, 417, 423, 425, 455, 494, 501
Gesheva, V., 431
Ghabbour, EA.,
Ghibaudi, E., 461
Ghisalba, O., 494
Ghisalberti, EL., 463
Ghiviriga, I., 498
Ghorbel, A., 451
Ghosh, AC., 426
Giacobbe, RA., 398
Giacomini, C., 497
Giannakakou, P., 429

Giannini, C., 427
Gibb, SW., 393
Gibbons, MJ., 506
Gibbs, JB., 471
Gibson, DT., 497
Giese, B., 394
Giese, NA., 398
Giese, RA., 408
Gil, RR., 412
Gilbert, JR., 455
Gilbert, OL., 469
Gill, JH., 431, 436, 440, 456
Gille, JC., 388
Giniès, C., 494
Ginsburg, DW., 400
Giovenzana, GB., 446
Girijavallabhan, VM., 475
Girolami, RL., 454
Gitler, AD., 400
Giudice, G., 387
Giuliano, L., 503
Giuseppone, N., 429
Givskov, M., 421, 500, 507
Gize, AP., 388
Glas, B., 476
Gledhill, JR., 476
Gloer, JB., 426, 453, 507
Glombitza, K-W., 459, 464
Glotch, TD., 506
Goeger, DE., 417, 425
Goericke, R., 457
Goerke, H., 463
Goetz, G., 503
Goetz, GH., 424, 438
Goetz, MA., 398, 471
Goff, F., 387
Goffredi, SK., 385
Goins, DK., 502
Goins, K., 406
Golakoti, T., 430
Gold, A., 487
Gold, MH., 488
Goldberg, ED., 381
Goldfarb, TD., 476
Goldmacher, VS., 452, 502
Goldstein, BP., 454, 474
Golen, J., 456
Gollmick, F., 436

Golovleva, LA., 495
Golstein, PE., 424
Golz, G., 444
Gomez-Paloma, L., 424, 427, 442
Gompel, M., 441
Gong, SL., 391
Gontang, E., 385
Gonzalez, AC., 392
González, AG., 401, 411, 414
Gonzalez, D., 497, 498
González, IC., 411
González, J., 410
González, MC., 444
González, N., 409
González-Coloma, A., 400, 501
Gooday, GW., 466
Goodfellow, M., 385, 399
Goodwin, KD., 394, 495
Göransson, U., 441, 499
Gordon, DP., 503
Gordon, H., 497
Gordon, MP., 495, 496
Gore, R., 443
Gören, AC., 403
Goren-Inbar, N., 388
Görls, H., 435
Goss, RJM., 420
Gossauer, A., 507
Goto, K., 463
Gouda, H., 474
Gould, SJ., 474
Govindan, M., 460
Gowd, KH., 473
Grabley, S., 407
Grace, EJ., 398
Graedel, TE., 382, 391, 501
Gräfe, U., 436
Graham, J., 411
Gram, L., 421
Granato, AC., 468
Grasset, M., 500
Graupner, PR., 425, 455
Gray, D., 406
Gray, GN., 399, 403
Gray, PJ., 498
Gray, WR., 428
Grebenjuk, VA., 383
Grebnev, BB., 444

Grechkin, AN., 420
Green, III FR., 425
Green, NJL., 476, 478
Green, P., 491
Green, PS., 491
Greene, CH., 408
Greengard, P., 443
Greenplate, JT., 502
Greenstein, M., 456
Gregg, K., 420
Greif, G., 453
Grever, MR., 427
Greyerz, E., 475
Grgurina, I., 432
Gribbestad, IS., 457
Gribble, GW., vii, 379–381, 383, 430, 435, 461, 479, 489, 492–494, 501, 503
Griesser, H., 441
Griffith, GS., 469
Griggs, CC., 408
Grimaldi, FS., 386
Grimsrud, EP., 395
Grimvall, A., 14, 381, 385, 389, 391, 392, 394, 478, 479, 485
Grimwade, J., 445
Grisham, MB., 490
Grkovic, T., 445
Grob, F., 485
Gröger, D., 448
Groll, M., 432
Grøn, C., 385, 392
Gronewold, TMA., 427
Gross, F., 422
Gross, H., 425, 507
Gross, HJ., 499
Grossi, G., 476
Grossi, V., 395
Großmann, J., 496
Groszko, W., 391, 395
Groth, I., 436, 449
Groves, JT., 488
Grube, A., 437–440
Gruber, K., 495
Gruijters, BWT., 449
Grün-Wollny, I., 443
Grüschow, S., 398
Gryndler, M., 390
Gu, Y., 455

Guadaño, A., 501
Guan, Z., 470
Guardaño, A., 400
Gudder, DA., 417
Guella, G., 403–405, 407, 414, 437, 442
Guénard, D., 442
Guerriero, A., 445
Guh, J-H., 420
Guimarâes, SMP., 461
Guinaudeau, H., 433
Guiraud, P., 422
Guiry, MD., 414
Guiry, PJ., 472
Guiso, M., 416
Gujer, E., 477
Gulavita, NK., 431, 460
Gullo, V., 475
Gullo, VP., 471, 474
Gullotti, M., 482, 484
Gulyas, J., 428, 500
Gunasekera, SP., 441, 445
Gunatilaka, AAL., 400, 450
Gundersen, H., 400
Gündisch, D., 440
Gunther, FA., 382
Guo, S-P., 436
Guo, Y-W., 406, 408, 409, 414
Guo, Z., 494
Gurjar, MK., 418, 474
Gurrath, M., 422, 502
Gurrieri, S., 387
Gürtler, H., 434
Guschina, IA., 469
Gust, B., 449
Gustafson, K., 500
Gustafson, KR., 405, 431, 447
Gustafsson, Ö., 463, 478
Gustavson, DR., 436
Gut, A., 390
Gutiérrez, M., 471
Guyot, M., 407
Guz, NR., 450
Gwo, H-H., 408

H
Haag, T., 486
Habermehl, GG., 433

Hacker, J., 384, 502
Hagen, WR., 496
Hagenbuch, P., 430
Hagenmaier, H., 476
Hager, LP., 391, 480–484
Hager, MH., 452
Haggart, DA., 457
Häggblom, MM., 380, 495, 496
Hagiwara, H., 414, 415
Haglund, P., 475
Hahn, DR., 455
Hahn, F., 474
Hahn, ME., 435
Hahn, S., 430
Haiber, G., 396
Haider, S., 503
Hajdu, E., 466, 468
Hajek, M., 398
Hale, KJ., 439
Hall, A., 435
Hall, JG., 413
Hall, PA., 388
Halliwell, B., 448
Hallock, YF., 445
Haltiwanger, RC., 400
Hamada, M., 425, 434, 436, 445, 472, 489
Hamada, N., 450, 475, 476
Hamamoto, Y., 461
Hamann, MT., 406, 409, 411, 437,
    445, 466, 468, 499, 501, 502
Hambe, B., 473
Hambsch, B., 479
Hamburger, M., 401
Hamdorf, B., 420
Hamel, E., 425, 429
Hämeri, K., 383
Hamilton, JTG., 380, 390, 391, 395,
    419, 420, 492, 495, 501
Hamilton, VE., 506
Hammel, KE., 479, 485, 487
Hammer, E., 496
Hammer, PE., 434, 488
Han, B., 425
Han, C-C., 408
Han, K., 496
Han, L., 406, 407
Han, LJ., 405, 443, 460, 461
Han, MY., 398

Han, Y-J., 481
Hanada, M., 452
Hancock, J., 470
Hancock, REW., 456
Handa, H., 432
Handa, S., 432
Handayani, D., 462
Hanefeld, U., 493
Haneishi, T., 396
Hanessian, S., 426
Hangsterfer, AN., 436
Hanif, N., 462
Hansen, K., 476
Hansen, TV., 497
Hanson, AD., 489
Hanson, JR., 406
Hanson, KM., 460
Hanss, J., 486
Hanus, LO., 386, 455
Hao, J., 429
Hara, H., 503
Harada, N., 447
Haraguchi, K., 435, 461
Harayama, Y., 444, 445
Harburn, JJ., 468
Hardcastle, KI., 408
Hardt, IH., 456
Harfoot, G., 497
Harfoot, GJ., 497
Hargreaves, J., 463
Harigara, Y., 415
Harms, A., 494
Harnisch, J., 387, 388, 395
Harper, DB., 380–382, 389–391, 394, 395, 419, 492, 495, 498, 501
Harper, J., 455
Harper, JL., 422
Harper, MK., 383, 418, 424, 440
Harput, US., 415
Harran, PG., 429
Harrigan, GG., 424, 438, 454, 503
Harris, GW., 391
Harrison, KN., 409
Harrowven, DC., 406
Hartman, NT., 457
Hartmann, A., 390
Hartog, AF., 482
Hartung, IV., 454

Hartung, J., 423
Haruna, M., 401
Harvala, C., 414
Harvell, CD., 408
Harvey, BMR., 390
Hasan, AFMF., 456, 470
Hasan, Z., 482
Hasegawa, K., 448
Hasegawa, M., 456
Haseley, SR., 383
Haselkorn, R., 431
Haselmann, KF., 392
Hashima, H., 446
Hashimoto, A., 480
Hashimoto, M., 432
Hashimoto, S., 477
Hashimoto, T., 403, 459, 471
Hashizume, H., 493
Hashizume, M., 445
Hasler, P., 476, 477
Hassall, CH., 427
Hata, Y., 402
Hatanaka, S., 423
Hatano, H., 453
Hatano, S., 452
Hatori, H., 432
Hatscher, C., 488
Hatton, AD., 393
Hattori, S., 489
Hattori, T., 462
Hauchecorne, A., 506
Haumann, M., 486
Hauperich, S., 464
Haupt, C., 488
Hausner, P., 430
Haustedt, LO., 454
Hauze, DB., 426
Hawkes, DB., 484
Hawkins, CL., 489–492
Hawksworth, DL., 386
Hawser, SP., 471
Hay, MB., 386
Hay, ME., 408, 498, 500
Hayakawa, I., 433, 446, 467
Hayakawa, Y., 425, 474
Hayashi, I., 461
Hayashi, M., 397, 398, 446, 471
Hayashi, Y., 433

Hayat, S., 449
Hayatsu, R., 388, 395
Haydock, SF., 420
Hayes, JM., 388
Hayes, MA., 398, 434
Haygood, MG., 424
Haynes, D., 435, 478
Hayward, DG., 477
Hayward, S., 463
Haywood, MG., 383
Hazell, LJ., 490, 491
Hazen, SL., 413, 464, 491, 492
Hazuda, D., 474
He, H., 384, 456
He, J., 422, 496
He, J-Y., 423
He, L., 406, 447, 460, 461
Heal, MR., 390
Healy, PC., 422, 458
Heard, DE., 501
Hebestreit, K., 383, 393
Hecht, HJ., 486
Hecht, SM., 506
Heck, JV., 474
Heckman, DS., 386
Hedges, SB., 386
Hedman, R., 455
Hedner, E., 441, 499
Hegde, VR., 474
Heide, L., 449
Heiland, K., 431, 440
Heilbrun, LK., 446
Heimann, AC., 454
Heimbuch, B., 474
Heimer de la Cotera, EP., 473
Heinecke, JW., 380, 413, 448, 464, 489–492
Heinzel, N., 440
Heinzle, E., 459
Heisel, R., 459
Heitto, LV., 385
Helas, G., 391
Helidoniotis, F., 461, 485
Helliwell, M., 497
Hellwig, V., 453, 458, 473
Heltzel, CE., 430
Hemenway, MS., 433
Hempenstall, F., 498

Hemrika, W., 480, 482, 486, 489
Hemscheidt, T., 430
Henderson, GB., 396
Henderson, JP., 380, 448, 489, 490
Henderson, K., 420
Hendricks, DT., 400
Hénichart, J-P., 410
Henkel, TP., 499
Henkelmann, B., 476
Henry, CM., 503
Hensens, OD., 398
Hentschel, U., 383, 384, 496, 502, 506
Hentzer, M., 421, 500
Herald, CL., 464
Herald, DL., 438
Herath, KB., 470
Heraty, LJ., 481
Herdman, M., 431
Hernández Franco, L., 441
Hernandez, J., 484
Hernández, LR., 401
Herrera, J., 404
Herrera, JS., 414
Herrmann, H., 382
Herrmann, M., 417
Hertewich, UM., 459
Hertweck, C., 386
Hervey, A., 422
Herz, W., 401, 467
Herz, W., 468
Herzke, D., 400
Hesler, GA., 436
Hess, PG., 388
Hess, WM., 455
Hettich, RL., 422
Heumann, KG., 393, 394
Heydor, A., 500
Heydorn, A., 421
Heygster, G., 383
Heykants, J., 503
Hickford, SJH., 501
Hickman, J., 446
Hidalgo, HG., 388
Hideshima, T., 432
Hiebl, J., 400
Higa, T., 400, 403, 405, 407, 410, 412, 418, 428, 429, 448, 453, 462
Higashi, T., 389

Higuchi, H., 503
Higuchi, K., 412, 441
Higuchi, R., 414
Hill, DS., 434, 488
Hill, RA., 396, 450
Hill, RT., 384, 502
Hillyard, DR., 428
Himmelfarb, J., 464
Hinde, R., 384
Hino, M., 432
Hino, T., 447
Hinterding, K., 430
Hiraga, Y., 447
Hirai, H., 451
Hiraide, H., 446
Hirama, M., 399, 452
Hiramatsu, A., 469
Hirano, K., 466
Hirano, S., 472
Hiraoka, M.,
Hirata, E., 415
Hirayama, M., 470, 472
Hirono, S., 474
Hirosawa, S., 436
Hirota, H., 403, 408, 437, 438, 465, 467, 499
Hirth, KC., 485
Hisamichi, Y., 435
Hisatomi, G., 389
Hites, RA., 462
Hitotsuyanagi, Y., 407
Hjelm, O., 458, 459, 478
Ho, JZ., 470
Hobbs, JN. Jr., 456
Hoberg, M., 416
Hobson, KA., 435
Hochgürtel, M., 437, 499
Hochlowski, JE., 454
Hodges, TW., 406
Hodin, F., 485
Hodzic, S., 499
Hoekstra, E., 396
Hoekstra, EJ., 381, 392, 393, 459, 479
Hoekstra, P., 435
Hoepker, AC., 466
Hoffman, BM., 481
Hoffmann, D., 494
Hoffmann, H., 436, 499

Hoffmann, HMR., 454
Hoffmann, L., 444, 454
Hoffmann, R., 388
Hoffmann, RW., 453
Hoffmann, T., 383
Hoffmann-Röder, A., 413
Höfle, G., 417, 418, 427
Hofmann, D., 434
Hofmann, G., 438
Hofrichter, M., 488
Hofstede, CM., 389
Hogan, PC., 452
Hohaus, K., 488
Høiby, N., 421, 500
Hokama, Y., 417, 453
Holden, JL., 416
Holdenreider, O., 450
Holder, JSE., 396
Holland, EA., 394
Holland, HL., 482, 483
Holland, ND., 431, 503
Hollemeyer, K., 459
Höller, U., 396
Hollmann, F., 488
Holloway, J., 398
Holloway, MK., 474
Hollwedel, J., 383
Holm, M., 391
Holman, TR., 439, 462, 502
Holm-Hansen, O., 420
Holmstrand, H., 478
Holmström, C., 421, 499
Holness, NJ., 412
Holt, JA., 393
Holzenkämpfer, M., 449
Hölzer, M., 434, 488
Homnick, CF., 474
Honda, G., 449
Honda, K., 489
Hong, CS., 387
Hong, EP., 411
Hong, J., 442, 467
Hong, JK., 397
Hong, TW., 439
Honma, K., 465
Hönninger, G., 382, 387
Hoogsteen, K., 471
Hook, DF., 453

Hooper, D., 429
Hooper, GJ., 445
Hooper, JNA., 383, 416, 424, 433, 439–441, 445, 447, 468
Hope, H., 406, 410, 424
Hopke, J., 384
Hopkins, JR., 501
Hopping, G., 428
Horgen, FD., 503
Hori, Y., 451
Horibe, Y., 397
Horii, Y., 478
Horikawa, M., 503
Horikoshi, H., 395
Horn, M., 384
Horne, DA., 438, 439, 442
Horton, PA., 460
Hoshi, T., 414, 415
Hoshino, Y., 425
Hosoe, T., 436
Hosokawa, N., 472
Hosokawa, S., 446
Hossian, MB., 462
Hostettmann, K., 401, 472
Hotchkiss, RS., 490
Houck, D., 426
Hout, S., 402
Hoveyda, AH., 474
Howald, WN., 495
Howard, J., 434
Howard, JAK., 419
Howard, JC., 481
Howard-Jones, AR., 446, 493
Howell, F., 477
Howell, SB., 429
Howes, PD., 406
Hozumi, Y., 460
Hsu, FF., 413, 464, 490, 491
Hu, J., 463
Hu, J-F., 437, 445
Hu, JY., 422
Hu, L-H., 470
Hu, S., 483, 484
Hu, W-P., 379
Hu, Y., 455, 473
Huang, F., 420
Huang, FL., 420
Huang, H., 409

Huang, H-C., 405
Huang, J-S., 409
Huang, L., 398
Huang, WH., 386
Huang, W-Y., 450
Huang, X., 407, 429
Huang, Y-L., 409
Hubbard, BK., 473
Hubbe, JM., 382
Huber, R., 432
Huber, U., 444
Hubschwerlen, C., 447
Hudlicky, T., 368, 497, 498
Huebner, A., 503
Huesing, J., 502
Huff, T., 458
Hughes, TV., 442
Huhn, T., 426
Hühnerfuss, H., 440
Hui, J., 469
Hulme, AN., 443
Hulsebosch, T., 387
Humanes, MM., 381, 487, 488
Hummel, RH., 476
Humphreys, JL., 497
Hunek, S., 386
Hunt, SW., 382
Hunter, IS., 484
Hunter, PR., 384
Huo, J., 402
Hupe, D., 434
Hupe, DJ., 398
Huq, NP., 491
Hur, S., 495
Hurd, AR., 452
Hurst, JK., 413
Husebo, TL., 430
Hussain, M., 446
Hussein, TA., 403
Ali, AT., 401
Hutzinger, O., 476
Huwe, JK., 476
Hwang, BY., 463
Hwang, HJ., 406
Hylin, JW., 382
Hynninen, PH., 470
Hywel-Jones, NL., 386
Hyytiäinen, H., 431

I

Ianaro, A., 413, 416, 418
Iavarone, C., 416
Ichiba, T., 410, 416, 418, 448
Ichihara, Y., 445, 446
Ichikawa, H., 397
Ichikawa, Y., 482
Ichino, T., 396
Ichiyama, S., 495
Ide, T., 415
Igarashi, M., 425
Igarashi, T., 456
Igarashi, Y., 446
Iguchi, K., 410, 413, 420, 421
Iida, K., 452
Iinuma, H., 436, 472
Iitaka, Y., 426
Iizuka, T., 429
Ikeda, D., 412, 436
Ikeda, H., 398
Ikeda, M., 390
Ikeda, T., 425
Ikeda, Y., 428
Ikonomou, MG., 388
Ikura, T., 397
Ilan, M., 466
Iliopoulou, D., 405, 407, 414
Ilyin, SG., 404, 444, 460
Im, KS., 442, 467
Imajo, S., 452
Imamura, N., 431, 448, 454
Imanishi, T., 397, 456
Imperial, J., 500
Imre, S., 403, 404, 414, 468
Inaba, K., 437, 438
Inagaki, M., 469
Inanaga, S., 433
Ingram, B., 386
Inman, WD., 427
Innocenti, G., 470
Inokoshi, J., 398
Inoue, H., 485
Inoue, K., 398
Inoue, M., 452
Inukai, M., 396
Inukai, Y., 429
Inuzuka, Y., 389
Inuzuka, Y., 390

Ireland, CM., 386, 418, 421, 422, 426, 440, 443–445, 448, 456, 465
Irita, H., 459
Iritani, M., 397
Irschik, H., 417
Irvine, R., 478
Irvine, RL., 477
Irving, E., 439
Irwin, A., 491
Isaka, M., 386, 471
Isenbeck-Schröter, M., 386
Ishibashi, M., 416, 428
Ishida, K., 426, 431
Ishigaki, C., 501
Ishiguro, M., 452
Ishiguro, Y., 457
Ishihama, M., 461
Ishihara, J., 487
Ishii, M., 452
Ishii, T., 407
Ishikawa, J., 425
Ishikawa, Y., 468
Ishiwata, H., 424
Ishiyama, A., 431
Ishiyama, D., 470
Ishiyama, H., 428
Ishizuka, M., 412, 472
Isidorov, V., 390
Isidorov, VA., 15, 388, 393
Islam, K., 471
Ismailoglu, UB., 459
Ismailov, Z., 434
Isnansetyo, A., 461
Isobe, M., 425, 461
Isosaari, P., 478
Israel, V., 430
Issa, HH., 412, 453
Isshiki, K., 412, 434, 445
Isski, K., 434
Isupov, MN., 486
Itabashi, T., 472
Itagaki, F., 428
Itaya, T., 460
Ito, A., 475
Ito, H., 410
Ito, K., 401
Ito, M., 449
Ito, S., 437

Ito, T., 397, 417
Itoh, H., 394
Itoh, M., 455
Itoh, N., 389, 485, 487
Itoh, T., 405
Itoh, Y., 438
Itokawa, H., 426, 446, 464
Itou, Y., 418
Ivy, P., 446
Iwagawa, T., 409, 436, 442
Iwai, Y., 431, 454
Iwamoto, H., 409, 443
Iwasa, K., 433
Iwasaki, I., 387
Iwasaki, J., 410
Iwashima, M., 413
Iwata, M., 445, 470
Iwata, R., 454
Iwatani, W., 470
Iwatsuki, C., 440
Izaki, K., 418
Iznaguen, H., 440
Izumi, AK., 417
Izumi, Y., 394, 485, 486, 489
Izumida, H., 440

J
Jackson, M., 454
Jacob, DJ., 382
Jacob, G., 396
Jacob, JS., 464
Jacobs, RS., 411
Jacobsen, R., 500
Jacobsen, RB., 429
Jacoby, C., 419
Jain, V., 446
Jakobsson, E., 463
Jakupovic, J., 401
Jalil, S., 458
Jamieson, C., 415
Janda, KD., 500
Jang, J-H., 465
Jannun, R., 485
Janosik, T., 442
Janota, K., 456
Jansen, R., 418, 427
Janssen, DB., 494, 495

Janssen, MAC., 503
Janssen, PAJ., 503
Jansson, B., 462
Janus, SA., 502
Janussen, D., 396
Jao, E., 475
Jarman, WM., 435
Jasinski, JP., 435
Jaspars, M., 417, 427, 466, 503
Jaun, B., 426, 427
Jaya, A., 388
Jayaprakasam, B., 412
Jayasuriya, H., 470, 474
Jdanova, M., 390
Jeffe, PR., 394
Jeffers, MR., 390
Jeffers, PM., 393
Jefferson, MM., 490
Jefford, CW., 403, 407
Jelakovic, S., 431
Jenh, C-H., 474
Jenkins, J., 475
Jenkins, JK., 475
Jenkins, RG., 471
Jenner, A., 448
Jennings, SG., 383
Jensen, CM., 430
Jensen, KA. Jr., 487
Jensen, P., 499
Jensen, PR., 385, 412, 432, 433, 453,
   456, 457, 469, 470, 499, 506
Jensen, R., 457
Jensen, S., 475
Jeong, JH., 405
Jeong, S., 429
Jerlich, A., 489, 490
Jessiman, AS., 453
Jha, RK., 501
Ji, N-Y., 440, 507
Jia, Y., 506
Jia, Z-J., 402, 405
Jiang, B., 436, 438
Jiang, H., 469
Jiang, J., 426
Jiang, Q., 491
Jiang, S-H., 433
Jiang, Z., 460
Jimenez, B., 463

Jiménez, C., 409, 424
Jímenez, DR., 439
Jimenez, EC., 428, 473
Jiménez, JI., 383, 424, 425, 444
Jin, N., 488
Jobst, H., 478
Jockusch, RA., 441
Johansen, S., 457
Johansson, C., 478, 479
Johansson, E., 385, 459, 479
Johansson, G., 392
Johansson, M-B., 478
Johansson, T., 441
Johjima, T., 487
John, S., 457
Johns, WD., 386
Johnsen, S., 457
Johnson, A-L., 442, 499
Johnson, CG., 463
Johnson, JD., 396
Johnson, R., 430
Johnson, RA., 497
Johnson, RK., 438
Johnson, TA., 447, 462, 502
Johnson, TR., 445
Johnston, AE., 478
Jojima, Y., 429
Jolliffe, KA., 497
Jompa, J., 422, 465
Jonas, U., 496
Jones, AD., 457
Jones, AE., 501
Jones, AL., 399
Jones, G., 384
Jones, GB., 451
Jones, JL., 478
Jones, KC., 380, 476, 478
Jones, PG., 397, 472
Jones, T., 398
Jongaramruong, J., 399, 404
Jonsson, P., 441
Jonsson, PR., 499
Jonsson, S., 385, 478
Joo, B., 447
Jordan, A., 380, 388, 395
Jordan, MA., 430
Jørgensen, BB., 385
Jos, A., 420

Joshi, BK., 453
Joullié, MM., 426
Ju, J., 452
Juagdan, EG., 404
Julius, D., 500
Jun, W., 435, 463
Jung, G., 474, 501
Jung, HA., 460
Jung, J., 415
Jung, JH., 442, 460, 467
Junk, PC., 462
Jun-ying, L., 470
Juranic, Z., 397, 402
Jurchen, JC., 441
Jurek, J., 399
Jüttner, F., 447, 501

K
Kabuto, M., 487
Kadali, SS., 450
Kadima, TA., 481
Kagel, RO., 476
Kahn, K., 495
Kahne, D., 473
Kai, N., 407
Kaiser, CR., 404
Kaji, A., 503
Kaji, H., 503
Kajiura, T., 452
Kakkar, P., 503
Kaleschke, L., 383
Kalidindi, R., 404
Kalin, RM., 390, 391
Kalinovsky, AI., 405, 407, 413
Kalkoff, W-D., 477
Kallenborn, R., 400, 435
Kallmeyer, J., 385
Kalume, DE., 473
Kamada, N., 404
Kamano, Y., 445, 446, 464
Kameda, M., 425
Kamei, H., 452
Kamei, K., 453
Kamei, Y., 461, 503
Kamel, MS., 447
Kamenarska, Z., 485
Kamerling, JP., 383

Kameyama, T., 452
Kamijima, C., 410
Kamimura, D., 444, 445
Kaminski, JJ., 470
Kamisaka, S., 423
Kamiya, S., 470
Kammann, U., 440
Kammerer, B., 429
Kamo, S., 464
Kan, Y., 417
Kanai, M., 432
Kanai, T., 460
Kanai, Y., 470
Kanakubo, A., 461
Kanazawa, S., 470
Kanbe, K., 489
Kanchanapoom, T., 447
Kaneko, M., 436
Kanematsu, K., 465
Kanerva, LT., 484
Kang, H., 447, 460
Kang, JS., 396, 460
Kang, N., 492
Kang, S., 433
Kankaanpää, HT., 385
Kannan, K., 478
Kanoh, N., 487
Kao, J., 491
Kaouadji, M., 422
Kaptein, B., 446
Karagouni, AD., 399
Karaguni, I-M., 430
Karchesy, J., 402
Kardos, NL., 386
Karlaganis, G., 477
Karlöf, L., 389
Karlsson, S., 485
Karplus, M., 453
Karpov, AS., 441
Karpov, GA., 393
Karuso, P., 438
Karwowski, JP., 454
Kasai, R., 415, 447
Kashem, MA., 481
Kashman, Y., 410, 429, 436, 466, 467
Kaspars, KA., 389
Kassel, S., 456
Kästner, M., 496

Katagiri, M., 415
Katano, T., 463
Kataoka, K., 440
Kataoka, M., 489
Katapodis, P., 497
Kather, N., 459
Kato, H., 437, 438, 465, 467
Kato, M., 433
Kato, MJ., 403
Kato, S., 468
Kato, T., 426, 440
Kato, Y., 399, 451
Kato-Noguchi, H., 448
Katsifas, EA., 399
Katsumata, R., 456
Katz, B., 426
Katzer, W., 398
Kau, DA., 434
Kau, TR., 466
Kauffman, C., 469
Kauffman, CA., 432, 433, 453, 457, 506
Kaugare, S., 478
Kaur, B., 445
Kaur, G., 427
Kautsky, L., 462, 475
Kavuru, MS., 464
Kavvadias, D., 448, 449
Kawabata, J., 438, 462
Kawabata, S., 401
Kawada, T., 471
Kawagishi, H., 457
Kawaguchi, H., 452
Kawaguchi, S., 405
Kawahara, S., 453
Kawai, K., 436, 472
Kawai, M., 412
Kawai, N., 447
Kawakubo, T., 398
Kawamoto, T., 405, 407, 507
Kawamura, N., 434, 493
Kawamura, T., 411, 412
Kawamura, Y., 470
Kawano, K., 450
Kawasaki, I., 439
Kawasaki, M., 425
Kawasaki, T., 442, 444
Kawashima, A., 470
Kawashima, K., 471

Kawashima, Y., 463
Kawata, S., 452
Kawuloková, L., 486
Kaya, K., 423, 507
Kazi, AB., 437
Kazunga, C., 487
Kearns, PS., 468
Keasling, JD., 470
Keder, NL., 486
Kedersha, NL., 502
Keene, J., 388
Keene, WC., 382, 389, 391, 501
Kehraus, S., 403, 440, 499
Kekos, D., 497
Kelleher, NL., 452, 493
Keller, S., 488, 489, 501
Kelley, JA., 447
Kelley, WP., 441
Kellogg, RM., 495
Kelly, M., 437, 445, 455, 461, 468, 506
Kelly, RF., 500
Kelly, SR., 499
Kelly-Borges, M., 406, 416, 418, 424, 437, 445, 448, 467
Kempf, H-J., 434
Kenigsberg, P., 480
Kenji, T., 476
Kennard, CHL., 424
Kennedy, JT., 390, 498
Kennedy, MC., 420
Kenyon, VA., 462
Keppler, F., 11, 380, 386, 390, 395, 396, 457, 492, 501
Keppler, H., 387
Kerhoas, L., 415, 470
Kerkhof, LJ., 496
Kerkman, R., 482
Kernan, MR., 406, 462
Kerr, RG., 409, 493
Kertesz, M., 495
Keseru, GM., 459
Kesingland, AC., 433
Kessler, JH., 503
Ketola, RA., 392
Kettle, AJ., 413, 464, 490
Kettrup, A., 476
Keusgen, M., 464
Keyzers, RA., 501

Khadeer, A., 439
Khain, A., 501
Khalil, AT., 409
Khalil, MAK., 389, 391, 393, 395
Khalil, Z., 428
Khan, AM., 449
Khatvon, R., 402
Khlifi, S., 415
Kho, YH., 397
Khosla, C., 423, 488
Khosrovi, B., 490
Khoury, Ael, 451
Kicklighter, CE., 500
Kido, M., 475
Kieb, M., 486
Kierkegaard, A., 462
Kigoshi, H., 385, 396, 402, 411, 454, 455, 465, 467
Kihara, T., 434
Kijjoa, A., 467, 468
Kikuchi, K., 411, 412, 439
Kikuchi, M., 415
Kilian, G., 395
Kiljunen, E., 484
Killham, K., 478
Killmer, L., 438
Kim, C-G., 494
Kim, C-J., 447, 474
Kim, D., 414, 415
Kim, D-K., 467
Kim, D-S., 460
Kim, G., 433
Kim, H., 414, 415
Kim, H-K., 398
Kim, HS., 463
Kim, JC., 460
Kim, J-G., 442
Kim, J-P., 447
Kim, J-S., 425, 461, 474
Kim, J-Y., 442
Kim, S., 414, 415, 442
Kim, S-G., 506
Kim, SH., 481, 502
Kim, S-K., 396, 449
Kim, S-U., 449
Kim, W-G., 447
Kim, YJ., 387, 473
Kim, Y-K., 449

Kim, Y-P., 397, 398
Kimblin, C., 481, 486
Kimura, H., 408
Kimura, J., 400, 404, 421
King, A., 471
King, DB., 393
King, GM., 461
King, RJ., 500
King, RM., 401
Kinghorn, AD., 412
Kingston, DGI., 400, 403
Kinnel, RB., 440
Kinoshita, N., 425, 434, 436, 472
Kinoshita, T., 455
Kinter, M., 491
Kintzinger, JP., 412
Kinzyo, Z., 475
Kiprof, P., 506
Kirchmann, H., 459
Kirchner, M., 499
Kirschning, A., 494
Kirst, GO., 393
Kirtikara, K., 386
Kiryu, M., 424
Kishore, KH., 448
Kislev, ME., 388
Kiss, E., 486
Kissau, L., 430
Kita, K., 398
Kita, M., 411, 413, 455, 467
Kita, Y., 444, 445, 507
Kitagawa, I., 412, 415, 441
Kitahara, T., 425, 450
Kitamura, H., 410
Kitamura, K., 440
Kitamura, M., 462
Kittakoop, P., 386, 471
Kiuchi, F., 449
Kiyosawa, S., 407
Kiyota, H., 418
Kizu, H., 445, 446, 464
Kjaer, A., 434
Kjelleberg, S., 421, 422, 499, 500, 507
Kjeller, L-O., 476, 477
Kladi, M., 379, 405, 406
Klausmeyer, P., 471
Kleiman, G., 391
Klein, D., 444, 454

Klein, DM., 462
Klein, G., 397
Klein, R., 490
Kleinschnitz, M., 448
Kleymann, G., 453
Kliche-Spory, C., 450
Klimm, C., 476
Kline, DL., 506
Kline, PC., 488
Kloareg, B., 485, 487, 500
Klohr, SE., 436, 452
Klöser, H., 385, 394
Klostermeyer, D., 458
Knapp, LW., 481
Knauth, LP., 382
Knipping, EM., 382
Knockaert, M., 443
Knott, MG., 400
Knust, H., 453
Ko, C-L., 410, 420
Ko, MKW., 383
Kobayashi, H., 422, 425, 440, 462, 467, 474
Kobayashi, J., 416, 428, 437–440, 442, 443, 465, 466, 507
Kobayashi, K., 407
Kobayashi, M., 412, 431, 432, 441, 469
Kobayashi, S., 446, 452
Kobayashi, Y., 409
Kocak, G., 502
Köck, M., 437–440, 499
Kodali, S., 470
Kodama, K., 396
Kodama, M., 402, 440
Koehn, FE., 417
Koenigsberg, SS., 496
Kogame, K., 402
Kogler, H., 474
Koh, Y., 439
Kohda, S., 434
Koizumi, N., 469
Kojima, Y., 433, 451, 473
Kojiri, K., 470, 472
Kokas, OJ., 498
Kokubo, S., 410
Kolesnikova, SA., 407, 413
Kollmann, A., 415
Kolz, J., 459, 471
Komander, D., 466

Komiya, T., 397, 398
Komiyama, K., 397, 398, 445, 446, 471
Kondapi, AK., 468
Kondo, H., 470, 472, 482
Kondo, S., 428
Kondo, T., 424
Kondoh, H., 472
Kondru, RK., 417
Kong, C-H., 448
Kong, F., 429, 454
Kong, YL., 398
König, GM., 396, 399, 403, 404, 421, 440, 457, 468, 499, 502, 507
König, WA., 431
Konishi, M., 452
Konishi, T., 407
Konivalinka, G., 401
Konno, A., 462
Konopelski, JP., 506
Konoshima, T., 407
Konya, K., 472
Kop, T., 397
Kopecny, J., 420
Koppmann, AKR., 390
Koppmann, R., 391, 395
Korablev, O., 506
Korn, ED., 427
Kornprobst, J-M., 407
Koropalov, V., 389, 391
Korshalla, JD., 456
Kosaka, T., 465
Kosemura, S., 448
Koshino, H., 397, 455
Kotake, A., 446
Kotaki, Y., 465
Köth, A., 496
Kothéus, M., 459
Kotiw, M., 398, 430
Kotoku, N., 469
Kotte, K., 392
Kotzsch, A., 434, 488
Kouko, T., 442
Koumaglo, KH., 450
Kouno, K., 457
Kousaka, K., 421
Kouzai, T., 485
Kovacs, KM., 435
Kovalenko, LJ., 383

Kovtun, Y., 452
Koyama, H., 491
Koyama, T., 413, 462
Kozhushkov, SI., 431
Kraal, B., 419
Krab, IM., 418
Kragl, U., 484
Krahn, MM., 435
Krajick, K., 501
Kramer, RA., 444
Krantz-Rülcker, C., 479
Krasokhin, VB., 475
Kratky, C., 495
Krause, N., 413
Krauß, P., 476
Krauß, Th., 476
Kraut, L., 459
Kräutler, B., 495
Krebs, HC., 433
Kreis, P., 502
Křen, V., 486
Krishnan, KS., 473
Krishnaswamy, NR., 401
Kristoffersden, P., 421
Krivobok, S., 451
Krohn, K., 397, 402, 450, 456, 472, 507
Krueger, G., 392
Krüger, U., 382
Krummel, PB., 389
Krummel, PD., 389
Krupa, D., 502
Kuball, HG., 458
Kubanek, J., 408, 410, 500
Kubista, Ev
Kubo, A., 433
Kubo, O., 507
Kubo, Y., 402
Kuboki, H., 472
Kubota, NK., 409, 443
Kubota, T., 466, 507
Kubota, Y., 493
Kuenen, JG., 459
Kuga, H., 401, 454
Kuhlmann, J., 430
Kuiper-Goodman, T., 423
Kukla, MJ., 503
Kulakov, L., 495
Kulatheeswaran, R., 408

Kulcharyk, PA., 491
Kullmann, Y., 476
Kulmala, M., 383
Kulp, SE., 477
Kumagai, H., 412, 448
Kumagi, H., 412
Kumar, N., 421, 422, 507
Kunimoto, S., 436
Kuniyoshi, M., 400, 403, 407
Kuntiyong, P., 455
Kunze, A., 499
Kunze, B., 418, 427
Kunzendorf, A., 488
Kunzmann, A., 462
Kuo, R-Y., 433
Kuo, Y-H., 410
Küpper, FC., 393, 487, 489
Küppers, M., 389
Kura, F., 491
Kuramoto, M., 410, 433
Kurata, K., 461, 500
Kurihara, H., 462
Kurihara, T., 495
Kuroki, O., 426
Kurotaki, M., 455
Kuruto-Niwa, R., 413
Kuruüzüm, A., 459
Kuwahara, S., 471
Kuwamoto, S., 485
Kuwano, M., 469
Kuwano, T., 469
Kuyper, TW., 386
Kuypers, FA., 413
Kuzovkina, IN., 448
Kwak, JH., 409
Kwon, B-M., 398
Kwon, HC., 456
Kwon, J-Y., 398

L

Laatsch, H., 418, 443, 484, 493
Labadie, M-D., 460
LaBarbera, DV., 426
Lacey, E., 431, 436, 440, 456
Lachance, N., 422
Lackner, H., 431
LaCount, MW., 496

Lacy, C., 417, 466
Ladbury, JE., 453
Ladino, CA., 502
Lago, JHG., 403
Lahav, R., 501
Lahtiperä, M., 463
Laine, W., 446
Laing, C., 427
Lakin, MJ., 382
Lakner, FJ., 483
Lal, AR., 406
Lallemand, J-Y., 470
Lam, KS., 432, 436, 446
Lam, MHW., 478
Lam, PKS., 478
Lam, TH., 468
Lamarque, J-F., 388
Lamb, C., 390
Lambeir, A-M., 481
Lambert, G., 441
Lambert, JM., 502
Lamparski, LL., 476, 478
Lampert, MB., 490
Lampis, G., 434
Lan, Y-H., 449
Laney, M., 460
Lang, G., 436, 445, 472
Lang, M., 458
Lange, AJ., 489
Lange, CA., 392
Langley, DR., 452
Laniewski, K., 478
Lansakara-P, DSP., 482
Lanzalunga, O., 483
Lanzirotti, A., 386
Lapadatescu, C., 494
Lapawa, M., 454
Laprevote, O., 442
Largent, DL., 422, 440
Larkin, MJ., 391, 495
Larsen, LK., 430
Larsen, RM., 416
Larsen, TO., 450
Larson, GM., 453
Larter, M., 406
Lary, DJ., 382
Laskin, AI., 381
Lasram, S., 451

Lassen, P., 479
Lassøe, T., 463
Lassota, P., 456
Laturnus, F., 385, 390–394, 485
Lau, C., 477
Lauble, H., 420
Laurén, MA., 385
Laurenti, E., 461, 496
Lauritsen, FR., 392, 458
Laus, G., 380
Lautens, M., 430
Lavayre, J., 442
Lawrence, MG., 382
Lawrence, NJ., 441
Lawton, LA., 503
Lazo, JS., 447
Lee Guern, F., 388
Lee Pennec, G., 383
Lee Quéré, J-L., 494
Leadlay, PF., 420
Leahy, JW., 454
Lear, MJ., 452
Lebeuf, M., 435
Lebioda, L., 496
Leblanc, C., 485, 489
Leboeuf, M., 433
Lebrihi, A., 451
Lebrilla, CB., 468
Lechevalier, HA., 381
LeCloarec, M-F., 387
Lee, AY., 442, 464
Lee, BH., 406
Lee, C., 387
Lee, CH., 397
Lee, C-O., 442, 467
Lee, DS., 492
Lee, D-U., 433
Lee, G., 406
Lee, H., 414
Lee, HJ., 397, 463
Lee, H-S., 466, 503
Lee, JD., 501
Lee, JH., 463, 503
Lee, JJ., 463
Lee, K-H., 447
Lee, MD., 494
Lee, MS., 436, 452
Lee, NH., 406

Lee, N-K., 441
Lee, PE., 394
Lee, RT., 479
Lee, S-J., 449
Lee, SY., 449, 507
Lee, T-H., 442
Lee, TH., 455, 503
Lee, UC., 398
Leece, BA., 452
Lee-Lu, R., 455
Leese, RM., 461
Lee-Taylor, J., 390
Lee-Taylor, JM., 391, 394
Lefevre, CK., 458
Legendre, J-M., 487
Legg, M., 466
Lehnert, H., 383
Lehrer, RI., 429
Leimkuhler, C., 473
Leisch, H., 497
Leisinger, T., 495
Leistner, E., 436, 452, 494
Leize-Wagner, E., 485
Lellouch, E., 389
Lemieux, PM., 477
Lenoir, D., 435
Leon, F., 402
Leonard, A., 445
Léonce, S., 446
Leong, CY., 445
Leost, M., 441, 443
Leri, AC., 386, 394
Leslie, P., 423
Letai, A., 432
Levenfors, J., 428
Levine, JS., 388
Lewandowicz, A., 495
Lewis, AC., 501
Lewis, EA., 422
Ley, SV., 453
Li, C., 435
Li, C-J., 437
Li, D-R., 455
Li, F., 418
Li, G., 405, 432
Li, H., 431
Li, H-J., 389
Li, J., 429

Li, J-J., 408
Li, J-Y., 455
Li, K., 439, 507
Li, MH., 457
Li, M-Y., 409
Li, P., 415
Li, Q., 429
Li, Q-X., 409
Li, S., 405, 406, 443, 447, 460, 461
Li, S-M., 449
Li, T., 430
Li, T-L., 420
Li, W., 413
Li, X., 396, 402
Li, X-M., 440, 507
Li, Y., 416, 428
Li, YF., 391
Li, Y-M., 404, 433, 448
Liang, AP., 417
Liang, J., 430
Liang, XT., 472
Liao, J-H., 405
Liao, Q., 489
Liaud, M-F., 486
Libby, RD., 481
Lichte, E., 437, 499
Lichtig, J., 476
Lidgren, G., 441
Lidstrom, ME., 394, 495
Lieberknecht, A., 441
Liebisch, P., 486
Lievens, SC., 410
Lightstone, FC., 495
Ligon, JM., 434, 488
Lii, Y., 461
Likos, J., 424, 438
Lim, LL., 464
Lim, TK., 441
Limbert, M., 474
Limin, S., 388
Limosani, P., 451
Lin, C-J., 448
Lin, G-Q., 455
Lin, L-P., 412
Lin, M-R., 408, 409
Lin, S., 494
Lin, W., 407
Lin, Y-C., 409, 420, 450

Lincoln, DE., 395, 459, 481, 496
Lindel, T., 436, 437, 439, 444, 499
Linden, A., 399
Lindhorst, TK., 486
Lindquist, N., 441, 464
Lindsay, BS., 465
Lindsay, RC., 460, 461
Ling, T., 429
Lingham, RB., 474
Linington, RG., 438
Liou, C-S., 459
Lipscomb, JC., 495
Lipson, DA., 386
Lira, SP., 426
Lira-Rocha, A., 444
Lirazan, MB., 428, 429
Lis, AW., 448
Lisitzyna, L., 392
Liss, PS., 392, 393, 395
Little, RD., 411, 487
Littlechild, J., 479, 489
Littlechild, JA., 486
Liu, B., 455
Liu, C., 502
Liu, C-M., 402
Liu, DQ., 454
Liu, H., 462, 463
Liu, H-Y., 409
Liu, J-F., 436, 438
Liu, J-J., 447
Liu, K., 423
Liu, L., 480
Liu, P., 448
Liu, Q-W., 461
Liu, S., 467
Liu, W., 452
Liu, Y., 411, 467
Liu, Y-H., 475
Liu, Y-T., 475
Liu, Z., 402
Liukkonen, M., 386
Livett, BG., 428
Ljungquist, P., 478
Ljungström, E., 382
Llamas, NM., 498
Lobert, JM., 389, 391
Lobkovsky, E., 432, 445, 456, 469, 502
Locher, HH., 447

Lockhart, A., 492, 493
Loebenberg, D., 470
Löffler, FE., 496
Logan, JA., 391
Lohmann, S., 436
Lokker, NA., 398
Lolas, M., 471
Loll, PJ., 473
Lonergan, GC., 458, 475
Long, BH., 446
Long, C., 402
Long, GC., 423
Longley, RE., 441, 445
Longmore, R., 497
Longmore, RW., 497
Longoria, A., 484
Loo, CP., 502
Look, GC., 482
Loong, DTJ., 497
Lopez, S., 447
Lopez-Artíguez, M., 420
López-Canet, M., 426
López-Vera, E., 473
Lorber, M., 476
Lorenz, P., 450
Lorenzo, M., 414
Loria, P., 502
LoRusso, PM., 446
Los, G., 429
Loscha, K., 431
Loseto, G., 464
Losey, HC., 493
Lou, L-G., 402
Lough, AJ., 410, 429
Loughlin, WA., 484
Love, SP., 387
Lovell, CR., 460, 481, 496, 499, 500
Lovett, AA., 389
Low, JC., 390
Lowery, CA., 500
Lowes, DJ., 497
Lowy, FD., 503
Lozada, D., 483
Lteif, R., 451
Lu, C., 453, 455
Lu, CC., 441
Lu, J-H., 404, 448
Lu, W., 473

Lu, X., 471
Lü, Y., 460
Lucas, MC., 406
Lucas-Meunier, E., 473
Luckas, B., 435
Lücking, R., 470
Luckman, SW., 456
Ludewig, K., 397, 450
Luesch, H., 424, 425, 454, 503
Luft, T., 449
Luksemburg, W., 476
Lull, C., 444
Lumbsch, HT., 469, 470
Luna-Herrera, J., 468
Lunding, A., 458
Lund-Pero, M., 490
Luo, X-D., 412
Luoma, EV., 476
Luppi, G., 446
Luria, M., 383
Lusby, WR., 457
Lusis, AJ., 464, 490
Lutes, CC., 477
Luther, RR., 428
Lutje Spelberg, JH., 495
Lutz, RJ., 452
Lyakhova, EG., 407, 413
Lydersen, C., 435
Lymar, SV., 413
Lyssenko, KA., 444

M
Ma, D., 425
Ma, M., 406, 447, 460, 461
Ma, WCJ., 461
Ma, Y-A., 494
Ma, Y-B., 412
Maas, EW., 422
Mabe, J., 456
Mabry, TJ., 401
Mabuchi, I., 499
Macaulay, JB., 458, 494
MacBeth, JL., 436
MacDonald, S., 451
Macedo-Ribeiro, S., 480, 482
Macherla, VR., 432
Machida, K., 415

Machida, T., 389
Macías, FA., 401
Maciejewski, AJH., 387
Macielag, M., 503
Mackay, AC., 454
Mackay, MA., 384
Macko, V., 426, 427
MacMillan, DWC., 506
MacMillan, JB., 417, 424, 454
MacNaul, K., 470
MacNeil, S., 415
Macura, S., 402
Maddess, ML., 430
Madsen, H., 476
Maeda, N., 491
Maenhaut, W., 391
Mafnas, C., 385
Magae, J., 471
Magalhâes, A., 468
Magarvey, N., 422
Magiatis, P., 443
Magno, S., 416, 465–467
Maguire, N., 496
Magyarosy, A., 470
Mahajan, AS., 501
Mahandru, MM., 469
Maharaj, D., 410
Mahieux, A., 506
Mahindaratne, MPD., 405
Mahmoud, AA., 401, 402
Mahmud, T., 493, 494
Mahuteau, J., 433
Mai, B., 432
Maia, MF., 488
Maier, A., 418
Maiese, WM., 444
Maitland, DJ., 463
Majdik, Z., 382
Majerus, P., 450
Makarieva, TN., 444
Maki, Y., 434
Makino, B., 412
Maksimenka, K., 459
Mal, D., 456, 470
Malabarba, A., 473
Malchow, R., 484
Maldonado, LA., 385
Maliakal, S., 400, 485

Malin, G., 392, 393
Mallais, T., 396
Malle, E., 491
Malloy, TA., 476
Malmstrøm, J., 467
Malmvärn, A., 462, 475
Malochet-Grivois, C., 410
Malone, JF., 498
Maloney, D., 475
Malpass, JR., 432
Malspeis, L., 427
Mamada, H., 431
Manam, RR., 432
Mancini, I., 403–405, 414, 437, 442
Manda, T., 451
Mandalakis, M., 463, 478
Mandelkow, E-M., 502
Manduchi, L., 483
Manefield, M., 421
Manes, LV., 399
Manfredi, A., 482
Mangalindan, GC., 440
Mangani, F., 388
Mangieri, EA., 428
Mangoni, A., 437, 459, 466
Manley, SL., 390, 480
Mann, M., 473
Manø, S., 391
Manohar, RN., 434
Manoj, KM., 481, 483
Manríquez, CP., 411
Manta, E., 403, 404
Manzo, E., 408, 414, 433
Mao, S-C., 406
Maoka, T., 459
Maoufoud, S., 446
Mar, W., 442
Marante, FJT., 458
Marchesini, A., 436, 484
Marchetti, F., 407
Marcinkiewicz, J., 490
Marco-Urrea, E., 496
Marcourt, L., 402
Marelich, GP., 464
Mariezcurrena, RA., 404
Marín, S., 451
Marma, MS., 403, 407
Marminon, C., 446

Márquez, B., 417
Márquez, BL., 417, 423, 425
Marquez, LA., 490
Marquez-Rocha, FJ., 484, 486
Marriott, G., 410, 429
Marsh, G., 461–463
Marshall, AG., 473
Marshall, KM., 445
Marshall, RA., 394, 395
Martarello, L., 492, 493
Martens, D., 476
Martí R., 476
Martin, AP., 386
Martin, GDA., 470
Martin, I., 398
Martín, JD., 406, 411
Martín, MJ., 406
Martin, S., 398, 399
Martin, SM., 434
Martínez, E., 414
Martinez, JS., 487
Martínez, M., 476
Martinez-Cortizas, A., 386
Martínez-Luis, S., 444, 502
Martini, M., 388
Martini, U., 459
Martinot, TA., 498
Märtlbauer, E., 451
Marty, D., 503
Maruya, KA., 440
Maruyama, S., 455
Marzorati, L., 482
Masaoud, M., 396
Maskey, RP., 418, 443
Mason, G., 475
Massa, S., 434
Massanet, GM., 401
Massey, R., 451
Mastalerz, P., 381
Masuda, H., 471
Masuda, M., 402, 404, 405, 407, 411, 413, 414, 491
Masuda, Y., 465
Masugi, T., 434
Masuko, K., 428
Masuma, R., 397, 398, 431, 454, 471
Masuno, MN., 454, 466, 468
Mata, R., 444, 502

Matainaho, L., 433
Maté, JL., 413, 421
Matern, U., 431
Mathela, CS., 406
Mathur, DB., 476
Matias, PM., 397
Matlock, DB., 400
Matoso, CM., 381
Matrai, PA., 393
Matson, JA., 436, 446, 452
Matsson, O., 495
Matsuda, A., 411
Matsuda, F., 447
Matsuda, H., 415, 426, 431
Matsuda, M., 423, 464
Matsuda, Y., 398, 463
Matsudo, T., 398
Matsui, J., 433
Matsui, S., 472
Matsui, S-I., 452
Matsumoto, GK., 408, 462
Matsumoto, N., 472
Matsumoto, S., 451
Matsumoto, SS., 445
Matsumura, E., 397
Matsumura, K., 442
Matsumura, R., 415
Matsunaga, S., 418, 431, 438–440, 455, 465, 467
Matsuo, Y., 402, 411–414
Matsushima, C., 398
Matsuzaki, K., 398, 474
Matsuzawa, S., 411, 412
Matthiesen, U., 458
Mattia, CA., 414
Mattocks, KM., 502
Matucha, M., 390, 485
Matulova, M., 446
Matveev, V., 383
Mauger, AB., 431
Mauro, JM., 480
Maximilien, R., 421
Maxwell, AR., 410
May, HD., 496
Maycock, CD., 397
Mayer, AMS., 501
Mayer, P., 439
Mayer-Bartschmid, A., 453, 473

Mayne, B., 483
Mayne, CL., 443
Mayo, M., 428
Mayrhofer, H., 469
Mayumi, T., 469
Mazzarella, L., 414
McAlpine, JB., 454
McAnulla, C., 495
McCarthy, PJ., 445
McCarty, LS., 381
McCauley, JA., 474
McClintock, JB., 400, 445, 498
McCloud, TG., 471
McCloy, S., 499
McConnell, JC., 382
McCook, A., 393
McCormick, JL., 475
McCormick, ML., 448
McCoy, WF., 500
McCubbin, JA., 430
McCulloch, A., 389, 391, 392, 395, 396
McDonald, IR., 495
McDonald, LA., 456, 465
McDougald, D., 422
McElroy, CT., 382
McEwan, AR., 420, 492
McGee, DI., 396
McGeehin, PKM., 498
McGenity, 503
McGlinchey, RP., 420, 493
McGuire, D., 428
McGuire, M., 399
McHugh, MM., 452
McIntosh, JM., 428
McIntyre, HP., 393
McKee, TC., 440, 443
McLachlan, M., 478
McLaughlin, DI., 448
McLaughlin, RK., 448
McLean, S., 410, 429
McLeod, MD., 497
McLinden, CA., 382, 383
McMahon, JB., 447
McMenamin, ME., 464
McMillan, A., 390
McMurry, J., 507
McNab, H., 443
McNicholas, C., 398

McNulty, J., 438
McPhail, AT., 475
McPhail, K., 417, 501
McPhail, KL., 403, 414, 423, 425, 494
McQuaid, JB., 501
McRae, KJ., 497
McRoberts, CW., 492
McRoberts, WC., 390, 394, 419, 495, 498
Mearns-Spragg, A., 460
Mecham, RP., 491
Meddour, A., 420
Medina, JR., 416
Meehan, T., 457
Meguro, S., 462
Meharg, AA., 478, 496
Mehrtens, G., 392, 485
Mehta, AC., 464
Meier, GP., 496
Meijer, HAJ., 389
Meijer, L., 441, 443, 502
Meili, J., 426, 427
Meintanis, C., 399
Meinwald, J., 426
Meketa, ML., 438
Melamed, Y., 388
Melcher, J., 400, 463
Melkani, AB., 406
Melo, R., 487, 488
Melton, DF., 490
Mendes, L., 501
Méndez, C., 446, 488, 493
Menez, E., 399
Menke, JG., 470
Menkis, A., 428
Menna, M., 438, 467
Menon, K., 430
Menzel, LP., 429
Meragelman, KM., 440
Merentes, E., 467
Merkel, P., 426
Merkert, H., 384
Merkul, E., 441
Merlet, P., 391
Merrick, WC., 411
Mertz, FP., 456
Messerschmidt, A., 480, 482
Mester, T., 458
Metcalf, JS., 503, 507

Métrich, N., 387
Meurer, K., 436
Meyer-Klaucke, W., 489
Meyers, CLF., 493
Meyers, RA., 381
Michael, AP., 398
Michael, PJ., 387
Michaelis, W., 479
Michel, A., 450, 456
Michel, G., 489
Michels, JJ., 500
Midgley, PM., 391
Mierau, V., 397
Mierzwa, R., 470, 471, 475
Mignerey, AC., 480
Mihara, A., 398
Mihopoulos, N., 407, 414
Mikami, Y., 425, 439, 443, 507
Miki, C., 495
Miki, K., 488
Mikros, E., 407
Milanowski, DJ., 447
Milat, M-L., 470
Milic, DR., 397
Millar, PD., 406
Miller, BR., 389, 390
Miller, D., 436
Miller, DL., 476
Miller, VP., 483
Milligan, KE., 417, 425
Milliken, CE., 496
Millington, DS., 396
Milne, BF., 427
Milosavljevic, S., 402
Minale, L., 429, 454
Minami, Y., 452
Minassian, F., 438
Mincer, TJ., 385, 432, 506
Minette, HP., 417
Minguez, S., 451
Minoura, K., 397, 417, 418
Miono, T., 400, 403
Mirailles, J., 459
Mircea, M., 383
Mironenko, T., 420
Misico, RI., 412
Mitani, T., 462
Mitchell, HJ., 474, 475

Mitchell, SS., 432, 448
Mitome, H., 408
Mitome, Y., 405
Mitra, SN., 492
Mitsiades, C., 432
Mitsiades, N., 432
Mitsos, C., 439
Mitsos, CA., 444
Mitsuhashi, S., 411, 412, 439
Mittelman, S., 471
Miura, I., 475
Miwa, H., 507
Miwa, K., 452
Miyagawa, N., 452
Miyahara, K., 407
Miyaji, M., 436
Miyake, FY., 442
Miyaki, T., 452
Miyamae, 434
Miyamoto, T., 414
Miyaoka, H., 408
Miyashita, K., 397, 456
Miyatake, M., 406
Miyawaki, H., 450
Miyazaki, M., 442
Mizuno, Y., 411, 412, 414
Mkivananzi, H., 400
Mocek, U., 471
Mochida, K., 418
Mogg, A., 432
Moghaddam, AP., 495
Mogollon, L., 484
Mogridge, N., 464
Mohamed, NM., 384
Mohn, WW., 494
Mohou, H., 446
Mojena, M., 398
Moka, W., 438
Moldanová, J., 382
Molin, S., 421, 500
Molinski, TF., 410, 416, 417, 424, 439,
    448, 454, 466, 468
Mollenhauer, D., 417
Moller, SJ., 501
Mollicone, D., 388
Möllmann, U., 436
Mollo, E., 408, 409, 414
Moloney, JM., 419, 434

Mombrú, AW., 403, 404, 498
Momose, I., 436, 472
Momose, Y., 436
Monari, M., 446
Monde, K., 262, 458
Moni, RW., 416
Monks, PS., 382
Montalvo, NF., 384
Montanari, AM., 464
Monte, FJQ., 412
Monti, F., 471
Montmessin, F., 506
Montzka, SA., 389
Monzani, E., 482
Mooberry, SL., 424, 425, 430
Moon, S-S., 449
Moore, BS., 384, 493
Moore, CA., 485
Moore, DL., 427
Moore, M., 398
Moore, RE., 395, 408, 417, 424, 425, 430, 444, 454, 507
Moore, RM., 380, 385, 389–395, 487, 495, 501
Moos, HW., 389
Mootoo, BS., 410
Morales, Z., 459
Mörck, R., 478, 479
Moreau, D., 406
Moreau, P., 446
Morelli, I.,
Morgenstern, T., 401
Mori, C., 469
Mori, H., 475
Mori, K., 503
Mori, T., 387
Morikawa, T., 415
Morimoto, K., 476
Morimoto, Y., 409, 411, 447
Morinaga, N., 485
Morishita, M., 507
Morita, H., 426
Moriyasu, M., 433
Moro, S., 486
Morris, BD., 447, 464
Morris, DR., 480
Morris, LA., 427
Morrison, LF., 503

Mortby, E., 451
Mortimer, M., 478
Morton, GO., 444
Morton, J., 475
Morton, JB., 475
Morton, RB., 427
Moser, U., 477
Moser-Thier, K., 412
Moses, JI., 389
Moss, SJ., 419, 494
Motoyama, T., 400
Moujir, L., 405
Moulin, E., 453
Mourabit, AA., 493
Moutsos, VI., 469
Moya, P., 444
Moyer, CL., 506
Mozurkewich, M., 382
Mroczko, M., 481
Mshicileli, N., 451
Mtolera, MSP., 392, 480
Mu, H., 416
Mueller, DM., 413, 448, 464, 491
Mues, R., 459
Muhammad, I., 448
Mühlbacher, J., 471
Muijsers, AO., 480
Muir, D., 435
Mukai, A., 425
Mukhopadhyay, T., 418
Mulet, J., 502
Mulholland, DA., 397
Mulholland, NP., 507
Mullally, JE., 421
Mullaney, JT., 429
Muller, C., 506
Müller, D., 499
Müller, G., 385, 386, 388, 456
Müller, H., 453
Müller, IM., 383, 506
Müller, JF., 435, 463, 478
Müller, O., 430
Müller, TJJ., 441
Müller, WEG., 383, 384, 438, 506
Mulvaney, R., 393
Mundt, S., 457
Munetomo, E., 470
Munro, MHG., 379, 436, 445, 472, 501

Murai, A., 415, 487
Murai, K., 507
Murakami, E., 463
Murakami, M., 426, 431, 449
Murakami, S., 470
Muralikrishna, M., 462
Muramatsu, Y., 395
Muraoka, O., 415
Murch, SJ., 503, 507
Murni, A., 462
Murooka, K., 472
Murphy, B., 475
Murphy, CD., 379, 380, 392, 419, 420, 479, 487, 492
Murphy, PT., 416
Murray, LM., 441, 442
Murrell, JC., 495
Murshudov, GN., 486
Murthy, USN., 448
Murthy, VS., 414
Musacchio, A., 443
Musman, M., 403
Muto, N., 431
Myers, AG., 452
Myers, RA., 428
Mynderse, JS., 456
Myneni, SCB, 11, 386, 394

N
Nachtigall, K., 478
Nagai, H., 417, 422, 453, 462
Nagai, K., 440, 474
Nagaki, H., 395, 396
Nagakura, N., 450, 475
Naganawa, H., 412, 425, 428, 434, 436, 445, 472, 489, 493
Nagaoka, M., 452
Nagaoka, T., 457
Nagasawa, E., 423
Nagashima, S., 426
Nagatsu, A., 415
Nageswara, Rao, K., 462
Nagle, DG., 406, 417
Nagraj, N., 482
Nagy, I., 488
Nahrwold, M., 430
Naidoo, D., 397

Naidu, R., 469
Naik, CG., 466
Nair, MG., 412
Nair, MSR., 422
Nair, SS., 473
Naismith, J., 479
Naismith, JH., 354, 420, 434, 480, 488, 492
Naithani, V., 503
Najafi, R., 490
Nakagawa, M., 447, 461
Nakagawa, S., 413, 469
Nakagawa, T., 487
Nakahara, S., 433
Nakajima, H., 432, 451, 455
Nakajima, M., 396
Nakajima, S., 423, 470, 472
Nakamichi, Y., 413
Nakamura, H., 413, 425, 434, 436, 472, 499
Nakamura, KT., 472, 489
Nakamura, M., 458, 487
Nakamura, S., 415
Nakamura, T., 446
Nakanishi, K., 452, 456
Nakanishi, M., 412
Nakano, H., 453
Nakano, S., 394, 407, 414
Nakano, T., 456, 475
Nakano, Y., 462
Nakao, S., 469
Nakao, Y., 438, 440, 448, 465
Nakashima, K., 401
Nakatani, M., 409, 436, 442
Nakatani, R., 507
Nakatsukasa, WM., 456
Nakayama, Y., 475
Nakazaki, A., 446
Nakazawa, T., 389
Nam, J-Y., 398
Nam, KW., 460
Namatame, I., 398, 453
Namikoshi, M., 384, 422, 462
Nandiraju, S., 400
Nara, K., 413
Naranjo, S., 422
Narasimhulu, M., 468
Narina, SV., 422
Narkowicz, CK., 440
Nascimento, MSJ., 467

Nasini, B., 472
Nasini, G., 453, 454, 458, 472
Nass, R., 415
Nasu, SS., 406
Naumann, K., 379, 381, 503
Nauseef, WM., 491
Navarre, DA., 427
Naylor, S., 399, 431
Nedachi, M., 387
Neefs, E., 506
Negishi, R., 422
Negrete, GR., 457
Neidigh, JW., 448
Neild, GH., 503
Neilson, AH., 379, 494
Nelson, L., 428
Nelson, RK., 436
Neretin, LN., 385
Nestrick, TJ., 476, 478
Neu, J., 473
Neubert, REM., 389
Neufeld, DA., 388
Neuman, DJ., 399
Neumann, G., 502
Neuteboom, STC., 432
Nevejans, D., 506
Newcomb, M., 482
Newcombe, NJ., 497
Newman, A., 470
Newman, DJ., 379, 429, 501
Newman, LA., 496
Newton, GL., 466
Ng, SB., 445
Ngo, A., 433
Nguyen, BV., 498
Nguyen, MN., 464
Nguyen, T., 447
Nguyen, V-A., 423
Nhan, TQ., 448
Ni, X., 391
Nicas, TI., 473
Nicholas, GM., 466
Nicholson, B., 432
Nicholson, G., 399, 474
Nicholson, GJ., 429
Nickless, G., 390, 395
Nicolaou, KC., 406, 429, 451, 473–475
Nicoletti, M., 415

Nicolosi, G., 483
Nicotra, VE., 412
Niedan, V., 386, 457, 492
Niedan, VW., 388, 396, 456, 457
Niehaus, J., 399
Nielsen, DB., 428
Nielsen, J., 469
Nielsen, JS., 429
Nielsen, PH., 385, 443, 469
Nielsen, SE., 434
Nieschalk, J., 419
Nieto, RM., 424
Nietsche, J., 454
Nightingale, PB., 392
Nightingale, PD., 389, 393, 394
Niimura, Y., 423
Nijenhuis, I., 496
Nikiforov, V., 463
Nikolaev, SV., 431
Nikolausz, M., 496
Nikolic, M., 502
Nilsson, CL., 473
Ning, S., 494
Niranjan Reddy, VL., 465
Nishi, M., 407
Nishida, F., 418
Nishida, H., 448, 451, 473
Nishiguchi, GA., 411
Nishikawa, Y., 411
Nishimura, S., 439
Nishimura, T., 471
Nishimura, Y., 422
Nishino, H., 491
Nishio, M., 450
Nishiyama, S., 468
Nitanda, N., 413
Niven, JS., 391
Nkusi, G., 385, 388, 396, 456
Noda, T., 399
Noger, D., 477
Nogle, LM., 417
Nogradi, M., 459
Noijiri, Y., 389
Nold, SC., 500
Nonaka, K., 452
Nordlund, E., 479
Noro, T., 503
Norstrom, R., 436

Norstrom, RJ., 435
Norte, M., 411, 414
North, WJ., 394
Northcote, P., 444
Northcote, PT., 379, 440
Northrup, AB., 506
Nortrup, D., 477
Norval, RAI., 457
Norwood, DL., 396
Notholt, J., 383
Nowak, T., 430
Nozawa, K., 472
Nozoe, S., 423, 457, 464
Nuell, MJ., 480
Numata, A., 397, 417, 418
Nunozawa, T., 457, 464
Nüske, J., 488
Nussbaumer, T., 476
Nyberg, F., 441
Nygård, T., 400
Nyui, S-I., 491
Nzengung, V., 393

O
O'Brate, A., 429
O'Brien, R., 453
O'Connell, SJ., 423, 494
O'Connor, FM., 383
O'Connor, SE., 488
O'Doherty, S., 389
O'Dowd, CD., 383
O'Dowd, CR., 498
O'Dwyer, PJ., 430
O'Hagan, D., 361, 380, 381, 419, 420, 492, 493
O'Hara, T., 435
O'Malley, DP., 439
O'Neil, GW., 436
Obata, R., 468
Oberer, L., 426, 431
Öberg, G., 11, 381, 385, 391, 392, 396, 458, 459, 479
Öberg, GM., 380
Öberg, LG., 343, 476
Öberg-Asplund, G., 478
Oberti, JC., 412
Obringer, C., 487

Obuchi, T., 472
Ocaña, JM., 422
Ochi, M., 440, 475
Oda, MN., 491
Oda, T., 471
Odom, JM., 494
Oelschläger, H., 435
Oetjen, H., 501
Ogamino, T., 468
Ogasawara, N., 472
Ogata, Y., 394, 487
Ogawa, A., 444
Ogawa, M., 415
Ogawa, S., 429, 455
Ogawa, T., 463
Ogi, N., 421
Ogihara, Y., 415, 427
Ogino, H., 470
Ogino, J., 430
Ogita, T., 396
Ogura, T., 412
Oh, D-C., 457
Oh, K-B., 442, 503
Oh, M-N., 442
Ohashi, K., 449
Ohashi, Y., 401
Ohba, S., 412
Ohbayashi, M., 452
Ohfune, Y., 425
Ohi, K., 449
Ohizumi, Y., 499
Ohkuma, H., 452
Ohnuki, T., 450
Ohr, HD., 393
Ohsawa, N., 394, 485, 487
Ohshima, H., 490
Ohshiro, T., 394, 485, 486
Ohta, S., 439
Ohta, T., 406, 418, 423, 439, 457, 460, 464
Ohtake, N., 448
Ohtani, II., 418
Ohtani, K., 415
Ohtsu, Y., 432
Ohzono, M., 444
Oikawa, H., 411
Oiwa, R., 431, 454
Ojika, M., 402, 413, 418, 424, 429, 449, 468
Ojima, F., 457

Okabe, T., 470
Okada, K., 423
Okada, M., 445
Okada, N., 394, 487
Okada, T., 439
Okada, Y., 461
Okami, Y., 489
Okamoto, K., 402, 457
Okamoto, Y., 413, 449, 468
Okamura, H., 409, 436, 442
Okamura, M., 489
Okazaki, T., 470
Okino, T., 403, 408, 426, 499
Okita, T., 411
Okita, Y., 426
Öksüz, S., 401
Okuda, RK., 485
Okuda, T., 450
Okue, M., 425
Okuhara, M., 451
Okumura, Y., 449
Okuyama, A., 470
Okuyama, T., 461
Olaniran, AO., 495, 496
Olbrich, D., 436, 463
Oldham, NJ., 400
Oliver, MP., 425
Olivera, BM., 428, 429, 473, 500
Olivo, HF., 433
Olmstead, MM., 468
Olsen, CE., 434
Olson, ER., 434
Olson, M., 455
Olson, TM., 392
Olsson, A., 490
Olsson, M., 435, 462
Olsthoorn-Tieleman, LN., 419
Omae, I., 499
Omori, AT., 497
Omura, S., 397, 398, 431, 448, 453, 454, 471, 474
Onaka, H., 446
Onega, M., 493
Ongania, K-H., 495
Ono, M., 469
Onodera, N., 414
Oppenheimer, C., 387
Oppentocht, JE., 494

Oppong, KA., 498
Oram, JF., 491
Orcutt, BN., 506
Oremland, RS., 495
Oritami, T., 418
Orjala, J., 423
Orozco-Cardenas, M., 480
Orsini, MA., 396
Orsini, P., 453
Ortea, J., 414
Ortega, MJ., 399
Ortega, MJ., 422
Ortega-Barria, E., 410
Ortiz de Montellano, PR., 483
Ortiz, E., 473
Ortiz-Bermúdez, P., 479, 485
Osada, H., 418
Osawa, T., 401
Osborne, NJT., 384
Osborne, RL., 481
Oshiro, N., 400
Oshiro, T., 489
Osipov, AN., 490
Oskarsson, N., 387
Oskarsson, N., 387
Osmulski, PA., 480
Osorio, C., 450
Osterloo, MM., 506
Osterman, RM., 475
Ostermann, S., 495
Osthoff, HD., 501
Otani, T., 452
Otani, Y., 433
Otero-Quintero, S., 445
Otoguro, K., 431, 453
Otsuka, H., 415
Otsuka, M., 410
Otsuki, A., 394
Ott, J., 456
Otteneder, H., 450
Ouellet, M., 422
Oum, KW., 382
Ouye, A., 443
Ovaa, H., 432
Ovenden, SPB., 442
Overman, LE., 414
Oza, VB., 494
Ozawa, S., 471

Ozawa, T., 387
Öztunc, A., 404, 414

P

Paasivirta, J., 463
Paepke, O., 478
Page, MJ., 422, 441
Page, SE., 388
Paik, S., 430
Pairet, L., 398, 434
Pais, M., 424, 465
Paiz, N., 405
Pajunen, H., 478
Palacios, Y., 399
Palavinskas, R., 435
Palermo, JA., 406, 427, 441
Palladino, MA., 432
Pallanza, R., 474
Palmisano, F., 451
Palmisano, G., 446
Paloma, LG., 429
Palomar, AJ., 411
Pan, Y-L., 410
Panasenko, OM., 489, 490
Panayotou, G., 497
Panda, D., 471
Panesar, MS., 433
Paneth, P., 495
Pang, C-H., 473
Pannecouque, C., 405
Panneels, V., 419
Pannell, L., 429
Pannell, LK., 396, 431, 440, 443
Pansini, M., 465–467
Pant, AK., 406
Paonita, A., 387
Papazafiri, P., 405, 407
Papendorf, O., 457
Papeo, G., 438
Papiernik, SK., 393
Papke, O., 478
Päpke, O., 478
Parameswaran, PS., 466
Paranagama, PA., 450
Paré, PW., 402
Parella, T., 496
Parenti, F., 434

Parikka, M., 391
Park, HJ., 429, 460
Park, H-S., 407
Park, J., 463
Park, M., 429
Park, PU., 400, 417
Park, SH., 406
Park, Y., 467, 492
Park, YC., 400
Parker, JC., 473
Parkes, RJ., 385
Parkin, S., 424
Parkinson, T., 451
ParksWC., 491
Parmeggiani, A., 418, 419
Parrish, JD., 487
Parsek, MR., 421, 500
Partida-Martinez, LP., 386
Parvez, M., 401
Pascale, M., 451
Pascoe, KO., 410
Pass, D., 444
Passreiter, CM., 458
Pasta, P., 482, 483
Patallo, EP., 479, 488
Patane, MA., 474
Patel, J., 438
Patel, M., 470, 474, 475
Patel, MG., 471
Paterson, I., 453–455, 506
Pathom-aree, W., 385
Patil, AD., 438
Patnam, R., 450
Patra, A., 408
Patricelli, MP., 396
Patrick, BO., 437
Pattenden, G., 507
Patterson, GML., 430, 444, 507
Patterson, JE., 396
Patti, A., 483
Pattisina, LA., 438, 439
Pattison, DI., 489–492
Paubert, G., 389
Paul, BJ., 498
Paul, DC., 430
Paul, GK., 441
Paul, KG., 476
Paul, NA., 500

Paul, VJ., 384, 400, 405, 408, 417, 424, 425, 454, 466, 498, 500
Paulin, L., 431
Paulus, E., 474
Paustenbach, DJ., 476
Pauwels, R., 503
Pavasars, I., 396, 479
Pavone, V., 407
Pawlik, JR., 108, 306, 372, 408, 498, 499
Pawson, DL., 503
Pearce, AN., 422
Pearce, C., 426
Pearce, GL., 491
Pearl, L., 443
Pearl, LH., 453
Pec, MK., 412
Pecoraino, G., 387
Pecoraro, VL., 489
Pedersen, JR., 388
Pedersén, M., 392, 394, 395, 480
Pedler, BE., 436
Pedro, JR., 401
Pedro, M., 467
Peev, DR., 401
Peipp, H., 450, 471
Peixinho, S., 466
Pelaez, F., 398
Peleg, M., 383
Pelizaeus, B., 484
Pelletier, E., 436
Pelletier, I., 485, 488
Pelletier, J., 411
Pelloux-L,éon, N., 438
Pelter, Av
Pelzer, S., 474
Peng, J., 437, 499, 502
Peng, J-N., 445
Penkett, SA., 389, 393, 394
Penn, MS., 491
Penn, SG., 468
Pennaka, HK., 499
Pennathur, S., 448
Pennisi, M., 387
Penoni, A., 446
Pereira, A., 419
Pereira, M., 385
Pereira, RC., 500
Perera, R., 481

Pérez, C., 406
Perez, J., 429
Perez, LB., 465
Pérez-Vásquez, A., 502
Pero, RW., 490
Perović-Ottstadt, S., 438
Perry, NB., 458
Perry, TL., 502
Perzanowski, HP., 448
Peschke, JD., 493
Pessah, IN., 466, 468
Peter, T., 457
Peters, L., 440, 502
Peters, S., 458
Petersen, BO., 443
Petigara, BR., 480
Petit, G., 402
Petra, F., 405
Petri, A., 481
Petrick, G., 395, 478
Petri, M., 386
Pétron, G., 388
Petroutsos, D., 497
Petrucci, F., 413
Pettersson, A., 463
Pettit, GR., 429, 438, 445, 446, 455, 464, 466, 468, 497, 502
Pettit, RK., 466, 468
Pezzuto, JM., 403, 412, 422
Pfefferle, W., 457
Pfeifer, O., 485, 486
Pfeiffer, B., 446
Pfenningsdorff, A., 392
Pfister, G., 388
Pham, HTL., 457
Pham, NB., 416, 466
Phares, V., 430
Phelps, CD., 496
Phillips, AJ., 413
Phillips, DAS., 427
Phillips, T., 499
Phillips, TG., 388
Phipps, AK., 481
Phipps, RK., 458
Pi, S., 399
Pica-Granados, Y., 484
Piceno, Y., 460
Picheansoonthon, C., 447

Pickard, M., 481
Pickard, MA., 480, 484
Piel, J., 384
Pierré, A., 446
Pieters, RJ., 495
Pietra, F., 403–405, 407, 414, 437, 442, 445, 501
Pietraszkiewicz, H., 447
Pietruszka, J., 453
Pignatelli, B., 491
Pika, J., 408
Pilati, T., 446
Pilinis, C., 393
Pillai, R., 448
Pillay, B., 495, 496
Pillay, D., 495, 496
Pilling, MJ., 501
Piña, IC., 465, 467
Pinckney, J., 460
Pinheiro, US., 468
Pinkert, A., 445
Pinsky, M., 501
Pinto, AC., 404, 406, 446
Piovan, A., 470
Piozzi, F., 406
Piper, PW., 453
Piraee, M., 422
Pirjola, L., 383
Pironet, F., 384
Pironti, V., 483
Pisano, B., 418
Pitombo, LF., 404
Pitsinos, EN., 469
Pitt, AR., 489, 490
Pittet, A., 450
Pitzenberger, SM., 474
Plane, JMC., 501
Plastridge, RA., 382
Platt, U., 382, 383, 387, 393, 501
Pochetti, G., 483
Podar, K., 432
Poelarends, GJ., 494, 495
Pohanka, A., 428
Pohnert, G., 489, 500
Polášek, M., 486
Poletti, R., 416
Poli, S., 484
Pollock, WH., 391

Polo, E., 401
Polonsky, J., 471
Polychronopoulos, P., 443
Polzin, JJ., 400
Pomerantz, SC., 448
Pompei, R., 434
Pomponi, SA., 431, 441, 445, 460, 493, 503
Ponglimanont, C., 424
Pons, A., 451
Poon, GK., 503
Poon, KF., 478
Poore, AGB., 421
Pootoolal, J., 473
Popkirova, B., 430
Poplawski, J., 401
Popov, AM., 475
Popov, DY., 405, 475
Popp, P., 477
Porco, JA. Jr., 468
Porter, J., 428
Porter, LW., 390
Portmann, RW., 382
Porzel, A., 396
Possani, LD., 473
Posteri, H., 438
Potier, P., 493
Potin, P., 485, 489, 500
Potts, BCM., 432
Poulet, G., 382
Poulin, M., 385
Pouliot, M., 415
Poulos, TL., 480, 481
Pouria, S., 503
Power, GM., 386, 387
Prabhakar, 465
Pracht, J., 386, 395, 492
Prade, L., 480
Pramanik, B., 474, 475
Prange, JA., 478
Prangé, T., 470
Pranschke, J., 435
Pratilas, CA., 453
Preeprame, S., 397
Preiser, PR., 419
Preiser, W., 503
Prepas, E., 388
Presley, RW., 428
Preston, CM., 416

Price, IR., 500
Price, N., 394
Prieto-Davo, A., 456
Prilepskii, EB., 393
Primo, J., 444
Prince, D., 388
Prins, JM., 395
Prinsen, M., 502
Prinsep, MR., 379, 447, 464
Pritchard, DK., 448
Prodromou, C., 453
Prokofeva, NG., 475
Proksch, P., 384, 438, 439, 462, 467, 468, 499, 502
Proost, P., 488
Protais, P., 433
Prudhomme, M., 446
Prusak, AC., 408
Pszenny, AAP., 382
Puar, MS., 470, 471, 474, 475
Pucci, P., 432
Pudleiner, H., 484
Puglisi, MP., 400, 498
Puliti, R., 414
Pullen, C., 436
Punt, PJ., 481
Purewal, R., 398
Puricelli, L., 441
Pusecker, K., 429
Putaud, J-P., 383
Putschew, A., 386
Putz, E., 392
Puyana, M., 499
Pyle, JA., 383

Q
Qi, S-H., 409
Qian, FG., 457
Qian, P., 402
Qian, P-Y., 409
Qiao, W., 422
Qin, J., 492
Qin, S., 418
Qiu, F., 418
Qu, X-D., 464
Quack, B., 395
Quaghebeur, K., 449

Quartino, ML., 385
Queiroz, PPS., 406
Quémerais, E., 506
Quinn, RJ., 416, 422, 433, 439, 440, 444, 447, 466
Quiñoá, E., 400, 441
Quintero, R., 484
Qureshi, A., 440

R
Rachmat, R., 410, 418, 453
Radchenko, OS.,
Radhika, P., 462
Radisky, DC., 444
Radke, L., 476
Radu, I-I., 414
Rae, AD., 497
Raeymaeckers, A., 503
Raffel, RJ., 406
Raftery, J., 498
Raggatt, ME., 493
Rahbaek, L., 443
Rahkonen, P., 477
Rahm, S., 463
Rai, GP., 484
Räisänen, R., 470
Rajbhandari, I., 406
Raju, TV., 466
Ramgopal, M., 426
Ramíerz, C., 409
Ramírez, MR., 468
Ramji, N., 401
Ramo, Rao, M., 465
Ramos, AJ., 451
Ramsewak, RS., 410
Randrianarivelojosia, M., 397
Raner, GM., 481
Ranfranz, LM., 454
Rangachari, K., 419
Rangel, HR., 467
Raniga, P., 403
Rankin, AM., 383
Rantalainen, A-L., 463
Rao, AM., 459
Rao, AS., 474
Rao, AVR., 474
Rao, MR., 465

Rao, PB., 429
Rao, T., 465
Rao, VK., 459
Rao, VL., 407, 409
Rao, W., 386
Raphel, D., 395
Rapoport, H., 470
Rappe, C., 343, 476, 477
Raptis, RG., 410
Rashid, MA., 405, 431, 447
Rasmussen, RA., 389, 393, 395
Rasmussen, TB., 500
Rasmussen, U., 503
Rassias, G., 429
Rasumssen, TB., 421
Rateri, DL., 491
Rathinasamy, K., 471
Ratnakumar, S., 410
Ratnapala, L., 405
Ratnayake, AS., 418, 421
Ratnayake, R., 456
Ratola, N., 451
Rav, NSK., 409
Ravanat, J-L., 491
Raventos-Suarez, C., 408, 470
Ravinder, K., 466, 468
Ravindranath, N., 448
Ravishankara, AR., 501
Rawat, DS., 383
Rayner, ADM., 469
Read, KA., 501
Read, R., 421
Rebérioux, D., 410
Redburn, J., 433
Reddy, AV., 440, 468
Reddy, CA., 496
Reddy, CM., 436, 461, 463, 481, 506
Reddy, KL., 474
Reddy, KP., 502
Reddy, MR., 448
Reddy, MV., 429
Reddy, NS., 437
Reddy, RE., 426
Reddy, RS., 452
Redeker, KR., 390
Reed, C., 503
Reed, KA., 432
Reeder, MR., 415

Rees, MD., 489
Reeves, CE., 389, 393, 394
Rehder, D., 479, 486, 489
Reichenbach, H., 417, 418, 427
Reichert, M., 471
Reid, KA., 419
Reimann, A., 478, 479
Reina, RG., 386
Reineke, N., 440
Reinhardt, K., 435
Reinhardt, TE., 390
Reisch, J., 448
Reiss, JA., 413
Reissig, H-U., 434
Remiszewski, SW., 465
Remkes, IJ., 482
Ren, D., 421
Renard, P., 446
Renganathan, V., 488
Renhowe, PA., 414
Renirie, R., 480, 482
Renner, MK., 412
Repeta, DJ., 457
Repetto, G., 420
Reshetnyak, MV., 404
Retailleau, P., 424
Rétey, J., 430
Rether, J., 397
Rew, Y., 473
Reyes, F., 406, 443
Reyes, M., 416
Reynolds, EE., 398
Reynolds, WF., 429
Reynolds, WJ., 410
Rezanka, T., 244, 247, 253, 317, 386, 416, 421, 455, 469
Rhee, S-K., 496
Rheinberger, S., 395
Rheingold, AL., 414
Rheinheimer, J., 402
Rhew, RC., 389, 390
Rho, J-R., 442, 466
Rho, M-C., 471
Ribe, S., 417
Ribechini, E., 443
Riccio, R., 427
Ricco, R., 442
Rice, SA., 421, 422, 500

Richards-Gross, SE., 442
Richardson, P., 432
Richelmi, C., 482
Richnow, HH., 479, 496
Richter, A., 383
Richter, GM., 490
Richter, JM., 444
Rideout, JA., 468
Ridley, CP., 424
Riedel, K., 421, 500
Riederer, P., 448, 449
Rieley, JO., 388
Rigby, AC., 473
Riguera, R., 400, 441
Rimpler, H., 415
Rinehart, KL., 384, 494
Ringelberg, DB., 500
Rinner, U., 497
Riou, D., 410
Ripamonti, F., 471, 474
Riquelme, A., 400, 501
Ritalahti, KM., 496
Ritson-Williams, R., 498
Ritter, T., 429
Rittschof, D., 467
Rivier, J., 500
Rivier, JE., 428
Rizk, AFM., 407
Rizk, T., 451
Roach, MP., 481
Roberge, M., 410, 433, 437, 454
Robert, F., 411
Roberts, JM., 501
Roberts, MA., 417, 423, 425, 455
Roberts, S., 483
Robertson, F., 470
Robertson, LW., 485
Robichaud, J., 415
Robillard, N., 410
Robinson, DA., 420, 492
Robinson, SJ., 447
Robledo, NR., 484
Rocha, JHC., 463
Rochfort, SJ., 407
Röckmann, T., 390, 501
Rodin, A., 506
Rodrigues, C., 466
Rodríguez, A., 443

Rodríguez, AD., 409, 410
Rodríguez, B., 406
Rodríguez, J., 409, 424, 427, 471
Rodriguez, JM., 394
Rodríguez, R., 444
Rodríguez, RM., 474, 475
Rodríguez-Larena, MC., 476
Rodríguez-Luis, F., 401
Rodstedth, M., 385
Roe, SM., 443, 453
Roepstorff, P., 473
Roeselová, M., 382
Rogers, CN., 422, 500
Rogers, EW., 468
Rohr, J., 446, 493
Roll, DM., 408, 456
Roller, PP., 395
Rominger, F., 441
Rooney, F., 442
Roque, NF., 403, 461
Rorrer, GL., 400, 485
Rosas, FE., 458
Roscoe, HK., 383, 501
Rose, WC., 446
Roseman, JB., 414
Rosen, D., 383
Rosen, N., 453
Rosenfeld, D., 501
Rosenfeld, L., 503
Ross, C., 498
Ross, J., 436
Ross, P., 435
Ross, PS., 436
Ross, SA., 466
Rossi, J., 423
Rossman, AY., 386
Rössner, E., 431
Rostas, JAP., 442
Rothenberg, G., 486
Rouhi, AM., 506
Rouhiainen, L., 431
Rouse, GW., 385
Rousis, V., 406
Roussakis, C., 406, 410, 427, 437, 454
Roussis, V., 379, 405, 407, 408, 414
Rovirosa, J., 395, 399–401, 404, 414, 501
Row, LR., 459
Rowland, FS., 388, 391

Roy, K., 418
Roy, R., 450
Ruano, JZ., 411
Rubnov, S., 442
Ruby, C., 470
Rudi, A., 410, 436, 466, 467
Rudolph, J., 390, 391, 395
Rudzinski, J., 495
Ruesink, JL., 408
Ruijssenaars, HJ., 482
Rumpf, J., 488
Rundberget, T., 444
Running, SW., 388
Ruokojarvi, P., 477
Ruppe, S., 435
Rusaouen-Innocent, M., 500
Rush, C., 483, 489
Rutledge, PS., 406
Ruuskanen, J., 477
Ryan, CA., 480
Ryan, JJ., 463
Ryan, XP., 443
Ryberg, P., 495
Ryoo, I-J., 474
Ryu, G., 406

S
Saares, RJ., 385
Sachse, GW., 388, 391
Sacks, JT., 441
Sadler, R., 478
Saeki, BM., 468
Saemundsdóttir, S., 393
Safe, L., 475
Safe, S., 475
Sage, L., 451
Sahin-Erdemli, I., 459
Sahnow, DJ., 389
Said, SA., 447
Saida, K., 471
Sailler, B., 464
Saini, HS., 489
Saint-Louis, R., 436
Sainz, E., 427
Saisho, T., 433
Saito, T., 389, 451, 473, 501
Saiz-Lopez, A., 501

Sakagami, Y., 413, 418, 449, 468
Sakaguchi, N., 439
Sakai, S., 484
Sakai, T., 456, 457
Sakakibara, Y., 401
Sakamoto, B., 417
Sakamoto, H., 457
Sakata, T., 503
Sakemi, S., 473
Sako, M., 434
Saksena, AK., 475
Sakuma, Y., 466
Sakurai, M., 450
Sala, F., 502
Sala, S., 502
Sala, T., 469
Salamonczyk, GM., 494
Salas, AP., 446, 493
Salas, JA., 446, 488, 493
Salaün, J-P., 500
Salawitch, RJ., 383
Saldaña, J., 403, 404
Saleem, M., 401
Salguero, M., 420
Salietti Vinué, JM., 381
Salinas, PC., 502
Salituro, GM., 474
Salkinoja-Salonen, M., 386
Salmon, RA., 501
Salmond, GPC., 421
Salomon, CE., 383, 417
Saltzman, ES., 389, 495
Saltzmann, ES., 393
Salvá, J., 399, 422
Salvadori, P., 481
Samaai, T., 444, 445, 506
Samb, A., 459, 466
Sambucetti, LC., 465
Sameshima, H., 425
Samiotaki, M., 497
Samoszuk, MK., 464
Samreen, 412
Samuelsson, B., 385
San Martín, A., 401, 406
Sancelme, M., 446
Sánchez, C., 446, 493
Sanchez, I., 399
Sánchez, JA., 410

Sanchis, V., 451
Sand, P., 448, 449
Sandar, V., 468
Sandén, P., 385, 391, 392
Sander, R., 382
Sanders, ML., 465
Sanders, SC., 502
Sandler, JS., 447, 455
Sandoval, E., 467
Sandoval-Villasana, AM., 484
Sanfilippo, C., 483
San-Martín, A., 395, 399, 400, 404, 413, 414, 501
Sano, T., 423
Santacroce, C., 407
Santana, O., 400, 501
Santelli, CM., 506
Santo, RD., 434
Santoni, G., 486
Santra, MK., 471
Sapidou, E., 471
Saracoglu, I., 415, 459
Sarakinos, G., 469
Sargent, MV., 469
Sarre, O., 475
Sasada, Y., 401
Sasaki, D., 452
Sasaki, K., 401
Sasaki, T., 416, 452, 465
Sasamura, H., 432
Sass, R., 390
Sasse, F., 427
Sastry, VG., 409
Sata, N., 461
Sata, NU., 418, 448
Satake, M., 456
Sato, A., 395
Sato, B., 451, 455
Sato, H., 438, 442, 455
Sato, M., 428
Sato, S., 402
Sato, Y., 410
Satoh, H., 458
Satoh, Y., 398, 469
Satomi, T., 401
Sattler, I., 407, 418
Satyanarayana, N., 468
Sauerbrei, B., 494

Sauleau, P., 424
Saunders, JC., 439
Sausville, EA., 427
Sauter, H., 471
Savage, NH., 383
Savino, M., 451
Savona, G., 406
Sawa, R., 434, 472, 493
Sawa, T., 434, 445, 472, 493
Sawada, T., 475
Sawangwong, P., 467
Saxena, G., 456
Scaloni, A., 432
Scarratt, MG., 389, 392
Schadt, CW., 386
Schaefer, S., 419
Schäfer, M., 474
Schaffrath, C., 419, 420, 492
Schall, C., 393, 394
Schaneberg, BT., 453
Schapiro, V., 498
Scharffe, D., 391
Schatowitz, B., 476
Schatz, PF., 443
Schatzman, RC., 460
Schaub, TM., 473
Schauer, F., 496
Schauffler, S., 395
Schaur, JR., 489
Schaur, RJ., 489, 490
Schebeske, G., 391
Schefer, AB., 468
Scheibner, K., 488
Scheinmann, F., 439
Schellekens, K., 503
Schembri, MA., 421
Scher, JM., 459
Scherer, G., 435
Schetz, JA., 445
Scheuer, P., 404
Scheuer, PJ., 383, 399, 405, 406, 408, 409, 416, 417, 424, 425, 440, 448, 461, 466
Schilke, P., 388
Schiller, J., 490
Schilling, J-G., 387
Schimana, J., 449
Schinazi, RF., 466
Schippers, A., 385

Schlabach, M., 400, 435
Schleberger, C., 431
Schlegel, B., 436
Schleyer, M., 429
Schlingmann, G., 434, 493
Schlumpf, E., 476
Schmeda-Hirschmann, G., 471
Schmid, A., 488
Schmid, M., 384, 502
Schmidt, A., 459, 464
Schmidt, EW., 383, 428, 431
Schmidt, J., 396
Schmidt, JM., 438, 455, 468
Schmidt, SK., 386
Schmidt, U., 428
Schmiedel, U., 479
Schmitt, D., 491
Schmitt, DC., 468
Schmitz, FJ., 5, 74, 221, 273, 297, 370, 408, 409, 411, 418, 424, 437, 447, 460, 462, 465–467
Schmitz, P., 436
Schmitz, W., 385
Schnabel, PC., 445
Schnarr, NA., 423, 488
Schneekloth, JS. Jr., 435
Schneider, K., 399
Schneider, NM., 389
Schneider, TR., 474
Schnerr, H., 488, 489
Schoemaker, HE., 484, 489
Schoenfeld, RC., 467
Schöler, HF., 380, 385, 386, 388, 390, 395, 396, 456, 457, 492
Scholokova, OV., 475
Schols, D., 503
Scholz, E., 435
Schomburg, D., 486
Schoofs, G., 488
Schopflocher, D., 388
Schopohl, D., 419
Schotterer, U., 395
Schraa, G., 496
Schramm, K-W., 435, 476
Schreck, CE., 506
Schreier, P., 448, 449
Schrey, P., 476
Schröder, HC., 383, 438, 506

Schroeder, DR., 436, 446, 452
Schroeder, FC., 466
Schroll, R., 390
Schuchter, L., 430
Schulman, M., 470
Schulte, GR., 383
Schultz, A., 496
Schultz, PG., 452
Schultz-Tokos, JJ., 382
Schulz, B., 397, 402, 450, 456, 472
Schulz, GE., 431
Schulzke, C., 486
Schumacher, D., 475
Schumacher, KK., 426
Schumacher, RW., 447
Schwaiger, N., 401
Schwandner, FM., 7, 388
Schwartz, GG., 450
Schwartz, SM., 448
Schwarz, B., 438
Schwarz, JB., 433
Schweigert, N., 487
Scippa, S., 429
Scognamiglio, G., 411
Scolaro, A., 453
Scoullos, M., 407, 414
Searle, PA., 448, 454
Sedmera, P., 486
Seelbach, K., 484
Seephonkai, P., 471
Seeram, NP., 412
Seff, K., 405
Seffaj, T., 446
Segraves, NL., 441, 447, 462
Seibert, SF., 499
Seibl, J., 426, 427
Seibold, C., 488
Seifert, R., 479
Seigle-Murandi, F., 422, 451
Seiki, M., 438
Seinfeld, JH., 383
Seki, T., 470
Sekiguchi, M., 438
Sekine, M., 420, 421
Sekizawa, R., 472
Seldes, AM., 406, 427, 441
Seleghim, MHR., 426
Selenka, F., 476

Self, S., 387, 506
Sellström, U., 462
Selover, SJ., 409
Selva, E., 474
Selvaggini, C., 484
Semesi, AK., 392, 480
Senda, H., 470
Senderowicz, AMJ., 427
Senn, HM., 420
Senthilmohan, R., 464
Seo, Y., 442, 466
Seoane, G., 497, 498
Seoane, GA., 498
Sephton, S., 506
Setiawan, A., 462
Seto, H., 418, 425, 474
Setoi, H., 432
Sette, LD., 426
Settle, M., 491
Seuzaret, C., 391
Sévenet, T., 442
Severinghaus, JP., 389
Severino, N., 488
Sewald, N., 430
Seward, TM., 388
Sewell, MA., 417
Sewram, V., 451
Seyjagat, J., 395
Shadoff, LA., 476
Shaffer, PL., 499
Shafiee, A., 474
Shah, RR., 462
Shah, SAA., 412
Shang, G., 494
Shang, S., 460
Shang, TQ., 495, 496
Shanu-Wilson, P., 434
Shao, B., 491
Shao, N., 468
Sharma, K., 506
Sharma, N., 470
Sharma, ND., 498
Sharma, R., 410
Sharma, RP., 401
Sharma, SV., 453
Sharma, VK., 406
Sharp, K., 498
Shaw, G., 479

Shaw, GR., 384, 478
Shaw, KJ., 461
Shearer, MJ., 395
Shedd, AL., 481
Sheldon, RA., 480, 482, 484, 487, 489
Sheldrake, GN., 497
Sheldrake, HM., 415
Sheldrick, GM., 474
Shelton, KD., 481
Shen, B., 452, 494
Shen, G., 453
Shen, X., 502
Shen, Y., 453, 455
Shen, Y-C., 409, 410, 420
Shen, Z., 492
Sheng, D., 488
Sheng, Y., 490
Shengule, SR., 438
Shenvi, RA., 444
Shephard, GS., 451
Shepherd, MJ., 451
Shepson, PB., 382
Sherman, DH., 423, 494
Shetty, R., 428, 429
Sheu, J-H., 405, 408, 409
Shi, D., 406
Shi, DY., 407, 443
Shi, H., 461
Shi, J., 406, 447, 460, 461
Shi, JG., 405, 443, 460
Shi, W., 464
Shi, Y-P., 405
Shibasaki, M., 432
Shibata, K., 423
Shibata, T., 411, 432
Shibazaki, M., 439
Shibuya, M., 469
Shida, H., 408
Shida, N., 402
Shie, J-J., 498
Shields, L., 463
Shigematsu, N., 431, 432
Shigemori, H., 428
Shigenaga, MK., 464
Shih, C., 430
Shih, DM., 464
Shikano, M., 433
Shima, H., 411, 439

Shima, J., 418
Shimada, A., 457
Shimada, Y., 449, 487
Shimizu, I., 471
Shimizu, S., 489
Shimogawa, H., 455
Shimomura, K., 451
Shimomura, M., 408
Shimonishi, M., 485
Shimonishi, Y., 428
Shin, D., 442, 473
Shin, D-S., 503
Shin, HJ., 426
Shin, J., 442, 466, 503
Shinada, M., 444
Shinada, T., 425
Shinnar, AE., 429
Shinoda, K., 463
Shinohara, H., 387, 388
Shinohara, T., 474
Shin-ya, K., 425, 474
Shinya, M., 394
Shiomi, K., 398, 431, 453
Shiori, T., 425
Shirahama, H., 447
Shirai, R., 457
Shiraki, T., 395
Shiro, M., 409
Shirokawa, S., 446
Shirota, O., 456
Shizuri, Y., 440, 462
Shmul, G., 410
Shoeb, M., 417
Shoemaker, RH., 399, 471
Shrestha, KL., 403
Shubina, LK., 405
Sibanda, L., 451
Sicinska, D., 495
Siebe, C., 387
Siebert, A., 495
Siebert, U.,
Siegel, MM., 444
Siegert, F., 388
Siems, ST., 389, 390
Sierra-Alvarez, R., 497
Sievering, H., 382
Signoretto, M., 483
Sigvaldason, GE., 387

Sih, CJ., 494
Silk, P., 340
Silk, PJ., 386, 458, 475, 494
Sills, MA., 431, 460
Silva, AMS., 467, 468
Silva, CL., 466
Silva, GL., 412
Silva, JA., 487
Silver, J., 474
Silver, PA., 466
Silverman, KC., 474
Sim, ATR., 442
Sim, CJ., 442
Sim, K-Y., 470
Simchoni, O., 388
Simmonds, PG., 390, 395
Simon, M., 435
Simons, BH., 480
Simons, TL., 501
Simpson, JS., 403, 405, 424
Simpson, MT., 486
Simpson, TJ., 493
Sims, JJ., 393
Sims, JK., 417
Sincich, C., 445
Singh, IP., 464
Singh, SB., 470, 471, 474
Sinkkonen, S., 463
Sinnwell, V., 431
Sioris, CE., 383
Sirirath, S., 418
Sisti, M., 446
Sitachitta, N., 384, 417, 423, 461
Sive, BC., 391
Sivonen, K., 431
Siwu, ERO., 411
Sjödin, A., 462
Sjögren, M., 441, 499
Skaff, O., 492
Skalicky, JJ., 421
Skaltsounis, A-L., 443
Skelton, BW., 404, 463
Skepper, CK., 454
Skoropowski, G., 440
Skoumbourdis, AP., 439
Skropeta, D., 403, 405
Skulberg, OV., 384
Slivka, A., 490

Smidt, H., 495, 496
Smilda, T., 495
Smit, AJ., 502
Smith, CD., 427, 444
Smith, CJ.,
Smith, D., 426
Smith, III AB., 474
Smith, J., 439
Smith, JD., 491
Smith, TW. Jr., 388
Smith, KM., 421
Smith, S., 440
Smith, SK., 474
Smonou, I., 483
Smythe-Wright, D., 395
Snader, KM., 399
Snapper, ML., 474
Sneden, AT., 453
Snell, TW., 408
Snider, BB., 455
Snieckus, V., 415
Snipes, CE., 455
Snoeijs, P., 392
Snyder, GT., 387
Snyder, JK., 485
Snyder, SA., 429
Sobarzo-Sánchez, EM., 433
Sobek, H., 486
Soda, K., 419, 495
Soedarsono, 462
Soedjak, HS., 486
Soga, K., 423
Sogin, ML., 506
Sohal, J., 434
Solaja, BA., 397
Solit, D., 453
Solomon, S., 382
Sölter, S., 441
Solyanikova, IP., 495
Someno, T., 412
Son, BW., 396, 460
Son, K-H., 398, 449
Sonchaeng, P., 468
Sone, H., 424, 454
Sonenshine, DE., 457
Song, F., 460
Song, FH., 460
Song, LL., 412

Song, Q., 457
Song, Y-C., 450
Song, Z., 421
Sonnenbichler, J., 450, 471
Sono, M., 401
Sorek, H., 467
Soria-Mercado, IE., 456
Sorokin, D Yu., 459
Sosa, ACB., 438
Soto, H., 404, 414
Sotokawa, T., 399
Souto, ML., 411, 412
Sowers, KR., 496
Sowers, LC., 448
Spagnuolo, C., 406
Spalteholz, H., 490
Spantzel, J., 488
Speicher, A., 459, 471
Spencer, JB., 420
Spenger, RE., 382
Spicer, CW., 382
Spickett, CM., 489, 490
Spilling, CD., 468
Spiteller, D., 420
Spiteller, P., 458, 494
Spitthoff, B., 388, 456
Spízek, J., 386
Sponchiado, SRP., 426
Sprecher, DL., 491
Sprengeler, PA., 474
Srebnik, M., 380, 416
Srebotnik, E., 479, 485, 487
Sridevi, A., 468
Sridevi, KV., 440
Srinivasulu, M., 465
Srujan Kumar, D., 465
Staats, PS., 428
Stach, JEM., 385
Stacpoole, PW., 503
Stadler, M., 397, 450, 453, 473
Ståhlberg, C., 385
Stams, AJM., 496
Stanley, AE., 493
Stapleton, BL., 424
Starks, CM., 432
Starmer, JA., 500
Starnes, S., 384
Staudigel, H., 506

Stauffer, RL., 386
Stavrou, S., 495
Stazi, F., 446
Stchedrin, AP., 404
Stearns, GW., 500
Steel, D., 429
Steelink, C., 507
Steen, H., 473
Steeves, RM., 452
Steglich, W., 261, 457, 458, 471, 472
Stehl, RH., 476
Steinberg, P., 421
Steinberg, PD., 421, 422, 500, 507
Steiner, JR., 399
Steinert, M., 384
Steinman, R., 422
Stelmaszyńska, T., 489
Stenflo, J., 473
Stenger, VA., 478
Stenstrom, Y., 497
Stenutz, R., 463
Stepanov, A., 506
Stermitz, FR., 450
Stern, G., 435
Sterner, O., 396, 397, 449, 450, 458
Stessman, CC., 502
Steube, K., 438
Stevens, AA., 396
Stevens, J., 476
Stevens-Miles, S., 398
Stevenson, JP., 430
Steward, CC., 496, 499, 500
Stewart, CC., 460
Stewart, SG., 497
Stien, D., 439
Stierle, DB., 409
Stijve, T., 382
Stimmel, MB., 427
Stockenström, S., 451
Stocker, R., 490, 491
Stocks, BJ., 476
Stoll, E., 463
Stoltz, BM., 442
Stoncius, A., 430
Stonik, VA., 404, 405, 407, 413, 444
Stout, CD., 420
Strand, SE., 496
Stratford, IJ., 498

Strath, M., 419
Stratmann, K., 444
Streitwieser, A. Jr., 391
Strobel, DF., 389
Strobel, G., 455
Ströbele, M., 399
Strömberg, B., 388
Stromquist, ER., 506
Stroud, V., 393, 395
Struchkov, YuT., 404
Strukul, G., 483
Strunz, GM., 381, 396
Stucky, GD., 481
Studer, A., 495
Studer, C., 477
Studier, MH., 388, 395
Stuiber, DA., 460, 461
Stuppner, H., 401
Sturchio, NC., 481
Sturges, WT., 389, 391, 393, 394
Sturrock, GA., 389, 390
Sturtz, G., 402
Stuta, J., 383
Su, C., 455
Su, E., 494
Su, J-H., 409
Su, J-Y., 403, 404, 424, 448
Su, YL., 472
Suárez, AI., 467
Subbaraju, GV., 430
Subrahmanyam, C., 410
Subrahmanyan, C., 408
Sucher, CM., 389
Sucker, J., 397
Suckling, CJ., 484
Suda, H., 470, 472
Sudalai, A., 422
Sudarsono, 439
Sudarsono Ferdinandus, E., 438
Sudo, H., 415
Suenaga, K., 396, 410, 411, 413, 455, 467
Suescun, L., 403, 404
Suffert, J., 452
Suga, H., 421
Sugano, M., 395
Sugawara, F., 455
Sugier, J., 389
Sugimoto, M., 415

Sugimoto, T., 463
Sugimoto, Y., 433
Sugiura, Y., 452, 470
Sugiyama, J., 423
Sugiyama, T., 418
Suh, JH., 406
Sui, JK., 462
Sukopp, I., 459
Sulake, RS., 467
Sullivan, AM., 480
Sumaryono, W., 438
Summers, RS., 506
Sun, C-Y., 455
Sun, H., 402
Sun, J., 405, 406, 443
Sun, Q., 442
Sun, W., 430, 481, 487
Sunazuka, T., 398
Sundaramoorthy, M., 480, 481
Sundell, S., 486
Sundin, P., 416
Sung, P-J., 408, 409
Sunga, GN., 454
Sunnenberg, G., 389
Suntornchashwej, S., 425
Suomalainen, S., 431
Suominen, KP., 386
Supekova, L., 452
Supriyono, A., 438
Surico, MTJ., 476
Suryanaramyana Murty, U., 465
Süssmuth, R., 449
Süssmuth, RD., 399, 474, 488, 501
Suursaar, UK., 385
Suwanborirux, K., 425
Suzuki, A., 401
Suzuki, H., 470, 472, 474, 475
Suzuki, K., 439, 474, 491
Suzuki, M., 394, 402, 404, 405, 407, 411–414, 440, 461, 500, 507
Suzuki, S., 449
Suzuki, T., 402, 411, 412, 414, 415, 418, 490, 491
Suzuki, Y., 398
Svensmark, B., 392
Svenson, A., 476
Svensson, T., 385, 390–392
Svoronos, D., 461

Swaffar, DS., 444
Swali, R., 440
Swanson, SE., 476
Swanson, SJ., 454
Swarts, HJ., 386, 449, 457, 458
Swasithamparam, K., 463
Sweedler, JV., 441
Swersey, JC., 465
Swetnam, TW., 388
Synstad, B., 466
Szerminski, WR., 483
Szewczyk, J., 426

T

Tabacchi, R., 457
Tabata, N., 398
Taboada, S., 443
Tabudravu, JN., 427, 466
Tackett, LP., 438, 468
Tadayoni, BM., 502
Tae, J., 406
Tagle, B., 409
Taglialatela-Scafati, O., 410, 413, 418, 437, 441
Tahara, H., 398
Tahir, A., 422, 438, 465
Tait, VK., 385
Tajbakhsh, A., 469
Takada, N., 396, 411, 455, 467
Takada, Y., 400, 421
Takahashi, A., 457, 464
Takahashi, H., 407, 420, 470
Takahashi, K., 387, 462
Takahashi, M., 418, 475
Takahashi, T., 423
Takahashi, Y., 394, 402, 405, 407, 411, 413, 414, 431, 434, 445, 452, 454, 493
Takahasi, Y., 443
Takaichi, S., 459
Takaishi, M., 411
Takaishi, Y., 422, 450
Takamatsu, S., 397, 398, 406, 471
Takao, T., 428
Takaoka, I., 397
Takaoka, M., 477
Takaoka, S., 459
Takase, S., 451, 455

Takashima, A., 441
Takashima, K., 461
Takasugi, M., 458
Takasugi, Y., 487
Takeda, N., 477
Takeda, Y., 415
Takegami, K., 423
Takeichi, Y., 401
Takemura, K., 409, 442
Takenaka, Y., 450, 475, 476
Takeshita, J., 448
Takeuchi, M., 396
Takeuchi, R., 418
Takeuchi, T., 422, 425, 434, 436, 445, 472, 489, 493
Takeya, K., 407, 426
Takeyama, H., 477
Takishima, K., 423
Takita, T., 445
Takizawa, K., 436
Takushi, A., 415
Talbot, AC., 453
Talbot, RW., 390
Talpir, R., 429
Tamai, M., 472
Tamburini, C., 503
Tamura, G., 471
Tamura, M., 458, 487
Tamura, S., 401
Tamura, T., 419
Tan, C-H., 433, 461
Tan, J-J., 433
Tan, JS., 470
Tan, LT., 384, 470, 503
Tan, R-X., 450
Tanabe, S., 435
Tanabe, T., 485, 486
Tanahashi, T., 450, 475, 476
Tanaka, A., 503
Tanaka, C., 410, 429, 462
Tanaka, H., 397, 398, 451, 471, 474, 487
Tanaka, J., 56, 81, 82, 121, 159, 239, 274, 403, 405, 410, 412, 418, 428, 429, 453, 462
Tanaka, M., 459, 471
Tanaka, N., 482, 489
Tanaka, O., 415

Tanaka, T., 452, 463
Tanaka, Y., 441, 453
Tane, K., 439
Tang, L., 495
Tang, T., 382
Tangen, J., 496
Tanhua, T., 394
Taniguchi, K., 500
Taniguchi, S., 446
Taniguchi, T., 485
Taniguchii, K., 461
Tanimoto, H., 387
Tanisaki, M., 434
Taniyama, T., 415
Tanner, HR., 453
Tao, D., 402
Tapia, CA., 502
Targett, NM., 434
Tarhanen, J., 477
Tarling, T., 454
Tarricone, C., 443
Tasdemir, D., 440
Tashiro, C., 476
Taskova, RM., 401
Tatian, M., 441
Tatko, CD., 503
Tatli, II., 459
Tato, M., 445
Tatsukawa, R., 477
Taylor, D., 457
Taylor, MW., 506
Taylor, SW., 424, 429
Taylor, TN., 386
Teeyapant, R., 499, 502
Teicher, BA., 430
Teisan, S., 432
Teixeira, VL., 500
Ten Brink, HB., 484, 489
Teng, C-M., 420, 433
Tennant, S., 436, 456
Teran, A., 474
Terano, H., 451, 455
Terao, Y., 413
Terlau, H., 440
Terner, J., 480, 481
Terracciano, J., 470, 471
Terracciano, S., 427
Terui, Y., 470

Teruya, T., 413, 455, 467
Tesevic, V., 402
Teshima, Y., 470
Test, ST., 490
Teste, J., 465
Tettamanzi, MC., 412
Teunissen, PJM., 458
Teuten, EL., 436, 461, 463
Teuten, EM., 463
Thacker, RW., 384
Thakur, SS., 473
Thaning, C., 455
Tharnish, P., 466
Thebtaranonth, Y., 386, 471
Theobald, N., 440
Theoduloz, C., 471
Theriault, RJ., 454
Thief, W., 420, 482
Thirionet, I., 465
Thoison, O., 442
Thomas, EL., 490
Thomas, M., 484
Thomas, SM., 421
Thomassen, MJ., 464
Thomasson, F., 422
Thomas, WA., 427
Thompson, BL., 422
Thompson, JE., 467, 499
Thompson, RC., 473
Thoms, C., 384, 499
Thomson, J., 503
Thor, G., 469
Thordarson, Th., 387
Thornton, D., 430
Thorson, JS., 452
Thukkani, AK., 490
Thunnissen, A-MWH., 502
Thuß, U., 477
Thutewohl, M., 430
Tichy, H-V., 453
Tiedje, JM., 494
Tien, G., 503
Tiernan, TO., 476
Tijerina, T., 491
Tillekeratne, LMV., 405
Tilley, L., 502
Tilotta, MC., 440
Tilvi, S., 466

Timmis, KN., 503
Tincu, JA., 429
Tinoco, R., 484
Tinto, WF., 410, 429
Tischler, M., 454, 456
Tittlemier, SA., 435, 436, 463
Tius, MA., 430
Tizio, SC., 488
Tkacz, JS., 398
Tobias, DJ., 382
Tobita, Y., 400
Toda, S., 419
Todd, JS., 400, 417
Todorova, MN., 401
Toelzer, S., 494
Togashi, M., 471
Tohma, H., 445
Tokarczyk, R., 385, 393, 394, 495
Toki, S., 398
Tokimura, F., 487
Tokumasu, S., 470
Tokuyama, H., 447
Tolstikov, 399
Tomasini, C., 446
Tomishima, Y., 470
Tomita, K., 436, 452
Tomoda, H., 397, 398, 448, 453, 468
Tomono, Y., 408, 499
Tomy, G., 435
Tonai-Kachi, H., 473
Tong, C., 433
Toom-Sauntry, D., 389
Topcu, G., 401, 403
Topol, EJ., 491
Toppet, S., 449
Tori, M., 401, 471
Torii, S., 396
Torkewitz, NR., 434
Tornabene, LL., 506
Tóth, G., 401
Totzke, F., 453
Toupet, L., 407, 410
Toussaint, D., 452
Tov-Alper, DS., 383
Towers, GHN., 456, 470, 485
Towerzey, L., 462
Toyota, M., 459, 471
Tran, HD., 490

Tranter, CJ., 422
Trauner, D., 433
Travert, N., 493
Treseder, KK., 390
Trevisan, V., 483
Trew, SJ., 434
Trianto, A., 462
Trifunovic, S., 402
Tringali, C., 502
Trinh, KQ., 494
Trogolo, C., 416
Tronstad, L., 495
Trotta, F., 496
Trousdale, EK., 424
Truchet, M., 500
Trudeau, M., 422
Trudinger, CM., 389
Truumees, I., 470
Truumees, I., 471
Tsai, L-H., 502
Tsankova, ET., 401
Tsao, R., 444
Tschirret-Guth, RA., 483, 487
Tsuchida, T., 472
Tsuda, M., 428, 437–440, 442, 443, 465, 466, 507
Tsujimoto, Y., 404
Tsujita, H., 469
Tsujita, M., 389
Tsukamoto, M., 470, 472
Tsukamoto, S., 406, 437–439, 465, 467
Tsukuda, E., 463
Tsuruga, M., 471
Tsurumi, Y., 432
Tsuruo, T., 397
Tsuruoka, A., 447
Tuan, G., 434
Tucker, KD., 471
Tucker, MP., 485
Tulkens, PM., 473
Turk, J., 413
Turnbull, MM., 404
Turuo, T., 433
Tuynman, A., 489
Tyler, SC., 390
Tysklind, M., 477, 478
Tzakou, OA., 401

U

Ubukata, M., 471
Uchida, K., 450
Uchida, M., 413
Uchida, R., 397
Uchida, S., 470
Uchihata, Y., 428
Uchio, Y., 409, 443
Uchiumi, Y., 426
Uddin, J., 411
Uddin, MJ., 410
Udell, KS., 496
Ueda, K., 410, 411, 418, 455, 467
Ueda, Y., 397
Uemoto, H., 428, 437, 438
Uemura, D., 396, 410, 411, 413, 433, 455, 462, 465, 467
Uemura, T., 414
Ui, H., 431
Ullrich, R., 488
Umekita, M., 472
Umeyama, A., 437
Umezawa, H., 445
Umezawa, K., 428, 472
Umezu, K., 457
Unger, M., 463
Ungur, N., 408
Uno, J., 425
Unson, MD., 384
Unversucht, S., 479, 488
Uosaki, Y., 398
Urban, L., 433
Urban, S., 405, 406, 439, 501
Urhahn, T., 391
Usami, Y., 397
Usher, KM., 506
Usleber, E., 451
Usui, T., 418
Usuki, T., 452
Utkina, N., 338
Utkina, NK., 460, 475

V

Vaccarini, C., 412
Vacelet, J., 383, 424
Vagias, C., 379, 405–407, 414
Vaillancourt, FH., 423, 432, 488, 489
Vairappan, CS., 404, 405, 407, 414, 507

Vaishampayan, U., 446
Vajs, V., 402
Valenza, M., 387
Valeriote, F., 427
Valeriote, FA., 411, 430, 447, 507
Valinluck, V., 448
Vallée, Y., 438
Vallefuoco, T., 437
Vallribera, M., 426
Valls, N., 426
Valor, LM., 502
Van Aalten DMF., 466
Van Bambeke, F., 473
Van Bavel, B., 463, 478
Van de Pas, BA., 496
Van de Velde, F., 480, 482, 484
Van de Wal, RSW., 389
Van den Berg, JJM., 413, 492
Van den Bergen, H., 419
Van den Brink, 392
Van den Broeke, MR., 389
Van denVeen, C., 389
Van der Helm, D., 462
Van der Oost, J., 496
Van der Sar, SA., 472
Van der Vliet, A., 464
Van der Wielen, PWJJ., 503
Van Deurzen, MPJ., 482, 484, 487, 489
Van Dorsselaer, A., 485
Van Duyne, GD., 408
Van Gelder, J., 503
Van Herk, T., 350
Van Hylckama Vlieg, JET., 495
Van Laethem, Y., 473
Van Lanen, SG., 452, 494
Van Leeuwen, JGE., 479
Van Merode, AEJ., 495
Van Pée, K-H., 354, 380, 422, 434, 479, 481, 484–486, 488, 489, 497
Van Peteghem, C., 451
Van Rantwijk, F., 480, 482, 484, 487, 489
Van Sande, J., 419, 424
Van Soest, R., 429, 433, 445, 454
Van Soest, RWM., 383, 418, 436–440, 462, 465, 467, 468
Van Veldhuizen, A., 449
Van Vliet, AJ., 483
Van Volkenberg, MA., 473

Van Wagoner, RM., 465
Vandaele, A-C., 506
Vanderleyden, J., 488
Vannelli, T., 495
Varasi, M., 438
Vargas, RR., 482
Varnavas, SP., 503
Varner, RK., 390
Vartiainen, T., 478
Vasiliauskas, R., 428
Vaskovsky, VE., 407, 413
Vasquez, VR., 392
Vatinno, R., 451
Vatter, S., 435
Vaughn, D., 430
Vázquez, MG., 441
Vázquez-Duhalt, F., 484
Vecchione, M., 503
Veitch, JM., 446
Veitch, NC., 487
Velankar, M., 432
Velasco, L., 484
Veleiro, AS., 412
Veltri, CA., 418, 421
Venables, D., 426
Venables, DA., 445, 469
Venãncio, A., 451
Venkatesham, U., 440, 465
Venkateshwar Goud, T., 465
Venkateswarlu, Y., 437, 440, 460, 465, 466, 468
Venoyama, T., 463
Ventura, MR., 397
Venugopal, MJRV., 409
Verbist, J-F., 410
Verbitski, SM., 421, 443
Verdel, EF., 488
Verdier-Pinard, P., 425
Verhagen, FJM., 381, 386, 393, 457, 458
Verigin, V., 463
Vermeer, HJ., 383
Vernier, J-M., 433
Vértesy, L., 474
Vervoort, HC., 429, 442
Vetter, W., 379, 380, 396, 400, 435, 436, 463
Vicent, T., 496
Vijayakumar, EKS., 418
Vijender Reddy, A., 465, 466

Vikelsøe, J., 476
Vilella, D., 474
Villar, J., 412
Villard, E., 506
Vilter, H., 486, 487
Vining, LC., 422, 423, 485
Vinogradov, I., 506
Viola, G., 470
Violante, F de A., 446
Virovaya, MV., 475
Visconti, A., 451
Viswanathan, PN., 503
Vita-Marques, A., 426
Vitinius, U., 484
Vliegenthart, JFG., 383
Vobach, M., 440
Vogt, R., 382
Vögtle, M., 497
Vojtko, CM., 454
Voldoire, A., 446
Vollenbroek, EGM., 480
Vollmer, MK., 390
Volpe, G., 434
Von Czapiewski, K., 395
Von Elert, E., 426
Von Glasow, R., 382
Von Kuhlmann, R., 382
Vonica, CA., 443
Vosburg, DA., 488, 489
Voskoboev, AJ., 481
Voss, JC., 491
Vovelle, J., 500
Vrabcheva, T., 451
Vrijenhoek, RC., 385
Vrijmoed, LLP., 450
Vroye, L., 424
Vulpanovici, A., 455
Vyas, DM., 446

W
Wada, M., 419
Wada, S., 418
Wada, Y., 445
Wadey, P., 479
Waeg, G., 491
Wage, T., 479, 488, 489
Wagenknecht, H-A., 481, 484

Wagner, E., 485
Wagner, H-C., 476
Wagner, M., 384, 502
Wagner, MM., 430
Wagner, T., 501
Wagner Tacobi, H-W., 383
Wahidhulla, S., 466
Wahidulla, S., 411
Wahome, PG., 403
Waikedre, J., 442
Waikins, SF., 401
Waite, TD., 492
Wakabayashi, K., 423
Wakimoto, T., 455, 477
Waldmann, H., 430
Walk, T., 429, 474
Walker, CS., 428, 429
Walker, JV., 486
Walker, K., 433
Walker, M., 449
Walker, S., 473
Walla, MC., 459
Walla, MD., 395
Wallace, DWR., 395
Wallace, MS., 428
Waller, MP., 482
Walser, M., 390
Walsh, A., 395
Walsh, C., 473, 493
Walsh, CT., 423, 432, 446, 488, 489, 493
Walshe, ND., 433
Walter, J., 502
Walters, IAS., 507
Walton, JD., 426
Walton, LJ., 493
Wan, J., 450
Wang, B-G., 440, 507
Wang, D., 482
Wang, F-D., 412
Wang, G-H., 409
Wang, G-Y-S., 410, 465
Wang, H., 415, 502
Wang, H-B., 433
Wang, J., 433, 470
Wang, L., 452, 490
Wang, M., 453
Wang, N-J., 431
Wang, N-Y., 390

Wang, S., 406, 460, 461
Wang, SJ., 405, 443
Wang, T., 456
Wang, W., 382, 461
Wang, W-L., 440
Wang, X., 448, 464, 479, 481
Wang, X-N., 412
Wang, Y., 391, 402, 461, 491
Wang, YL., 472
Wani, S., 488
Ward, AC., 385, 399
Ward, CJ., 503, 507
Ward, DE., 390
Ward, P., 476
Ward, RS., 408, 410, 459
Wardlaw, JH., 469, 470
Wariishi, H., 487
Waring, AJ., 429
Waring, MJ., 497
Warner, KL., 495
Warters, RL., 444
Warwick, NJ., 383
Watabe, S., 418
Watanabe, CMH., 452
Watanabe, H., 425, 449, 450
Watanabe, K., 420, 421, 442, 466
Watanabe, M., 426
Watanabe, MF., 426
Watanabe, O., 389
Watanabe, R., 467
Watanabe, S., 451
Watanabe, T., 418
Watanadilok, R., 468
Waters, ML., 503
Watkins, M., 428, 473
Watling, R., 390
Watson, H., 489
Watson, JT., 481
Watson, K., 497
Wattanadilok, R., 467
Watters, D., 458
Watts, JEM., 496
Watts, KS., 425
Wayman, K., 440
Weaver, HA., 389
Webb, M., 393
Webb, PM., 384
Webb, RI., 424

Webb, VL., 422, 445
Weber, K., 463, 499
Weber, R., 478
Webster, JD., 387
Webster, NS., 384
Weckesser, J., 431
Weete, JD., 466
Wehrl, M., 384
Wei, Q., 429
Wei, X., 410
Weichbrodt, M., 435
Weidler, M., 397
Weidner-Wells, M., 503
Weijers, CAGM., 449
Weinbrenner, S., 428
Weinreb, SM., 434, 438, 439
Weise, D., 382
Weisenstein, DK., 383
Weiss, B., 499
Weiss, H., 496
Weiss, RF., 389, 390
Weiss, SJ., 490
Weissflog, L., 392
Welch, CJ., 455
Welch, M., 421
Welch, TR., 453
Weldon, PJ., 395
Welling, M., 391
Wellington, EMH., 434
Wellington, KD., 406
Wen, Y-M., 409
Wennberg, PO., 383
Werk, TL., 455
Werker, E., 388
Werner, P., 479
Wesén, C., 416
Wessels, M., 403
Wesson, KA., 497
Wesson, KJ., 502
West, LM., 440
Westergreen, AD., 496
Westerling, AL., 388
Westwell, MS., 474
Wever, R., 393, 480, 482, 484–489
Weyand, M., 486
Whalley, AJS., 463
Whipple, WJ., 388
Whitby, HL., 498

White, AH., 404, 463
White, AW., 502
White, DC., 500
White, J., 388
White, JD., 455
White, R., 432
White, RL., 422, 423, 487
White, RM., 392
Whitehead, RC., 497, 498
Whiteman, KR., 452
Whiteman, M., 448
Whitfield, FB., 460, 461, 485, 487
Whitman, S., 439, 462, 502
Whitten, DN., 454
Whitton, MM., 496
Wiberg, K., 475
Wichard, T., 489
Wicklow, DT., 453
Widdison, WC., 452, 502
Widdowson, M., 506
Wideman, MA., 502
Widersten, M., 495
Wie, M., 490
Wiedenfeld, H., 459
Wiedinmyer, C., 388
Wiegrebe, W., 433
Wiencke, C., 385, 391, 393, 394
Wiener, JJM., 506
Wiese, B., 473
Wiese, KJ., 439
Wijeratne, EMK., 450
Wijnberg, JBPA., 380, 386, 449, 457, 458, 493
Wikman, P., 502
Wildman, HG., 469
Wilhelm, M., 476
Wilhelm, SD., 452
Wilkins, AL., 444
Wilkins, SP., 445, 502
Will, S., 478
Willemsen, P., 465, 500
Willetts, A., 483, 486, 489
Williams, DC., 430
Williams, DE., 437, 438, 454
Williams, DH., 398, 399, 417, 438, 439, 473, 474
Williams, DR., 444
Williams, ER., 441

Williams, GC., 409
Williams, J., 492
Williams, MD., 429, 466, 468
Williams, MV., 448
Williams, PG., 408, 425, 432, 453, 456, 457
Williamson, RT., 417, 423, 425, 455
Willis, AC., 470, 497, 498
Willis, CL., 423, 494
Willis, J., 498
Willoughby, R., 493
Willuhn, G., 458
Wilquet, V., 506
Wilson, AM., 495, 496
Wilson, DM., 499
Wilson, JA., 445
Wilson, KJ., 384
Wilson, L., 430, 471
Wilson, P., 451
Wilson, RJM., 419
Win, H., 400
Wind, NS., 498
Wingen, LM., 382
Winkelmann, G., 434
Winssinger, N., 453, 473
Winstead, EL., 388
Winterbourn, CC., 413, 464, 490, 492
Winterton, N., 361, 381, 492
Winther, J-G., 389
Wipf, P., 417, 447
Wippel, A., 469
Wirtz, SS., 466
Wiryowidagdo, S., 438
Wise, ML., 400
Wishkerman, A., 492
Witt, KL., 393
Witte, L., 438, 462, 502
Witter, JB., 387
Wittsiepe, J., 476
Wladislaw, B., 482
Woerdenbag, HJ., 502
Woestenborghs, 503
Woggon, W-D., 481, 484
Wohlfarth, G., 496
Wohlleben, W., 474
Wolf, D., 408, 418
Wolf, H., 431
Wolfe, AR., 457
Wolfe, NL., 393

Wolfender, J-L., 401, 472
Wolfensberger, M., 477
Wolpert, TJ., 426, 427
Wolters, AM., 441
Wong, C., 420
Wong, C-H., 482, 498
Wood, TK., 421
Wood, WF., 133, 134, 395, 422, 440, 458
Woodall, CA., 495
Woodin, SA., 395, 459, 481, 496
Woodruff, WH., 481
Woods, AE., 481, 488
Woods, JS., 476
Woolard, FX., 395
Woolery, M., 433
Wosniok, W., 440
Wray, V., 404, 438, 439, 462, 467, 468, 502
Wright, A 5, 200
Wright, AD., 396, 399, 403, 404, 421, 422, 440, 457, 502
Wright, AE., 445, 460
Wright, AF., 431
Wright, GD., 473
Wright, JT., 421, 500
Wright, PC., 384, 460
Wrigley, SK., 398, 399, 434, 493
Wrobleski, ML., 439
Wu, C-L., 459
Wu, D-G., 412
Wu, F-H., 415
Wu, H., 421
Wu, J., 435
Wu, J-J., 415
Wu, M., 417
Wu, P-H., 433
Wu, Q-H., 402
Wu, Q-X., 405
Wu, S-H., 412
Wu, S-L., 408, 409
Wu, W., 448, 464, 492
Wu, X-Y., 450
Wu, Y-C., 405, 433, 449
Wu, Y-Z., 502
Wunder, A., 458
Wunderli, S., 477
Wyatt, D., 406
Wynands, I., 488, 489

X
Xenaki, H., 405
Xiao, F., 460
Xiao, S., 386
Xiao, Z-H., 409
Xie, H., 415, 452
Xiong, F., 415
Xu, G., 402
Xu, GY., 457
Xu, J., 432, 494
Xu, J-P., 408, 455
Xu, L., 436, 463, 481, 506
Xu, N., 460, 461
Xu, NJ., 460
Xu, S., 394
Xu, X., 460
Xu, X-H., 403, 404, 448
Xu, XL., 460
Xu, Z., 501

Y
Yabannavar, AV., 484
Yagura, T., 449
Yahiro, Y., 428
Yaipakdee, P., 485
Yakushijin, K., 438, 439, 442
Yamada, A., 396, 410
Yamada, H., 431
Yamada, K., 385, 396, 402, 411, 424, 433, 454, 455, 467
Yamada, N., 408
Yamada, T., 397, 437
Yamada, Y., 408
Yamaguchi, A., 398
Yamaguchi, K., 410, 425, 426, 431
Yamamoto, H., 394, 470
Yamamoto, K., 387, 450, 453, 470
Yamamoto, Y., 410, 421
Yamamura, H., 412
Yamamura, S., 448
Yamanaka, S., 429
Yamanoi, S., 453
Yamasaki, K., 415, 447
Yamasaki, M., 398
Yamashita, J., 438
Yamashita, K., 418
Yamashita, M., 439
Yamashita, N., 478

Yamashita, T., 423, 447
Yamashita, Y., 406
Yamauchi, Y., 433
Yamazaki, M., 469
Yamazaki, R., 463
Yamdagni, R., 401
Yamori, T., 397, 433
Yan, F., 498
Yanagisawa, I., 396
Yang, A., 445
Yang, C-R., 415
Yang, DJ., 398
Yang, HY., 456
Yang, J., 425
Yang, JS., 472
Yang, K., 485
Yang, K-L., 496
Yang, L., 402, 455, 460
Yang, RSH., 393
Yang, S., 424
Yang, S-P., 402
Yang, X., 383
Yang, Y., 406, 447, 460, 461, 492
Yang, YC., 405, 443, 460
Yang, Y-L., 449
Yang, ZP., 481
Yang, Z-Q., 453, 502
Yantosca, R., 391
Yao, G., 468
Yao, G-M., 404, 448
Yao, RC., 456
Yao, X., 442, 462
Yaosaka, M., 418
Yarnell, A., 488
Yarwood, D., 431
Yasuda, T., 438
Yasui, H., 432
Yasumoto, T., 453, 456, 465
Yata, H., 411
Yates, SR., 393
Yazawa, K., 410, 425, 433
Ye, B-C., 507
Ye, RW., 421
Ye, W-C., 415
Ye, Y., 441
Yearwood, T., 428
Yeh, E., 446, 488, 489
Yeh, GC., 490

Yeh, VSC., 433
Yen, H-F., 433
Yeung, BKS., 406
Yeung, K-S., 453
Yevich, R., 391
Yi, KY., 406
Yi, X., 481
Yin, J., 432
Yin, S., 412
Yip, S-C., 470
Yiwen, C., 470
Yoda, N., 407
Yokochi, A., 425
Yokokawa, F., 425
Yokota, E., 499
Yokota, K., 465
Yokouchi, Y., 389–391, 394, 501
Yokoyama, A., 440
Yokoyama, M., 403
Yonehara, N., 387
Yonemoto, T., 452
Yoneshigue-Valentin, Y., 500
Yoo, I-D., 447, 474
Yoo, JS., 397
Yoo, S., 406
Yoon, J., 492
Yoon, KS., 481
Yoon, YJ., 383
York, C., 498
Yoshida, E., 398, 469
Yoshida, K., 452
Yoshida, M., 387, 398, 445, 463
Yoshida, S., 395, 457
Yoshida, WY., 408, 424, 425, 430, 454
Yoshida, Y., 391
Yoshikawa, H., 396
Yoshikawa, K., 418
Yoshikawa, M., 415
Yoshikawa, N., 451, 473
Yoshimura, E., 403, 408, 499
Yoshimura, H., 423
Yoshimura, I., 386
Yoshimura, S., 432, 455
Yoshioka, S., 395
Yoshioka, T., 472
Yost, RA., 506
Young, K., 396
Young, LY., 496

Young, PR., 389
Yourkevich, O., 481
Youssef, DTA., 401
Yousuf, S., 412
Yu, DQ., 472
Yu, J-H., 449
Yu, J-M., 402
Yu, M., 429, 433
Yu, T-W., 494
Yuan, X., 386
Yuasa, E., 437
Yue, J-M., 402, 412
Yúfera, EP., 444
Yukawa, T., 390
Yun, B-S., 474
Yunker, CE., 457
Yvon-Lewis, SA., 495

Z

Zacharewiski, T., 475
Zähner, H., 431
Zaks, A., 483, 484
Zambonin, CG., 451
Zampella, A., 427, 429, 454
Zampella, G., 489
Zapp, J., 459
Zaylskie, RG., 476
Zea, S., 437
Zebühr, Y., 475
Zeeck, A., 418, 431, 449, 459
Zehner, S., 380, 479, 488, 489
Zeid, IF., 402
Zein, N., 452
Zeller, G., 423, 494
Zeng, L-M., 403, 404, 424, 448
Zeng, T., 391
Zenkevich, IG., 393
Zenkoh, T., 432
Zennegg, M., 477
Zerikly, MK., 493
Zetzsch, C., 382
Zhan, Z-J., 402
Zhang, BX., 479
Zhang, D., 507
Zhang, D-H., 455
Zhang, E., 496
Zhang, H., 445, 446, 464

Zhang, H-P., 446
Zhang, H-W., 450
Zhang, J., 415
Zhang, J-W., 455
Zhang, L-Y., 415
Zhang, R., 482, 491
Zhang, S., 409, 411
Zhang, S-J., 461
Zhang, T., 461
Zhang, W., 409
Zhang, X., 444
Zhang, XW., 402
Zhang, Y., 412
Zhao, J., 460, 461
Zhao, JL., 460
Zhao, Q., 402
Zhao, R., 452
Zhao, S., 447
Zhao, S-X., 415
Zhao, S-Y., 438
Zhao, Y., 402
Zheng, L., 491
Zheng, Y-J., 495
Zhong, YL., 424
Zhou, G-X., 454
Zhou, LM., 429
Zhou, P., 494
Zhou, S-N., 450
Zhou, W-S., 455
Zhu, D-Y., 433, 461
Zhu, J., 506
Zhu, L., 446, 493
Zhu, LY., 462
Zhu, S., 420
Zhu, X., 492
Zhu, Y., 402, 417
Zhu, ZN., 469
Zidorn, C., 401
Zilfou, JT., 444
Zimmerli, B., 450
Zimmermann, MP., 416
Zink, DL., 474
Zinsmeister, HD., 386
Zitzmann, W., 453
Zizak, Z., 402
Zlatopolskiy, BD., 431
Zmuidzinas, J., 388
Zoete, V., 453

Zöllinger, M., 439
Zolotov My, 389
Zong, Q., 480, 484
Zubía, E., 399, 422
Zuleta, IA., 445
Zuo, P., 507
Zurita, JL., 420
Zuurmond, AM., 419

Subject Index

A
A11-99-1, 27
A-40926-A/B, 328
A-40926-PA, 328
A-40926-PB, 328
A-80915-A–D, 249
AA-57, 38, 39
AB0022A, 341, 342
Abalones, 374
Acacia confusa, chlorotryptamine alkaloid, 213
Acanthella sp., Δ^9-kalihinol Y, 68
– kalihinanes, 68
Acanthella carteri, hanishin, 184
Acanthella cavernosa, 369
– kalihinenes, 68
Acanthostigmella sp., CJ-19,784 (bromine analog of chlorflavonin), 231
Acanthus ebracteatus, benzoxazinoid glucosides, 220
Acanthus ilicifolius, benzoxazinoid glucosides, 220
Acarogobiens, 316, 317
Acarospora gobiensis, acarogobiens, 317
Acetamides, polychlorinated, 26
Acetogenins, C_{15}, 96
5-Acetoxy-2,10-dibromo-3-chloro-7, 8-epoxy-α-chamigrene, 53
9α-Acetoxyanadalucin, 43
5-Acetoxyoxachamigrene, 53
8-Acetylcaespitol, 55, 57
3'-O-Acetylchloramphenicol, 134
Acetyldeschloroelatol, 54

Acetylelatol, 54
Acetylenes, 22
Achillea clavennae, chlorine-containing guaianolides, 43
Achillea clusiana, 2-epi-chloroklotzchin, 40
Achillea depressa, bibsanin, 42
Achillea ligustica, seco-tanapartholide, 41
Aconitase, 126
Acorn worms, bromophenols, 373
Acorospora gobiensis, bromo acids, 109
Acridone alkaloid A6, 222
Acrodontiolamide, 134
Acrodontium salmoneum, acrodontiolamide, 134
Actin cytoskeleton disruptor, 150
Actinomadura sp., A-40926-PA
328– decatromicins, 183
Actinomadura madurae, maduropeptin chromophore, 233
Actinomycin G_2 171, 172
Actinomycins Z_3/Z_5, 171
Actinoplanes sp., BE-40665D, 325
– SCH, 54445 (polycyclic xanthone), 318
Actinoplanes deccanensis, lipiarmycins, 236
Actinosynnema pretiosum ssp. *auranticum*, ansamitocinoside P-2, 234
Acutumine, chlorine-containing, 174
Acyclovir vs. topopyrone B, 320
Acyl homoserine lactone, 373
Adenocarpus foliolosus, sesquiterpene acid, 44
S-Adenosine-L-methionine (SAM) 12, 125
Aegle marmelos, chloromarmin, 228

Aeromonas hydrophila, growth inhibition, 299
Aerophobin, 371
Aeroplysinin, 371, 375
Aerothionin, 300, 307, 372
Aeruginosins, 144–146
Agallochins, 61
Agaricic ester, 326, 327
Agaricoglycerides, 326
Agaricus arvensis, CCl$_3$, 14
Agaricus macrosporus, agaricoglycerides, 326
Agarwood, 227
Ageladine A, 187, 188
Agelas spp. 371, 372
Agelas sp., (–)-7-N-methyldibromophakellin, 189
– agesamides, 188
– brominated fatty acids, 106
– monobromoisophakellin, 189
– nagelamides, 191, 195
Agelas ceylonica, longamides, 184
Agelas clathrodes, dispacamide, 185
Agelas conifera, 372
– bromosceptrin, 190
– dispacamide, 185
Agelas dispar, clathramides, 185
– dispacamide, 185
– longamides, 184
Agelas longissima, (+)-(S)-longamide, 184
– agelongine, 185
Agelas mauritiana, mauritiamine, 187
Agelas nakamurai, ageladine A, 187
– mukanadin A, 185
– mukanadin C/debromolongamide, 184
– pyrroles, 183
– sceptrin-related nakamuric acid, 190
– slagenins, 187
Agelas oroides, cyclooroidin, 187
– taurodispacamide A, 186
Agelas sventres, sventrin (N-methyloroidin), 186
Agelas wiedenmayeri, bromopyrrole homoarginine, 186
Agelastatins, 191
Ageliferin, 372
Agelongine, 185
Agesamides, 188

Aguilaria sinensis ("agarwood"), chromone, 227
AI-2 bacterial quorum sensing system, 373
Aiolochroia crassa, N-methylaerophobin-2, araplysillin III, hexadellin C, 307
Ajuga parviflora, *neo*-clerodane ajugarin-I, 60
Akashins, 211
Alcalinaphenols, 258, 259
Alcyonium paessleri, illudalane sesquiterpenoids, 60
Alcyopterosins, chlorinated, 60
Aldingenins, 50, 51
Algicides, 219
Algoane, 50
Alkaloids, acridone A6, 222
– agelongine, 185
– aporphine, 175
– frogs, 174
– halogenated, 174
– pyrroles, brominated, 183, 371
– tetracyclic brominated, 175
Alkanes, 9
Alkenes, chlorinated/brominated, 21
Alkyl iodides, 19
Alkylating agent, 134
Alkynes, 22
– chlorinated/brominated, 23
Aloe sabaea, N-4'-chlorobutylbutyramide, 26
Alternatamides, 201, 202
Altohyrtins, 236
Alzheimer's disease, 44, 219
Amanita castanopsidis, 135
Amanita gymnopus, (2S)-2-amino-5-chloro-4-hydroxy-5-hexenoic acid, 135
Amanita miculifera, 2-amino-5-chloro-5-hexenoic acid, 135
Amanita vergineoides, (2S,4Z)-2-amino-5-chloro-4-pentenoic acid, 134
Amathamides, 282
Amathaspiramides, 282, 284
Amathia spp., amathamides, convolutamines, convolutamine H, 282
Amathia alternata, alternatamides A-D, 201
Amathia convoluta, amathamide G, volutamides, 282

- convolutamydines, 216
- convolutindole A, 198
Amathia wilsoni, amathaspiramides, 282
Ambigol C, 278, 279
Ambiguine G nitrile, 213
Ambiguines, 213
Amblyomma, tick control, 256, 257
Ambrosia maritima, 11β-hydroxy-13-chloro-11,13-dihydrohymenin, 40
Amino acids, halogenated, 134
3-Amino-10-chloro-2-hydroxydecanoic acid, 135
(2S)-2-Amino-5-chloro-4-hydroxy-5-hexenoic acid, 135
(2S,4Z)-2-Amino-5-chloro-4-pentenoic acid, 134
2-Amino-5-chloro-5-hexenoic acid, 135
Aminochlorophenol, 259
Aminocoumarins, 228
2-Aminoimidazole alkaloids, 193
Amoora yunnanensis, dammaranes, 91
Amphimedon terpenensis, 6-bromo-(5E,9Z)-tetracosadienoic acid, 106
Amycolatopsis sp., 4-oxovancosamine-containing glycopeptides, 329
- ansacarbamitocins, 243
- epoxyquinomicins, 325
Amycolatopsis mediterranei, SP-969/SP-1134, 332
Amycolatopsis orientalis, monodechlorovancomycin, 332
Anabaena strain, 90, anabaenopeptilide 90B, 167
Anabaenopeptilide, 90B, cyclic depsipeptide, 167
Ancorina sp., ancorinolates, 197
Ancorinolates, 197, 198
Andalucin, 38, 39
Androstanes, polychlorinated, 94
Angucycline antibiotic, 250
Anhydroaplysiadiol, 64
Anisyl metabolites, chlorinated, 258
Annona purpurea, romucosine F, 175
Ansacarbamitocins, 243
Ansamitocinoside P-2, 234
Ansamitocins, 242
Ansamycin, 242
Antazirine, 111

Anthocidaris crassispina, purealin, 370
Anthraquinones, chlorinated, 319
- tetracyclic, 319
Anthrasesamone C, 320, 321
Antibacterial activity, nagelamides, 191
- TAN-876 A, 178
Antibiotics, 322
- chlorophenol-oligosaccharide (orthosomycins), 333
- prenyl-phenol, 322
Anticancer agents, diazonamides, 156, 157
Antifeedants, 35, 132, 373, 374
- bromopyrrole alkaloid, stevensine, 371, 372
Antifouling activity, 369
- barnacles, 68, 194
Antifungal activity (*Aspergillus fumigatus, Candida albicans*), 118
Antihistaminic activity, taurodispacamide A, 186
Antiinflammatory activity, 112
Antimalarial activity, kalihinols, 68
Antimicrobial activity, chloptosin, 154
Antineoplastic compounds, dibromophakellstatin, 189
Antioxidative activity, 197
Antiplatelet aggregation activity, 175
Antiserotonergic activity, agelongine, 185
Antitumor activity, (+)-bromoxone, 28
Antiviral compounds, 370
Anverene, 34
Apakaochtodenes, 35, 36
Aphid repellents, 374
Aplicyanins, 209, 210
Aplidiamine, 265, 266
Aplidiopsis sp., aplidiamine, 265
Aplidium sp., floresolide C, 236
- iodinated tyrosine alkaloids, 284
Aplidium cyaneum, aplicyanins, 209, 210
Aplidium longithorax, polybrominated diphenyl ethers, 273
Aplidium meridianum, 2-bromopyrimidine, 203
Aplydactone, 53
Aplyparvunin, 96, 97
Aplysamine, 296, 297
Aplysia dactylomela, aplydactone, 53
- caespitenone, 55

- dactylallene, 100
- isopinnatol B, 62
- lankalapuols, 54
- (−)-(3E,6R,7R)-pinnatifidenyne, 97
- puertitol-B acetate, 50

Aplysia kurodai, 28
- aplysiallene, 96
- *ent*-isoconcinndiol, 63
- laurinterol acetate

Aplysia parvula, 373
- 5-acetoxy-2,10-dibromo-3-chloro-7, 8-epoxy-α-chamigrene, 53
- aplyparvunin, 96
- (3Z)-bromofucin, 100
- halogenated furanones, 130

Aplysia punctata, acetates, 33
- chemical defense, 35
- neopargueroldione, 62
- perforatol, 54
- punctatol, 55

Aplysia sp., 10-bromo-β-chamigren-4-one, 51

Aplysiadiol, 63
Aplysiallene, 96
Aplysiatoxins, 112, 235
Aplysillin A, 265
Aplysin, 58

Aplysina spp., aplyzanzine A, 293
- bromotyrosines, 295
- wounding, 370

Aplysina aerophoba 4, 371
- antimicrobial activity, 375

Aplysina archeri, bromotyrosine alkaloids, archerine, 304, 306

Aplysina caissara, caissarines, 307

Aplysina cauliformis, isomeric carbamates, 304

Aplysina cavernicola, 4
- antimicrobial activity, 375
- oxohomoaerothionin, hydroxyfistularin, 304

Aplysina fistularis, aerothionin, homoaerothionin, 300
- bromophenols/bromotyrosines, 370, 372

Aplysina fistularis fulva, aplysillin A, 265

Aplysina gerardogreeni, calafianin, aerothionin, phenylacetic acid, 307

Aplysina insularis, deoxyfistularin, oxoaerophobin, 304

Aplysinal, 58
Aplysinamisin, 371

Aplysinella sp., aplysinillin, 308
- hydroxyceratinamine, dibromotyramine, 293
- psammaplysin F, 303

Aplysinella rhax, psammaplins, bisaprasin, N,N-dimethylguanidium, aplysinellins, 293

Aplysinellins, 293, 295
Aplysinillin, 308
Aplysinopsin, 198, 200
Aplysins, hydroxylated, 57
Aplyzanzine A, 293
Apoptosis, chloptosin, 154
Aporphine alkaloids, 175
Aquariolides, 79, 80
Aranochlors, 116, 117
Araplysillin III, 307, 308
Archerine, 304–306

Arenicola cristata, bromophenols, 265

Ariolimax columbianus, clitolactone 132–134

Aristolane, 46
Armatols, 88, 90
Armellide B, 324
Armentomycin (2-amino-4, 4-dichlorobutyric acid), 135

Armillaria cepestipes, arnamiol, 324
Armillaria gallica, arnamiol, 324
Armillaria mellea, arnamiol, 324
- melleolide J/armillarikin, 324

Armillaria monadelpha, arnamiol, 324
Armillaria novae-zelandiae, 6′-chloro-10α-hydroxymelleolide
- melleolides, 324

Armillaria ostoyae, arnamiol, 324
Armillaria tabescens, arnamiol, 324
Armillaridin, 324
Armillariella mellea, melleolides, 324
Armillarikin, 324
Arnamiol, 324

Arnica sachalinensis, chlorinated phenol, 260

Aromadendrane, 46
Aromatics, halogenated, 254–256

Artemisia dracunculus, benzodiazepines, 225
Artemisia lanata, guaianolide andalucin, 38
Artemisia suksdorfii, chlorinated sesquiterpene lactones, 41
Artemisina apollinis, 1,1,2-tribromooct-1-en-3-one, 25
Ascochlorin, antidiabetes, 322
– LL-Z1272g/ilicicolin D, 322
Ascophyllum nodosum, trichloromethane, 13
Asparagopsis armata ("limu kohu"), 20
Asparagopsis taxiformis 19, 20
– halogenated carboxylic acids, 25
Aspergillus sp., chlorocarolides, 132
– ICM0301C/ICM0301D, 90, 91
– methyl (di)chloroasterrates, 277
– SCH, 202596, 322
Aspergillus candidus F1484, antifungal compound F1484, 227
Aspergillus fischeri var. *thermomutatus*, CJ-12662/UK-88051, 175
Aspergillus flavus, host, 235
– tetrathioaspirochlorine, 323
Aspergillus fumigatus, growth inhibition, 231
Aspergillus ochraceus, ochratoxin A, 229
Aspergillus ostianus, chlorinated furanones, 132
Aspergillus parasiticus, parasitenone, 266
Aspirochlorine (A30641 = oryzachlorine), 323
Aster tataricus, astin I, 143
Asteronotus cespitosus, pyrrolidone, 139
Astins, 143, 144
Astrosclera willeyana, *N*-methylageliferins, 190
Asystasioside E, 104
Atherosclerotic tissues, 281
Atlantic hagfish (*Myxine glutinosa*), cathelicidins, 156
Atpenins, 223
Auletta cf. *constricta*, jasplakinolides, 150
Aurantosides, 118, 119
Aurilol, 89, 90
Aurisides, 239, 241
Aurones, 226
Auxarconjugatins, 182

Auxarthron conjugatum, auxarconjugatins, 182
Avicennin A, 229
Axinella sp., axinellamines, 193
Axinella brevistyla, pyrrole-derived alkaloids, 189
Axinella carteri, 3-bromohymenialdisine, 188
– debromolongamide, 184
– ugibohlin, 187
Axinella corrugata, stevensine, 371, 372
Axinella tenuidigitata, longamides, 184
Axinella verrucosa, bromopyrroles, 186
Axinellamines, 193, 194
Axinohydantoins, 188, 189
Axinyssa sp., axinyssimides A–C, 47
Axinyssimides A–C, 47, 48
Azamerone, 250
Azaphilones, 28, 29, 325
AZT, HIV-I, 376

B

((–)-BABX), 320
Bacillus sp., chlorxanthomycin, 321
Bacillus subtilis, growth inhibition, 274
Bacteria, terrestrial, 6
Bacterial "cleansing", 369
Bactobolin, 134
Balaenoptera acutorostrata, dimethoxytetrabromobiphenyl, 272
Balanus amphitrite 47, 68, 187, 369
– antifouling spermidine, 185
Balanus improvisus, 369
Balhimycins, 329, 330
Barbaleucamides, 139
Barbamide, 135, 139
Barettin, 201, 369
– dihydrobarettin, 201
Barnacles, 68, 369
– antifouling compounds, 194
– deterrents, 33
– inhibition, 47, 187
Bastadins, 282, 295, 297, 312–314
– antiinflammatory activity, 375
Batis maritima, atmospheric CH_3Cl, 11, 12
– chloromethane, 10
Batzella sp., discorhabdins, 214

– secobatzelline A, 216
Batzellines, 214, 216
Bazzania trilobata, bazzanins, 262
Bazzanins, 262–264
BE-19412A, antitumor antibiotic, 321
BE-23254 (antibiotic), 250
BE-40665D, 325, 326
Beauveria felina, [β-Me-Pro] destruxin E chlorohydrin, 143
Beech, chloromethane, 11
Bejaranoa balansae, furanoheliangolides, 40
Bejaranoa semistriata, 40
Benzastatin C, 222
Benzene, halogenated, 255
Benzo[*a*]pyrene, 256
Benzo[*a*]pyrene diol epoxide, 256
Benzodiazepines, 225
Benzofurans, halogenated, 226–227
Benzoic acids, chlorinated, 254
Benzoquinones, 28
Benzoxazolinones, 223
Benzoxepins, chlorinated, 226
Berardius bairdii, dimethoxytetrabromobiphenyl, 272
Beticolins, 318, 319
Bibsanin, 42
Biofouling, 369, 374
Biomass combustion 7, 10, 11
– bromomethane, 15–17
– chloromethane, 11
Biotranshalogenation, 12
Biphenyls, polybrominated (PBBs), 272
– polychlorinated (PCBs), 277
Bisabolanes, chlorinated, 44
Bisabolenes, 46, 50
– γ-Bisabolenes, 53
Bisaprasin, 293, 294
Bis-benzyl ether, 274
Bis-bibenzyls, chlorinated, 262
– "ring-opened" 262
Bischloroanthrabenzoxocinone (BABX) 320, 321
Bis(deacetyl)solenolide D, 81
Biselides, 244, 246
Bisezakynes, 99
Bispsammaplin A, 297, 299
Bis-spiroketals, tetraphenolic

Bitungolides, 121, 122
Bjerkandera spp., chlorinated anisyl metabolites, 258
Bjerkandera sp. BOS55, CCl_3, 14
– chlorinated benzoic acids, veratryl chloride, 258
Bjerkandera adusta, chlorometabolite production of, 258
Bjerkandera fumosa, chlorinated phenols, 257, 258
Blue mussel (*Mytilus edulis*), 369
– dioxins, 338
Boldine, 175
Bonfires, PCDDs/PCDFs, 344
Botryllamides, 284, 285
Botryllus sp., botryllamides, 284
– cadiolides, 131
Bracken fern (*Pteridium aquilinum*), carcinogen ptaquiloside, 44
Brasilanes, 55
Brasilenane, 46
Briaexcavatolide, 73
Briantheins, 72
Briaranes, 76
Briareins, 70, 71, 82
Briareum sp., briviolides, 74
Briareum asbestinum, briareins, 70–72
– 11-hydroxybriantheins, 71
Briareum stechei, milolides, 73, 74
Briviolides, 74, 75
Bromide, biotranshalogenation, 12
Brominated biphenyls, 272
Brominated dibenzofuran, 274
Brominated diphenyl ethers, 273–275, 277, 378
Bromine oxide 3, 374
Bromine-ozone chemistry, stratospheric, 3
5-Bromo-4,7-dihydroxyindole, 197
6-Bromo-4,5-dihydroxyindole, 197
6-Bromo-(5E,9Z)-docosadienoic acid, 106
4-Bromo-1,1-epoxylaur-11-ene, 57, 58
6-Bromo-5-hydroxyindole, 197
2-Bromo-5-hydroxytryptophan, 153
(6*R*,9*R*,10*S*)-10-Bromo-9-hydroxychamigra-2,7(14)-diene, 52
6-Bromo-2-mercaptotryptamine, 199
6-Bromo-(5E,9Z)-pentacosadienoic acid, 106

5-Bromoabrine, 198, 199
Bromoacids, 106
Bromoageliferin, 372
Bromoallene, 96
– C_{15} acetogenins, 96
Bromoallenic fatty acids, 109
Bromoanisole, 266
3-Bromobarekoxide, 64, 65
Bromobenzaldehyde, 269
Bromobenzene, 254, 257
Bromobenzoquinone, 268
Bromobutane, 18
8-Bromo-chamigren-1-en, 53
Bromochloro nor-steroids, 93
Bromochloromethanes, 18
Bromochlorovinyl, 36
Bromocontryphan, 156
Bromocyclococanol, 49
Bromodeoxytopsentin, 204, 206
Bromodichloromethane ($CHBrCl_2$), 18
Bromoditerpenoids, 63
Bromoester A (2-octyl 4-bromo-3-oxobutanoate), 26
Bromoethane (CH_3CH_2Br), 18
Bromoform, 16, 17, 374
(3Z)-Bromofucin, 100
4-Bromohamigeran B, 58, 59
Bromohydroquinone, 268
3-Bromohymenialdisine, 189
5-Bromohypaphorine, 198, 200
5-Bromohypaphorine, 200
5-Bromoindole-3-acetic acid, 198
Bromoiodomethane ($BrCH_2I$), 20
10-Bromoisoaplysin, 57, 58
3-Bromomaleimide, 189
Bromomesitylene, 254
Bromomethane, 15–17
Bromomethane, soil pesticide, 16
Bromomethane, sources, 16
Bromomethylidene-10β-bromo-β-chamigrene, 53
5-Bromo-*N,N*-dimethyltryptophan, 198
4-Bromopalau'amine, 194
Bromoperoxidase, 337
Bromoperoxidase, 265
4-Bromophenol, 265
Bromophenols, 370, 378
Bromophenols, fungicides, 375
Bromophenols, simple, 265, 266, 268–270

Bromophycolides, 67
Bromopsammaplin A, 297, 299
2-Bromopyrimidine, 203
Bromopyrroles, 185
Bromopyrroles, phakellin-type, latonduines, 189
Bromosceptrin, 190, 191, 372
Bromosphaerone, 67
3-Bromostyloguanidine, 194
Bromotopsentin, antiinflammatory activity, 375
Bromotryptamine, 200
Bromotryptamine peptides, 201
Bromotryptamines, 198, 200
5-Bromotryptophan, 198
6-Bromotryptophan, 373
6-Bromotryptophan amino acids, 155, 156
Bromotubercidin, 224
Bromotyramines, 282
Bromotyrosine spiroisoxazolines, 312
Bromotyrosines, 282, 286, 289, 293, 295–297, 300
Bromotyrosines, 370
5-Bromouracil, 224
Bromoverongamine, 286, 287
Bromovulones, 128
Bromoxone, 28
(+)-(10S)-10-Bromo-β-chamigrene, 51
10-Bromo-β-chamigren-4-one, 51, 52
Bruguiera gymnorrhiza, *ent*-kaurane, 61
Bryozoans, 198
Bugula dentata, bipyrroles tambjamines, 179
Burkholderia cepacia, growth inhibition, 266
Bursatella leachii, deacetylhectochlorin, 141
– malyngamides, 112
Byrsonima microphylla, chlorinated diphenyl ether, 277
Byssoloma subdiscordans, dichloro-*O*-methylnorlichexanthone, 318

C

C-1027 chromophore, 232
C_{15} acetogenins, 96
Cacospongia, brominated cacoxanthenes, 58
Cacoxanthenes, brominated, 58

Cadinane, 46
Cadiolides, 131, 132
Caespitenone, 55, 57
Caissarines, 307
Calafianin, 307
Calalpae fructus, 4-hydroxybenzoyl esters, 104
Caldariomyces fumago, CCl$_3$, 14
Calenzanane, 54
Calenzanol, 54, 55
Calicheamicin, 232
Callicladol, 86, 87
Callinectes similis (blue crab), 81
Calliostoma canaticulatum, 6-bromo-2-mercaptotryptamine, 199
Callipelta sp., callipeltosides, 239
Callipeltosides, 239, 240
Callophycus serratus, bromophycolides, 67
Callyspongia bilamellata, phoriospongins, 167
Calmodulin inhibitors, eudistomidines, 375
Calyculin J, 247
Camouflage, chemical, 373
Camptothecin vs. topopyrone B, 320
Canadian peat bogs, dioxins/dibenzofurans, 339
Cancer, 375
– halomon, 32
Candida albicans, growth inhibition, 231, 252, 282
– hectochlorin, 141
– massadines, 193
Canthigaster solandri, 373
Carbamates, 304
Carbamidocyclophanes, 255, 256
Carbazoles, halogenated, 197, 217
Carbohydrates, halogenated, 231
Carbolines, halogenated, 197, 218–220
Carbonaceous black shales (Central Asia), CCl$_4$, 15
Carbonimidic dichlorides, 55
Carboxylic acids, fluorine-containing, 124
– halogenated, 25
Carcinogens, ptaquiloside, 44
Carijenone, 128, 130
Carijoa multiflora, carijenone, 128
– chlorinated pregnanes, 94
Carpopeltis crispata, ochtodenes, 35

Carpophyllum angustifolium, phloroglucinols, 279
Carteramine A, 195, 196
Cartilagineol (allo-isoobtusol) 51, 52
Catarraphia dictyoplaca, cyclographin, 316
Catechols, abiotic formation of organochlorines, 22
– brominated, 268
Cathelicidins, HFIAP-1/2/3, 156
Caulibugula intermis, caulibugulones, 220
Caulibugulones, 221
Celastramycins, 182, 320, 321
Celenamides, 150
Centaurea acaulis, 14-chloro-10β-hydroxy-10(14)-dihydrozaluzanin, 42
Centaurea conifera, chlorohyssopifolin A (centaurepensin), 40
Centaurea glatifolia, epicebellin J, 40
Centaurea scoparia, 40
Cephalosporium acremonium strain IFB-E007, graphislactone G, 230
Ceramium tenuicorne, brominated diphenyl ether, 275
Ceratamines, 175, 176
Ceratinamine, 286, 288
– *N*-methyl- 292
Ceratostoma erinaceum, brominated imidazole, 223
Cercospora beticola, beticolin, 318
CF$_3$CF$_3$CF$_2$H, 23
Chaetoceros calcitrans, chloromethane, 9
Chaetochiversins, 230
Chaetomium chiversii, chaetochiversins 230
Chaetomorpha basiretorsa, halogenated biindole, 209
Chagosensine, 243, 244
Chamigrenes, halogenated, 46, 51, 63
Chamonixia pachydermis, pachydermin, 182, 326
Charles W. Morgan, 180
Chartelline C, 212
Chelonin B, 201
Chemical camouflage, 373
Chemical waste, PCDDs/PCDFs, 344
Cherry, chloromethane, 11
Chimaphilin, 8-chlorochimaphilin, 250
Chinikomycins, 122

Chinzallene, 96, 97
Chloptosin, 154, 155
Chloramphenicol, 134
Chlorflavonin, bromine analog, 231
Chloride, methylation, 11, 374
Chloridium sp., CJ-21,164, 326
Chlorinase, 125
Chlorinated aromatics, 7
Chlorinated benzoic acids, 254
Chlorinated bridged biphenyls, 262
Chlorinated depsides, 314–315
Chlorinated lactone, 257
Chlorinated orcinols, 259
Chlorinated polycyclic aromatic hydrocarbons (PAHs), 256
Chlorinated xanthones, 317–319
Chlorination of aromatics, 254
Chlorine oxide, 3
α-Chloro divinyl ethers, 114
(2R)-12-Chloro-2,3-dihydroillicinone E, 44
13-Chloro-3-*O*-β-D-glucopyranosylsolstitialin, 41
1-Chloro-5-heptadecyne, 26
12-Chloro-11-hydroxydibromoisophakellin, 189
14-Chloro-10β-hydroxy-10(14)-dihydrozaluzanin, 42
6-Chloro-5-hydroxy-*N*-methyltryptophan, 153
1-Chloro-3-methyl-2-butene, 21
5-Chloro-1-*O*-methyl-*o*-hydroxyemodin, 319
7-Chloro-1-*O*-methyl-*o*-hydroxyemodin, 319
Chloroacetic acids (MCA) 26, 345, 347
– abiotic formation, 347
Chloroacetylphosphonic acid, 26
8-Chloroadenine, 224
Chloroalkanes, long-chain, 24
Chloroanisoles, 258
Chloroasterrate, methyl, 277, 278
Chloroatranol, 259, 260
Chlorobenzenes, 254
Chlorobenzoic acids, 254
Chlorobifuhalol, 279
Chlorobiocin (clorobiocin, RP, 18,631), 228
Chlorobisfucopentaphlorethol-A, 279
5-Chlorobohemamine C, 175, 176 5-

Chlorobromomethane (CH_2BrCl), 18
8-Chlorocannabiorcichromenic acid, 227
Chlorocarolides, 132
6-Chlorocatechin, 231
8-Chlorochimaphilin, 250
Chlorochrymorin, 38, 39
Chlorocinnamic acids, 339
Chlorocresol, 339
5-Chlorocytosine, 224
5-Chlorodeoxycytidine, 224
(*E*)-Chlorodeoxyspongiaquinol, 55
(*E*)-Chlorodeoxyspongiaquinone, 55
Chlorodesnkolbisine, 223, 224
7-Chlorodeutziol, 104
Chlorodibromomethane ($CHBr_2Cl$), 18
Chlorodibromophenol, 266
Chlorodifucol, 279
Chlorodimethoxybenzoic acid, 258
Chlorodivaricatic acid, 314, 315
Chlorodysinosin A, 147
5-Chloroemodin, 319
Chloroenone quassinoid eurycolactone B, 61
Chloroepicatechin, 260, 261
Chloroethyne (chloroacetylene), 22
Chlorofluorocarbons (CFCs), 23
Chloroform, 339, 340, 345, 347, 378
Chlorofusin, 31, 32, 100
Chlorogentisylquinone, 28
Chlorohematommic acids, 259, 260
Chlorohydrin sesquiterpenes, 38, 39
Chlorohydrins, 44, 60, 94, 104, 143, 256
Chlorohydroxybenzoic acid, 258
Chlorohydroxyphenylacetamide, 258
2-Chlorohyperzine E, 175, 176
Chlorohyssopifolin A (centaurepensin), 40
Chlorohyssopifolins, 38
12-Chloroillicinone, 44
12-Chloroillifunone, 44, 61
3-Chloroindole, 197
Chloroiodomethane, 20
Chloroisoplagiochin D, 262, 263
5-Chloroisorotiorin, 29, 30
Chlorojanerin, 40
Chlorolecideoidin, 316
Chlorolichexanthone, 318
Chlorolissoclimides, 84, 85
Chloromarmin, 228
Chloromethane 9, 374

– methyl donor, 369
– sources, 10
Chloromethoxybenzaldehyde, 265
Chloromethoxybenzoic acids, 258, 339
Chloromethoxyphenols
16-Chloromilolide B, 73, 74
Chloromycorrhizin A, 28
Chloronaphthalene, 254
Chloronatronochrome, 260, 261
Chloronectrin, 322
Chloronectrin, cylindrol A_4, 322
Chloronitrophenol, 259
5-Chloro-*o*-hydroxyemodin, 319
Chloro-*O*-methylnorlichexanthone, 318
3-Chloro-*N*-methyltyrosine, 169
3-Chloro-*O*-methyltyrosine, 159
Chloropeptin 1, 330
Chloroperoxidase 6, 26, 254, 262, 340
Chloroperoxidase-promoted halogenation, 345
Chlorophenol-oligosaccharide antibiotics (orthosomycins), 333
Chlorophenols, 257–265, 340, 378
– peroxidase-catalyzed transformation, 343, 346
Chlorophenylacetic acid, 254
Chlorophenylcarboxylic acids, 254
21-Chloropuupehenol, 60
Chloropuupehenone, 375
Chloroquinocin, 250
Chlororesorcinol, 257
2-Chlorosamaderine A, 27
Chlorosesamone, 250
Chlorostyrenes, 254, 260
Chlorosulfolipids, 105
4-Chlorothreonine, 171
Chlorotoluenes, 254
Chlorotriol, 50
6-Chlorotryptophans, 152
3-Chlorotyrosine, 281
5-Chlorouracil, 224, 225
Chlorovulone II, 127
Chloroxanthone, 317–319
Chlorxanthomycin, 321
Chlovalicin, 28
Cholesterol, 94
Cholesteryl ester transfer protein (CETP), isochromophilones, 29

Choline acetyltransferase, 44
Chondramides, 150, 151
Chondria armata, armatols, 89
Chondrochlorens, 122, 123
Chondromyces crocatus, 150
– chondrochlorens, 122
Chromeno[2,3-β]pyrrole ring, 178
Chromodoris hamiltoni, hamiltonins, 69
Chromones, halogenated, 226
Chromophores, 231–233
– C-1027, 231
– kedarcidin, 232
– maduropeptin, 232, 233
– N1999A2, 233, 234
Chroococcus turgidus, 5
Chrysanthemum morfolium, chlorochrymorin, 38
Chymotrypsin, 156
Chymotrypsin inhibitors, 147
Cigarette smoke, 8
Cistanche tubulosa, kankanoside C, 104
CJ-19,784 (bromine analog of chlorflavonin), 230
CJ-21,164 326, 327
Cladonia furcata, brominated fatty acids, 109
Cladophora albida, trichloromethane, 13
Cladosporium, growth inhibition, 259, 274
Clathramides, 185
Clathrynamides, 121, 122
Clavidol, 86, 89
Clavinflol B, cytotoxicity, 83, 84
Claviol, 51, 52
Clavularia inflata, clavinflol B, 83
Clavularia viridis, chlorovulone II 127– yonarasterols, 93
Cleansing agent, air pollution, 374
neo-Clerodane ajugarin-I, 60
Cliona chilensis, celenamides, 150
Cliona nigricans, clionastatins, 94, 95
Clionastatins, 94, 95
Clitocybe elegans, melledonal D, 324
Clitocybe flaccida, clitolactone, 132, 133
Clitolactone, 132–134
Clolimalongine, 174
Clorobiocin (chlorobiocin, RP, 18,631), 227
Clostomicins, 236

Cnemidocarpa bicornuta, brominated tyramines, 284
Coastal wetlands, atmospheric CH$_3$Cl, 10
Cochliobolus victoriae (*Helminthosporium victoriae*), victorins, 148
Colubricidin A, 238
Complestatins (chloropeptin II) 330, 331
Compost, chloromethoxybenzaldehyde, 265
– polychlorinated dioxins, 343
Condylactis gigantea, 6-bromo-(5E,9Z)-eicosadienoic acid, 106
Cone snails, toxin, 155, 373
Coniine, 26
Coniochaeta tetraspora, compound CT-1, 316
Coniothyriomycin, 325, 326
Coniothyrium sp., coniothyriomycin, 325
– cryptosporiopsinol, 27
– palmarumycins, 252
ω-Conopeptide MVIIA, 155
Conotoxins, 156
– GVIIIA, 373
Conus delessertii, conotoxins, 156
Conus imperialis, heptapeptide, 156
Conus monile, conotoxins, 156
Conus peptides, 155
Conus radiatus, 6-bromotryptophan, 156
Conus textile, conotoxins, 156
Convolutamines, 282, 283
Convolutamydines, 216, 217
Convolutindole A, 198, 199
Coptotermes lacteus, CCl$_3$, 14
Corallina officinalis, chloromethane, 9
– trichloromethane, 13
Corallina pilulifera, 374
Corallinaether, 274
– polybrominated dibenzofuran, 341
Corallinafuran, polybrominated dibenzofuran, 341, 342
Coscinamides, 208, 209
Coscinoderma sp., coscinamides, 207
Coumarins, halogenated, 226–229
Crassostrea virginica, dibromoindoles/tribromoindole, 197
Cremanthodium discoideum, 44
Crematories, PCDDs/PCDFs, 344
Crustose coralline red alga, corallinafuran/corallinaether, 341

Cryptococcus neoformans, growth inhibition, 230
Cryptophycins, 159–163
Cryptosporiopsin, 27
Cryptosporiopsinol, 27
CT-1, 316
Cuparanes, 46
– halogenated, 57
Cyanobacteria, 135, 376
– halogenated fatty acids, 111
– symbiosis, 4
Cyanobacterial blooms, 376
Cyanobacterial microcystins, 376
Cyanopeptolin, 954, 147
Cyanosporasides, 255
Cyclin-dependent kinase, 194
Cyclitols, 28
Cyclobutane-containing metabolites, 324
Cyclochlorotine, 143
Cyclocinamide A, 151, 152
Cyclodepsipeptides, antibiotic, 149
6,8-Cycloeudesmanes, 54
Cyclographin, 316
Cyclohexadienones, 36
Cyclohexapeptide, chlorinated, 154
Cyclohexene, 28
Cyclolaurane, 57
Cyclolithistide A, 167
Cyclooroidins, 187, 188
Cyclopentanes, 27
Cyclopentapeptides, proline-containing, 143
Cyclopentenones, 27
Cyclopeptide, 369
Cyclopropane-cyclopentane ring, 49
Cyclopropane fatty acid, 111
Cylindricines, 176
Cylindrocarpon lucidum, cylindrol A4, 322
Cylindrocarpon olidum, 8-chlorocannabiorcichromenic acid, 226
Cylindrochlorin (ilicicolin E/8′, 9′-dehydroascochlorin), 322
Cylindrol A$_4$ 322, 323
Cymbastela sp., agelastatins, 191
– geodiamolides, 157
Cymopol, 373
Cymopolia barbata, 3′-methoxy-7-hydroxycymopol, 58

Cynara scolymus, cynarinin B, 42
Cynarinin B, 42, 43
Cynthia savignyi, cynthichlorine, 216
Cynthichlorine, 216, 217
Cystic fibrosis, 3-chlorotyrosine, 281
– control of, 266
Cystophora retroflexa, halogenated phlorethols and fucophlorethols, peracetates, 279
Cystophora torulosa, fucophlorethols, 279
Cytospora sp., cytosporin B, 28
Cytosporin B, 28
Cytostatic compounds
Cytotoxic activity, 112
Cytotoxicity, lyngbyabellins, 141, 142

D

(Z)-Dactomelyne, 96
Dactylallene, 100
Dactylopyranoid, 62, 63
Dactylosporangium aurantiacum ssp. *hamdenensis*, tiacumicins, clostomicin A, 236
Dakaramine, 295, 296
Dasyphila plumariodes, isolaurefucin methyl ether, 99
Dasyscyphus sp. A47-98, VM, 4798-1a/1b, 27
Dauricumidine, 174
Dauricumine, 174
DDT/DDE, 180, 277
1-Deacetoxyalgoane, 50
1-Deacetoxy-8-deoxyalgoane, 50
Deacetoxystylocheilamide, 112, 113
Deacetylhectochlorin, 141
Deacetyljunceellin, 78
Deacetylparguerol, 62, 63
Dead Sea, 3
2-Debromohymenin, 188, 189
Debromolongamide, 184
2-Debromotaurodispacamide A, 186
Decatromicins, 183
Dechlorobarbamide, 135
Dechlorogangaleoidin, 316
Defense function, 369
Deformylflustrabromine, 198, 199
Degluco-balhimycin, 329, 330

1,2-Dehydro-3,4-epoxypalisadin B, 50
8′,9′-Dehydroascochlorin, 322
Dehydromicrosclerodermins, 152
10-epi-15,16-Dehydrothyrsiferol, 86, 87
3-epi-Dehydrothyrsiferol, 86, 89
Dehydrovenustatriol, 86, 87
Dehydroxylinarioside, 104
Delisea fimbriata, 1,1,2-tribromooct-1-en-3-one, 25
Delisea pulchra, brominated furanones, 373
– furanone, 130
– pulchralides, 130
Delorazepam, 224
Demethylbalhimycin, 329, 330
Demethylchodatin, 318
Dendridine A, 209, 210
Dendrilla cactos, bastadins-22/-23, 313
Dendronephthya sp., brominated oxylipins, 128
Dendrophyllia sp., brominated oxylipins, 128
9-Deoxyelatol, 52
Deoxyfistularin, 304
Deoxyisoreticulidine B, 57
Deoxyprepacifenol, 47
Depsides, chlorinated, 314, 315
Depsidones, halogenated, 315–317
Dermacentor variabilis, tick control, 257
Dermatitis ("swimmer's itch"), 111
Dermatophytosis, pyrrolnitrin, 177
Dermocybe sanguinea, 5, 7-dichloroendocrocin, 319
Deschlorobromocaespitol, 55, 57
Deschloroelatol, 53
Desmarestia antarctica, CH_2Cl_2, 12–13
Destruxin-A4 chlorohydrin, 143, 144
Deterrents, 179
– *Fundulus heteroclitus* (mummichog fish), 81
Devancosamine-vancomycin, 329, 330
Diadema antillarum, 373
Diazepam, 224
Diazona chinensis/*Diazona angulata*, diazonamides, 156, 157
Diazonamides, 156, 157
Dibenzofurans, 337
Dibenzo-*p*-dioxins, 337
– polybrominated, 337

5,6-Dibromoabrine, 198, 199
Dibromoacetaldehyde, 374
Dibromoageliferin, 372
2,10-Dibromo-3-chloro-7-chamigrene, 53
5,6-Dibromo-2′-demethylaplysinopsin, 201
Dibromodeoxytopsentin, 204, 206
1,2-Dibromoethane, 18
Dibromoethylene, 20
Dibromohydroxyphenols, 265
Dibromoindoles, 197
Dibromoiodomethane ($CHBr_2I$), 20
Dibromoisophakellin, 189
3,4-Dibromomaleimide, 189
Dibromomethane (CH_2Br_2), 17
Dibromomethoxybenzoic acid, 266, 267
4,5-Dibromopalau'amine, 194
Dibromophakellstatin, antineoplastic, 189, 190
2,6-Dibromophenol, 265
3,5-Dibromo-3-(2′,4′-dibromophenoxy) phenol, 373
4,5-Dibromopyrrole-2-carboxamide, 371
4,5-Dibromopyrrole-2-carboxylic acid, 373
Dibromosceptrin, 372
2,3-Dibromostyloguanidine, 194
Dibromotyramines, 370
Dichapetalum toxicarium, 16-fluoropalmitoleic acid, 124
Dichloroacetate, lactic acidosis, 375
Dichloroacetic acid (DCA), 26
Dichloroasterrate, methyl, 277, 278
Dichlorodiaportin, 229
5,7-Dichloroendocrocin, 319, 320
6,7-Dichlorohexahydropyrrolo[2,3-β] indole, 154
7,7′-Dichlorohypericin, 319, 320
Dichloroimine sesquiterpenes, 47
Dichloroiodomethane, 20
Dichloroisoplagiochins, 262, 264
Dichlorolissoclimide, 84
Dichloromethane, 12
Dichloromethoxybenzaldehyde, 257
Dichloromethoxybenzyl alcohol, 257
Dichloromethylbenzoate, 257
Dichloro-*O*-methylnorlichexanthone, 318
2,4-Dichlorophenol, 339, 340
2,6-Dichlorophenol (tick sex pheromone), 256

3,4-Dichlorophenylacetic acid, 256
Dichloroproline, 143
Dichloropyruvate, 135
Dictyodendrilla sp., dendridine A, 209
Dictyostelium purpureum K1001, AB0022A, 341
Didemnidae (ascidian family), biselides, 244
Didemnimides, 203, 203
Didemnolines, 218
Didemnum conchyliatum, didemnimide B and D, 203, 203
Didemnum sp., didemnolines, 218
Didemnum sp., fascaplysin, reticulatine, 219
Didemnum voeltzkowi, 5′-deoxy-3-bromotubercidin, 224
Dideoxyagelorins, 310, 311
Didiscus oxeata, mukanadin D, 185
Didiscus sp., dibromomethoxybenzoic acid, 265
Dieckol, nonhalogenated dioxin, 338
8,9-Dihydrobarettin, 369
3,4-Dihydro-11-hydroxybrianthein, 71
9,10-Dihydrokeramadine, 191, 192
Dihydromaldoxin, 277
Dihydrophenanthrene, 265
(E)-Dihydrorhodophytin, 97, 98
3,4-Dihydroxybenzoic acid, 347, 348
3,7-Dihydroxycymopolone, 58
Diiodomethane, 19
Diketopiperazines, polychlorinated, 136
Dimelaena cf. *radiata*, chlorodivaricatic acid, 314
Dimethoxytetrabromobiphenyl, 272
N,N-Dimethylguanidium, 293
Dimethylketals, 93
Dimethyloctadiene, halogenated, 33
Dinitrogen pentoxide (N_2O_5), 374
Dioxapyrrolomycin, 177
Dioxepandehydrothyrsiferol, 86, 89
Dioxins, 337
Diphenyl ethers, halogenated, 273–281
– polybrominated, 337, 338
Diphenylmethanes, polybrominated 270–273
Discodermia calyx, calyculin J, 247
Discodermia polydiscus, 6-hydroxydiscodermindole, 203
Discorhabdins, 214, 215

Dispacamide, 371
Distaplia regina, 3,6-dibromoindole, 197
Disulfatobastadin, 312
Diterpenes, 60
– gorgonian, 70
– iodinated, 67
– marine, 62
– neo-clerodane, 60
– neo-irieane, 65
– sponges, 68
Diterpenoids, dolabellanes, 83
DNA polymerases, inhibition, 183
Dogs, 2-iodohexadecanal, 123
Dolabella auricularia, aurilol, 89
– aurisides, 239
– dolabellin, 140
– dolastatin, 19, 241
Dolabellanes, 83
Dolabellin, 140, 141
Dolastatin, 19, 241
Dolphins, Q1, 180
Domesite lignite, PCDDs/PCDFs, 344
Doris verrucosa, diterpene isocopalane verrucosins, 69
Douglas fir forest, chlorinated phenols, dibenzo-*p*-dioxins, dibenzofurans, 340, 342
Doxorubicin, 83
Dragmacidins, 206
Dragmacidon sp., nortopsentin D, 206
Drimane-phenolic, 60
Druinella sp., purealidin S, 310
– purpuramine J, 297
Drupella fragum, brominated hydroxyindoles, 197
Dual-specificity phosphatase VHR, 121
Dysamides, 136, 137
Dysidamides, 140
Dysidea sp., chlorodysinosin A, 147
– dibromohydroxyphenols, 265
– dysithiazolamide, 139
– (–)-(S)-4,4,4-trichloro-3-methylbutanoic acid, 137
Dysidea chlorea, dysamides, 136
Dysidea dendyi, spongiadioxins, 337
Dysidea fragilis, (4E)-(S)-antazirine/(4Z)-antazirine, 110
– dysamide D

Dysidea granulosa 4, 5, 373
Dysidea herbacea, 4
– (–)-neodysidenin, 137
– herbacic acid, 137
– herbamide A, 135
– polybrominated diphenyl ethers, 273, 274
Dysideaprolines, 139
Dysidenamide, 138
Dysidenin, 137
– iodide transport inhibitor, dog thyroid, 140
Dysithiazolamide, 139

E
Echinodictyum sp., echinosulfonic acids, 208
Echinosulfonic acids, 207–209
Ecklonia kurome, nonhalogenated dioxins, 338
Eckol, 338, 339
Ectyplasia perox, 27
Eisenia bicyclis, eiseniachlorides, 128
Eiseniaiodides, 128–130
Elatenyne, 96, 102, 104
Elatol, 373
Eleutherobia sp., briareins, 82
Ellisella sp., briaranes, 80, 81
Emericella falconensis, falconensins, 325
Emericella fruticulosa, falconensins, 325
Emericella unguis, guisinol, 314
Emiliana huxleyi, chloromethane, 9
Enacyloxin, 123, 124
Endocladia muricata, chloromethane, 12
Enduracididine, 201
Enediynes, halogenated, 232–234, 255
– N1999A2, 233
Enhydra fluctuans, chlorinated melampolides, 40
Enshuol, 86, 87
Enterobacter agglomerans, pyrrolnitrin, 177
Enterococcus, vancomycin-resistant, growth inhibition, 213, 249, 312, 314
Enteromorpha sp., trichloromethane, 13
Enteromorpha compressa, chloromethane, 9
Eosinophil peroxidase (EPO), 282
Epibatidine, frogs, 174

Epicebellin J, 39, 40
Epicentaurepensin, 39, 40
2-Epi-chloroklotzchin, 40, 41
10-Epidehydrothyrsiferol, 86, 87
12-Epi-hapalindole G, 213
10-Epikalihinol I, 68, 69
Epilaurallene, 96
13-Epilaurencienyne, 97, 98
Epinardins, 216
(3E)-13-Epipinnatifidenyne, 97, 98
EPO (eosinophil peroxidase), 282
promoted bromination, 282
Epoxymonoterpene, alicyclic polyhalogenated, 35
3,4-Epoxypalisadin B, 50
Epoxyquinomicins, 325, 326
Erythrolides, 79, 80
Erythropodium caribaeorum, chlorinated diterpenes, 78
Erythropoietin gene expression, destruxin chlorohydrin, 143
Esophageal cancer cells, plocoralides, 34
Estrones, chlorinated, 94
Eucalyptus leaves, CCl_3, 14
Eucalyptus sp., chloromethane, 11
Eucheuma denticulatum, tetrachloromethane, 15
– trichloromethane, 13
Eudesmanes, 46
– halogenated sesquiterpenes, 54
Eudistoma sp., brominated tyramines, 284
– halogenated carbolines, eudistomins, 218
– pibocins, 213
Eudistomins, 218
Eunicea sp. cembrane, 83
Eupachifolin D, 38, 39
Eupachinilides, 41
Eupaglehnins, 40, 41
Eupalinilides, 41, 42
Eupatorium chinense, sesquiterpenoids, 41
– var. *simplicifolium*, eupachifolin D, 38
Eupatorium glehni, 41
Eupatorium lindleyanum, chlorinated guaianes eupalinilides, 41
Eupenicillium shearii, kaitocephalin, 143
Eurycolactone B, 61, 62
Eurycoma longifolia, halogenated quassinoid, 61

Eurypon laughlini, laughine
Euryspongia sp., sesquiterpene quinone (E)-chlorodeoxyspongiaquinone, 55
Euthyroideones, 222, 223
Euthyroides episcopalis, euthyroideones, 222
Evernia prunastri, bromobenzene, chlororesorcinol, 257
Everninomicins, 333–336
– 13,382-1 (ziracin; SCH, 27899), 333
Excavatolide A, 73
Excoecaria agallocha, labdane-type diterpenes, 61
Excoercarin F, 61

F
F1484, antifungal compound, 227
Falconensins, 325
Falkenbergia hillebrandii, trichloromethane, 13
Farnesyl isocyanide, 55
Fascaplysin, 219, 220
Fascaplysinopsis reticulata, fascaplysin, reticulatine, 219
Fatty acids, brominated, 108
– chlorinated, 105
– cyclopropane, 111
Fe(III) oxidation, 6
– natural organic phenols, bromination, 17
Feeding deterrence, furanones, 130
Feeding deterrents, 204, 214, 270
– brominated sesquiterpenes, 54
– *Callinectes similis* (blue crab), 81
– golden fish (*Carassius auratus*), 100
– gorgonian diterpenes, reef fishes, 70
– herbivorous reef fish, 35
– sventrin, *Thalassoma bifasciatum*, 186
– *Thalassoma bifasciatum* 186, 189
Fires 7, 10
– chloromethane, 11
Fireworks, PCDDs/PCDFs, 344
Fischerella ambigua, ambigol C, 278
– 2,4-dichloro-benzoic acid, 254
Fischerella ATCC, 43239, 374
Fischerella muscicola, welwitindolinone, 213
Fischerindoles, 213

Fish toxicity, aplyparvunin, 96
Fistularines, 307–309
Flabellazoles, 195
Flavan-3-ol, halogenated, 230
Flavones, halogenated, 230
Floresolide C, 236
Fluorescent pyrene, chlorxanthomycin, 321
Fluorinase, 125, 126
o-Fluorinated fatty acids, 124
Fluorites, 23
5′-Fluoro-5′-deoxyadenosine, 125
5′-Fluoro-5′-deoxy-D-ribose-1-phosphate, 125
18-Fluoro-9,10-epoxystearic acid, 124, 125
Fluoroacetaldehyde, 125
Fluoroacetaldehyde dehydrogenase, 125, 126
Fluoroacetate, 125, 126
– poisoning, 126
Fluoroacetate dehalogenase, 126
Fluoroacetic acid, 124–126
20-Fluoroarachidic acid, 124, 125
Fluorobenzene, 254
Fluorochlorobenzene, 254
Fluorocitrate, 126
20-Fluoroeicosenoic acid, 124, 125
(Z)-16-Fluorohexadec-7-enoic acid, 124, 125
(Z)-20-Fluoroicos-9-enoic acid, 124, 125
18-Fluorolinoleic acid, 124, 125
(Z)-18-Fluorooctadec-9-enoic acid, 124, 125
16-Fluoropalmitoleic acid, 124
Fluororibulose-1-phosphate, 125
18-Fluorostearic acid, 124, 125
4-Fluorothreonine, 124–126
Fluostatins, 31, 32
Flustra foliacea, bromotryptamines, 200
– deformylflustrabromines, 198, 200, 375
– hexahydropyrrolo[2,3-b]indole, 212
Flustramides, 212
Flustramines, 212
Flying fox (*Pteropus giganteus*), 1-chloro-3-methyl-2-butene, 21
FOM-8108, chlorogentisylquinone, 28
Fomes annosus (*Heterobasidion annosus*), isocoumarins, 229

Forest fires, 7
Forest leaf litter, atmospheric CH_3Cl, 10
Forest soil, CCl_4, 15
10β-Formamido-5-isocyanatokalihinol-A, 68
10β-Formamido-5β-isothiocyanatokalihinol-A, 68
10β-Formamidokalihinol-A, 68
10β-Formamidokalihinol-E, 68
Fosfonochlorin (chloroacetylphosphonic acid), 26
Fouling, 373
FR177391, 244, 246
FR225659/FR225656, gluconeogenesis inhibitors, 173
FR901463, 230
Frankenia grandifolia, chloromethane, 10
Frateuria sp. W-315, enacyloxin, 123
Frog alkaloid, 174
Fucophlorethols, 279
Fucus serratus, trichloromethane, 13
Fucus vesiculosus, chloromethane, 9
Fulgensia canariensis, isofulgidin, 316
Fulgensia fulgida, fulgoicin, fulgidin, 316
Fulgidin, 315, 316
Fulgoicin, 315, 316
Fulvic acids, 345
– chloroperoxidase-induced chlorination, 346
Fungal strain B, 90911, methyl(di)chloroasterrates, 277
Fungi, organohalogen compounds, 322
Furanoheliangolides, 40
Furanones, brominated, 130, 373
Furo[3,2-β]pyranyl framework, 96
Furocaespitanelactol, 55, 57
Furoplocamioid C, 374
Furoplocamioids, 37
Fusarium sp., chlorofusin, 31
– neomangicols, 90
Fusarium avenaceum, fosfonochlorin, 26
Fusarium heterosporum, 90
Fusarium oxysporum, 369
– fosfonochlorin, 26
Fusarium oxysporum f. sp. *lilii*, plant defense, 259, 262
Fusarium tricinctum, fosfonochlorin, 26

G

Galanin receptor GALR1, inhibition, 322
Galanthamine, 219
Gambusia affinis, dactylallene toxicity, 100
Gastropterids, 373
Gelliusines, 207, 208
Gemmacolides A/B, 76
Geodia sp., geodiamolides, 157
Geodia barretti, 369
– barettin, 201
Geodiamolides, sponge cyclic peptides 157, 158
Geranium pratense subsp. *finitimum*, chloroepicatechin, 260
Geranylgeranyltransferase type I, massadines, 193
Germacrane, 46
Germacranolides, 40
Gigartina skottsbergii, chloromethane, 9
Gigartina stellata, trichloromethane, 13
Globularia alypum, globularioside, 104
Globularioside, 104, 105
Gluconeogenesis inhibitors, FR225659/FR225656, 173
D-Glucose-6-phosphate phosphohydrolase inhibitor, 326
Glucosylmentzefoliol, 104
Glutamate receptor antagonist, 143
Glycopeptide antibiotics, 328
Glycopeptides, chlorinated, 328
Gmelina arborea, bromine-containing lignan, 260
Gomisin J, HIV-1 reverse transcriptase inhibitors, 376
Goniothalamus amuyon, pyrone 8-chlorogoniodiol, 227
Gonorrhea, 328
Gorgonella umbraculum, diterpenes, 70
– umbraculolides, 80
Gorgonian corals, 70
Gorgonian diterpenes, 70
Gracilaria spp. 374
Gracilara cornea, trichloromethane, 13
Gracilaria coronopifolia, aplysiatoxin, malyngamides, 112
– manauealides, 236
Gracilaria verrucosa, chlorohydrins, 114
Graminichlorin, 38, 39

Graphis sp., methylated diaportins, 229
Graphislactone G, 230
Grb2-SH2 domain, antagonist, 31
Great Salt Lake, 3
Grenadadiene, 111, 112
Griseofulvin, 322
Guaiane, 46
Guaianolide andalucin, 38
Guaianolides, 40, 41, 43
Guimarane, 46
Guisinol, 314, 315
Gymnasella dankaliensis, 116
Gymnastatins, 116, 117
Gymnoascus reessii, (12E)-isorumbrin 182
Gymnogongrus antarcticus, chloromethane, 9

H

Hafellia parastata, isofulgidin, 315
Halichlorine, 176
Halichondria sp., 1,1,2-tribromooct-1-en-3-one, 25
Halichondria cylindrata, halicylindramides, 163
Halichondria japonica, gymnastatins, 116
Halichondria okadai, halichlorine, 176
– trichodenones B/C, 27
Halicortex sp., dragmacidin F, 206
Halicylindramides, 163–165
Halides, 3
– distribution 4, 16
Halimione portulacoides, long-chain chloroalkanes, 24
Haloalkanes, 9
Halogenase (non-haem) SyrB2, 171
Halogenated aromatics, 254–256
Halogenated benzene, 254–255
Halogenated benzofurans, 226–227
Halogenated carbohydrates, 231
Halogenated carbolines, 218–220
Halogenated chromones, 227
Halogenated complex phenols, 270–328
Halogenated coumarins, 227–230
Halogenated depsidones, 315–317
Halogenated diphenyl ethers, 273–281
Halogenated enediynes, 232–234, 255

Halogenated flavan-3-ol, 231
Halogenated flavones, 231
Halogenated isocoumarins, 227–230
Halogenated isoflavones, 231
Halogenated macrolides, 234–249
Halogenated naphthoquinones, 249–253
Halogenated nucleic acid bases, 224
Halogenated polyacetylenes, 231
Halogenated polyethers, 234–249
Halogenated pyrones, 227
Halogenated quinolines, 220–225
Halogenated quinones, 249–253
Halogenated simple phenols, 256–270
Halogenated tetracyclines, 253
Halogenated tyrosines, 281–314
Halogens, reactive, formation, 3
Halomon, 32, 33
Haloperoxidase-promoted halogenation, 345
Haloperoxidases, 373, 374
Halopytis incurvus, brominated phenols, 286
Hamacanthins, 204–206
Hamigera tarangaensis, hamigerans, 58
Hamigerans, 58, 59
Hamiltonins, 69, 70
Hanishin, 184
Hapalindoles, 213, 374
Hapalosiphon delicatulus, ambiguine G nitrile, 213
Hapalosiphon laingii, 12-epi-hapalindole G, 213
Harbouria trachypleura, trachypleuranin B, 22
Hasubanan type alkaloid, 174
Haterumaimides, 84, 85
Haterumalides, 243, 244
HCl, gaseous, 6
Hectochlorin, 141
Helicobacter pylori inhibition, axinellamines, 193
Helicomyces sp., FR225659/FR225656, 173
Helicusins, 28
Hemiasterella minor, geodiamolides, 157
Hemibastadinols, 295, 296
Hemibastadins, 295, 296
Hemichordata, 372
Heptachlorodibenzo-*p*-dioxin, 345
Heptachloro-*p*-dioxins, 343
Heptachloro-1′-methyl-1,2′-bipyrrole (Q1), 180

Heptatoxins, 135
Herbacic acid, 137
Herbamide A, 135, 136
Herbertus sakuraii, dichloroisoplagiochins, 262
Hericium erinaceus, chlorodimethoxybenzoic acid, orcinols, 257
Herpes virus VZV, topopyrone B, 320
Heterobasidion annosum (*Fomes annosus*), isocoumarins, 227
Hexachlorodibenzo-*p*-dioxin, 344, 345
Hexachloroisoperrottetin, 262
Hexadella sp., moloka'iamines, kuchinoenamine, 299
Hexadellin C, 307, 308
Hexadepsipeptides, polychlorinated cyclic, 154
Hexahydropyrrolo[2,3-β]indole cyclohexapeptide, chlorinated, 154
HF, gaseous, 6
Hiburipyranone, 229
Himerometra magnipinna, hydroxyhomoaerothionin, 310
Histidine decarboxylase, inhibition, 325
HIV-I, gomisin J, 376
– imidazobenzodiazepinone, 376
HIV-inhibitory activity, 376
– ancorinolates, 197
– cyclic depsipeptide, 167, 168
Homaxinella sp., longamides, 184
Homoaerothionin, 300, 372
Homoarginine, 186
Homophymia conferta, aurantosides, 118
Homosesquiterpenic fatty acids, 110
Hormaomycin, 171
Horse, 2-iodohexadecanal, 123
Horse chestnut, chloromethane, 11
Horseradish peroxidase, dioxins/dibenzofurans, 343
Host defense, 369
Household waste, PCDDs/PCDFs, 344
5-HT$_3$ serotonin receptor, inhiobition, 373
Human leukocyte myeloperoxidase, dioxins/dibenzofurans, 343
Humic acids, 345
– chlorination, 348
Humic soil layer, chloroperoxidase, dioxins, 340

Humicola sp. FO-2942, monorden, pochonins, 235
Humicola fuscoatra, radicicol, 235
Humulane, 46
Huperzia serrata, 2-chlorohyperzine E, 175
Hydrogen peroxide, 374
Hydrothermal vents, Kamchatka, CCl_4, 15
4-Hydroxy(E)-aconitate, 126
11β-Hydroxy-13-chloro-11,13-dihydrohymenin, 40
10-Hydroxyaplysin, 57, 58
10-Hydroxyaplysin, 58
3β-Hydroxyaplysin, 58
8'-Hydroxyascochlorin, 322, 323
4-Hydroxybenzoic acid, 369
Hydroxyceratinamine, 293, 294
3β-Hydroxychlorolissoclimide, 84
7-Hydroxycymopochromanone, 58
Hydroxycymopochromenols, 58
7-Hydroxycymopol, 58
3-Hydroxycymopolone, 58
16-Hydroxydehydrothyrsiferol, 86, 87, 89
6-Hydroxydiscodermindole, 203
10-Hydroxyepiaplysin, 57, 58
Hydroxyfistularin, 305
Hydroxyhomoaerothionin, 310, 311
Hydroxyindoles, 197
4-Hydroxymilolide C, 73, 74
9-Hydroxymukanadin B, 186
15-Hydroxypalisadin A, 50, 51
11-Hydroxyptilosarcenone, 82
5-Hydroxytryptamine, 373
Hygrophorus paupertinus, 3-chloroindole, 197
Hymeniacidon sp., konbu'acidins, 194
– spongiacidins, 188
– tauroacidins, 185, 186
Hymenialdisine-axinohydantoin bromopyrrole, 188
Hymenialdisines, 188, 189
– antiinflammatory activity, 375
– cyclin-dependent kinase inhibitor (GSK-3β-/CK1), 375
Hymenidin, 186
Hypericins, 319
Hypericum ascyron, vinetorin, 318
Hypholoma spp., chlorinated anisyl metabolites, 258

Hypholoma elongatum, trichloro(di) methoxyphenols, 257
Hypholoma fasciculare, chlorinated depside, 314
Hypholoma subviride, dichloromethoxybenzaldehyde, 257
Hypnea spinella, trichloromethane, 13
Hyrtios sp., 21-chloropuupehenol, 60
– poipuol, 269
Hyrtios erecta, 5,6-dibromo-2'-demethylaplysinopsin, 201
– quinolones, 220

I

Ianthella sp., iantherans, 226, 309
Ianthella basta, bastadins, disulfatobastadin, sulfatobastadin, 312, 313
– hemibastadins, hemibastadinols, 295, 296
Ianthella quadrangulata, bastadin-21, 312
Ianthellin, bromine-containing, 369
Iantherans, 226, 227
Ianthesines, 309, 310
Ibhayinol, 50
Ice plant (*Mesembryanthemum crystallium*), chloromethane, 12
ICM0301C/ICM0301D, 90, 91
Ilicicolins, 322
Illicium tashiroi, 12-chloroillifunone, 44
Imidazobenzodiazepinone, chlorinated, HIV-1, 376
Imidazoles, 286
– brominated and chlorinated, 223
Indanone sesquiterpenes, 44
Indisocin, 216, 217
Indoles, brominated, 198, 200
– halogenated, 197
– iodine-containing, 199
– sulfate-sulfamate, 197
Indolocarbazoles, 217
Inflammation, 375
Insect repellents, 35
Insecticidal activity, 97
– agelastatins, 191
Insecticides, 374, 375
Interleukin 6, 28
Io, chlorine, 8

Iodine 6, 373
– cyclic mastalgia, 376
Iodine-containing indoles, 199
Iodine oxide, 374
1-Iodobutane ($CH_3(CH_2)_3I$), 20
2-Iodobutane ($CH_3CH_2CH(CH_3)I$), 20
Iododiphlorethol, halogenated, 279
Iodoethane (CH_3CH_2I), 19
Iodolactones, 123
Iodomethane (CH_3I, methyl iodide), 19
– sources, 19
1-Iodo-2-methylpropane, 20
1-Iodopropane ($CH_3CH_2CH_2I$), 19
2-Iodopropane (($CH_3)_2CHI$), 19
Iodovulones, 127
Iotrochota purpurea, itampolins, 299
– matemone, 217
Ircinia sp., furanoses-terterpenes, 90
– haterumalide B, 243
Iridoids, chlorinated, 104
Iron-oxidizing bacteria, thermal vents, 377
Islanditoxin, 143
Isobromodeoxytopsentin, 204
Isobromotopsentin, 203, 204
Isochromophilones, 29, 30
Isochrysis sp., chloromethane, 9
Isoconcinndiol, 63
Isocopalane verrucosins, 69
Isocoumarins, halogenated, 227–230
Isocyanide diterpenes, 69
Isocyanokalihinanes, 68
Isocyanoterpenoids, chlorinated, 369
Isodehydrothrysiferol, 86, 87
Isodihydromaldoxin, 277, 278
Isofistularin, 371
Isoflavones, halogenated, 231
Isofulgidin, 315, 316
Isohalomon, 32, 33
Isokibdelones, 251
Isolaurallene, 102, 103
Isolaurefucin methyl ether, 99
Isolaurepinnacin, 102
Isolaurinterol, 58, 373
Isomalyngamide A, 114
Isomaneonene A, 100
Isomarinone, 250
Isonitriles, 374
Isoobtusol, 51, 52

Isopalisol, 49
Isoparguerol, 62
Isopinnatol B, 63
(3Z)-Isoprelaurefucin, 100
Isorigidol, 52, 53
Isorogiolal, 65, 66
(12E)-Isorumbrin, 182
Isothiocyanate, 55
Isovanillic acid, 369
Isoxazoline alkaloids, brominated, 370
Itampolins, 299, 300
Itomanallenes, 96, 97
Itomanol, 54

J
Jaborosa bergii, chlorohydrins, 92
Jaborosa odonelliana, jaborosalactone 92
Jaborosa runcinata, jaborosalactones, 92
Jaborosa sativa, 92
Jaborosalactol, 93
Jaborosalactones, 92
Jamaicamides, 114, 115
Jamesoniella colorata, hexachloroisoperrottetin, 262
Japonenynes, 96
Jaspis cf. *coriacea*, chlorocarolides 132
Jaspis splendans, jasplakinolides, 150
Jaspis wondoensis, bromopsammaplin A, bispsammaplin A, 297
Jasplakinolides (jaspamides) 150, 151
Jatropha curcas, chlorinated imidazole 223
Johnstonol, 51
Junceella fragilis, diterpenes, 70
– (+)-junceelloide A, 77
– junceellonoids, 77
Junceella juncea, chlorinated briarane diterpenoids, 75
– diterpenes, 70
– juncenolides, 76
– juncins, 76
Junceellin, 70, 71
Junceelloide A, 77
Juncenolides, 76, 77
Juncins, 74–76

K

Kaitocephalin, 143
Kalihinenes, 68, 69
Kalihinols, 68, 69
Kalihipyran B, 68
Kankanol, 104, 105
Kankanoside C, 104, 105
Kappaphycus alvarezii, 374
KB cells, 93
Kedarcidin chromophore, 233
Keramadine, 371
Keramamides, 153, 154
Keto esters, brominated, 25
Kibdelones, 251
Kibdelosporangium sp., kibdelones, isokibdelones, 250
Kilauea, acidic plumes ("acid rain"), 7
Kirkpatrickia variolosa, 1,1,2-tribromooct-1-en-3-one, 25
Konbamide, 153, 375
Konbu'acidins, 194–196
Korormicin, 121
Kottamides, 201, 202
KS-504 compounds, 375
Kuchinoenamine, 299
Kuehneromyces mutabilis, methyl dichloromethylbenzoate, 257
Kumausallene, 102
Kumausyne, 102
Kumusine , 224, 225
Kutzneria sp. 744, 154
Kutznerides, 154, 155

L

Labdane bromoditerpenoids, 63
Labdanes, 61
– ent-labdanes
Lachnum papyraceum, chloromycorrhizin A, 28
– isocoumarins, 227–230
Lachnumons, 28
Lactarius spp., 1-chloro-5-heptadecyne, 26
Lactone, chlorinated, 257
Lactoperoxidase, dioxins/dibenzofurans, 343
Lambia antarctica, CH_2Cl_2, 13
Lamellodysidea herbacea, diphenyl ethers, 274
– dysidamides, 140

Laminaria digitata, trichloromethane, 13
Laminaria saccharina, CH_2Cl_2, 13
– trichloromethane, 13
Landfill fires, PCDDs/PCDFs, 344
Lankalapuol A, 54
Lanosol, 274
Latonduines, 189, 190
Latrunculia sp., discorhabdin W, 215
Latrunculia apicalis, discorhabdin G, 214
Latrunculia bellae, discorhabdins, 214
Latrunculia corticata, latrunculinosides, 247
Latrunculia purpurea, discorhabdin Q, 214
Latrunculinosides, 247
Laughine, 186
Laurallene, 96
Laureatin, 102, 103
Laurefucin type halogenated bicyclic acetogenins, 99
Laurencia spp. (Rhodomelaceae, Ceramiales), 46
– bisezakynes, 99
– brominated diterpenes, 62
– chamigrenes, 51
– diterpenes, 63
– labdane bromoditerpenoids, 63
– paniculatol, 63
– polyether terpenoids, 90
– tribrominated ma'iliohydrin, 52
Laurencia aldingensis, aldingenins, 50
Laurencia brongniartii, polybromoindoles, 197, 209
Laurencia calliclada, callicladol, 86
Laurencia caraibica, 47
Laurencia cartilaginea, 52
– ma'ilione, 51
Laurencia claviformis, claviol, 51
– (3Z)-13-epipinnatifidenyne, 97
Laurencia concreta, 52
Laurencia elata, elatenyne, 96
Laurencia filiformis, 5-acetoxy-2,10-dibromo-3-chloro-7,8-epoxy-α-chamigrene, 53
– parguerenes, 63
Laurencia glandulifera, laurencin, 96
Laurencia intricata, itomanallenes, 96
– itomanol, 54

Laurencia japonensis, anhydroaplysiadiol, 63
– japonenynes, 96
Laurencia luzonensis, 50
– 3-bromobarekoxide, 64
– isopalisol, 49
– luzodiol, 65
Laurencia majuscula, halogenated sesquiterpenes, 47
– 8-bromo-chamigren-1-en, 53
– (6*R*,9*R*,10*S*)-10-bromo-9-hydroxychamigra-2,7(14)-diene, 52
– cedrene-type sesquiterpene majusin, 47
Laurencia mariannensis, 9-deoxyelatol, 52
– (12E)-lembyne A, 100
– pacifenol/deoxyprepacifenol, 47
Laurencia microcladia, 6,8-cycloeudesmanes calenzanol/calenzanane, 54
– lung cancer toxicity, 57
– rogioldiols, 65
Laurencia nidifica, halogenated chamigranes, 51
Laurencia nipponica, neoisoprelaurefucin, 99
– nipponallene, 96
– pargueranes, 63
Laurencia obtusa 97, 100, 374
– brasilanes, 55
– bromocyclococanol, 49
– chlorotriol, 50
– (3Z)-13-epilaurencienyne, 97
– labdanes, 64
– neoisoprelaurefucin, 100
– oxachamigrene, 53
– perforatone analogs, 54
– prevezols, 64
– scanlonenyne, 99
– β-snyderol, 49
Laurencia okamurai, laureperoxide, 57
Laurencia omaezakiana, enshuol, 86
Laurencia paniculata, *ent*-labdane paniculatol, 63
Laurencia pannosa, (3Z)-chlorofucin, 100
– pannosallene, 96
– pannosanol/pannosane, 53
Laurencia pinnatifida, 97
– dehydrothyrsiferol, 86

Laurencia rigida, (–)-10α-bromo-9β-hydroxy-α-chamigrene, 51
Laurencia saitoi, 374
Laurencia scoparia, β-bisabolenes, 49
– isorigidol, 53
– mailione/isorigidol, 52
Laurencia similis, bromoindoles, 197
Laurencia subopposita, 45
Laurencia tristicha, hydroxylated aplysins, 10-hydroxyepiaplysin/10-hydroxyaplysin, 57
Laurencia viridis, 86
– brominated polyether squalene-derived metabolites, 86
Laurencia yonaguniensis, neoirietetraol, 65
Laurencienynes, 97, 98
Laurencin, 96, 102, 103
Laurenes, halogenated, 57
Laurenyne, 102, 103
Laureperoxide, 57, 58
Laurinterol, 58
– acetate, 58
Laurokamurene A, 58
Lazy slugs, 69
LDL (low-density lipoprotein), 281
Leaf litter, 378
Leathesia nana, bromophenols, 268, 270
– phlorotannins, 338
– polybrominated diphenyl ethers, 274
Lecanora argentata, dechlorogangaleoidin, 316
Lecanora broccha, dichloro-*O*-methylnorlichexanthone, 318
Lecanora californica, dechlorogangaleoidin, 316
Lecanora chlarotera, norgangaleoidin, 315
Lecanora cinereocarnea, dibenzofurans, 341
Lecanora fructulosa, brominated fatty acids, 109
Lecanora iseana, dibenzofurans, 341
Lecanora jamesii, *O*-methylsulphurellin, 314
Lecanora leprosa, chlorolecideoidin, 316
Lecanora lividocinerea, *O*-methylanziaic acid, 314
– *O*-methylnorhyperlatolic acid, 314
– *O*-methylnorstenosporic acid, 314

Lecanora pachysoma, demethylchodatin, 318
Lecanora sulphurescens, chlorolecideoidin, 316
Leiodelide B, 241
Leiodermatium sp., leiodelide B, 241
Leishmanicidal activity, 92
Lembynes, 100
Lentinellus cochleatus, chlorostyrene, 260
Leontodon palisae, 13-chloro-3-*O*-β-D-glucopyranosylsolstitialin, 41
Lepidozia incurvata, bazzanins L-R/S, 262
Lepiota sp., lepiochlorin, 132
Lepista nuda, chlorophenols, 258, 265
Leptoclinides debius, amino acid derivatives, 201
Leptogium saturninum, brominated fatty acids, 109
Lethariella canariensis, chloroatranol, chlorohematommic acids, 259
Leucetta antarctica, 1,1,2-tribromooct-1-en-3-one, 25
Leucetta chagosensis, chagosensine, 243
Leucine aminopeptidase inhibitors, 144
Leucoagaricus carneifolia, chlorinated lactone, 257
Leukemia P388, 27
Liatris graminifolia, graminichlorin, 38
Lichens, 109, 341
– anthraquinones, 319
– chlorinated phenolics, 6
Lignan, bromine-containing, 260
Lignin degradation, 369
Lignin peroxidase, 369
Ligularia cymbulifera, bisabolanes, 44
Lilium maximowiczii, chlorinated fungicides, 369
– chlorinated orcinols, 259, 262
Limacia oblonga, acutumine, 174
Lindtneria trachyspora, chloronitrophenol, 259
Lipiarmycins, 236, 237
Lipids, halogenated, 105
Lipopurealins, 289, 290
Lipoxygenase inhibitors, sponge-derived terpenoids, 375
Lissoclimides, food poisoning, oysters, 84

Lissoclinum sp., chlorinated lissoclimide-type diterpenoids, 84
– haterumalide B, 84, 243
Lissoclinum voeltzkowi, labdane diterpenes, 84
LL-37H248 (fungus), spiroxins, 252
Loligo pealei, Q1, 180
Longamide B methyl ester, 184
Longamides, 184
Longissiminone B, 259, 260
Low-density lipoprotein (LDL), 281
Luteusins, 28, 30
Luzodiol, 65
Luzofuran, 50
Luzonenone, 50
Luzonensin, 49
Luzonensol, 50
– acetate, 49
Lyngbya bouillonii, lyngbyaloside, 239
Lyngbya majuscula 36, 111, 112
– barbamide, 135
– dysidenamide, 138
– hectochlorin, 141
– isomalyngamides, 112
– jamaicamides, 114
– lyngbyabellins, 141
– nordysidenin, 138
– pitiamide A, 113, 114
– pseudodysidenin, 138
– tetrahydroquinolines, 220
Lyngbyabellins, 141, 142
Lyngbyalosides, 239, 240
Lysolipins, 318
Lytocarpus philippinus, lytophilippines 243
Lytophilippines, 243, 245

M
M43C, 329, 330
Ma'edamines, 296, 298
Ma'iliohydrin, tribrominated, 52
Ma'ilione, 51, 52
Maaliane, 46
Macrolides, halogenated, 234–348
Maduropeptin chromophore, 233
Magnaporthe grisea, bromophenols, 375
Mailione, 52

Majapolenes A, 47, 48
Majapols, 48
Majusculoic acid, 111
Majusin, 47, 48
Makalika ester, 114, 115
Makalikone ester, 114, 115
Makaluvamines, 214, 216
Malaria, 375
– inhibition of malaria parasite, 235
Malbranchea aurantiaca,
   malbrancheamide, 213
Malbrancheamide, 213, 375
Maldoxin, 322
Maldoxone, 316, 317
Malyngamides, 112, 113
Manauealides, 236
cis-Maneonene C, 100
Maracens, antimycobacterial, 114–116
Marenzellaria viridis, alkyl/alkenyl
   halides, 24
Marine polybrominated diphenyl ethers,
   273–277
Marinone, 250
Marmycin B, angucycline-type, 320
Mars, chloride salts, 378
Massadines, 191, 193, 194
Mastigophola diclados, (di)
   chloroisoplagiochins, 262
Matemone, 217
Maui acorn worm (*Ptychodera* sp.), (+)-
   bromoxone, 28
Mauritiamine, 187, 188
Maytanbicyclinol, 234
Maytanbutine, 234
Maytansinoids, 234
– antitumor activity, 375
Maytenus aquifolia, celastramycin A, 182
Maytenus buchananii, 2'-*N*-demethyl-
   maytanbutine, maytanbicyclinol, 234
Maytenus hookeri, host plant, 242
MC21-A, 272
Melampolides, chlorinated, 40
Melanin synthesis inhibitor, 27
Melledonals, 324
Mellein, 229
Melleolides, 324, 325
– melleolide J/armillarikin, 324
Meloidogyne incognita (nematode host), 227

Menispermum dauricum, dauricumine, 174
Mentha longifolia, chlorinated menthone
   longifone, 37
Mentha villosa, oleanane, 91
Mentzelia cordifolia, mentzefoliol, 104
Meridianins, 209
Mertensene, 35
Mesembryanthemum crystallium,
   chloromethane, 12
Metalloproteinases, inhibition, 187
Meteorites, 254
Meteorites, organochlorines, 8
9-Methoxydispacamide B, 186
8-Methoxysaxalin, 228
Methyl 3',5'-dichloro-4,4'-di-*O*-
   methylatromentate, 326
Methyl 6-bromoindole-3-carboxylate, 198
Methyl bromide *see* Bromomethane
Methyl chloride transferase, 12
1-Methyl-2,3,5-tribromoindole, 197
N(1')-Methyl-2-bromoageliferin, 190
1-Methyl-3,5,6-tribromoindole, 197
N-Methylaerophobin-2 307, 308
O-Methylanziaic acid, 314
Methylbalhimycin, 329, 330
N-Methyl-ceratinamine, 292
N-Methyldibromoisophakellin, 189, 190
Methyldibromophakellin, 189
N-Methylmanzacidin C, 189, 190
N-Methylmonobromophakellin, 189
O-Methylnorhyperlatolic acid, 314
O-Methylnorstenosporic acid, 314
8-*O*-Methylsclerotiorinamine, 31, 32
O-Methylsulphurellin, 314
Microcoleus sp., pitiamide A, 113, 114
Microcoleus lyngbyaceus, polychlorinated
   acetamides, 26
Microcystins, 135, 376
Microcystis aeruginosa, plasmin inhibitors
   micropeptins, 476-A/478-B, 169
– (NIES-98), aeruginosin, 98-A, 144
– (NIES-299), microginins, 299-A/299-B,
   144
– (NIES-478), micropeptins, 476-A/478-B,
   144
– NIVA Cya, 43, cyanopeptolin, 954, 147
Microginins, 144–147
Micromonospora sp., pyrrolosporin A, 183

Subject Index

Micromonospora carbonacea, everninomicin-6, 335
– var. *africana*, everninomicin, 13,382-1 (ziracin; SCH, 27899), 333
Micromonospora echinospora subsp. *armeniaca*, clostomicins, 236
Micropeptins, 144, 145
– 476-A/478-B, 169, 170
Microsclerodermins, 152
Microspinosamide, HIV-inhibitory, 167, 168
Microtetraspora (*Actinomadura*) *spiralis*, pyralomicins, 178
Midpacamide, 371
Milolides, 73
Minabein-4, 82
Minabien-6 (11-hydroxyptilosarcenone), 82
Miuraenamides, 158, 159
Mollisia melaleuca, 27
Moloka'iamines, 292, 299, 300, 370
Molokinenone, 60
Mololipids, 292
Monamycins, 149
Mondia whitei, 5-chloropropacin, 228
Moneses uniflora, 8-chlorochimaphilin, 250
Monk seal blubber, brominated cacoxanthenes, 58
Monoacetylagaricoglycerides, 326, 328
Monoamine oxidase inhibition, 28
Monobromoisophakellin, 189, 190
Monodechlorovancomycin, 332, 333
Monomethyl ether, 258
Monordens, 235
Monoterpene ethers, 36
Monoterpenes, 32
– acyclic, 32
– alicyclic, 35
Morulin Pm, 156
Mosquito fish (*Gambusia affinis*), 100
MPO (myeloperoxidase) 94, 224, 281
MR566A, 27
Mukanadin C/debromolongamide, 184
Mukanadin D, 185
Mussels, 105
Mycena spp., chlorinated anisyl metabolites, 258
– tetrachlorocatechol, monomethyl ether, 258
Mycena alcalina, alcalinaphenols, 258

Mycena galopus, chlorinated benzoxepins, 226
Mycena metata, CCl_3, 14
Mycobacteria, 114
Mycobacterium tuberculosis, growth inhibition, 307
– pyrrolnitrin, 177
Mycorrhizin A, 28
Mycothiol-*S*-conjugate amidase, inhibition, 297, 300
Myeloperoxidase (MPO) 94, 224, 281, 282
– dioxins/dibenzofurans, 343
Myeloperoxidase–H_2O_2– chloride, 94
Myrcene, 32
Myriastra clavosa, myriastramide B, 169
Myriastramide B, 169, 170
Mytilus edulis, 369
– brominated diphenyl ether, 277
– 1,3,8-tribromodibenzo-*p*-dioxin, 338
Mytilus galloprovincialis, chlorosulfolipids, 105
Myxine glutinosa (Atlantic hagfish), cathelicidins, 156
Myxobacteria, SMH-27-4, 158

N
N1999A2 (enediyne), 233
Nagelamides, 191, 192, 195, 196
Nakamuric acid, 190, 191
Nakirodin A, 295, 296
Nakiterpiosin, 93, 94
Nakiterpiosinone, 93, 94
Naphthalene glycosides, 260
Naphthols, halogenated, 181
Naphthomycin K, 242
Naphthoquinones, halogenated, 249–253
Napyradiomycins, 249
Nectria galligena, ilicicolins, 322
Neisseria gonorrhoeae, gonorrhea, 328
NeoC-1027 chromophore, 232, 233
Neodysidenin, 137
Neoirietetraol, 65
Neoisoprelaurefucin, 99, 100
Neolaurallene, 96
Neomangicols, 91
Neomeris annulata, brominated sesquiterpenes, 54

Neonipponallene, 96, 97
Neopargueroldione, 63
Neopicrorhiza scrophulariiflora, piscroside A, 104
Neorogioldiol, 64–66
Neosiphonia superstes, neosiphoniamolide A, 157
Neosiphoniamolide A, iodinated, 157
Nephroma laevigatum, 7-chloro-1-*O*-methyl-*o*-hydroxyemodin, 319
Netted barrel sponge (*Verongula gigantea*), 295
Neuroglossum ligulatum, CH_2Cl_2, 13
Neuronal nicotinic receptor, 375
Neuropathic pain, ω-conopeptide MVIIA, 155
Neuroprotectins, 330
Neurotoxin, *β*-*N*-methylamino-L-alanine, 376
Niphogeton ternata, polyacetylene, 231
– psoralen (8-methoxysaxalin), 228
Nipponallene, 96, 97
Nitophyllum marginata, 1-methyl-2,3,5-tribromoindole, 197
Nitryl chloride (NO_2Cl), 374
Nocardia blackwellii, indisocin, 216
Nocardia transvalensis, transvalencin A, 143
Nonachloro-2-phenoxyphenol, 343
Norcembrane sinularectin, 83
Norditerpene dilactones, halogenated, 61
Nordysidenin, 138
Norgangaleoidin, 315, 316
Norsphaerol, 67
Norte's obtusenynes, 102
Nortopsentin D, 206
Norway maple, chloromethane, 11
Norway maple, chloromethane, 11
Nosporamide A, 377
Nostoc sp. 111
– carbamidocyclophanes, 255
– cryptophycins, 159–163
– nostocarboline, 219
Nostocarboline, 219, 220
Nostocyclophanes, 255
Notomastus lobatus, bromophenols, 265, 373
NPI-0052, 172
Nucleic acid bases, halogenated, 224
Nui-inoalides, 74, 75, 82

O

Oak, chloromethane, 11
Obtusadiol, 64
Obtusallenes, 100
Obtusenynes, 97, 99, 102
OCDD, 343, 344
Ocean, salinity, 3
Oceanapia sp. 309
– bromo acids, 106
– bromotyrosine derivate, 297
– petrosamines B, 175
Ochratoxin A, 229, 230
Octachlorodibenzo-*p*-dioxins (OCDD) 343, 344
Odonthalia corymbifera, bromophenols, 375
– diphenylmethane, 270
– lanosol, 274
Old man's beard (*Usnea longissima*), 259
Oleanane, 91
Oocydin A (haterumalid NA), 243
Oppositane, 46
Orcinols, 257
Organobromines, 18
Organochlorines, meteorites, 8
– peatlands, 6
Organofluorines, 23
Organohalogens, terrestrial environments, 5
Organoiodines, humification in peatlands, 6
– marine biogenic, 3
Orina sp., gelliusines, 207
Oroidin, 185, 371
– derivatives, 186
– dimer, 187
Orthosomycins, 333
Oscillaginin A, 135
Oscillariolide, 243
Oscillatoria agardhii, aeruginosins, 205A/205B, 144
– oscillaginin A, 135
Oscillatoria spongeliae, 4
– polychlorinated amino acid metabolites, 135
– polychlorinated metabolites/polybrominated diphenyl ethers, 140
Osedax spp. 4
Osmundaria obtusilobu, sulfated oligobromophenols, 268

Oudemansin B, 323
Oxachamigrene, 53
Oxoaerophobin, 304
Oxocyclostylidol, 188
Oxohomoaerothionin, 304, 305
4-Oxovancosamine, 329
Oxylipins, brominated, 128
Ozone 3, 374

P

$P2X_7$ antagonists, 195
p53-MDM2 antagonist, 31
Pachychalina sp., bromotyrosines, 293
Pachyclavularia violacea, pachyclavulide D, 81
Pachyclavulide D, 81, 83
Pachydermin, 182, 326
Pacifenediol, 51
Pacifenol, 47, 51
Pacifidiene, 51
PAHs (polycyclic aromatic hydrocarbons), 256
Palau'amine, 194, 195
Palisadin B, 50, 51
Palmarumycins, 252
Palythoa caribaeorum, 6-bromo-(5E,9Z)-eicosadienoic acid, 106
Panacene, 102, 103
Pancreatic elastase, 169
Paniculatol, 64
Pannosallene, 96, 97
Pannosanol/pannosane, 53
Pantherinine, 175
Pantofuranoids, 36, 37
Pantoisofuranoids, 36, 37
Pantoneura plocamioides, 36
– pantoneurotriols, 33
– pantopyranoids, 37
Pantoneurines A, 37, 38
Pantoneurotriols, 33
Pantopyranoids, 37, 38
Paralemnalia thyrsoides, chlorinated norsesquiterpenoid paralemnolin A, 55
Paralemnolin A, 56
Parasitenone, 266, 267
Parguerenes, 63

Parmelia comtseliadalis, brominated fatty acids, 109
Parmelia linctina, brominated fatty acids, 109
Patientosides, 260, 261
PBBs (polybrominated biphenyls), 272
PCBs (polychlorinated biphenyls), 180, 277, 344
PCDDs, 344, 345
PCDFs, 344, 345
PDGF, RP-1551s, 30
Peatbogs/peatlands, atmospheric CH_3Cl, 10
– dioxins/dibenzofurans, 339, 340
– organohalogens, 11
– organochlorines, 6
Peltigera canina, brominated fatty acids, 109
Pelvetia canalicuta, trichloromethane, 13
Penicillium sp., atpenins A4 and A5, 223
– SPC-21609, RP-1551s, 30
– topopyrones, 320
Penicillium citrinum, methyl (di) chloroasterric acids, 277
Penicillium crustosum, thomitrem A, 213
Penicillium multicolor, 31
– isochromophilones, 29
Penicillium nalgiovense, dichlorodiaportin, 229
Penicillium sclerotiorum, 5-chloroisorotiorin, 29
Peniophora spp., chlorinated anisyl metabolites, 258
Peniophora pseudopini, CCl_3, 14
Pentabromopseudilin, 177
Pentachlorophenol-treated wood, 343
Pentachloropseudilin, 177
Pentalenolactone, 38
Pepticinnamin E, farnesyl-protein transferase inhibition, 158
Peptide metabolites, polychlorinated, cysteine-derived, 140
Peptides, halogenated, 134
– tryptophan-derived, 150
Peracetates, 279
Perforane, 46
Perforatol, 54, 55
Perforatone analogs, 54, 55
Perforenol B, 54, 55

Periconia byssoides, (+)-pericosine A, 28
Periconia circinata, peritoxins, 148
Periconia macrospinosa, isocoumarins, 229
Pericosine A, 28
Peritoxins, 148, 149
Perknaster fuscus (feeding deterrent), 214
Perophora namei, perophoramidine, 209
Perophoramidine, 209, 210
Perthamide B, cyclic octapeptide, 163
Pestalone, 314, 315
Pestalotia sp., pestalone, 314
Pestalotiopsis sp., compound RES-1214-2, 277
Pesticides, bromomethane, 15
Petrosamines B, 175
Pezicula carpinea, mycorrhizin A, 28
Pezicula livida, (+)-cryptosporiopsin, 27
– mycorrhizin A
Phaeocystis sp., chloromethane, 9
Phaeodactylum tricornutum, chloromethane, 9
Phakellia fusca, 5-fluorouracil derivatives, 224
Phakellia mauritiana, dibromophakellstatin, 189
Phakellia pulcherrima, kalihinols, 68, 69
Phakellin alkaloids, 189
Phakellin-type bromopyrroles, latonduines, 189
Phallusia mammillata, morulin Pm, 156
Phellinus spp., chlorinated anisyl metabolites, 258
Phellinus pini, CCl_3, 14
Phellinus pomaceus, chloromethane, 12
Phenethylamines, brominated, 282–300
Phenols, chlorinated, 257–265
– complex, halogenated, 270–328
– simple, brominated, 265–270
– halogenated, 256–270
Phenylacetic acid, 307
Phenylacetonitrile, 286
Phlomis younghusbandii, phloyoside II, 104
Phlorethols, halogenated, 279
2-Phloroeckol, 338, 339
Phloroglucinols, 279
Phloyoside II, 104
Pholiota spp., chlorinated anisyl metabolites, 258

Pholiota destruens, dichloromethoxybenzaldehyde, 257
– dichloromethoxybenzyl alcohol, 257
Phoma sp., topopyrones, 320
– TC, 1674, TMC-264, 230
Phomopsis sp., sesquiterpene acid, 44
Phorbas sp., phorbasides, 239
– phorboxazoles, 241
Phorbas glaberrima, bromophenols, 266
– 1,1,2-tribromooct-1-en-3-one, 25
Phorbasides, 239, 240
Phorboxazoles, 241*Phoriospongia* sp., phoriospongins, 167
Phoriospongins, 167, 169
Phormidium sp., phormidolide, 243
Phormidolide, 243, 244
Phospholipids, 106
Phyllidiella pustulosa, kalihinols, 69
Phylloporia spp., chlorinated anisyl metabolites, 258
Phyllospongia sp., tribrominated diphenyl ether, 275
Phyllospongia dendyi, polybrominated diphenyl ethers, 274
Physalin H, 92
Physalis angulata, physalin H, 92
Physostegia virginiana ssp. *virginiana*, stegioside I, 104
Phytophthora sp. 158
Pibocins, 213
Picea abies, 6,7-dichlorohexahydropyrrolo [2,3-β]indole, 154
Pinicoloform, 26
Pinna muricata, pinnaic acid, 176
Pinnaic acid, 176
Pinnatifidenyne, 102
Piscroside A, 104, 105
Pitiamide A/B, 113, 114
Plagiochila sp., chloroisoplagiochin D, 262
Plagiochila peculiaris, bazzanin J, chloroisoplagiochin D, 263
Plakohypaphorines, 199, 200
Plakoris simplex, plakortether C, 121
Plakortamines, 218, 219
Plakortether C, 121, 122
Plakortis nigra, plakortamines, 218
Plakortis simplex, plakohypaphorines, 199
Plasmin inhibitors, 144

Subject Index

Plasmodium falciparum, inhibition of, 235
Plectophomella sp., mellein derivatives, 229
Pleurobranchus albiguttatus, 3β-hydroxychlorolissoclimide, 84
Pleurobranchus forskalii, 3β-hydroxychlorolissoclimide, 84
Plexaureides praelonga, praelolide, 70
Plocamenols, 34
Plocamenone, 32, 33
Plocamiopyranoids, 37, 38
Plocamium cartilagineum 32, 33
– anverene, 34
– epoxides, 35
– furoplocamioids, 36
– halogenated homosesquiterpenic fatty acids, 110
– insecticides, 35
– plocamiopyranoid, 37
– prefuroplocamioid, 34
Plocamium corallorhiza, plocoralides
Plocamium costatum, halogenated monoterpenes, 33
Plocamium hamatum, 33
– chloromertensene, 373
Plocoralides, 34
Pluchea arguta, 3,4-di-epi-3′-chloro-2′-hydroxyarguticinin, 44
Pluchea carolonesis, eudesmane, 44
Plutonites, 23
Pochonia chlamydosporia var. *catenulata*, pochonins, 235
– tetrahydromonorden, 235
Pochonins, 235
Podocarpus macrophyllus, rakanmakilactones, 61
Poecillastra wondoensis, bromopsammaplin A, bispsammaplin A, 297
Poipuol, 269
Poitane, 46
Polyacetylenes, halogenated, 231, 232
Polybrominated biphenyls (PBBs), 272
Polybrominated dibenzo-*p*-dioxins, 337
Polybrominated diphenyl ethers, 337
Polybromoindoles, 197
Polychaeta, 372
Polychlorinated biphenyls (PCBs), 277

Polycitones, retroviral reverse transcriptases, inhibition, 183
Polycitor africanus, polycitone B, 183
Polycitrin B, 183
Polycyclic aromatic hydrocarbons (PAH), 256
Polydora socialis, alkyl/alkenyl halides, 24
Polyether triterpenes, marine, 86
Polyethers, halogenated, 234–248
Polygala vulgaris, chloroxanthone, 318
Polyketides, 121
Polysiphenol, 265
Polysiphonia ferulacea, polysiphenol, 265
Polysiphonia lanosa, rhodomelol, 265
– trichloromethane, 13
Polysiphonia sphaerocarpa, simple bromophenols, 266
Polysiphonia urceolata, urceolatol, 269
Porphyridium sp., chloromethane, 9
Porpoise, 179
Portieria hornemannii 32–34
– halomon, 32, 35
Potato (*Solanum tuberosum*), atmospheric CH_3Cl, 11
Praelolide, 70, 71
Predator deterrents, 373
Pre-dioxin nonachloro-2-phenoxyphenol, 343
Prefuroplocamioid, 34
Pregnanes, 94
Premia subscandens, asystasioside E, 104
Prepacifenol epoxide, 51
Prevezols, 64, 65
Prialt, 155
Prinsepia utilis, lactones, 132
Prorocentrum sp., chloromethane, 9
Prostaglandins, halogenated, 127
Prostanoids, 127
Protein phosphatase 2A, nagelamides, 191
Protoberberine alkaloids, 175
Prunolides, 131, 132
Prymnesins, 247–248
Prymnesium parvum, prymnesins, 247
Psammaplins, 286–289, 293–295
Psammaplysenes, 292, 293
Psammaplysilla sp., *N*-methylceratinamine, 292
– wai'anaeamines, 292

- psammaplysenes, 292
Psammaplysilla purea, purealin, 370
 - lipopurealins, 289
 - purealidins, 300
Psammaplysilla purpurea, 370
 - bis(deacetyl)solenolide D, 81
 - dibromomethoxybenzoic acid, 266
 - purpuramines, purpurealidins, 289
 - purpurealidins, 300
Psammaplysin A, acyl derivatives of, 303
Psammaplysins, 303
Psammocinia aff. *bulbosa*, cyclocinamide A, 151, 152
Psammoclemma sp., (6E)-clathrynamide A
 - echinosulfonic acids, 207–208
Psammopemma sp., psammopemmins, 203
Psammopemmins, 203
Pseudoalteromonas spp. 369
 - F-418, korormicin, 121
Pseudoalteromonas luteoviolacea, chlorodibromophenol, 266
Pseudoalteromonas phenolica, compound MC21-A, 272
Pseudoarachniotus roseus, aranochlors, 116
Pseudoceratidine, 185, 186
Pseudoceratina sp., bromoverongamine, 286
 - ceratamines, 175
 - *N*-methyl-ceratinamine, wai'anaeamines, 292
 - simple carboxylic acid, 300
Pseudoceratina crassa, brominated phenylacetonitrile, imidazole, 286
Pseudoceratina purpurea, ceratinamides, 303
 - ceratinamine, 286, 287
 - psammaplins, 286, 287
 - psammaplysin A, 303
 - pseudoceratins, 286, 287
 - purpuroceratic acids, 303
 - tokaradines, 286, 287
 - zamamistatin, 303
Pseudoceratina purpurea, pseudoceratidine, 185
Pseudoceratina verrucosa, bromotyrosines, 286, 300
Pseudoceratinines, 286, 287, 300, 302
Pseudoceratins, 287, 289
Pseudodistoma aureum, eudistomin V, 218

Pseudodysidenin, 138
Pseudomonas spp., FR, 901463, nonchlorinated epoxides, 231
 - pyrrolnitrin, 177
Psoralen, 228
Ptaquiloside, carcinogen, 44
Pteridium aquilinum, ptaquiloside, 44
Pteroeides sp. 80
 - diterpenes, 82
Pteropus giganteus, 1-chloro-3-methyl-2-butene, 21
Pterula sp. pterulinic acids, pterulone, 226
Pterulinic acids, 226
Pterulone, 226
Ptilocaulis spiculifer, dakaramine, 295
Ptilonia magellanica, pyranosylmagellanicus, 99
Ptychodera flava, bromohydroquinones, bromobenzoquinone, 268
Ptychodera sp., (+)-bromoxone, 28
Puertitol-B acetate, 50
Pufferfish (*Canthigaster solandri*), 373
Pulchralides, 130
Punaglandin 8, 127
Punctatol, 55, 57
Purealidins, 289, 290, 300, 301, 302, 310, 311
Purealin, 370
Purpuramines, 289, 296, 297–298
Purpurealidins, 289, 291, 300, 302
Putterlickia retrospinosa, celastramycin A, 182
Putterlickia verrucosa, celastramycin A, 182
Puupehenone, 60
Pycnoclavella kottae, kottamides, 201
Pycnopodia helianthoides, repellants, 199
Pyoluteorin, 177
Pyralomicins, 178
Pyrano[3,2-β]pyrans, 102
Pyrano[3,2-β]pyranyl vinyl acetylene, 96
Pyranosylmagellanicus, 99
Pyrazinone bromotyrosines, 296
Pyrone 8-chlorogoniodiol, 227
Pyrroindomycin B, 213, 214
Pyrrole alkaloids, brominated, 183, 371
Pyrroles, halogenated, 177
Pyrrolidone, 139
Pyrrolizidine alkaloids, *Senecio selloi*, 174

Pyrrolizidines, 5-chlorobohemamine C, 175
Pyrrolnitrin, 177
Pyrrolomycins, 177
Pyrrolosporin, 183

Q
Q1 180, 181
– human milk, 180
– polyhalogenated, 180
Quassinoids, 61
Quinolines, halogenated, 220–225
Quinone reductase, 92
Quinones, halogenated, 249–253

R
Radicicol, 235
Radiciol, anticancer agents, 375
Rakanmakilactones, 61, 62
Rat thyroid, 2-iodohexadecanal, 123
Rebeccamycin, 217, 218
Red tides, 376
Renilla reniformis, 81
Renillins, 80, 83
RES-1214-2 (dihydromaldoxin), 277, 278
Resinicium pinicola, pinicoloform, 26
Resormycin, 143
Reticulatine, 219, 220
Reticulidia fungia, reticulidins, 55
Reticulidins, 55, 56, 57
Rhamnosyl-balhimycin, 329, 330
Rhaphisia lacazei, topsentin and hamacanthin, 204
Rhaphisia pallida, cyclic *N*-bromoimide, 223
Rhizoplaca peltata, 109
Rhodomela confervoides, brominated catechols, 268
– brominated diphenylmethanes, 270
– bromophenols, 266
– bromotyrosine, 282
– lanosol-purine metabolites, 269
– tetrahydroquinolines, 220
Rhodomelol, 265
Rhodospirillum salexigens, 25
Rhopaladins, 204, 205
Rhopalaea sp., rhopaladins, 204
Rhopaloeides odorabile, 4

Rhyncholacis pedicillata, host plant, 243
Riccardia marginata, chlorinated bibenzyls, 259
Rice paddies, atmospheric CH_3Cl, 10
Richibucto (New Brunswick), dioxins/chlorinated furans, 339
Rigidol, 51, 52
Rinodina dissa, isofulgidin, 315
Ritterella, halogenated carbolines, 218
Rogiolal, 65, 66
Rogioldiols, 64–66
Rogioloxepane A, 102, 103
Rollinia mucosa, romucosine B, 175
Romucosines, 175
Root fungi, atmospheric CH_3Cl, 10
RP, 18,631 (clorobiocin/chlorobiocin), 228
RP-1551s, 30, 31
Rubrolides, halogenated, 131
Rubrosides, 118, 120
Rumbrin, 182
Rumex patientia, 6-chlorocatechin, 231
– patientosides, 260
Russula roscea, 133
Russula subnigricans, dichloromethoxyphenol, 257
Russuphelins, 278
Russuphelol, 278, 279
Ruta spp., acridone alkaloid A6, 222

S
Sabella pavonia, 373
Saccoglossus bromophenolosus, dibromophenol, 269
Saccoglossus kowalevskii, bromophenols, 265
– 2,3,4-tribromopyrrole, 179, 373
Sagaminopteron bilealbum, brominated diphenyl ether, 273
Sagaminopteron nigropunctatum, 373
Sagaminopteron psychedelicum, 373
Salicornia sp., chloromethane, 10
Salinispora pacifica, cyanosporasides, 255
Salinispora tropica, salinosporamides, 172
– sporolides, 235
Salinity, early ocean, 3
Salinosporamides, 172, 173
Salt lakes, halobacteria, trichloromethane, 14

Salt marshes, bromomethane, 16, 17
Saltwort (*Batis maritima*), atmospheric CH_3Cl, 11
Samadera madagascariensis, quassinoid 2-chlorosamaderine A, 27
Sarcocornia fruticosa, long-chain chloroalkanes, 24
Sargassum spinuligerum, fucophlorethols, 279
Scanlonenyne, 99
Sceptrin, 372
Sceptrins, 190, 191
Sch, 27899, 333
Sch, 27900, 333, 334
Sch, 49088, 333, 335
Sch, 54445 (polycyclic xanthone) 318, 319
Sch, 58761, 333, 334
Sch, 58769, 335, 336
Sch, 58771, 335, 336
Sch, 58773, 335, 336
Sch, 58775, 335, 337
Sch, 202596, 322
Sch, 204698 331Sch, 212394, 331
Scleroderma sp. (poison puff ball), methyl 3',5'-dichloro-4,4'-di-*O*-methylatromentate, 326
Scleroderma sinnamariense, methyl 2',5'-dichloro-4,4'-di-*O*-methylatromentate, 326
Sclerotiorin, 28
Scolelepsis squamata, alkyl/alkenyl halides, 24
Scyptolins, 169, 170
Scytonema hofmanni PCC, 7110, scyptolins, 169
Sea pens, 82
Sea-salt spray/aerosols, chlorine/bromine 3, 374
Sea urchins, 374
Seabird eggs, 179
Secobatzelline A, 216
Securamines, 212
Securiflustra securifrons, securamines, 212
Sediments/clays, PCDDs/PCDFs, 344
Senecio selloi, 18-hydroxyjaconine, 174
Seragamides, 158, 159
Serratia spp., oocydin A (haterumalid NA), 243

Serratia liquefaciens, FR177391, 244
Sesamum indicum, anthrasesamone C, 320
– chlorosesamone, 250
Sesquiterpene chlorohydrins, 55
Sesquiterpene dichloroimines, 55
Sesquiterpene lactone glucoside, chlorinated, 41
Sesquiterpene lactones, chlorinated, 38
Sesquiterpenes, halogenated, 38
– marine, 46
Sesterterpenes, 90
Sewage sludge, polychlorinated dioxins 343
Sex pheromones, control of ticks, 256
Shrublands, atmospheric CH_3Cl, 10
Sidonops microspinosa, microspinosamide, 167
Siliquariaspongia japonica, aurantosides, 118
– rubrosides, 118
Simocyclinone D8, 228
Sinularectin, 83, 84
Sinularia cruciata, 373
Sinularia erecta, norcembrane sinularectin, 83
Slagenins, L1210 murine leukemia, 187
Smenospongia sp., 5-bromotryptophan, 198
– methyl 6-bromoindole-3-carboxylate, 198
Smenospongia aurea, makaluvamine O, 216
Soil pesticide, bromomethane, 16
Solanum tuberosum, atmospheric CH_3Cl, 11
– benzodiazepines, 225
Solenolides, 70, 71, 74
Solfataras, CCl_4, 15
Sorangium cellulosum, maracens, 114
SP-969, 332
SP-1134, 332
Spatoglossum variabile, aurones, 226
Spermidine derivative, antifouling, 185
Sphaerococcus coronopifolius, bromine-containing diterpenes, 67
Sphaerolabdadiene-3,14-diol, 67
Spiophanes bombyx, alkyl/alkenyl halides, 24
Spirastrella coccinea, spirastrellolide A, 238
Spirastrella hartmani, halogenated heliananes, 58
Spirastrellolide A, 238, 239

Spiro-chamigrene, halogenated, 51
Spirocyclohexadienone, maldoxin, 322
Spirocyclohexadienyl isoxazolines, 282, 300
Spirographis spallanzanii, iodination of tyrosine, 373
Spiroisoxazolines bromotyrosine, 300–312
Spiroxins, 252
Sponge diterpenes, 68
Sponge-bacteria symbiosis, 4
Spongiacidins, 188, 189
Spongiadioxins, 337–339
Spongistatins, 241*Spongosorites* sp., dragmacidin E, 206
– hamacanthins, deoxytopsentin, spongotine, 204
– isobromotopsentin, 203
Spongosorites genitrix, bromodeoxytopsentin, 204
– isobromodeoxytopsentin, 204
Spongotines, 204, 206
Sporolides, 235
Sporopodium citrinum, chlorolichexanthone, 318
– chloro-*O*-methylnorlichexanthone, 318
Sporothrix sp. 28
Staphylococcus aureus, methicillin-resistant, 154, 183, 376
– growth inhibition, 204, 213, 249, 266, 272, 312, 314
Stegioside I, 104
Stenella coeruleoalba, dimethoxytetrabromobiphenyl, 272
Stephanospora caroticolor, aminochlorophenol, 259, 261
– chloronitrophenol, 259, 261
– stephanosporin, 259, 261
Stephanosporin, 259, 260
Stereobalanus canadensis, bromophenols, 373
Steroids, 92
– halogenated marine, 93
Stevensine, 371, 372
Stevia sanguinea, 40
Stoichactis helianthus, 6-bromo-(5E,9Z)-heneicosadienoic acid, 106
Streblospio benedicti, alkyl/alkenyl halides, 24

Streptomyces sp., azamerone, 250
– akashins, 211
– bischloroanthrabenzoxocinone, 320
– celastramycins, 182, 320
– chloptosin, 154
– clorobiocin, 228
– colubricidin A, 238
– dechloromarmycin, 321
– hormaomycin, 171
– MA7-234, complestatins, 330
– manumycin antibiotics chinikomycins, 122
– naphthomycin K, 242
– pentalenolactone, 38
– pepticinnamin E, 158
– pyrrolizidine 5-chlorobohemamine C 175
– Q27107, neuroprotectins, 330
– SCH, 212394, 331
– TAN-876 A/TAN-876 B, 178
– xantholipin, 318
Streptomyces aculeolatus, compounds A80915-A–D, 249
Streptomyces antibioticus Tii, 6040, simocyclinone D8, 228
Streptomyces armeniacus, streptopyrrole, 178
Streptomyces armentosus var. *armentosus*, armentomycin (2-amino-4, 4-dichlorobutyric acid), 135
Streptomyces aureofaciens, gene for chlorination of tetracycline, 253
Streptomyces cattleya, 4-fluorothreonine, 124
– organofluorines, 125
Streptomyces fradiae, actinomycins Z_3/Z_5, 171
Streptomyces fumanus, pyrrolomycins, 177
Streptomyces globisporus, neoC-1027 chromophore, 232
Streptomyces griseoflavus, hormaomycin, 171
– 593A, 223
Streptomyces iakyrus, actinomycin G_2, 171
Streptomyces jamaicensis, monamycins, 149
Streptomyces nitrosoreus, benzastatin C, 222

Streptomyces platensis, resormycin, 143
Streptomyces rimosus, streptopyrroles, 178
Streptomyces rugosporus,
    pyrroindomycin B, 213
Streptomyces strain A, 23254 (angucycline
    antibiotic), 250
Streptomyces strain AJ, 9493, enediyne
    N1999A2, 233
Streptomyces strain BE-23254 (angucycline
    antibiotic), 250
Streptomyces strain LL-A9227,
    chloroquinocin, 250
Streptomyces venezuelae, 3′-*O*-
    acetylchloramphenicol, 134
Streptomyces violaceoniger, lysolipins, 318
Streptopyrroles, 178
Strobilurin B, 323
Strongylodesma algoensis, discorhabdins,
    214
Stropharia sp., dichloromethoxybenzyl
    alcohol, 257
Styela clava, styelin D, 156
Stylissa aff. *massa*, massadines, 193
Stylissa caribica, homoarginine, 186
– *N*-methyldibromoisophakellin, 189
– oxocyclostylidol, 188
– stylissadines, 191
– tetrabromostyloguanidine, 195
Stylissa carteri, carteramine A, 195
– 2-debromostevensine, 188
– latonduines, 189
Stylissa flabellata, stylissadines A, 195
Stylissadines, 191, 193, 195
Stylocheilamide, 112
Stylocheilus longicauda, makalika ester,
    114
– malyngamides, 112
Styloguanidine, 194, 195
Stylotella agminata (*Stylotella aurantium*),
    palau'amine, 194
Stylotella aurantium, axinohydantoin, 188
– carbonimidic dichlorides, 55
– dichloroimine stylotellane A, 47
– hymenialdisine, (10E)-diastereomer, 188
– isopalau'amines, 194
– sesquiterpene chlorohydrins, 55
Stylotellanes, 47–49
Suaeda vera, long-chain chloroalkanes, 24

Suberea aff. *praetensa*, dideoxyagelorins,
    310
Suberea sp., aplysamine-2, 296
– ma'edamines, 296
– purpuramine H, 296
Suberedamines, 296, 298
Suberites japonicus, seragamides, 158, 159
Sulfatobastadin, 312, 313
Sulfatohemibastadins, 295, 296
Sventrin (*N*-methyloroidin) 186, 187
Swimmer's itch, 235
Symphyocladia latiuscula, bis-benzyl ether,
    274
– polybrominated phenols, 272
– symphyoketone, 266
– tribromophenols, 269–270
Symphyoketone, 266, 267*Symploca* sp.,
    tasihalides, 67
Synechococcus sp., chloromethane, 9
Synoicum blochmanni, rubrolides, 131
Synoicum prunum, prunolides, 131
Syringomycins, 171
Syringtoxins, 171

T
Tafricanin A, 60
Takaokamycin/hormaomycin, 171
Talaromyces flavus, fosfonochlorin, 26
Talaromyces helicus, helicusins, 28
Talaromyces luteus, luteusins, 28
Tambjamines, 179
TAN-876 A/B, 178
Tasihalides, 67
Tauroacidins, 185, 186
Taurodispacamide A, 186
Tauropinnaic acid, 176
TCDD, 344
Teclea nobilis, chlorodesnkolbisine, 222
Temazepam, 225
Termites, CCl_3, 14
Terpene isonitriles, 375
Terpenes, higher, 86
Terpios hoshinota, nakiterpiosin, 93
Tetrabromoethane, 18
Tetrabromostyloguanidine, 195, 196
Tetrachlorocatechol, 258
1,3,7,9-Tetrachlorodibenzo-*p*-dioxin, 339

2,3,7,8-Tetrachlorodibenzo-*p*-dioxin (TCDD), 344
2,4,6,8-Tetrachlorodibenzofuran, 339
Tetrachloroethylene (PERC), 20
Tetrachlorohypericin, 319, 320
Tetrachloromethane (carbon tetrachloride, CCl_4), 15
Tetracycline chlorination gene, 253
Tetracyclines, halogenated, 253
Tetrafluoroethylene, 23
Tetrafluoromethane, 23
Tetrahydromonorden (tetrahydroradicicol), 235
Tetraselmis sp., chloromethane 9
Tetrathioaspirochlorine, 323
Teucrium racemosum, 60
Teuracemin, 60
Thalassia testudinium, 373
Thalassiosira weissflogii, chloromethane 9
Thelephenol, 270
Thelepus sp., bis-benzyl ether, 274
– bromophenols, 373
Theonegramide, 166
Theonella sp., halogenated keramamides, 153
– kumusine, 224
– microsclerodermins, 152
– perthamide B, 163
– theonellamides, 166
Theonella cupola, dehydromicrosclerodermins, 152
Theonella swinhoei, 4
– aurantosides G/H/I, 118
– bitungolides, 121
– chloroleucine-containing cyclolithistide A, 167
– theopalauamide, 166
Theonellamides, 166
Theopalauamide, 166
Thermal springs (Ashkhabad, Turkmenia), CCl_4, 15
Thialkalivibrio versutus, chloronatronochrome, 261
Thiazoles, 138
Thomitrem A, 214
Thorectandra sp., 5-bromo-*N*, *N*-dimethyltryptophan, 198
Threonine, biological chlorination, 171

Threonine transaldolase-PLP, 125, 126
Thrombin inhibitors, 144
Thyroid hormone, 376
Thyroxine, 281, 282
Thyrsenols, 86, 87
Thyrsiferol, 90
Thyrsiferyl, 23-acetate, 90
Tiacumicins, 236, 237
Tinea pedis (athlete's foot), 322
TMC-264, (compound), 230
TNF-α promoter activity/synthesis, inhibition, 27
Tokaradines, 286, 288
Topoisomerase I inhibitors, 320
Topopyrones, 320
Topsentins, 203–207
– bromodeoxytopsentin, 204, 206
– dibromodeoxytopsentin, 204, 206
– isobromodeoxytopsentin, 204
– isobromotopsentin, 203, 204
– nortopsentin D, 206, 207
Trachelospermum jasminoides (plant host), 230
Trachycladine A, 224
Trachycladus laevispirulifer, kumusine (trachycladine A), 224
Trachypleuranin B, 228
Trametes sp., trametol, 258, 259
Transvalencin A, zinc-containing antibiotic, 143
Tribrominated ma'iliohydrin, 52
Tribromoacetamide, 25
1,3,8-Tribromodibenzo-*p*-dioxin, 338
Tribromoethylene, 20
Tribromoindoles, 197
Tribromomethane (bromoform, $CHBr_3$), 17
1,1,2-Tribromooct-1-en-3-one, 25
2,4,6-Tribromophenol, 266, 372
2,3,4-Tribromopyrrole, 179
(–)-(S)-4,4,4-Trichloro-3-methylbutanoic acid, 137
Trichloroacetic acid (TCA), 26
1,1,1-Trichloroethane (methyl chloroform, CH_3CCl_3), 24
Trichloroethylene (TCE), 20
Trichloroleucine amino acid, non-*N*-methylated, 138
Trichloromethane (chloroform), 13

– biomass combustion, 11, 15
– volcanic emissions, 15
Trichloromethyl metabolites, 137
Trichodenones B/C, 27
Trichoderma harzianum, MR566A, 27
Trichoderma virens, trichodermamide B, 173
Trichodermamides, 173
Tricholoma magnivelare, dichloromethoxybenzaldehyde, 257
– dichloromethoxybenzyl alcohol, 257
Trichophyton mentagrophytes, growth inhibition, 259
Triene bromohydrin, 49
Trifluoroacetic acid, 127
– salts, 193
Triiodothyroxine, 376
Triterpenes, heterocyclic, 86
– pentacyclic, squalene-derived, 86
Trophon geversianus, brominated imidazole, 223
Tropical islands, chloromethane, 10
Tropical plants, atmospheric CH_3Cl, 10
Trunculariopsis trunculus, brominated imidazole, 223
Trypsin inhibitors, 144
Tryptamines brominated, 198, 200
Tryptophan amino acids, 153
Tryptophan 7-halogenase, 177
Tryptophans, brominated, 198
Tsitsikamma spp., discorhabdins, 214
Tskhaltubo (Georgia), CCl_4, 15
Tubastraea sp., tubastrindole A, 206
Tubastrindole A, 206, 207
Tuberculosis, 375
Tubipora musica, brominated oxylipins, 128
Turbo marmorata, turbotoxins, 286
Turbotoxins, 286
Turraea pubescens, turrapubesin, 90
Turrapubesin, 90, 91
Tursiops truncatus, dimethoxytetrabromobiphenyl, 271
Tyramines, brominated, 282–300
Tyrian purple, 203, 210
Tyrosine kinase inhibition, 185
Tyrosines, brominated, 282–300
– halogenated, 281–314
– multiple, halogenated, 300

U

Ubiquitin isopeptidase activity, 128
Ugibohlin, 187, 188
Ulosa spongia, carbonimide dichlorides, 47
Ulosins, 47–49
Ulva lactuca, chloromethane, 9
– dibromomethoxybenzoic acid, 266
– trichloromethane, 13
Umbraculolides, 80, 82
Urceolatol, 269
Ureido-balhimycin, 329, 330
Urphoside B, 104, 105
Usnea longissima, longissiminone B, 259

V

Vancomycin, 328
Vancomycin-type glycopeptide antibiotics, methicillin-resistant bacteria, 378
Vascular cell adhesion molecule-1, inhibition, 176
Vellozia bicolor, isopimarane diterpene, 12-chloroillifunone, 60
Venus, HCl/HF, 380
Veratryl alcohol, 371
Veratryl chloride (dimethoxybenzyl chloride), 258
Vernchinilides, cytotoxic activity, 43
Vernolide C, 43
Vernonia chinensis, chlorinated sesquiterpene lactones vernchinilides, 43
Vernonia cinera, 43
Verongamine, 286
Verongia aerophoba, antibiotic/cytotoxic activity, 377
– fistularin-3, 309
Verongid sponge, mololipids, moloka'iamines, 292
– nakirodin A, 296
Verongida, 372
Verongula gigantea, bromotyrosine, 295
Veronica pectinata var. *glandulosa*, urphoside B, 104
Verrucosins, chlorinated, 69
Verticillium sp., 8′,9′-dehydroascochlorin, 322
Verticillium hemipterigenum, vertihemipterin, 322
Vertihemipterin, 322, 323

Victoricine, 148
Victorins, 148
Vinetorin, 318
Vinyl chloride, 21
Violacene, 35
Virantmycin, 222
VM, 4798-1a/b, 27
Volcanic emissions, bromomethane, 18
– CCl$_4$, 15
– ethyl iodide, 20
– halogenated alkynes, 23
– vinyl chloride, 21
Volcanoes, 6
– chloromethane, 12
Volutamides, 282–284

W
Wai'anaeamines, 292
Welwitindolinones, 213
Wetlands, atmospheric CH$_3$Cl, 10
– bromomethane, 16Whale blubber, 179
White rot fungi (*Phellinus pomaceus*), chloromethane, 12
Wíthanolides, 92
Woodrot fungi, atmospheric CH$_3$Cl, 10
Wrangelia sp., 25

X
Xantholipin, 318, 319
Xanthones, chlorinated, 317–319
Xanthoparmelia camtschadalis, 109
Xanthoparmelia tinctina, 109

Xanthoria sp., brominated fatty acids, 109
Xanthoria elegans, 109
Xerula longipes, oudemansin B, 323
Xerula melantricha, oudemansin B, 323
Xestoquinones, 250
Xestospongia sp., aragusteroketal C, 93
– brominated fatty acids, 108
– xestoquinones, 250
Xestospongia muta, 4
Xestospongia testudinaria 4, 108
Xestosterol, 108
– esters, 108
Xylaria sp. 325
– chlorohydroxyphenylacetamide, 258
– dihydromaldoxin, isodihydromaldoxin, 277
– maldoxin, 322
– maldoxone, 317
Xylariamide A, 163

Y
Yonarasterols, 93, 94

Z
Zamamistatin, 303
Zea mays, benzoxazolinones, 223
Zebra mussel antifouling, 372
Ziconotide (ω-conopeptide MVIIA), 155
Ziracin (SCH, 27899)333*Zyzzya* spp., discorhabdin Q, 214
Zyzzya fuliginosa, makaluvamines, batzellines, 216

Printing and Binding: Stürtz GmbH, Würzburg